D0773494

TELEVISION ELECTRONICS:

Theory and Servicing

eighth edition

Milton S. Kiver
Milton Kaufman

VNR VAN NOSTRAND REINHOLD COMPANY
New York

ISBN: 0-442-24871-7

Originally published under the title *Television Simplified*.

Manufactured in the United States of America

Published by Van Nostrand Reinhold Company Inc.
135 West 50th Street
New York, New York 10020

Van Nostrand Reinhold Company Limited
Molly Millars Lane
Workingham, Berkshire RG11 2PY, England

Van Nostrand Reinhold
480 Latrobe Street
Melbourne, Victoria 3000, Australia

Macmillan of Canada
Division of Gage Publishing Limited
164 Commander Boulevard
Agincourt, Ontario M1S 3C7, Canada

15 14 13 12 11 10 9 8 7 6 5 4 3

PREFACE

The eighth edition of *Television Electronics: Theory and Servicing* (formerly *Television Simplified*), has been completely redesigned and updated to the current state of the art. The book is designed as a text for television electronics courses that follow a course in basic electronics or basic radio.

The purpose of the book is to prepare electronics technicians and engineers for a career in some phase of the television industry. Every effort has been made to ensure the book's usefulness to instructors, their students, and to self-study students.

This book covers in detail the operation, circuitry, and trouble-shooting of solid-state color and monochrome television receivers. Some coverage of vacuum-tube television receivers is also presented. Other current and important topics that are covered include (1) cable television, (2) video-tape and video-cassette recorders, (3) video games, (4) integrated circuits, (5) communications satellites, (6) color television signal generation, (7) digital circuitry, and (8) closed-circuit television.

NEW CHAPTERS AND CHAPTER REORGANIZATION

The eighth edition contains 27 chapters. The previous edition contained 24 chapters. The new chapters are

- *Chapter 2: Television System Applications.* This chapter includes topics such as cable television, video-tape and video-disc recording, video games, and closed-circuit television.
- *Chapter 7: Principles of Monochrome Television Receivers.* This is a basic block diagram explanation of the operation of monochrome television receivers. It also includes the test equipment and tests used for monochrome television receivers.
- *Chapter 10: Frequency Synthesis, Automatic Fine Tuning, and Remote Control.* This chapter includes discussions of binary numbers, digital frequency dividers, the phase-locked loop, microcomputers, and frequency synthesis push-button tuning. The topics of Automatic Fine Tuning and Remote Control are also updated and covered in detail.

The material on vertical deflection oscillators has been placed in a separate chapter (Chapter 21), which also includes a digital IC vertical countdown circuit. Horizontal oscillators and horizontal AFC appear in Chapter 22, which also includes a digital IC horizontal countdown circuit. All chapters in the eighth edition have been reorganized and updated. Obsolete material has been removed.

CHAPTER FORMAT

The opening format of each chapter has been improved to increase the usefulness of this edition. The beginning of each chapter includes

- A chapter outline
- An introduction
- Chapter objectives

The format at the end of most chapters includes

- A summary of chapter highlights.
- Examination questions with answers. These questions are multiple choice, true-false, and fill-in blank types.
- Review essay questions. These require the application of text information, drawing diagrams, answering questions by referring to diagrams, solving troubleshooting problems, providing definitions, and giving the functions of circuit components.
- Examination problems with selected answers. These are generally numerical problems related to the chapter contents.

A comprehensive summary is presented at the end of each chapter. This provides students with a review of the chapter material. Reference figure numbers or chapter section numbers are given for many items in each summary. This makes it easier for students to obtain specific additional information.

The self-testing aids in the text can be very helpful in improving the student's understanding of the material in each chapter. They can also be helpful to instructors in providing class assignments and to self-study students in checking their progress.

NEW CIRCUITRY AND DIAGRAMS ADDED

The very latest types of circuitry have been added to the eighth edition. Among these are frequency synthesis tuning, more integrated circuits, noise cancellers, horizontal and vertical digital countdown circuits, and new FM demodulators. The latest remote control techniques are described in detail. Also described, is the latest color television receiver automatic circuitry, including automatic hue and saturation control, and automatic brightness and contrast control.

Many new circuit schematic and block diagrams have been added to enhance the reader's understanding of the material. Among these are a number of simplified diagrams presented wherever a more detailed analysis of the information is required.

TEXT ORGANIZATION

The eighth edition begins with two introductory chapters. These chapters explain the basic concept of the television system. They also describe the various applications of television and its associated equipment. Chapters 3 and 4 describe video signals and the principles of scanning and synchronizing a television picture. This is followed in Chapter 5 by discussions of television camera tubes and camera systems. Next we explain television frequency radio-wave propagation, and various indoor and outdoor television receiving antennas.

Following this, the principle of monochrome and color television receivers are described in Chapters 7, 8, and 9. Beginning with Chapter 10, the individual sections of television receivers are discussed in detail. This material begins with tuners and progresses through various signal circuits. This is covered in Chapters 10 through 17. Monochrome and color picture tubes are covered in Chapter 18.

A synchronized scanning raster is produced on a picture tube by the action of synchronizing and deflection circuits. These circuits are described in Chapters 20 through 24. Following the discussion of the synchronizing and deflection circuits, the FM sound system is explained in Chapter 25. This chapter also includes discussions of the latest types of FM demodulators.

Chapters 26 and 27 provide details on the operation of a complete solid-state color television receiver. Information is also provided on the alignment, adjustment, and troubleshooting of television receivers. In addition, various types of test equipment are described.

TROUBLESHOOTING AND ALIGNMENT

Each chapter that describes a section of a television receiver ends with a section describing common troubles that might occur in that section. This information is enhanced by the use of photographs and drawings. Chapter 26 presents a detailed explanation of the operation of a solid-state color television receiver. Chapter 26 also includes detailed instructions for the alignment and adjustment of color television receivers. A new type of oscilloscope, a color-bar generator and a color-pattern generator are described in Chapter 26. Their uses are also discussed there.

Chapter 27 presents important information on the techniques of troubleshooting television receivers. The methods of signal tracing and signal injection are covered in detail and many explanatory diagrams are presented. Many types of television receiver faults are described with the aid of photographs and drawings. A number of full-color photographs help to clarify possible troubles in color circuitry. Additional types of test equipment and their uses are covered in Chapter 27. These include transistor testers, color picture tube brighteners, and a color picture tube analyzer and restorer.

END OF BOOK MATERIALS

The end of this book contains a comprehensive glossary of terms and abbreviations. These will prove most helpful to students desiring definitions relating to television electronics. Four useful appendices are also provided: (1) U.S. television channels and related frequencies, (2) Frequently used metric units, (3) A table of binary numbers with a method of conversion from decimal to binary numbers, and (4) SI unit prefixes.

ACKNOWLEDGMENTS

The authors wish to gratefully acknowledge all the individuals and companies who were instrumental in providing helpful assistance and information. Credits are given in the text for photographs and drawings furnished by the various companies. Particular appreciation is extended to Mr. El Mueller of Quasar Electronics Corp., Mr. John Shouse of General Electric Co., Mr. Greg Carey of Sencore, GTE Sylvania Inc., Heath Co., Sony Inc., Zenith Radio Corp., Radio Corporation of America, B&K-Precision Dynascan Corp., and Westinghouse Electric Corp.

We also wish to cite Mr. David F. Stout, Mr. Burton Santee, and Mr. Bob Barkley for their valuable cooperation and Mr. Dave Favin for his conscientious review of the manuscript.

In addition, we wish to gratefully acknowledge the considerable efforts of Marjorie Bruce of Delmar Publishers and of Maryanne Miller and the rest of the staff of Editing, Design & Production, Inc., in editing and producing the eighth edition.

MILTON S. KIVER

MILTON KAUFMAN

TABLE OF CONTENTS

chapter 1 TELEVISION SYSTEM CONCEPTS **1**

Introduction 1.1 Digital techniques 1.2 Modern television receivers 1.3 Use of semiconductors in television receivers 1.4 Summary of desirable image characteristics 1.5 Picture detail 1.6 Block diagrams of the television system Summary *14*

chapter 2 TELEVISION SYSTEM APPLICATIONS **16**

Introduction 2.1 Television systems that use cables 2.2 Extending the coverage of television 2.3 Videotape and video-disc recording 2.4 Video games 2.5 Projection television 2.6 Alpha-numeric and graphic displays 2.7 Summary of U.S. television standards 2.8 Foreign television standards Summary *43*

chapter 3 PRINCIPLES OF SCANNING, SYNCHRONIZING, AND VIDEO SIGNALS **46**

Introduction 3.1 Basic television system 3.2 Introduction to scanning 3.3 Simplified scanning principles 3.4 Scanning rate 3.5 Flicker 3.6 The complete scanning process 3.7 Overview of blanking and synchronizing signals 3.8 The composite monochrome video signal 3.9 The horizontal sync and blanking pulses 3.10 Vertical sync, vertical blanking, and equalizing pulses 3.11 The color-synchronizing signal Summary *81*

chapter 4 VIDEO SIGNALS AND PICTURE QUALITIES **68**

Introduction 4.1 Negative and positive picture phase 4.2 Why television requires wide frequency bands 4.3 Effect of loss of low and high video frequencies 4.4 Desirable picture qualities 4.5 Components of the color video signal 4.6 Normal and special video signals Summary *81*

chapter 5 TELEVISION CAMERA TUBES AND CAMERA SYSTEMS **85**

Introduction 5.1 Basic TV camera fundamentals 5.2 Camera tube characteristics 5.3 Operation of the vidicon 5.4 Operation of the plumbicon 5.5 Silicon diode array vidicon 5.6 Silicon imaging device (SID) 5.7 Single-tube color camera 5.8 Color television camera principles 5.9 Studio color television camera 5.10 Portable color TV camera 5.11 The minicam system 5.12 Special purpose TV cameras 5.13 Studio color video-tape recorder (VTR) Summary *112*

chapter 6 TV WAVE PROPAGATION AND TV ANTENNA SYSTEMS **117**

Introduction 6.1 Radio wave propagation 6.2 Line-of-sight distance 6.3 Unwanted signal paths 6.4 Antenna characteristics 6.5 Tuned antennas 6.6 UHF antennas 6.7 General-purpose antennas 6.8 Combination antennas 6.9 RCA ministate antenna system 6.10 Indoor antennas 6.11 Antenna rotators 6.12 Transmission lines 6.13 Amplifying the antenna signal 6.14 Antenna-system accessories 6.15 Master-antenna distribution systems 6.16 Community-antenna television systems 6.17 Grounding and lightning protection 6.18 Trouble shooting antenna systems Summary *155*

chapter 7 PRINCIPLES OF MONOCHROME TELEVISION RECEIVERS **159**

Introduction 7.1 Some TV receiver fundamentals 7.2 Block diagram discussion 7.3 Monochrome tele-

vision receiver controls 7.4 Use of PCs, ICs, FETs, MOSFETs and modular construction 7.5 Use of nuvistor, novar, and compactron vacuum tubes 7.6 Types of test equipment used for monochrome TV receivers 7.7 General types of tests required for monochrome TV receivers 7.8 Some common monochrome TV receiver troubles Summary 174

chapter 8 PRINCIPLES OF COLOR TELEVISION **178**

Introduction 8.1 Elements of color 8.2 Chromaticity chart 8.3 The NTSC color television system 8.4 I and Q signals 8.5 Color-signal components 8.6 The color subcarrier 8.7 The color burst signal 8.8 Derivation of the color subcarrier frequency 8.9 Composite colorpexed video wave forms 8.10 How compatibility is achieved Summary 198

chapter 9 PRINCIPLES OF COLOR TV RECEIVERS **202**

Introduction 9.1 RF tuner 9.2 Video-IF system 9.3 Sound IF, FM detector, and the audio system 9.4 Video detector and video amplifiers 9.5 Chrominance section 9.6 Color sync section 9.7 Sync separators and AGC 9.8 Horizontal- and vertical-deflection systems 9.9 High-voltage circuits 9.10 Color picture tube and convergence circuits 9.11 Processing color sidebands 9.12 Automatic circuits 9.13 Some common color TV receiver troubles Summary 218

chapter 10 TV-RECEIVER TUNERS **222**

Introduction 10.1 Types of TV tuners for VHF and UHF reception 10.2 Electrical characteristics of tuners 10.3 Interference, stability, and noise problems in tuners 10.4 Channel allocations 10.5 Characteristics of tuned circuits 10.6 Vacuum tubes for tuners 10.7 Typical tube-type RF amplifiers 10.8 Transistor RF amplifiers 10.9 Field-effect transistors and RF amplifiers 10.10 Mixers and mixer circuit operation 10.11 Local oscillator operation 10.12 A typical nuvistor tuner 10.13 A typical transistor tuner 10.14 Varactor tuning and tuners 10.15 UHF solid-state tuners 10.16 Sources of trouble in VHF and UHF tuners Summary 268

chapter 11 FREQUENCY SYNTHESIS, AFT AND REMOTE CONTROL **272**

Introduction 11.1 The binary number system 11.2 Dividers and prescalers 11.3 The phase-locked loop (PLL) 11.4 The microcomputer (μC) 11.5 The Quasar compu-matic touch-tuning system 11.6 Compu-matic circuit analysis 11.7 Automatic fine tuning (AFT) 11.8 Functions of (AFT) systems 11.9 Functions and types of remote control devices 11.10 Ultrasonic remote control systems 11.11 Electronically generated sound signals 11.12 An electronic remote control system 11.13 A direct-access remote control system 11.14 Troubles in compu-matic tuning systems 11.15 Troubles in AFT circuits 11.16 Troubles in remote control systems Summary 320

chapter 12 VIDEO IF AMPLIFIERS **325**

Introduction 12.1 Major video IF functions 12.2 Frequency response curves 12.3 Comparison of vacuum tube and solid-state video IF amplifiers 12.4 Types of video IF amplifiers 12.5 Intermediate frequencies 12.6 Amplifier bandwidth 12.7 Stagger tuning 12.8 Interstage coupling 12.9 Wave traps 12.10 Sound IF frequency separation 12.11 Typical video IF amplifiers 12.12 An IC video IF amplifier 12.13 Surface-acoustic wave filter (SAWF) 12.14 Video-IF amplifier troubles Summary 361

chapter 13 VIDEO DETECTORS **367**

Introduction 13.1 Positive- and negative-picture phases 13.2 Video detector filtering and high-frequency compensation 13.3 Shunt-video detectors 13.4 Video detectors in color television receivers 13.5 Synchronous (linear) video detectors 13.6 Synchronous video detector operation 13.7 Troubles in the video detector stage Summary 379

chapter 14 AUTOMATIC GAIN CONTROL (AGC) CIRCUITS **383**

Introduction 14.1 General types of AGC circuits 14.2 Peak-AGC systems 14.3 Disadvantages of peak AGC 14.4 Keyed-AGC systems 14.5 Delayed AGC and diode clamping in AGC circuits 14.6 AGC systems in solid-state receivers 14.7 Noise cancellation circuits 14.8 Solid-state AGC system 14.9 AGC system in an RCA IC 14.10 Expanded block diagram of RCA IC noise processor 14.11 Expanded block diagram of RCA IC AGC processor 14.12 Troubles in AGC circuits Summary 412

chapter 15 VIDEO AMPLIFIERS 416

Introduction 15.1 Video signal requirements of picture tubes 15.2 Video signal amplitude 15.3 Eye resolving power 15.4 Effects of loss of low- and high-video frequencies 15.5 The DC component of a video signal 15.6 Phase distortion 15.7 Square-wave response of video amplifiers 15.8 Video requirements for color TV 15.9 Types of video amplifiers for monochrome and color sets 15.10 Comparison of tube and solid-state video amplifiers 15.11 Contrast controls in video amplifiers 15.12 Automatic contrast control 15.13 Automatic brightness and contrast control 15.14 Video peaking 15.15 Improved video peaking system 15.16 The comb filter 15.17 Elementary comb filter 15.18 Comb filter block diagram Summary 444

chapter 16 VIDEO AMPLIFIER DESIGN 449

Introduction 16.1 The gain of a pentode-video amplifier 16.2 The gain of a transistor video amplifier 16.3 High-frequency behavior 16.4 Shunt peaking 16.5 Series peaking 16.6 Series-shunt peaking 16.7 Degenerative high-frequency compensation 16.8 Low-frequency compensation 16.9 The selection of tubes and transistors for video amplifiers 16.10 Typical transistor video amplifier circuits 16.11 Integrated circuits 16.12 Troubles in video-amplifier circuits Summary 475

chapter 17 DC REINSERTION 481

Introduction 17.1 The DC component of video signals 17.2 Reinserting the DC component 17.3 DC reinsertion with a diode 17.4 Television receivers that do not employ DC restoration 17.5 DC restorers in color receivers 17.6 Troubles in DC restorer circuits Summary 493

chapter 18 TV PICTURE TUBES 497

Introduction 18.1 Picture-tube specificatons 18.2 The electron gun 18.3 Electron-beam deflection 18.4 Electromagnetic deflection with electrostatic focus 18.5 Ion spots 18.6 Problems in obtaining brightness and contrast 18.7 Rectangular screens and tube-safety shields 18.8 Color-picture tube with a delta gun and color-dot triads 18.9 RCA in-line gun, color-picture tube 18.10 Trinitron-picture tube 18.11 Tri-potential color tube 18.12 Picture-tube protection considerations 18.13 Troubles in monochrome and color-picture tubes Summary 537

chapter 19 LOW VOLTAGE TV POWER SUPPLIES 542

Introduction 19.1 Types of power supplies 19.2 Rectifiers 19.3 Filters 19.4 Transformer and transformerless power supplies 19.5 Voltage multipliers 19.6 Voltage regulators 19.7 The zener diode regulator 19.8 The series-pass transistor regulator 19.9 Three-terminal voltage regulators 19.10 Switching regulators 19.11 Solid-state monochrome power supplies 19.12 A hybrid color TV power supply 19.13 Horizontal frequency power supply 19.14 Troubles in low-voltage TV power supplies Summary 569

chapter 20 SYNCHRONIZING CIRCUITS 574

Introduction 20.1 Effects of loss of synchronization 20.2 Syncronizing pulse and video signal separation 20.3 Syncronizing pulse separator circuits 20.4 Reduction of noise to improve syncrhonization 20.5 Transistor noise-cancellation circuits 20.6 Syncronizing pulse separator in an RCA IC 20.7 Vertical and horizontal pulse separation 20.8 Function of the equalizing pulses 20.9 Vertical syncronizing pulse and equalizing pulse display 20.10 Syncrhonizing troubles Summary 597

chapter 21 VERTICAL OSCILLATORS AND DIGITAL COUNTDOWN 601

Introduction 21.1 Requirement for sawtooth current 21.2 Need for trapezoidal wave forms 21.3 The vacuum tube blocking oscillator 21.4 Vacuum tube blocking oscillator, sawtooth generator 21.5 Transistor blocking oscillators 21.6 A transistor-blocking oscillator, vertical deflection system 21.7 Vacuum tube plate-coupled multivibrators 21.8 Plate-coupled, multivibrator sawtooth generator 21.9 Combination vacuum tube multivibrator and output stage 21.10 Vac-

uum tube cathode-coupled multivibrators 21.11 Cathode-coupled, multivibrator-sawtooth generator 21.12 Synchronizing the multivibrator 21.13 Transistor multivibrator oscillators 21.14 Transistor, combination multivibrator and output stage 21.15 Miller integrator, vertical-deflection circuit 21.16 Digital vertical countdown circuits 21.17 Troubles in verticle deflection oscillator systems Summary 636

chapter 22 HORIZONTAL OSCILLATORS AND HORIZONTAL AFC **640**

Introduction 22.1 Fundamentals of horizontal AFC 22.2 Vacuum tube AFC and multivibrator circuits 22.3 Multivibrator stabilization 22.4 Solid-state AFC and Hartley oscillator circuits 22.5 Solid-state AFC and blocking oscillator 22.6 Reactance-controlled, sinusoidal oscillator 22.7 Digital horizontal countdown 22.8 Troubles in horizontal oscillator and AFC systems Summary 665

chapter 23 HORIZONTAL OUTPUT DEFLECTION CIRCUITS AND HIGH VOLTAGE **669**

Introduction 23.1 Horizontal output block diagram 23.2 Vacuum tube monochrome circuits 23.3 Flyback high voltage 23.4 Solid-state monochrome TV circuits 23.5 Transistor horizontal output circuits 23.6 Solid-state dampers 23.7 NPN and PNP horizontal output stages 23.8 Adjustments in monochrome TV receivers 23.9 Vacuum tube color TV horizontal output circuits 23.10 Solid-state color TV horizontal output circuits 23.11 Pincushion (PIN) distortion 23.12 Silicon-controlled rectifier (SCR) horizontal output circuits 23.13 Horizontal output transformers and special assemblies 23.14 Deflection yokes 23.15 Horizontal high-voltage components 23.16 Adjustments in color-TV horizontal systems 23.17 X-ray emission 23.18 X-ray protection circuits 23.19 Troubles in horizontal-deflection circuits 23.20 High-voltage troubles 23.21 Relationship between horizontal and high-voltage problems Summary 716

chapter 24 VERTICAL OUTPUT DEFLECTION CIRCUITS **721**

Introduction 24.1 Operation of vertical-deflection circuits 24.2 Special requirements for monochrome TV transistor vertical-deflection circuits 24.3 Special requirements for color transistor and vacuum tube vertical output circuits 24.4 Vertical dynamic convergence 24.5 Vertical blanking 24.6 Monochrome TV vacuum tube vertical output circuit 24.7 Monochrome TV transistor vertical output circuits 24.8 Color transistor vertical output circuit 24.9 Vertical output transformers 24.10 Vertical yokes 24.11 Troubles in vertical output circuits Summary 745

chapter 25 THE FM SOUND SYSTEM **750**

25.1 Review of AM and FM systems 25.2 Glossary of FM terminology 25.3 Properties of FM waves 25.4 Advantages and disadvantages of FM 25.5 Preemphasis and deemphasis 25.6 The TV receiver FM sound section 25.7 The 4.5-MHz sound-IF amplifier 25.8 Sound-IF limiting 25.9 A basic FM discriminator 25.10 The Foster-Seeley discriminator 25.11 The ratio detector 25.12 The quadrature detector 25.13 The transistor quadrature detector 25.14 The phase-locked loop (PLL) FM detector 25.15 A digital FM demodulator 25.16 The FM differential peak detector 25.17 A complete vacuum tube sound section 25.18 A complete transistor sound section 25.19A partial IC sound section 25.20 A complete IC sound section 25.21 Complementary-symmetry audio push-pull output stage Summary 788

chapter 26 COLOR TV RECEIVER CIRCUIT ANALYSIS, TEST EQUIPMENT AND ALIGNMENT **793**

Introduction 26.1 The monochrome section 26.2 The color section 26.3 Equipment for TV receiver alignment 26.4 The oscilloscope 26.5 Sweep alignment generators 26.6 RF signal generators 26.7 Marker signals 26.8 Color pattern generators 26.9 Solid state (analog and digital) multimeters 26.10 Alignment using frequency-sweep generator 26.11 Alignment using bar sweep generator 26.12 Horizontal AFC and anode voltage adjustments 26.13 Color picture tube adjustments Summary 856

chapter 27 SERVICING AND TROUBLESHOOTING TELEVISION RECEIVERS **862**

Introduction 27.1 Color-picture tube test jig 27.2 Transistor tester 27.3 Picture-tube analyzer and restorer 27.4 Picture-tube brighteners 27.5 High-voltage probes 27.6 Substitute tuner (subber) 27.7 Horizontal-output transformers, horizontal yoke and vertical-yoke testing 27.8 Analysis and isolation of TV-receiver malfunctions 27.9 Isolation by observation 27.10 Localizing color troubles by observation 27.11 Signal injection 27.12 Signal tracing 27.13 Troubleshooting the tuner section 27.14 Troubleshooting the video-IF and video-detector circuits 27.15 Troubleshooting the AGC section 27.16 Troubleshooting the video amplifiers 27.17 Troubleshooting picture tubes and associated circuits 27.18 Troubleshooting low-voltage power supplies 27.19 Troubleshooting sync-separator stages 27.20 Troubleshooting the vertical-deflection system 27.21 Troubleshooting the horizontal-deflection system 27.22 Guidepoints for troubleshooting color-TV receivers 27.23 Color-TV receiver troubles in monochrome circuits 27.24 Color-TV receiver troubles in color circuits Summary 919

APPENDICES 921

Appendix I: U.S. Television Channels and Related Frequencies Appendix II: Frequently Used Metric Units Appendix III: SI Unit Prefixes Appendix IV: Binary Numbers

ANSWERS TO SELECTED QUESTIONS AND PROBLEMS 929

GLOSSARY 937

INDEX 951

Chapter 1

1.1 Digital techniques
1.2 Modern television receivers
1.3 Use of semiconductors in television receivers
1.4 Summary of desirable image characteristics
1.5 Picture detail
1.6 Block diagrams of the television system
Summary

TELEVISION SYSTEM CONCEPTS

INTRODUCTION

Television is the science of transmitting rapidly changing pictures from one place to another. Radio-frequency waves are usually used for the transmission of television pictures. However, in some applications, such as closed-circuit television (CCTV) or cable television (CATV), coaxial cables carry the signal from one point to another. Television dominates the home-entertainment industry. It is also used widely in science, industry, education, and military applications.

New types of components, new circuits, and new concepts are continually being introduced in the television industry. Therefore, it is crucial that the TV technician know about the modern technical aspects of television systems. A number of advancements have been made since 1941, when the complete television receiver consisted only of **monochrome,** or black and white, circuitry. These include color television, solid-state television receivers, satellite television relays, video-tape and video-disc recorders and players, and closed-circuit television. Other changes include home video and sound cameras, video games, automatic time-programmed channel selection, push-button channel selection for VHF and UHF channels, cable television, and large-screen projection television.

When you have completed the reading and work assignments for Chapter 1, you should be able to:

- Define the following terms: random access, monochrome, integrated circuit, scanning, synchronizing pulse, vestigial sideband transmission, and line-of-sight broadcasting distance.
- Describe how modern television receivers differ from those manufactured in 1941 and explain briefly how digital techniques are used in television receivers.
- Describe the function of VIR circuits.

- Discuss the advantage of using ICs in television receivers and list several applications.
- Summarize the qualities of a desirable television image.
- Describe the different sizes and styles of television receivers.
- Understand the basic operation of television transmitters and receivers.

MODERN TELEVISION RECEIVERS

Early television receivers contained as many as 30 vacuum tubes. However, almost all modern receivers use many integrated circuits (ICs) and are completely solid state. Digital techniques are used in some receivers for electronic channel selection, both at the receiver and also with remote control systems. The tuners (VHF and UHF) on these sets have no moving parts. All tuning is done electronically. On some sets a digital readout of the channel number being viewed and the time in hours, minutes, and seconds, may also be shown on the screen whenever desired. Home computers are now quite popular. The television receiver screen may be used for the computer readout, if desired.

This brief introduction, shows that the television system accommodates many different applications. In this chapter and the next, the basic concepts and some of the more important applications of the television system are described.

1.1 DIGITAL TECHNIQUES

Digital techniques[1] have replaced analog techniques in many applications, including the digital multimeter. Other examples are microprocessors,[2] high-fidelity audio tape recording, frequency counters, timers, and satellite communications. Some advantages of digital techniques are greater accuracy, faster response, less noise and distortion, less drift, fewer errors, and more automatic operation.

The binary number scheme is used in digital electronics. Two discrete voltage levels represent an "on" or "off" condition. The voltage levels that are commonly used are 0 volts and +5 volts. The use of standard integrated circuits (ICs) is crucial in the practical application of digital devices. An example is the popular TTL (transistor-transistor logic) family of IC logic devices. These devices "recognize" the 0-volt and 5-volt conditions. The use of standard, mass-produced ICs greatly reduces the cost of digital devices. It also substantially decreases space requirements, weight, and power consumption.

DIGITAL USES IN TELEVISION RECEIVERS

Some of the uses of digital techniques in color television receivers are as follows:

1. Push-button channel selection (including remote control). In this case, the selected channel and correct time may be shown momentarily as a digital display on the screen (see Figure 1.1A).
2. Digital techniques can eliminate some conventional television receiver controls. These include the fine-tuning control and the vertical and horizontal hold controls.

---

[1]Digital circuits operate on the basis of pulses, or "on-off" conditions. This type of operation differs from analog circuits where the information is present as continuous data, such as voltage or current.

[2]A microprocessor (μP) is one part, the central processing unit, of a microcomputer (μC). The μP may be manufactured on a single integrated circuit (IC) chip or on several chips.

3. Digital techniques make time-programmed channel selection possible.
4. Microprocessors and other computer techniques used in automatic timers and automatic channel selection.
5. Vertical-interval reference (VIR) circuits. These enable automatic color corrections to be made in both hue and saturation.

DIGITAL USES IN TELEVISION TRANSMISSION

A proposal has been made to transmit television video signals by digital techniques. Such techniques are currently used in transmitting television from space vehicles. The major advantage is the almost total elimination of "snow" (noise) from the picture. However, this digital system requires extremely wide bandwidths which are not available with the present television transmission frequencies. One possible solution to this problem is to originate the signal from a space satellite, with a carrier frequency of 12 000 MHz (12 gigahertz). Then, the increased bandwidth can be accommodated. It is also anticipated that bandwidth-compression schemes may reduce the required bandwidth. (The use of digital techniques in television receivers is covered later.)

1.2 MODERN TELEVISION RECEIVERS

A highly sophisticated, completely solid-state color television receiver is shown in Figure 1.1A. Figure 1.1B shows a front view, with the computer-type section pulled forward and opened up in the servicing position. Figure 1.1C shows the rear of the receiver and the TV circuit modules that can be removed for convenient servicing.

MAJOR FEATURES

This receiver can be programmed to provide up to 32 automatic channel changes within two 12- or 24-hour periods. This is accomplished by operating the programmer keyboard (8 buttons) and the random access[1] keyboard (12 buttons) shown in Figure 1.1B. The tuners (VHF and UHF) are completely electronic and contain no moving parts.

Another feature of this receiver is the automatic antenna rotor control. When using an antenna with a motorized rotor, the automatic rotor control turns the antenna to the correct azimuth heading each time a channel is changed. Up to eight separate headings can be selected, with up to three stations for each individual heading. This feature is very useful in reception areas where it is necessary or desirable to receive television stations from widely different directions.

The horizontal and vertical oscillators are connected in phase-locked loop circuits. This eliminates conventional horizontal and vertical hold controls.

A wireless remote control device permits the viewer to turn the set "on" or "off," adjust the volume, adjust the tint, scan up or down through the channels, and turn the automatic programmer "on" or "off."

DIGITAL SCREEN DISPLAY

In this receiver, the digital readout is used as a computer readout for the time-programmable system. The first entry is the

[1]*Random access* means that channels may be chosen in any order.

A

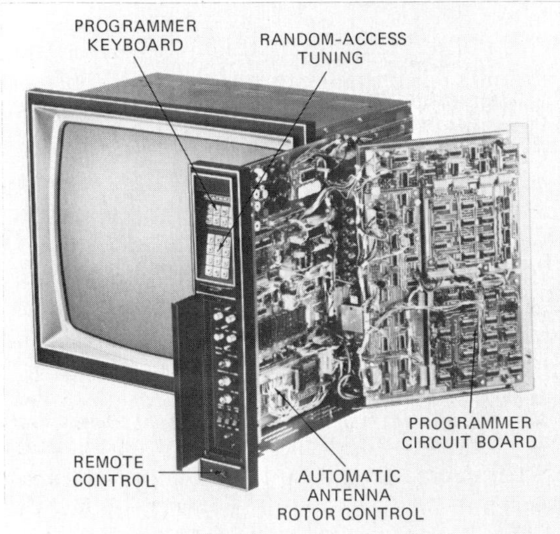

B

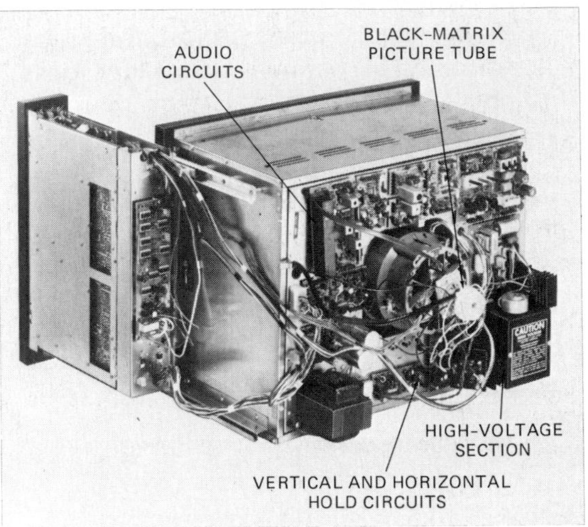

C

Figure 1.1A A 25-in (63.5-cm) solid-state color television receiver using digital techniques. The channel number in use and the correct time are displayed briefly. **B.** An open computer section, showing the large number of ICs in the receiver of Figure 1.1A. **C.** The individual plug-in circuit modules, the audio section, and the high-voltage section of the receiver in Figure 1.1A are shown in this rear view. *(Courtesy of Heath Company.)*

time the viewer wants to see a program. The second entry is the channel number. This information appears individually on the screen for each selection but briefly, to avoid interfering with the desired program.

The receiver is completely solid-state and examination of Figure 1.1B reveals a large number of integrated circuits (ICs), in the opened-up computer section. Also note the individual, solid-state, plug-in circuit modules, the separate high-fidelity audio section, and the high-voltage section.

STYLES OF TELEVISION RECEIVERS

Modern television receivers come in a variety of styles and sizes ranging from tiny monochrome receivers with 3-in (7.5-cm) screens to large console, color receivers with 25-in (63.5-cm) screens. More sophisticated features, such as time-programmed channel selection, are available only in the "top-of-the-line" receivers. Three other styles of television receivers are shown in Figure 1.2.

A

C

B

Figure 1.2A A 9-in (22.9-cm) solid-state portable monochrome television receiver. This receiver is only 7 in (17.8 cm) deep and can be operated from 120 V AC, a battery pack, or an auto cigarette lighter socket. *(Courtesy of Magnavox Consumer Electronics Company.)* B. A 19-in (48.3-cm) solid-state portable color television receiver with automatic picture brightness and VIR IC circuitry. *(Courtesy of Panasonic Company.)* C. A 25-in (63.5-cm) solid-state console color television receiver with push-button electronic tuning. Note that the receiver has a digital channel number display. *(Courtesy of Quasar Electronics Corp.)*

9-in Monochrome Portable A 9-in (22.9-cm) portable monochrome receiver is shown in Figure 1.2A. This receiver features three-way power. That is, it may be operated from 120 V AC, from a battery pack, or from a car battery using an optional adapter. The receiver is 100% solid-state, except for the picture tube.

19-in Color Portable Figure 1.2B shows a portable 19-in (48.3-cm) color receiver that is for AC operation only. The receiver is solid-state and has an automatic room-light sensing device which adjusts the picture brightness to ambient room-light conditions. An important feature is the VIR (vertical-interval reference) circuitry.[1] (The VIR system is briefly discussed later in this chapter.) VIR circuitry automatically maintains color inten-

[1]See Section 1.3 for a discussion of the VIR system.

sity and hue by reference to a special signal broadcast by the television station.

25-in Color Console Figure 1.2C shows a 25-in (63.5-cm) console color receiver that is all solid-state, including push-button, solid-state (varactor) VHF and UHF tuners. A newly developed "tri-potential" picture tube, with in-line guns and a "black matrix" screen provides excellent picture quality. The same type of picture tube is also used in many other receivers.

The receiver features a "super modular" chassis containing a large percentage of the receiver circuitry. A color correction system, called "dynacolor" automatically corrects errors in flesh tones and color intensity. A "sharpness" control allows the user to regulate the degree of picture detail. An "over-voltage shutdown" circuit assures X-ray protection. An over-voltage sensor turns off the horizontal oscillator (and thus, the high voltage), if the high voltage exceeds a predetermined level.

1.3 USE OF SEMICONDUCTORS IN TELEVISION RECEIVERS

There are a number of reasons why semiconductor devices have replaced tubes in the design of television receivers. For signal application, the transistor is more efficient than a tube because it does not require a heated filament. The fact that there are no filaments also means fewer cooling problems with solid-state operation. Since filament burn-out is the most frequent cause of tube failure, solid-state circuitry is more reliable. Solid-state sets are also smaller and lighter than comparable tube sets. Solid-state components do not require warmup time.

The smaller physical size of semiconductor circuits and the lower cost of printed circuit fabrication makes it possible to use more automatic circuitry to simplify the operation of television receivers. Figure 1.1B shows how modern transistor circuits are fabricated with compact circuit boards. A service kit of television transistors is shown in Figure 1.3A.

INTEGRATED CIRCUITS

One of the important developments in the use of semiconductors is the *integrated circuit* (IC). This is a method of making a number of circuits on a single semiconductor unit called a *chip.* Figure 1.3B shows an integrated circuit. The inset of this illustration shows the chip with connections to the various individual circuits. One chip may contain a complete audio section, and an IF section, or other sections in the receiver. The chip shown in Figure 1.3B contains a complete color demodulator. Other ICs may be used for automatic fine tuning, video signal processing (brightness and sharpness), volume control, and color signal processing.

VIR ICs

Another important use of ICs is for VIR television receiver operation. A non-IC VIR circuit may require as many as 180 components and needs adjustment. The VIR IC designed for the receiver illustrated in Figure 1.2B reduces the number of components by two-thirds and is integrated on an 0.366-in^2 (9.3-cm^2) chip. This IC requires no adjustment and is fabricated in a single-chip, 24-pin, dual-in-line package.

ADVANTAGES OF ICs

ICs are more reliable than transistor circuits because fewer soldered connections are needed for a complete circuit. Their reliability had been proven many times in computer applications before they were used in

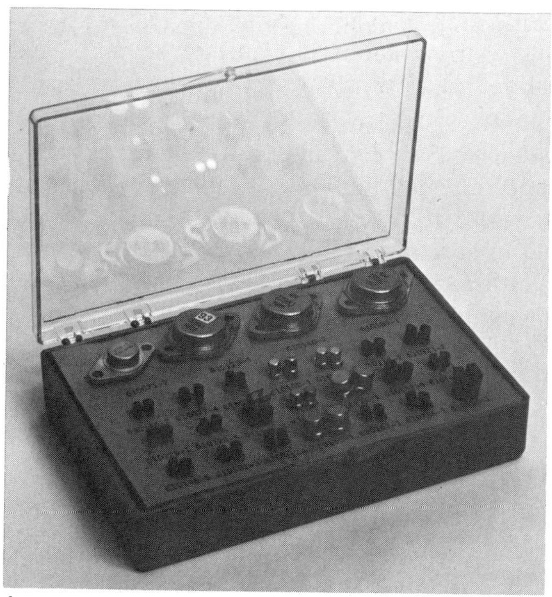

A

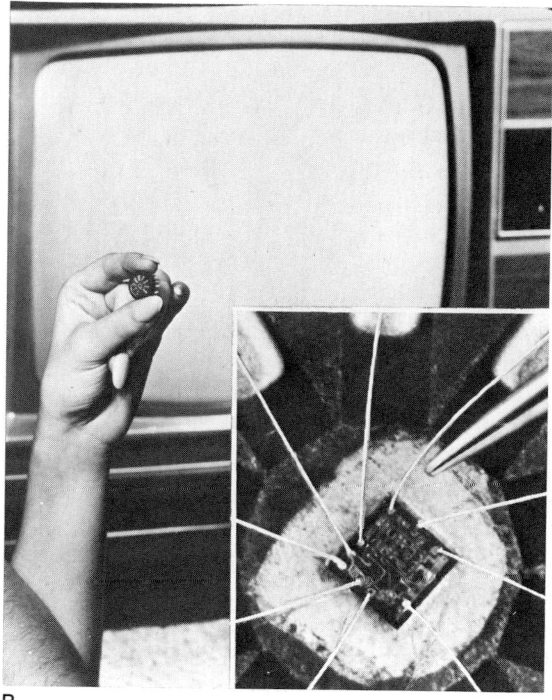

B

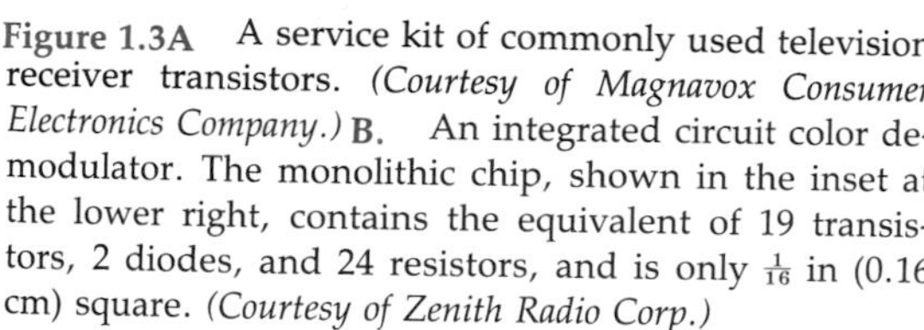

Figure 1.3A A service kit of commonly used television receiver transistors. *(Courtesy of Magnavox Consumer Electronics Company.)* **B.** An integrated circuit color demodulator. The monolithic chip, shown in the inset at the lower right, contains the equivalent of 19 transistors, 2 diodes, and 24 resistors, and is only $\frac{1}{16}$ in (0.16 cm) square. *(Courtesy of Zenith Radio Corp.)*

television receivers. ICs are also cheaper to construct than transistor circuits and are more compact.

1.4 SUMMARY OF DESIRABLE IMAGE CHARACTERISTICS

The image is the final product of the television system and everything centers around its production. The following summarizes the minimum requirements for a satisfactory picture:

1. The composition of the image should be such that none of the elements that go into its make-up are visible from ordinary viewing distances. The image should have the fine, smooth appearance of a good photograph.
2. If a color picture is being displayed, the colors should be realistic. There are many different shades of blue, but most of them would not be suitable for the color of the sky. A good picture will be a reproduction of the colors as they exist in real life.
3. The eye must not be able to perceive a flicker in the reproduced picture. To accomplish this, the television receiver displays sixty fields (30 frames) per second. An advantage of using a high number of fields per second is that the motion on the screen appears to be smooth and continuous.
4. If the picture is to be viewed by more than one person, it should be large

enough so that it can be viewed comfortably by everyone. On the other hand, the picture must not be so large that the viewer can see the lines that make it up.

5. Enough light must come from the picture tube to view the screen by day or by night. Less image brightness is necessary when the room illumination is low than when it is high. Therefore, a brightness control is needed.
6. An effective contrast range is desirable. Contrast is the ratio of maximum to minimum brightness on the same screen. In broad daylight, the contrast ratio between areas in bright sunlight to shaded areas may run as high as 10 000:1. Fluorescent screens normally emit only a limited amount of light. Only contrast ratios up to 100:1 are obtainable but these prove quite satisfactory. To take advantage of the maximum range of contrast values, a contrast control is needed.

1.5 PICTURE DETAIL

Consider the photograph shown in Figure 1.4. This picture was obtained from a negative that contained a large number of chemical grains which were originally sensitive to light. When a photograph is printed in a newspaper or book, it is *half-toned* or divided into very small picture elements. This procedure permits one shade of ink (usually black) to be used to produce the shades of black, gray, and white shown in the picture. When a picture is divided into tiny picture elements it is said to be *half-toned.*

The picture elements of a photograph, or a half-toned illustration, should be so small that they cannot be seen with the naked eye. A fine-grain illustration has many picture elements per unit of area. It can be viewed more closely than a coarse-grain picture before these elements are detected. You can see the picture elements in Figure 1.4 by using a strong magnifying glass.

THE PICTURE TUBE

With television images, the same kind of situation prevails. The television picture is displayed on the screen of a picture tube like the one shown in Figure 1.5. This is a cathode-ray tube somewhat similar to the ones used in oscilloscopes. The electron beam from the gun produces light when it strikes the screen. The amount of light depends

Figure 1.4 Photograph of a color picture tube assembly line developed from a negative. This picture has been half-toned. *(Courtesy of RCA.)*

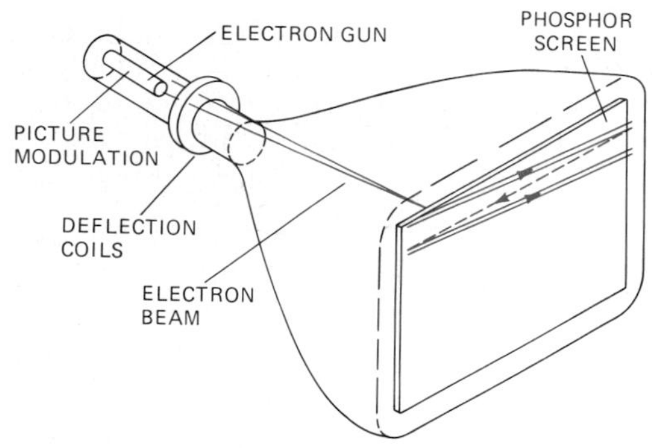

Figure 1.5 Basic method of displaying a television picture on a picture tube.

upon the strength of the beam current. The beam can be moved to any point on the screen by using the magnetic field produced by a deflection current.

In the receiver, each picture element is as large as the area of the circular electron beam striking the fluorescent screen of the picture tube. The light that is seen when observing a picture-tube screen comes from the energy given off by particles of the fluorescent coating on the inner face of the tube, when it is struck by the electron beam. The eye of the observer will merge the points of light if they are closely spaced. Therefore, they will not appear as separate points. An electron beam with a small diameter is required for a detailed television picture to be reproduced. This is required in the receiver picture tube and in the camera tube.

SCANNING LINES

The television picture is produced by moving the electron beam rapidly back and forth across the screen. This produces 525 lines per frame, of which 483 lines are used for picture display information. This procedure is called *scanning.* As the lines are scanned, the beam is made brighter and dimmer in order to "paint" the "active" picture. Ideally, the individual lines making up the picture should not be visible. When they are seen, they distract the viewer and blur the picture.

When 483 lines are crowded together to make a picture that is 8 in high, the lines are not seen easily. But if the same number of lines make up a picture 20 in high, the lines are easily detected unless the viewer moves back some distance from the screen. If the number of lines could be increased in the larger picture, the individual lines would not be seen. However, the Federal Communications Commission (FCC) establishes how many lines can be scanned and this limits the size of the picture.

1.6 BLOCK DIAGRAMS OF THE TELEVISION SYSTEM

Before discussing television circuitry in detail, an overall description of the television system will be given.

THE TELEVISION TRANSMITTER

Figure 1.6 shows a simplified block diagram of a television transmitter. This system involves two transmitters. One is used for transmitting the picture, or *video,* signal.

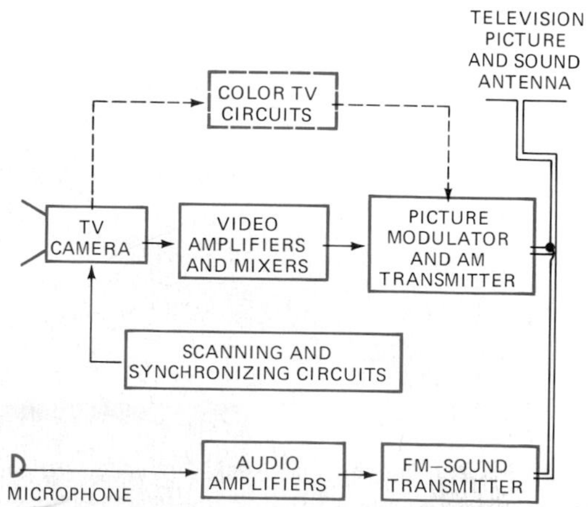

Figure 1.6 Simplified block diagram of a television transmitter.

The other is used for transmitting the sound, or *audio,* signal.

The TV camera converts the scene being televised into electrical impulses which are amplified and transmitted. When transmitting monochrome television only, the monochrome camera(s) utilize one monochrome camera tube in each camera. There are two basic types of color cameras for color television. One type uses a single-color camera tube in each camera. The other type has three pick-up tubes in each camera: one each for the red, green, and blue colors.

Figure 1.7 shows a modern broadcast color camera tube designed for compact studio or hand-held cameras.

TELEVISION TRANSMISSION

The camera tube scans the scene being televised in the same way the picture is scanned in the picture tube of the receiver. Therefore, scanning and synchronizing voltages must be delivered to the camera. A *synchronizing signal* must also be transmitted so that the scene in the receiver can keep in step with the scene at the transmitter. The required scanning and synchronizing signals are also delivered to the video amplifiers and mixers in the transmitter.

The output signals from the video amplifier and mixers are delivered to an amplitude-modulation system. This is an AM transmitter that differs somewhat from the transmitters used in AM broadcast radio. Instead of both sidebands being transmitted, only one full sideband and a part of the other is contained in the output signal. This is called vestigial sideband transmission. By eliminating most of one sideband the electromagnetic spectrum is conserved. Therefore, more stations can transmit in a given range of frequencies. Vestigial sideband transmission should not be confused with single sideband transmission, in which one sideband is completely removed.

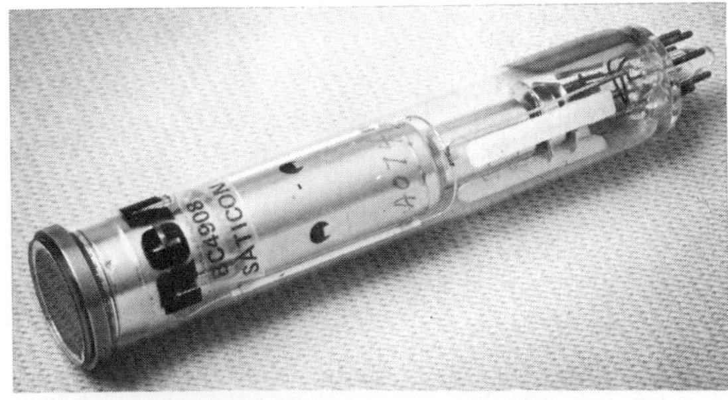

Figure 1.7 A broadcast color camera tube, $\frac{2}{3}$ in (18 mm) in diameter and about 4 in (10.2 cm) long. *(Courtesy of RCA Closed-Circuit Video Equipment.)*

SYSTEM OPERATION

All television broadcast channels, whether VHF (54 to 216 MHz) or UHF (470 to 890 MHz), occupy a bandwidth of 6 MHz. Within this band are broadcast the monochrome video signal, the color video signal, the synchronizing pulses, and the FM sound signal.[1] The "line-of-sight" broadcast distance is about 80 miles (128.7 km) for VHF channels and about 35 miles (56.3 km) for UHF channels.

The studio color cameras and the picture tubes of color receivers can produce red, green and blue colors. However, these are *primary* colors. Thus, by proper combinations, all the other desired colors and white are produced.

Transmission Characteristics

If a color television scene is being transmitted, the output from the color TV camera goes through special color processing circuits. This is shown in dotted lines in Figure 1.6. When the color or "chrominance" signal

[1]For a complete listing of channel frequencies, see "Television Standards," p. 40, and Table 2.2, p. 41.

is being transmitted, a monochrome signal is also being transmitted. This is a requirement which was made when color television was first introduced to the American public. It was necessary that the existing (monochrome) receivers receive and display the signal from a color transmission with no change in quality. In other words, it was necessary that the color and monochrome systems be compatible. Color transmission uses special color TV circuits to superimpose the color signal onto the existing monochrome signal.

Of the 6 MHz-channel, the luminance, or "Y" video signal occupies a total bandwidth of about 5.45 MHz, including the vestigial sideband (1.25-MHz wide). The upper video sideband is 4.2-MHz wide. The sound signal bandwidth is 50 kHz. The balance of the 6 MHz is taken up by "guard bands" to prevent adjacent channel interference. The chrominance-video signal does not take up any significant additional bandwidth. It consists of a chrominance-modulated subcarrier of 3.58 MHz. This, in turn, modulates the picture carrier of each channel. By a clever scheme, the chrominance signal is actually fitted into "spaces" in the frequency spectrum of the luminance signal. Interleaving the two signals in this way is called frequency multiplexing.

THE AUDIO SYSTEM

The audio system in a TV transmitter is completely separate from the video system. It consists of a microphone, audio amplifiers, and an FM transmitter. Both the video and the audio signals are transmitted from the same antenna by using a special coupling device called a *diplexer.*

THE STUDIO CONSOLE

The control console of a modern color television studio is shown in Figure 1.8.

Figure 1.8 Control console of a modern color television studio. Note the video waveforms displayed on the two lower, centrally located monitors. *(Courtesy of RCA Broadcast Systems.)*

Note the convenient location of the various operating controls and the signal and picture monitors.

THE TELEVISION RECEIVER

Figure 1.9 shows a simplified block diagram of a television receiver. The blocks drawn with the solid lines represent the monochrome circuitry necessary for reproducing the picture in black and white. In a color receiver, a special color demodulator (shown in dotted lines) is used. Of course, the picture tube used must also be capable of reproducing the colors.

A modern color receiver picture tube is shown in Figure 1.10. This tube has a screen with a diagonal measurement of 25 in (63.5-cm). In-line electron guns simplify tube adjustments. Note that each color dot is surrounded by a black matrix to improve picture contrast and detail.

There are four sections common to all receivers. These sections are present in the block diagram of Figure 1.9. The four basic sections are an antenna system, a method of selection, a method of detection, and a device for reproducing the intelligence.

The antenna for television receivers is much more elaborate than that used for radio receivers. This is because the television signal is more affected by reflections from large objects, and also because the signal has a more limited distance over which it can travel.

Receiver Operation

The selection function of the receiver is accomplished in the tuner. The tuner makes it possible to select one channel and reject all others. All television receivers used today are superheterodyne types. That is, the tuner section includes a local oscillator and mixer for converting the RF signal to an intermediate frequency (IF).

The output of the tuner is delivered to the IF amplifiers and then to the video detector. Detection in the television receiver actually involves two steps: the video signal is amplitude modulated (requiring an AM detector); the audio signal is frequency modulated (requiring an FM detector). The video and audio signals are amplified by the video and audio amplifiers, respectively. The picture tube reproduces the video signals,

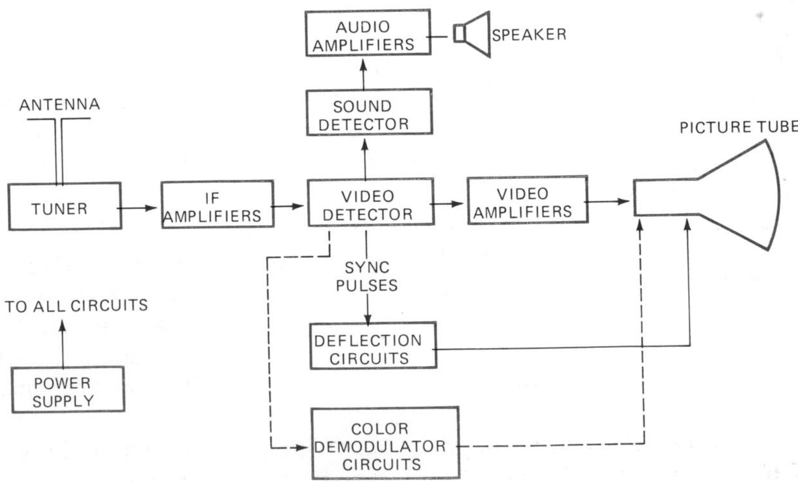

Figure 1.9 Simplified block diagram of a television receiver.

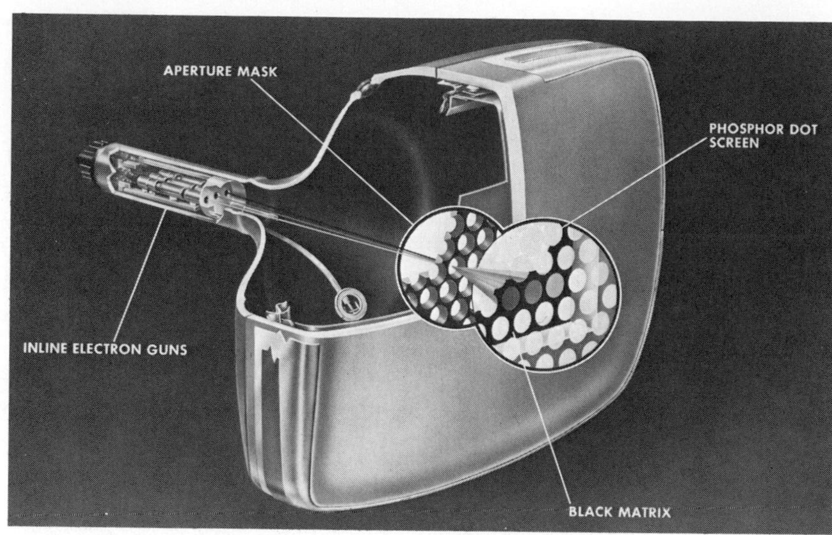

Figure 1.10 Color television picture tube (25-in or 63.5-cm diagonal size) with in-line electron guns. *(Courtesy of General Electric Company.)*

the receiver speaker reproduces the audio signals.

Recall that special signals are combined with the television signal so that the received picture can be synchronized with the one that is transmitted. These special signals are called *synchronizing pulses or sync pulses.* Sync pulses control the frequency and phase of the deflection circuit currents. These currents cause the electron beam to scan the picture tube screen.

If a color signal is present, a special color circuit demodulator is actuated. (This demodulator is shown in dotted lines in Figure 1.9.) An automatic circuit (color killer) in a color television receiver keeps the color demodulators from operating when no color signal is present. The transmitter may switch back and forth between color pictures and monochrome pictures. The receiver control circuits automatically choose the circuitry to produce the monochrome or color picture.

THE PICTURE TUBE

Picture tubes vary in size from 3 in (7.62-cm), to 25 in (63.5-cm) and may be of the monochrome or color type. A typical, large screen, color picture tube is shown in Figure 1.10. This tube has a 25-in (63.5-cm) diagonal measurement. It features "in-line" electron guns, that enable simpler color convergence of the three guns, as well as simpler convergence circuitry.

The color dots are arranged in trios of red, green, and blue. Each of the three guns illuminates one of these primary colors. Other types of color tubes have thin vertical stripes of color, alternating in the order of red, green and blue. These may also utilize the in-line gun principle. White is produced by combining red, green, and blue in the proper proportions. All other colors are reproduced by combining two or three of these primary colors in the correct proportions.

A monochrome picture tube has only one electron gun. Therefore, it produces a black and white picture. The inside of the screen is coated with a uniform phosphor that emits light when the electron beam hits it. The amount of light produced depends on the strength of the beam. This is controlled by the video signal voltage on its control grid.

SUMMARY OF CHAPTER HIGHLIGHTS

1. Television pictures are normally transmitted by radio-frequency waves. In some applications, coaxial cables are used for transmission. (Introduction)
2. Television innovations include color television, all solid-state receivers, satellite television relays, videotape and video-disc players, closed-circuit television, and projection television. (Introduction)
3. Digital techniques are used in color television receivers. The binary number scheme is used with 0 volts and +5 volts as the two "states" of "on" and "off." This technique permits the elimination of the fine-tuning and vertical and horizontal hold controls. It is also used with the vertical-interval reference (VIR) signal to provide color corrections. (Section 1.1)
4. Most modern television receivers are completely solid state. Integrated circuits (ICs) are also widely used. (Section 1.1)
5. Computer-type techniques are used in some receivers for electronic channel selection. The tuners on such receivers have no moving parts. (Sections 1.1 and 1.2)
6. An integrated circuit (IC) chip in a color receiver may contain an audio section, an IF section, a VIR section, or a color demodulator section, among other possible receiver sections. (Section 1.3)
7. The television receiver displays 60 fields (30 frames) per second. This field rate is necessary to prevent flicker from appearing in the picture. (Section 1.4)
8. The amount of picture detail is a function of the area of the electron beam striking the fluorescent screen of the picture tube. It is also a function of the number of lines per frame. However, this is fixed by the FCC at 483 *active* (picture) lines, (525 total lines per frame).
9. All television channels (VHF and UHF) are 6 MHz wide. The luminance, or "Y" video signal, occupies a total bandwidth of 5.45 MHz. The "Y" video frequencies in the upper sideband extend to 4.2 MHz. The color, or chrominance, signal is a modulated subcarrier of 3.58 MHz. Combining red, blue, and green colors in the correct proportions can produce white and all other desired colors. (Section 1.6)
10. The four sections common to all receivers are (1) an antenna system, (2) a method of selection, (3) a method of detection, and (4) a device for reproducing the intelligence. (Section 1.6)

EXAMINATION QUESTIONS

(Answers provided at the back of the text)

The following items are to be answered true (T) or false (F).

1. The abbreviation IC stands for integrated circuit.
2. Digital techniques in television receivers are made practical by the use of printed circuit boards.
3. Push-button tuning is made possible through

the use of computer-type techniques.
4. On some television receivers, the time in hours, minutes, and seconds may be displayed.
5. Conventional television receiver screens measure diagonally from 1 in (2.54 cm) to 25 in (63.5 cm).
6. Digital techniques in television receivers make it possible to eliminate some controls.
7. An integrated circuit chip may contain no more than one circuit stage.
8. To eliminate flicker, there are 30 fields per second.
9. The AM video signal is transmitted with both complete sidebands. However, the FM audio signal is transmitted with vestigial-sideband transmission.
10. All television broadcast channels have a bandwidth of 6 MHz. Of this, the "Y" signal occupies a total bandwidth of about 5.45 MHz.

REVIEW ESSAY QUESTIONS

1. Describe briefly how modern television receivers differ from those made in 1941.
2. What digital information may be displayed on the screen of some television receivers? Is this a continuous display?
3. What is the meaning of "random access"?
4. Describe briefly how digital techniques are used in television receivers.
5. Define a binary number.
6. What is the range of screen sizes on modern television receivers?
7. Describe briefly the function of VIR circuits.
8. Discuss the advantages of using ICs in television receivers. Give at least five applications.
9. On a picture tube, what is the general size of a picture element?
10. What is the general function of a camera tube? Why are three tubes used for studio color television cameras?
11. What is the approximate line-of-sight broadcast distance for VHF television stations?
12. What is meant by "multiplexing", when referring to the luminance and chrominance signals?
13. A color picture tube shows basically red, blue, and green colors. How then is a full color picture displayed?

Chapter 2

2.1 Television systems that use cables
2.2 Extending the coverage of television
2.3 Videotape and video-disc recording
2.4 Video games
2.5 Projection television
2.6 Alpha-numeric and graphic displays
2.7 Summary of U.S. television standards
2.8 Foreign television standards
Summary

TELEVISION SYSTEM APPLICATIONS

INTRODUCTION

Chapter 1 presented the basic concepts of the television system. In addition, some of the popular applications were listed. In this chapter, the more common television system applications will be described in greater detail. In addition, a summary of U.S. and foreign television standards is presented at the end of the chapter.

When you have completed the reading and work assignments for Chapter 2, you should be able to:

- Define the following terms: cable television, closed-circuit television, master-antenna television, community-antenna television, directional-coupler multitap, microwave relay station, and premodulator processing.
- List the major components of the basic cable television system.
- Describe the components of the closed-circuit television system and discuss two applications of the system.
- Explain briefly how satellites produce wide area television coverage.
- Summarize the major problems associated with video-tape recording.
- List two video-disc systems.
- Describe the function of the "game chip" and digital video pulses in video games.
- List the major subsections of a video game and briefly describe the function of each.
- Describe the characteristics of one-tube and three-tube projection television systems.
- Summarize the major U.S. and foreign television standards.

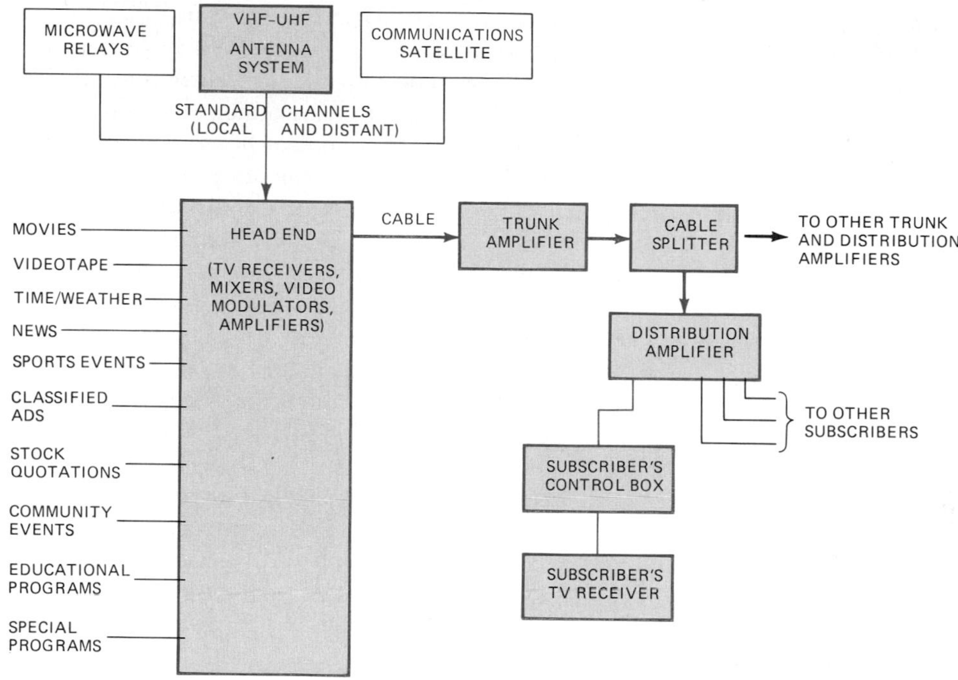

Figure 2.1 Simplified block diagram of the cable television (CATV) system.

2.1 CABLE TELEVISION SYSTEMS

Cable television systems fall into the following four categories:

1. Cable television (CATV)–a combination of radio-wave transmission and reception and a radio-frequency cable distribution system.
2. Closed-circuit television (CCTV)–operates through cables only; no radiowave transmission is involved. There are only video signals here.
3. Master-antenna television (MATV)–this system permits the operation of a large number of television receivers from a single antenna system.
4. Community-antenna television–as in CATV (Cable), this is a combination of radio-wave transmission and reception and a cable-distribution system. This system is useful for special applications, as in the case of a community located behind a mountain or in a valley. In such cases, very poor signals would otherwise be available to viewers.

CABLE TELEVISION (CATV)

This system is also known as "Cablevision." The basic system is shown in the block diagram of Figure 2.1. It consists of four main components (1) the antenna system, (2) the head-end system, (3) the trunk-amplifier system, and (4) the distribution-amplifier system.

The Antenna System An elaborate antenna system is erected in a ghostfree area of good signal strength. This system picks up local VHF and UHF channels. In addition, by means of microwave relays and communication satellites, the antenna receives selected channels from distant parts of the country.

At a number of cable-vision stations, the antenna system is located at a high elevation. Here, the received signals are beamed to a central distribution station by means of microwaves.

The Head-End System The head-end system receives the signals from the antenna system. A number of other locally generated signals, as shown in Figure 2.1, are also fed into the head-end system. Within this system, VHF channels (2 to 13) are processed and sent out on a cable at their original VHF frequencies. Selected UHF channels are heterodyned down to a designated cable VHF or UHF frequency. Other program inputs modulate specific cable frequencies. All of the above channels are combined in a mixer system and are fed to the first trunk amplifier.

The Trunk-Amplifier System

The trunk-amplifier system carries all cable channels through the coaxial cable to the various areas served by the cable system. Amplifiers and equalizers maintain the signal level and proper frequency response. Trunk amplifiers are spaced from 1 400 to 2 000 feet apart. It is possible to have up to 50 trunk amplifiers in cascade, depending upon the number of channels being carried.

The Distribution-Amplifier System

The output of a trunk amplifier feeds a distribution amplifier (bridging amplifier). Up to four coaxial lines may be tapped off from a distribution amplifier to serve individual subscribers. A commonly used tapoff device is the *directional-coupler multitap*. The output of each tap feeds a 75-ohm coaxial cable which goes to the subscriber's house. With TV receivers having a 300-ohm antenna input, a transformer (at the receiver) is used to convert the unbalanced 75-ohm cable impedance to a balanced 300-ohm antenna input.

Heterodyning Down At the TV Receiver The CATV channels arrive at the subscriber's home with a variety of VHF and UHF frequencies. However, the subscriber can only tune the TV set to Channel 3 or Channel 4. Consequently, equipment is provided at each TV set to heterodyne every CATV channel to Channels 3 or 4. The cable company selects whichever of these two channels is not active in the particular area as the output channel in each home. The subscriber then tunes the TV set to the selected channel and leaves it there for all CATV reception.

System Matching Proper "impedance" matching is required throughout the CATV system to avoid line reflections which impair picture quality. A system such as the one shown in Figure 2.1 supplies each subscriber with strong, ghostfree signals.

Remote-Control Boxes A cable television installation may make available as many as 66 or more channels. This requires a 450-MHz bandwidth system. Table 2.1 lists the frequency assignments for cable television channels. A remote-control box with push buttons or digital-control buttons for each channel is provided for each subscriber television receiver. This box may be wired or wireless. The subscriber receiver is tuned to either Channel 3 or Channel 4. All channel selecting is then done through the remote-control box.

Typical remote-control boxes are shown in Figure 2.2. Figure 2.2A shows a 36-channel, 300-MHz bandwidth, wired, push-button, remote-control box. Figure 2.2B shows a 58-channel, digital set-top, remote-control box, which can also be operated from a wireless, infrared transmission, hand-held control box.

Some models of TV receivers have cable TV tuning capability. This eliminates the necessity of separate cable TV channel-selecting equipment. For example, one Panasonic TV receiver has a 105-channel CATV push-button tuner. Other TV receivers with this facility may have the capability of tuning fewer CATV channels.

In addition to providing more channels of interest, each subscriber receives every channel clearly. Unedited, first-run movies are sent over some channels without commercial interruption. These channels may be used solely for this purpose, or for special entertainment programs as well.

A

B

Figure 2.2 Remote control boxes used with home cable television installations. **A.** A 36-channel, 300-MHz bandwidth, set-top wired unit. **B.** A 58-channel, digital set-top wired unit and hand-held wireless remote-control unit. *(Courtesy of Jerrold Division, General Instrument Corp.)*

Other Uses For Cable Television The uses for cable television systems are always expanding. The following describes some new applications.

With a two-way CATV system, the subscriber may request or provide various types of information. Two-way communications can be used for taking polls on desired topics. Other services include home security, home banking, credit card transactions and home shopping.

Subscribers can obtain information and entertainment (movies, sports, etc.) services on request. Microcircuit technology makes possible electronic text-delivery services, such as electronic newspapers, classified ads, and complete programmed instruction courses.

One use for two-way CATV systems is for subscriber-callable video games. Games such as NFL Football, Major League Baseball and NBA Basketball are available in some areas. Cable system operators store the programs in minicomputers and transmit them to subscribers on request. A special terminal connected to the subscriber's TV set is required for all two-way cable television services.

Table 2.1 **FREQUENCY ASSIGNMENT FOR CABLE TELEVISION CHANNELS.**

35-CHANNEL FREQUENCY ASSIGNMENT		61-CHANNEL INCREMENTAL FREQUENCY ASSIGNMENT				69-CHANNEL HARMONIC FREQUENCY ASSIGNMENT			
CHANNEL DESIGNATION	*FREQUENCY RANGE MHz*	*CHANNEL DESIGNATION*	*FREQUENCY RANGE MHz*	*PICTURE CARRIER MHz*	*SOUND CARRIER MHz*	*CHANNEL DESIGNATION*	*FREQUENCY RANGE MHz*	*PICTURE CARRIER MHz*	*SOUND CARRIER MHz*
2	54–60	2	54–60	55.25	59.75	2H	52.75–58.75	54	58.5
3	60–66	3	60–66	61.25	65.75	3H	58.75–64.75	60	64.5
4	66–72	4	66–72	67.25	71.75	4H	64.75–70.75	66	70.5
5	76–82	5*	76–82	77.25	81.75	5H**	76.75–82.75	78	82.5
6	82–88	6*	82–88	83.25	87.75	6H**	82.75–88.75	84	88.5
A	120–126	14I	120–126	121.25	125.75	14H	118.75–124.75	120	124.5
B	126–132	15I	126–132	127.25	131.75	15H	124.75–130.75	126	130.5
C	132–138	16I	132–138	133.25	137.75	16H	130.75–136.75	132	136.5
D	138–144	17I	138–144	139.25	143.75	17H	136.75–142.75	138	142.5
E	144–150	18I	144–150	145.25	149.75	18H	142.75–148.75	144	148.5
F	150–156	19I	150–156	151.25	155.75	19H	148.75–154.75	150	154.5
G	156–162	20I	156–162	157.25	161.75	20H	154.75–160.75	156	160.5
H	162–168	21I	162–168	163.25	167.75	21H	160.75–166.75	162	166.5
I	168–174	22I	168–174	169.25	173.75	22H	166.75–172.75	168	172.5
7	174–180	7	174–180	175.25	179.75	7H	172.75–178.75	174	178.5
8	180–186	8	180–186	181.25	185.75	8H	178.75–184.75	180	184.5
9	186–192	9	186–192	187.25	191.75	9H	184.75–190.75	186	190.5
10	192–198	10	192–198	193.25	197.75	10H	190.75–196.75	192	196.5
11	198–204	11	198–204	199.25	203.75	11H	196.75–202.75	198	202.5
12	204–210	12	204–210	205.25	209.75	12H	202.75–208.75	204	208.5
13	210–216	13	210–216	211.25	215.75	13H	208.75–214.75	210	214.5
J	216–222	23I	216–222	217.25	221.75	23H	214.75–220.75	216	220.5
K	222–228	24I	222–228	223.25	227.75	24H	220.75–226.75	222	226.5
L	228–234	25I	228–234	229.25	233.75	25H	226.75–232.75	228	232.5
M	234–240	26I	234–240	235.25	239.75	26H	232.75–238.75	234	238.5
N	240–246	27I	240–246	241.25	245.75	27H	238.75–244.75	240	244.5
O	246–252	28I	246–252	247.25	251.75	28H	244.75–250.75	246	250.5
P	252–258	29I	252–258	253.75	257.75	29H	250.75–256.75	252	256.5
Q	258–264	30I	258–264	259.25	263.75	30H	256.75–262.75	258	262.5
R	264–270	31I	264–270	265.25	269.75	31H	262.75–268.75	264	268.5
S	270–276	32I	270–276	271.25	275.75	32H	268.75–274.75	270	274.5
T	276–282	33I	276–282	277.25	281.75	33H	274.75–280.75	276	280.5
U	282–288	34I	282–288	283.25	287.75	34H	280.75–286.75	282	286.5
V	288–294	35I	228–234	289.25	293.75	35H	286.75–292.75	288	292.5
W	294–300	36I	294–300	295.25	299.75	36H	292.75–298.75	294	298.5

CLOSED-CIRCUIT TELEVISION (CCTV)

Another system that uses cables is called **closed-circuit television** (CCTV). In this system, the signal is transmitted along a coaxial cable from the camera to one or more monitors. The general public cannot receive the signal. In some applications, the camera is mounted at some point where it would be inconvenient or impossible for a human to be, such as near an atomic energy experiment or at great depths in the ocean.

The signal for closed-circuit television is not radiated into space, therefore, it does not come under the rules and regulations of the FCC. Much less power is required to operate a closed-circuit television system than a cable television system. Large RF power amplifiers are not required. Therefore, the system has a lower initial cost and lower operating cost. Another advantage is the privacy of the system. A set that is not connected to the cable cannot receive the signal.

CCTV Components

The composite photo of Figure 2.3 shows a group of typical CCTV components. These are described as follows (the letters identify the individual components in the photo):

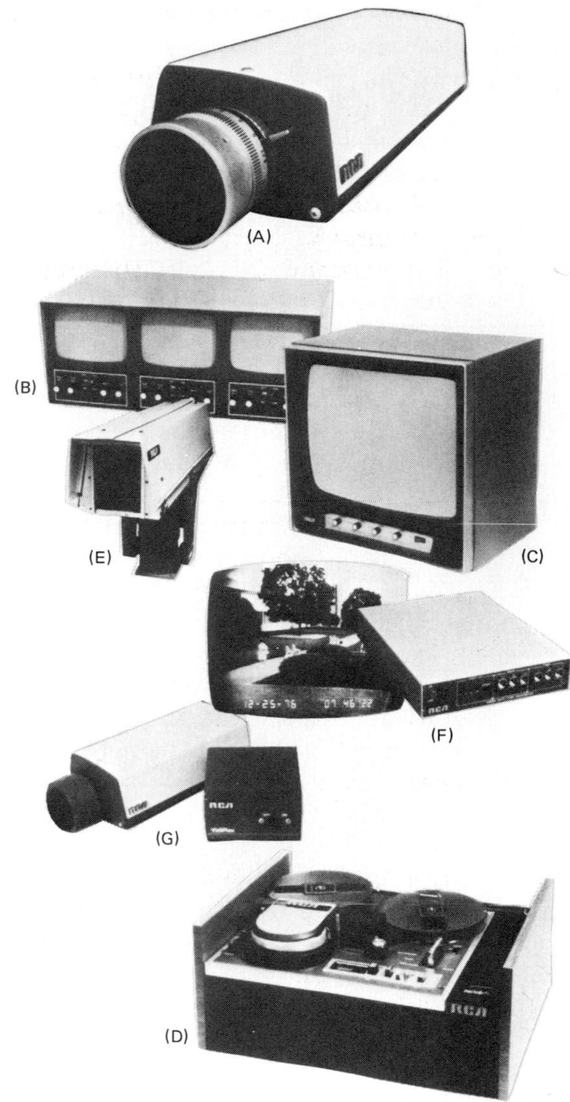

Figure 2.3 A group of typical closed-circuit television (CCTV) components. *(Courtesy of RCA Closed Circuit Video Equipment.)*

A. **CCTV camera** can be equipped with either a "silicon target" vidicon, or a "Newvicon" camera tube. These tubes provide useable pictures at fairly low light levels. Another camera, utilizing an RCA silicon intensifier target (SIT) tube can provide useable pictures with the illumination of only one-quarter moonlight.

B. and C. **CCTV Monitors.** Item B is a triple set of monitors with a 6-in (15.24-cm) picture. This triple set can be used for convenient viewing of the pictures from three dif cameras, each at a differe tion. Item C has a 17-in picture tube and pro picture detail than th cm) monitors. Othe

CHANNEL DESIGNATION	*FREQUENCY RANGE MHz*	*PICTURE CARRIER MHz*	*SOUND CARRIER MHz*
37I	300–306	301.25	305.75
38I	306–312	307.25	311.75
39I	312–318	313.25	317.75
40I	318–324	319.25	323.75
41I	324–330	325.25	329.75
42I	330–336	331.25	335.75
43I	336–342	337.25	341.75
44I	342–348	343.25	347.75
45I	348–354	349.25	353.75
46I	354–360	355.25	359.75
47I	360–366	361.25	365.75
48I	366–372	367.25	371.75
49I	372–378	373.25	377.75
50I	378–384	379.25	383.75
51I	384–390	385.25	389.75
52I	390–396	391.25	395.75
53I	396–402	397.25	401.75
54I*	72–78	73.25	77.75
55I*	78–84	79.25	83.75
56I*	84–90	85.25	89.75
57I	90–96	91.25	95.75
58I	96–102	97.25	101.75
59I	102–108	103.25	107.75
60I	108–114	109.25	113.75
61I	114–120	115.25	119.75

*Use either 5 and 6 or 54I, 55I, and 56I.

CHANNEL DESIGNATION	*FREQUENCY RANGE MHz*	*PICTURE CARRIER MHz*	*SOUND CARRIER MHz*
37H	298.75–304.75	300	304.5
38H	304.75–310.75	306	310.5
39H	310.75–316.75	312	316.5
40H	316.75–322.75	318	322.5
41H	322.75–328.75	324	328.5
42H	328.75–334.75	330	334.5
43H	334.75–340.75	336	340.5
44H	340.75–346.75	342	346.5
45H	346.75–352.75	348	352.5
46H	352.75–358.75	354	358.5
47H	358.75–364.75	360	364.5
48H	364.75–370.75	366	370.5
49H	370.75–376.75	372	376.5
50H	376.75–382.75	378	382.5
51H	382.75–388.75	384	388.5
52H	388.75–394.75	390	394.5
53H	394.75–400.75	396	400.5
54H	70.75–76.75	72	76.5
55H**	76.75–82.75	78	82.5
56H**	82.75–88.75	84	88.5
57H	88.75–94.75	90	94.5
58H	94.75–100.75	96	100.5
59H	100.75–106.75	102	106.5
60H	106.75–112.75	108	112.5
61H	112.75–118.75	114	118.5
62H	400.75–406.75	402	406.5
63H	406.75–412.75	408	412.5
64H	412.75–418.75	414	418.5
65H	418.75–424.75	420	424.5
66H	424.75–430.75	426	430.5
67H	430.75–436.75	432	436.5
68H	436.75–442.75	438	442.5
69H	442.75–448.75	442	448.5

**5H and 6H are same as 55H and 56H.

The 35-channel frequencies at the left are being superceded by the other frequencies shown

different picture tube sizes are also available.

D. **The time-lapse videotape recorder** can provide time-lapse, video recordings for periods ranging from 11 to 99 hours. This component is valuable for viewing events over long periods of time.

E. **The pan and tilt unit** is a heavy-duty model for outdoor use. The CCTV camera is firmly mounted inside the unit. It can then be panned through 360 degrees and tilted up and down through large angles.

F. **The time and date generator** has a built-in calendar and tracks days, months and years automatically. The date and the time in hours, minutes, and seconds are displayed in digital form on the monitor.

G. **One-cable Camera.** This view shows the camera and its special power supply. The units are designed for reduced installation cost and elimination of conduit and utility wiring. This camera is ideal for use in hard-to-reach places. Only one cable—the video cable—connects the power supply to the camera. Power is also fed through this same cable. The units may be separated by as much as 3,000 feet.

Uses of CCTV

The use of videotape recordings in conjunction with many CCTV installations means that permanent records can be kept of crimes, medical operations, training material, and any other occurrences requiring review. The development of transistor and IC video circuits has led to widespread educational, industrial, and private use of CCTV systems. The following discusses some representative ones.

Educational Uses A small closed-circuit camera may be located close to a laboratory experiment, so that a large number of students can view it simultaneously. A lecturer, school principal, or board president may address a large number of people in many different locations simultaneously. An important area of application is in vocational and technical schools. Here, the use of expensive and complex equipment may be demonstrated to many students, simultaneously. Students may also practice television acting, programming, and directing with a low-budget television system.

Industrial Uses Closed-circuit television cameras can be mounted to monitor many meters and gauges from a single remote position. Close-up views of hazardous operations, such as very high-temperature or very low-temperature experiments and machine handling of radioactive materials, can be monitored continuously from a remote position. Detection of burglary, fires starting in remote positions, assembly line failure, and so on, is an easy matter with strategically located closed-circuit cameras.

Private Uses As in industry, protection of personal property against burglary and trespassing can justify the relatively low cost of a closed-circuit television system. A camera near a baby's crib will allow a parent in another part of the house to keep an eye on the child. When a camera is mounted outside the door of a house or an apartment, the resident can identify callers before opening the door.

Crime Prevention CCTV is widely used to assist in crime prevention. Installations may be found in banks, department stores, jewelry stores, apartment house lobbys, and various other sensitive locations.

Medical Uses The field of medicine finds

important uses for CCTV installations. Operations may be shown in color to a large number of students or interested physicians. The observer can be located at some distance from the operating room. Patients in intensive care units or other locations can also be monitored using closed-circuit television.

Other CCTV Uses Other closed-circuit television uses are as follows: 1. Live and prerecorded programs can be used in business to train personnel and observe their performance. 2. In freight yards CCTV is used to observe and assist personnel in making up and routing trains. 3. CCTV permits the exploration of normally inaccessible places.

The almost unlimited number of applications of closed-circuit television means additional employment for television technicians. The circuitry for both the camera and the monitor is less complicated than the systems used in television broadcasting.

Figure 2.4 Microwave relays are used to transmit television signals over long distances. *(Courtesy of Western Electric.)*

2.2 EXTENDING THE COVERAGE OF TELEVISION

Television signals are transmitted in the range of frequencies between 54 and 890 MHz. This range includes frequencies in the very high frequencies (VHF) and in the ultra high frequencies (UHF).

Unlike the lower frequencies used for AM broadcast radio, these frequencies are not reflected from the ionosphere. Their transmission is limited to *line-of-sight distances.* The *average* distance from the transmitter to the receiver antennas is about 45 miles (72.4 km). This is an average value because the terrain between the transmitter and the receiver antennas, the frequency used, and the heights of the antennas must be taken into consideration.

The use of Cablevision and community antenna systems can achieve transmissions well beyond the line-of-sight. Both of these systems use coaxial transmission lines. There are also methods of greatly extending the coverage of television, other than by the use of cables. These will now be discussed.

MICROWAVE RELAYS

Microwave relay stations are used to transmit signals coast-to-coast and throughout the United States, (Figure 2.4). These stations receive the signal, amplify it, and then retransmit it. They are located at intervals of 50 to 100 miles (80.5 to 161 km). They operate in the range of 4 GHz to 11 GHz. In addition, domestic satellites are also used to

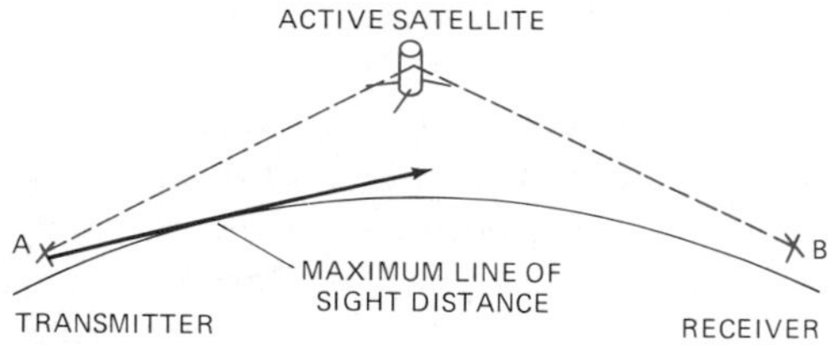

Figure 2.5 Active satellites are used to extend television coverage. Note that the distances involved are far greater than the line of sight.

provide television coverage for the entire United States, Puerto Rico, Alaska and Hawaii.

USE OF SATELLITES

To greatly extend the range of television programs (as well as voice and data communications), active satellites are used. They permit domestic transmissions and worldwide transmissions. An active satellite receives a signal, amplifies it, and then retransmits it. It differs from a passive satellite that echoes the signal by reflecting it back to Earth.

Figure 2.5 shows how the active satellite works. The television signal that is transmitted from point A is received by the satellite and retransmitted to the station at point B. Note the line-of-sight distance over which transmission is normally achieved.

Domestic Satellites

Both the United States and Canada now employ domestic satellites to provide television, communications, and data coverage from their territories. For the United States, this coverage includes Puerto Rico, Hawaii and Alaska. The United States domestic communications satellite system is known as Westar and is operated by Western Union. (The satellites are built by the Hughes Aircraft Company.) The satellite coverage is illustrated in Figure 2.6. Actually, two satellites are used to provide increased capacity and also backup capability in case of failure.

Each Westar (12-transponder) satellite has a capacity of 7 000 voice circuits, or of 12 simultaneous color television channels. It can also relay private messages and data communications. The Westar satellites are in

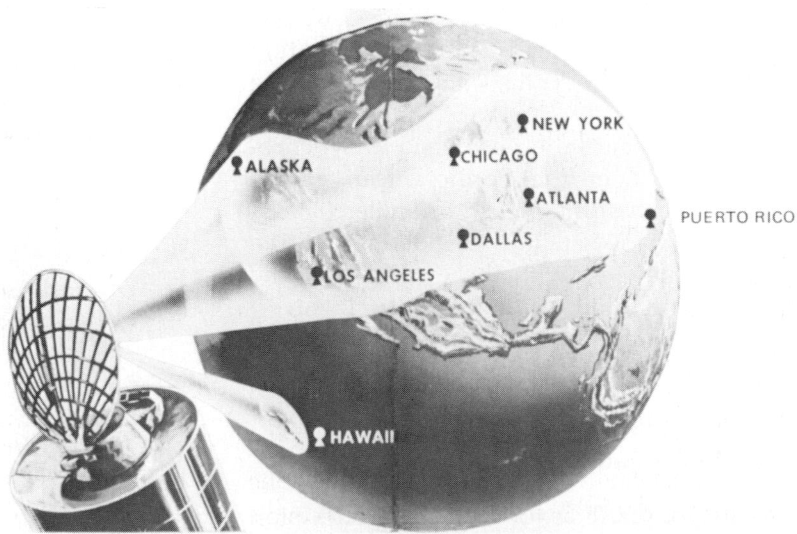

Figure 2.6 The area covered by the Westar Domestic Satellite (the satellite is 11.6 ft (3.54 m) in height and 6.25 ft (1.91 m) in diameter). *(Courtesy of Hughes Aircraft Company.)*

"synchronous orbit" over the Pacific Ocean, at an altitude of 22 300 miles (35 881 km) above the equator. The synchronous orbit means the satellites are positioned (and readjusted by earth commands) in a fixed location in space relative to the earth.

Westar Earth Stations

The ground network of the Westar system consists of five strategically located earth stations. These are located near the cities of New York, Atlanta, Chicago, Dallas, and Los Angeles. The earth stations relay satellite traffic to regional Western Union facilities in each serving city. In addition, they will interconnect with the company's 8 000-mile (12 872-km) ground microwave network to distribute voice, video, and data traffic.

Canadian Domestic Satellites

The Canadian Satellite System consists of three Anik (Eskimo for "brother") synchronous satellites and more than 20 Canadian earth stations spread across Canada. Transmission is limited to Canada. The system is operated by Telesat Canada, a private company. Interconnection to Telesat earth stations is provided by user systems, such as the Canadian Broadcasting Corporation (CBC) and the Trans-Canada Telephone System.

The satellites are positioned in synchronous orbit over the Pacific Ocean at the equator at an altitude of 22 300 miles (35 881 km). The Anik satellites are identical to the Westar satellites, with the exception of the antenna feedline configuration. However, they have the same communications capabilities. Both types operate on solar power (300 W).

Worldwide Satellites

The use of satellites for worldwide television (and other data) transmission began with the medium altitude repeater, Telestar, in 1962. This was followed in 1963 by Syncom, the world's first synchronous satellite, which was positioned over the Atlantic Ocean. This satellite had a capacity for only 50 two-way voice channels or one TV channel. Syncom was followed by a series of more sophisticated satellites having much greater communications capabilities. The more recent worldwide satellites with color television capability are the Intelsat IV (Figure 2.7) and the Intelsat IVA series. These satellites are positioned in synchronous orbit over the Atlantic, Pacific, and Indian Oceans.

Each Intelsat IVA satellite has a capacity for 11 000 voice channels or 20 simultaneous color TV channels. All of the worldwide satellites described to this point were built by the Hughes Aircraft Company for

Figure 2.7 Intelsat IV worldwide communications satellite. It is about 18 ft (5.5 m) high and 8 ft (2.44 m) in diameter. The spotbeam antennas can aim signals into high communications traffic areas. *(Courtesy of Hughes Aircraft Company.)*

the International Telecommunications Satellite Organization (INTELSAT), consisting of 82 nations and directed by the Communications Satellite Corporation (COMSAT). The satellites relay TV or other data to 111 ground stations strategically situated around the world.

TV transmission standards and color TV standards differ in the United States and Europe. However, the ground stations convert from one standard to the other, as required. A summary of these standards is given at the end of this chapter.

2.3 VIDEO-TAPE AND VIDEO-DISC RECORDING

The television video signal may be recorded and played back by a magnetic tape system. The basic principle is similar to that used to record and play back audio frequencies. Video-tape recorder (VTR) and playback units are used for home entertainment and for industrial, educational, and broadcast functions.

The problems associated with video recording are much greater than for audio recording. The tape must not only record the television FM sound, but also the video luminance and chrominance signals. In addition, a control signal is needed to achieve correct playback.

PROBLEMS OF VIDEO-TAPE RECORDINGS

Three major problems of video-tape recording arise from (1) the high ratio of the lowest to the highest frequencies; (2) the high upper frequency limit, and (3) the noise and amplitude distortion which may result.

Audio frequencies cover a range of about 20 to 20 000 Hz, or about 10 octaves (an octave is double or half frequency). This range can be readily equalized to provide a flat-frequency response on the tape. However, the video range of about 30 Hz to 4 MHz, is almost 18 octaves. This range cannot be equalized in a practical manner.

RECORDING VIDEO FREQUENCIES

The high video frequencies (to 4 MHz in studio recorders and to 2.5 MHz in portable or home recorders) can be recorded by using a very narrow video-head gap and high effective head-to-tape speeds (writing speeds). Head gaps of 2×10^{-5} in (5.1×10^{-5} cm) or less are employed. One method of reducing tape consumption consists of relatively slow-moving tape passing over a rapidly rotating video head. In broadcast studios, fast writing speeds are obtained by the use of either four rotating heads or a large-sized helical drum. With the helical drum, the tape records diagonally across the width of the 1-in (2.54-cm) studio tape. These machines can record and play back the NTSC color-luminance video signal directly.

RECORDING SCHEME FOR PORTABLE OR HOME RECORDERS

A different recording scheme is used in portable or home recorders, which must use somewhat lower writing speeds. Here, the luminance bandwidth is limited to about 2.5 MHz. The chroma component (at 3.58 MHz) is separated and reduced in frequency to 629 kHz or 688 kHz, by heterodyning with a 4.209-MHz or 4.268-MHz crystal oscillator. Both luminance and chrominance signals are recorded in the same tape track. The tape used in these recorders is $\frac{1}{2}$ in (1.27 cm) wide.

The problems of video-luminance frequency equalization and of noise and ampli-

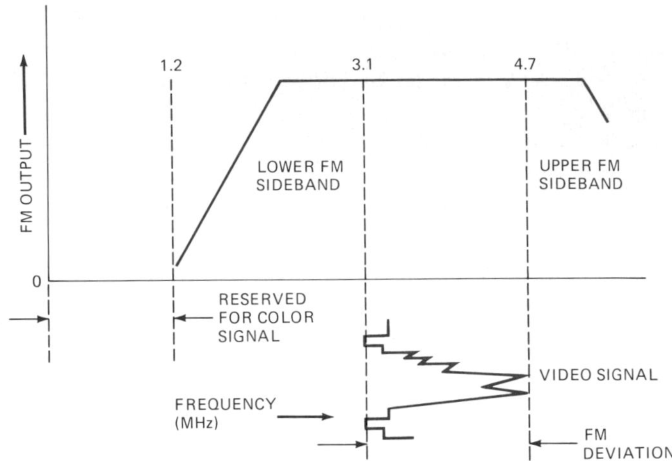

Figure 2.8 In a video-cassette recorder, the AM video luminance signal is converted to an FM signal; the chroma signal is reduced to 629 kHz or 688 kHz.

tude distortion are solved by changing the luminance signal (30 Hz to 2.5 MHz) into an FM signal.

Luminance Signal Converted To FM

A typical FM conversion for a home videocassette recorder is illustrated in Figure 2.8. Here the FM deviation is from 3.1 to 4.7 MHz. Note that each level of the video signal is converted to a specific frequency. Thus, the tip of sync is at 3.1 MHz and the peak white level is at 4.7 MHz. Intermediate amplitude levels produce corresponding frequencies. In the playback process, the converted FM-luminance signal is demodulated and reverts to the original amplitude-varying signal.

This FM signal is applied to the videotape head. The bandwidth is now less than one octave and is equalized readily. (An octave is one-half or twice the original frequency.) The normal characteristic of an FM signal (constant amplitude) makes possible low-noise and low-amplitude distortion of the recording.

HOME VIDEO RECORDER

Home entertainment video recorders can record television programs off the air in monochrome or color. These programs can be played back through any standard television receiver. In addition, combination video and sound cameras (color or monochrome) can be attached to the recorder to produce original video and sound recordings. Figure 2.9A shows a home video recorder of the Video Home System (VHS) type. A combination video and sound camera for use with a home video recorder appears in Figure 2.9B. This type of recorder features up to 6-hour record/play capability on a cassette. It has built-in VHF and UHF tuners and a built-in electronic timer. The tuners permit the user to view one program on a television receiver, while recording another program on a different channel. The automatic timer provides for unattended recording.

BLOCK DIAGRAM OF THE RECORD-PLAYBACK SYSTEM

The block diagram is shown in Figure 2.10. The composite-video signal (including chroma) is sent to four paths. In the record condition, the upper path sends the composite video signal directly to the TV receiver. This enables the user to monitor the program being recorded.

A

B

Figure 2.9A. A VHS (video home system) video-tape recorder/player (VTR) that uses cassettes and features four-hour capability, integral tuners, and a built-in electronic timer. *(Courtesy of Magnavox Consumer Electronics.)* B. Combination monochrome, video, and sound camera for use with a video-tape recorder (VTR). Color cameras are available also. *(Courtesy of Magnavox Consumer Electronics.)*

Premodulator Processing

The path immediately below the one just described enters the *premodulator-processing circuits.* This is the luminance-signal path. The first group of circuits encountered provides *premodulator processing.* Here, corrections are applied to the composite-video signal prior to the FM modulator. This processing includes (1) reducing the video bandwidth below 3.5 MHz, (2) AGC for the video signal, (3) pre-emphasis to improve signal-to-noise ratio, and (4) FM-deviation clamping.

FM Modulator

Following premodulator processing, the luminance signal (only) is applied to the FM modulator. The FM modulators used in all VTRs are voltage-controlled oscillators (VCOs). For home VTRs, the VCOs are either astable or free-running multivibrators. Waveshape is unimportant here since the tape is driven to magnetic saturation in both directions. The VCOs are part of IC chips.

The output of the VCO is an RF square wave which varies in frequency in accordance with the luminance-video signal. This FM wave is then applied to the record amplifier.

Record Amplifier

The record amplifier is a power amplifier. Its function is to supply sufficient current to the video-head to saturate the tape magnetically. The record amplifier also provides some frequency equalization. A high-pass filter is placed between the FM modulator and the record amplifier. (This filter is not shown in Figure 2.10.) This prevents FM sidebands from existing below about 1 MHz where the chroma signals (reduced in carrier frequency) exist.

Record-Chroma Processing

As shown in Figure 2.10, the processed chroma signal is fed to the video head together with the processed luminance signal. The combination signal forms a composite video signal suitable for tape recording. However, as mentioned above, the color

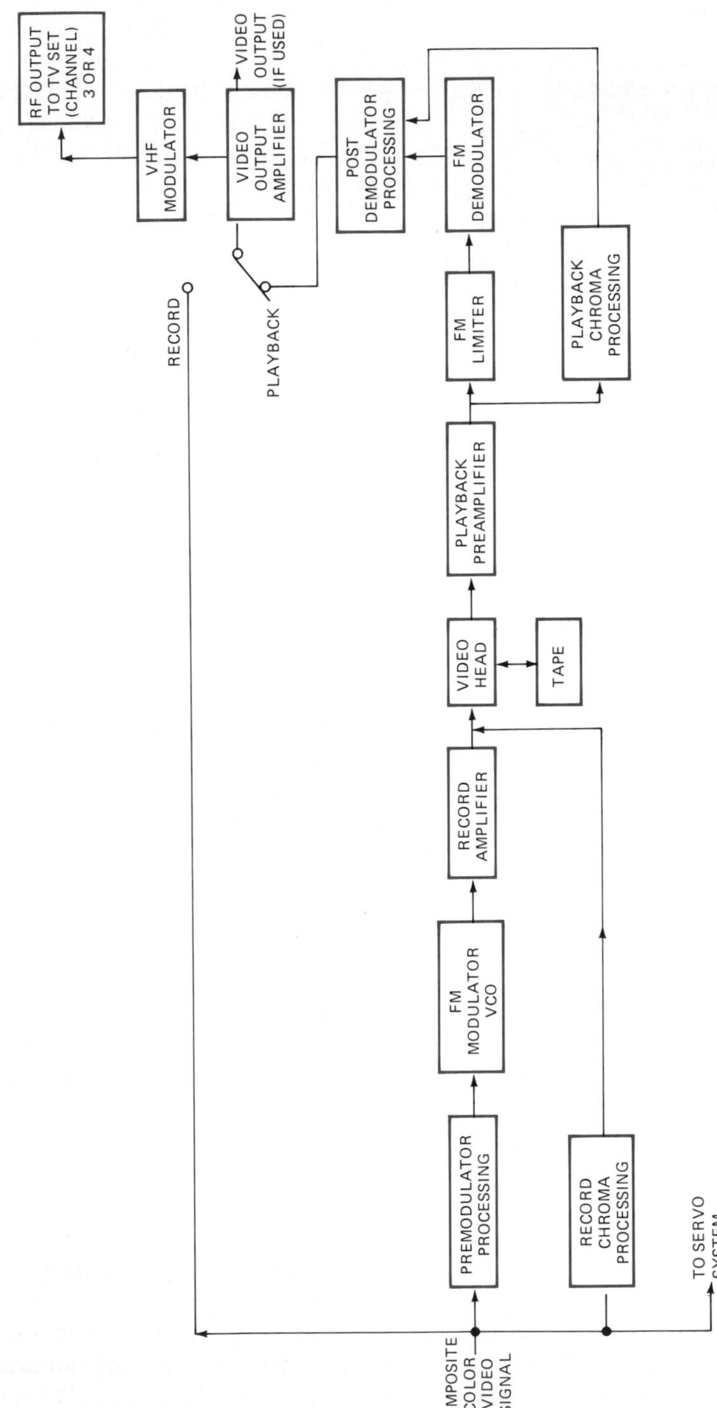

Figure 2.10 Simplified block diagram of the luminance record-playback system of a videocassette recorder.

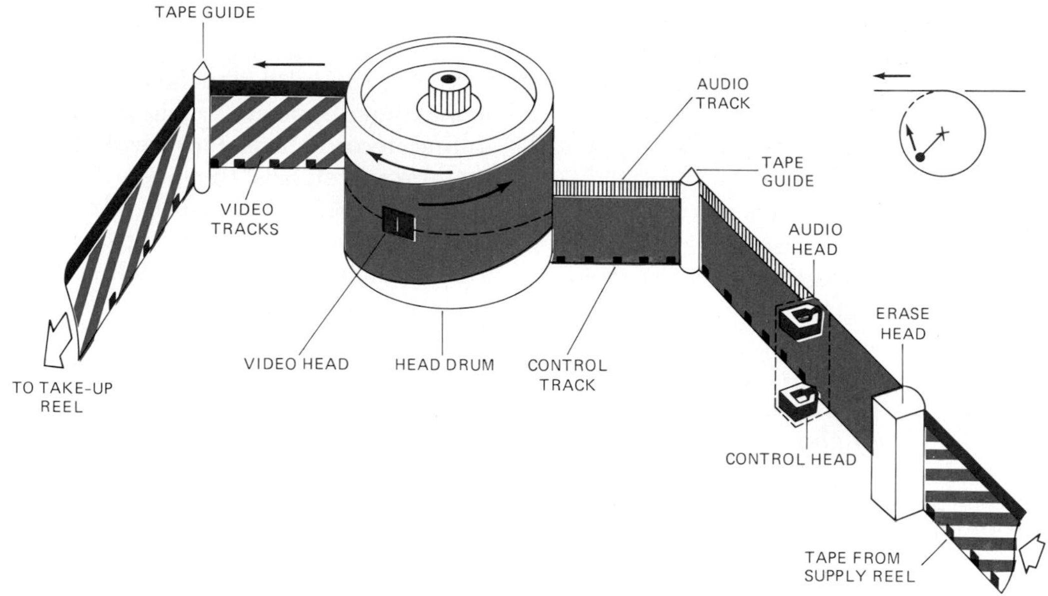

Figure 2.11 Simplified diagram of the Quasar "Alpha-Wrap" $\frac{1}{2}$-in (1.27-cm) videotape system.

sidebands which originally were on a 3.58-MHz subcarrier are now heterodyned down and exist around a much lower subcarrier carrier frequency. For Beta-max systems, this lower frequency is 688 kHz and for VHS systems, it is 629 kHz. Thus, the same color information is transferred to a new subcarrier frequency. Both the luminance and the color signals are recorded on the same tape track.

Video-Head and Tape Threading

Figure 2.11 shows the "Alpha-Wrap" tape system used by Quasar in a video cassette recorder (VCR). The same path is followed for recording and playback. The tape, which is $\frac{1}{2}$ in-(1.27 cm) wide, leaves the supply reel and moves past the erase head. This head is activated only for recording. Next, the tape passes the audio and control track heads. The audio head receives and plays back only audio frequencies on a separate track. The control head records amplified vertical-sync pulses. These pulses are used on playback, through a servo system, to synchronize the head-drum position and speed.

Next, the tape passes around the head drum in a helical fashion. Because of the rotating video-head, tape speeds in the order of 0.8 in/s (2.03 cm/s) can produce writing speeds of about 130 in/s (330.2 cm/s). This results in cassette playing times of up to eight hours. Note in Figure 2.11 that the video tracks on the tape are recorded diagonally across the width of the tape. Each track represents a complete television vertical field.

Playback Circuits

The playback circuits perform the following basic functions: (1) amplify the relatively weak signals picked up from the tape, (2) de-emphasis, (3) time-delay equalization between chrominance and luminance signals, (4) noise cancellation, (5) FM demodulation, (6) restoration of 3.58-MHz chrominance signal subcarrier and sidebands, (7) combining of restored chrominance signal and demodulated FM-luminance signal, and (8) place the signals on a Channel 3 or 4 carrier.

Playback Preamplifier

This amplifier has uniform response throughout the luminance signal FM range. It provides an RF output of several volts. At the output of the playback preamplifier, there is a high-pass filter. This filter passes the luminance-FM frequencies, but rejects the lower frequency chrominance signals.

FM Limiter

The FM limiter functions in the same manner as those used in FM receiver circuits. It removes amplitude variations from the FM-luminance signal. A limiter that is used frequently consists of two diodes in parallel, but connected with opposite polarities. This limiter is frequently included as part of an IC chip. The output of the FM limiter feeds the FM demodulator.

FM Demodulator

The luminance-FM signal is a square wave. The FM demodulator operates on a different principle than the FM receiver. One demodulator has an output consisting of positive pulses, whose number (per second) varies in accordance with the FM-square wave deviation frequency. The average value of the positive pulses, therefore, varies in accordance with the deviation frequency. The changes in the average value represent the shape of the original video-luminance signal. The output of the FM demodulator feeds into the **post demodulator processing** circuits, together with the output of the **playback chroma processing** circuits.

Playback-Chroma Processing

Basically, the color signals are now handled in a manner opposite to that which occurred during **record-chroma processing.** That is, the chroma signals that were heterodyned down from 3.58 MHz to 688 kHz or 629 kHz are now heterodyned back up to 3.58 MHz. This is done by the **playback-chroma processing** circuits.

When the composite FM-video signal from the playback preamplifier is applied to the playback-chroma processing circuits, it first encounters a low-pass filter. The filter passes only the 688-kHz or 629-kHz chrominance signals. It rejects the FM-luminance signal.

Next, the chrominance signal is heterodyned back up to 3.58 MHz. The 3.58-MHz chrominance signal next passes through a band-pass filter. The band-pass filter passes only the desired 3.58-MHz subcarrier and its sidebands to the **post-demodulator processing** circuits.

Post-Demodulator Processing

The outputs from the FM demodulator and from the playback-chroma processing circuits are fed to the post-demodulator processing circuits. Here they are added together to form a composite color-video signal. This signal is now theoretically identical with the original signal that was applied to the input-record circuits.

Video-Output Amplifier

A record-playback switch connects the post-demodulator processing output signal to the video-output amplifier when it is in the **playback position.** The video-output amplifier is a power amplifier. Its function is to amplify the level of the composite color video signal to a level suitable for application to the VHF modulator.

VHF-Modulator Circuits

The VHF-modulator circuits include (1) a switchable crystal oscillator, (2) a frequency doubler, (3) an AM modulator, and (4) mixer circuits.

The crystal oscillator is a switchable type. It operates on one of two selectable frequencies, which are one-half of the Channel 3 video-carrier frequency (30.625 kHz) and one-half of the Channel 4 video-carrier frequency (33.625 kHz). One-half frequencies are generated in order to obtain optimum oscillator stability. The output from the crystal oscillator is then applied to a frequency doubler. The output of this latter stage is the actual Channel 3 video-carrier frequency (61.25 MHz) or Channel 4 video-carrier frequency (67.25 MHz).

The video-carrier frequency selected is applied to an AM modulator together with the composite color-video signal. Here the video signal amplitude modulates the Channel 3 or 4 video carrier. However, although the modulated video carrier is now suitable for application to the antenna terminals of a TV set, the audio signal is not yet present.

In the VHF-modulator circuits, the audio signal is applied to a varactor which determines the frequency of a 4.5-MHz oscillator. The audio signal produces frequency deviation of the 4.5-MHz oscillator. Thus, the output of the oscillator is an audio-modulated FM signal with a carrier frequency of 4.5 MHz.

This FM-sound signal is next applied to a mixer. Here, it heterodynes with the video-carrier signal. As a result, the FM-sound signal is moved up in frequency to the normal FM-sound carrier of Channel 3 (65.75 MHz) or Channel 4 (71.75 MHz).

The composite-video signal at the correct RF video-carrier frequency and the FM-sound signal at the correct RF sound-carrier frequency are now added together. The result is a composite RF video and sound signal which is equivalent to that transmitted by a TV station.

This composite RF signal is fed to the antenna terminals of a TV receiver tuned to either Channel 3 or 4, whichever is not active. The TV receiver then processes this input signal in the identical manner as it would a TV signal transmitted over the air.

VIDEO-DISC RECORDING (VDR)

Video signals can also be recorded on a video disc. Video discs look like 12-in (30.5-cm) LP audio records. The techniques for recording and playback of video discs are much more complex than those for audio recording. The video-disc system is used for playback only. The discs are purchased prerecorded. When video discs are played on both sides, they can provide a color television program lasting up to two hours. A wide variety of programs are available, including motion pictures, plays, operas, and educational programs. A laser-optical, video-disc player, and typical video discs are shown in Figure 2.12. Video discs and players are considerably more economical than

Figure 2.12 Laser-optical video-disc player and some typical prerecorded 12-in (30.5-cm) diameter discs. *(Courtesy of Magnavox Consumer Electronics.)*

comparable videotapes and videotape-recorders.

RCA Disc Systems The RCA SelectaVision 12-in (30.5-cm) video discs have a metallic layer sandwiched between two uncoated plastic layers. Masters for the discs are cut mechanically. This is a relatively simple process, compared to the former electron-beam recording process. The playback head operates by a variable-capacitance system. The video information is recorded in the form of grooves on both plastic surfaces. A sapphire stylus is equipped with a metal electrode which measures by capacitance the distance from the stylus tip to the metal layer. Capacitance variations are translated into voltage variations, which constitute the color video signal. The disc revolves at 450 r/min and plays a color television program for up to 60 min per side.

Phillips-MCA System The North American Phillips-MCA system (player shown in Figure 2.9) is a laser-read, optical video-disc system. A precisely controlled laser beam is used for recording. It cuts minute, oblong depressions representing sound, color, and brightness information. The depressions are about 27.6×10^{-6} in (0.7 micron or μ) wide and vary in length from about 31.5×10^{-6} to 98.4×10^{-6} in (0.8 to 2.5 microns).[1] Tracks are spaced less than 78.7×10^{-6} in (2 microns) apart. In the player, the audio and video information is recovered by means of a laser beam and optical system. Light from a helium-neon laser is reflected back from the depressions in the aluminum coated video disc. This light varies in accordance with the depressions, and thus, with the sound and video signals. The light variations are focused onto a photo detector. Its output is an electrical sound and color television picture signal. The pickup never touches the disc. Both sides of the disc can be played to provide up to two hours of program time.

The laser-optical disc plays from the inside to the outside. It does not revolve at a constant speed. Rather, it works on the "variable angular velocity" principle. The speed begins at 1 800 r/min when playing the inside, and decreases to 600 r/min at the outside. In this way, the speed of the track being played is always constant in relation to the laser-beam pickup and longer playing time is achieved.

[1]A micron is 10^{-6} meter.

2.4 VIDEO GAMES

A wide variety of electronic games may be played on the screens of TV receivers. Electronic pulses simulate playing field lines, players, and game balls. The video pulses are added to horizontal and vertical sync and blanking pulses to make up a composite video signal. This composite video signal modulates an oscillator for either Channel 3 or 4. The modulated carrier wave is fed to the VHF antenna terminals of any television receiver. No modifications of the receiver are required. Some models provide a digital readout of the score and a full-color presentation. The game units are also known as home TV programmers.

TV GAME BLOCK DIAGRAM

A block diagram of a TV scoreboard game manufactured by Radio Shack is shown in Figure 2.13. Four games can be played on this unit (1) practice, (2) handball (smash), (3) hockey, and (4) tennis. Paddle controls are located on a remote box. The player has several options which include (1) ball speed, (2) slice angle, (3) paddle size.

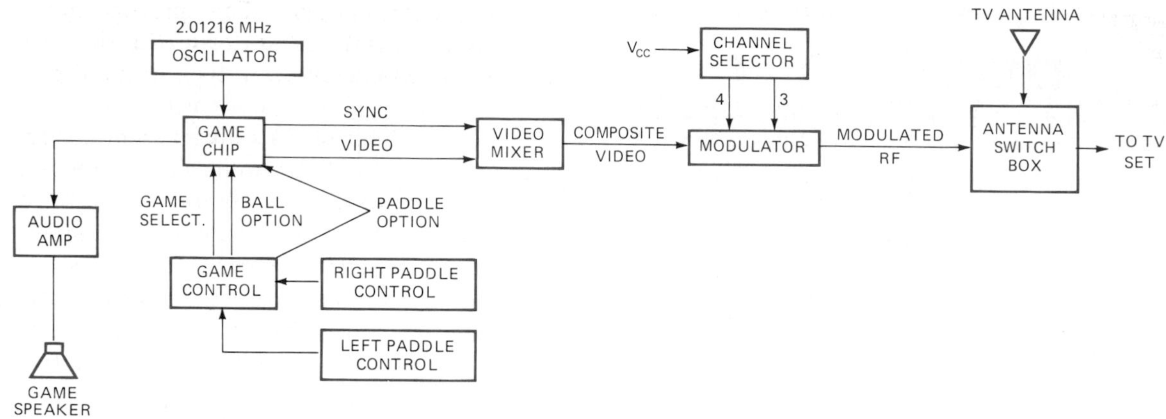

Figure 2.13 Simplified block diagram of a Radio Shack video game. Four different games are available.

Although this particular remote box is connected to the game unit by wires, other game units are wireless.

The Six Subsections

As shown in the block diagram, the scoreboard game is divided into six major subsections. These are (1) game chip, (2) oscillator, (3) video mixer, (4) modulator, (5) audio amplifier and (6) power supply.

The game-IC chip generates all of the sync, video, and logic signals. It also processes the ball speed and paddle size options. In addition, it receives game selection information from the game-selector switch. All of the other five sections support the game-IC chip.

The crystal oscillator operates at a frequency of 2.01216 MHz and is a CMOS type. It supplies the master-reference frequency to the game chip. All generated pulses are related to the master-reference frequency.

The video mixer (an IC) receives the horizontal and vertical-sync signals, the left- and right-paddle video, the ball video, and the score and field-video signals. The combined signals are then fed to the modulator.

The modulator contains two separate oscillators. One or the other is activated by switching power to it. The oscillators are tuned to the picture-carrier frequency of Channel 3 (61.25 MHz) and Channel 4 (67.25 MHz). The user selects the channel that is not broadcasting in the local area. The video and sync signals modulate the selected channel video carrier, similar to a normal TV broadcast signal. The output signal of the modulator can be connected directly to the antenna terminals of a TV receiver.

For the convenience of switching to normal TV reception or to game reception, an antenna (switch) box is provided. The game-modulator output and the normal TV antenna are connected to the antenna box. A switch selects either signal to be fed to the TV receiver.

Sound is provided for the game to make it more realistic. "Pong" sounds occur each time a ball is struck by a paddle or hits a wall. Audio is provided by an output from the game chip to an audio amplifier and game speaker.

Other Video Games

Other video games use plug-in cartridges which provide a wide variety of games. For

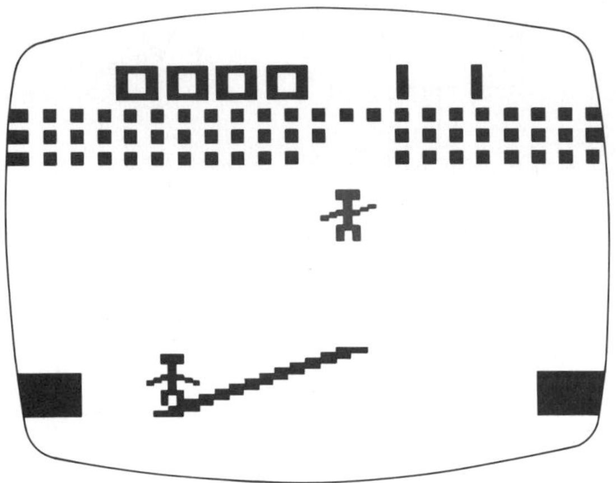

Figure 2.14 The Circus Atari video game as it appears on a TV screen.

example, Atari offers 43 different cartridges. Compatible cartridges are also offered from another company.

One Atari game, known as Circus Atari, is shown in Figure 2.14 as it appears on the TV screen. Here two circus clowns try to propel each other on a see-saw to hit and puncture three rows of balloons overhead. (It's not as easy as it appears.)

Several examples of other video games are given in Figure 2.15.

RCA HOME TV PROGRAMMER

The RCA Studio II home TV programmer is shown in Figure 2.16. This unit has

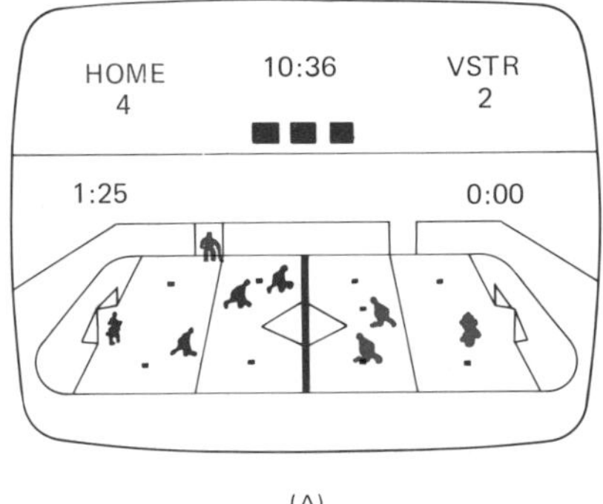

(A)

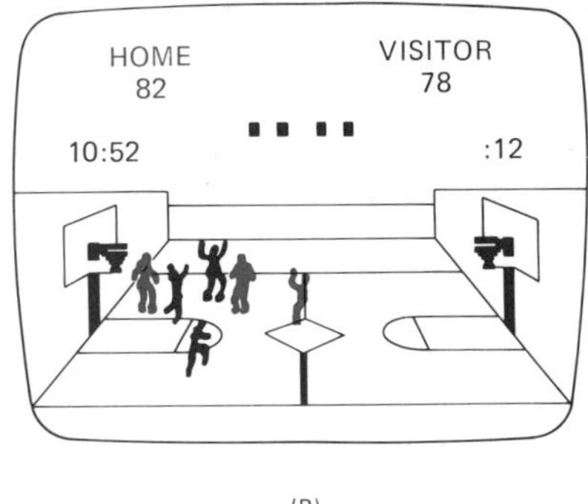

(B)

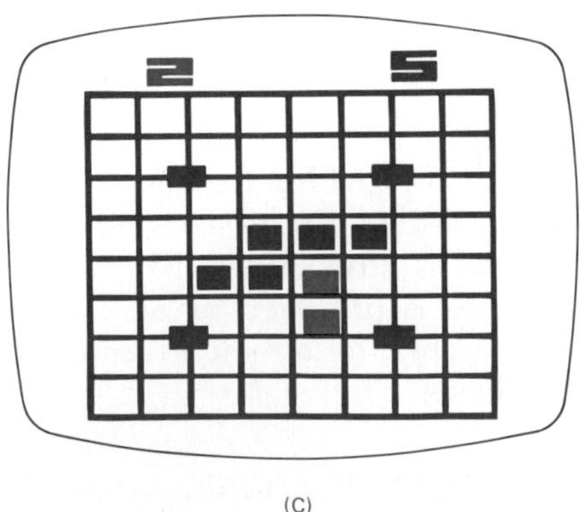

(C)

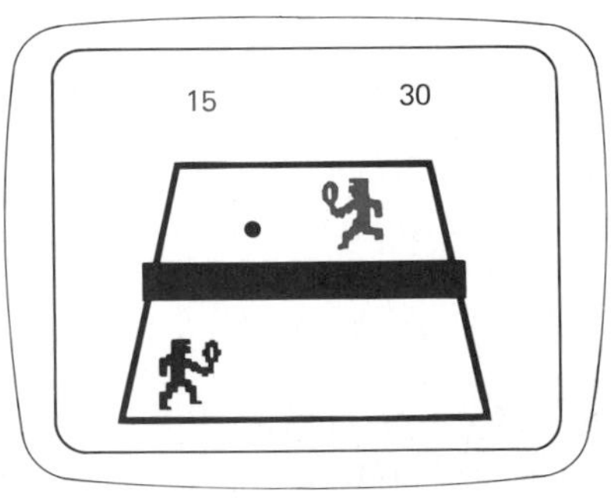

(D)

Figure 2.15 Some representative video games. **A.** NHL Hockey **B.** NBA Basketball. **C.** Othello game **D.** Tennis game.

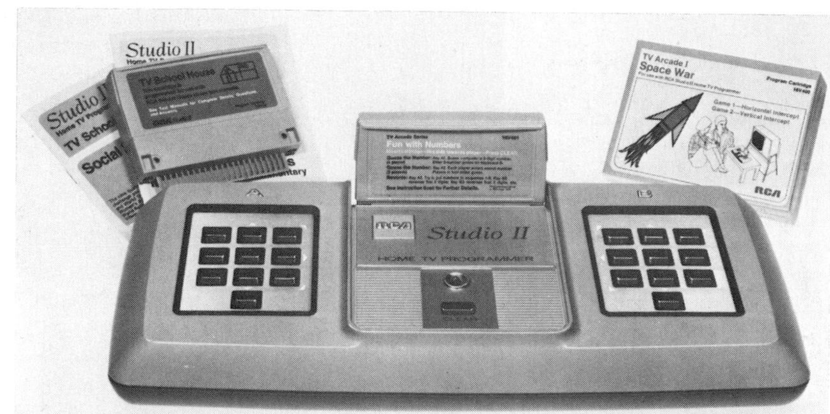

Figure 2.16 Home TV programmer (video game unit) that features large-scale integration (LSI) circuits and microprocessor control and uses plug-in cartridges. *(Courtesy of RCA.)*

five built-in programs. Plug-in cartridges are used to add other programs. A microprocessor (a limited computer) enables the user to select from a wide variety of games by plugging in the desired cartridge. There are two sets of digital keyboards for two players (Figure 2.16). A sound system adds realism to the games.

LSI CIRCUITS

The use of digital LSI (large-scale integration) ICs simplifies the wiring of the game units. They contain hundreds of diodes and transistors. These ICs are used in the RCA unit shown in Figure 2.16 and by most other manufacturers. For example, in the simplest Magnavox Model only four LSI-type ICs are used. More complex models have additional ICs and show the games in full color on a color television receiver.

2.5 PROJECTION TELEVISION

In projection television, the image is optically projected so that it appears greatly enlarged upon a special screen. This scheme is similar to the projection of still slides or motion pictures upon a screen. Projection television is used for special theatre programs, for industrial and educational purposes, and for home entertainment. Except for home entertainment purposes and certain special situations, the projection units are employed as part of a CCTV system. Here the video and sound signals are transmitted by coaxial cable.

FRONT PROJECTION

Except for home entertainment units, projection systems are of the front-projection type, similar to movie projection. This type of system is shown in Figure 2.17. The projected image appears upon a Kodak Ektalite screen, measuring 40 in × 40 in (101.6 cm × 101.6 cm). This special screen has a finely grained, specially treated, aluminum foil surface which is extremely reflective. The screen is curved to concentrate the reflected light and serves an area approximately 60 degrees wide by 30 degrees high. Viewing may be done in the presence of ambient light, which is properly located.

Home entertainment units may be either front projection or rear projection units.

REAR PROJECTION

With rear projection, the image is projected through the rear of a special flat

Figure 2.17 A front projection color television system for home use. The special, high-reflective curved screen measures 40 in × 40 in (101.6 cm × 101.6 cm). Horizontal viewing angle is about 60°. *(Courtesy of Eastman Kodak Company.)*

screen. An optical system within the cabinet focuses the image from a single, special, color picture tube upon the rear of the screen. One General Electric unit has a 45.7-in (116.1-cm) diagonal screen. The screen is made of three plastic layers. The rear layer is molded into a fresnel lens. This lens concentrates the light in forward directions to improve picture brightness. The middle layer is a diffused surface upon which the picture is formed. The outer surface is treated to minimize the effect of room lighting on picture contrast. The normal viewing angle (horizontally) is about 30°.

BRIGHTNESS CONSIDERATIONS

The color picture tube image on a conventional color television receiver is not suitable for projection. The basic reason is that the expanded image does not have adequate brightness. To obtain the necessary brightness, special one-tube and three-tube systems are used to achieve adequate brightness. A specially constructed color picture tube is used with the one-tube system. The General Electric receiver mentioned previously employs a single 13-inch color picture tube operating at a much higher voltage than the conventional 30-kV value. This color tube is almost five times as bright as a conventional, direct-viewing color tube. Other home projection systems and many larger screen commercial systems use a three-tube configuration.

THREE-TUBE SYSTEM

With the three-tube system, a separate projection picture tube is used for each of the three primary colors: red, blue and green. For a home projection system using three tubes, high-efficiency screens are used with diagonal measurements of 5 feet (1.52 m), 6 feet (1.82 m), or 7 feet (2.13 m).

All high-efficiency screens have restricted viewing angles for maximum brightness. Typically, a home screen may have a horizontal viewing angle of 30° to 60°. The images from the three color tubes are superimposed on the screen to produce the full color range. For a home system, the high voltage on each picture tube need be no greater than about 30 kV. Some of these three-tube systems employ the Schmidt optical system to reflect and enlarge the pro-

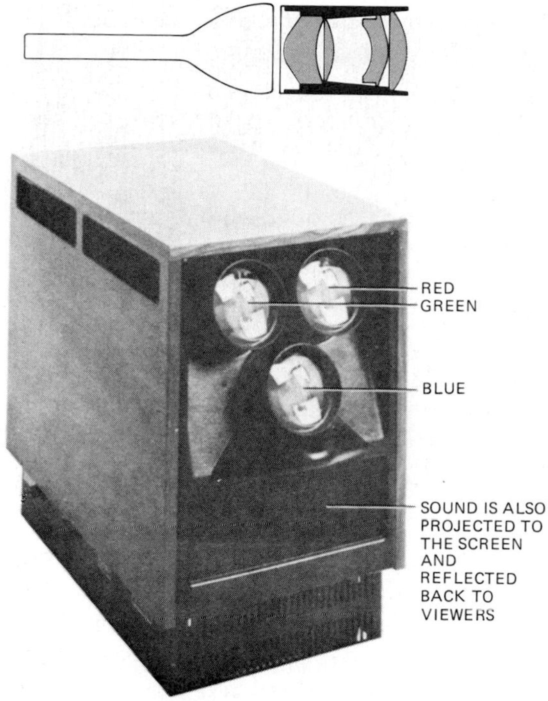

Figure 2.18 Example of a three-tube color television projection unit for home use. Each color tube has a 5-in (12.7-cm) diameter. The image of each tube is projected by a four-element, *f*1.3 lens (see inset). *(Courtesy of Advent Corp.)*

jected image. This scheme employs a spherical mirror and has high light efficiency. Schmidt optics are also used in some industrial and educational television projectors with larger screens. The three-tube system, in general, produces brighter pictures, with greater contrast and detail than the one-tube system. However, they are more expensive than one-tube systems.

A three-tube system manufactured by the Advent Corp. is shown in Figure 2.18.

LARGER SCREEN SIZES

Other types of installations use screens that are much larger than home-unit screens. These include installations in theatres, educational institutions, television studios, military centers, and hospitals. Screen sizes vary according to need and may be as large as 42 ft × 55 ft (13 m × 17 m). To achieve the required brightness, the three-tube system using very high color-picture anode voltages (80 kV to 100 kV) may be employed.

Eidophor System A different scheme known as the Eidophor system can project a 42 ft × 55 ft (13 m × 17 m) picture. Here, the light emitted by a 2.5 kW Xenon lamp is modulated by the video signal. This is done with the aid of a spherical mirror coated with a very thin layer of oil. The electrical charges of the video signal distort the oil layer as it is scanned by an electron beam. This distortion causes the Xenon light to be reflected in a standard television scanning pattern. Three such units are used: one each for the red, blue, and green colors.

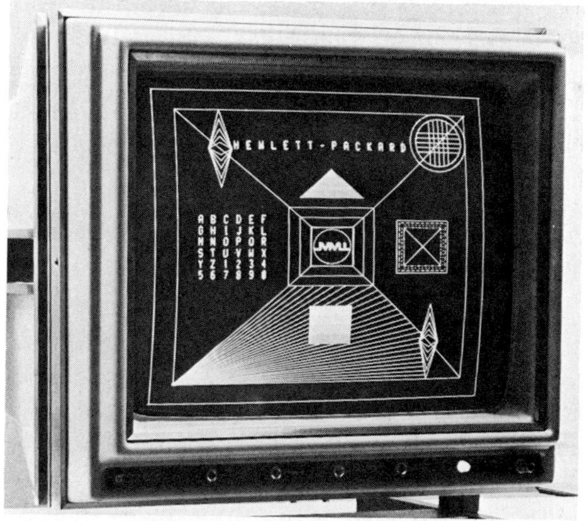

Figure 2.19 A composite display of alphanumerics and graphics. The diagonal screen measurement is 19 in (48 cm). The accelerating potential is 28.5 kV. *(Courtesy of Hewlett-Packard.)*

2.6 ALPHA-NUMERIC AND GRAPHIC DISPLAYS

Figure 2.19 illustrates how alpha-numeric and graphical displays may be presented on a large screen cathode-ray tube. The Hewlett-Packard unit illustrated has a 19-in (48-cm) diagonal screen. The display shown is for illustration only. It can be produced by programming a computer. Producing a display like this one requires video, scanning, and synchronizing signals. A special cathode-ray tube is used, with an accelerating potential of 28.5 kV and electrostatic deflection. The high accelerating potential, coupled with the small size of the electron beam assures a bright, crisp, and clear image over the active viewing area.

TYPES OF DISPLAYS

An almost limitless number of displays can be presented on an appropriate unit. Typical displays include: computer graphics, radar, stock quotations, computer-stored information, medical patient monitoring, and instrumentation readouts. For permanent records, excellent photos can be taken of a display, using a CU-5 or CU-9 Polaroid Land camera.

2.7 SUMMARY OF U.S. TELEVISION STANDARDS

The transmitted television signal must comply with strict standards established by the Federal Communications Commission (FCC). These standards are detailed in their Rules and Regulations, Part 73. The standards insure that uniformly high quality monochrome and color picture and sound signals will be transmitted. Standards are also required so that television receivers can be designed to receive these standard signals, and thus provide a correct reproduction of the original picture and sound.

IMPORTANT TELEVISION STANDARDS

The standards of greatest interest for the purpose of this book, are given below:

1. Each VHF and UHF television station is assigned a channel that is 6 MHz wide. The complete or composite signal, must fit into this 6-MHz bandwidth. The composite signal includes: the video carrier, one vestigial sideband for the video signal, one complete sideband for the video signal, the color signals, the synchronizing and blanking pulses, and the FM sound signal.
2. The visual (picture) carrier frequency is 1.25 MHz above the low end of each channel.
3. The aural (sound) carrier frequency is 4.5 MHz above the visual carrier frequency.
4. The chrominance (color) subcarrier frequency is 3.579545 MHz above the picture carrier frequency. For practical purposes, the subcarrier frequency is expressed as 3.58 MHz.
5. One complete picture frame consists of 525 scanning lines (483 "active" lines actually produce the picture). The remaining lines are blanked out during the vertical retrace. There are 30 frames per second.
6. Each frame is divided into two fields; thus, there are 60 fields per second. Each field contains $262\frac{1}{2}$ lines, interlaced with the preceding field.
7. The scene is scanned from left to right horizontally at uniform velocities, progressing downward with

Table 2.2 **FREQUENCIES OF TELEVISION CHANNELS**

VHF BAND			UHF BAND					
CHANNEL NO.	*CHANNEL FREQ. (MHz)*		*CHANNEL NO.*	*CHANNEL FREQ. (MHz)*	*CHANNEL NO.*	*CHANNEL FREQ. (MHz)*	*CHANNEL NO.*	*CHANNEL FREQ. (MHz)*
2	54 – 60		14	470 – 476	38	614 – 620	62	758 – 764
3	60 – 66		15	476 – 482	39	620 – 626	63	764 – 770
4	66 – 72	lower VHF band	16	482 – 488	40	626 – 632	64	770 – 776
5	76 – 82		17	488 – 494	41	632 – 638	65	776 – 782
6	82 – 88		18	494 – 500	42	638 – 644	66	782 – 788
			19	500 – 506	43	644 – 650	67	788 – 794
7	174 – 180		20	506 – 512	44	650 – 656	68	794 – 800
8	180 – 186		21	512 – 518	45	656 – 662	69	800 – 806
9	186 – 192		22	518 – 524	46	662 – 668	70	806 – 812
10	192 – 198	upper VHF band	23	524 – 530	47	668 – 674	71	812 – 818
11	198 – 204		24	530 – 536	48	674 – 680	72	818 – 824
12	204 – 210		25	536 – 542	49	680 – 686	73	824 – 830
13	210 – 216		26	542 – 548	50	686 – 692	74	830 – 836
			27	548 – 554	51	692 – 698	75	836 – 842
			28	554 – 560	52	698 – 704	76	842 – 848
			29	560 – 566	53	704 – 710	77	848 – 854
			30	566 – 572	54	710 – 716	78	854 – 860
			31	572 – 578	55	716 – 722	79	860 – 866
			32	578 – 584	56	722 – 728	80	866 – 872
			33	584 – 590	57	728 – 734	81	872 – 878
			34	590 – 596	58	734 – 740	82	878 – 884
			35	596 – 602	59	740 – 746	83	884 – 890
			36	602 – 608	60	746 – 752		
			37	608 – 614	61	752 – 758		

each additional scanning line. The scene is retraced rapidly (blanked out), from the bottom to the top, at the end of each field.

8. The "aspect ratio" of the picture is four units horizontally to three units vertically.
9. At the transmitter, the equipment is so arranged that a decrease in picture light intensity during scanning causes an increase in radiated power. This is known as "negative picture transmission."
10. The video part of the composite signal is amplitude modulated. The synchronizing and blanking pulses are also transmitted by amplitude modulation and these pulses are added to the composite video signal.
11. The color signal is transmitted as a pair of amplitude modulation sidebands. These sidebands effectively combine to produce a chrominance signal varying in hue, or tint (phase angle of the signal) and saturation (color vividness). Saturation corresponds to the amplitude of the color signal. Color information is transmitted by slipping the color signal frequencies between spaces in the monochrome video signals. This process is called **frequency interleaving** and is discussed in more detail in Chapter 8.

Table 2.3 **SUMMARY OF TELEVISION STANDARDS**[1]

	U.S., MEXICO, CANADA, JAPAN	FRANCE	ENGLAND	ITALY, SPAIN, GERMANY	RUSSIA
No. of frames per second	30	25	25	25	25
No. of fields per second	60	50	50	50	50
No. of lines per frame	525	625	625	625	625
No. of lines per field	262½	312½	312½	312½	312½
Horizontal scan frequency, Hz	15 750	15 625	15 625	15 625	15 625
Color subcarrier, MHz	3.58	4.43	4.43	4.43	4.43
Video signal bandwidth, MHz	4.2	6.0	5.5	6.0	6.0
Channel bandwidth, MHz	6.0	8.0	8.0	8.0	8.0
Picture carrier modulation	AM	AM	AM	AM	AM
Picture transmission	Negative	Positive	Negative	Negative	Negative

[1]The scanning frequencies shown in Table 2.3 are for monochrome transmission only. However, those frequencies used for color transmission are very close to the values for monochrome transmission. For example, the U.S. Color Standards are: 29.97 frames/s, 59.94 fields/s, and 15 734.264 horizontal lines/s. This difference occurs because in a color transmission, the preceding rates are derived from the color subcarrier frequency, instead of the power line frequency.

12. The sound signal is frequency modulated. As in FM broadcast systems, the sound signal may also be produced by the **indirect FM method.** The maximum deviation of TV is ±25 kHz.
13. Table 2.2 shows the frequencies assigned to the United States VHF and UHF channels.
14. Vertical-interval reference (VIR) signal. According to FCC Rules and Regulations, line 19 of the vertical blanking interval of each field is reserved for a special signal. This is the VIR signal. The VIR signal is known as the **chrominance reference.** While originally intended for station use, it is also utilized by television receivers. At the station, the VIR signal is used as a reference to assure the correct transmitted hue and saturation of the color signal. At the receiver, the signal and automatic circuits help to provide accurate color reproduction. When this method is used it eliminates the need for manual color control adjustment.

2.8 FOREIGN TELEVISION STANDARDS

An abbreviated table of foreign television standards and United States Standards is shown in Table 2.3. Note that in countries using a 50-Hz power line frequency, the 50-Hz field rate is established. Also note that these countries use a number of lines (625) and a horizontal scanning frequency (15 625), which are exact multiples of the frame rate of 25 Hz. All countries use the system of interlaced scanning, with two fields per frame.

SUMMARY OF CHAPTER HIGHLIGHTS

1. Cable television (CATV), also called Cablevision, combines radio-wave transmission and reception and a radio-frequency cable distribution system.
2. The basic CATV system consists of four main sections (1) the antenna system, (2) the head-end system, (3) the trunk-amplifier system and, (4) the distribution-amplifier system. (Figure 2.1)
3. Distant stations are received by a CATV system via microwaves and communications satellites.
4. CATV signals at the subscriber's TV receiver are fed in on either Channel 3 or 4.
5. With a two-way CATV system, the subscriber may request or provide information. In some CATV systems the subscriber may also engage in subscriber-callable video games.
6. Closed-circuit television (CCTV) transmits only the video and audio signals (no RF) through cables to the receiver(s).
7. Master-antenna television (MATV) can operate many television receivers from a single antenna system.
8. Community-antenna television (CATV) is a combination of radio-wave transmission and reception and a cable-distribution system. It is used in such cases as a community located behind a mountain, or in a valley.
9. Microwave relays and synchronous-orbit communications satellites enable television signals to be transmitted over

extremely long ranges. The "Westar" domestic satellite provides a 12-channel, color television coverage for the United States, Puerto Rico, Alaska, and Hawaii. There are also worldwide communication satellites in operation.

10. Color television programs can be recorded on magnetic tape using basic principles similar to audio-tape recording. But the much higher effective-tape speed (writing speed) is obtained by the use of rotating tape heads. On home-video-cassette recorders (VCRs or VTRs), the luminance signal (30 Hz to 2.5 MHz) is changed to an FM signal by an FM modulator. The 3.58 MHz chrominance signal is reduced to 688 kHz or 629 kHz and recorded on the same track as the luminance signal. The chrominance signal is heterodyned back up to 3.58 MHz in playback circuits.
11. The FM modulator used in all video-tape recorders (VTRs) are voltage-controlled oscillators (VCOs). A typical FM deviation is 3.1 MHz to 4.7 MHz.
12. With the "Alpha-Wrap" video-tape system, four magnetic heads are used. These are the (1) erase head, (2) control-track head, (3) audio head and, (4) video-head.
13. The output signals from a VTR are heterodyned up to Channels 3 and 4 and fed with the appropriate channel frequencies to the antenna input of a TV set.
14. One type of 12-in (30.5 cm) video disc, revolving at 450 r/min, plays a color television program for up to 60 min on each side.
15. Video discs are prerecorded and can be used for playback only.
16. Two types of video-disc players are the variable-capacitance type and the laser-read, optical video-disc type.
17. Television video games use digital pulses to simulate playing field lines, players, and game balls. They utilize microprocessors and send out their signals on Channel 3 or 4.
18. There is a great variety of video games. Some units employ plug-in cartridges to increase the number of available programs.
19. Many video game units employ an IC "game chip." This chip generates all of the sync, video, and logic signals.
20. A stable crystal oscillator generates a master reference frequency for the game circuits.
21. Projection television systems are of the front projection and rear projection types. Home units use either one or three color-projection tubes and special, highly reflective directional screens. Commercial units may project an image as large as 42 ft × 55 ft (13 m × 17 m). Home unit screens are in the order of 5 ft (1.52 m) measured diagonally.
22. An important use of television-type displays is to show alpha-numerics and computer-produced graphs on a large screen cathode-ray tube.
23. For a summary of important U.S. television standards, refer to pp. 20–21, 41, 42, and Tables 2.1, 2.2, and 2.3.
24. The vertical-interval reference (VIR) is placed on Line 19 of the vertical-blanking interval of each field. It is used as a color reference at television transmitters. It is also used in television receivers to provide automatic correction of the hue and saturation of colors.
25. For a summary of foreign television standards, refer to Table 2.3.

EXAMINATION QUESTIONS

(Answers provided at the back of the text)

Answer true (T) or false (F).

1. Closed-circuit television is abbreviated as CATV.
2. In a cable-vision system, distant TV stations may be received by the system via microwaves or communications satellites.
3. In a CATV system, the output of a trunk amplifier goes directly to a subscriber.
4. CATV signals at the subscriber are fed to the TV set on either Channel 3 or 4.
5. With a two-way CATV system the subscriber may request or provide information.
6. A domestic satellite can retransmit television programs to the entire United States.
7. A videotape must record the luminance and chrominance signals, as well as the FM sound signal and a control signal.
8. In a home-video-cassette recorder, the AM video signal is changed to an FM signal.
9. In Question 8, the chrominance signal center frequency is reduced to 629 kHz or 688 kHz.
10. In Question 8, a typical FM deviation of the video signal is ± 150 kHz.
11. With video games, the composite RF signal is usually sent to the TV set on Channel 5 or 7.
12. In a video game, a "game chip" generates the sync, video, and logic signals.
13. In video games, a plug-in cartridge is used to supply only realistic audio signals.
14. The maximum diagonal screen size for any projection television system is 7 ft. (2.134 m).
15. Alpha-numeric displays require a video signal plus scanning and synchronizing signals.
16. The chrominance subcarrier frequency is exactly 3.579545 MHz.
17. The "aspect ratio" is 4 units vertically and 3 units horizontally.
18. In the United States, the television channels cover a range of 54 MHz to 890 MHz.
19. European television stations have a horizontal-scan frequency of 15 750 Hz.

REVIEW ESSAY QUESTIONS

1. What is the basic equipment used for closed-circuit television (CCTV)? Briefly discuss two applications.
2. Briefly discuss the basic operation of a cable-vision system.
3. Define "microwave relay station."
4. What is the meaning of a satellite being in "synchronous orbit"?
5. Explain briefly how a satellite produces wide area television coverage.
6. Describe briefly how a video-tape recorder is able to record frequencies up to about 5 MHz.
7. How is it possible for a video-tape recorder to record independently of a television receiver?
8. Discuss briefly how the RCA SelectaVision video discs are played to reproduce a television program.
9. What is the function of the digital-video pulses in video games?
10. Briefly discuss the function of the "game chip" in video games.
11. Discuss briefly the advantages and disadvantages (if any) of a projection television system.
12. List at least four uses of alpha-numeric and graphic displays.
13. What are the frequency ranges for Channels 5, 13, 29, and 54?
14. Describe briefly the function of the VIR signal in a television receiver.

3.1 Basic television system
3.2 Introduction to scanning
3.3 Simplified scanning principles
3.4 Scanning rate
3.5 Flicker
3.6 The complete scanning process
3.7 Overview of blanking and synchronizing signals
3.8 The composite monochrome video signal
3.9 The horizontal sync and blanking pulses
3.10 Vertical sync, vertical blanking, and equalizing pulses
3.11 The color-synchronizing signal
Summary

PRINCIPLES OF SCANNING, SYNCHRONIZING, AND VIDEO SIGNALS

INTRODUCTION

A television picture is composed of many tiny pieces. The picture is "disassembled" at the camera tube and "reassembled" at the picture tube. In order that this overall process may be successful, both the "disassembly" and "reassembly" must be accomplished in a synchronized and orderly fashion. The composite video signal must be made in such a way that it will disassemble and reassemble the picture, with the aid of the transmitting and receiving circuitry.

This chapter investigates the principles of scanning, synchronizing, and video signals which make it possible to transmit and receive television pictures.

After you have completed the reading and work assignments for Chapter 3, you should be able to:

- Define the following terms: blanking signals, synchronizing signals, raster, aspect ratio, trace and retrace, persistence of vision, front porch, back porch, serrated pulse, color burst, and interlaced scanning.
- Discuss the principles of scanning an entire television frame, including interlacing.
- Discuss the function of horizontal and vertical synchronizing and blanking signals and their rates.
- List the components of a monochrome and color, composite video signal and know their functions.

- Discuss the problem of flicker and how it is overcome.
- Summarize the technical reasons for choosing to divide each television picture frame into 525 horizontal lines and two fields.
- Explain why equalizing pulses are used.
- Discuss the details of a complete vertical retrace, including the vertical sync, equalizing, and blanking pulses.
- Discuss the differences in scanning monochrome and color-television pictures.

3.1 BASIC TELEVISION SYSTEM

Figure 3.1 shows the basic components of a system for transmitting and receiving television programs. The sections related to scanning and synchronizing are shown in heavy outlines.

Scanning and **synchronizing** are needed to reproduce the picture and keep the reproduced picture in step with the picture originated at the transmitter. The synchronizing signals, or more simply, the **sync signals,** are generated at the transmitter, where, together with the video signal, they amplitude-modulate the transmitted signal. The receiver uses the sync signals to synchronize the scanning of the electron beam of the picture tube with that of the transmitter camera tube. As will be shown later, the sync signals control the frequency and phase of the oscillators in the scanning and synchronizing section of the receiver.

Another function of the scanning and synchronizing sections of the transmitter is to generate **blanking signals.** During retrace periods it is desirable to turn off the electron beam in the camera and in the receiver picture tube. This allows the beam to be repositioned for the start of a new line or new field.

3.2 INTRODUCTION TO SCANNING

Figure 3.2 illustrates the basic scanning elements in the transmitter and receiver systems. The scene to be televised is focused by a lens system onto a light-sensitive plate located in the camera tube. This plate is made in such a way that the electric charge at any point on the plate depends upon the amount of light falling on that point. The video signal output current of the camera tube depends directly upon the amount of light falling on the plate at each point. As the electron beam in the camera tube scans the plate, a video (picture) signal current is generated. The video signal consists of variations of current which correspond to varia-

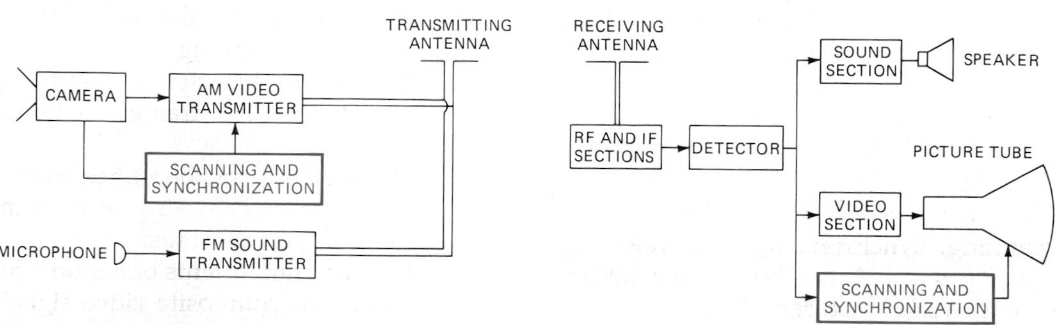

Figure 3.1 Simplified block diagram of a television transmitter and receiver.

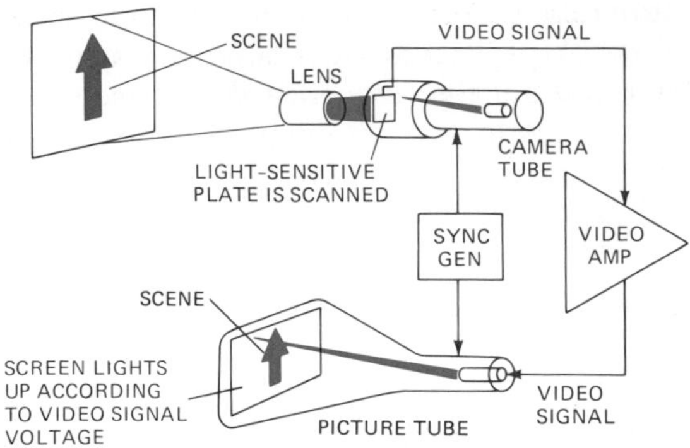

Figure 3.2 The scene is focused on a light-sensitive plate in the camera tube that is scanned by an electron beam. The same scanning process is used in the receiver to reconstruct the scene.

tions in brightness on the light-sensitive plate. This video signal is amplified and delivered to the gun of the receiver picture tube. (Not shown in Figure 3.2 are the intervening transmitter, radio-frequency wave, and front portion of the receiver.)

SYNCHRONIZATION

The beam is swept back and forth across the screen in step with the beam that is sweeping the light-sensitive plate in the camera tube. This permits the video signal from the camera tube to reproduce the picture at the receiver picture tube. The transmitter/synchronizing generator provides the signal that controls the motion of the electron beam within the camera tube. This signal also controls the beam in the picture tube. As the picture tube electron beam scans any one point, its strength (or the number of electrons contained in its beam) is controlled by the amplitude of the video signal.

Identical scanning is employed at both the transmitter and the receiver. Synchronizing signals must be used to assure that the position of the electron beam in the receiver picture tube corresponds with the same position on the light-sensitive plate of the camera tube, at any given time.

3.3 SIMPLIFIED SCANNING PRINCIPLES

In Chapter 1, it was mentioned that the picture on the picture tube screen is "painted" by moving an electron beam back and forth as it scans individual lines. Each line is separated from the next one by a very narrow space.

Figure 3.3 shows the basic principle involved in scanning a picture. Starting at the top left side, the electron beam moves from left to right across the screen along line 1–1. During this time, picture information is being displayed. (In this illustration, it is assumed that the viewer is facing the screen of the picture tube.) After the beam reaches the right side of the screen at point 1, it is rapidly moved to the left side of the screen along the dotted line 1–2. No picture information is transmitted while the beam moves from right to left. The line marked 1–1 is

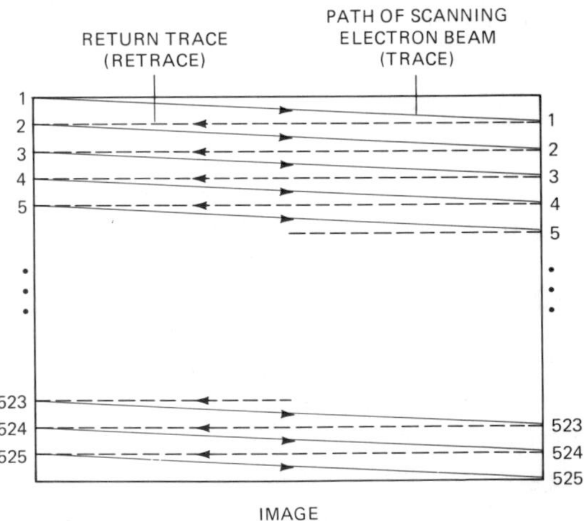

Figure 3.3 The procedure for scanning a television picture (somewhat simplified); solid line 1-1 is called a *trace*, and the dotted line 1-2 is called a *retrace*.

called a **trace,** and the dotted line marked 1–2 is called the **return trace,** or **retrace.**

After retrace, the beam is at the left side of the screen at point 2. It then moves from left to right across the screen for the next line, indicated as line 2–2. This procedure is repeated throughout the picture.

This is a simplified and slightly modified illustration of the scanning procedure. The scanning procedure is the same as that for reading a line on a page. The reader's eyes start at the top left side and move across a line of type. After reaching the right side of the page, the eyes quickly move back to the left side and begin the next line.

THE RASTER

The electron beam moves back and forth across the screen of the picture tube whether or not there is a picture displayed. When there is no picture, the beam will trace out a white rectangle called a **raster** on the screen. When the standards for television were first established, it was decided that a rectangular picture (or rectangular raster) would be most desirable, and the relationship between the height and width of the rectangle was established.

ASPECT RATIO

The width of the rectangle divided by its height is called the **aspect ratio.** In television, the aspect ratio of the picture (and the raster) is four to three. Therefore, if the picture is four inches wide, it must be three inches high. If the picture is twelve inches wide, it must be nine inches high to obtain the correct aspect ratio.

By adjusting the voltages to the picture tube circuitry, a rectangle with different aspect ratios is produced. But, the picture is **transmitted** with an aspect ratio of four to three. Therefore, the transmitted picture is reproduced with minimal distortion when the receiver picture also has an aspect ratio of four to three.

3.4 SCANNING RATE

When the trace reaches the bottom of the raster, it is quickly moved back to the top of the screen where it begins to scan lines again. The television screen actually displays a number of complete rasters (or fields) each second.

When the standards for television were first established, it was necessary to decide on a repetition rate for showing complete pictures. How many complete frames were to be shown per second? The choice of frame frequency involves a tradeoff. Ideally, a high frame frequency should be used so that the eye cannot perceive the individual pictures making up the continuous motion. However, a high frame frequency produces electronic problems in the scanning. For one thing, the amount of brightness on the screen is definitely affected by the rate at which the beam is moved across the screen. Specifically the faster the beam is moved, the more difficult it is to get sufficient brightness. In addition, a higher frame rate requires a wider transmission bandwidth.

FRAME AND FIELD RATES

To obtain the desired amount of brightness, to utilize a reasonable transmission bandwidth, and to achieve the effect of continuous motion, a compromise was needed. It was finally decided that for monochrome transmission, thirty complete pictures would be displayed on the screen each second. These complete pictures are known as **frames.** A frame frequency of only thirty pictures per second yields a flicker which is discernible to the eye. Therefore, each picture is divided into two parts called

fields. The field frequency is sixty fields per second and the frame frequency is thirty frames per second. Each field contains one-half of the total picture elements. The above rates are very slightly different for color transmissions.

CHOICE OF RATES

Orginally, the values of sixty fields per second and thirty frames per second were selected so that the television picture could be synchronized to the standard United States power line frequency of 60 hertz (Hz). For a monochrome transmission, all scanning frequencies were previously[1] derived from the 60-Hz line frequency. The receivers are automatically synchronized to the transmitted synchronizing pulses. As a result of this power-line frequency synchronization, the effects of hum on the picture will be stationary on the TV receiver screen. Hum may be caused by imperfect power supply filtering. If there were no synchronization, the hum effects would cause vertically moving patterns to pass through the picture. TV transmitters in adjacent cities frequently use synchronized power line frequencies to prevent this unwanted type of interference.

COLOR FIELD RATE

When color television programs are broadcast, the field frequency is reduced slightly from 60 Hz to 59.94 Hz. Thus, hum interference patterns are not perfectly synchronized for color broadcasts. However, they move vertically at the rate of only 0.06 Hz, which is a slow rate and not readily visible.

3.5 FLICKER

If related still films follow each other fairly rapidly on a screen, the human eye combines them and the motion appears continuous. The eye can do this because of a phenomenon called **persistence of vision.** Due to this characteristic of the eye, visual images do not disappear as soon as their stimulus is removed. Instead, the light appears to diminish gradually. On the average, about 1/50 second is required before the light disappears entirely. If this did not occur, motion picture and television entertainment might be impossible.

MOTION PICTURE FRAME RATE

It has been found that the action appears continuous when theater films are presented at a rate of 15 stills per second. But flicker is still detected at this speed, and detracts from the complete enjoyment of the film. The flicker is due to the picture impression in the viewer's mind decreasing to too low a value before the next film is presented on the screen. Increasing the rate at which the stills are presented will gradually cause the flicker to disappear. At 50 frames per second—that is, 50 complete pictures per second—there is no trace of flicker, even under adverse conditions. The rate is not absolute, however, but depends greatly upon the brightness of the picture. With average illumination, lower frame rates prove satisfactory.

In the motion-picture theater, 24 individual still films (or frames) are flashed onto the screen each second. Flicker is still notice-

[1]Currently, most TV stations utilize the color TV, vertical synchronizing rate of 59.94 Hz for both monochrome and color broadcasts. The slight difference in frequency from 60 Hz has no significant effect on TV receiver performance.

able at this rate. Therefore, a shutter in the projection camera breaks up the presentation of each frame into two equal periods. This increases the fundamental rate to an effective rate of 48 frames per second. A shutter moving across the film while it is projected onto the screen accomplishes this. Thus, each picture is actually being viewed twice and all traces of flicker are eliminated.

THE TELEVISION FRAME RATE

In television, a fundamental rate of 30 images (or frames) per second was chosen. As previously noted, this frequency and the effective rate are related to the frequency of the AC power lines. This choice of frame-sequence rate requires less filtering to eliminate an AC ripple. In audio systems, this AC ripple is known as *hum*. At a rate of 24 frames per second, any ripple not eliminated by filtering produces a weaving motion in the reproduced image. AC ripple is less of a problem when the rate is 30 frames per second.

INTERLACED SCANNING

All traces of flicker are eliminated by using an effective rate of 60 frames per second. This is accomplished by increasing the downward rate of travel of the scanning electron beam so that every other line is sent, rather than every successive line. When the bottom of the image is reached, the beam is sent back to the top of the image. Then, the lines that were skipped in the previous scanning are sent. The initial set of scanning lines is assumed to start at the upper left portion of the camera-image plate. This is point A of Figure 3.4. Starting at point A with line 1, every other line is scanned. Lines 1, 3, 5, 7, 9, and so on are

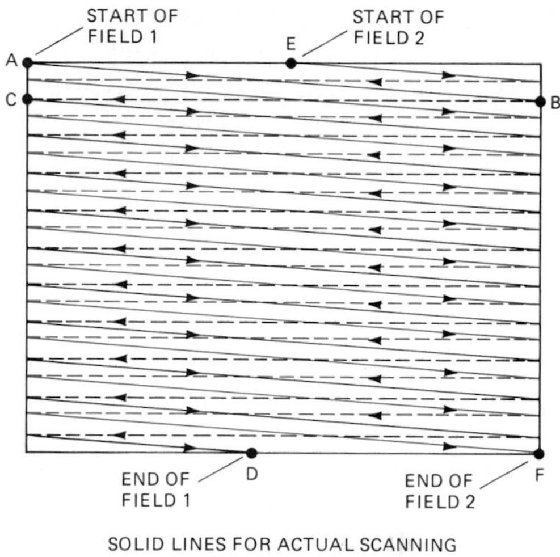

Figure 3.4 The path of the electron beam in interlaced scanning.

scanned. The field produced is known as the "odd-line field." When this field is completed, the beam retraces rapidly to point E in Figure 3.4. From point E, the "even-line field" is developed by scanning lines 2, 4, 6, 8, 10, and so on. Both of these operations, the odd- and even-line scanning, take 1/30 second. Therefore, 30 frames per second is still the fundamental rate. However, since all the even lines are transmitted in 1/60 second and the same is true of the odd lines, the total scanning time adds up to 1/30 second. The human eye cannot separate the two. Therefore, the *effective* rate is now 60 frames per second, and no flicker is noticeable. This method of sending television images shown in Figure 3.4 is known as **interlaced scanning.**

With each frame divided into two parts each field will have one-half of 525 lines ($262\frac{1}{2}$ lines) from its beginning to the start of the next field. With interlaced scanning, each frame is broken up into an even-line field and an odd-line field. Each field contains $262\frac{1}{2}$ lines. Each frame has a total of 525 lines.

CHOICE OF 525 LINES

Each complete scene is sent at a rate of 30 frames per second. The desired amount of picture detail is obtained by dividing the picture into a total of 525 horizontal lines. Five-hundred twenty-five lines were chosen for the following technical reasons:

1. The frequency bandwidth available to the transmission of the television signals. As will be shown later, the required bandwidth increases with the number of lines.
2. The amount of detail required for a well-produced image.
3. The ease with which the synchronizing (and blanking) signals can be generated.

3.6 THE COMPLETE SCANNING PROCESS

From the foregoing discussion, it is possible to reconstruct the entire scanning process. Only the movement of the electron beam at the television camera will be considered. However, this is identical to the motion that exists at the receiver screen.

THE ODD FIELD

At the start of the scanning motion at the camera-tube image plate, the electron beam is at the upper lefthand corner, (point A of Figure 3.4). Under the combined influence of the two sets of deflection coils, the beam then moves at a small angle downward to the right. When point B is reached, the blanking signal acts while the beam moves rapidly back to point C. At this point, the third line begins, as is required for interlaced scanning. The blanking signal then ends and the electron beam again begins its left-to-right motion. In this manner, every odd line is scanned.

THE EVEN FIELD

When the end of the bottom odd line is reached (point D), the blanking signals are applied as the beam is brought up to point E. Point E is above the first odd line of field 1 by a distance approximately equal to the thickness of one line. The beam is brought here as a result of the odd number of total lines used (525). Recall that each field has $262\frac{1}{2}$ lines from its beginning to the start of the next field. Therefore, when the beam reaches point E, it has moved through the necessary $262\frac{1}{2}$ lines from its starting point A. From here, the beam again starts its left-to-right motion, moving in between the previously scanned lines, as shown in Figure 3.4. The beam continues until it reaches point F. From there it is brought back to point A. From point A, the entire sequence repeats itself.

SUMMARY

The electron beam moves back and forth across the width of the image $262\frac{1}{2}$ times in going from point A to point D to point E. The remaining $262\frac{1}{2}$ lines needed to equal the total of 525 lines are obtained when the beam moves from point E to point F back to point A. This process occurs readily and accurately at the transmitter and receiver. A more detailed analysis, including the number of horizontal lines that are lost when vertical blanking interval occurs, is given later in this chapter.

3.7 OVERVIEW OF BLANKING AND SYNCHRONIZING SIGNALS

The cathode-ray beam at the receiver picture tube must follow the camera tube scanning beam at every point. Each time the camera-tube beam is blanked out, the same process must occur at the receiver and at the proper place on the screen. For this reason, blanking-pulse signals are sent along with the video signals containing the image details. When these blanking pulses are applied to the control grid of a cathode-ray tube, the grid is biased to a large negative value that prevents any electrons from passing through the grid to the fluorescent screen. Vertical and horizontal blanking pulses prevent the retrace lines from being visible on the picture tube.

However, blanking voltages do not cause the movement of the beam from the right-hand to the left-hand side of the screen, or from the bottom to the top. Another set of pulses (called **synchronizing pulses**) are superimposed on the blanking signals to control the deflection oscillators at the receiver. These oscillators, in turn, control the position of the beam. A horizontal sync pulse at the end of each line causes the beam to be brought back to the left-hand side, in position for the next line. Vertical sync pulses occur at the end of each field to bring the beam back to the top of the image.

3.8 THE COMPOSITE MONOCHROME VIDEO SIGNAL

Figure 3.5A shows how the picture detail, blanking signals, and synchronizing pulses combine to form the composite monochrome video signal. The figure covers the scan of three complete lines.

HORIZONTAL BLANKING AND SYNC PULSES

At the end of each line, the horizontal blanking signal is imposed on the beam and automatically prevents the electron beam from reaching the image plate at the camera tube or the fluorescent screen at the receiver. With the horizontal blanking signal ON, a horizontal synchronizing pulse causes the horizontal deflection coils to move the position of the electron beam from the right side of the picture to the left side. The sync pulse controls a deflection oscillator circuit which provides the proper deflection coil currents needed for the retrace. The job of the synchronizing pulse is finished once the retrace is complete. A fraction of a second later, the horizontal blanking pulse releases its bias on the grid of the cathode-ray tube and the electron beam begins to scan again. This process continues until all the lines (odd or even) in one field are scanned. Additional details of the composite signal appear in Figure 3.5B. Details involving the horizontal blanking pulse and the sync pulse are given later.

VERTICAL BLANKING AND SYNC PULSES

The vertical motion ceases at the bottom of the field. It is then necessary to bring the beam quickly to the top of the image so that the next field can be traced. Since the vertical sync pulse and the retrace require a longer period of time than the horizontal sync pulse and retrace, a longer blanking signal (vertical) is inserted. As soon as the vertical blanking signal takes hold, the vertical-sync pulse is sent. The form of this pulse is shown in Figure 3.6. The horizontal-synchronizing pulses cannot be interrupted, even while the vertical deflection coils bring the electron beam to the top of the field.

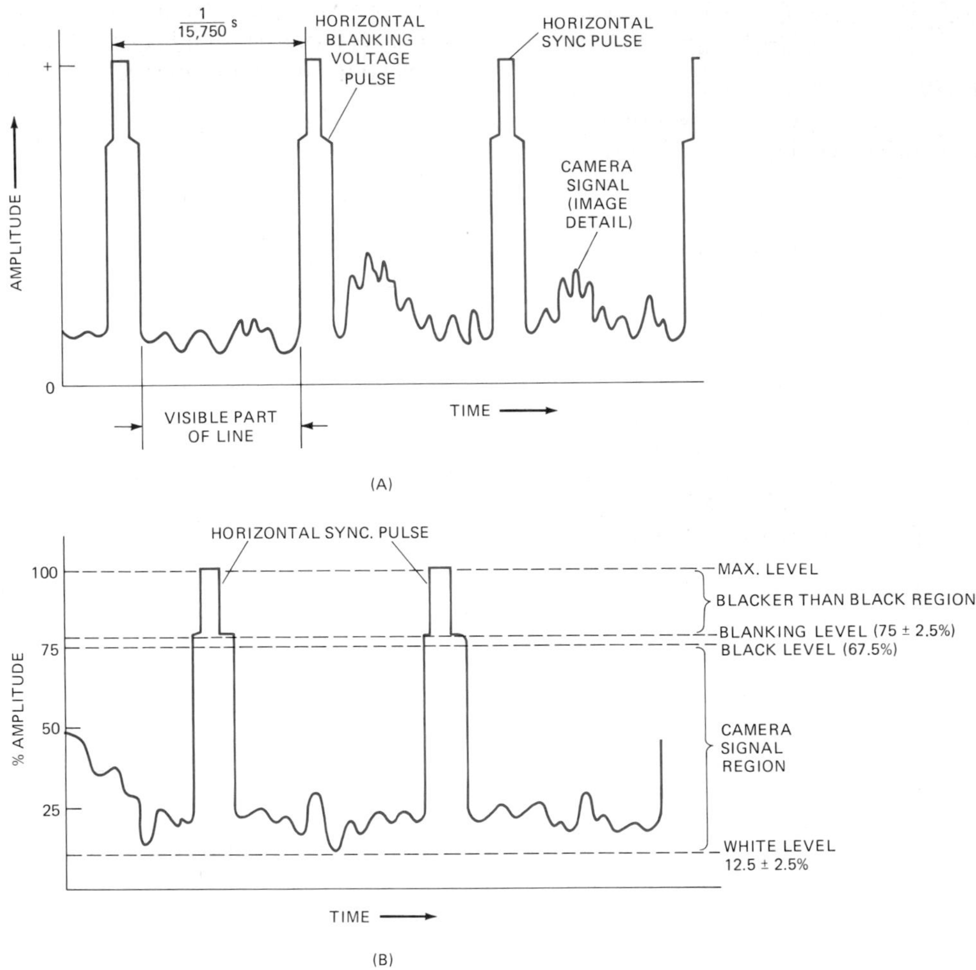

Figure 3.5A. The composite monochrome video signal for three scanned lines. **B.** Details of the amplitude levels for the composite monochrome video signal.

Therefore, the long vertical pulse is broken into appropriate intervals. In this way, both horizontal and vertical pulses can be sent at the same time. Each type of pulse is accurately separated at the receiver and transferred to the proper deflection system.

SERRATED VERTICAL PULSES

The vertical deflection coils bring the electron beam back to either point A or point E (Figure 3.4). Then, the scanning of the next field begins. **Serrated vertical pulses** are the series of synchronizing pulses that combine to make up the total vertical synchronizing signal.

At the bottom horizontal line, a vertical sync pulse is inserted to bring the electron beam back to the top of the screen again. During the period that the vertical sync pulse is active, the horizontal deflection oscillator must not be neglected. This would cause the horizontal oscillator to slip out of

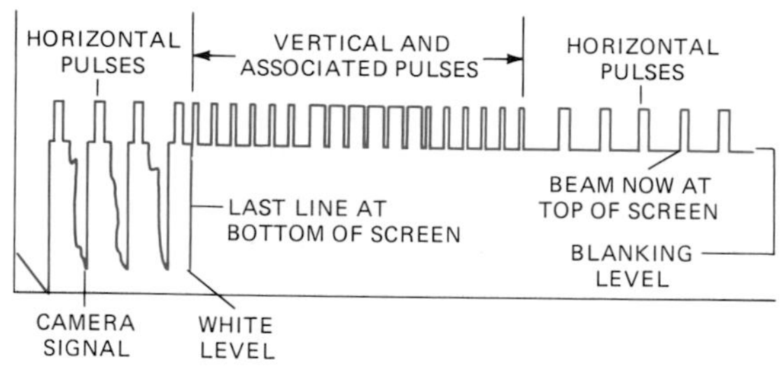

Figure 3.6 The form of the vertical-synchronizing pulses (vertical serrated pulses).

synchronization. To avoid this, the vertical sync pulses are arranged in serrated form to accomplish vertical and horizontal synchronization simultaneously. Figure 3.7 illustrates the composite video and sync waveform at the beginning of each field. These two waveforms are slightly different so that interlacing can be performed. The first horizontal line of field 1 and the last horizontal line of field 2 will be complete lines. The last horizontal line of field 1 and the first horizontal line of field 2 will be half lines. The serrated sync pulses and the equalizing pulses will be discussed later.

SYNC AND VIDEO SIGNAL SEPARATION

The pulses of a video wave must be separated from the other portions of the signal before they can be used. The separation may

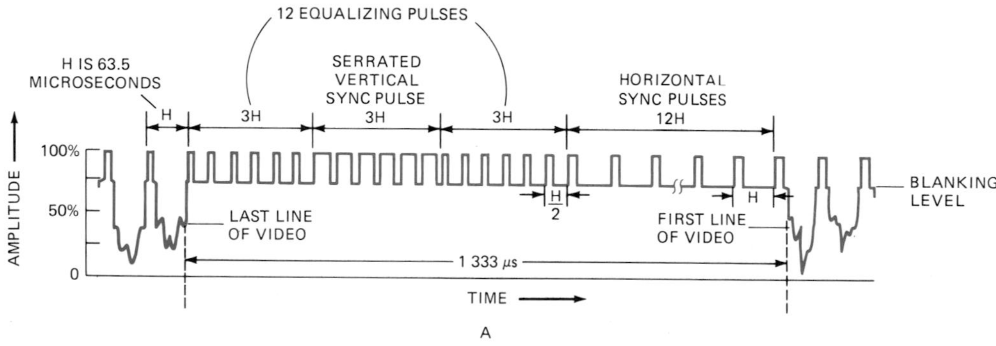

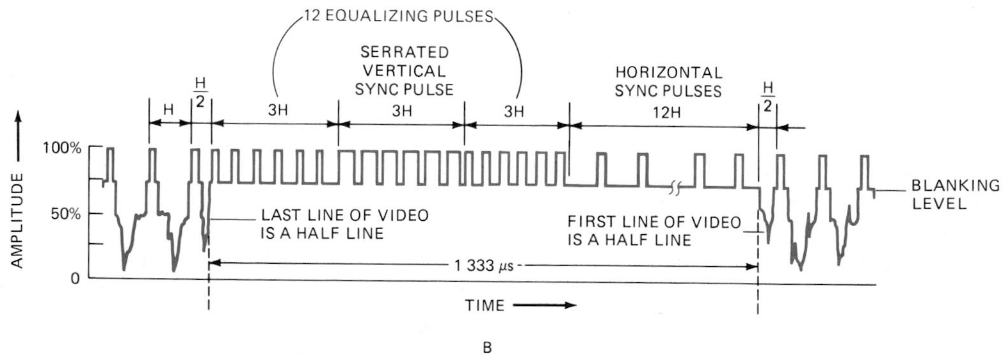

Figure 3.7A. Video and sync waveforms at the beginning of the first field. H is the time of one complete horizontal cycle, or 63.5 microseconds. **B.** Video and sync waveforms at the beginning of the second field.

occur anywhere from the video detector to the last video stage before the cathode-ray tube. In practice, receiver designers obtain the input for the synchronizing stages from a point beyond the video detector, usually at the output of one of the video amplifiers. The signal has sufficient amplitude at such points and is in the proper form for controlling horizontal- and vertical-deflection oscillators with a minimum of additional stages. For example, circuit designers often do not apply the video signal to the sync separator until it has passed through the first video amplifier. In this way, an extra pulse amplifier is eliminated.

3.9 THE HORIZONTAL SYNC AND BLANKING PULSES

The basic functions of the horizontal sync and blanking pulses were described previously. Recall that the horizontal blanking pulses extinguish the electron beam of the picture tube (and camera tube) during the horizontal retrace. In this way, horizontal retrace lines do not appear in the picture at the time when no picture information is present. The horizontal sync pulses perform the all important task of synchronizing each scanning line in the picture tube with its corresponding line in the camera tube.

DETAILS OF THE SYNC PULSES AND HORIZONTAL BLANKING

There are 15 750 scanning lines per second. A horizontal sync pulse controls the timing of each line. Thus, there are 15 750 horizontal sync pulses and horizontal blanking pulses per second. Details of the horizontal sync and blanking pulses are shown in Figure 3.8A. The position of these pulses in the composite video screen was shown in Figure 3.5. As mentioned earlier, the horizontal blanking pulses make the horizontal retrace lines invisible. This is accomplished by having the blanking pulses at the "black level." In other words, these pulses cut off the electron beams of the camera and picture tubes for the pulse duration.

As shown in Figure 3.8A, the duration of each horizontal blanking pulse is 0.16 H. Since 1 H is the time from one line to the next, or 63.5μs, the pulse (in microseconds) is 0.16 × H (63.5μs) = 10.16μs.

The actual horizontal retrace time is somewhat less than the blanking time and is in the order of 7μs.

THE FRONT PORCH

In Figure 3.8A you will notice that the horizontal sync pulse that initiates the horizontal retrace actually begins 0.02 H, or 1.27μs, after the start of the horizontal blanking pulse. This 1.27μs period composes the *front porch*. This delay in the start of the sync pulse insures that the end of each line is blanked before the retrace begins. The front porch blanking interval produces a vertical black bar at the right-hand edge of the picture that is usually off the viewing portion of the screen.

THE HORIZONTAL SYNC PULSE, RETRACE, AND BLANKING

The horizontal sync pulse follows the front porch. This pulse has a duration of 0.08 H, or 5.08μs. Its function was described previously.

Retrace time is about 7μs. Blanking after the start of the horizontal sync pulse continues for 8.89μs. Thus, the left side of the raster is blanked for 1.89μs to produce a vertical black bar on the left side of the picture. This bar is also normally off the screen, but can be seen by reducing the width of the picture.

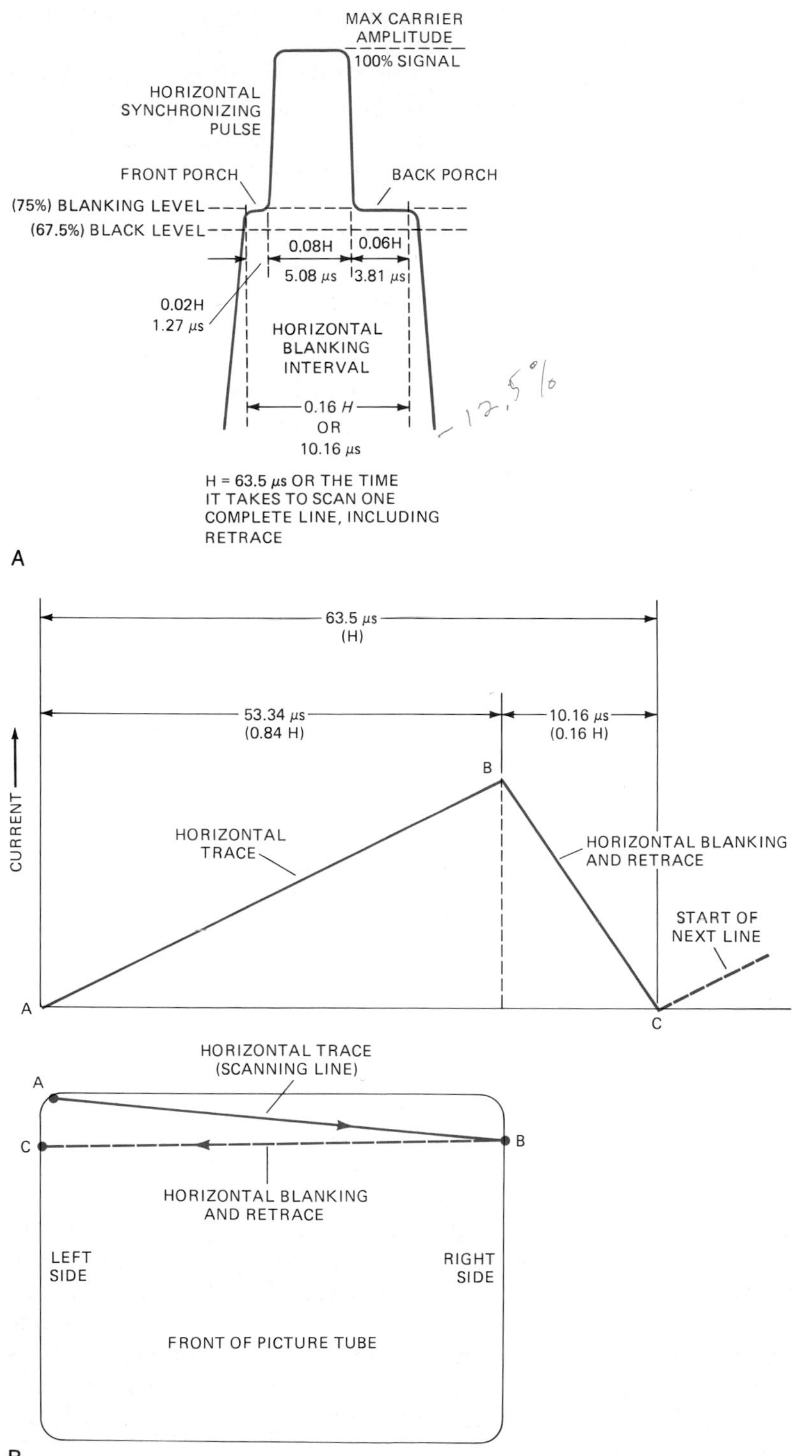

Figure 3.8A. A monochrome horizontal blanking pedestal, showing the horizontal blanking interval and sync pulse. The total horizontal blanking time of 10.16 μs is precisely controlled. See Table 3.3 for more information. **B.** The horizontal trace-and-retrace current related to the front of a picture tube. This sawtooth wave of current passes through the horizontal deflection coils at the rate of 15 750 Hz (for monochrome transmission).

THE BACK PORCH

A back porch exists, following the completion of the horizontal sync pulse. This porch has a duration of 0.06 H, or 3.81μs. The back porch is exactly three times as long as the front porch. This is necessary to accommodate the so-called "color-burst" sync signal. The color burst is transmitted with all color programs. It will be discussed briefly later in this chapter and in more detail in a chapter on color television. The color burst consists of a maximum of 11 cycles of a 3.58-MHz signal, which is about 3.07μs long. The color-burst signal is superimposed on the back porch and is used to synchronize the color portion of a color television program.

HORIZONTAL SAWTOOTH CURRENT ACTION

Figure 3.8B illustrates the horizontal trace and retrace with respect to the front face of a picture tube. Note that the waveform is a sawtooth of current. This waveform is passed through a pair of horizontal deflection coils mounted around the neck of the picture tube. The magnetic field from these coils causes the electron beam to move in a linear fashion from the left side (A) to the right side (B) of the picture tube. This process takes $0.84 \times H$ (63.5μs) = 53.54μs. When the beam reaches the right side, the horizontal sync pulse occurs and the retrace is initiated. As a result, the beam is driven back to the left side (C) to start the next line. For simplicity at this time, Figure 3.8B shows the retrace taking 10.16μs. However, it was stated earlier that the actual retrace takes only about 7μs. The total horizontal blanking period is, of course, 10.16μs. These details will be explained more fully in a later chapter.

3.10 VERTICAL SYNC, VERTICAL BLANKING, AND EQUALIZING PULSES

This section describes in more detail the other parts of the sync signals shown in Figure 3.7. Consider the basic form of the vertical sync pulse shown in Figure 3.9. At the bottom of each field this basic vertical sync pulse is inserted into the signal. This pulse controls the vertical deflection oscillator and forces the beam to be brought back to the top of the screen. No provision is made in the signal, at this point, for horizontal deflection oscillator control as the vertical pulse is acting. Such a condition is undesirable as it permits the horizontal oscillator to slip out of control. To prevent this, the vertical pulse is broken into smaller intervals, so that both actions can occur simultaneously. The vertical synchronizing pulse, in its modified form, is shown in Figure 3.10.

Breaking up the vertical pulse permits the horizontal synchronizing voltages to continue without interruption. However, the action of the vertical pulse is substantially unchanged. The vertical pulse remains above the blanking voltage level practically all of the time it is acting. The serrated pulse width is appreciably longer than the preceding horizontal pulse width. The two types of pulses are capable of separation because their waveforms are different, as is evident from Figure 3.10.

THE VERTICAL BLANKING INTERVAL

When the beam is blanked out at the bottom of an image and returned to the top, it does not move straight up. That is, it moves from side to side during its upward swing. This movement is due to the rapidity with which a horizontal line is traced out as

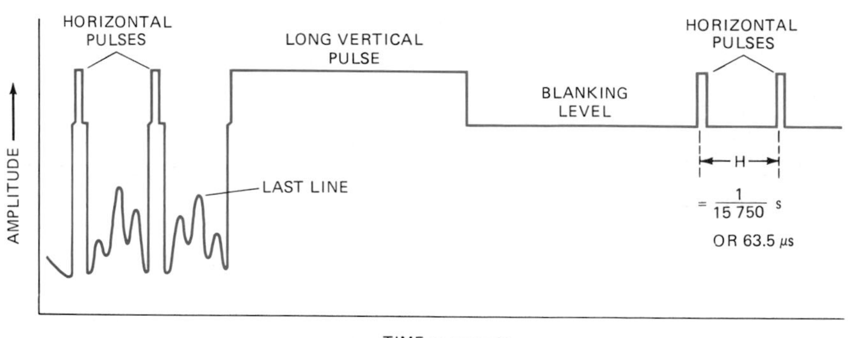

Figure 3.9 The basic form of the vertical synchronizing pulse.

compared with the vertical retrace period. In fact, there are approximately 21 horizontal lines traced out while the beam is brought back to the top of the picture. Of these 21 lines, the distribution is as follows:

1. Approximately four lines are scanned at the bottom of the raster.
2. Five lines are scanned while the beam is returning to the top of the picture. The number may vary with different receivers.
3. The balance, in this case 12 lines, is scanned at the top of the raster. Thus, in each field, 21 horizontal lines are lost in the blanking interval between fields. Of the 525 lines which are sent out, only 525 − 2(21) or 483 lines are actually effective in forming the visible image.

WHY 21 LINES ARE BLANKED

The method of arriving at 21 horizontal lines is quite simple. The electron beam is blanked out for approximately 1 333 μs between fields while the beam is being shifted from the bottom to the top of the image. During this interval, the horizontal-deflection oscillator is still active. Thus, as the beam moves up under the influence of the vertical-deflection current, it also moves back and forth because of the horizontal-de-

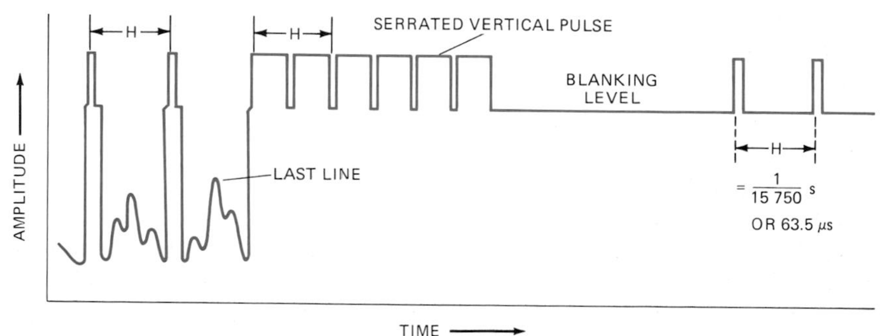

Figure 3.10 The serrated vertical synchronizing pulse allows horizontal synchronization to be maintained during the vertical blanking interval.

flection current. One horizontal line requires $\frac{1}{15}$ 750 s, or 63.5μs. The number of horizontal lines traced out is approximately equal to 21 lines. In a frame containing two fields, 42 lines are lost. It is possible to see these vertical retrace lines by turning up the brightness control on a television receiver when no station is being received and only the scanning raster is visible.

EQUALIZING PULSES

The vertical sync pulse is inserted once into the video signal when a horizontal line is half completed, and once into the video signal at the end of a complete line. (See Figure 3.7). This occurs because each field consists of 262½ lines (the number required to produce interlaced scanning).

This condition is shown in a simplified form in Figure 3.11. Here a video signal for each field of one frame is shown. Note again that there is a ½-line difference just prior to the start of the serrated vertical pulse. This ½-line difference will not affect the horizontal deflection synchronization. But it can affect the timing of the vertical synchronization, and thus the interlacing.

To ensure that the vertical deflection oscillator receives the necessary triggering voltage at the same time after every field, a series of six equalizing pulses is inserted into the signal immediately before and after the vertical synchronizing pulses. These equalizing pulses (Shown in Figure 3.12) do not disturb the operation of either oscillator (as will be shown in the chapter on Synchronizing Circuit Fundamentals). Yet they do permit the vertical sync pulse to occur at the correct time after every field.

VERTICAL SYNC AND EQUALIZING PULSE DETAILS

The total vertical blanking interval (Figure 3.7) occupies 1 333μs (21 lines). This interval contains 12 equalizing pulses, the six serrations of the vertical sync pulse, and 12 horizontal sync pulses. The basic functions of these pulses were discussed previously. To review their basic functions briefly: during the vertical blanking interval, the electron beams of the camera and picture tubes are extinguished for 1 333μs. This vertical blanking begins when the beams approach the bottom of the screen. The vertical blanking remains on, while the electron beams retrace their paths to the top of the screen.

During the vertical blanking interval, the initial equalizing pulses are followed by the serrated vertical sync pulse, which triggers the start of each vertical retrace. These pulses are followed by more equalizing pulses and by horizontal sync pulses.

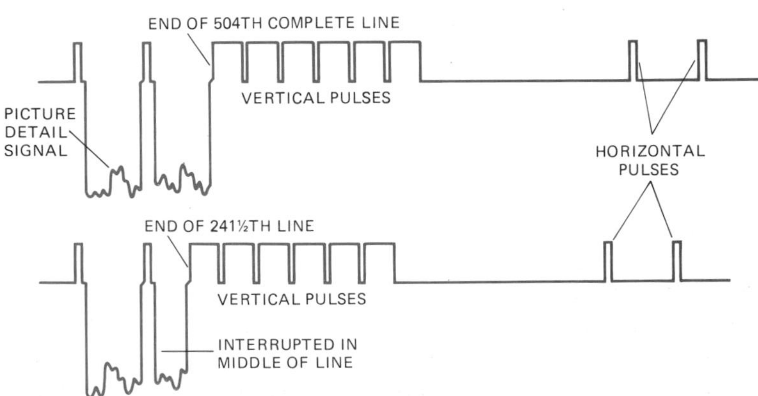

Figure 3.11 The form of the video signal at the end of 241½ and 504 lines (equalizing pulses are not shown).

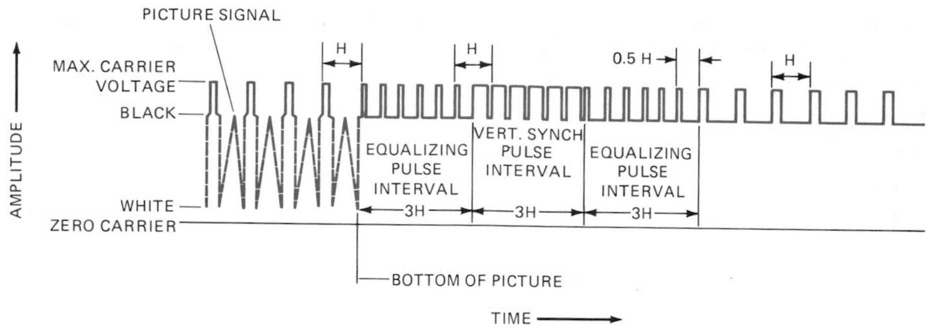

Figure 3.12 The position of the equalizing pulses in the video signal (H = 63.5 microseconds).

Throughout the entire vertical blanking interval, horizontal synchronization is continuously maintained by the action of all three of these pulses. However, note that equalizing pulses and the vertical serrations are only spaced $\frac{1}{2}$ H or 31.75μs apart. Thus, horizontal sync is performed only by those pulses which occur a full H, or 63.5μs apart.

Details of the vertical sync and equalizing pulses are shown in Figure 3.13A. Here we see that the width of a typical equalizing pulse is 0.04 × H (63.5 s) = 2.54μs. The spacing from the trailing edge of the last equalizing pulse of the first group to the leading edge of the first vertical pulse segment is 0.5 × H (63.5μs) = 31.75μs. Each vertical pulse segment has a width of 0.43 × H (63.5μs) = 27.31μs. Note that the value is more than five times wider than a horizontal sync pulse. The spacing between vertical pulse segments is 0.07 × H (63.5μs) = 4.45μs.

VERTICAL SAWTOOTH CURRENT ACTION

Figure 3.13B illustrates the vertical trace and retrace with respect to the front face of the picture tube. As with horizontal deflection, the current waveform is also a sawtooth. This vertical sawtooth of current is passed through a pair of vertical deflection coils which are arranged in a package with the horizontal deflection coils. These coils are located on the neck of the picture tube. The resultant magnetic fields (vertical and horizontal) cause the horizontal scanning lines to move downward in a linear fashion until an entire field is scanned. When the bottom of the screen is reached, the vertical sync pulse initiates the vertical retrace.

The time of one complete field is designated by the letter "V." V = $\frac{1}{60}$ s 16 667μs. Note that this period includes the vertical trace, plus the vertical blanking period. The vertical blanking period equals 1 333μs (21 lines). Therefore, the vertical trace period is 16 667μs − 1 333μs = 15 333μs (241 $\frac{1}{2}$ lines). For simplicity, the vertical retrace in Figure 3.13B is indicated as taking 1 333μs. However, the actual retrace from the bottom to the top of the screen takes about the time of five lines, or about 317μs. The balance of the 1 333μs (vertical blanking time) is taken up by about four lines (254μs) at the bottom of the raster and about 12 lines (762μs) at the top of the raster.

3.11 THE COLOR SYNCHRONIZING SIGNAL

When color is being transmitted (most programs), a special color sync burst signal is added to the back porch of all horizontal

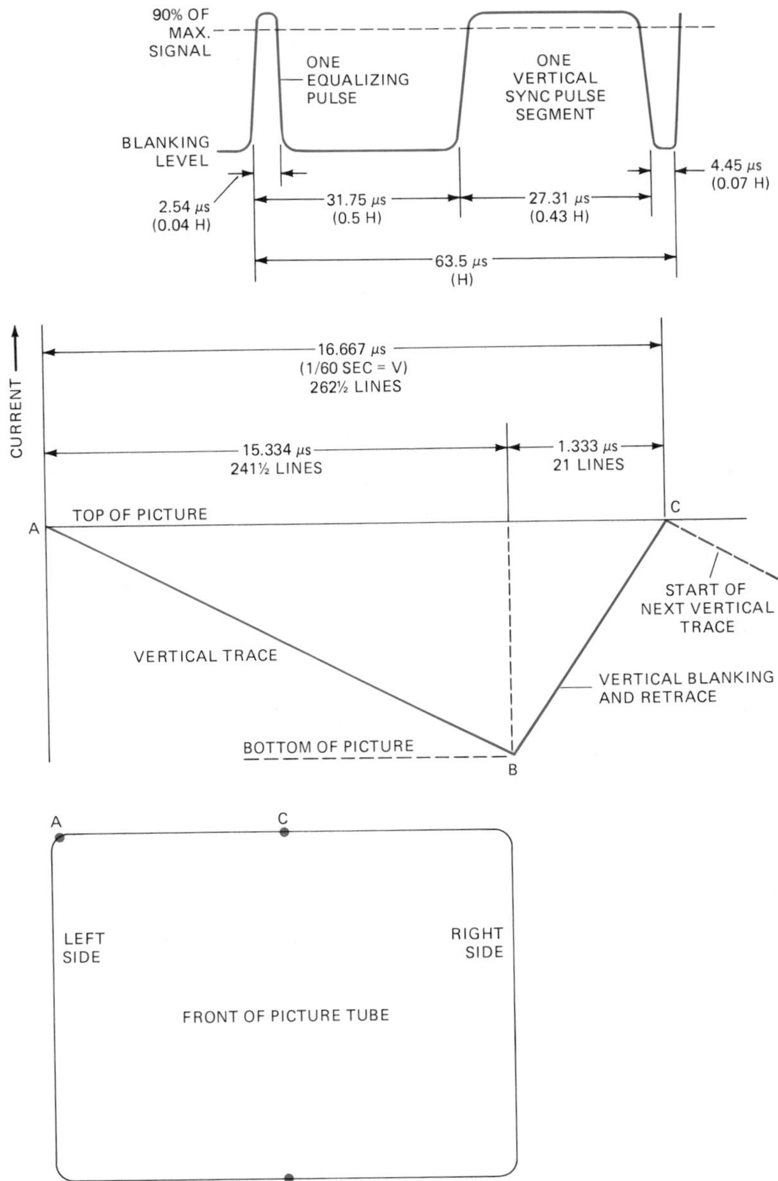

Figure 3.13A. Details of an equalizing pulse and a vertical pulse segment. The last equalizing pulse preceding the first vertical pulse segment is shown. **B.** The vertical trace-and-retrace current related to the front of a picture tube. This sawtooth of current passes through the vertical deflection coils at the rate of 60 Hz (for monochrome transmission).

blanking pulses, shown in Figure 3.14. (Compare this figure with Figure 3.8A). The dimensions of Figure 3.14 are identical with those of Figure 3.8A. The two figures differ only in the addition of the color burst. The burst consists of 8 to 11 cycles of 3.58 MHz. As mentioned in Chapter 1, this is the color subcarrier frequency. The burst has the same phase as the transmitter 3.58-MHz subcarrier.

The function of the burst in a color television receiver, when fed to the appropriate circuitry, is to insure that the colors at the receiver picture tube match those of the orig-

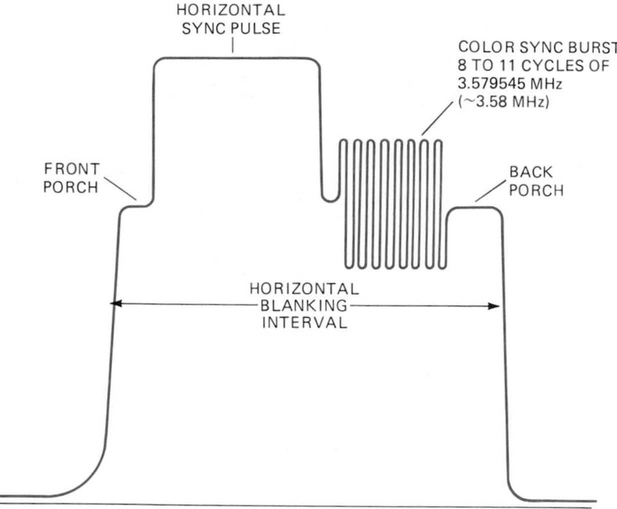

Figure 3.14 The color sync burst signal is present on the back porch of every horizontal blanking pulse, only when color is being transmitted.

inal scene at the studio. Since the burst is transmitted only with color programs, this provides a means for a color receiver to determine whether or not color is being received. In the absence of a color program (and burst), a color "killer" circuit deactivates the receiver color circuits. This prevents the appearance of colored "confetti" on the screen during monochrome broadcasts. If such confetti were present, it would be very annoying to the viewer.

The burst does not interfere with horizontal deflection synchronization. Note in Figure 3.14 that it follows the horizontal sync pulse. The burst has only half the peak amplitude of the horizontal sync pulse and does not affect horizontal synchronization. The burst occurs during the horizontal blanking and retrace time and does not affect the picture.

The manner in which the burst is used in a color television receiver is described in detail in the chapter entitled, "Principles of Color TV" (Chapter 9). Basic information regarding the color picture signal is presented in Chapter 4.

SUMMARY OF CHAPTER HIGHLIGHTS

1. Scanning and synchronizing are needed to keep the reproduced picture in step with the picture at the studio. (Figure 3.2)
2. The picture on the picture tube is "painted" by an electron beam. The beam moves back and forth scanning individual lines, which are very close together. (Figures 3.3 and 3.4)
3. The motion of the picture tube beam, from left to right, is called the trace. The motion of the return beam, from right to left, is called the retrace. (Figures 3.3 and 3.4)
4. In the absence of a picture, the picture tube beam will trace out the area contained by a white rectangle. This rectangle is called the raster. The width divided by the height is 4 to 3 and is called the aspect ratio. (Figure 3.4)

5. Each complete picture is known as a frame. There are 30 frames per second. Each frame is divided into two halves. Each half is called a field and there are 60 fields per second. (For color transmissions, the rates are slightly different.) (Section 3.4)
6. The frequency of 60 fields per second was selected to coincide with the standard United States power-line frequency of 60 hertz. (Section 3.4)
7. The use of 60 fields per second eliminates flicker in the picture.
8. Each television frame consists of 525 scanning lines. Each field has $262\frac{1}{2}$ scanning lines. (Section 3.5)
9. A system of interlaced scanning is used for television pictures. One field consists of lines 1, 3, 5, and so on. This is the odd-line field. The next field consists of lines 2, 4, 6, and so on. This is the even-line field. These two sets of lines fit in between each other, making up a complete frame. (Figure 3.4, Section 3.5)
10. Blanking pulses prevent the electron beam at the picture tube from lighting the screen during retrace periods. Horizontal sync pulses initiate the return of the electron beam from the right side to the left. Vertical sync pulses initiate the return of the beam from the bottom to the top. (Figures 3.5 and 3.6)
11. The composite monochrome video signal consists of picture detail, horizontal and vertical sync and blanking pulses, and vertical equalizing pulses. (Figure 3.7)
12. The sync pulses must be separated from the composite video signal before they can be used. This is done by a sync separator circuit. (Section 3.7)
13. There are 15 750 scanning lines per second (slightly less for color). (Section 3.9)
14. The color burst consists of a maximum of 11 cycles of a 3.58-MHz signal. It is superimposed on the back porch of the horizontal blanking signal. (Figure 3.14)
15. A sawtooth waveform of current passes through the horizontal and vertical deflection coils. This waveform is needed to provide linear scanning and rapid retraces. (Figures 3.8B and 3.13B)
16. The vertical sync pulse is serrated to permit continuous horizontal synchronization. (Figures 3.10, 3.11 and 3.12)
17. The equalizing pulses maintain accurate interlacing of the odd- and even-line fields.
18. The color sync burst signal has the same frequency (3.58 MHz) and the same phase as the transmitter 3.58-MHz subcarrier. The function of the burst is to aid in insuring that the colors at the receiver will match those at the transmitter. (Figure 3.14)
19. Tables 3.1, 3.2, 3.3 and 3.4 summarize important information presented in this chapter.

Table 3.1 **SCANNING FREQUENCIES**

ITEM	FREQUENCY, Hz
V field (monochrome)	60
V field (color)	59.94
H lines (monochrome)	15 750
H lines (color)	15 734.26

Table 3.2 **SYNC AND BLANKING PULSE FREQUENCIES**

ITEM	FREQUENCY, Hz
V sync and blanking (monochrome)	60
V sync and blanking (color)	59.94
H sync and blanking (monochrome)	15 750
H sync and blanking (color)	15 734.26
Equalizing pulses (monochrome)	31 500
Equalizing pulses (color)	31 468.52

Table 3.3 **DETAILS OF THE HORIZONTAL PERIOD**

ITEM	DURATION, μS
H (one horizontal cycle)	63.5 $\frac{(1.)}{(15\ 750)}$
Active picture time	53.34 (0.84 H)
Horizontal blanking	10.16 (0.16 H)
Horizontal sync pulse	5.08 (0.08 H)
Front porch	1.27 (0.02 H)
Back porch	3.81 (0.06 H)
Horizontal retrace	(approx.) 7.0 (0.11 H)
Color burst (11 cycles)	(approx.) 3.07 (0.05 H)

Table 3.4 **DETAILS OF THE VERTICAL PERIOD**

ITEM	DURATION, μS
V (one vertical field)	16 667 $\left(\frac{1}{60}\right)$
Active picture time ($241\frac{1}{2}$ lines)	15 334 (0.92 V)
Vertical blanking (21 lines)	1 333 (0.08 V)
Vertical-pulse segment width	27.31 (0.043 H)
Vertical-pulse segment space	4.45 (0.07 H)
Equalizing-pulse width	2.54 (0.04 H)
Equalizing-pulse space	29.21 (0.46 H)
6 Vertical sync pulses	190.5 (3 H)
6 Equalizing pulses	190.5 (3 H)
Vertical retrace (typical)	317.5 (5 H)

EXAMINATION QUESTIONS

(Answers provided at the back of the text)

Supply the missing word(s) or number(s).

1. The picture tube beam is kept in step with the camera tube by ______ signals.
2. The process of producing a raster is called ______ .
3. The aspect ratio is ______ to ______ , horizontal to vertical.
4. Horizontal retrace takes place from ______ to ______ .
5. Vertical retrace takes place from ______ to ______ .
6. The monochrome frame rate is ______ Hz. The monochrome field rate is ______ Hz.
7. In television, the power line frequency is the same as the ______ .
8. In interlaced scanning, one frame is made up of an ______ line field and an ______ line field.
9. There are ______ lines in one frame.
10. There are ______ lines in one field.
11. 15 750 Hz is the ______ scanning frequency.
12. 60 Hz is the ______ scanning frequency.
13. The color horizontal scanning frequency is ______ . The color vertical scanning frequency is ______ .
14. The composite monochrome video signal consists of horizontal and vertical sync and blanking pulses, equalizing pulses, and the signal.
15. At the end of each field, a ______ sync pulse is sent.
16. Horizontal retrace takes about ______ μs.
17. 5.08 μs is the duration of the ______ pulse.
18. Linear scanning requires a current wave having a ______ shape.
19. Equalizing pulses are required for proper ______ of the fields.
20. 1 333 μs represents the ______ time. This is equal to ______ scanning lines.
21. One complete field takes ______ μs.
22. The color burst frequency is ______ MHz. The burst consists of ______ to ______ cycles.
23. The color burst is positioned on the ______ of the horizontal blanking pulse.
24. Vertical serration is done to permit continuous synchronization.
25. Correct scanning timing is done by ______ and ______ sync pulses.
26. There is no picture tube beam current during ______ pulses.

REVIEW ESSAY QUESTIONS

1. Briefly discuss the need for vertical synchronizing and blanking signals in monochrome television.
2. Define: raster, aspect ratio, trace, and retrace.
3. Name all the basic components of a color, composite video signal.
4. Prepare a scale drawing of a simple, interlaced raster having two fields and not more than 10 lines in each field. (Do not show vertical retraces.)
5. Discuss briefly the problem of flicker and how it is overcome in television.
6. Prepare a scale drawing of the composite color video signal for two complete lines. Indicate the duration of the various parts. (See Figures 3.5, 3.14, and Table 3.3 in the Summary.)
7. Why are equalizing pulses used? What is their spacing in μs?
8. Refer to Figure 3.12. What would happen if the vertical sync pulse were not serrated?

9. Refer to Figures 3.8B and 3.13B. Why is it necessary to use the sawtooth waveforms?
10. Why are the vertical blanking pulses so much longer than the horizontal blanking pulses?
11. Discuss the details of a complete vertical retrace.
12. Define: front porch, back porch, serrated pulse, and color burst.

EXAMINATION PROBLEMS

(Selected problem answers are provided at the back of the text.)

1. Calculate the following as a portion of H: horizontal sync pulse, horizontal blanking pulse, active picture time, and back porch.
2. How many lines are in one field if the color horizontal frequency is 15 734.26 Hz and the color vertical frequency is 59.94 Hz?
3. Calculate the time in μs for one complete horizontal cycle (trace and retrace) for the English TV Standard. (Hint: horizontal frequency is 15 625 Hz.)
4. If the actual vertical flyback occurs in five lines, calculate the time in μs taken up by the remaining lines within the vertical blanking period.
5. Referring to Figure 3.7A, what is the time in μs, from the beginning of the first equalizing pulse to the beginning of the 5th vertical serration?
6. Identify the following periods or frequencies relating to scanning: 16 667 μs, 15 734.26 Hz, 63.5 μs, 31.75 μs, 31 468.52 Hz, 33 333 μs.
7. Prepare a table listing the following values: V frequency (monochrome), H frequency (color), equalizing pulse frequency (monochrome), H times (μs), V times (μs), color burst frequency (MHz), field active picture time (μs), and horizontal blanking time (μs).
8. Calculate the time in μs of one cycle of the color burst.
9. Draw two cycles of the horizontal sawtooth current wave. Identify the portions and relate them to the scanning lines.
10. In Figures 3.7A and B, identify the exact portions of the waves used for horizontal synchronization.

Chapter 4

4.1 Negative and positive picture phase
4.2 Why television requires wide frequency bands
4.3 Effect of loss of low and high video frequencies
4.4 Desirable picture qualities
4.5 Components of the color video signal
4.6 Normal and special video signals
Summary

VIDEO SIGNALS AND PICTURE QUALITY

INTRODUCTION

This chapter discusses negative and positive picture transmission. It also explains why wide bands are required for television. Important picture qualities, such as detail, contrast and brightness are considered. The composition of the color video signal is also explained. Special video signals, such as multiburst and VIR signals are also covered.

By the time you have completed the reading and work assignments for Chapter 4, you should be able to:

- Define the following terms: negative picture transmission, vestigial-sideband transmission, resolution, contrast, brightness, chrominance saturation, and hue.
- Discuss the difference between negative and positive picture transmission and list one important advantage of negative picture transmission.
- Know the effects of low and high video frequency on the TV picture.
- Discuss why it is necessary to apply the video signal with only positive picture phase to the picture tube grid.
- Discuss the need for vestigial sideband transmission.
- Discuss the need for wide frequency bands in television.
- List the desirable picture qualities.
- Know what the color subcarrier frequency is and define the range of the upper and lower color sideband frequencies.
- Discuss the luminance and chrominance portions of a color television signal.
- List the VITS and briefly describe the function of each one.

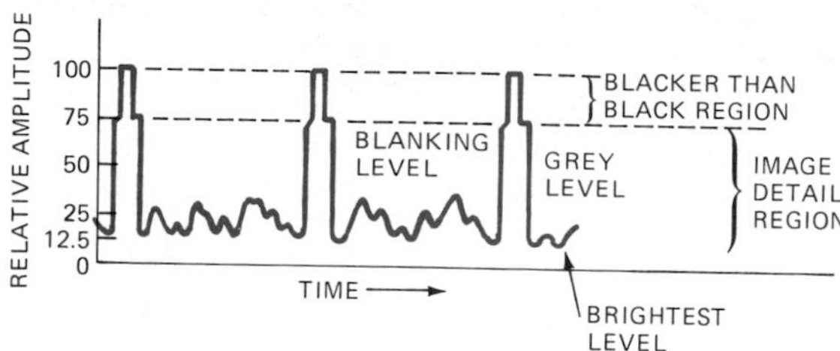

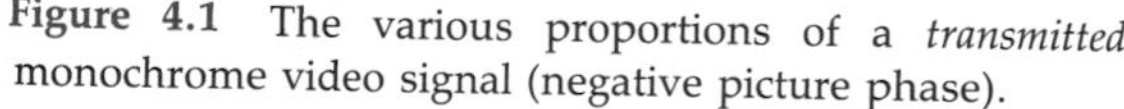

Figure 4.1 The various proportions of a *transmitted* monochrome video signal (negative picture phase).

4.1 NEGATIVE AND POSITIVE PICTURE PHASE

In the composite video signal shown in Figure 4.1, 75% of the total amplitude is reserved for picture signal variations. The remaining 25% is used for the horizontal and vertical synchronizing pulses. However, the full 75% available for the picture signal is not used for this purpose. Note that the brightest (maximum white) level is at 12.5%, not at zero. Zero level represents zero amplitude modulation. This could result in distortion of bright picture values.

NEGATIVE PICTURE PHASE

The wave in Figure 4.1 has "negative picture phase." The brightest parts of the picture produce the lowest amplitude video signals. Conversely, the darker and black parts of the picture produce higher amplitude video signals. The "blacker than black" signals (sync pulses), when added to the video signals, produce the highest amplitudes, reaching a maximum of 100%. This is the signal polarity which is used to amplitude modulate the picture carrier wave.

Negative Picture Transmission "Negative picture transmission" means that the video modulating signal has negative picture phase. It is a method of modulation that represents double-sideband amplitude modulation of the signal shown in Figure 4.1. Note in Figure 4.2 that the tips of the sync pulses produce the highest percentage of AM. This is the 100 percent RF carrier level. The RF carrier level for blanking pulses is at 75%. Below 75 percent, extending to 12.5 percent, are the picture brightness values. The 75% value represents black. The 12.5% value represents white. Various shades of grey exist between these two limits.

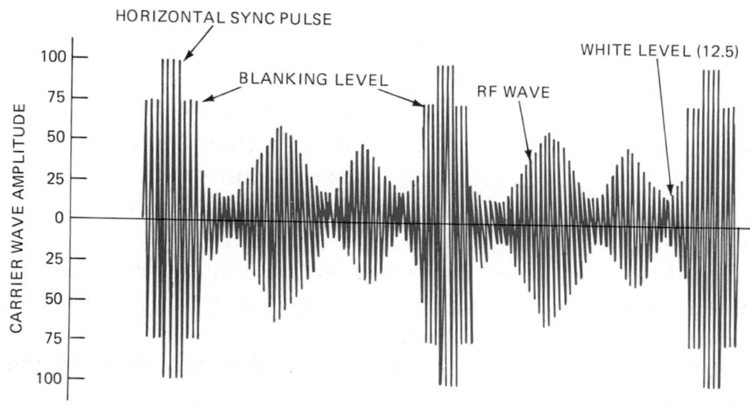

Figure 4.2 The amplitude-modulated video carrier wave. This wave results from modulation by a video signal similar to that in Figure 4.1, but simplified here for clarity.

Advantages of Negative Picture Transmission

AM negative picture transmission has two important advantages.

1. Some noise pulses add to the carrier amplitude. However, note in Figure 4.2 that any increase in carrier amplitude is always toward black. Thus, noise pulses are less apparent in the picture.
2. Pictures having a high brightness (white) content will produce a lower average percentage of modulation of the RF carrier wave. This results in a saving of transmitter power and produces more efficiënt operation.

RECEIVER VIDEO SIGNAL

In the receiver, before the video signal is applied to the control grid of the picture tube, the signal must possess positive picture phase (See Figure 4.3). The grid of the picture tube is then biased by enough negative voltage that the electron beam is automatically prevented from reaching the fluorescent screen when the blanking voltage portion of the signal is present. With the positive picture phase, the camera signal voltages are all more positive than the blanking pulse. On these portions of the video signal, the electron beam is permitted to impinge on the screen with varying amounts of electrons. A bright spot in the received image causes the grid to become more positive than when the voltage of a darker spot is applied. More electrons in the beam mean that more light is emitted at the screen when the beam has more electrons. Thus, different grid voltages produce images of various shades and light gradations.

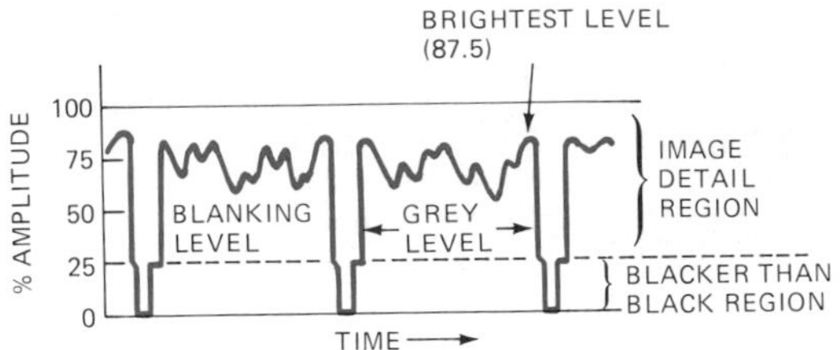

Figure 4.3 The form of the composite video signal as applied to the grid of a TV receiver picture tube *(positive picture phase)*.

The blanking voltage in the video signal prevents the electron beam from reaching the fluorescent screen. The blanking signal occurs at the point in the video wave where the darkest positions of the image are found. The beam is entirely cut off when the blanking voltage acts at the control grid of the picture tube. Nothing appears on the screen. The blanking level is then properly called the *black level,* because no light at all appears on the fluorescent screen.

Consider the video signal of Figure 4.3. The synchronizing pulses are found beyond the blanking level. When applied to the picture tube control grid along with the rest of the wave, the pulses drive the grid to a negative voltage even greater than the cutoff voltage. This pulse region is labeled *blacker than black,* because the position of the blanking signal was labeled black. The synchronizing pulses that pass through the video amplifiers with the video signal do not interfere in any way with the action of the control grid. Therefore, they are not removed at the picture tube. After the detector, the complete video wave is applied to the sychronizing- and video-amplifier circuits at the same time. The synchronizing clipper tube permits only the pulses to pass through to the sync circuits. On the other hand, the video amplifiers allow the entire signal to pass on to the picture tube.

4.2 WHY TELEVISION REQUIRES WIDE FREQUENCY BANDS

Wide-band amplifiers are used extensively in television receivers. These amplifiers are designed to amplify signals with spectrums extending from 4 to 6 MHz. The different forms which these amplifiers may assume and their characteristics are discussed in later chapters; however, the reason for the extremely wide bandwidth is considered now.

The more elements in a picture, the finer is the detail that is portrayed. The picture can also stand closer inspection before its smooth, continuous appearance is lost. Each $\frac{1}{30}$ second, 525 lines are scanned, or a total of 15 750 lines each second.

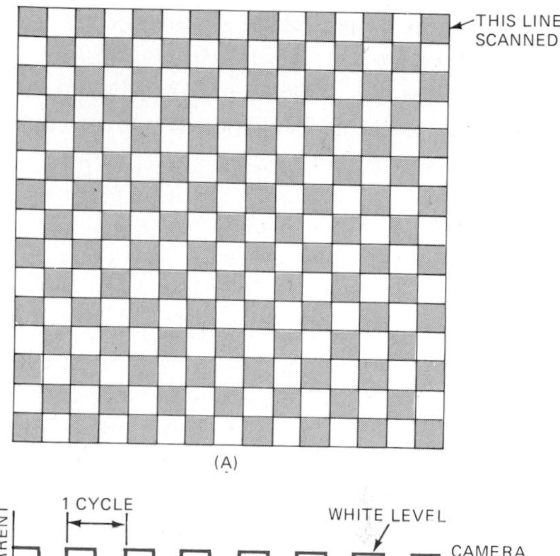

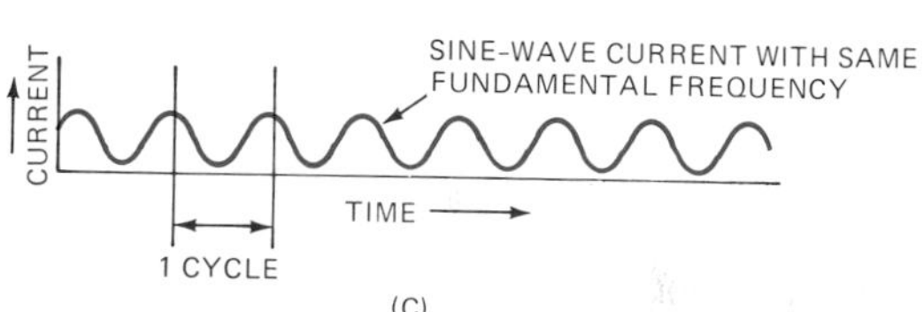

Figure 4.4 The number of elements in an image determines the width of the frequency band required to transmit it. (Only a small portion of a complete picture is shown here.)

ELEMENTS IN A PICTURE

Assume that a horizontal scanning line is 20 in (50.8 cm) long and that one picture element is .037 in (0.094 cm) long. Thus, there are 540 horizontal picture elements on one line. There are 15 750 lines per second. Therefore, (for monochrome TV) there are 15 750 × 540 = 8 505 000 elements transmitted per second. To use this number of elements most advantageously, the relationship between the number of elements and the bandwidth must be determined.

Imagine that the image plate in the camera tube is broken up into a series of black and white squares. Each square represents one element. A portion of the resulting pattern is shown in Figure 4.4A. As the scanning beam passes over each element in turn, a pulse of current flows every time a white square is reached. At the next (black) element, the current drops to zero. As one horizontal line is scanned, the electric pulses of current have the shape shown in Figure 4.4B.

Taking the total number of white and black squares on a line and dividing their sum by two gives the number of cycles the current goes through when one horizontal line is scanned. With 540 elements on a line, a fundamental frequency of 270 Hz per line is generated.

Under present standards, 525 lines are scanned in $\frac{1}{30}$ seconds, or a total of 15 750 in one second. Employing 540 elements per line, 8 505 000 picture elements are sent each second. For the present analysis, this results in a frequency of $\frac{8\,505\,000}{2}$ Hz, or 4.25 MHz. In actual practice, a bandwidth of about 4.2 MHz is allowed. This very large bandwidth must theoretically be passed by the video amplifier section alone.

SIDEBANDS

About 4.2 MHz are required to accommodate the video information alone. However, the *transmitted* bandwidth is 6 MHz. Of the extra 2 MHz, the FM audio signal uses 50 kHz (± 25 kHz). Not all of the bandwidth is utilized. The reason that extra space exists is explained in the process for generating the television video signal.

Double Sidebands

On ordinary AM radio broadcast frequencies, most stations occupy a 10-kHz bandwidth, or 5 kHz on each side of the carrier position. If a station is assigned the frequency of 700 kHz, it transmits a signal that occupies just as much frequency space on one side of 700 kHz as on the other. With a modulating frequency of 5 kHz, the deviation is 5 kHz (or 5,000 Hertz) on either side of the carrier position of 700 kHz. These side frequencies are known as *sidebands.* For the present illustration, each sideband may have a maximum deviation of 5 kHz about the mean (or carrier) position. The signal information is contained in the sidebands, since they are not generated until sounds are projected into the microphone. At the radio receiver, the variations in *one* of the sidebands are transformed into audible sounds and heard by the radio listener.

The sidebands with frequencies that are higher than the carrier frequency contain the same information as the sidebands with frequencies that are lower than the carrier. If one set of sidebands is eliminated above or below the carrier, the receiver still obtains all of the necessary information. While single sideband transmission does exist for certain communication facilities, there is an economic reason for not eliminating a sideband. A transmitter naturally generates both sidebands. It is often cheaper to transmit both than it is to try to eliminate one, using expensive and complicated filters.

Video Signal

Now, consider the video signal. It is generated by the same type of circuitry that is used at the AM broadcast frequencies. Since 4.2 MHz is needed for the picture detail, a signal generated would extend 8.4 MHz, or 4.2 MHz on either side of the carrier. This spectrum does not include the sound sidebands. An 8.4-MHz band is undesirable because of the bandwidth occupied and the difficulties of transmitting a signal of this wide bandwidth. In situations like this, it is necessary to remove most of one sideband.

Vestigial Sidebands

The undesired sideband is largely removed by filters that follow the last RF amplifier of the television transmitter. However, it is not easy to construct filters that will cut off one sideband completely and leave only the desired sideband. Furthermore, nothing must occur in the process of eliminating one sideband that changes the amplitude or the phase of any of the components in the desired sideband. A suitable compromise is to remove most of one sideband. In this way, the remaining sideband is not affected by the filtering. Part of the transmitted 6-MHz bandwidth is occupied by what may be called the remnants of the undesired sideband. This method is known as a *quasi-single-sideband or vestigial-sideband* operation.

Figure 4.5A shows a television video signal with both sidebands. Figure 4.5B shows a signal that has been filtered to par-

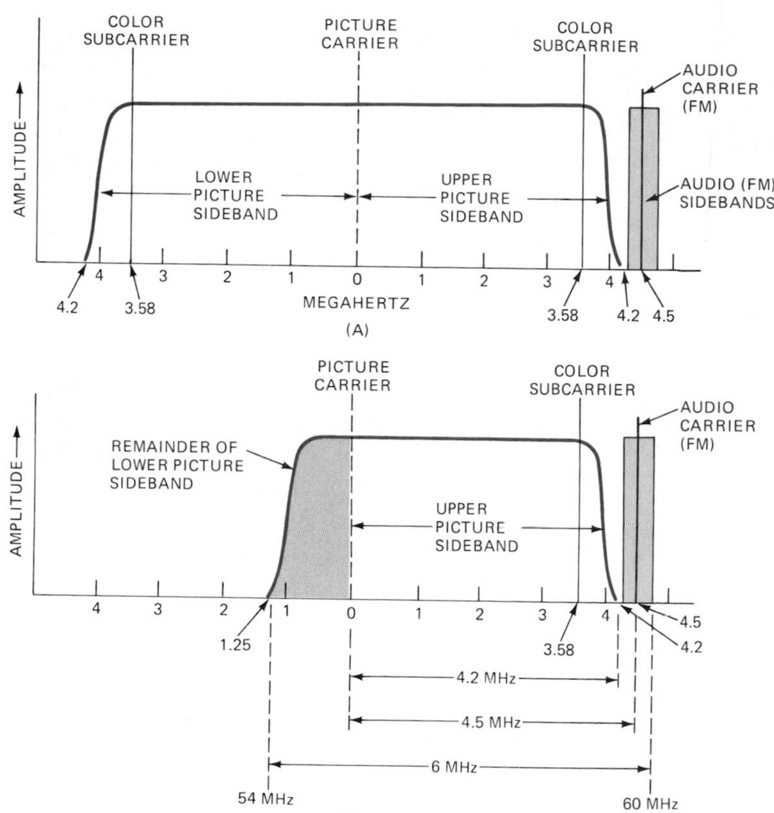

Figure 4.5A. The TV transmitted frequency spectrum as it would appear if both complete picture sidebands were sent. **B.** The actual TV frequency spectrum. Note that the lower picture sideband is reduced in bandwidth (vestigial-sideband transmission).

tially remove one sideband. The frequency of the picture carrier is found 1.25 MHz above the low-frequency edge of the television channel. Then for 4.2 MHz above the picture carriers, there is the television video signal with the desired picture and sync information. Note in Figure 4.5B that the color subcarrier is 3.58 MHz above the picture carrier. The FM sound carrier is 4.5 MHz above, with a maximum deviation of ± 25 kHz. The color video sidebands are on both sides of the 3.58-MHz subcarrier. These sidebands extend about 0.6 MHz above and 1.5 MHz below 3.58 MHz. There is no interference between the color sidebands and the monochrome upper sideband. The reason for this will be explained in the chapter entitled, "Principles of Color TV" (Chapter 9). A 0.3-MHz bandwidth separates the high-frequency edge of the video signal and the FM carrier. In this manner, the allotted 6 MHz are distributed.

4.3 EFFECT OF LOSS OF LOW AND HIGH VIDEO FREQUENCIES

Uniform response over a 4.2-MHz band is desired in the picture IF and video amplifiers. However, special circuit designs must be used to achieve this. (These circuits are fully explained in later chapters of this book.) For the moment, it is only necessary to point out the effects of poor response at the high- or low-frequency ends of the band.

HIGH-FREQUENCY LOSS

It has been shown that a greater number of picture elements require a greater bandwidth. Since picture detail is determined mainly by the number of very small elements, any decrease in the response at the higher frequencies will result in less fine detail available at the receiver picture tube screen. The picture will lose some of its sharpness and may even appear somewhat blurred if the high-frequency response is degraded enough. In commercial monochrome television receivers, a video passband of 3.3 MHz is generally considered good. Anything below 2.5 MHz is undesirable. In home video-tape recorders and in some portable television receivers the video response is limited to 2.5 MHz. This value provides adequate detail, particularly when a color picture is present. The color seems to make the decreased detail less noticeable. The color portion of the signal will be affected if the frequency response of some of the circuits in a color television receiver falls below about 4.2 MHz. This bandwidth is required to pass the 3.58-MHz color subcarrier and upper sideband.

LOW-FREQUENCY LOSS

"Low" frequencies in television are generally considered to extend from about 30 Hz to 100 kHz. Within this range, the band from approximately 30 Hz to 10 kHz is responsible for correct background shading. (Frame to frame changes in background shading occur at the frame rate of 30 Hz.) Amplitude or phase distortion within this range will result in incorrect shading of large picture areas. In addition, slow changes or background shading may be lost or distorted.

The band from approximately 10 kHz to 100 kHz is responsible for the reproduction of relatively large horizontal areas, such as lettering and other large, horizontal details. Loss or distortion of these frequencies may result in changes of brightness values of large horizontal details.

WHY FREQUENCIES ARE LOST

For all practical purposes, all frequencies are correctly transmitted by television stations. Therefore, any loss is confined to receiver defects or design. Some high frequencies in the range of 100 kHz to 4.2 MHz may be beyond the scope of some television equipment that is deliberately limited to a top of 2.5 MHz. However, a "loss" of high frequencies can also occur due to a deficiency in a television receiver caused by misalignment of wideband-tuned circuits, or by a defective component in a tuned circuit or a wideband video amplifier. A "loss" of low frequencies (30 Hz to 100 kHz) is often caused by a defective component in the video amplifier circuits.

The above points will be elaborated upon in later chapters.

4.4 DESIRABLE PICTURE QUALITIES

The picture must have certain qualities if the viewer is to consider it pleasing. The more important of these qualities are discussed in the following paragraphs.

PICTURE DETAIL

The picture should resolve enough small picture elements to bring out fine details clearly, including vertical edges. (The terms "resolution" and "definition" are also used to describe fine picture detail.)

The picture on a small screen television receiver may appear sharper and clearer than the picture on a large screen set. However, both have approximately the same number of picture elements. The difference is caused by the spreading out of picture elements and scanning lines on the larger screen. The detail on a television screen is about equal to that of 16-mm film. This is roughly 160,000 picture elements per frame.

PICTURE CONTRAST

Television receivers have a user-operated "contrast" control. This control adjusts the amplitude of the video signal applied to the control grid of the picture tube. Adjusting this amplitude sets the ratio of the dark-to-light portions of the picture. Once set, the contrast control is rarely changed. The setting depends upon (a) viewer preference, (b) the particular receiver, and (c) ambient light falling upon the picture tube. Occasionally, the differences in specific programs may require temporary readjustment. Higher contrast settings are more desirable in greater ambient light. A low-contrast picture appears "washed out" when viewed with fairly high ambient light.

PICTURE BRIGHTNESS

Picture brightness is closely associated with contrast. Brightness and contrast should be adjusted together to suit the receiver, the viewer, and the ambient lighting conditions. A "brightness" control will vary the overall illumination. Some television receivers have an automatic brightness control feature. This automatically adjusts picture-tube brightness to compensate for different room-ambient lighting conditions. In these receivers, a manual brightness control is provided to set the desired range.

Brightness refers to the average intensity of the overall picture. Unlike contrast, it does not vary from scene to scene. Once set, the brightness level remains constant. (Except for automatic control, of course.) The brightness control varies the DC bias between the control grid and cathode of the picture tube. Lesser values of negative bias produce a brighter picture.

Other factors affecting brightness are (1) the type of picture tube, and (2) the final accelerating potential at the picture tube (high voltage). To improve the apparent brightness (and contrast), many picture tubes have a special external coating. This coating tends to absorb, rather than reflect, the ambient light.

COLOR SATURATION

Color saturation refers to the amount of color in a picture. It may also be described as color intensity, color amplitude, or color level. It is a function of the amplitude of the 3.58-MHz color signal. When this is added to a monochrome picture, a complete color picture is formed. Controlling the amount of added color is somewhat similar to controlling contrast. They both represent adjustment of signal amplitudes. The control which adjusts the color saturation is called the **color control** or the **chroma control.** When it is turned fully counterclockwise, the picture appears in black and white. As the control is turned clockwise, colors begin to appear. At the extreme clockwise position, the color intensity is probably too strong for correct viewing. In practice, the monochrome picture is first correctly adjusted for brightness and contrast. Then the color control is slowly advanced until the desired effect is reached.

COLOR TINT

Color tint (also known as *hue*) refers to the actual color of objects. The control which affects this quality is the hue or tint control. The most accurate way to set this control is to adjust it for correct flesh tones of the face. If this is done, the remaining colors will be correct. Automatic circuits in many television receivers provide correct colors without the need for manual control.

One such circuit, described briefly in Chapter 1, utilizes the VIR (Vertical Interval Reference) signal. This circuit automatically provides the correct tint and saturation of colors. There are other types of automatic color control (ACC), also called automatic tint control (ATC). The principles of these controls and VIR control will be described in a later chapter.

Unlike the contrast and color saturation controls, which control the signal amplitude, the hue (or tint) control actually varies the phase of the 3.58-MHz color signal.

GAMMA

This term comes from the Greek letter gamma (γ). In the television system, the measurement of gamma is applied to camera tubes and to picture tubes. Gamma is a number which expresses the compression or expansion of original light values. Such variations (if present) are inherent in the operation of the camera tube or picture tube.

With camera tubes, the gamma value is generally 1. This value represents a linear characteristic that does not change the light values from the original scene when they are translated into electronic impulses, (see Figure 4.6A). However, the situation is different in the case of picture tubes which have a gamma of approximately 3. The number varies slightly for different types of tubes and for different manufacturers. For picture tubes it is desirable to provide improved contrast. Emphasizing the bright values to a greater degree than the darker values accomplishes this. Figure 4.6B is a typical, picture tube, gamma characteristic. Note that the "bright" portions of the signal operate on the steepest portion of the gamma curve. Conversely, the "darker" signal portions operate on a lesser slope. Thus, it can be seen that the bright picture portions will be emphasized to a greater degree than the darker picture portions.

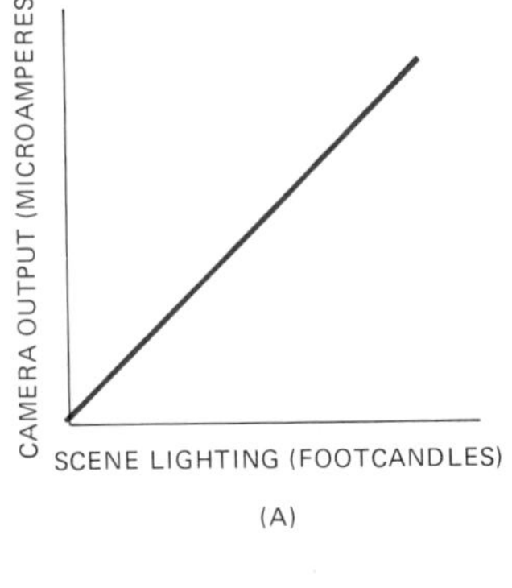

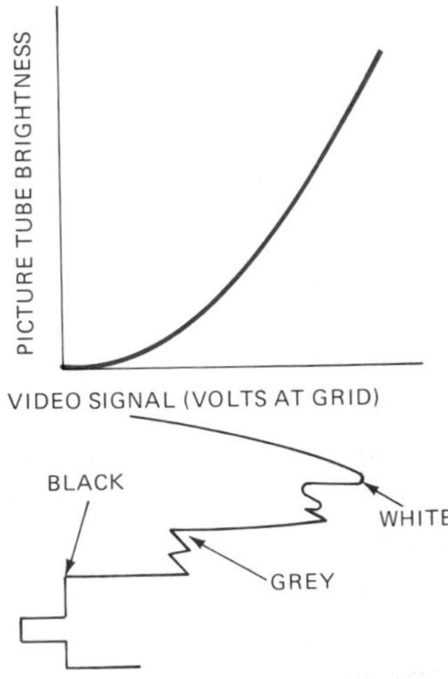

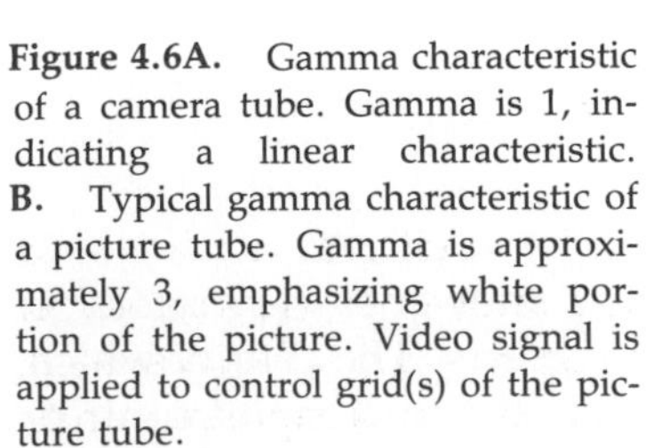

Figure 4.6A. Gamma characteristic of a camera tube. Gamma is 1, indicating a linear characteristic. **B.** Typical gamma characteristic of a picture tube. Gamma is approximately 3, emphasizing white portion of the picture. Video signal is applied to control grid(s) of the picture tube.

4.5 COMPONENTS OF THE COLOR VIDEO SIGNAL

The NTSC color television system is a compatible one. That is, the complete color signal produces a normal black and white picture on a monochrome receiver. In addition, a black and white television broadcast is correctly reproduced in monochrome on both color and monochrome receivers.

The color portion of the composite video signal consists of sidebands of the 3.58-MHz color subcarrier. When viewing a color broadcast on a monochrome receiver, it is undesirable to have the color signal components appear on the picture tube. These components will cause an objectionable interference pattern on the scene. Therefore, the 3.58-MHz signal is not permitted to pass through the video amplifier.

THE COMPOSITE COLOR VIDEO SIGNAL

The composite color video signal is the combination of two types of signals. It contains the monochrome, or "Y," *(luminance)* signal and the color *(chrominance)* signal.

The "Y" *(luminance)* signal is a complete monochrome television signal. By itself, it is capable of reproducing a black and white picture on either a monochrome or a color television receiver. The Y signal is also the only one supplying information relating to fine picture detail. As shown in Figure 4.5B, the luminance signal sidebands occupy a bandwidth extending from about 1.25 MHz below the picture carrier to about 4.2 MHz above the picture carrier. Only the upper sideband carries the fine detail information.

By itself, the chrominance signal is not capable of reproducing a useable picture. It represents in full color only the medium to large areas of the picture. The chrominance signal modulates the 3.58-MHz color subcarrier. The color sidebands extend from about 1.5 MHz below 3.58 MHz to about 0.6 MHz above 3.58 MHz. (This signal is more fully explained in a later chapter.)

In the color television receiver circuits, the luminance and chrominance signals are combined in such a way that the original red, blue, and green color signals are recovered. When these are fed to the color picture tube together with the luminance signals, a full color picture is displayed on the screen.

PROPERTIES OF THE EYE

The human eye sees a full color range only when the object is relatively large. When the detail becomes small, the eye can only discern changes in brightness. These properties of the eye are utilized in the NTSC color television system. Only the relatively large and medium-sized areas are sent in color. The fine detail is sent only in black and white. (Actually, the "Y" signal represents the entire picture, in monochrome.)

4.6 NORMAL AND SPECIAL VIDEO SIGNALS

The appearance of normal composite, color video signals on an oscilloscope is described first. Then, the uses of some special video signals are described.

OSCILLOSCOPE PHOTOS

Figure 4.7 shows oscilloscope photos of two scanned horizontal lines (Figure 4.7A) and two fields (Figure 4.7B). Figures 4.7A and 4.7B show the entire composite, color

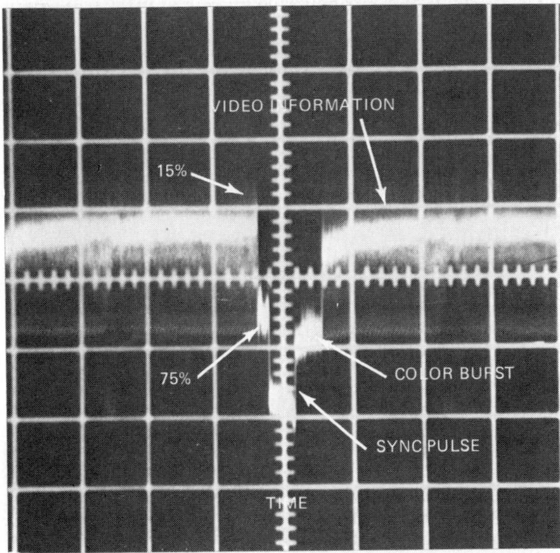

A

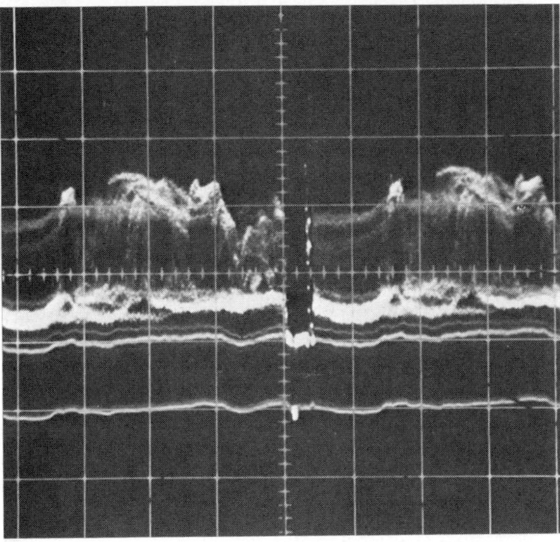

B

Figure 4.7 Oscilloscope photos of the composite color video signal. **A.** Two horizontal *lines* of video information, including horizontal sync and blanking pulses and the color bursts (positive picture phase). **B.** Two complete *fields* of video information. The vertical blanking and synchronizing pulses are at the center of the display. *(Courtesy of Sencore.)*

video signal. Note the video information, the horizontal blanking and sync pulses, and the color burst. To obtain this display, the oscilloscope sweep frequency is operated at one-half the horizontal scanning frequency, or 7875 Hz.

Two complete fields are shown in Figure 4.7B, with the vertical sync and blanking pulses at the center of the display. The two horizontal lines at the bottom of this photo are caused by the "sweeping by" of horizontal sync pulses that are not locked in. This happens because the oscilloscope sweep frequency is now set at 30 Hz (the horizontal rate is 15 734.26 Hz for a color signal). Because of this "sweeping by," it is difficult to see the vertical sync and blanking pulses clearly. However, by stripping the sync pulses from the picture information and then separating the horizontal and vertical sync pulses, each type can be seen clearly. The techniques for accomplishing this will be described in a later chapter. Actually, these techniques are a normal part of television receiver operation.

SPECIAL VIDEO SIGNALS

In accordance with FCC Rules and Regulations, (Part 73.682), lines 17 through 21 of the vertical blanking interval may be used for the transmission of test-, cue-, control-, and program-related data signals. Except for program-related data signals (line 21), the special signals on lines 17 through 20 are known as Vertical Interval Test Signals (VITS). Test signals may include a signal to supply reference modulation levels. This signal transmits light-intensity variations viewed by the camera. Other test signals are designed to check the performance of the overall television transmission system, or of its individual components. Cue and control signals are related only to the operation of

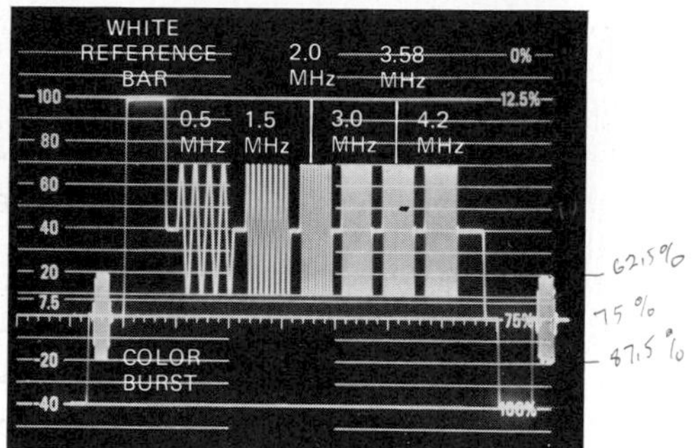

Figure 4.8 The multiburst test signal that is inserted in odd fields on line 17 of the vertical blanking interval. IRE units are at the left and percentages of amplitude at the right. *(Courtesy of Tektronix, Inc.)*

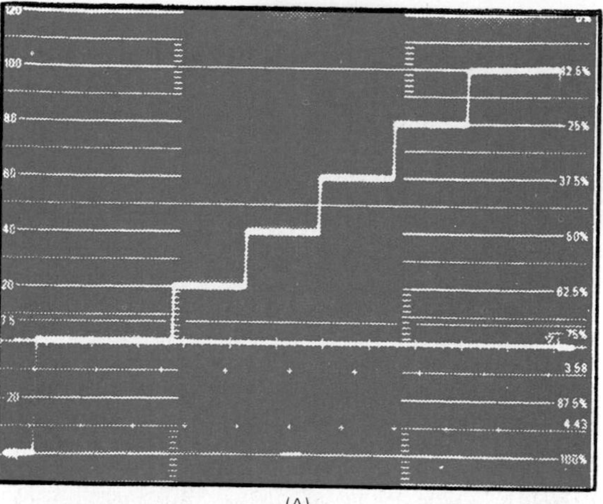

(A)

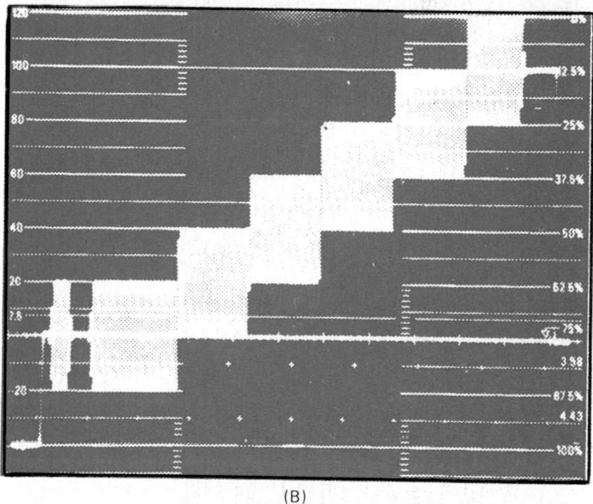

(B)

Figure 4.9 Five-step linearity test signals. A. Unmodulated staircase signal used for luminance greyscale checks. B. Modulated (3.58-MHz) staircase signal. May be keyed on during a selected line in the vertical blanking interval.

the individual television broadcast station. These special signals are valuable to engineers at the stations.

Multiburst Test Signal

(Per FCC Part 73.699) The multiburst-test signal is illustrated in Figure 4.8. It may be inserted only in the odd field, on line 17 of the vertical blanking interval. The numbers on the left side of the figure are IRE units. The numbers on the right are percentages. At the left of the figure is the color burst. This is followed by a white reference bar at 100% IRE, or 12.5% amplitude. Next are the multiburst frequencies of 0.5 MHz, 1.5 MHz, 2.0 MHz, 3.0 MHz, 3.58 MHz and 4.2 MHz. These frequencies are useful in checking the amplitude and frequency response of the television system.

Staircase Test Signal

Figure 4.9 illustrates the staircase test signal. In Figure 4.9A, the signal is unmodulated and there are equal steps of luminance changes from black to white. These steps progress through varying values of gray (gray scale). This signal is used to evaluate gray scale tracking. This signal and the modulated one following may be keyed on during a selected line of the vertical blanking interval. Lines 11 through 22 in the odd field, or lines 324 through 335 in the even field are available to carry the staircase test signal.

Figure 4.9B is the same staircase signal after it is modulated at 3.58 MHz (burst phase). Some of the applications of this modulated signal include measurements of differential gain, luminance signal linearity,

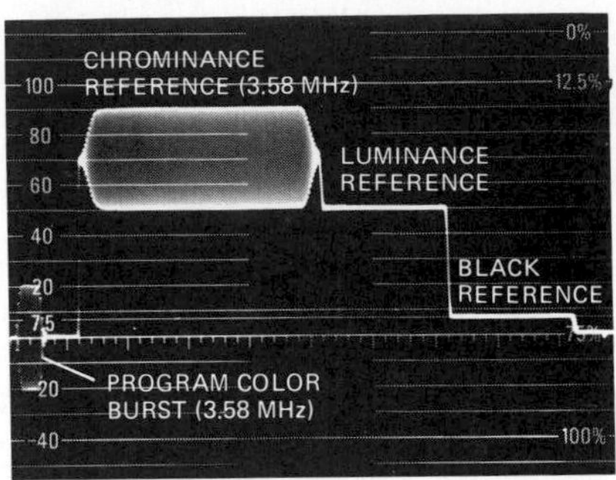

Figure 4.10 The vertical interval reference (VIR) signal is transmitted in odd and even fields, on line 19. The chrominance reference signal has the same phase as the program-color burst signal, at 3.58 MHz.

dynamic gain, luminance signal distortion caused by chrominance signal nonlinearity, and burst-phase errors.

Vertical Interval Reference (VIR) Signal

The VIR signal was discussed previously, mainly with regard to its function of maintaining correct hue and saturation values in color television receivers. The VIR signal can also be used in the adjustment of the playback conditions of a VTR. It also has a very important function in television color transmission. When combined with suitable circuitry, it enables automatic correction of the following: (1) video gain, (2) chrominance to luminance gain ratio, (3) black level, and (4) chroma phase, color-burst amplitude, and sync amplitude errors.

The VIR signal is illustrated in Figure 4.10. Note the program color burst (extreme left), the chrominance reference, and the luminance and black reference levels. (This signal is per FCC Part 73.699.) It is sent on line 19 of both fields.

Program Related Data Signal

In the vertical blanking interval shown in Figure 4.11, all of line 21 of the odd fields and the first half of line 21 of the even fields may be used to transmit a program-related digital data signal (per FCC Part 73.699). The first half of line 21 of the even fields is used for a framing code to be used by the data decoder in the receiver.

A program-related data signal will provide a visual readout on the receiver picture tube of information which is being transmitted at the same time as the audio signal. Non-program-related data signals may also be transmitted, if the information to be dis-

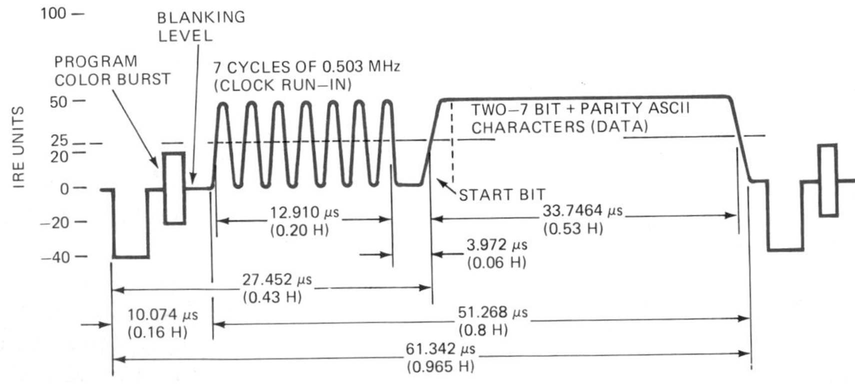

Figure 4.11 The data signal format on line 21 in odd fields of the vertical blanking interval. *(FCC, Part 73.699)*

played is of a broadcast nature. Some of the types of information which may be transmitted by data signals and displayed visually are (1) text for deaf viewers, (2) emergency messages, (3) other broadcast messages, (4) correct time, and (5) channel number.

SUMMARY OF CHAPTER HIGHLIGHTS

1. 75% of the total composite video signal is reserved for the picture signal. The remaining 25% is occupied by the horizontal and vertical synchronizing pulses. (Figures 4.1 and 4.3)
2. In negative picture transmission (U.S. Standard), the blackest portions of the picture produce the highest percentage of AM. An important advantage of this system is the reduction of the appearance of noise pulses in the picture. (Figures 4.1 and 4.3)
3. At the receiver picture tube, the video signal is applied to the grid with positive-picture phase. (Section 4.1, Figures 4.1 and 4.3)
4. Wide-frequency bands are an inherent requirement of the television system. These bands are needed to transmit and reproduce a finely detailed picture. (Section 4.2, Figures 4.4 and 4.5)
5. The television channel occupies a 6-MHz bandwidth. Of this, 5.45 MHz is the picture signal bandwidth. The remainder is taken up by the FM sound signal and a guardband. (Figure 4.5)
6. The color subcarrier has a frequency of 3.58 MHz. Its sidebands extend from 0.6 MHz above to 1.5 MHz below 3.58 MHz. (Section 4.2, Figure 4.5)
7. In vestigial-sideband transmission, a portion of one sideband is not transmitted. (Figure 4.5)
8. A loss of high video frequencies will degrade fine picture detail. A loss of low video frequencies will result in incorrect background shading. (Section 4.3)
9. The television picture qualities are (1) Picture detail, (2) Picture contrast, (3) Picture brightness, (4) Color saturation, (5) Color tint, (6) Gamma. (Section 4.4)
10. The NTSC color television system is a compatible one. A color TV signal can produce a monochrome picture on a black and white receiver. Also, a monochrome transmission will be reproduced in black and white on a color TV receiver. (Section 4.5)
11. Only relatively large- and medium-sized areas are reproduced in color. (Section 4.5)
12. The composite color video signal is the combination of the luminance and chrominance signals. (Section 4.5)
13. In Figure 4.7B the two horizontal lines at the bottom of the signal are caused by the "sweeping by" of horizontal sync pulses.
14. The multiburst test signal is used to check the frequency response of the television system. It is inserted in odd fields on line 17 of the vertical blanking interval. (Figure 4.8)
15. The staircase, unmodulated signal is used to evaluate gray scale tracking. It may be inserted in lines 11-22 of odd fields, or lines 324-335 of even fields, during the vertical blanking interval. (Figure 4.9)
16. With proper circuitry, the VIR signal can be used to provide automatic correction of various transmission factors. It is sent on line 19 of both fields. (Figure 4.10)
17. The program-related data signal may be

inserted in all of line 21 of odd fields, plus the first half of line 21 of even fields. One use is to display written text for deaf viewers, simultaneously with the picture and audio message. (Figure 4.11)

EXAMINATION QUESTIONS

(Answers are provided at the back of the text.)

Part A Select correct multiple choice answer.

1. The correct meaning of "negative picture transmission" is
   a. the picture is transmitted similarly to a photo negative.
   b. the AM of the picture carrier increases for dark picture areas.
   c. the AM of the picture carrier increases for bright picture areas.
   d. the lower sideband of picture transmission is partially filtered.
2. In the composite video signal, the picture signal
   a. extends from 0 to 75%.
   b. has a permissible amplitude from 12.5% to 100%.
   c. may extend from 12.5% to 75%.
   d. is transmitted as a single sideband.
3. The picture AM, lower and upper sideband frequencies, may extend from
   a. 1.25 MHz to 4.2 MHz.
   b. 4.0 kHz to 4.0 MHz.
   c. 0.25 MHz to 4.5 MHz.
   d. 1.25 MHz to 3.58 MHz.
4. Refer to Figure 4.7. The arrow pointing to the 75% level refers to
   a. the sync pulse level.
   b. the color level.
   c. the blanking level.
   d. the luminance level.
5. The multiburst signal is inserted during the vertical blanking interval
   a. in even fields on line 17.
   b. in line 17, during odd fields.
   c. in all fields, during line 17.
   d. in line 18, during odd fields.
6. The vertical interval reference (VIR) signal is used to
   a. check the TV transmitter frequency response.
   b. control automatic color circuits in the TV receiver.
   c. check the TV receiver frequency response.
   d. check the amplitude and phase of the picture carrier.
7. The data signal format may be inserted
   a. on the color subcarrier.
   b. in the vestigial sideband.
   c. on line 20, odd fields, of the vertical blanking interval.
   d. on line 21, odd fields, of the vertical blanking interval.

Part B Supply the missing word(s) or number(s).

1. In the composite video signal, the maximum white signal is at ______ percent.
2. With negative picture phase, 100% amplitude is reached by the tip of the ______.
3. The effect of noise pulses on the picture is reduced by ______ picture transmission.
4. Blanking pulses cut off the ______ of the picture tube.
5. The maximum deviation of the TV FM sound signal is ______ kHz.
6. To reproduce fine picture detail, the maximum video frequencies may extend up to about ______ MHz.
7. The color video sidebands extend to about ______ MHz above 3.58 MHz and to about ______ MHz below 3.58 MHz.
8. With vestigial-sideband transmission, the lower frequency sideband is reduced to a maximum of ______ MHz.
9. A loss of high frequencies will result in a loss of ______ in the television picture.
10. Picture contrast is the ratio of the ______ to ______ portions.

11. Color intensity is more accurately called color ________.
12. The actual color of objects is called the ________ or ________ of the objects.
13. The number which expresses the compression or expansion of light values is known as ________.
14. A color video signal is composed of the ________ and ________ components.
15. By itself, the luminance signal is capable of producing a ________ picture.
16. The three primary colors in a color television picture are ________, ________, and ________.
17. The abbreviation VITS stands for ________.
18. The staircase test signal may be used to check the gray ________.

REVIEW ESSAY QUESTIONS

1. Discuss briefly the difference between negative and positive picture transmission. What is one important advantage of negative picture transmission?
2. List the amplitude percentages in the composite video signal of: (a) Brightest picture signal, (b) Blanking level, (c) Tip of sync pulses, and (d) Darkest picture signal.
3. Refer to Figure 4.2. Why is the brightest white level at 12.5% not at 0% AM?
4. Draw to scale a composite monochrome video signal, which at the grid of a picture tube produces three equally spaced white vertical bars. (Hint: Separate with black bars.)
5. Why is it necessary to apply the video signal with only positive picture phase to the picture tube grid?
6. What is meant by the "blacker-than-black" region of the video signal?
7. Refer to Figure 4.4. Discuss briefly the need for wide frequency bands in television.
8. Name two television-related devices which may employ a video bandwidth of approximately 2.5 MHz.
9. Draw to approximate scale a simple diagram of the 6-MHz wide TV channel (video frequency band), labeling all sections.
10. What is the color subcarrier frequency? What is the range of the upper and lower color sideband frequencies?
11. Define vestigial-sideband transmission. Why is it used in television transmission?
12. Describe the appearance of the picture on a TV receiver whose upper video frequency limit is 1.0 MHz.
13. In a TV picture, what are the frequencies between 30 Hz and 10 kHz responsible for reproducing? From 10 kHz to 100 kHz?
14. Define the following terms: (a) Resolution, (b) Contrast, (c) Brightness, (d) Saturation, (e) Hue, and (f) Gamma.
15. Explain what is meant by a "compatible" color television system.
16. Is color sent in fine detail? Explain the reason for your answer.
17. Discuss briefly the luminance and chrominance portions of a color television signal.
18. Refer to Figure 4.7. Explain why the "burst" signal does not interfere with normal horizontal synchronization.
19. List the VITS. Briefly describe the function of each one.
20. Is the program-related data signal one of the VITS? What is its general purpose?

EXAMINATION PROBLEMS

(Selected problem answers are given at the back of the book.)

1. Make up a simple table showing the amplitude levels of the various parts of a composite, color video signal.
2. Refer to Figure 4.4. If each horizontal scanning line contains 350 picture elements, what video frequency does this represent?
3. In Problem 2 above, if the scanning lines are 20 in (50.8 cm) long, what is the width

of each picture element, in inches and centimeters?

4. List the frequency ranges responsible for: (a) picture shading, (b) picture lettering, and (c) fine detail.
5. Complete the following table:

ITEM	FREQUENCY
Picture carrier	
Sound carrier	
Color subcarrier	
Width of picture vestigial sideband	
Width of full picture sideband	
Total audio bandwidth	
Difference between picture and sound carriers	
Difference between picture carrier and color subcarrier	
Total TV channel bandwidth	

6. Identify these TV sidebands by their bandwidth: (a) 1.5 MHz, (b) 0.6 MHz, (c) 1.25 MHz, and (d) 4.2 MHz.
7. What VITS may be found in odd fields on Line 17? On Line 19 of both fields?

Chapter 5

5.1 Basic TV camera fundamentals
5.2 Camera tube characteristics
5.3 Operation of the vidicon
5.4 Operation of the plumbicon
5.5 Silicon diode array vidicon
5.6 Silicon imaging device (SID)
5.7 Single-tube color camera
5.8 Color television camera principles
5.9 Studio color TV camera
5.10 Portable color TV camera
5.11 The minicam system
5.12 Special purpose TV cameras
5.13 Studio color video-tape recorder (VTR)
Summary

TELEVISION CAMERA TUBES AND CAMERA SYSTEMS

INTRODUCTION

It was stated in a previous chapter that in the telecasting process an optical image is converted into an electrical image for transmitting the picture to the receiver. This important conversion is done in the TV camera by an image pickup tube. What this tube "sees" and converts into equivalent electrical impulses determines the form of the image finally reproduced at the receiver.

This chapter describes the operation of the TV camera tube in greater detail. It discusses conventional types of camera tubes, special types of tubes, and describes their particular applications. Finally, this chapter covers various types of television camera systems.

By the time you have completed the reading and work assignments for Chapter 5, you should be able to:

- Define the following terms: gamma, light-transfer characteristic, lag, dark current, CCD, CCU, and spectral response.
- Draw the components of the standard vidicon target plate.
- Describe the operation of a standard vidicon.
- Describe briefly the difference between the Plumbicon and standard vidicon.
- Describe briefly how the silicon diode array vidicon differs from the standard vidicon.
- Describe briefly the difference between a single-tube and a three-tube color camera.
- Describe what a beamsplitter is and how it is used.
- Explain briefly how motion picture film is televised.
- Discuss the function of the CCU (and multiple CCUs) in the studio TV color camera.
- Explain three different methods of

powering portable color TV cameras.
- Describe briefly how the "Minicam" system operates.
- Describe briefly the operation of studio VTRs.
- Describe briefly the characteristics of TV cameras used for education, medicine, and industry.

5.1 BASIC TV CAMERA FUNDAMENTALS

For high quality images at the receiver, the camera tube must resolve the scene being televised into as many basic picture elements as possible. The greater the number of these elements, the higher is the quality of detail in the reproduced picture. The scanning beam in the pickup tube must also produce electrical signals that faithfully represent each of these picture elements. To do this, the optical to electrical conversion must achieve a high enough signal-to-noise ratio to provide proper pickup sensitivity when low light-level scenes are being telecast. In other words, when there is no incident light on the face of the pickup tube, there must be little, if any, output signal. Because of these requirements, the camera-tube characteristics and its electrical operation are important considerations.

MONOCHROME TV CAMERA

Figure 5.1 is a simplified, block diagram of a typical monochrome-TV camera. An optical system focuses light reflected by the scene onto the faceplate of the camera tube. A photoelectric process then transforms the light image into a virtual electronic replica in which each picture element is represented by a voltage. A scanning beam in the pickup tube next converts the picture, element by element, into electrical impulses. At the out-

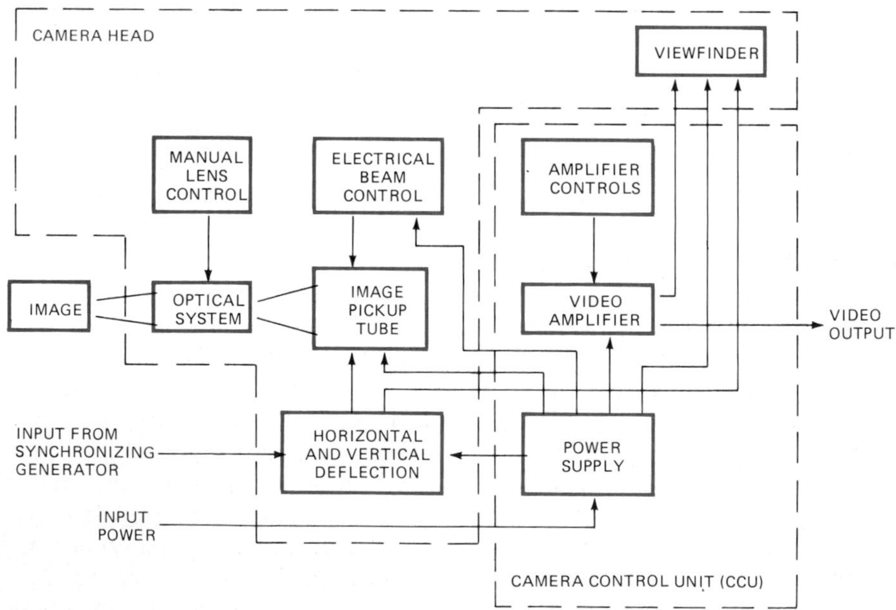

Figure 5.1 Simplified diagram of a typical monochrome television camera.

put, an electrical sequence develops that represents the original scene. The output of the camera tube is then amplified to provide the video signal for the transmitter. A sample of the video signal is also provided for observation in a CRT viewfinder mounted on the camera housing.

Electronic circuits that provide the necessary control, synchronization, and power supply voltages operate the TV camera tube. A deflection system is included in the TV camera to control the movement of the camera tube scanning beam. Many TV cameras receive synchronizing pulses from a studio-control unit. This unit also provides the sync pulses that synchronize the receiver with the camera.

Some TV cameras, however, generate their own control signals. In turn, they provide output pulses to synchronize the control unit. Manual controls are also provided at the rear of the camera for setting the optical lens and for zooming. Because of the complex electronic circuits and controls, early TV cameras were rather large and awkward to handle. Recent innovations have revolutionized the camera's construction.

APPLICATIONS

Television has become the entertainment medium of the major population areas of the world. However, television is not limited to entertainment. Television is in widespread use in education, medicine, industry, aerospace, and oceanography. Many of these applications require special camera tubes. Some of these tubes have widely varying sensitivities to light of different wavelengths throughout the spectrum. Much effort has been spent on image detection devices. Today, there is a TV camera tube that can do the job, whatever the need. Some of these special purpose camera tubes are discussed later in this chapter.

SOLID-STATE CAMERA CIRCUITS

Formerly, vacuum tubes were used in the electronic circuits of the TV camera. These tubes consumed a great deal of power in heating the filaments. They also produced a large amount of heat. The combined heat from the vacuum tubes and the heat generated by studio lighting raised the camera temperature to a rather high level. As a result of this high-heat environment, the electronic circuits in the camera had to be aligned several times each day during regular program production. In general, the TV camera circuits were not stable and required at least one operator for each camera. Also, the vacuum tubes required spacious compartments. This resulted in a large, awkward camera.

Today, transistors, integrated circuits, and modular construction have made dramatic improvements in television broadcasting. Transistors have replaced the vacuum tube. This eliminates the excessive power consumption and most of the heat. Integrated circuits have combined into one tiny chip many of the circuits that previously included numerous capacitors, resistors, and transistors. These are all mounted on a large-area circuit board. The elimination of the vacuum tube and its heat made modular construction feasible. This allowed a remarkable reduction in the overall size of the TV camera. Modular construction has virtually eliminated troubleshooting at the component level. Now, the trouble is traced to a particular module. Then, the module is simply replaced. The defective unit can be repaired and returned to service when needed. These advantages, together with temperature compensated deflection circuits, the use of feedback, and better regulated power supplies, result in a hundred-to-one stability improvement. In fact, it is now a common practice for one person to operate several TV cameras.

COLOR TV CAMERAS

There are two general categories of color TV cameras: one group uses a single (special) camera tube and the other group uses three camera tubes.

In the single tube color camera the tube faceplate has a vertical stripe filter that separates the colors prior to the scanning beam area. This principle is discussed in more detail later in this chapter. When three tubes are used, a separate tube is employed for each of the three primary colors: red, green, and blue. A color filter system separates the incoming light from the image into the three colors and focuses each color onto the faceplate of an appropriate camera tube. The three camera tubes are identical except that the photo-sensitive material in each tube is more responsive to the particular color for which the tube is used.

Each color is first determined by the filter system. The output from each tube consists of voltages that represent the elements of the particular primary color reflected by the scene. These outputs are then amplified to provide three channels of video signals, one channel for each color. Color cameras will be discussed in more detail, after examining the camera tubes which are in general use in the television industry.

5.2 CAMERA TUBE CHARACTERISTICS

One important characteristic of all camera tubes is their *light-transfer characteristic.* This is the ratio of the faceplate illumination in footcandles to the output signal current in nanoamperes (nA). This characteristic may be considered as a measure of the "efficiency" of a camera tube. Typical values of output current range from 200 nA to 400 nA.

GAMMA

This characteristic was described in Chapter 4. It is also known as the dynamic range, or linearity of the faceplate illumination versus the output signal. As mentioned before, camera tubes are normally operated with unity gamma. (Refer to Figure 4.6A.)

SPECTRAL RESPONSE

For a proper color reproduction, the spectral response of a camera tube is an important parameter. As nearly as possible, the tube should have the same spectral response as the human eye. This is necessary to render colors in their proper tones. It is also important in reproducing black and white pictures, thereby producing the proper gray scale. Tubes designed to operate in a color camera have a greater response to each of the primary colors. Today, spectral response distribution has made possible the manufacture of camera tubes that are sensitive to the infrared, the ultraviolet, and even the X-rays. But variations in spectral response have had little effect on the other operating characteristics of the tube.

If the photosensitive material in a camera tube was able to emit an electron for each photon of light focused upon the material, the quantum efficiency of the material would be 100 percent. The formula for quantum efficiency is:

$$\text{Qeff} = \frac{\text{electrons}}{\text{photons}}$$

A quantum efficiency of 100% is almost impossible. However, quantum efficiency is a practical way to compare photosensitive surfaces in the camera tube. In this comparison, photocurrent per lumen is measured, using a standard light source. (A lumen is the amount of light that produces an illumination of one footcandle over an area of one square foot.) The source adopted for the measurement is a tungsten-filament light operating at a color temperature of 2870°K. Since the lumen is actually a measure of brightness stimulation to the human eye, quantum efficiency is a convenient way to express the sensitivity of the image-pickup tube.

A graph of the spectral response of a vidicon camera tube is shown in Figure 5.2.

LAG

This term refers to the time lag during which the image on a camera tube decays to an unnoticeable value. All camera tubes have a tendency to retain images for short periods after the image is removed. Some types do this more than others. Lag on a television picture causes smear (comet trails) to appear following rapidly moving objects.

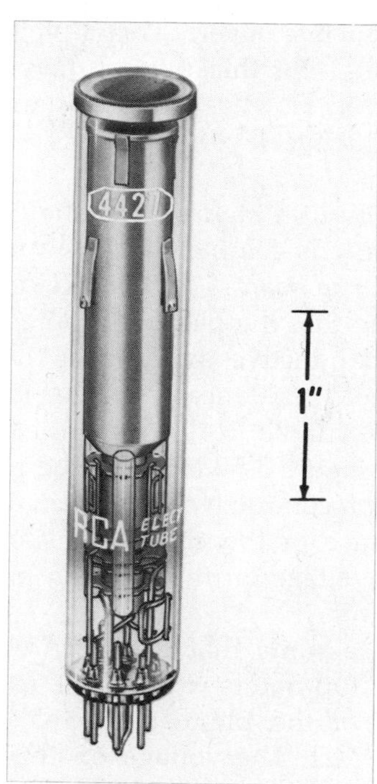

Figure 5.2 A vidicon camera tube, 0.5 in (1.27 cm) in diameter and 3.5 in (8.9 cm) long. *(Courtesy of RCA Electronic Components.)*

Lag may be expressed as a percentage of the initial value of the signal current remaining 1/20s after the illumination is removed. For the type of tube shown in Figure 5.2, using an initial signal output current of 250 nA, the lag is 5%. Typical lag values for vidicons range from 1.5% to 5%. Other types of camera tubes (such as the "Plumbicon") have lag values as low as 1.5%.

DARK CURRENT

This term is applied to all types of camera tubes and other photoelectric devices. It refers to the current that flows through the device even in total darkness. For camera tubes, a low value of dark current is desirable.

5.3 OPERATION OF THE STANDARD VIDICON

The standard vidicon is a relatively simple and compact camera tube. A typical standard vidicon is shown in Figure 5.3. This particular tube is 0.5 in (1.27 cm) in diameter and 3.5 in (8.9 cm) long. Other vidicons may vary in diameter from 0.5 in (1.27 cm) to 1.5 in (3.8 cm), and in length from 3.5 in (8.9 cm) to 8 in (20.3 cm). Vidicons are widely used for closed-circuit television (CCTV) and for TV studio and film cameras. Some vidicons can produce useable pictures operating in near-total darkness, or in near-direct sunlight.

VIDICON COMPONENTS

The external and internal components of a vidicon are shown in Figure 5.4. The vidicon tube proper consists basically of a glass faceplate and target, and a special electron

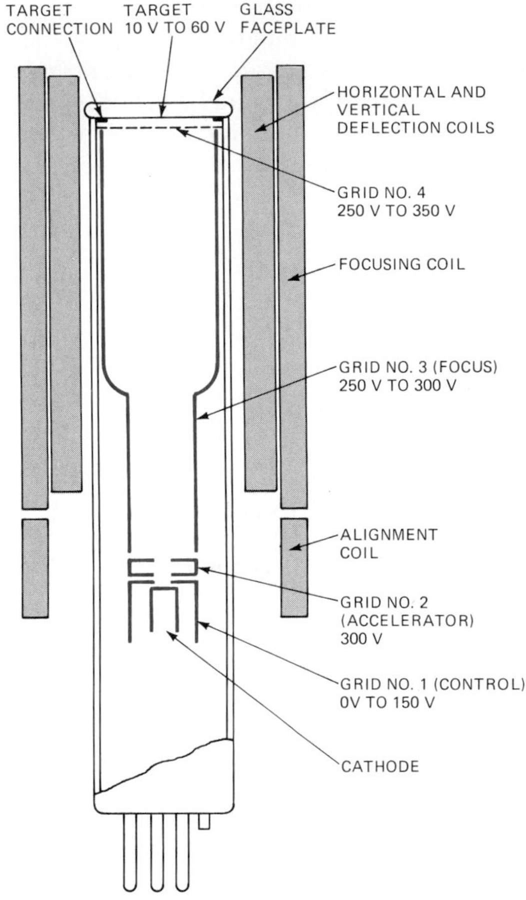

Figure 5.3 Internal construction and external components of a vidicon camera tube. Voltages shown are typical.

gun. The external components consist of the horizontal and vertical deflection coils (yoke), the beam-alignment coil, and the focus coil.

Target Construction The target consists of a transparent conducting film (the signal electrode) deposited on the inner surface of the faceplate. A very thin photoconductive layer is also deposited over the signal electrode (Figure 5.5A).

The conducting film acts as a signal electrode for the electrical output signals of the camera. It has an applied voltage of 10 V to 60 V, depending upon the tube type. The photoconductive layer has a thickness of 30 $\times 10^{-6}$ in (76.2×10^{-6} cm). Materials used for this layer include antimony or selenium compounds. One important characteristic of a photoconductive material is that the resistance decreases as the amount of light falling upon it increases.

Target Operation The operation of the vidicon target can be explained with the aid of Figure 5.5B. In total darkness, the thin photoconductive layer has a resistance of about 20 MΩ across its thickness. In the presence of bright light, this resistance decreases to about 2 MΩ. The change in resistance is sharply localized to individual picture elements. Thus, it can be said that a "picture" of various resistances is "painted" on the photoconductive layer. This "picture" appears through the thickness of the photoconductive layer. However, the resistance of the photoconductive layer across its surface is quite low.

As shown in Figure 5.5B, the circuit for electron flow is completed by the resistance of the electron beam. For the sake of simplicity, assume that the picture impinging upon the photoconductive layer causes the pattern of different resistances shown in Figure 5.5B. For example, if the picture element at the top of Figure 5.5B is black, the resistance of the photoconductive layer remains at 20 MΩ. Because of the circuit resistances, assume the voltage on the electron beam side falls to +6 V.

Now, assume that the picture element below the top one is white. This causes the resistance of the photoconductive layer to drop to 2 MΩ. The voltage on the electron beam side is now assumed to be 9.5 V. Notice in Figure 5.5B that dark picture elements result in lower positive charge voltages. Conversely, light picture elements result in

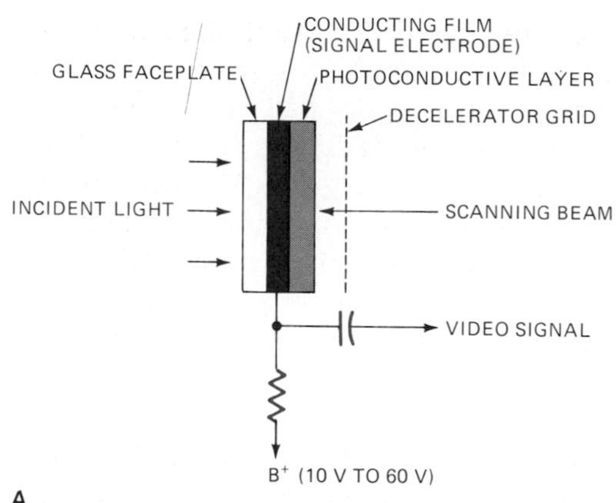

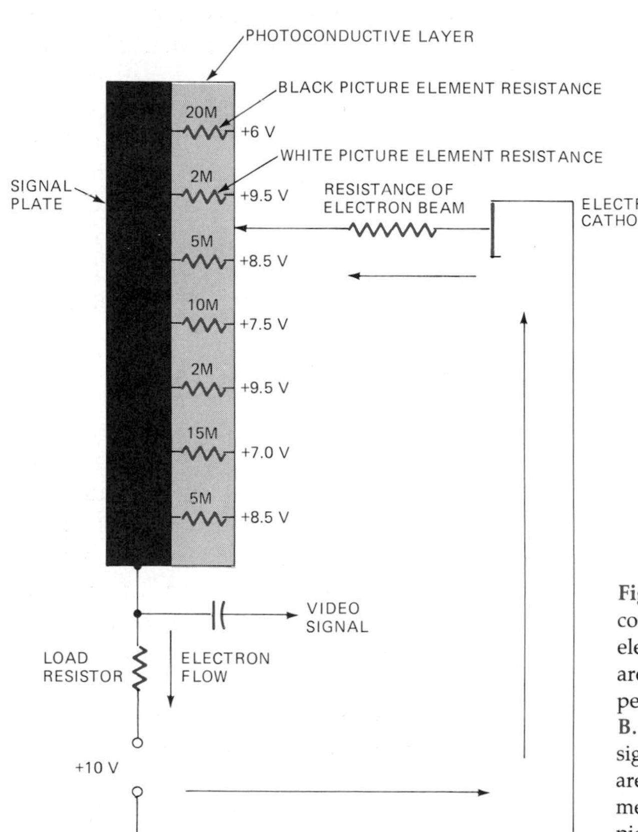

Figure 5.4A. Simplified diagram of the standard vidicon target plate. Output signal is taken from the signal electrode. Conducting film and photoconductive layer are very thin (not shown to scale). The B+ value depends upon the specific type of vidicon tube. **B.** Simplified diagram showing how the camera video signal is developed in a standard vidicon. Values shown are for explanatory purposes only. Black picture elements result in the highest resistance (20 MΩ); white picture elements result in the lowest resistance (2 MΩ).

higher positive element charge voltages. Of course, intermediate light values result in intermediate positive element charge voltages appearing on the electron beam side of the photoconductive layer.

The above process creates a charge pattern of different positive voltages facing the electron gun. This charge pattern corresponds to the light values in the optical image.

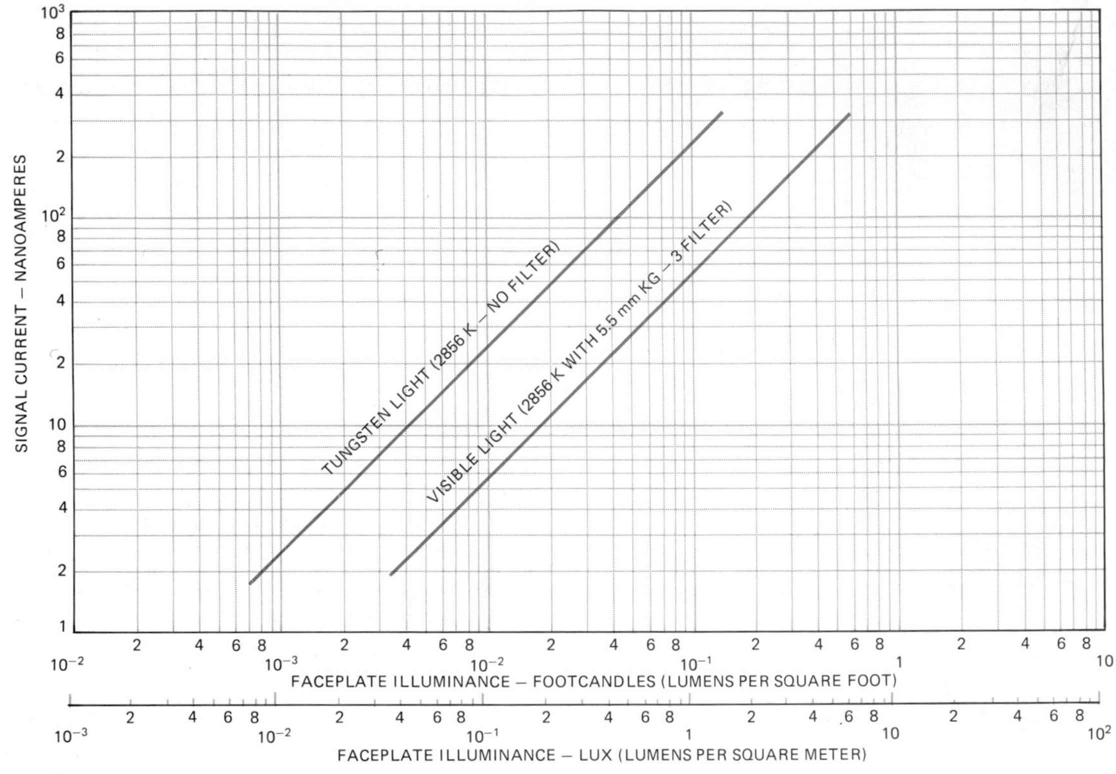

Figure 5.5 Light-transfer characteristics of a vidicon camera tube. *(Courtesy of RCA Electronic Components.)*

VIDEO SIGNAL

The photoconductive layer facing the electron beam has a positive voltage charge pattern. The electron beam scans the photoconductive layer in the standard 525 lines-per-frame pattern. As the beam contacts each positive voltage charge element, it gives up electrons in proportion to the actual voltage present at the point. The beam tends to drop the voltage of each charge element toward zero. As a result, a varying electron current is produced which flows through the load resistor, as shown in Figure 5.5B. This current produces the video-signal output of the camera tube. Note that the resultant video signal currents (and voltages) have the greatest amplitude for white picture elements and the least amplitude for black picture elements.

LIGHT TRANSFER CHARACTERISTIC

This is a measure of the video output signal current versus faceplate illumination and is shown in the graph of Figure 5.6. The performance is shown for two different qualities of light. For example, a faceplate illumination of 10^{-1} footcandle with tungsten light produces a signal current of about 260

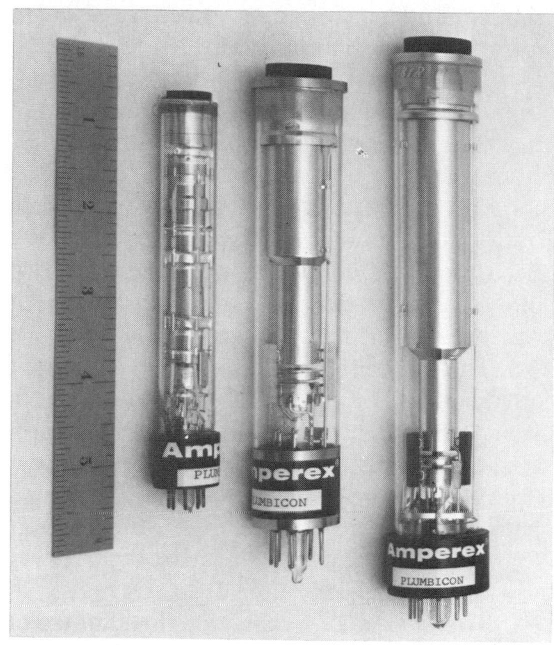

Figure 5.6 Typical Plumbicon tubes. *(Courtesy of Amperex Electronic Corp.)*

nA, for a target voltage of 8 V. The graph is for a silicon diode-array vidicon and is discussed later in this chapter.

THE ELECTRON BEAM

The electron beam is formed by the "electron gun." The electron gun consists of the cathode, the No.1 control grid, the No.2 accelerator grid, and the No.3 focus grid, as shown in Figure 5.4. The focus grid acts with the external focus coil to provide a combination of electrostatic and electromagnetic fields to produce a finely focused beam. The No.4 grid is a fine wire mesh just preceding the target (photoconductive layer). This grid is physically and electrically connected to grid No.3. The No.4 grid provides a uniform electrostatic field in the vicinity of the target. The target is at a potential of 10 V to 60 V, and the No.4 grid is at 250 V to 300 V. Because of this large potential gradient, the beam electrons are slowed down appreciably before they contact the target. If the electrons contacted the target with their initial velocity, they would cause secondary emission from the target. If this happened, incorrect signal currents would result.

5.4 OPERATION OF THE PLUMBICON

The Plumbicon, developed by Philips of Holland, is a small, lightweight television camera tube that has fast response and produces high quality pictures at low light levels. Its small size and low-power operating characteristics make it an ideal tube for transistorized television cameras designed to serve a particular purpose. Modern color television cameras are making widespread use of the Plumbicon because of its simplicity and spectral response to the primary colors. Typical Plumbicon camera tubes are shown in Figure 5.7.

Functionally, the Plumbicon is very similar to the standard vidicon. Focus and deflection are both accomplished magnetically. The main difference between the Plumbicon and the standard vidicon is in the target.

PLUMBICON TARGET

Figure 5.8 is a simplified diagram of the Plumbicon target. As shown in part (A) of this figure, the inner surface of the glass faceplate is coated with a thin transparent conductive layer of tin oxide (SnO_2). This layer forms the signal plate of the target. A photoconductive layer of lead monoxide (PbO) is deposited on the scanning side of the signal plate. These layers are specially

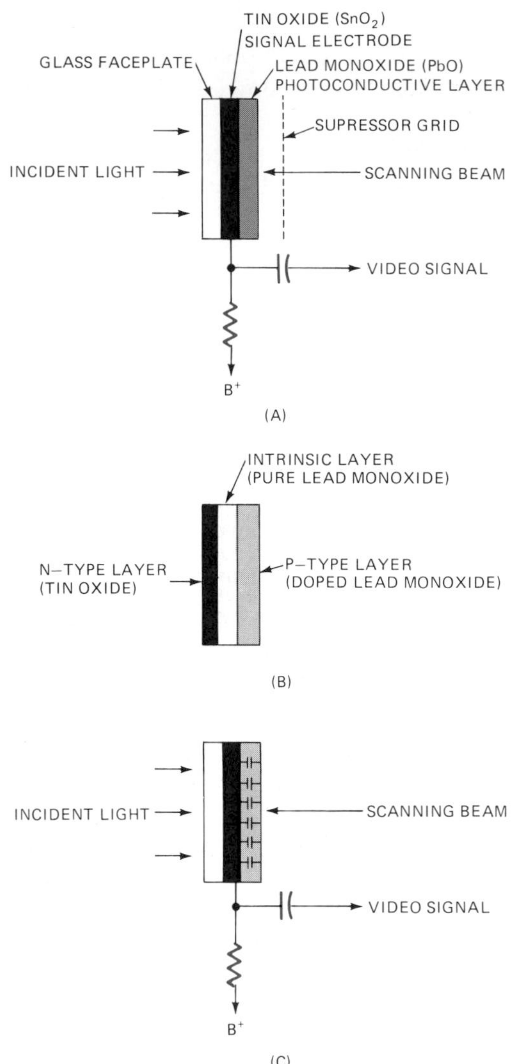

Figure 5.7A-C. Operation of the Plumbicon target.

prepared to function as three sublayers. Each layer has a different conduction mode.

The tin oxide layer on the inner side of the faceplate is a strong N-type semiconductor, commonly found in transistors. Next to this N-type region is a layer consisting of almost pure lead monoxide. This is an intrinsic semiconductor. The scanning side of the lead monoxide is doped to form a P-type semiconductor. Together, these three layers form a P-I-N junction diode.

The photoconductive target of the Plumbicon functions much like the photoconductive target in the standard vidicon. Light from the scene being televised is focused through the transparent layer of tin oxide onto the photoconductive lead monoxide. Each picture element charge takes the form of a small capacitor with its positive plate toward the scanning beam, as shown in Figure 5.8. The target signal plate becomes the negative side of the capacitor. When the low-velocity scanning beam lands on the charge element, it releases enough electrons to neutralize the charge built up on the element capacitor. The scanning-beam current through the external signal-plate load resistor develops the video-signal output.

The spectral response of the Plumbicon can be varied while it is being manufactured to suit almost any application. Since the tube has gained wide popularity in color television cameras, it is available with a spectral response suitable for any of the primary colors. The particular color response of the tube is designated by the letter R (red), G (green), or B (blue) following its type number. For example, a Plumbicon for use in a red channel is designated as a type XQ1427R. When this type of tube is intended for monochrome use, no letter follows the type number.

PLUMBICON SPECIFICATIONS

The specifications listed in Tables 5.1 and 5.2 (courtesy of Ampex Electronic Corp.) are for the type XQ1427 (R, G, B) Plumbicon, Figure 5.9. This is a $\frac{2}{3}$-in (18 mm) diameter tube using a high-resolution, lead-oxide photoconductive target. The tube is for use in broadcast studios, news gathering, and industrial applications. A comparison of

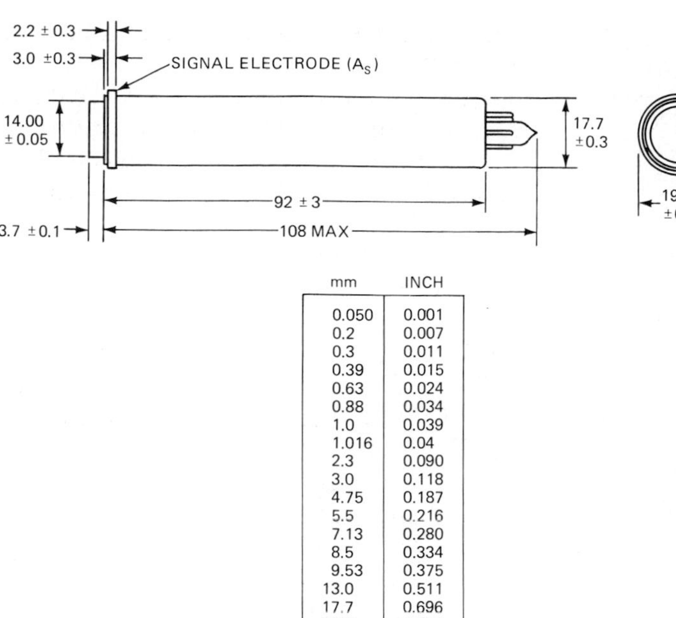

mm	INCH
0.050	0.001
0.2	0.007
0.3	0.011
0.39	0.015
0.63	0.024
0.88	0.034
1.0	0.039
1.016	0.04
2.3	0.090
3.0	0.118
4.75	0.187
5.5	0.216
7.13	0.280
8.5	0.334
9.53	0.375
13.0	0.511
17.7	0.696
18.0	0.708
19.6	0.771
92.0	3.622
108.0	4.252

Figure 5.8 Outline drawing for Plumbicon XQ1427 (R, B, G). All dimensions are in millimeters. *(Courtesy of Amperex Electronic Corp.)*

the resolution characteristics of Plumbicons XQ1427, XQ1427R, XQ1427G, and XQ1427B can be found in Table 5.3.

1600 TV LINE PLUMBICON

A more recent development in Plumbicon camera tubes is the type 45 X Q. This tube has a limiting resolution of 1600 TV lines, as compared with 600 to 750 TV lines for other types of Plumbicons. This tube is shown in Figure 5.10. It has a 1.18-in (3-cm) diameter, with an effective target diameter of 1.02 in (2.6 cm).

A major use for this tube is expected to be in the field of electronic cinematography. This is motion picture production on video tape. Earlier tube designs did not have adequate resolution for this function. In addition to TV broadcasting and motion pictures, other applications include flight simulation, TV fluoroscopy, and dynamic film and document scanning. A feature of the tube is very low lag (below 5%) and an integral bias-light system. The bias lighting is provided by an LED array and helps to reduce the "build up" lag.

5.5 SILICON DIODE ARRAY VIDICON

This is a vidicon camera tube with a special type of target. The target is actually constructed on a very thin slice of N-type silicon. A typical silicon diode array vidicon is shown in Figure 5.11A. Its outline drawing is shown in Figure 5.11B. The tube is designed to accept an optical input image with a diagonal measurement of 0.433 in (1.1 cm). Because of the small size of the tube, it can be used in compact, hand-held surveillance cameras. Some of the advantages of this tube over conventional vidicons are (1) high sensitivity, (2) broad spectral response, (3)

Table 5.1 **GENERAL PLUMBICON SPECIFICATIONS**

MECHANICAL	
Focusing Method	Magnetic
Deflection Method	Magnetic
Dimensions and basing	See outline drawing (Figure 5.9)
Mounting Position	any
Weight	0.8 oz.
Base	JEDEC E7–1, with pumping stem
Accessories	
Socket	56049
Mask for reduction of flare	56033
Deflection and Focus Coil Assembly	
monochrome	KV12 or equivalent
color	AT1105
OPTICAL	
Dimensions of Quality Area of Target (.26″ × .35″)	6.6mm × 8.8mm
Image Orientation	Horizontal scan parallel to plane of tube axis and gap between pins 1 and 7.
Faceplate	
Refractive index	n = 1.49
Refractive index of antihalation glass disc	n = 1.52
Sensitivity at color temperature of illumination—2856°K	
XQ1427	375 A/1m
XQ1427R	100 A/1m
XQ1427G	140 A/1m
XQ1427B	32 A/1m

low lag, (4) good resolution of 450 TV lines, and (5) low dark current. The silicon target is highly resistant to image burn-in. It has appreciable sensitivity to red and near infrared radiation.

TARGET OPERATION

Figure 5.12 is a greatly enlarged cross section of a small portion of a silicon diode array target. The tiny P-type silicon buttons in contact with the N-type silicon layer form photodiodes. Approximately 30 000 photodiodes are formed on the target. Each photodiode has a diameter of about 275×10^{-6} in (699×10^{-6} cm). It is composed of a P-type button in contact with the N-type silicon. Photodiodes (photoconductive mode) are normally operated with *reverse* bias provided by the +8 V applied to the N-type silicon. The path for current flow is completed by the scanning electron beam.

Table 5.2 **TYPICAL PLUMBICON OPERATING CONDITIONS AND PERFORMANCE**

OPERATING CONDITIONS (using coil unit AT1105)	
Cathode Voltage	0 V
Grid No. 2 Voltage	300 V
Signal Electrode Voltage	45V
Grid No. 4 Voltage	500 V
Grid No. 3 Voltage	300 V
Beam Current	150 – 300 nA
Focusing and Deflection Coil Current	
Monochrome coil assembly KV12	
Focus Current	120 mA
Line Current (P–P)	160 mA
Frame Current (P–P)	25 mA
Color coil assembly AT1105	
Focus Current	40 mA
Line Current (P–P)	320 mA
Frame Current (P–P)	120 mA
Faceplate Temperature	20 to 45 °C (68 to 113 °F)
Blank voltage, peak-to-peak, grid No. 1	50 V
PERFORMANCE	
Dark Current	– 1.5 nA
Gamma of Transfer Characteristic (Gamma stretching circuitry is recommended)	0.95 ± 0.05
Spectral Response, max	500 nm
cut-off, XQ1427 G, B	650 to 850 nm
XQ1427 R	850 nm
Limiting Resolution	600 TV Lines

Table 5.3 **COMPARISON OF PLUMBICON RESOLUTION CHARACTERISTICS**

CHARACTERISTIC	PLUMBICON TYPE			
	XQ1427	*XQ1427R*	*XQ1427G*	*XQ1427B*
Highlight Signal Current I_S	150 nA	75 nA	150 nA	75 nA
Beam Current I_b	300 nA	150 nA	300 nA	150 nA
Modulation Depth				
At 320 TV Lines	48%	42%	48%	55%
At 400 TV Lines	30%	25%	30%	35%

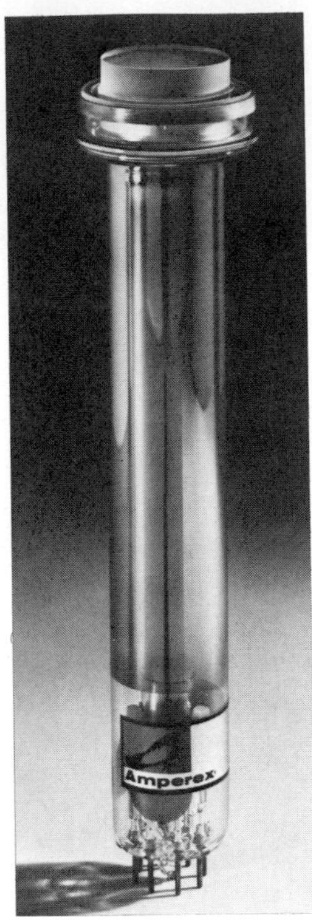

Figure 5.9 A 1 600 line TV Plumbicon. This camera tube is suitable for use in telecinematography. *(Courtesy of Amperex Electronic Corp.)*

The conduction of a photodiode increases with the light intensity. Thus, as the light from the optical image is focused upon the target, a pattern of photodiodes in various states of possible conduction is formed. As the scanning electron beam reaches each photodiode, a current passes through the load resistor according to the light then impinging upon the particular photodiode. The variations of this current form the video-signal output from the camera tube. The signal-output current varies between 4 nA (dark current) and 200 nA (bright light current).

5.6 SILICON IMAGING DEVICE (SID)

The silicon imaging device in Figure 5.13 is a completely solid-state device. Its overall picture performance is comparable with that of $\frac{2}{3}$-inch vidicon camera tube. Note that the SID (without its socket) measures 1.2 in (30.48 mm) long by 0.79 in (20.07 mm) wide. This is a self-scanned device which is hermetically sealed. It does not utilize a scanning electron beam.

CHARACTERISTICS

The SID contains 512 vertical and 320 horizontal picture elements, for a total of 163 840 picture elements. The image diagonal measures 0.48 in (12.2 mm). This device features high resolution and ultra-low blooming characteristics. The SID is highly resistant to image burn-in. Undesirable vidicon characteristics, such as lag and microphonics, are not present in the SID. Further, the unit has low voltage and power requirements. The maximum voltage required is +20 V.

OPERATION

The operation of the SID is described with the aid of the block diagram shown in Figure 5.14. An optical image is focused upon the image area and suitable waveforms are applied through the socket connections. The image area is an array containing 320 parallel vertical columns of 256 sensing cells. During the normal active TV field display time, the optical image produces a corresponding charge pattern of electrons. This charge pattern is transferred to the storage area during the vertical blanking interval.

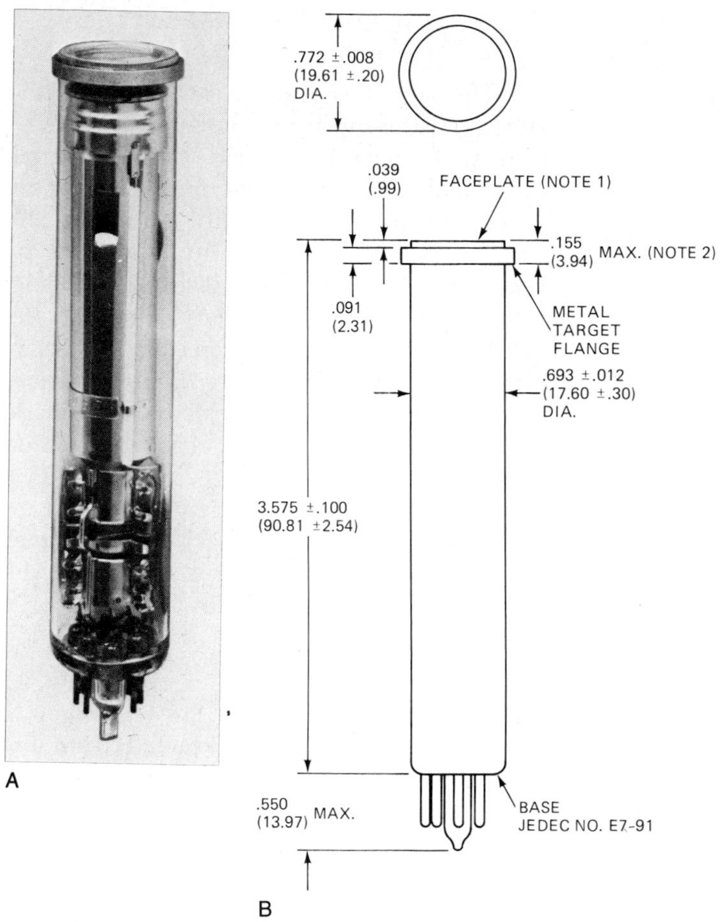

NOTE 1 – FACEPLATE GLASS IS CORNING NO. 7056 HAVING A THICKNESS OF 0.067 ±0.005 (1.70 ±0.13) AND A REFRACTIVE INDEX (N_D) = 1.49 AT 589.3 NANOMETERS.
NOTE 2 – OPTICAL DISTANCE FROM OUTER SURFACE OF FACEPLATE TO TARGET SURFACE.

DIMENSIONS ARE IN INCHES UNLESS OTHERWISE STATED. DIMENSIONS IN PARENTHESES ARE IN MILLIMETERS AND ARE DERIVED FROM THE BASIC INCH DIMENSION (1 INCH = 25.4 mm).

Figure 5.10A. A type 4833 silicon diode array vidicon. **B.** Outline showing all important dimensions of a silicon diode array vidicon. *(Courtesy of RCA Electronic Components.)*

STORAGE AREA

The storage area has the same construction as the image area. It also contains 320 parallel vertical columns of 256 sensing cells. These line up with the image area columns. The storage area serves as a temporary storage site for the previous TV picture field. This function allows for conversion of the charge pattern image into a sequential horizontal readout (similar to horizontal scanning).

HORIZONTAL REGISTER

The horizontal register receives one complete horizontal line of picture informa-

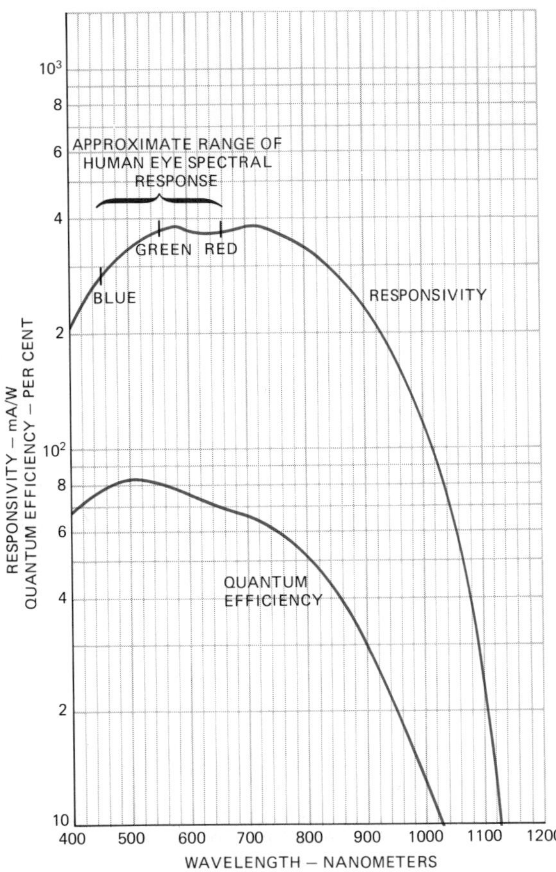

Figure 5.11 The silicon diode array target. Approximately 30 000 photodiodes are formed on the target. This illustration shows a small section of the target (greatly enlarged). The inset shows a simplified equivalent circuit of a small section of the target.

tion from the storage area during each horizontal blanking interval. The register contains 320 cells corresponding to the 320 columns in the image and storage areas. The picture elements are read at a 6.1-MHz picture element rate. Thus, the 320 active picture elements are read out in the active horizontal line time of 53.34 μs (nominal).

CCD SIGNAL

The signal appearing in the image and storage areas and in the horizontal register, is a CCD (charge-coupled device) signal. This means that light-generated charges are caused to move sequentially along the silicon chip(s). This movement is caused by low-voltage (0 V to 20 V) pulses applied successively to electrodes adjacent to the elements. The charges move because the electrode next to each charge is made more positive and the charge electrode is made less positive. A series of accurately timed pulses make these voltage changes. This process is similar to horizontal-line scanning. When the charges reach the particular output electrode involved, they form the output signal current.

OUTPUT CIRCUIT

The CCD signal is extracted from the horizontal register by the *output circuit* (Figure 5.14). The output of this circuit is the video camera signal. The test point shown in the block diagram is used to make measurements. At this point, the photo-current response to a light stimulus can be measured.

5.7 SINGLE-TUBE COLOR CAMERA

As previously mentioned, some color TV cameras employ a single pickup tube. One such camera is shown in Figure 5.15. This is a high quality camera. More compact, portable TV cameras are also available with a single pickup tube. The camera shown in Figure 5.15 is manufactured by the Sony Corp. and the pickup tube used is called the "Trinicon"®. Use of a single, stable tube minimizes maintenance and required operator skill, as compared with a three-tube camera.

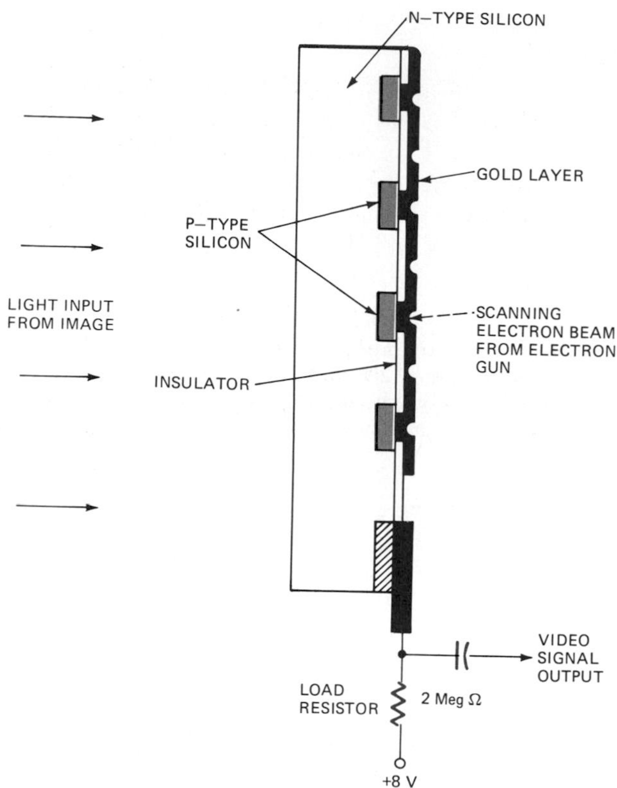

Figure 5.12 The silicon diode array target. Approximately 30 000 photodiodes are formed on the target. This illustration shows a small section of the target (greatly enlarged).

TRINICON® TUBE

The Trinicon® tube is a specially designed, magnetic deflection vidicon. The Trinicon® delivers two signal outputs: a luminance (Y) signal and a dot sequential color signal. The dot sequential signal is processed into a standard NTSC color video signal.

Description The Trinicon® is a 1-in (25 mm) diameter pickup tube. An outline drawing of the tube is shown in Figure 5.16A. Details of the target of this tube are shown in Figure 5.16B. The important differences from a standard vidicon tube are in the target assembly. The components of the target are (1) faceplate (for protection only), (2) a three-color filter, (3) a mask, (4) an index electrode, and (5) a photo-conductive layer. There are 280 sets of red, green, and blue filters arranged vertically. The most important parts of the target are the red-, green-, and blue-stripe filters and the index electrode.

Operation The operation of the Trinicon® is explained with the aid of Figure 5.17. Part A shows the three-color filter which resolves the incoming light image into the three primary colors (red, green, and blue). For simplicity, assume that the subject is completely green. As a result, the light passes through the green filter stripes alone to produce output signal pulses, as shown in Figure 5.17B. Output pulses are produced in a similar manner for red and blue objects. For white light, as in Figure 5.17C, an almost continuous signal is produced.

Figure 5.13A. A silicon-imaging device. **B.** Dimensional outline of a silicon-imaging device as mounted in its socket. The image size is 288 mils × 384 mils (7.31 mm × 9.75 mm). Dimensions are given in inches, with millimeters in parentheses. *(Courtesy of RCA Solid-State Division.)*

A

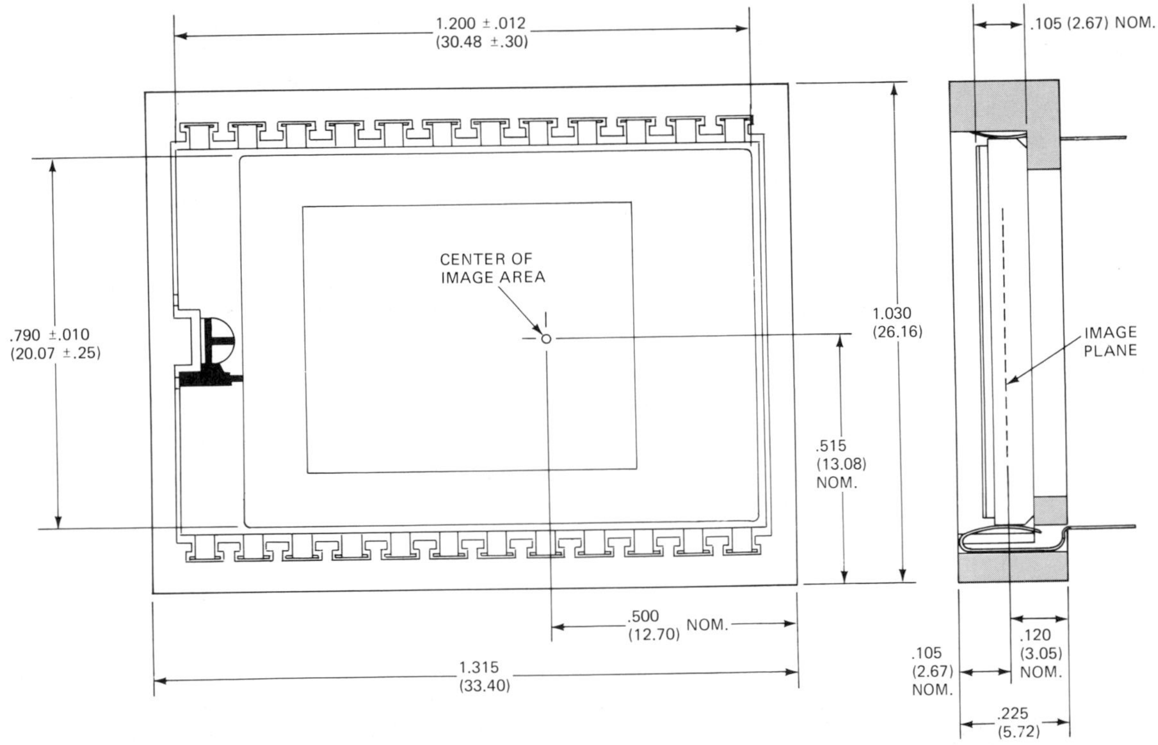

1.200 ±.012 (30.48 ±.30)
.105 (2.67) NOM.
CENTER OF IMAGE AREA
.790 ±.010 (20.07 ±.25)
1.030 (26.16)
IMAGE PLANE
.515 (13.08) NOM.
.500 (12.70) NOM.
1.315 (33.40)
.105 (2.67) NOM.
.120 (3.05) NOM.
.225 (5.72)

B

Subcarrier The time required to scan one trio of red, green and blue filter stripes is 0.22 μs. This period corresponds to a frequency of 4.5 MHz, or the *subcarrier* frequency for the Trinicon®. This subcarrier signal amplitude varies with the intensity of the various colors during image scanning. A subcarrier is present only when there is scanned color information (suppressed type of subcarrier).

Indexing A method of indexing is needed

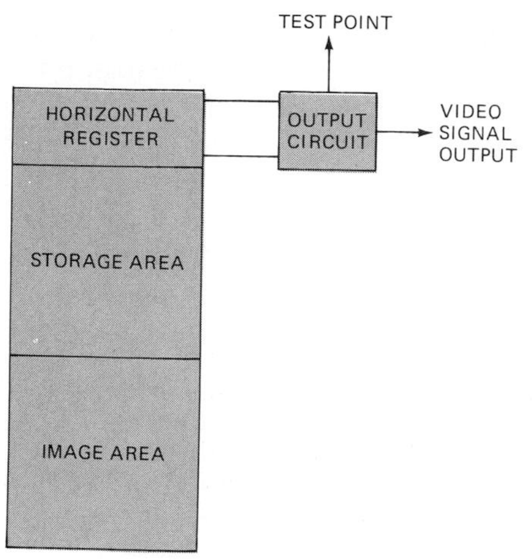

Figure 5.14 Simplified block diagram of the silicon-imaging device.

to determine if a signal pulse is from a red, green, or blue source. Otherwise, the colors are indistinguishable from one another. In other words, a reference phase must be established. In the Trinicon®, this reference phase is established by applying a square wave to the indexing electrode. This square wave reverses each 1 H line. The net effect is to cause the three-color signals to combine into the same format as the NTSC

Figure 5.15 A studio color TV camera employing a single pickup tube. The camera features high-quality pictures. It has a 5-in (12.7-cm) viewfinder, which is also used to check the video signals. *(Courtesy of Sony Corp.)*

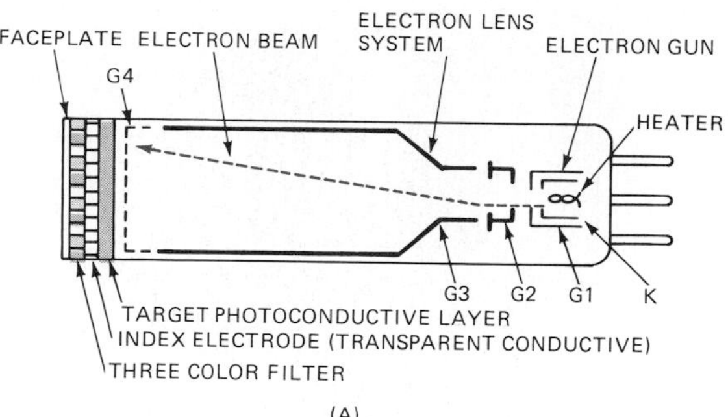

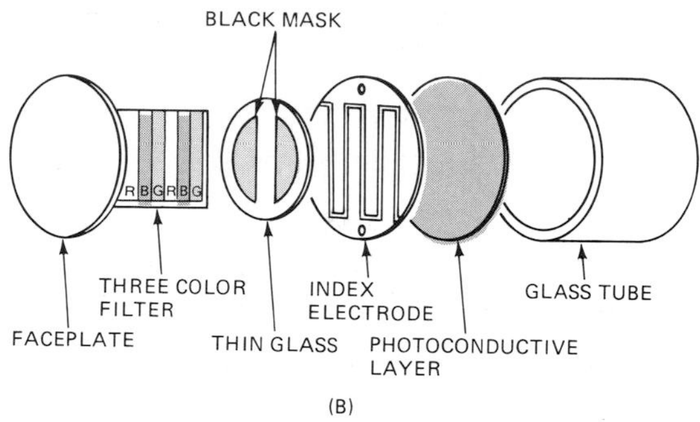

Figure 5.16A. The major components of the Trinicon. The tube is 1 in (2.5 cm) in diameter. **B.** Details of the Trinicon target. *(Courtesy of Sony Corp.)*

color video signal. This signal includes both the luminance (Y) and chrominance components.

5.8 COLOR TELEVISION CAMERA PRINCIPLES

To televise a color picture using the method of separating the three primary colors, three individual color tubes are required. In some color cameras, a fourth tube is used to provide an improved monochrome picture and better detail for the color picture. This fourth tube is described later in this chapter.

Except for the number of camera tubes used, the distinguishing feature of the color camera is its optical system. It is here that all

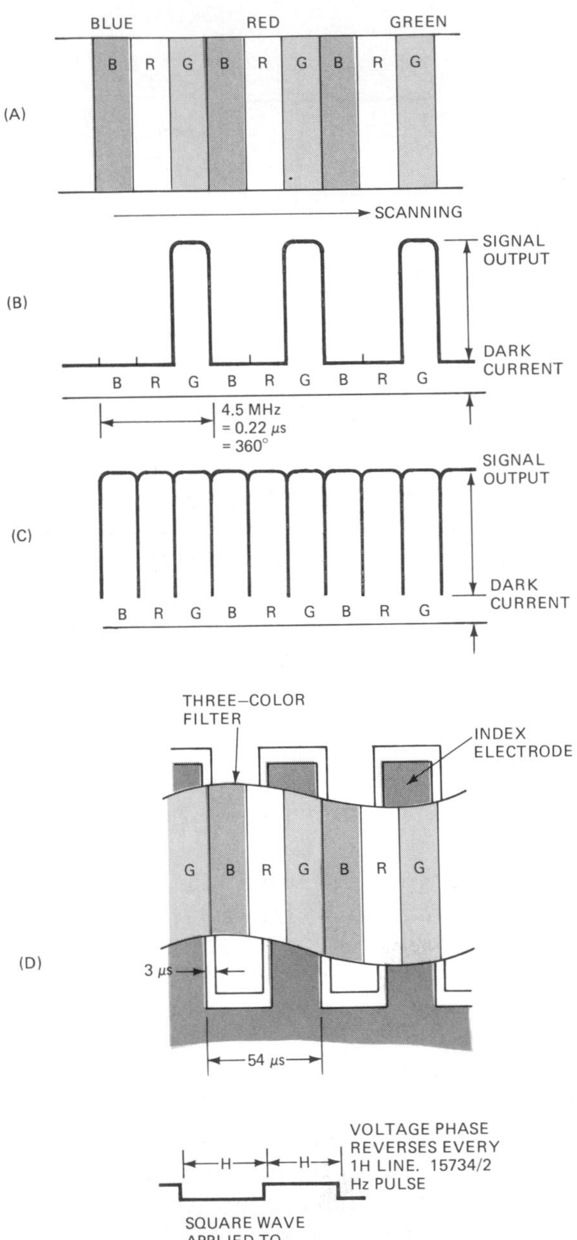

Figure 5.17A. The Trinicon three-color filter. **B.** Output signal for green light only. **C.** Output signal for white light. **D.** Positioning of the indexing electrode and three-color filter. *(Courtesy of Sony Corp.)*

light entering the camera lens is divided, or "split," into separate color beams: one beam for each camera tube. This beam-splitting is accomplished by mirrors, prisms, and lenses arranged in a unit called a beam-splitter. In the four-tube camera, four-way beam-splitters do the dividing, while three-way units do the job in the three-tube cameras.

FOUR-WAY BEAM-SPLITTER

A simplified diagram of a four-way beam-splitter used in a typical four-tube camera is shown in Figure 5.18. Incident light from the scene being televised is divided into two parts. One part of the split beam is focused onto the faceplate of the fourth camera tube, in the luminance channel, to provide the "Y" voltage (monochrome signal) of the video signal. The other part of the split beam is reflected by a front-surface mirror through a relay lens into a system of dichroic mirrors. These mirrors are designed to reflect light of only one particular color and to allow light of all the other colors to pass through. In Figure 5.18, the first dichroic mirror (1) reflects only the red light from the scene onto the faceplate of the red-channel tube. This provides the R voltage of the video signal. Green and blue light pass through the mirror and strike the front surface of the second mirror (2). This second mirror reflects the blue light into the blue channel and passes the green light on into the green channel. These two latter channels provide the B and G voltages of the video signal.

THREE-WAY BEAM-SPLITTER

The optical system in the three-tube camera is somewhat simpler than the system found in the four-tube camera. Figure 5.19 shows a simplified diagram of the compact

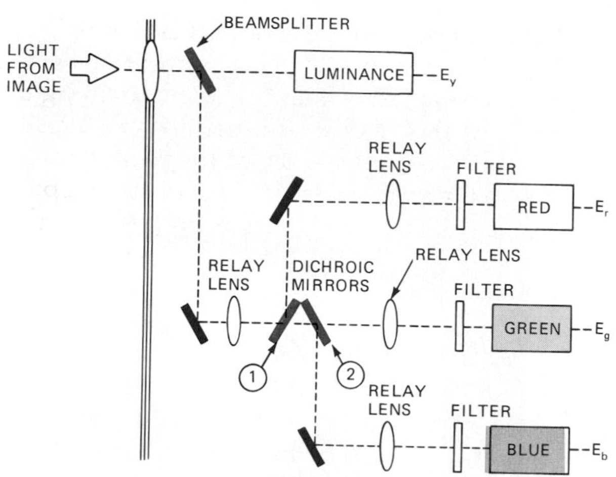

Figure 5.18 Simplified diagram of a typical four-way beam-splitter.

three-way beam-splitter in a Norelco three-tube color camera. In this color-splitting system, dichroic layers are integral with, and enclosed within, a single-sealed assembly of small prisms. Color-trimming filters on the exit faces of the prisms provide suitable color characteristics, in conjunction with the spectral separation of the dichroic layers of the prisms and the spectral response of the Plumbicon tubes used in the camera.

The Plumbicon tube is positioned at the red, green, and blue exits of the beam-splitting optical system. The tube has an antihalation faceplate. This prevents light reflected back from the sensitive layer in the tube from generating an unwanted signal and impairing image quality. Each Plumbicon is mounted in an identical yoke assembly which is independently adjustable to center and align the tube on its optical axis. Each yoke also provides the precise adjustment for rotating the scanned area of the tube for angular registry.

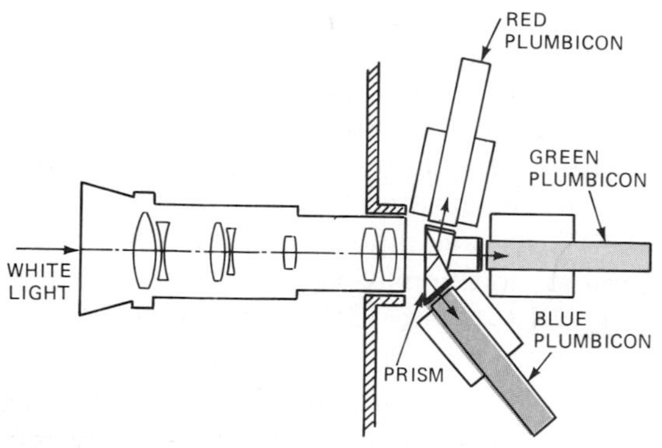

Figure 5.19 Simplified diagram of the Norelco optical system.

5.9 STUDIO COLOR TV CAMERA

A color television camera that is suitable for either studio or field use is shown in Figure 5.20. It uses three $\frac{2}{3}$-in (17 mm) diameter Plumbicons and a three-way beam-splitter. The physical properties are noted in the figure legend.

This camera is extremely stable, reliable, and it can withstand appreciable shock and vibration. It can also operate in a wide range of temperature and humidity environments. It utilizes state of the art metal film resistors, ICs, and LSI components.

FEATURES

The synchronizing signal generator is built into the camera head. No external generator is used. The camera head also con-

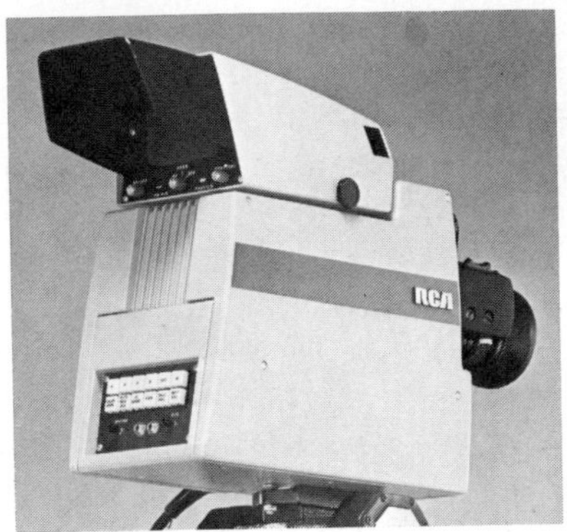

Figure 5.20 A studio color television camera. Camera dimensions (less lens) are: height $17\frac{1}{2}$ in (444 mm), width $8\frac{3}{4}$ in (213 mm), length 18 in (457 mm). Weight is 38 lb (17.2 kg). *(Courtesy of RCA Broadcast Systems.)*

Figure 5.21 The camera control unit (CCU) for the camera of Figure 5.20. The unit weighs only 6 lb and measures $8\frac{1}{2}$ in (21.6 cm) × $10\frac{1}{2}$ in (26.7 cm) × $3\frac{3}{4}$ in (9.5 cm). *(Courtesy of RCA Broadcast Systems.)*

tains a high quality audio amplifier. This amplifier drives a 600-ohm line at 10 dBm. Also featured are automatic iris control, automatic white balance control, automatic flare correction, and automatic cable equalization. The camera has fast warmup with useable pictures available in 5 to 7 seconds. The camera has a built-in color filter wheel.

The camera has a single 200-mm lens, of the motorized (servo) type. The operator's viewfinder can be seen in Figure 5.20. This is a monochrome picture tube with a diagonal dimension of 5 in (127 mm). It has a resolution of 600 TV lines at the center of the screen. The operator is provided with contrast and brightness controls for the viewfinder. Video-level indication is part of the viewfinder display. If the 100% level is exceeded, the peaks will go negative. By means of push buttons the operator can view the following waveforms: (1) luminance, (2) red, (3) green, (4) blue, and (5) external video (composite).

Typical camera operation is at 125 footcandles, with a lens opening of f/2.8. For extremely low light levels, the operator can switch in an additional +9 dB of video gain. The camera has low power consumption. The total power consumption (90 V to 130 V and 180 V to 260 V at 47–63 Hz) is approximately 100 W. Provision is made for interphone communications. The system is designed for use with dynamic, microphone-type headsets.

CAMERA CONTROL UNIT (CCU)

This unit is shown in Figure 5.21. It makes possible multi-CCU installations in various studio and van locations. Such installations save time in setup when changing camera locations. The camera is simply plugged into the CCU and is ready for operation in a matter of seconds. Cable compensation is automatic. It should be noted that, if desired, this camera can be operated without the CCU. The camera is completely self-contained.

Figure 5.22 A portable color TV camera. Camera dimensions (less lens and shoulder support) are: height 10.5 in (267 mm), width 4.3 in (109 mm), depth 16.0 in (406 mm). Weight is 17.3 lb (7.9 kg). *(Courtesy of RCA Broadcast Systems.)*

5.10 PORTABLE COLOR TV CAMERA

A color TV camera especially designed for portable operation is shown in Figure 5.22. Unlike some previous designs, this camera requires no backpack and it is completely self-contained. A CCU (camera control unit) is not used. The camera is easy and convenient to use for newsgathering, documentaries, and on-site commercials. It provides high-quality, full bandwidth pictures and operates from 12 V DC sources, including automobile batteries. In addition, the camera can be operated for 75 min from a rechargeable nickel-cadmium battery belt (weighing 6 lb) which is worn around the operator's waist. The camera can also be operated from an AC adapter.

FEATURES

The camera is enclosed in a sealed, rugged case which protects it from dust, moisture, and RFI (radio frequency interference). It features automatic iris, white balance, and flare control. Picture sharpness and colorimetry compare favorably with large studio cameras. The unit contains special preamplifiers, which permit use in low-light conditions. It is designed to be operated without highly trained personnel and requires infrequent maintenance.

The camera features solid-state construction, with most components mounted on circuit board modules. As shown in Figure 5.23A, the modules slide into connectors on a "mother" board, where they are locked into place. The camera can provide usable pictures within 5 to 7s after it is turned on. It will meet specifications under temperatures ranging from −4 °F to +122 °F (−20 °C to +50 °C).

TUBES AND VIEWFINDER

A three-way color tube system is used, as illustrated in Figure 5.23B. Three $\frac{2}{3}$-in (1.70-cm) Plumbicons are employed. The lens, prism, yokes, camera tubes, and preamplifiers are mounted as a single unit to maintain precise registration of the system. The viewfinder incorporates a 1.5-in (3.8-cm) diagonal cathode ray tube. It is provided with LED (light-emitting diode) indicators to provide the operator with necessary technical information. The camera has high sensitivity: 45 footcandles at f/1.6 (or higher by switching in additional preamplifiers). The power requirement at 11 V to 14 V is 42 W.

PORTABLE COLOR TV CAMERA WITH VCR

Figure 5.23C illustrates another portable, color TV camera. This camera has a built-on miniature VCR. Thus, the program recording is made directly on the camera assembly without the need for external cables or equipment. One-half inch wide recording

A

C

B

Figure 5.23 *(Legend on facing page)*

Figure 5.24 A portable color TV camera with an attached microwave transmitting-receiving system. The microwave assembly can be seen at the left of photo. The entire unit weighs 26 lb (11.8 kg). *(Courtesy of Ikegami Electronics (USA), Inc.)*

tape in a cassette is used. The VCR is seen at the left in the figure. (RCA supplies a similar camera using one-quarter inch wide recording tape on a cassette.)

5.11 THE MINICAM SYSTEM

The audio and video output of portable, color TV cameras used in the field may be transmitted by microwaves to a base station. This arrangement is often referred to as the "Minicam" system.

One such camera is pictured in Figure 5.24. This camera can function up to almost 5 000 ft (1 500 m) from the base station antenna. The microwave link permits the camera to be operated in crowded areas or in situations where it is not possible to use cables. The camera can be used indoors and outdoors.

Bidirectional operation is used. The audio and video signals are multiplexed and transmitted directly from the camera transmitter and antenna on the 2-GHz frequency band. Control signals at 950 MHz from the base station are received by a whip antenna on the microwave assembly. The assembly includes a two-way intercom circuit.

The transmitting antenna at the camera is equipped with an automatic 360° homing mechanism. This keeps the microwave antenna directed at the base station antenna.

This system provides up to 13 different channels. This will allow different cameras to broadcast from the same area.

5.12 SPECIAL PURPOSE TV CAMERAS

Over the past several decades, television has completely revolutionized home entertainment. Now, it is making great inroads in education, medicine, industry, aerospace, and oceanography. In fact, there is hardly a field of endeavor that does not use television in some particular application to perform tasks that are unsafe, inconvenient, or impossible. However, not all of these tasks can be performed by television equipment that is designed primarily for use in the broadcasting studio. For this reason, many special purpose cameras have been developed.

Figure 5.23A. Side view of opened portable color TV camera. Note the plug-in, solid-state circuit module boards. *(Courtesy of RCA Broadcast Systems.)* **B.** The portable color TV camera, showing the three-way pickup tube arrangement, lens, and viewfinder. The viewfinder, located directly above the lens, is shown in the "storage" position for transporting. *(Courtesy of RCA Broadcast Systems.)* **C.** A portable color TV camera that includes an attached VCR. The VCR cassette uses one-half inch wide recording tape. *(Courtesy of Ikegami Electronics (USA), Inc.)*

EDUCATION, MEDICINE, AND INDUSTRY

Television cameras that are designed specifically for education, medicine, and industry are typically small, simple to set up and operate, and are much less expensive than their studio counterparts. They can be operated by students or other personnel who are not specifically trained or experienced in television program production. Therefore, operator controls are reduced to a minimum. In many cases, the camera has only an on-off power switch. The method of mounting the camera is determined by the purpose for which the camera is designed. Simple roll-around pedestals are often used for classroom or studio-type operation. Photographic-type tripods and simple shelf-type brackets serve in field or fixed applications.

Figure 5.25 shows a color camera that is designed primarily for educational, medical, and industrial uses. It is 19½ in long, 8 in wide, 13 in high, and weighs less than 50 lb. This camera achieves full fidelity color reproduction through the use of a single vidicon, a special color-detecting optical filter, and innovative electronics. No revolving color wheels or other moving parts are used. All circuit components are completely solid state. In addition, a zoom lens and a 5-inch viewfinder are included in the camera. The operation of this camera is based upon the sequential, rather than on the simultaneous color system.

Figure 5.25 A single vidicon color camera. *(Courtesy of RCA Broadcast Systems.)*

CONSIDERATIONS FOR TELEVISING FILMS

Many TV programs consist of motion picture films. However, as mentioned in a prior chapter, the basic film frame rate is 24/s, while the TV frame rate is 30/s. Obviously, if no accommodation is made, these rates are incompatible. It is not possible to run the film sound track at any other rate because sound changes would occur. The problem is solved by running the film continuously at 24 frames per second and operating the shutter at the TV field rate of 60 frames per second.

The shutter is synchronized to the TV field rate in the following manner. Odd numbered film frames are projected during the time of two television fields. Even numbered film frames are projected during the time of three television fields. When four film frames are completed ($\frac{1}{6}$s), ten television fields or five TV frames are completed ($\frac{1}{6}$s). Therefore, 24 film frames in one second coincide with the scanning of 30 TV frames at the same time.

COLOR FILM (TELECINE) CAMERA

Projection of 16-mm and 35-mm film, as well as slides, is performed at TV studios. The projector focuses the optical image upon the camera tube targets via the optical system. A three-tube telecine camera is shown in Figure 5.26. This unit can be equipped with either three vidicons or three Plumbicons. The camera has a number of automatic features, including film density and contrast

Figure 5.26 An RCA TK28 color film (telecine) camera. Note the position of the three pickup tubes. The unit may be powered from 100 V AC to 130 V AC or 200 V AC to 260 V AC, at 47 to 62 Hz. Power consumption is 250 W. *(Courtesy of RCA Broadcast Systems.)*

correction, film-base color error correction, color balance, and white level and black level control. Resolution with either type of pickup tube is 600 TV lines (measured center screen).

Figure 5.26 clearly shows the location of the three-pickup tubes. The optical image is focused upon the lens located at the upper left side of the cabinet. The panel at the top contains most of the required electronics and controls. Seven plug-in solid-state modules are employed. The power supply panel is visible near the bottom of the cabinet.

5.13 STUDIO COLOR VIDEO-TAPE RECORDER (VTR)

A video-tape recorder designed for studio use is shown in Figure 5.27. The tape used in this VTR is 2 in (5.08 cm) wide. Unlike home or portable VTRs studio VTRs record and play back the full bandwidth of the composite NTSC color video signal. The video frequency bandwidth response is flat from 30 Hz to 4.5 MHz. Audio frequency response (on a separate track) is flat from 50 Hz to 16 kHz.

Figure 5.27 A studio video-tape recorder (VTR). The unit uses 2-in (5.08-cm) wide magnetic tape. It can provide up to 250 minutes of recording time. *(Courtesy of Ampex Corp.)*

The VTR shown in Figure 5.27 can accommodate tape reels from 7 in to 16 in (17.8 cm to 40.6 cm) in diameter. Tape pulling speeds are either $7\frac{1}{2}$ in/s (19.05 cm/s) or 15 in/s (38.1 cm/s). Recording times are 250 min for the slower speed and 125 min for the higher speed. These values are based on a 9400-ft (2866-m) reel of 2-in (5.08-cm) tape. The effective or "writing" tape speeds are very much higher than the tape pulling speeds so that the high video frequencies can be recorded. In this type of design, four record-playback heads are mounted 90° apart on a drum. This drum rotates at a constant speed of 14 400 r/min. The speed and phasing of the tape writing are closely controlled by servo amplifiers. Unlike the helical recording method, the video signal is recorded as a series of transverse tracks across the 2-in (5.08-cm) tape. Separate tracks are used for the control and audio signals. To reduce maintenance time, LED status indicators are located on the control console and printed wiring board.

SUMMARY OF CHAPTER HIGHLIGHTS

1. The camera pickup tube converts light rays received from the scene being televised into electrical impulses.
2. The electrical impulses represent basic picture elements. The greater the number of these elements, the greater the picture resolution.
3. Light rays incident on the faceplate of the camera pickup tube produce an electron-charge replica of the scene on a target within the tube. Each electron charge represents a picture element. The intensity of the charges is determined by the bright or dark elements in the scene.
4. A scanning beam that is developed by an electron gun in the neck of the tube converts each picture element charge into an impulse of current. The magnitude of the impulses is proportional to the intensity of the picture-element charges in the electron replica.
5. Scanning in the camera-pickup tube is identical to scanning in the TV receiver picture tube. External deflection coils move the electron beam in the standard scanning pattern.
6. Modern TV cameras feature solid-state circuitry, resulting in reduced size, weight, power consumption, and improved reliability.
7. Color TV cameras may have either three pickup tubes or a single pickup tube.
8. Some important characteristics of camera tubes are (a) light-transfer characteristics, (b) gamma, (c) spectral response, (d) lag, and (e) dark current.
9. The vidicon is a simple and compact camera tube. Its basic components are an electron gun and a target (Figure 5.4). The diameter of the vidicon varies from $\frac{1}{2}$ in (13 mm) to $1\frac{1}{2}$ in 38 mm).
10. The operation of the vidicon is based upon the properties of a photoconductive layer (Figure 5.5B). The electron beam causes output currents to flow in proportion to the light intensities of various parts of the image.
11. The Plumbicon camera tube is widely used in color television studio and field cameras. It differs from the vidicon in its target construction (Figure 5.8). The spectral color response of the tube is

designated by the letter R, G, or B, following its type number, such as XQ1427R.

12. A 1600 TV line Plumbicon was developed for use in the field of electronic cinematography.
13. The silicon diode array vidicon has a target consisting of about 30 000 photodiodes. The photodiode conduction increases with light intensity so that it is possible to establish a photodiode "image" in various states of conduction (Figure 5.12).
14. The silicon-imaging device (SID) (Figure 5.13), is a solid-state, self-scanned type of camera pickup device. It does not use a scanning beam. It is a **charge-coupled device.** The effect of scanning is performed by electronic circuitry.
15. One type of single color camera pickup tube is the Trinicon®. The important difference between a Trinicon® and a vidicon is in the target assembly (Figure 5.16). An NTSC color video signal is produced by a three-color filter, an index electrode with suitable electronics circuitry, and the scanning beam. (Figures 5.16 and 5.25)
16. The light transfer capability of a camera pickup tube is expressed as the gamma of the tube. Gamma is the ratio of the brightness variation in the reproduced image to the brightness variation in the original scene. Normally, camera pickup tubes are operated, as nearly as possible, at unity gamma.
17. For proper tone rendition, the spectral response of the camera pickup tube should be the same as that of the human eye.
18. Quantum efficiency is a means of comparing photosurface sensitivities. A perfect photosurface would emit an electron for each photon of light focused upon the surface.
19. In some color cameras, a separate camera pickup tube is used for each of the primary colors: red, green, and blue. Some cameras use a fourth tube for a luminance channel to provide better detail and sharper monochrome pictures.
20. Beam-splitters in a color camera optical system divide the light reflected from the scene into the three primary colors. Each portion of the light is then directed onto the faceplate of the respective camera pickup tube.
21. In some color TV cameras, the synchronizing signal generator and an audio amplifier are built into the camera head. Cameras may also use a motorized zoom lens and automatic controls.
22. Synchronization, power supply, and control voltages for some cameras are provided by a camera control unit in the studio. (Section 5.9 and Figure 5.21)
23. The purpose of the camera control unit is to provide remote control and to relieve the camera operator of many of the technical functions required in reproducing high quality pictures during a telecast. (Section 5.9 and Figure 5.21)
24. The camera operator views the scene that is being telecast on a TV receiver-type viewfinder that is mounted on the camera. The viewfinder also provides important operational waveforms to the operator.
25. Some portable color TV cameras are completely self-contained. These cameras do not require any backpack or external CCU (camera control unit). They may be operated from 12 V dc sources, a belt-type battery pack, or an AC adapter. They are available as three-tube or single-tube cameras. (Figures 5.22, 5.23 and 5.24)
26. In the "Minicam" system, portable color TV cameras are connected by ca-

bles or microwave links to the studio broadcast equipment. (Figure 5.24)

27. When televising motion picture films, the film runs continuously at 24 frames per second, while a shutter operates at 60 TV fields per second. When four film frames are completed ($\frac{1}{6}$s), five TV frames are completed ($\frac{1}{6}$s). (Figure 5.26)
28. Studio VTRs use 2 in (5.08 cm) wide magnetic tape and record the full bandwidth of the NTSC color-video signal. A 9 400-ft (2 866-m) reel of tape can provide up to 250 minutes of recording or playing time. (Figure 5.27)
29. Table 5.4 compares the characteristics of a variety of camera tubes. The Newvicon tube employs new photoconductive materials (zinc telluride and zinc selinide) which give it a light sensitivity 20 times greater than vidicons. The Cosvicon is used in single-tube color cameras.

EXAMINATION QUESTIONS

(Answers are provided at the back of the text.)

The following items are to be answered true (T) or false (F).

1. When low light-level scenes are telecast, it is essential that the TV camera provide a low signal-to-noise ratio.
2. All types of TV cameras generate their own synchronizing and control signals.
3. Multitube color TV cameras must employ a minimum of four pickup tubes.
4. The light-transfer characteristic is the ratio of the faceplate illumination to the output signal current of a camera tube.
5. Some camera tubes are sensitive to infrared and X-rays.
6. Dark current is the current flowing in a camera tube in the total absence of infrared radiation.
7. Vidicons operate by the principle of photoemission of the target.
8. In the vidicon target, the conducting layer acts as the signal electrode.
9. In camera tubes, the optical image is transferred into an electrical charge image.
10. Most camera tubes employ magnetic deflection and either magnetic or electrostatic focusing.
11. Color TV cameras use Plumbicon tubes because of their low lag, simplicity, and excellent spectral response to the primary colors.
12. The Plumbicon is one of a class of SID (silicon diode array) camera tubes.
13. The 1600 TV line Plumbicon is specifically designed for portable field cameras.
14. The type of camera tube employing photodiodes is the silicon diode array vidicon.
15. The SID pickup device works by means of a very fine scanning beam.
16. In a single TV color pickup tube (Trinicon®), the tube delivers both the "Y" signal and the chrominance signal.
17. Basically, the Trinicon® differs from the vidicon by its three-color filter and index electrode.
18. In a three-way beam-splitter, the first dichroic mirror focuses the image onto the faceplate of the "Y" signal camera tube.
19. The viewfinder on a TV camera is also used as an indicator for video signals.
20. Lens openings for TV cameras are commonly about f/28.
21. Some portable TV cameras can be operated from a rechargeable lead-acid battery belt.
22. The term "Minicam system" refers to an extremely compact TV color camera.
23. To televise motion picture film, a high intensity cathode-ray tube is used in the motion picture camera.
24. A studio color VTR uses 2 in (5.08 cm) wide magnetic tape and records the full bandwidth of the NTSC signal.

Table 5.4 **TYPICAL CHARACTERISTICS OF TV CAMERA TUBES**
(Courtesy of Panasonic Company)

	TYPE NO.	APPLICATION	FEATURES	HEATER CURRENT (mA)	TUBE DIAMETER (inch)	MAX. LENGTH (mm)	FOCUSING METHOD (NOTE 1)	DEFLECTION METHOD (NOTE 2)	MESH	SENSITIVITY		TARGET VOLTAGE	DARK CURRENTS	GAMMA	LAG (NOTE 4) (%)	LIMITING RESOLUTION AT CENTER (TV LINES)
										FACEPLATE ILLUMINATION (NOTE 3) (lX)	SIGNAL CURRENT (nA)					
	S4075	Compact TV camera	Ultra-sensitive in visible spectrum No burning-in	95	2/3	108	M	M	Separate	1	260	10	25	3	1	10	650
	S4092	Compact TV camera	Ultra-sensitive in visible spectrum No burning-in	95	2/3	108	E	M	Separate	1	260	10	25	3	1	12	600
	S4102	Compact TV camera	Ultra-sensitive in visible spectrum No burning-in	95	2/3	108	E	M	Separate	1	260	10	25	3	1	12	600
Newvicon™	S4113	TV camera for near infrared light	Ultra-high spectral response extended to near infrared region	95	2/3	108	M	M	Separate	1	320	10	25	7	1	10	650
	S4076	High-resolution TV camera X-ray TV camera	Ultra-sensitive in visible spectrum High resolution, No burning-in	95	1	162	M	M	Separate	0.5	240	10	25	6	1	20	750
	S4093	LLL TV camera	Fiber-optic faceplate	95	1	162	M	M	Separate	0.5	170	10	25	7	1	20	650
	20PE11	Home use TV camera	Compact, Light-weight	110	2/3	108	M	M	Integral	10	200	10	80	20	0.74	20	500
	20PE13A	Compact TV camera	Compact, High-resolution	95	2/3	108	M	M	Separate	10	220	10	80	20	0.74	20	650
	20PE14	Mini-type TV camera Handheld VTR camera	Compact, Light-weight	95	2/3	108	E	M	Separate	10	220	10	80	20	0.74	20	600
	S4097	Mini-type TV camera Handheld VTR camera	Compact, Light-weight	95	2/3	108	E	M	Separate	10	220	10	80	20	0.74	20	600
Vidicon	7262A	Standard TV camera	Standard	95	1	133	M	M	Integral	10	300	10	100	20	0.74	25	600
	7735A	Standard TV camera	Standard	600	1	162	M	M	Integral	10	300	10	100	20	0.74	25	600
	8507	High-resolution TV camera "Tele-cine" camera	High-resolution	600	1	162	M	M	Separate	10	300	10	100	20	0.74	25	750
Cosvicon	8541	High-resolution TV camera "Tele-cine" camera	High-resolution	95	1	162	M	M	Separate	10	300	10	100	20	0.74	25	750
	S4089	Single-tube color camera	Built-in stripe (cyan. Yellow)	95	1	162	E	M	Separate	20	300	10	100	25	0.74	25	250
	S4110	Single-tube color camera	Built-in stripe filter (cyan. Yellow)	95	1	162	M	M	Separate	20	300	10	100	25	0.74	25	250

Note 1. M: Magnetic focusing
E: Electrostatic focusing

Note 2. M: Magnetic deflection
Note 3. 2856 K Tungsten
Note 4. Lag at 50 m sec. after light-off

REVIEW ESSAY QUESTIONS

1. Discuss briefly the *basic* manner in which an optical image is transformed into an electrical signal.
2. Refer to Table 5.1. Which 1-in (25.4 mm) diameter camera tube(s) is suitable for (1) X-ray TV camera, (2) studio TV camera, (3) telecine camera, (4) single-tube color camera.
3. Define: gamma, light-transfer characteristic, lag, dark current.
4. Make a simple drawing of the components of the standard vidicon target plate.
5. In Question 4, explain briefly how the output signal is developed.
6. List all of the different types of target plates mentioned in this chapter.
7. In Figure 5.4, what is the function of grid No. 4? Of the target connection?
8. Explain how the Plumbicon differs physically from the standard vidicon. (Refer to Figure 5.8.)
9. Draw a simple diagram of the Plumbicon target and explain briefly how it works.
10. What is a photoconductive layer?
11. Describe briefly how the silicon diode array vidicon differs from the standard vidicon. (Refer to Figure 5.12.)
12. In Question 11, how does the target produce an output signal? (Refer to Figure 5.12.)
13. Describe, in a general way, the construction of the SID and how it operates. (Refer to Figures 5.13 and 5.14.)
14. What is meant by CCD, CCU?
15. What is a Trinicon®? Briefly describe its operation. (Refer to Figures 5.16 and 5.17.)
16. What is a beamsplitter and how is it used? (Refer to Figure 5.18.)
17. In a studio TV color camera, what is the function of the CCU? Of multiple CCUs? (Refer to Figure 5.21.)
18. Explain three different methods of powering portable color TV cameras. What automatic features might they have? What is the function of the LED indicators?
19. Describe briefly how the "Minicam" system operates. (Refer to Figure 5.25.)
20. The normal motion picture frame rate is 24/s. The TV frame rate is 30/s. How is it possible to transmit motion pictures over television?
21. How is it possible for a studio VTR to record directly the full bandwidth of the NTSC color TV signal?

EXAMINATION PROBLEMS

(Selected problem answers appear at the back of the text.)

1. In Figure 5.2, what is the quantum efficiency if the wavelength is 660 nanometers? 900 nanometers?
2. Refer to Figure 5.6. If the faceplate illuminance is 10^{-2} footcandles for tungsten light, what is the signal current? What is the signal current for visible light at 10^{-1} footcandles?
3. Refer to Figure 5.12. The load resistor is 50 kilohms. The bright light signal current is 250 nA and the grey light signal current is 50 nA. Calculate the peak-to-peak camera output signal voltage.
4. In Figure 5.17B show mathematically why 0.22 μs corresponds to 4.5 MHz.
5. A vidicon has an initial signal-output current of 100 nA. The illumination is removed and $\frac{1}{20}$s later the signal current has dropped to 7.5 nA. What is the lag? What is the lag if the signal current drops to 1.0 nA?
6. Show mathematically how it is possible for a 24/s motion picture frame to be shown on a 30/s frame television system.
7. A studio VTR is using 2-in (5.08-cm) wide tape on a 7000-foot (2134-m) reel. If it operates at $7\frac{1}{2}$ in/s (19.05 cm/s), what is the recording time in minutes?

Chapter 6

6.1 Radio wave propagation
6.2 Line-of-sight distance
6.3 Unwanted signal paths
6.4 Antenna characteristics
6.5 Tuned antennas
6.6 UHF antennas
6.7 General-purpose antennas
6.8 Combination antennas
6.9 RCA ministate antenna system
6.10 Indoor antennas
6.11 Antenna rotators
6.12 Transmission lines
6.13 Amplifying the antenna signal
6.14 Antenna-system accessories
6.15 Master-antenna distribution systems
6.16 Community-antenna television systems
6.17 Grounding and lightning protection
6.18 Trouble shooting antenna systems
Summary

WAVE PROPAGATION AND TV ANTENNA SYSTEMS

INTRODUCTION

The antenna of a television receiver requires much more attention than the antenna of a sound receiver. This is especially true as far as placement is concerned. In order to obtain a clear, well-formed image, the following requirements must be met:

1. Sufficient signal strength must be developed at the receiving antenna.
2. The signal must be received from one source, not several.
3. The receiving antenna must be placed well away from man-made interference sources.

In sound receivers, a certain amount of interference and distortion is permissible. For television, however, the standards are stricter and so elaborate antenna systems are needed to guard against many types of interference and distortion.

GHOSTS

The antenna must be positioned carefully, not only to provide the strongest signal to the receiver but also to avoid so-called **ghosts** on the screen. Ghosts are multiple images caused by the almost simultaneous reception of the same signal from two or more directions. Figure 6.1 shows a television antenna receiving two signals. The first signal comes directly from the transmitting tower. The second signal, however, strikes the antenna only after following a (longer) indirect path caused by reflection from tall buildings, water towers, etc. Because the reflected signal travels farther, it will arrive at the receiver antenna a fraction of a second later than the direct signal.

With sound receivers, the ear cannot detect the small difference in time between the direct and indirect signals. On a television screen, however, the scanning beam has traveled a short distance between the time

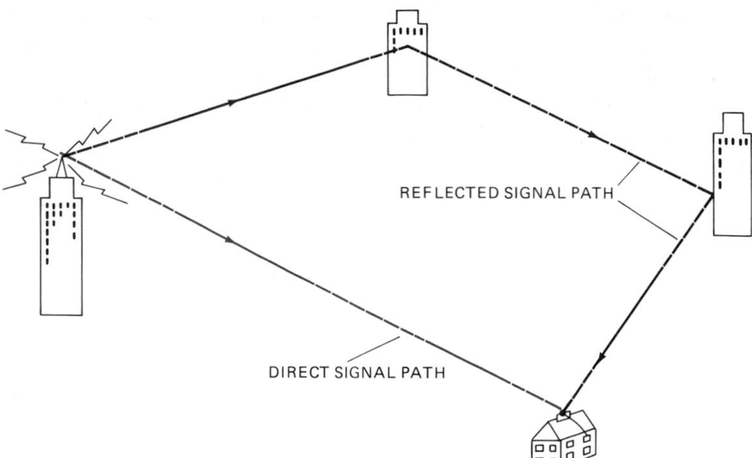

Figure 6.1 The reflected signal and the direct signal arrive at the receiving antenna and form double images, or ghosts.

the direct signal arrives and the time the reflected signal arrives at the receiver. Hence, the image contained in the reflected signal appears on the screen displaced a short distance from similar detail contained in the direct signal. The result is shown in Figure 6.2. When the effect is pronounced, there is a distinct double image and the picture is blurred. To correct this condition, it is necessary to make changes in the antenna system so that only one signal is received. This is usually done with a highly **directional antenna,** that is, one that will receive a signal from one direction only. The directional antenna is positioned to minimize or eliminate the undesired signal. The antenna should not favor the reflected signal unless it is impossible to obtain a clear image from the direct signal. The properties of reflecting surfaces change with time, and so there is no certainty that a good reflected signal will always be received.

Figure 6.2 A ghost image on a television screen. Note that the ghost is displaced to the right.

The placement of the antenna is one of the most important requirements of a television installation. To obtain the best results, you must understand the behavior of radio waves at the high frequencies used for television.

When you have completed the reading and work assignments for Chapter 6, you should be able to:

- Define the following terms: ghosts, directional antenna, bidirectional antenna, nondirectional antenna, sky wave, ground wave, line-of-sight distance, gain, bandwidth, impedance, resonant frequency, half-wave antenna, dipole antenna, directivity, reflector, driven element, director element, log-periodic antenna, parallel-wire transmission line, coaxial cable, booster, preamplifier, band separator, balun unit, MATV, CATV.
- Describe the basic antenna systems

commonly used, both indoor and outdoor.

- Explain antenna resonant frequency and calculate suitable half-wave rod lengths for various transmission frequencies.
- Design lightning protection for an antenna system.
- Discuss the various points to inspect when servicing an antenna system.
- Describe the different kinds of transmission wires and the advantages and disadvantages of each.

6.1 RADIO WAVE PROPAGATION

Transmitted radio waves can be characterized by their direction of travel. A wave that closely follows the surface of the earth is known as a **ground wave.** A wave that travels upward, at an angle determined by the position of the transmitting antenna, is called a **sky wave.** At low frequencies—up to approximately 1.5 MHz—the attenuation of ground waves is low and signals travel for long distances before they disappear. Above the broadcast band, ground wave attenuation increases rapidly and long-distance communication is carried on mostly by sky waves.

SKY WAVES

Sky waves leave the earth at an angle that may have any value from 3 to 90°. They travel in an almost straight line until they reach the ionosphere. This region, which begins about 70 miles (112 m) above the earth's surface, contains large concentrations of charged gaseous ions, free electrons, and uncharged, or neutral, molecules. The ions and free electrons bend all passing electromagnetic waves back toward earth. Whether the bending is complete (and the wave does return to earth) or only partial depends on several factors:

1. The frequency of the radio wave
2. The angle at which the wave enters the ionosphere
3. The density of charged particles (ions and electrons) in the ionosphere at that particular moment
4. The thickness of the ionosphere at that moment

Experiments show that as the frequency of a wave increases, a smaller entering angle is necessary for complete bending. To illustrate this, consider Waves A and B in Figure 6.3. Wave A enters the ionosphere at a small angle (ϕ). Thus, little bending is required to return it to earth. Wave B, subject to the same amount of bending, does not return to earth because its entering angle (θ) is too large. Naturally, Wave B would not be useful for communicating between points on Earth.

NON-RETURN OF WAVES

By raising the frequency still higher, the maximum allowable incident angle necessary at the ionosphere becomes smaller. Finally, a frequency is reached where it becomes impossible to bend the wave back to Earth no matter what angle is used. For ordinary ionospheric conditions, this frequency is about 35 or 40 MHz. Above this frequency, sky waves cannot be used for radio communication between distant points on earth. Only **direct waves** are of any use. Television bands starting above 40 MHz fall into this category. By *direct waves* we mean the radio waves that travel in a straight line from transmitter to receiver. Ordinarily, because their frequencies are low enough, radio waves are sent to the ionosphere and from there to a distant receiver. At television frequencies, the ionosphere is no longer use-

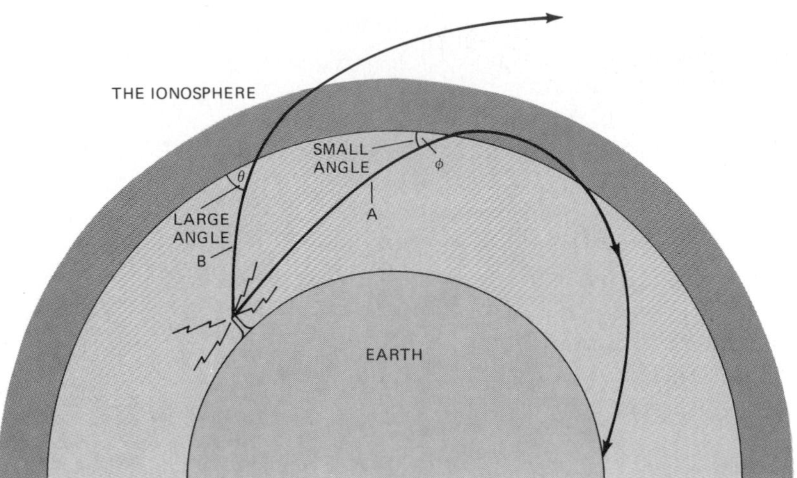

Figure 6.3 A radio wave must enter the ionosphere at small angles if it is to be returned to Earth.

ful, and so the waves being transmitted must be concentrated into a path leading directly to the receiver. It is this restriction—that direct rays must be used—that limits the distance for effective high-frequency communication.

Weather conditions sometimes cause the concentration of charged particles in the ionosphere to increase sharply. At these times, it is possible to bend radio waves of frequencies up to 60 MHz. The exact time and place of these conditions cannot be predicted, and so they are of little value for commercial operation. They do explain, however, why high-frequency signals can sometimes be received over very long distances.

6.2 LINE-OF-SIGHT DISTANCE

At the frequencies used for television, reception is possible only when the receiver antenna directly intercepts the signal as it travels away from the transmitter. The farthest distance from a transmitter that a receiver can be placed and still have its antenna intercept the signal is called the **line-of-sight distance.** It is computed as follows.

In Figure 6.4, the height of the transmitting antenna is h_t, the radius of the earth is R, and the distance from the top of the transmitting antenna to the horizon is d. These distances form a right triangle. The Pythagorean Theorem states that *the sum of the squares of the sides of a right triangle equals the square of the hypotenuse.* Using this relationship for the distances shown in Figure 6.4 gives the following equation:

$$R^2 + d^2 = (R + h_t)^2$$

By expanding,

$$R^2 + d^2 = R^2 + 2\,Rh_t + h_t^2$$

The value of h_t is very small relative to R, and so the h_t^2 term can be dropped without seriously affecting accuracy. Also, R^2 can be subtracted from both sides of the equation. The equation then becomes

$$d^2 = 2\,Rh_t$$

The radius of the earth is 4,000 miles, or 21,120,000 ft. Substituting this value for R into the equation and taking the square root of both sides, we get

$$d = \sqrt{42{,}240{,}000\,h_t}$$
$$= 6{,}499\,\sqrt{h_t}$$

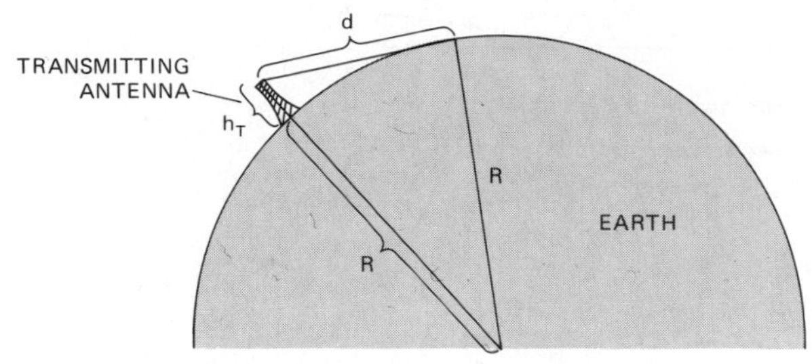

Figure 6.4 Computation of the line-of-sight distance for high-frequency radio waves.

Dividing by 5,280 to convert the distance to miles gives

$$d = 1.23\sqrt{h_t}$$

where

- d is the line-of-sight distance, in miles, from the top of the transmitting antenna
- h_t is the height, in feet, of the transmitting antenna

The relationship between d and h_t for various values of h_t is shown in Figure 6.5.

The ground coverage for any transmitting antenna will increase with its height.

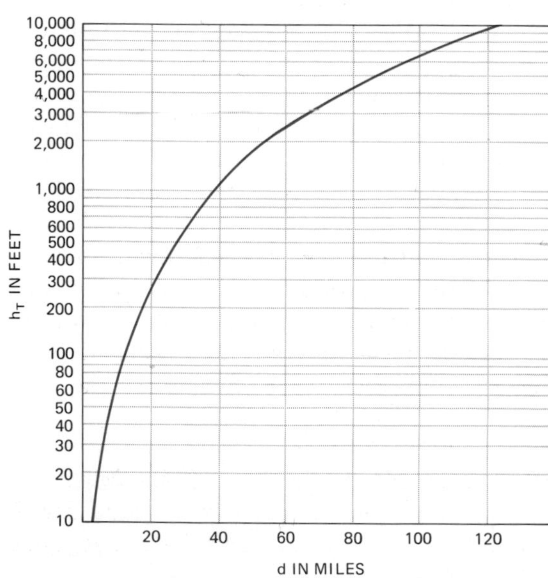

Figure 6.5 The relationship between the height of the transmitting antenna in feet and the distance in miles from the antenna that the ray may be received.

Likewise, the number of receivers capable of receiving the signal will increase. These facts account for the placement of television antennas atop tall buildings and on high plateaus.

HORIZON DISTANCE

If a receiving antenna is mounted some distance in the air, the line-of-sight distance between it and the transmitter will be increased. This condition is depicted in Figure 6.6. By geometric reasoning, the line-of-sight distance between the two antennas can be determined from the distances shown in Figure 6.6:

$d_1 = 1.23\sqrt{h_t}$ = distance from transmitting antenna to horizon

$d_2 = 1.23\sqrt{h_r}$ = distance from receiving antenna to horizon

$d = d_1 + d_2$ = maximum distance from transmitting antenna to receiving antenna

$$d = 1.23\sqrt{h_t} + 1.23\sqrt{h_r}$$
$$= 1.23(\sqrt{h_t} + \sqrt{h_r})$$

where

- h_t is the height, in feet, of the transmitting antenna
- h_r is the height, in feet, of the receiving antenna
- d is the maximum line-of-sight distance, in miles, between the two

These equations are for the geometric line-of-sight distance. In reality, electromag-

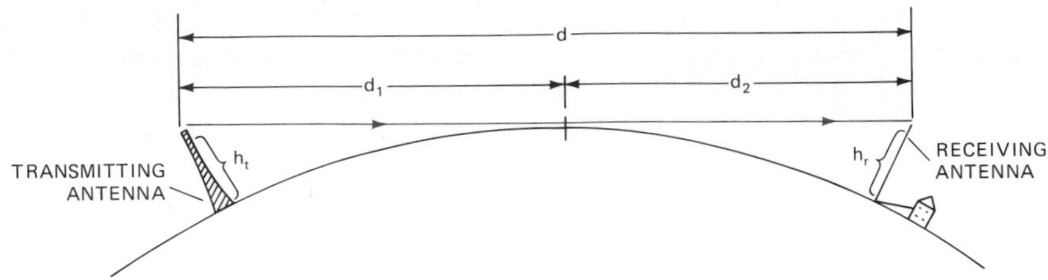

Figure 6.6 The increase in the line-of-sight distance from the receiving antenna to the transmitter achieved by raising both structures as high as possible.

netic waves are bent slightly as they move across the contact point at the horizon. As a result, television line-of-sight distance is about 15% greater than geometric line-of-sight distance. Thus, for a geometric line-of-sight distance of 30 miles (48 km), the television line-of-sight distance is 30 + (0.15 × 30) = 34.5 miles (55.5 km).

The strength of the received signal increases with the height of either antenna or of both. For television signals, this increase is most important. Proper placement of the antenna and utilization of its directive properties help to decrease (and many times to eliminate) reception of all but the desired direct wave.

6.3 UNWANTED SIGNAL PATHS

The distances just calculated apply to a direct signal. However, there are other paths that waves may follow from the transmitting antenna to the receiving antenna. Signals that follow these other paths are undesirable because they interfere with the direct-signal image on the screen. One type of indirect signal, the one caused by reflection from surrounding objects, has already been discussed. Another type of indirect signal is one that arrives at the receiver after being reflected from the earth's surface. This path is shown in Figure 6.7. At the point where the wave strikes the earth, a phase shift of up to 180° may occur. This phase shift places a reflected wave at the receiving antenna. This reflected wave generally acts against the direct wave. The overall effect is a general weakening of the signal and the appearance of ghost images.

There are two compensating conditions that reduce the problem of ground reflection. One is that the signal is weakened when it grazes the earth. The second condition results from the added phase shift caused by the fact that the path of the reflected signal is longer than that of the direct signal. In other words, two phase shifts affect the reflected signal: (1) one at the point of reflection from the earth and (2) one that is the result of the longer signal path. These two phase shifts are additive, so the total phase shift can be nearly 360°. Since 360° is equivalent to no phase shift, the problem of signal cancellation is reduced considerably.

The worst possible phase shift is 180° because then the direct and reflected waves are subtractive. Of course, the direct wave is stronger than the reflected wave, so the two

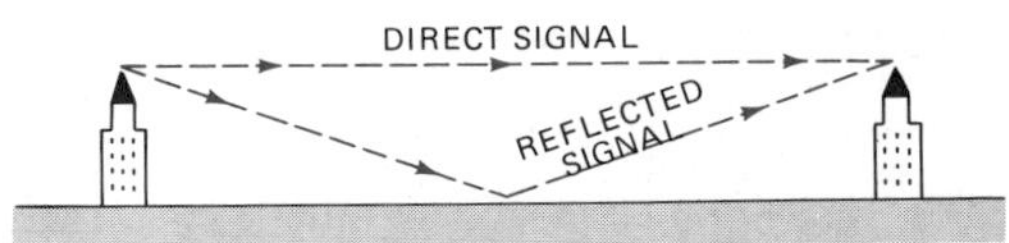

Figure 6.7 The reflected radio wave, arriving at the receiving antenna after being reflected from the Earth, may reduce the strength of the direct ray considerably.

do not cancel completely, but the reduction in signal strength may lessen receiver picture quality.

6.4 ANTENNA CHARACTERISTICS

POLARIZATION

The height of the receiving antenna is one important factor that determines the quality of the reproduced picture. Another factor is the way the antenna is placed, vertically or horizontally. The position of the antenna is determined by the nature of the electromagnetic wave.

The energy of all electromagnetic waves is divided between an electric field and a magnetic field. In free space these fields are at right angles to each other. Thus, when these fields are visualized and represented by their lines of force, the wave front appears as shown in Figure 6.8. The two rectangles represent wave fronts, and the arrows inside the rectangles represent the direction of the fields. The direction of travel of these waves in free space is always at right angles to both fields. If the direction of the electric field is vertically upward and that of the magnetic field is horizontally to the right, as in Figure 6.8B, then the direction of wave travel is as indicated in the figure.

In radio, the polarity of a radio wave is taken to be the same as the direction of the *electric* field. Hence, a vertical transmitting antenna radiates a vertical electric field (the lines of force are perpendicular to the ground), and the wave is said to be vertically polarized. A horizontal antenna radiates a horizontally polarized wave. In most cases, the signal that is induced in the receiving antenna is greatest if this antenna has the same polarization as the transmitting antenna.

Polarization Characteristics Horizontally and vertically polarized waves have different characteristics. For antennas located close to the earth, vertically polarized rays yield a better signal. When the receiving antenna is raised about one wavelength above the ground, however, either vertical or horizontal antennas may be used. When the antenna is at least several wavelengths above

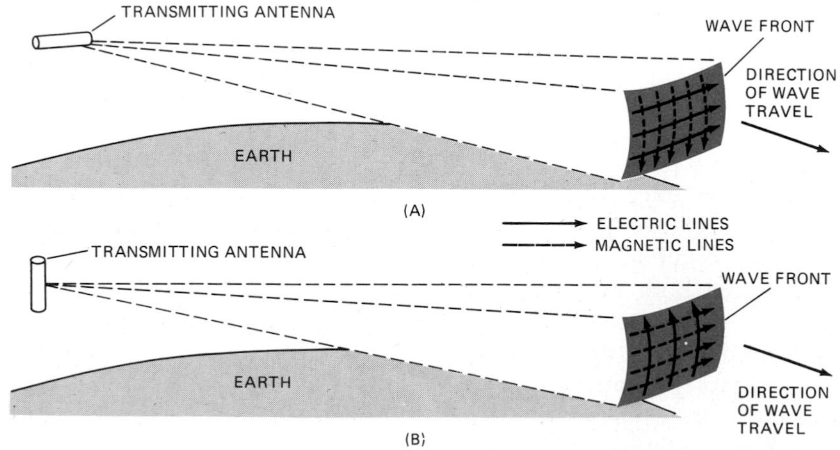

Figure 6.8 The components of an electromagnetic wave, showing their relationship with the direction of travel of the wave front.

the ground, horizontally polarized waves give a more favorable signal-to-noise ratio. In television, the wavelengths are short and so antennas can be placed several wavelengths in the air. Horizontally polarized waves are standard for the television industry, and so all television receiving antennas are mounted horizontally.

ANTENNA DIRECTIVITY

Antennas are either **directional** (unidirectional) or **nondirectional** (omnidirectional). A nondirectional antenna is one that radiates (or receives) equally well in all directions. Television transmitting antennas are usually nondirectional. On the other hand, television receiving antennas are usually unidirectional because they must pick up the maximum signal from the direction of the TV transmitter(s). Receiving antennas must also reject unwanted signals from other directions. Thus, a unidirectional antenna receives the strongest signals from only one general direction. In practice, the receiving antenna is usually "pointed" toward the transmitting antenna.

Directivity patterns are shown in Figures 6.10B, 6.12, 6.21B, 6.21D, and 6.22B.

ANTENNA GAIN

Antenna gain is the effectiveness of a unidirectional antenna relative to the effectiveness of a half-wave dipole having the same polarization when both antennas are oriented for maximum pickup. Gain is commonly expressed in decibels (dB) or as a voltage ratio. For example, an antenna with a voltage gain of 6 dB has a voltage pickup twice that of a half-wave dipole. Antennas also have increased gain when they have sharper directivity. Antenna gain graphs are shown in Figures 6.21B, 6.21D, and 6.22B. Unidirectional antennas are composed of a number of elements as compared to a simple half-wave dipole. These will be described later in this chapter.

ANTENNA BANDWIDTH

As a general rule in conventional antenna design, the higher the gain and directivity, the narrower the **bandwidth.** Although television receiving antennas must have high gain and directivity, they must also have adequate bandwidth to cover all VHF and UHF channels. Special antenna design is one method of achieving optimum television antenna performance. One popular design is the so-called log-periodic type, which is discussed later in this chapter. This design provides the desired combination of high gain, high directivity, and adequate bandwidth.

The bandwidth of any tuned antenna (Section 6.5) is also a function of the diameter of the antenna conductors. The greater the conductor diameter, the greater the antenna bandwidth. For this reason and for mechanical strength, antenna elements are made of tubing with a diameter of about $\frac{1}{4}$ in (0.64 cm).

ANTENNA IMPEDANCE

Antenna impedance is generally referred to the point at which the transmission line is attached. For television receiving antennas, this point is at the center of the connected element. The impedance is nearly equal to the $\frac{V}{I}$ ratio at the center, which is the ratio of received signal voltage to the signal current at the center. For the simple half-wave dipole shown in Figure 6.9, the impedance is 72 ohms. For more elaborate

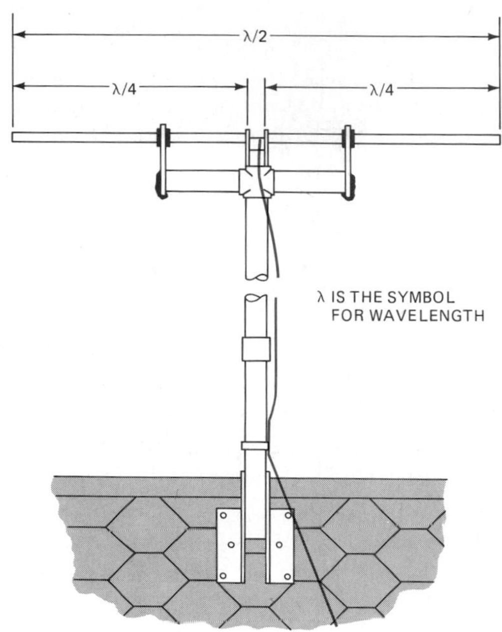

Figure 6.9 A half-wave dipole-antenna assembly, which may be suitable for some ideal locations. The feed-point impedance is 72 ohms.

antennas, the impedance may be 300 ohms. These two values are standard in the television industry.

6.5 TUNED ANTENNAS

Television antennas are made of wires or rods cut to a specific length. This length is the determining factor in the **resonant frequency** of the antenna. An antenna has inductance and capacitance; therefore it has a resonant frequency. Electromagnetic waves induce voltage in the antenna. The closer the resonant frequency is to the frequency of the electromagnetic wave, the greater is the signal voltage generated in the antenna. For example, a 50-MHz wave induces more voltage in an antenna that is resonant at 50 MHz than in one that is resonant at 60 MHz. The greater the signal induced in the antenna, the greater the signal-to-noise ratio that is possible at the output of the receiver.

HALF-WAVE ANTENNAS

An ungrounded wire or rod whose length is one half the wavelength of the signal to be received is called a **half-wave antenna** or a **Hertz antenna.** This type of antenna is very popular because it is one of the smallest antennas for its frequency and consequently requires little space. In many locations, however, it is necessary to have antennas with greater gain and directivity than the simple half-wave antenna.

A simple half-wave antenna is shown in Figure 6.9. Two metallic rods are used for the antenna itself. They are mounted horizontally on a supporting structure. Each of the rods is one quarter of a wavelength. Thus, the total length of the two rods is the necessary half-wavelength. This arrangement is a **dipole antenna.** The transmission lead-in wires are connected one to each rod. The two-wire line then extends to the receiver. The line must be fastened at several points to the supporting mast with standoff insulators so that it does not swing in the wind. Any such motion may weaken the electrical connections at the rods.

An important property of dipole antennas is that they receive signals with the greatest intensity when the rods are at right angles to the approaching signal. This arrangement is illustrated in Figure 6.10A. Signals approaching the antenna from either end are very poorly received. Figure 6.10B shows how waves at any angle are received. This graph is an overall response curve for a horizontally held dipole antenna in the horizontal plane.

With the antenna positioned as shown in Figure 6.10A, a strong signal will be received from Direction A because this signal hits the antenna at a right angle. As the angle the signal makes with the rod decreases, the strength of the received signal decreases. For a signal arriving from Direction B, therefore, the voltage is at a minimum (or zero).

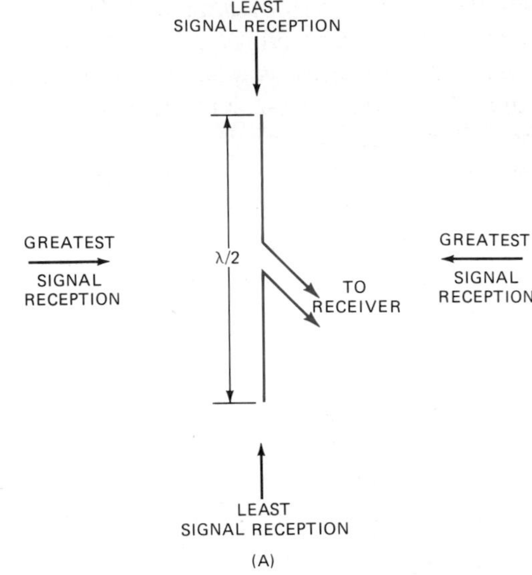

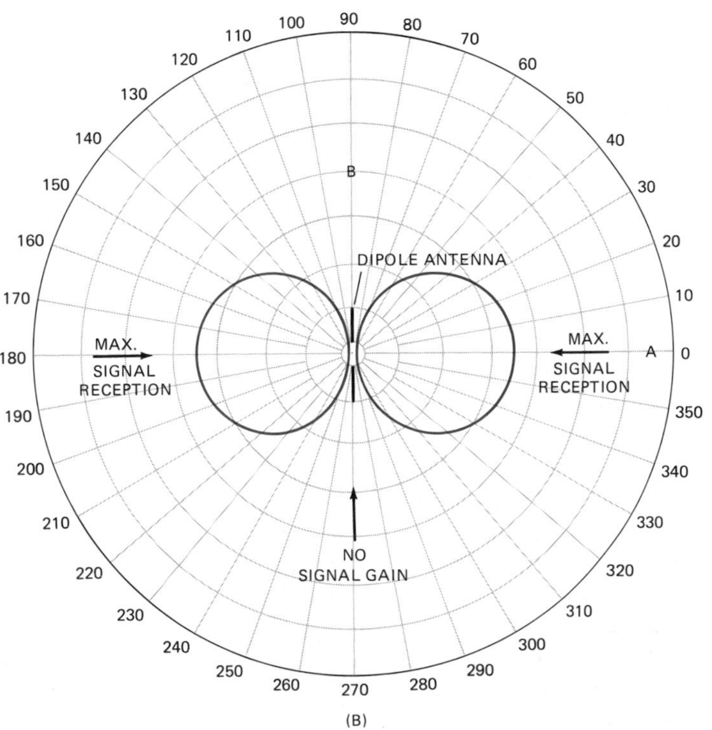

Figure 6.10A. Dipole antennas of the type shown here receive signals best from the directions indicated. **B.** The directional response curve of a dipole antenna. Note the bidirectional characteristic.

The reception for waves coming in at other angles can be determined from the graph. Note that the strongest signal is obtained from two directions. Therefore, the dipole may be called **bidirectional.** Other systems can be devised that are unidirectional or nondirectional or that have almost any desired properties. The response curve for each system quickly indicates its properties in any direction.

TUNING THE ANTENNA

An antenna must be tuned in order to have the strongest signal develop along its length. Tuning is usually accomplished by cutting the wires or rods to a specific length. The length is determined by the frequency of the signal; it is longer at the lower frequencies and shorter at the higher frequencies. It might be supposed, then, that a television receiver capable of receiving signals with frequencies ranging from 54 MHz to 88 MHz would need several antennas, one for each channel. However, this is not the case.

Antennas are similar to other tuned circuits in that they have a specific bandwidth. Thus bandwidth is related to the effective "Q" of the antenna. Thus, TV antennas have a bandwidth wide enough to accommodate all VHF or UHF channels. Separate antennas are used for the VHF and UHF channels because of the great difference in frequencies.

Dipole Length Computation For frequencies between 54 MHz and 88 MHz, a resonant frequency of 65 MHz may be chosen for the antenna design. Although an antenna cut to this frequency will not give optimum results at other frequencies, reception will be satisfactory in most cases.

To compute the length of dipole needed for the 65-MHz half-wave antenna, the following formula is used:

$$L = \frac{468}{f}$$

where

- L is the length, in feet, of the dipole
- f is the frequency, in megahertz, at which the dipole resonates

When f is 65 MHz, the correct dipole length is 468/65, or 7.2 ft (2.2 m). Practically, 7 ft (2.1 m) may be used, and so each half of the half-wave dipole will be 3.5 ft (1.1 m) long. For a full-wave antenna, approximately 14 ft (4.3 m) will be needed.

Because half-wave antennas are smaller than full-wave antennas, they are easier to make. As a result, they can be made to withstand high winds and the weight of ice or snow.

HALF-WAVE DIPOLE WITH REFLECTOR

The simple half-wave antenna may provide satisfactory reception in ideal locations. However, problems with ghosts or insufficient signal strength may require more elaborate antenna arrays. Such arrays are commonly used and provide greater antenna gain and directivity. The **gain** of a receiving antenna is a ratio: its effectiveness in receiving a signal divided by the effectiveness of a dipole receiving the same signal at the same location. The gain is usually expressed in decibels (dB). An antenna with a positive gain will deliver a better signal to a receiver than will a dipole. The gain of the antenna of Figure 6.11 is about 5 dB above the gain of a simple dipole.

The **directivity** of an antenna refers to its ability to receive signals from one direction and reject signals from all other directions. Antennas with high directivity are often used to eliminate or reduce ghosts.

The antenna shown in Figure 6.11 is a dipole with an additional rod called a **re-**

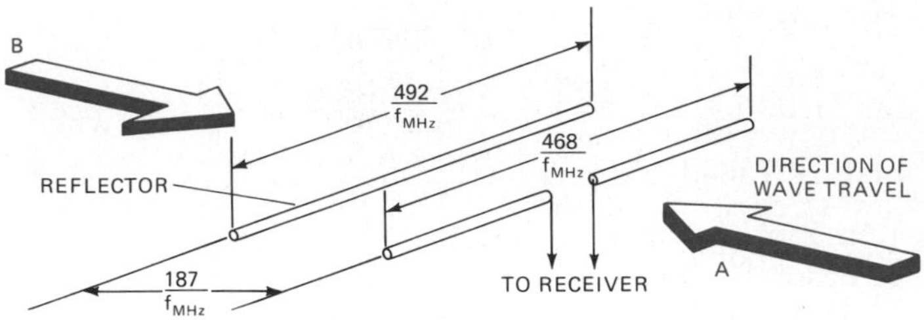

Figure 6.11 Dipole receiving antenna and reflector. The antenna will receive best from direction A.

flector. There is no electrical connection between the dipole, also called the **driven element,** and the reflector, or **parasitic element.** The space between the two is 0.15 to 0.25 wavelengths of the signal the dipole is resonant to. This is an important factor in antenna operation.

Direction A Let us assume that a signal arrives in the direction shown by Arrow A in Figure 6.11 and that the dipole (driven element) is resonant to this signal. The signal generates a voltage in the driven element and also in the reflector. As a result, signal current flows in the reflector and the signal is reradiated. Now, the reradiated signal from the reflector also induces a signal voltage in the driven element. If the reflector and dipole are properly spaced, the reradiated signal will be in phase with, and therefore will reinforce, the original signal. This accounts for the gain of this antenna over a simple dipole.

If a signal approaches from Direction B in Figure 6.11, it, too, will be reradiated by the reflector. However, this reradiated signal will arrive at the driven element out of phase with the original signal from Direction B.

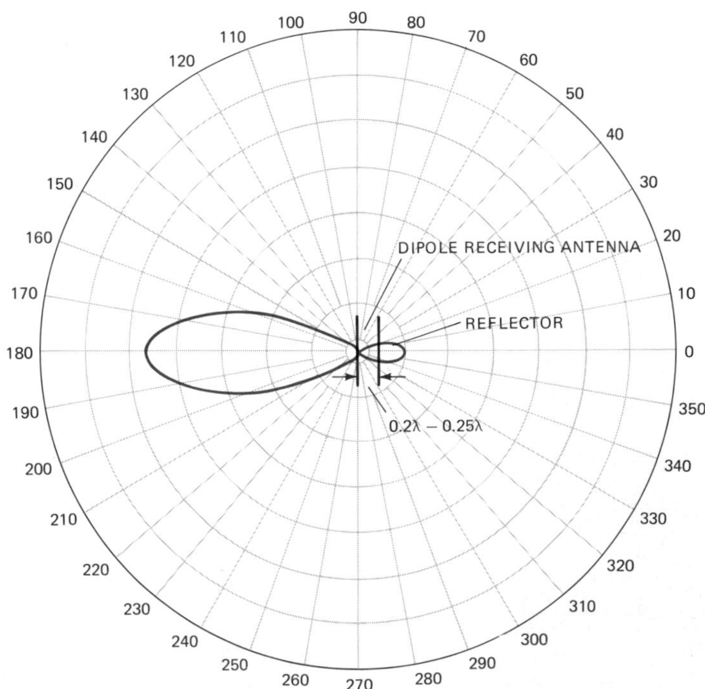

Figure 6.12 The directional response curve for a half-wave dipole antenna with a reflector. Note the basically unidirectional characteristic.

The overall result is a large decrease in the strength of the received signal. Note that the time required for the wave from Direction A to travel from the dipole to the reflector and back to the dipole is longer than the time required for the wave from Direction B and the Direction B reradiated signal from the reflector to arrive at the dipole. This accounts for the difference in phase between the original and reradiated waves for the two cases.

Increased gain is only one advantage of a reflector system. Figure 6.12 shows that the angle at which a strong signal may be received is narrower in a reflector system. This reduces the number of reflected signals, which produce ghosts. Also, partial or complete discrimination is possible against any interference, man-made or otherwise.

The curve in Figure 6.12 is idealized. The response curve of an actual array has small lobes extending in the direction of the reflector, indicating that signals approaching from the rear can be received. This is understandable, since the waves do not arrive at the driven element exactly 180° out of phase. The voltage induced in the driven element by waves approaching from the reflector side is much less than that induced by waves arriving from the front.

DIPOLE WITH REFLECTOR AND DIRECTOR

Additional gain and directivity can also be obtained from an antenna by also adding a **director element.** A director is a wire or rod whose length is slightly less than one half wavelength. Like the reflector, it is a parasitic element. Normally, the reflector is slightly longer than the driven element and the director is slightly shorter. When the dipole is placed perpendicular to the direction of the incoming signal, the director is the first element to intercept the signal. Figure 6.13 shows an antenna with a director and

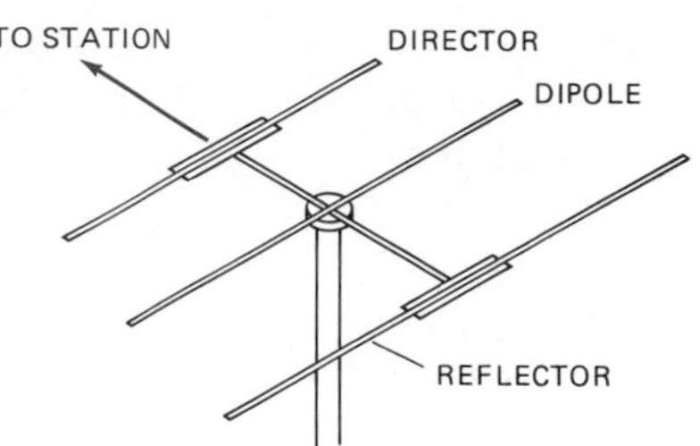

Figure 6.13 A dipole antenna with reflector and director. The reflector and director are electrically insulated from the dipole.

reflector. The arrow points in the direction of greatest directivity. The director picks up part of the signal and then reradiates it with a phase relationship that strengthens the signal arriving at the driven element. This increases the directivity of the dipole and reduces the ability of the array to pick up signals arriving from the rear.

Some antennas have more than one director. The Yagi antenna shown in Figure 6.14 is an example. This type of antenna has very high gain and a highly directional response. Some Yagi antennas have more directors than shown, which increases the gain and directivity even more. A disadvantage of the Yagi antenna is that it responds only to a relatively narrow range of frequen-

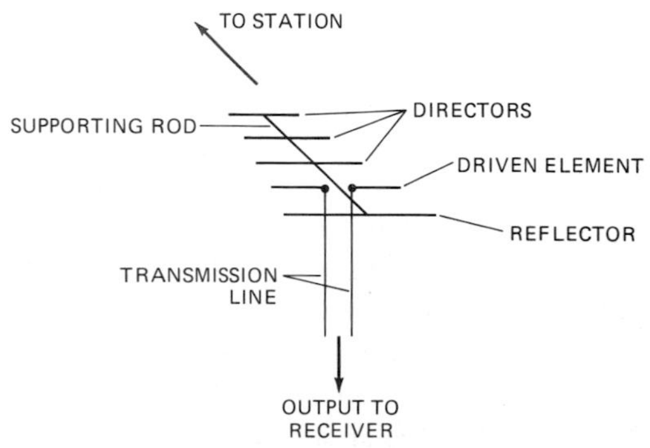

Figure 6.14 A Yagi antenna with three directors and one reflector. This antenna has high gain and sharp directivity.

cies. Therefore, it cannot be used as an all-purpose antenna in areas where there are a number of stations.

6.6 UHF ANTENNAS

In its original allocation plan, issued in 1946, the Federal Communications Commission set aside 12 channels for commercial television broadcasting. At first 13 channels were allocated, but Channel 1 (44 to 50 MHz) was subsequently dropped. It did not take long to realize that 12 channels were far too few for extensive nationwide coverage, and so an additional 70 channels in the UHF band (470 to 890 MHz) were added in 1952.

The problems facing the television technicians working in the UHF band differ in principle from those inherent in the UHF band. The problems, however, do differ in degree. The technician must still erect an antenna system to capture as much signal as possible. However, UHF signals are often weaker than VHF signals. Furthermore, losses in the transmission lines are greater for UHF.

To send as much UHF signal to the receiver as possible, the antenna must be erected carefully. This means that the technician must find the best horizontal spot and the best height for the antenna. There are numerous antenna designs to choose from. Fortunately, high-gain arrays are feasible for the UHF signal because antenna dimensions are smaller. A half-wave dipole at 550 MHz will be roughly one tenth the size of a half-wave dipole at 55 MHz. This means that more elements can be added to the UHF array without causing it to become unwieldy. Since the gain of an antenna generally increases with the number of elements, higher gain can be expected from the UHF arrays.

We shall now discuss a few representative designs.

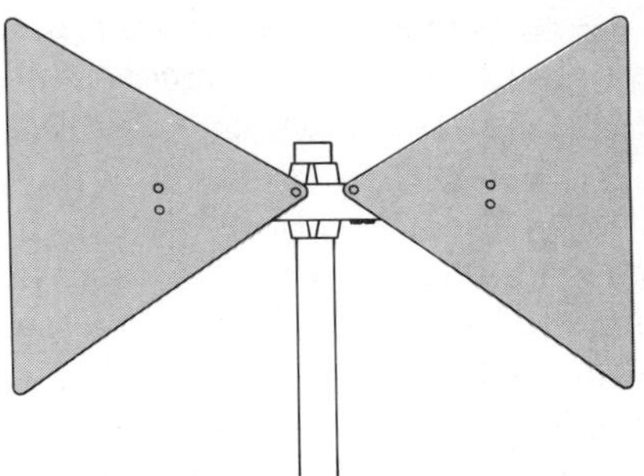

Figure 6.15 A UHF fan dipole antenna. The directional pattern is basically the same as that shown in Figure 7.10B.

FAN DIPOLE

The half-wave dipole, also known as the **bow tie** or **di-fan** antenna, is the simplest type of VHF antenna. It is also the simplest UHF antenna. Figure 6.15 shows an example of a dipole antenna with specially shaped dipoles. By using triangular sheets of metal (fans) instead of rods, the unit becomes a broadband affair capable of receiving all those signals within the UHF band. For maximum gain, a fan dipole should be slightly longer than a rod dipole.

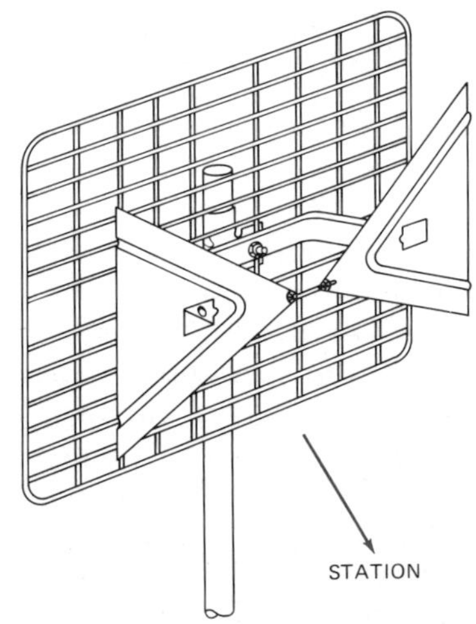

Figure 6.16 A fan dipole antenna with a screen reflector. This antenna is basically unidirectional.

The response pattern of a fan dipole is a "figure eight" unless a screen reflector is placed behind the array as shown in Figure 6.16. In this case, the pattern becomes unidirectional. Even with a reflector, these antennas are not highly directional and so they provide satisfactory reception only in strong signal areas that have few ghost problems. For greater gain, fan dipoles can be stacked two high and four high. This arrangement also provides better discrimination against ground-reflected signals and reduces the number of interfering signals.

MESH REFLECTORS

Note that the di-fan antenna in Figure 6.16 has a **mesh reflector** instead of the rod used in VHF antennas. Mesh screens are considerably more efficient reflectors than rods. The only reason screens are not used extensively for VHF is that they are too bulky and their wind resistance is too great. Mesh screens are as effective as solid metallic sheets when the mesh openings are no more than 0.2 wavelength or less at the highest operating frequency. Reflector dimensions are not critical, but the edges should extend a little beyond the dipole elements.

PARABOLIC REFLECTORS

The headlights of a car have parabolic reflectors to concentrate the light. In much the same fashion, **parabolic reflectors** are used to receive and transmit radio waves with a high gain and a sharp directivity.

Instead of an entire parabolic reflector, it is possible to use only a section of one, as shown in Figure 6.17. The gain of a parabolic reflector can be 8 dB greater than the gain of a resonant half-wave dipole. The vertical directivity of this antenna is sharp, but the horizontal directivity is somewhat broad. Where high gain is desired and the ghost problem is not serious, this array will provide excellent results.

Figure 6.17 A UHF antenna consisting of a folded dipole and a section of a parabolic reflector. (*Courtesy of Taco, Inc.*)

CORNER REFLECTORS

Instead of curved surfaces, as reflectors it is possible to use two flat surfaces that intersect to form an angle. This type of reflector, depicted in Figure 6.18, is known as a **corner reflector.** The driven element, usually a dipole antenna, is placed at the center of the corner angle and at some distance from the vertex of the angle.

The response pattern of this antenna depends not only on the corner angle but also on the distance between the dipole and the vertex. If the dipole is too far from the vertex, the response pattern will have several lobes. If the dipole is too close, the vertical response will be broadened and susceptibil-

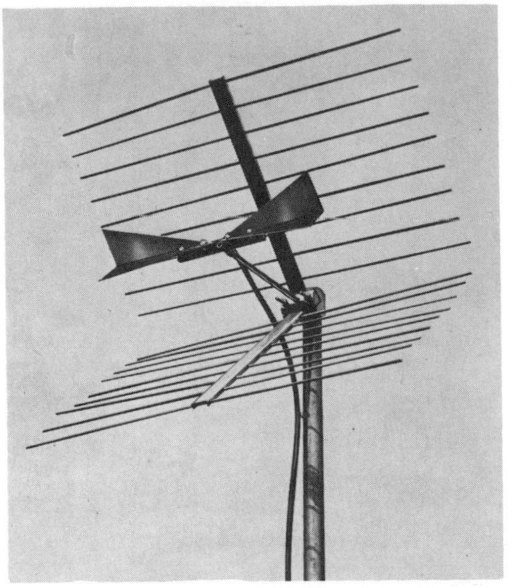

Figure 6.18 A corner reflector for a UHF antenna. Note that the shape of the driven fan element follows that of the reflector. *(Courtesy of JFD Electronics Corp.)*

ity to ground-reflected signals will increase. The corner angle in Figure 6.18 is 90°. There is a similar angle in the two triangular pieces of the fan dipole. Gain over the entire UHF-TV band is high, ranging from about 7 dB at 500 MHz to 13 dB at 900 MHz.

6.7 GENERAL-PURPOSE ANTENNAS

FOLDED DIPOLES

An antenna widely used in the past is the **folded dipole,** shown in Figure 6.19A. This antenna consists of two dipole antennas connected in parallel. The separation between the two sections is from 3 to 5 in (7.62 to 12.70 cm). The folded dipole has the same bidirectional pattern as that shown for the simple dipole in Figure 6.10B. Dipole antennas and folded dipole antennas have approximately the same gain. However, the response of a folded dipole is more uniform over a given band of frequencies than that of the simple dipole.

The directivity of the folded dipole can be increased by adding a reflector. Figure 6.19B shows a folded dipole with a reflector. The directional pattern of this antenna is the same as that of the simple dipole with a reflector, as illustrated in Figure 6.12. The best length and spacing for the reflector can be determined by the relationships given previously for a simple dipole and reflector. Figure 6.19C shows the correct relationships for the length of the elements and the spacing between elements for a folded dipole with a director and a reflector.

Other popular types of television antennas are also shown in Figure 6.19. These antennas are variations of the basic dipole or folded dipole. Figure 6.19D shows a stacked-dipole array with reflectors. This antenna is sometimes referred to as a "lazy H" because it looks like an *H* lying on its side. It consists of two half-wave dipoles placed at the front of the assembly, mounted one above the other. The center terminals of the dipoles are connected by a parallel-wire line. Each conductor of the lead-in line to the television receiver is attached to a conductor of this connecting line at a point midway between the dipoles. A reflector is mounted behind each dipole.

In Figure 6.19E, two folded dipoles (with reflectors) are mounted one above the other. The upper dipole is cut for a resonant frequency approximately in the center of the upper VHF band (174 to 216 MHz). The longer folded dipole is resonated at the center frequency of the lower television band. A short two-wire conductor connects the two dipoles. From the lower dipole, a two-wire conductor feeds the signal to the receiver. With this assembly, each antenna can be oriented independently for best reception from stations within its band, thus providing good coverage on both bands.

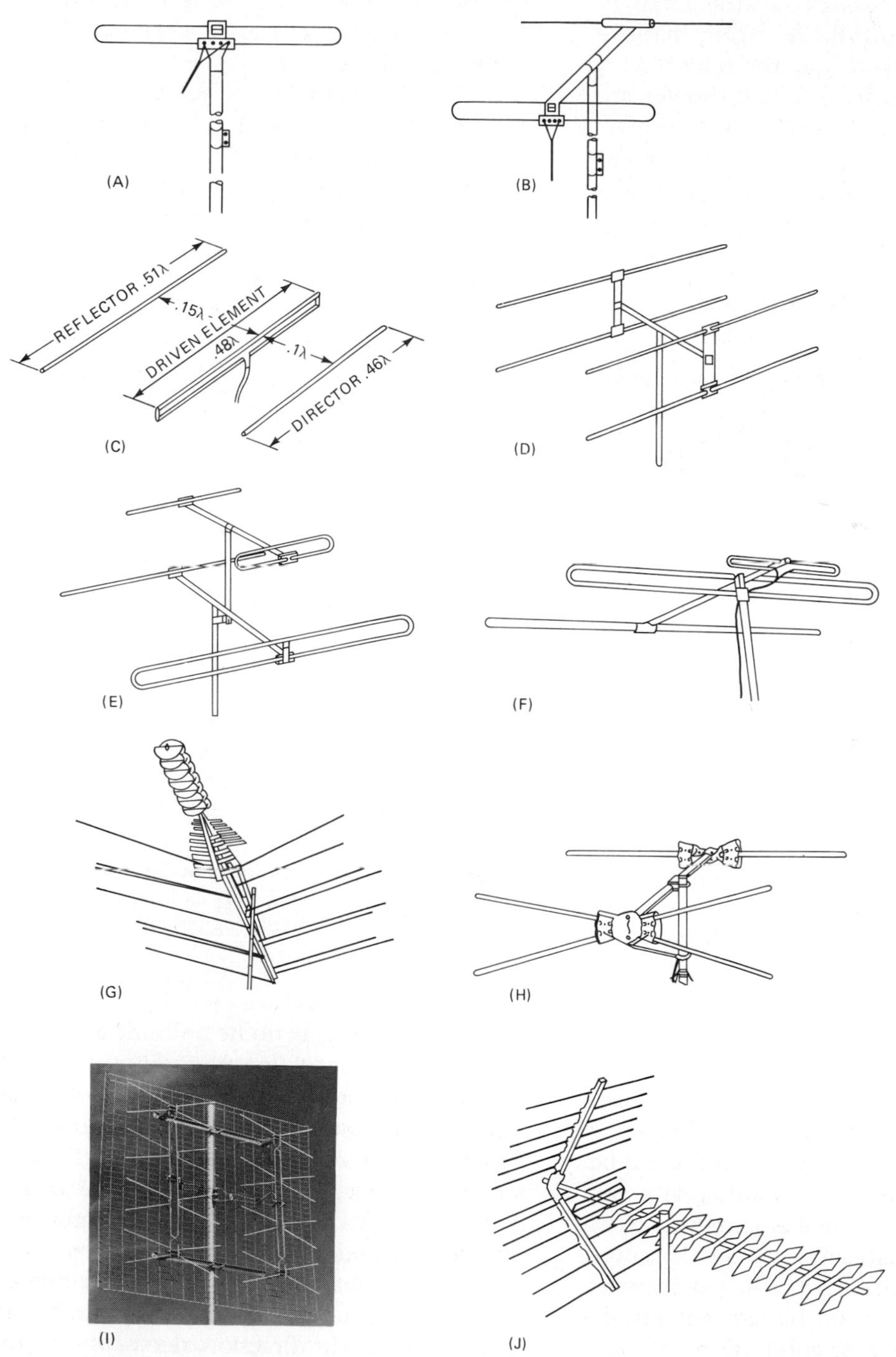

Figure 6.19 Some types of television receiving antennas. *(**G** courtesy of JFD Electronics Corp.; **I** and **J** courtesy of Channel Master.)*

Figure 6.19F is essentially the arrangement shown in Figure 6.19E except that the longer folded dipole acts as the reflector for the shorter one. The two folded dipoles are connected in the same manner as in the array in Figure 6.19E. However, independent orientation of each folded dipole is not possible in the array shown in Figure 6.19F.

LOG-PERIODIC ANTENNAS

A type of antenna used extensively for color TV signal reception is shown in Figure 6.19G. This style, known as a **log-periodic antenna,** has a very wide bandwidth. In some versions, the elements are bent forward and a UHF section is added in front of the longer VHF section.

There is no single style of log-periodic antenna. The term applies to a variety of antennas having the same characteristics. Two antennas may be completely different in appearance, but both may be log-periodic antennas.

The term *log-periodic* does not refer to the relative lengths of the elements or to the distances between them. Instead, it refers to their electrical characteristics. A graph of antenna impedance versus the logarithm of the frequency is shown in Figure 6.20A. Note that the maximum value of impedance repeats periodically on the logarithmic scale. This feature gives rise to the name *log-periodic.* This repetition is true not only for the impedance of a log-periodic antenna but also for other electrical characteristics.

An important feature of log-periodic antennas is the geometric relationship between the spacing of the elements and their length. Characteristically, these spacings and lengths vary in a geometric progression that results in a larger and larger antenna as the distance from a theoretical point, the apex, increases. In a similar manner, the length of the elements also increases, as

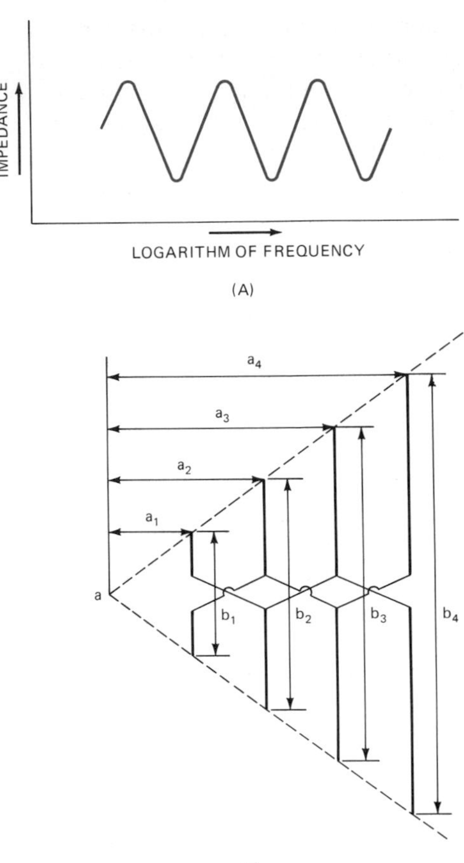

Figure 6.20 A. The periodic nature of antenna impedance when plotted on a logarithmic scale. **B.** Illustrating the element lengths and spacings of a log-periodic antenna.

shown by the log periodic antenna in Figure 6.20B. This type of log-periodic antenna is also known as a dipole array log-periodic antenna because it consists of a number of dipoles which vary in spacing and length. The antenna is designed so that the ratio of the distances between the dipoles and the lengths of the dipoles is constant.

Assume that a television station broadcasts a signal and that a log-periodic antenna is pointed in the direction necessary to receive this signal. Only one or two of the dipole elements in the antenna will react to the frequency. All other elements are inactive at that particular frequency. At some other frequency, however, other elements will be active. In other words, for any particular frequency being received, only one or two log-periodic elements are active.

CONICAL ARRAY ANTENNAS

The conical array shown in Figure 6.19H is used extensively because it can receive low and high VHF signals. Note that the front elements are bent forward. The rear elements, the reflectors, generally extend straight out. The response pattern of this antenna contains only one major lobe on all channels (Figure 6.12). This is an improvement over the conventional dipole, where an element cut for low frequencies will have a multilobed pattern on high channels. Similarly an element cut for high frequencies will have poor response on low channels. With the conical antenna, one array may be adequate for all VHF channels.

Conical antennas may be installed singly or stacked two or four high. The same is true of most other antennas.

EIGHT-BAY BOW-TIE ANTENNAS

An example of a stacked antenna is the eight-bay bow-tie UHF antenna illustrated in Figure 6.19I. The UHF bow-tie elements are stacked four high and two across. The steel screen behind these elements acts as a reflector. This antenna provides high gain and sharp directivity and covers the entire UHF band. It does have some drawbacks, however. It is cumbersome to work with and difficult to install. Also, because it is made of steel, it can rust.

Figure 6.19J shows an antenna that provides better UHF performance without the disadvantages of the eight-bay bow-tie antenna. Note the folded dipole driven element just forward of the corner reflector. The 14 elements forward of the folded dipole are diamond-shaped directors.

6.8 COMBINATION ANTENNAS

In many locations, there are both VHF and UHF stations. In these areas, combination antennas simplify the installation problems. Combination antennas are very popular for either new or renewed installations. This is particularly true when color TV receivers are involved. Combination antennas provide high gain, good directivity, broad bandwidth, and have excellent mechanical characteristics. In addition, in areas of intense and varied types of interference, the narrow beam width and high forward directivity of these antennas provides clearer pictures. Antenna pattern front-to-back ratios as high as 35 dB are available. In addition to VHF and UHF reception, some antennas also supply FM reception.

Two examples of VHF-UHF-FM combination antennas, their bandwidth characteristics, and directional patterns are shown in Figure 6.21. These antennas are designed for monochrome and color TV reception. A three-way signal splitter is used with each of these antennas. The splitter separates the VHF, UHF, and FM signals so that they can be fed to their appropriate antenna terminals.

The antenna in Figure 6.21A has 17 electrical elements, divided as follows:

- Low VHF (Channels 2 to 6, FM)—3 driven, 1 parasitic
- High VHF (Channels 7 to 13)—3 driven, 7 parasitic
- UHF (Channels 14 to 83)—1 driven, 12 parasitic

(Some elements are used for more than one band.)

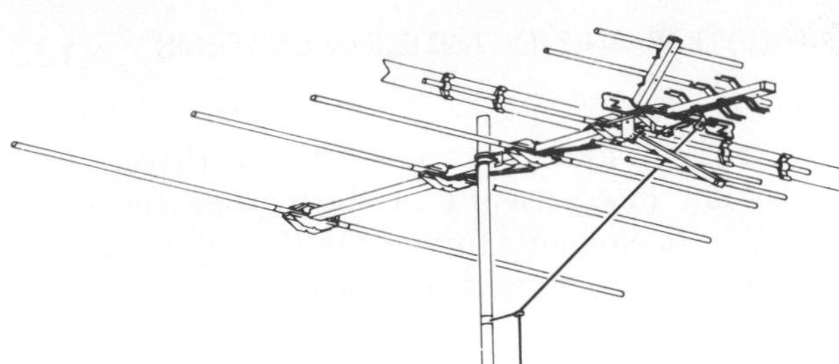

Figure 6.21 A. VHF-UHF-FM antenna for metro reception areas. B. Bandwidth and directional patterns for A. *(Courtesy of Zenith Radio Corp.)* C. VHF-UHF-FM antenna for far suburban reception areas. D. Bandwidth and directional patterns for C. *(Courtesy of Zenith Radio Corp.)*

A

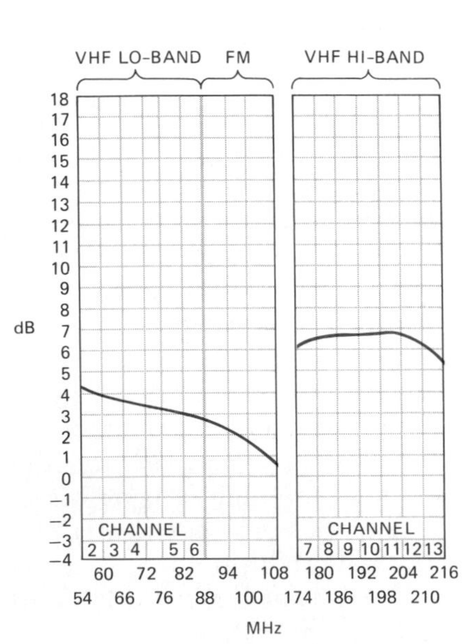

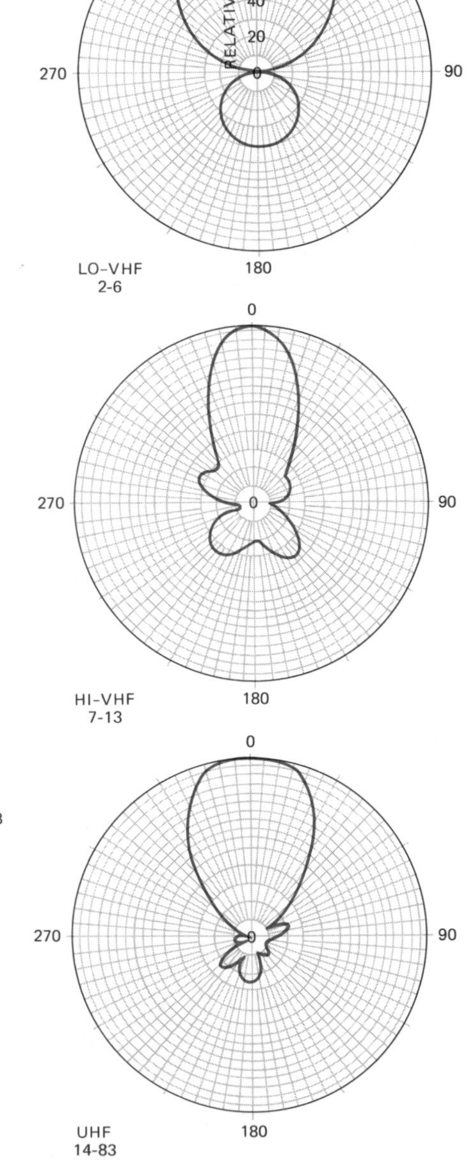

B

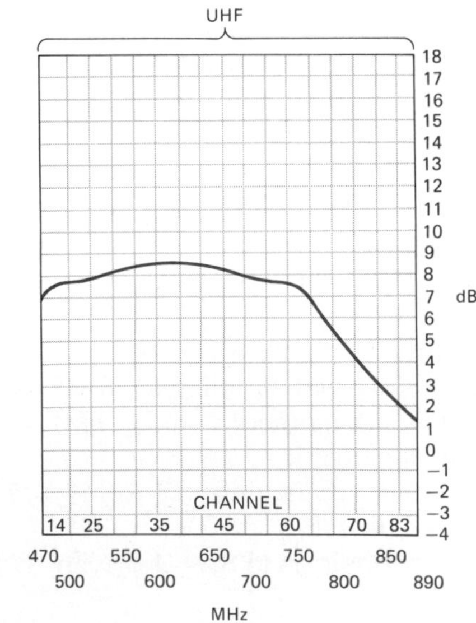

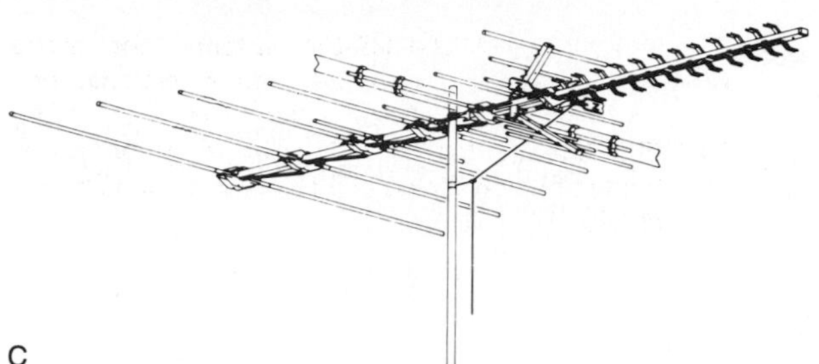

C

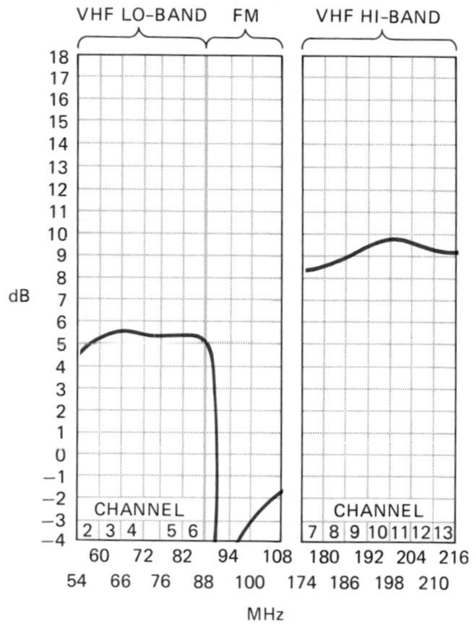

VHF LO-BAND
FM
VHF HI-BAND
dB
18 17 16 15 14 13 12 11 10 9 8 7 6 5 4 3 2 1 0 −1 −2 −3 −4
CHANNEL
2 3 4 5 6
CHANNEL
7 8 9 10 11 12 13
60 72 82 94 108 180 192 204 216
54 66 76 88 100 174 186 198 210
MHz

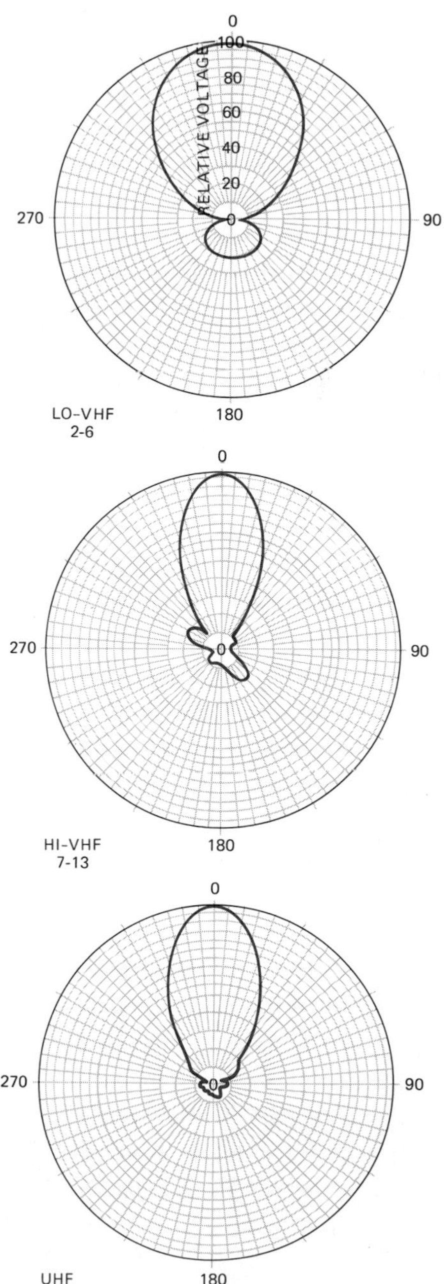

0
RELATIVE VOLTAGE
100
80
60
40
20
0
270
90
180
LO-VHF
2-6
0
270
90
180
HI-VHF
7-13
0
270
90
180
UHF
14-83

D

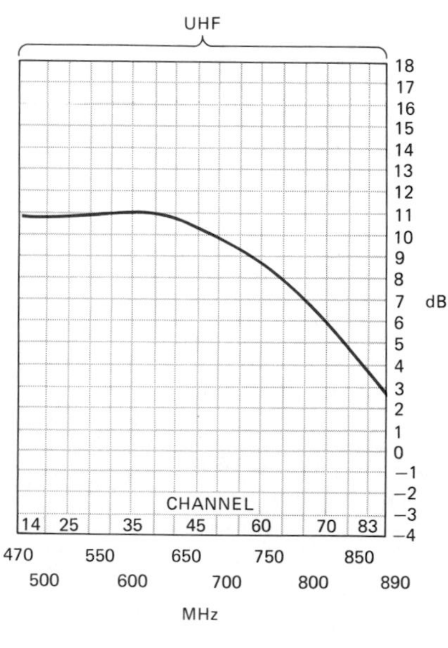

UHF
18 17 16 15 14 13 12 11 10 9 8 7 6 5 4 3 2 1 0 −1 −2 −3 −4
dB
CHANNEL
14 25 35 45 60 70 83
470 550 650 750 850
500 600 700 800 890
MHz

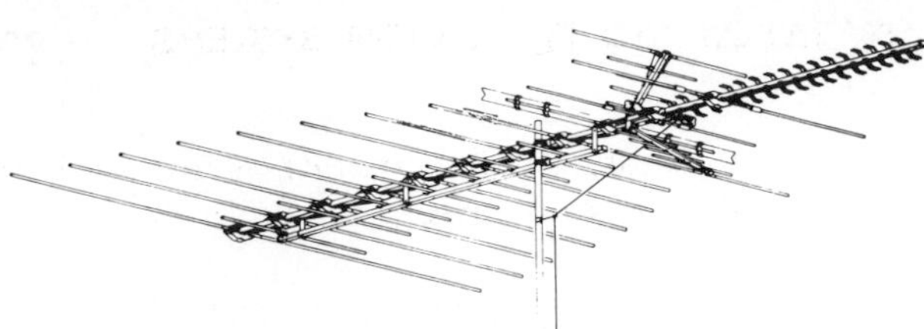

Figure 6.22A. VHF-UHF-FM antenna for deep fringe areas. **B.** Bandwidth and directional patterns for A. *(Courtesy of Zenith Radio Corp.)*

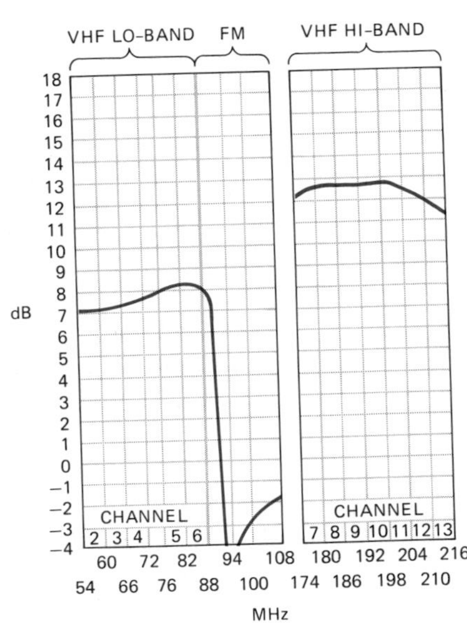

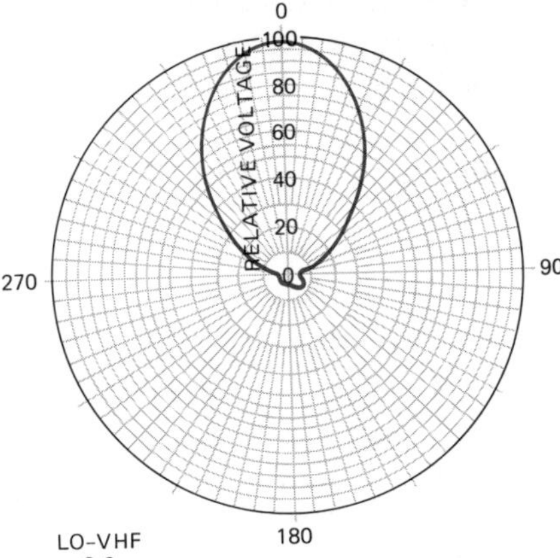

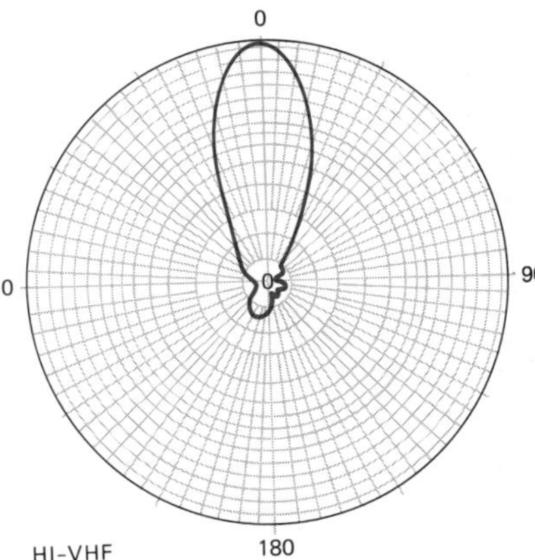

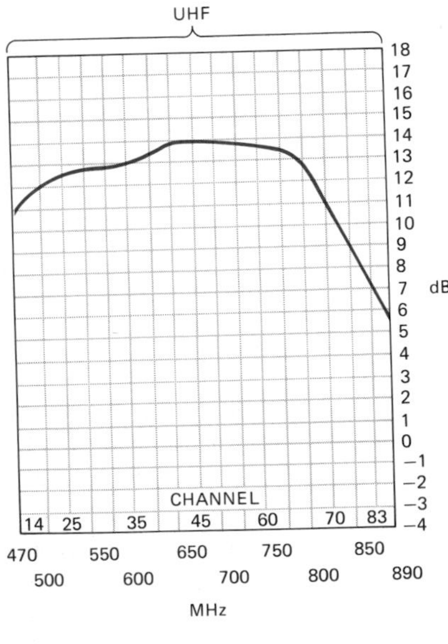

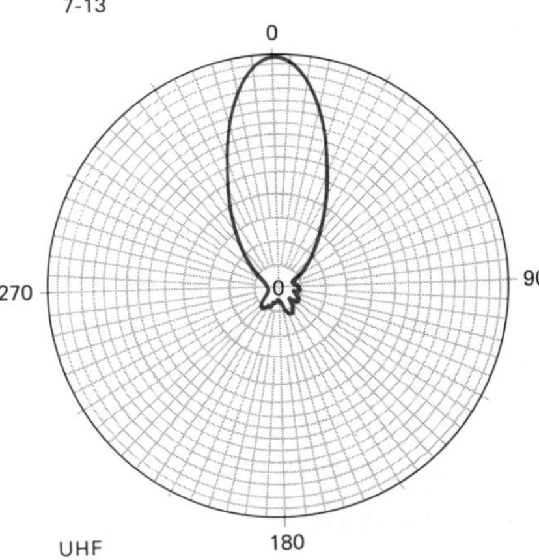

The antenna in Figure 6.21C has 42 electrical elements, divided as follows:

- Low VHF (Channels 2 to 6, FM)—5 driven, 1 parasitic
- High VHF (Channels 7 to 13)—5 driven, 8 parasitic
- UHF (Channels 14 to 83)—1 driven, 31 parasitic

(Some elements are used for more than one band.)

FRINGE-AREA ANTENNAS

In fringe areas where the signal level is very low, high-gain antennas are needed. The gain of an antenna increases with the number of elements it possesses. Therefore, fringe-area arrays have many more elements than antennas used where the signal is strong. A gain of about 10 dB is easily obtained. With some, the gain may reach 15 dB or more. Most high-gain combinations are sharply directional. Such antennas must be carefully aimed; otherwise the captured signal will be much lower than it should be. Figure 6.22A shows a high-gain, high-directivity antenna suitable for fringe-area reception, and Figure 6.22B shows the bandwidths and directivity graphs. This antenna receives VHF, UHF, and FM. As before, a three-way signal splitter is required.

By comparing Figures 6.21A, 6.21C, and 6.22B, you can see the increasing gains and narrowing directional patterns. All these antennas have good bandwidth, however.

The antenna in Figure 6.22A has 70 electrical elements, divided as follows:

- Low VHF (Channels 2 to 6, FM)—9 driven, 2 parasitic
- High VHF (Channels 7 to 13)—9 driven, 2 parasitic (4 colinear)
- UHF (Channels 14 to 83)—1 driven, 45 parasitic

Fringe Antenna Installation Hints A number of points should be kept in mind when installing a fringe-area antenna:

1. It is usually true that the higher the antenna, the stronger the signal received. This is not *always* true, however, because there are heights at which cancellation due to ground-reflected waves will occur. It is always a good idea to try changes in height in weak signal areas.
2. If possible, the antenna should be connected to its receiver before the supports are fixed in place permanently.
3. When stations are to be received from different directions, the final placement of the antenna must be a compromise. In extreme cases, it may be desirable, or even necessary, to erect several antennas with each oriented to a different compass point.

6.9 RCA MINISTATE ANTENNA SYSTEM

A recent innovation in outdoor (or indoor) TV antenna technology is shown in Figure 6.23. This is the RCA ministate antenna, a miniaturized VHF-UHF rotating antenna system. It is suitable for monochrome or color. This unit is used for TV reception in metropolitan and suburban areas up to 35 miles (56 km) from a transmitter. It is very light and compact.

The antenna system consists of the radome-enclosed, unidirectional antenna and a hand-held remote control unit (Figure 6.23A). The remote control unit turns the antenna for best reception on any channel. A direction-indicator light shows the direction in which the antenna is aimed.

The heart of this system is shown in Figure 6.23B. It consists of the unidirectional antenna, an integrated solid-state amplifier,

A

B

Figure 6.23 A. The RCA Mini-State TV antenna system. The hand-held remote control device is shown in the inset. The antenna is 21 in (53.34 cm) in diameter, 7 in (17.78 cm) in height, and weighs 6 lb. **B.** The antenna with radome removed. (See text for details.) Note bow-tie driven element. *(Courtesy of RCA.)*

interference filters, and a motor-driven antenna rotator unit. These components are all enclosed in a weatherproof plastic radome.

6.10 INDOOR ANTENNAS

In areas that receive fairly strong signals and are also relatively ghost-free, it is often possible to use an indoor antenna. Indoor antenna designs vary from simple "rabbit ears" to much more elaborate models. Figure 6.24A shows the basic rabbit ears antenna for VHF-FM only. It has no tuning features, but its telescoping rods can be extended to a length of 35 in (89 cm). The entire assembly can be rotated for best picture.

A somewhat more elaborate model is shown in Figure 6.24B. This is a VHF-UHF-FM antenna with a double-tuned UHF loop. The full-swivel telescoping rods extend to 45 in (114 cm). A 12-position tuning switch is provided. This switch can modify the response pattern and raises the resonant frequency of the antenna so as to minimize interference and ghost signals. Generally, the switch is rotated (with the antenna connected to the receiver and with a signal being received) until the best picture is obtained. The swivel rods are also adjusted for best picture.

An elaborate and highly efficient VHF-UFH-FM indoor antenna is shown in Figure 6.24C. It has a high-efficiency UHF antenna with a director and a reflector. The rods can be extended to 46 in (117 cm). Each rod has an axial inductor near its lower end. These provide improved VHF and FM reception. The VHF, FM, and UHF antennas are mounted on separate counterrotating decks which are controlled by the rotator control. This control is located at the lower right of the unit. A tuning dial and two push buttons are provided to obtain the best picture.

A simple indoor antenna is supplied as an integral part of some television receivers. One such arrangement is shown in Figures 6.24D and 6.24E. The antenna is mounted directly on the back of the receiver. Its swivel base permits movement to capture the greatest amount of signal. When not in use, the arms are telescoped inward and turned down behind the receiver cabinet, as shown in Figure 6.24E. Another arrangement is to have two swivel rods that can telescope completely into the rear of the receiver.

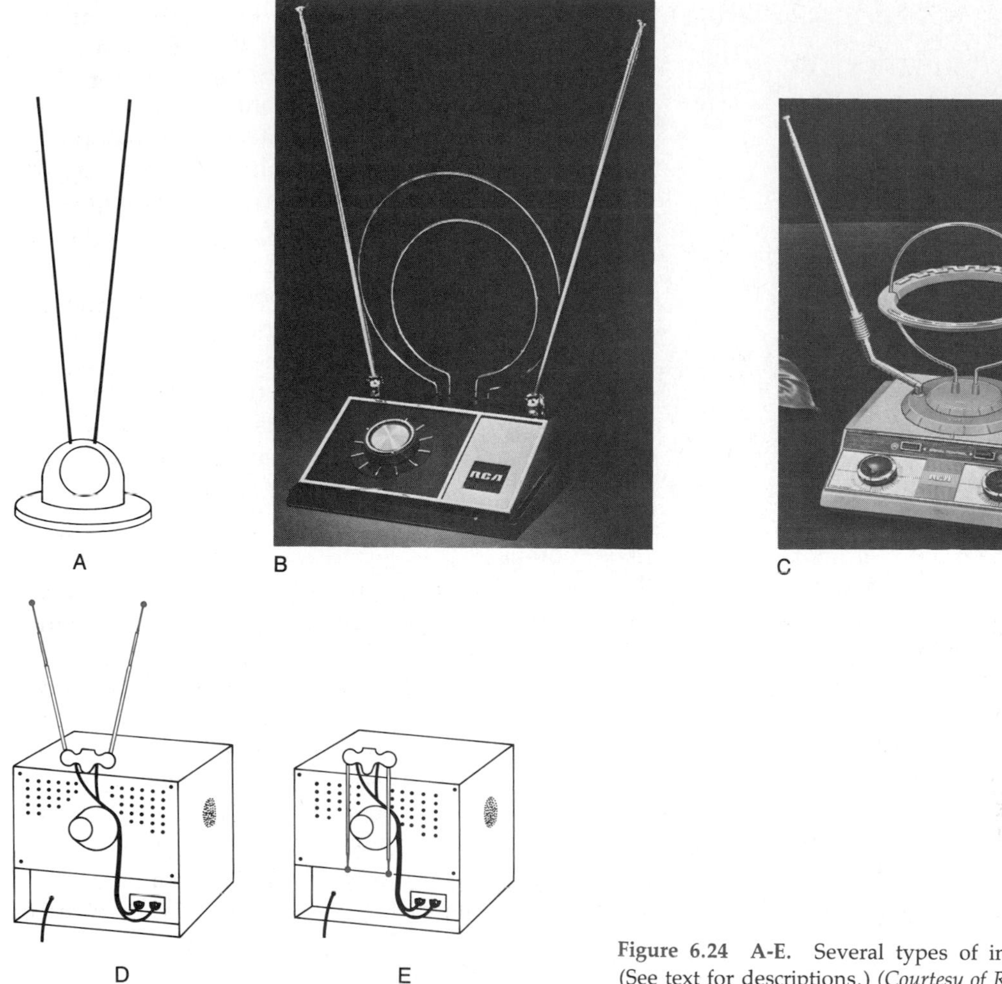

Figure 6.24 A-E. Several types of indoor antennas. (See text for descriptions.) *(Courtesy of RCA.)*

6.11 ANTENNA ROTATORS

In some areas, the TV broadcast stations are located at widely different points of the compass. One way to solve this difficult reception problem is to install several antennas. A more practical solution is to install one suitable antenna mounted on a TV antenna rotator. Figure 6.25 shows a rotator and its control unit. The rotator is mounted on a mast fastened to a roof or tower. The TV antenna mast is then secured in the top of the rotator. The control unit is then used to quickly and easily point the antenna in any direction. The control unit can be placed in any convenient location near the TV receiver. The rotator is equipped with automatic brake pads. These will hold the antenna in position in winds up to 70 mph (112.7 km/h).

6.12 TRANSMISSION LINES

The energy intercepted by the antenna's driven element must be delivered by a **trans-**

Figure 6.25 A TV antenna rotator and control box. This unit can accommodate masts up to 2 in (5.08 cm) diameter. *(Courtesy of Channel Master.)*

mission line to the receiver with the least possible loss. The receiver may be located far from the antenna. Since it is a conductor, it may act like an antenna and pick up signals of its own. These signals may be electrical "noise" impulses, such as those generated by automobile ignition systems and machinery, or they may be video signals from the transmitter. In any event, they are undesirable.

TYPES OF TRANSMISSION LINES FOR VHF

Although there are many different designs for transmission lines, two general types are used extensively in television installations: **parallel-wire** lines and **concentric,** or **coaxial cable,** lines.

For convenience and economy, it is desirable to use only one antenna that is capable of receiving all VHF television stations. Such an antenna should have a fairly uniform response from 54 MHz to 216 MHz (the log-periodic antenna, for example). A resonant dipole antenna presents an impedance at its center of about 72 ohms. For maximum power transfer, the connecting transmission line should *match* this value. In other words, the impedance of the transmission line should also be 72 ohms. When we use the same dipole for a band of frequencies, we find that the 72-ohm dipole impedance value is no longer valid. A dipole cut for 50 MHz presents a 72-ohm impedance at that frequency. At 100 MHz, the impedance has risen to 2 000 ohms. It is clear that the best transmission-line impedance is no longer 72 ohms. A higher value is needed, one that is a compromise between 72 ohms and 2 000 ohms.

It is desirable to use as high an impedance as possible because line loss is inversely proportional to the characteristic impedance. On the other hand, such factors as the length of the line and the wire gauge must also be considered. It is common practice to design the input circuit of a television receiver to be used with either a 72-ohm or a 300-ohm transmission line. A 300-ohm line used with a half-wave dipole produces a broad frequency response without too great a loss due to mismatching. Satisfactory monochrome reception is achieved this way. A folded dipole has an impedance close to 300 ohms at its resonant frequency, and a much more uniform response is obtained with this type of antenna.

Parallel-Wire Transmission Lines The flat parallel-wire transmission line shown in Figure 6.26A is one of the most popular transmission lines for VHF television. The wires are encased in a plastic ribbon of polyethylene, which is strong, flexible, and unaffected by sunlight, water, cold, acids, or alkalis. At 100 MHz, the line loss is about 1.2 dB per 100 ft (30.5 m) of line. Characteristic impedance ranges from 75 ohms to 300 ohms are

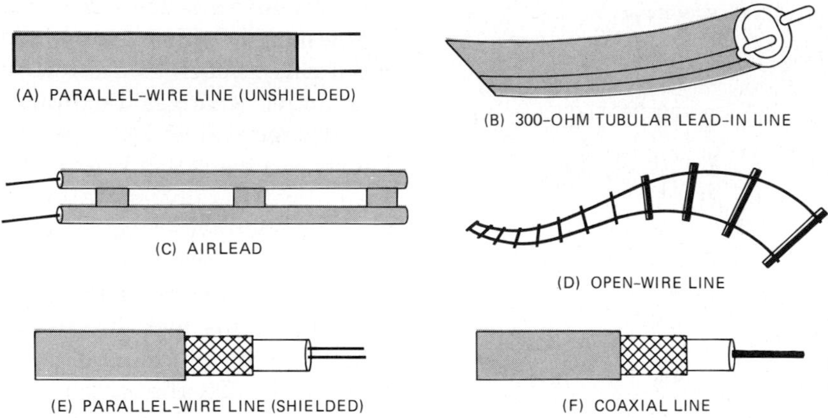

Figure 6.26 A-F. Types of popular transmission lines used for television and FM installations.

obtainable. The line is balanced, which means that both wires have the same average potential with respect to ground. It is, however, unshielded and therefore not recommended for use in locations where there is much electrical noise.

A tubular twin-lead line is pictured in Figure 6.26B. Although this version is more expensive than the flat twin-lead line, it is less affected by weather. Rain, sleet, and snow do not physically affect the flat line, but they do cause signal loss. At 100 MHz, flat and tubular lines both have an attenuation of 1.2 dB per 100 ft (30.5 m) under dry conditions. When wet, however, the loss in a flat line rises to 7.3 dB per 100 ft (30.5 m), whereas that in a tubular line is only 2.5 dB per 100 ft. In strong signal areas this loss may not be important, but it is excessive for a weak signal area.

A third type of parallel-wire line, known commercially as an airlead, is shown in Figure 6.26C. It has 80% of the polyethylene webbing removed, which reduces loss (dB attenuation per 100 ft) by at least 50%. The line impedance is still 300 ohms.

A fourth parallel-wire line, the one in Figure 6.26D, is almost completely open. It is held together by small polystyrene spacers placed approximately 6 in (15.2 cm) apart. The attenuation of this line is only about 0.35 dB per 100 ft (30.5 m) at 100 MHz. It is relatively unaffected by changes in weather. The impedance of this line is 450 ohms.

A completely shielded parallel-wire transmission line is pictured in Figure 6.26E. This is called a "twin axial" cable. The two wires are enclosed in a dielectric, possibly polyethylene, and the entire unit is shielded by a copper-braid covering. An outer rubber covering provides protection against the elements. Grounding the copper braid converts it into an electrostatic shield, which prevents any stray interference from reaching the inner conductors. Furthermore, the line is balanced with respect to ground. It can have any impedance from 50 ohms up, but a 225-ohm line is generally used in television installations. The attenuation of this line is 3.4 dB per 100 ft (30.5 m) at 100 MHz, considerably higher than the attenuation of any of the unshielded lines. Because of this, and because of its greater cost, the shielded line is used only where electrical noise is particularly severe.

Coaxial Cables A coaxial, or concentric, cable is illustrated in Figure 6.26F. It is made up of an insulated center wire enclosed by a concentric metallic covering, which generally

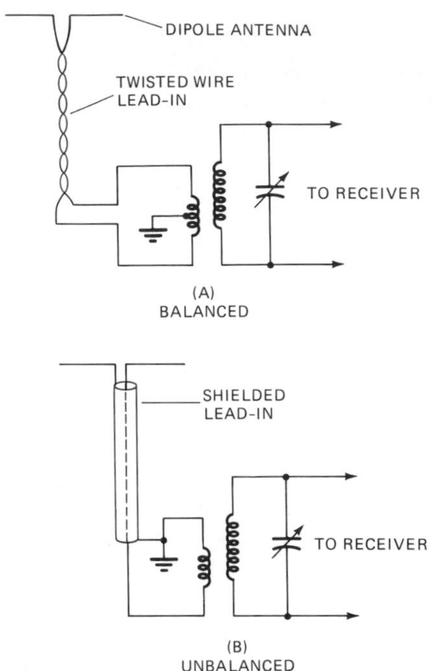

Figure 6.27A,B. Methods of connecting lead-in wires to the input coil of a receiver.

is a flexible copper braid. The inner wire is kept in position by a solid dielectric chosen for its low loss properties. In some applications, the signal carried by the coaxial line is confined to the inner conductor and the outer copper-braid conductor is grounded to shield against stray electromagnetic fields. This arrangement causes the line to be unbalanced with respect to ground, and the input circuit of the receiver must be connected accordingly. Coaxial cables are available in a range of impedances from 10 to 150 ohms.

At the receiver, the connections for balanced and unbalanced lines differ, as shown in Figure 6.27. For a balanced line, the input coil is center-tapped and grounded at this center terminal. Stray fields, since they cut across both wires of a balanced line, induce equal voltages in each line. The current that flows because of the induced voltages is traveling in the same direction in both conductors and so is balanced. Many receivers are designed so that a simple change in the input circuit, such as the removal or adding of a jumper, changes the input impedance to the desired value. A type of transformer knows as a balun can be used to connect balanced lines to unbalanced lines.

The open-wire line has by far the lowest attenuation. Thus, it is frequently used in fringe areas. Its characteristic impedance is 450 ohms. This high impedance sometimes requires a matching network between the line and the 300-ohm receiver. This is not always necessary, however, because receiver impedances may vary considerably from their stated value of 300 ohms. The 300-ohm line is the one used most in VHF because it is economical, it matches receiver input impedances directly, and its attenuation is low. However, in areas where electrical noise is high, unshielded open-wire or 300-ohm lines become impractical and one of the coaxial cables must be used.

TRANSMISSION LINES AT UHF

With an increase in frequency, line attenuation also rises. Thus, at 500 MHz the open-wire loss for 100 ft (30.5 m) is 0.78 dB, the 300-ohm line loss becomes 3.2 dB, RG-11U attenuation increases to 5.0 dB, and RG-59U loss reaches a rather high 9.4 dB. Comparative figures at 100 MHz and 1 000 MHz are given in Table 6.1. It can well be understood why the amount of line needed should be figured closely so that no more than absolutely necessary is used.

An interesting sidelight on line attenuation is its rapid rise in unshielded lines when they become wet. The 300-ohm flat line is most vulnerable, jumping from 3.2 dB (at 500 MHz) when the line is dry to 20.0 dB when it is wet. What this rise can do even to a strong signal is obvious, and this important point should be kept in mind.

Table 6.1 **TRANSMISSION LINE LOSS**

	100 MHz		500 MHz		1,000 MHz	
TYPE OF LINE	*WET*	*DRY*	*WET*	*DRY*	*WET*	*DRY*
450-ohm open-wire[a]	—	0.35	—	0.78	—	1.1
300-ohm flat	7.3	1.2	20.0	3.2	30.0	5.0
300-ohm tubular	2.5	1.1	6.8	3.0	10.0	4.6
RG-11U	—	1.8	—	5.0	—	7.6
RG-59U	—	3.8	—	9.4	—	14.2

[a]Estimated dry values; wet values unknown. All values are dB loss per 100 ft (30.5 m).

Guidelines to be followed when installing an antenna are:

1. Do not let the transmission line lie on the roof, where it can become covered by ice, snow, or water.
2. Do not have long horizontal runs exposed to the elements, which would permit buildup of ice or moisture.
3. Keep the line away from gutters, pipes, or other metal objects.
4. Avoid sharp bends in the line. If a bend must be made, make it gradual.
5. Secure the line tightly with standoff insulators so that it does not sway in the wind or otherwise alter its position.
6. Observe these precautions in both VHF and UHF installations.

The 300-ohm tubular line is considerably less affected by moisture than the other types of line. Thus, this line is better for some installations. No data are available on the attenuation increase in wet open-wire lines, although it is not considered to be appreciable. Shielded cables, such as the RG-11U and RG-59U are practically unaffected by weather.

Impedance matching at the antenna and, more important, at the receiver requires careful attention in UHF installations. If there is a mismatch at the receiver, energy will be reflected back along the line and will cause standing waves. The attenuation of a line may increase by as much as 2 dB over its normal rating when standing waves are present. In strong signal locations the additional loss may not be serious, but in moderate and weak signal areas it can mean the difference between usable and unusable signals.

6.13 AMPLIFYING THE ANTENNA SIGNAL

There are two basic devices for amplifying the antenna signal: (1) a signal booster or a combination booster and multiple-set coupler and (2) an antenna preamplifier. A booster-coupler is shown in Figure 6.28. This unit can feed four television receivers.

SIGNAL BOOSTERS

Owners of television receivers located in weak signal areas often try to improve picture quality by adding an external booster to their sets. **Boosters** are basically nothing more than RF amplifiers. When one is attached to a set, it means, in effect, that one or more RF amplifiers have been added to that already in the receiver.

Figure 6.28 A booster-coupler that can feed amplified TV signals to four TV receivers. *(Courtesy of Winegard Co.)*

A booster strengthens the incoming signal so that it produces a picture having the full contrast range. At the same time, the booster improves the signal-to-noise ratio so that the picture will be clear and free of noise spots. Of these two objectives, the improvement in the signal-to-noise ratio is the more difficult to attain, but it is the most important.

A booster capable of high gain but incapable of providing a good signal-to-noise ratio will give a picture filled with disturbing noise spots. A booster with little internal noise but capable of little gain will not amplify the signal enough to permit it to override the noise of the set. Again the picture will be covered with noise spots. The booster must have both attributes or it might as well have none.

Before we leave the subject of noise, it should be pointed out that some noise can be generated outside the set or the booster. This noise, if present, comes down the transmission line with the signal and is indistinguishable from the signal so far as the booster is concerned. To overcome this noise, it must be eliminated at the source. If that is not practical, as much of it as possible must be kept from reaching the signal through the antenna or the lead-in line. Standard methods of attack include increasing the antenna height, antenna replacement, and using a shielded lead-in line. It also helps to position the booster at the antenna (or at least as close to the antenna as possible). In this way, the booster will strengthen the signal before it is subjected to the noise. This enables the signal to overcome the adverse effects of the noise. In this way, the signal-to-noise ratio is also improved.

ANTENNA PREAMPLIFIERS

There are certain disadvantages to the use of a booster. One of the most important is that an additional unit must be connected to the television set. Another is their rather unsightly appearance.

Raising the antenna, using an antenna with a higher gain, and using two or more antennas in combination may eliminate the need for a booster, but not in all cases. An-

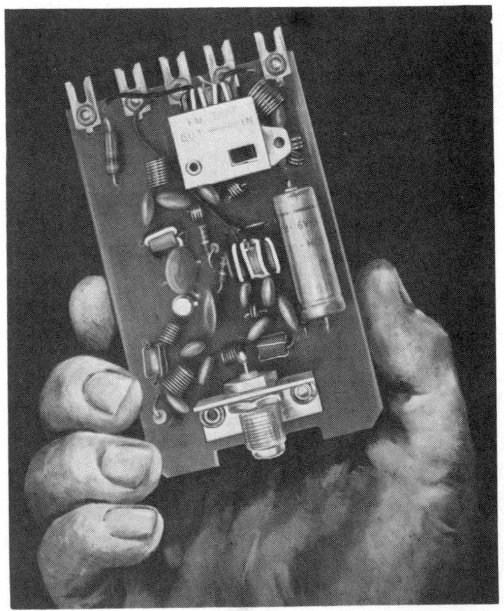

Figure 6.29 An antenna preamplifier. Lightning protection and strong signal overload protection are provided. *(Courtesy of Winegard Co.)*

other solution is to use an antenna **preamplifier.** This is simply a broad-band transistor RF amplifier mounted on the antenna or antenna mast (near the antenna terminals). Figure 6.29 shows such a preamplifier. Because transistors are more reliable than tubes and operate at lower voltages, a preamplifier can be mounted on the antenna mast.

Either a grounded-emitter or grounded-base circuit configuration may be used for the antenna preamplifier. The grounded-emitter circuit will give a somewhat higher gain, but replacement transistors must be selected carefully to avoid oscillation caused by a change in the transistor parameter. The grounded-base configuration is often used because it is more stable at the wide temperature ranges that can be expected with a mast-mounted amplifier. But it does have a lower gain.

Since the preamplifier is untuned, it can amplify any frequency in the television VHF, UHF, and FM broadcast range. Negative feedback is usually used for broadening the response and increasing the stability. All such preamplifiers are mounted in a weatherproof enclosure.

Because a transistor preamplifier requires such a small amount of power, the power can be fed through the transmission

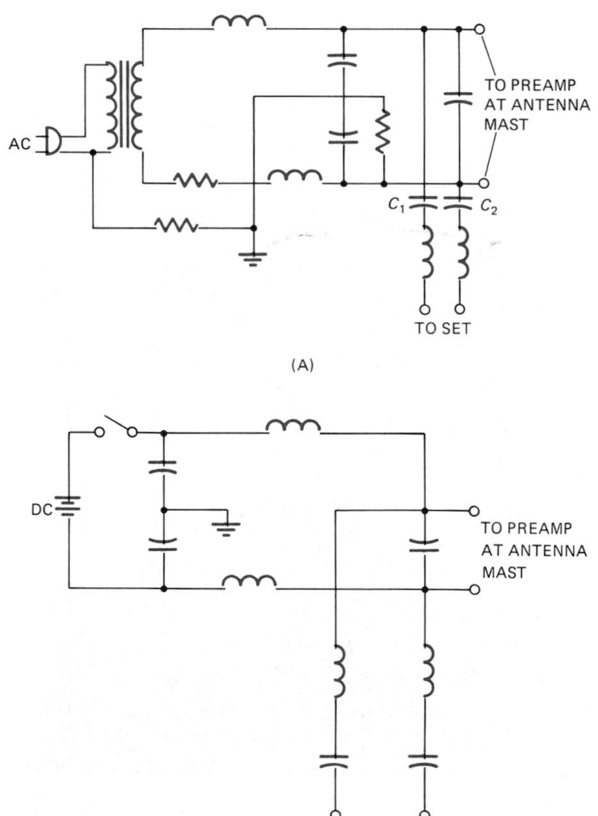

Figure 6.30A,B. Examples of power supplies used for operating preamplifiers. Power is obtained through the transmission lines.

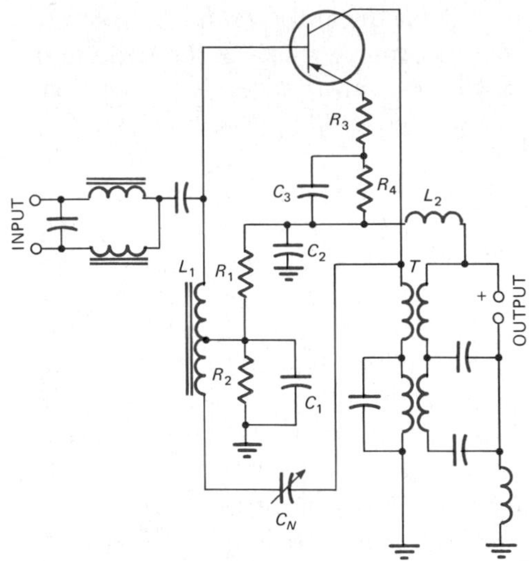

Figure 6.31 Circuit of an antenna preamplifier.

line that couples the antenna to the receiver. In this type of operation, an ac power supply can be located in the receiver cabinet and the preamplifier is energized whenever the receiver is being used. Figure 6.30A shows how the AC line is coupled to the antenna transmission line. A rectifier, located in the preamplifier at the antenna, converts the AC to DC for operating the transistor. In other versions, the rectifier is located in the AC supply.

Batteries may also be used for operating preamplifiers. If the battery was mounted on the mast, replacement would be inconvenient. Therefore, such batteries are usually mounted in the house or some other convenient place. Figure 6.30B shows a battery-powered preamplifier.

Figure 6.31 is a typical preamplifier circuit. The transistor is operated in the grounded-emitter configuration. The collector is at DC ground potential, and a positive voltage is supplied to the emitter and the base. The voltage dividers R_1 and R_2 establish the bias, and C_1, C_2, and L_2 filter the transistor-operating voltages. Neutralization is provided by C_N.

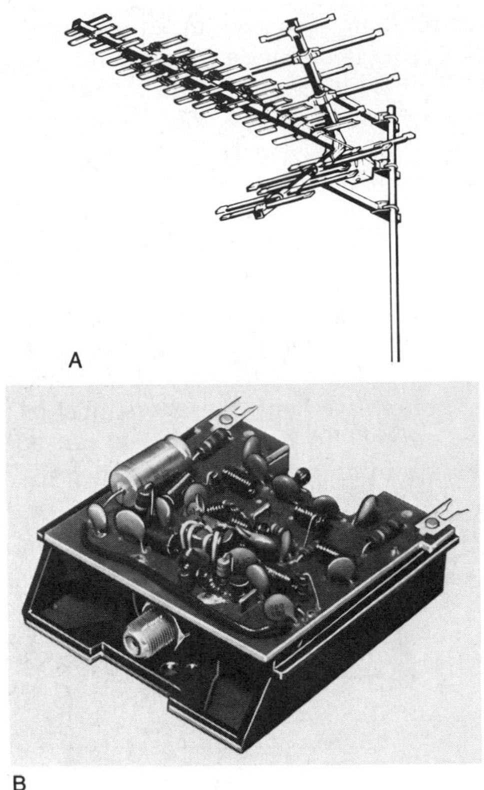

Figure 6.32 **A.** High-gain UHF antenna. **B.** High-gain UHF antenna preamplifier. *(Courtesy of Winegard Co.)*

UHF-ONLY ANTENNA PREAMPLIFIER

In many areas, it is difficult to get good UHF reception. In such areas, the combination of a high-gain UHF antenna and an antenna preamplifier may add up to 30 miles (48.3 km) to UHF reception distance. These units are shown in Figure 6.32. The UHF antenna is the corner reflector type. The UHF preamplifier accepts a maximum input signal voltage of 0.13 V. It provides a maximum signal output voltage of 0.9 V.

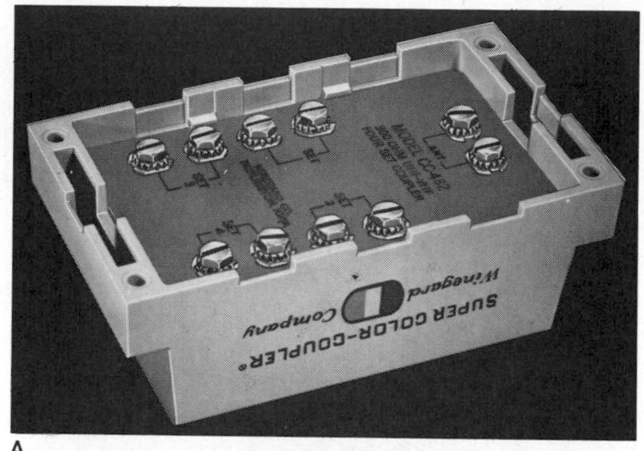

A

B

Figure 6.33 TV receiver splitters. A. This unit feeds four 300-ohm receiver inputs from a 300-ohm line. B. This unit feeds four 75-ohm receiver inputs from a 75-ohm line. *(Courtesy of Winegard Co.)*

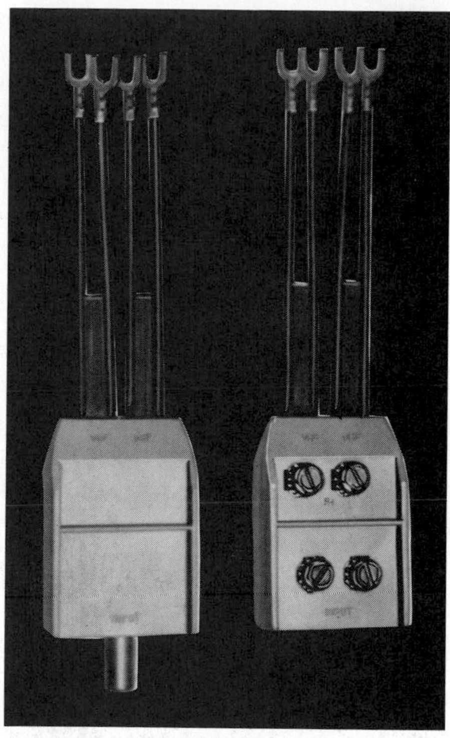

Figure 6.34 VHF-UHF band separators. A. Accepts a 75-ohm coaxial input and has separate VHF and UHF 300-ohm outputs. B. Accepts a 300-ohm input (bottom terminals) and provides separate VHF, UHF, and FM (upper terminals) 300-ohm outputs. *(Courtesy of Winegard Co.)*

6.14 ANTENNA-SYSTEM ACCESSORIES

SPLITTERS

Two types of **splitters,** also known as couplers, are shown in Figure 6.33. In Figure 6.33A is an 82-channel (VHF-UHF-FM), four-set splitter, designed to accept a 300-ohm line and feed it to as many as four 300-ohm receiver inputs. These splitters have no amplification (unlike the one in Figure 6.28), and thus a fairly strong incoming signal is required. This unit may be mounted outdoors or indoors.

The unit in Figure 6.31B performs the same basic function as the one in Figure 6.33A. However, this unit accepts the signal from a 75-ohm coaxial cable and distributes it over four 75-ohm coaxial cables to as many as four 75-ohm receivers.

BAND SEPARATORS

Band separators are mounted in back of television receivers. They are designed to accept VHF, UHF, and FM output and feed it to appropriate receiver terminals. Two models are shown in Figure 6.34. The separator shown in Figure 6.34A converts the signal from a 75-ohm coaxial cable downlead to separate VHF and UHF signals. These are fed out on individual 300-ohm lines. In the unit shown in Figure 6.34B, the input is from a 300-ohm line. Outputs at 300-ohm are separate VHF, UHF, and FM signals.

MATCHING TRANSFORMERS

Matching transformers, also known as **balun units,** match a 300-ohm VHF-UHF-FM antenna or a 300-ohm input television receiver to a 75-ohm coaxial cable downlead. Outdoor or indoor mounting is possible.

INTERFERENCE FILTERS AND TRAPS

Interference filters and **traps** eliminate various types of interference encountered in TV reception. They are installed on the back of the set, in series with the antenna transmission line. Some types of interference which are thus filtered are (1) amateur radio, (2) neon lights, (3) medical diathermy, (4) strong local FM stations, and (5) strong local TV stations.

6.15 MASTER-ANTENNA DISTRIBUTION SYSTEMS

When many sets are to be operated from a single antenna system, a **master-antenna television (MATV) system** is used. Figure 6.35 illustrates a typical MATV system. It is made up of three main sections:

1. The antenna section
2. The distribution amplifier
3. The signal distribution section

ANTENNA SECTION

In Figure 6.35, the antenna system consists of (1) the VHF-UHF-FM antenna and (2)

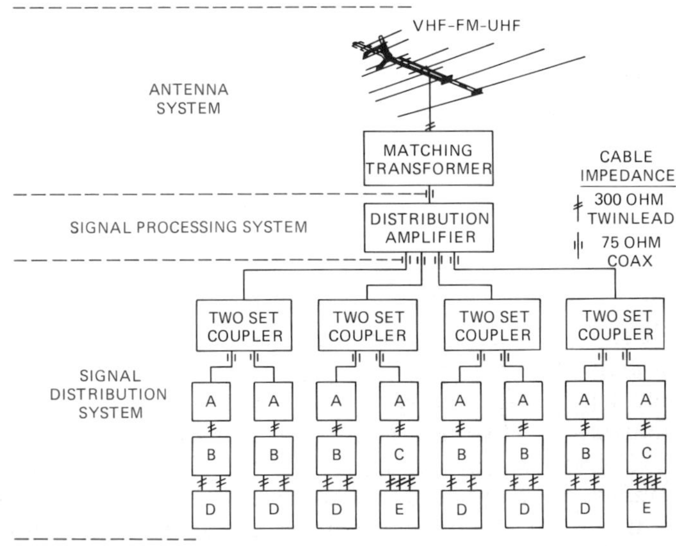

Figure 6.35 A typical MATV signal distribution system. This is basically a 75-ohm coaxial cable system, feeding eight outlets.

the 300-ohm-to-75-ohm antenna matching transformer. The problems encountered in mounting an antenna for an MATV system are similar to the problems for other antenna systems. One significant difference is that in an MATV installation, a lower gain can be tolerated in favor of a greater bandwidth. This is because an amplifier system is used in the signal-processing section. When the signal amplitude is very low, a preamplifier may be mounted on the mast near the antenna terminals.

Coaxial cable is used almost exclusively in MATV systems because it rejects noise signals and does not reradiate signals. If twin-lead (balanced) line is used for the connection to the driven element, it must be converted to an unbalanced line. A special transformer, called a **balun,** is used for this purpose. The antenna lead-in may also contain traps or filters for eliminating interference at certain frequencies.

If a broad-band antenna is used, one channel may be very much stronger than the others. In such cases, a tuned-circuit **attenuator** may be needed to reduce the strong signal so that all frequencies presented to the receiver will be of the same strength. The antenna and distribution amplifier (see below) are referred to as the **head end.**

DISTRIBUTION AMPLIFIER

As shown in Figure 6.35, the antenna signal, at a 75-ohm impedance, is fed through coaxial cable to the **distribution amplifier.** A distribution amplifier suitable for use with the system shown in Figure 6.35 is illustrated in Figure 6.36. This is a VHF-UHF-FM amplifier designed for small and medium systems. FM signals may be either amplified or attenuated by actuating a selector switch. The average gain per output terminal is 9 dB.

Other types of distribution amplifiers of-

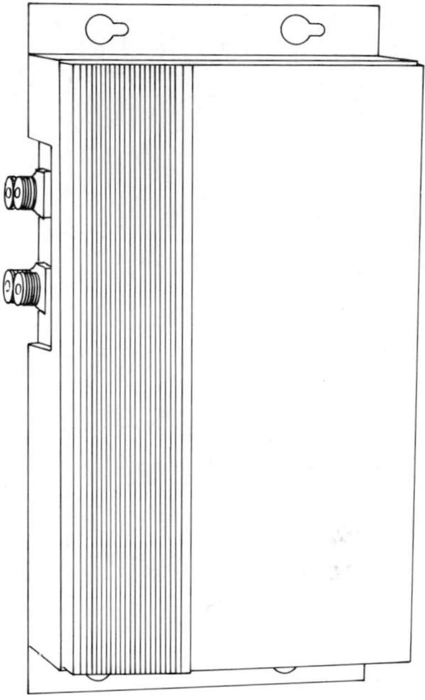

Figure 6.36 A distribution amplifier suitable for use with the eight-outlet system of Figure 6.35. The unit consumes 3.6 W at 120 V AC, 60Hz. *(Courtesy of RCA.)*

fer additional features. Some of these are (1) separate low-band and high-band VHF channel gain controls, (2) separate UHF channel gain control, (3) switchable FM trap to eliminate undesired FM signals, (4) higher gain to accommodate additional outlets, and (5) built-in features which eliminate the need for external FM traps, attenuators, and band separators.

SIGNAL DISTRIBUTION SECTION

As shown in Figure 6.35, a **signal distribution** system includes (1) four two-set couplers, (2) light-duty wall outlets (the squares marked A), (3) six VHF-UHF band separators (B), (4) two VHF-UHF-FM band separators (C), (5) six TV sets (D), and (6) two TV-FM combination sets, or two TV and two FM sets (E).

In this particular system, a single VHF-

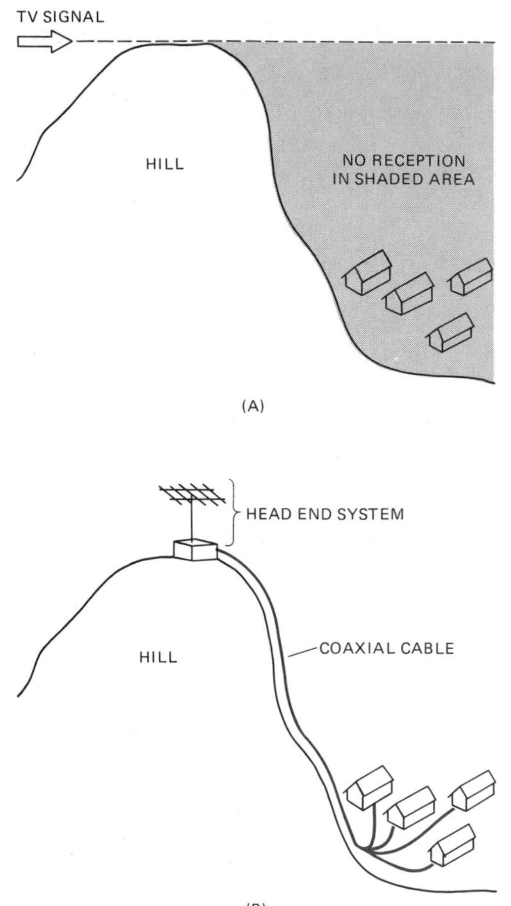

Figure 6.37 A. A situation whers CATV is needed. B. How a CATV system delivers the signal to the community.

UHF-FM antenna provides strong, clear signals capable of operating eight television sets and two FM sets. Many other combinations and total numbers of outlets are possible. For example, an entire hotel, motel, or apartment house can be serviced by a rather elaborate MATV system. The manufacturers of MATV systems provide technical assistance in the planning of such systems.

6.16 COMMUNITY-ANTENNA TELEVISION SYSTEMS

Very often there are locations within the line-of-sight distance from the transmitter where the television signal cannot be received because of hilly terrain. Figure 6.37A, shows such a situation. The houses in the valley cannot receive the line-of-sight television signal. This is the type of situation that leads to a **community-antenna television system (CATV)**, shown in Figure 6.37B. It consists of a receiving antenna mounted on the hill and a signal-processing system similar to the ones used in the MATV system. As in the MATV systems, the antenna and distribution amplifier together are called the **head end** of the system. The television signal is delivered to the town via a coaxial cable. In the town, tap-off points are used for each house that desires reception. The houses that are connected to the system pay a monthly fee.

Not only does the CATV system find applications where reception is normally very poor, but it is also used in locations where the viewer is usually limited to one or two channels. CATV systems with elaborate antennas and signal-processing equipment can pick up distant stations and deliver them to the customer, thereby increasing the number of channels to choose from.

In some cases, microwave links are used to transport a signal from one point to another more distant one. Suppose a local UHF station is at Point A in Figure 6.38 and it is desired to receive this signal at Point B, which is more than 30 miles (48 km) away. The signal could be transported via coaxial cable, but a less expensive system is illustrated in Figure 6.38. The first step is to convert the signal to a microwave frequency. This microwave signal is then transmitted from Point A to Point B. Repeaters are used to receive and retransmit the signal at points

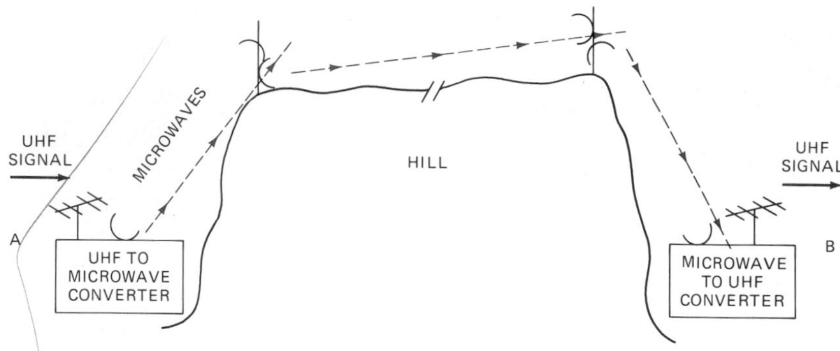

Figure 6.38 Use of microwaves in a CATV system.

along the way. At Point B the signal is converted from microwave back to a UHF signal and delivered to the head end of the CATV system. In addition to saving miles of coaxial cable (and therefore reducing the cost), this system has the advantage of making use of the relatively uncrowded microwave band.

6.17 GROUNDING AND LIGHTNING PROTECTION

GROUNDING THE ANTENNA MAST

For the protection of the people using the TV receiver and service personnel, the antenna mast must be properly grounded. In a number of areas of the United States, mast grounding is required by local ordinance. Grounding the mast prevents the accumulation of static charges on it. It also prevents the mast from accidentally assuming a voltage above ground from contact with voltage-carrying base wires or connectors.

Suitable grounding, which is also a part of lightning protection, can be achieved by connecting a heavy-gauge wire from the mast to a cold-water pipe or to a metal stake driven at least 3 ft (0.93 m) into the earth.

LIGHTNING PROTECTION

A **lightning arrestor** has two protective functions. When correctly installed, it provides a high-resistance path to ground to prevent the accumulation of static charges on the antenna. Secondly, if lightning strikes the antenna, the charge is shorted to ground. This protects both the television receiver and its user.

For outdoor installations, the arrestor can be installed on a mast grounded as described above. The arrestor is connected to the transmission line and to the grounded mast. For indoor installations, the arrestor can be mounted on a cold-water pipe, which constitutes an effective ground.

6.18 TROUBLE-SHOOTING ANTENNA SYSTEMS

Since the antenna is exposed to rain, snow, wind, and smog, it deteriorates over a period of time. It is important to know the symptoms of a defective antenna. Obviously, if there is insufficient signal delivered to the receiver by a poor antenna system, no amount of work on the set will cure the trouble. A good technician will first carefully observe the antenna system when making a service call. Before he enters the house, the technician will know whether or not (1) the antenna has been damaged by high winds, (2) the antenna is pointed in the right direction, and (3) the transmission line is securely in place. Any of these factors can cause insufficient signal strength at the receiver.

Unshielded transmission lines can become short-circuited even though nothing is touching the wires. You can demonstrate this by wrapping a piece of metal foil around the line and moving it with your hand. In most cases, the picture will be affected even though the foil is isolated from the wires. House painters sometimes get paint on an unshielded line. If the paint has a lead base, as many of them do, the paint can effectively short-circuit the line. (You can think of the paint as being a capacitive coupling between the lines, which is the same thing as a short circuit.) Always inspect the transmission line in both new and old installations. Unshielded line should not be run close to rain gutters, aluminum siding, conduit, water pipes, furnace ducts, etc. All of these things can seriously affect picture quality.

INSPECTING OLDER ANTENNA INSTALLATIONS

If the antenna has been in service for some time, there are a number of features that should be inspected. We shall mention these as a checkoff list. Experienced technicians will train themselves to look for these things in a quick glance.

1. Note the orientation of the antenna. Strong winds can change it. Make a quick comparison of the direction in which the other antennas in the neighborhood are pointing to see if the antenna being worked on is properly oriented.
2. Check the transmission line. Make sure that it is fastened at the antenna and that there are no visible breaks. Make sure that the line has not been painted and that it has not come loose from its fastenings. Watch for a fluttering transmission line. It may be a source of trouble.
3. Find out if the customer has recently switched from a monochrome receiver to a color receiver. If this is the case, the old antenna may not be adequate. The only solution here is to replace the old antenna.
4. Watch for white flashes in the picture—usually accompanied by static in the sound. They are an indication that there is an intermittent break in the transmission line. The break may not be visible.
5. Check the ground connections for the mast and lightning arrestors. Improper grounding will not affect the picture, but it represents a serious electric-storm hazard.
6. Look for excessive snow in the picture or for a very weak picture. This indicates that the antenna lead may be open or disconnected. Of course, the trouble could be in the receiver. A small portable receiver known to be in operating condition is useful for checking the signal from the antenna. If the portable can produce a satisfactory picture when connected to the antenna but the customer's receiver cannot, then the trouble is in the customer's receiver rather than in the antenna. If a satisfactory picture cannot be obtained on the portable, then the antenna should be suspected. A field-strength meter is useful if a large amount of antenna work is to be done, especially if you are going to work on MATV and CATV systems.

SUMMARY OF CHAPTER HIGHLIGHTS

1. The requirements for television antennas are more stringent than those for radio antennas. Television antennas must be designed for correct gain, bandwidth, directivity, and impedance.
2. Television signals are broadcast at line-of-sight frequencies. This means that they cannot ordinarily be received satisfactorily at a distance greater than about 75 miles (120 km) from the transmitter. (Figures 6.4, 6.5, and 6.6)
3. The line-of-sight distance between two antennas on earth is given by the equation $d = 1.23\ (\sqrt{h_t} + \sqrt{h_r})$, where h_t is the height, in feet, of the transmitting antenna above ground, h_r is the height, in feet, of the receiving antenna above ground, and d is the distance, in miles, between the antennas.
4. Multipath distortion occurs when a television signal arrives at the antenna from more than one direction. This is the cause of ghosts. (Figure 6.7)
5. Television signals are horizontally polarized. This means that the electric field is horizontal, that is, parallel to the surface of the earth. (Figure 6.8)
6. An antenna, like an *LC* tuned circuit, has a resonant frequency. This is the frequency at which the antenna will receive maximum energy.
7. Half-wave antennas (often called Hertz antennas) are used as the driven elements for television receiving antennas. The driven element is the one that captures the signal and delivers it to the transmission line.
8. Half-wave dipoles and folded dipoles are the most frequently used driven elements for television receiving antennas. (Figure 6.9)
9. Two types of *parasitic elements* are used with television receiving antennas: directors and reflectors. These elements intercept the signal and reradiate it to the driven element. (Figures 6.10, 6.11, 6.12, and 6.13)
10. In general, the greater the number of parasitic elements, the greater the directivity and the gain of the antenna. (Figures 6.14 and 6.21)
11. The *directivity* of an antenna is its ability to receive signals from one general direction and to reject signals from all other directions. (Figures 6.12, 6.21, and 6.22)
12. The *gain* of a receiving antenna is a measure of its effectiveness in receiving a signal, relative to the effectiveness of a dipole in receiving the same signal. (Figures 6.12, 6.21, and 6.22)
13. Because of the ultrahigh frequencies and shorter transmission ranges involved, some UHF antennas may be more elaborate than comparable VHF antennas. (Figures 6.18 and 6.19)
14. The bandwidth of an antenna—that is, its ability to receive a range of frequencies—is increased by shaping the driven element. Fan shapes and conical shapes are used to obtain a broadband antenna. (Figures 6.15, 6.16, 6.17, 6.18, and 6.19)
15. The log-periodic antenna is very common. Its major advantage lies in its very wide bandwidth. Such antennas may also have high gain and directivity. (Figure 6.20)
16. Combination television receiving antennas are the most common ones used. They offer high gain, high directivity, and a wide bandwidth for VHF and

UHF channels. They are largely based on log-periodic design. (Figures 6.21 and 6.22)

17. Fringe-area antennas are designed to provide the highest gain and directivity. They must be capable of receiving very low signal levels while providing acceptable pictures. (Figure 6.22)
18. Indoor antennas vary from simple rabbit ears to more elaborate, tunable devices, which can receive VHF, UHF, and FM channels. (Figures 6.23 and 6.24)
19. For maximum transfer of power, the impedance of the antenna, the transmission line, and receiver should all be the same. (Figure 6.26)
20. Coaxial cable (72 ohms) has the advantage that the inner conductor is shielded by the outer conductor and therefore does not have electrical noise voltages injected into it. Another advantage is that a signal on the inner conductor will not radiate waves into space. (Figure 6.26)
21. Twin-lead transmission line (300 ohms) has a lower loss than coaxial cable but does not have the advantage of shielding. (Figure 6.26)
22. Coaxial cable is called an *unbalanced transmission line* because one conductor is grounded.
23. Twin-lead transmission line is said to be a *balanced line* because both conductors are at the same potential with respect to ground.
24. Special transformers called *baluns* are used to connect balanced lines (300 ohms) to unbalanced lines (72 ohms).
25. Two basic devices for amplifying the antenna signal are (1) boosters and (2) antenna preamplifiers. Antenna preamplifiers provide a better signal-to-noise ratio. (Figures 6.28, 6.29, 6.30, and 6.31)
26. Antenna signal splitters, or couplers, feed two or more television (and FM) receivers from a single antenna. They do not amplify the signal, which is divided equally among the receivers. (Figure 6.33)
27. A master-antenna television (MATV) system is used to operate a number of sets from a single antenna system. The basic system consists of (1) the antenna section, (2) the distribution amplifier, and (3) the signal-distribution section. (Figure 6.35)
28. A community-antenna television system (CATV) is used where the terrain interferes with good television reception.
29. Television antennas may be struck by lightning. Protection should be provided so that the lightning does not travel into the house.

EXAMINATION QUESTIONS

(Answers are provided at the back of the text.)

Answer true (T) or false (F).

1. Ghosts are additional images produced by an antenna having excessive gain.
2. The sky wave will not return to earth at frequencies greater than about 15 MHz.
3. Line-of-sight transmission may extend beyond the horizon.
4. The polarization of an antenna is determined by the direction of its magnetic lines of force.
5. Antenna gain refers to an antenna's effectiveness relative to a half-wave dipole.
6. Antenna impedance is measured at the

point of connection of the transmission line.

7. The equation commonly used to determine the length of a dipole in feet is $L = \frac{468}{f\ (\text{MHz})}$.
8. When a reflector is added to a dipole, the antenna pattern becomes basically unidirectional.
9. When a director is added to a dipole, the antenna bandwidth is increased.
10. The UHF fan dipole is also known as a bow-tie or a di-fan antenna.
11. A folded dipole is made by bending a half-wave dipole in two.
12. The term *log-periodic* refers to the relative antenna element lengths.
13. The pattern shown in Figure 6.10B is called omnidirectional.
14. In a stacked antenna array, the elements may be positioned vertically.
15. A combination antenna may efficiently receive VHF, UHF, and FM.
16. A fringe-area antenna has the highest gain and sharpest directivity.
17. The RCA ministate antenna has a bidirectional pattern for improved color reception.
18. Some indoor antennas can receive VHF, UHF, and FM signals.
19. Coaxial lines for television have an impedance of 300 ohms.
20. To obtain optimum power transfer, the transmission line impedance should be within 25% of the antenna impedance.
21. The 300-ohm flat transmission line will directly match the impedance of some antennas.
22. The transmission line with the lowest attenuation is the shielded parallel-wire line.
23. An antenna preamplifier provides a better signal-to-noise ratio than a booster.
24. Band separators provide separate VHF, UHF, and FM signals.
25. In a MATV system, separate antennas for VHF, UHF, and FM are required.

REVIEW ESSAY QUESTIONS

1. Discuss how a ghost is produced on a television screen. Is the ghost displaced to the right or left of the normal image? Why?
2. Explain briefly how a sky wave may return to earth. What is the approximate maximum frequency at which this may occur?
3. What is meant by line-of-sight distance? What is the formula for calculating it?
4. Define polarization, antenna, directivity, antenna gain, and antenna impedance.
5. Why are tuned antennas used?
6. What are parasitic elements? What types are there and where is each situated?
7. In Figure 6.12, explain how a reflector causes the pattern to become unidirectional.
8. Is a UHF antenna fundamentally different from a VHF antenna? Explain.
9. In Figure 6.18, which is the driven element?
10. List all the basic antenna types discussed in this chapter.
11. Draw to approximate scale, on polar coordinate paper, the antenna pattern of a high-gain combination antenna.
12. Discuss briefly the use of a corner reflector. In what television band is it used?
13. Explain how the log-periodic antenna acquired its name. Are all such antennas physically identical? Why?
14. What do we mean by a combination antenna? List the desirable characteristics of this antenna.
15. Discuss briefly the requirements of a fringe-area antenna.
16. Describe briefly the various types of indoor antennas and their general characteristics.
17. What are the different types of transmission lines?
18. For the transmission lines listed for Question 17, briefly state the advantages and disadvantages of each.
19. When would you use a booster-coupler? An antenna preamplifier?

20. Explain the function of a band separator.
21. Explain the use of a balun.
22. Discuss briefly the purpose of and basic components used in a MATV system.
23. Why are CATV systems used? What are their basic components?

EXAMINATION PROBLEMS

(Selected answers are provided at the back of the text.)

1. Compute the distance to the horizon from the top of a 1 000-ft (305-m) antenna. (Section 6.2)
2. A television transmitter tower places the antenna 800 ft (244 m) above the surface of the earth. A receiving antenna is mounted on a 60-ft (18-m) mast. What is the maximum theoretical line-of-sight distance that the receiving antenna can be located from the transmitter? (Do not take into account the 15% due to bending of the waves.) (Section 6.2)
3. The geometric line-of-sight distance between two points is 45.8 miles (73.3 km). What is the television line-of-sight distance? (Section 6.2)
4. Which of the following lengths would be suitable for a half-wave antenna at a frequency of 58.5 MHz? (a) 12 ft (3.7 m), (b) 10 ft (3.0 m), (c) 8 ft (2.4 m), (d) 6 ft (1.8 m).
5. Calculate the gain of an antenna that radiates 10 mW of power when under identical conditions a standard dipole radiates 1 mW.
6. Draw a simple block diagram of a MATV system that uses a single antenna and feeds four television receivers.
7. In Figure 6.30A, what is the function of capacitors C_1 and C_2?
8. In Figure 6.22B, compare the antenna gain and directivity of the low-VHF and high-VHF bands.
9. Draw a dipole with director and reflector. Show the lengths for these elements if they are to be used at 60 MHz.

Chapter 7

7.1 Some TV receiver fundamentals
7.2 Block diagram discussion
7.3 Monochrome television receiver controls
7.4 Use of PCs, ICs, FETs, MOSFETs, and modular construction
7.5 Use of nuvistor, novar, and compactron vacuum tubes
7.6 Types of test equipment used for monochrome TV receivers
7.7 General types of tests required for monochrome TV receivers
7.8 Some common monochrome TV receiver troubles
Summary

PRINCIPLES OF MONOCHROME TELEVISION RECEIVER OPERATION

INTRODUCTION

A television receiver is basically a superheterodyne. It is more complex than an AM or FM radio receiver, but the two are similar in basic operating principle. Color television receivers are more complex than their monochrome counterpart. This is because of the additional functions required to process and reproduce the color signals. These additional functions include the detection and conversion of the color signal sidebands into voltages appropriate to drive the red, green and blue guns of the color picture tube to produce a full-color picture. As a first step in understanding the operation of either monochrome or color television receivers, they are considered in the form of a simplified block diagram.

In this chapter, we shall discuss, in a simplified manner, the major sections of monochrome TV receivers. (Color TV receivers will be covered in later chapters.) We shall also discuss the use of solid-state components and several special tubes. In addition, some general types of TV test equipment and tests will be mentioned, as well as some basic TV receiver troubles.

When you have completed the reading and work assignments for Chapter 7, you should be able to:

- Define the following terms: PC, IC, FET, MOSFET, high voltage, solid state, common IF amplifiers, intercarrier IF, AFC, AGC channel selectors, modular construction, hybrid TV receiver, and video peaking.
- Draw a simple block diagram of a monochrome TV receiver. Explain briefly the function of each block.

- Name all the external controls on a monochrome TV receiver. Briefly describe the function of each one.
- Describe the advantages of using PCs, ICs, and modular construction in TV receivers.
- Briefly discuss the action of an AGC circuit.
- Know which TV channels are tuned by the VHF and UHF tuners. Also what are the frequency ranges of the VHF and UHF TV channels.
- Know the approximate range of frequencies passed by the common-IF amplifiers.
- Understand the importance of the correct video-signal polarity applied to the picture tube.
- Describe briefly how the picture tube high voltage is produced.
- Know how to adjust an AGC control.
- Know how to use an RF sweep generator and oscilloscope in connection with a TV receiver.
- Know how the observation of visual and aural symptoms of a TV receiver can help you to determine the source of trouble.

7.1 SOME TV RECEIVER FUNDAMENTALS

The complete signal from a monochrome TV transmitter includes not only video voltages but also horizontal and vertical blanking and synchronizing voltages and sound information. The sound is transmitted on its own carrier, which is frequency-modulated. The video signals required for the reproduction of the picture are amplitude-modulated and are on a separate picture carrier.

Because of the complexity of TV transmission and the need to reproduce both picture and sound correctly, a TV set must include not only an AM receiver to process picture signals, but also a portion of an FM receiver to process sound signals. In addition, the synchronizing pulses must be separated from the composite video signal. These are fed to the circuits that control the scanning of the electron beam in the picture tube.

All TV receivers (as is the case with all types of electronic equipment) require power supplies to provide the various operating voltages. Generally, a monochrome TV receiver will contain a low-voltage power supply to operate the various transistor or vacuum tube circuits and a high-voltage power supply to operate the picture tube. These same supplies are also found in color TV sets, but the high-voltage supplies are more complex because of the requirements of the color picture tube.

Most modern color TV receivers are fully solid-state. However, some monochrome receivers are still being built with vacuum tubes. In addition, many millions of older vacuum tube receivers are in use and will be for a number of years. Television receivers in current use fall into one of the following three general categories of construction:

1. All solid-state construction, with widespread use of integrated circuits and digital-computer techniques
2. Hybrid construction, with both solid-state construction and vacuum tubes; the deflection and power supply circuits in such receivers may use vacuum tubes
3. All vacuum tube receivers. This applies mainly to some current monochrome receivers and to older monochrome and color receivers

The construction of TV sets using plug-in or wired-in modular circuit boards is popular and is designed to simplify the servicing of solid-state circuits. Although the replacement cost of a complete plug-in circuit board is greater than that of, say, a single transistor, this may be offset by the savings in time required to troubleshoot and replace discrete components. Figure 7.1 shows a monochrome TV receiver that has modular solid-state construction.

Figure 7.1 Rear view of a monochrome TV receiver. The circuitry is in the form of modules, built on printed circuit boards. This is a solid-state receiver. *(Courtesy of Magnavox Consumer Electronics Co.)*

7.2 BLOCK DIAGRAM DISCUSSION

A monochrome TV receiver can be divided into ten basic sections, as shown in Figure 7.2. These sections are:

1. Tuner section
2. Common video and sound IF amplifiers and automatic gain control circuit
3. Video detector
4. Sound IF amplifier, detector, and audio amplifiers
5. Video amplifier
6. Sync-pulse separator
7. Vertical-deflection section (integrator, oscillator, and amplifier)
8. Horizontal-deflection section (automatic frequency control, oscillator, and amplifier)
9. High-voltage power supply
10. Low-voltage power supply

Each of these ten basic sections will be discussed briefly.

TUNER SECTION

The **tuner section,** also known as the **RF section** or the **front end,** includes both the VHF and UHF tuners, which cover all the television broadcast channels, from Channel 2 to Channel 83. The function of the tuner is to select the desired channel, amplify the selected band of RF frequencies, and reduce these frequencies to those amplified by the IF amplifiers (41 to 47 MHz). The bandwidth of the tuner can pass only the frequencies of each selected channel. RF frequencies are changed to IF frequencies by heterodyning against the local oscillator frequency, as in any superheterodyne receiver.

UHF Operation

The VHF-UHF selector switch is an integral part of the VHF channel selector switch. When the VHF channel selector is in the UHF position, the UHF oscillator and mixer convert the UHF signals to a lower frequency in the IF range. This IF signal is amplified in the VHF tuner, which now functions only as an IF amplifier, and then fed to the IF amplifiers. The UHF tuner does not have an RF amplifier because of the difficulty in producing low-noise, high-gain amplification in the UHF range.

The UHF tuner consists basically of a transistor oscillator and a crystal mixer stage.

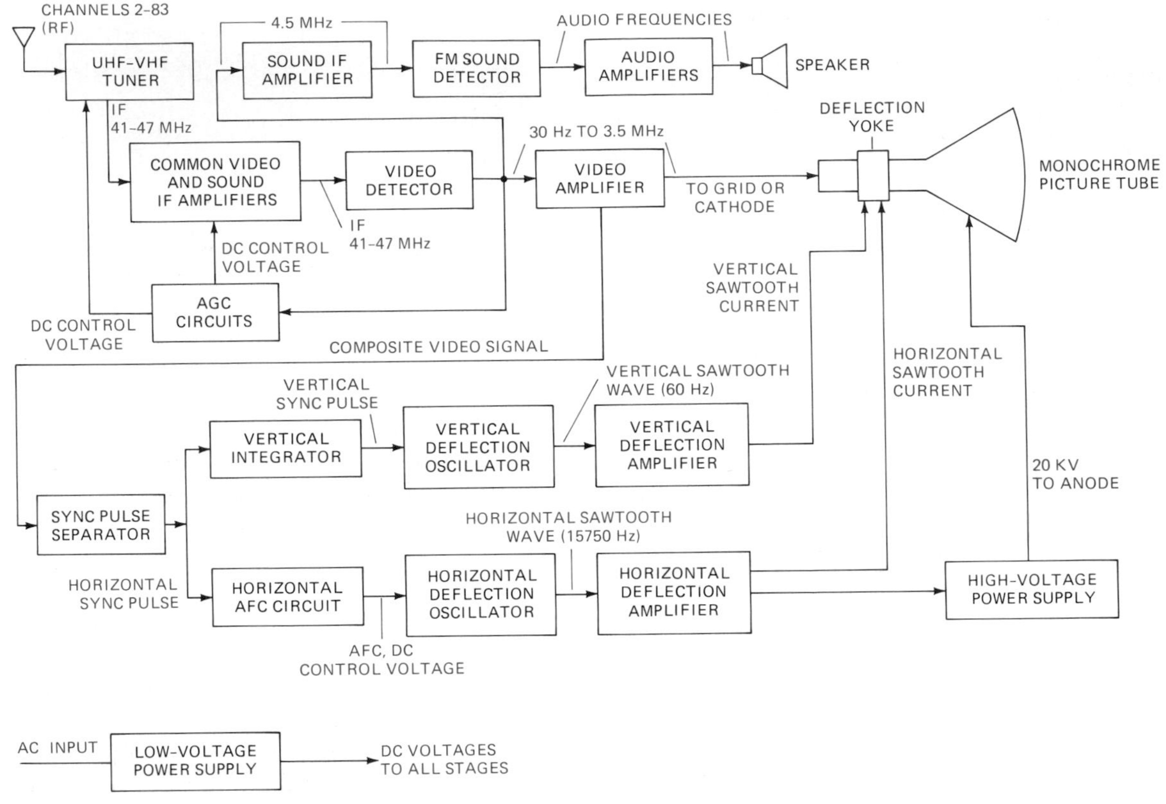

Figure 7.2 Simplified block diagram of a monochrome television receiver.

The UHF channel and oscillator signals heterodyne in the crystal mixer, whose usable output is in the IF range (41 to 47 MHz). Within this range (Figure 7.3) we have the picture IF carrier frequency of 45.75 MHz and the sound IF carrier frequency of 41.25 MHz. Note that the *fixed* difference between these two frequencies is 4.5 MHz. This 4.5-MHz frequency will be used as the FM sound IF frequency in later stages of the television receiver. The UHF tuner selects Channels 14 to 83 (470 to 890 MHz) either by the detent action of a rotary switch or by push buttons in more sophisticated receivers.

VHF Operation

The VHF tuner (in its normal VHF function) consists of an RF amplifier, mixer, and local oscillator. The RF amplifier increases the strength of the signal from the antenna. The gain of the RF amplifier is automatically controlled by a voltage from the automatic gain control (AGC) circuit. The RF amplifier also acts as a buffer to prevent undesirable coupling of the VHF oscillator signal to the antenna. The VHF tuner selects Channels 2 to 13 (54 to 216 MHz). Again, this is accomplished by detent action or push buttons. When VHF channels are being tuned, the UHF tuner is disabled.

As in the case of the UHF tuner, each VHF channel signal is heterodyned with the correct oscillator frequency in the mixer stage (transistor or vacuum tube). The usable mixer output for each channel is in the IF range, with the picture IF carrier frequency always 45.75 MHz and the sound IF

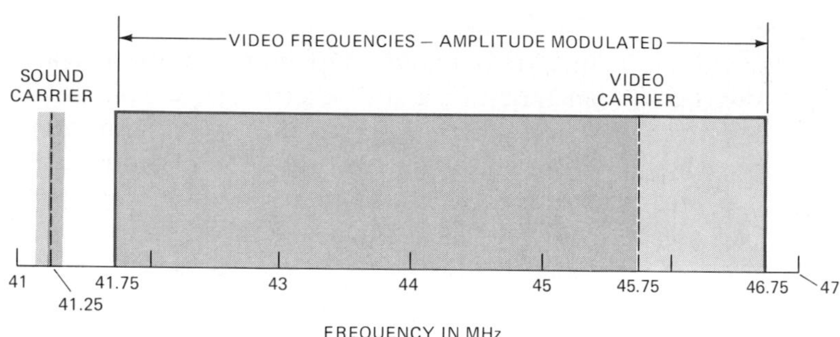

Figure 7.3 The distribution of sound and video IF frequencies.

carrier frequency always 41.25 MHz. Many receivers have a fine-tuning control associated with the tuner. This control permits limited manual adjustment of the oscillator frequency for best picture. With many other receivers, this is done automatically by AFT (automatic fine-tuning) circuits.

COMMON VIDEO AND SOUND IF AMPLIFIERS

Monochrome TV receivers may have two or three common IF amplifiers. Both the video and the sound IF frequency are passed by these amplifiers. This includes the video IF carrier and its sidebands plus the sound IF carrier and its sidebands. The required bandwidth may be obtained by "stagger-tuning" the frequencies of each IF amplifier. It may also be obtained by "broad-banding" when only two IF stages are used. The details of the IF response curve will be covered in Chapter 12. However, for the present, it is sufficient to say that the IF response is *not* flat over this range.

As shown in the block diagram of Figure 7.2, the gain of the common IF stages (and the tuner) is controlled by the AGC circuit. The voltage to operate this circuit originates at the output of the video detector. The AGC circuit varies the gain of these stages automatically, in accordance with the received signal strength. The output of the common IF stages is one or higher and is fed to the video detector.

AGC CIRCUITS

Superheterodyne AM radios use a volume control circuit to provide automatic adjustment of the output sound volume against variations in RF signal strength. In television receivers, a similar circuit is used but it is called **automatic gain control,** or **AGC.** The term *automatic volume control* is not appropriate for television because both the sound and the picture signal are stabilized against the signal strength changes by this circuit. The AGC circuit in a television receiver is more complex than the one in a radio receiver. The strength of a television video signal changes according to variations in the brightness of the scene being televised, but the height of the synchronizing pulses depends only on the strength of the RF signal. For this reason, television receiver AGC circuits use the amplitude of the synchronizing pulses rather than the overall video signal to develop a DC control voltage. This control voltage is then applied to the RF and IF amplifiers to reduce their gain when strong signals are received and to increase their gain for weak signals.

VIDEO DETECTOR

The output of the last common IF amplifier is coupled into the **video detector.** Here, the video carrier is heterodyned against its sidebands to provide different frequencies that constitute the original modulating video frequencies, thus recovering the original video information. Also in the video detector, the video carrier is heterodyned against the sound carrier to produce the 4.5-MHz **intercarrier** sound IF frequency. The sound IF frequency is then applied to the sound IF amplifier, and the video signals are applied to the video amplifier. A portion of the video information is also fed into the AGC and sync-pulse separator circuits. As mentioned before, signal strength variations in the IF amplifiers are stabilized by feedback DC voltage from the AGC circuits. A portion of this AGC voltage is also used to adjust the signal output of the tuner.

Note: In color broadcasts and for color TV receivers, the output of the video detector will also contain a 3.58-MHz color subcarrier and its sidebands. This signal is fed to special color circuits. In monochrome receivers, however, the demodulated color signals are not permitted to pass through the video amplifier. This is done to avoid an interference pattern on the screen of monochrome receivers.

AUDIO SECTION

The FM sound signal is amplified in the 4.5-MHz sound IF amplifier and then fed into an FM detector. In some receivers, an FM **discriminator** is used as the FM detector. This circuit requires an additional stage, a **limiter,** to provide adequate noise reduction. Because of this, many receivers use a ratio detector or other single-stage detector. After detection, the sound signals are amplified to a level suitable for driving a loudspeaker.

VIDEO AMPLIFIER

Figure 7.2 shows that the detected picture signal is fed into the **video amplifier.** The polarity of this signal can be either positive or negative, depending on the particular type of picture detector used. This polarity is important because it, and the number of signal inversions during amplification, determine which element of the picture tube the amplified signal is applied to. Negative video output signals are connected to the picture tube control grid. (Positive signals go to the cathode.) Thus, a positive signal from a video detector, which is inverted one time in a single-stage video amplifier, is applied as a negative signal to the picture tube control grid. This signal can be made positive and then applied to the cathode by adding a second video amplifier. Most monochrome vacuum tube TV receivers have a single-stage video amplifier. Solid-state monochrome receivers usually have two video amplifier stages.

The basic function of the video amplifier is to provide sufficient drive signal to the picture tube. The signal must be capable of driving the tube from darkest black to brightest white. This will make it possible to reproduce all picture elements in their proper shades. The signal at this point, of course, is a composite video signal consisting of picture information and blanking and synchronizing pulses. The blanking pulses extinguish all retrace lines. The synchronizing pulses are not used at this point. A typical value of peak-to-peak video signal applied to the picture tube is about 100 V. The bandwidth of the video amplifier may vary from about 2.5 MHz to about 3.25 MHz.

SYNC-PULSE SEPARATOR

The **sync-pulse separator** removes the synchronizing pulses from the detected video signal, amplifies them to the proper level, and then feeds them into the vertical- and horizontal-deflection sections. These pulses keep the vertical and horizontal oscillators synchronized, or in step with, the scanning beam in the TV camera.

VERTICAL-DEFLECTION SECTION

Figure 7.2 shows that the amplified synchronizing pulses are coupled through an integrator to the **vertical oscillator.** This integrator filters and shapes the pulses so that only the vertical pulses affect the vertical oscillator. In this way, the vertical oscillator is properly controlled in producing the required 60-Hz, modified sawtooth wave form that drives the vertical output amplifier.

The **vertical output amplifier** is a power amplifier that drives the vertical windings of the deflection yoke mounted on the neck of the picture tube. This vertical (magnetic field) drive moves the electron beam in the picture tube from the top of the screen to the bottom at a 60-Hz rate. After each excursion (downward sweep of the beam), the beam is moved rapidly (or retraced) to the top of the screen to start the next downward sweep. During the retrace interval, the vertical blanking pulse blanks out the picture tube screen until the scanning beam is in the proper position to begin the next downward sweep. Each excursion develops one field. Two successive fields make up one complete frame (or one complete picture).

HORIZONTAL-DEFLECTION SECTION

The output of the sync-pulse separator is also applied to the **horizontal-deflection section.** The function of this section is similar to that of the vertical-deflection section. However, the electron beam in the picture tube is moved from left to right as the vertical deflection moves the beam from top to bottom. To produce a sufficient number of horizontal scanning lines for each complete field, the horizontal oscillator frequency is maintained at 15.750 Hz by the **automatic frequency control,** or **AFC,** circuit. This circuit prevents noise or other interference pulses from triggering the horizontal oscillator and keeps the horizontal oscillator locked in with the scanning beam in the TV camera.

Similar to the vertical-deflection oscillator, the **horizontal oscillator** generates a modified sawtooth wave form that is coupled to the **horizontal output amplifier.** This power amplifier increases the level of the horizontal sawtooth wave and shapes it properly for driving the horizontal windings of the deflection yoke on the neck of the picture tube. At the end of each horizontal line, the beam is rapidly retraced to the left-hand side of the screen to start the next line. During the retrace interval, a horizontal blanking pulse blanks out the screen until the scanning beam is in the proper position to begin the next horizontal line. A retrace pulse, produced by a winding on the horizontal output transformer, develops the high voltage for the picture tube.

HIGH-VOLTAGE POWER SUPPLY

At the end of each horizontal scanning line, the drive current in the horizontal windings of the yoke suddenly changes in order to retrace the scanning beam. This sudden current change produces a high-voltage pulse across a winding of the horizontal output transformer. Here, by transformer action, the pulse is increased to about 20 000 V. After this increase, the pulse is rectified, filtered, and applied to the high-voltage anode on the picture tube. It is this high voltage that accelerates the electron beam, causing it to strike the screen at a very high speed.

LOW-VOLTAGE POWER SUPPLY

The low-voltage power supply shown in Figure 7.2 provides the necessary voltages for all of the circuits in the receiver. For vacuum tube receivers, the low-voltage power supply also provides the power to heat the tube filaments. Solid-state receivers need filament power only for the picture tube. Regulation is seldom used in monochrome vacuum tube receivers but may be used in solid-state receivers. These power supplies use multiple-winding transformers and a full-wave rectifier with filters. Some manufacturers use a simple AC power supply without a transformer. Batteries provide the power for portable TV receivers.

The DC voltage available from a low-voltage power supply depends upon whether the supply is used for a vacuum tube or a solid-state TV receiver. For tube-type receivers, a range of 125 to 300 V DC is common. For solid-state receivers, the range varies from 10 to 100 V.

7.3 MONOCHROME TELEVISION RECEIVER CONTROLS

Monochrome TV receivers have a number of controls. These are necessary to adjust the sound level as well as the picture size and quality. In addition, some type of control is necessary to select channels.

VOLUME CONTROL

For the sound section of the television receiver, a conventional on-off **volume control,** supplemented occasionally by a **tone control,** is used. Instead of turning the volume control knob fully counterclockwise to turn the set off, some manufacturers prefer to use a push-pull control. With this control, the receiver is energized when the knob is pulled out and de-energized when it is pushed in. Volume variation is obtained by turning the knob clockwise or counterclockwise. This arrangement has the advantage that the same volume control setting can be maintained when the receiver is turned off and then turned on again.

CHANNEL SELECTOR

For the video section of the television receiver, additional controls are required. These are few in number and easily manipulated. A primary control is the **channel selector.** It contains two parts: one for selecting VHF stations (Channels 2 to 13) and the other for selecting UHF stations (Channels 14 to 83). In the VHF range, there may be as many as seven channels in one community, but no more. If there are more than seven stations in a community, the remainder will be in the UHF range. The VHF channel selector is usually a 12-position switch with an associated **fine-tuning control.** This control is basically a vernier-tuning adjustment. The VHF tuning circuits used in television receivers are relatively fixed, and the desired station is obtained with the selector switch.

In addition to other circuits, the proper oscillator components are selected when the channel selector switch is moved into position. If a change should occur in the resonant properties of the oscillator circuits (due to heating or aging) and no fine-tuning adjustment were provided, the sound would possibly become distorted and the images would not be faithfully reproduced. To prevent this, a fine-tuning control is placed on the front panel. Within limits, this control permits the user to tune in the signal so that the proper frequencies are delivered to the receiver video and audio IF amplifiers.

CONTRAST CONTROL

Another control associated with the picture is the **contrast control,** which adjusts the relationship between the light and dark areas on the screen. Turning the control clockwise causes the darker areas of the picture to turn from gray to black. Simultaneously, the lighter areas will achieve maximum brightness. Rotating it counterclockwise causes the dark areas to lighten from black to gray. At the same time, the bright areas will become dimmer. The contrast control is also called the **picture control** or **pix control.**

BRIGHTNESS CONTROL

The **brightness control** regulates the overall shading of the picture and establishes the brightness of an image; that is, whether the overall image is light, medium, or dark. The brightness control adjusts the DC bias of the picture tube. This in turn determines the number of electrons in the scanning beam and consequently the screen brightness. The grid-to-cathode bias for normal viewing is in the order of 35 to 45 V.

BUZZ CONTROL

The **buzz control** is frequently a screwdriver adjustment on the rear of the chassis. Some receivers are susceptible to a 60-Hz buzz in the sound. The control is simply adjusted for minimum aural buzz.

AGC CONTROL

Some receivers use this control. Generally it is also a screwdriver adjustment on the rear of the chassis. It usually requires only an initial adjustment, achieved as follows:

1. With the brightness control at a normal setting, turn the contrast control maximum clockwise.
2. Determine the strongest channel and tune it in for best picture.
3. Adjust the AGC for high contrast, short of overloading. If overloaded, picture brightness values are abnormal. In this condition, the black and white elements of the image may be reversed. If excessive buzz is noted, back off the AGC control until buzz is minimized.

VIDEO-PEAKING CONTROL

The **video peaking control** is also called the **sharpness control.** This is not an essential control and is found only in some receivers. It is used to adjust the high-frequency response of the video amplifier. As noted previously, the greater the high-frequency response, the more detail in the picture. Hence, the "sharpness" of the picture will be enhanced. Unfortunately, this condition will also bring up any noise (snow) present, as well as some interference patterns. Thus, this control must be adjusted for each individual installation. In some cases, a lower high-frequency response setting will produce a more pleasing picture.

VERTICAL HOLD CONTROL

When available, the **vertical hold control** is generally mounted on the front panel. It is used to adjust the free-running frequency of the vertical-deflection oscillator close to 60 Hz. When this is achieved, the vertical synchronizing pulses will lock in the oscillator at exactly 60 Hz (monochrome transmission). At lock-in, the picture will be stationary and correctly centered vertically. If the control is incorrectly adjusted, the picture will roll vertically, either up or down. In some late-model receivers, automatic circuits make it unnecessary to have a vertical hold control.

HORIZONTAL HOLD CONTROL

Horizontal hold control is most often a rear-chassis, screwdriver adjustment. Once set, this control seldom requires attention. The adjustment generally varies the depth of a tuning slug within the horizontal oscillator coil. This changes the resonant frequency of the horizontal oscillator. As the control is turned slowly from its proper setting (normal picture), the picture will first shift horizontally and then break up into slanting segments. When setting this control, you should switch channels to ensure no picture breakup at different signal strengths. Because of the horizontal AFC circuit, this control setting will hold unless some defect occurs in the horizontal AFC circuit or in the horizontal-deflection oscillator circuit.

SETUP CONTROLS

The **vertical height** and **vertical linearity** controls are rear-chassis, screwdriver controls. They are adjusted together to fill the screen vertically and to provide the correct vertical linearity. However, you will find that the height control affects mostly the bottom area. Conversely, the linearity control affects mostly the top area.

The **horizontal width** (or drive) **control** adjusts the width of the picture. It, too, is a rear-chassis control. This control (if present) is set so that the picture goes just beyond the vertical edges of the screen. This is necessary to prevent the appearance of black vertical (horizontal blanking) bars on both sides of the picture.

7.4 USE OF PCs, ICs, FETs, MOSFETs, AND MODULAR CONSTRUCTION

PCs, ICs, FETs, MOSFETs, and modular construction are innovations that have increased TV receiver reliability; decreased its size, weight, and power dissipation; and reduced manufacturing costs and service time. Without these innovations, the small, battery-operated portable receiver would not be practical.

PRINTED CIRCUITS

Printed circuits (PCs), play an important role in TV receivers. The PC technique enables many components to be mounted on a single flat board, with a separate connection for each component. The circuit diagram and component identification are often printed right on the board to make troubleshooting and parts location simpler. Also, PCs have greatly increased the reliability of TV receivers by eliminating assembly-line wiring errors and by providing a firmer support for the components. PCs have also eliminated many of the critical lead-dress problems of the hand-wired chassis. On a PC board, all leads are fixed and never disturbed during component replacement. Figure 7.4 is a top view of a TV receiver PC. Note the clear markings of all parts.

INTEGRATED CIRCUITS

The importance and uses of **integrated circuits** (ICs) were discussed in Chapter 1. For many years, ICs were used for the inte-

Figure 7.4 A TV receiver printed circuit board. Each part and control is clearly labeled. Components are grouped by function to minimize trouble shooting time. *(Courtesy of Sony Corp. of America)*

grator in the sync-input circuit of the vertical oscillator. Although considerably larger than the newer ICs, that vertical integrator contained three or four capacitors and as many resistors. Many of the newer ICs contain several amplifier stages in an envelope the size of a small transistor. In fact, many ICs contain several dozen transistors and diodes, a dozen or more resistors, and many capacitors. All of these components are "grown" on a tiny substrate and then encapsulated in a small envelope. In troubleshooting, the overall operation of the IC is checked, and if found unsatisfactory, the IC is replaced. Trouble shooting an IC section can be as simple as determining whether or not all of the inputs are present and if there is a satisfactory output. ICs have contributed greatly to the compactness and reliability of present-day TV receivers. A ceramic, flat-pack IC is shown in Figure 7.5. Several complete TV receiver stages may be included in this IC.

Figure 7.5 A ceramic, flat-pack integrated circuit. *(Courtesy of RCA Solid-State Components)*

FIELD-EFFECT TRANSISTORS

The **field-effect transistor,** or **FET,** is finding ever-increasing use in TV receivers. It has a very high input impedance and operating characteristics that are very similar to those of a triode vacuum tube. Because of its square-law characteristics when biased near cutoff, the FET is often used as the mixer in TV tuners. Another desirable characteristic of the FET is that its noise level is much lower than that of other transistors, making it useful in RF stages, where low noise levels are necessary to reduce picture "snow." FETs are also used extensively in audio amplifier circuits. Because of its high input impedance, strong electric fields or static electrical charges can develop sufficient voltage to damage an FET when it is not connected into a circuit. For this reason, the leads of an FET are connected to each other until the unit is installed in the circuit. (See Chapter 10 for a basic explanation of transistor operation, including FETs and MOSFETs.)

MOSFET

The **MOSFET,** or **metal oxide semiconductor field-effect transistor,** is similar to the FET except that its input impedance is even higher and its gate electrode is insulated (Chapter 10). This insulated gate renders the input impedance of the MOSFET virtually impervious to normal signal magnitude and to the polarity of the input gate voltage. To compare this feature to the grid of a vacuum tube: a positive voltage on the grid of the tube causes current flow, whereas a positive voltage on the gate of a P-channel MOSFET causes no current flow. The MOSFET has four electrodes, whereas the FET has three. This fourth electrode is called the substrate and is usually connected to ground.

MODULAR CONSTRUCTION

In modular construction, various sections, such as the IF amplifier, AGC, and synchronizing circuits, are each mounted separately or in combination on a printed wiring board. These modules are then plugged or wired into a master (or mother) chassis, drawer, or rack (Figure 7.1). The chief advantages of this type of construction are ease of manufacture, at-home servicing, and improvements in performance, dependability, and reliability.

During assembly, each module is completed and tested as though it were an individual unit. After final acceptance at the factory, the modules are delivered to the main TV receiver assembly area, where they become a working part of the receiver. In the home, the service technician isolates receiver troubles to a particular section and then merely puts in a replacement module. In the shop, the modules are easier to repair than a complete chassis. Thus, the manufacturer provides a more precise unit and repair bills may be lower because less time is spent in troubleshooting and repair.

7.5 USE OF NUVISTOR, NOVAR, AND COMPACTRON VACUUM TUBES

An innovation in television has been the development of special-purpose vacuum tubes. These tubes have better operating characteristics at high RF frequencies and offer several additional advantages because of their small size. Many of the present-day hybrid TV receivers have from one to three of these special-purpose tubes. The most common are the nuvistor, the novar, and the compactron.

NUVISTOR

The **Nuvistor** is a very small triode vacuum tube made of metal and ceramic (Figure 7.6). Instead of having the usual planar configuration, the nuvistor is a cantilevered coaxial device. Its elements are attached to cones supported on the internal ends of its connecting pins. The nuvistor has a high transconductance, or low noise figure, and very low dielectric losses. These tubes are used mostly in tuners.

NOVAR

The **novar** is a beam-pentode vacuum tube with a glass-button base. Although it is larger than most miniature tubes, it offers the advantage of better heat dissipation. A popular novar tube is the 6JE6, which is often used in the horizontal output circuit. The main attributes of the novar are its small size and high voltage-breakdown characteristics.

COMPACTRON

As its name implies, the **compactron** is a compact vacuum tube. It has an unconventional outward appearance but conventional

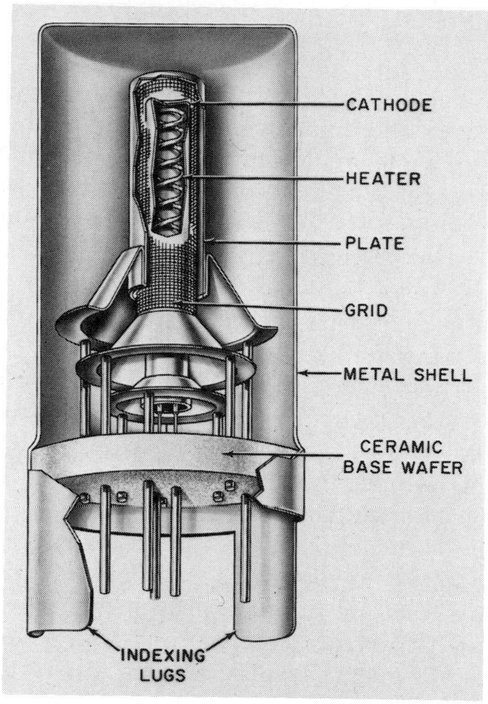

Figure 7.6 A cutaway view of a Nuvistor tube. The overall length is 0.8 in (2.03 cm). The diameter is 0.435 in (1.1 cm). *(Courtesy of RCA.)*

internal construction. Compactrons differ from conventional tubes in that they have a 12-pin, glass-bottom base. This allows from one to four vacuum-tube functions to be included in the single envelope. Among the types of tubes commonly included in the compactron are triple triodes, dissimilar double-triode pentodes, twin pentodes, and dissimilar double pentodes. The compactron is used in audio and other multitube circuits. Its chief advantage is that it saves space.

7.6 TYPES OF TEST EQUIPMENT USED FOR MONOCHROME TV RECEIVERS

Over the years, test equipment manufacturers have developed numerous pieces of equipment that combine many test functions in one unit. Equipment of this nature is convenient and easy to use, and it seldom offers the maze of cables and test leads often encountered when individual pieces of test equipment are used. Individual test equipment is quite popular, however, because it is less complex to service and less prone to malfunctions that would affect several test functions.

TV receiver test equipment can be divided into three categories according to test functions: general measurements, wave form observations, and signal injection. Often several of the categories must be combined to obtain the desired test results.

GENERAL MEASUREMENTS

General measurements include voltage, current, resistance, and capacitance as well as the characteristics of vacuum tubes, transistors, flyback transformers, yokes, and coils. Test equipment in this category is as follows:

1. VTVM (vacuum tube voltmeter or solid-state equivalent)
2. VOM (volt-ohm-milliammeter)
3. Resistance-capacitance comparator bridge
4. Tube tester
5. Transistor tester (in-circuit and out-of-circuit)
6. Flyback transformer and yoke tester

WAVE FORM OBSERVATION

The second test equipment category is wave form observation. This is performed with an oscilloscope. Various types of oscilloscopes may be used:

1. Conventionally synchronized oscilloscopes

2. Triggered-sweep oscilloscopes
3. Dual-trace oscilloscopes
4. Wide-band oscilloscopes

For monochrome receiver servicing, a relatively inexpensive oscilloscope will generally suffice. However, more elaborate units are desirable for color receiver servicing. This will be discussed in Chapter 27.

SIGNAL INJECTION

The third test equipment category includes all equipment that is used to inject a signal into the receiver circuits. This test equipment must be used in conjunction with equipment from the other two categories. Equipment in this category is as follows:

1. RF signal generator
2. RF sweep generator, or bar-sweep generator
3. RF crystal marker generator (may be built into sweep generator)
4. Audio signal generator

The uses of the above-mentioned test equipment as well as special color TV receiver test equipment will be described in Chapter 27.

7.7 GENERAL TYPES OF TESTS REQUIRED FOR MONOCHROME TV RECEIVERS

There are a number of tests that are identical for monochrome and color receivers. Some tests are performed only on color receivers. There are no tests that are performed only on monochrome receivers. Some common tests performed on monochrome receivers are described in the following paragraphs. Additional information concerning these tests is given in later chapters.

GENERAL MEASUREMENTS

Vacuum tube and transistor testers are used to determine whether either of these components are defective. A word of caution is necessary at this point because tube and transistor tests are not always conclusive. They can sometimes be misleading. It is better, if possible, when testing these components to compare their operation with tubes and transistors known to be good.

The VTVM is used for voltage measurements in all circuits where a low-impedance VOM might load the circuit — for example, in high-impedance and AGC circuits. VOMs are used mainly for resistance measurements and for checking voltages in power supplies and other low-impedance circuits. Capacitance tests are normally made with a resistance-capacitance comparator bridge. There are, however, several capacitor testers that can measure circuit capacitance without the need to remove the capacitors from the circuit. With the comparator bridge, though, resistors and capacitors must be removed from the circuit, but they can then be compared with known standard units for precise evaluation.

All of these measurements are identical for monochrome and color receivers.

WAVE FORM OBSERVATIONS

Often, voltage and resistance measurements alone cannot isolate the trouble to a particular circuit or stage. In this case, it is desirable to use an oscilloscope to observe the wave forms developed at various test points in the circuit. For these observations, the receiver may be set up for normal operation and the actual TV signals from the antenna are used. If a portion of the set is inoperative, then signal injection may be used to produce the desired wave forms.

SIGNAL INJECTION

Tuned-circuit alignment, or bandwidth checking, requires an RF sweep (or bar sweep) generator for injecting the band-sweep frequencies and an oscilloscope for observing the wave forms developed across the various circuits. In addition, a marker generator is often used to identify particular frequencies on the observed wave form, such as band edges, IF frequencies, and video and sound carriers. These tests are performed in both color and monochrome receivers.

7.8 SOME COMMON MONOCHROME TV RECEIVER TROUBLES

Both monochrome and color TV receiver troubles and troubleshooting techniques will be covered in subsequent chapters. At this point, we shall only introduce some basic, common troubles you may encounter in monochrome receivers. (You should refer to Figure 7.2 as necessary.)

NO RASTER, NO SOUND

Here, the screen is dark and there is also no sound. The problem must be in a section that affects both the audio and the deflection section. The problem is most likely a lack of DC operating voltages. This indicates that the trouble is probably (l) a defective low-voltage power supply, (2) a blown AC fuse, or (3) an open heater in a series filament receiver.

NO RASTER, BUT NORMAL SOUND

The sound is normal, therefore the low-voltage power supply is functioning. The condition of no raster could be caused by (1) a defective picture tube, (2) a defective high-voltage power supply, or (3) a defective horizontal-deflection section. In order to produce high-voltage for the picture tube, it is necessary for the horizontal-deflection section to be operating.

NORMAL PICTURE, NO SOUND

With a normal picture but no sound, the low-voltage power supply must be functioning. In addition, most of the signal circuits as well as the deflection circuits are operating. This problem, then, is localized to thc sound section of the receiver.

NORMAL RASTER, NORMAL SOUND, NO PICTURE

In order for this condition to occur, there must be signal at the output of the video detector (See Figure 7.2). However, since there is no picture, the problem must be (1) a defective video amplifier or (2) a defective picture tube.

NORMAL RASTER, NO SOUND OR PICTURE

The low-voltage power supply is functioning, as are the deflection circuits. Since there is neither sound nor picture, the trouble is in the signal circuits, which include (1) the tuner, (2) the common IF amplifiers, and (3) the video detector.

PICTURE SHORT VERTICALLY

If the only problem is a vertically short picture, the trouble could be (1) misadjustment of the vertical height and/or vertical linearity controls or (2) a defect in the vertical-deflection section, probably in the vertical-deflection amplifier. (A similar symptom in the horizontal direction, except to a minor extent, may not appear. Any trouble that causes a considerable decrease of width will probably also result in no high voltage, and thus no raster.)

NO VERTICAL DEFLECTION

A thin, bright, horizontal line across the center of the screen means that there is no vertical deflection. Such a line indicates that the horizontal section is functioning properly. Thus, the problem is localized to the vertical-deflection section. The defect is most likely in the vertical-deflection oscillator or the vertical-deflection amplifier. Other possibilities are the vertical output transformer and the vertical-deflection yoke windings.

SUMMARY OF CHAPTER HIGHLIGHTS

1. A television receiver is basically a superheterodyne. However, it is more complex than an AM or FM radio receiver. (Figure 7.2)
2. A television receiver must include an AM portion to process the video signals. It must also possess an FM portion to process the sound signals. (Figure 7.2)
3. All television receivers contain a low-voltage power supply to operate the transistor or vacuum tube circuits. Receivers also contain a high-voltage power supply for the picture tube. (Figure 7.2)
4. Most modern television receivers are fully solid-state. (Figure 7.1)
5. A monochrome television receiver is madeup of ten basic sections. (Figure 7.2)
6. The tuner selects and amplifies the RF channels. It also converts each of the RF signals to an IF signal. (Figure 7.2)
7. The common IF amplifiers amplify both sound and video signal IF carriers and their sidebands. Their approximate bandwidth extends from 41 to 47 MHz. (Figure 7.2)
8. The video detector recovers the original video information and applies it to the video amplifier. It also produces at its output the 4.5-MHz FM sound carrier and sidebands. (Figure 7.2)
9. The video amplifier accepts the detected video signal. It amplifies this signal to a level suitable for driving the picture tube grid or cathode. (Figure 7.2)
10. The AGC system maintains a relatively constant sound and video level in the receiver, regardless of reasonable signal variations. (Figure 7.2)
11. The sync-pulse separator removes the synchronizing pulses from the composite video signal. It amplifies and transfers these pulses to the horizontal- and vertical-deflection sections. (Figure 7.2)
12. The vertical-deflection section develops a current sawtooth wave form at 60 Hz. This produces the vertical, picture tube beam deflection. (Figure 7.2)
13. The horizontal-deflection section has two major functions: (1) it produces horizontal beam deflection at 15 750 Hz and (2) it produces the picture tube high voltage (up to about 20 kV). (Figure 7.2)

14. The channel selector and fine-tuning controls are responsible for changing channels and tuning them in properly.
15. The contrast and brightness controls adjust the overall dark-to-light picture element relationship.
16. The vertical and horizontal hold controls adjust their respective deflective oscillator frequencies for lock-in.
17. PCs and solid-state components have made it possible to manufacture small, lightweight TV receivers that require little power. They also have greater reliability than most older models. (Figures 7.3 and 7.4)
18. A small IC package may contain several complete TV receiver stages. (Figure 7.4)
19. Modular construction simplifies home TV repair because an entire module can be changed easily. (Figure 7.3 and Section 7.4)
20. The nuvistor is an extremely small vacuum tube used in some tuners. (Figure 7.5)
21. Monochrome receiver test equipment may be divided into three categories: (1) general measurements, (2) wave form observation, and (3) signal injection. (Section 7.6)
22. When troubleshooting a TV receiver, it is necessary to observe the symptoms carefully. An analysis of these symptoms will enable you to better pinpoint the source of trouble. (Section 7.7)

EXAMINATION QUESTIONS

(Answers are provided at the back of the text.)

Part A Select the correct multiple choice answer.

1. A TV receiver is basically a
   a. superregenerative receiver.
   b. single-sideband receiver.
   c. superheterodyne receiver.
   d. narrow-band receiver.
2. Most modern TV receivers are
   a. hybrid type.
   b. all solid-state.
   c. all vacuum tube type.
   d. hand-wired.
3. The common IF stages amplify
   a. only video signals.
   b. only synchronizing signals.
   c. only video and synchronizing signals.
   d. video and sound signals.
4. The output of the video amplifier feeds
   a. the picture tube grid or cathode.
   b. the synchronizing circuits.
   c. the two deflection AFC circuits.
   d. the 4.5-MHz sound IF amplifier.
5. The operating frequency of the vertical-deflection amplifier for a monochrome TV receiver is
   a. 59.94 Hz.
   b. 15 750 Hz.
   c. 60 Hz.
   d. 60.94 Hz.
6. The brightness control adjusts
   a. the overall picture intensity.
   b. the ratio of light-to-dark picture elements.
   c. the picture tube gamma.
   d. the transfer ratio.
7. The video peaking control can adjust
   a. high-frequency response.
   b. mid-frequency response.
   c. AGC voltage.
   d. common IF amplifier bandwidth.
8. The nuvistor is
   a. a new type of transistor.
   b. a negative-temperature-coefficient resistor.

   c. used in ICs.
   d. a special vacuum tube.
9. A typical piece of signal injection test equipment is the
   a. oscilloscope.
   b. bar-sweep generator.
   c. frequency counter.
   d. digital VOM.
10. The symptom is no raster, normal sound. One likely cause is
   a. defective horizontal-deflection oscillator.
   b. defective vertical-deflection amplifier.
   c. defective low-voltage power supply.
   d. excessively high voltage.

Part B Supply the missing word(s) or number(s).

1. An additional function performed by a color television receiver but not by a monochrome receiver is the processing of the ______.
2. A television receiver must include an AM receiver to process the ______ signals.
3. Most modern television receivers are built with ______ components.
4. The tuner section changes the RF channel frequency to an ______ frequency.
5. The common IF amplifier passes the ______ and ______ signals.
6. The 4.5-MHz FM sound IF is produced in the ______ stage.
7. Whether the video signal is applied to the grid or cathode of the picture tube depends upon its ______.
8. The horizontal-deflection oscillator is stabilized by the ______ circuit.
9. A monochrome TV receiver may generate high voltage as high as ______ V.
10. An overloaded signal condition may be corrected by adjustment of the ______ control.
11. The horizontal-deflection oscillator is adjusted to 15 750 Hz by means of the ______ control.
12. Assembly-line wiring errors are eliminated by the use of ______ boards.
13. Replaceable major circuits are mounted on ______.
14. The three general categories of TV receiver test equipment are ______, ______, and ______.
15. If the symptom is no raster and no sound, a probable defective section is the ______.
16. If the symptom is a bright horizontal line on the screen, with normal sound, a probable defective stage is the ______ oscillator.

REVIEW ESSAY QUESTIONS

1. Explain what is meant by the term *hybrid TV receiver*.
2. List the major sections of a monochrome TV receiver.
3. What are the shape and frequency of the wave fed to the input of the horizontal-deflection amplifier? Why is this waveshape required?
4. Draw a simple block diagram of a monochrome TV receiver. Label all blocks.
5. Discuss briefly the action of the AGC circuit. Which stages does it control?
6. What are some advantages of modular construction?
7. Which channels are tuned by the UHF tuner? What is the frequency range?
8. In Figure 7.2, why is an FM sound detector necessary?
9. What is the approximate range of frequencies passed by the common IF amplifiers?
10. Name the circuits which make up the horizontal-deflection section. Briefly give the function of each stage.
11. What is an IC? What are some advantages of using ICs in TV receivers?
12. When receiving UHF channels, is the VHF tuner functioning? Explain your answer.
13. What is the approximate peak-to-peak video voltage applied to the picture tube? Why is the polarity of this signal important?
14. Discuss briefly the operation of the vertical integrator. Between what two stages is it connected?
15. Describe briefly how the picture tube high voltage is produced.

16. In your own words, tell how you would properly adjust an AGC control.

17. When is a VTVM rather than a VOM used?

18. Describe briefly the use of an RF sweep generator and oscilloscope.

19. Explain how observing visual and aural symptoms of a TV receiver aids in troubleshooting.

20. Complete the following table by giving the function in a TV receiver for each of the following frequencies, or frequency range.

FREQUENCY	FUNCTION
41.25 MHz	________
45.75 MHz	________
4.5 MHz	________
3.58 MHz	________
30 Hz to 3.25 MHz	________

8.1 Elements of color
8.2 Chromaticity chart
8.3 The NTSC color television system
8.4 I and Q signals
8.5 Color-signal components
8.6 The color subcarrier
8.7 The color burst signal
8.8 Derivation of the color subcarrier frequency
8.9 Composite colorplexed video wave forms
8.10 How compatibility is achieved
Summary

PRINCIPLES OF COLOR TELEVISION

INTRODUCTION

Emphasis throughout the preceding chapters was largely on the underlying principles of the transmission and reception of monochrome signals. In such a system, only black, white, and various shades of gray appear on the TV receiver screen. The result is similar in all respects to a black-and-white motion picture. Although the reproduced image is certainly far from being an exact duplicate of the full-color scene it represents, it imparts sufficient information to be entertaining. The public was long accustomed to monochrome images in the motion pictures and therefore accepted with little or no objection the same type of image in a television receiver.

The appeal of color television lies in its greater naturalness. We live in an environment that contains many shades of color, and to desire the same lifelike qualities in television is quite understandable. Color in an image heightens the contrast between elements, brightens the highlights, deepens the shadows, and appears to add a third dimension to a flat reproduction. More detail appears to be present in colored images containing fewer scanning lines than corresponding monochrome pictures.

In this chapter, we are going to study the elements of color and of the NTSC (National Television System Committee) color television system.

When you have completed the reading and work assignments for Chapter 8, you should be able to:

- Define the following terms: true color, primary color, chromaticity chart, illuminant C, saturation, hue, NTSC system, interleaving, multiplexing, Y signal, luminance signal, color subcarriers, chrominance, I signal, Q signal, color burst signal, and compatibility.

- Know the exact frequency of the color subcarrier and understand the reasons for its choice.
- Discuss the elements of color and tell how to use a chromaticity chart. Understand color mixing.
- Briefly explain the principles of the NTSC color TV system.
- Explain the differences between monochrome and color transmission and reception.
- Discuss the interleaving of color and monochrome signals. Know the equation for the Y signal.
- Describe the two components of the color signal.
- Discuss the two wave forms that should be examined when servicing a color TV receiver.
- Discuss the colors and bandwidths represented by the I and Q signals.
- Tell how a color transmission is received on a monochrome receiver and how a monochrome transmission is received (in monochrome) on a color receiver.
- Discuss how the color subcarrier is reinserted at a color TV receiver.
- Explain the function and composition of the color burst.
- Understand the phasor color diagrams.
- Understand why three colors are used for large areas and two colors for medium size areas.

8.1 ELEMENTS OF COLOR

Color, physicists tell us, is a property of light. If we pass sunlight through a prism, a variety of colors are produced. Sunlight contains all colors, but, owing to the limitations of the human eye and to the fact that the colors visible through a prism blend into one another, we can count only six fairly distinct colors (red, orange, yellow, green, blue, and violet). Upon closer inspection of this color distribution, numerous fine gradations can be distinguished, both between different colors and within any one color. For example, the orange that borders the red is a different shade than the orange that borders the yellow.

TRUE COLOR

It is a common experience to find that objects that look to be one color under an electric light may be a considerably different color when examined in sunlight. This is because the color of an object is a function of the wavelengths of light that the object does not absorb. Thus, if we shine white light on an object and none of it is absorbed, we see a white object. If under the same white light the object appears blue, however, then the object is absorbing all the other components of white light and reflecting blue.

To see the **true color** of an object, we must examine it under a light that contains all the wavelengths of visible light. Thus, a blue object appears much darker under an incandescent lamp than it does in sunlight. This is because the lamp has an excess of red light and a deficiency of blue. Since a blue object will reflect only blue light, it will reflect less light under an incandescent lamp and thus appear darker. In sunlight, all colors are present to the same extent and so the object assumes its true color.

With **transparent objects,** color is determined by which light is transmitted through the object. Thus, in a green piece of glass, green is permitted to pass through while the other colors are absorbed.

COMBINING COLORS

Anyone who has ever experimented with projector lamps has discovered that

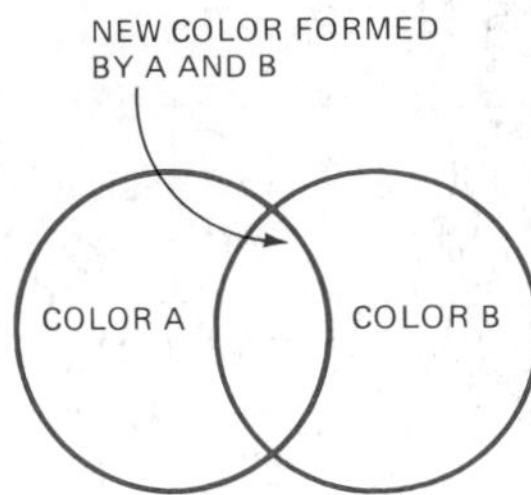

Figure 8.1 Two circles of light, A and B. Where they overlap, they form a third color different from either A or B.

when differently colored lights from several projectors are combined, the resultant color is different from that of any of the projected lights. Thus, for example, yellow can be formed by combining red and green light, and white can be produced by combining red, green, and blue. The color of the mixed light will appear as one color because the eye cannot distinguish the various components of the mixture that produced the color.

This method of **color formation** is illustrated in Figure 8.1. Two circles of colored light are projected onto a screen and positioned so that they overlap to some extent. In the overlap region, a new color is produced by the combination of Color A and Color B. Where the circles of light do not overlap, each light retains its original color. If a third circle of light is added, as shown in Figure 8.2, then additional colors are obtained. These colors are

• Color A (blue)	Color D (formed from A and B): cyan
• Color B (green)	Color E (formed from A and C): magenta
• Color C (red)	Color F (formed from B and C): yellow
	Color G (formed from A, B, and C): white

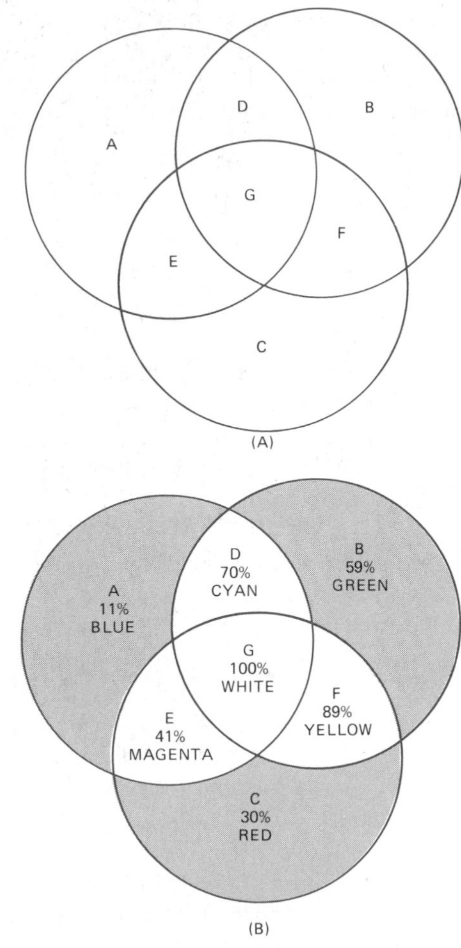

Figure 8.2A-B. The mixing of three colors—A, B, and C—results in four new ones: D, E, F, and G.

In the overlap areas, the eye is not able to distinguish each of the colors forming the mixture but instead sees only a new color.

Figure 8.2B shows that the brightness of the color in any overlap area is the sum of the brightnesses of the contributing colors. Yellow, for example, appears to be 89% as bright as the reference white because it is a combination of a green having a brightness of 59% and a red having a brightness of 30%.

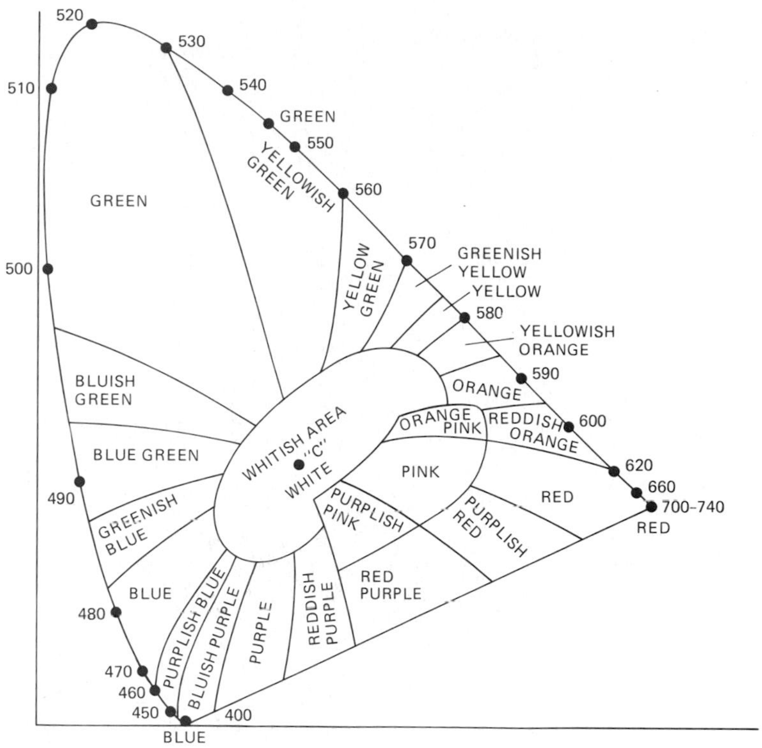

Figure 8.3 A chromaticity diagram. The numbers listed around the perimeter of the chart represent the wavelengths of the various colors in millimicrons.

THE PRIMARY COLORS

The number of different colors that can be formed with three colored lights depends upon the initial colors chosen. Red, blue, and green, when combined with each other in various proportions, produce a wider range of colors than any other combination of three colors. Four different colors produce an even wider range of colors. With the addition of more and more colors to the mixing scheme, the range of possible colors keeps increasing. Obviously, though, a line must be drawn somewhere, and therefore the use of three colors has become standard. The three colors chosen—red, green, and blue — are referred to in color television technology as the **primary colors.**

8.2 CHROMATICITY CHART

A convenient diagram for color mixing is the tongue-shaped (or horseshoe-shaped) chart shown in Figure 8.3. This is known as a **chromaticity chart.** The positions of the various colors from blue at one end to red at the other are indicated around the curve. Any point not actually on the tongue-shaped curved line but within the area enclosed by it represents a mixture of colors. Since white is such a mixture, it, too, lies in this area, specifically at Point C. This particular point was chosen at an international convention in England and is generally referred to as **illuminant C.** Actually, of course, there is no one white light; sunlight, skylight, and daylight are all forms of white light, and yet

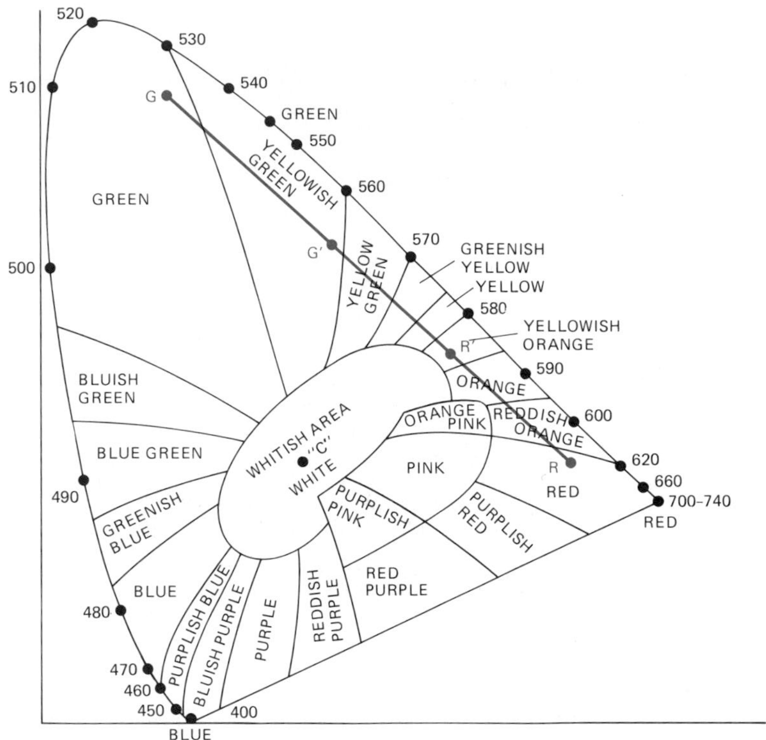

Figure 8.4 The line drawn between points R and G passes through all the colors that can be obtained by mixing these two shades of the red and green hues.

their components differ considerably. The standard for the white of a monochrome television receiver is represented by a point in the central region of the diagram, approximately at Point C.

If Figure 8.3 were reproduced in full color, you would see that the colors change gradually from point to point. The deepest, most intense colors are at the outer edge of the diagram. Here we find the really deep red, deep blue, and deep green shades very seldom seen in everyday life. More familiar are the lighter colors, appearing as we move toward the center. These are the pastels, such as pink, light green, and pale blue. Finally at the center come the whites, with Point C as the reference white or, for our purposes here, the "whitest" white. Actually this is a rather nebulous shade, entirely arbitrary in value and chosen simply for convenience.

COLOR MIXING

The chromaticity chart lends itself readily to **color mixing.** A straight line joining any two points inside the area defined by the curve indicates all the variations that can be obtained by combining, in varying amounts, the two colors represented by the two starting points. Thus, in Figure 8.4 the line connecting Points R and G represents various shades of red and green. If there is more red than green in the mixed color, the point representing it will be closer to R than

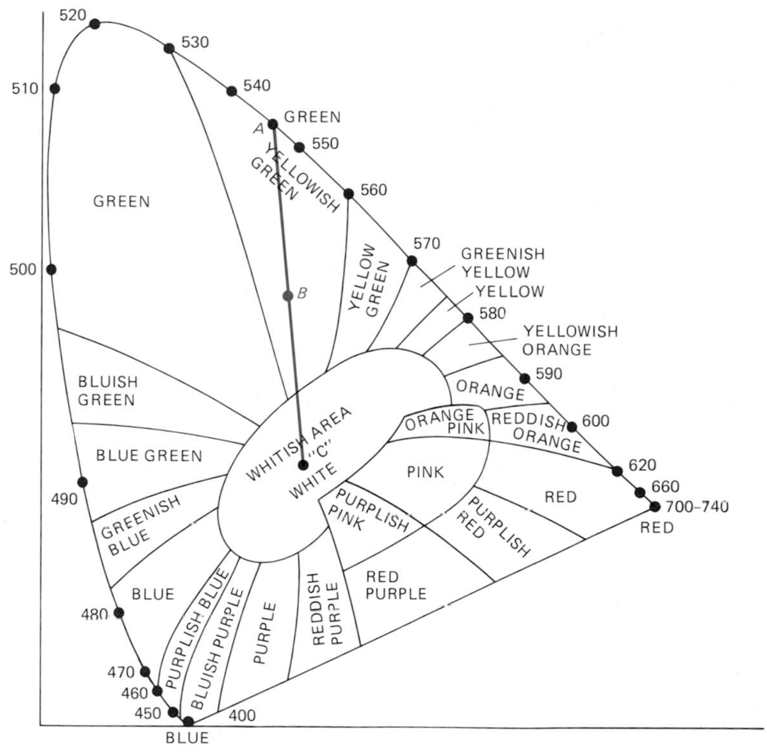

Figure 8.5 When moving from A to C, the green becomes less and less saturated, or lighter in intensity.

to G; Point R′ is such a point. If a greater percentage of green is used in the mixture, the point representing the new color will be closer to G than to R; Point G′ is such a point. This method of combining colors can be carried out for any two colors on the chart.

COLOR PURITY

It is possible to specify the **purity** of a color by its distance from Point C. Consider Point B in Figure 8.5. It is exactly halfway along the line between Point C (white) and Point A (green). Hence, Point B represents a mixture of 50% green and 50% white, and we say that the purity of this color is 50%. If the distance between Point C and Point B was 75% of the total distance from Point C to Point A, the purity of the color at Point B would be 75%. As Point B moves closer and closer to the outer curved line, the purity of the color increases until it becomes 100% at the outside curve, that is, Point A. As Point B moves closer to Point C, color purity decreases. At Point C, the purity is said to be zero.

SATURATION AND HUE

Saturation is frequently used in place of purity. Any color located on the tongue-shaped curved line of the chromaticity chart is said to be completely saturated. As we leave the curve and approach Point C, more and more white light is added to the color and it becomes less saturated, or more desaturated. At Point C, the saturation is zero. If a color is highly saturated, we say that it is a *deep* color, for example, deep red or deep green. If it contains a considerable amount of white light, we say it appears faded, as faded red or faded green.

The word *hue* is frequently heard in connection with saturation. **Hue** means color, such as red, green, and orange. The term is associated with wavelength. When we call a certain color green, or orange, or red, we are specifying its hue. Thus, hue refers to the color as it appears to us, whereas saturation tells us how deep the color is.

8.3 THE NTSC COLOR TELEVISION SYSTEM

We are now ready to study the NTSC color television system, the system officially adopted by the FCC in 1953. This system has been designed so that its signal occupies no more than 6 MHz (video and sound) and so that it carries not only the full monochrome signal but color information as well.

The question is; How is all this information compressed into a 6-MHz bandwidth? The answer is to be found in the nature of a television signal. It was shown as far back as 1929 that a monochrome 4-MHz video signal does not occupy every cycle of the 4 MHz assigned to it. Rather, this signal appears in the form of *clusters of energy* located at the harmonics of the 15 750-Hz line-scanning frequency. The monochrome signal energy is grouped around these points, with relatively wide gaps between the clusters (Figure 8.6). Since these empty spaces are not being used, they can be employed for the transmission of additional information. This is where the color information of the NTSC color television signal is placed (Figure 8.7). The practice of placing the information of one signal between the energy clusters of another signal is known as **interleaving** or **multiplexing** (see below).

Thus, a complete color signal consists of two components: a monochrome (Y) signal and a signal that carries color information. Let us examine each component separately.

THE MONOCHROME SIGNAL

The monochrome portion of the total color signal is equivalent in all respects to the signal from a conventional monochrome transmitter. It is formed by combining the red, green, and blue signals from their respective color cameras in the following proportions (Figure 8.2):

$$Y = 0.59G + 0.30R + 0.11B$$

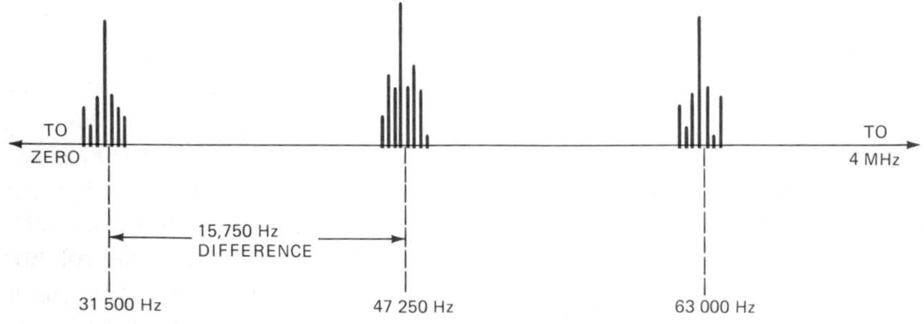

Figure 8.6 Part of the spectrum distribution of a monochrome signal.

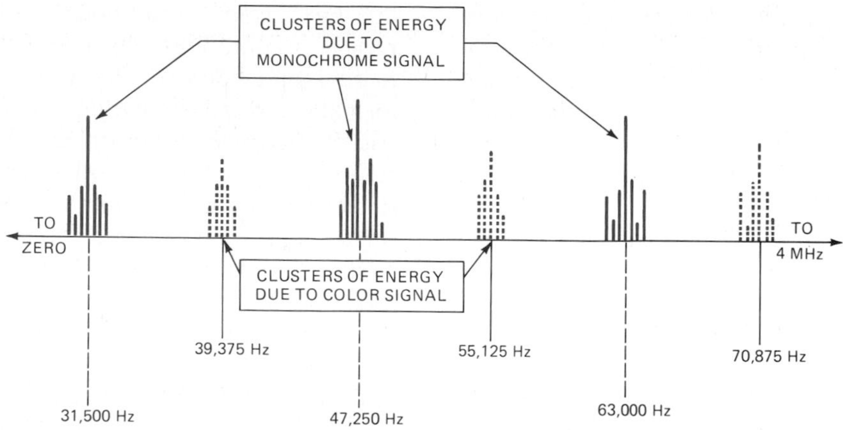

Figure 8.7 The information of the color signal is inserted in gaps between the clusters of energy of the monochrome signal.

where

- Y = monochrome signal[1]
- G = green signal
- R = red signal
- B = blue signal

This particular proportion was chosen because it closely follows the color sensitivity of the human eye. That is, if you combine equal amounts of green light, red light, and blue light on a viewing screen, the color you see will be white. However, if you then look at each light separately, the green will appear to be twice as bright as the red and from six to ten times as bright as the blue. This is because the eye is more sensitive to green than to red and more sensitive to red than to blue. It is because of this that the proportions given above were chosen.

Thus, the monochrome signal is composed of 59% green signal (that is, 59% of the output of the green camera), 30% red signal, and 11% blue signal. It contains frequencies from 0 to 4 MHz.

Other names for this monochrome signal are **luminance signal** and **brightness signal.** These terms were chosen because they clearly indicate the action of this signal. Every monochrome video signal contains only the variations in amplitude of the picture signal, and these amplitude variations are what produce the changes in light intensity on the picture tube screen.

THE COLOR SIGNAL

The second component of the color television signal is the color signal itself. This, as we just saw, is interleaved with the monochrome signal. To determine what information this portion of the total signal must carry, let us first see how the eye reacts to color. The color signal is also known as the **chrominance** or **chroma signal.**

Color Versus Area Many people have investigated the color-discerning characteristics of the human eye. Briefly, here is what they have found. The typical human eye sees a full color range only when the area or object being viewed is relatively large. As the size of this area or object decreases, it becomes more difficult for the eye to distinguish between colors. Thus, the eye requires

[1]The use of the letter Y to denote the monochrome portion of a color signal is common practice and should become familiar to you.

three primary colors for viewing large objects or areas, but it can get along very well with only two when viewing a medium-size area. That is, two colors in different combinations will provide the limited range of colors that the eye can see in medium-size areas.

When the area being viewed is very small, all that the eye can discern are changes in brightness. Colors cannot be distinguished from gray, and, in effect, the eye is color-blind.

These properties of the eye are put to use in the NTSC color system. First, only large and medium-size areas are colored; the fine detail is rendered in monochrome. Second, as we shall see later, the color information is regulated according to bandwidth. That is, the large objects receive more of the green, red, and blue than the medium size objects.

Interleaving The color signal takes the form of a subcarrier and an associated set of sidebands. The subcarrier frequency is approximately 3.58 MHz. This figure is the product (approximately) of 7 875 Hz multiplied by 455 (explained later). The 7 875 is one half of 15 750, and if we use an odd multiple (1, 3, 5, etc.) of 7 875 as a carrier, then the frequency will fall midway between the harmonics of the 15 750-Hz line-scanning frequency. If we used even multiples of 7 875, we would end up with 15 750 Hz or one of its harmonics, and this would place the color signal at the points (throughout the band) already occupied by the monochrome signal (refer back to Figure 8.7). By taking an odd multiple of 7 875, we cause the second signal to fall between (**interleave** with) the clusters of energy produced by the monochrome signal, and the two do not interfere with each other.

The Modulating Signal Now that we have a color carrier (or subcarrier, as it is known), the next step is to modulate it with the proper signal. This will enable the receiver to develop a color picture. Ordinarily, the information required would consist of R, G, and B[1] because these are the three primary colors, from which all other colors are derived. This means modulating the color subcarrier with three different voltages (signals). Actually, however, we can do the same job using only two signals if we resort to the following modification.

Take the R, G, and B signals and combine each with a portion of the monochrome signal that has been inverted 180°. This inversion is accomplished in the following way. First we pass a portion of the brightness signal (Y signal) through a low-pass filter (Figure 8.8). This permits only the lower-frequency components (0 to 1.5 MHz) to get through, which is satisfactory since the color signals are concerned only with these lower frequencies. Then the brightness signal is passed through an amplifier, where it is inverted. This gives us the desired −Y. This is then added to each of the three color signals to produce a G−Y, an R−Y, and a B−Y signal. These are called color-difference signals.

At the receiver, the original R, G, and B can be obtained by adding noninverted Y to G−Y to get G, to R−Y to get R, and to B−Y to get B.

B−Y, R−Y Thus far, we have only exchanged R, G, and B for R−Y, G−Y, and B−Y. However, once this is done, it turns out that, instead of requiring three color-difference signals, all we really need are two, say R−Y and B−Y. This is because the G information is already present in the Y, or brightness, signal since the latter contains voltages from all three colors ($Y = 0.59G + 0.30R + 0.11B$). Hence, if we send only R−Y and B−Y in the color signal to the re-

[1]From now on, we shall use R, G, and B instead of the words *red, green,* and *blue.*

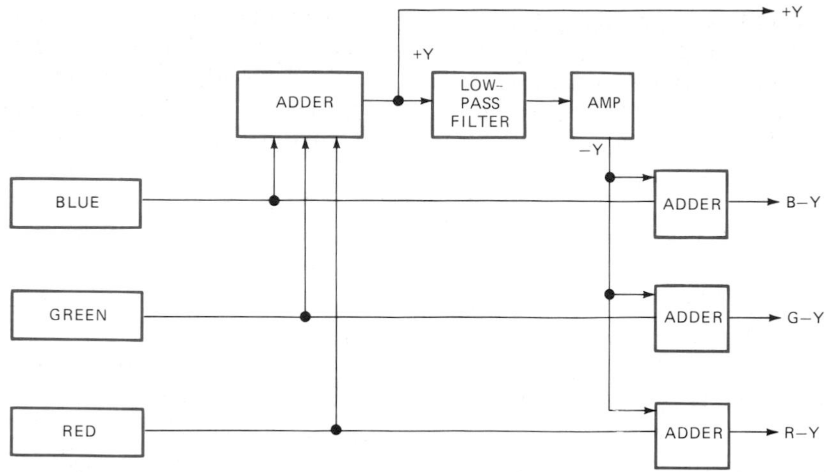

Figure 8.8 Block diagram illustrating how color signals minus brightness signals are formed.

ceiver, we can use these to obtain the G − Y information that we need.

We now have only two pieces of color information to send. Somehow the 3.58-MHz color subcarrier frequency must be modulated by R − Y and B − Y voltages without the two conflicting with each other. The best solution to this problem, designers have found, is to take the B − Y and R − Y signals and apply each to a separate modulator. At the same time, 3.58-MHz carriers are also applied to the two modulators. These two 3.58-MHz carriers are 90° out of phase with each other. After the carriers are amplitude-modulated, they are combined to form a resultant carrier. This is best illustrated by means of phasors.

Phase and Amplitude In Figure 8.9A, the B − Y phasor represents the B − Y modulated carrier and the R − Y phasor represents the carrier modulated by the R − Y voltage. When these voltages, or signals, are combined, a **resultant** is formed. If the R − Y and B − Y signals are equally strong, the resultant will occupy the position shown in Figure 8.9B. If the B − Y signal is dominant, the resultant will be closer to it (Figure 8.9C). If the R − Y signal is the stronger, the resultant phasor will shift toward it (Figure 8.9D).

Thus, the **phase angle** of the resultant will be governed by the color, or hue, of the picture, whereas the **amplitude,** or length, of the resultant phasor will determine the saturation of the colors. This particular fact is of great importance in the receiver. If we change the phase angle of the resultant, with respect to B − Y or R − Y, then the colors reproduced on the screen will be incorrect. Hence, circuit designs incorporate a special phasing control that enables us to compensate for any phase shift that may occur. The position of this control in the circuit will be discussed in the next chapter.

Note that the B − Y and R − Y signals amplitude-modulate their separate carriers prior to the addition, and that therefore each modulated signal possesses a 3.58-MHz carrier and a series of sidebands (like every AM signal). When the resultant is formed, the sidebands are brought along with it.

The Total Color Signal If we pause and reconstruct our total color signal, here is what we find. First, there is the Y, or monochrome, signal, which extends over the en-

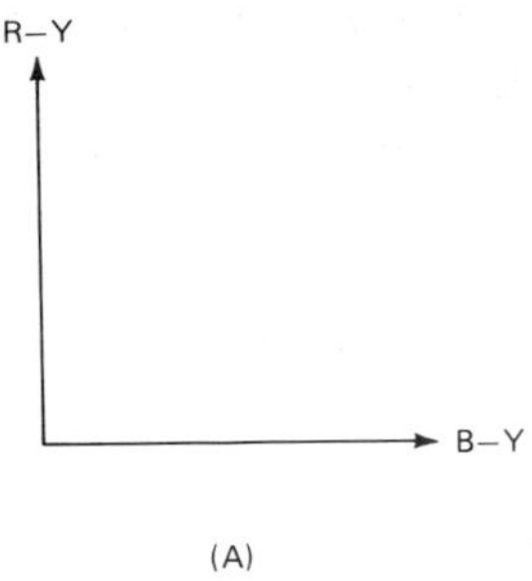

(A)

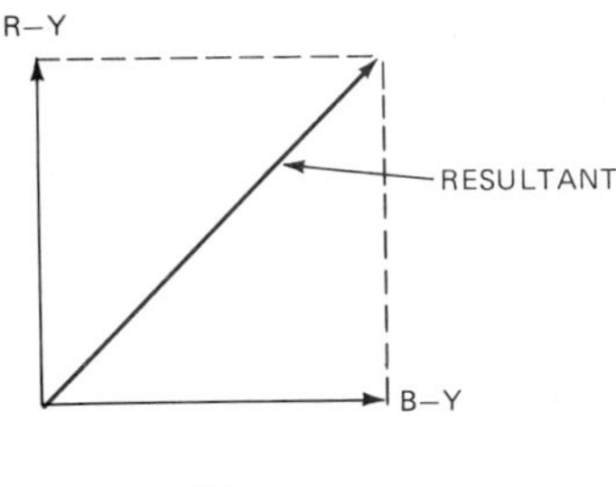

(B)

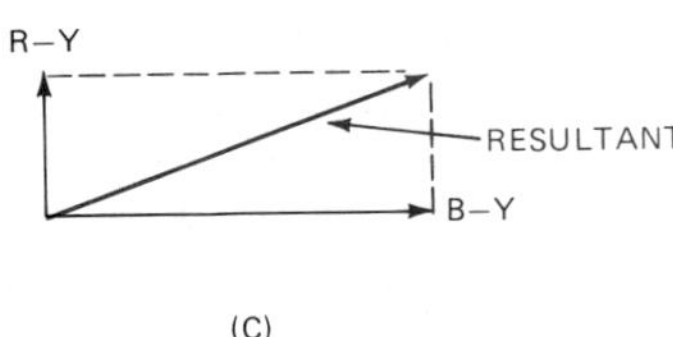

(C)

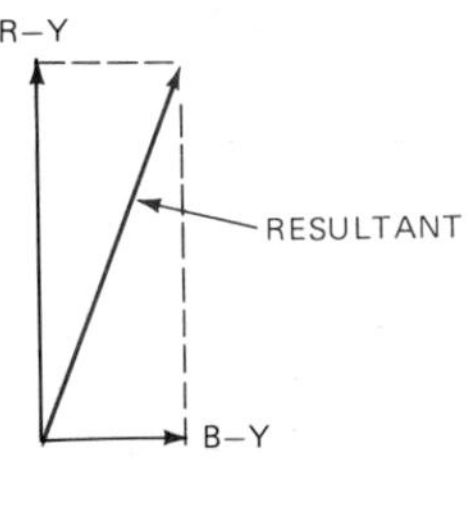

(D)

Figure 8.9A-D. The angular position and amplitude of the resultant carrier for various amplitudes of B−Y and R−Y.

A. The R−Y and B−Y vectors.
B. The resultant when R−Y and B−Y are equal.
C. The resultant when B−Y is stronger than R−Y.
D. The resultant when R−Y is stronger than B−Y.

tire video frequency range from 0 to 4.0 MHz. Second, there is a color subcarrier, with a frequency of 3.58 MHz. This carrier is modulated by the R−Y and B−Y signals, and the modulation information is contained in a series of sidebands that extend above and below 3.58 MHz. Just how far above and below depends on the band of frequencies contained in the R−Y and B−Y modulating voltages. It has been determined that the eye is satisfied by the color image produced if we include color information requiring frequencies only up to 1.5 MHz. The portion of the image from 1.5 to 4.0 MHz is transmitted in black and white. Hence the sideband frequencies of the color modulating voltages (R−Y and B−Y) need extend only from 0 to 1.5 MHz.

We can even modify this set of conditions somewhat because the *three* primary colors are required only for *large* objects or areas, say, those produced by video frequencies up to 0.5 MHz. For *medium-size* objects, say, those produced by video frequencies from 0.5 to 1.5 MHz, only two primary colors are needed. To take advantage of this situation, we need only two color signals: one that has a bandpass only up to 0.5 MHz and one that has a bandpass from 0 to 1.5 MHz. The next problem, then, is to determine the composition of these two signals.

Composition of the Two Color Signals To understand the answer to this problem, let us return to the phasor diagram (Figure 8.9) which shows the R−Y and B−Y signals. This diagram is redrawn in Figure 8.10A, and we have added the equivalent equation for Y.

$$Y = 0.59G + 0.30R + 0.11B$$

For R−Y, then, we have

$$R-Y = R - 0.59G - 0.30R - 0.11B$$

or

$$R-Y = 0.70R - 0.59G - 0.11B$$

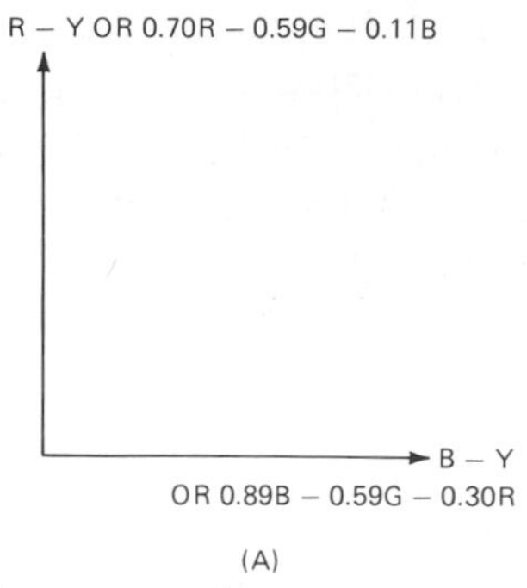

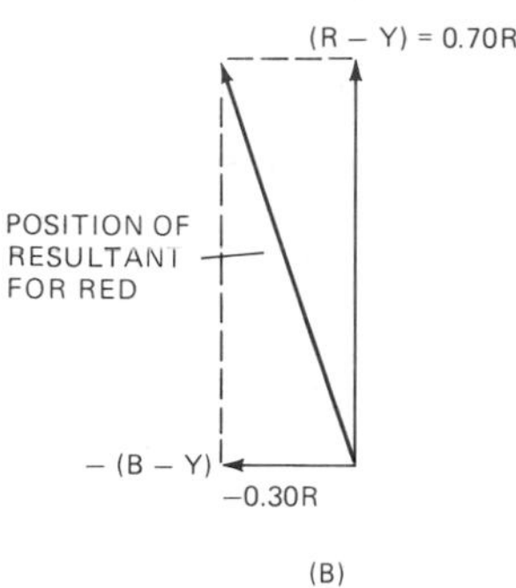

Figure 8.10A-B. How color determines the position of a resultant vector. **A.** Equations showing the composition of B−Y and R−Y in terms of R, G, and B. **B.** Position of the signal vector when a scene containing only red is being scanned.

For B−Y we obtain

$$B-Y = B - 0.59G - 0.30R - 0.11B$$

or

$$B-Y = 0.89B - 0.59G - 0.30R$$

This means that the R−Y and B−Y phasors contain R, G, and B voltages in these proportions.

Now, let us suppose that the color camera is scanning a scene containing only red. Then, no green or blue voltages will be present and the R−Y signal becomes simply 0.70R and the B−Y signal is reduced to −0.30R. This set of conditions is shown in Figure 8.10B, along with the position of the resultant phasor. In other words, this is the position of the resultant phasor when only red signal is being sent.

By following the same process, we can obtain the position of the resultant phasor when only green is being sent, or blue, or any other color formed by mixing the three primary colors in any combination. A number of colors are shown in Figure 8.11. We see how the phase of the color subcarrier changes as the color to be transmitted varies. To repeat, the phase angle of the resultant phasor is governed by the hue of the picture, whereas the amplitude (length of the phasor) determines the saturation of the colors.

8.4 I AND Q SIGNALS

The designers of the NTSC system found that, although they could use R−Y and B−Y for the color signals, better results are obtained with two other signals situated not far from the R−Y and B−Y signals. These two other signals were labeled I and Q (meaning *in phase* and *quadrature*, respectively). Their position with respect to R−Y and B−Y is shown in Figure 8.12.

Thus, where before we had R−Y and

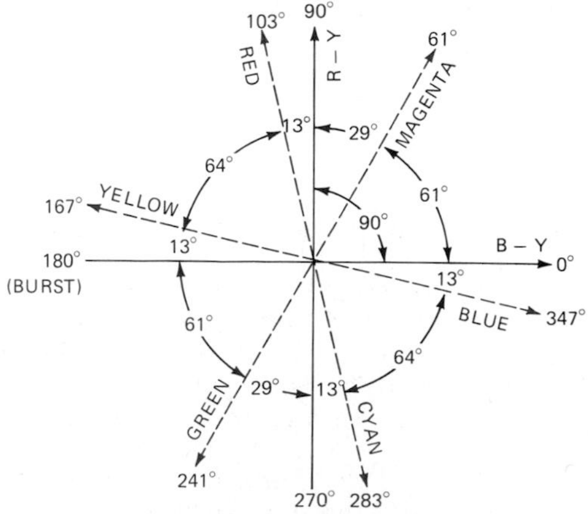

Figure 8.11 The phase of the color subcarrier depends upon the color to be transmitted.

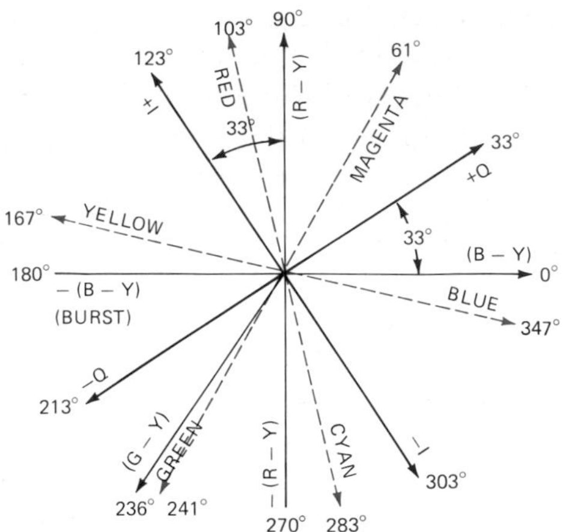

Figure 8.12 The positions of the I and Q signals with respect to R−Y and B−Y. Note that burst phase is −(B−Y), or yellow-green.

B−Y signals modulating the 3.58-MHz color subcarrier, we now substitute **I and Q signals.** The Q signal has sidebands up to 0.5 MHz, and the I signal is permitted to have sidebands up to 1.5 MHz.

What do we gain from this? For all color-signal frequencies up to 0.5 MHz, both I and Q are active. Since they are 90° apart, as were R−Y and B−Y, they will act just the way R−Y and B−Y acted. That is, they will produce, in combination with each other, all of the colors shown in Figure 8.12. Hence, whether we use I and Q or R−Y and B−Y as our modulating voltages for color-signal frequencies up to 0.5 MHz, we obtain precisely the same results.

Consider, however, the situation for color-signal frequencies from 0.5 MHz to 1.5 MHz. The Q signal drops out, and only the I signal produces color on the picture-tube screen. From Figure 8.12, we see that positive values of the I signal will produce colors between yellow and red (actually, a reddish orange). Negative values of I will produce colors between blue and cyan (in general, in the bluish-green range). Hence, when only the I signal is active, the colors produced on the screen will run the gamut from reddish orange to bluish green.

Why do we want this arrangement? Recall that, for medium-size objects (say, those produced by video signals from 0.5 MHz to 1.5 MHz), the sensitivity of the eye for color is reduced. Actually, for medium-size objects, the eye is sensitive principally to the bluish greens and the reddish oranges. The NTSC signal (via its I component) takes advantage of this fact by producing only bluish greens and reddish oranges for medium-size objects.

8.5 COLOR-SIGNAL COMPONENTS

We are now in a position to consider the color signal in all its aspects:

1. There is a monochrome signal with components that extend from 0 to 4.2 MHz. This is the Y signal.
2. The color subcarrier frequency is set at 3.58 MHz (actually 3.579 545 MHz).
3. This color subcarrier is modulated by two color signals called the I and Q signals, with $I = 0.60R - 0.28G - 0.32B$ and $Q = 0.21R - 0.52G + 0.31B$.
4. The Q signal has color frequencies that extend from 0 to 0.5 MHz. This means that the upper Q sideband extends from 3.58 MHz to 3.58 + 0.5, or 4.08, MHz. The lower Q sideband goes from 3.58 MHz down to 3.58 − 0.5, or 3.08, MHz.
5. The I signal has color frequencies that extend from 0 to 1.5 MHz. When this modulates the color subcarrier, upper and lower sidebands are formed. The lower sideband extends from 3.58 MHz down to 3.58 − 1.5, or 2.08 MHz. If there were a full upper sideband, it would extend all the way up

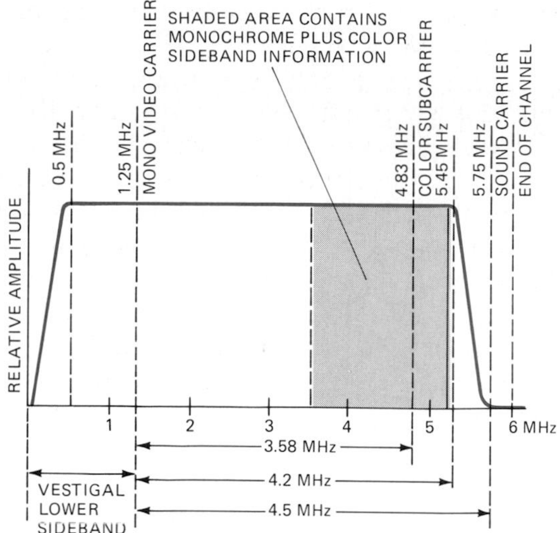

Figure 8.13 The distribution of the full color signal within its allotted band.

to 3.58 + 1.5, or 5.08 MHz (upper video limit is 4.2 MHz). This would prevent the use of a 6.0-MHz overall band for the television signal (video and sound). To avoid this spilling over beyond the limits of the already established channels, the upper sideband of the I signal is limited to about 0.6 MHz. This brings the upper sideband of the I signal to 4.18 MHz. The video bandpass then ends rather sharply just before 4.5 MHz (Figure 8.13).

8.6 THE COLOR SUBCARRIER

There is one more important fact about the makeup of a color television signal, and this concerns the color subcarrier. We know that the 3.58-MHz carrier is modulated by the I and Q color signals. In conventional modulation, (as in standard AM broadcasting) both the carrier and the sidebands are present when the signal is finally sent out over the air. The modulation is contained in the sidebands and actually is all that is necessary for signal transmission. However, the carrier is sent along because it is required in the receiver in order to reverse the modulation process and recreate the original modulating voltages.

In the NTSC system, the color subcarrier is not sent along with its sidebands (after the latter have been formed). Instead, it is suppressed by means of a balanced modulator. This particular practice is followed for two reasons. First, by suppressing the color subcarrier, (at 3.58 MHz) we reduce the formation of a 0.92-MHz beat note between it and the 4.5-MHz sound IF carrier. This 0.92-MHz beat note would appear as a series of interference lines on the picture tube. It is true that the color sidebands are present and that they can (and do) beat with the 4.5-MHz sound IF carrier to produce similar low-frequency beat notes. In any signal, however, the carrier usually contains far more energy than any of its sidebands. Hence, when we suppress the carrier, we are in effect suppressing the chief source of this interference. Other interference produced by some of the stronger sidebands near 3.58 MHz can be easily dealt with by using traps in the IF system. (This will be seen when we examine the circuitry of a receiver.)

The second reason for using the suppressed-carrier method is that it leads to the automatic removal of the entire color signal when the televised scene is to be sent wholly as a monochrome signal. When this occurs, I and Q become zero and, since the balanced modulator suppresses the carrier, no color signal is developed.

With these advantages of carrier suppression noted above comes one disadvantage. When the color sidebands reach the color section of the receiver, a carrier must be reinserted to permit detection. Offhand,

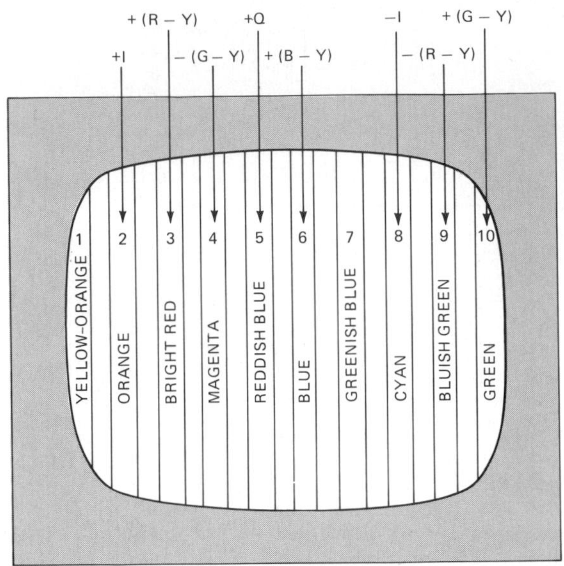

Figure 8.14 The bar pattern developed by a color-bar generator. See Figure 8.12 for corresponding phase angles.

one might suppose that an oscillator operating at 3.58 MHz is needed. This is one requirement. A second and vitally important consideration is the phase of this reinserted carrier. Remember that, at the transmitter, I and Q were 90° out of phase as they were introduced into the modulator. If this phase difference is not maintained in the reinserted carrier, the final colors will not have the proper hue. Figure 8.14 shows the phase and color relationship as seen on a picture tube screen. Compare this to Figure 8.12.

8.7 THE COLOR BURST SIGNAL

The proper detection of the color subcarrier sidebands is not possible if the correct phase relationship is not maintained between the 3.579 545-MHz subcarrier oscillators located at the transmitter and receiver. Since the 3.579 545-MHz signal is suppressed at the transmitter, a short, separate **burst** sample of the 3.579 545-MHz transmitter oscillator is sent (Figure 8.15.)

To provide information about the frequency and phase of the missing color subcarrier, a **color burst** is sent along with the signal. This burst follows each horizontal pulse and is located on the back porch of each blanking pedestal. It contains a minimum of eight cycles of the subcarrier and is in phase with the color subcarrier used at the transmitter. In the receiver, this burst is used to lock in the frequency and phase of a 3.58-MHz oscillator. Thus, we are assured at all times that the reinserted carrier will do its job correctly when it recombines with the color sidebands.

The color burst does not interfere with horizontal synchronization because it is lower in amplitude and follows the horizontal synchronizing pulse.

8.8 DERIVATION OF THE COLOR SUBCARRIER FREQUENCY

It was mentioned in preceding pages that the color subcarrier frequency is approximately 3.58 MHz. This frequency was se-

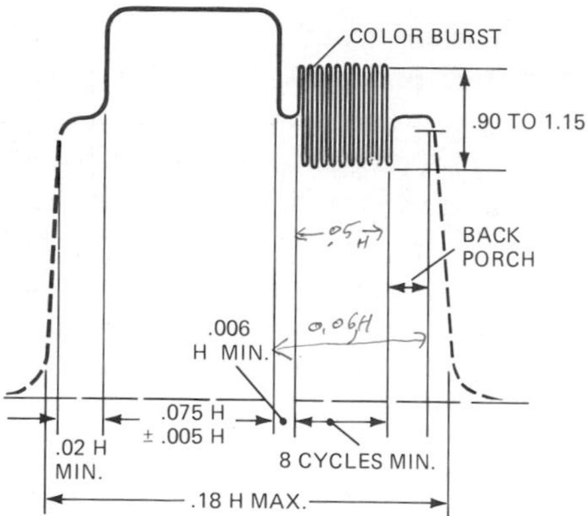

Figure 8.15 The position of the color burst for the subcarrier oscillator sync: on the back porch of a horizontal-sync pulse.

lected so that the chrominance (color) video information will be nearly zero when a color program is being received on a monochrome receiver. The color subcarrier frequency was chosen to be an odd multiple of one half the horizontal-scan frequency. More detail will follow on this choice.

COLOR-SIGNAL CANCELLATION

Color signals do not follow the brightness (monochrome) signal and, if not minimized, would appear on a monochrome screen as a spurious, unrelated dot pattern. Having a color subcarrier frequency that is an odd multiple of one half the horizontal-scan frequency causes the color video information on any scan line to be 180° out of phase with the next scan of that line on the next frame.

The color signal, which varies about the brightness signal and modulates each horizontal-scan line, is inverted 180° each frame. This prevents the color signal from producing an objectionable black-and-white interference pattern on the screen. The color video signal is inverted 180° because the number of cycles the signal passes through each frame is equal to a whole number of cycles plus one half of a cycle.

Figure 8.16A shows a portion of one horizontal-scan line for a monochrome video signal. A portion of one modulated horizontal-scan line is shown in Figure 8.16B. One frame (525 lines) later, the modulation is inverted 180°, as in Figure 8.16C. Since the two signals are approximately equal in amplitude but opposite in polarity, they cancel to nearly zero (Figure 8.16D). The apparent cancellation appears to be a continuous process because of the human eye's **persistence of vision.** The eye retains the brightness signal of the horizontal line from frame to frame.

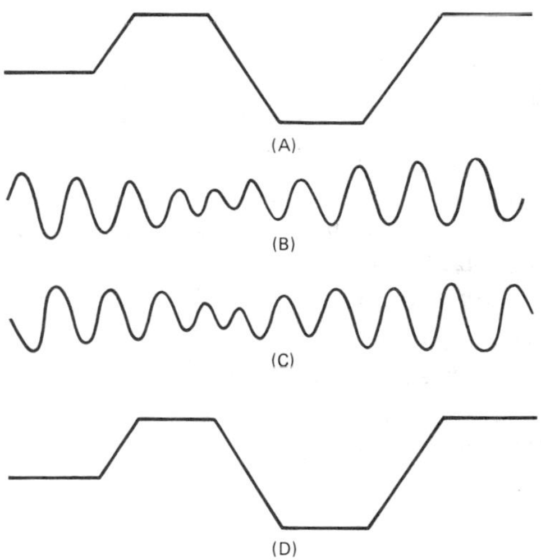

Figure 8.16A-D. Chroma interference cancellation in frequency interleaving.

A. Black and white video during one scan line.
B. Chroma video signal during one scan line.
C. Same scan line, chroma video signal, exactly one frame later. (Note that signal is 180° out of phase with part B.)
D. Visual addition of waveforms A, B, and C. (Note elimination of chroma video signals.)

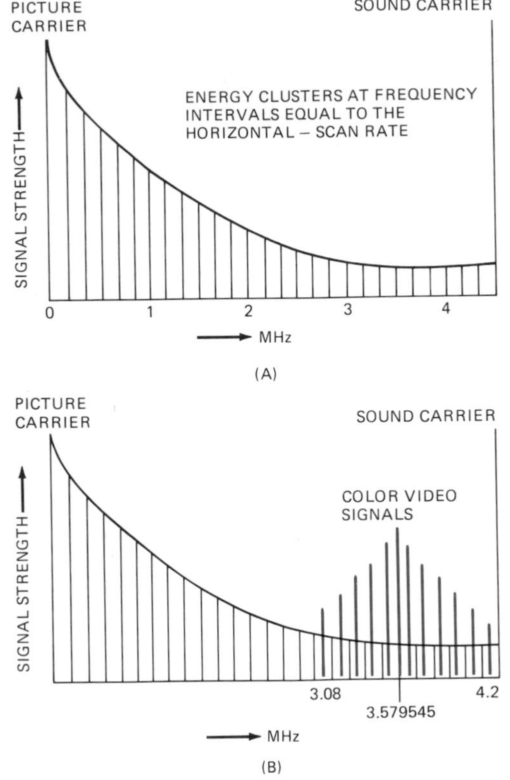

Figure 8.17 Black and white video frequency energy distribution.

CHOICE OF SUBCARRIER FREQUENCY

It is highly desirable to have the color subcarrier frequency as high above the picture carrier frequency as possible in order to minimize interference with the monochrome video information, Figure 8.17A. The average energy level of the monochrome video signal falls very rapidly with increasing video frequencies (Figure 8.17A). By placing the color subcarrier as high as possible, as in Figure 8.17B, the difference in energy levels minimizes possible interference. Practical considerations set the upper limit at 3.6 MHz.

There is a very objectionable 0.92-MHz signal generated from the beat between the sound carrier (4.5 MHz) and the color subcarrier (4.5 − 3.58 = 0.92). This beat is much less objectionable if the difference between the sound carrier frequency and the video carrier frequency is some multiple of the horizontal-scan frequency. In standard monochrome television, the 285th and 286th harmonics of 15 750 Hz are at 4.488 75 MHz and 4.504 50 MHz.

The line frequency whose 286th harmonic is 4.5 MHz is 15 743.26 Hz [F horizontal = (4.5 MHz/286) = 15 734.26 Hz]. This frequency is within the deviation limit set by the NTSC monochrome standards. With the horizontal-scan frequency changed, the color subcarrier must be chosen to interleave. It must be an odd multiple of one half the horizontal-scan frequency. With 3.6 MHz as an upper limit, it was found that the 455th harmonic of one half the horizontal-scan frequency becomes

$$\text{F color subcarrier} = 455 \times \frac{15\,734.26}{2}$$
$$= 3.579\,545 \text{ MHz}$$

Since there are 525 lines and a 2:1 interlace is used, the new vertical-scan frequency becomes

$$\text{F vertical} = \frac{2}{525} \times \frac{15\,734.26}{1} = 59.94 \text{ Hz}$$

8.9 COMPOSITE COLORPLEXED VIDEO WAVE FORMS

The generation of a complete video signal, monochrome and color signals combined, is shown in Figure 8.18. One horizontal-scan line crosses the test pattern in Figure 8.18A to produce the video signals shown in Parts B through I. The pattern consists of the three primary colors (red, green, and blue), the three two-color combinations of the primary colors (cyan, yellow, and magenta),

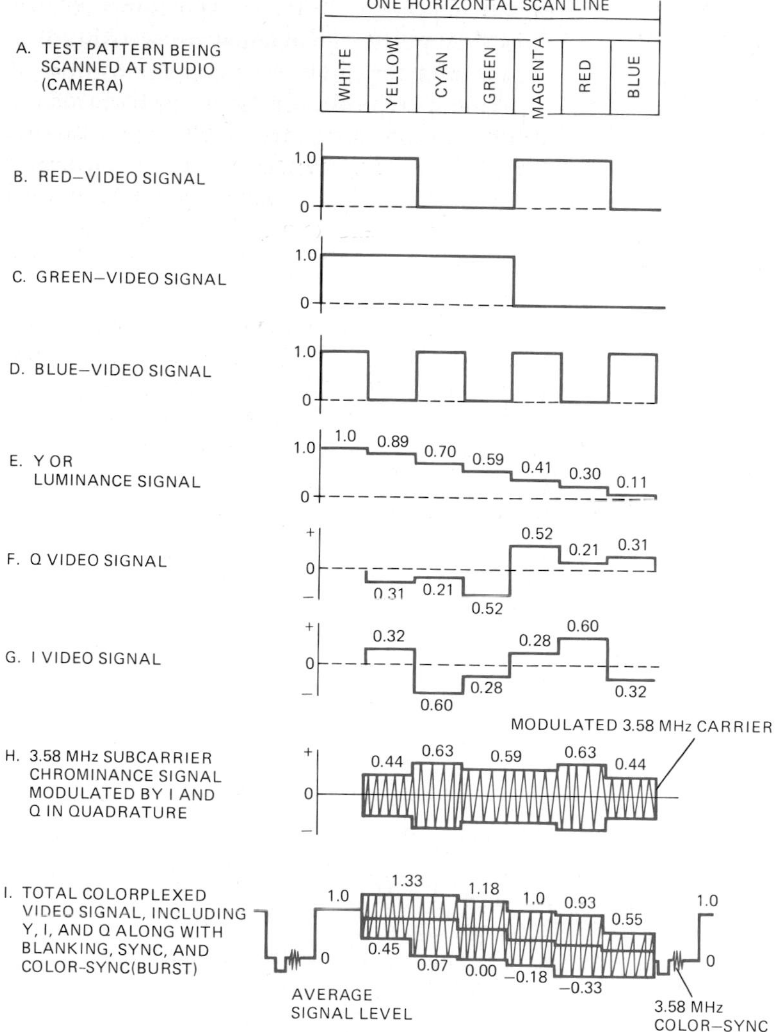

Figure 8.18 Colorplexed composite video signal.

and the one three-color combination of the three primary colors (white).

The wave forms shown in Parts B, C, and D give the primary-color wave forms with an amplitude of 1, that is, 100% of fully saturated color. The luminance (that is, the Y signal) in Part E shows the relative brightness for each color bar in the pattern (compare this with Figure 8.2). The I and Q signals are illustrated in Parts F and G. Note that the I and Q signals can have positive or negative values. (Figure 8.19 is alternate representation of Figure 8.18. The Y, I, and Q video signals are formed from the basic R, G, and B camera outputs.)

The I and Q video signals are passed to the modulators to generate the 3.58-MHz sidebands, as shown in Figure 8.18H. The amplitude of the 3.58-MHz signal in Part H is formed by the vector addition of the quadrature I and Q signals in quadrature. For example, the blue color bar I value is −0.32

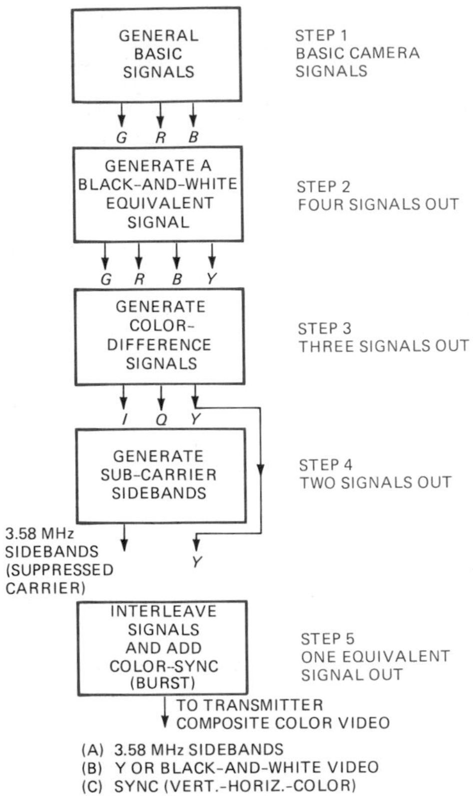

Figure 8.19 Summary of color processing prior to the actual transmission.

and Q is 0.31, and so the vector addition in Part H for blue is $(-0.32)^2 + (0.31)^2 = 0.1985 = (0.44)^2$. The complementary colors, such as yellow and blue, have the same peak amplitude but are 180° out of phase in the 3.58-MHz subcarrier signals. Refer to Figure 8.12 for verification and comparison.

At the transmitter, the 3.58-MHz chrominance information is combined with the luminance signal, and with the color synchronizing, and horizontal and vertical synchronizing and blanking pulses to produce the composite **colorplexed signal** shown in Figure 8.18I. Adding the Y luminance signal in Part E causes a shift of the 3.58-MHz chrominance information from a zero reference to a new level equal to the luminance value of the corresponding colors. This average DC level must be recovered at the receiver for the correct color renditions.

The preceding sections introduced a number of diagrams and wave forms. In servicing a TV receiver, however, only certain wave forms are of practical value.

IMPORTANT WAVE FORMS

Two wave forms of particular interest can be easily viewed on a wide-band service oscilloscope. The first is transmitted from the TV studio whenever color is being broadcast. The color sidebands vary and present a problem because they are not repetitious. A service oscilloscope must have a repeating wave form to give a stationary pattern. The horizontal synchronizing pulses with the color synchronizing burst on the blanking pedestal back porch are repetitious (Figure 8.20). Note that the oscilloscope sweep is set to view two horizontal synchronizing pulses. Close observation will show several distinct signals. The blanking pulses, horizontal synchronizing pulses, color burst, and video modulation (Y and chroma) are all illustrated. These signals can be traced through the Y or video section of color TV receivers up to the color picture tube.

Another important wave form can be observed if a color-bar generator is on hand.

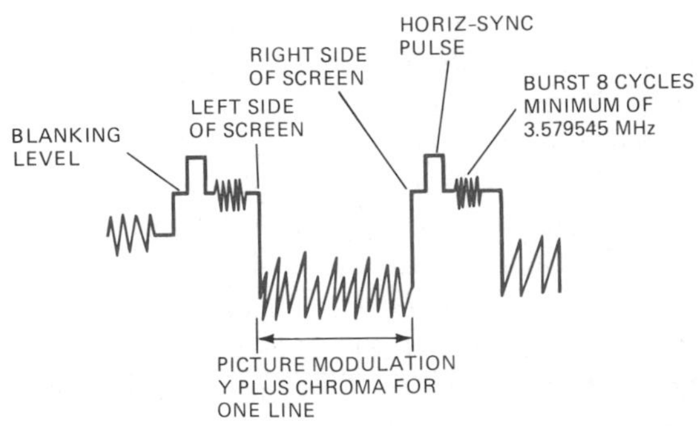

Figure 8.20 Composite video signal waveform.

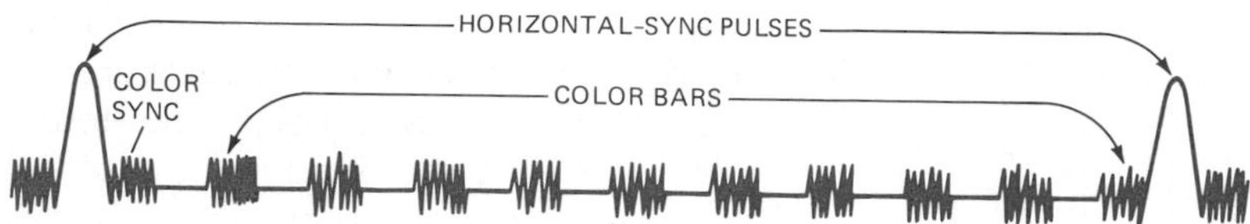

Figure 8.21 An example of a color-bar generator signal.

A group of color burst signals are produced in the color-bar generator and will produce color bars on a receiver. Figure 8.21 is a typical example of the output from such a generator. Each burst block contains roughly eight cycles of subcarrier signal. Only the burst block immediately following the horizontal synchronizing pulse is used by the receiver for color synchronization. The remaining ten burst blocks (10) generate a series of color bars, as shown in Figure 8.14. An oscilloscope can be used to trace this wave form through both the Y and the color sections of color TV receivers. Many color-bar generators have peculiarly shaped horizontal synchronizing pulses in the output signal, but the burst blocks are the key signals to observe. Use the manufacturer's service data shown on each chassis for the gain in each section of the receiver and compare the data with your particular wave forms for future reference.

8.10 HOW COMPATIBILITY IS ACHIEVED

To properly reproduce color broadcasts in monochrome, on monochrome receivers, the color broadcast signals must be specially constituted to produce a Y signal. This signal is used to produce the monochrome picture. The process of transmitting the color broadcast signals so that they are usable on a monochrome receiver is called **compatibility.** (Technically, compatibility also includes the fact that a color broadcast includes the normal monochrome synchronizing and blanking signals.)

Compatibility is illustrated in Figure 8.22. Note that the red, green, and blue outputs of the color camera are 30%, 59%, and 11%, respectively, to produce the Y signal. (This composition of the Y signal is explained in Section 8.3.) The additional color signals that are required to produce color pictures on a color set are accepted by the tuner of a monochrome set. However, monochrome sets will not pass the 3.58-MHz color subcarrier and its adjacent sidebands through the video amplifier. In addition, the monochrome set does not possess the various color-processing circuits of a color set, and the special color signals simply do not affect the monochrome set.

MONOCHROME SIGNAL IN COLOR TV RECEIVERS

Compatibility means not only that a color transmission can be reproduced in black and white on a monochrome set but also that a monochrome broadcast can be reproduced in monochrome on a color set. This is illustrated in Figure 8.23. For the color picture tube to reproduce either a correct color picture or a correct monochrome picture, certain static adjustments must be made to the picture tube circuitry. Among these are **screen adjustments.** The color picture tube includes a red, green, and blue screen grid (Chapter 14). The screen voltages are adjusted (generally without a picture) so that the resulting light background on the color picture tube is white. Since black is simply the absence of any light, the resulting picture on a color tube from a monochrome

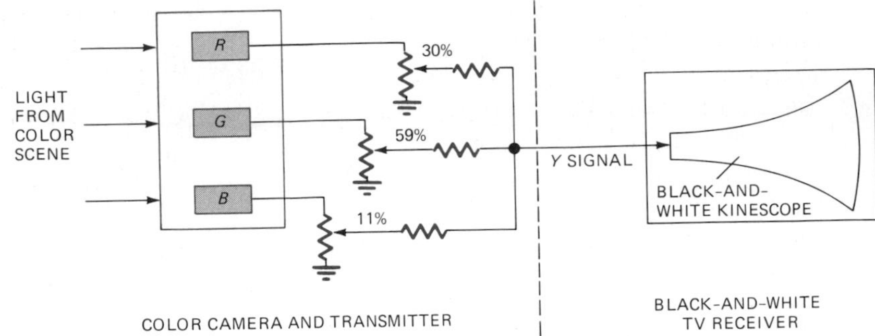

Figure 8.23 Black and white broadcast TV signal compatibility on a color TV receiver.

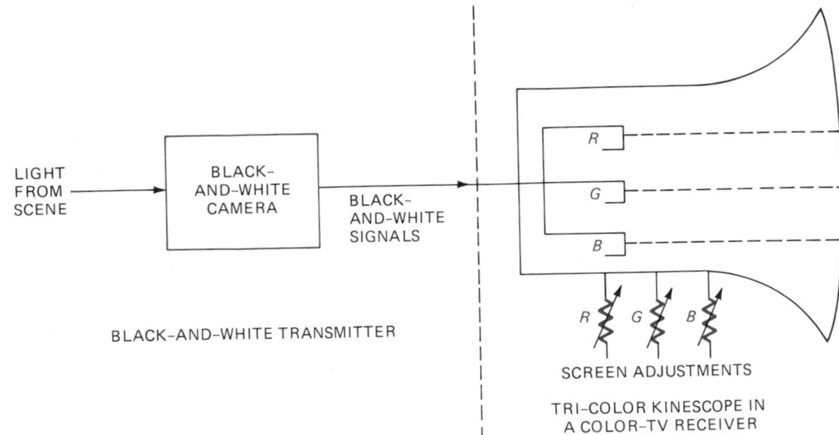

Figure 8.23 Black and white broadcast TV signal compatibility on a color TV receiver.

broadcast will then be black and white only.

During a monochrome broadcast, the color circuits of a color receiver are disabled by a circuit called the color killer, which is described in Chapter 9. If they were not so disabled, colored snow, or confetti, might appear on the screen during monochrome reception.

SUMMARY OF CHAPTER HIGHLIGHTS

1. Color is a property of light. To see the true color of an object, the object must be examined under a light containing all the wavelengths of visible light. (Section 8.1)
2. White can be produced by combining the primary colors, which are red, green, and blue. (Section 8.1 and Figure 8.2)
3. The NTSC color system was adopted by the FCC in 1953. It provides compatibility so that color transmissions can be received in monochrome on monochrome receivers. (Section 8.3)

4. A chromaticity chart is a two-dimensional view of basic color mixing plus brightness. Hue describes the wavelength of a particular color, and saturation tells how deep a color appears. (Section 8.2 and Figures 8.3, 8.4, and 8.5)
5. A monochrome TV video signal does not occupy the entire space of its 4-MHz frequency band. It appears in the form of clusters of energy. (Figure 8.6)
6. The color information of the NTSC color signal is placed between the *clusters* of the monochrome video signal. This scheme is known as *interleaving* or *multiplexing*. (Figure 8.7)
7. The monochrome signal is generated by adding red, green, and blue camera signals in the ratio of 30%, 59%, and 11%, respectively. This signal is also called the *luminance* or *brightness* or Y signal. The monochrome signal of a total color signal is formed as follows: $Y = 0.59G + 0.30R + 0.11B$. (Section 8.3)
8. The NTSC system takes full advantage of how the human eye views color. Large objects can be reproduced by three colors; medium-size objects can be reproduced by orange-red and blue-green colors; small objects are seen only as variations in brightness (monochrome). (Section 8.3)
9. The primary colors—red, green, and blue—are used in the NTSC system. They will produce a wider range of color combinations on the color picture tube than any other three colors. (Section 8.3)
10. Color camera signals are adjusted to produce a response closely matching the color brightness characteristics of the human eye. Red, green, and blue outputs are adjusted to 30%, 59%, and 11% to produce white on the receiver color tube. (Section 8.3)
11. The total color signal is composed of the Y (monochrome) signal and the color video signal. (Section 8.3)
12. Color information is transmitted as amplitude-modulated and phase-modulated sidebands of the 3.579 545-MHz subcarrier. The choice of 3.579 545 MHz reduces the interference produced on a monochrome receiver when color is being transmitted and makes frequency multiplexing possible. (Section 8.3 and Figures 8.9, 8.10, and 8.11)
13. The two color signals that modulate the 3.579 545-MHz subcarrier are called I and Q signals. (Figure 8.12)
14. The I signal has sidebands up to 1.5 MHz and will produce colors (acting alone) from reddish orange to bluish green (cyan). (Section 8.4)
15. The Q signal has sidebands up to 0.5 MHz and will produce colors (acting alone) from magenta to yellowish green. (Section 8.4)
16. In the NTSC color TV system the color subcarrier is suppressed during transmission and is reinserted in color receivers. (Section 8.6)
17. Transmitter modulators eliminate the color subcarrier.
18. A burst of color subcarrier is transmitted on the back porch of each horizontal synchronizing pulse. The burst synchronizes the color receiver with the color camera at the studio, insofar as color reproduction is concerned. (Section 8.6)
19. The phase and frequency of the reinserted color subcarrier are locked in correctly by the color burst. (Section 8.7 and Figure 8.15)
20. Two important wave forms can be viewed on color receivers: (1) the composite color video signal and (2) color-bar signals from a color-bar generator. (Figures 8.18 and 8.21)

EXAMINATION QUESTIONS

(Answers are provided at the back of the text.)

Part A Select the correct multiple choice answer.

1. To see the true color of an object, we must examine it
   a. under an ordinary light bulb.
   b. in sunlight.
   c. under ultraviolet light.
   d. under argon light.
2. The primary colors used for color television are
   a. red, blue, and yellow.
   b. red, blue, and green.
   c. red, blue, and cyan.
   d. red, blue, and magenta.
3. The color having the greatest brightness is
   a. yellow.
   b. green.
   c. cyan.
   d. magenta.
4. The NTSC color TV system was adopted in
   a. 1963.
   b. 1973.
   c. 1947.
   d. 1953.
5. Monochrome and color video signals are compressed into a standard 6-MHz channel by a method known as
   a. interlocking.
   b. multiplexing.
   c. multispectruming.
   d. interweaving.
6. The signals used to modulate the color subcarrier at the transmitter are the
   a. B − Y and G − Y.
   b. R − Y and G − Y.
   c. I and Q.
   d. X and Y.
7. The exact color subcarrier frequency is
   a. 3.59 MHz.
   b. 3.579 545 MHz.
   c. 4.579 45 MHz.
   d. 3.795 45 MHz.
8. Luminance information is sent by the following signal:
   a. Y.
   b. G − Y.
   c. picture carrier.
   d. 3.58-MHz.
9. The phase angle of the 3.58-MHz signal represents
   a. saturation.
   b. percent modulation.
   c. hue.
   d. carrier suppression.
10. Color burst phase is
    a. −(B − Y).
    b. B − Y.
    c. 90°.
    d. −(R − Y).

Part B Supply the missing word(s) or number(s).

1. If white light shines on a body and none is absorbed, the body appears to be ______.
2. The color ______ can be formed by combining red and green.
3. A diagram used for color mixing is called a ______ chart.
4. If less white light is added to a color, it is more ______.
5. The monochrome video signal appears in clusters around harmonics of ______ Hz.
6. The total color signal consists of a ______ signal and a ______ signal.
7. The monochrome signal (Y) is expressed as: Y = ______ G + ______ R + ______ B.
8. The color signal is ______ with the monochrome signal.
9. In viewing medium-size objects, the eye is satisfied with ______ colors.
10. Concerning the modulated color subcarrier, the amplitude represents the ______ of a color.
11. For large objects requiring all three primary colors, the color video bandwidth is 0 to ______ MHz.
12. The two signals that modulate the color subcarrier are the ______ and ______ signals.
13. ______ MHz is the exact color subcarrier frequency.

14. The ________-MHz beat would occur between the color subcarrier IF and the 4.5-MHz sound carrier IF, if the color subcarrier were not suppressed.

15. The color burst contains information regarding the ________ and ________ of the missing color subcarrier.

REVIEW ESSAY QUESTIONS

1. Discuss briefly why red, green, and blue were selected as the NTSC primary colors.
2. In each case, name the two colors which may produce the following: (a) yellow, (b) magenta, (c) cyan, (d) orange.
3. If blue is subtracted from white light, what color results?
4. Define hue, saturation, interleaving, and color subcarrier.
5. Explain briefly the use of a chromaticity chart.
6. Describe briefly the NTSC system of multiplexing (or interleaving).
7. What is the Y signal? How is it composed?
8. What are the two major components of the total color signal?
9. The color subcarrier frequency is exactly 3.579 545 MHz. Briefly explain the reason(s) for this choice.
10. What is the function of the color burst in a color TV receiver? What effect does the burst have on a monochrome receiver?
11. Explain briefly why color video signals do not affect the picture on a monochrome receiver.
12. Why is the color subcarrier wave suppressed at the transmitter?
13. What parts of a color television picture are reproduced only in monochrome?
14. What two color signals modulate the color subcarrier? What colors do each represent?
15. What characteristic of the color subcarrier represents hue? Saturation?
16. Briefly discuss the two important wave forms that can be viewed on a wide-band service oscilloscope.
17. Explain briefly how compatibility is achieved.
18. Explain the significance of Figure 8.7.
19. List the phase angles of the following colors: (a) blue, (b) yellow, (c) green, (d) red, (e) cyan, (f) magenta.
20. Discuss the process of chroma interference cancellation (Figure 8.16).

EXAMINATION PROBLEMS

(Selected answers are provided at the back of the text.)

1. Taking white as 100%, give the luminance (Y) percentages of (a) red, (b) green, (c) blue, (d) yellow.
2. Use Figure 8.4 to determine all the colors that can be obtained by mixing red and blue and by mixing green and blue.
3. Give the phase difference in degrees between (a) blue and magenta, (b) red and magenta, (c) blue and cyan, (d) yellow and burst, (e) burst and green.
4. What is the beat frequency between the highest I sideband frequency and the 4.5-MHz sound IF.
5. Write equations showing the three-color composition of the I and Q signals (individually).
6. Show mathematically why the color subcarrier frequency must be an odd multiple of one half of the horizontal scan rate.

Chapter 9

9.1 RF tuner.
9.2 Video-IF system.
9.3 Sound IF, FM detector, and the audio system.
9.4 Video detector and video amplifiers.
9.5 Chrominance section.
9.6 Color sync section.
9.7 Sync separators and AGC.
9.8 Horizontal- and vertical-deflection systems.
9.9 High-voltage circuits.
9.10 Color picture tube and convergence circuits.
9.11 Processing color sidebands.
9.12 Automatic circuits.
9.13 Some common color TV receiver troubles.
Summary

PRINCIPLES OF COLOR TV RECEIVERS

INTRODUCTION

In Chapter 7 we covered the principles of monochrome TV receivers. It was shown that a color TV receiver must not only perform all the functions of a monochrome TV set, but additional functions as well. These include the detection and processing of the color signal sidebands into voltages that produce a color picture. Also, color synchronization must be accomplished with great accuracy. There are other considerations to be investigated in connection with producing a color picture on the color picture tube. In this chapter, we will discuss the basic requirements of the color TV receiver.

Many of the monochrome circuits in a color TV receiver are slightly different from those of black-and-white sets. We will begin this chapter by briefly discussing the entire color receiver. Then we will examine the content of the color sync and chrominance sections. We will do this beginning with the block diagram of Figure 9.1. However, please note that topics common to monochrome receivers, such as the RF tuner and the video IF system, are treated here specifically with regard to their basic requirements for a color TV receiver. These topics will also be treated in detail, both with regard to their monochrome and color functions, in following chapters.

By the time you have completed the reading and work assignments for Chapter 9, you should be able to:

- Define the following terms: AFPC, ATC, ABC, ABL, AFT, and VTR.
- Draw a block diagram of a color television receiver, including the color picture tube.

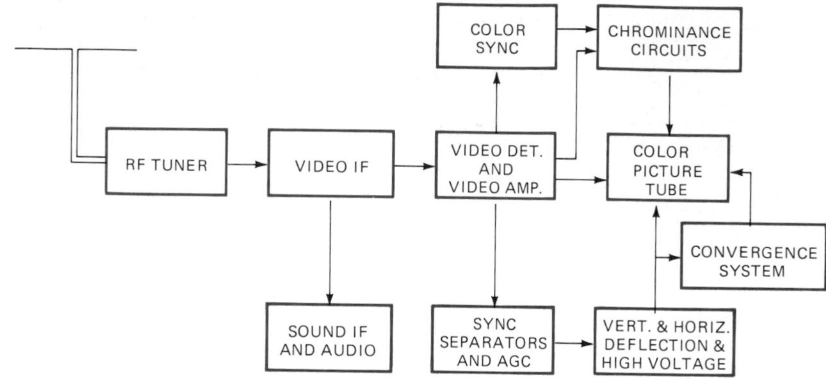

Figure 9.1 Master block diagram of a color television receiver.

- Explain the function of the delay hue in a color-TV receiver.
- List the sections of a color television receiver that are not found in the monochrome television receiver.
- Draw and label a block diagram to show the major stages of the chrominance section of a television receiver. Briefly explain the purpose of each stage. Indicate signal flow.
- List and explain at least four automatic circuits which may be in a color receiver.
- Discuss why the I, Q demodulation system provides better color pictures than the other demodulation systems.
- Discuss what might cause abnormal color intensity in a color-TV receiver.
- Discuss some symptoms of a defective color picture tube.
- Discuss the picture tube, high voltage power supply requirements.
- Discuss the requirements of the video-IF section in a color-TV receiver.

9.1 RF TUNER

The RF tuner consists of an RF amplifier, an oscillator, and a mixer stage. This section of the receiver is similar to that employed in the black-and-white receivers, since the RF requirements of both types of sets are alike. However, the allowable tolerance of the RF frequency responses is more critical in color receivers. A dip of 30% in the center part of the response curve might be satisfactory for monochrome reception. For color TV reception, however, it would cause

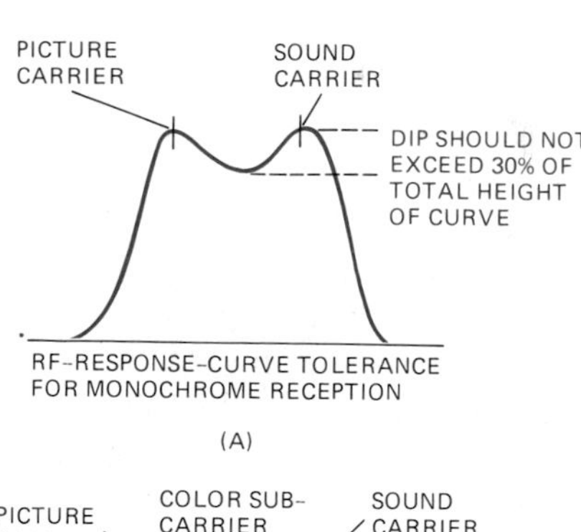

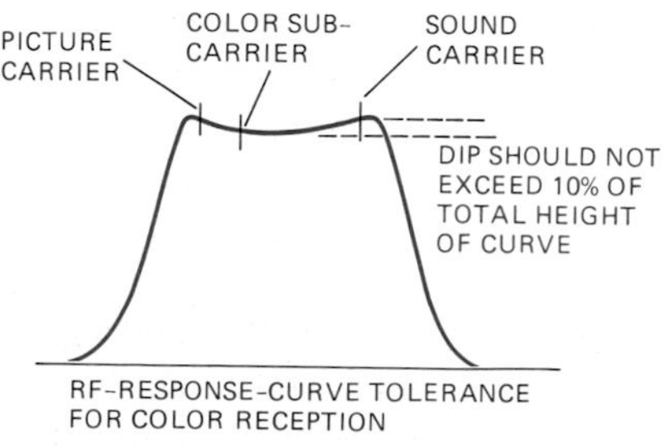

Figure 9.2A-B. RF response curves: **(A)** Suitable for monochrome reception; **(B)** required for color reception.

a degradation in the picture quality due to the unequal amplification. As shown in Figure 9.2, the RF response curve of a tuner for color reception must have a more uniform characteristic within the signal portion of the band.

If a tuner possessing the response of Figure 9.2A is employed for color reception, the color subcarrier will be attenuated as much as 20% to 30% compared with the picture and sound carrier. The response of Figure 9.2B is required on all channels.

9.2 VIDEO-IF SYSTEM

The video-IF system in color television receivers, in general, contains three or four separate amplifiers, see Figure 9.3. In form, these amplifiers closely resemble the IF section of a monochrome receiver. The change to color reception in no way alters the basic function of this section of the receiver. That function is to establish the overall bandpass and sensitivity of the receiver.

The stages in the video-IF system are stagger-tuned, in the 41 to 46 MHz range, with suitable traps for the accompanying sound (41.25 MHz), for the sound carrier of the adjacent lower channel (47.25 MHz), and for the video carrier of the adjacent higher channel (39.75 MHz). The adjacent channel traps generally have an attenuation of 55 to 60 dB. The sound carrier of the same channel (41.25 MHz) may have more than one trap to insure that this signal is kept down at the proper level. Failure to observe this precaution will tend to produce a noticeable 920-kHz beat on the picture tube screen, especially where the colors are highly saturated. This 920-kHz beat is the difference between the color subcarrier (3.58 MHz) and the sound carrier (4.5 MHz).

Typical IF-response curves for color reception are shown in Figure 9.4. These curves differ mainly in the location of the color subcarrier. In Figure 9.4A, the color subcarrier is shown near the top of the curve with 41.6 MHz at the "knee," or limit of the response. This curve is typical of many early receivers.

The overall IF-response curve extends to approximately 4.2 MHz (at the 50% points) in order to include the color subcarrier and all its sidebands. It will be remembered that the upper sidebands of the color subcarrier extend about 0.6 MHz above 3.58 MHz or up to 4.2 MHz. Hence, to reproduce the picture in its full and true color, it is necessary that the upper color sidebands be permitted to pass. Beyond 4.2 MHz, the IF response drops sharply to the level of the sound carrier.

The IF response shown in Figure 9.4B is typical of modern receivers. The same general shape of the response curve is maintained except for a slight decrease in the

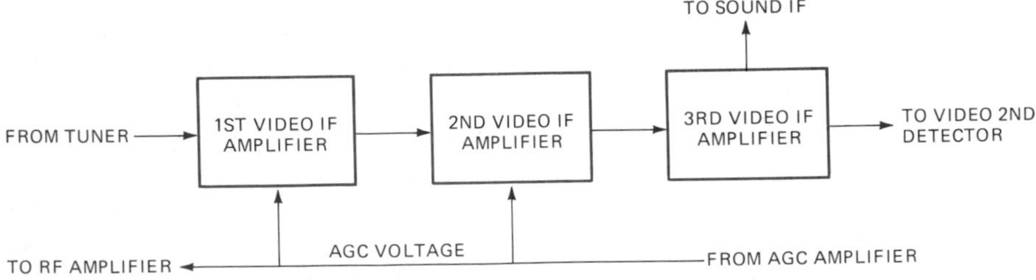

Figure 9.3 A typical video-IF system. Note the similarity to the video system of monochrome receivers. In some color receivers one or two additional IF stages are used.

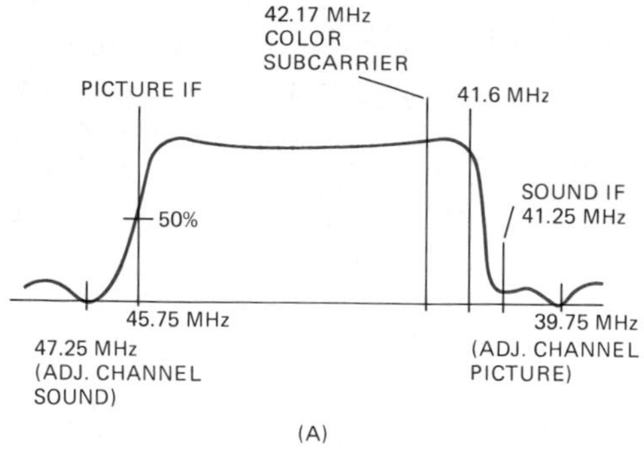

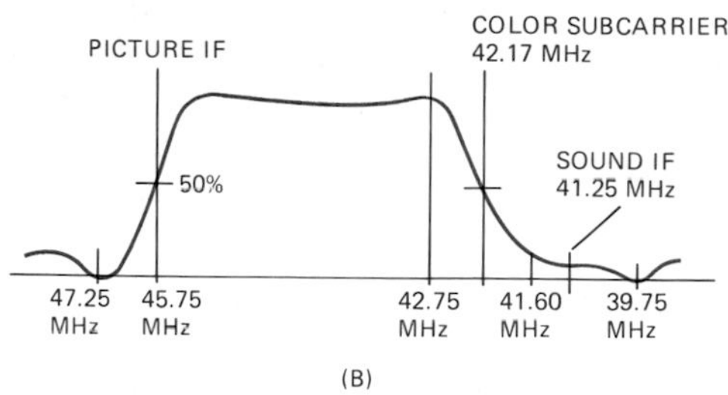

Figure 9.4A-B. Typical IF-response curves for color receivers: **(A)** early receivers; **(B)** recent receivers.

sharpness of the slope on the color-subcarrier side. Also, the color-subcarrier signal is carried through the IF stages at an amplification level equal to the picture IF (50% or 6 dB). This design further reduces the tendency toward a 920-kHz beat pattern on the picture tube screen. However, other effects created by this approach are compensated for in the video amplifier and chroma-bandpass stages discussed later.

AGC voltage is applied to all the IF amplifiers, except the last, in order to control their gain in accordance with the level of the received signal. If the last IF stage is AGC-fed, excessive attenuation will result. A portion of the same voltage is also fed to the RF amplifier for similar control.

From the video-IF system, the signal is applied to two points: the video-second detector and the sound system. Let us consider the latter first.

9.3 SOUND IF, FM DETECTOR, AND THE AUDIO SYSTEM

It is the practice in color television receivers to separate the sound and video carriers just as soon as it becomes feasible to do so. This generally occurs at the last IF stage. The purpose, of course, is to keep the amount of 920-kHz beat interference voltage as small as possible so that its effect on the screen will be negligible.

The practice generally followed is to use a separate detector (usually a solid-state diode) in which the sound and video carriers are mixed to produce the 4.5-MHz sound-IF signal. This signal is then amplified by one or two sound IF-amplifier stages and applied to a sound-detector stage. From the sound detector, the original audio intelligence is recovered. It is then fed to one or two audio-amplifier stages. Finally, it is fed to the speaker. The system is identical in all respects to the sound section of many modern monochrome receivers.

9.4 VIDEO DETECTOR AND VIDEO AMPLIFIERS

Returning to the video system, we find that the signal enters the video-second detector after leaving the IF stages. Here the signal is demodulated, giving back the 0 to 4.2-MHz monochrome or luminance signal and the color sidebands. The various synchronizing pulses plus the color burst are also present. The latter, it will be recalled, is needed to reestablish the proper frequency and phase of the missing color subcarrier.

At the video-detector output, a number of things must occur. First, the brightness portion of the total signal must be fed to a separate amplifier. Second, the color sidebands must be separated from the full signal and transferred to a separate chrominance section. Third, the color burst must be made available to the color-sync circuits. Finally, the sync separator and AGC system must also be tied into the signal path.

BLOCK DIAGRAM

There are a number of different ways in which all the foregoing functions can be carried out. The block diagram in Figure 9.5 illustrates one method. The signal from the video-second detector is fed to a video amplifier. Here both the chroma and monochrome signals are amplified. The monochrome signal is then transferred to a video-second amplifier. From this stage it goes either to a matrix network or to the cathodes of the picture tube. The chroma signal is taken from the first-video amplifier and coupled to a bandpass amplifier in the chrominance section. A color receiver may employ one or two bandpass amplifiers. The chroma signal is also fed to a burst amplifier. By means of accurate gating, the burst amplifier separates the burst signal from the chroma signal. The brightness and contrast controls are also associated with the video-output stage just as in monochrome receivers. A delay line (approximately 1.0 microsecond) is inserted between the first and the second video amplifiers. The need for the delay line arises from the fact that the color signal passes through a rather narrow bandpass filter in its system. This filter acts to slow down its passage. To insure the simultaneous arrival of the Y (or brightness) signal with the color signal at the matrix (or picture tube grids), an artificial delay line having a delay of from 0.6 to 1.0 microsecond (depending on the receiver design) is inserted in the Y channel.

It is interesting to note that the bandpass of the Y channel beyond the color take-off point is reduced below 3.58 MHz (to about 3.25 MHz). This is done to further minimize any visual dot pattern which the 3.58-MHz color subcarrier signal may develop on the face of the picture tube screen.

9.5 CHROMINANCE SECTION

We come now, for the first time, to a section of the receiver which has no counterpart in any monochrome receiver. This is the chrominance section and covers several stages. The number of stages used in this section and the type of circuitry employed to accomplish its function vary considerably

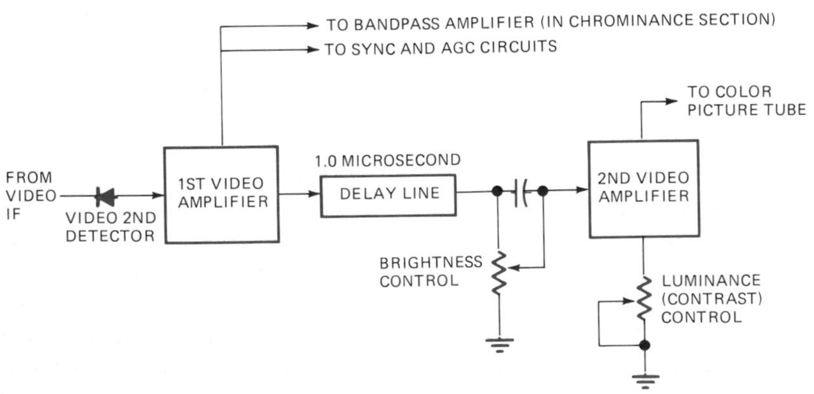

Figure 9.5 Block diagram of video-second detector and video amplifier (luminance channel). Color receivers may use two to four video amplifiers.

among TV sets. However, the function of the chrominance section is the same, regardless of the type of circuitry employed. The section must demodulate the color signal in such a way that the original red, green, and blue components of the chroma signal are recovered as originally seen by the color camera tube. Essentially, the chrominance section separates and amplifies the chrominance signal and demodulates (detects) the individual red, green, and blue signal components. Then, by a matrixing, or signal adding, network, it couples the correct portions of these signal components to the tricolor picture tube.

The chrominance section will be further analyzed later in this chapter.

9.6 COLOR SYNC SECTION

The stability of the 3.58-MHz carrier signal reinserted into the chroma demodulators is an important factor in the reproduction of the original red, green, and blue voltages. It is the function of the color sync section to develop a stable 3.58-MHz signal, and to make certain it possesses the proper frequency and the proper phase. If the phase is wrong, the reproduced color will also be wrong. In a color television system, this is a very noticeable form of distortion.

9.7 SYNC SEPARATORS AND AGC

The sync separators and the AGC section of a color television receiver do not differ in any important aspects from the same stages in monochrome receivers. Thus, the function of the sync separators is to separate the horizontal and vertical sync pulses from the rest of the video signal. Once this is accomplished, the pulses are applied to the horizontal and vertical sweep systems through appropriate integrating and differentiating networks. A noise inverter is linked with the sync separator to prevent noise pulses from affecting the vertical and horizontal sweep systems.

For the AGC section, any method used in monochrome sets may be used in color receivers. At the present time, keyed AGC is favored. This preference is not because color television is being used, but because of the inherent characteristics of keyed AGC.

The stages controlled by the AGC voltage include the RF amplifier and one or more video-IF stages. Clamping of the AGC voltage fed to the RF amplifier may also be employed.

9.8 HORIZONTAL- AND VERTICAL-DEFLECTION SYSTEMS

In the horizontal- and vertical-deflection circuits we again encounter circuits similar to those found in monochrome receivers (Figure 9.6). The vertical system usually consists of a multivibrator with an output amplifier. In the horizontal section, the oscillator is preceded by an automatic frequency-control system. Beyond the oscillator, there is an output amplifier and the horizontal-output transformer. A damper diode is connected across the horizontal windings of the deflection yoke to eliminate oscillations which may occur during beam retrace. The energy absorbed by the damper in this process is converted into an additional voltage which results in a boosted B+ when added to the normal B+ voltage.

A horizontal-retrace blanking circuit is included in color receivers to prevent the 3.58-MHz burst from reaching the grids of the color picture tube. This burst follows the horizontal-sync pulse. If it were permitted to reach the picture tube, it would develop a yellow strip during the horizontal blanking interval.

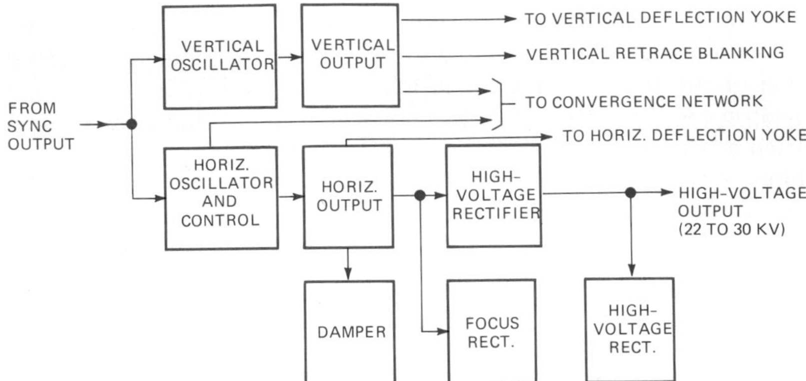

Figure 9.6 Block diagram of the vertical and horizontal deflection systems of a typical color receiver.

9.9 HIGH-VOLTAGE CIRCUITS

The high-voltage requirements of the color picture tube are considerably more critical than they are for a conventional black-and-white picture tube. The color tube requires up to 30 000 volts at a maximum current drain of 1.5 milliamperes (mA). In addition, a focus voltage of 5 to 8 kV must be available to the tube.

The heavy requirements of the picture tube in regard to beam current result in some serious problems in the designing of a combination deflection and high-voltage system. The power used by the high-voltage circuit is an appreciable portion of the total power. Therefore, changes in the beam current due to changes in picture brightness can cause variations in scanning linearity and in the various operating potentials of the tube itself. To avoid such variation, it is necessary to maintain the high-voltage load constant whether the picture is bright or dim. This requires the use of a special high-voltage regulator.

9.10 COLOR PICTURE TUBE AND CONVERGENCE CIRCUITS

Color receivers employ a tricolor picture tube. The tube possesses three electron beams. The conventional black-and-white tube employs only one electron beam.

The screen of the tube possesses three different color-emitting phosphors. This, of course, is basic to the entire color television system, since we employ the three primary colors (red, green, and blue) to synthesize the wide range of hues and tints required for the satisfactory presentation of a color picture. These color-emitting phosphors are arranged in a dot or line pattern on the picture tube screen.

Convergence circuits and convergence magnets are required to insure that the beams strike the proper points on the screen. The magnets take care of any permanent misalignment of the electron beams or the offset caused by the earth's field. The convergence circuits are commonly called the dynamic convergence circuits. They modify the instantaneous alignment of the beams to make further corrections to the convergence.

9.11 PROCESSING COLOR SIDEBANDS

Several ways of processing the color sidebands are possible. We will first briefly discuss the original system which utilizes the complete I and Q signals (see Figure 9.7). We will then discuss the three systems which are in common use today.

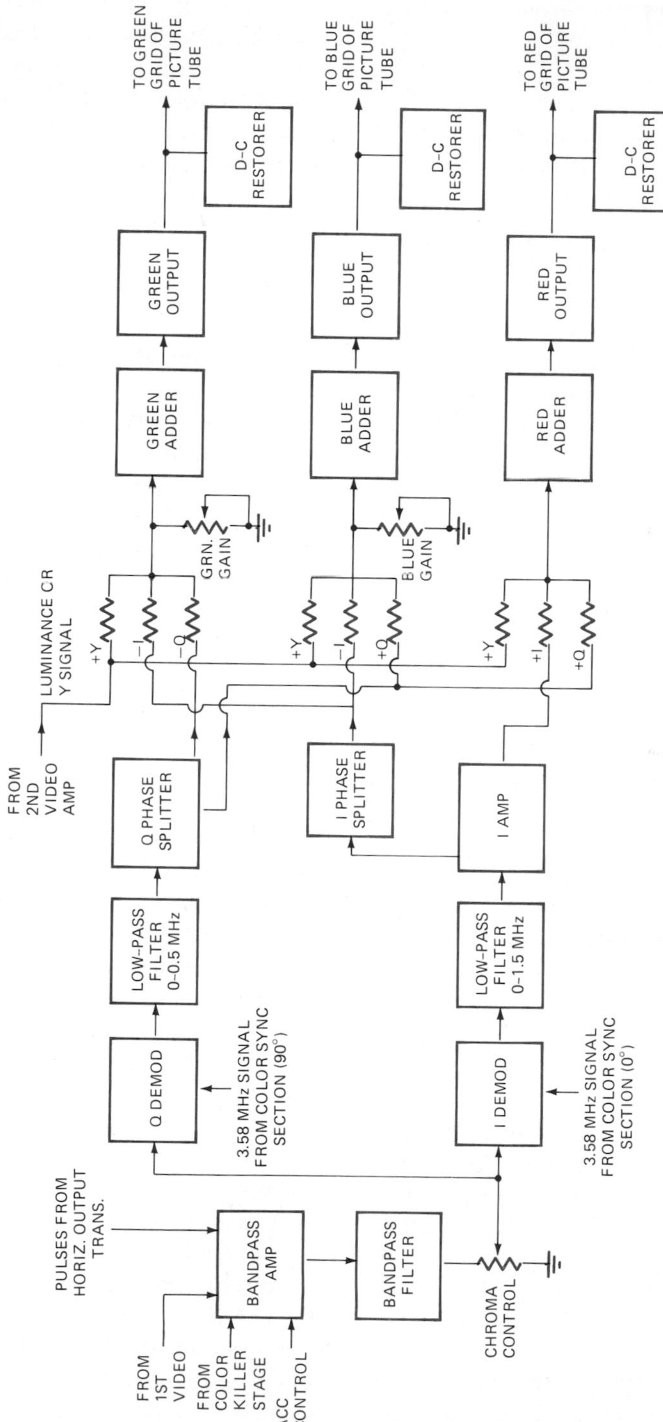

Figure 9.7 The chrominance section of a color receiver employing I and Q demodulators. No special delay is shown for the I stages, because the required delay is in distributed form.

I AND Q SYSTEM

Bandpass Amplifier

The full color signal is obtained from the video system and fed to a bandpass amplifier. Also applied to the bandpass amplifier is a gating pulse which keys off the amplifier by applying a pulse derived from the horizontal-deflection output transformer. The pulse arrives only during the horizontal-retrace interval when the color burst is passing through the system. By gating out the burst from the bandpass amplifier, we prevent color background unbalance in the picture tube. This unbalance may arise if the DC restorers in the chrominance channel clamp onto the color bursts rather than the normal chrominance signal.

The bandpass amplifier also receives a DC biasing voltage from a stage known as a "color killer." This stage is located in the color-sync section of the receiver. Its purpose is to bias the bandpass amplifier to cutoff in the absence of a color signal (that is, when black-and-white signals *only* are being received). This precaution is taken to insure that no random color appears on the picture tube screen during a monochrome transmission. Such random color would be produced by noise or monochrome signals reaching the I and Q demodulators.

In the output circuit of the bandpass amplifier there is a bandpass filter which permits signals from 2.1 to 4.2 MHz to pass, but strongly attenuates all others. This filter thus separates that portion of the signal containing the color sidebands from the section of the signal containing only monochrome information.

Chroma Control

The bandpass filter is terminated in a color-intensity control or chroma control. From this point we can take as much of the color signal as we feel is required and feed it to the following demodulators. This control actually determines how saturated (deep) the colors appear on the screen. It is a front panel control which the user of the set can adjust as necessary. (It may also be labeled "chroma," "color intensity," "color saturation," or "color." Regardless of the name used, its function is the same in any color receiver.)

Color Demodulators

After this control, the color signal is fed in equal measure to the I and Q color demodulators. Also arriving at these demodulator stages is a 3.58-MHz signal. This signal represents the missing color subcarrier. It must be recombined with the color signal of the demodulators so that the original I and Q signals can be detected. Both I and Q stages receive a 3.58-MHz voltage, the only difference being that one 3.58-MHz voltage lags behind the other 3.58-MHz voltage by 90 degrees. This particular phase relationship is required because the color signal was modulated this way at the transmitter and demodulation is the reverse process.

With a color signal being received, the entire chrominance channel is operative. The output of the two demodulators represents the original I and Q color signals that were originally developed at the transmitter. The I signal, then, is passed through a 0 to 1.5-MHz bandpass filter and a special 0.5-μs delay line. It may receive additional amplification before being made available to the adding, or matrix, network in positive and negative polarity. The double-polarity I signals are required in the final mixing process from which red, green, and blue voltages are recreated. A single phase splitter provides the positive and negative I signals.

The use of a 0.5-μs delay network in the I channel again stems from the narrow 0 to 0.5-MHz bandpass filter through which the

Q signal is sent. The Y signal, it will be remembered, had to be delayed 1 μs for the same reason. The difference in delay between the Y and I signals arises from the different characteristics of their respective networks. In the Y channel, the bandpass of the circuits extends from 0 to 3.5 or 4.0 MHz. In the I channel, the bandpass extends only from 0 to 1.5 MHz. The narrower bandpass introduces some delay, requiring less additional delay in order to slow the I signal down to the Q signal.

In the Q channel, the demodulated Q signal passes through a 0 to 0.5-MHz bandpass filter and reaches a phase splitter from which positive and negative Q signals are made available to the matrix.

I, Q and Y Signals

We now have at the matrix the I, Q, and Y signals. By properly combining them, we can reobtain the red, green, and blue voltages that were originally combined to form the I, Q, and Y signals. The addition is carried out in rather simple fashion by using a series of resistors connected as shown in Figure 9.7. At the output of the matrix section, each of the three color voltages is separately amplified and then transferred via separate DC restorers to the appropriate control grid of the tri-gun picture tube.

The demodulators in color receivers which reproduce the original I and Q components from the chroma signal are usually low-level demodulators. Thus, demodulation is performed at a low color signal level. By necessity, the signals must be further amplified following demodulation.

OTHER SYSTEMS

The I, Q system just described, when properly designed, provides excellent color reproduction. However, price is an important aspect of receiver sales. If it is possible to achieve acceptable results at lower cost, some circuit designers will use a more economical system. Several other systems have been developed. These other methods are the R − Y and B − Y system, the X and Z system, and the R − Y, B − Y, G − Y system.

The R − Y and B − Y System

In this commonly used system, both channels possess the same bandwidth, generally 0 to 0.5 MHz. While this method does not color as much of the picture as the I, Q system, the visual results are acceptable. In this narrower system, only the larger objects are in color. Medium and small detail are rendered in black and white only. By this modification, no time-delay networks are needed in the chrominance section. However, the time-delay filter in the Y section is still retained. The change also permits other simplifications which are economically advantageous.

R − Y and B − Y systems have been used with low-level and high-level demodulators. In high-level demodulation, the G − Y signal is often formed in a cathode circuit which is common to both R − Y and B − Y demodulators, Figure 9.8. The detected R − Y, B − Y, and G − Y color signals are then fed directly to the grids of the picture tube. High-level demodulation means that no additional amplifiers are needed beyond the demodulators. In such a system, the 3.58-MHz carrier that is reinserted with the two color sidebands does not possess the same 90-degree relationship previously indicated for the I, Q system, even when low-level R − Y, B − Y detection is employed. This stems from the fact that a common cathode is being utilized at the demodulators to provide the G − Y signal. If a 90-degree phase separation between the 3.58-MHz carriers being applied to each demodulator is main-

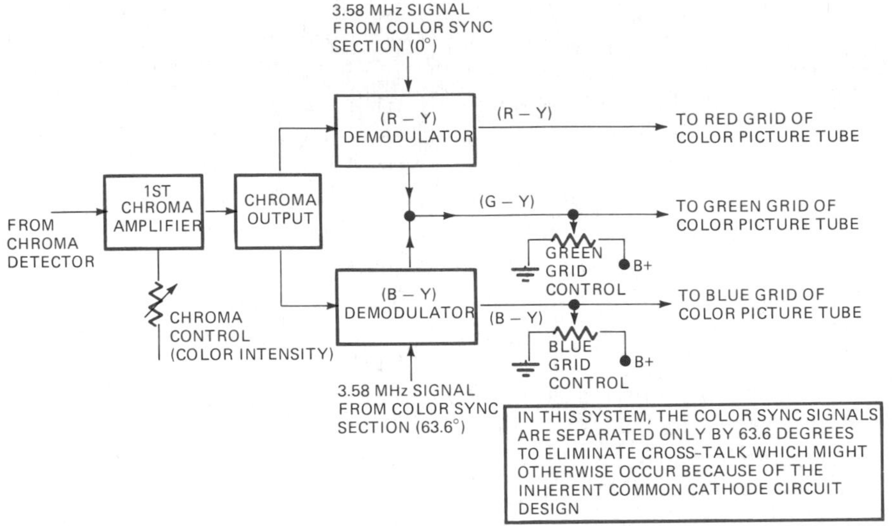

Figure 9.8 The chrominance section of a color receiver employing R – Y and B – Y demodulators. Note that bandwidths for the two demodulators are equal and that no delay line is required.

tained, cross talk will occur and cause improper coloring of the image. That is, the colors sent to each grid of the picture tube will not be pure as they could be.

When low-level R – Y and B – Y demodulation is carried out and the G – Y signal is formed in a separate circuit, the reinserted 3.58-MHz carriers more nearly possess a 90-degree phase relationship.

Obtaining B – Y

In some receivers, R – Y and G – Y are demodulated and a portion of each is added to obtain B – Y. This is feasible because if we can add R – Y and B – Y to obtain G – Y, we can also use R – Y and G – Y to derive B – Y. It should be noted, however, that if we employ this method, a different phase relationship is required between the 3.58-MHz signals sent to the demodulators. This can be seen from Figure 9.9.

The three color difference signals are then applied to the grids of the picture tube. Matrixing is accomplished in the picture tube, as the –Y luminance signal is coupled to the three picture tube cathodes. A –Y signal applied to the cathodes is effectively the same as a +Y applied to the grids. Thus:

$$(R - Y) + Y = R \text{ (red)}$$
$$(G - Y) + Y = G \text{ (green)}$$
$$(B - Y) + Y = B \text{ (blue)}$$

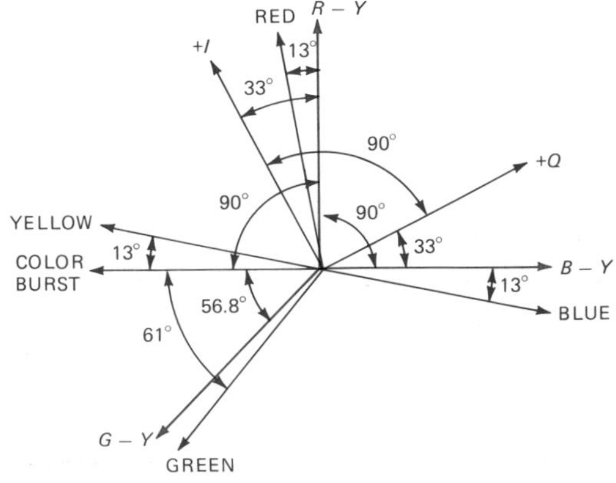

Figure 9.9 The phase relationships between the various color signals, including R – Y, B – Y, and G – Y.

Color-Tube Matrixing

As previously mentioned, the three color-difference signals are applied to the grids of the tricolor picture tube at the same time that the luminance (or Y) signal is fed to the three cathodes of the picture tube. Thus, matrixing is performed within the tricolor tube itself. When the Y signal (appearing on the "red" cathode) is added to the R − Y signal appearing on the "red" grid, a voltage (bias) results that represents R alone. The B − Y and G − Y signals are added in the same manner to the Y signal, resulting in the correct voltage and beam current for B and G. Traps in the output circuits of the demodulators are resonant to 3.58 MHz, preventing this frequency from reaching the picture tube grids.

X and Z System

Still another low-level modulation system uses what are called X and Z demodulators, Figure 9.10. This is basically an R − Y, B − Y system, but the angular separation of the reinserted 3.58-MHz carrier signals is 57.5° and not 90°. The output from the X demodulator is R − Y, and the output from the Z demodulator is B − Y. From the common emitters or the common cathodes of the following amplifier, the G − Y signal is obtained, amplified, and then transferred to the appropriate grid of the color picture tube. Notice that the terms X and Z have been selected arbitrarily to identify the demodulators. They have no other significance.

This matter of color demodulation in which phase angles other than 90° are employed in the reinserted 3.58-MHz carrier often puzzles the service technician working with color circuits. The precise angle selected depends upon the way the circuits have been designed to handle the color signals. Demodulation can be performed in a variety of ways, each with its own phase requirements. The point to remember in all these systems is that pure R − Y, B − Y, and G − Y signals are provided at their respective grids at the color picture tube.

The R − Y, B − Y, and G − Y System

Demodulators can be designed to obtain all three color signals directly. This is evident

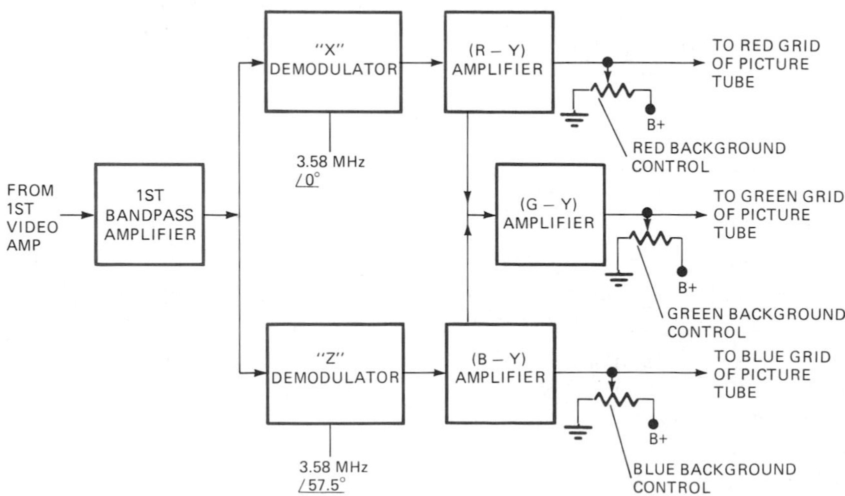

Figure 9.10 The chrominance section of a color receiver employing X and Z demodulators.

from Figure 9.9 which shows that the G − Y color signal is only 57° ahead of the 3.58-MHz carrier. One method of demodulation accomplishes this by delaying the color signal 57°. The 3.58-MHz carrier is then used to obtain the G − Y signal directly, Figure 9.11. The B − Y signal is obtained by shifting the 3.58-MHz carrier 57° and the color signal 180°. The R − Y demodulator uses the same color signal as the G − Y demodulator. However, the 3.58-MHz carrier must be shifted 147° for proper demodulation.

Note: The foregoing discussions of the processing of the color sidebands will be further emphasized with reference to specific schematics, in Chapter 10.

9.12 AUTOMATIC CIRCUITS

The automatic circuits mentioned below are discussed briefly, for general information. Most color TV sets have some or all of these automatic circuits.

Automatic Color Control (ACC) This is also known as "automatic chroma control." The function is similar to AGC as applied to the tuner and IF stages. In this case, the gain of a chroma amplifier is automatically controlled. This provides a constant color amplitude in spite of signal amplitude variations.

Automatic Fine Tuning (AFT) This is also used in some monochrome receivers. In this

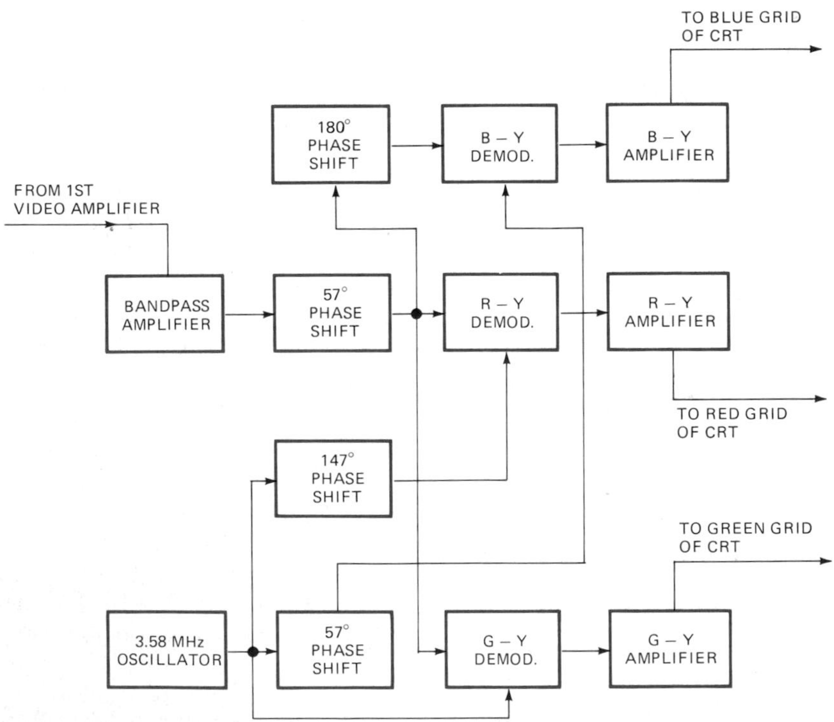

Figure 9.11 The chrominance section of a receiver employing R−Y, B−Y, and G−Y demodulators. *(Courtesy of RCA.)*

system, the video IF carrier frequency is fed to the AFT circuit. If the frequency is incorrect, a correction voltage is generated and fed to the tuner oscillator. The oscillator frequency is then adjusted to provide the correct video IF carrier frequency. This results in the production of the best color (and monochrome) picture.

Automatic Frequency and Phase Control (AFPC) All color TV receivers include some type of AFPC circuit. The function of AFPC is to make the frequency and phase of the 3.58-MHz oscillator output the same as that of the color burst. This will insure the correct hues being displayed.

Automatic Tint Control (ATC) When switched on, this circuit causes flesh tones to be emphasized for a more consistently eye-pleasing picture. This is particularly useful when switching channels, or when a channel's program content varies considerably.

Automatic Brightness Control (ABC) This circuit automatically adjusts the picture tube brightness in accordance with room ambient light.

Automatic Brightness Limiter (ABL) As the name implies, this circuit limits the maximum picture tube brightness. It is useful in preventing excessive brightness due to incorrect manual adjustment.

Vertical Interval Reference (VIR) This circuit function has been discussed in a previous chapter. When used, it provides completely automatic control of the hue and saturation of a color picture. It has been recently introduced and furnishes perhaps the most consistently accurate color renditions.

9.13 SOME COMMON COLOR TV RECEIVER TROUBLES

Color TV receiver troubles and troubleshooting techniques will be covered in Chapter 27. The following sections introduce some common troubles you may encounter in color TV receivers. Keep in mind that a color TV receiver also contains the circuits of a monochrome receiver. Thus, for such circuits, the monochrome receiver troubles discussed in Chapter 7 will also apply here. Before attempting to localize a problem to the color circuits, first be certain the monochrome circuits are functioning properly. By turning off the color control, are you getting a normal monochrome picture? If not, the trouble is in the monochrome circuits. An exception to this rule is a defective color picture tube or defective circuits supplying operating voltages to the tube.

1. Symptoms Of A Defective Color Picture Tube
   (a) A defective gun (or beam) will make it impossible to see a picture in monochrome. One of the primary colors will be missing.
   (b) Flashing in the picture, particularly if this changes when the set is struck, is often caused by an internally shorted color picture tube.
   (c) A "washed out" monochrome and color picture.
   (d) Change of color when set is vibrated or struck.
2. Normal monochrome picture, no color (See Figure 9.7.) The following section(s) may be defective:
   (a) color killer
   (b) 3.58-MHz subcarrier oscillator
   (c) chroma-bandpass amplifier.

   Chart 9.1 illustrates a trouble shoot-

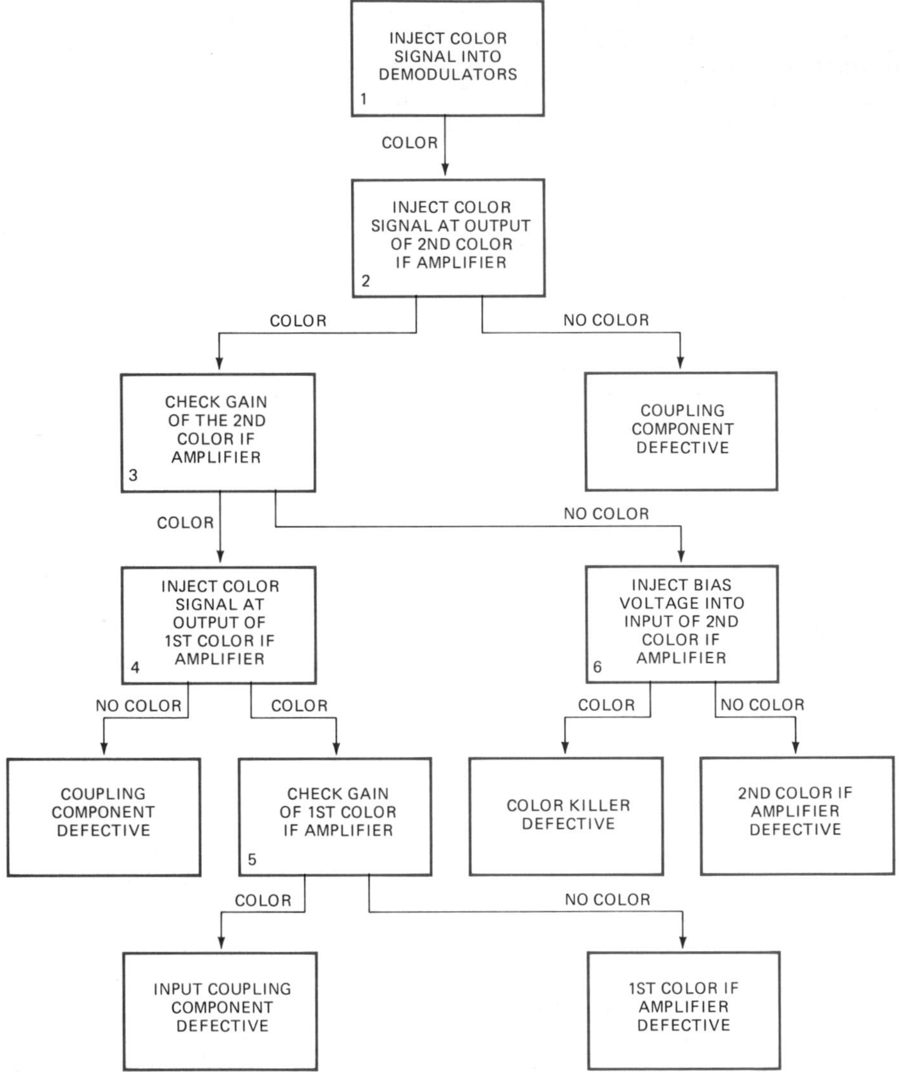

Chart 9.1 Illustrating a trouble shooting procedure to help localize the problem of normal monochrome picture, no color. *(Courtesy of B&K-Precision, Dynascan Corp.)*

ing procedure to help localize this problem.

3. Abnormal color intensity (saturation), see Figure 9.7. Assuming adequate signals from the first video amplifier, the trouble is likely to be in the bandpass amplifier or in its ACC control voltage.
4. No color sync, monochrome sync normal. This is most likely caused by a defect in the 3.58-MHz subcarrier oscillator or in its AFPC circuits. The appearance on the screen is color bars.
5. All hues incorrect, good monochrome picture. First check the setting of the hue, or tint, control. Other possibilities are defects in the bandpass amplifier, or the AFPC system (including the 3.58-MHz oscillator).
6. One hue missing, monochrome pic-

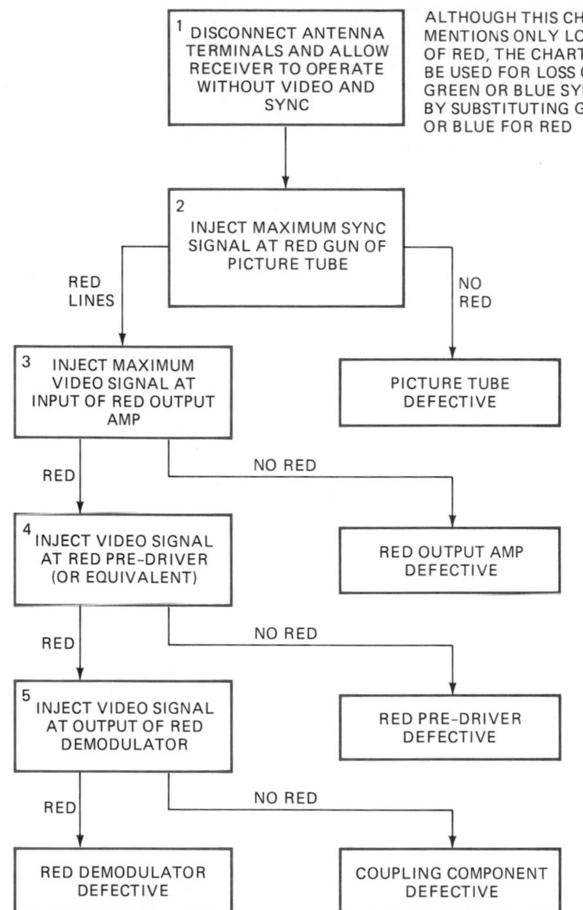
1 DISCONNECT ANTENNA TERMINALS AND ALLOW RECEIVER TO OPERATE WITHOUT VIDEO AND SYNC
ALTHOUGH THIS CHART MENTIONS ONLY LOSS OF RED, THE CHART CAN BE USED FOR LOSS OF GREEN OR BLUE SYMPTOMS BY SUBSTITUTING GREEN OR BLUE FOR RED
2 INJECT MAXIMUM SYNC SIGNAL AT RED GUN OF PICTURE TUBE
RED LINES
NO RED
3 INJECT MAXIMUM VIDEO SIGNAL AT INPUT OF RED OUTPUT AMP
PICTURE TUBE DEFECTIVE
RED
NO RED
4 INJECT VIDEO SIGNAL AT RED PRE-DRIVER (OR EQUIVALENT)
RED OUTPUT AMP DEFECTIVE
RED
NO RED
5 INJECT VIDEO SIGNAL AT OUTPUT OF RED DEMODULATOR
RED PRE-DRIVER DEFECTIVE
RED
NO RED
RED DEMODULATOR DEFECTIVE
COUPLING COMPONENT DEFECTIVE

Chart 9.2 Illustrating a practical troubleshooting procedure for the condition of one hue missing, monochrome picture normal. *(Courtesy of B&K-Precision, Dynascan Corp.)*

ture normal. (See Figures 9.7, 9.8, 9.10, and 9.11.) By observation, the missing hue can be easily identified. The trouble is most likely to be in an individually defective red, green, or blue channel. Chart 9.2 illustrates a troubleshooting procedure to help localize this problem.

7. Hues changing, normal monochrome picture. This generally indicates weak color sync amplitude. The trouble may be in the 3.58-MHz subcarrier oscillator or in its associated AFPC circuit.

Note: For additional troubleshooting procedures relating to color-TV receivers, see Chapter 27.

SUMMARY OF CHAPTER HIGHLIGHTS

1. A color TV receiver must not only perform all the functions of a monochrome TV receiver, but also all the additional functions required to add color to the picture.
2. The tuner of a color TV set should not have a dip in its RF-response curve greater than 10%. (Figure 9.2)
3. The overall response curve of a color set IF section extends to 4.2 MHz. This is necessary to amplify the color subcarrier frequency, plus all of its sidebands. (Figure 9.4)
4. The sound and video-IF carriers in a color set are generally separated after the last common IF stage. This reduces possible 920-kHz beat interference. (Figure 9.1)
5. After the video detector output of a color TV receiver, (1) The monochrome portion of the total color signal is fed to a separate Y-video amplifier; (2) The color sidebands must be separated from the total color signal and fed to the chrominance section; or (3) The color burst must be fed to the AFPC circuits.
6. An artificial delay line (0.6 to 1.0 μs) is inserted into the Y channel. This insures that the Y and color signals arrive simultaneously at the color picture tube. (Figure 9.5)
7. The chrominance section has no counterpart in a monochrome receiver. Basically, it separates and amplifies the chrominance signal, detects the red, green, and blue signal components, and couples these components to the color picture tube. (Figure 9.7)
8. The subcarrier which was suppressed at the transmitter is reinserted at the color receiver. This is done by the 3.58-MHz oscillator which is frequency and phase controlled with reference to the color burst. (Section 9.6)
9. The high-voltage requirements for the color picture tube are rather critical. A high voltage of 30 000 V is required, at 1 500 μA. The voltage must be regulated. (Section 9.9)
10. Convergence circuits and magnets are required to insure that the color tube beams strike the proper points on the tube screen. (Section 9.10)
11. There are several methods of processing the color sidebands. These are known as (1) I, Q system, (2) R − Y, B − Y system, (3) X, Z system, and (4) R − Y, B − Y, G − Y system. (Section 9.11)
12. The I, Q color signal processing system provides the best color rendition. However, it is also the most expensive and many color TV sets use one of the other systems. (Figure 9.7)
13. The automatic circuits relating to good color rendition are (1) Automatic Color Control (ACC), (2) Automatic Fine Tuning (AFT), (3) Automatic Frequency and Phase Control (AFPC), (4) Automatic Tint Control (ATC), (5) Automatic Brightness Control (ABC), (6) Automatic Brightness Limiter (ABL), (7) Vertical Interval Reference (VIR). (Section 9.12)
14. When troubleshooting a color-TV receiver, first establish if the monochrome portions are operating correctly. If they are, then carefully observe the symptoms. This will help greatly in localizing the trouble. (Section 9.13)

EXAMINATION QUESTIONS

(Answers are provided at the back of the text.)

The following items are to be answered true (T) or false (F).

1. Monochrome circuits in a color TV receiver are exactly the same as in a monochrome TV receiver.
2. The RF amplifier (tuner) in a color TV receiver can have a response dip up to 30%.
3. The adjacent, lower channel sound IF trap is tuned to 47.25 MHz.
4. The sound IF carrier level is kept at the required level by the AGC system.
5. For a color TV receiver, the common IF response curve must extend to about 4.5 MHz.
6. The color subcarrier signal is passed through the common IF stages at the 50% amplitude, response curve level.
7. The 920-kHz beat interference signal can occur between the picture carrier and the color subcarrier.
8. The color signals are fed to the chrominance section from the output of the last common IF stage.
9. The Y-signal delay line delays the high-frequency monochrome signals to coincide with the high-frequency color signals.
10. To minimize a possible dot pattern on the color picture tube, the bandpass of the Y channel is reduced below 3.58 MHz.
11. The 3.58-MHz oscillator and AFPC system generates I and Q signals.
12. Color TV receiver high-voltage power supplies deliver 30 000 V and also 5 000 to 8 000 V.
13. The I, Q system of color processing at the receiver produces the best color picture, compared to other available systems.
14. In the absence of a Y signal, the color killer disables the bandpass amplifier.
15. The bandpass filter passes frequencies from 0.6 to 1.5 MHz.
16. The chroma control can adjust the color saturation.
17. In the R − Y, B − Y system, both channels utilize a 0.5-MHz bandwidth.
18. In some receivers, B − Y is obtained by combining a portion of the R − Y and G − Y signals.
19. It is possible to perform color matrixing within the color picture tube.
20. ACC stands for "Automatic Contrast Control."
21. The Automatic Tint Control (ATC) controls the gain of the color subcarrier oscillator.
22. The symptom is no color, but a normal monochrome picture. A possible defective section is the I signal demodulator.
23. A condition of no color sync can be caused by a defective 3.58-MHz oscillator.
24. If only one hue is missing, but the monochrome picture is normal, one of the color channels is defective.

REVIEW ESSAY QUESTIONS

1. Draw a block diagram of a color television receiver including the color picture tube. Label all blocks and indicate signal flow.
2. Which sections of the color block diagram are not found in a monochrome receiver? Explain why in each instance.
3. What precautions must be observed with respect to sound takeoff in a color receiver?
4. With which portion of the incoming signal is the color sync section specifically concerned? How does it use this information?
5. Trace the path of the monochrome (Y signal)

of a color transmission, from the second detector to the picture tube. Draw a simple block diagram to illustrate.

6. Draw and label a simple block diagram to show the important stages of the chrominance section. Indicate signal flow.
7. List the basic functions of the chrominance section. Briefly, explain the purpose of each stage.
8. Why can the I, Q system provide better color pictures than the other demodulation systems?
9. What are the manual controls that are unique to a color TV receiver? Explain what each one does.
10. In Figure 9.2, why is curve (A) not suitable for color reception?
11. Refer to Figure 9.3. How does this section differ from that of a monochrome TV receiver?
12. Discuss the purpose of the delay line in Figure 9.5.
13. In the block diagram of Figure 9.6, what blocks are not found in a monochrome TV receiver? Why?
14. Briefly explain the function of the "I-phase splitter" shown in Figure 9.7.
15. In Figure 9.8, how is the G − Y signal derived?
16. If the chroma-bandpass amplifier is inoperative, what happens to the picture? Why?
17. In an I, Q demodulation system, if the I demodulator is inoperative, what will the picture look like?
18. Give the phase difference between the following signals (Figure 9.9): (a) I, Q; (b) Q, blue; (c) I, red; (d) I, burst; and (e) green, burst.
19. In Figure 9.10, how is the G − Y signal derived?
20. Describe briefly four differences of the monochrome sections in a color TV receiver.
21. What is the "color killer"? What does it do?
22. What is the approximate pass band for the chroma bandpass amplifier? Explain the reason for this.
23. List four automatic circuits which may be in a color receiver. Give a brief explanation of each.
24. What might cause abnormal color intensity in a color TV receiver? Why?
25. What might cause all hues to be incorrect (monochrome picture normal)?
26. The colors keep changing in a random manner. What do you suspect is wrong?

EXAMINATION PROBLEMS

(Selected problem answers appear at the end of the text.)

1. Draw a typical IF response curve for current color TV receivers. Indicate all important frequency points. Label traps as such.
2. Draw a color signal vector diagram, showing the following: (1) B − Y, (2) burst, (3) I, Q, (4) R − Y, (5) G − Y. Label the correct angle for each.
3. In the table below, identify the frequencies shown:

FREQUENCY (MHz)	APPLICATION
55.25	
59.75	
58.83	
45.75	
45.25	
42.17	
41.25	
39.75	

4. For Channel 4 (66 to 72 MHz), calculate: (1) picture carrier frequency, (2) color subcarrier frequency, and (3) sound carrier frequency.

5. What blocks are missing (if any) from the color TV receiver partial block diagram shown below?

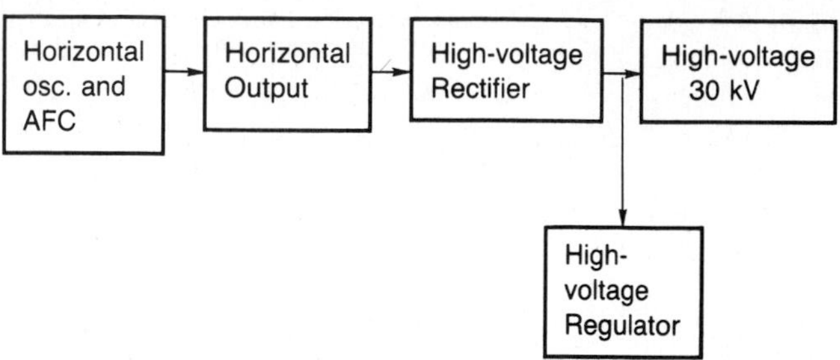

Chapter 10

10.1 Types of TV tuners for VHF and UHF reception
10.2 Electrical characteristics of tuners
10.3 Interference, stability, and noise problems in tuners
10.4 Channel allocations
10.5 Characteristics of tuned circuits
10.6 Vacuum tubes for tuners
10.7 Typical tube-type RF amplifiers
10.8 Transistor RF amplifiers
10.9 Field-effect transistors and RF amplifiers
10.10 Mixers and mixer circuit operation
10.11 Local oscillator operation
10.12 A typical nuvistor tuner
10.13 A typical transistor tuner
10.14 Varactor tuning and tuners
10.15 UHF solid-state tuners
10.16 Sources of trouble in VHF and UHF tuners
Summary

TELEVISION RECEIVER TUNERS

INTRODUCTION

All receiving systems have at least four basic sections: an antenna system for converting electromagnetic signals to electric signals; circuitry for selecting one station and rejecting all others; a detector for converting the RF (or IF) signal into signals that can operate a transducer; and a transducer to convert the electric signals into some other form of energy. Examples of transducers are loudspeakers that convert electric energy to sound energy and picture tubes that convert electric energy to light energy.

The **tuner**—sometimes called the **front end**—selects the desired station and rejects all others. In addition, it does the following:

1. It terminates the transmission line between the antenna and the receiver in the proper impedance and presents a proper impedance match to the receiver IF system.
2. It amplifies the RF signal to present an acceptable signal-to-noise ratio.
3. It isolates the local oscillator signal from the antenna, thus preventing the oscillator signal from radiating and interfering with other receivers.
4. It converts the RF signal to an IF signal as required in a superheterodyne receiver. All television receivers use the superheterodyne principle.

Modern television receiver tuners may be divided into two major categories: (1) mechanical tuners and (2) electronic tuners. With mechanical tuners, channel selection is accomplished by switching in different sets of tuned coils for each channel. Electronic tuning is accomplished with **varactors,** which are silicon diodes that, when reverse-biased, act as voltage-variable capacitors. Varactors used with coils form the required tuned circuits. Tuning is accomplished by applying the correct DC reverse bias to the varactors. In sophisticated designs, digital techniques are combined with varactor tuners. In this chapter, we shall discuss the various types of tuner circuits and complete tuners.

When you have completed the reading and work assignments for Chapter 10, you should be able to:

- Define the following terms: tuner, front end, hybrid tuner, random access, balun, thermal noise, cascode amplifier, nuvistor, masking voltage, neutralizing network, FET, JFET, ultra-audion, drift, AFT, image frequency, mixer, varactor, frequency-synthesized tuning, digital tuning, and switching diode.
- Discuss briefly how varactor tuning works.
- Discuss briefly the types of resonant circuits in UHF tuners.
- Explain the main advantages and disadvantages of vacuum tubes and transistors in tuners.
- Know what types of tuner-RF amplifiers need neutralization and explain two methods of neutralization.
- Discuss briefly at least two types of mixers.
- Explain how a tube or transistor RF cascode amplifier works.
- List the RF frequencies for the picture carrier, color subcarrier, and sound carrier for Channels 2, 7, and 13. Know how to calculate these frequencies for any channel.
- Know how to localize some basic tuner troubles.
- Know how tuning voltages are supplied to a varactor tuner.
- Explain briefly how to calculate image frequencies.
- Discuss briefly AFT operation.
- Understand the requirements for the RF amplifier and mixer bandpass.
- Know how a dual-gate FET is used as an RF amplifier.
- Explain briefly how diode-band switching works.
- Know where the UHF tuner output is connected.
- Know how to calculate tuner oscillator frequencies for any channel.

10.1 TYPES OF TV TUNERS FOR VHF AND UHF RECEPTION

Television tuners can be classified in more than one way. There are **VHF tuners** and **UHF tuners, vacuum tube tuners** and **solid-state tuners.** There are tuners with both tubes and semiconductors, which are called **hybrid tuners.** In some of the earlier receivers, multichannel VHF tuning was achieved with a ganged set of spiral wound inductors and sliding contacts. These were called **continuous tuners.** Tuning of VHF stations is now almost always accomplished by incremental steps for tuning from station to station. These tuners are sometimes called **step tuners** or **incremental tuners.**

WAFER-TYPE INCREMENTAL TUNERS

A vacuum tube type of incremental VHF tuner is illustrated in Figure 10.1A. This tuner uses frame grid tubes and printed circuit (PC) wafer sections. It is designed specifically for color sets. It also features preset oscillator frequencies.

The tuner shown in Figure 10.1B is a completely transistorized VHF tuner. It also uses PC wafer sections.

Tuners incorporating incremental inductors are called **continuous tuners.** They

(A)

(B)

(C)

Figure 10.1A-C. Three examples of VHF tuners. **A.** A vacuum-tube type, using frame grid tubes and printed circuit wafer sections. **B.** A transistor type, using printed circuit wafer sections. **C.** Exploded view of transistor VHF tuner. *(**A** and **B** courtesy of Oak Manufacturing Co.; **C** courtesy of Zenith Radio Corp.)*

change channels by a progressive shorting out of sections of the total inductance. Connections for the various sections are brought out at wafer contacts. In such layouts it is important to keep the length of wiring leads to the barest minimum. This will avoid bringing in undetermined amounts of inductance and unwanted coupling effects to neighboring circuit elements. The latter condition takes on an added importance when one recalls that modern tuners have a high component density relative to their prede-

(A)

(B)

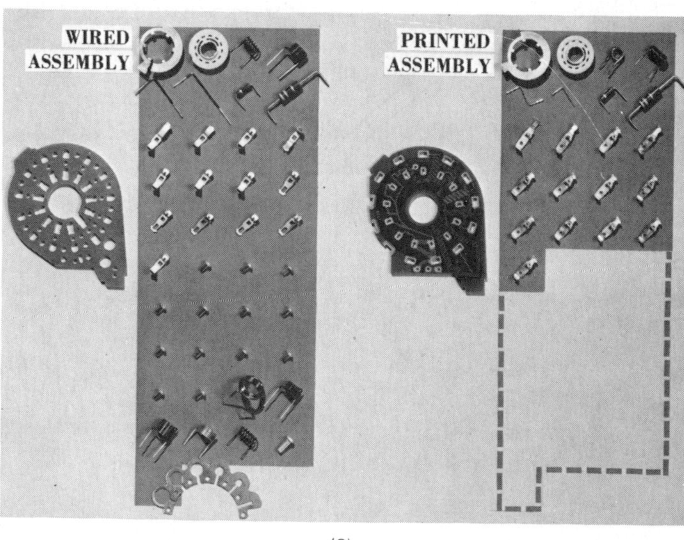

(C)

Figure 10.2A-C. Wafer construction for VHF TV tuners. **A.** Conventional wafer construction. **B.** Printed circuit wafer construction. **C.** Comparison of number of parts in conventional and printed circuit wafer construction. *(Courtesy of Oak Manufacturing Co.)*

cessors of the late 1940s. (*Component density* refers to the compactness of electronic circuits.)

Figure 10.1C is an exploded view of the important details of another transistorized VHF tuner.

Typical Tuner Wafer The construction of a typical tuner wafer is shown in Figure 10.2A. Note that for Channels 2 through 6, individual coils are connected in series and are progressively shorted out as the channel frequency increases. For Channels 7 through 13, sections of a stamped inductance ring are shorted out for higher channel frequencies. Oscillator wafers are generally tuned by individual screw adjustments accessible through the front panel of the set. There is a separate adjustment for each channel. These generally function by varying the effective inductance for each channel. RF and mixer wafers are not tuned for each channel but have an overall tuning adjustment that is preset at the factory. However, the tuning can be readjusted by a technician if necessary.

PC Wafer A newer type of tuner uses a PC wafer (Figure 10.2B). In this tuner, all of the inductances except those used for Channels 6 and 13 have been replaced by PC inductances. This method of construction has several advantages over that shown in Figure 10.2A: (1) greater reliability, (2) greater uniformity of alignment, (3) lower cost, and (4) greater versatility with regard to accommodating future designs.

A comparison between the parts used in the wired assembly of Figure 10.2A and the PC version of Figure 10.2B is made in Figure 10.2C. Note the simplicity of the PC design. The PC wafer is tuned by using fixed capacitors and by varying the inductances (coils) used for Channels 6 and 13. The tuner shown in Figure 10.1A is constructed with PC wafers.

TURRET-TYPE TUNERS

Turret VHF tuners are made up of a bank of precisely wound coils with tight tolerances. The coils are switched into or out of the tuner circuit as the channel selector knob is turned from channel to channel. Both wafer switching and turret tuning are widely used and work remarkably well.

Resetability is the ability of a tuner to repeatedly retune to exactly the same frequency after being switched. Resetability is better for turret tuners than for their wafer counterparts. For this reason, color receivers often have turret tuners. Figure 10.3 illustrates a turret tuner.

UHF TUNERS

All television sets are required by FCC regulations to have both a VHF and a UHF tuner. The UHF tuner is separate from the VHF unit and is tuned by a separate control knob. Whereas VHF tuners are generally of the incremental type, many UHF tuners are of the continuous-tuning type. It has been found, however, that most viewers find it difficult to tune in UHF stations with a continuous tuner, and so in more recent models UHF stations are tuned either by switching or by push buttons. Many of these tuners use varactors for channel switching. (Chapter 11 discusses the use of varactors in automatic fine tuning circuits.)

ELECTRONIC TUNING

Electronic tuning of all VHF and UHF channels is accomplished with the aid of varactors. The desired varactor tuning voltages are obtained by one of two basic methods: mechanical switching or digital switching.

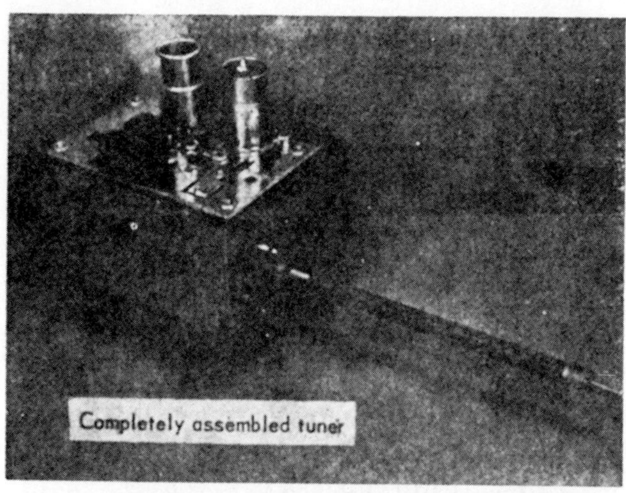

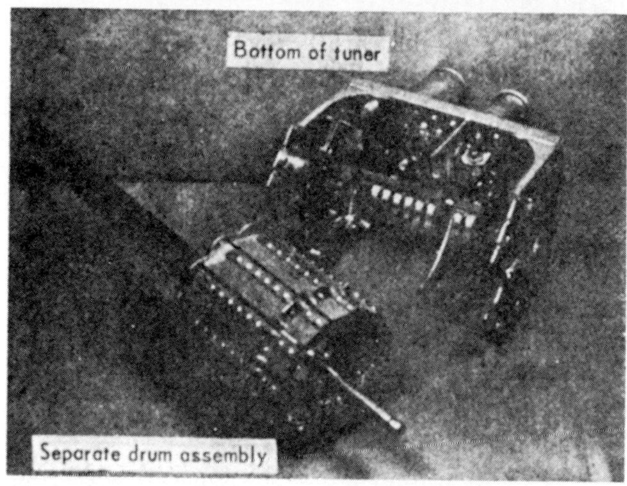

Figure 10.3 A VHF turret tuner. Each strip holds the RF, mixer and oscillator coils for one VHF channel. *(Courtesy of Magnavox Consumer Electronics Co.)*

Mechanical Switching The simplest method of electronic tuning is by **mechanical switching.** In this system, an array of 16 or 20 potentiometers are connected between a source of constant DC voltage and ground. Each potentiometer is preset to develop a voltage that can tune the front end to a VHF or UHF channel. The appropriate tuning voltage for each channel is then selectively routed to the varactors by means of a rotary switch or labeled push buttons. In some television receivers, digital circuits instead of potentiometers generate the tuning voltages.

The VHF frequency range (54 to 216 MHz) has a ratio of 1 to 4. This is too great for capacitive tuning by a varactor diode. Consequently, part of the tuning inductances are shorted out to reduce inductance for Channels 7 through 13. This shorting is accomplished by energizing switching diodes that are connected across a portion of the inductances (see Figure 10.38).

Disadvantages of Mechanical Switching Although mechanical switching is widely used, it does have some disadvantages.

These include (1) mechanical wear and tear, (2) dirty or loose contacts, and (3) the inconvenience of switching through a number of channels to reach the desired one.

Digital Switching The disadvantages of mechanical switching can be overcome with a **digital switching system.** Such systems have no moving parts, and channels are selected by push button. In addition, **random access** digital systems make it possible to switch from any VHF or UHF channel to any other channel in a fraction of a second. This makes UHF channel selection as simple as VHF channel selection.

In many of these systems, the number of the selected channel appears digitally by means of LEDs (light-emitting diodes). The number may appear as a front-panel display or momentarily on the picture tube.

Omega System In the Omega digital tuning system, potentiometers are not used to obtain tuning voltages. Instead, a computerized, solid-state memory stores digital equivalents of the required tuning voltages for each channel. As needed, this data is retrieved from the memory, processed, and applied to the varactor tuner as tuning voltages. (Digital tuning will be described further in Chapter 11.)

RCA Frequency-Synthesized (FS) Tuning In the RCA frequency-synthesized system, fine tuning is eliminated by the use of a digital and phase-locked loop (PLL) arrangement. This scheme comes in three variations: (1) keyboard control, (2) manual scan, and (3) remote scan.

The keyboard system is the most basic. It includes a ten-button keyboard entry panel and an LED channel number display. To select a channel, two numbered buttons are pressed. For example, for Channel 4, the buttons marked 0 and 4 are pressed.

In the manual scan system, a channel up/down selector button is provided. Holding this button in selects channels in numerical order, at one-half-second intervals. A memory device permits the system to skip nonactive channels.

The remote scan system works essentially the same as manual scan. The basic difference is that channel changes (and other functions) are made from a position remote from the receiver. Momentary on-screen display of channel number and time of day is provided.

10.2 ELECTRICAL CHARACTERISTICS OF TUNERS

To understand the electrical functions of the individual components of a television tuner, it is essential to divide the tuner into convenient sections. These subsections of a VHF tuner are shown in the functional block diagram of Figure 10.4. This diagram shows a balun and an IF trap at the input. The **balun** matches the 300-ohm transmission line (commonly used to bring the antenna signal to the receiver) to the 75-ohm input impedance level. Input impedances of most RF amplifiers are frequency-dependent, but a 75-ohm impedance is a representative figure for the mid-VHF band.

The **IF trap** keeps any frequencies transmitted in the IF band from being injected into the mainstream signal flow. The balun and IF trap are normally placed side by side on a separate mount and are located outside of the encased tuning mechanism of the tuner. This arrangement allows for easy access to the traps for alignment purposes. Often the mount also carries a pair of capacitors, one in each lead, to prevent any damage to the receiver by lightning, etc. (C_1 and C_2 in Figure 10.5A). Figure 10.5 shows the input circuitry of two VHF tuners.

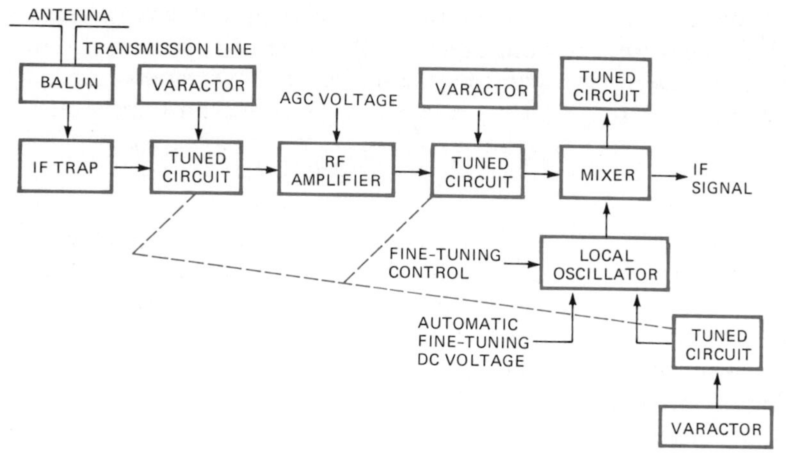

Figure 10.4 Block diagram of a VHF tuner. The tuned circuits connected with a dashed line are varied simultaneously.

THE BALUN

The **balun** is a matching transformer with a ferromagnetic core. On this core have been wound four tightly coupled and evenly spaced windings. These windings provide an impedance transformation from a 300-ohm balanced source to either a 75-ohm or a 300-ohm unbalanced load. Since the resistive component of the input impedance of most RF devices at VHF is close to 75 ohms, the balun is normally used to provide a 75-ohm

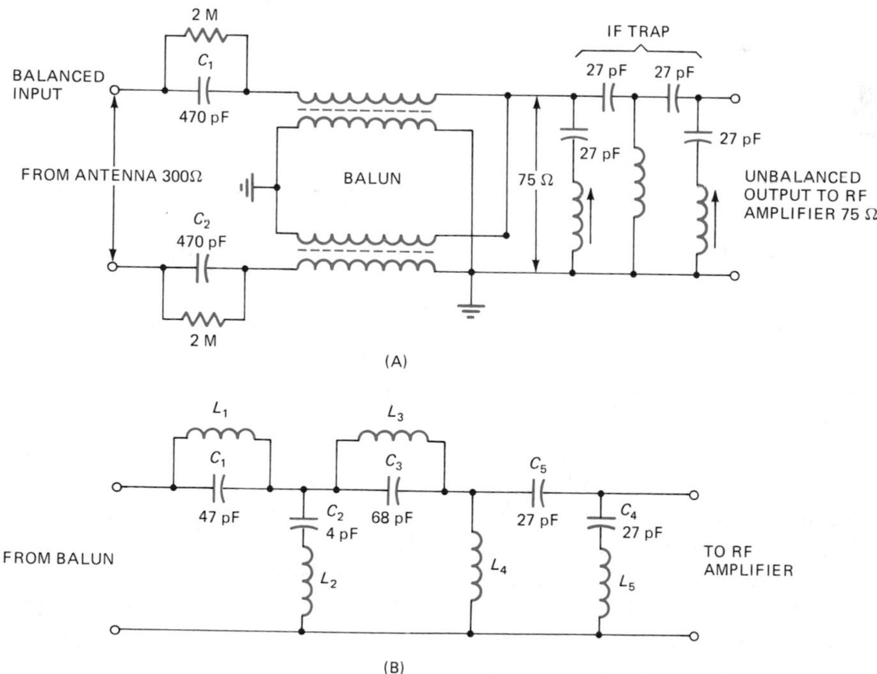

Figure 10.5 Typical input circuitry for VHF tuners.

output transformation. Figure 10.5A shows this configuration. The ferromagnetic core helps to keep the impedance transformation uniform throughout the range of frequencies of interest.

IF TRAP

One of the sources of interference outside the signal passband occurs in the IF band when the receiver is tuned to the lowest channels. The video and sound carriers of Channel 2 are at 55.25 MHz and 59.75 MHz, respectively. This is so close to the IF band pass of TV receivers (41 to 46 MHz) that the natural attenuation of the tuner signal circuits may not adequately reject these RF frequencies. In the **IF trap,** which follows the balun, a number of series-tuned or parallel-tuned subsections prevent interfering signals near the IF range from penetrating the RF amplifier and reaching the mixer. (Undesired signals that reach the mixer, or IF system, are more difficult to reject.)

Some of the inductors that make up the traps are accessible and can be adjusted during alignment to give a proper IF response at the input of the IF amplifier. Two versions of the trap are shown in Figure 10.5. One acts like a high-pass filter for frequencies beyond about 40 MHz (Figure 10.5A). In the trap shown in Figure 10.5B, additional tuned networks (L_1 and C_1; L_2 and C_2) are built in to reject FM broadcast signals, which occupy the range from 88 to 108 MHz.

The RF amplifier (Figure 10.4) boosts the incoming VHF signal before it is mixed with the local oscillator to produce an IF signal. The design of this amplifier is one of the key factors in the overall performance of the tuner, since the amplifier handles the signals at their weakest level. Thus, in addition to magnifying the signal, the amplifier must not degrade the noise performance of the tuner. Also, the response should be as uniform as possible over the entire VHF range from Channels 2 through 13. Tuning is done with four varactors.

THE UHF TUNER

The input requirements of UHF tuners are quite different from those of VHF units. Figure 10.6 shows the UHF tuner in block diagram form. IF traps are almost never used, primarily because the frequencies of the lowest UHF channel (Channel 14) are well above the highest IF channel frequency. Secondly, any passive network, such as a trap, reduces signal strength because of its insertion loss. This adversely affects the signal-to-noise ratio.

The UHF antenna signal input is coupled to the MOSFET RF amplifier. (Earlier models did not use RF amplifiers in the UHF tuner because suitable transistors or tubes were not available.) The RF amplifier is gain-controlled by automatic gain control (AGC) voltage. The circuits are tuned by three tracking varactors. The output of the RF amplifier, together with the UHF oscillator signal, is applied to the diode mixer. Then, the IF output of the mixer is applied to the IF buffer amplifier. This latter stage helps to stabilize the mixer frequency response across the UHF band. From the IF buffer amplifier, the signal goes to the input of the VHF mixer, which functions only as an IF amplifier when receiving UHF channels. In this case, the VHF oscillator and RF amplifier are inoperative.

10.3 INTERFERENCE, STABILITY, AND NOISE PROBLEMS IN TUNERS

In addition to converting an RF signal to an IF signal, the television tuner must provide immunity against undesired signals and

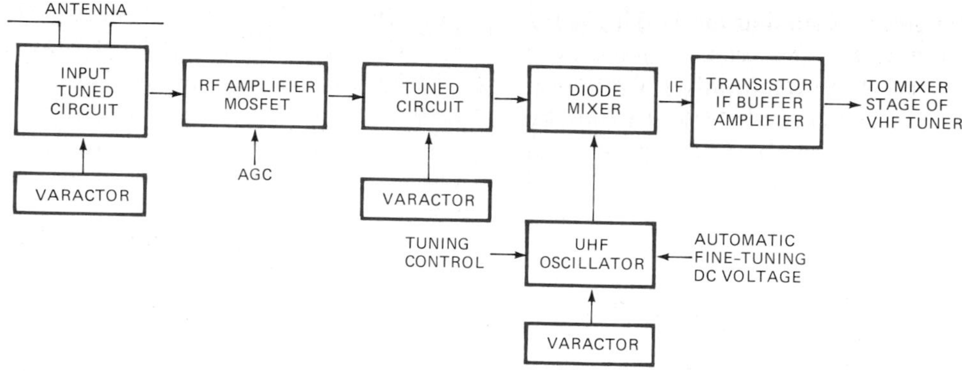

Figure 10.6 Simplified block diagram of a varactor-tuned UHF tuner.

must keep the overall noise level as low as possible. Interference problems will be discussed first.

IMAGE REJECTION

Normally, the local oscillator frequency is above the incoming RF frequency by an amount equal to the IF frequency of the receiver. In other words, *oscillator frequency minus RF frequency equals IF frequency*. With all superheterodyne receivers, there is a problem with image frequencies. The image frequency is above the oscillator frequency by an amount equal to the IF frequency of the receiver. Thus, *image frequency minus oscillator frequency equals IF frequency*. We see, then, that there are two different signals that can heterodyne (mix) with the local oscillator frequency to produce frequencies in the IF range. These two signals are the desired RF frequency from the station being tuned in and an *undesired image frequency*, which comes from a station whose carrier frequency is above the local oscillator frequency by an amount equal to the IF frequency.

The combined selectivity of the RF input coupling circuits and output circuits must be adequate to reduce tuner response to image frequencies to a level that is below the visible and audible levels. For IF frequencies of 41.25 MHz (sound) and 45.75 MHz (video), the corresponding image frequencies are 82.50 MHz (sound) and 91.50 MHz (video) away from the desired signal inputs. Since the television broadcast spectrum is quite wide, it is possible that these image frequencies represent another channel operating nearby. However, frequency separation is large enough to ensure that the images are rejected, to the extent of 60 dB or better, by the tuned circuits of the RF amplifier. In fact, one of the reasons for using higher IF frequencies is to achieve better image rejection when resonant circuits of moderate Q values are used.

IF-BAND INTERFERENCE

Another source of interference is the IF band itself. IF frequencies generated outside the television receiver can find an easy passage through the tuner to the IF amplifier. This is especially true when the tuner is tuned to a low channel. As already mentioned, special filter traps tuned to the sound and video IF frequencies are placed at the input of the RF amplifier to reduce this interference. An overall IF attenuation of 60 dB is adequate for acceptable tuner performance.

LOCAL OSCILLATOR RADIATION

Electromagnetic radiation from the local oscillator of one television receiver can be the source of considerable interference at a neighboring receiver. In the early years of television receiver development, this fact was not fully understood. After a time, though, the FCC set limits on the allowable field strength of radiation for both VHF and UHF tuners. The RF amplifier in a VHF tuner prevents this radiation from traveling to the antenna. Proper cable and case shielding cuts down on interference radiation by the local oscillator.

OSCILLATOR STABILITY

The need for a stable oscillator cannot be overemphasized. Any frequency changes that result from the inability of the step-tuning mechanism to resonate the circuits to produce the IF from exact channel frequencies, can be taken care of by the fine-tuning adjustment. What is usually more of a problem is the oscillator drift that takes place when the temperature of the tuner changes. At 25 °C (ordinary room temperature), the local oscillator can drift by as much as 1 MHz as a result of temperature changes in the tuner. This can throw off the sound and video performance and result in a complete loss of the color signal.

Fortunately, adequate design compensates for such drift in both vacuum tube and solid-state tuners. For instance, automatic fine tuning (AFT) circuits lock the local oscillator into the desired frequency so that temperature changes cannot cause drift. (These circuits are called automatic frequency control (AFC). However, there is also an AFC circuit in the horizontal sweep circuit, and so AFT is preferred here to avoid confusion.)

SUPPLY VOLTAGE CHANGES

Another thing that can cause oscillator drift is a change in the DC supply voltage resulting from poor regulation. A transistor oscillator can be particularly sensitive to such effects. For example, when the DC voltage changes from 7 to 10 V, the oscillator frequency changes from 250 to 251 MHz. Many receivers now have regulated power supplies, which deliver a constant DC voltage despite changes in load and changes in the line voltage input. This eliminates the problem of oscillator drift caused by voltage changes.

NOISE PROBLEMS

Electrical noise may originate from such external sources as poorly maintained electric machinery in the vicinity of the receiver or from sunspots. Noise may also be produced in the circuit elements themselves. For instance, **thermal noise** is generated in resistors because their atomic components are affected by temperature. **Shot noise,** which is a serious problem in vacuum tubes, is caused by the random fluctuation of the plate current. (Transistors also produce electrical noise.) Thermal noise and shot noise are not frequency-dependent.

Since the tuner handles the RF signal at its lowest amplitude, the input stage, the RF amplifier (if one is used), must be designed so that the signal-to-noise ratio is optimum. A poor signal-to-noise ratio produces snow on the picture tube screen.

TUNER SPECIFICATIONS

Having introduced the various elements that affect the performance of a television tuner, we can now list the "specifications"

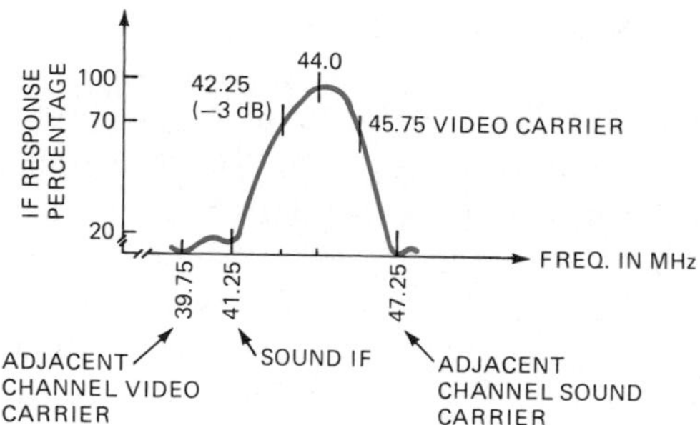

Figure 10.7 Typical monochrome response curve for a tuner plus IF amplifier.

for one. The overall response for a monochrome TV receiver is sketched in Figure 10.7. The shape of the response curve depends a great deal on the load presented to the tuner. For this reason, receiver manufacturers almost always specify the IF amplifier input conditions that the tuner must match.

The IF frequency ranges in Table 10.1 have been given as 41.25 to 45.75 MHz. All recently manufactured receivers have IF frequencies in this range, but there may be some older sets in operation that have lower IF ranges—usually in the range from 20 to 30 MHz. It is necessary to consult the manufacturer's literature to determine IF values for these older receivers.

10.4 CHANNEL ALLOCATIONS

Table 10.2 shows the frequency allotments for the various television channels in the United States. The entire spectrum allocated for television transmission is divided into 12 channels in the VHF region and 70 channels in the UHF region. The lower and upper limits of each channel are 6 MHz apart, and this value is commonly referred to as the **bandwidth of the channel.** This should not be confused with the formal definition of the term *bandwidth,* which is the frequency region where the voltage (or current) response is equal to or greater than 0.707 times the peak value.

Table 10.2 also lists the picture and sound carrier frequencies specified for all

Table 10.1 **GENERAL SPECIFICATIONS FOR A TELEVISION TUNER**

Video IF	45.75 MHz
Sound IF	41.25 MHz
Relative response	See Figure 10-7. Depends on load conditions
Image rejection	VHF—Better than 60 dB minimum
	UHF—Better than 70 dB minimum
Adjacent channel rejection	Better than 60 dB (min.)
Noise figure	VHF Tuners—5 to 8 dB (typical)
	UHF Tuners—7 dB (typical minimum)
	11.5 dB (typical maximum)
RF amplifier gain	Ch. 2-6—28 dB minimum
	Ch. 7-13—22 dB minimum
Oscillator frequency resetability	Better than ±400 kHz for black and white units
	Better than ±100 kHz for color units
Oscillator frequency drift	For a temperature change of 22°C from ambient:
	125 kHz maximum
	For a 10% change in the DC supply voltage:
	50 kHz maximum
Fine-tuning range	±0.6 MHz (Average)

Table 10.2 **U.S. TELEVISION CHANNEL FREQUENCIES (MHz)**

SOUND IF 41.25 PICTURE IF 45.75

	CHANNEL NO.	*FREQ. LIMITS*	*CENTER FREQ.*	*PICTURE CARRIER*	*SOUND CARRIER*	*OSC. FREQ.*	*PICTURE IMAGE FREQ.*	2 × *OSC.* + *PIX IF*	2 × *OSC.* − *PIX IF*	*CHANNEL NO.*
VHF	2	54–60	57	55.25	59.75	101	146.75	247.75	156.25	2 VHF
	3	60–66	63	61.25	65.75	107	152.75	259.75	168.25	3
	4	66–72	69	67.25	71.75	113	158.75	271.75	180.25	4
	5	76–82	79	77.25	81.75	123	168.75	291.75	200.25	5
	6	82–88	85	83.25	87.75	129	174.75	303.75	212.25	6
	7	174–180	177	175.25	179.75	221	266.75	487.75	396.25	7
	8	180–186	183	181.25	185.75	227	272.75	499.75	408.25	8
	9	186–192	189	187.25	191.75	233	278.75	511.75	420.25	9
	10	192–198	195	193.25	197.75	239	284.75	523.75	432.75	10
	11	198–204	201	199.25	203.75	245	290.75	533.75	444.25	11
	12	204–210	207	205.25	209.75	251	296.75	547.75	456.25	12
	13	210–216	213	211.25	215.75	257	302.75	559.75	468.25	13
UHF	14	470–476	473	471.25	475.75	517	562.75	1079.75	988.25	14 UHF
(INCOMPLETE)	15	476–482	479	477.25	481.75	523	568.75	1091.75	1000.25	15
	54	710–716	713	711.25	715.75	797	802.75	1559.75	1468.25	54
	55	716–722	719	717.25	721.75	803	808.75	1571.75	1480.25	55
	56	722–728	725	723.25	727.75	809	814.75	1583.75	1492.25	56
	82	878–884	881	879.25	883.75	965	970.75	1895.75	1804.25	82
	83	884–890	887	885.25	889.75	971	976.75	1907.75	1816.25	83

Note: All frequencies are in megaHertz.

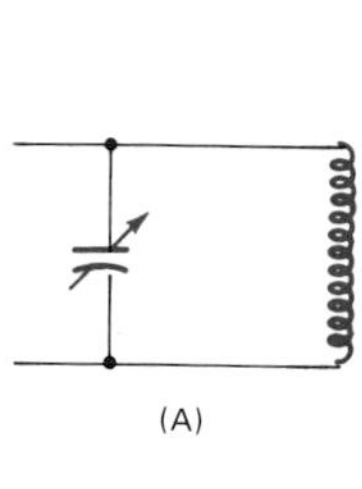

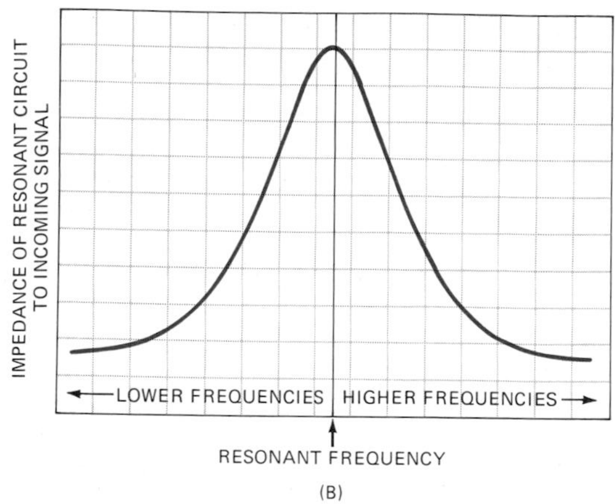

Figure 10.8A-B. A parallel tuning circuit and its response curve.

tabulated channels, the picture image frequencies (oscillator frequency plus picture IF frequency), and the location of spurious responses.

The television sound signal is transmitted by frequency modulation, and the video signal amplitude modulates the channel picture carrier. The video IF is 45.75 MHz, and the corresponding sound IF is 41.25 MHz. Thus, the IF picture and sound carrier spacing is 4.5 MHz. In order to meet the requirements of this system, the receiver IF section must pass the entire 4.5-MHz band plus the lower picture IF sideband and the upper sound sideband.

The reader is encouraged to complete the information in Table 10.2 for the missing UHF channels as an exercise. The UHF spectrum is continuous, but the VHF allocation has two frequency gaps.

10.5 CHARACTERISTICS OF TUNED CIRCUITS

The television signal occupies a 6-MHz band in the radio spectrum. This range is far greater than anything we receive with an ordinary radio set. This problem is solved at the television receiver in the RF and mixer stages. The response of the tuned receiving circuit should be uniform throughout the 6-MHz band. Yet it should also be selective enough to discriminate against unwanted image frequencies or stations on adjacent bands. Before the circuits of the RF and mixer stages are considered, it will be helpful to discuss wide-band tuning circuits.

PARALLEL-TUNED CIRCUITS

A single coil and capacitor, connected as shown in Figure 10.8A, form a parallel-tuning circuit. Figure 10.8B shows the variation of impedance that this combination presents at or near the resonant frequency. At frequencies below resonance, the parallel combination acts as an inductor with a lagging current, and impedance drops off rapidly to a fairly low value. Above resonance, the effect is capacitive with a leading current, and again impedance decreases quite rapidly. At the resonant point, capacitive and inductance reactances cancel each other and the impedance becomes high and wholly resistive.

EFFECT OF Q

Sometimes more specific information than that shown in Figure 10.8B is neces-

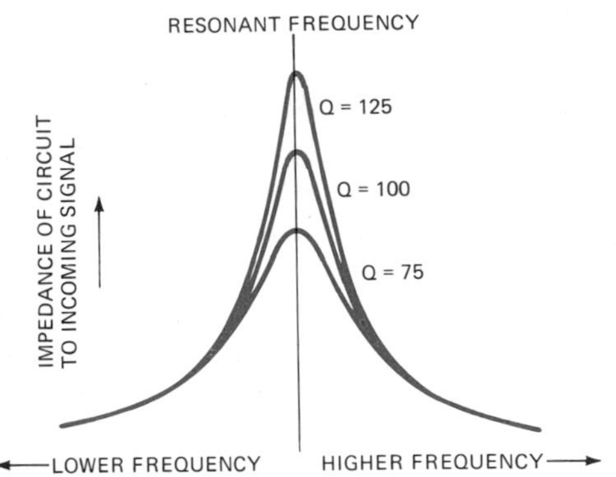

Figure 10.9 The variation in the response curve with different values of Q.

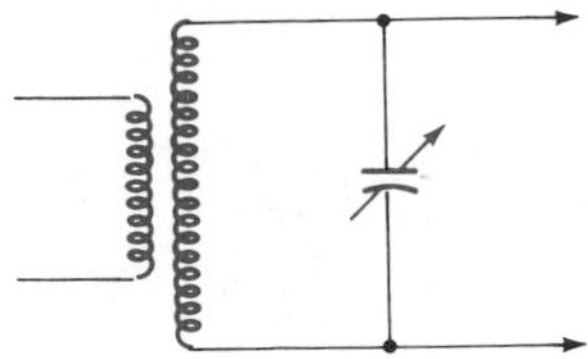

Figure 10.10 A common form of transformer-coupled tuning circuit used in receivers.

sary. Hence, in Figure 10.9 several resonant curves have been drawn, each for a circuit having a different value of Q, which is the ratio of inductive reactance to coil resistance. The value of Q indicates two things:

1. The sharpness of the resonant curve near the resonant frequency (in other words, the selectivity of the tuning circuit)
2. The amount of voltage developed by the incoming signal across the resonant circuit at resonance

For any given circuit, the greater its Q value, the more selective the response and the greater the voltage developed. While these factors may be highly desirable, they are useful only if they do not interfere with the reception of radio signals.

At broadcast frequencies, each station occupies a band 10 kHz wide. Within this region, uniform response is desirable. However, the sharply peaked curve of Figure 10.8B shows that response is not equal at all points within this region. The portion of the signal exactly at the resonant frequency, for example, develops a greater voltage across the resonant circuit than those portions at the outer fringes, plus and minus 5 kHz away. A coil and capacitor combination having a lower Q would give a more uniform response and might be chosen over one with a higher Q. It is true that less voltage would result from this change. With the advent of high-gain tubes and transistors, however, amplification is not too serious a problem. Emphasis today is on fidelity, which is also necessary for the reproduction of images in television receivers.

TRANSFORMER COUPLING

The simple circuit just described is sometimes used by itself for tuning. A more usual combination, though, is an untuned primary coil inductively coupled to a tuned secondary coil (Figure 10.10). With this form of coupling, additional gain may result from having more turns in the secondary coil than in the primary coil. The stepped-up voltage applied to the grid of the next stage is greater than that obtained with the single coil-capacitor combination. Just how big the voltage increase is depends upon the design of the coils.

The shape of the response curve of the primary circuit depends to a great extent upon the degree of coupling between the coils. When the coefficient of coupling k is low (that is, when the coils are relatively far apart), the interaction between the coils is small. In this case, the secondary response curve retains the shape shown in Figure 10.8B.

INCREASED K

As the coupling coefficient *k* increases, the secondary circuit reflects a larger impedance into the primary and the primary current is affected more by variations in the tuning of the secondary capacitor. This, in turn, changes the number of flux lines cutting across the secondary coil. The end result is a gradual broadening of both primary and secondary response curves. With very close coupling, the secondary response curve may continue to broaden and even develop a slight dip at the center. The dip, however, will never become too pronounced.

Both Circuits Tuned The discussion so far has dealt with coupled circuits where the primary is untuned. Hence, no matter how close the coupling is, the secondary curve will essentially look like the curve in Figure 10.8B. With *two* tuned circuits coupled together, however, such as IF transformers, the effect of each circuit on the other becomes more pronounced. With close coupling of two tuned circuits, the familiar double-humped curve of Figure 10.11 is obtained. The closer the coupling, the broader the curve and the greater the dip at the center.

INCREASING BANDWIDTH

For television reception, none of the coupling combinations just described provide the necessary uniform bandwidth. Loose coupling produces a response curve that is too sharp, and tight coupling decreases the voltage of the frequencies near resonance because of the dip. Between these two extremes, there is a semblance of uniform response about the center point of the curve but never across the entire 6-MHz spread. By shunting a low-valued resistor across the coil and capacitor, however, we can artificially flatten the curve to receive the necessary 6 MHz. The extent of the flat portion of the response curve depends inversely on the value of the shunting resistor. The higher the resistance, the narrower the uniform section (flat portion) of the curve. Hence, what we could not accomplish with a coil or a capacitor, we can do with a combination of the two plus a resistor.

One undesirable result of increasing the width of a response curve with resistors is the lower Q value that is obtained. As the value of Q decreases, the voltage developed across the tuned circuit becomes smaller for the same input. An inevitable reduction in output results. There are many ways of combining the tuned circuits and loading resis-

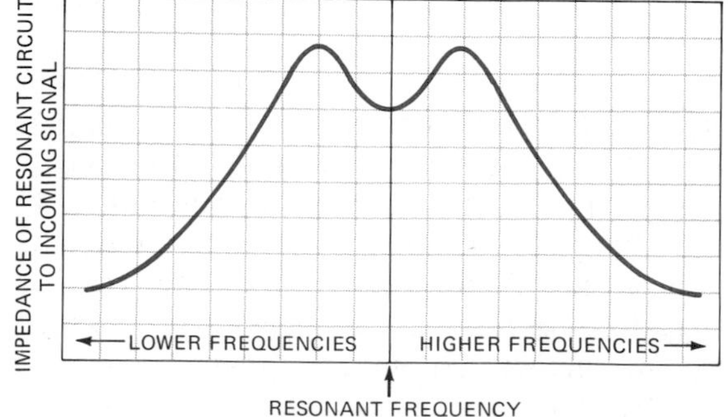

Figure 10.11 Close coupling between two tuned circuits produces this type of response curve.

tors to achieve optimum gain and selectivity. Several of the more widely used circuits will be discussed in the following section on RF amplifiers.

10.6 VACUUM TUBES FOR TUNERS

TRIODES

The ability of a receiver to amplify a signal is governed not only by the amplification obtained from the tubes but also by the noise generated in the tubes and in the associated receiver networks. The noise developed by the first stage (the RF amplifier) is the most significant because at this point the incoming-signal level is closer to the noise level than at any other point in the receiver. Once the signal becomes much larger than the noise, it can easily override the noise and hence mask its presence. Whatever noise voltage appears at the input of the RF amplifier is amplified together with the signal. To obtain a picture as free of noise spots as possible, there must be as much signal and as little noise as possible at the front end of the set.

The best choice for a low-noise tube is a **triode RF amplifier.** This is because noise originating in a tube varies directly with the number of positive elements in the tube and a triode has fewer such elements than a pentode. Hence, a triode is more desirable from a noise standpoint than a pentode. It is for this reason that high-frequency triodes are used in RF amplifiers.

A triode normally provides less gain than a well-constructed high-frequency pentode. As a result, special dual triodes have been developed for use in a circuit known as a **cascode amplifier,** in which two triodes are connected in series. In this arrangement the triodes provide about as much gain as a pentode. This gives the desired amplification at a lower noise level. Cascode amplifiers will be discussed at length later in the chapter.

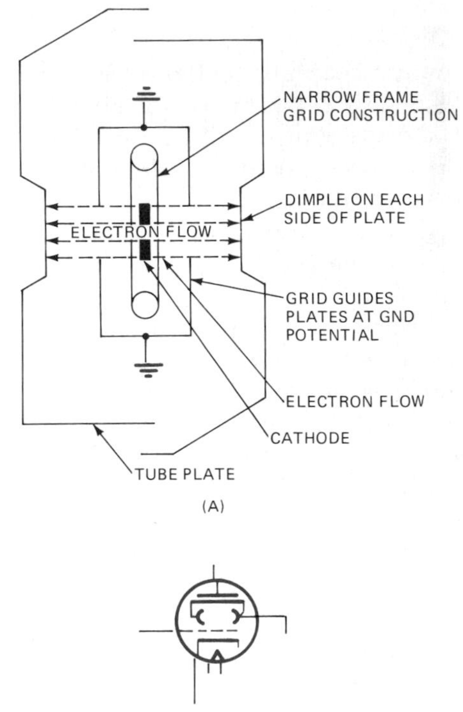

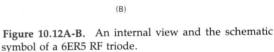

Figure 10.12A-B. An internal view and the schematic symbol of a 6ER5 RF triode.

Grid Guides A significant change has taken place in high-frequency triode construction. In an effort to achieve higher efficiency and reduced plate-to-grid capacity, extra elements called **grid guides** or **shield plates** have been inserted in the region between the grid and the plate. An internal view of the 6ER5, a tube of this group, is pictured in Figure 10.12. The grid guides are U-shaped plates that surround the grid. When the plates are grounded, a shield is inserted between the anode and the grid, reducing the capacity between these elements. In addition, the electron flow from cathode

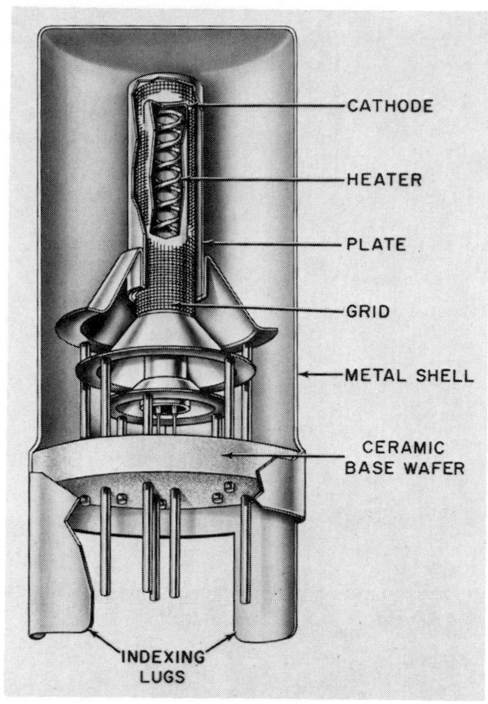

Figure 10.13 A cutaway view of a nuvistor tube. *(Courtesy of RCA Consumer Electronics.)*

to plate is concentrated into a smaller area, and the electrons strike the plate only at an indented section called a **dimple.** This concentration discourages sideway or random travel of electrons and reduces the noise produced by such flow. Note that the grid guides do not themselves intercept any electrons traveling from cathode to plate. Hence they are not another element, which is why these tubes are still considered triodes. Because of the dimples on the plate, this element can be placed closer to the grid. This reduces electron transit time, a desirable feature at VHF.

Nuvistor Tubes Figure 10.13 shows a cross section of a **nuvistor tube** which is a miniaturized triode. The overall length of this tube is only about 0.8 in (2.0 cm), and its diameter is 0.435 in (1.1 cm) at the widest point. All of the nuvistor electrodes are proportionately smaller than those of a conventional triode. This permits extremely close grid-to-cathode spacing, with a high mutual conductance. A typical value of mutual conductance for a nuvistor tube is 15 000. Additional advantages are a very low filament current (typical value, 135 mA) and a low plate voltage resulting in low tube noise.

Although the interelectrode capacitance is very small in a nuvistor, it must still be neutralized when used as an RF amplifier in tuner circuits. This is true of all triode amplifiers. The neutralizing circuits will be discussed later in this chapter.

TETRODES

High-frequency **tetrode tubes,** for example, the 6CY5, have also been developed specifically for use as RF amplifiers in television receivers. These tubes have high mutual conductance (on the order of 8 000 to 10 000), and they do not develop much more internal noise than triodes. The high mutual conductance is achieved by using fine (8-mil) wire for the control grid. The low noise level is partly due to the fact that the plate takes considerably more current than the screen grid (the ratio is approximately 7:1). Since current division is one of the governing factors of internal noise in a tube, the relatively small percentage of the total current captured by the screen grid helps to keep the noise level low.

10.7 TYPICAL TUBE-TYPE RF AMPLIFIERS

The typical television receiver RF stage is very similar to the same stage in AM broadcast receivers. It has three functions. First, it amplifies the signal in that part of the set where the signal is at its lowest value. In outlying reception areas or noisy locations, this extra amplification may be the deciding factor in obtaining satisfactory reception. Second, it discriminates against signals in adjacent bands. This is especially applicable to image frequencies. A properly designed RF stage will help the signal to override any small disturbances produced in the tubes themselves. (The internal tube disturbance is known as **noise.**) In television receivers these internal disturbances are amplified along with the video signal. If they are stronger than the received signal, they will appear as small black or white spots on the screen. These spots are often referred to as snow, or **masking voltages.** Finally, the RF amplifier reduces local oscillator radiation, which can be quite offensive to neighboring receivers.

The tube used in the RF stage, besides having low noise and high mutual conductance, should also have a remote cutoff characteristic. With remote cutoff properties, the stage does not distort as readily when strong input signals are received. Furthermore, **automatic gain control** (AGC) voltages can be applied to the tube. This makes the amplifier more stable and helps maintain a steady signal output. (AGC in a television receiver is similar to AVC in a radio receiver.)

TRIODE RF AMPLIFIERS

The low-noise qualities of triodes make them attractive for the RF amplifier stage, and a number of circuits that use these tubes have been designed. The simplest uses a single triode (Figure 10.14A). The low plate resistance loads the tuning circuits sufficiently to achieve the desired bandwidth without the need for any external shunting resistors. A high mutual conductance helps to achieve fairly good gain even at these VHF frequencies.

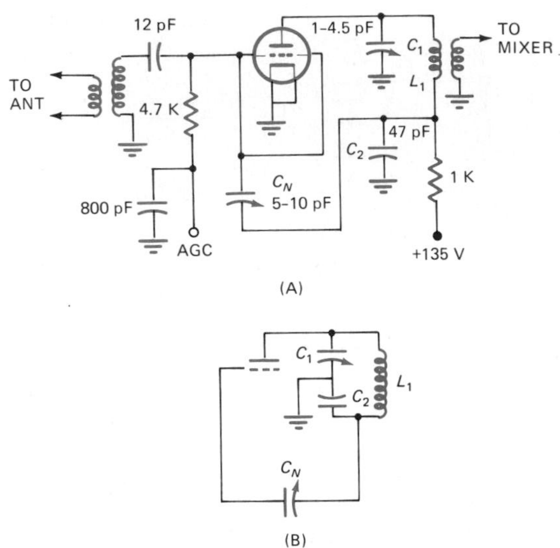

Figure 10.14A. A triode RF amplifier. **B.** Part of the circuit rearranged to better reveal out-of-phase voltage applied to grid to prevent oscillation.

When used as shown in Figure 10.14A, a triode tube requires a **neutralizing network** to prevent oscillation at high frequencies. This need stems from the relatively large capacitance that exists between the plate and grid elements inside the tube. At sufficiently high frequencies, the output signal can use this capacitance to travel from the plate to the grid, and this can cause oscillation. A neutralizing network feeds an out-of-phase signal of the same amplitude back to the grid. In this way, the first signal is effectively neutralized and oscillation avoided.

Neutralizing Signal The required out-of-phase voltage for the grid is obtained at the bottom end of the plate coil (L_1) across C_2 (Figure 10.14). The RF voltage at the plate end of the coil is 180° out of phase with the voltage at the other end. The capacitor C_N feeds back to the grid as much out-of-phase voltage as necessary to neutralize the signal

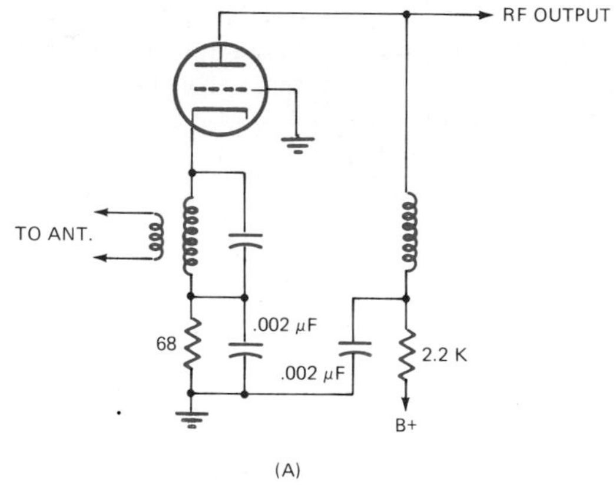

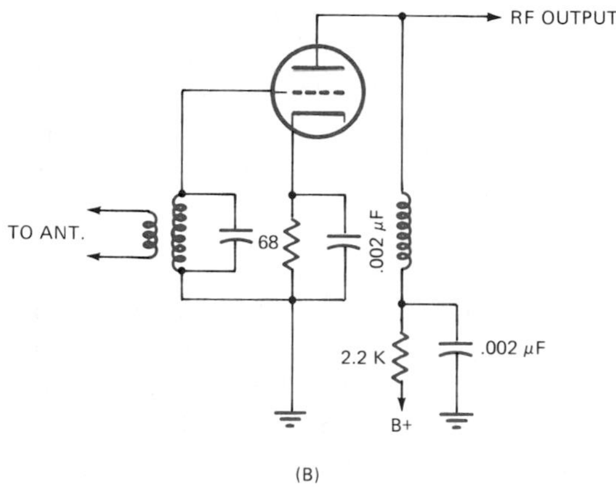

Figure 10.15 A comparison between **A.** grounded-grid and **B.** conventional RF amplifiers.

voltage that reaches the grid by way of the plate-to-grid interelectrode capacitance.

The tube actually has two grid leads and two cathode leads. This minimizes lead inductance, which in turn permits a single neutralization adjustment to be effective throughout the entire VHF band. If such inductances were permitted to become high enough, operation on Channels 7 to 13 would require one neutralizing adjustment and Channels 2 to 6 would require another.

Grounded-Grid Amplifiers Triode RF amplifiers are often used in an arrangement known as a **grounded-grid amplifier.** This type of amplifier is contrasted with a conventional amplifier in Figure 10.15. Note that the grid of the tube is at RF ground potential and that the signal is fed to the cathode. The tube still functions as an amplifier because the flow of the plate current is controlled by the grid-to-cathode potential. Instead of the grid potential being varied and the cathode held fixed, the grid is fixed and the cathode potential is varied. The net result is the same. In addition, the grid being grounded acts as a shield between the input and output circuits. This prevents feedback of energy that is essential to the development of oscillations. Thus, neutralization is not needed.

The grounded-grid amplifier also offers low input impedance, enabling the amplifier to match the antenna transmission-line impedance. The low impedance provides a broader band-pass response curve, which is particularly desirable for 6-MHz television signals.

Cascode Amplifiers Still another RF amplifier arrangement that uses triodes is the **cascode amplifier** (Figure 10.16). Here, two triodes are connected in series, that is, the plate of the first section goes directly to the cathode of the second section. The same current flows through both tubes, and the amplitude of this current is controlled by the AGC bias on the first triode.

The input-tuned circuit of this series amplifier connects to the control grid of the first triode. The output-tuned circuit is in the plate lead of the second triode. The first stage is operated as a conventional amplifier, that is, with the input signal applied to the grid and the output signal obtained from the plate. The second stage is used as a grounded-grid amplifier. The inductor L_1 between the two stages helps to neutralize the grid-to-plate capacitance of the first triode

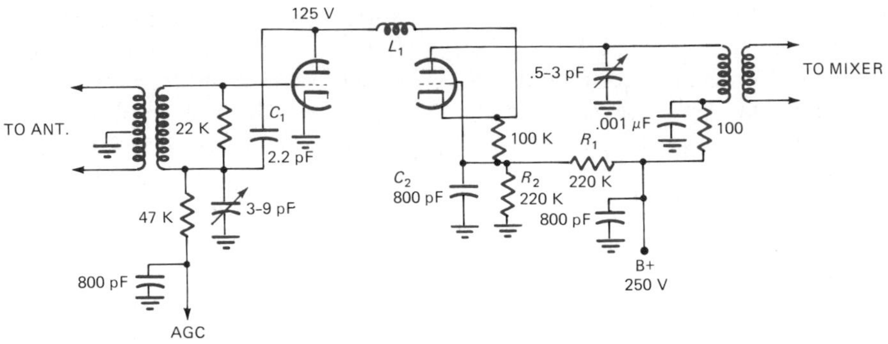

Figure 10.16 A cascode RF amplifier.

(with help from capacitor C_1). It is designed to resonate with the grid-cathode capacitance of the second section on the high VHF channels. In addition to contributing to the stability of this combination, inductor L_1 is also largely responsible for the low-noise qualities of the cascode circuit.

The role of capacitor C_1 in neutralizing the input triode to prevent it from oscillating can be understood by noting that this capacitor connects from the plate of the first triode to the bottom end of the coil in the grid circuit. Thus, it feeds its signal to the bottom end of this coil at the same time that the grid-to-plate capacitance in the first triode feeds back its signal to the top of the coil. In this way we achieve the 180° phase reversal required for the two voltages to counterbalance and neutralize each other.

Direct coupling is used between the first triode plate and the second triode cathode. With cathode feed to the second triode, C_2 is used to place the grid at RF ground potential. Since the two triodes are in series across a common plate supply, the cathode of the second triode may be assumed to be 125 V positive with respect to chassis ground. A divider across the plate supply consisting of R_1 and R_2 places the grid of the second triode at a sufficiently positive potential (with respect to its cathode) for proper operating bias.

The cascode circuit is widely used, and a number of special twin-triode tubes have been developed for this purpose. All have basically similar electrical characteristics but different internal connections to facilitate the placement of components and the layouts found in various television tuners. The cascode arrangement gives an overall gain that is somewhat less than that obtainable from a well-designed high-frequency pentode. However, the noise figure of a cascode combination is considerably better than that of a pentode.

10.8 TRANSISTOR RF AMPLIFIERS

Triode RF amplifiers may be operated as either grounded-cathode or grounded-grid amplifiers. The grounded-cathode amplifiers provide a somewhat higher gain but have the disadvantage that neutralization is required to offset the plate-to-grid capacitive feedback. With grounded-grid amplifiers, the grid acts as a shield between the input signal at the cathode and the output signal at the plate and neutralization is not necessary.

Transistor RF amplifiers may be operated as grounded-emitter amplifiers or as grounded-base amplifiers. In this respect they are similar to the triode circuits. The grounded-emitter circuit (comparable to the grounded-cathode triode configuration) provides the higher gain and requires neutralization. The grounded-base circuit (comparable to the grounded-grid triode configuration) does not need neutralization.

NEUTRALIZING SIGNALS

The signal fed from collector to base via the collector-base junction capacitance may be either regenerative or degenerative, depending upon the frequency of the signal and the nature of the collector load. If **regenerative,** the feedback signal *reinforces* the signal at the base. If **degenerative,** the feedback signal partially cancels the signal at the base. In either case, the feedback signal may be highly undesirable in RF amplifiers. In the examples of neutralized RF amplifiers given here, both regenerative and degenerative feedback circuits are represented.

GROUNDED-EMITTER AMPLIFIERS

Figure 10.17 shows a typical grounded-emitter amplifier. Since the collector signal voltage is 180° out of phase with the base signal voltage, the collector-to-base capacitance causes degenerative feedback. In other words, the collector voltage, when fed back to the base via the collector-to-base capacitance, partially cancels the input signal and causes a large reduction in gain. To prevent this, an in-phase signal must be fed to the base to cancel the out-of-phase feedback.

The feedback voltage from the transformer secondary is 180° out of phase with the collector voltage at the point of takeoff. This feedback voltage is fed to the base via the neutralizing capacitor C_N. A *positive* AGC voltage is also fed to the base of the NPN transistor. (A *negative* voltage would be needed if a PNP transistor were used.) As with any AGC system, the AGC voltage is more effective if it is fed to the RF stage rather than to the IF stages only. However, the input and output impedances of a transistor vary considerably with changes in the operating voltages, and this is a decided disadvantage. Thus, in some transistor-TV receivers, there is no AGC voltage fed to the RF amplifier.

Tuning the RF Stage The primary winding of transformer (T) in Figure 10.17 is changed whenever the channel selector knob of the

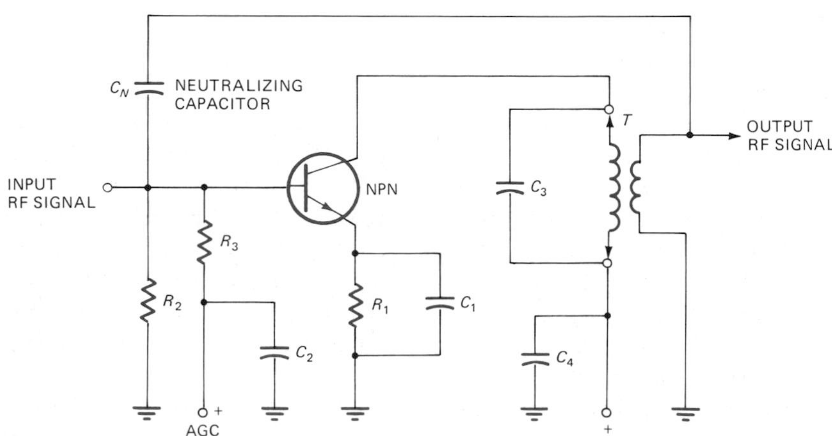

Figure 10.17 A grounded-emitter RF amplifier with neutralization.

receiver is turned. This tunes the RF stage to the carrier frequency of the input signal. The primary is tuned by C_3, and C_2 and C_4 decouple the RF amplifier circuit from the other stages.

PHASE REVERSAL ACROSS A TRANSFORMER PRIMARY

Another method of neutralizing a transistor RF amplifier is shown in Figure 10.18. The basic circuit, shown in Part A, is an NPN conventional amplifier biased with a voltage divider made up of R_1 and R_2. The neutralizing capacitor C_N supplies the feedback-signal voltage to the base of the transistor. A decoupling filter (R_3 and C_4) keeps signal voltage variations from the other stages from affecting the RF amplifier.

The undesired feedback signal from the collector to the base via the base-collector capacitance is degenerative. To cancel the effects of this feedback signal, the feedback signal via the neutralizing capacitor C_N must be regenerative. In other words, the neutralizing capacitor must be connected at a point where the signal is 180° out of phase with the collector signal. To show how the signal phase reversal occurs between the collector of the transistor and Point A, the circuit has been redrawn in Figure 10.18B. Note that the junction of C_1 and C_2 is grounded. Therefore, the voltage at one side of the primary is 180° out of phase with the signal at the other end.

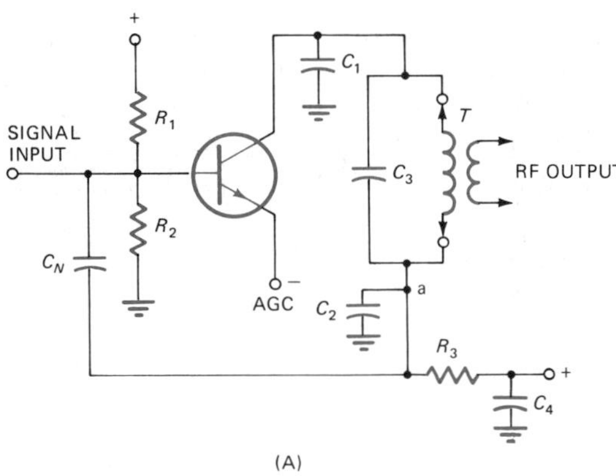

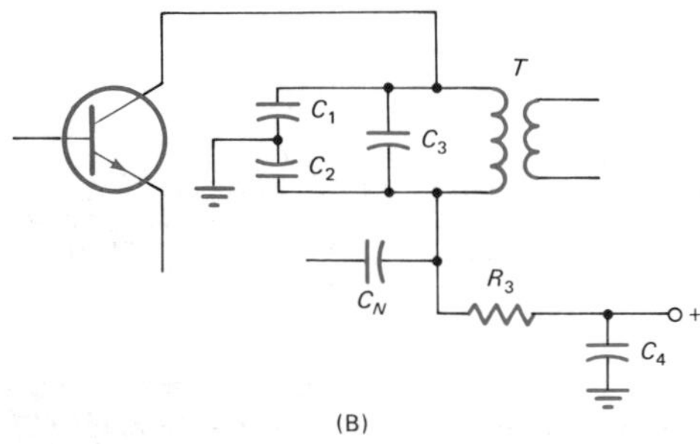

Figure 10.18A. The basic neutralized RF amplifier circuit. **B.** Circuit redrawn to show how phase reversal across a transformer primary is obtained.

TAPPING THE PRIMARY WINDING

A third method of neutralizing a transistor RF amplifier is shown in Figure 10.19. In this case, the primary winding of the transformer is tapped and grounded (for RF). The voltage across the lower tap is of course, 180° out of phase with the voltage at the collector side. Therefore the proper feedback voltage polarity is obtained. This circuit is more frequently used with IF amplifiers.

10.9 FIELD-EFFECT TRANSISTORS AND RF AMPLIFIERS

In modern solid-state television receivers, **field-effect transistors** (FETs) are used in a number of different circuits. Here we are especially interested in the FET as an RF amplifier. Before discussing a typical RF FET amplifier, we shall review the basic theory of FETs.

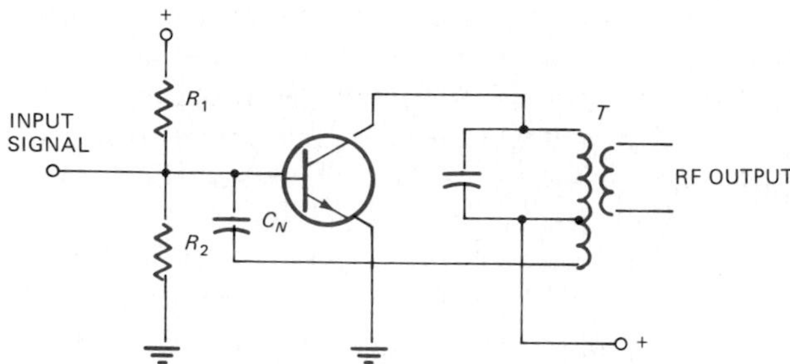

Figure 10.19 A third method of neutralization.

TYPES OF FETS

Two types of FETs are in popular use. One is the **junction FET** (JFET), and the other is the **insulated-gate FET** (IGFET), which are now called MOSFETs (metal oxide semiconductor field-effect transistors). The newer term comes from the fact that a metal oxide is used for the gate insulation.

Figure 10.20 shows the theory of operation for the JFET. The device depicted has an N channel through which current flows. There are two types of JFETs: the **N-channel** type and the **P-channel** type. They are named according to whether N material or P material is used for a conducting path through the device.

N-Channel JFETs In the N-channel JFET, current flow is by electrons. These electrons leave the source and are attracted to a positive voltage on the drain. Moving between the source and the drain, they pass through the gate area. The dotted lines around the gate in Figure 10.20A represent the boundaries of a depletion region at the PN junction. This depletion region limits the number of electrons that can flow from the source to the drain. When no voltage is applied to the gate, the depletion region is small. Under this condition, there is little opposition to current flow through the device and the drain current is high.

When a negative voltage is applied to the gate, the PN junction is reverse-biased. This increases the size of the depletion region, as shown in Figure 10.20B. The result is a decrease in the drain current. The more negative the gate voltage for this type of FET, the greater the depletion region and the smaller the amount of drain current. Ultimately, a point is reached where the depletion region is so large that it prevents any

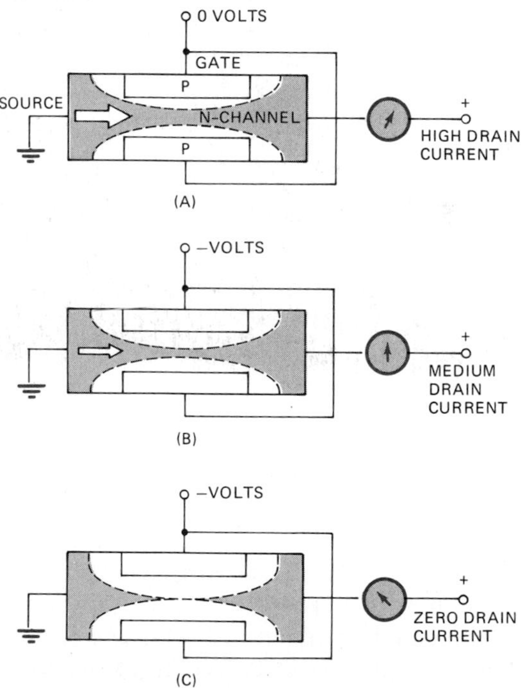

Figure 10.20A-C. The voltage and current relationships in an *N*-channel JFET. **A.** Current through FET with zero volts on the gate. The arrow indicates the electron flow. **B.** Negative gate voltage reduces drain current. **C.** More negative pinch-off voltage reduces drain current to zero.

current flow through the channel. The negative voltage on the gate required to produce this situation is called the **pinch-off voltage** (Figure 10.20C.)

P-Channel JFETs A P-channel JFET operates in the same manner just described except that the voltages on the gate and drain are reversed. The more *positive* the voltage on the gate, the lower the current flow in a P-channel JFET. Current flow through a P-channel JFET is due to "hole" flow and is toward a negative voltage on the source.

JFET Symbols Figure 10.21 shows the two symbols used for N-channel and P-channel JFETs. Typical voltage polarities are shown on the symbols.

The N-channel JFET is often likened to a vacuum tube for the following reason. In the JFET drain voltage is positive and gate voltage is negative, whereas in the vacuum tube plate voltage is positive and grid voltage is negative.

Depletion-Mode JFETs and Enhancement-Mode JFETs The JFET of Figure 10.20 operates by controlling the size of the depletion region with a gate voltage. The more negative the gate voltage, the larger the depletion region. The converse is also true. This type of FET is often referred to as a **depletion-mode JFET.**

Another type of operation is possible. The gate-channel junction can be made sufficiently large and a sufficient amount of doping can be applied to cut off the current through the channel whenever the voltage on the gate is zero. In order to make current flow through this FET, it is necessary to *forward-bias* the junction to reduce the size of the depletion region around the gate. This type of FET is referred to as an **enhancement-mode JFET.**

The important difference between depletion-mode and enhancement-mode FETs is that current flows through the depletion-mode FET when the gate voltage is zero, but no current flows through the enhancement-mode FET when the gate voltage is zero.

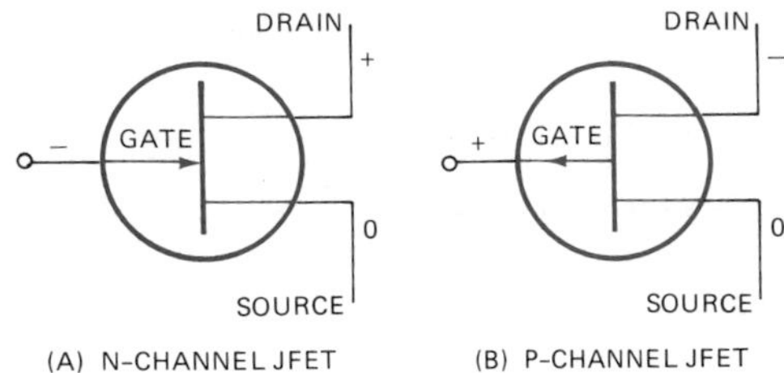

Figure 10.21A-B. Symbols for JFETs, showing typical operating voltage polarities.

The MOSFET The disadvantage of JFETs is that a current will flow between the gate and the channel with a very small amount of forward bias. Even with reverse bias, the minority charge carriers flowing across the junction can result in an appreciable input current relative to that of a vacuum tube. To get around this problem, an insulated gate is sometimes placed at the junction between the gate and the channel. Figure 10.22 shows a cross section of a MOSFET. As with

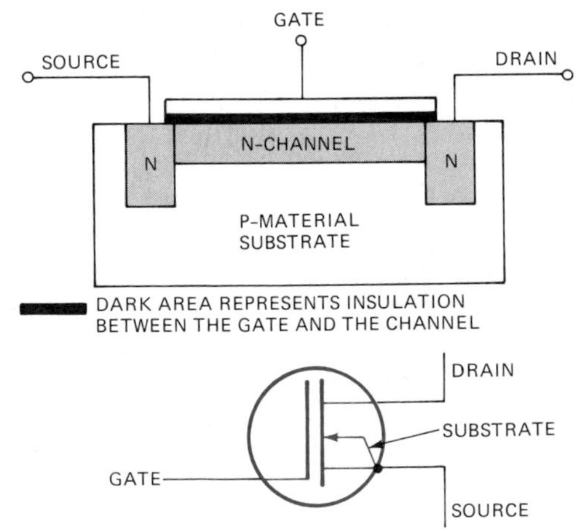

Figure 10.22 A MOSFET in cross section and its symbol.

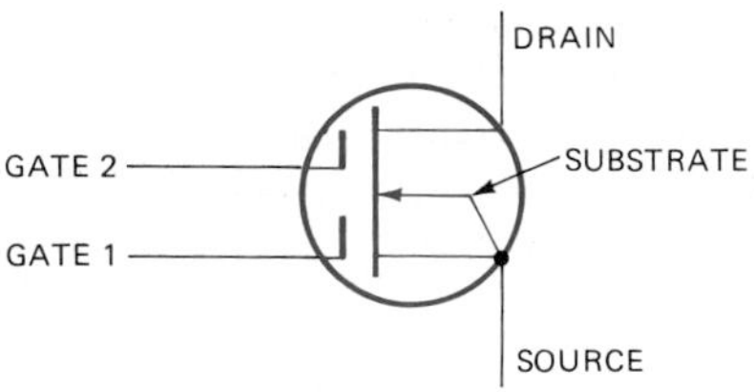

Figure 10.23 The symbol for a dual-gate MOSFET.

the JFETs, MOSFETs may be either N-channel or P-channel types. Also, either type may be designed for operation in the depletion or the enhancement mode. The insulating layer—usually silicon dioxide—between the gate and the channel prevents current flow in the gate circuit since neither majority or minority charge carriers can move across the dielectric layer. Therefore, the MOSFET has a very high input impedance.

The Insulated Gate The insulated gate layer is made very thin to allow the gate voltage to exercise maximum control over the current in the conducting channel. Because the layer is so thin, it can be easily punctured by any excessive voltage. For example, an electrostatic charge can destroy an IGFET. It is common practice to ship MOSFETs with their leads tied together to prevent accidental static charges from destroying them before they can be placed in a circuit. Remember that a static charge from a person's hand or from the probe of a meter can easily destroy the very thin insulating layer, and special care must be taken when handling or measuring in circuits containing MOSFETs. For example, grounding a finger ring or bracelet on the hand is an acceptable precautionary measure.

Dual-Gate FETs Some FETs have more than one gate. One important example is the **dual-gate,** or **tetrode, FET.** The symbol for this type of FET is shown in Figure 10.23. Both of the gates control the current flow through the FET. The characteristics of a dual-gate FET are similar to those of a tetrode. The dual-gate FET is very useful as a mixer for the oscillator and RF input signals. It is also useful as an RF amplifier because the AGC and RF voltages can be applied to the device at separate electrodes.

ADVANTAGES OF FETS

In an ordinary transistor, current flow through the device is dependent upon both hole flow and electron flow. Since both polarities of the charges are required for this operation, ordinary transistors are sometimes referred to as **bipolar devices.** By comparison an FET is a **unipolar device,** because the flow of current through it is either by hole flow or by electron flow, but not by

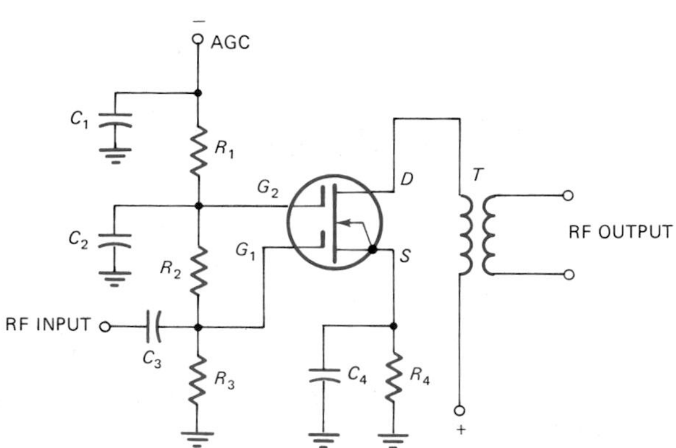

Figure 10.24 A dual-gate MOSFET RF amplifier.

both. Since an electron-hole combination causes noise, a unipolar device (FET) generates less noise. This is an advantage in high-frequency operation.

Another advantage of the FET is its high input impedance. In this regard it is similar in operation to the vacuum tube. Unlike transistor amplifiers, which require current flow in the base circuit and therefore require an input power, an FET circuit may be designed to have virtually no current flow in the gate circuit. By way of comparison, vacuum tube amplifiers also have virtually no current flow in their grid circuits.

Another advantage of the FET is its **square-law operation.** Current flowing through the FET is directly proportional to the square of the voltage on the gate, which means that it is a square-law device rather than a linear device. With proper operating voltages, the output of an FET amplifier is rich in second harmonics. Transistors and vacuum tube amplifiers have third and higher harmonics in addition to the second harmonic. This results in cross-modulation distortion of the output signal. Thus, the FET has an advantage over both tubes and transistors when it is used as a mixer.

A VHF MOSFET RF AMPLIFIER

Figure 10.24 shows a simplified circuit with a dual-gate MOSFET used as a VHF RF amplifier. An AGC voltage is developed across a voltage divider made up of R_1, R_2, and R_3. The capacitors C_1 and C_2 filter the AGC voltage so that only a pure DC operating voltage is applied to the gates. Most of the AGC voltage is applied to Gate 2 (G_2).

The RF signal is applied through C_3 and developed across R_3. This signal voltage appears on the first gate (G_1). Thus, the amount of current through the FET is con-

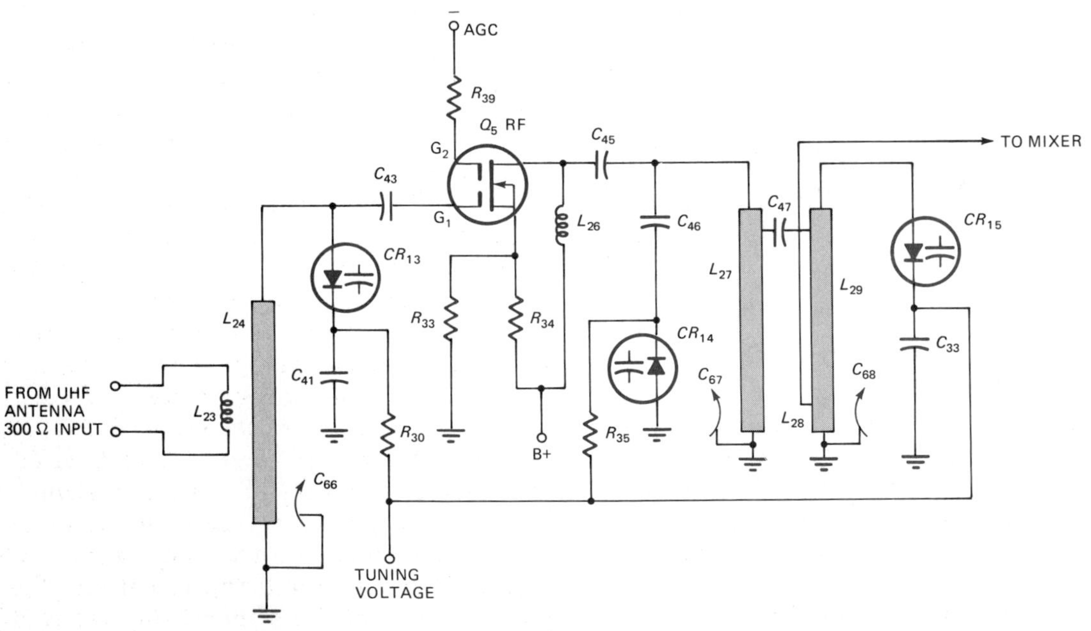

Figure 10.25 Simplified diagram of a dual-gate MOSFET UHF-RF amplifier. CR_{13}, CR_{14}, and CR_{15} are varactors, which tune inductances L_{24}, L_{27}, and L_{29} respectively. *(Courtesy of RCA Consumer Electronics.)*

trolled by two voltages: the AGC voltage and the signal voltage.

The source bias is supplied by R_4, and C_4 is used to prevent degeneration. The output signal is taken from the secondary of the transformer T in the drain circuit.

A UHF MOSFET RF AMPLIFIER

A simplified schematic diagram of a dual-gate MOSFET UHF RF amplifier is shown in Figure 10.25. AGC voltage is applied to Gate 2 (G_2). After being tuned by inductor L_{24} and varactor CR_{13}, the UHF signal from the antenna is applied to G_1 through C_{43}. The antenna input in L_{23} is inductively coupled to L_{24}. The interstage (between RF amplifier and mixer) inductors L_{27} and L_{29} are inductively and capacitively (C_{47}) coupled. These inductors are tuned by varactors CR_{14} and CR_{15}, respectively.

Inductors L_{24}, L_{27}, and L_{29} are made of printed copper patterns. The tuned circuits are initially adjusted by moving the copper tabs labeled C_{67}, C_{68}, and C_{69} relative to the inductors. The RF output to the mixer (Schottky barrier diode) is taken from a tap on L_{29}. The tap is labeled L_{28}. All tuning inductors are made of copper patterns printed on the circuit board.

10.10 MIXERS AND MIXER CIRCUIT OPERATION

The tuner has an RF amplifier, a mixer circuit, and an oscillator. RF amplifiers were covered in previous sections. We shall now discuss mixer circuit configurations. (Oscillator circuits will be discussed in Section 10.11.)

As shown in the block diagram of Figure 10.4, the mixer receives signals from both the RF and the oscillator circuit and converts them to a difference-frequency output, called the **intermediate frequency** (IF) **signal.** This conversion is achieved by applying both the RF and oscillator voltages to a nonlinear vacuum-tube, a transistor, or a diode (UHF). If the device is nonlinear, heterodyning will take place and the output signal will consist of both the sum and difference frequencies. Assuming that the frequency difference between the RF and oscillator signals is appreciable, a selective network tuned to the lower-frequency component of the output will pass the desired IF only.

VACUUM TUBE MIXER CIRCUITS

In vacuum tube tuners, energy from the oscillator may be either capacitively or inductively coupled to the mixer. Two frequently used methods are pictured in Figure 10.26. Capacitive coupling is used in the circuit of Part A, and inductive coupling is used in the circuit of Part B of Figure 10.26. Interaction between the input signal and the oscillator outside the mixer tube is kept as low as possible. This prevents changes of the oscillator frequency and minimizes the level of oscillator radiation at the antenna.

In addition to the fact that there are FCC regulations covering radiated signals, such signals are troublesome because any considerable amount of radiated signal can produce a complete loss of contrast or even a negative picture in nearby television receivers. When the interfering frequency is close to the picture carrier of the station being received by the other sets, the **beat interference** produces vertical, horizontal, or slanted stripes across the screen. (*Beat interference* refers to the difference-frequency signal obtained when the interfering signal and the picture carrier of the station being received mix in the second detector stage of the receiver.)

In the vacuum tube mixer of a television

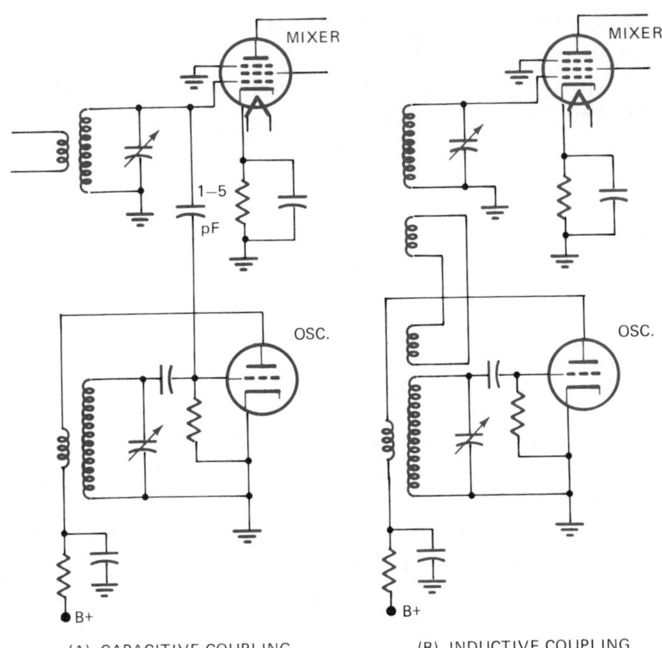

Figure 10.26A-B. Two types of circuits for coupling oscillator signals to vacuum tube mixers.

receiver, the received signal and the oscillator voltage both modulate the electron stream to form the desired audio and video IF voltages. In nearly all tube-type receivers, the signal from the RF amplifier is transformer-coupled or impedance-coupled to the mixer. The oscillator voltage is transferred to the mixer tube either capacitively or inductively.

In vacuum tube TV receivers, the mixing for a UHF tuner is done by a semiconductor diode. This device requires less oscillator voltage than a vacuum tube. As an additional advantage, mixer noise is reduced when diodes are used. (A diode-mixer circuit is shown in Figure 10.30.) Diode-type UHF mixers are also frequently used in solid-state receivers.

VHF TRANSISTOR MIXER CIRCUITS

In a transistor mixer, the amplifier configuration, the DC biasing and coupling to the RF oscillator, and the IF amplifier load should be considered. Adequate precautions must be taken against the generation of IF frequencies from image frequencies and oscillator second-harmonic frequencies. Also, since the mixer amplifies IF frequencies, it must be stabilized in this range. This is done either by neutralization or by the insertion of a series-resonant IF trap from input to ground.

Common-Base Mixers There are many ways to inject the RF and oscillator signals into a transistor mixer. The common-base configuration has found wide acceptance because of its stable operation, relative constancy of input impedance with input frequency, and immunity against cross-modulation. The choice of transistor type is governed by considerations of high gain at the IF frequency, base-emitter characteristics, and low base-emitter capacitance. The cutoff frequency need not exceed 50 MHz. The AGC voltage is usually applied to the RF amplifier. Thus, the transistor character-

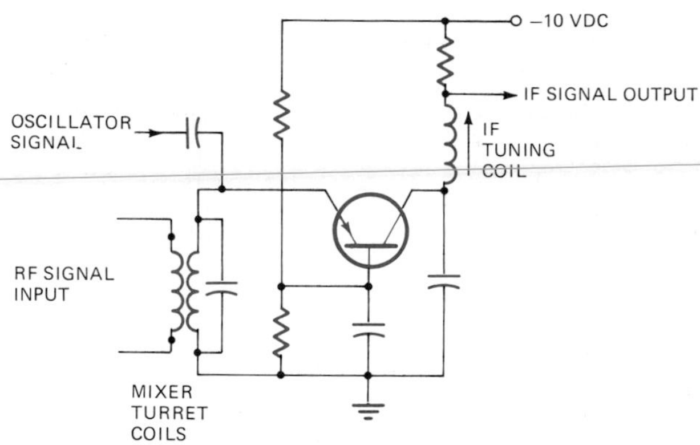

Figure 10.27 A common-base mixer circuit with the RF and local oscillator signals injected at the emitter.

istics in this respect are not critical. A common-base mixer with oscillator injection at the emitter is shown in Figure 10.27.

Common-Emitter Mixers Common-emitter transistor amplifier circuits are also used for the mixer stage. The RF and oscillator signals may both be injected at the base or at the emitter. In some cases, one signal is injected at the base and the other at the emitter. In any case, heterodyning will take place, provided, of course, that the stage is not linear.

Figure 10.28 shows a common-emitter mixer stage. The RF and oscillator signals are fed to the base via C_1 and C_2, respectively. Emitter stabilization is provided by R_1 and C_4. The base is biased by the voltage divider R_2 and R_3. Normally, this bias voltage is very low to ensure that the transistor operates in the nonlinear range. A filter comprising C_7 and the RFC (radio-frequency choke) prevents the IF signal at the collector from entering the power supply.

Neutralization A small IF signal voltage developed across the RFC, is fed back to the base via C_6. This represents an important difference between common-base and common-emitter mixers, since neutralization is not required for the common-base configuration. This is especially true since the input

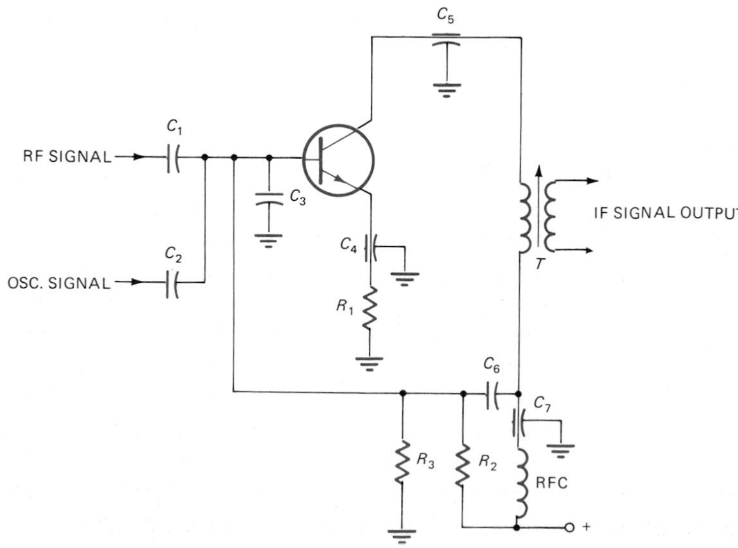

Figure 10.28 A common-emitter mixer stage.

and output circuits are tuned to different frequencies. In the common-emitter circuit, however, the stage is operating as a common-emitter IF amplifier and the IF signal fed back from collector to base (via the collector-base junction capacitance) must be neutralized. The IF transformer (T) has two purposes: it transfers only the IF signal and rejects all other signals from the mixer, and it matches the output impedance of the common-emitter mixer to the input of the first IF amplifier stage.

Cascode Circuits The cascode configuration for two triodes was discussed in Section 10.7. It consists of a common-cathode circuit followed by a common-grid circuit. The plate of the common-cathode circuit is tied directly to the cathode of the common-grid circuit. Transistor amplifiers are also connected in a cascode configuration. Cascode transistor circuits have a common-emitter circuit followed by a common-base circuit, with direct coupling between the collector and emitter. Because of their high gain, good isolation between input and output, and low noise characteristics, transistor cascode amplifiers are used for both RF and mixer amplifiers in tuners.

Figure 10.29 shows a cascode mixer stage. The oscillator and RF signals are applied to the base of the common-emitter transistor (Q_2), and the IF signal is taken from the collector of Q_1. When the tuner is switched to the UHF position, the IF signal from the UHF tuner is fed to Q_2 through the isolating diode D_1. Thus, the cascode amplifier provides an extra stage of IF amplification for UHF stations. This partially offsets the fact that there may be no RF amplifier in the UHF tuner.

The base of Q_1 is grounded by C_1, and the base bias for Q_1 and Q_2 is developed by the voltage dividers R_1, R_2, and R_3. Emitter stabilization is obtained with R_4 and C_2. Note that neutralization is not required for this circuit because of the low input impedance of the grounded base stage.

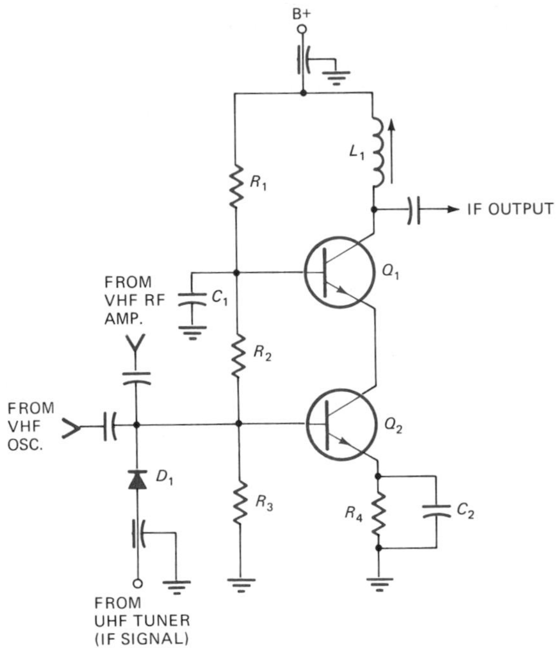

Figure 10.29 A cascode mixer circuit. (On UHF operation, the VHF RF amplifier and VHF oscillator are disabled.)

MOSFET Mixers The dual-gate MOSFET (Figure 10.23) makes an ideal mixer amplifier. One of the gates can be used for the RF input and the other for the oscillator input. Alternately, both inputs can be delivered to one gate. In this case, the other gate is used as a screen to isolate the input and output stages of the mixer.

UHF DIODE MIXERS

A simplified diagram of a UHF diode mixer is shown in Figure 10.30. This circuit is from the same UHF tuner shown in Figure 10.26. Note that the UHF RF signal is applied to the mixer diode CR_1 from L_{28}. In addition, the oscillator signal is coupled to CR_1

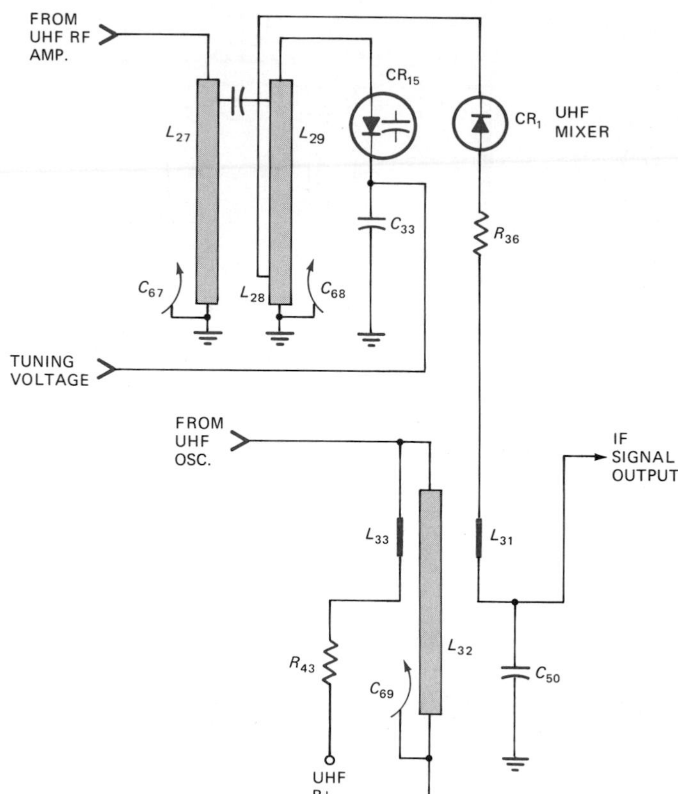

Figure 10.30 Simplified diagram of a UHF diode (CR_1) mixer. (See also Figure 10.25, from same UHF tuner.) CR_1 is a Schottky-barrier diode. *(Courtesy of RCA Consumer Electronics.)*

via L_{31}. Since the diode is a nonlinear device, heterodyning takes place in it and the difference, or IF, frequency is produced. The diode used in this tuner is the Schottky barrier diode, which has excellent characteristics as a mixer in the UHF band.

10.11 LOCAL OSCILLATOR OPERATION

The local oscillator provides a signal that heterodynes in the mixer with the RF signal. As in radio superheterodyne receivers, the local oscillator frequency is normally *above* the incoming RF frequency by an amount equal to the receiver IF frequency.

An important characteristic of the local oscillator used in tuners is its freedom from **drift.** The term *drift,* as applied to oscillator circuits, means any undesired change in oscillator frequency caused by temperature change, aging of the circuit components, change in the DC operating voltage, or change in the oscillator load. In monochrome receivers, a small amount of oscillator drift can be tolerated. In color television receivers, however, the color portion of the signal can be completely lost if the oscillator drifts too much. For this reason, automatic fine tuning (AFT) circuits may be used to maintain oscillator frequency at a fixed value.

The local oscillator frequency is changed whenever the receiver is tuned from one station to another. The **fine-tuning control** of the receiver allows the oscillator frequency to be varied over a narrow range. The setting of this control is very important if a satisfactory color picture is to be obtained. For this reason, manufacturers may include some kind of fine-tuning indicator (FTI) circuit to simplify adjustment of the fine-tuning control. The AFT circuit is also an aid to proper adjustment of the fine-tuning control. It will lock the oscillator onto the correct frequency whenever the control is within range of the AFT circuit limits.

VACUUM TUBE LOCAL OSCILLATOR CIRCUITS

Figure 10.31 shows what is perhaps the most frequently used oscillator circuit in present-day television receivers. The circuit in Figure 10.31A is known as the **ultraudion.** Figure 10.31B shows the ultraudion oscillator circuit as it is often drawn on schematic drawings. It is equivalent in its operation to the well-known Colpitts circuit (Figure 10.31C). In the ultraudion, the voltage division across the tank circuit is accomplished through the grid-to-cathode (C_{gk}) and the plate-to-cathode (C_{pk}) capacitances in the tube. The feedback voltage that sustains the oscillations is developed across C_{gk}.

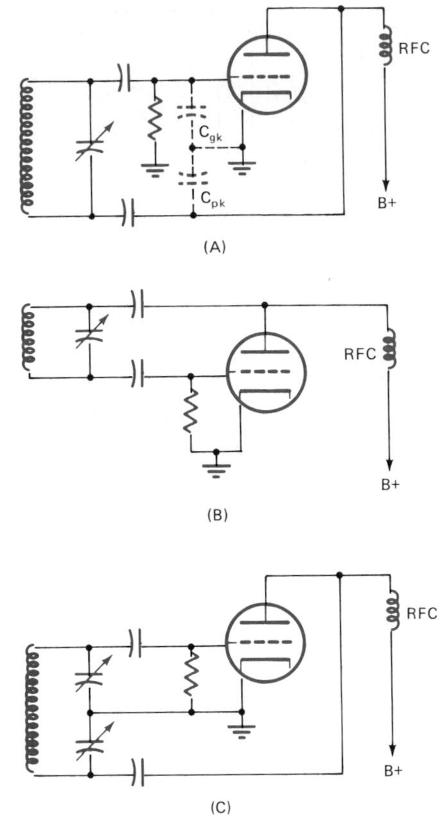

Figure 10.31A-C. The ultraudion oscillator A. and B. compared to a Colpitts oscillator C.

Oscillator Operation An ultraudion oscillator circuit used in television receivers is shown in Figure 10.32. As shown in Figure 10.32A, the inductance of the resonant circuit is changed as the receiver is tuned from station to station. The equivalent circuit is pictured in Figure 10.32B. The voltage-dividing capacitance network consists of the effective capacitance of C_{gk} in series with the parallel combination of C_{gk} and C_1. Capacitor C_t represents the combination of grid-plate capacitance, distributed capacitance, and C_2. Capacitor C_2 is a temperature-compensating capacitor that helps reduce oscillator drift. In spite of this, drift does occur, and C_1 is provided to permit the user of the set to adjust the oscillator frequency to the best picture. Because C_1 is actually a vernier adjustment, it is labeled *fine-tuning control* and placed on the front panel. Any shift in oscillator frequency immediately alters the IF produced as a result of the mixing action. The effect is the same as detuning the receiver. By means of the fine-tuning control, the oscillator frequency can be readjusted to its proper value. Capacitor C_3 keeps the DC plate voltage from the exposed coils. Capacitor C_4 allows the oscillator to develop grid-leak bias across R_1.

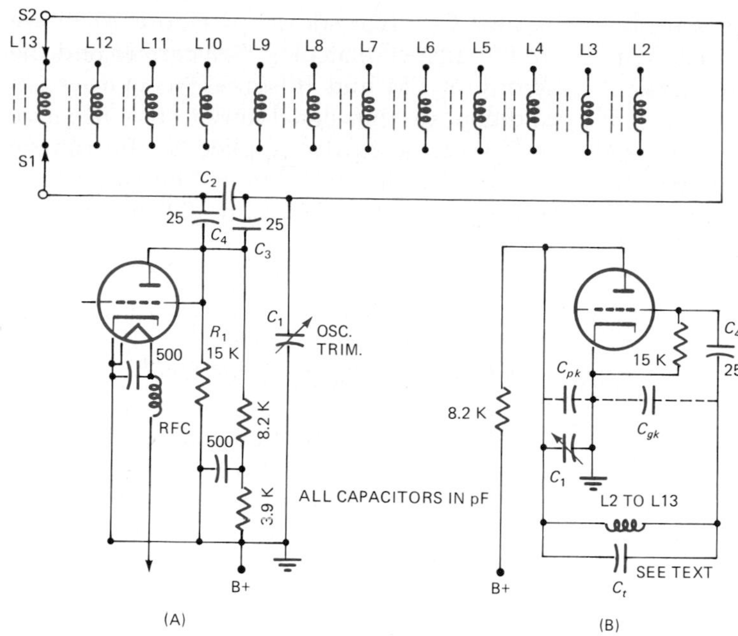

Figure 10.32A. A typical ultraudion oscillator, showing switched inductances for VHF channels. **B.** Equivalent circuit.

TRANSISTOR LOCAL OSCILLATOR CIRCUITS

Figure 10.33 shows a Colpitts oscillator circuit with a transistor amplifier. This circuit, including variations, is usually used as a local oscillator in transistor television tuners. The transistor amplifier is Q_1. Its base is biased for conduction by a voltage divider made up of R_1 and R_2. The junction of these resistors places a positive voltage on the base of the NPN transistor, as required for conduction. Inductor L_1 acts as a load for the collector. This load permits the oscillations to reach the desired amplitude.

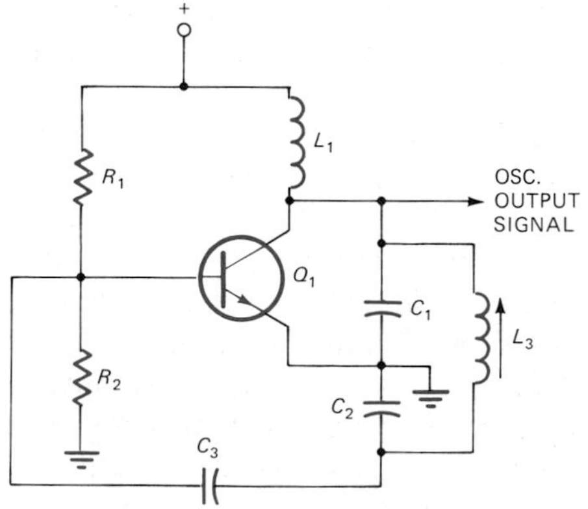

Figure 10.33 The transistor Colpitts oscillator is frequently used as a tuner local-oscillator circuit. Coil L_3 is changed each time a different channel is selected.

The frequency-determining network in the circuit of Figure 10.33 is made up of C_1, C_2, and L_3. Often, one of these capacitors is made variable and serves as the fine-tuning control of the receiver. Note that capacitor C_1 is connected directly from the collector to the emitter. In most transistor oscillator circuits, this collector-to-emitter capacitor is present in one form or another. Capacitor C_3 is necessary to prevent a DC short-circuit path from the collector to the base through coil L_3. It also serves to deliver the regenerative feedback signal required from the collector to the base circuit.

In one important variation of the circuit shown in Figure 10.33, the junction capacitance within the transistor for feedback, and therefore the external feedback capacitance network, is not visible. In another version, to be shown later, a common-base transistor is used as an oscillator in a transistor tuner.

UHF TRANSISTOR OSCILLATORS

A simplified diagram of a UHF transistor oscillator is shown in Figure 10.34. This circuit is from the tuner that Figures 10.26 and 10.30 were taken from. It is a Colpitts-type oscillator that is tuned by varactor CR_{16}. The base of Q_7 is forward-biased by R_{47} and R_{48} to enable oscillations to start. The tuning inductor is L_{32}. This inductor is a copper pattern on the circuit board, as was the case for inductors in Figures 10.25 and 10.30.

10.12 A TYPICAL NUVISTOR TUNER

The advantages of a nuvistor triode have already been explained (Section 10.6). Figure 10.35 is the schematic diagram of a tuner that uses a nuvistor for an RF amplifier.

RF amplifier V_1 is a conventional (that is, grounded-cathode) amplifier that uses a nuvistor tube. The signal for the amplifier arrives at the tuner via the feedthrough capacitor C_{22}. Not shown in Figure 10.35 are an impedance-matching transformer and the filters for FM and IF signal frequencies. In addition to the signal input, the negative AGC voltage is also applied to the control grid of V_1. The AGC line is filtered by R_{1051} and C_{1050}, and a decoupling filter comprises C_{20} and R_3.

The output signal of V_1 is electromagnetically coupled from L_6 to L_7. Capacitors C_8 and C_{12} tune these coils. The RF amplifier circuit is neutralized via the feedback capacitor C_{21}.

There are two input signals to the mixer (V_2-A). The RF signal from L_7 is applied to the control grid at Pin 9, and the oscillator signal from V_2-B is applied to the screen grid at Pin 7. The difference signal—which is the IF frequency—is taken from the plate at Pin 6.

10.13 A TYPICAL TRANSISTOR TUNER

Figure 10.36 pictures a typical VHF transistor tuner. There are separate antenna input circuits for the VHF and UHF tuners. (The UHF tuner is not shown.) The RF am-

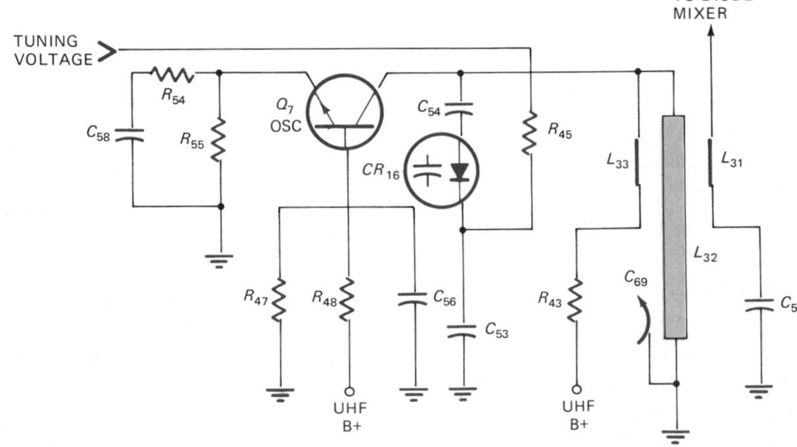

Figure 10.34 Simplified diagram of a UHF transistor oscillator. (See also Figures 10.25 and 10.30 from same tuner.) CR_{16} is a varactor to tune the oscillator. *(Courtesy of RCA Consumer Electronics.)*

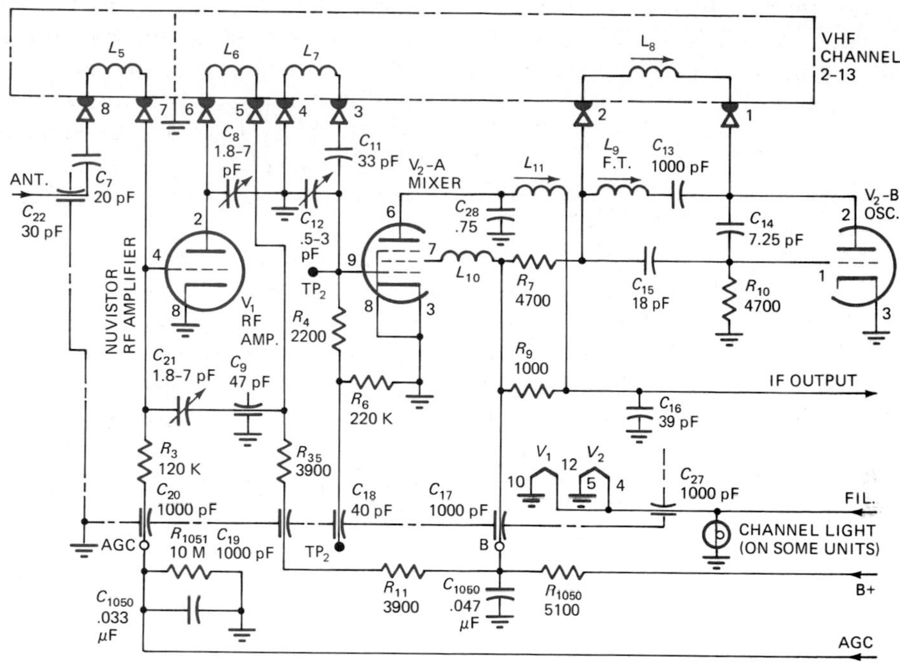

Figure 10.35 A nuvistor tuner. The nuvistor is connected as a neutrode RF amplifier. (Neutrode means neutralized triode.) Unless otherwise stated, capacitor values are in μF.

plifier is an NPN common-emitter circuit. The input to the RF amplifier is tuned by Tuned Circuit 1. The output of the RF amplifier is tuned by Tuned Circuit 2, which is inductively coupled to Tuned Circuit 3. Thus, the RF signal is coupled to the mixer inductively and the oscillator signal is injected into the mixer capacitively via C_{225}.

The mixer is a common-emitter circuit with a base bias established by a voltage divider consisting of R_{206} and R_{208} across the B+ source. Emitter stabilization is established by R_{207}, and C_{217} maintains the emitter at RF ground potential.

The oscillator is a common-base circuit. Capacitor C_{230} grounds the base. When the tuner is set for a VHF station, the oscillator base is biased by R_{212} and R_{213} across the B+ source. Collector voltage for the oscillator is obtained through R_{211}. Regenerative feedback from the collector to the emitter is established by C_{219}. Tuned Circuit 4 sets the oscillator frequency. The four inductively tuned circuits are adjusted simultaneously when the channel selector is moved from station to station.

10.14 VARACTOR TUNING AND TUNERS

One of the important developments in television tuners has been the solid-state tuning system. This system is based on the use of the so-called varactor diode. This diode is also known as a voltage-variable capacitor and by the trade names Varicap and Selicap. The varactor is a special solid-state diode that acts as a capacitor whose capacitance varies inversely with the amount of reverse bias applied across it. All varactors operate only with reverse bias. In tuned circuits that use them, the resonant frequency is changed merely by changing the

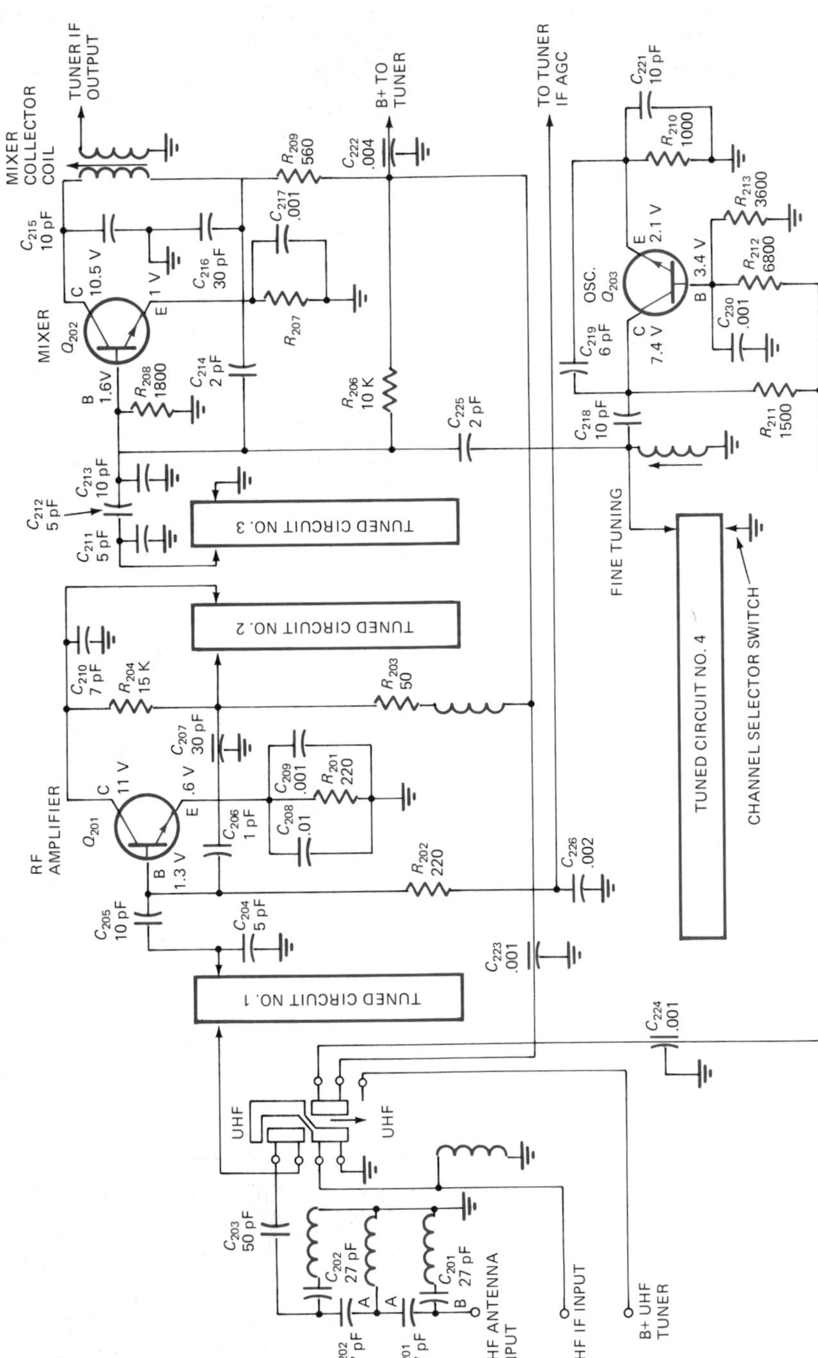

Figure 10.36 A typical VHF transistor tuner circuit. Unless otherwise stated, capacitor values are in μF.

amount of reverse bias across the varactor. The greater the reverse bias across the varactor, the lower the varactor capacitance and the higher the tuned-circuit resonant frequency.

VARACTOR CIRCUITS

In order to control the frequency of a tuned circuit, the varactor diode is placed across the tuning capacitor as shown in Figure 10.37. In Figure 10.37A the amount of reverse voltage on the varactor depends on the settings of resistors R_1 and R_2 and, of course, on the position of the switch (SW). When the switch is in Position 1, the voltage at the arm of R_1 is picked off. This voltage establishes a certain amount of varactor capacitance. Capacitor C_1 is needed for isolating the varactor circuit so that the reverse-bias voltage is grounded. In practice, the reactance of C_1 is negligible for the frequencies involved. Therefore the varactor can be considered to be connected directly across tuning capacitor C_2.

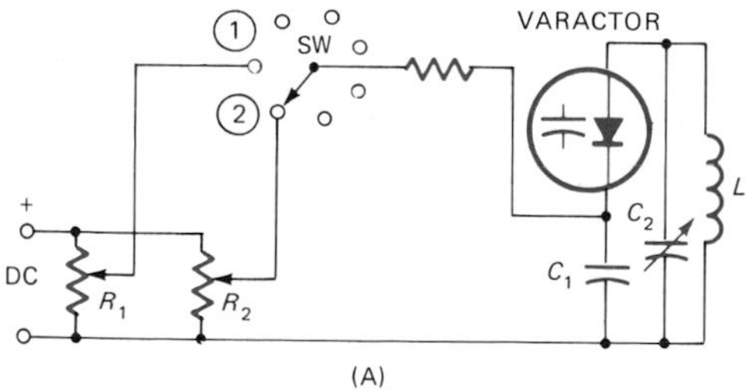

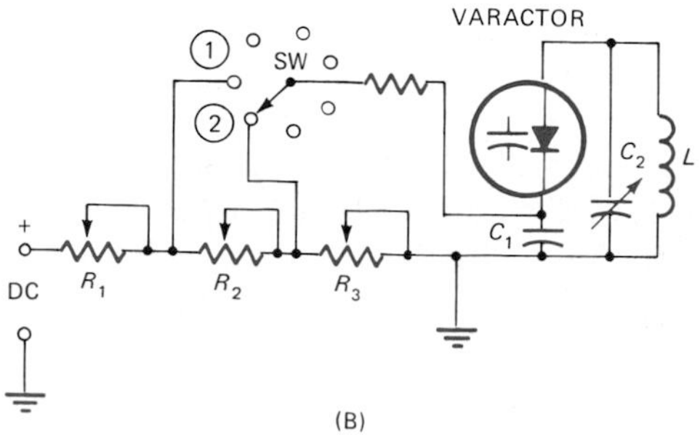

Figure 10.37 Two ways of using varactors in detent tuners.

When the switch is turned to Position 2, the voltage at the arm of resistor R_2 determines the capacitance of the varactor. The varactor, in parallel with the tuning capacitance, determines the frequency of the tuned circuit comprising L and C_2. Only two positions are shown on the switch in Figure 10.37A, but it is presumed that all of the channel positions contain variable resistors that pick off different amounts of voltage from the applied DC. This type of tuning system has the advantage that only one inductance and capacitance is needed for all channel frequencies.

Another variation of the circuit in Figure 10.37A uses push button switches (rather than a rotating switch) to set the voltage on the voltage-variable capacitor. Both arrangements are useful because they provide channel selection by switching. This is especially important in the UHF band because the FCC requires step tuning of stations in this range.

The circuit of Figure 10.37B is slightly different from that in Figure 10.37A. Here the voltage picked off for the varactor is established by a number of resistors in series. Only three of the resistors are shown, but it is presumed that there is a variable resistor for each position on the switch. A single variable resistor in series with the resistor lineup can be used for fine tuning in this arrangement.

Switching Diodes As previously mentioned, the VHF frequencies extend from 54 to 216 MHz. This is a 4:1 tuning ratio, which cannot be accommodated by simple varactor tuning. To solve this problem, switching diodes are used, as illustrated in Figure 10.38. In Figure 10.38A, we see that for the low-VHF band (Channels 2–6), two inductors in series are tuned by the varactor diode

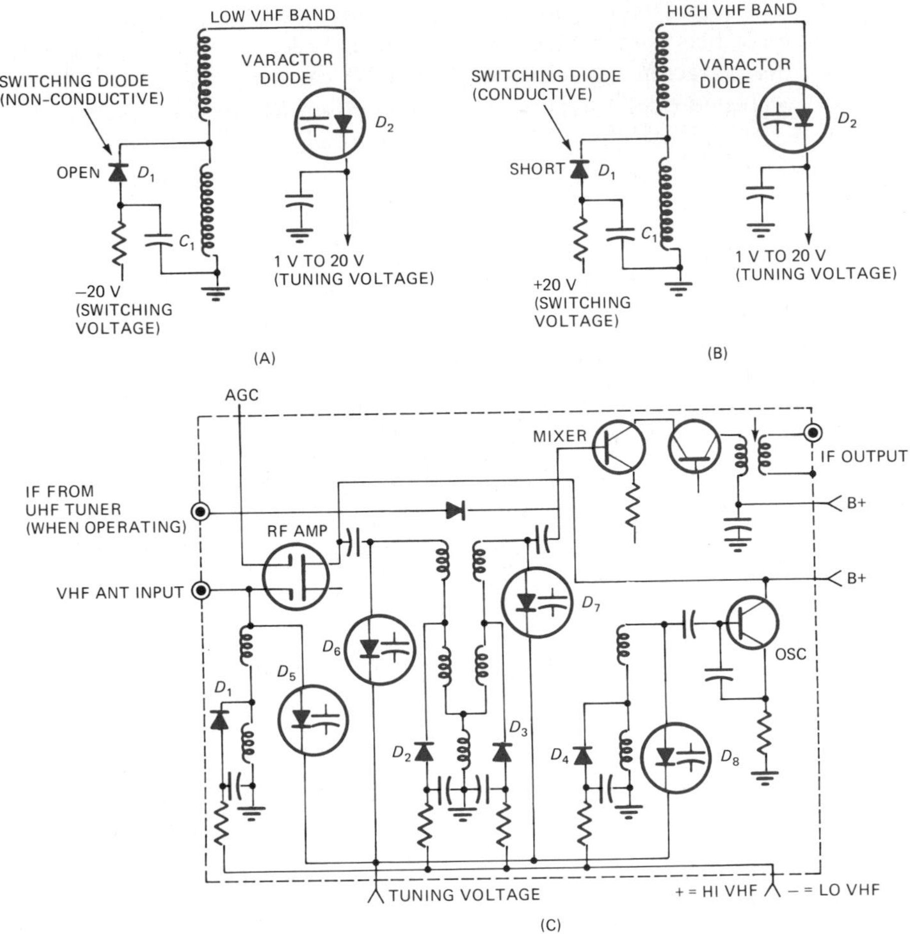

Figure 10.38A-C. Simplified drawings illustrating VHF varactor tuning and switching-diode operation. **A.** Switching diode is non-conductive for low VHF band. **B.** Switching diode is conductive for high VHF band. **C.** Simplified diagram of varactor VHF tuner, using four switching diodes and four varactors. *(Courtesy of Quasar Electronics Co.)*

D_2. At this point, the switching voltage of −20 V cuts off the switching diode D_1. However, for Channels 7 to 13, the switching diode D_1 is made to conduct by changing the −20 V to +20 V through a switching arrangement. The conduction of D_1, in conjunction with the low impedance of capacitor C_1, effectively shorts out the lower inductance. The reduced inductance can now be tuned to Channels 7 through 13 by the varactor diode.

Four Tuned Circuits For simplicity, only one tuned circuit is shown in Figure 10.38. However, many VHF tuners have four varactor-tuned circuits: (1) antenna circuit, (2) RF amplifier circuit, (3) mixer circuit, and (4) oscillator circuit. Thus, four switching diodes as well as four varactor diodes are used.

Simplified Varactor Tuner A simplified diagram of a VHF varactor tuner is shown in

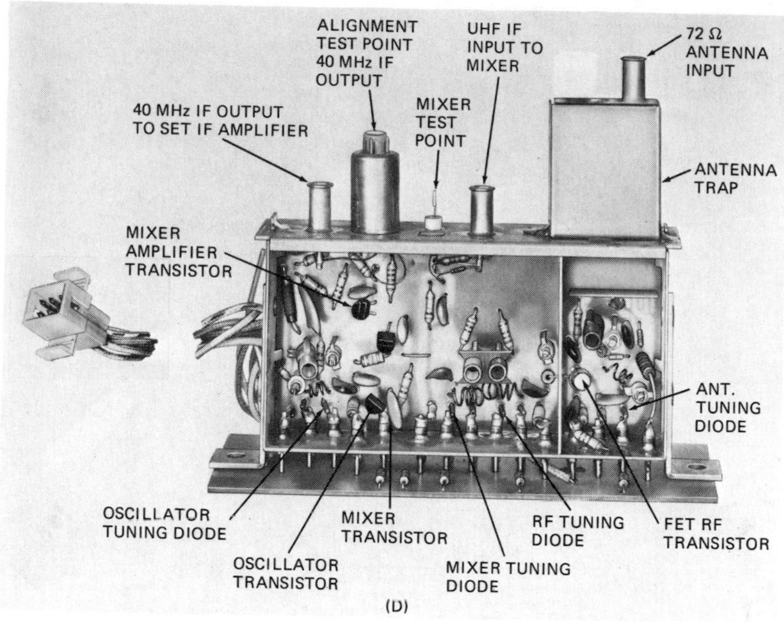

Figure 10.38D. Bottom view of a varactor-tuned VHF tuner, using four varactors. *(Courtesy of Zenith Radio Corp.)*

Figure 10.38C; this unit has four varactor-tuned circuits. Each varactor, labeled D_5, D_6, D_7, and D_8, is controlled by a common DC tuning voltage. Each tuning inductor has a switching diode, labeled D_1, D_2, D_3, and D_4, across a portion of it to permit operation at the high and low VHF bands.

For low-band VHF operation, the switching diodes are reverse-biased by a negative voltage and the entire inductor is used. For high-band VHF operation, the switching diodes are forward-biased by a positive voltage. This shorts out the lower portion of the inductor.

10.15 UHF SOLID-STATE TUNERS

Most UHF tuners fall into two general categories, those tuned by parallel-line circuits and those tuned by coaxial-tuned circuits (sometimes called cavities). In addition, there also are varactor-tuned UHF tuners. However, in these cases, the varactor varies the total tuning capacitance of the tuned circuit, as noted. The major advantage of varactor tuning is that it permits easy selection of UHF stations by rotary switch or push buttons. It also allows UHF stations to be tuned by remote control (Chapter 11).

UHF COAXIAL-TANK TUNERS

Figure 10.39 shows one version of a UHF tuner that uses a popular tuning arrangement. This circuit uses a coaxial tank with capacitor-end tuning. Antenna coil L_1 couples the incoming RF signal to a coaxial tank, which in turn transfers the signal to the mixer diode (CR_1) through a suitably positioned window. The oscillator delivers its signal through the pickup coil (L_3), and the

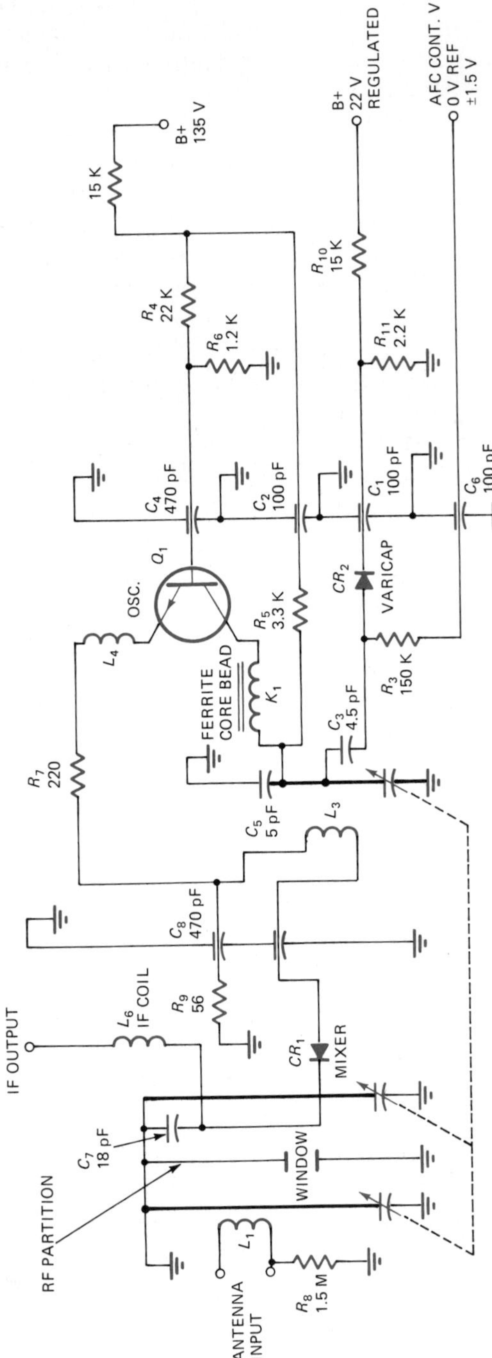

Figure 10.39 A UHF tuner that uses coaxial tank circuits with capacitor-end tuning. *(Courtesy of General Electric Co.)*

resultant IF becomes available at L_6. This IF signal can be fed to the VHF tuner, which then acts as an IF amplifier.

The circuit in Figure 10.39 features a special AFC arrangement that helps to stabilize the oscillator frequency. At exact resonance, varactor diode CR_2 is reverse-biased by 2.8 V across its terminals furnished by a 22-V regulated supply. Resistors R_{10} and R_{11} are used as a voltage divider. Should the oscillator frequency drift, a voltage proportional to the drift will be generated, increasing or decreasing the reverse bias for the varactor. Since the varactor is a *voltage-variable capacitor,* the circuit elements can be arranged so that the varactor offers a higher capacitance across the coaxial tuning circuit when the effective oscillator tank capacitance decreases, causing a frequency increase. The effect is to counteract the decrease in tank capacitance. This arrangement checks the oscillator drift caused by the DC supply variations, as well as any temperature rise occasioned by the transfer of heat from the hotter parts of the receiver. A typical varactor is a silicon planar diode with a Q of 200 at 50 MHz and a reverse-breakdown voltage of 20 V. The rest of the circuitry is conventional.

Another coaxial-tank UHF tuner with capacitor-end tuning is shown in Figure 10.40. There are three coaxial tanks (or cavities), one each for the RF, mixer, and oscillator sections. In each case, the inner walls of the metal partitions act as the outer conductors of the coaxial line and the wound coils act as the inner conductors. The three end-tuning capacitors are ganged on a single shaft for ease of tuning. As with the circuit in Figure 10.39, this tuner also uses a varactor circuit for the oscillator to provide the AFT. (See Chapter 11 for further details of AFT.) The antenna input signal is inductively coupled to the inner RF conductor L_2 by the antenna coupling coil L_8. Coupling between the RF and mixer inner lines is accomplished through a window cut in the intervening partition. Note the bimetal strip in

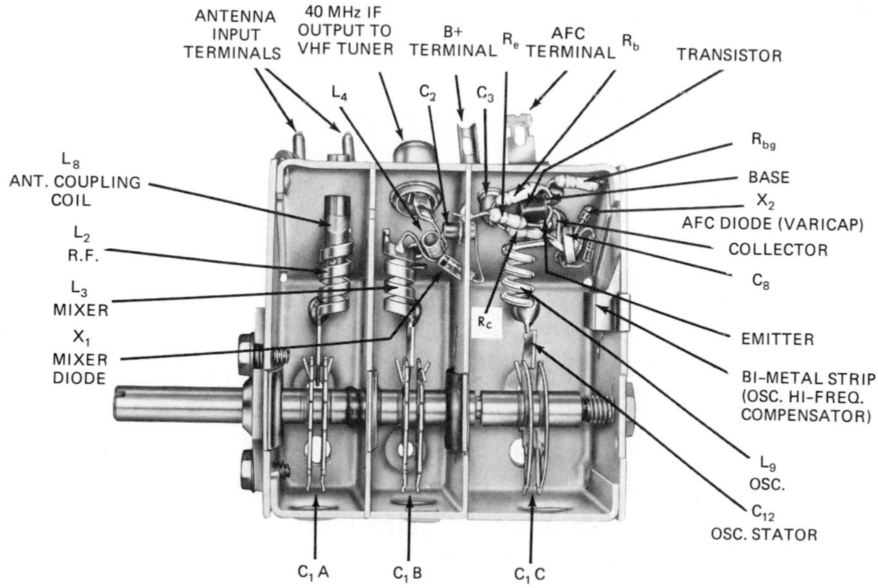

Figure 10.40 A transistor, coaxial-tank UHF tuner. Note the individual coaxial cavities for the RF, mixer, and oscillator sections. *(Courtesy of Zenith Radio Corp.)*

the oscillator section; this strip provides temperature-frequency compensation, thereby improving oscillator stability. The strip also acts as a capacitor that varies with temperature. This capacitor is formed between the strip and the oscillator inner conductor L_9.

UHF VARACTOR TUNERS

A complete schematic diagram of a varactor-tuned UHF tuner is shown in Figure 10.41. This tuner is represented by the block diagram of Figure 10.6. Individual circuit diagrams of the RF amplifier, mixer, and oscillator sections are shown in Figures 10.25, 10.30, and 10.34, respectively. (Circuit descriptions were presented when these diagrams were discussed, as well as general circuit functions when Figure 10.26 was analyzed. A few additional comments should be noted.)

The range of varactor DC tuning voltage for all UHF channels is from 1 to 27.5 V. This voltage is used to tune the antenna, RF, mixer, and oscillator circuits. Note that a +18-V line to Q_5, Q_6, and Q_7 is switchable. In VHF operation, the UHF section is inoperative. In UHF operation, the UHF tuner IF output is fed to the VHF mixer, which functions as an amplifier. During UHF operation, the VHF oscillator and mixer are inoperative. AGC voltage in the range from 0 to 12 V is fed to G_2 of the UHF RF amplifier (Q_5).

The bottom view of a UHF varactor tuner is illustrated in Figure 10.42. Note the absence of any variable-tuning capacitors. The tuning is accomplished by preset trimmers in conjunction with the variable capacity of the varactors. In the Zenith tuner shown in this figure, four varactor-tuning diodes are used, one each in the antenna, RF, mixer, and oscillator sections. The varactor diodes are supplied in matched sets of four for proper tracking correlation. If it becomes necessary to replace one of these, a new matched set of all four must be installed.

CHANNEL SELECTION

In the unit shown in Figure 10.42, any combination of VHF and UHF channels can be set up, for example, Channels 2, 32, 5, 7, 44, 9, etc. All tuning is accomplished by means of a 3-position band-selector switch and a 14-position channel-selector drum. The band-selector switch is adjusted to either VHF-LOW (Channels 2 to 6), VHF-HIGH (Channels 7 to 13), or UHF. The exact channel desired is then obtained by turning the VHF-UHF fine-tuning control until the channel number appears in a window. Varying this latter control changes the resistance of a master potentiometer in accordance with each of the 14 detented positions of the channel-selector drum. Each potentiometer resistance change produces a corresponding voltage change, which is applied to the varactor tuning diodes. They in turn tune the resonant circuits by a capacitance variation.

VHF-UHF COMBINATIONS

At this point it is appropriate to label various sections of the VHF-UHF signal flow in a receiver. Figure 10.43 gives an idea of the connections, both mechanical and electronic, in a typical all-channel receiver. The VHF tuner in its UHF setting acts as an IF amplifier to boost the input to the video IF amplifier, which thus receives nearly equal signal strengths for VHF and UHF channels.

DIGITAL TUNING TECHNIQUES

Digital tuning, including the use of phase-locked loops (PLL) and synthesized frequencies, is very important. The prin-

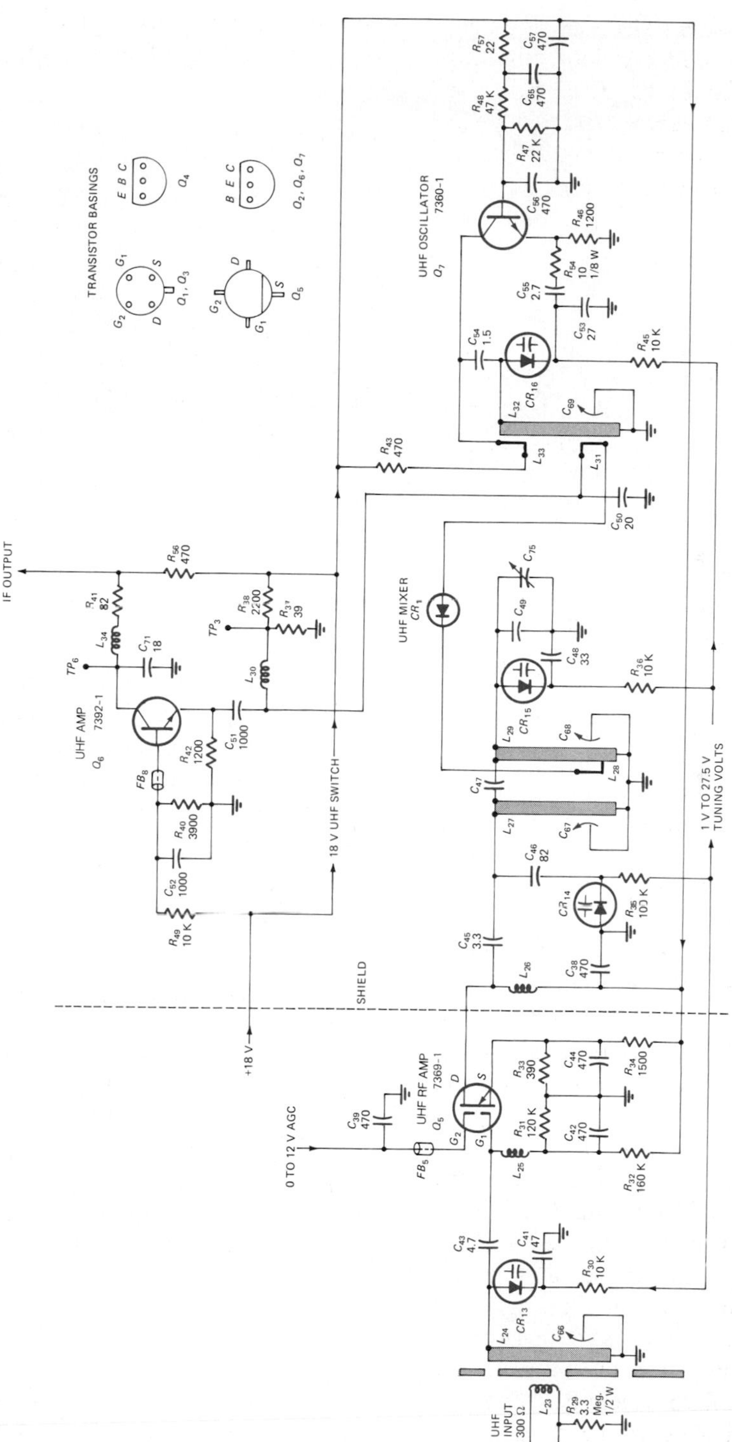

Figure 10.41 Complete schematic diagram of a varactor-tuned UHF tuner. (Individual sections are shown schematically in Figures 10.25, 10.30, 10.34.) Note the addition of a UHF IF (buffer) amplifier. Unless otherwise stated, capacitor values are in pF. *(Courtesy of RCA Consumer Electronics.)*

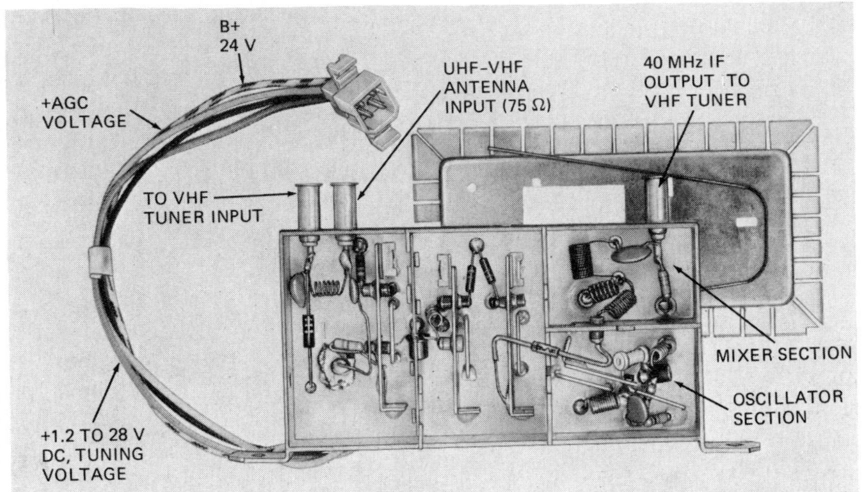

Figure 10.42 The bottom view of a UHF varactor tuner. Note the absence of tuning capacitors. *(Courtesy of Zenith Radio Corp.)*

ciples of these topics, as well as AFT and remote control operation, are covered in Chapter 11.

ALIGNMENT AND ADJUSTMENT

The alignment and adjustment of tuners and the type of test equipment used are described in Chapter 26.

10.16 SOURCES OF TROUBLE IN VHF AND UHF TUNERS

The identification of trouble areas in a television tuner becomes easier if one recalls the following major differences between transistor and vacuum tube characteristics.

1. Transistors used as voltage amplifiers are normally cool when in use, and unless they are working at abnormal operating points, they will not get warm enough to indicate faulty performance. Transistor burnouts in equipment are relatively rare. Vacuum tubes, on the other hand, tend to get hot, even in normal operation.
2. Vacuum tube control grid bias values are significantly higher than base or gate bias values for transistors.
3. Transistor input impedances are markedly lower, and a 20 000 ohms-

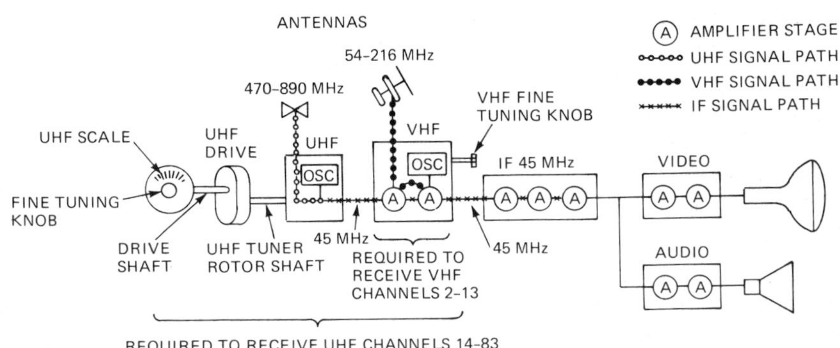

Figure 10.43 A block diagram for UHF-VHF tuning. *(Courtesy of Zenith Radio Corp.)*

volt voltmeter is adequate for measuring in most transistor circuits. With tubes, a VTVM (Vacuum Tube Voltmeter) or FET meter is often required.

4. Large changes in ambient temperature affect transistor operation much more than they do vacuum tube operation.

When the receiver is not functioning normally, the source of the problem can sometimes be determined from the sound or picture symptoms. The tuner is often to blame when one of the following conditions is present:

1. There is a good raster but no sound or picture;
2. Good sound and picture quality is attained at markedly different settings of the fine-tuning control;
3. A picture is available on only some of the channels operating in the vicinity; or
4. The picture is snowy or intermittent.

A step-by-step troubleshooting procedure is followed when one or more of these symptoms appear. Tuner contacts may require careful cleaning and lubrication if the picture is snowy or intermittent. Any broken coils on the turret should be carefully replaced and, of course, realigned. However, a complete realignment of the unit should be undertaken only after a visual examination and voltage checks have revealed no abnormality.

In vacuum tube tuners, a change of tubes may sometimes highlight a faulty area. Transistors are in general more reliable and have a very long life. Thus, before a transistor replacement, other possible causes of the trouble should be carefully considered. A faulty circuit element may be isolated by injecting a suitable sweep voltage at its input and observing the response on an oscilloscope. An *in-circuit* transistor tester can be used to check the transistors, without removing them from the tuner. A faulty oscillator may be isolated if, on application of the proper supply voltage, it yields no output. Poor picture and sound are indications of a faulty RF amplifier, mixer, or faulty common IF amplifier circuitry.

UHF TUNER PROBLEMS

A UHF tuner in its *simplest* form has only two active elements: the **diode detector** and the **local oscillator**. The transmission lines used for tuning are rigidly fixed to their supports, and the end-tuning variable air capacitors are almost always trouble-free. It is no surprise therefore that these units give the least trouble. When problems do occur, they are often attributable to mechanical gearing arrangements or to an accumulation of dust on the capacitor plates. In replacing any components in the UHF circuitry, care should be exercised since a small amount of excessive lead length can misalign the tuner.

Should the picture become noisy on a specific UHF channel, reduced oscillator injection voltage may be a cause. Coupling from the oscillator to the mixer may need attention. If the tuner should require complete alignment, it is advisable to follow the manufacturer's instructions as detailed in his product manuals. This is particularly true in UHF tuners, where a slight twist on the end-capacitors may offset the proper working of the entire unit. The importance of this fact cannot be overemphasized.

VARACTOR TUNER PROBLEMS

In varactor tuners, there are no mechanical tuning devices, and so symptoms of problems are different from those found in other tuners. The most common symptoms are (1) erratic or unstable tuning, (2) no tuning capability, or (3) AFT (voltage) drift.

For erratic or unstable tuning, the cause may be (1) incorrect DC power supply voltage, (2) defective tuning-voltage potentiometer or incorrect potentiometer calibration, or (3) incorrect VHF tuner mixer and high-low band-switching voltages. For no tuning capability, the possible sources of trouble are basically the same as for erratic or unstable tuning. For AFT drift, the symptom is drifting of the tuning voltage as seen on a DC voltmeter. If this occurs, the AFT system should be isolated from the tuner and its individual circuits checked.

More detailed troubleshooting procedures are given in Chapter 27.

PRECAUTIONS

Precautions necessary while servicing any electronic product are in order when servicing tuners. In this case, however, such precautions are to be followed even more rigorously. For instance, the transistor heat sinks should be adequate, soldering guns must not overheat the transistors, and the solder joints must be clean.

SUMMARY OF CHAPTER HIGHLIGHTS

1. The tuner, also called the front end, selects the desired station and rejects all others. It converts the RF signal to an IF signal.
2. Television receiver tuners may be classified as (1) mechanical tuners and (2) electronic tuners. Electronic tuners use varactors as tuning devices.
3. Mechanical-switch tuners are also known as incremental or step tuners. These have detents to select either VHF or UHF channels.
4. Electronic (varactor) tuners select channels by means of specific applied DC voltages. These are frequently push-button tuned and may also have digital circuits to select VHF or UHF channels.
5. Varactor tuners use switching diodes to go from the low VHF channels (2 to 6) to the high VHF channels (7 to 13).
6. Digital channel-switching systems have no moving parts and enable switching between VHF and UHF channels in a fraction of a second.
7. A random access digital tuner system makes it possible to quickly select any VHF or UHF channel regardless of the numerical sequence.
8. Frequency-synthesized tuning uses digital and PLL circuits. This eliminates the fine-tuning control.
9. The subsections of a typical VHF tuner are shown in Figure 10.4. Those of a typical UHF tuner are shown in Figure 10.6.
10. The television tuner must be able to reject undesired signals, such as image signals, and to provide a high signal-to-noise ratio.
11. A television tuner must have a highly stable oscillator. Excessive drift can degrade the TV picture. An AFT circuit helps to stabilize the oscillator.
12. Table 10.1 gives the general specifications for a television tuner. Television channel frequencies used in the United States are given in Table 10.2.
13. The response of the RF and mixer stages should be uniform throughout the 6-MHz band. However, it must also reject adjacent bands and image frequencies. (Figure 10.7)
14. The RF amplifier must have a good signal-to-noise ratio to help prevent the appearance of snow on the TV screen.
15. Transistor RF amplifiers may be oper-

ated as grounded-emitted or grounded-base amplifiers. The grounded-emitter circuit requires neutralization. (Figure 10.17)

16. The dual-gate FET is frequently used as an RF amplifier. The RF signal is applied to one gate and the AGC voltage to the other gate. (Figures 10.24 and 10.26)
17. The mixer receives signals from the RF amplifier and oscillator and converts them to the IF. The mixer must be nonlinear.
18. Common-emitter circuits are used as mixers. The circuit is operated in a nonlinear manner. (Figure 6.10)
19. UHF mixers are generally of the diode type. The Schottky-barrier diode is one type used. (Figure 10.30)
20. The tuner oscillator generally employs a Colpitts-type oscillator, which is also known as an ultraudion oscillator. (Figures 10.31, 10.33, 10.34)
21. Varactor tuners have no moving parts. The varactors tune the tuner circuits by applying specific voltages. (Figure 10.38)
22. Varactor tuners are used for VHF and UHF channels. They simplify channel selection. (Figures 10.38, 10.41)
23. Some troubles that may occur in varactor tuners are (1) erratic or unstable tuning, (2) no tuning capacity, or (3) AFT (voltage) drift.
24. In mechanical tuners, troubles frequently encountered are (1) dirty or loose switch contacts and (2) a weak detent spring.

EXAMINATION QUESTIONS

(Answers are provided at the back of the text.)

Part A Answer true (T) or false (F).

1. AGC is applied to the RF amplifier.
2. Varactors act as voltage-variable capacitors and thus can tune circuits.
3. Digital techniques are used in conjunction with varactor tuners to select VHF and UHF channels.
4. A hybrid tuner incorporates both vacuum tubes and semiconductors.
5. Switching diodes in varactor tuners are used only on UHF channels.
6. A random access system makes it possible to select channels in any desired order.
7. In frequency-synthesized tuning, a fine-tuning control is not used.
8. The image frequencies are rejected by the tuned circuits of the mixer.
9. To broaden the frequency response of a tuned circuit, a shunting resistor may be used.
10. Triode tubes are used as RF amplifiers in some tuners because of their low noise characteristics.
11. A grounded-emitter RF amplifier requires neutralization.
12. A FET is a bipolar device.
13. Many VHF mixers are diodes because of their low noise characteristics.
14. Varactor tuners may have as many as four varactor-tuned circuits.
15. One type of possible trouble in varactor tuners is AFC drift.

Part B Supply the missing word(s) or number(s).

1. In UHF operation, the VHF mixer functions as an ________ amplifier.
2. The tuner is also known as the ________.
3. Electronic tuners have ________ as the tuning elements.

4. In electronic tuners, channels are changed by applying different ______.
5. In electronic tuners ______ diodes are used to extend the tuning range.
6. A digital system that can select channels in any order is the ______ system.
7. The frequency-synthesis tuning system makes it possible to eliminate the ______ control.
8. A balun unit matches a balanced to an ______ impedance.
9. The RF amplifier must have a good ______ to ______ ratio.
10. Image frequencies are rejected by the tuned circuits of the ______ amplifier.
11. The tuner oscillator must have excellent ______ to prevent picture degradation during operation.
12. In Figure 10.14, the neutralizing capacitor is labeled ______.
13. In Figure 10.41, AGC is applied to gate ______.
14. Two tubes or transistors connected in series constitute a ______ amplifier.
15. Neutralization is accomplished with the proper phase of ______ voltage.
16. One advantage of a FET is that it generates less ______.
17. An excellent mixer amplifier is the dual-gate ______.
18. The automatic circuit that locks the oscillator on the correct ______ frequency, is abbreviated ______.
19. A popular oscillator circuit is the ______ oscillator.
20. A four-varactor tuner will tune the antenna circuit, the ______ circuit, the mixer circuit, and the oscillator circuit.
21. For tuning voltage drift in a varactor tuner, you should check the ______ system.

REVIEW ESSAY QUESTIONS

1. Draw a simple block diagram of a four-varactor UHF tuner. Label all sections.
2. In Question 1, briefly give the basic function(s) of each block.
3. Draw the response curve for a monochrome tuner plus IF amplifier. Label all important frequencies.
4. Why is the response at 41.25 MHz not zero? (In Question 3.)
5. Briefly discuss the function of the AFT circuit.
6. Give at least three functions of the RF amplifier.
7. In Question 6, what important characteristics should the transistor or tube have?
8. In Figure 10.5, what is the function of C_1 and C_2?
9. Explain the practical significance of Figure 10.9 in television receivers.
10. In Figure 10.14B, briefly explain the significance of the connections of C_1 and C_2.
11. Where is the fine-tuning control connected? Briefly discuss its function.
12. What is the effect on a color TV picture of excessive oscillator drift?
13. To which stage is the output of the VHF mixer connected? What is the approximate (usable) frequency range of the output?
14. To which stage is the output of the UHF mixer connected? Why is this done?
15. Why is it necessary to neutralize some types of RF amplifiers?
16. Briefly explain, with the aid of a simple schematic diagram, how neutralization takes place.
17. What is the connection between tuner operation and snow on the screen?
18. Refer to Figure 10.26 and give the basic function of the following components: (1) L_{24}, (2) CR_{13}, (3) C_{41}, (4) C_{67}.
19. In Figure 10.28, what is the function of R_2 and R_3? Of C_7 and RFC?
20. In Figure 10.29, how does Q_2 obtain its correct collector voltage?
21. In Figure 10.30, explain briefly how RF and oscillator signals arrive at the mixer (CR_1).
22. Refer to Figure 10.34. What would be the effect on oscillator operation (if any) if resistor R_{48} were to open?
23. Describe briefly two types of troubles that may occur in varactor tuners.

24. In Question 23, where might the cause of the trouble lie in each case?
25. What is the function of switching diodes in a varactor tuner? Are these used in a VHF or a UHF tuner? Why?
26. Explain briefly, at least two methods of supplying tuning voltages to a varactor tuner.
27. In the RCA frequency-synthesized tuning system, explain how the manual scan system operates.
28. Briefly describe the Omega digital-tuning system.
29. In Figure 10.6, what is the basic function of the IF buffer amplifier?
30. Explain briefly how an image frequency might pass through the common IF amplifiers.
31. Name the traps used in the common IF stages and their frequencies.
32. Make up a simple table showing the RF frequencies for the picture carrier, color subcarrier, and sound carrier for Channels 2, 5, 14, 55, and 83.
33. What is a cascode amplifier? What are its advantages?

EXAMINATION PROBLEMS

(Selected problem answers are provided at the back of the text.)

1. Give the frequency ranges for Channels 2, 3, 5, 13, 16, 27, and 80.
2. List the various schemes for selecting channels.
3. Identify the circuit shown below. Give the function of each labeled part.
4. In Figure 10.26, identify the parts making up the antenna and RF amplifier tuned circuits.
5. In Figure 10.30, identify the parts making up the mixer tuned circuit.
6. With a video IF of 45.75 MHz, calculate the local oscillator frequencies for Channels 2, 4, 9, 14, 55, and 82.
7. What are the picture image frequencies when tuned to Channels 3, 5, 15, and 83?
8. Make up a simple table showing the sound-image frequencies for Channels 4, 11, 13, 40, 60, and 70. (Note that sound IF = 41.25 MHz.)

11.1 The binary number system
11.2 Dividers and prescalers
11.3 The phase-locked loop (PLL)
11.4 The microcomputer (μC)
11.5 The Quasar compu-matic touch-tuning system
11.6 Compu-matic circuit analysis
11.7 Automatic fine tuning (AFT)
11.8 Functions of (AFT) systems
11.9 Functions and types of remote control devices
11.10 Ultrasonic remote control systems
11.11 Electronically generated sound signals
11.12 An electronic remote control system
11.13 A direct-access remote control system
11.14 Troubles in compu-matic tuning systems
11.15 Troubles in AFT circuits
11.16 Troubles in remote control systems
Summary

FREQUENCY SYNTHESIS, AFT, AND REMOTE CONTROL

INTRODUCTION

Many changes have taken place in electronics in a relatively short time. In recent years, we have seen the transition from vacuum tubes to transistors and then to integrated circuits and specialized devices. In many instances, we find that these new developments are quite complex. At first glance they may appear confusing and difficult to understand. In most cases, though, the more complex devices (such as microprocessors) can be considered to be "black boxes." Since a microprocessor may contain 11 000 transistors, it is better to concentrate on general functional concepts rather than on a detailed analysis of internal IC circuitry. This approach greatly simplifies the understanding of such devices.

A good understanding of the functional concepts and of the new terms accompanying new developments will provide techniques of troubleshooting which will remain useful for a long period of time. This is true for even the more complex electronic tuning systems. In order to understand such systems, however, it is necessary to first discuss basic concepts and the terms that relate to them.

All electronic tuning systems are similar in many respects. They all rely on varactor tuners for channel selection. Further, they all use electronic circuitry to control the varactor tuner. In this chapter, we shall discuss the principle of frequency synthesis (FS) in electronic tuning systems. Frequency synthesis (FS) is an electronic system, which can generate a number of *crystal stable* frequencies, while using only *one* crystal. In an electronic tuner, frequency synthesis is used to generate crystal stable frequencies for the tuner oscillator, for all VHF and UHF channels. This is explained in Section 11.5. Frequency synthesis involves the use of a highly stable, crystal reference oscillator, a phase-locked loop (PLL), and digital circuits that include a microcomputer. A microcomputer (μC) is a complete, self-contained computer on a single IC chip (or on several chips). This type of system enables a user to select any channel by direct access. It also eliminates the need for a fine-tuning control.

In addition, we shall discuss automatic fine tuning (AFT) and remote control.

When you have completed the reading and work assignments for Chapter 11, you should be able to:

- Define the following terms: frequency synthesis, binary number, bit, byte, logic high, logic low, phase comparator, VCO, CPU, radix, ROM, programmable divider, prescaler, counter, digital phase-locked loop, microcomputer, automatic fine tuning, chip, and latch circuit.
- Explain the general operation of a frequency synthesis system.
- Describe the three sections of a microcomputer and the function of each section.
- Discuss the function of a microcomputer in frequency synthesis.
- Discuss the function of varactors in electronic tuners.
- Understand how *direct access* tuning works.
- Explain the type of signals processed by digital systems.
- Explain how an AFT circuit works.
- Know how to convert decimal numbers to binary numbers.
- Describe briefly the basic differences of the various remote control systems discussed in the chapter.
- Understand how the division factor of a programmable divider is established.
- Explain the operation of a PLL.
- Understand how to troubleshoot computerized tuners, AFT circuits, and remote-control systems.

11.1 THE BINARY NUMBER SYSTEM

Semiconductor devices can be made to operate as *switches*. This makes them ideal for use in digital circuits. Such a switch is either ON or OFF and thus has two stable states. This type of switch is also called a **bistable device.**

For such switches to perform arithmetic and/or logic operations, it is necessary to represent numbers by only two digits, or **bits:** 0 and 1. A group of bits is called a **byte.** A 1 may represent a closed switch (ON), and a 0 may represent an open switch (OFF). The number system that comprises only the digits 0 and 1 is called the **binary number system** and is used in digital systems. For example, all signals into and out of, as well as inside, a microprocessor must be in binary form.

Digital systems process ON or OFF signals that are represented by two discrete voltage levels. These are generally 0 V and +5 V. In a positive logic system, 0 V is represented by a 0 and +5 V is represented by a 1. In a negative logic system, the opposite applies. The 0-V and +5-V levels are practically an industry standard.

An ON signal is referred to as a "high" and is represented by binary number 1. Similarly, an OFF signal is referred to as a "low" and is represented by binary number 0.

The **radix** of a number system is the number of symbols required to write any number of the system. The decimal number system has a radix of 10. The binary number system has a radix of 2. Two other number systems are (1) the octal number system, which has a radix of 8, and (2) the hexadecimal number system, which has a radix of 16.

Both the octal and hexadecimal number systems are used to program microprocessors. However, special circuits convert these number codes into eight-bit binary bytes that are then sent to a microprocessor. Recall that a microprocessor can function only with binary numbers.

BINARY NUMBERS

The decimal system is incompatible with switching circuits. Thus, decimal numbers must be converted to equivalent binary numbers. Such a number can then be applied to computer and control circuits. The binary system may seem confusing at first, but it is actually quite simple. We must get accustomed to working with only two digits (0 and 1) instead of ten digits.

In comparing numbers having different bases, it is customary to show the base as a subscript after the number; thus,

$$13_{10} = 1101_2$$

The binary number is read not as *one thousand one hundred and one* but rather as *one one zero one*.

DECIMAL TO BINARY CONVERSION

The following simple procedure illustrates how to convert a decimal number to a binary number. For our example, we shall convert decimal number 13 to a binary number.

1. Divide the decimal number by 2. The remainder is always either 0 or 1. Thus, $\frac{13}{2} = 6$ with remainder 1.
2. Place the remainder to the right of the partial quotient obtained in Step 1. (The 6 is the partial quotient.)
3. Divide the partial quotient of Step 1 by 2: $\frac{6}{2} = 3$, remainder 0. Place this remainder to the right of the *new* partial quotient (the 3).
4. Repeat this process until a quotient of zero is obtained.
5. The equivalent binary number is equal to the remainders (the 0s and 1s) arranged so that the first remainder obtained is the rightmost binary digit (this is called the least significant bit) (LSB) and the last remainder obtained is the leftmost binary digit (called the most significant bit, MSB) of the binary number. Thus, correctly arranging the remainders, we have

$$13_{10} = 1101_2$$

A shortened version of what we just calculated is as follows:

2)13		
2)6	1	least significant bit (rightmost digit in the binary number)
2)3	0	
2)1	1	
0	1	most significant bit (leftmost digit in the binary number)

To convert decimal 103 to a binary number:

2)103		
2) 51	1	least significant bit
2) 25	1	
2) 12	1	
2) 6	0	
2) 3	0	
2) 1	1	
0	1	most significant bit

This is correctly written as $103_{10} = 1100111_2$.

The following simple procedure illustrates how to convert a binary number to a decimal number.

1. Multiply the most significant bit (leftmost digit–MSB) by 2.
2. If the bit next to the MSB is 1, add 1 to the partial product obtained in Step 1. If this bit is 0, add 0.
3. Multiply the result obtained in Step 2 by 2 and add 1 or 0 to the partial product, depending on what the next bit is.
4. Continue this process until the least significant bit (rightmost digit–LSB) is included in the conversion.

The following two examples illustrate this process.

Convert binary number 11010 to a decimal number:

$$(1 \times 2) + 1 = 3$$
$$(3 \times 2) + 0 = 6$$
$$(6 \times 2) + 1 = 13$$
$$(13 \times 2) + 0 = 26$$

Therefore, $11010_2 = 26_{10}$

Convert binary number 110101 to a decimal number.

$$(1 \times 2) + 1 = 3$$
$$(3 \times 2) + 0 = 6$$
$$(6 \times 2) + 1 = 13$$
$$(13 \times 2) + 0 = 26$$
$$(26 \times 2) + 1 = 53$$

Therefore,

$$110101_2 = 53_{10}$$

For your convenience, Table 11.1 shows the binary number equivalents for decimal numbers from 1 through 25.

Table 11.1. **DECIMAL TO BINARY NUMBER CONVERSION TABLE**

DECIMAL	BINARY	DECIMAL	BINARY
0	0000	14	1110
1	0001	15	1111
2	0010	16	10000
3	0011	17	10001
4	0100	18	10010
5	0101	19	10011
6	0110	20	10100
7	0111	21	10101
8	1000	22	10110
9	1001	23	10111
10	1010	24	11000
11	1011	25	11001
12	1100		
13	1101		

11.2 DIVIDERS AND PRESCALERS

Dividers and **prescalers** perform the same basic function. A prescaler, however, is a high-frequency divider that is often used to provide lower-frequency signals for succeeding dividers. Dividers may be either fixed or programmable. When fixed, they divide the input signal by a specific factor, such as 2, 5, 75, 181, or 256. When programmable, the division factor can be altered by changing the input logic of the divider. This type of divider is generally shown as a divide by N. Generally speaking, ICs perform these divider functions. Their design may be N type, P type, or C (complementary), in one of the following formats:

- RTL (resistive transistor logic)
- DTL (diode transistor logic)
- TTL (transistor transistor logic)
- ECL (emitter coupled logic)

Logic circuitry may be encapsulated in a chip or may be included as part of a large-scale integrated (LSI) package that includes many other devices and functions. Usually there are several inputs to these dividers including the signal to be divided, the binary number, the enable (start) pulse, and the reset pulse.

The binary inputs are the logic highs or lows (voltages) that establish the divide ratio. Logic highs are normally represented by 1 and logic lows by 0. The binary input shown in Figure 11.1 consists of two logic highs and two logic lows (1010).

Usually, the input signal is a continuous train of pulses. Thus, to accomplish the desired function, an enable pulse begins the function and a reset pulse re-establishes the starting point. If, for example, a 10-kHz signal is applied to a divide-by-10 divider, the enable pulse begins the counting function, which is completed at the end of the 10 000th pulse. At this time, a reset pulse starts a new count, which is also divided by 10. The output, therefore, is 1 kHz.

Many applications require continuous output from a divider until it is cancelled or reprogrammed. This is accomplished by applying the binary information to the divider inputs through **latch circuits.** These circuits are activated by closing switches, connecting wires together, rotating controls, or even by the output of a computer.

A latch circuit connected to a divide-by-10 divider is illustrated in Figure 11.2

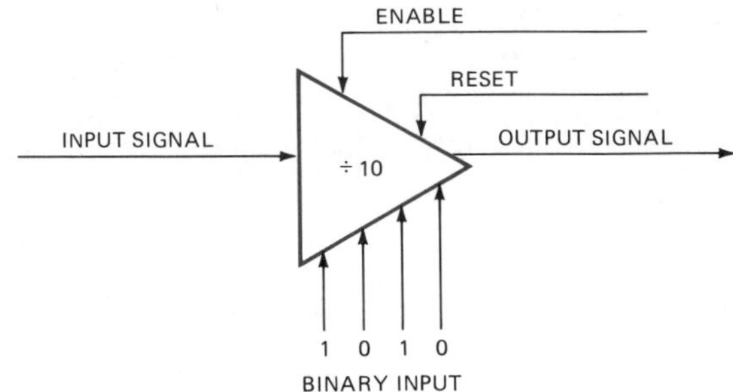

Figure 11.1 The binary input to a divider determines the factor by which it divides. In this case the binary input is 1010_2, so the divider divides by 10_{10}.

THE DIVISION FACTOR OF A DIVIDER

The division factor of a divider is determined by the binary number applied to the input and is limited by the number of bits of information the divider will accept. This is determined by the input ports. As an example, consider the four-bit programs in Figure 11.3. In the first example, all input ports are high and thus the division factor is 15. In the second example, only the second and fourth input ports are high. Thus, the division factor is 10. Any division factor from 1 through 15 can be obtained by properly programming the input ports of this divider.

As another example, consider an eight-bit divider, as illustrated in Figure 11.4A. If all input ports are a logic high, the division factor is 255. Thus, proper programming can provide any desired division factor from 1 through 255. Actually, factors of several thousand are not uncommon. This becomes evident if four more bits are added to the above, as shown in Figure 11.4B. In this case, we now have a 12-bit divider with all input ports a logic high. Proper programming can provide any desired division factor from 1 through 4095 for the 12-bit divider.

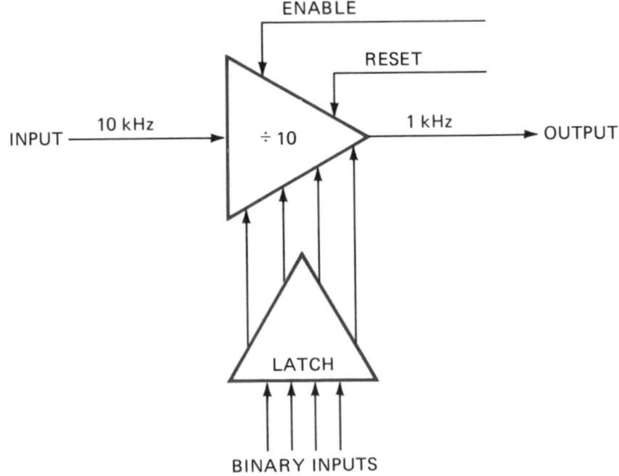

Figure 11.2 A latch circuit continuously supplies the binary input to a divider until the latch is reset.

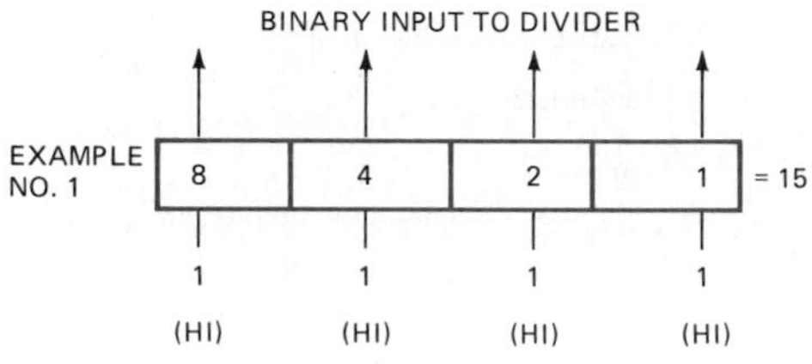

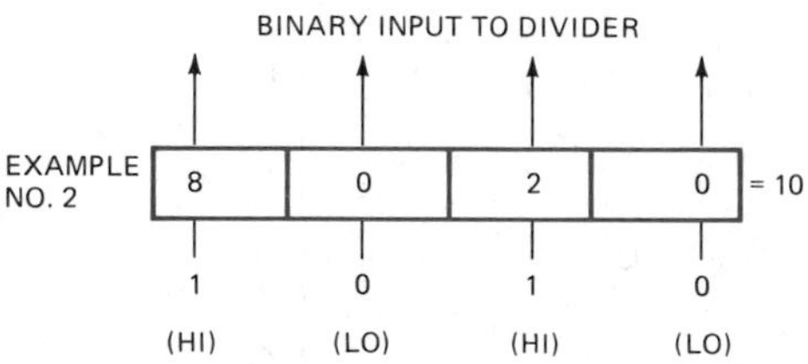

Figure 11.3 Examples illustrating division by 15 and by 10, represented by binary inputs 1111_2 and 1010_2.

Hence, if we have an N-bit divider, we can divide by any number from 1 to $2^N - 1$.

A PRACTICAL DIVIDER APPLICATION

Divider applications are many and varied. As an example, assume that a 450-kHz signal is required for a specific purpose and a stable 9-MHz signal is readily available. The desired signal may be obtained as shown in Figure 11.5. The 9-MHz signal is applied to a divide-by-10 divider, and the output is applied to a divide-by-2 divider. This provides the desired 450-kHz signal. (Note that this could not be done in a single divider since the maximum division available $= 2^N - 1 = 2^4 - 1 = 15$.)

COUNTERS

Counters work like dividers. They are logic blocks that produce an output pulse for N input pulses. As an example, digital clocks divide the power line frequency by 60 to obtain one pulse each second, and then divide this pulse by 60 and its output by 60 to obtain minutes and hours. This device counts input pulses for specific time intervals.

11.3 THE PHASE-LOCKED LOOP (PLL)

The basic **phase-locked loop** (PLL) **circuit** is not new. It has been in use for many years. Two common examples in television receivers are (1) the horizontal AFC and horizontal oscillator and (2) the AFPC (automatic frequency phase control) and 3.58-MHz oscillator. In recent television receiver designs, PLL circuits have been combined in the digital frequency dividers to stabilize the wide-range frequency-control circuits found in modern VHF/UHF TV tuners.

Essentially, PLL circuits compare two

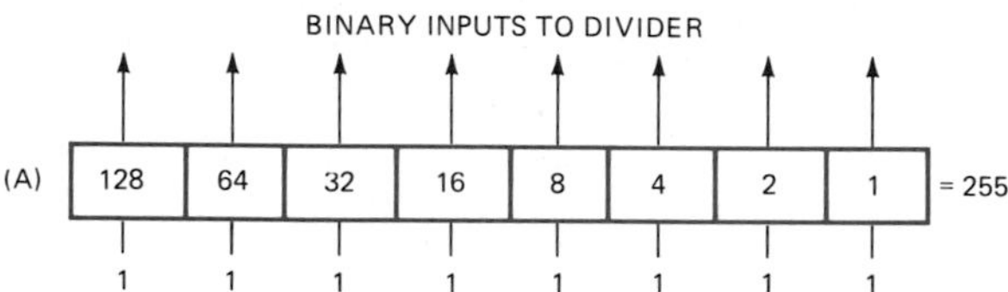

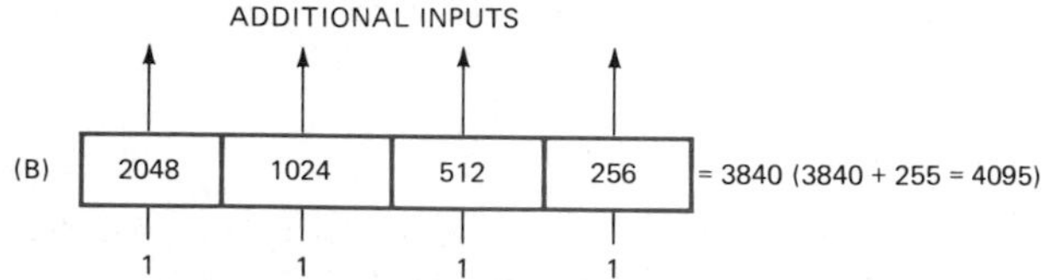

Figure 11.4A. Illustrating how an 8-bit divider can divide by 255. (All input ports are a logic high.) B. If 4 more bits are added to (A), the resulting 12-bit divider can now divide by 4095. (All input ports are a logic high.)

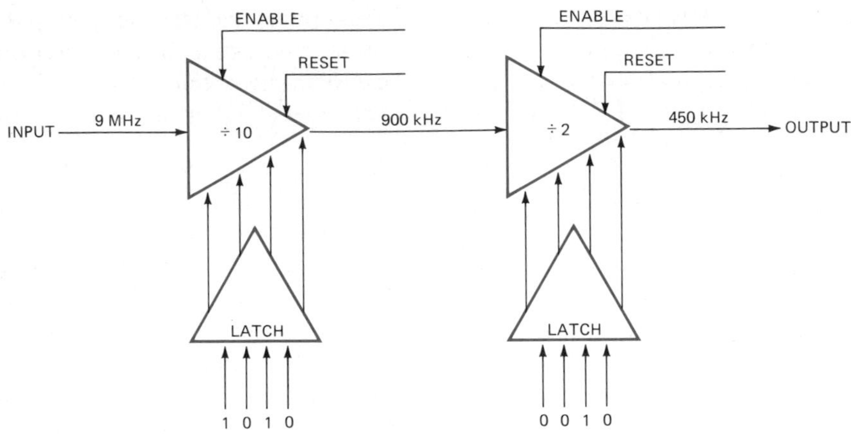

Figure 11.5 Method of obtaining a stable 450-kHz output from a stable 9-MHz source.

signals and develop a correction voltage proportional to the frequency difference or phase difference between them. This voltage then alters the frequency or phase of one of the signals to bring it into step with the other signal. The two signals are generally a voltage-controlled oscillator (VCO) signal and a stable, crystal-oscillator reference signal. The PLL circuit develops a voltage proportional to the difference in phase or frequency of the VCO and the reference crystal oscillator and then uses this voltage (called a correction or error voltage) to alter the output signal of the VCO and bring it into step with the crystal oscillator.

BASIC PLL OPERATION

In the operation of a PLL, a crystal oscillator provides a stable reference signal to a phase comparator (Figure 11.6). The output of a voltage-controlled oscillator (VCO) is also applied to the same phase comparator. If the two signals are synchronized, no correction is required. However, if the VCO drifts off frequency, the phase comparator develops a correction voltage proportional to the phase difference of its input signals. The output of the phase comparator is filtered and applied to the VCO. This output is a DC error voltage. The voltage, when applied to the VCO, causes it to assume the correct frequency and phase. The VCO is thus able to provide crystal-oscillator accuracy at the desired frequency (or frequencies).

High-Frequency Crystal Controls Lower Frequency VCO Assume that crystal-oscillator accuracy is required for a 10-kHz signal. Precision crystals at 10 kHz are not read-

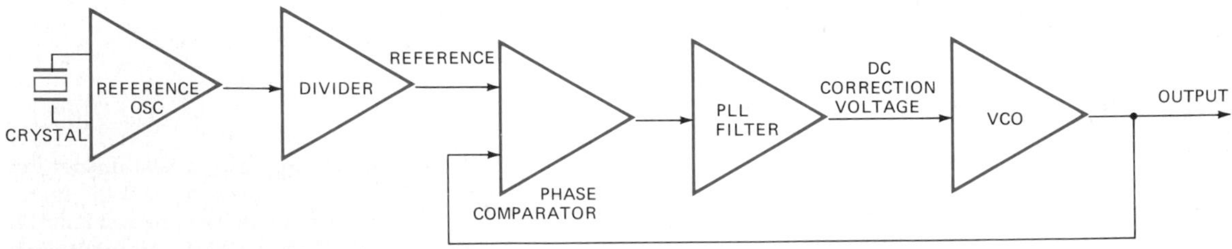

Figure 11.6 A basic phase-lock loop (PLL) circuit. A highly stable crystal oscillator generates the reference frequency.

ily available and, in any case, are quite expensive. However, high-frequency crystals are readily available at low cost. Thus, it is desirable to use a 20-MHz crystal oscillator to stabilize a 10-kHz VCO. This scheme is illustrated in Figure 11.7. The highly stable 20-MHz reference crystal oscillator is fed to a divide-by-2000 digital-frequency divider. The output of the divider is a crystal-stable 10-kHz signal. This new signal is now used as the reference to stabilize the 10-kHz VCO.

PLL IN TV TUNERS

PLL systems in combination with a microprocessor (MPU, or μP) can provide electronic selection and proper tuning of all VHF and UHF channels. With this system, no fine-tuning control is needed because the correct tuning for each channel is accurate and automatic. How any one of the 82 TV channels is tuned in will be discussed later in this chapter.

11.4 THE MICROCOMPUTER (μC)

As mentioned previously, a microcomputer is a complete computer on a single μC chip or on several chips. Similar to a microcomputer, the human brain is also a computer. It stores information and data in a memory, selects specific data from the memory when needed, interprets the data selected, and is capable of performing various operations. It can receive and store new data in its memory and even revise data already stored in memory.

COMPUTER SECTIONS

All computers contain three fundamental sections: memory, central processing unit (CPU) and input-output (I/O). The memory stores information and data required to satisfy the computer capabilities. There are, however, two types of memories. A "read only memory" (ROM), stores specific predetermined data which remains unchanged. A "read and write memory" (RAM), is capable of receiving and storing data which can be changed as required.

The "central processing unit" (CPU) performs a sequence of step-by-step operations determined by various programs stored in the memory, thus it controls the computer. This is the *microprocessor*. The "input" section funnels information into the CPU. The output of the CPU is fed out through the "output" section.

COMPUTER CAPABILITIES

Basically, a computer performs very simple operations one at a time, but its advantage lies in the fact that it is capable of performing each operation in an extremely short time. As an example, a program of 10 000 individual instructions appears quite complex to most people, but a computer can execute all of the instructions in a few thou-

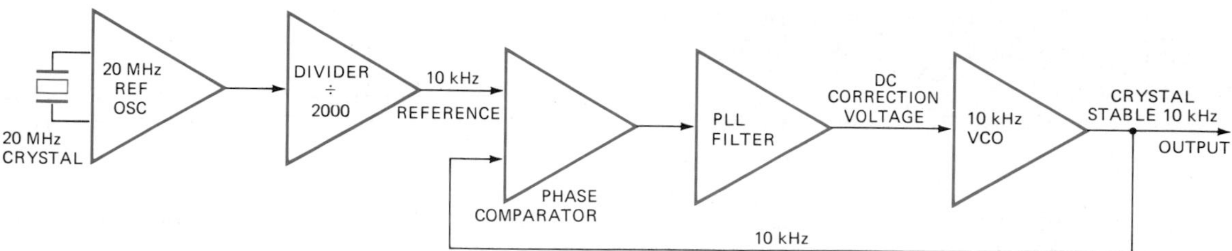

Figure 11.7 Using a highly stable 20-MHz crystal-reference oscillator to stabilize a 10-kHz VCO.

sandths of a second. As a result of this fantastic speed, most individuals conclude that computers are unlimited in their capabilities. Generally speaking, though, they are limited by programming techniques, capabilities as designed and, of course, cost.

A SIMPLE PROGRAM

The following simple program illustrates the basic operation of a computer. Assume that a group of pigeonholes represents the computer memory and that each hole contains a simple directive.

A1	A2	A3	A4	A5	A6	A7	A8
B1	B2	B3	B4	B5	B6	B7	B8
C1	C2	C3	C4	C5	C6	C7	C8
D1	D2	D3	D4	D5	D6	D7	D8
E1	E2	E3	E4	E5	E6	E7	E8
F1	F2	F3	F4	F5	F6	F7	F8
G1	G2	G3	G4	G5	G6	G7	G8
H1	H2	H3	H4	H5	H6	H7	H8

An individual—you, for example—can represent the CPU of the computer. Assume you are instructed to begin at pigeon hole A1 and follow each directive explicitly to determine whether $1 + X$ is greater than, less than, or equal to $6 + Y$.

- "A1—record the mathematical value 1"

You write the figure 1 on a scratch pad, which represents the accumulator in the computer, and then proceed to the next pigeonhole.

- "A2—add 2 to the existing value"

$$1 + 2 = 3$$

- "A3—place this value in H8"
- "A4—record the mathematical value 6"
- "A5—add 3 to the existing value"

$$6 + 3 = 9$$

- "A6—place this value in H7"
- "A7—retrieve and record the value in H8"

The instruction at A3 directed you to place the value 3 in H8. Thus you now write 3 on the scratch pad and proceed to A8.

- "A8—subtract the existing value from the value in H7"

The instruction at A6 directed you to place the value 9 in H7. Thus $9 - 3 = 6$.

- "B1—if the existing value is not zero proceed to B3"
- "B3—if the existing value is a negative number proceed to B5"
- "B4—the value in H7 is greater than the value in H8; stop"

Notice that you begin at A1 and progress in an orderly sequence—A2, A3, A4, and so on—until you are directed to skip a pigeonhole. This represents a very basic computer program and illustrates how simple operations are performed one at a time. Computers, however, are capable of performing thousands of sequential operations in only a few hundred milliseconds. Thus, the application of microcomputers in TV tuning systems is not surprising.

Digital electronics, microcomputers, counters, and dividers are fascinating and interesting, which is true of all sciences. Obviously, these devices will be used in many home-entertainment products in the future. Thus, you should become thoroughly familiar with their basic concepts. A number of excellent publications are available to assist in this study.

11.5 THE QUASAR COMPU-MATIC TOUCH-TUNING SYSTEM

This system has been chosen for explanation because it is an excellent example of frequency-synthesized tuning.

BASIC OPERATION

The control panel for this receiver is shown in Figure 11.8 and the remote control unit in Figure 11.9. At the control panel, the user may choose either direct-access channel selection or the "seek" function. For direct access, the desired channel number is simply entered on the keyboard. This may be done in any desired order. The selected channel number is shown in a digital readout (using LEDs) at the top of the panel. For direct access, two digits must be entered for each channel. For example, "02" has to be entered for Channel 2. When the first button is pressed, the "units" digit will light up. The second entry lights up the "tens" digit, and the channel changes. If the second entry is not made within 3 to 4 seconds following the first entry, the set remains on the original channel and the original channel number reappears in the display. If an inactive channel is selected and the "seek" switch is ON, the set will seek the next active channel.

Alternately, the user may operate the "channel up" or "channel down" buttons. The receiver will then automatically seek the next active channel and lock onto it.

Note: The receiver must be allowed to stop on a channel before a new channel entry is made, for the following reason. If an invalid channel has been entered, 00, 01, and 84 through 99, the set will search for an active channel. If the channel-select buttons are pressed while a search is being made,

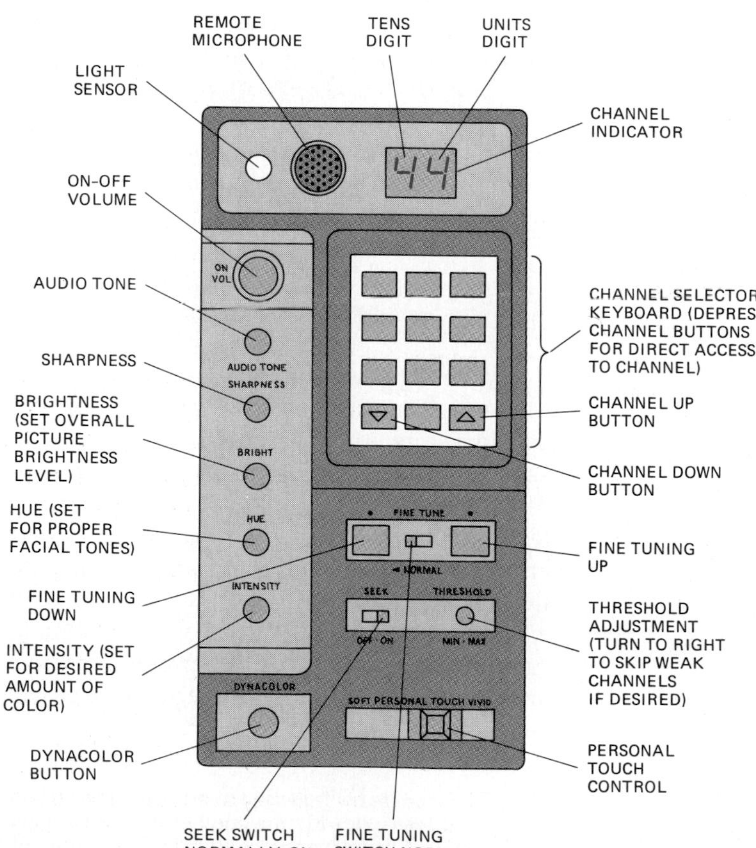

Figure 11.8 Control panel of the Quasar Compu-Matic color TV receiver. *(Courtesy of Quasar Electronics Co.)*

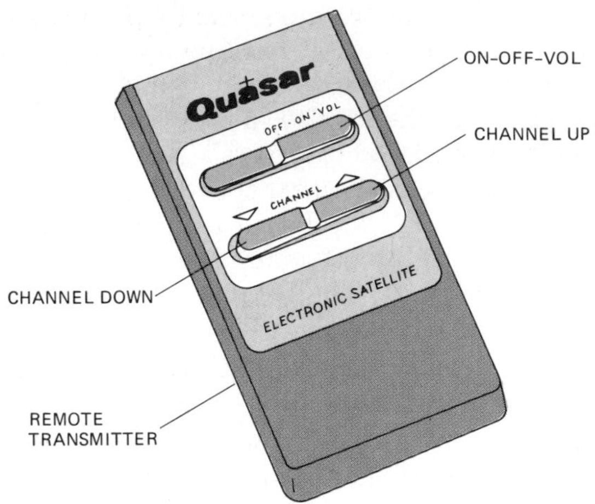

Figure 11.9 A basic remote control unit for the Quasar Compu-Matic color TV receiver. *(Courtesy of Quasar Electronics Co.)*

the last number may be held in memory and cause the next selection to be erroneous. If this occurs, press zero (0). The set then selects a station, and then two digits can be entered for the desired station.

REMOTE OPERATION

One type of remote unit for this receiver is shown in Figure 11.9. This unit provides control of ON, OFF, and volume, in three steps. Depressing the "channel up" or "channel down" buttons provides the seek function described above. A direct-access remote unit is available on some TV receiver models. It incorporates an infrared transmitter and receiver. Such a unit is described later in this chapter.

SUMMARY OF OPERATING MODES

The following is a brief description of the three operating modes:

1. Seek Off, Fine Tune Normal
   System tunes to channel selected by digit entry and remains on that channel whether it is active or not. Thus, the system can be tuned to unused channels for use with VTR, video games or for troubleshooting. This mode also provides for manual fine tuning.
2. Seek On, Fine Tune Normal
   Permits channel selection by digit entry or "channel up" and "channel down" buttons. If the channel selected is inactive, the system performs a fine-tune search for sensory signals that indicate the presence of a channel signal (a frequency offset may occur in some cable systems). If the sensor circuits are satisfied, the system will remain at that channel location. If not satisfied, the system automatically steps through each channel in sequence to the next active channel. In this mode, the channel may be manually fine-tuned.
3. Seek On, Automatic Fine Tune
   Functions are same as in Mode 2 except that system automatically compensates (fine-tunes) for frequency offsets[1] on any VHF channel. This mode defeats manual fine tuning.

When the seek switch is on (Mode 2 or Mode 3), unusable signals can be bypassed by adjusting the threshold control. With seek OFF, invalid digit entry, such as 00 or 01, causes the system to tune to Channel 83. If the seek switch is ON and channel 83 is an unused channel, the system will step to the lowest active VHF channel. Invalid digit entries 84 through 99 cause the system to tune to Channel 2. If this channel is inactive and the seek switch is ON, the system will step to the lowest active VHF channel.

---

[1]Some number channels located in adjacent areas where they might interfere with one another have their video carriers shifted in frequency by 10 kHz, in opposite directions.

FINE-TUNE BUTTONS

Fine tuning is normally *not* required because the set tunes precisely to the assigned channel frequency. However, fine-tune buttons are provided to compensate for unusual signal conditions, interference, tuning for games, and so on on a specific channel. To achieve the desired effect, one of the fine-tune buttons is pressed and held. If color fades or sound interference is noted in the picture, the first fine-tune button is released and the other button is pressed and held until the desired tuning is achieved. The receiver reverts to the precise channel frequency upon re-entry of the same or another channel.

OVERVIEW OF SYSTEM OPERATION

The Quasar electronic-tuning system uses varactor tuners in conjunction with a PLL to maintain crystal accuracy and stability of selected channels. A microcomputer provides the intelligence, decision-making, and control for the system. A brief overview of the system is presented with the aid of the simplified block diagram of Figure 11.10.

PLL An extremely accurate and stable crystal oscillator and dividers provide a reference signal to a phase comparator, which is a key element in the PLL system. The tuner oscillator signal is also divided down and applied to the phase comparator. The oscillator division factor must result in the same frequency as the reference signal. When this occurs, the channel is tuned to its precise nominal frequency. When there is a difference between these two input signals, the phase comparator develops a voltage that automatically corrects the tuner oscillator frequency until the phase comparator input signals once again coincide. As a result, the tuners assume the accuracy and stability of the reference oscillator.

Tuning Voltage VHF and UHF tuning voltage is obtained by clamping +130 V to +33 V. Output pulses from the phase comparator are integrated to control transistor conduction, which establishes the correct tuning voltage for each channel.

Changing Channels by Direct Access Touch pads on a front-panel keyboard provide direct entry of two digits to the microcomputer. When the user enters the desired channel digits (05, 09, 12, 35, and so on) several things occur:

1. A microcomputer output to the LED driver activates the proper LED segments to display the channel number selected.
2. The microcomputer loads the correct division factor for the selected channel into the tuner oscillator divider chain. As a result, the input frequency to the phase comparator is altered.
3. Unequal frequency inputs to the phase comparator produce an output that adjusts the tuning voltage to the level required for the selected channel.
4. The microcomputer determines if the channel selected is low VHF, high VHF, or UHF and applies the proper outputs to the VHF-UHF B+ switch and VHF high/low band switch.
5. An output from the microcomputer activates the audio-mute switch in the absence of a station signal.

Sensory Circuits The sensory circuits receive several signals (45.75 MHz, 41.25 MHz, vertical synchronizing pulses, composite video, and AFT voltage) and perform logic functions that provide drive levels for

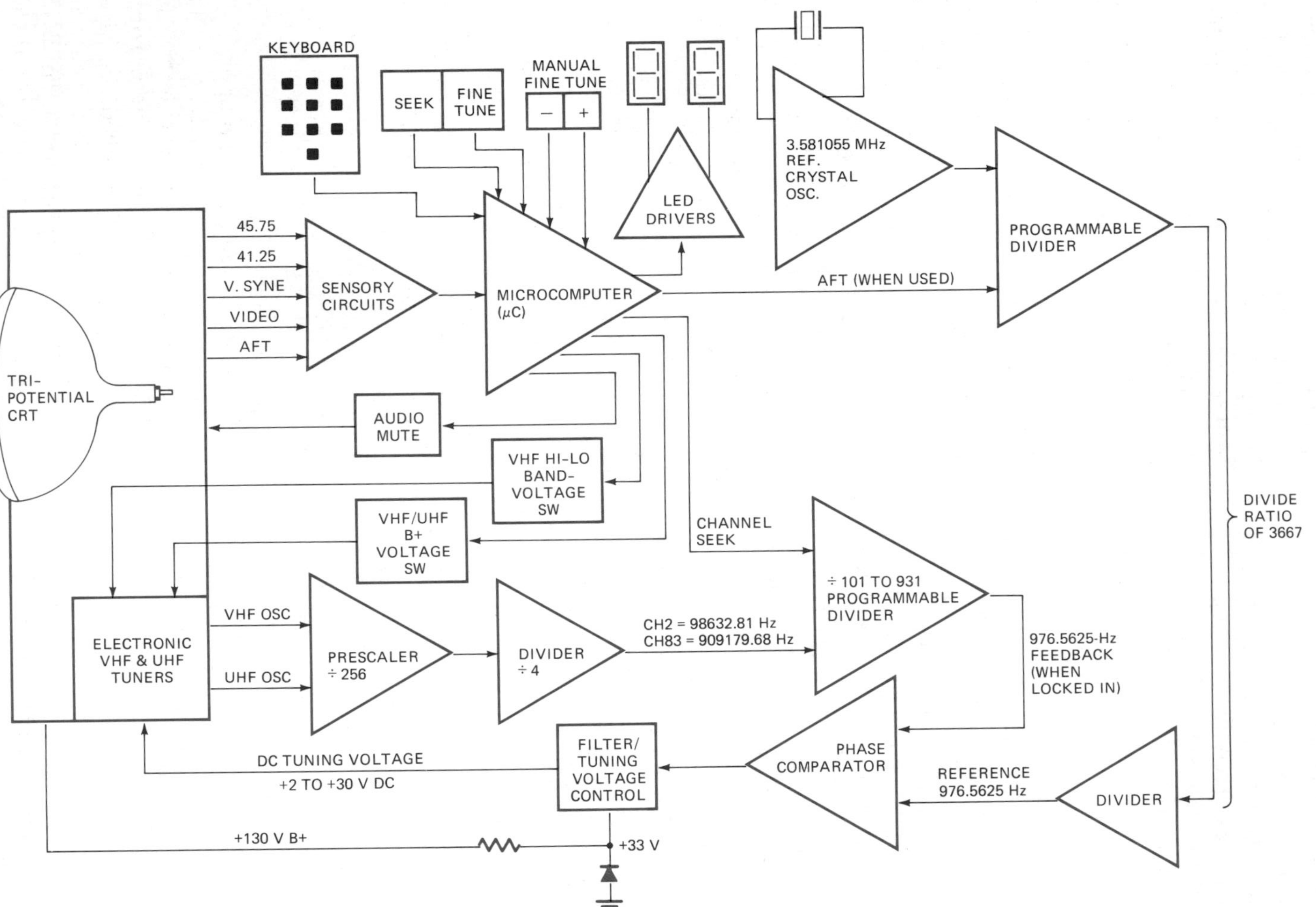

Figure 11.10 Simplified block diagram of the Quasar Compu-Matic electronic tuning system. The Channel 2 and Channel 83 frequencies shown above are derived from the tuner oscillator frequencies of 101 MHz (Channel 2) and 931 MHz (Channel 83). *(Courtesy of Quasar Electronics Co.)*

the microcomputer. The input signals tell the microcomputer whether or not a station signal is present and how accurately it is tuned. In the seek mode, the system steps to the next channel if a station signal is not present. When seek is off, the microcomputer disregards the sensor drive levels. Thus, it remains tuned to an unused channel when seek is off.

Seek and Fine Tune Seek and fine-tune switches on the front panel permit the viewer to establish the various modes listed above. With seek ON, the system bypasses unused channels. When the seek switch is in the OFF position, the viewer can select unused channels for operation with a video tape recorder or video games.

The fine-tune switch provides for manual or automatic fine tuning. In the "normal" position, the viewer can fine-tune manually by depressing one of the fine-tune buttons. In the other switch position, the system automatically fine-tunes when required. Thus, it compensates for frequency errors that may exist in cable systems.

11.6 COMPU-MATIC CIRCUIT ANALYSIS

Since the PLL IC contains 1 153 transistors and the microcomputer IC contains 11 029 transistors, the major portion of this section is devoted to functional concepts rather than to an analysis of internal IC circuitry. (The discrete switching circuits, audio mute, VHF-UHF B+ and VHF high/low band, and sensory circuits will be presented in greater detail.)

PLL

The PLL is a key element in the tuning system because it assures full-time crystal accuracy and stability of all VHF and UHF channel frequencies. A detailed block diagram of the PLL is given in Figure 11.11. The output of a stable 3.581 055-MHz crystal oscillator is applied to dividers that establish a **reference frequency** input to the phase comparator. Two dividers between the reference oscillator and phase comparator are illustrated in Figure 11.11 to emphasize that a portion of the divider chain is programmable. Actually, it is a 14-stage divider with some of its input ports fixed and others variable. Binary inputs from the microcomputer to this 14-stage divider establish a division factor of 3667. Thus, 3.581 055-MHz divided by 3667 provides a crystal-stable 976.5625-Hz reference frequency to the phase comparator. The local (tuner) oscillator signal is frequency-divided and compared to the crystal-controlled reference frequency. If the frequency of these signals is not exactly the same, the phase comparator develops a correction voltage. This voltage is used to adjust the frequency of the tuner oscillator so that the two inputs to the phase comparator are exactly the same frequency.

Tuner Oscillator Division The crystal-controlled reference oscillator frequency is always 3.581 055-MHz, but the tuner oscillator frequency differs for each channel. Obviously, the system must provide, for each of the 82 channels, the proper frequency division between the local oscillator in the tuner and the comparator. This is accomplished by using a programmable divider.

The local oscillator signal from the VHF or UHF tuner is amplified and applied to a prescaler (high-frequency divider). As various channels are selected, switching circuits apply B+ to the proper tuner and preamplifier. The prescaler is a fixed divide-by-256 divider. Its output (local oscillator frequency divided by 256) is applied to a fixed divide-by-4 divider and a programmable divider network, which completes the required frequency division.

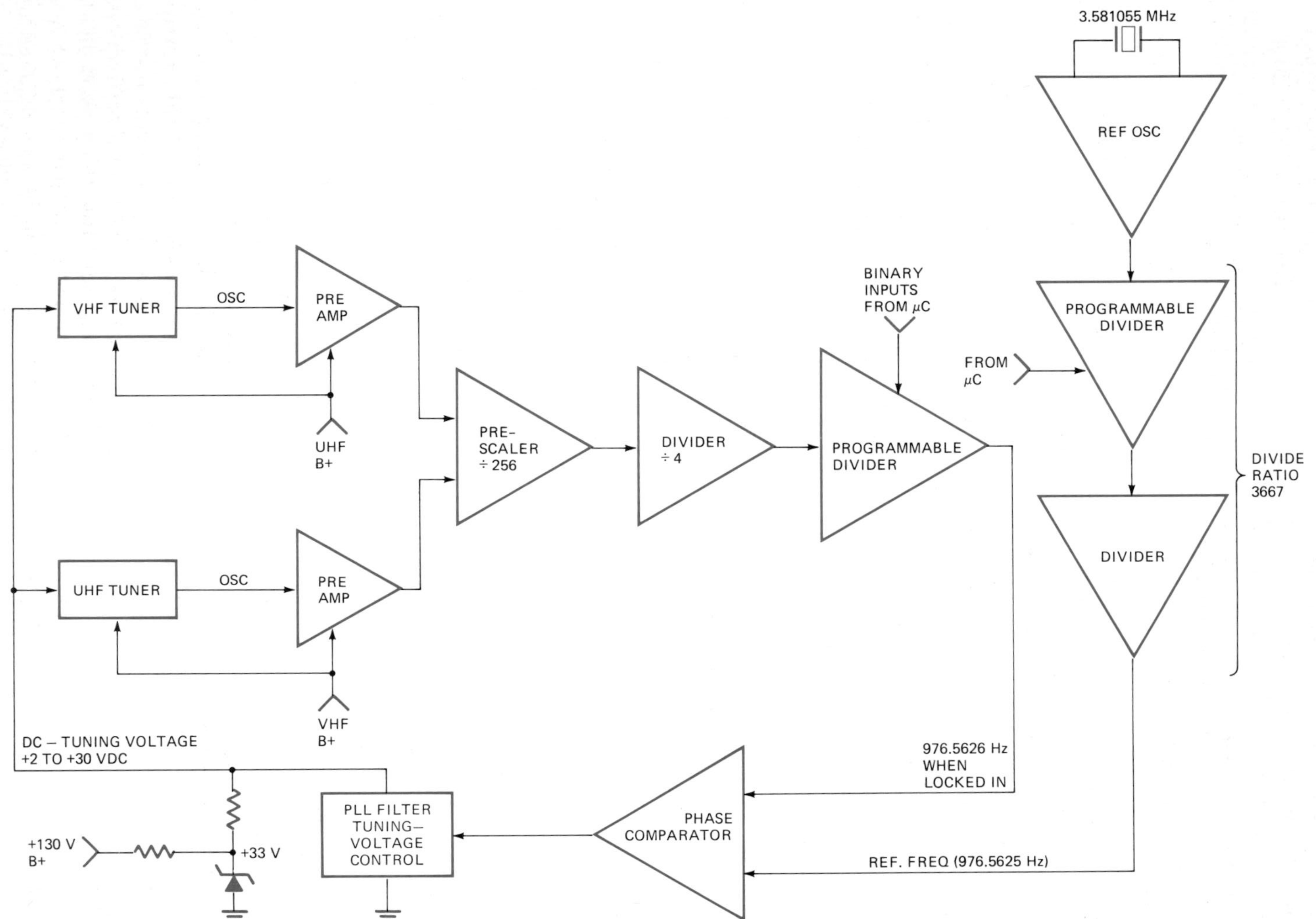

Figure 11.11 Block diagram of the PLL of the Quasar Compu-Matic electronic-tuning system. *(Courtesy of Quasar Electronics Co.)*

The signal obtained by frequency-dividing the local oscillator signal must be exactly the same as the reference frequency. However, the frequency of the local oscillator is different for each channel, and two of the three dividers in the chain are fixed (Figure 11.11). The third divider is programmable and receives binary inputs from the microcomputer, which provides the proper frequency division for each channel. Note in Figure 11.11 that the reference frequency is exactly 976.5625 Hz. For the selected channel to be correctly locked in, the tuner oscillator frequency must have a divided value of 976.5625 Hz. This value is achieved as follows:

$$\frac{\text{tuner oscillator frequency}}{256 \times 4 \times N} = 976.5625 \text{ Hz} \qquad (11.1)$$

Notice that the tuner oscillator signal is divided by 256, by 4, and by N. The variable N represents the binary inputs to the programmable divider. As an example, assume that the system is tuned to Channel 2. The tuner oscillator frequency for Channel 2 is 101 MHz. Therefore:

$$\frac{101\,000\,000}{256 \times 4 \times N} = 976.5625 \text{ Hz}$$

The value of N for Channel 2 is determined as follows:

$$\begin{aligned} N &= \frac{101\,000\,000}{256 \times 4 \times 976.5625} \\ &= \frac{101\,000\,000}{1\,000\,000} \\ &= 101 \text{ (same as tuner oscillator frequency in MHz)} \end{aligned}$$

Substituting 101 for N in Equation 11.1:

$$\frac{101 \times 10^6}{256 \times 4 \times 101} = 976.5625 \text{ Hz}$$

It follows then that the value of N for *any* channel is:

$$N = \frac{\text{local oscillator frequency}}{256 \times 4 \times 976.5625}$$

The tuner oscillator frequency for Channel 83 is 931 MHz (931×10^6). Therefore:

$$N = \frac{931 \times 10^6}{256 \times 4 \times 976.5625} = 931$$

(same as tuner oscillator frequency in MHz)

or

$$\frac{931 \times 10^6}{256 \times 4 \times 931} = 976.5625 \text{ Hz}$$

Note: In these two examples, the value for N is the same as the tuner oscillator frequency in MHz. This is true for all 82 channels.

Two additional examples:

Channel 6 (tuner oscillator frequency = 129 MHz)

$$\frac{129\,000\,000}{129 \times 256 \times 4} = 976.5625 \text{ Hz}$$

Channel 13 (tuner oscillator frequency = 257 MHz)

$$\frac{257\,000\,000}{257 \times 256 \times 4} = 976.5625 \text{ Hz}$$

Remember that the microcomputer always loads the programmable divider (following prescaler) with a binary number that is equivalent to the required value of N.

As long as both inputs to the phase comparator are equal (at 976.5625 Hz), no correction voltage is developed by the comparator. However, if the tuner oscillator frequency drifts, its input to the phase comparator changes (no longer 976.5625 Hz). In this case, the phase comparator develops a DC correction voltage (Figure 11.11). This voltage is fed to the VHF or UHF tuner oscillator (VCO) to correct the frequency. Actually, the output of the phase comparator controls another circuit, which in turn supplies the DC tuning voltage (to be described shortly).

Changing Channels Channels are changed by altering the frequency division resultant between the tuner oscillator and the phase comparator. As an example, assume that the system is operating on Channel 2, whose tuner oscillator frequency is 101 MHz. An input from the microcomputer to the divider chain produces the following ratio:

$$\frac{101 \times 10^6}{101 \times 256 \times 4} = 976.5625 \text{ Hz}$$

The viewer depresses the "0" touch pad and then the "6" touch pad to obtain Channel 6. This produces a new binary input from the microcomputer. The ratio then becomes:

$$\frac{101 \times 10^6}{129 \times 256 \times 4} = 764.5654$$

Thus, a frequency difference exists between the two comparator inputs. This difference develops a correction voltage, which is used to adjust the tuning voltage to the level required to tune the local oscillator to 129 MHz. When this occurs, the two inputs to the comparator will once again be exactly the same frequency:

$$\frac{129 \times 10^6}{129 \times 256 \times 4} = 976.5625 \text{ Hz}$$

Any frequency difference between the comparator inputs produces a series of pulses at its output. These pulses must be filtered to produce the control voltage that adjusts the tuning voltage, as described below.

PLL Filter and Tuning-Voltage Control Refer to the schematic diagram of Figure 11.12. The +130-V DC input is clamped to about +33 V DC by zoner diode D_{32}. This establishes the maximum varactor tuning voltage. If there is a frequency difference between the two phase comparator input sig-

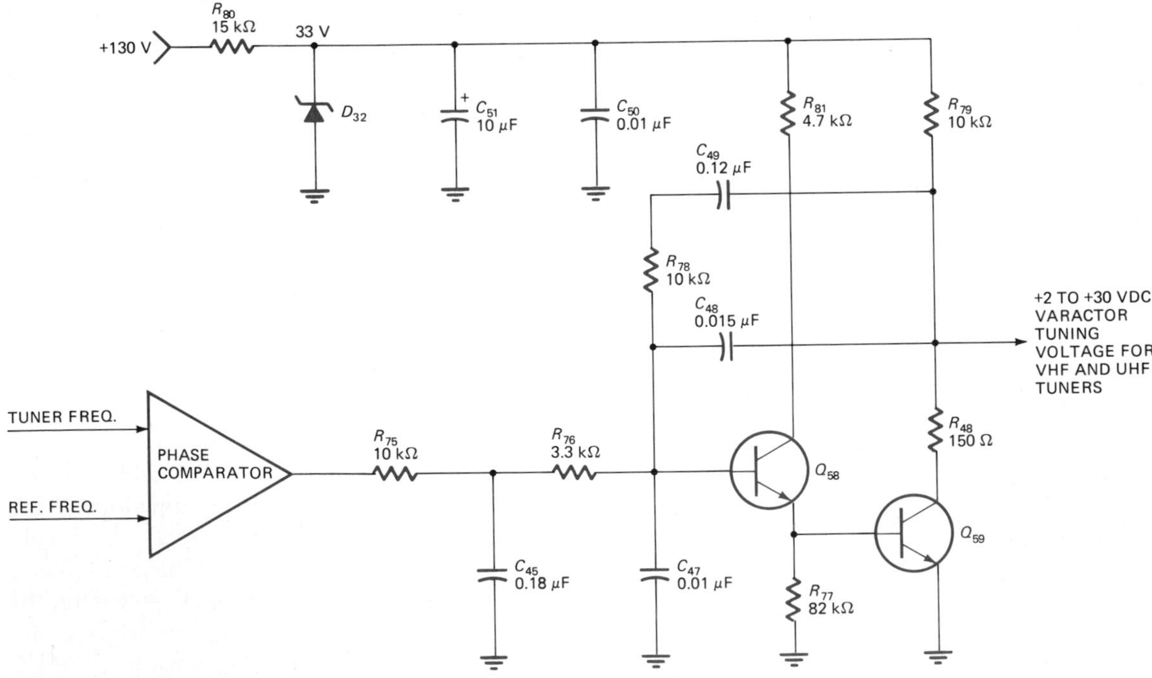

Figure 11.12 Schematic diagram of the PLL filter and tuning voltage control circuits.

nals, the comparator produces a series of DC pulses at its output. These are integrated (filtered) into the base bias voltage for transistor Q_{58}. The integration is performed by R_{75}, C_{45}, R_{76}, and C_{47}.

Transistors Q_{58} and Q_{59} are connected so that conduction of Q_{58} provides base drive for Q_{59}. The latter (Q_{59}) in conjunction with R_{48}, functions as a variable resistance between the varactor tuning voltage and ground. *Increasing* its drive *decreases* the tuning voltage, and *decreasing* base drive *increases* the tuning voltage. This voltage normally varies between +2 and 30 V DC.

A static phase difference between the two comparator input signals (identical frequencies but a slight phase displacement) provides a small Q_{58} base current to maintain the desired tuning voltage.

MICROCOMPUTER

The microcomputer is a complete, self-contained computer that provides intelligence and control for the entire system. Its design includes the following features:

1. Tuning adjustments are not required if the set is moved from one city to another.
2. Programming power and standby power (batteries) are not required.
3. Channels may be selected by direct access or by seek up or down to next higher or lower active channel.
4. Automatic-fine tuning (AFT) on all VHF channels compensates for frequency errors in cable systems.
5. Manual fine tuning on all channels and last-channel memory are provided.

A simplified diagram illustrating the basic inputs and outputs of the microcomputer is given in Figure 11.13. The microcomputer controls the system by continuously scanning its various inputs and making many computations based on details stored in a read only memory (ROM). The computer program is contained in the metallic part of the IC. Thus, it does not require standby batteries when power is removed. There are 629 steps in the program, occupying 629 of the possible 768 ROM locations. A portion of the memory is used for routine self-testing.

The microcomputer scans the keyboard 195 times per second and examines the seek and fine-tune switch positions, looking for new command signals. It senses closure of the seek switch, fine-tune switch, 0 through 9 keys, channel up/down keys, and manual fine-tune up/down keys. It also receives inputs from the sensor circuits that determine whether or not a station signal is present and, if so, how accurately the set is tuned.

When a new channel is selected by digit entry, the microcomputer loads the correct binary information to the tuner oscillator and reference oscillator dividers and the LED drivers. It determines whether the new channel is the low VHF, high VHF, or UHF band. It applies the correct outputs to the VHF/UHF B+ voltage switch, VHF high/low band switch, and audio-mute switch. All of these functions occur instantaneously.

If seek and automatic fine tune are active when the new channel is tuned in, inputs from the sensor indicate the presence or absence of a station signal. If a signal is present, the sensor inputs also indicate tuning accuracy. The microcomputer reacts to this information as follows:

1. If the new channel is a VHF channel and the sensor inputs to the microcomputer indicate a no-signal condition, the microcomputer begins a localized search (±1.5 to ±2.0 MHz). If it does not find and properly tune a station in a specific time interval, the

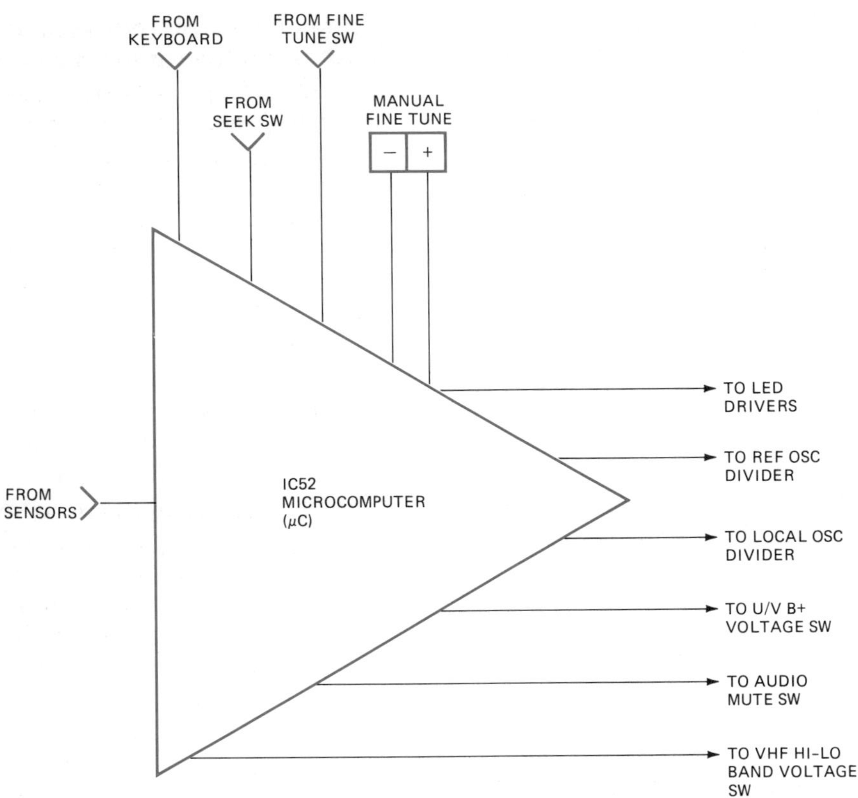

Figure 11.13 Simplified diagram showing the basic inputs and outputs of the microcomputer.

system will *advance* to the next channel and repeat the process. The computer program allows 300 milliseconds to determine if a channel is present and an additional 328 milliseconds to perform a localized search. AFT (localized search) compensates for frequency errors in systems. Thus, this function is isolated to VHF channels only. The seek function (advancing to the next channel in the absence of station signals) applies to both VHF and UHF.

2. When seek is off, the microcomputer disregards the sensor inputs. Thus the set tunes to the exact frequency of the selected channel. It remains on the channel whether a signal is present or not, which permits use of a VTR or video game. Under these conditions, the viewer can manually fine-tune by placing the fine-tune switch in "normal" and holding down one of the fine-tune buttons. (UP or DOWN).
3. Seek (advancing to the next channel) is accomplished by pressing the UP or DOWN button, which loads the proper number for the next channel into the tuner oscillator divider chain. This produces unequal inputs at the PLL comparator, which changes the channel by shifting the tuning voltage.
4. Automatic or manual fine tuning is accomplished by changing the binary input to the reference oscillator di-

vider. Thus, the reference frequency at the comparator either increases or decreases. This produces a proportional change in tuning voltage, which changes the tuner oscillator frequency in a relationship. When the desired fine-tune correction is completed, the reference oscillator maintains the tuning position, ensuring crystal accuracy and stability.

SENSORY CIRCUITRY

Two comparator ICs are used to determine whether a station signal is present and if so, how accurately it is tuned. A simplified schematic of the sensor circuits is given in Figure 11.14

Five different signals from the super module (vertical synchronization, AFT voltage, 45.75-MHz picture IF carrier, composite video, and 41.25-MHz sound IF carrier) are applied to the individual comparator. The comparator reacts to these signals and applies logic signals to the microcomputer (Figure 11.13).

The output of each comparator is either a logic high or a logic low, which is determined by the relationship between its inputs, as follows:

1. When the positive (+) input is high relative to the negative (−) input, the output is a logic high.
2. When the positive (+) input is low relative to the negative (−) input, the output is a logic low.

For seek and AFT, the sensor inputs to the microcomputer indicate whether or not a signal is present and how accurately it is tuned. When a properly tuned signal is present, all five sensor outputs are a logic high. The sensors function irrespective of operating mode, but the microcomputer disregards inputs from the sensors when in the manual operating mode (seek off).

Signals from the super module are converted to proportional DC voltages, which are applied to the appropriate sensor input terminals. As an example, about 8.5 V derived from vertical synchronization pulses is applied to the negative terminal of Comparator 1 and to the positive terminal of Comparator 2. A properly tuned signal establishes the proper voltage levels to produce a logic high at the output of both comparators. If the received signal is badly mistuned, the vertical synchronizing pulses voltage could drop to a level that is below the negative terminal of Comparator 2. Thus, the polarity is reversed and the output becomes a logic low. In the absence of vertical synchronization, the input voltage to both comparators increases and the output of Comparator 1 switches to a logic low.

Notice that the AFT voltage is applied to Comparators 3, 4, and 5. The output of Comparators 3 and 4 in conjunction with the output of Comparator 5 determines whether a frequency offset is too high or too low. The computer reacts to these sensor signals and provides fine-tune correction in the proper direction. If either Comparator 3 or 4 switches to a logic low, the microcomputer input is low. Since the output of these comparators (Discriminator 1) is common, either increasing or decreasing AFT voltage results in a logic low at the microcomputer. Only an increasing AFT voltage results in a logic low from Comparator 5 (Discriminator 2). When the AFT voltage increases, both Discriminator 1 and Discriminator 2 are logic lows. When the AFT voltage decreases, only Discriminator 1 is a logic low and thus the microcomputer reacts to these inputs by fine-tuning in the proper direction.

The outputs of Comparators 6, 7, and 8 are common. Thus the absence of one or all of the input signals results in a logic low at

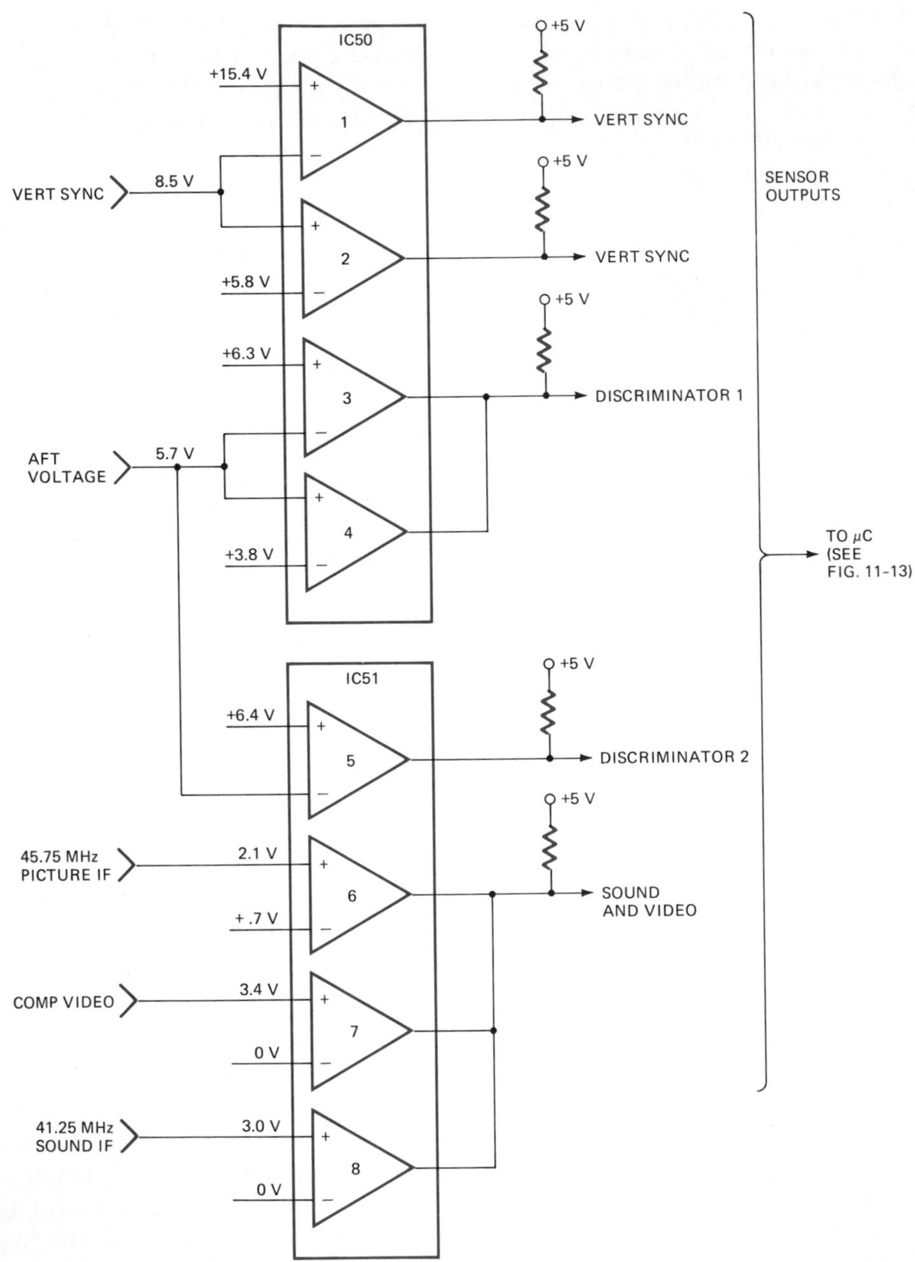

Figure 11.14 Simplified diagram of the sensor circuits.

the microcomputer. The output of these comparators in conjunction with Comparator No. 1 determines whether or not a signal is present.

CHANNEL INDICATORS

The channel indicator assembly contains two gold (amber), seven-segment LEDs and latch/decoder/driver ICs. When activated, the left LED displays a "tens" digit and the right LED displays a "units" digit. A simplified schematic diagram of the channel indicators circuitry is shown in Figure 11.15.

The two latch/decoder/driver ICs receive a binary number at inputs A–B–C–D (connected in parallel) and an enable pulse at terminal "EL." To change channels by direct entry, the viewer must depress two touch pads. When the first number is depressed, a binary number from the microcomputer is

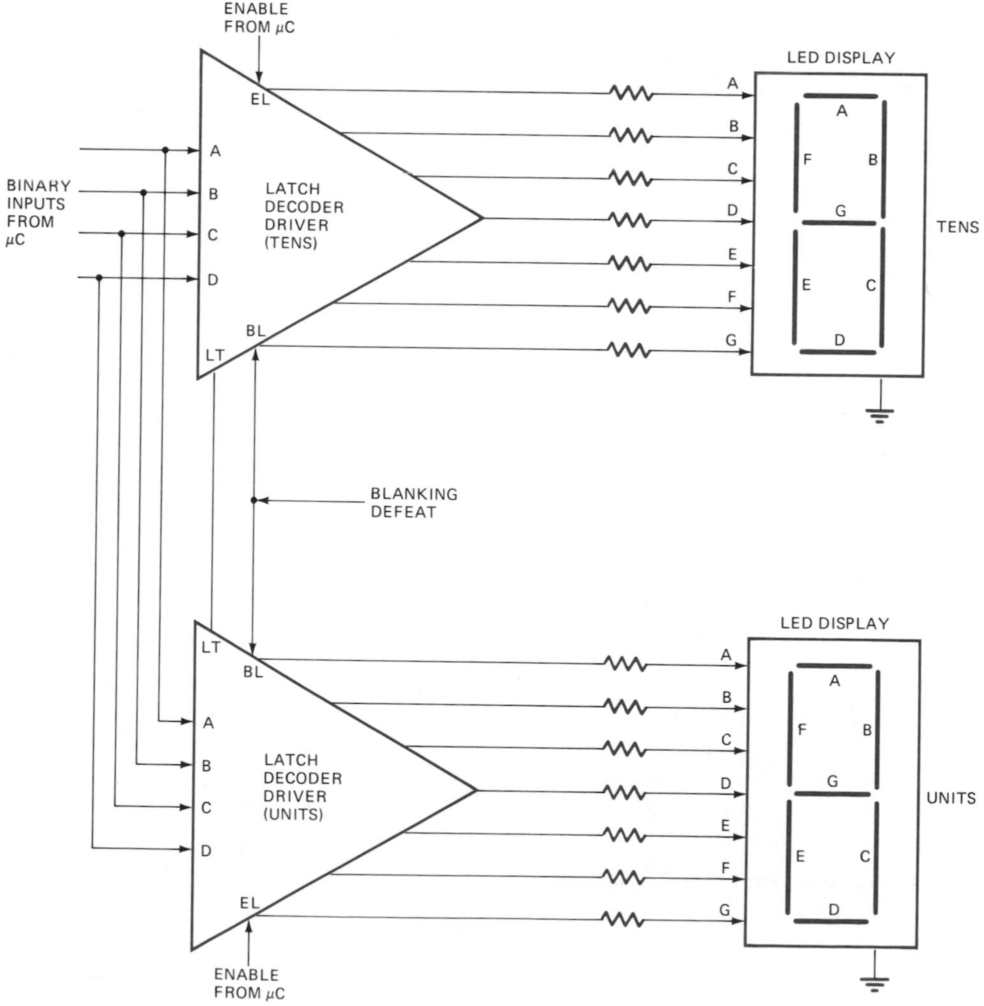

Figure 11.15 Simplified diagram of the channel indicator LED display circuitry.

presented to both ICs but only the "tens" IC receives an enable pulse. This pulse activates its latch and accepts the binary input. Since the "units" IC did not receive an enable pulse, it ignores the binary input. When the viewer depresses the second number, the microcomputer presents a binary input at both ICs but only the "units" IC receives an enable pulse. Thus, it accepts the binary input, which is ignored by the "tens" IC.

The binary number loaded into each IC is maintained by internal latch circuitry until it is changed or cancelled. A decoder section in each IC converts the binary number to its equivalent decimal number and activates appropriate internal driver stages. Driver outputs apply a positive voltage through current-limiting resistors to the LED. Each LED receives up to seven positive voltages from its latch/decoder/driver IC, depending on the decimal number to be displayed. As an example, to display Channel 12, positive voltage from the IC drivers is applied to Elements B and C of the "tens" LED and to Elements A, B, G, E, and D of the "units" LED.

IC terminals "LT" (connected in parallel) provide a quick check of the LED segments, when grounded. Both LEDs will display the numeral 8 if all segments and circuit components are good. A fixed DC voltage is applied to the "BL" terminal of both ICs. This terminal is normally used for multiplexing applications.

SWITCHING CIRCUITS

Outputs from the microcomputer control the VHF and UHF tuner B+ voltages, the VHF high/low band-switching voltage, and the audio-mute switch.

VHF B+ and High/Low Band Switching In Figure 11.16, transistor Q_{55} provides the VHF tuner B+ voltage and transistors Q_{52} and Q_{54} provide the voltage required for VHF high-band operation. When the set is

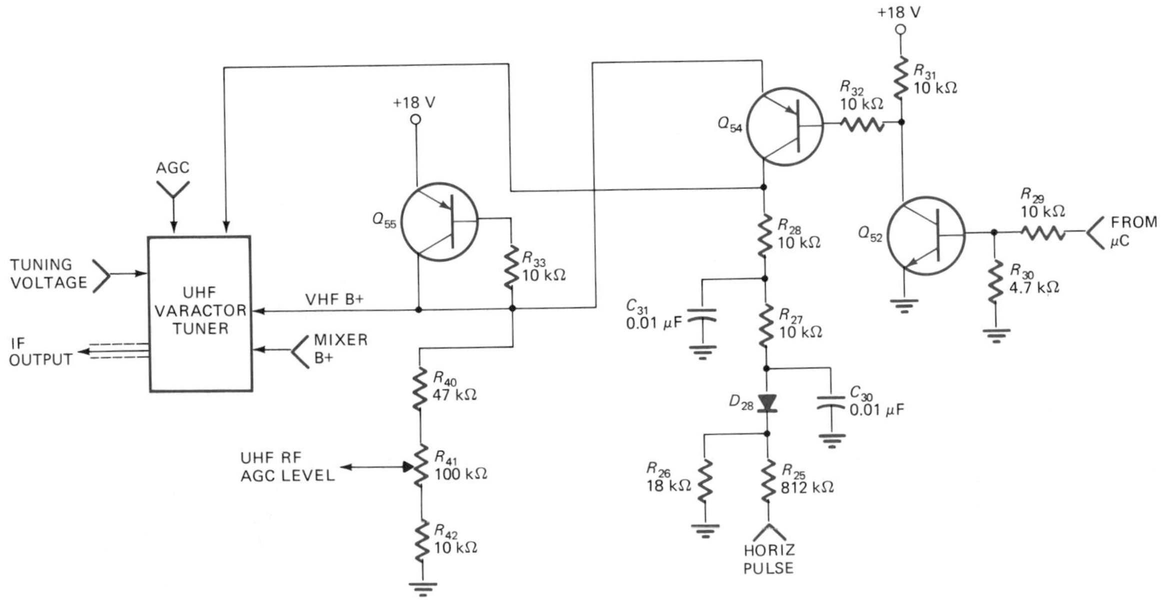

Figure 11.16 Simplified diagram of VHF B+ switching and VHF hi/lo-band switching.

tuned to a VHF channel, Q_{55} conducts and applies B+ to the VHF tuner. If the channel is in the low band, Q_{52} does not receive an input voltage from the computer and Q_{54} cannot conduct. Negative going retrace pulses from the pin cushion circuit board are rectified by D_{28} and charge capacitors C_{30} and C_{31}. This negative voltage is applied to the tuner through R_{28} to reverse-bias its band-switching diodes for low-band VHF operation.

For a high-band channel, Q_{52} receives a positive voltage from the microcomputer and it saturates. This saturates Q_{54}, which applies B+ from the collector of Q_{55} to the band-switching diodes in the tuner and discharges capacitor C_{31}.

UHF Band Switching Refer to Figure 11.17. For UHF channels, the switching circuits remove B+ from the VHF tuner and apply it to the UHF tuner. A positive voltage from the microcomputer saturates transistor Q_{53}, which in turn saturates Q_{56}. B+ couples through this transistor to the UHF tuner and the base circuit of Q_{55}. This voltage turns off Q_{55} and removes B+ from the VHF tuner. This same voltage is applied through R_{40}, R_{41}, and D_{29} to the AGC circuitry of the super module. Adjustment of R_{41} establishes the proper UHF RF AGC level.

Audio-Mute Switching An output from the microcomputer mutes audio in the absence of a station signal (Figure 11.18).

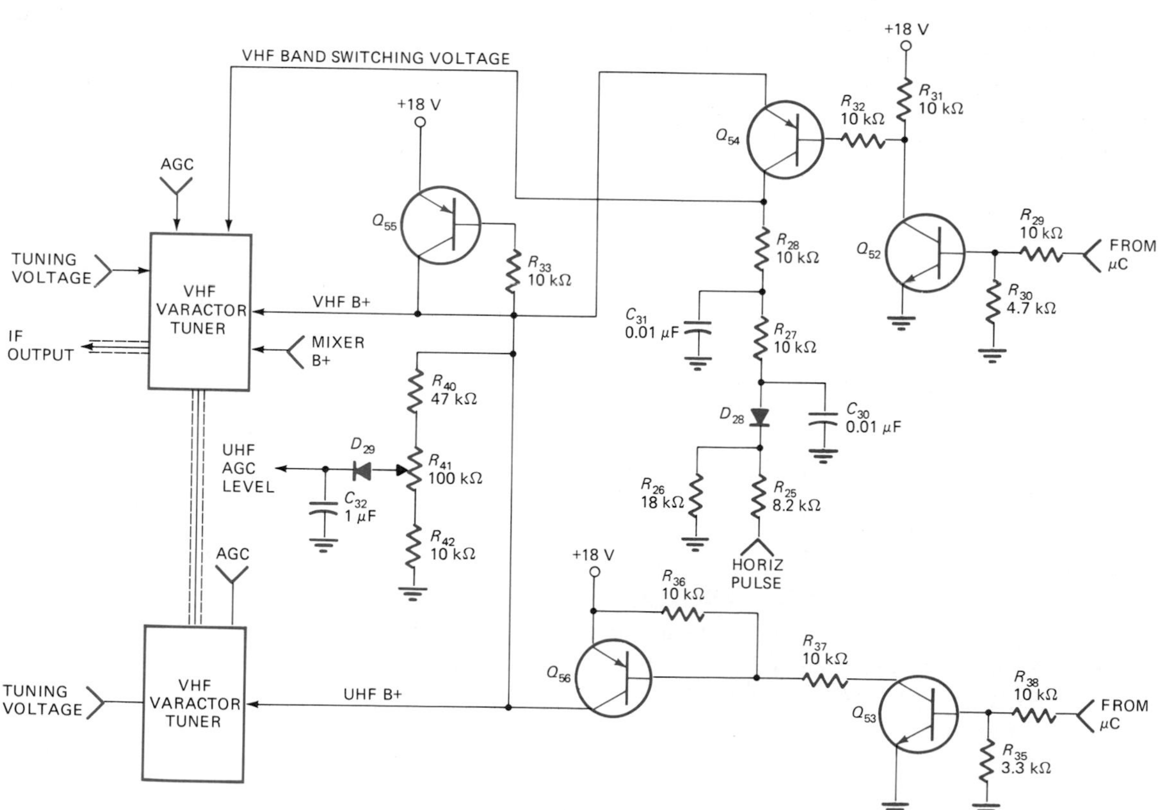

Figure 11.17 Simplified diagram of UHF-band switching.

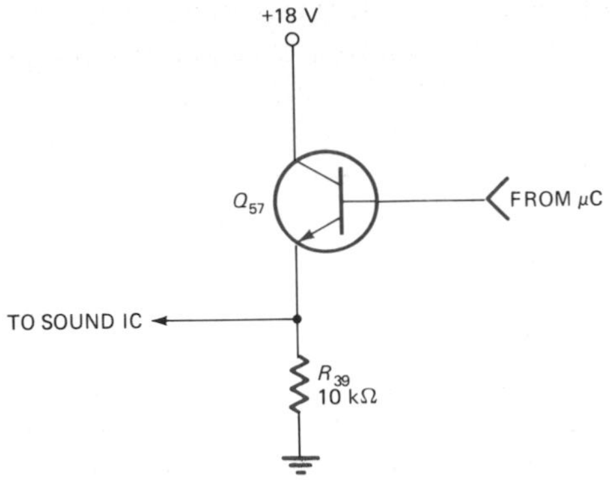

Figure 11.18 Simplified diagram of the audio-mute switch.

When there is no station signal, positive voltage from the computer provides base drive to transistor Q_{57}, which conducts and applies a dc voltage to Pin 6 of sound IC 201. This voltage drives the sound IC into the cutoff and mutes the audio. When a signal is present, the microcomputer removes the bias to Q_{57}; thus, the audio circuit functions normally.

11.7 AUTOMATIC FINE TUNING

When frequency synthesis is used, there is no need for a fine-tuning control (or for AFT). However, many television receivers do not use frequency synthesis. In such cases, AFT is very helpful in achieving proper picture quality and correct color reproduction. Receivers with frequency synthesis may also include AFT. In such receivers, AFT can correct for the effects of inaccurate cable frequencies or for channel frequency offsets.

The fine-tuning control of a TV set adjusts the frequency of the local oscillator in the tuner. For a monochrome set, a moderate amount of misadjustment of this control can be tolerated. In color receivers, however, even a small deviation of the oscillator frequency from its correct setting may result in serious deterioration or complete loss of color.

Even if the fine-tuning control is initially set correctly, the oscillator frequency may drift from its nominal frequency and cause color deterioration. Oscillator drift is caused mainly by the change in oscillator circuit component values due to temperature variations.

AFT circuits are designed so that any adjustment of the fine-tuning control within a predetermined range locks the tuner oscillator onto the correct frequency. Tuning indicators may be used to adjust the manual fine-tuning control to the required range. In their simplest form, tuning indicators are lights that are lit whenever the control is not correctly adjusted.

11.8 FUNCTIONS OF AFT SYSTEMS

As pointed out in Chapter 10, the local oscillator for the tuner is tuned to a frequency that is above the carrier frequencies of the VHF and UHF channels by an amount equal to the IF. (The stability of sound and picture carrier frequencies is ensured by elaborate circuitry in the transmitter.) The AFT circuits allow the tuner oscillator frequency to be set at the proper value. This will ensure that picture and sound IF frequencies from the tuner output appear at the proper points of the IF amplifier response curve.

It has also been pointed out that a stable local oscillator frequency is much more important in a color receiver than in a monochrome receiver. In fact, special design efforts are made so that in changing channels the oscillator is set precisely to the required frequency.

OSCILLATOR DRIFT

Oscillator drift is caused by such factors as ambient temperature change, component aging, and power supply voltage fluctuations. A number of remedial measures may improve the stability of an oscillator circuit. The frequency-determining elements, such as inductors and capacitors, should be chosen so that their temperature coefficient is negligible or so that a change in one is counterbalanced by a compensating change in the other. Clearly, it is difficult to make such provisions in television tuners, where a portion of the tuning components are often composed of stray capacitances and inductances. It must be noted, however, that capacitors having well-controlled positive and negative temperature coefficients are available.

A power supply that is regulated to within a very small percentage of the nominal value will eliminate voltage change as a serious cause of oscillator frequency drift in some receivers. The expense involved in incorporating such regulation is, of course, a concern of the TV receiver manufacturers.

The factors that contribute to inherent oscillator drift can be compensated by the addition of AFT circuitry. This ensures that the local oscillator frequency is always at the correct value. The basic operating principle of AFT circuitry has been well known for many years. In fact, such circuitry appeared in monochrome TV sets decades ago. The AFT circuit is basically an automatic frequency control circuit similar to those used in FM and other receivers. The basic principle is illustrated with the aid of the block diagram of Figure 11.19.

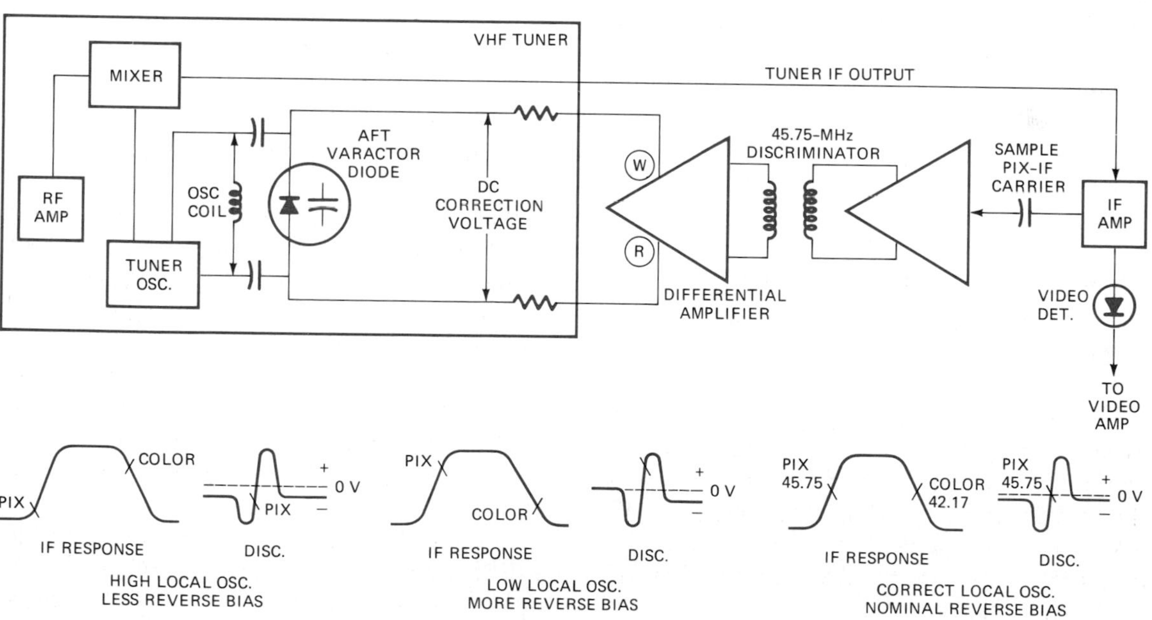

Figure 11.19 Simplified block diagram and waveforms, illustrating the basic principle of AFT. All frequencies are in MHz. 45.75 MHz is the picture-IF carrier and 42.17 MHz is the color-IF subcarrier, when the tuner oscillator frequency is correct. Off-frequency conditions are exaggerated here for clarity.

BASIC AFT OPERATION

When the tuner oscillator frequency is correct, the IF response is illustrated by the wave form in the lower right portion of Figure 11.19. Note that this condition results in practically equal amplitude for the picture IF (45.75 MHz) and color IF (42.17 MHz) carriers.

These IF carrier frequencies are dependent upon the difference frequencies between the incoming RF carrier frequencies and the tuner oscillator frequency. The difference frequencies should be above the picture RF carrier frequency by exactly 45.75 MHz. If the tuner oscillator frequency differs from this amount, either higher or lower, the frequency-sensitive detector (discriminator) provides a DC output having either a positive or a negative polarity, depending upon whether the oscillator is above or below its required frequency. In addition, the magnitude of the DC voltage will be proportional to the amount of error of the oscillator frequency. The AFT circuit makes it possible to mistune the fine-tuning control somewhat without any deterioration of picture quality. It also automatically compensates for any oscillator drift.

Oscillator On Frequency When the tuner oscillator is exactly on frequency (wave forms on the right in Figure 11.19), the sample picture IF carrier is at exactly 45.75 MHz. When this frequency is applied to the 45.75-MHz discriminator and differential amplifier, the outputs at W and R consist of equal DC voltages. Under this condition, the *differential* voltage applied to the varactor has a nominal value that is needed to produce the value of capacitance required to maintain the correct frequency.

Oscillator Frequency Too Low When the tuner oscillator frequency is too low (center wave forms of Figure 11.19), the picture and color IF carrier frequencies are decreased. Note the unequal position of the IF response. The lower frequency picture IF carrier has moved to a higher amplitude point. The lower color IF carrier frequency has moved to a lower amplitude point. This condition causes weak color or loss of color.

Because the picture IF carrier frequency is now *lower* than 45.75 MHz, the discriminator-differential amplifier *increases* the positive voltage at *R* and *decreases* it at *W*. This applies more reverse bias to the varactor, which *reduces* its capacitance. In turn, this increases the oscillator frequency and cancels the original error (too low frequency).

Oscillator Frequency Too High When the tuner oscillator frequency is too high (wave forms at the left in Figure 11.19), the picture and color IF carrier frequencies are increased. The picture IF carrier has moved to a low amplitude point, and the color IF carrier has moved to a high amplitude point on the IF response curve. This condition results in a weak picture, perhaps with a loss of synchronization.

The picture IF carrier frequency is now *higher* than 45.75 MHz. The output of the differential amplifier is such that the positive voltage at *R decreases* and that at *W increases*. This *reduces* the reverse bias across the varactor, which increases its capacitance. As a result, the tuner oscillator frequency is reduced, thus cancelling the error (too high frequency).

The off-frequency conditions shown in Figure 11.19 are exaggerated for the sake of clarity.

Pull-In and Hold-In Range All AFC circuits (including AFT) have a "pull-in" range and a "hold-in" range. The **pull-in range** is the frequency farthest from the nominal frequency that the AFT circuit can acquire and lock into when the AFT circuit is originally unlocked. An example of this is when we are switching channels. The **hold-in range** is that frequency range, starting from the locked AFT condition, in which the AFT circuit can stay locked even if there is oscillator drift or if the fine-tuning control is changed.

The pull-in range of the circuit in Figure 11.19 is usually about ±50 kHz. The hold-in range is about ±1 MHz.

An AFT Frequency-Sensitive Detector The AFT detector circuit shown in Figure 11.20 uses a transistor (Q_1) in a common-emitter configuration for the AFT frequency-sensitive detector. The circuit uses a second common-emitter transistor (Q_2) as an AFT amplifier. The input to the detector is taken from the output of the third IF amplifier. This input is applied to the base of Q_1. The basic circuit of the detector is that of a Foster-Seeley FM discriminator, which is explained in Chapter 22. The detector transformer T_1 is tuned to the picture IF carrier frequency of 45.75 MHz. The action of the detector is such that when the incoming carrier frequency is exactly 45.75 MHz, the output as applied to the base of amplifier Q_2 is 0 V. If

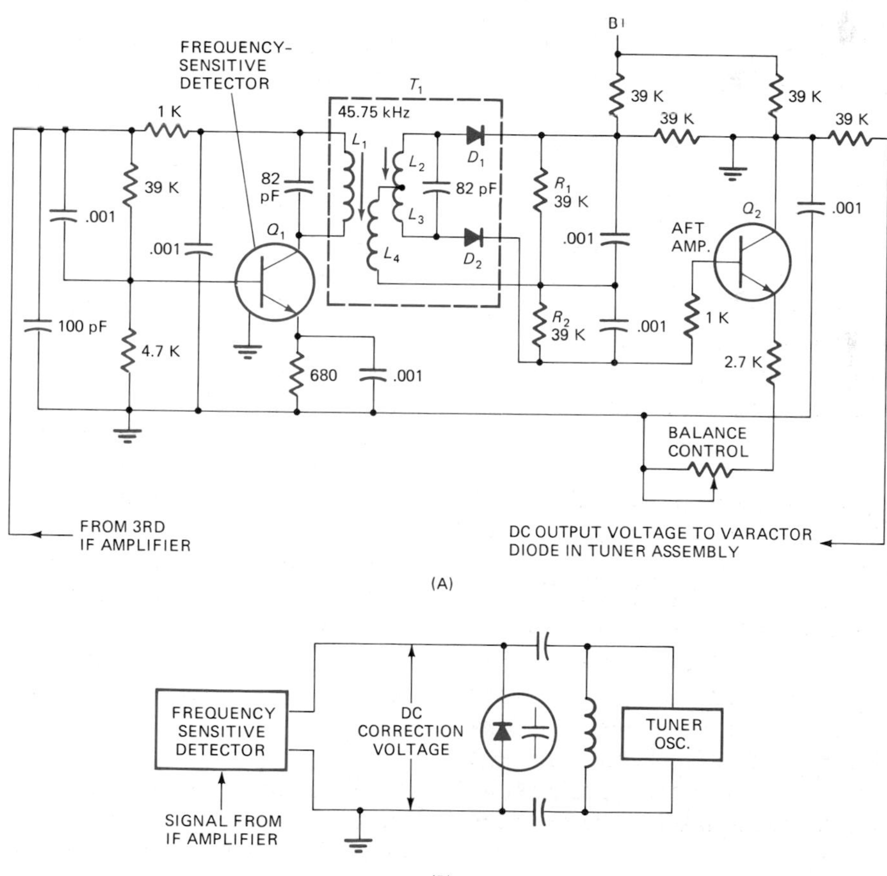

Figure 11.20A. An AFT frequency-sensitive detector and amplifier circuit. (Capacitances in μF unless otherwise noted.) **B.** The detector circuit connected to the varactor in the tuner assembly. Unless otherwise stated, capacitor values are in μF.

the local (tuner) oscillator is not at its nominal frequency for the channel selected, the picture IF carrier frequency will be either above or below 45.75 MHz. In this event, the detector output is either a more positive or a less positive voltage. In addition, within limits, the farther from nominal frequency the local oscillator is, the greater the departure from nominal DC output voltage. In the circuit of Figure 11.20, the DC detector output voltage is amplified by transistor Q_2 and is then applied to a varactor in the tuner assembly, as pictured in Figure 11.20B.

Detector transistor Q_1 is normally operated at a very low quiescent emitter current (usually less than 50 μA) to improve detection efficiency. This also provides amplification in the collector circuit.

11.9 FUNCTIONS AND TYPES OF REMOTE CONTROL DEVICES

The complexity of a TV remote control device depends on the number of things it does. Its most important function is station changing, since this is the major reason why most viewers leave their chairs and go to the receiver. Next, it is useful to be able to alter the sound level, preferably in a series of small steps. Turning the receiver on and off is also a desirable control function. Finally, if not too costly, the control of contrast and/or brightness is useful. For some color receivers, the tint and color circuits may be adjusted by remote control. Obviously, the more functions to be altered remotely, the more complex the circuitry of the control device. This, in turn, directly affects the cost and reliability of both the unit and the television receiver.

Two popular schemes of remote control signaling to the television receiver are:

1. Ultrasonic waves (34 to 54 kHz range), mechanically or electronically generated
2. Infrared wave transmission and reception, using an infrared transmitting LED and a receiving sensor.

11.10 ULTRASONIC REMOTE CONTROL SYSTEMS

An ultrasonic remote control system generates several frequencies in the range from 34 to 54 kHz. Each frequency represents a specific function, such as set on-off, volume up-down, and channel changing. The number of frequencies used depends upon the design and complexity of the unit. Some units may have only three ultrasonic frequencies: for example, on-off and volume, 38.2 kHz; channel seek up, 41.8 kHz; and channel seek down, 40.0 kHz. (Control frequencies are not standard and are selected by each manufacturer.)

More complex remote units may generate 14 or more control signals. These generally control direct-access tuning (described previously). (Additional information is presented later in this chapter.) These control frequencies are electronically generated.

A MECHANICAL ULTRASONIC REMOTE CONTROL SYSTEM

Ultrasonic signals can be generated in a number of ways, but the overall principle is always the same. The sound waves, at various frequencies, are picked up by a unit in the receiver and used for controlling several receiver functions. A different sound wave frequency is used for each function.

Ultrasonic Transducers The operation of an ultrasonic remote control system depends upon the use of transducers. A transducer in the transmitter (the hand-held remote control unit) converts a mechanical vibration to ultrasonic waves. A transducer in the receiver converts the ultrasonic waves to electric signals, which operate a control in the receiver.

In the systems to be described, the actuating transmitted signal is an ultrasonic wave produced by vibrating cylindrical rods. There are other methods for producing ultrasonic sounds for remote control. For example, a vibrating column of air in a whistle, or a mechanical device, may be used. In any event, the systems are quite similar despite differences in the method used for transmitting and receiving the sounds. The waves produced in the systems to be described are picked up by a microphone, converted to equivalent electric signals, amplified, and then used to actuate relays which perform the desired functions.

Cylindrical Vibrating Tubes In the Zenith remote control system of Figure 11.21, and in several other systems as well, the transmitter is a cylindrical aluminum tube that vibrates longitudinally. If such a tube is struck on one end by a hammer moving along the tube axis, it emits a sustained note that has a definite frequency. For example, an aluminum rod 2.5 in (6.4 cm) long has a fundamental resonant frequency of about 50 kHz. The internal damping of the aluminum is so slight that a large part of the vibrational energy stored in the rod after the original blow of the hammer is radiated.

A single tube of a specified length will produce a certain resonant frequency. Therefore, to control three or four functions within a television receiver, three or four rods of slightly different lengths are used (Figure 11.21). In the illustration we see the working parts of the transmitter. A hammer, which is a steel cylinder weighing about

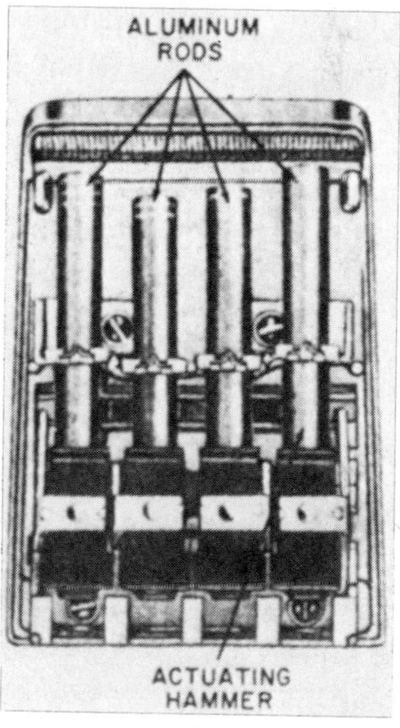

Figure 11.21 The external appearance and internal construction of an ultrasonic transmitter for remote control of a TV receiver. *(Courtesy of Zenith Radio Corp.)*

0.1 oz (2.5 g), is located at one end of the rod. When a button is pressed, the hammer is pushed away from the rod by the force of a spring. As the button is further advanced, the spring is suddenly released and the hammer strikes the rod. Generally, if there are several rods, the frequency difference of the various rods may be on the order of 1 000 Hz or so.

Once a rod has been struck and the energy transmitted, it is then desirable to dampen the remaining energy as quickly as possible. Mechanical damping is used for this purpose. In the unit shown in Figure 11.21, damping is achieved by a small piece of spring wire covered by a plastic sleeve. This wire protrudes through the mounting and touches the rod. When the button is pressed, the damper is withdrawn; when the button is released, the sleeve again makes contact with the rod.

Microphone Pickup At the receiver, the ultrasonic energy is picked up by a crystal microphone with a barium titanate crystal element. This material, when cut in the form of a small bar or plate and placed between two conducting electrodes, will generate a voltage when the bar or plate is mechanically strained by ultrasonic waves. Conversely, when a voltage is applied across the two electrodes, the barium titanate will be mechanically strained. This is the well-known **piezoelectric effect.**

Figure 11.22 shows a typical microphone. Two thin, rectangular wafers of barium titanate are combined, as shown in Part A. Silver (the conducting electrode) is applied over a small section of each end of each wafer. These electrodes are indicated in Part A by the dark segments at each end of the assembly. Between the two wafers, at the nodal points of vibration, two thin metal strips are cemented. These strips serve as electrical contacts and as a mechanical suspension system.

To broaden the response of this transducer, a small U-shaped piece of aluminum is added to the assembly, as shown in Part B. This makes the microphone behave like two tuned circuits closely coupled. It enables ultrasonic frequencies from 37.75 kHz to 41.25 kHz to be picked up and converted to their equivalent electric signals.

Figure 11.22C shows the entire microphone in cross section. From left to right, there is first a supporting piece, which carries the barium titanate wafers by means of the two thin metal strips mentioned previously. Next is the aluminum bridge to which the wafers are cemented, together with a plastic piece and a rectangular window that fits closely around the bridge. Beyond the bridge, there is a space equal in length to a one-quarter wavelength at about 40 kHz. Beyond this, there is a rectangular horn 2 in (5.1 cm) long. Both the one-quarter-wavelength space and the horn serve to match the impedance of the barium titanate assembly to the air. In other words, the horn and the one-quarter-wavelength space help to couple the wafers to the air.

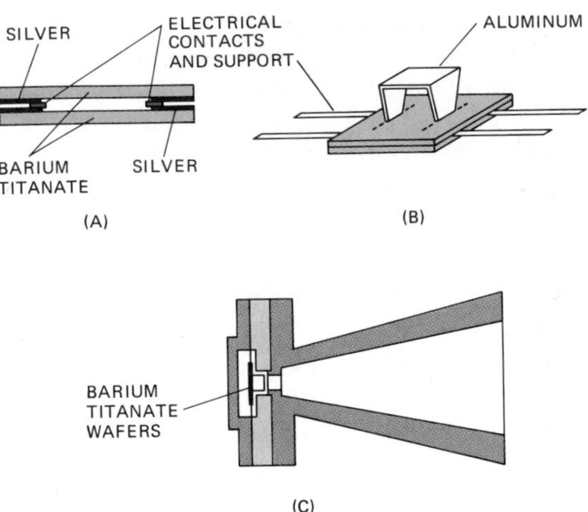

Figure 11.22 The microphone utilized by Zenith to pick up the ultrasonic sound of the transmitter shown in Figure 11.21.

The combination of mechanical transmitter, microphone, and amplifier in this system provides sufficient sensitivity to make the aiming of the transmitter quite unnecessary in most homes. Sound reflected from the floor, walls, ceiling, or furniture makes it possible to operate the receiver controls with the transmitter in almost any position. The line of sight and the approximate aiming of the transmitter become important only at maximum range—a distance of 40 ft (12.2 m)—which is rarely encountered.

11.11 ELECTRONICALLY GENERATED SOUND SIGNALS

Instead of using a mechanical device such as a tube of aluminum, the ultrasonic generator may comprise an electronic oscillator and a speaker. Such a system is shown in Figure 11.23A. A typical remote control panel is illustrated in Figure 11.23B.

THE OSCILLATOR

The oscillator Q_1 is a Hartley type. The primary winding of transformer T is tapped to provide the regenerative feedback from collector to base by way of a feedback capacitor C_1. Base bias is established by the resis-

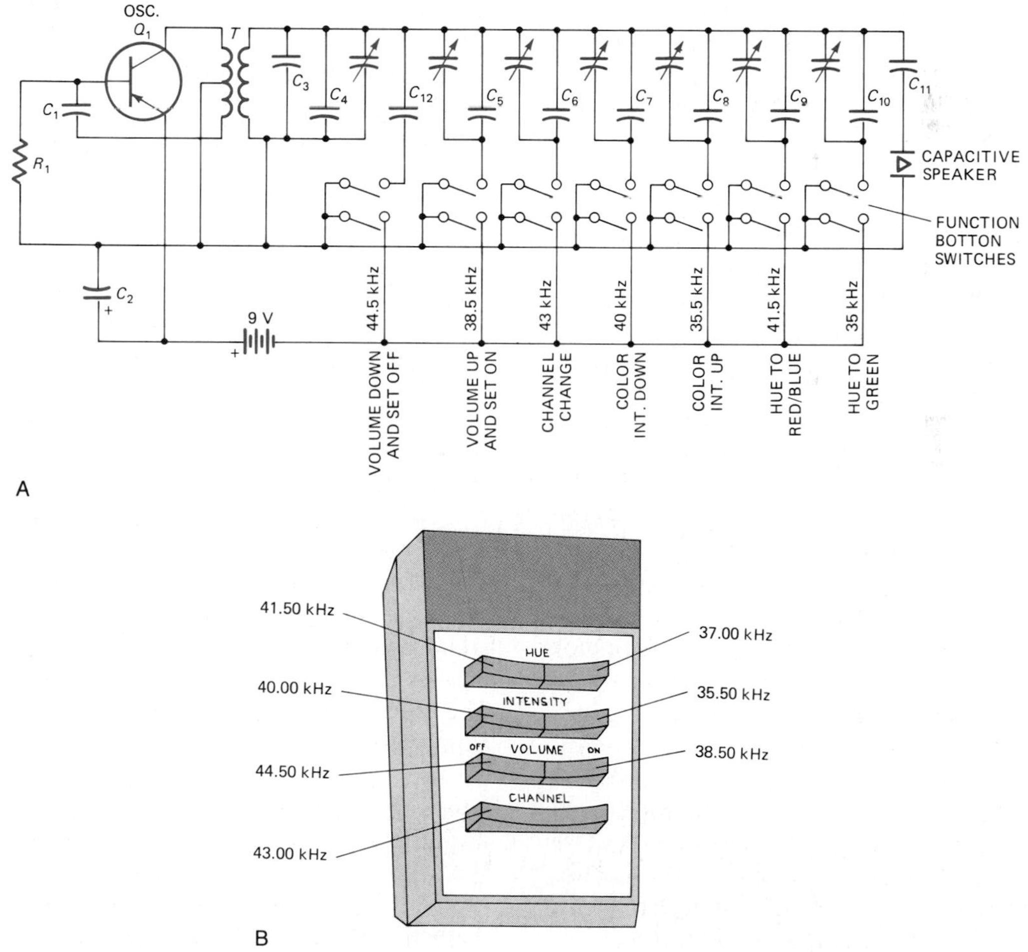

Figure 11.23A. With this type of remote transmitter, the ultrasonic signals are developed by an oscillator and converted to air waves by a capacitive speaker. **B.** A typical remote control panel for a color TV receiver. Ultrasonic frequencies are transmitted to the receiver to provide the various indicated functions. *(Courtesy of Quasar Electronics Co.)*

tor R_1, connected to the negative side of the supply. The collector is connected directly to the negative side of the 9-V battery when one of the function switches is closed. Note that the oscillator base and collector circuits are open when all of the switches are open.

Each of the function switches connects a different capacitor combination across the secondary winding of transformer T. Capacitors C_3, C_4, and C_{13} are permanently connected across this winding. The oscillator frequency is controlled by the total amount of capacitance across the secondary winding. Closing a function switch places another capacitor combination across the C_3–C_4–C_{13} connection, thus increasing the capacitance and lowering the resonant frequency. The oscillator signal is delivered to a capacitive speaker that radiates the ultrasonic sound.

Capacitor C_2 is an electrolytic capacitor placed across the 9-V supply when the circuit is in operation. This prevents degeneration due to the internal resistance of the battery.

THE FUNCTION SWITCHES

Let us assume that the *channel change* button is pushed on the remote control transmitter. (All of the "switches" are push buttons that are normally open.) This completes both the base and collector circuits for transistor Q_1 and so the circuit oscillates. At the same time, capacitors C_{15} and C_{16} are connected in parallel with the secondary winding of transformer T. This increases the secondary capacitance and decreases the oscillator frequency. This arrangement is quite common in remote control circuits that use electronically generated ultrasonic signals. In the receiver, the circuit actuated is the one that is resonant to the 43-kHz signal. This in turn operates a motor that turns the channel selector switch from station to station.

Note that there are two function switches for controlling the volume. One is for turning the volume up, and the other is for turning the volume down. When the volume is turned down all the way, the set is turned off, but it can be turned on again remotely by pushing the "volume up" button. Two color intensity buttons turn the color intensity up or down. Color intensity is sometimes referred to as color control or color amplitude. It sets the amount of color in the picture, but not the hue. In other words, varying the color control should change the greens, for example, from light green to dark green but should not change them from green to some other color. The hue control, which on some receivers is called the tint control, sets the receiver for the proper phase relationship of the color signals. The hue control should be set for the proper flesh tones in the reproduced color picture. Operating the hue control can change flesh tones either to a greenish color or to a reddish blue, depending upon which of the function switches is depressed.

RECEIVERS FOR ULTRASONIC CONTROL SYSTEMS

Three or four transistors are generally used for amplifying the ultrasonic signal from the microphone. This amplifier system is usually referred to as the preamplifier. Figure 11.24 shows a transistor preamplifier. The signal from the microphone is coupled from J_1 to Q_1 through the coupling capacitor C_1. Resistance-capacitance coupling is used between stages. The gain of Q_3 is set by R_{12}, and the setting of this resistor determines the sensitivity of the system.

Typical Receiver Figure 11.25 shows a block diagram of a typical receiver for an ultrasonic remote control circuit. The ultrasonic transmitter, which is sometimes referred to as a bonger, produces waves at

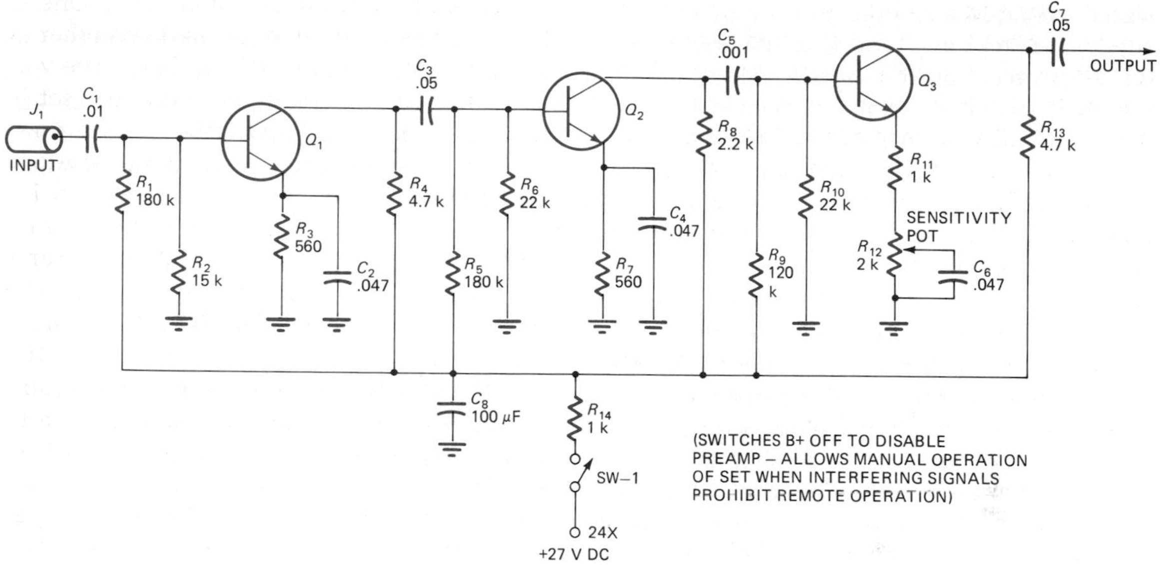

Figure 11.24 This receiver amplifier is typical of those used with ultrasonic remote-control systems. Capacitor values are in μF.

various ultrasonic frequencies according to the control being actuated. Different frequencies are used for the different functions of the hand-held transmitter.

The microphone receives the sound and converts it to an electric signal that can be amplified. In most receivers, there are a number of amplifier stages for this purpose. The output of the amplifier is delivered to a line that is used as the input to a number of *LC* tuned circuits. Each tuned circuit is resonant to a particular frequency and will not respond to any of the other frequencies delivered by the ultrasonic transmitter. Suppose, for instance, that the transmitter is sending a 37-kHz ultrasonic frequency. This signal is converted to an electrical signal, amplified, and then fed to all of the tuned circuits shown in the block diagram. However, only the tuned circuit marked 37 kHz will respond to this signal.

When there is an input of 37 kHz, the

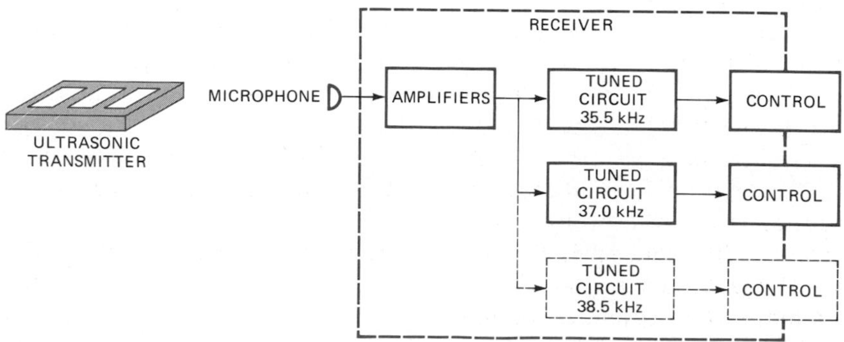

Figure 11.25 Block diagram of a receiver used with an ultrasonic control system.

signal operates a control circuit that may be a relay, a stepping switch, a DC motor, or an electronic control circuit. Motors and stepping switches are often used to turn or vary the controls. Although there are only three tuned circuits shown in Figure 11.25, it is possible to have as many as eight or ten, depending upon the number of functions being controlled.

Control Circuits Figure 11.26 illustrates typical control circuits for an ultrasonic receiver. The input to the circuits comes from the amplifier and is delivered simultaneously to a number of tuned circuits. (Only two of the tuned circuits are shown in Figure 11.26.) The transistor bases are grounded through coils L_1 and L_2, and there is no DC base bias present. When a transistor is not forward-biased, it is in the cutoff condition. Therefore, all of the transistors in this line are normally at cutoff.

Let us assume that there is an input signal of 37 kHz. The tuned circuit in the base of Q_2 is resonant to this frequency, and a signal will be delivered to the base. Whenever this signal goes in a positive direction, the base is forward-biased and the transistor conducts. The output signal of the transistor is a series of pulses that are smoothed by the capacitor C and the coil of the relay. The relay is actuated, and the relay contacts are closed. These relay contacts are used to complete the circuit of a motor or a stepping switch.

11.12 AN ELECTRONIC REMOTE CONTROL SYSTEM

The Quasar remote control system described here is an example of a solid-state system. The only electromechanical devices are an on-off relay in the power supply circuit and a relay and a motor in the channel selector. The remote transmitter for this system is the one shown in Figure 11.23. In the receiver, the electric signal from the transducer is fed to a three-stage preamplifier similar to the one shown in Figure 11.24.

The preamplifier feeds the all-function driver illustrated in Figure 11.27. The all-function driver, in turn, drives the function output stage and the channel selection discriminator coil. The function output stage is a tuned-input amplifier that accepts signals

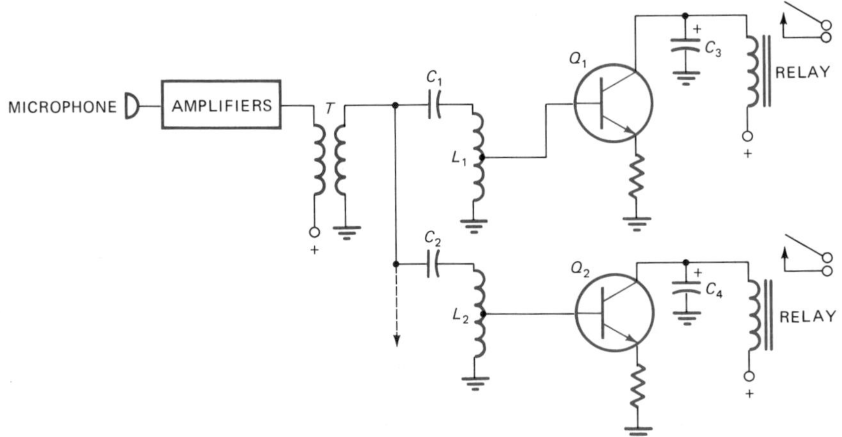

Figure 11.26 Typical control circuits for an ultrasonic remote-control receiver.

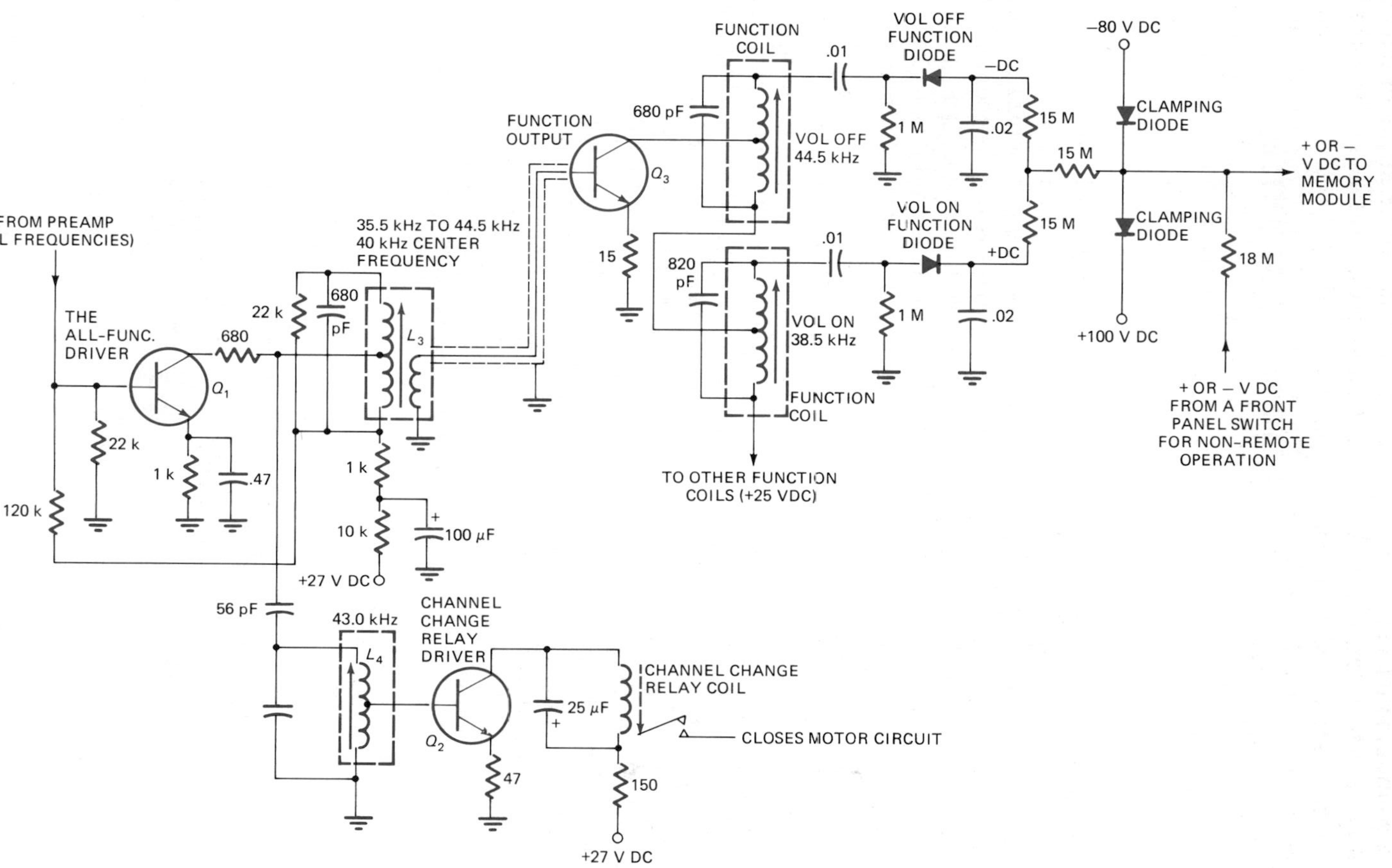

Figure 11.27 The all-function driver and the function output stages of the Quasar remote-control system. Unless otherwise stated, capacitor values are in μF.

with frequencies ranging from 35.5 to 44.5 kHz. Up to this stage the response is wideband, but from this point on the receiver is very selective.

FUNCTION DISCRIMINATORS

The output stage feeds the function discriminators. The name *function discriminator* is somewhat misleading because this is not the type of discriminator used for demodulating the FM signals. Instead, it is a frequency-selective system that accepts one frequency and rejects all others. Thus, it discriminates against the unwanted frequencies and is sensitive to only one frequency for each function.

The function discriminator consists of three pairs of coils and two function diodes. The six coils (two for each function) form a series load for the function output collector. Figure 11.27 shows the function discriminator circuitry for the volume off-on functions, along with the clamping diodes for the memory module. (The operation of the memory module is explained later in this section.) Each of the function coils is a high-*Q* coil with a bandwidth between 500 and 1000 Hz. Under these conditions, the coil appears as a virtual short unless the frequency of the incoming signal is resonant with the coil.

At resonance, the coil develops a maximum AC voltage, which is then rectified by one of the function diodes. In Figure 11.27, if a 38.5-kHz amplified signal is fed to L_1, it is passed because this high-Q coil is tuned resonant at 44.5 kHz. The system develops a maximum voltage across L_2, which is tuned to 38.5 kHz. The AC voltage from L_2 is half-wave rectified so that a positive voltage appears at the clamping diode of the memory module. The operation of the system beyond this stage depends on the working of three memory modules for intensity, volume, and hue functional controls.

MEMORY MODULES

The memory module in Figure 11.28 is a simple device composed of a neon bulb, a low-leakage capacitor, and a MOSFET.

Since a MOSFET has a large input impedance (of the order of 10^{14} ohms) and a neon bulb will not conduct until the voltage across it reaches the ionization potential, both appear to be open circuited. Once the memory capacitor is charged, it will remain so for a long time (more than 1 000 hours). The charge on the capacitor holds a fixed potential on the gate of the MOSFET. Thus, the drain and source currents are fixed. Voltages used for control purposes in later stages of the system are taken from a 10 000-ohm resistor in a source follower arrangement.

Charging current for the capacitor goes through the neon bulb. Voltages applied to the bulb are clamped by the diodes outside the module. If the function signal exceeds +100 V, the 100-V DC clamping diode conducts and this limits the applied voltage to +100 V DC. Since the ionization potential of the bulb is approximately 80 V, the remaining 20 V is used to charge the memory capacitor. The 80-V clamp works the same way

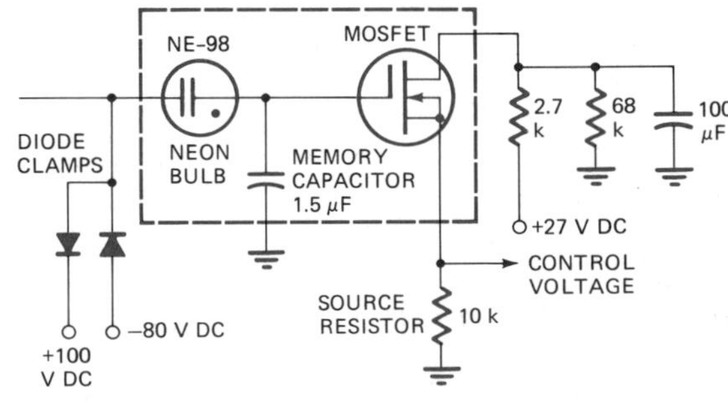

Figure 11.28 A memory module circuit used in the Quasar remote-control system.

except that the current flows in the opposite direction. The negative clamp is lower than the positive clamp, so that the MOSFET gate voltage will not exceed its cutoff (or pinch-off) voltage.

USE OF CONTROL VOLTAGE

The operation of all the circuits in the remote control system is the same up to this point. That is, all functions have their signals generated, received, amplified, discriminated, and stored in exactly the same manner. However, the way in which the control voltage taken from the source resistor (10 000 ohms) is used varies from function to function.

In the ON-OFF function, for instance, the control voltage operates a one-shot flip-flop that energizes a relay. This circuit is shown in Figure 11.29. The audio is controlled by changing the bias of a two-stage audio amplifier, as shown in Figure 11.30. The amplitude of one of the two 3.58-MHz, continuous-wave color reference signals (180° out of phase) is varied by a controlled amplifier. Also, a single-stage controlled resistor decides the intensity. Similarly, a driver transistor used as a switch operates a relay to pull in the channel selection motor.

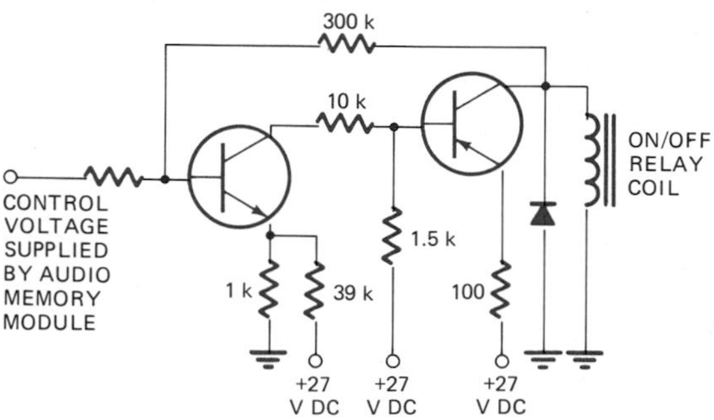

Figure 11.29 Solid-state ON-OFF switch (single-shot flip-flop).

Each function will now be discussed in detail, in the order in which the discriminator coils are arranged. The seven functions of this remote system and their corresponding control-signal frequencies are summarized in Table 11.2.

Volume-On Function The remote transmitter has two volume control buttons: volume off and volume on (Figure 11.23). Pressing the "on" button charges the capacitor in the volume memory module (Figure 11.28). The potential at the gate of the FET rises above the pinchoff, and the FET conducts. A control voltage is thus developed across the source resistor. This control voltage splits between the audio and on-off controls and is thus used for two operations. One is to supply the turn-on voltage of a single-shot flip-flop, which Quasar calls a solid-state switch (Figure 11.29). The load for the switch is a relay coil. When this is energized, the relay contacts that are in series with the main off-on switch are closed (Figure 11.31). The closing of these two sets of contacts applies power to the power transformer primary, the CRT filament transformer, and the power transformer. The skip and hold switches, along with the tuner motor and relay contacts shown in Figure 11.31, will be explained later. Since the Quasar chassis has the quick-on capability, a quick-on defeat switch is in series with the CRT filament transformer.

Audio Function The other path for the control voltage is to an audio preamplifier. An audio signal is received from the chassis, controlled in the remote control unit, and returned to the chassis for further amplification. Figure 11.30 illustrates the two-stage audio preamplifier located in the remote unit. The variable DC voltage received from the memory module shifts the bias of both transistors. Increasing the forward bias in-

Table 11.2 **REMOTE-CONTROL TRANSMITTER FUNCTIONS AND FREQUENCIES FOR THE QUASAR SYSTEM.**

NO.	FUNCTION	OSCILLATOR/TRANSMITTER FREQUENCY kHz
1	Volume-up, on	38.50
2	Volume-down, off	44.50
3	Intensity-up	35.50
4	Intensity-down	40.00
5	Hue-Red-Blue	41.50
6	Hue-Green	37.00
7	Channel change	43.00

creases the volume. Pressing the volume-off control reverses the above process. Decreasing the forward bias for the two amplifiers decreases the volume level. When the control voltage falls below the firing point of the one-shot flip-flop, the switch snaps back to the off position. The holding current for the relay (Figure 11.31) is removed, and the contacts open. The set is now off, but the filament voltage is still applied to the CRT through the quick-on switch.

Hue Function The hue reference phase is controlled by adding together two continuous-wave, 3.58-MHz signals that are initially 180° out of phase. These signals are generated in the appropriate circuitry in the receiver. Varying the amplitude of one of the signals controls picture hue. In the remote system, pressing the hue-control button changes the voltage in the hue memory module. This hue-control voltage is applied to a two-stage amplifier (Figure 11.32).

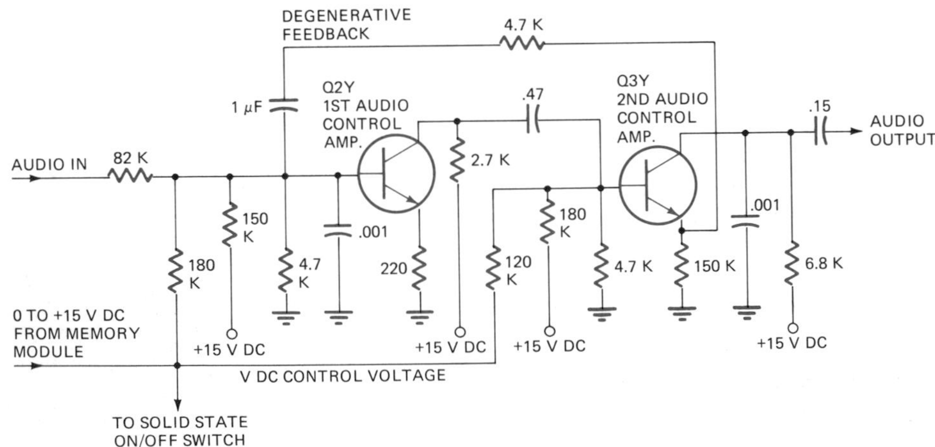

Figure 11.30 Two-stage audio amplifier with degenerative feedback. The control voltage from the memory module sets the gain by controlling the base bias voltage. (This amplifier is used in the Quasar remote-control system.) Unless otherwise stated, capacitor values are in μF.

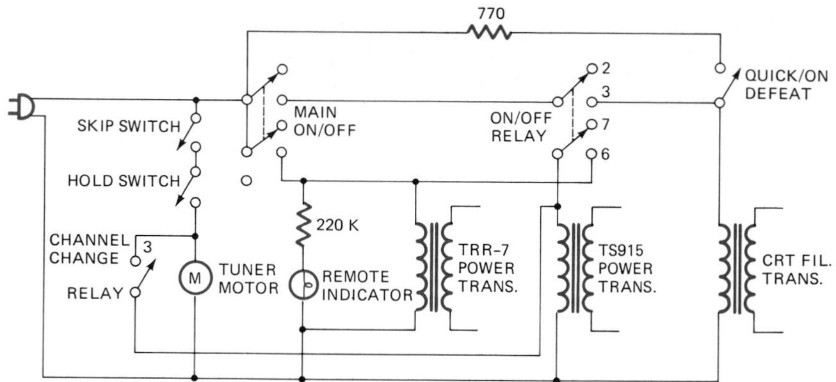

Figure 11.31 This circuit shows how the on-off relay in the solid-state switch controls power to the receiver. It is used in the Quasar remote-control system.

The signal to be controlled is accepted from the main receiver chassis and fed to a common-base transistor amplifier. From there it passes through an emitter follower stage. This results in a power gain without any phase inversion or losses due to impedance mismatching. If the receiver did not have the remote control, the hue could be controlled with a potentiometer. The components in the circuit in Figure 11.32 that are not involved in biasing and load are used to eliminate amplified transients and harmonics, which would interfere with good picture reproduction. The more positive the voltage out of the function discriminator, the more positive the voltage on the FET gate. This causes an increase in FET conduction and an increase in the control voltage. Increasing the control voltage increases the bias on the hue-control amplifier. This results in an increase in the signal amplitude and a hue shift to red-blue. Decreasing the gate voltage has the opposite effect, i.e., toward green.

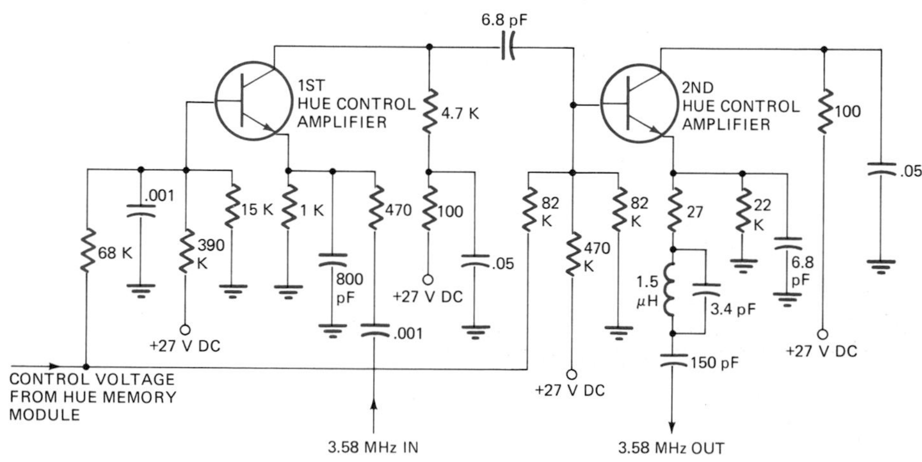

Figure 11.32 The hue-control amplifier used in the Quasar remote-control system. Unless otherwise stated, capacitor values are in μF.

Intensity Function The color intensity control operates in a similar manner. As the control voltage from the module increases, the conduction of the color intensity amplifier is increased. The word *amplifier* is a misnomer in this application. The transistor is used as a variable resistor which simulates the action of a control potentiometer. High color intensity is achieved by an increased (positive-going) control voltage. Figure 11.33 shows the intensity variable resistor, which controls the gain of the second color IF amplifier.

Channel Selection Function Channel selection is the only remote function that does not use a memory module. The all-function driver (Q_1) is loaded by two coils (Figure 11.27): the low-Q coil L_3 (35.5 to 44.5 kHz) and the narrow-band, high-Q coil L_4 (43.00 kHz). The voltage developed turns the channel change relay driver transistor fully on. The collector current through transistor Q_2 pulls in the relay to apply power to the tuner motor. The location of the relay relative to the power supply is shown in Figure 11.31. When the motor starts to turn, it acts as a solenoid. The forward motion toward the

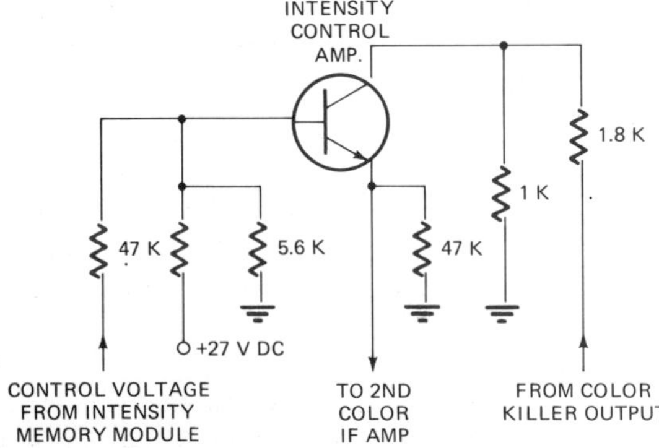

Figure 11.33 The color-intensity control amplifier (which is really a transistor that behaves as a variable resistor). It is used in the Quasar remote-control system.

front of the set activates a **hold switch**. A **skip switch** is in series with the hold switch. The combination is in parallel with the relay contacts.

Skipping of channels is accomplished by detuning the fine-tuning coils.

Rotating the fine-tuning knob counterclockwise at least ten turns brings the tuning slug out so that its head activates the skip switch (Figure 11.31). The initial rotation of the motor is enough to carry the tuner drum past the channel that is to be changed. Each channel that has its slug out keeps the skip switch closed. The hold switch is closed. Therefore, the motor keeps turning until a channel is in place. This allows the motor to stop because the skip switch is now open. Programming the tuner enables the operator to select one channel by pressing the button only once and then releasing it.

11.13 A DIRECT-ACCESS REMOTE CONTROL SYSTEM

The system to be described here is the Quasar system used in conjunction with the Compu-Matic (direct access) tuning system already described in this chapter. Remote control is accomplished by means of infrared transmission and reception. A sketch of the remote control transmitter unit is shown in Figure 11.34. This unit has 16 buttons (functions) and is battery-powered (1.5 V and 3.0 V). Buttons 0 through 9 are for channel selection (direct access).

There are also channel seek up and down buttons, volume up and down buttons, an audio mute, and a power on-off button. A function indicator (LED) lights whenever any button is pressed, showing that a function button has been activated.

Channels are selected as described in Section 11.6. The mute button instantly cuts off sound, which is returned to the previous level by again pressing the mute button.

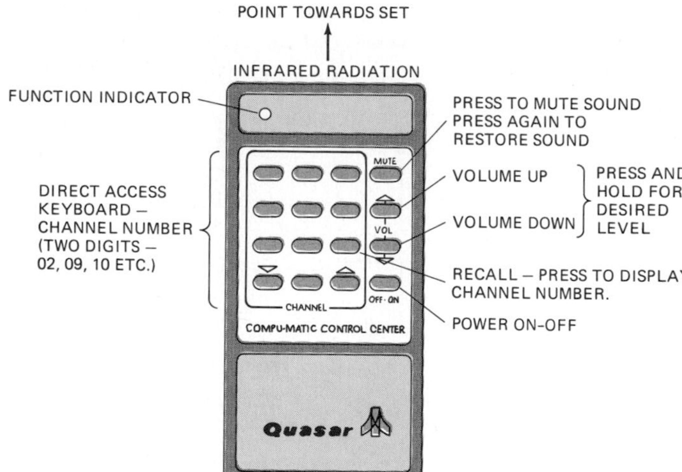

Figure 11.34 The direct-access, 16-button, remote transmitter used with the Compu-Matic television receiver. The function indicator (LED) lights when any button is pressed. *(Courtesy of Quasar Electronics Co.)*

The channel in use can be recalled by pressing the 9 button. The channel number will be displayed on the screen for about six seconds.

DIRECTIVITY

Since the transmission to the television receiver is by infrared waves, it is somewhat directional. However, this makes the system less susceptible to interference. Note in Figure 11.34 that for the most effective range, the function-indicator end of the transmitter should be pointed toward the set.

THE REMOTE TRANSMITTER

A simplified schematic diagram of the Quasar remote transmitter is shown in Figure 11.35. When a button is pushed, the infrared waves are modulated via a 42-kHz carrier by one of 16 pulse patterns.

Activating the Circuitry When a function button is pushed, the remote transmitter circuitry is activated and a train of pulses from encoder IC_{1001} is applied to the appropriate encoder input port. Encoder IC_{1001} develops a distinctive output pulse pattern for each of the 16 buttons. One pulse pattern at a time is applied to the base of modulation transistor Q_{1001}. IC_{1001} also applies DC voltage (+1.5 V) to the collector of modulation transistor Q_{1001} and to the base of gate transistor Q_{1003}.

Infrared Modulation When gate transistor Q_{1003} receives the DC base voltage from encoder IC_{1001}, it activates the 42-kHz oscillator (Q_{1002}). However, when any transmitter button is depressed, an individual pulse pattern is also fed to the base of gate transistor Q_{1003}.

The effect of the pulse pattern is to turn Q_{1003} off and on in accordance with the individual pattern sequence. In turn, Q_{1003} modulates the 42-kHz oscillator (Q_{1002}) in accordance with the same pattern. We now have a pulse-modulated, 42-kHz carrier wave. This modulated carrier is coupled through tuned-modulation transformer T_{1001} to the infrared driver Q_{1004}.

Infrared Output The modulated 42-kHz signal, via the infrared driver Q_{1004}, causes the three infrared LEDs to transmit infrared light waves modulated by the individual pulse pattern. In addition, the modulated 42-kHz output from T_{1001} is rectified by

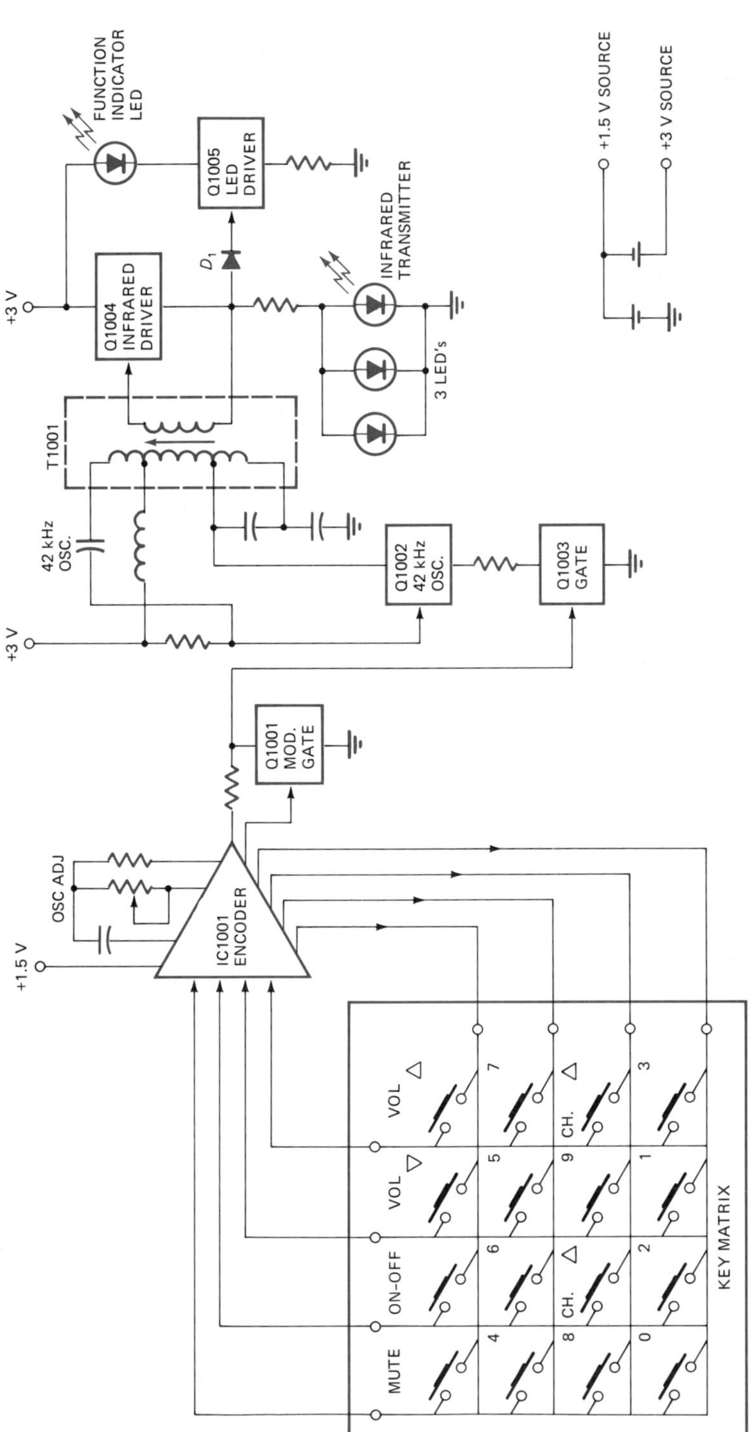

Figure 11.35 A simplified diagram of the Quasar direct-access, remote-infrared transmitter (16 button). Battery voltages of +1.5 V and +3.0 V are utilized. (*Courtesy of Quasar Electronics Co.*)

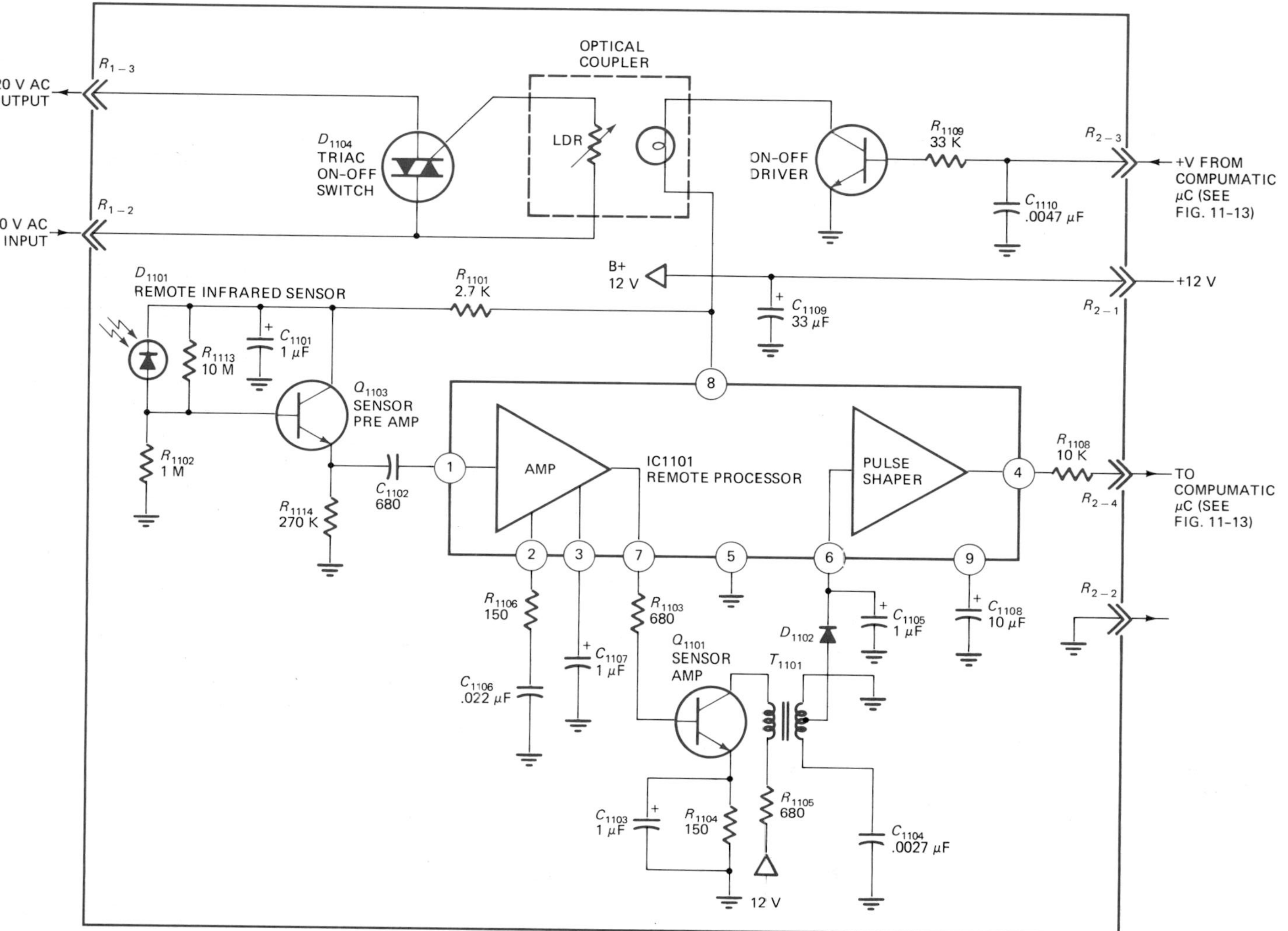

Figure 11.36 Schematic diagram of the Quasar direct-access remote-infrared receiver. Unless otherwise stated, capacitor values are in μF. *(Courtesy of Quasar Electronics Co.)*

diode D_1 and lights the function indicator LED (Figure 11.34) on the transmitter panel. This LED shows that a function button has been pressed. If it fails to light, this may be a sign that the battery needs to be replaced.

Volume Up, Volume Down All functions except volume up and volume down react only to the 42-kHz carrier *modulation* patterns. However, for volume up and volume down the pulse patterns only *initiate* the function. The function then reacts continuously (within its maximum or minimum limits) to the carrier.

THE REMOTE CONTROL RECEIVER

A schematic diagram of a remote control receiver is shown in Figure 11.36.

Signal Path The pulse-modulated infrared signals from the remote infrared transmitter are picked up by the remote infrared sensor D_{1101}. These signals are first amplified by the sensor transistor preamplifier, Q_{1103} and then coupled to IC_{1101}.

In IC_{1101}, the modulated signals are further amplified. They then pass through Terminal 7, DC-coupled, to the sensor transistor amplifier Q_{1101}. The signals then pass through transformer T_{1101} and detector diode D_{1102}, through Terminal 6 and to the pulse shaper in IC_{1101}. Only the positive half of the amplified signal is applied to the pulse shaper. The shaped, individual, pulse pattern is then DC-coupled to the Compu-Matic microcomputer (Figure 11.13). The microcomputer then produces the appropriate outputs to perform the ordered function.

Power On To turn the power on, the ON-OFF button (Figure 11.34) is pressed. This action produces a positive voltage at Pin 2 of the Compu-Matic microcomputer. This positive voltage is fed through connector $R_{2\text{-}3}$ (Figure 11.36) to the base of ON-OFF transistor driver Q_{1102}. The driver now conducts and causes the lamp to light in the optic coupler E_{1101}. (The optic coupler isolates the microcomputer from the 120-V AC line.) When the lamp lights, it reduces the resistance of the LDR (light-dependent resistor).

The low resistance of the LDR now applies 120 V AC to the gate of the triac D_{1104}. (The triac is an AC semiconductor switch. It is a bidirectional, gate-controlled thyristor (semiconductor switch) that provides full-wave control of AC power.) Since the triac conducts on both the positive and the negative half cycles of the AC line voltage, it acts as an AC power switch.

Power Off To turn the television receiver off, the ON-OFF button on the remote transmitter is pressed again. This removes the positive voltage from the base of on-off driver Q_{1102}. The lamp in the optic coupler E1101 is extinguished, the LDR assumes a high resistance, and the triac ceases to conduct. This removes the 120-V AC from the receiver.

11.14 TROUBLES IN COMPU-MATIC TUNING SYSTEMS

Troubleshooting the Compu-Matic tuning system requires a logical technique that separates normal and abnormal functions. Begin with a visual check to make certain that all plugs are on the proper terminals and are making good contact. Since symptom analysis can provide valuable clues and reduce diagnostic time, all functions should be checked.

NO PICTURE, NO SOUND

This condition may be caused by a defect either in the tuning system or in the video IF circuitry. One way to isolate between these two sections is to substitute a **tuner subber** for the receiver tuning system. If the tuner subber produces picture and sound, the fault lies with the tuning system. If no picture and sound appear even with the subber, the fault is with the IF circuitry.

If no tuner subber is available, the following simple procedure may be used. Disconnect the IF cable that connects the tuner to the IF board. Connect a jumper to the center terminal of the cable connection on the IF board. Scratch the other end of the jumper on the metal frame of the IF board. If flashes appear on the screen and static is heard on the speaker, the IF circuitry is probably OK and the problem is in the tuning system. If there are no flashes or static; the trouble is most likely in the tuning system.

TUNING BAND(S) INOPERATIVE

If all tuning bands (low VHF, high VHF, and UHF) are inoperative, check voltages to the tuners and prescaler. If the tuning voltage is either high or low and does not change when channels are selected, suspect (1) the prescaler or tuning-voltage control transistors (high tuning voltage), (2) the reference oscillator crystal, or (3) PLL IC (low tuning voltage). If only one tuning band is inoperative, check the discrete switching circuits and microcomputer outputs to these circuits (Figures 11.10, 11.11, 11.13, 11.16, 11.17).

SEEK FUNCTION INCORRECT

Seek function is active normally only when channel or seek buttons are depressed. Continuous seek may be caused by a defective sensor IC_{51} (Figure 11.14), misadjusted threshold control, or the microcomputer. Select a channel by direct entry. If seek stops, the microcomputer is probably functioning properly. Check threshold adjustment first. If adjustment has no effect, check voltages and signal inputs to IC_{51}.

INCORRECT CHANNEL DISPLAY

If all functions are normal except channel display, suspect a defect on the channel display board. A tuning problem accompanying erroneous channel display indicates a defective microcomputer.

REMOTE FUNCTIONS INACTIVE

If the remote transmitter does not activate the system but all functions are normal using the keyboard on the receiver, suspect the remote transmitter, infrared receiver, or the sensory or remote control receiver. If some but not all remote functions seem normal, the remote transmitter may be at fault (Figures 11.34 to 11.36).

11.15 TROUBLES IN AFT CIRCUITS

In addition to Figure 11.37, discussed below, refer also to Figures 11.19 and 11.20.

AFC, AGC, and AFT circuits are closed-loop systems. This means that part of the output signal is fed back (sometimes after it has been modified in some way) to the input of the circuit. Locating problems in this type of system is difficult because the output signal depends upon the input signal and the input signal depends in part upon the output signal.

Suppose, for example, that a receiver with an AFT circuit drifts off frequency after it is tuned to a channel. This could be because the AFT control is not working, or it

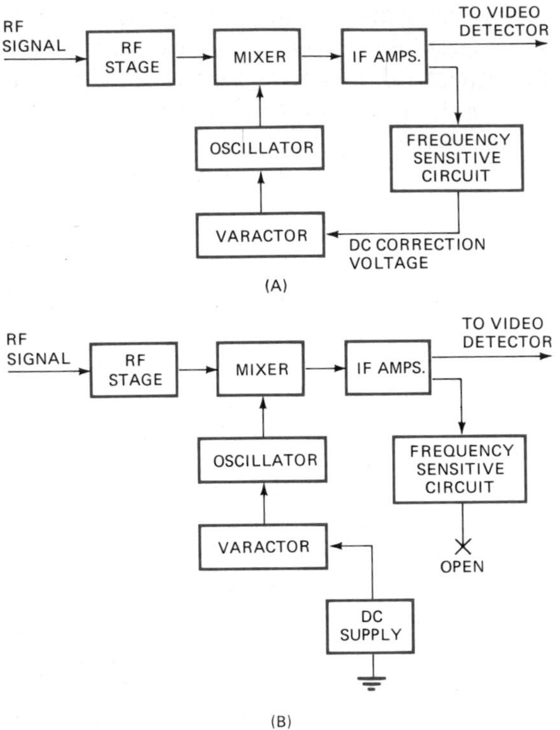

Figure 11.37A-B. The AFT circuit is a closed-loop system that presents a difficult troubleshooting problem. **A.** The closed-loop system. **B.** In order to troubleshoot a closed-loop system, some part of the system must be opened.

could be because the oscillator is drifting beyond the range of frequencies that the AFT circuit can control.

Manufacturers usually give typical operating voltages and test points for measuring these voltages. This is the first step in looking for a fault in an AFT circuit. For other than routine troubles, the only sure way to isolate a fault in a closed-loop system is to open the feedback loop. In some systems, it may be necessary to substitute a signal or a voltage to compensate for the open circuit. In an AFT system, the procedure is to open the part of the circuit that delivers a DC voltage to the oscillator control, that is, to the varactor diode. Very often, receivers have an AFC ON-OFF troubleshooting switch to make it a simple matter to open the loop. A DC voltage from a battery or a bias pack is substituted for the control DC. This procedure is illustrated in Figure 11.37 (also see Figures 11.19 and 11.20).

AFT BLOCK DIAGRAM

Figure 11.37A displays the AFT system in block diagram form. There are several points in this closed-loop system that could be open-circuited. The input to the frequency-sensitive circuit could be opened, but this is likely to affect the performance of the IF amplifier due to the change in its load. The varactor circuit could be opened, but this would probably put the oscillator completely off frequency. Therefore, the most likely place to open the circuit is in the DC line. This is indicated in Figure 11.37B. The DC supply shown replaces the normal DC feedback voltage.

Once the feedback loop is opened, standard troubleshooting techniques can be used.

11.16 TROUBLES IN REMOTE CONTROL SYSTEMS

ULTRASONIC-WAVE SYSTEMS

The ultrasonic waves from the remote control transmitter may be produced mechanically. In that case, the transmitter is not likely to be a source of trouble. If the ultrasonic waves are produced by a transistor oscillator, the most likely problem will be a depleted battery. Usually the customer can replace this battery, but it should not be overlooked as a source of trouble. If the remote transmitter is the type that radiates a modulated RF signal, any point in the circuitry, including the battery, can be at fault. Voltage measurements are the first step.

One useful technique for determining if a remote control transmitter is working properly is to tune an AM radio to a harmonic frequency of the transmitter's radiation. (This technique will work, of course, only for transmitters that radiate electromagnetic waves.) If the transmitter is working, there will be a buzzing sound in the radio loudspeaker.

One Function Out If one of the remote control functions is not working but all others are, then the transmitter and receiver circuitry is the least likely source of trouble. Instead, one of the components in the inoperative section is probably at fault.

All Functions Out If none of the functions work, then the trouble may be in either the transmitter or the receiver. The first step is to check the power supplies and the operating voltages. The ultrasonic receivers can be checked by the same techniques used for troubleshooting audio amplifiers. Receivers designed for remote control are usually of the TRF (tuned radio frequency) type, and troubleshooting is accomplished with RF and audio generators.

INFRARED-WAVE SYSTEMS

General The infrared remote unit transmits a pulse-modulated 42-kHz carrier to the receiver. An infrared sensor receives the transmitted signal, which is then amplified by several stages and applied to the Compu-Matic tuning system.

A remote control problem can be caused by a defect in either the transmitter or the receiver. Generally speaking, an oscilloscope is most useful in accurately troubleshooting the system.

If all functions are inoperative, replace the batteries in the hand unit. If still inoperative, connect an oscilloscope to the infrared sensor output lead (C_{1101} in Figure 11.36). Use the 10-to-1 probe, 20-μs horizontal, 50-mV vertical. Position the hand unit about 10 ft (3m) from the infrared sensor and press the 9-R button (Figure 11.34). A signal of approximately 1 V peak-to-peak (CP-P) 42 kHz signal should appear on the scope. The transmitter should be oriented for maximum signal on the oscilloscope. If this signal is present, troubleshoot the remote receiver. If not present, troubleshoot the transmitter (Figure 11.35) and/or the infrared sensor assembly (Figure 11.36).

If one or more, but not all, functions are inoperative, the transmitter is probably at fault. Follow the transmitter troubleshooting procedures given below.

Troubleshooting the Remote Transmitter If all functions are inoperative (battery is okay), run through the following sequence. Connect the 10-to-1 oscilloscope probe to the anode of the infrared transmitting diode. Set horizontal speed at 20 ms and vertical deflection at 50 mV. Depress the 9-R button. The signal at this point should be 1 V P-P at 42-kHz (one complete cycle = 24 μs).

If the carrier frequency and amplitude are correct, change the oscilloscope horizontal setting to 10 ms. Depress the 9-R button several times. Modulating pulses should be observed when the 9-R button is depressed. If so, the infrared sensor assembly at the receiver or the remote receiver circuitry is at fault.

Troubleshooting the Remote Receiver If a 1-V P-P 42-kHz signal is observed at the infrared sensor output (as outlined above), move the oscilloscope probe to the anode of D_{1101} (Figure 11.36). Horizontal remains the same (20 ms), but vertical should be set to 0.2 V. Position the transmitter about 1 ft (0.3m) from the infrared sensor and press the 9-R button. The 42-kHz signal at this point should be 6 V P-P.

If one or more, but not all, functions are inoperative, connect the 10-to-1 oscilloscope probe to the base of Q_{1101} (Figure 11.36), set the horizontal at 10 ms, and the vertical at 50 mV. A single pattern of pulses should appear as each transmitter button is depressed. If a pulse pattern does not appear for one or more buttons, remove the battery and check for continuity through the button contracts. If there is continuity when the buttons are pressed, suspect IC_{1001} (Figure 11.35) or its associated oscillator components.

SUMMARY OF CHAPTER HIGHLIGHTS

1. Electronic tuning systems all use varactors for channel selection. They all use electronic circuitry to control the varactor tuner.
2. Frequency synthesis is an electronic system which can generate a number of *crystal stable* frequencies, while using only *one* crystal. Frequency synthesis involves the use of a crystal oscillator, a phase-locked loop (PLL), and digital circuits, which include a microcomputer (μC). (Figure 11.6)
3. A microcomputer is a complete computer on a chip (or several chips). (Figure 11.13)
4. Frequency-synthesized tuning enables the user to select channels by *direct access*. It also eliminates the need for a fine-tuning control. (Figure 11.10)
5. A microcomputer may contain 11 000 transistors or more. Thus, concentration should be on functional concepts and not on internal circuit analysis.
6. The binary number system uses only the digits 0 and 1. The digits are called bits. A group of bits is called a byte. (Section 11.1)
7. All signals into, out of, and inside a microcomputer must be in digital form. (Section 11.4)
8. Digital systems process ON or OFF signals represented by 0 V and +5 V. In a positive logic system, an ON signal is a logic high (binary 1) and an OFF signal is a logic low (binary 0). (Section 11.1)
9. The radix is the number of different symbols required in a numbering system. The binary number system has a radix of 2. (Section 11.1)
10. A table giving decimal-to-binary number equivalents is presented in Table 11.1.
11. Dividers and prescalers are frequency dividers. A prescaler is a high-frequency divider. Dividers may be fixed or programmable. Programmable divider ratios are controlled by binary input to logic circuitry (Figure 11.10).
12. Any desired division factor can be obtained by correctly programming the input ports of a programmable divider. (Section 11.2)
13. Phase-locked loops (PLLs) compare two signals. They develop a correction voltage in a phase comparator proportional to the frequency and phase difference between them. (Figure 11.6)
14. The PLL correction voltage is fed to a voltage-controlled oscillator (VCO) to correct the frequency and phase difference. (Figure 11.6)
15. In the frequency-synthesis system, accurate crystal stable frequencies are developed because one input to the phase comparator is divided down from a highly stable, reference crystal oscillator. (Figure 11.10)

16. The three fundamental sections of a computer are (1) memory, (2) central processing unit (CPU), and (3) input-output (I/O).
17. A computer performs simple operations, one at a time, but at great speed. (Section 11.4)
18. With "direct access" channel selection, simply enter the desired channel number on the keyboard. For example, enter 02, 11, or 77.
19. In the Quasar frequency-synthesis system, the reference frequency is supplied by a stable 3.581 055-MHz crystal oscillator. 3.581055 MHz − 3667 = 976.562 Hz. This is the reference frequency fed to the phase comparator. (Figure 11.10)
20. In frequency-synthesis tuning, the tuner oscillator frequency is divided by fixed and programmable dividers to equal 976.562 Hz on all channels. This signal is also fed to the phase comparator. (Figure 11.10) This is the frequency as corrected by the phase comparator output to the VCO (tuner oscillator). (See Section 11.7)
21. In frequency synthesis, channels are selected by changing the frequency division between the tuner oscillator and the phase comparator. (Figure 11.11)
22. The microcomputer (μC) IC is a complete computer. It controls the entire frequency synthesis tuning system. (Figure 11.13)
23. Automatic fine tuning (AFT) maintains the tuner oscillator at a precise frequency in spite of oscillator drift or moderately incorrect fine-tuning control settings. It is particularly useful in color television receivers to maintain correct color rendition. (Section 11.7)
24. In AFT, a sample of the picture IF carrier is fed to a 45.75-MHz discriminator. The discriminator provides a DC correction voltage (when required) that is fed to a varactor to correct the tuner oscillator frequency. (Figure 11.19)
25. All AFT circuits have a *pull-in* and a *hold-in* range. Typical values are 50 kHz for pull-in range and 1 MHz for hold-in range.
26. Two types of remote control transmission signaling are (1) mechanically or electronically generated ultrasonic waves, and (2) infrared waves generated by infrared LEDs.
27. Ultrasonic remote control systems transmit in the range from 34 to 54 kHz. Individual frequencies in this range are used for specific functions. (Figures 11.21 to 11.33)
28. Direct-access remote control systems may have 14 or more individual transmitting frequencies. (Figures 11.34 to 11.36)
29. Ultrasonic remote control waves are transmitted through the air and picked up by a special crystal microphone at the television receiver. From the microphone, they are passed through an amplifier and then to the control circuits. (Figures 11.21, 11.24 to 11.26)
30. The Quasar direct-access remote control transmitter has 16 buttons (functions) and transmits by means of infrared waves. Transmission is somewhat directional, but relatively immune to interference. (Figures 11.34 and 11.35)
31. The infrared waves in the Quasar remote transmitter are modulated by a 42-kHz carrier. This carrier in turn is modulated by 16 individual-pulse patterns (one at a time). Each function button produces, when depressed, its own individual pulse pattern, which corresponds to a specific function. (Figure 11.35)
32. At the television receiver, an infrared sensor picks up the modulated infrared waves. These waves are amplified and

shaped. They are then sent to the Compu-Matic microcomputer, which orders the required function. (Figures 11.13 and 11.36)

33. In troubleshooting a Compu-Matic tuning system, it is useful to isolate the area of trouble by using a tuner subber.
34. AFT circuits are closed-loop systems. In troubleshooting such a system, it is useful to open the loop (after voltage checks). This is done at the circuit that delivers a DC voltage to the varactor. (Figure 11.37)
35. In a remote control electronic oscillator system, a frequent cause of trouble is a depleted battery.
36. Ultrasonic receivers can be checked by the techniques used for audio amplifiers.
37. In troubleshooting infrared remote control systems, the use of an oscilloscope is helpful. (Section 11.17)

EXAMINATION QUESTIONS

(Answers are provided at the back of the text.)

Part A Supply the missing word(s) or number(s).

1. Electronic tuning systems must all use ______ tuners.
2. A computer on a chip is called a ______.
3. A switch that is either ON or OFF is called a ______ device.
4. The binary number system uses only ______ and ______.
5. The decimal equivalent of binary number 010001 is ______.
6. A high-frequency frequency divider is called a ______.
7. The binary ______ applied to its inputs determines the division factor of a divider.
8. A 500-kHz signal is required from a 10-MHz crystal oscillator. This may be obtained by dividing by ______ and by 4.
9. The PPL compares ______ and develops a correction voltage from them.
10. In a frequency synthesis tuning system, the ______ is not required.
11. A ______ provides the overall control for a frequency synthesis tuning system.
12. Tuning voltages are applied to the ______.
13. The phase comparator receives signals from the ______ oscillator and the VDO oscillator.
14. In a *direct-access* tuning system, channels are selected in ______ order.
15. A ______ is the term used for a group of bits.
16. Inside a microcomputer, all signals must be in ______ form.
17. In frequency synthesis tuning, fixed and ______ dividers are used to divide the tuner oscillator frequency.
18. A sample of the ______ carrier is fed to a discriminator in the AFT system.
19. An AFT system has a ______ range and a ______ range.
20. The transmitting frequency range of ultrasonic remote control systems is from ______ to ______.
21. In the infrared remote control system, ______ generate the infrared waves.
22. Each of the 16 function buttons produces an individual ______ in the infrared remote control system.
23. After the modulated infrared signal is processed by the remote control receiver, it is sent to the ______, which orders the required function.
24. In troubleshooting an AFT system, the closed ______ must sometimes be opened.
25. It is very useful to use a (an) ______ when troubleshooting infrared remote control systems.

B Answer true (T) or false (F).

1. In frequency synthesis, a discriminator develops the correction voltage.
2. The term *radix* refers to the physical dimensions of a chip.
3. Since a microcomputer may contain 11 000 transistors, it is essential to analyze its internal circuitry.
4. The binary number system is used in digital systems.
5. The decimal numbering system is not compatible with electronic switching circuits.
6. Dividers and prescalers are basically different types of circuits.
7. The binary inputs to a divider establish its division factor.
8. Division factors for binary-controlled dividers on the order of several million are not uncommon.
9. Counters function in a manner similar to dividers.
10. In frequency synthesis, a highly stable crystal oscillator has its frequency divided to provide the tuner oscillator frequency for each channel.
11. It is difficult and expensive to obtain crystals in the range of 3 to 4 MHz.
12. All computers contain three sections: memory, erasability, and accumulator.
13. ROM stands for read-only memory.
14. *Direct access* means that channels can be selected in any desired order.
15. Television receivers using frequency synthesis never have AFT circuitry.
16. AFT circuits have phase comparators to develop the required correction (error) voltages.
17. A typical hold-in range for an AFT circuit is 1 MHz.
18. Ultrasonic remote control systems transmit in the range from 34 to 54 kHz.
19. When ultrasonic signals are electronically generated, they are transmitted by a small, nondirectional antenna.
20. In the infrared remote control system, wave transmission is nondirectional.
21. In the Quasar infrared remote control system, the carrier frequency is 42 kHz.
22. A problem of no picture, no sound cannot occur in a television receiver that uses frequency synthesis.

REVIEW ESSAY QUESTIONS

1. Draw a simple block diagram showing the essential sections of a frequency synthesis system.
2. Explain briefly how the system represented by the block diagram in Question 1 works.
3. Define: FS, PLL, μC, direct access. (Note: do not merely supply names for abbreviations.)
4. Why is it necessary for each function in Table 11.2 to use a different ultrasonic frequency?
5. Explain how it is possible to eliminate the fine-tuning control in a frequency synthesis system.
6. Discuss the need for binary inputs to a frequency divider. (Section 11.3)
7. Explain the operation of a latch circuit. (Figure 11.2)
8. Discuss how the tuner oscillator frequency of the unit shown in Figure 11.11 is produced. Why is it crystal-stable?
9. What is (are) the basic function(s) of the microcomputer in a frequency synthesis system? (Figure 11.13)
10. Discuss the need for VHF high/low band switching and describe the general scheme used for it. (Figure 11.16)
11. Explain how the binary number system works. Why is it used in digital circuits?
12. Define binary, bit, byte, radix, bistable.
13. Tell how to convert a decimal number to its binary equivalent. (Section 11.2)
14. Give the names for these abbreviations: RTL, DTL, TTL, ECL.
15. In a microcomputer, briefly describe the function of (a) the memory, (b) the central processing unit, and (c) the input/output.

16. Explain what the seek function does in a tuning system.
17. Draw a simple block diagram of an AFT system.
18. For the block diagram drawn in Question 17, explain its operation.
19. Referring to Figure 11.10, explain how it is possible for the tuner oscillator frequency to become 976.5625 Hz at the phase-comparator input for all channels.
20. Name two types of ultrasonic remote control devices.
21. Briefly describe the operation of each type of device named in Question 20.
22. Draw a simple block diagram of the transmitting and receiving portions of the Quasar infrared remote control system.
23. Briefly describe the operation of the system whose block diagram was drawn for Question 22.
24. What is a tuner subber? How is it used?
25. Discuss, in your own words, the value of an oscilloscope in troubleshooting an infrared remote control system.

EXAMINATION PROBLEMS

(Selected answers are provided at the back of the text.)

1. Construct a table (similar to Table 11.1) of the binary equivalent of the decimal numbers 30, 45, 60, and 100.
2. Convert binary number 11 to a decimal number. Show your calculations.
3. In Figure 11.3, program the input ports for a division factor of 7 (refer to Table 11.1).
4. Show how a digital clock obtains hours, minutes, and seconds from a 60-Hz power line frequency.
5. Draw a simple block diagram of a PLL system that shows how to obtain a crystal-stable 15-kHz signal from a 30-MHz reference crystal.
6. Referring to Figure 11.10, explain the steps involved in obtaining a divided 976.5625-Hz tuner oscillator signal when tuned to Channel 2.
7. In Figure 11.38, fill in the three items requested.
8. If the pull-in range of an AFT system is ±50 kHz (±1 MHz), what are the extremes of the tuner oscillator frequency on Channel 2 that are within this range?
9. The hold-in range of an AFT system is ±1 MHz. What are the extremes allowed for the tuner oscillator frequency on Channel 6?
10. In an AFT system with a hold-in range of ±1 MHz, what are the extremes allowed for the color IF carrier?

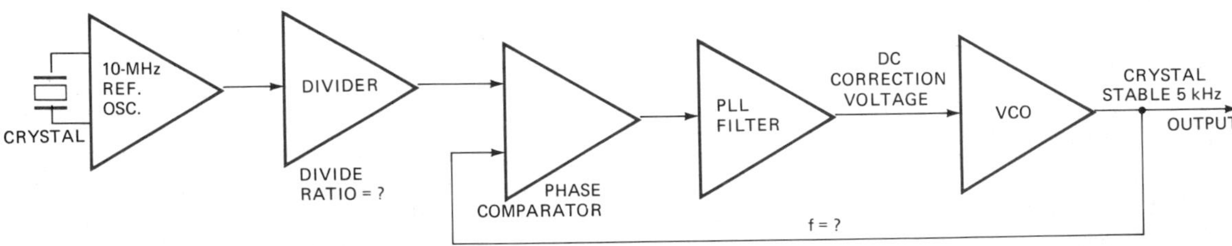

Figure 11.38 Figure for Problem 7.

Chapter 12

12.1 Major video IF functions
12.2 Frequency response curves
12.3 Comparison of vacuum tube and solid-state video IF amplifiers
12.4 Types of video IF amplifiers
12.5 Intermediate frequencies
12.6 Amplifier bandwidth
12.7 Stagger tuning
12.8 Interstage coupling
12.9 IF Wave traps
12.10 Sound IF-frequency separation
12.11 Typical video IF amplifiers
12.12 An IC video IF amplifier
12.13 Surface-acoustic wave filter (SAWF)
12.14 Video-IF amplifier troubles
Summary

VIDEO IF AMPLIFIERS

INTRODUCTION

Only two sections of a TV receiver handle the entire band of frequencies required for the reproduction of the picture and sound. These sections are the tuner and the common video IF amplifiers. As discussed in Chapter 10, the tuner selects the desired TV channel and then converts the band of frequencies of the channel selected to the IF band of the video IF amplifiers. These IF amplifiers then amplify the signal enough to drive the video detector and the sound and video amplifiers. In this process, the quality of the picture and sound depends on how well the RF tuner and the IF amplifiers can select and amplify only the desired frequencies. All other frequencies must be attenuated to a very low level. In color TV receivers, the video IF amplifiers must also remove the sound IF frequencies before they reach the video detector. Figure 12.1 shows the relationship of the video IF amplifiers to the tuner and the video detector, for both monochrome and color receivers.

In addition to driving the video detector, video IF amplifiers must compensate for vestigial sideband transmission. This compensation is required output at the video detector for frequencies near the video carrier as well as those in the higher-frequency portion of the IF band. The video IF amplifiers must also provide special tuned circuits, or traps, to prevent interference from the associated-channel sound carrier, the adjacent channel sound carrier, and, in some cases, the adjacent-channel picture carrier. In addition, a color video IF amplifier must separate the sound frequencies from the picture frequencies before they arrive at the video detector. This reduces interference patterns on the picture tube caused by heterodyning between the sound IF frequencies and the 3.58-MHz color subcarrier (Figure 12.1B). In monochrome receivers, the video IF amplifiers must attenuate the associated sound carrier so that its amplitude will be low at the output to the video detector. This prevents sound bar patterns on the picture tube.

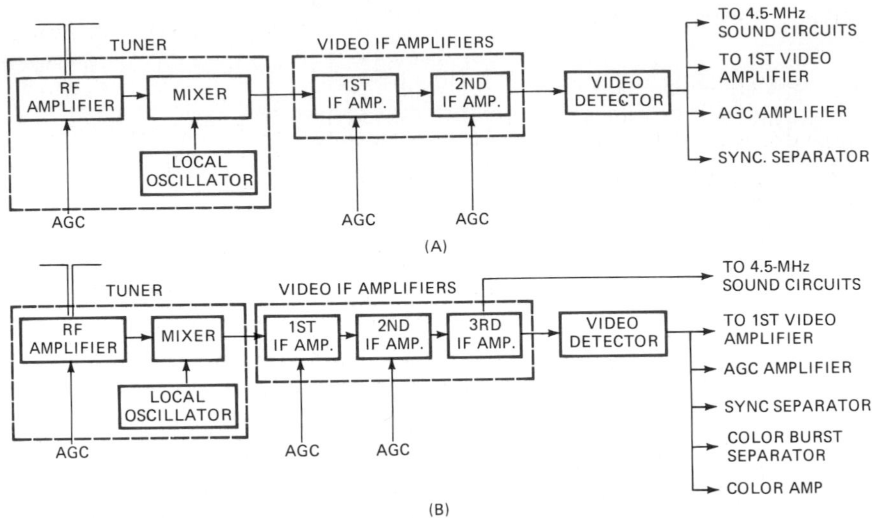

Figure 12.1A-B. The relationship of the tuner, video-IF amplifiers, and video detector for **A.** monochrome and **B.** color sets.

To sum up, the most important functions of video IF amplifiers are selectivity, gain, and elimination of unwanted frequencies.

TV receiver video-IF amplifiers may be found in the form of (1) vacuum tube, wired or printed circuits, (2) transistor printed circuits, or (3) transistor and integrated circuits. The subject of video-IF amplifiers will be discussed under the following headings:

When you have completed the reading and work assignments for Chapter 12, you should be able to:

- Define the following terms: common IF amplifier, IF gain, IF selectivity, bifilar coil, stagger tuning, marker frequency, bandwidth, piezoelectric effect, surface acoustic wave, half-power point, trap circuits, absorption trap.
- Draw and label a block diagram of a color TV receiver, common IF section. Include all traps and their frequencies.
- Explain how the 920 kHz beat frequency is prevented from reaching a color TV picture tube.
- List the number of common IF stages generally used in monochrome and color TV receivers. Explain why this is done.
- List all specific frequencies normally found in a color TV receiver, common IF section. Identify each one.
- Explain why the position of the sound carrier with respect to the video carrier, is reversed after the mixer stage.
- Draw the overall response curve of a color TV receiver. Label all the important frequencies, including trap frequencies at the correct locations. Tell why the video carrier frequency is at the 50% point.
- Describe the bandwidths occupied by the various components of the IF, color video signal.
- Discuss the most troublesome, undesired responses of the common video IF system.
- Explain the use of *markers*.
- Explain why some transistor IF amplifiers must be neutralized.

- Know what sections of a TV receiver ICs are used for.
- Discuss the construction of a SWAF.
- Explain the use of and the operating principle of a SWAF.
- Discuss the various troubles that may occur in a common IF section and their effects.

12.1 MAJOR VIDEO IF FUNCTIONS

A brief summary of the major functions of the common video IF amplifiers in both monochrome and color receivers is as follows:

1. They provide most of the RF gain in the receiver.
2. They provide most of the RF selectivity in the receiver.
3. They reduce accompanying sound interference.
4. They reduce adjacent channel interference.
5. They separate the sound and picture IF frequencies.
6. They supply IF signals of sufficient amplitude to drive the AM video detector (and video amplifiers) and the FM sound detector (and audio amplifiers).

The circuitry in video IF amplifiers used in color sets is more complicated than in monochrome receivers. Also, amplifier alignment is more critical in color sets. In general, color IF amplifiers may have to pass a wider bandwidth, without attenuation, than the amplifiers used for many monochrome sets. This is necessary because the color IF sideband frequencies must not be attenuated.

12.2 FREQUENCY RESPONSE CURVES

In any discussion of selectivity and gain, it is important to speak in terms of bandwidth, or frequency response. Anyone studying video IF amplifiers must understand **frequency response curves** and how they relate to the operation of the amplifiers. Figure 12.2 presents a typical response curve for an arbitrary amplifier. This curve represents the output voltage of the amplifier versus the frequency of a constant-input amplitude signal. In solid-state circuits, the curve may represent the output current, or voltage, developed by the transistor amplifiers. Note that as the frequency increases, the voltage remains low until it is near 2 MHz. At that point, further increases in frequency cause the voltage to increase. Around 5 MHz, the voltage starts to decrease. Finally, at about 8 MHz, the voltage again is at its original level. This curve then represents the response of an arbitrary amplifier.

The bandwidth of the amplifier in Figure 12.2 is approximately the band of frequencies between the lowest and highest frequencies on the response curve that produces a specific output. This particular output is specified for video IF amplifiers and is discussed in greater detail in Section 12.5. The selectivity of the amplifier is determined by the steepness of the slopes, or skirts, of

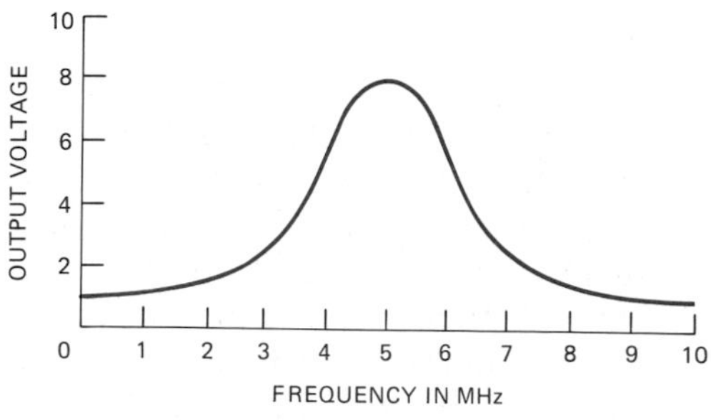

Figure 12.2 The typical response curve of an arbitrary amplifier.

the response curve. The steeper the slope, the greater the selectivity because unwanted frequencies outside the curve will develop a much lower output voltage. The voltage gain of the amplifier is represented by the highest point, or peak, on the response curve. This gain is usually expressed in decibels.

You should review parts of Chapter 10 for additional basic information on tuned circuits and response curves.

12.3 COMPARISON OF VACUUM TUBE AND SOLID-STATE VIDEO IF AMPLIFIERS

Functionally, vacuum tube and solid-state video IF amplifiers are identical. In construction, though, they are quite different. Vacuum tube amplifiers are either hand-wired on a metal chassis or mounted on PC boards. Solid-state IF amplifiers are invariably constructed on PC boards. Many recent solid-state receivers use IC IF amplifiers mounted between tiny IF transformers on a PC board. In modular construction, both vacuum tube and solid-state circuits have been used.

NEUTRALIZATION

There are several distinguishing features of transistor amplifiers that should be noted. Pentodes are used extensively in vacuum tube IF amplifiers, and therefore neutralization is not needed. The transistor is more comparable to a triode. The feedback signal between the collector and base (through the junction capacitance) must often be neutralized. A few examples of **neutralizing circuits** are shown in Figure 12.3. Figure 12.3A presents the neutralization that takes a voltage from the tap on the IF transformer. Another method of neutralization takes a voltage of the proper phase from the secondary winding of the transformer (Figure 12.3B). This method has the advantage of not requiring a tap on a transformer winding. In a third method (Figure 12.3C), a separate winding on the transformer L_N provides the feedback signal required for the neutralization. All solid-state video IF amplifiers are not neutralized in TV receivers. If the amplifier has low gain and low impedance, and if it is a broad-band amplifier, then neutralization may not be required.

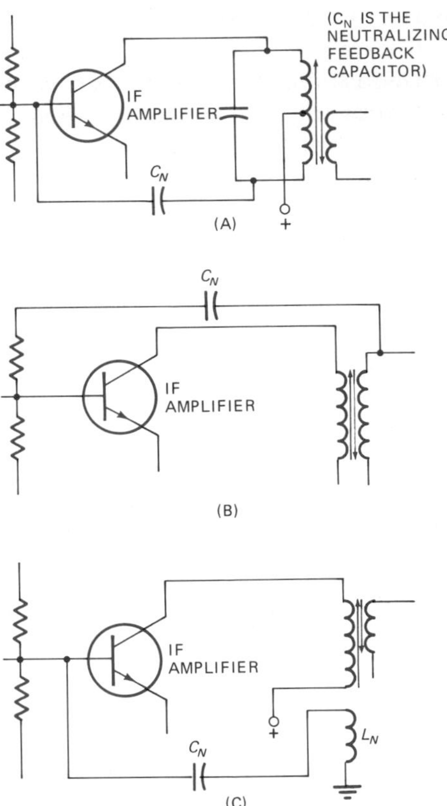

Figure 12.3A-C. Several neutralizing circuits for transistor-IF amplifiers. **A.** An IF amplifier neutralized by using a tapped transformer winding. **B.** The neutralization process without the need for a tap on the transformer winding. **C.** Obtaining a neutralizing signal with a separate transformer winding.

IMPEDANCE MATCHING

Another important feature of transistor IF amplifiers is the need for **impedance matching** between the collector circuit of one amplifier and the base circuit of the following stage. The input impedance of a transistor is quite low in comparison to that of a vacuum tube. This assumes, of course, that a common-emitter circuit is used. This type of circuit is by far the most popular one used for IF amplifiers. Impedance matching can, of course, be accomplished by selecting the proper turn ratios for the transformers used in the IF stages. There is, however, some tendency to avoid the use of transformers in low-cost TV receivers. Instead, impedance coupling is used extensively. When impedance coupling is used, a special impedance-matching network is needed, such as the one illustrated in Figure 12.4.

AGC VOLTAGE

In vacuum tube circuits, it is common practice to deliver a negative AGC voltage to one or more of the IF amplifier stages. In transistor IF amplifiers, either NPN or PNP transistors may be used. Therefore, the AGC voltages may be either positive or negative. (AGC circuits are discussed in Chapter 14.)

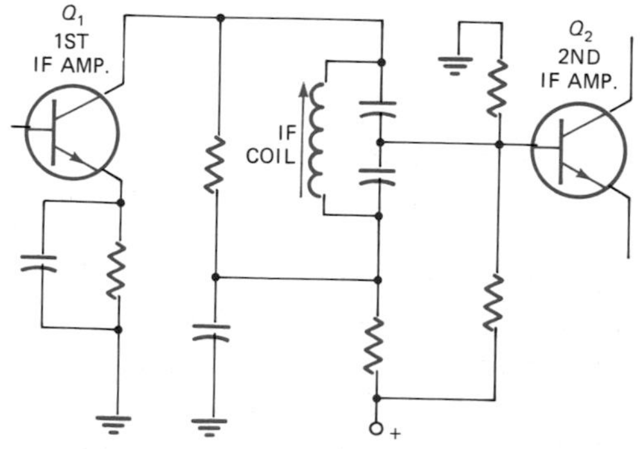

Figure 12.4 A circuit used for impedance matching between IF amplifiers. Many variations of this circuit are possible.

PRINTED CIRCUIT BOARDS

Printed circuit (PC) boards are used in both vacuum tube and solid-state video IF amplifier circuits. The boards used with vacuum tubes are usually thicker than their solid-state counterparts. The extra thickness is required to support the heavier vacuum tubes and their associated sockets. Vacuum tube PCs require an extra printed circuit for the tube filaments.

Solid-state PCs sometimes have plug-in sockets for transistors. On many boards, however, the transistors are soldered directly to the circuits printed on the board. Even though this is less convenient when servicing is necessary, it is by far the most efficient and reliable method. This is because the transistors are not connected into the circuit by spring contacts.

Another feature of the PC is its adaptability to modular construction. For this application, contacts are mounted along the edges of the board for easy insertion into mating connectors on the master chassis or rack.

INTEGRATED CIRCUITS

Integrated circuits have been developed for both monochrome and color video IF amplifiers. These components are about the same size as the transistors normally used for the IF amplifier. However, they contain all of the components, including transistors, for an entire IF section. The ICs do not contain the IF transformers or the coupling-impedance devices. Each IC unit has from 12 to 20 base pins. An IC plugs into a socket the way vacuum tubes and transistors do. ICs are used in both hybrid and all-solid-state receivers. They are particularly useful in modular construction.

IF AMPLIFIER SIZE

A major difference between vacuum tube and solid-state video IF amplifiers is size. Even though vacuum tubes are used in modular construction, the all-solid-state circuits using transistors or ICs are more suited for this technique. Most solid-state video IF modules are no larger than a playing card. A vacuum-tube module is two to three times that size. Solid-state circuits require no filament voltage and produce far less heat.

12.4 TYPES OF VIDEO IF AMPLIFIERS

Even though all video IF amplifiers perform basically the same function, there are several different types that must be considered. First, video IF amplifiers are designated as monochrome or color types, depending on the TV receiver they are intended for. Second, they contain either vacuum tube or solid-state circuits. Third, all video IF amplifiers are classified according to their number of stages, from one to four. Two- and three-stage amplifier systems are the most common. Finally, all video IF amplifiers are classified according to the method used for interstage coupling. In the succeeding discussion, all of the classifications are presented separately.

MONOCHROME VIDEO IF AMPLIFIERS

Most monochrome TV receivers have either a two-stage or a three-stage video IF amplifier. An impedance-matching device is used to couple the signal from the tuner to the first IF stage. A sound trap is used to reduce the sound carrier amplitude to a low level. All associated sound signals are passed on through the system to the detector, as shown in Figure 12.1A. Traps may also be provided to eliminate the adjacent channel sound. Some manufacturers also include traps to prevent adjacent channel picture interference. In monochrome IF amplifier systems, AGC is normally applied to two stages to provide a gain variation of at least 60 dB. This value is required to compensate for the wide range of signal strengths passed on from the tuner. Such broad control is difficult to achieve when only one stage is controlled. Monochrome video IF amplifiers generally do not have sufficient bandwidth for all color picture IF sideband frequencies.

COLOR VIDEO IF AMPLIFIERS

Color TV receivers usually have a three-stage video IF amplifier system. Input to the color IF amplifiers is similar to that of the monochrome amplifiers. An impedance-matching device is used to couple the first IF stage to the tuner coupling link, which is usually a length of coaxial cable. Sound traps are provided in the color IF system to reduce the amplitude of the associated sound carrier and to eliminate adjacent channel sound frequencies. Since the associated sound carrier is now allowed to reach the video detector, a tuned circuit in the third stage removes the sound carrier (Figure 12.1B). Another sound trap is then placed between the third stage and the video detector to prevent associated sound interference from reaching the video detector. As in the monochrome video IF system, AGC is generally applied to the first and second stages. Because of the color side-bands, the bandwidth and selectivity requirements of color-IF amplifiers are more stringent than those of monochrome systems.

VACUUM TUBE VIDEO IF AMPLIFIERS

Vacuum tube video IF amplifiers can be classified as either monochrome or color. In these systems, pentodes are normally used for the amplifiers and transformers provide the interstage coupling. Two vacuum tube stages are frequently used for monochrome systems and three for many color applications. In the early days of color TV, some four-stage vacuum tube video IF systems were used.

SOLID-STATE VIDEO IF AMPLIFIERS

Transistor and IC video IF systems can also be classified as either monochrome or color. Even though the functions of these IF systems are identical to those of the vacuum tube systems, their circuits are different. The solid-state amplifiers have no filament circuits, and all components are considerably smaller. Transformers are normally used for interstage coupling, but impedance coupling may be used. The transistors are either NPN or PNP, as individual components or in IC units.

12.5 INTERMEDIATE FREQUENCIES

To understand the importance of selectivity and gain in video IF amplifiers, all of the frequencies handled in the video IF band must be considered, along with their amplitude and phase relationships. As shown in Figure 12.5A, there are two carriers involved in monochrome IF frequencies: an AM video carrier and an FM sound carrier. As shown in Figure 12.5B, there are three carriers for color transmission. The AM video and FM sound carriers are identical to those used for the monochrome IF. The third carrier is a phase AM-modulated color subcarrier. Each carrier has sidebands associated with it. Notice that in Figure 12.5 the carrier positions in the IF band are reversed from their normal RF transmitted order. The following paragraphs detail the reasons for this reversal. The various frequencies that make up the video IF band for both monochrome and color are also discussed.

FREQUENCY REVERSAL IN THE VIDEO IF BAND

When the composite video signal is initially transmitted, the sound carrier is 4.5 MHz above the video carrier. In the receiver IF stages, however, the sound carrier is 4.5 MHz below the video carrier, as shown in Figure 12.5. This signal reversal takes place in the mixer stage of the tuner. Figure 12.6 shows why the video and sound carriers exchange positions. (In color IF transmissions, the color subcarrier position is also reversed.) The frequencies shown in Figure 12.6A represent the video, color, and sound carrier frequencies for Channel 4. Also shown are the local oscillator frequency required to produce the standard 45.75-MHz video IF carrier, the sound carrier of 41.25 MHz, and the color subcarrier of 42.17 MHz. The transmitted sound carrier is shown at 71.75 MHz, which is 4.5 MHz above the 67.25-MHz video carrier. The color subcarrier is shown at 70.83 MHz, which is 3.58 MHz above the video carrier frequency. The exact frequency difference between the color subcarrier and the video carrier is 3.579545 MHz; however, to simplify the discussion, the value is rounded off to 3.58 MHz.

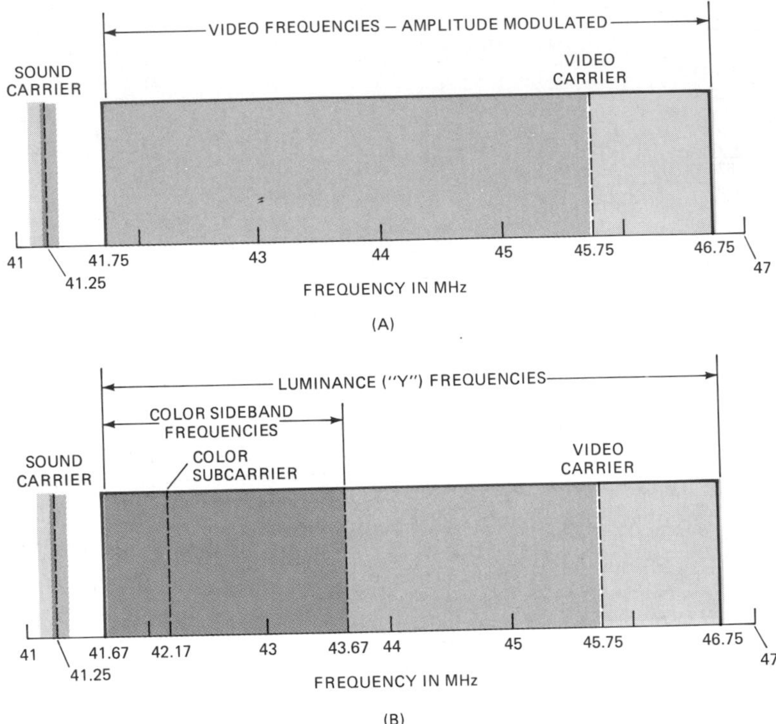

Figure 12.5 The distribution of sound- and video-IF frequencies for **A.** monochrome and **B.** color IF passbands in an ideal receiver.

Figure 12.6 indicates that the difference between the sound carrier and the local oscillator frequencies is smaller than the difference between the video carrier and local oscillator frequencies. This smaller difference, which also holds true for the color subcarrier, is why the sound carrier and color subcarrier are below the video carrier in the IF stages of the receiver.

The relationship between the sound and video carrier frequencies in the receiver IF stages is shown in Figure 12.6B. Here, because of the mixing action, the sound carrier and the color subcarrier are below the video carrier. This frequency reversal will not occur if the local oscillator frequency is below the RF frequency of the broadcast station. The normal procedure, however, is to make the local oscillator frequency higher than the received signal frequency.

MONOCHROME VIDEO IF FREQUENCIES

In the monochrome video IF stages, the sound carrier is established at 41.25 MHz and the video carrier is at 45.75 MHz. All of the frequencies between 41.75 and 45.75 MHz, as illustrated in Figure 12.5A, make up the upper sideband of the AM video carrier. Most of the lower sideband is eliminated in vestigial sideband transmission, but the frequencies between approximately 45.75 and 46.75 MHz of the lower sideband are included along with the carrier. These lower sideband frequencies require special compensation in the IF amplifiers, as discussed in Section 12.5. This frequency range ideally permits a 4-MHz-wide video IF band that is sufficient for high-quality picture reproduction. In practice, the video IF bandwidth of

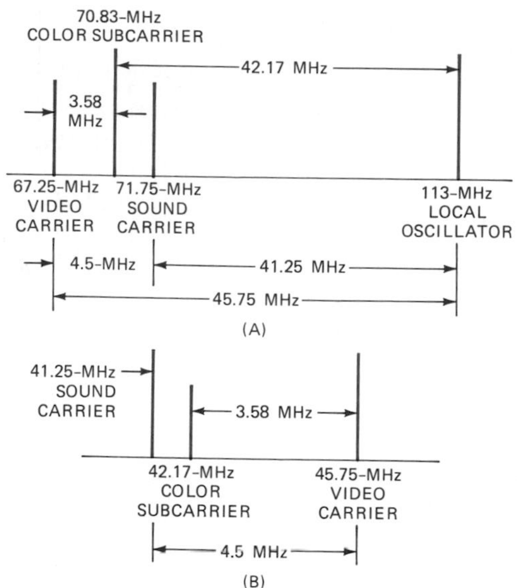

Figure 12.6A-B The reversal of carrier frequencies as they pass through the mixer into the IF stages. **A.** The distribution of frequencies for Channel 4 at the mixer stage input. **B.** The distribution of frequencies in the receiver IF stages.

monochrome receivers may vary between about 2.5 to 3.6 MHz. The wider bandwidth is used for receivers with larger screens and for more expensive receivers.

COLOR VIDEO IF FREQUENCIES

In the color video IF band, the FM sound and AM video carriers are identical to those in the monochrome IF band. However, chrominance, or color signals are superimposed on the upper video sideband. After the mixer stage of the tuner, the color IF sideband frequencies appear as shown in Figure 12.5B. In color receivers, monochrome video frequencies determine picture luminance, or brightness, and are referred to as the Y component.

SPURIOUS RESPONSES IN THE VIDEO IF BAND

The picture carrier video IF of 45.75 MHz is recommended by the Electronic Industries Association (EIA). This particular frequency was derived after many years of study and research and is currently the standard for both monochrome and color TV receivers. Because early TV receivers used lower video IF frequencies, they were prone to spurious responses. Even though these responses have been greatly reduced by making the 45.75-MHz picture carrier standard, it is important to understand how they may affect a TV receiver. The most troublesome of these spurious responses are (1) image response, (2) response to two stations separated in frequency by the IF value, and (3) direct IF response.

Image Response Image response occurs when an undesired signal is mixed with the local oscillator signal in the mixer stage. This produces a voltage at the IF, which will be passed by the IF amplifiers. As an illustration, a TV receiver with a video IF carrier value of 45.75 MHz has a video IF band extending from 45.75 MHz to approximately 41.75 MHz. If the receiver is tuned to Channel 2, which has a frequency range of 54 to 60 MHz, the local oscillator will operate at 101 MHz. Remember that the video carrier is 1.25 MHz above the low end of the channel frequency. This means that the frequency of the video carrier for Channel 2 is 55.25 MHz. The local oscillator frequency is then found by adding the picture IF carrier value of the receiver to the frequency of the picture carrier in the TV channel:

$$45.75 + (54 + 1.25) = 101.00 \text{ MHz}$$

From this discussion, it can also be seen that strong image-producing signals in the range from 142 to 148 MHz can beat with the 101-MHz local oscillator frequency to produce voltages within the video IF band of the receiver. These undesirable frequencies would, of course, have to get into the mixer stage of the tuner with enough strength to develop the spurious response. However, because of the high IF value of 45.75 MHz, the image frequency of 146.75 MHz is sufficiently removed in frequency from the desired signal (55.25 MHz) to be effectively eliminated from the mixer input by the RF resonant circuits in the tuner.

Stations Separated by the IF A second source of interference is caused by two or more stations separated by the IF value of the receiver. These stations can be any sources of RF energy. In this situation, one incoming signal acts as the mixing oscillator for the other signal (or signals when several sources are involved). The result is a difference frequency at the output of the mixer or converter stage that is equal to the IF value of the receiver.

Direct IF Response The third form of spurious response occurs when a receiver accepts a signal whose frequency is equal to the IF. To avoid this problem, 45.75 MHz is chosen as the IF because this frequency is not used to any appreciable extent for commercial or amateur transmissions. This eliminates the need to have special filters, wave traps, and shielding in television receivers.

12.6 AMPLIFIER BANDWIDTH

In the superheterodyne circuits of the TV receiver, a major portion of the overall gain and selectivity is provided by the IF amplifiers. For this reason, it is important for anyone working with TV receivers to understand the operation of the IF amplifiers, and especially their response characteristic curves. Figure 12.7 presents typical response curves for monochrome and color video IF amplifiers. The prime factors that determine the shape of the curves are vestigial sideband compensation and the type of IF am-

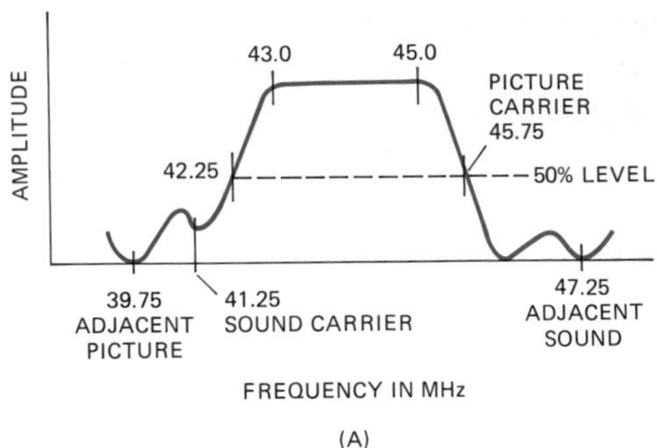

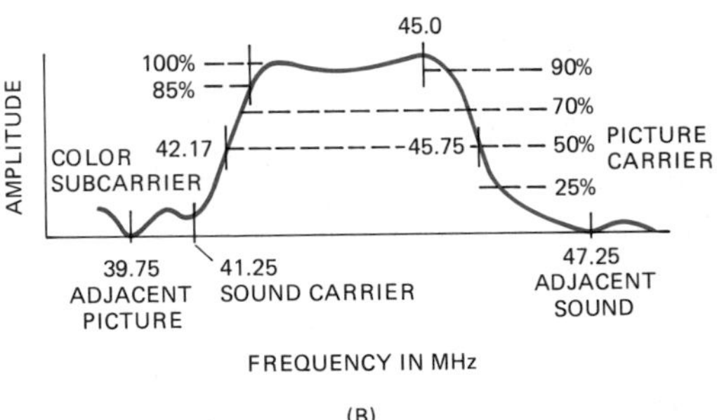

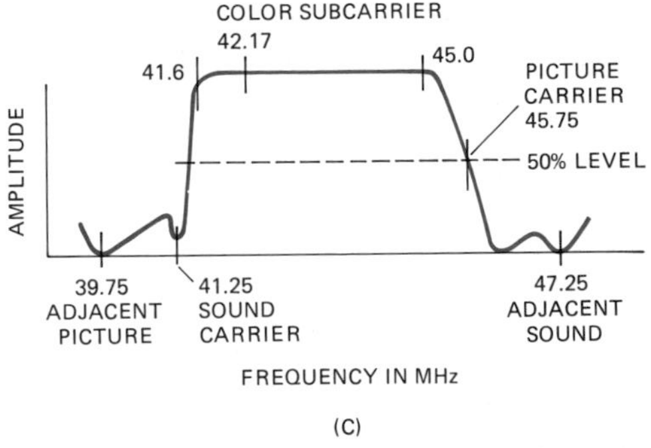

Figure 12.7A. The IF amplifier response curve of a typical monochrome TV receiver. **B.** The IF response curve of a typical color TV receiver. **C.** The IF amplifier response curve of the early color receivers.

plifier (monochrome or color). To achieve the desired bandwidth, the tuned circuits of the amplifier are usually **stagger-tuned**.

VESTIGIAL SIDEBAND TRANSMISSION COMPENSATION

Figure 12.5 indicates that the video carriers in both monochrome- and color-IF bands are amplitude modulated, but differ from conventional AM carriers by having only one complete sideband (Figure 12.5). The other sideband, of which remnants are still present, has been effectively suppressed. This is known as **vestigial sideband transmission** and is the standard for both monochrome- and color-TV broadcasting. When any carrier is amplitude-modulated, an upper and a lower sideband are generated; however, because identical information is contained in each sideband, only one sideband is required for demodulation in the receiver.

Complete suppression of the lower sideband is desirable, but it would be very expensive to do so. It is impossible for simple filters to completely eliminate one sideband without distorting portions of the other. As a compromise between economy and easily adjustable circuits on the one hand and minimum distortion and bandpass on the other, all but 1.25 MHz of the lower sideband of the video signal is removed. The transmitted video signal consists of this 1.25 MHz of the lower sideband plus the video carrier plus all of the upper sideband. With the addition of the associated sound carrier and its sidebands, the full 6 MHz allotted to each television station is obtained.

Method of Compensation Within the receiver, the IF amplifiers must use the upper sideband, together with the remnants of the lower sideband, to provide a response characteristic in which all sideband frequencies are amplified equally. In sound AM circuits this presents no particular problem because the two sidebands are identical. In TV receivers, however, the sidebands are different because of the vestigial lower sideband. The lower video frequencies (those close to the carrier) are contained both in the upper sideband and in the remnants of the lower sideband. All video frequencies above 1.25 MHz, however, are present only in the upper sideband because they have been suppressed in the lower sideband during transmission. If the low and the high video frequencies are amplified equally in the IF amplifiers, more low-frequency video voltage will be developed at the video detector output than high-frequency voltage. To prevent this, the IF amplifier response curve shown in Figure 12.7A is generally used. At the video carrier frequency (45.75 MHz), the response of the IF amplifiers is 50% of maximum. This response increases linearly to a maximum for the higher frequencies and decreases for the lower frequencies. Roughly speaking, the lower video frequencies, for which there are two sidebands, are amplified half as much as all video frequencies above 1.25 MHz. In this way, the response for the low and high video frequencies is equalized.

MONOCHROME VIDEO IF AMPLIFIER BANDWIDTH

Figure 12.8A illustrates the signal frequency distribution in a typical monochrome channel. The video carrier is set 1.25 MHz above the lower band edge, and the sound carrier is 4.5 MHz higher (that is, 0.25 MHz from the upper band edge). In the figure, the upper video sideband is 4 MHz wide. The width actually used, however, may vary from 2.5 to 4 MHz.

Marker Frequencies The frequencies that are labeled in Figure 12.7A are called **marker**

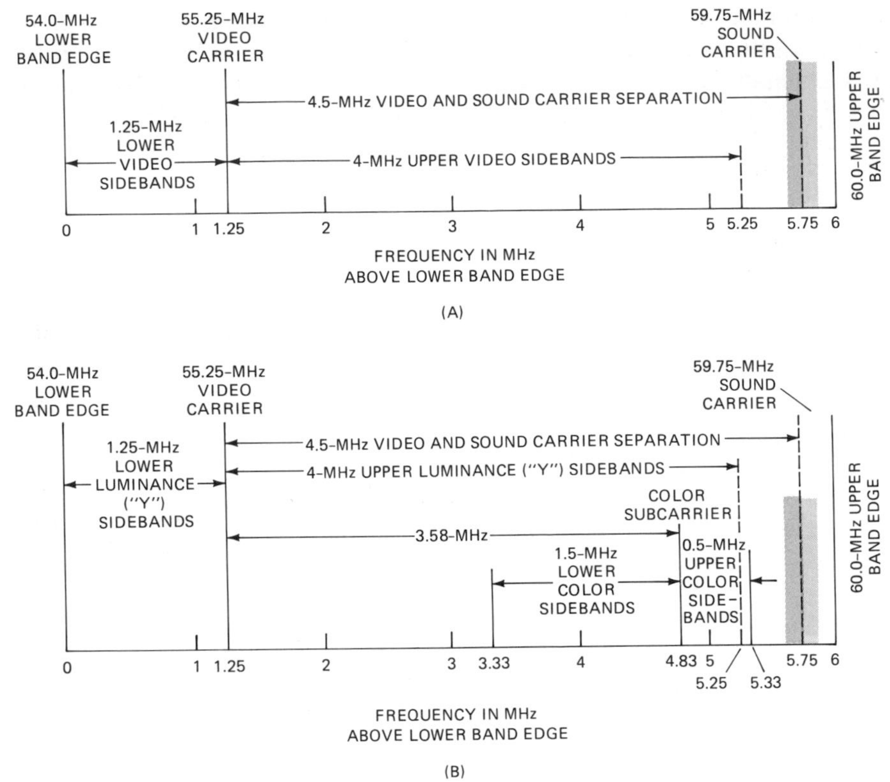

Figure 12.8 The signal frequency distribution for **A.** monochrome and **B.** color transmissions for a typical TV channel (Channel 2).

frequencies. These are the frequencies injected into the IF band during receiver alignment or testing. In this way, the response curve can be observed on an oscilloscope. The markers are used to spot particular frequencies along the curve. These markers are usually produced by crystal-controlled oscillators and are within a few hertz of their intended frequencies.

Aside from the standard frequencies for the sound, video, and color carriers, the marker frequencies can be any value specified by the manufacturer. In fact, there is a wide variation in the actual frequency of markers used, as well as the shape of the response curve. For that reason, for each receiver supplied by the manufacturer, there is a special service bulletin that describes the shape of the curve, how to obtain it, and the marker frequencies to use.

Figure 12.7 shows marker frequencies at 39.75 and 47.25 MHz. These frequencies represent the adjacent channel video and sound carriers, respectively. Special trap networks in the IF amplifier circuits attenuate these frequencies to a low level to prevent interference in the picture. Traps also reduce associated sound carrier to a level that is from 5 to 10% of the total response curve amplitude. At this level, the sound frequencies can be easily removed from the overall IF band and applied to the audio circuits. Also, this prevents these frequencies from reaching the video detector.

Response Curve Shape The shape of the

response curve is important in servicing television receivers. When the IF stages are being aligned, the video carrier must be placed close to the 50% point to compensate for vestigial sideband transmission. At the same time, the circuits must be tuned to provide maximum bandwidth. Detail in a television image is a function of the strength of the high video frequencies present. When the response drops at the upper end of the curve, the fine detail becomes fuzzy and indistinct. Poor low-frequency response gives rise to uneven shading of large areas on the screen, smearing, and a generally darker image.

COLOR VIDEO IF AMPLIFIER BANDWIDTH

The color IF amplifier response curve in Figure 12.7B includes all the frequencies in the monochrome response curve of Figure 12.7A, plus the frequencies associated with the color subcarrier and its sidebands. Note the similarity between the two curves. The critical difference between them is that the color IF subcarrier of 42.17 MHz must be set at the 50% point of the curve opposite the IF picture carrier. In early color sets, the color subcarrier and its sideband frequencies were positioned on the flat portion of the curve, as shown in Figure 12.7C. To accommodate this bandwidth with adequate gain, the early sets required four IF stages. The response curve of Figure 12.7B was adopted to reduce the number of IF stages. This narrower overall bandwidth made it possible to provide the required gain in only three IF stages. The effect of placing the color subcarrier and its sidebands on the slope of the response curve is compensated for in the Chroma amplifier (Chapter 9) by a tuned network that effectively places the color subcarrier and its sidebands on the flat portion of the Chroma amplifier response curve.

Trap networks are also provided in color IF amplifier circuits to suppress the adjacent video and sound carriers. These are shown at 39.75 and 47.25 MHz, respectively, in Figure 12.7B.

12.7 STAGGER TUNING

Achieving the necessary bandwidth in a video IF section with conventionally tuned amplifiers is a difficult and expensive task. A simpler and more economical way to increase the bandwidth of the amplifiers is to tune each stage to a slightly different frequency. This makes the overall response of the amplifiers wider than that of any individual amplifier. This method is known as **stagger tuning**. Because stagger tuning is used extensively in both monochrome and color video IF stages, it is important to understand this method and its effect on bandwidth. First, though, it is necessary to define bandwidth.

BANDWIDTH

A typical resonance curve for a parallel-tuned circuit is shown in Figure 12.9. The response is not uniform but varies from point

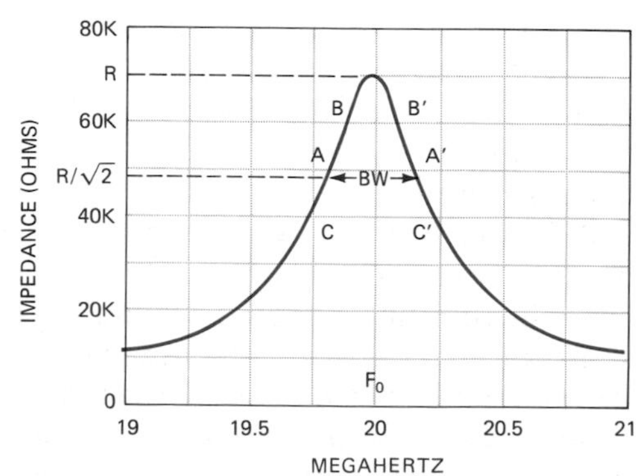

Figure 12.9 The accepted definition for the bandwidth of a tuned system.

to point. At the resonant frequency (labeled F_0 in the diagram), the response of the circuit is at its peak, or maximum impedance. From this point, the response tapers off in both directions until it becomes negligible. With this type of curve, the actual bandwidth is arbitrary. It could be said, for example, that all frequencies between Points B and B′ on the curve are within the bandpass of the circuit. Note that this does not prevent other frequencies—those that receive less amplification—from passing through the circuit.

The arbitrary definition generally accepted for bandwidth is illustrated in Figure 12.9. The **bandwidth** of a circuit is equal to the numerical difference, in hertz, between the two frequencies at which the impedance presented by the tuned circuit is 70.7% of the maximum impedance. Thus, in Figure 12.9, the impedance at Points A and A′ is 70.7% of the impedance at F_0. In this particular illustration, the bandwidth is approximately 0.4 MHz.

A further note of importance is that if the gain of the circuit is considered equal to one at F_0, it is down 3 dB at Points A and A′.

With this concept of bandwidth in mind, let us consider two single-tuned amplifiers, both tuned to the same frequency. If these two amplifiers are in **cascade** (i.e., connected so that the output signal of one is the input signal of the other), then the overall bandwidth is equal not to the bandwidth of either circuit, as might be expected, but to 64% of this value. The reason for the shrinkage in bandwidth will be apparent from the following discussion.

In the response curve of the first amplifier (Figure 12.10A), the amplification is 1 at F_0 and 0.707 at the ends of the bandpass. Assume that the mid-frequency is 45 MHz and that the end frequencies of the bandpass are 44 and 46 MHz. If each of these three frequencies has an amplitude of 1 V at the

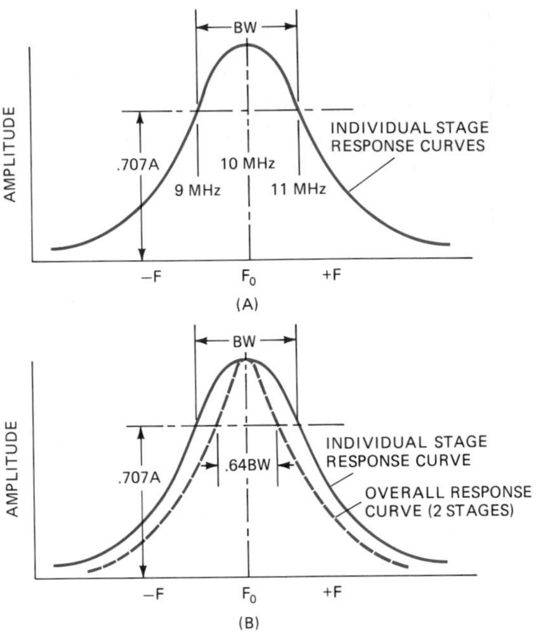

Figure 12.10A-B. Two tuned circuits, both peaked to the same frequency, produce an overall response in which the bandwidth is less than that of either curve taken separately.

input to this tuned stage, then at the output they would possess the following values: 1 × 0.707 = 0.707 V at 44 MHz; 1 × 1 = 1 V and at 45 MHz; and 1 × 0.707 = 0.707 V at 46 MHz.

These same three frequencies are now passed through the second tuned circuit. Since this second circuit possesses the same characteristics as its predecessors, the results at its output are 0.707 × 0.707 = 0.50 V at 44 MHz; 1 × 1 = 1 V at 45 MHz; and 0.707 × 0.707 = 0.50 V at 46 MHz. After passage through the two amplifiers, 44 and 46 MHz are no longer within the 70.7% region about the resonant frequency of 45 MHz. The result, of course, is a narrower bandpass. Actually, the bandpass is 36% narrower, as shown in Figure 12.10B.

Now consider two single-tuned amplifiers, each with the same bandwidth as

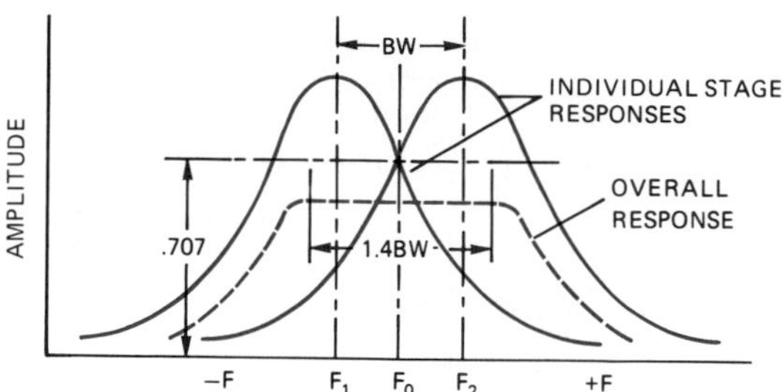

Figure 12.11 By stagger tuning two tuned circuits, we obtain a wider bandpass.

shown in Figure 12.11. The peaks for these two amplifiers are separated (or staggered) by an amount equal to their bandwidth. The result is a response in which the overall bandwidth (to the 70.7% point) is 1.4 times the bandwidth of a single stage. The overall gain, however, is now only one half that of the two stages tuned to the same frequency. This is because, at the center frequency of the overall response curve, the individual responses are only 70.7% of the peak response. The product of the stage gains is 0.50 ($0.707 \times 0.707 = 0.50$). This is what is meant by stagger tuning.

12.8 INTERSTAGE COUPLING

In modern TV receivers, either transformer or impedance coupling is used to connect two or three IF amplifiers to form the video IF system. For solid-state circuits, special lightweight, miniature transformers and coils have been developed for mounting directly on PC boards. These small components can also be used in modular construction and in small-screen, portable TV receivers. Other coupling methods have been tried, but only transformer and impedance coupling provide the tunability required in establishing the video IF bandwidth.

TRANSFORMER COUPLING

The response of a transformer-coupled IF stage depends largely upon the coefficient of coupling between the primary and secondary windings of the IF transformer. The coefficient of coupling of a transformer is a measure of how well it is able to transfer energy from the primary to the secondary. Thus, a coefficient of coupling with a value of 1 means that all of the flux lines of the primary link with all of the turns of the secondary. The equation usually given for the coefficient of coupling is:

$$\text{coefficient of coupling} = \frac{M}{L_1L_2}$$

where

M = mutual inductance between primary and secondary
L_1 = inductance of primary
L_2 = inductance of secondary

The value of the coefficient of coupling is always less than 1. In power transformers, it approaches 0.98 and 0.99. The windings of such a transformer are said to be closely coupled. As indicated by the equation, the greater the mutual inductance, the greater

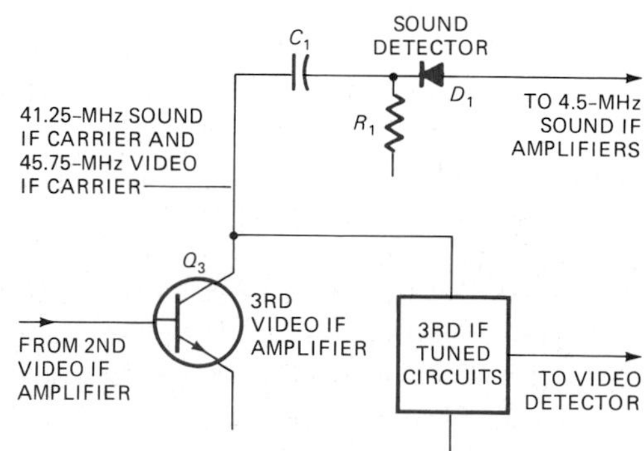

Figure 12.12 The sound separation at the third video-IF amplifier in a color TV receiver.

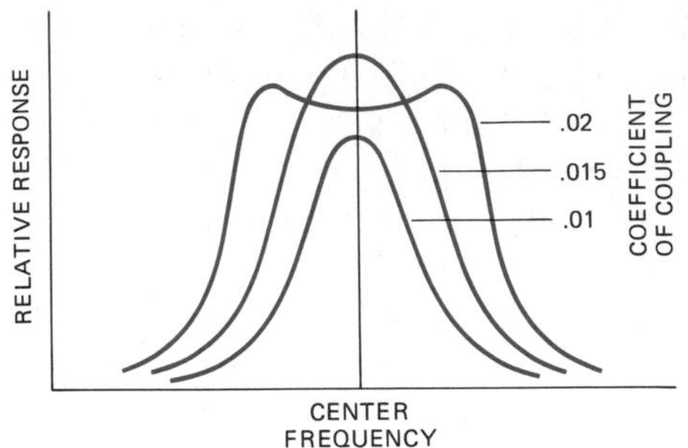

Figure 12.13 Relationship between the coefficient of coupling and the IF response.

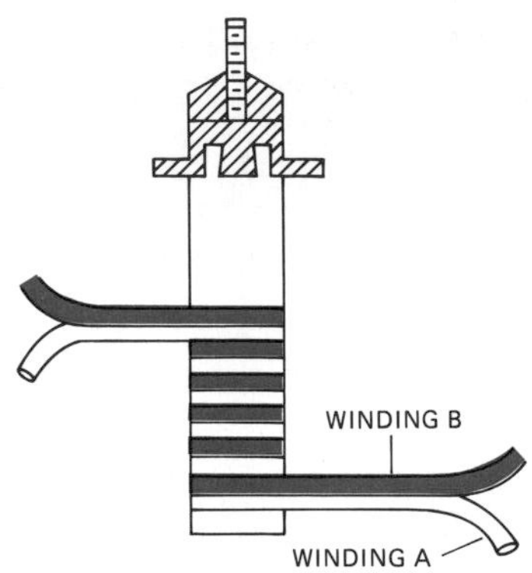

Figure 12.14 A bifilar coil.

the coefficient of coupling and hence the greater the numerical value of the coefficient. Figure 12.12 shows the relationship between the bandwidth and the coefficient of coupling. A coupling value of 0.02 is more desirable than one of 0.015 because the higher value allows a greater range of frequencies to pass.

Bifilar Coil A **bifilar-wound coil** is a type of transformer. It can be used to obtain the necessary bandwidth when transformer overcoupling is desired. As shown in Figure 12.13, a bifilar coil consists of two windings positioned so close together that the coefficient of coupling is nearly 1. The result is that practically all of the voltage developed across the primary is transferred to the secondary. A movable iron core inside the coil tunes both windings simultaneously. In most instances, the tuning rod extends above the top of the chassis so that it can be reached to align the circuit.

IMPEDANCE COUPLING

Figure 12.14 is a simplified diagram of an impedance-coupled vacuum tube IF stage. Impedance coupling is equally effective in solid-state circuits. In the diagram, the input signal is delivered to V_2 via C_1 and R_2. One side of R_2 is connected to the negative AGC source. In the absence of a signal, the AGC voltage drops to zero, so a small cathode resistor R_3 prevents the tube from being operated without bias. The plate load for V_1 consists of a tunable coil L_1. A decoupling filter (C_3 and R_4) prevents any interstage coupling that might occur as a result of the common power supply. The output signal, taken across coil L_2, is delivered to the next stage via C_2 and R_5.

There are many circuit variations possible with impedance coupling. For example, a tunable coil may be used in place of R_5 in the circuit shown in Figure 12.14. Also, the coupling capacitor may be replaced with an LC circuit to make the stage more selective.

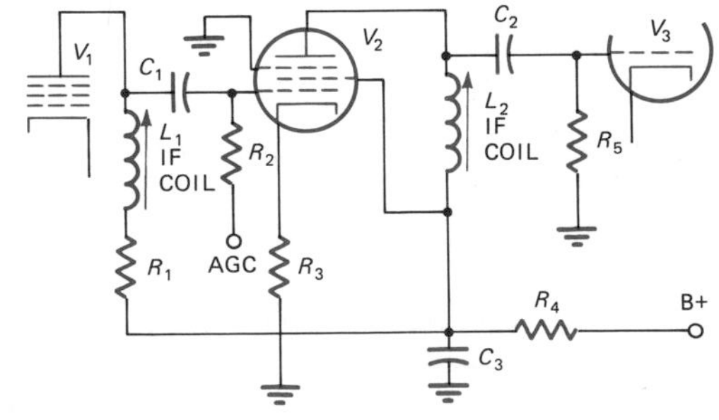

Figure 12.15 An example of impedance-coupled vacuum tube IF amplifier stages.

Figure 12.15 shows a typical impedance-coupling circuit used with solid-state IF systems.

12.9 IF WAVE TRAPS

The video IF band in a TV receiver is susceptible to interference signals. If these signals pass into the video amplifier section, they may distort or destroy the screen image. Fortunately, many of these signals never get beyond the RF tuner. Consequently, they are suppressed before they reach the first IF stage. Some undesirable signals, however, are so close to the selected channel frequency that they pass through the RF tuned circuits and reach the video IF system. The IF circuits present the greatest obstacle to undesired signals. Once these signals pass beyond this portion of the receiver, however, there is little that can be done to remove them. Every effort must be made to suppress all interference signals before they reach the video detector.

To receive a 6-MHz band of frequencies, the RF and mixer tuning circuits are designed with a low Q. This means that the sides of the input response curve are not very steep, but taper off gradually. Figure 12.16 illustrates the typical input response curve of a modern TV receiver. With this type of response curve, voltages at the frequency of the sound carrier of the next lower channel or the video carrier of the next higher channel may penetrate the RF stages and reach the video IF amplifiers. Unless trap circuits are inserted in the video IF system, these interference signals will affect picture quality.

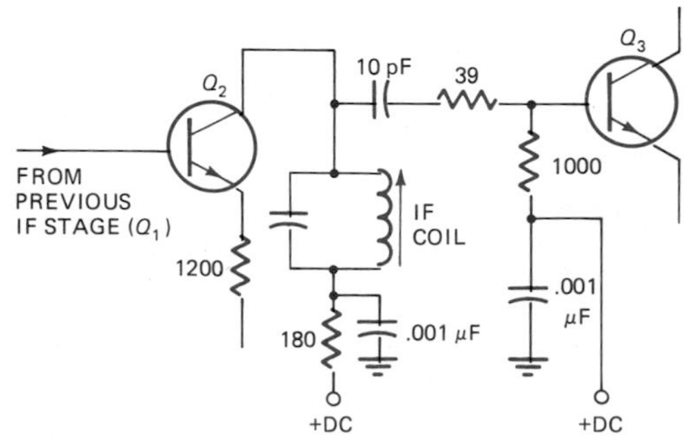

Figure 12.16 Typical impedance-coupled IF stages used in a solid-state color TV receiver.

INTERFERENCE FREQUENCIES

To eliminate interference signals (by filters or traps), it is first necessary to determine their frequencies. Sometimes this can be done by simple subtraction, based on the assumption that the video or sound carrier frequency of adjacent channels is causing the interference.

Suppose that the receiver is tuned to Channel 3 (60 to 66MHz) and the video carrier IF is 45.75 MHz. The sound carrier of the next lower channel (Channel 2, 54 to 60 MHz) is at 59.75 MHz. A signal at this frequency, when mixed with the local oscillator (107.00 for Channel 3), will develop a difference frequency of 47.25 MHz. This is the frequency of one interference signal (adjacent channel sound).

Another possible interference signal is the picture carrier of the next higher channel (in this example 66 to 72 MHz). When this signal beats with the local oscillator, a difference frequency of 107.00 − 67.25 = 39.75 MHz is produced (adjacent channel picture).

In all cases where adjacent channels exist, the two interfering frequencies will be 47.25 and 39.75 MHz for TV receiver with a 45.75-MHz video IF. There are, however, channels that are not subject to adjacent channel interference: Channel 2 (54 to 60 MHz) does not have an adjacent lower channel and Channel 4 does not have an adjacent higher channel. Remember that *adjacent* as used here means channels that follow each other without any frequency separation.

Channel 4 is followed by Channel 5, but the frequency of Channel 4 is 66 to 72 MHz, and that of Channel 5 is 76 to 82 MHz. The 4-MHz separation is sufficient to prevent any of the frequencies in Channel 5 from adversely affecting Channel 4. Channel 2, however, is closely followed by Channel 3, and so interference is possible. The same is true of many of the other channels. For this reason, the use of traps is important. The trap frequencies vary according to the video and audio IF values of the receiver. The purpose of the traps, however, remains unchanged.

Adjacent Channel Traps Adjacent channels are not assigned to TV stations broadcasting the same general area. There are cases, though, where adjacent channels are assigned to stations in nearby areas. As an example, consider New York and Philadelphia, which are only 90 miles (145 km) apart. New York is assigned VHF Channels 2, 4, 5, 7, 9, 11, and 13, and Philadelphia is assigned VHF Channels 3, 6, 10, and 12. Any TV receiver situated between these two cities is certainly subject to considerable interference and definitely requires trap circuits. This same situation occurs in many other parts of the country. Whether or not a receiver contains adjacent channel traps is largely a matter of design and economics. Some receivers have traps for only the signal from the lower or the higher adjacent channel; some have traps for both; and some have traps for neither.

Sound Traps for Color TV Receivers Sound traps are always used in color TV receivers because of the narrow separation between the sound carrier and the color subcarrier. This interference signal has a frequency of 920 kHz and is practically impossible to filter from the detected video signals. For that reason, sound traps in the video IF section reduce the sound carrier level to approximately 5% of the video IF amplitude. After the sound is separated from the video IF band, another sound IF trap further reduces the sound signals to a very low level. This prevents them from appearing in the video detector. In addition there may be a 4.5-MHz sound IF trap following the video detector.

TYPES OF TRAPS

Basically, there are five types of traps used with video IF amplifiers: series, parallel, absorption, degenerative, and bridged-T traps. For the most part, they are all used in both vacuum tube and solid-state circuits for either monochrome or color reception. The bridge-T trap is the most widely used circuit and is indispensable in color TV receivers.

Series Traps The **series trap** is a parallel, resonant-circuit configuration, as shown in Figure 12.17A. It is placed between two IF stages and is tuned to the frequency to be rejected. This type of trap circuit is a sharply tuned network designed to reject only one frequency or, at most, a narrow band of frequencies. When a signal voltage at the trap frequency appears at the input of the circuit, the impedance offered by the LC tank circuit is very high relative to the load impedance. Almost all of the undesired volt-

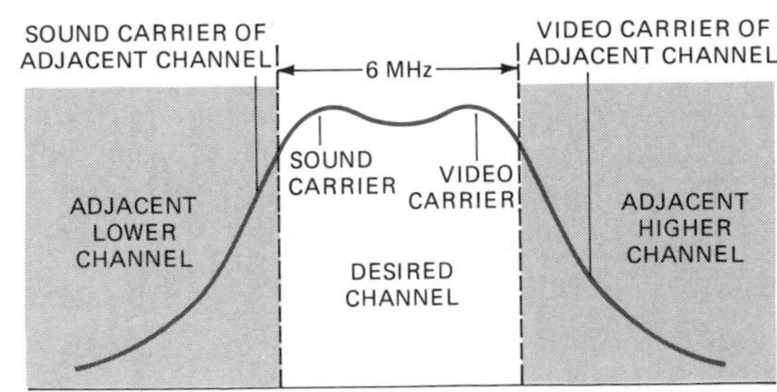

Figure 12.17 The RF response curve of most television receivers. Note that signals from adjacent channels can be received.

age is dropped across the trap network. A negligible amount of the voltage appears across the input circuit of the following IF amplifier. At all frequencies except the one it is tuned to, the trap circuit offers negligible impedance and the desired signals pass easily.

Parallel Traps **Parallel traps** are tuned circuits that are placed across, or in shunt with, the amplifier circuit. Figure 12.17B shows a series-resonant circuit used in this manner. At the frequency for which it is set, the trap acts as a short circuit, directing the resonant frequency to ground and preventing it from penetrating into the circuit. At other frequencies, the trap circuit presents a relatively high impedance, permitting these signals to proceed to the following stage. It is important that the parallel trap have a very high Q so that the circuit will bypass only a narrow range of frequencies.

A variation of the parallel trap is shown in Figure 12.17C. In this circuit, L_1 and C_2 form a parallel resonant circuit that is tuned to 42.25 MHz. The Q of the coil is 200. A fairly large voltage is developed across the circuit at this frequency. For all frequencies lower than its resonant frequency, a parallel resonant circuit appears inductive. At the resonant frequency, of course, it presents a purely resistive impedance. For frequencies above resonance, the impedance presented by L_1 and C_2 is capacitive. Note that for the higher frequencies, the parallel capacitor offers less impedance than the coil. Consequently, most of the current flows through the capacitor and the circuit current has a leading phase. Since the sound carrier, at 41.25 MHz, is below the 42.25-MHz resonant frequency of L_1 and C_2, the parallel combination appears inductive to the carrier. By resonating this inductance with C_1, a series-resonant path for the sound carrier is obtained and the carrier is bypassed to ground.

When a parallel resonant circuit (L_1 and C_2) is provided for 42.25 MHz, a sharp rise in voltage is obtained just beyond 41.25 MHz on the IF response curve. Since the 42.25-MHz value is included in the range of desired video frequencies (which extend from 45.75 MHz down to 41.75 MHz), all of the desired video frequencies are passed by the trap with negligible attenuation while at the same time the undesired sound IF carrier is suppressed.

Absorption Traps The **absorption trap** (Figure 12.18D) is a widely used type of rejection circuit. It consists of a coil (L_2) and a fixed capacitor (C_2) inductively coupled tightly to the load inductor (L_1) of an IF amplifier. When the IF amplifier receives a signal at the resonant frequency of the trap circuit, a high circulating current develops in the trap network as a result of the coupling between the trap and the load inductor. The voltage across load coil L_1 becomes quite low at the trap frequency. Consequently, very little of this interference voltage is permitted to reach the following stage. It is convenient to think of this kind of trap as being able to absorb all of the energy of the frequency to which it is tuned. Therefore, no energy at that frequency is left available to pass into the next stage. Absorption traps are also called suckout traps.

How absorption traps work can also be explained using fundamental transformer theory, with L_1 acting as the primary winding and the trap network (L_2 and C_1) acting as the secondary. Two resonant circuits closely coupled will produce a double-humped curve, such as the one shown in Figure 12.17E. Note the sharp decrease in the primary current at the center frequency. In the case of the two tuned circuits of Figure 12.17D, L_1 is tuned to the desired band of frequencies and the trap, or secondary, is tuned to the undesired frequency. Since the primary band coverage includes the undesired frequency to which the trap is tuned,

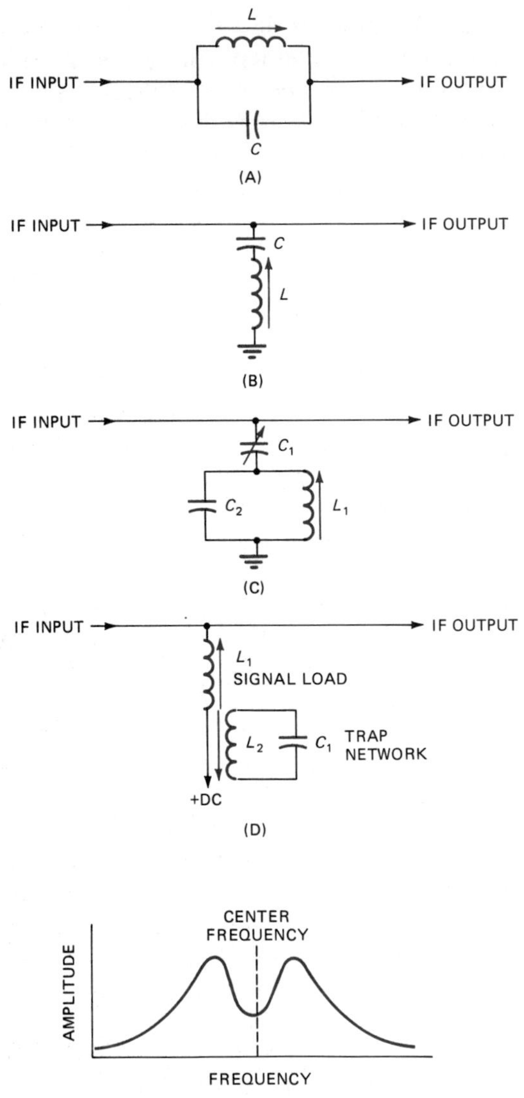

Figure 12.18A-E. Video IF-band traps. **A.** Series trap; **B.** parallel, or shunt trap; **C.** variation of the parallel trap; **D.** the absorption trap; **E.** closely-coupled resonant circuits' response curve.

there is a sharp drop in primary voltage at that frequency. It is this interaction between the primary and secondary coils that produces the marked decrease in voltage at the trap frequency. The other frequencies in the IF band are unaffected by the trap.

Degenerative Traps **Degenerative traps** (Figure 12.19) reduce the gain of an amplifier for those frequencies to which the trap is tuned. These traps are used in the emitter circuit of a solid-state amplifier or in the cathode leg of a vacuum tube circuit. In the latter application, the traps are often called *cathode traps.* The two types of traps normally used to provide degeneration are the absorption type and the series type. Figure 12.19A shows an absorption trap in which coil L_1 in series with C_1 forms a broadly tuned series-resonant circuit at the frequency to which the amplifier is tuned. This permits the amplifier to function normally for all signals within its frequency range. At the resonant frequency of the critically coupled trap, however, a high impedance is reflected into the emitter or cathode circuit by the trap and the gain of the stage is reduced by degeneration.

The series type of degenerative trap (Figure 12.18B) places a parallel circuit directly into the emitter or cathode leg. At the resonant frequency of the trap, the impedance in this part of the amplifier circuit is high. This produces a large degenerative voltage and thus reduces the gain of the amplifier. At all other frequencies, the imped-

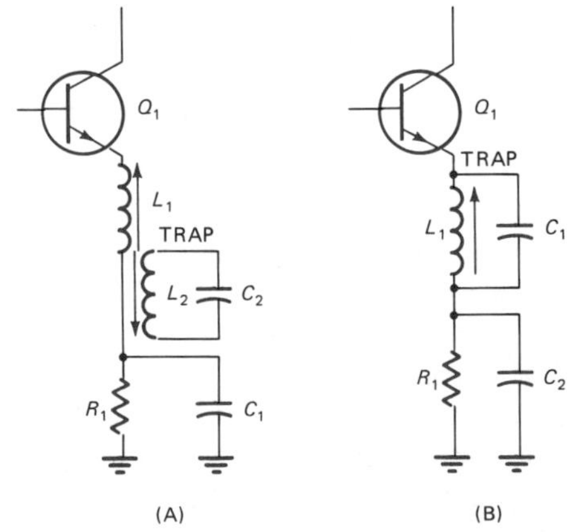

Figure 12.19 Degenerative traps. **A.** Coupled type; **B.** series type.

ance of this parallel network is low and there is only a small degenerative voltage. Therefore, only a slight loss in gain occurs except at the undesired frequency.

Bridged-T Traps A trap that is more complex than any of the foregoing, but also more effective, is the **bridged-T trap** (Figure 12.19A). In this circuit, L_1, C_1, and C_2 are resonated at the frequency of the signal to be rejected. Now, if the resistance of R is chosen properly, a large attenuation will be imposed upon a signal to which L_1, C_1, and C_2 are resonated. Ratios of 50 to 1 and 60 to 1 are easily attainable using standard components. This means that the strength of the desired signal at the output of the trap will be 50 to 60 times greater than the strength of the undesired signal.

Understanding the operation of this trap circuit can be simplified by transforming the bridged-T network shown in Figure 12.20B into the equivalent network shown in Figure 12.19C. This is called a **delta-wye transformation** and can be accomplished with well-known electrical theorems. If L_1, C_1, and C_2 of the bridged-T network are chosen properly, Z_1 will have a negative value. If R is made equal to Z_1, then the total impedance between Points 1 and 2 will become zero. This effectively short-circuits signals of the frequency to which the network is tuned. For all other frequencies, the bridged-T network offers negligible attenuation.

12.10 SOUND IF-FREQUENCY SEPARATION

Two basic methods have been used to separate the program sound from the video IF band. One method often used in early TV receivers was a split-sound system in which the sound was taken from the output of the mixer stage. Filters (or traps) then prevented the sound carrier from entering the video IF band. The disadvantages of this method are that critical adjustment of the fine-tuning control was required, sound degradation often resulted because of oscillator drift, and it was expensive.

The second method (the only one used today) is known as the **intercarrier sound system.** It has none of the disadvantages of the split-sound system. In the intercarrier method, the sound frequencies are allowed to pass through all the video amplifiers, but at a much reduced level. The program sound is separated in the video detector by beating the sound carrier against the video carrier, producing a 4.5-MHz sound IF. Filters, or traps, in the video amplifier section then separate the 4.5-MHz sound IF from the video signals. The sound IF can be filtered in the video amplifier section because the highest video frequency is 4.0 MHz, whereas the sound IF is 4.5 MHz.

For color TV, the intercarrier sound system had to be modified so that the sound IF carrier could be removed before it reached the video detector. If this carrier were allowed to reach the video detector, it would beat against both the video carrier and the color subcarrier. The result would be the development of two sound IF frequencies: one at 4.5 MHz and the other at 920 kHz. It would then be impossible to filter the 0.920-MHz component from the video signals. This would cause an interference pattern on the screen. Monochrome receivers use the intercarrier sound system and separate the sound at the video detector. Color TV receivers also use the intercarrier sound system, but remove the sound prior to the video detector.

Figure 12.20 is a simplified diagram of a typical sound-takeoff circuit in a color receiver. Even though the circuit is shown as part of a solid-state IF system, it is also used in vacuum tube systems. In the diagram, the 41.25-MHz sound carrier and the 45.75-MHz video carrier are coupled through C_1 to the

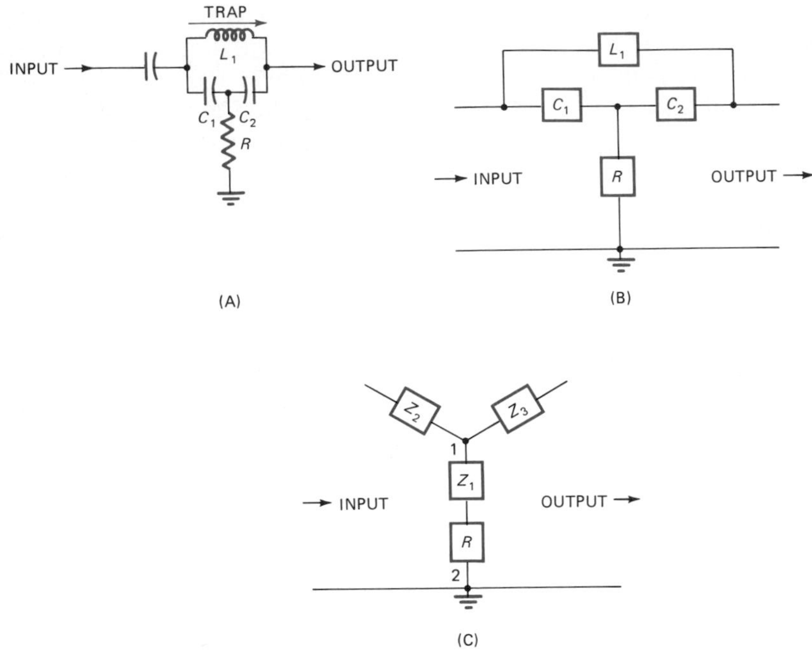

Figure 12.20A-C. The bridged-T trap. **A.** The schematic diagram; **B.** the bridged-T network in block form; **C.** the equivalent delta-wye transformation.

sound detector D_1 from the collector of the third IF amplifier Q_3. D_1 detects the 4.5-MHz difference between the two carriers and feeds it into the sound IF section.

12.11 TYPICAL VIDEO IF AMPLIFIERS

As stated previously, the main functions of the video IF amplifiers are to provide selectivity and gain. It was also pointed out that monochrome and color video IF amplifiers are the two basic types and that each of these types can have vacuum tubes or solid-state components. With these facts in mind, we can examine several typical amplifier circuits. There is actually very little basic difference in the various amplifier circuits. The most important fact to remember is that color IF circuits are more critical and therefore contain more traps or filters. Also, in color circuits the sound signals are removed prior to the video detector.

VACUUM TUBE COLOR IF AMPLIFIERS

Figure 12.21 presents a typical vacuum tube, transformer-coupled video IF system for a color TV receiver. In the diagram, each stage is shown as a separate tube. However, some manufacturers use two compactrons, with the first and second IF amplifiers in one and the third IF amplifier and a sound and synchronization amplifier in the other. Many of these circuits also use bifilar wound coils instead of transformers.

The video IF amplifier system shown in Figure 12.21 is coupled to the mixer stage of the tuner through a 41.25-MHz, series-reso-

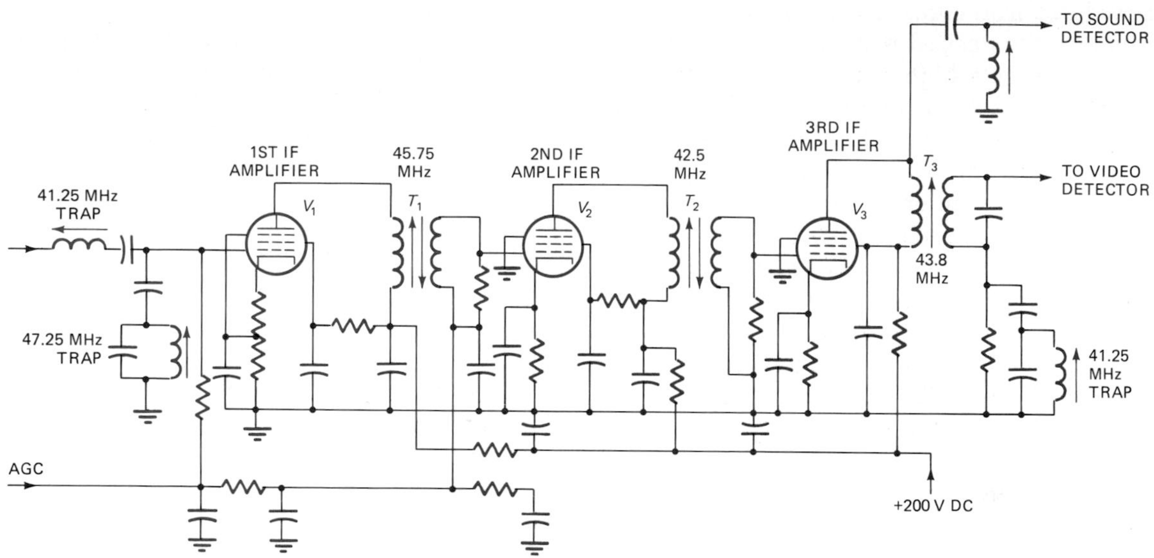

Figure 12.21 A vacuum-tube, transformer-coupled color-video IF system.

nant sound trap. This trap reduces the associated sound carrier to the required 5% level. This makes possible the intercarrier method of IF amplification. There is also a 47.25-MHz trap connected to the input of V_1 to reject the adjacent channel sound carrier. Some chassis also provide a 39.75-MHz trap at this point to reject the adjacent channel video carrier.

From the plate of V_1, the amplified video signals are coupled to the grid of V_2 via transformer T_1. This transformer is tuned to 45.75 MHz and helps place that frequency at the 50% point on the IF response curve. Transformer T_2 couples the output of V_2 to the input of V_3. By tuning T_2 to 42.5 MHz, the 50% point on the opposite side of the response curve is established and places the color IF subcarrier of 42.17 MHz at that point.

The plate circuit of V_3, the third video IF amplifier, feeds a portion of the entire IF band voltage into the sound detector circuit. Here, the sound IF carrier is beat against the video carrier to produce a 4.5-MHz sound carrier. Filters in the sound detector circuit then remove any color subcarrier frequencies, as well as the 920 kHz frequency that results when the sound carrier beats against the color subcarrier. Transformer T_3, also in the V_3 plate circuit, delivers the amplified IF voltage to the video detector. This transformer is then tuned to 43.8 MHz to flatten out the IF response curve near the center of the bandpass. A trap connected to the detector side of T_3 is tuned to 41.25 MHz to reject all traces of the associated sound carrier.

AGC is applied only to the first two stages in this circuit. There are, however, some circuits in which AGC is applied to all three stages, but this practice is not popular. About 60 dB of AGC control is usually desired. This value can be achieved easily with two stages.

Stagger-Tuned Circuits The interstage tuning circuits commonly used in stagger-tuned IF systems (Figure 12.21) are transformers, single coils, or bifilar-wound coils. Each transformer or coil contains an adjustable powdered iron core. Thus, the circuit can be tuned to the desired resonant frequency. Because relatively high frequencies are used in the IF band, actual capacitors are not always connected across each coil, although capacitance is present. This capacitance is a combination of the inherent capacitance between the turns of the coil plus the capacitance that the vacuum tubes (or transistors) and the connecting parts and wires contribute.

It is a common practice, when using single coils in a staggered arrangement, to place them in the plate circuit and use a coupling resistor in the following grid circuit. In short, each plate coil and the following grid resistor are in parallel. Low-valued resistors broaden the response of the tuning coils. This arrangement is needed to achieve an overall bandpass of 2.5 to 4 MHz when the individual responses of all the coils are combined.

SOLID-STATE COLOR IF AMPLIFIER

Figure 12.22 is a schematic diagram of a solid-state video IF system used in a color chassis. This system has three NPN transistor stages that are impedance-coupled and pass a band of frequencies centered at 43.8 MHz. Input to the IF system is through a plug-in coaxial link from the mixer stage in the tuner. The mixer output coil is tuned so that 42.17 MHz is at the 50% point on the IF response curve. The impedance of the coil is tapped to match the impedance of the coaxial link. The first IF amplifier (Q_1) is coupled to the input coaxial link through LA_2 and CA_1, and LA_2 is tuned so that 45.75 MHz is at the 50% point on the opposite side of the response curve from the 42.17 MHz point. Both 50% points are shown in Figure 12.7B. The mutual coupling of LA_2 and the mixer output coil provides a wide bandpass through the coaxial link.

Sound Traps Two sound traps are placed at the input to Q_1. One of these traps is tuned to reject the adjacent channel sound frequency at 47.25 MHz, and the other is tuned to suppress the 41.25-MHz associated sound carrier. The 47.25-MHz trap consists of RA_6, CA_3, CA_4 and LA_5. This trap is a bridged-T configuration and delivers to the base of Q_1 two 47.25-MHz voltages that are equal in amplitude but 180° out of phase. One of these voltages is developed across RA_6 and the other is developed across CA_3, CA_4, and LA_5. At the base of Q_1, the voltages cancel each other. This eliminates the adjacent channel sound carrier from the IF band. The 41.25-MHz trap is a series-resonant network consisting of CA_7, CA_{73}, and LA_9. This trap is loosely coupled to the input circuit. It improves skirt selectivity and provides better fine tuning.

AGC Operation In a solid-state color IF amplifier, AGC is applied only to the first IF amplifier stage. As the AGC voltage increases, forward bias is increased at the base of Q_1. The increased forward bias causes a reduction of amplifier gain. This effect is called **forward AGC** and is discussed further in Chapter 14.

Circuit Operation In the collector circuit of Q_1, it is CA_{21} and CA_{22} that divide the voltage developed across LA_{20} and couple it to the base of Q_2. The junction of the two capacitors matches the impedance of the coil to the base of Q_2. LA_{20} is tuned to 43.8 MHz, which is the center of the IF system passband. Coupling between Q_2 and Q_3 is identical to that between Q_1 and Q_2. Inductor LA_{28} is also tuned to 43.8 MHz. The associated sound carrier is tapped off at the collector of Q_3, the third IF amplifier.

The output circuit of the IF system is very similar to the input circuit. Here, LA_{39} is tuned so that 42.17 MHz is at the 50% point on the IF response curve (Figure 12.7B)

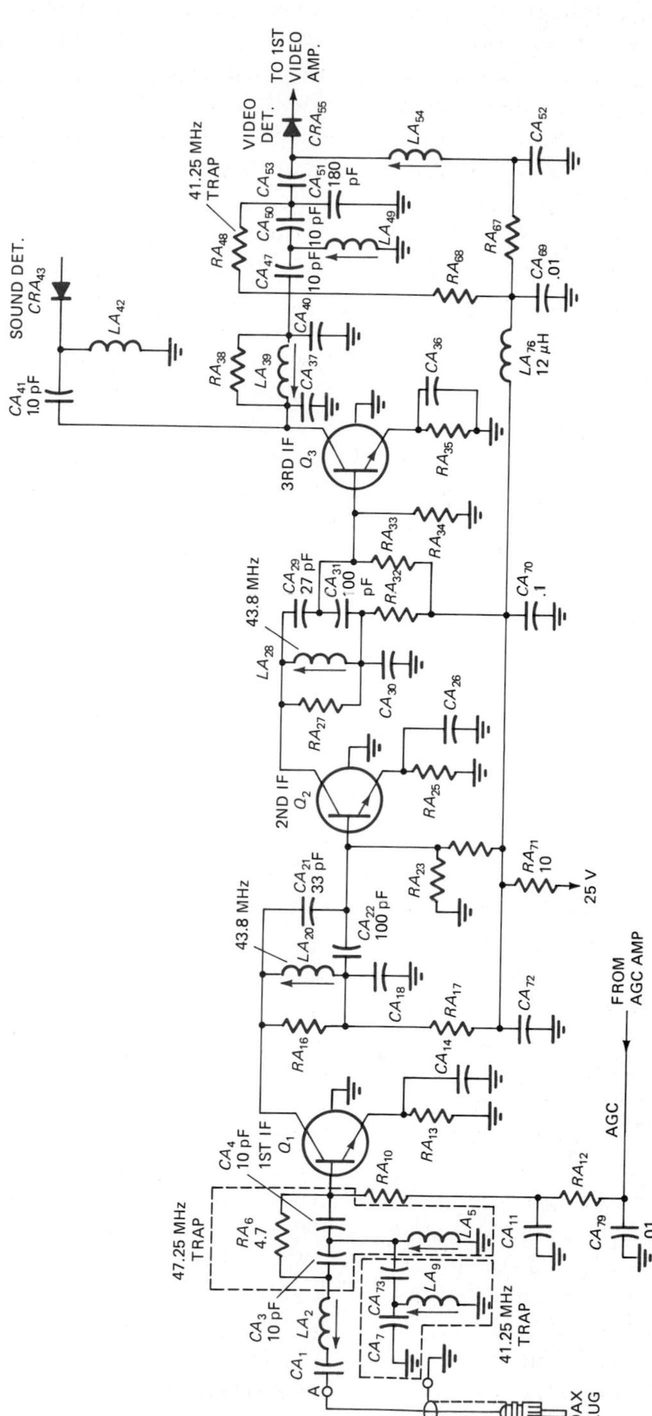

Figure 12.22 A solid-state video-IF system for a color TV receiver. Unless otherwise noted, capacitor values are in μF.

and LA_{54} positions 45.75 MHz at the same point on the opposite slope. A bridged-T trap consisting of RA_{48}, CA_{47}, CA_{50}, and LA_{49} rejects the 41.25-MHz associated sound carrier, and this prevents it from reaching the video detector. The three sound traps in this system, two at 41.25 MHz and one at 47.25 MHz, account for most of the selectivity in the IF system, in addition to suppressing interference.

12.12 AN IC VIDEO IF AMPLIFIER

Integrated circuits are widely used in television receivers. They form a complete section of the receiver in one compact package. Some typical IC sections are (1) picture IF, (2) sound IF, (3) AFT, and (4) chroma. A typical quad-in-line plastic package (QUIP) IC unit is shown in Figure 12.23.

BLOCK DIAGRAM

A block diagram of a single IC used for the complete IF system of a television receiver is shown in Figure 12.24. This unit can be used with either a color or a monochrome receiver. The IC is contained in a shielded, quad-formed, dual-in-line, 20-lead, plastic package similar to the one shown in Figure 12.23. The IC is divided into two sections, ICIA and ICIB. ICIA includes the first picture IF amplifier and the tuner AGC delay circuitry. The term *AGC delay* means that AGC action will not begin until a fixed DC bias is overcome by the AGC voltage. (For a discussion of AGC, see Chapter 14) ICIB includes all circuits not in ICIA.

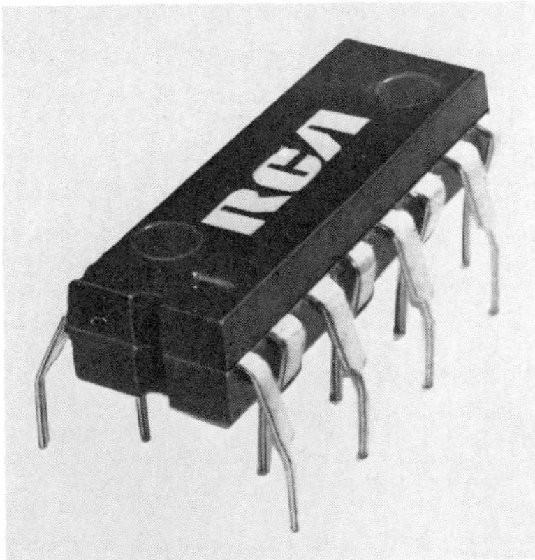

Figure 12.23 A typical IC unit used in television receivers. This unit is a Quad-In-Line Plastic Package (QUIP). The unit is approximately 1″ (2.54 cm) long and ¼″ (0.64 cm) thick. *(Courtesy of RCA Solid State Division.)*

Although this chapter is primarily concerned with picture IF amplifiers, it should be noted that the IC also contains some sound IF circuits, a video preamplifier, an AGC amplifier, and a zener voltage regulator. A complete list of the receiver functions performed by the IC is as follows:

1. Video IF amplification
2. Linear video detection
3. Video amplification
4. Detected video amplification
5. AGC circuits
6. AGC delay for the tuner RF amplifier
7. Output signal to drive AFT circuits
8. Sound IF carrier (4.5-MHz) detection
9. Sound IF carrier (4.5-MHz) amplification
10. Zener reference diode for voltage regulation

FIRST IF STAGE

In Figure 12.24, the IF input signal from the tuner passes through the input-tuned circuits (discussed later) and is applied via Terminal 6 to the first picture IF amplifier in ICIA. This is a cascode-type amplifier. The output of this cascode IF amplifier is fed through Terminal 9 to the interstage bandpass circuit. This circuit includes a 41.25-MHz sound trap.

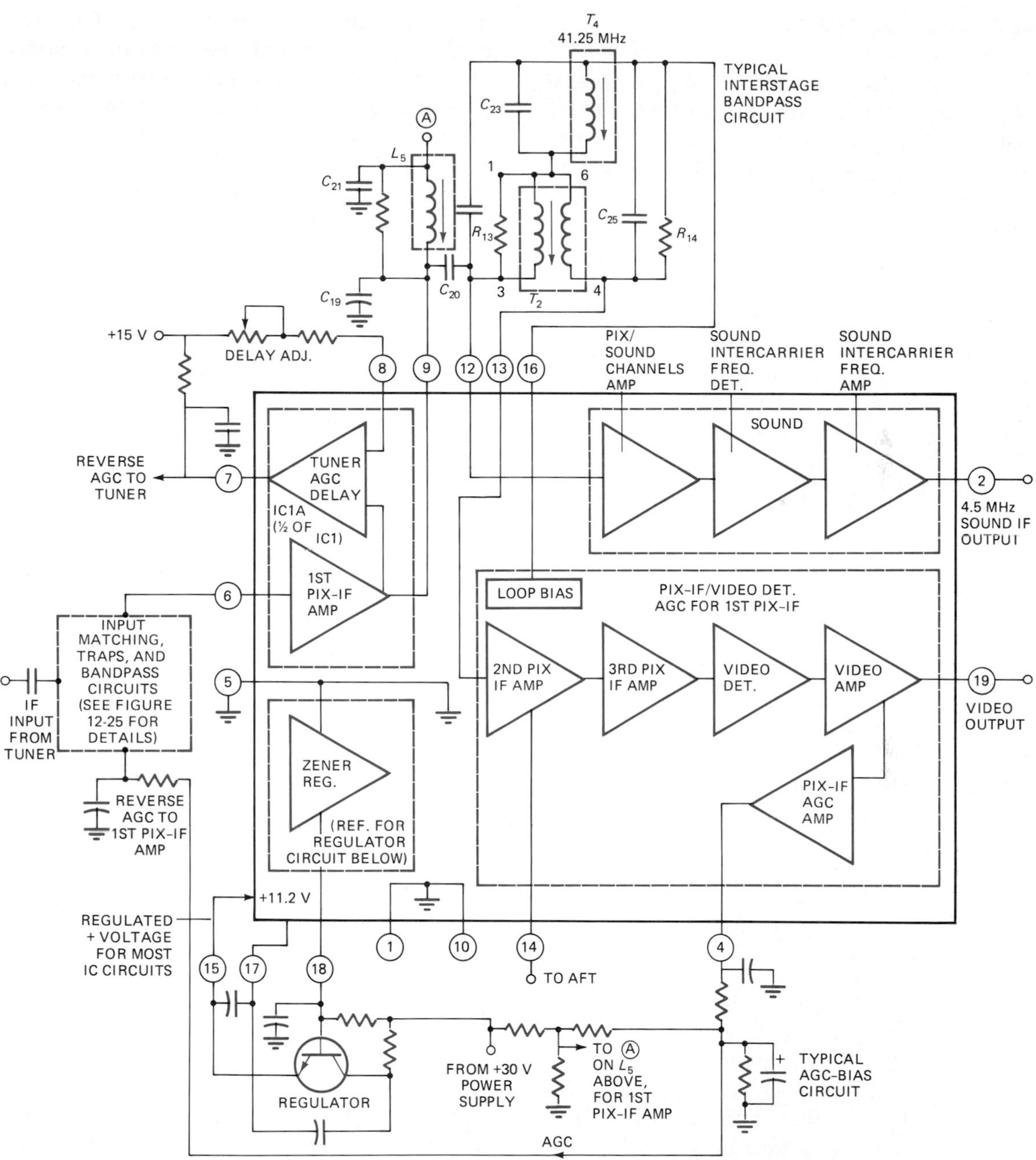

Figure 12.24 Block diagram of single IC used for the complete IF system. The IC is within the heavy border. The first picture-IF amplifier and the tuner-AGC delay are IC1A. The balance is in IC1B. Correlate with the schematic diagram of Figure 12.25. (See text.) Note that all tuned circuits, biasing networks, and the power supply are external to the IC. *(Courtesy of RCA Solid State Division.)*

SOUND TAKEOFF

Unlike the systems discussed previously, both the picture and sound IFs are picked off at the output of the first picture IF amplifier. They are fed via Terminal 12 to a separate wide-band (no tuned circuits) picture and sound channels amplifier. The 41.25-MHz bifilar-T trap removes this frequency at Terminal 13. This prevents it from passing through the second and third picture IF amplifiers and the video detector. This prevents the formation of a 920 kHz beat with the 42.17-MHz chroma subcarrier. This beat would cause an interference pattern on the color picture tube.

The amplified picture and sound IF are then fed to the sound intercarrier frequency detector. The output of this stage is the 4.5-MHz intercarrier sound IF. This output goes through a wide-band (no tuned circuits) sound intercarrier frequency amplifier. The amplified 4.5-MHz intercarrier sound IF is then fed to an FM detector, audio amplifiers, and the loudspeaker(s) (not shown).

SECOND AND THIRD IF STAGES

The output of the first picture IF stage, as mentioned previously, passes via Terminal 9 to the interstage bandpass circuit. This is a double-tuned bandpass circuit with the bifilar 41.25-MHz T-trap discussed above. At Terminal 13, the output, which consists of the picture carrier and chroma subcarrier, is amplified by two wide-band picture IF stages and the output signal is applied to the video detector. The maximum gain at picture IF frequencies is approximately 75 dB.

VIDEO DETECTOR AND VIDEO AMPLIFIER

The linear detector extracts the picture and chroma information from their respective IF carriers and feeds it to the video amplifier. This amplifier has a gain of about 12dB. The output of this amplifier is then fed to the video and chroma circuits of the television receiver.

BANDPASS SHAPING

The picture IF bandpass is shaped by two sets of tuned circuits (described in more detail later, in conjunction with Figure 12.25). These are (1) the input triple-tuned circuit, which feeds into Terminal 6 and the first picture IF amplifier, and (2) the interstage bandpass circuit between the first and second IF stages. As mentioned previously, the second and third picture IF amplifiers have no tuned circuits. They are wide-band amplifiers that pass, undistorted, the correctly shaped picture IF bandpass from the preceding tuned circuits. These proceding circuits include all necessary traps.

LOOP BIAS AND AGC

Loop bias is fed through Terminal 16, and through the bandpass circuit to the input stages of the amplifiers connected to Terminals 12 and 13.

The AGC voltage developed in the IC is applied to the first picture IF stage. This is done by an external path from Terminal 4 through the input networks to Terminal 6 and to the input of the first picture IF amplifier. An AGC output from the first picture IF stage is also fed to the tuner AGC delay circuit and to the tuner RF amplifier.

IC APPLICATION IN A COLOR TV IF SYSTEM

Figure 12.25 shows a schematic diagram of a typical application of the IC device in a color TV IF system. (This is the same device shown in Figure 12.24.)

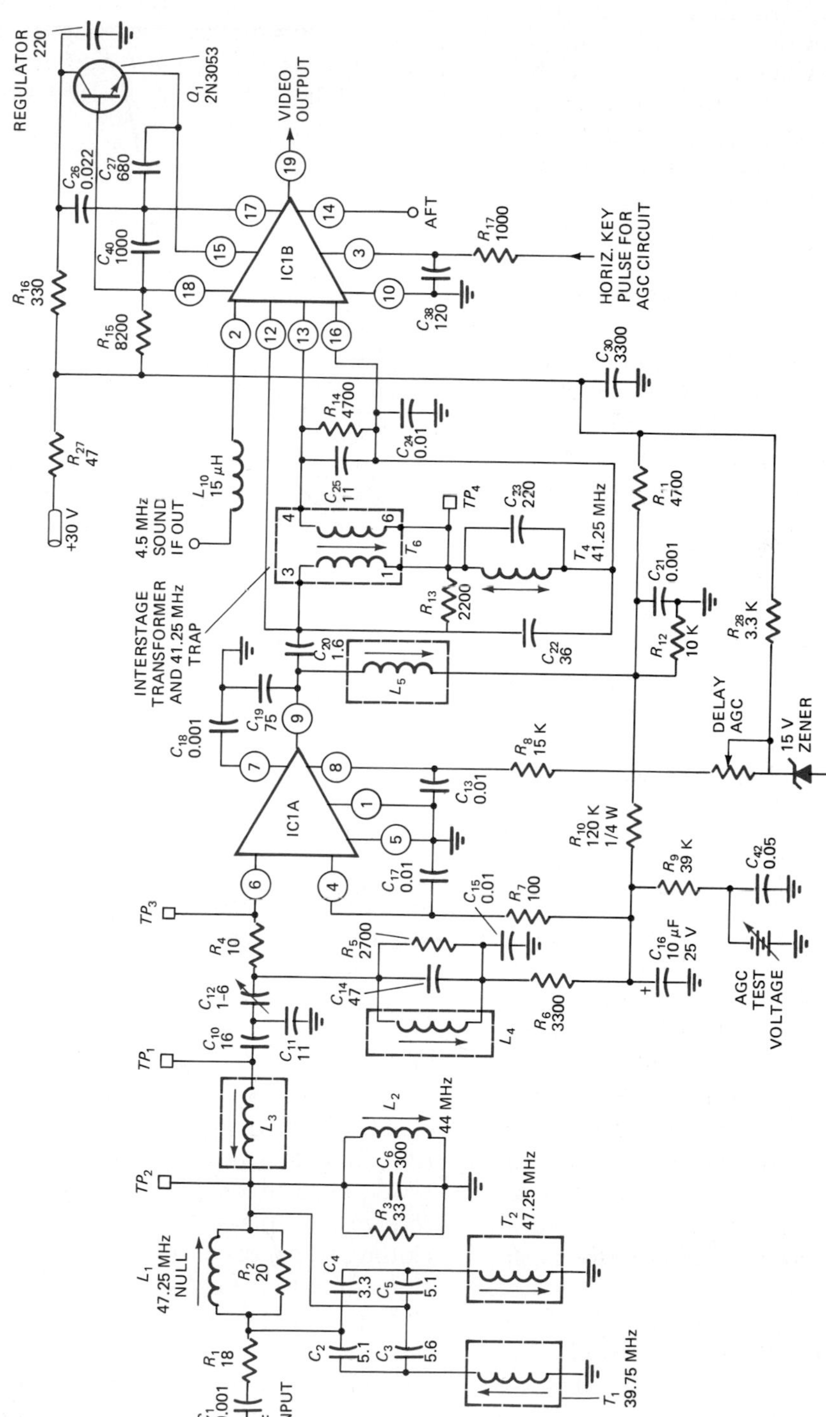

Figure 12.25 Schematic diagram showing a typical application of the IC unit of Figure 12.24 in a color TV IF system. Capacitor values less than 1 (except C_{16}) are in μF; those more than 1 are in pF. *(Courtesy of RCA Solid State Division.)*

The Input Circuit The input circuit plays an important part in IF bandpass shaping, as mentioned before. This is the circuit between the IF input from the tuner and Terminal 6. Two bridge-T traps at the beginning (left side) of the input circuit provide the desired attenuation of the adjacent channel picture carrier at 39.75 MHz and the adjacent channel sound carrier at 47.25 MHz. These traps include coils L_1, T_1, and T_2. The bridge impedance consisting of L_1 and R_2 is common to both traps. Coils L_2, L_3, and L_4 and trimmer capacitor C_{12} are adjusted both for proper bandwidth and to place the picture IF carrier (45.75 MHz) and the chroma IF carrier (42.17 MHz) at the 50% response points. Coupling from L_2 to L_4 is provided by coil L_3 and capacitors C_{10}, C_{11}, and C_{12}.

The Interstage Circuits The interstage circuit is composed of coil L_5, transformer T_6, and the 41.25-MHz associated channel sound trap. The function of the trap was explained previously. Coils L_5 and transformer L_6 are adjusted to provide a symmetrical IF response curve. The response of the interstage circuits alone is shown in Figure 12.26A. The overall IF response is shown in Figure 12.26B.

Video Detector The video detector detects video signals with a minimum of distortion. It contains no tuned circuits. The detector also has the advantage of detecting the chroma subcarrier (3.58 MHz) without introducing phase errors as a function of the video signal. This is a shortcoming of certain other detector designs.

Video Amplifier The video detector is direct coupled to the video amplifier. This provides excellent low-frequency and phase response. The circuit is designed to limit impulse noise, which could cause interference patterns on the picture tube. An output from the video amplifier goes to the picture IF AGC amplifier. The DC output of this amplifier goes via Terminal 4 to the first picture IF amplifier and the tuner RF amplifier. The total video amplifier output goes through Terminal 19 to the video and chroma circuits of the TV receiver.

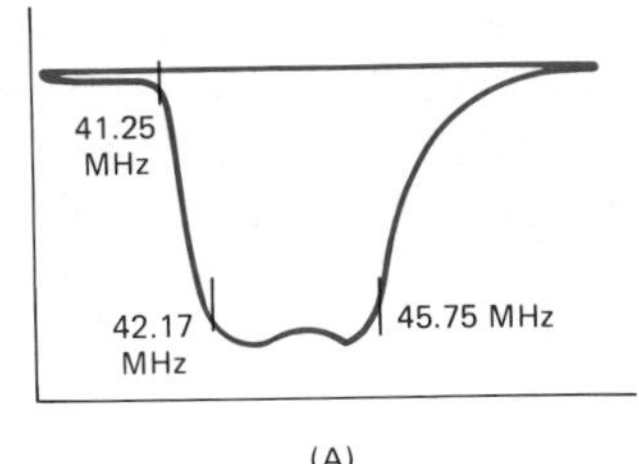

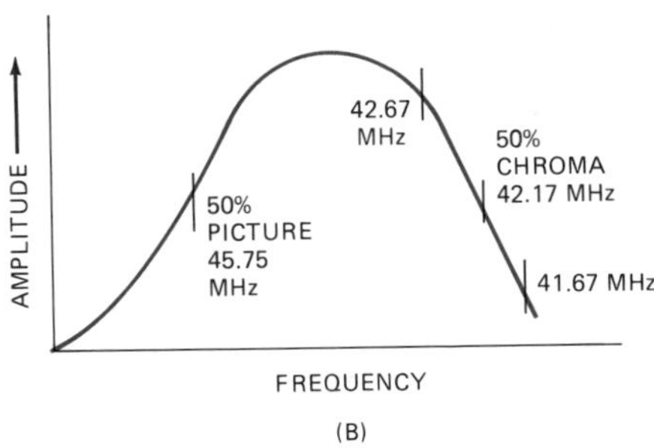

Figure 12.26A. Response of the interstage circuits alone. **B.** The overall IF response curve.

12.13 SURFACE-ACOUSTIC WAVE FILTER (SAWF)

The surface-acoustic wave filter (SAWF) is a surface-acoustic wave (SAW) device. When used in the IF section of a television receiver, it replaces the LC-tuned circuits (including traps) used in more conventional designs. In one application (Zenith), the SAWF is placed ahead of an IC three-stage broadband amplifier and provides the required bandpass shaping. (Schematic shown later.)

The operation of SAW devices is based on the functioning of a substance that exhibits the **piezoelectric effect.** This effect produces an electric potential across a piezoelectric substance when the substance is subjected to physical stress (such as bending). Conversely, applying an electric potential across a piezoelectric material causes it to undergo physical stress. This principle is widely used in crystal oscillators and filters, as well as in other devices.

The most suitable materials currently used for SAW devices are silicon oxide, thin-film zinc oxide, lithium tantalate, lithium niobate, bismuth germanium oxide, and quartz.

SURFACE ACOUSTIC WAVE

A SAW is a nonelectromagnetic wave that travels along the surface of a piezoelectric substance. Its frequency can be as high as several gigahertz. The piezoelectric material is known as the substrate, and its surface must be highly polished. The velocity of the SAW is only about 10^{-5} times that of electromagnetic waves. Thus, a SAW travels slowly, as sound waves do, but retains the frequency of its source. A surface acoustic-wave filter (SAWF) may also be called a surface-wave integrated filter (SWIF). SAW devices are used not only as filters, but also as delay lines, pulse processors and in various microwave devices and circuits.

SAW PROPAGATION AND DETECTION

Figure 12.27A shows the basic method of SAW propagation. The signal (here, in the TV IF band) is fed to the input transducer. (The transducers are known as **interdigital** (or interlocking) **transducers.**) The potential between the input transducer electrodes causes acoustic surface waves to be set up in the substrate. These waves travel to the output transducer, where they are transformed back into electric signals. In the process however specific bandwidth and frequency characteristics have been established, as described below. A basic SAWF is illustrated in Figure 12.27B. The interdigital transducers consist of periodic structures of film-coated metal electrodes. These are deposited on the surface of the piezoelectric substrate by a photolithographic process.

Bandwidth and Frequency Characteristics **SAWF bandwidth** and **frequency characteristics** are determined by the geometric structure of the transducers. In general, low-frequency SAWs are generated (or detected) at that portion of the transducer array that has the "fingers" relatively far apart. Conversely, high-frequency SAWs are generated (or detected) at that portion of the transducer that has the fingers relatively close together. A large number of uniform fingers results in restricted bandwidth operation. A uniform array with only a few fingers has a broad-frequency response.

For a specific frequency response with sharp cutoffs, the fingers are "amplitude-weighted" (Figure 12.27C). This is done by varying the amount of finger overlap in certain parts of the transducers. However, this scheme can produce phase distortion. One scheme that can provide specific frequency response and also reduce phase distortion is an array of "dummy" fingers (Figure 12.27D). Other methods of reducing phase distortion include the use of slant arrays and split fingers.

Transit Response It is possible for part of the SAW, when it reaches the output transducer, to be reflected back to the input transducer and then back again to the output transducer (double reflection). These reflected waves take three times as long to reach the output transducer as the incident (original) waves. If the SAWF is used as a

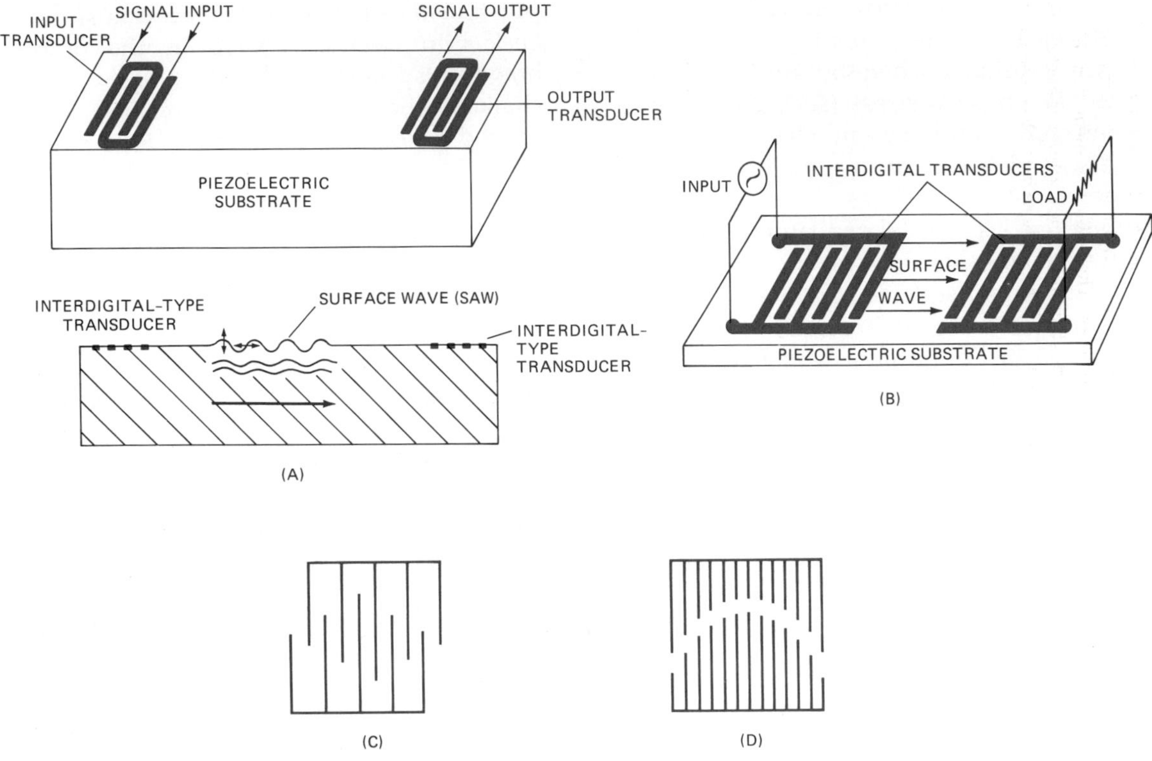

Figure 12.27A-D. Surface-acoustic-wave filter (SAWF). **A.** The signal is fed to the input transducer and generates surface waves on the piezoelectric substrate. The SAW is detected by the output transducer as electrical signals. **B.** Basic SAW filter. The geometric structure of the transducers determines the bandwidth and frequency characteristics of the SAWF. The configuration shown has uniform fingers, producing broad-band response. **C.** A "weighted" finger array. **D.** A "dummy" finger array.

TV-IF bandpass device, such reflections can result in ghosts in the picture.

One way to eliminate this is to use a **split-finger transducer** for the output array. Waves reflected from this array will have double the incident-wave frequency. Thus, they will be too far from the normal passband to cause any interference.

SAWF FOR TV-IF BANDPASS RESPONSE

An SAWF that provides the complete IF bandpass response for a Zenith color TV receiver is shown in Figure 12.28. Figure 12.28A shows a simplified diagram of the SAWF device, and Figure 12.28B shows the SAWF connected in the IF strip.

The SAWF Device The basic SAWF device consists of input and output interdigital transducers separated by a multistrap coupler. These units are mounted on a polished piezoelectric substrate. The substrate is made of lithium niobate and measures 0.5 in (1.3 cm) by 0.188 in (0.48 cm). The multistrap coupler (100 metal lines) prevents distortion of the SAW transmission caused by reflection of energy from the bottom to the

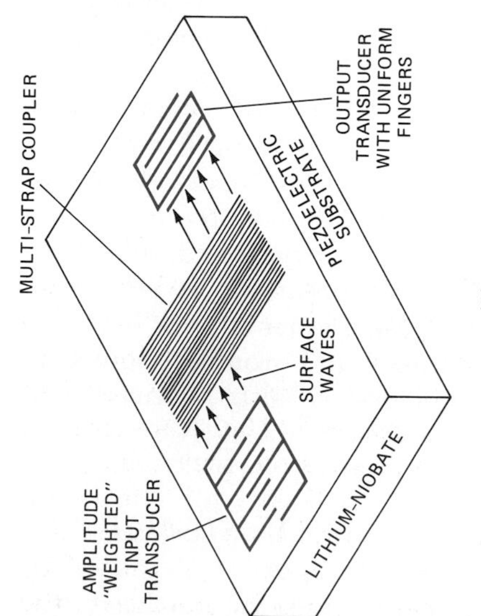

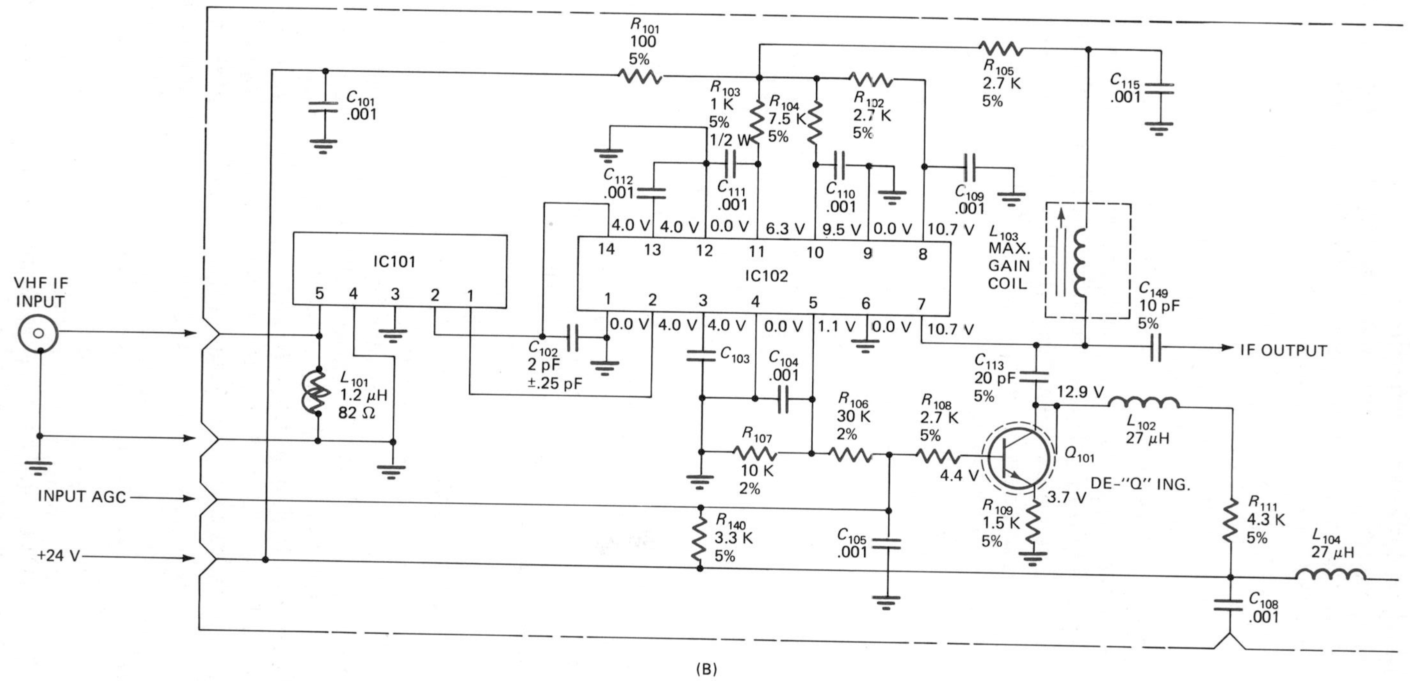

(B)

Figure 12.28A-B. A SWIF used to provide the IF bandpass characteristics for a color TV receiver. A. Simplified diagram of the SWIF. Note different input and output transducers. The multistrap coupler actually has 100 metal lines. The substrate measures 0.5 in (1.27 cm) by 0.188 in (0.48 cm). B. Partial schematic diagram showing the connection of the SWIF (IC101) in the IF section of a color TV receiver. Unless otherwise noted, capacitor values are in μF. (*Courtesy of Zenith Corp.*)

top of the substrate. This type of distortion is known as **bulk-wave reflection.**

Note in Figure 12.28A that the input transducer is *weighted* to provide the required selectivity for IF bandpass. The input transducer has very low signal loss for the IF passband centered at 44 MHz. It provides the required 41.25-MHz sound carrier attenuation as well as sharp trapping action at the 39.75-MHz (adjacent picture) and 47.25-MHz (adjacent sound) IF carriers. It also provides high attenuation for all signals outside the IF passband. The output transducer has only a few uniform fingers and therefore has broad-band response.

Mounting The substrate is mounted on the inside bottom of a ceramic package, the bottom of which has a deposited, thin-conductive coating. This coating is connected to the ground of the IF circuitry. For connections to the two transducers, four leads are brought out to package terminals. A ceramic lid is placed on top of the package, which is then sealed with epoxy.

Circuit Connections As shown in Figure 12.28B, the IF input from the tuner is connected to the input transducer of the SWIF (IC_{101}) at Terminals 4 and 5. Coil L_{101}) terminates the 75-ohm input IF cable and provides the correct impedance match to the SWIF. The output of the SWIF at Terminals 1 and 2 of IC_{101} is fed to Terminals 2 and 14 of IC_{102}, which is a three-stage IC IF broadband amplifier. The amplified IF output signal is taken from Pin 7, across coil L_{103}.

De-"Q"-ing Action Transistor Q_{101} is effectively connected (for signal) across output coil L_{103}. It acts as a variable shunt resistance across L_{103}, thus controlling its Q. By control of the Q (and impedance) of L_{103}, the video and sound IF output level is held constant (within design limits) even though signal strength may vary.

The IF AGC voltage from another module is fed to Terminals 5 and 10 of IC_{102} and to the base of Q_{101}. If the strength of the received signal increases, the AGC input voltage on the Q_{101} base also increases. This causes an increase in the collector current of Q_{101} and an effective decrease in its shunt resistance across L_{103}. With decreased Q, the IF output signal from L_{103} is proportionately decreased, tending to maintain the original level. The converse is also true. A decrease of received-signal strength results in a decreased Q_{101} base voltage. This decreases collector current and increases the transistor-shunt resistance. Now the Q and L_{103} increases and the signal output increases to the original level. Thus, Q_{101}, in conjunction with L_{103}, provides additional AGC action for greater signal level stability.

12.14 VIDEO IF AMPLIFIER TROUBLES

The video IF section of a television receiver must pass the video information, the color signals, the synchronizing and blanking signals, and the sound signals. If the receiver IF system is not operating properly, it cannot develop the voltages necessary for proper operation of the video, color, sweep, or sound sections.

Troubles in the video IF section can be classified in three categories: complete or partial loss of signal, improper alignment, and AGC problems. The symptoms are the same for both vacuum tube and solid-state receivers. The only difference is in voltage and resistance measurements.

The alignment and troubleshooting of television receivers is also discussed in Chapters 26 and 27.

COMPLETE LOSS OF SIGNAL

A complete loss of signal in the IF system can result from a lack of B+ or from a faulty vacuum tube, transistor, IC, or other component. The picture tube screen will have a raster, but there will be no sound or picture. The symptoms of an inoperative video IF stage are similar to those of a defective tuner, and so further tests are needed to isolate the defect.

Complete signal loss can also be produced by defects in the AGC circuits. The reason for this is that any IF amplifier that is AGC bias-controlled, either fully or partially, can by driven to cutoff by an incorrect AGC voltage. This is especially true in transistor IF amplifiers.

PARTIAL LOSS OF SIGNAL

Partial signal loss caused by a defect in the IF stages results in a picture of insufficient contrast. If the tuner is operating normally, there should be no snow in the picture. Snow can be produced by a malfunction or incorrect setting of the AGC system or by a defective RF amplifier in the tuner. The isolation of the defective area must be accomplished and this is discussed in Chapter 2. With partial signal loss caused by an IF defect, the sound may appear either quite normal or with a 60-Hz synchronizing pulse. In any case there will generally be sound when there is only a partial loss of the video IF signal.

IMPROPER ALIGNMENT

A television receiver IF system may become misaligned either during normal usage or when a component is replaced. In some cases the video IF section may have to be realigned whenever a tube, transistor, or IC is replaced. The aging of components and vibration or shock when the receiver is moved may also make alignment necessary.

Improper alignment of the video IF amplifier section can cause smearing of the picture. The adjustment of the fine-tuning control changes the amount of smear by varying the position of the video carrier on the IF response curve. The smear is due to a phase shift caused by excessive low-frequency response. Any phase shift in the video IF amplifier circuits can also upset the phase-modulated color hue signals. Troubles of this nature usually have little effect on the production of color, but the color appears in the wrong areas of the picture. If the smear is not tunable with the fine-tuning control, it is more likely caused by a defect in the video detector or video amplifier circuits.

Improper alignment of the video IF amplifiers may also be the cause of a loss of fine detail. This occurs when the higher video frequencies are attenuated. The color signal may be attenuated or completely lost because of an improperly aligned IF stage.

Sound interference can be caused by improper alignment in the video IF system. This problem is often traced to the traps used to obtaining the correct shape the IF response curve.

An excessive high-frequency response in the IF section can cause "ringing" in the amplifier circuits. Ringing is the appearance of a ghost-like region following the vertical lines of the picture. This problem is easily distinguished from the ghosts related to antenna problems because ringing can be varied with the fine-tuning control of the receiver.

Some transistor IF amplifiers use degenerative feedback to reduce the likelihood of oscillation and to broaden the frequency response of the amplifier. If oscillation occurs, the picture will show a pattern of light and dark areas and will often display a herringbone design. This pattern will remain on the screen even when the station selector is switched to an inactive channel. Also, the voltage across the detector load will be higher than normal and the voltage will not be affected, as it should be, by tuning to another station or to an inactive channel.

The obvious cure for alignment problems in the receiver is to follow the correct alignment procedures. Alignment should not be attempted without the test equipment recommended by the TV receiver manufacturer.

AGC TROUBLES

The symptoms described for a faulty video IF system also apply to trouble in the AGC circuits. Usually, AGC trouble appears on the picture tube CRT screen either as a washed-out picture with poor contrast and weak color or as an overloaded picture with too much contrast. The first condition is caused by too much AGC voltage, and the second by too little AGC voltage.

The difficulty in servicing AGC problems stems from the fact that this circuit is part of a feedback loop. The amount of AGC voltage depends upon proper operation of the IF system, and proper operation of the IF system depends upon the amount of AGC voltage applied to the controlled stages. Since the AGC and IF circuits are interdependent, it is difficult to separate the problems and symptoms in the two circuits.

The best way to determine whether the IF or the AGC circuit is at fault is to disconnect the AGC circuit and substitute a DC bias voltage to take its place. When the bias of the tuner is supplied by a substitute source, the circuit will operate properly, provided the trouble is in the AGC circuit. If the symptoms do not change when the AGC voltage is substituted for, then it follows that the trouble is in the tuner or the IF system. AGC troubles are discussed in Chapters 14 and 27.

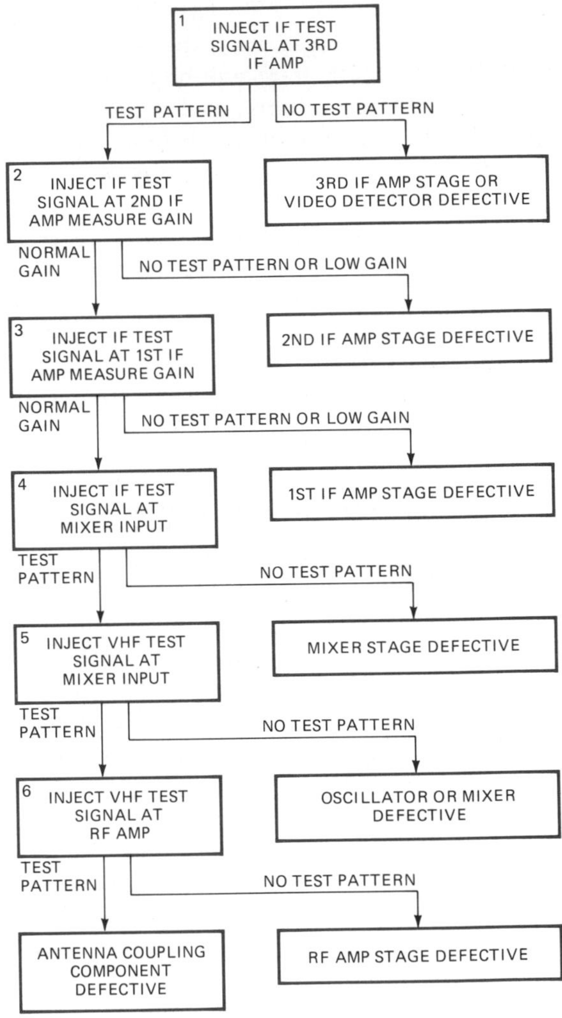

Chart 12.1 Illustrating a practical troubleshooting procedure for the condition of no picture, no sound. *(Courtesy of B & K-Precision, Dynascan Corp.)*

No Video, No Sound Chart 12.1 illustrates a practical troubleshooting procedure when there is no video and no sound.

SUMMARY OF CHAPTER HIGHLIGHTS

1. The RF tuner and the video IF amplifiers are the only sections of a TV receiver that handle the entire band of frequencies required for the reproduction of the picture and sound.
2. A major function of the video IF system is to drive the video detector and the sound and video amplifiers.
3. The video IF system provides most of the gain and selectivity in the TV receiver and provides equalization for vestigial sideband method of transmission. (Figure 12.7)
4. Functionally, vacuum tube and solid-state video IF amplifiers are the same.
5. Vacuum tube IF amplifiers are usually hand-wired onto metal chassis or mounted on PC boards. Solid-state IF amplifiers are always mounted on PC boards.
6. Many television receivers use IC IF amplifiers.
7. Both vacuum tube and solid-state circuits are used in modular construction.
8. Vacuum tube circuits supply a negative AGC voltage to one or more of the IF amplifier stages.
9. Either NPN or PNP transistors may be used in transistor IF amplifiers. Therefore, the AGC voltage may be either positive or negative.
10. There are two general types of video IF amplifiers: monochrome and color. Within the two general types, video amplifiers can be classified as either vacuum tube or solid-state.
11. To extend the classification, video IF amplifiers can be classified according to the number of stages: two, three, or four. Two-stage and three-stage systems are the most popular. Another classification is by the type of interstage coupling; either transformer-coupled or impedance-coupled. (Figures 12.15 and 12.16)
12. Monochrome TV receivers usually have a two-stage or three-stage video IF system. Color TV receivers generally have a three-stage video IF system to obtain the bandwidth required for the color signals. (Figure 12.10B)
13. The intermediate frequencies are 41.25 MHz for the sound carrier and 45.75 MHz for the video carrier. The color subcarrier is at 3.579545 MHz, which is 42.17 MHz in the IF band. (Figures 12.5 to 12.8)
14. Video transmission is by the vestigial sideband method. (Figure 12.5)
15. The video carrier is amplitude-modulated, and the sound carrier is frequency-modulated. The color subcarrier is both phase-modulated and amplitude-modulated.
16. When the composite video signal is transmitted, the sound carrier is 4.5 MHz above the video. Because of the action of the mixer stage in the tuner, however, this position is reversed in the receiver IF system and the sound carrier is then 4.5 MHz below the video carrier. (Figure 12.6)
17. In the IF system, the luminance video frequencies may occupy a 3.5-MHz band, the sound occupies 0.05 MHz, and color about 2.0 MHz. (Figure 12.5)
18. The most troublesome spurious responses of the video IF system are image response, response to two stations separated in frequency by the IF value, direct IF response, and response to signals reradiated from nearby television receivers.
19. To compensate for vestigial sideband

transmission, the video carrier is set at the 50% point on the video IF response curve. (Figure 12.7)

20. To prevent sound interference, the sound carrier is reduced to the 5% level in the common IF stages.
21. Fine detail in the picture depends on the high frequencies of the video signal. Shading of large areas on the screen depends on the lower video frequencies.
22. Marker frequencies are used to spot particular points on the IF response curve. (Figure 12.7)
23. Stagger tuning is used to achieve greater bandwidths. (Figure 12.11)
24. The bandwidth of an amplifier stage is defined as the width of the amplifier response curve at the half-power points. (Figure 12.10)
25. An important relationship that should be remembered about amplification systems is that the bandwidth of any amplifier is always inversely proportional to the gain: bandwidth × gain = constant. Expressed mathematically
26. In some monochrome TV receivers, the IF bandwidth may be as low as 2.5 MHz. The decreased amount of detail in the picture is hardly noticeable in a small-screen receiver.
27. In monochrome receivers the sound is separated after the video detector, but in color receivers the sound must be separated prior to the detector. If sound signals reach the detector television on a color receiver, they will beat with the color subcarrier and produce a 920 kHz interference signal on the picture tube. (Figures 12.1, 12.21, 12.22, 12.24, 12.25)
28. The reason for using transformer and impedance coupling for IF amplifiers is that both can be tuned to response to particular bands of frequencies. (Figures 12.15 and 12.16)
29. A bifilar-wound coil, which is a type of transformer, can be used to obtain the necessary bandwidth when transformer overcoupling is desired. (Figure 12.14)
30. Trap circuits are used in TV receivers to reduce the associated sound carrier (41.25 MHz) to the 5% level and to reject the adjacent channel sound carrier (47.25 MHz) and the adjacent channel video carrier (39.75 MHz). These traps also help shape the response curve.
31. Five types of traps are used with video. IF amplifiers: series, parallel, absorption, degenerative, and bridged-T. (Figure 12.18)
32. The bridged-T trap is the most effective type and is used in color video IF systems. A parallel trap offers a low impedance to ground to the undesired signal. An absorption trap removes the energy of the undesired frequency signal from the IF band. (Figure 12.18)
33. Color video IF amplifier circuits are more critical than monochrome circuits and so may contain more traps or filters. (Figures 12.21, 12.22, 12.24, and 12.25)
34. Some solid-state IF amplifiers must be neutralized to overcome the effects of high-frequency signal feedback from the collector circuit to the base circuit via the collector-to-base junction capacitance. (Figure 12.3)
35. ICs are widely used in television receivers. Some typical television IC sections are (1) picture IF, (2) sound IF, (3) AFT, and (4) chroma.
36. A single IC can contain the complete IF system (sound and picture) for a television receiver. However, the tuned circuits and power supply are external to the IC. The IC may also contain AGC circuitry and a zener reference for the external voltage regulator. (Figure 12.24)
37. In the block diagram of the IC of Figure

12.24, all tuned circuits precede and follow the first picture IF amplifier. The second and third picture IF amplifiers are broad-band types.

38. In Figure 12.24, AGC voltage is fed through the input-tuned circuits to the first-picture IF amplifier. From the first-picture IF amplifier, AGC voltage is fed via the tuner AGC delay to the tuner RF amplifier.
39. A surface acoustic wave filter can replace all the tuned IF circuits, including traps, in a TV receiver. (Figure 12.28)
40. The operation of a SAWF is based upon the functioning of a piezoelectric substance. The SAWF is capable of operating up to several gigahertz.
41. The SAW (surface acoustic wave) has the slow travel property of sound, but retains the frequency of its source. Propagation and detection of SAWs is by input and output metal-film transducers. These are called interdigital (interlocking) transducers.(Figure 12.27)
42. The bandwidth and frequency characteristics of a SAWF are determined by the geometric structure of the transducers. For a specific frequency response, the transducer fingers are *amplitude-weighted*. (Figure 12.27C)
43. A multistrap coupler on a SWIF is used to suppress bulk wave top-to-bottom reflections. (Figure 12.28A)
44. Trouble in the IF system can be classified in three categories: complete or partial loss of signal, improper alignment, and AGC trouble.
45. A defective IF amplifier can produce the same symptoms as a defective tuner or a defective AGC system.
46. Alignment of the IF system should never be attempted unless the proper equipment recommended by the receiver manufacturer is available.

EXAMINATION QUESTIONS

(Answers are provided at the back of the text.)

Part A Supply the missing word(s) or number(s).

1. Both _______ and _______ IF frequencies pass through the common IF amplifiers.
2. In a color IF section, the _______ IF frequencies must be removed before the video detector.
3. In question 2, if this is not done, there will be a _______ kHz interference pattern on the picture tube.
4. Important functions of the common IF amplifier are to supply most of the RF _______ and _______.
5. A single IC may contain an entire IF _______.
6. In a color TV IF section, the sound IF is tapped off _______ the video detector.
7. The sound, picture, and color carriers are _______ in position when changing from RF to IF frequencies.
8. The _______ subcarrier IF frequency is 42.17 MHz.
9. The exact value of the video color subcarrier frequency is _______ MHz.
10. _______ transmission is present when one complete and one partial sideband are transmitted.
11. The adjacent picture IF frequency is _______ MHz.
12. When two (or more) single-tuned amplifiers

are tuned to different frequencies within the desired band, this is called ________.

13. The most important trap frequencies are ________ MHz, ________ MHz, and ________ MHz.
14. A ________ trap is the type placed between two IF stages.
15. The ________ sound IF frequency is 4.5 MHz.
16. AGC voltage is generally applied to the ________ amplifier(s) and the ________ RF amplifier.
17. An amplifier that passes a number of frequencies is a ________ amplifier.
18. SAW stands for ________.
19. A SAW is a ________ wave.
20. SAW propagation occurs on the surface of a ________.
21. The ________ geometric structure determines the bandwidth of a SAWF.
22. Bulk wave reflection on a SAWF is prevented by a ________.
23. A defective IF amplifier may cause the same symptoms as a defective ________ or a defective ________ section.
24. Snow can be seen on the picture tube screen with a defective ________ amplifier.
25. The color signal may be lost because of a ________ IF stage.

Part B Answer true (T) or false (F).

1. In a color TV receiver, the sound IF frequencies are removed immediately before the first video amplifier.
2. The frequency of the color subcarrier wave is 920 kHz.
3. The sound intercarrier frequency is 4.5 MHz.
4. Monochrome IF circuits usually contain more traps and filters than color IF circuits.
5. All transistor IF amplifiers must be neutralized.
6. An IC for a television receiver may contain an entire IF section plus AGC circuitry.
7. In color TV receivers, an associated sound IF trap is usually placed between the last common IF stage and the video detector.
8. The IF carrier frequencies are reversed from those transmitted at RF frequencies.
9. For Channel 4, the RF sound carrier is at 71.75 MHz.
10. The color subcarrier frequency is 3.59 MHz (rounded off).
11. The Channel 4 RF color subcarrier frequency is 70.83 MHz.
12. The video IF bandwidth for inexpensive TV receivers may be 2.5 MHz.
13. The local oscillator frequency for Channel 2 is 107 MHz.
14. Image frequency response in television receivers is minimized by the choice of a high IF frequency band.
15. Vestigial sideband transmission is used for both monochrome and color transmission.
16. The picture carrier frequency is situated 1.5 MHz above the lower edge of the channel band.
17. The associated sound IF carrier is reduced by traps to about 5 to 10% of the maximum response curve amplitude.
18. The bandwidth of a tuned circuit is measured at the -3 dB points on the response curve.
19. Two common methods of coupling in the common IF stages are RC coupling and transformer coupling.
20. An absorption-type trap absorbs all frequencies except the trap-tuned frequency.
21. A common method of impedance matching in IF stages is to tap the tuned coil.
22. The IF band can be passed only by an amplifier with broadly tuned (stagger-tuned) stages.
23. An important feature of a bifilar transformer is that it is coupled at less than the critical value.
24. With SAWF devices, the required LC-tuned circuits are always placed ahead of the SAWF.
25. A SAW is a nonelectromagnetic wave that travels at about the speed of sound but retains the frequency of its source.
26. It is the structure of the interdigital (inter-

locking) transducers that establishes the bandwidth and frequency characteristics of a SAWF.

27. By "amplitude weighting" the SAWF transducer fingers, a specific frequency response with sharp cutoffs may be achieved.
28. One method of reducing transit response in a SAWF is to use a multistrap coupler.
29. Sound interference in the picture can be caused by incorrect common IF stage alignment.
30. A difficulty in servicing AGC problems arises from the fact that the circuitry is part of a closed loop.

REVIEW ESSAY QUESTIONS

1. Draw a simple block diagram of color TV common IF section, the tuner stages, and video detector. Show AGC inputs and all output connections.
2. In Question 1, explain briefly the operation of all stages shown.
3. Draw the color IF response curve, indicating all important frequency markers and their designation. Include traps and label marker amplitudes correctly.
4. Briefly explain why, in many color TV receivers, the sound IF must be separated prior to the video detector. (Figure 12.22)
5. Define: PC, IC, module, amplifier bandwidth, and vestigial sideband transmission. (Note: Do not merely supply names for the abbreviations.)
6. According to the standard definition, what is the bandwidth in Figure 12.2?
7. Explain how the circuit shown in Figure 12.4 provides impedance matching.
8. List some typical functions performed by an IC video IF amplifier.
9. In question 8, what circuits (or parts) are external to the IC? Why are coils not an integral part of the IC?
10. Explain how the correct phase of feedback voltage is obtained in each case shown in Figure 12.3.
11. In Question 10, why is capacitor C_N required?
12. Referring to Figure 12.7, explain the advantage, in the early color receivers, of having the 42.17-MHz marker on the flat portion of the response curve? Why is 42.17-MHz at the 50% point in current color TV receivers?
13. In Question 12, what is the disadvantage of having 42.17 MHz at the 50% point? How is this compensated for?
14. Draw a simple sketch of a SAWF. Label all parts.
15. In Question 14, explain in your own words the basic functioning of a SAWF.
16. In Figure 12.25, what is the function of the following: T_4, Q_1, L_5, T_1, and T_2?
17. In Figure 12.24, explain in your own words how AGC is applied to the IF and tuner sections.
18. In Figures 12.24 and 12.25, how is it possible for the second and third picture IF amplifiers to be untuned and still maintain the proper IF bandpass.
19. In Figure 12.22, Explain how impedance matching is achieved between Q_2 and Q_3.
20. What is the meaning of the following? (a) image response, (b) direct IF response, (c) intercarrier sound system, (d) impedance coupling.
21. Name three types of IF traps and briefly explain the operation of each.
22. Explain the function of 4.5-MHz traps.
23. What is the meaning of SAW, SAWF, SWIF?
24. What is the approximate upper frequency limit of a SAW?
25. In your own words, explain how the frequency response and bandwidth of a SAWF are achieved. What is the approximate velocity of propagation of a SAW?
26. List the problems that may occur as a result of improper IF alignment in a color TV receiver. Briefly explain your answers.

EXAMINATION PROBLEMS

(Selected problem answers are provided at the back of the text.)

1. Identify the following frequencies: 3.58 MHz, 4.5 MHz, 39.75 MHz, 41.25 MHz, 42.17 MHz, 45.75 MHz, 47.25 MHz, 107 MHz.
2. In Figure 12.5, calculate (a) upper and lower color sidebands, (b) upper and lower monochrome sidebands, (c) frequency difference between the video carrier and the color subcarrier, and (d) frequency difference between sound carrier and color subcarrier.
3. List the following frequencies for Channel 3: (a) upper and lower band limits, (b) video carrier, (c) color subcarrier, (d) sound carrier, and (e) tuner oscillator frequency.
4. In Figure 12.22, identify the input bridged-T trap and its trap frequency.
5. Calculate the RF color subcarrier frequencies and the tuner oscillator frequencies for Channels 5, 8, and 21.
6. What is the range of image frequency signals for a television receiver tuned to Channel 4?
7. Two single-tuned amplifiers each have a bandwidth of 2.2 MHz. Calculate the new bandwidth if these stages are stagger-tuned.
8. Show typical calculations to arrive at the figure of 39.75 MHz. Of 47.25 MHz.

Chapter 13

13.1 Positive- and negative-picture phases
13.2 Video detector filtering and high-frequency compensation
13.3 Shunt-video detectors
13.4 Video detectors in color television receivers
13.5 Synchronous (linear) video detectors
13.6 Synchronous video-detector operation
13.7 Troubles in the video detector stage
Summary

VIDEO DETECTORS

INTRODUCTION

The video detector for color and monochrome receivers is an AM detector since the video information is amplitude modulated. The video detector may be a nonlinear diode type or a linear-synchronous one. Both types will be described in this chapter. The diode detector will be discussed first.

A typical diode-detector circuit is shown in Figure 13.1. The demodulated video signal with its blanking and synchronizing pulses is developed across R_L. The form of the IF signal when it enters the second detector is illustrated on the left of the figure. The rectified resultant is illustrated at the right.

In diode operation, anode current flows only when the anode is positive with respect to the cathode. The effect of this action is to eliminate the negative portion of the incoming signal. Since the positive and negative sections of the modulated video signal are exact duplicates of each other, either one may be used. The signal polarity actually used depends upon the video circuitry following the video detector. This will be explained later in this chapter.

By the time you have completed the reading and work assignments for Chapter 13, you should be able to:

- Define the following terms: positive-picture phase, negative-picture phase, series diode-video detector, shunt diode-video detector, and synchronous-video detector.
- Draw a simple schematic diagram of a color TV receiver diode video-detector. Explain how the circuit operates. Describe some of the disadvantages of this type of circuit.
- Draw a simple sketch of a monochrome TV receiver diode-video detector. Explain briefly how it is possible to obtain either positive- or negative-picture phase.
- List three disadvantages of a diode-video detector.

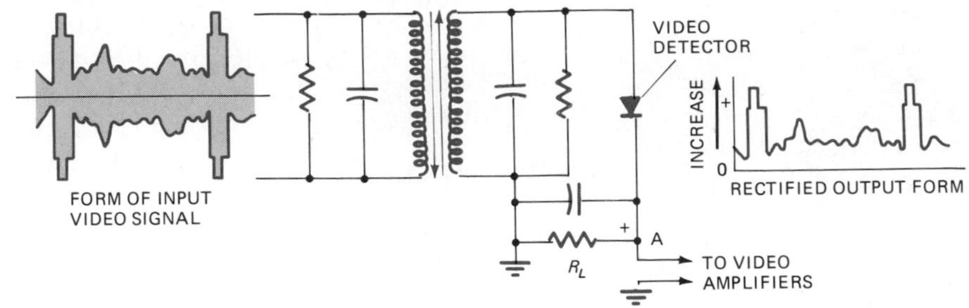

Figure 13.1 A diode detector circuit for a television receiver. Output is in negative picture phase.

- Briefly explain the operation of a synchronous-video detector.
- List four advantages of a synchronous-video detector.
- Explain why the word "synchronous" is used to describe the synchronous-video detector.
- Briefly explain why, in the synchronous-video detector, AM on the reference signal does not affect the detector operation.
- Name and describe two methods of generating the reference signal for a synchronous-video detector.
- Name three symptoms of a faulty video detector.

13.1 POSITIVE- AND NEGATIVE-PICTURE PHASES

At this point it is necessary to consider the polarity of the voltage drop across the load resistor R_L in Figure 13.1. For American television systems negative picture transmission is standard. This means that in the video detector, the brightest elements cause the least amount of diode current to flow. Maximum diode current is obtained when the blacker-than-black region of the synchronizing pulse is reached.

The signal in the negative-picture phase form, Figure 13.2A, cannot be applied to the grid of the picture tube. (This phase can be applied to the cathode of the picture tube and the effect will be the same as if a positive-picture phase signal is applied to the picture tube grid.) First, it must be reversed to the form shown in Figure 13.2B. This is necessary because when the blanking signals are applied to the control grid of a picture tube, they must bias it to cutoff. This objective can be attained only if the signal has the

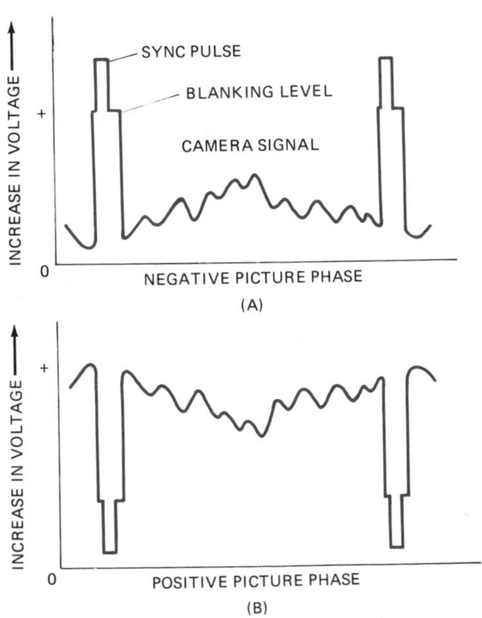

Figure 13.2A-B. Rectified video signals may be obtained from the output of the detector in either one of the two forms shown, depending upon how the detector is connected.

form given in Figure 13.2B. This latter form of the television signal is the positive-picture phase. It is interesting to note that if the negative phase of the signal is applied to the control grid of the picture tube, all the picture values will be reversed and the observed scene will be similar to a photographic negative. Another widely used method of distinguishing between the signal polarities is to employ the designation of **sync pulse positive** (Figure 13.2A) or **sync pulse negative** (Figure 13.2B).

In sound receivers, no attention is given to the relative phase of the audio signal because our ears are insensitive to all but gross phase differences. Television, on the other hand, deals with visual images. The reversal of phase produces noticeable effects. Possible ways of selecting the desired phase of the video signal are discussed in the following paragraphs.

NEGATIVE-PICTURE PHASE

Referring to the half-wave detector circuit of Figure 13.1, let us investigate the voltage developed across R_L. The incoming signal has the same form as at the antenna, with the synchronizing pulses giving rise to the greatest amplitudes. At the diode rectifier, these synchronizing signals cause the anode to become the most positive, resulting in a greater voltage drop across R_L and having the polarity as shown. On the other hand, those portions of the video signal representing the bright segments of the image will have the least positive (most negative) voltage at the diode anode. This results in a smaller voltage drop at R_L. Thus, point *A* of resistor R_L will still give rise to a large positive voltage for the synchronizing signal. This means that the signal is still in the negative picture phase. This phase of signal is unsuitable for application to the grid of the viewing tube.

POSITIVE-PICTURE PHASE

The direction of the current flow through R_L can be altered to give the opposite polarity by reversing the connections between the diode and the input transformer, Figure 13.3. Rectification eliminates the positive half of the modulated carrier and leaves the negative half. Nothing is lost since both halves contain the same information. Point A of Figure 13.1 becomes more negative for the blanking and synchronizing portion of the video signal. The bright elements cause the voltage at point A to become less negative. When the signal is applied in this form between the grid and the cathode of the image tube, the largest picture tube current will flow for the bright sections of the image. For the blanking and synchronizing parts of the signal, the voltage at the grid will be negative and the electron beam will be cut off as required.

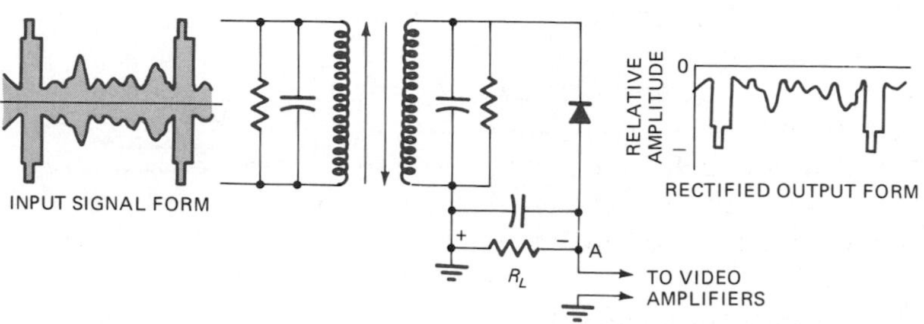

Figure 13.3 A diode detector connected to give a positive picture phase output signal. Note that, here, only the negative half of the input signal is rectified. In Figure 13.1 the opposite is true.

Video Signal Amplification

The signal that is developed at the diode-load resistor is not strong enough to be used directly at the picture tube. Further amplification is necessary. The following video amplifiers, which are generally of the *common cathode* or *common emitter* type, will reverse the polarity by 180 ° of any signal sent through them. Thus, if the video signal has a positive-picture phase at the diode-load resistor, it will have a negative-picture phase at the output of the first video amplifier. With another stage of amplification, the picture will be brought back to the positive phase again. As a general rule, an even number of video amplifiers is required (1) if the picture phase across R_L in the detector is positive; and (2) if the video signal is to be applied to the control grid of the picture tube. For a negative picture phase at R_L, an odd number of video amplifiers is needed. In this case, a positive-picture phase will appear at the grid of the image tube. These conditions are illustrated in block form in Figure 13.4.

Video Signal At Picture Tube Cathode

Figure 13.5 shows an example of the video signal being applied to the cathode of a picture tube. The output of the video detector is in the positive.picture phase. This phase is reversed 180 ° by the video amplifier. Thus, the video signal is in the negative-picture phase as it is applied to the cathode of the picture tube. The video signal applied to the **cathode** of a picture tube must always be opposite in polarity to the signal applied to the grid.

13.2 VIDEO DETECTOR FILTERING AND HIGH-FREQUENCY COMPENSATION

The frequencies present in the video detector circuit include the intermediate frequency values and the actual video signals themselves, from 30 Hz to 4 MHz. The latter signals are to be passed on to the video amplifiers and amplified to about 100 V to 125 V peak-to-peak. At these levels, they are able to modulate the electron current sufficiently in the picture tube to produce an image on the screen. At the detector output, the intermediate frequencies must be shunted around the load resistor to prevent their reaching the following video amplifiers. The use of high IF values makes the problem of filtering out the IF voltages comparatively simple. The rectified video signal has a maximum frequency of about 4 MHz. In early television receivers, the IF values ranged from 8.75 to 12.75 MHz. Considerable filtering was required because of the low order of separation between the desired frequencies (30 Hz to 4 MHz) and those which were to be by-passed (8.75 to 12.75 MHz). However, by increasing the separa-

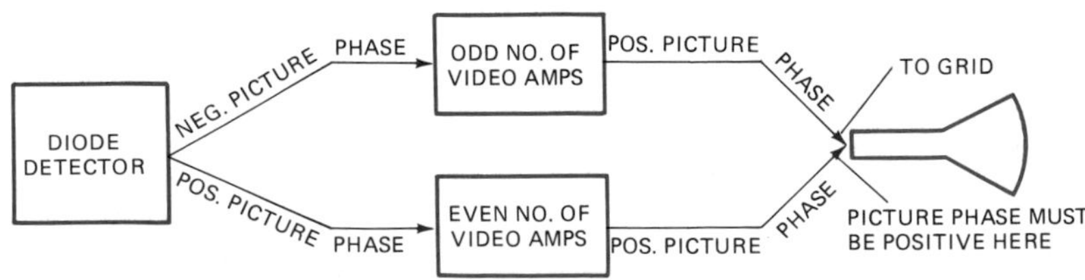

Figure 13.4 An illustration of why the number of video amplifiers after the detector is dependent upon the polarity of the signal obtained from the detector. Each video amplifier reverses the phase of the signal. Signals are being fed to the grid of the picture tube.

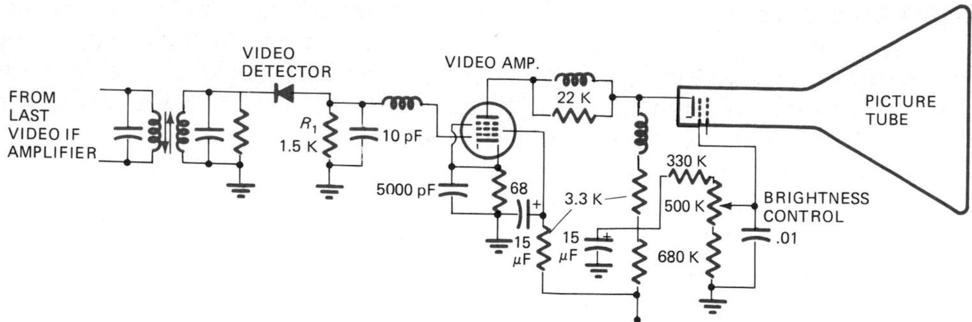

Figure 13.5 A video detector and a vacuum tube video amplifier feeding the video signal to the cathode of the picture tube.

tion between the two frequency ranges, the problem is simplified considerably. In modern television receivers, values for the video IF are in the range between 40 to 46 MHz.

FILTERING THE IF

Adequate filtering can be obtained using the arrangement shown in Figure 13.6. The rectified current passes through the low-pass filter composed of C_1, L_1, R_1, L_2, R_2, and C_2. Capacitor C_1 has a small fixed value of 5 pF. Actually, there is an additional capacitance across this point produced by the diode junction and the wiring. At the other end of the filter, C_2 is shown in a dotted form because no such component is inserted. However, the sum of the stray-wiring capacitance and the input capacitance of the following video amplifier produces the equivalent of an actual capacitor having a value of 10 to 15 pF. The two coils, L_1 and L_2, form part of the low-pass filter and also help to maintain a relatively flat frequency response up to 4 MHz. This counteracts any tendency of the circuit to attenuate these higher video frequencies. More will be noted on this point in the chapter on video amplifiers. The 22-kilohm (kΩ) resistor shunted across L_1 is used to prevent the response of the coil from rising abruptly at the higher video frequencies because of a natural resonant circuit formed by the coil and its inherent capacitance. The detector load resistor is R_2 (3 900 ohms).

13.3 SHUNT-VIDEO DETECTORS

All of the video detectors discussed thus far are of the series type. In other words, the rectifier (tube or semiconductor diode), the

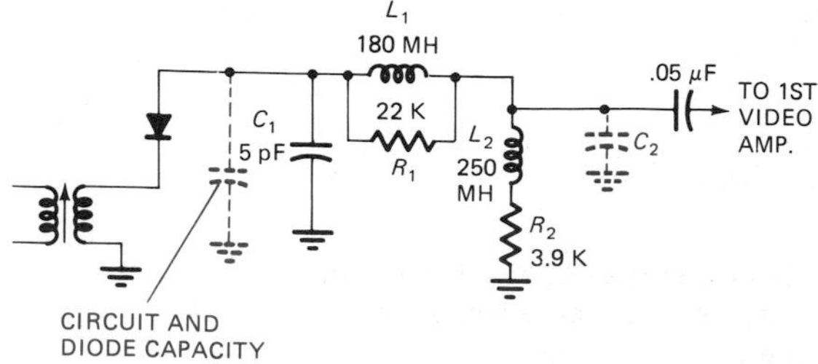

Figure 13.6 A video-detector circuit with a low-pass filter and load resistor.

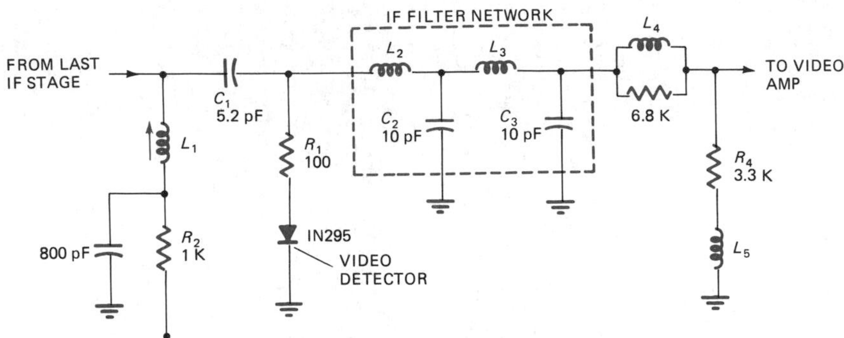

Figure 13.7 Schematic diagram of a shunt-type video detector.

input-tuned circuit, and the load resistor are all in series. It is also possible to achieve detection by placing the diode in shunt with the tuned circuit and the load resistor.

A shunt-detector circuit with a low-pass filter is shown in Figure 13.7. Coils L_2 and L_3 and capacitors C_2 and C_3 form the filter. Coils L_4 and L_5 serve as the video-peaking elements. Resistor R_1 is a current-limiting resistor designed to protect the diode. Note that the low-pass filter is placed before the load resistor, R_4. The filter may be placed after R_4, but the arrangement shown is more efficient.

THE VIDEO SIGNAL

It is instructive to see how a video signal at the detector output appears on the screen of an oscilloscope. A typical presentation is given in Figure 13.8. To obtain a stationary pattern, the sweep rate in the oscilloscope is either 15 750 Hz, or 7 875 Hz for two full lines, as shown in Figure 13.8. How clearly the sync pulses appear will depend on the overall response of the vertical amplifiers in the oscilloscope. If this response is too narrow, the sync pulses will appear with rounded corners. Even with the limited response of fairly low-cost oscilloscopes, the general form of the composite video signal will be clearly discernible.

The peak-to-peak amplitude at the output of the video detector is in the order of 1.0 V to 1.5 V for solid-state receivers. If a vacuum-tube video amplifier follows the video detector, a detector output of about 3.0 V peak-to-peak is required. In Figure 13.8, a light line can be seen at the level of the horizontal-blanking pulses. This line is caused by the unsynchronized vertical-blanking pulses.

13.4 VIDEO DETECTORS IN COLOR TELEVISION RECEIVERS

In monochrome receivers, it is a common practice to feed the video and sound

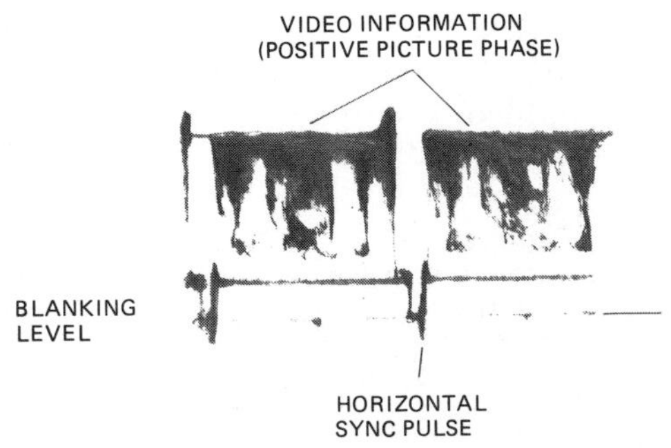

Figure 13.8 Two lines of a video signal, as observed on the screen of an oscilloscope. Note the horizontal sync and blanking pulses. The video information appears between the blanking intervals.

signals together through the video-IF amplifier stages and through the video-detector stage. In the detector, the video carrier and the sound carrier are heterodyned to produce the 4.5-MHz intercarrier sound-IF frequency.

920-kHz PATTERN

In color television receivers, it is undesirable to allow both the sound signal and video signal to pass through the video detector. The reason for this is that a 920-kHz difference signal will be generated as a result of heterodyning between the 3.58-MHz color carrier and the 4.5-MHz sound signal. The 920-kHz heterodyne signal will produce an interference pattern on the color screen.

To avoid the appearance of this pattern on the screen, the sound is obtained by tapping off the composite IF from the last video-IF amplifier stage and feeding it to a separate sound detector. The output of this sound detector is a 4.5-MHz FM sound signal which is then processed by the sound section of the TV set.

SOUND TRAPS

This method of obtaining the sound signal does not, in itself, eliminate the interference pattern due to the 920-kHz heterodyne signal. The pattern is suppressed by the use of two different traps, as indicated in Figure 13.9. A trap tuned to the sound IF of 41.25 MHz is inserted before the video detector. The trap greatly reduces the level of any sound-IF signal prior to its introduction to the video detector. Thus, the trap reduces the amplitude of the 4.5-MHz component contributing to the production of a 920-kHz signal. To reduce further the possibility of producing the interference pattern, a 4.5-MHz trap is placed after the video detector. This trap reduces any remaining heterodyned 4.5-MHz signal produced at the output of the video detector.

COLOR IF-RESPONSE CURVE

Figure 13.10 indicates a typical IF response curve for a color television receiver. This response curve is obtained by feeding a sweep generator into the tuner section and monitoring the output of the video detector. (The complete procedure for obtaining this pattern is covered in Chapter 27.) Since this response curve is obtained at the output of the video detector, the sound marker must be a point of minimum amplitude, as shown in Figure 13.10. The 41.25-MHz tuneable trap is adjusted to set the sound carrier at the minimum point. Note from the characteristic curve of Figure 13.10 that the receiver circuits, including the tuner, the video-IF amplifiers, and the video detector, are capa-

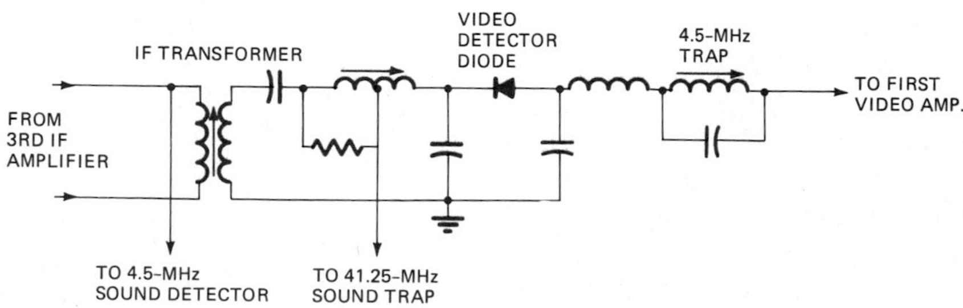

Figure 13.9 A video detector circuit for a color TV receiver.

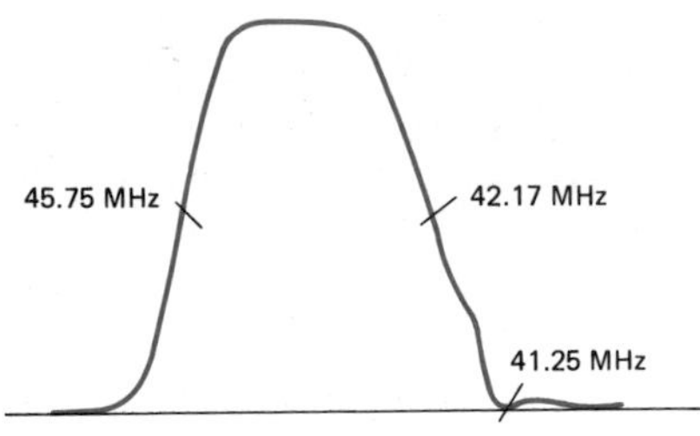

Figure 13.10 Characteristic curve of a color receiver. The audio marker shows that the response at 41.25 MHz is extremely low at the output of the video detector.

ble of passing all frequencies in the video range. This includes all frequencies in the color signal.

RCA COLOR TV VIDEO DETECTOR

Figure 13.11 illustrates the detector stage for an RCA solid-state color chassis. Capacitor C_1 at the output of the detector stage is used to attenuate the video-carrier frequency. A trap circuit, comprised of C_2, L_1, and R_1 is used to remove the 4.5-MHz beat frequency caused by the heterodyning of the audio- and video-carrier frequencies. Inductors L_2 and L_3 remove IF-harmonic frequencies generated in the nonlinear detector stage. Resistor R_2 is the resistance load for the detector, and L_4 is a peaking coil.

Note in the circuit of Figure 13.11 that a DC voltage is applied to the detector stage from a voltage divider comprised of R_3, R_4, and R_5. This DC voltage is used as the bias for the first video amplifier which follows the detector stage. It brings up a very important point. In vacuum tube receivers, the detector stages are normally at DC ground potential. However, in a solid-state color TV receiver, the detector stage often has a DC voltage present. Therefore, it is not true that you can completely ignore the DC voltage readings in the detector stage as far as the transistor receivers are concerned. If the DC voltage at this point is not correct, the first video amplifier will not be biased properly.

Detector In An IC. In many TV receivers, the video detector circuit is incorporated as part of an IC. Frequently the IC used is the video-IF IC, which may contain other circuits in addition to the video-IF and video-detector circuits. This was discussed in

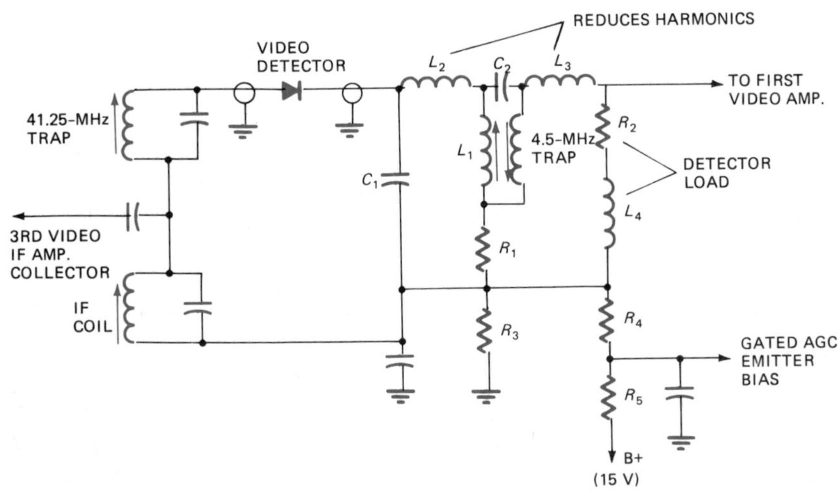

Figure 13.11 The video-detector stage of an RCA color TV receiver.

Chapter 12. The type of video detector mentioned in Chapter 12 is a linear detector. One type of linear detector is discussed in the following section.

13.5 SYNCHRONOUS (LINEAR) VIDEO DETECTOR

Diode-video detectors for color and monochrome TV receivers are simple, but have some important disadvantages. These disadvantages are overcome by the use of "synchronous" video detectors for color and monochrome TV receivers. Such detectors are incorporated as part of an IC. The use of a synchronous-video detector permits one video-IF amplifier to be eliminated and fewer IF traps are required.

DISADVANTAGES OF DIODE-VIDEO DETECTORS

The disadvantages of the diode-video detector are as follows:

1. Diodes have a barrier-threshold voltage which must be overcome before the diode conducts. This voltage is about 0.6 V for a silicon diode and about 0.2 V for a germanium diode. The diodes will not have full conduction until they receive about 0.5 V above the barrier voltage. This situation requires about 80 dB of IF-amplifier gain, with a minimum useable antenna signal of 100 μV.
2. With low-signal level diode operation, the diode operates on the nonlinear portion of its characteristic. A nonlinear device acts as a mixer (undesirable here). In this case, the various diode input frequencies, such as the picture-IF carrier, the sound-IF carrier, and the color-IF subcarrier, tend to beat together in the diode to produce sum and difference frequencies in its output. For example, the 41.25-MHz sound IF can beat with the 42.17-MHz color-IF subcarrier to produce a difference frequency of 920 kHz. This frequency will cause an interference pattern on the color tube. To prevent this interference, a 41.25-MHz trap in the IF strip *precedes* the diode-video detector. The 41.25-MHz signal is reduced to a low level prior to the detector to minimize the 920-kHz beat. In addition, a 4.5-MHz trap is placed at the output of the diode-video detector to eliminate the 4.5-MHz beat frequency at this point. This frequency is caused by the difference between the picture-IF carrier at 45.75 MHz and the sound-IF carrier at 41.25 MHz.

   However, if the video detector were a *linear* device (instead of a nonlinear device), such beat frequencies would *not* be produced and some of the traps could be eliminated.
3. A diode detector has no gain and actually has a signal loss. This loss must be made up by additional amplification in the TV receiver.

ADVANTAGES OF SYNCHRONOUS DETECTORS

The use of synchronous detectors is made practical because they are incorporated as part of an IC. The major advantages are as follows:

1. There is no barrier voltage to be overcome.
2. They have appreciable *gain.*
3. They are *linear* devices and therefore do not mix their input frequencies to produce sum and difference interfering output frequencies.

4. Because of the above factors, one IF stage and some IF traps can be eliminated.
5. Synchronous detectors do not produce the *high sum* frequencies common to diode detectors. Such frequencies can radiate into the tuner and antenna and may cause trouble with other TV receivers or even with other radio services. With diode detectors, FCC regulations require that the amount of such radiations be strictly limited. This results in additional cost and a larger receiver size.

13.6 SYNCHRONOUS VIDEO-DETECTOR OPERATION

The synchronous video detector requires two separate inputs, unlike the simple diode-video detector. These inputs are (1) the usual modulated-IF signal and, (2) a reference signal at the picture-IF carrier frequency and phase.

Note: The principle of using two inputs to a detector (demodulator) was described in Chapter 9. This was in connection with the color demodulators, each of which has a chroma signal input and a 3.58-MHz reference-frequency input.

SYNCHRONOUS VIDEO DETECTOR BLOCK DIAGRAM

A simplified block diagram of a synchronous video detector is shown in Figure 13.12. Basically, this diagram consists of two dual-differential amplifiers (shown in schematic form later) and two modulator transistors. The reference picture-IF carrier is fed in push-pull fashion to the differential amplifiers. The outputs of these amplifiers are connected in parallel. The result is a push-pull input and a parallel output.

The two modulator transistors have their inputs in push-pull fashion. These are each connected from the emitters of the differential amplifiers to ground. As a result, they can vary the output currents of the differential amplifiers according to the modulation of the IF signal. These current variations represent the original video modulation as it occurred at the transmitter.

This circuit is called a "synchronous" detector because it detects *only* IF modulation information that is synchronous in frequency and phase with the reference picture-IF carrier. When synchronization is correct, the only output from this detector will be the modulation frequencies. The IF picture-carrier frequency is cancelled in the differential amplifiers which act as *balanced* demodulators.

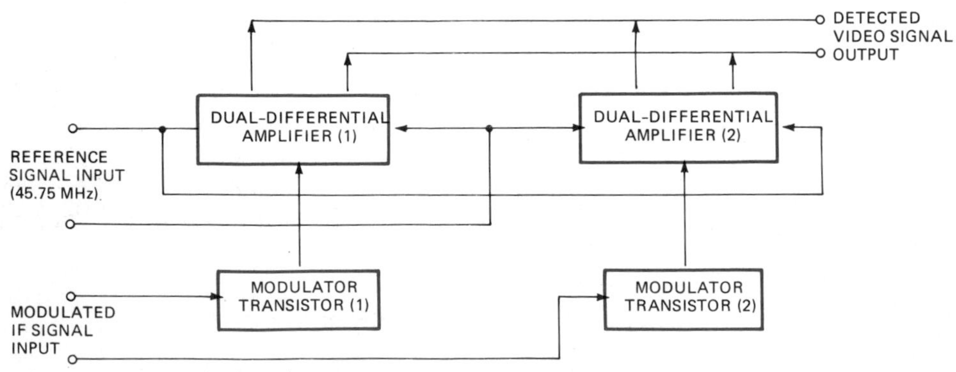

Figure 13.12 Simplified block diagram of a synchronous video detector. Note that two input signals are required.

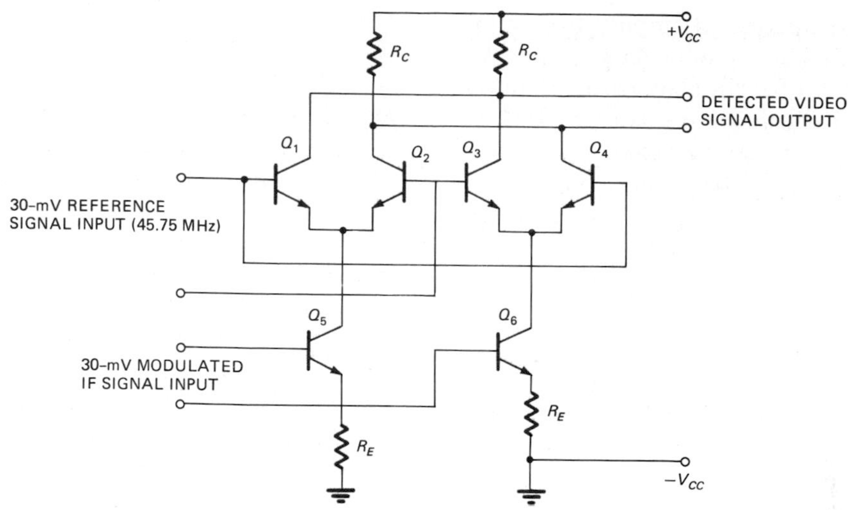

Figure 13.13 Schematic diagram of the synchronous video detector shown in Figure 13.12. The two dual-differential amplifiers consist of Q1, Q2, and Q3, Q4. The modulator transistors are Q5 and Q6.

SYNCHRONOUS VIDEO DETECTOR SCHEMATIC DIAGRAM

Figure 13.13 is the schematic diagram of the synchronous video detector shown in the block diagram of Figure 13.12. Differential amplifier (1) consists of Q_1 and Q_2. Differential amplifier (2) consists of Q_3 and Q_4. Q_5 and Q_6 are modulator transistors (1) and (2) respectively.

Reference Signal

The 45.75-MHz reference signal can be derived by either one of two basic methods. One method is to pass the composite IF signal through a narrow band amplifier which is tuned to 45.75 MHz. Thus, most IF-sideband frequencies are filtered out, leaving an amplified 45.75-MHz reference signal. Any remaining AM sidebands do not influence the operation of the detector for two reasons. The first reason is that the reference-signal amplitude into the detector is about 300 mV. This overdrives the differential amplifiers greatly, clipping off all amplitude variations. The second reason is that the reference-signal amplitude is about 10 times greater than that of the modulated-IF signal. This insures that the reference signal acts only as a carrier which is modulated by the modulated-IF signal.

The second method of generating the reference signal is by the use of a phase-locked loop circuit. (This circuit was described in a previous chapter.) A voltage-controlled oscillator (VCO) is set close to the reference-frequency (45.75 MHz) and is locked (in frequency and phase), to the picture-IF carrier (45.75 MHz) by the phase-locked loop. The output of the oscillator is fed to the reference-signal input of the synchronous detector. The VCO-tuned circuit may be labeled "Reference," or simply "Oscillator."

Added Amplifier

If an added amplifier is required it is included in the IC. Such an amplifier may be used as a narrow-band amplifier. Or an amplifier may be needed to amplify the output of the phase-locked loop circuit. However, the tuned circuit for the narrow-band amplifier will be mounted externally to the IC. This tuned circuit may be labeled "Limiter."

In Figure 13.13, note the complete absence of tuned circuits or traps. No traps follow this detector. Compared to a diode detector, fewer traps precede the detector. In addition, the synchronous detector provides appreciable amplification of the video signal, making it possible to eliminate one IF stage.

13.7 TROUBLES IN THE VIDEO-DETECTOR STAGE

In a monochrome receiver, both the sound and video signals pass through the video-detector stage. Therefore, a faulty detector can affect either or both of these signals. Since the sync pulses also pass through the detector stage, synchronization can be affected by a faulty component in the circuit. If a diode in the detector stage is not entirely defective, the circuit most affected by a weak video signal will show symptoms first. Thus, if a receiver cannot tolerate a reduction of the sync-pulse amplitude, there may be a partial loss of synchronization and a weak picture. However, the sound may be acceptable.

NO SOUND, NO PICTURE

The symptoms of no sound and no picture may be an indication of a faulty detector stage, if a raster is present. Semiconductor diodes are used for detectors in many receivers. When their front-to-back ratio decreases (that is, when the diode becomes faulty), a typical indication is a washed out picture. An image with good contrast cannot then be obtained with the contrast control.

OPEN PEAKING COIL

The diode, if used, is a likely source of trouble in the detector stage. However, it is also possible that one of the video peaking coils is open. The best way to check for an open video peaking coil in the detector stage, or in a video stage, is to short it temporarily while watching the picture. If there is no effect on the picture when the peaking coil is shorted, the peaking coil is not open. If a short-circuit test indicates that the peaking coil is faulty, the next step is to replace it with an identical coil.

DETECTOR IN COLOR SET

If the video-detector stage becomes faulty in a color receiver, it is necessary to use some common-sense troubleshooting. A look at the receiver schematic diagram will tell you which signals are present up to, and including, the detector stage. The sound is taken off at a point before the detector. Therefore, trouble with sound is not likely to be caused by the video-detector stage. Sync signals pass through the detector stage and may receive additional amplification in one or two video-amplifier (Y amplifier) stages. Therefore, a faulty video detector may cause problems with synchronization, a loss of picture contrast, or a total loss of the picture.

DETECTOR IN AN IC

When the video detector is part of an IC, the IC normally contains several other circuits. These include picture-IF amplifiers, sound-IF amplifiers, a video amplifier, and AGC circuits. If the symptom is a poor contrast picture (or none) with possibly poor synchronization, the defective portion of the IC can probably be identified by oscilloscope and/or voltage measurements. However, since the IC cannot be repaired, there is not much point in this procedure. In addition, some of the circuits in the IC, other than the detector, can cause the same symptoms as a defective detector. Therefore, if the trouble is identified as a defective circuit in the IC, it is preferable to replace the IC.

SUMMARY OF CHAPTER HIGHLIGHTS

1. Since the video information amplitude modulates the picture carrier, an AM detector must be used to demodulate this information (Figure 13.1).
2. Two general types of video detectors are used: (1) the nonlinear-diode detector and (2) the linear-synchronous detector. (Figures 13.1, 13.12, and 13.13)
3. Both types of video detectors are suitable for use in color and monochrome TV receivers.
4. The diode detector is a half-wave rectifier which passes either the positive- or the negative-video picture phase, but not both. (Figures 13.2 and 13.3)
5. In Question 4, the picture phase used depends upon the number of video amplifiers used. It also is a function of whether the video signal is sent to the grid or the cathode of the picture tube. (Figures 13.4 and 13.5)
6. The output circuit of a diode-video detector usually contains some form of low-pass filter and high-frequency compensation. (Figure 13.6)
7. In a positive-picture phase signal, the brightest picture information is in the positive direction. The sync pulses are in the negative direction. The opposite is true for a negative-picture phase signal. (Figures 13.1, 13.2, and 13.3)
8. Video amplifiers are generally of the type which causes a reversal of the picture phase. Therefore, depending upon whether the video signal is connected to the picture tube grid or cathode, the number of such amplifiers determines the desired final picture phase. (Figure 13.4)
9. At the output of a diode-video detector, the IF frequencies are filtered out. In addition, if this is a color TV receiver detector, a diode detector is generally followed by a 4.5-MHz trap. (Figures 13.9 and 13.11)
10. To view two full horizontal lines of information at the video detector output, set the oscilloscope sweep frequency to 7 875 Hz (one-half of the horizontal line frequency of 15 750 Hz). (Figure 13.8)
11. The peak-to-peak amplitude of the video signal at the output of a video detector is about 1.0 V to 1.5 V for solid-state receivers. It is about 3.0 V for vacuum tube receivers.
12. In color TV receivers, the sound is tapped off prior to the video detector and sent to a separate sound-IF detector. This helps to prevent the generation of a 920-kHz signal which will produce an interference pattern on the screen.
13. Video detectors must pass all frequencies in the video range, including those of the color signal. (Figure 13.10)
14. When the video detector is part of an IC, it is only one of several circuits in the IC. Other circuits may include picture-IF amplifiers, sound-IF amplifiers, and AGC circuits.
15. Some disadvantages of the diode-video detector are (1) it has barrier voltage to

overcome, (2) it operates as a mixer, because of nonlinearity characteristic, and (3) it has a signal loss through it.

16. Some advantages of the synchronous-video detector are (1) it has no barrier voltage, (2) it has appreciable gain, (3) it operates linearly and thus does not mix frequencies, and (4) it allows one IF stage and some IF traps to be eliminated.
17. The synchronous-video detector requires two inputs, (1) the usual modulated IF wave and (2) a reference frequency at 45.75 MHz.
18. The synchronous-video detector consists basically of two dual-differential amplifiers and two modulator transistors.
19. The only output from a synchronous-video detector are the video modulation frequencies.
20. The *synchronous*-video detector derives its name from the fact that it detects only IF modulation information that is *synchronous* in frequency and phase with the picture-IF carrier.
21. The synchronous-video detector reference signal can be derived directly from the picture IF signal, or by means of a phase-locked loop circuit.
22. There are no tuned circuits in the synchronous-video detector circuit.
23. A faulty video detector can cause the symptoms of (1) no picture, (2) a weak picture, and (3) poor picture synchronization.
24. If a synchronous-video detector becomes defective, the entire IC must be replaced.

EXAMINATION QUESTIONS

(Answers provided at the back of the text)

Part A Supply the missing word(s) or number(s).

1. The video detector may be either a (an) ________ type or a (an) ________ type.
2. Unlike the diode-video detector, the synchronous-video detector operates in a (an) ________ fashion.
3. In the positive-picture phase, the bright-picture information is in the ________ direction and the sync pulses are in the ________ direction.
4. Video information applied to the cathode of a picture tube must be ________ in polarity to that applied to the grid.
5. A ________ following a diode detector removes the IF carrier and high order harmonics.
6. A ________ -MHz trap is found in the output of a color TV video diode detector.
7. The use of a synchronous detector enables the elimination of some ________.
8. Because a synchronous detector has appreciable ________, its IF strip can delete one stage.
9. The two signal inputs to a synchronous detector are the ________ signal and the ________ signal.
10. The ________ is the only output from a synchronous detector.
11. Because it operates in a (an) ________ fashion, a diode detector generates unwanted frequencies.
12. The reference frequency for a synchronous detector is ________ MHz.
13. A symptom of poor synchronization can be caused by a defective ________ .

14. There is a signal ______ through a diode-video detector.

Part B Answer true (T) or false (F).

1. The synchronous detector demodulates only video information that is synchronous with the 42.17-MHz-IF carrier.
2. A diode detector can produce high-frequency interference signals.
3. The synchronous detector is a linear device.
4. Only the positive half of the modulated wave can be used to extract the video information.
5. A video signal having the negative-picture phase can be applied to the grid of the picture tube.
6. The grid and cathode of a picture tube require opposite phases of the video signal.
7. A high-pass filter will be found in the output circuit of a diode detector, together with a 4.5-MHz trap.
8. For a solid-state receiver, the peak-to-peak detector output video voltage is in the order of 1.0 V to 1.5 V.
9. A 920-kHz signal can be obtained by beating together the sound-IF carrier and the color-IF subcarrier.
10. A synchronous detector makes it possible to eliminate all IF traps.
11. The only harmonic output from a linear detector is even order harmonics.
12. An advantage of the diode detector is its low barrier-threshold voltage.
13. A diode detector can be made to operate as a linear detector by applying a forward bias which just overcomes the barrier-threshold voltage.
14. In color TV receivers, a 45.25-MHz trap always precedes a diode detector.
15. A synchronous detector has appreciable gain, making it possible to eliminate one IF stage.
16. The only output from a synchronous detector is the video signal.
17. One of the two signals applied to a synchronous detector is the 41.75-MHz reference signal.
18. In a synchronous detector, the modulator transistors vary the current through the differential amplifiers in accordance with the video modulation.
19. One scheme of obtaining the reference signal is to generate it by means of a VCO in a phase-locked loop circuit.
20. The only tuned circuit in a synchronous detector is the sound takeoff circuit.

REVIEW ESSAY QUESTIONS

1. Draw a simple schematic diagram of a color TV, diode video-detector stage. Include and label any necessary filters and traps.
2. For the diagram drawn in Question 1, briefly explain the operation of the circuit.
3. For the circuit drawn in Question 1, what are some disadvantages of this type of circuit.
4. For the circuit drawn in Question 1, where is the sound-IF takeoff point?
5. Draw a simple sketch of a monochrome TV receiver diode-video detector. Show the input and output waves to produce positive-picture phase.
6. For the detector drawn in Question 5, explain briefly how it is possible to obtain either positive- or negative-picture phase.
7. Define the terms: positive-picture phase, negative-picture phase, series diode-video detector, and shunt diode-video detector.
8. Refer to Figure 13.2A. Can this waveform be applied to the grid of a picture tube? Why?
9. In Figure 13.8, there is a light horizontal line at the blanking-pulse level. What causes this line? Is it synchronized? Why? Do you think it affects the waveform?
10. If the video signal is applied to the cathode rather than the grid of a picture tube, does this affect the reproduction on the screen? Briefly explain your answer.

11. Refer to Figure 13.12. In your own words, briefly explain the operation of the synchronous-video detector. (Also refer to Figure 13.13)
12. Refer to Figure 13.11. If R_1 opens, what effect will there be on the picture? Why?
13. In Figure 13.11, what is the effect of an open L_4 on the picture? Why?
14. List three disadvantages of a diode-video detector.
15. List four advantages of a synchronous-video detector.
16. State briefly why the word "synchronous" is used to describe the synchronous-video detector.
17. In the synchronous-video detector, briefly explain why AM on the reference signal does not affect the detector operation.
18. Name and describe two methods of generating the reference signal for a synchronous-video detector.

EXAMINATION PROBLEMS

(Answers to selected problems are provided at the back of the text.)

1. In Figure 13.6, identify the following: (1) low-pass filter, (2) load resistor, (3) stray capacities.
2. Refer to Figure 13.8. If you wanted to display three complete waveforms, what would be the oscilloscope horizontal sweep frequency?
3. Draw simple waveforms illustrating positive- and negative-picture phases.
4. In Figure 13.14, fill in the word(s) for the question marks. Also, what is missing from the diagram that is necessary for a color-TV receiver?
5. The frequencies of 41.25 MHz, 42.17 MHz, and 45.75 MHz are beating together in a nonlinear diode-video detector. Calculate all the possible sum and difference frequencies. Calculate the second and third harmonics of each of the three original frequencies.

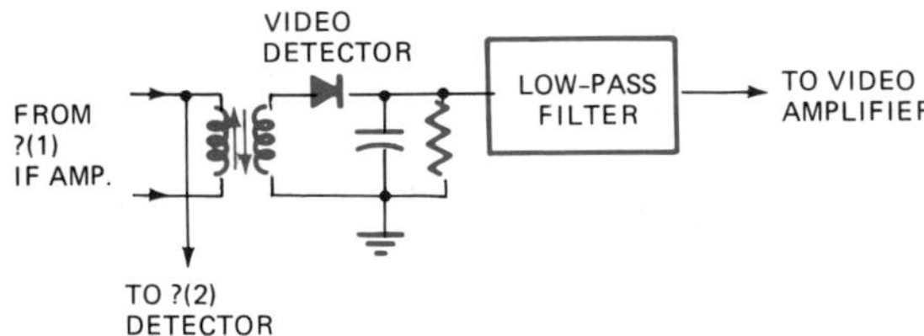

Figure 13.14 Illustration for Problem 4. This should be a diode-video detector for a color TV receiver.

Chapter 14

14.1 General types of AGC circuits
14.2 Peak-AGC systems
14.3 Disadvantages of peak AGC
14.4 Keyed-AGC systems
14.5 Delayed AGC and diode clamping in AGC circuits
14.6 AGC systems in solid-state receivers
14.7 Noise cancellation circuits
14.8 Solid-state AGC system
14.9 AGC system in an RCA IC
14.10 Expanded block diagram of RCA IC noise processor
14.11 Expanded block diagram of RCA IC AGC processor
14.12 Troubles in AGC circuits
Summary

AUTOMATIC GAIN CONTROL (AGC) CIRCUITS

INTRODUCTION

Automatic volume control (AVC), or more accurately, "automatic gain control," in a radio receiver keeps the output sound at a constant volume while wide variations occur in the strength of the input signal. Once the manual volume control is set to the output level that is desired, the AVC system tends to keep it there. In addition, when tuning to other stations, no adjustments are necessary to prevent a strong station from producing excessive volume. For television receivers, automatic gain control (AGC) is advantageous in keeping the picture contrast fixed at one level, even when the strength of the video signal at the input of the set may be varying. The eye is far more perceptive of changes than the ear, and anything that minimizes unwanted variations in image intensity is desirable. AGC is advantageous when switching from one station to another because input signal strengths may differ. Finally, more stable synchronizing is obtained if the signal fed to the synchronizing circuits is constant in amplitude.

One other important function of the AGC system is to prevent picture *overload* due to excessive RF amplifier and common-IF amplifier gain, for a strong signal. An example of picture overload is shown in Figure 14.1. Note the excessive contrast and poor horizontal synchronization.

AGC REFERENCE LEVEL

A study of the television modulated signal in Figure 14.2 reveals that, so far as AGC is concerned, the rapidly varying camera signal is of little use. We desire some point which will indicate the strength of the carrier and which changes only with the carrier amplitude.

Figure 14.1 An example of an overloaded picture, caused by improper AGC action. Note the washed out appearance and poor horizontal synchronization.

With the present system of transmission, the carrier is always brought to the same level when the synchronizing pulses are inserted. Thus, while the signal being received is constant in strength, the level of the synchronizing pulses will always reach the same value. If something affects the carrier level, the amplitude of the pulse level will also change. Because of the change, the gain of the set must be adjusted to maintain the previous level at the detector. Hence, the strength of the synchronizing pulses serves as a reference level for the AGC system.

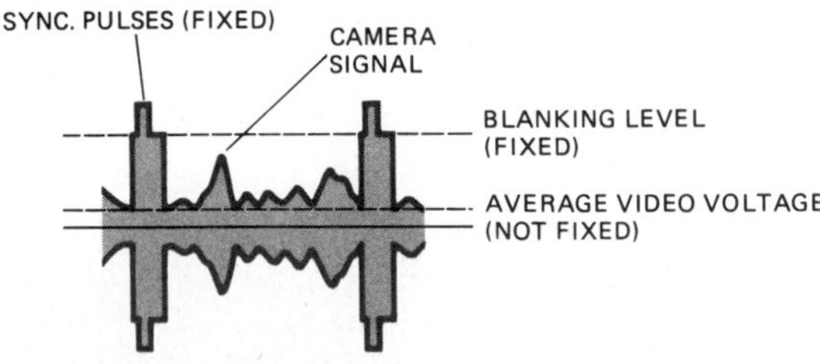

Figure 14.2 An amplitude-modulated television signal. The fixed voltage levels are suitable for AGC reference, since these voltages vary directly with the received signal strength.

AGC FUNCTION SUMMARY

The function of the AGC circuit in a television receiver is to stabilize the video signal against input signal variations. Both the video and audio signals are stabilized by using an AGC system. A voltage that depends upon the amplitude of the sync pulses is obtained and used to control the gain of the RF and IF stages. This is the essence of the AGC system in both monochrome and color receivers. However, there is a considerable difference between the method of controlling the gain of a vacuum tube amplifier and the method of controlling the gain of a transistor amplifier. Therefore, this chapter will treat tube and transistor AGC circuits separately, beginning with the tube circuits. Two general methods are employed to develop AGC voltages: *peak* and *keyed*. Each method will be considered in turn.

By the time you have completed the reading and work assignments for Chapter 14, you should be able to:

- Define the following terms: peak AGC, keyed AGC, coincidence circuit, delayed AGC, and noise cancelling.
- Briefly discuss the need for AGC in a television receiver.
- List the disadvantages of a peak-AGC system.
- Draw a simple schematic diagram showing how forward AGC is applied to an NPN-transistor amplifier.
- Explain briefly how positive or negative AGC-control voltage can be used with transistor amplifiers.
- Explain how a keyed-AGC system overcomes the disadvantages of the peak-AGC system.
- Explain how a 60-Hz ripple is developed in a peak-AGC system.
- Explain how 60-Hz intercarrier buzz may occur.

14.1 GENERAL TYPES OF AGC CIRCUITS

Before discussing the general types of AGC circuits, a brief review should be made of the means by which AGC can operate. The gain of a vacuum tube or FET can be varied by changing its bias, if the tube or FET is operating on a nonlinear portion of its characteristic. The portion of the characteristic used for AGC control is that approaching cutoff. Tubes having a remote-cutoff characteristic are generally more suitable for AGC control. For tubes or FETs, as the bias is *increased* and the cutoff characteristic is approached, the device will provide *less* amplification. The converse is also true. Thus, changing the bias voltage provides a means of controlling the gain of the controlled amplifiers.

The bias change is accomplished automatically in response to the incoming-signal strength. Thus, a weaker signal will cause less bias to be developed with a consequent increase in the RF- and IF-amplifier gain. On the other hand, a stronger signal will cause a greater bias to be developed, which will lower the RF- and IF-amplifier gain. The net result (within design limits) is to maintain a stable picture contrast regardless of the variations of the input signal strength.

AGC WITH TRANSISTOR AMPLIFIERS

When transistor amplifiers are controlled by AGC bias, the *basic* principle of operation is the same as for vacuum tubes. That is, the gain of the transistors is controlled by varying their bias automatically. However, since transistors (excepting FETs) have widely different characteristics than tubes, their applied voltages and methods of obtaining AGC action are somewhat different.

For example, either positive or negative AGC control voltages may be used with transistor RF- or IF-amplifiers because these amplifiers may be of either the NPN or PNP type. Additionally transistor amplifiers may use either forward or reverse AGC. These points will be explained in this chapter.

TYPES OF AGC CIRCUITS

There are several types of AGC circuits which will be described briefly here and in detail later in this chapter. These circuits are used with transistor and vacuum tube sets. Regardless of the method by which the AGC voltages are developed, they are fed to the RF amplifier (in the tuner) and also to the first (and sometimes also the second) common-IF amplifier. These voltages vary according to the strength of the received signal. They act to maintain a relatively constant video detector output over a wide range of antenna input signals (from about 75 μV to well over 100 000 μV). The AGC voltage fed to the RF amplifier is designated as "delayed." This term will be explained shortly.

Peak (or "simple") AGC This type of AGC circuit utilizes the peak amplitude of the sync pulses to develop the AGC voltage. The peak-amplitude voltage is rectified, filtered, and fed to the RF and IF amplifiers. AGC voltage is always a DC voltage.

Keyed AGC This circuit is widely used and is also known as "gated" AGC. With this system AGC voltage is developed only during the horizontal-retrace time. It has important advantages over peak AGC. These advantages are (1) faster response and (2) much better noise immunity.

Noise Cancelling Keyed AGC This system is similar to keyed AGC. However, it has additional circuitry to improve further the noise immunity of the system.

Delayed AGC This is generally used only for the RF amplifier. The "delay" is one of voltage, not time. There is no AGC action until the input signals exceed a certain minimum value. This precaution prevents any decrease in the gain of the RF amplifier when weak signals are being received.

Amplified AGC This is used to increase the sensitivity of the AGC system. It also functions to invert the polarity of the AGC voltage, as required in some circuits.

14.2 PEAK-AGC SYSTEMS

In peak AGC systems, the peak voltage, as represented by the tips of the sync pulses, is used to develop the AGC voltage. A separate diode rectifier is used to receive the same video-IF signal as the video detector. Diode conduction occurs for positive polarity signals. During this period a capacitor is charged to the peak voltage of the sync pulse. It is this voltage which is then filtered and employed for automatic gain control.

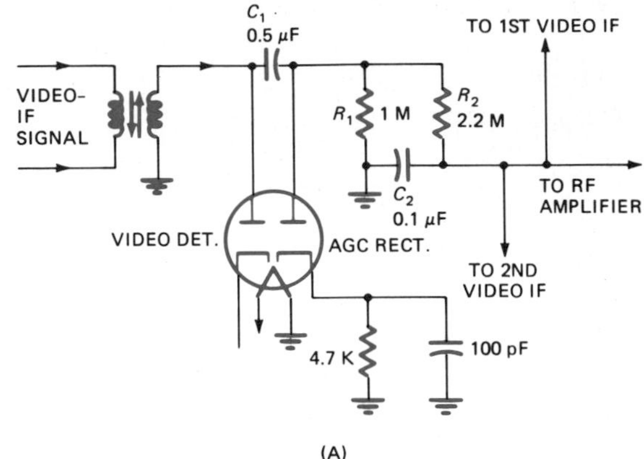

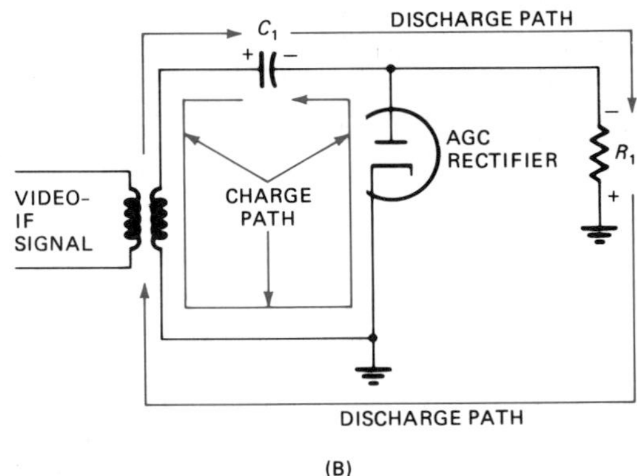

Figure 14.3A-B. A peak-AGC circuit. **A.** The schematic diagram. **B.** Simplified circuit showing the charge and discharge paths of capacitor C_1.

PEAK-AGC CIRCUIT

A typical peak-AGC circuit is shown in Figure 14.3A. The AGC diode is one section of a twin diode. (Although a twin diode vacuum tube is shown here, two semiconductor diodes may be used instead.) The diode receives the incoming signal from the video-IF system through capacitor C_1. The load for the AGC rectifier is R_1, a 1MΩ resistor. To understand the operation of this circuit, consider the equivalent diagram shown in Figure 14.3B. (The cathode resistor and capacitor are omitted from the equivalent diagram because they do not affect the AGC-voltage development.) The AGC rectifier will not conduct until its plate is driven positive with respect to its cathode. When this occurs, electrons flow from the cathode to the plate of the diode and into C_1, where the negative charge is stored. Because of the low impedance offered by the diode when it is conducting, C_1 charges to the peak of the applied voltage, which is the peak value of the synchronizing pulses.

Negative Signal Excursion

During the negative excursion of the incoming signal, the plate of the diode is driven negative with respect to its cathode. No conduction occurs through the tube. However, examine Figure 14.3B and note that a complete circuit exists with C_1, R_1, and the input coil all in series. Since a voltage exists across C_1 and a complete path is available, current will flow. This makes the upper end of R_1 negative with respect to ground. Because of the long time constant of R_1 and C_1, the charge accumulated across C_1 will discharge slowly through R_1. This discharge is so slow that only a small percentage of the voltage across C_1 will be lost during the interval when the tube is not conducting.

Positive Signal Excursion

When the incoming signal becomes positive again, the tube does not immediately conduct because the applied signal voltage must first overcome the voltage that remains across C_1. Since C_1 has lost little of its voltage, diode conduction will occur only at the very peak of the positive cycle. These peaks are the synchronizing pulses. Thus, the voltage across C_1 is governed entirely by the sync pulses, which is what we desire. The negative voltage across R_1 is filtered by R_2 and C_2 to remove the 15 750-Hz ripple of the horizontal sync pulse and the low-frequency video components. The rectified DC voltage is then fed to the RF and video-IF amplifiers as the AGC control voltage.

14.3 DISADVANTAGES OF PEAK AGC

The peak-AGC system is simple with regard to circuitry. However, it has several important disadvantages, as follows:

1. The peak system develops AGC voltages even for weak signals. These voltages will reduce RF amplifier gain on weak (as well as strong) signals. A poor signal-to-noise ratio results from weak signals.
2. Peak AGC has a *slow response*. This is necessitated by the requirement to remove the low-frequency video signals and vertical-sync pulse components at the output of the AGC rectifier. The removal is accomplished by the long-time constant filter (R_2, C_2) in Figure 14.3A, (0.22 sec). With a slow AGC response, the receiver gain cannot change rapidly enough to eliminate the effects of rapid signal strength changes. Such changes are caused by reflections from passing aircraft and also when switching channels.
3. Noise pulses are rectified by the AGC diodes along with the desired signal. If interference is present, a higher AGC voltage will be developed. This reduces the gain of the RF-amplifier and IF-amplifier(s), more than is desired. The result is a decrease in the signal-to-noise ratio of the receiver.
4. The control range of the simple, peak-AGC system is small. It cannot respond adequately to large changes of input-signal strength.

These disadvantages can be serious and may greatly reduce the viewer's enjoyment. However, they can be eliminated by the use of a "keyed" AGC system, which also incorporates *delayed* AGC for the RF amplifier.

14.4 KEYED-AGC SYSTEMS

One important advantage of keyed-AGC systems is the elimination of the picture intensity variations caused by rapid signal strength changes. Perhaps the most annoying of these variations is caused by passing aircraft. The effect is known as "airplane flutter."

AIRPLANE FLUTTER

Airplane flutter occurs whenever an airplane passes overhead or nearby. The picture intensity rises and falls, becoming light and dark in turn at a fairly rapid rate. This effect may last from 15 to 30 seconds, depending upon how long it takes the airplane to pass.

The intensity pulsation is caused by the airplane acting as a reflector. Some of the television signals striking the metallic surface of the airplane bounce off and reach the television antenna. When these reflected television signals arrive in phase with the normal signal that the antenna receives, they will *add* to the desired signal and strengthen it. When the reflected signals arrive out of phase with the normal signal, they subtract from the desired signal. Thus, the strength of the normal signal will be reduced. As the airplane moves, the reflected signal alternately adds to and then subtracts from the desired signal, depending upon the length of the reflected signal path. This rapid increase and decrease in the strength of the received signal cannot be counteracted by a slow-acting AGC filter. Therefore, the picture on the screen will vary in intensity.

A keyed-AGC system is able to overcome this flutter for two reasons. First, the AGC system is receptive to incoming signals only at certain specific times. Second, the resistance and capacitance that make up the AGC filter are lower in value than the corresponding components in conventional AGC systems. This means that the time constant of these components is shorter. Therefore, a keyed-AGC network can react quickly to a fairly rapid signal fluctuation, such as that produced by airplanes. The network can change the AGC bias fast enough to counteract this signal change.

SIMPLIFIED KEYED-AGC SYSTEM

A simplified keyed-AGC system is shown in Figure 14.4A. Here, a pentode is connected so that the detected video signal is applied to its control grid. The signal is in the negative picture phase, resulting in positive sync pulses. The plate of the pentode is connected to a winding on the horizontal-output transformer. The plate receives from this winding a positive pulse of voltage at the end of each horizontal line.

The pentode tube is biased so much that it will not conduct unless the grid and plate are positive simultaneously. If just one of these voltages is negative, the tube will not conduct.

The pulses applied to the grid are the horizontal-sync pulses. When these pulses arrive, the electron beam traveling across the face of the picture tube is about to start its *retrace*. As the retrace progresses, a large pulse of voltage is developed in the horizontal-output transformer and a portion of this pulse is fed to the plate of the tube. With both positive pulses of voltage present (one at the plate and one at the grid), the AGC tube is keyed into conduction and the AGC-bias voltage is developed. The coincidence of the two types of pulses is shown in Figure 14.4B. Note that the tube is turned on *only* for a short portion of each horizontal cycle.

Note the foregoing sequence of events carefully. This sequence contains the key to the operation of this system. Positive pulses must be present at both the control grid and the plate of the tube. Otherwise, they will not pass current and establish the proper AGC bias. The plate receives no positive

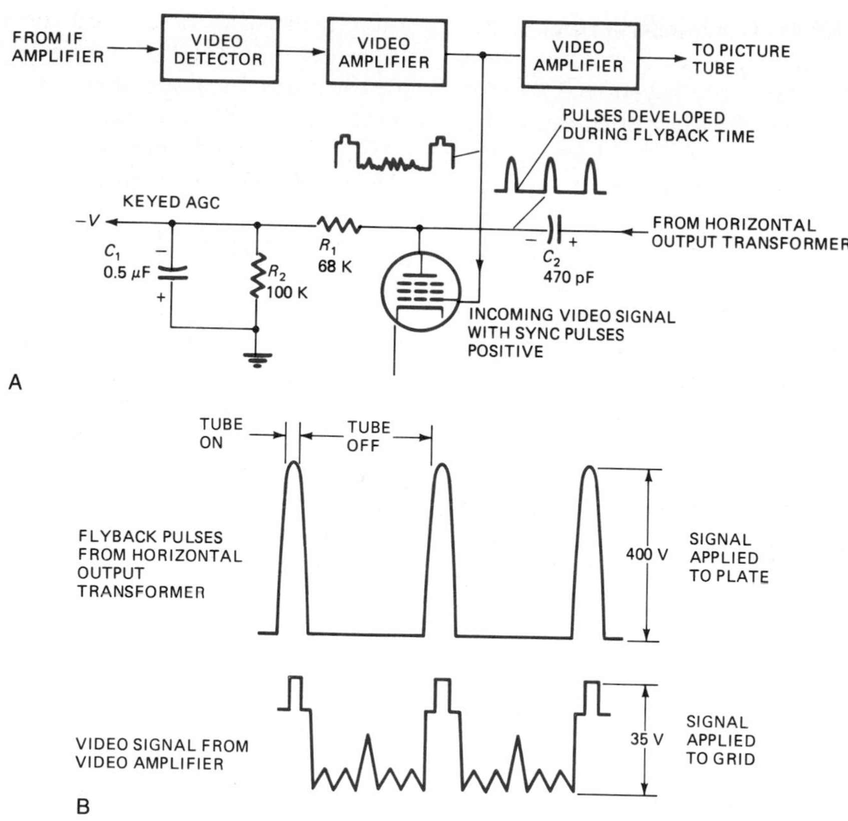

Figure 14.4A-B. **A.** The simplified diagram of the components of a keyed-AGC system. **B.** The flyback pulses and the horizontal-sync pulses coincide to turn on the keyed-AGC tube of part **A.** The flyback pulses supply "plate voltage" to the tube. The video signal overcomes the tube cutoff bias only at the sync-pulse level.

voltage other than that furnished by the horizontal-output transformer.

The tube conducts only when the sync pulses are active at its grid. It is inactive throughout the remainder of the video signal. Therefore, it is evident that the AGC tube (and consequently the AGC network) is responsive to undesirable noise pulses for only a very short period of time. Actually, the sync pulses occupy only 5% of the composite video signal. Therefore, only 5% of the total noise can be effective.

Contrast this with AGC systems other than the keyed type. They are supposed to be unresponsive to all but the sync-pulse tips. But this is true only if the amplitude of the sync pulses is greater than the amplitude of any of the noise pulses present. Any noise signal possessing a greater amplitude than the sync pulses will cause current to flow in the AGC circuit. Thus, noise pulses in such systems will develop a greater negative biasing voltage in the AGC network than that obtained from the sync pulses alone. Until this greater negative voltage diminishes and the normal sync pulses again resume control, the gain of the set will be reduced.

KEYED CIRCUIT OPERATION

When the tube in the keyed-AGC circuit of Figure 14.4A conducts, it charges the plate side of C_2 negatively. The *charging* current passes through the horizontal-output transformer to ground and back to the cathode of the keyed-AGC tube. Between pulses (when the tube is nonconducting), C_2 discharges through R_1 and R_2 and charges the AGC-filter capacitor C_1 with the negative polarity as shown in Figure 14.4A.

Note: A negative AGC voltage is required to provide AGC action for vacuum tube amplifiers. This is shown in the circuits for Figures 14.3 and 14.4A. However, for transistor amplifiers, the AGC voltage may be either negative or positive, as explained later in this chapter.

The AGC Filter

R_1, C_1, and R_2 form the AGC filter for this keyed system. The time constant of this filter (0.05 sec) is considerably shorter than that in the circuit of Figure 14.4A (0.22 sec). Thus, it can respond faster to rapid signal variations similar to those caused by airplane flutter. The shorter time constant is made possible because in this circuit there is no need to filter the 60-Hz vertical-sync pulses. These pulses cannot affect AGC circuit operation, as will be explained presently. The AGC filter need only be effective for the 15 750-Hz horizontal frequency.

Why No 60-Hz Ripple?

There is one question which will occur to many readers: Why do the peak-AGC networks tend to develop a 60-Hz ripple, while the keyed-AGC networks do not? To understand why, refer to Figure 14.5, where horizontal-sync and vertical-sync pulses are shown.

Note that the period between horizontal sync pulses is "H" (63.5 μs). Thus, the AGC filter network can discharge (partially) for this period. However, if you observe the equalizing pulses (c) and (d) and the vertical-sync pulse interval, note that the available discharge time between pulses is less than "H". Thus the DC-AGC voltage will tend to rise during these pulses. A long-time constant filter is required to filter the 60-Hz ripple.

In the keyed-AGC system shown in Figure 14.4A, the pentode is "fired" by the *combination* of a positive-plate pulse and a positive-grid pulse. The plate pulse, however, is

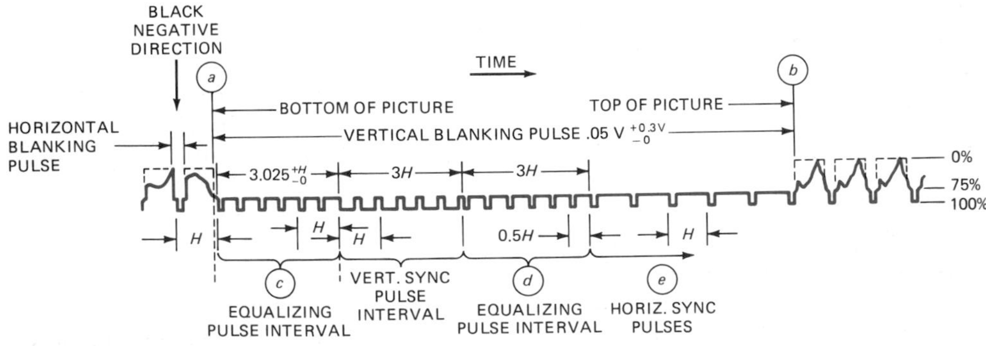

Figure 14.5 The relative time intervals for the horizontal-sync pulses and the vertical-sync pulse interval.

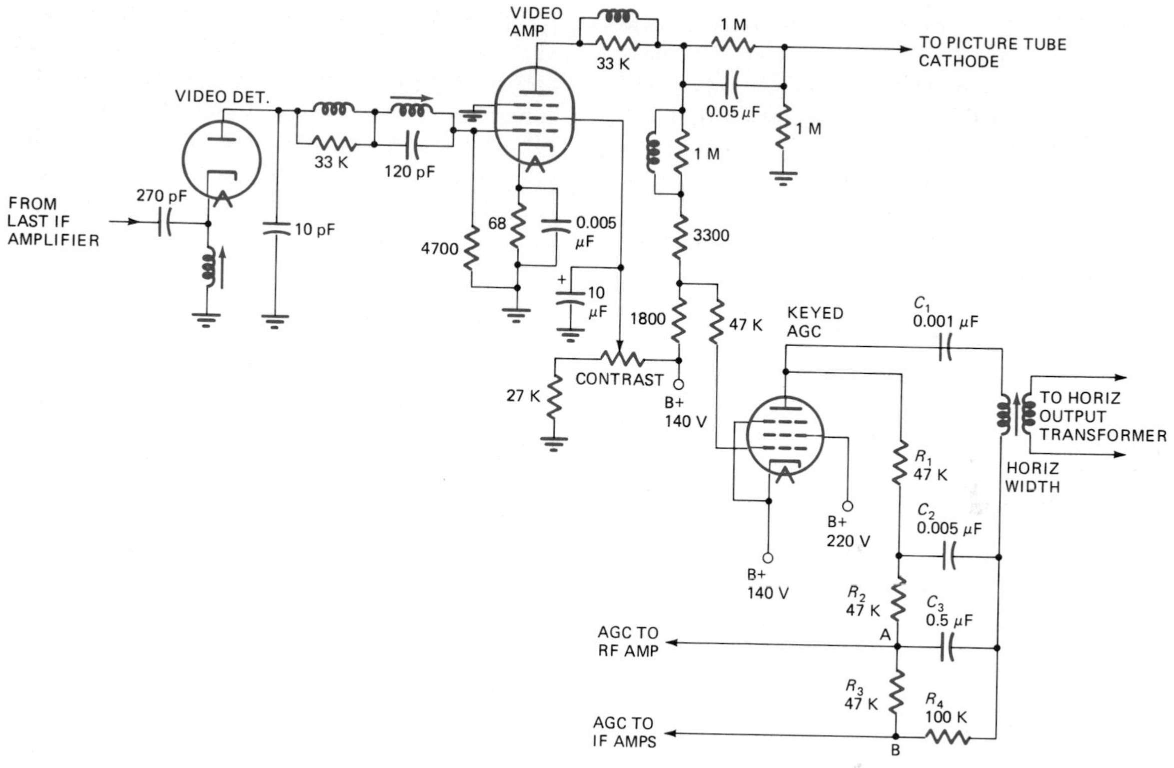

Figure 14.6 An example of a vacuum tube type of keyed-AGC system. Some systems use a triode for the keyed-AGC tube. Note the DC coupling to the grid of the keyer tube.

constant in duration since it is obtained from the horizontal-output transformer where the pulses do not change. Hence, whether the grid pulse is a horizontal-sync pulse or a vertical-sync pulse, the tube conducts for the same length of time. It is because of this behavior that only a 15 750-Hz ripple is present. It is this ripple frequency only that must be filtered in keyed-AGC systems.

Sync-Pulse Alignment

In applying the sync pulses to the grid of the AGC tube, care must be taken to see that they are all aligned to the same level. This alignment is necessary because the amount of current flowing through the AGC tube is determined in large measure by the amplitude of the sync pulses. If the sync-pulse amplitudes vary at the grid, the plate current and the AGC bias will also vary. To maintain the sync pulses at a constant level, DC coupling is used between the video amplifier output and the grid of the keyer tube. If AC coupling must be used, another method of maintaining sync pulse alignment is to use a diode-clamping circuit at the grid of the keying tube. (If the keyer tube draws grid current, this action will provide *clamping* of the sync pulses.)

COMPLETE VACUUM TUBE KEYED-AGC SYSTEM

A complete schematic diagram of a vacuum tube, keyed-AGC system is shown in Figure 14.6. Inspection of the diagram reveals that the video signal applied to the keyed-AGC tube is obtained from the plate circuit of the video amplifier. At this point the sync pulses are positive. (Note that the signal from this point is fed directly to the cathode of the picture tube.) A network consisting of a 3.3kΩ resistor, a 1.8 kΩ resistor, and a 47 kΩ resistor directs part of the video signal to the grid of the AGC tube. The network is designed to minimize the shunting effect of the AGC tube on the video-amplifier network. This upholds the high-frequency video response of the amplifier-coupling network.

The cathode of the AGC tube is connected directly to the 140-V B+ point. This is necessary because of the high-positive potential present on the control grid of the tube. Actually, with the AGC tube in operation, the control grid is approximately 25 volts less positive than the cathode. As a result, the tube does not conduct except when the horizontal-sync pulses are present.

AGC Voltage Development

A positive potential of 220 volts is applied to the screen grid of the AGC tube. The flyback pulse for the plate is taken from a special winding placed over the horizontal-width coil. This coil is connected across the secondary of the horizontal-output transformer. When the keyed-AGC tube conducts, capacitor C_1 (in the plate) charges negatively on the plate side. Between pulses, C_1 discharges through R_1, R_2, R_3, and R_4. The negative voltage drop across R_3 and R_4 (point A) is the RF amplifier AGC bias. The negative voltage drop across R_4 (point B) is fed to the IF amplifier(s) as the AGC bias. R_1, C_2, R_2, C_3, and R_4 form the keyed-AGC filter.

Under average signal conditions, the AGC voltage measured at the control grid of the controlled video-IF amplifier(s) is approximately −4.5 volts. This voltage will vary with signal strength and the contrast-control setting. The latter control adjusts the screen-grid voltage of the video-amplifier tube between the limits of +64 and +140 volts. At +64 volts the video signal is cut off completely; at +140 volts it receives its maximum amplification.

Use Of A Triode

Although a pentode vacuum tube has been shown as the keying tube in Figures 14.4A and 14.6, many vacuum tube receivers use triodes for this function. In this case, as with the pentode, the video signal is fed to the grid and the horizontal pulse from the flyback transformer is used as the plate voltage. The basic operation is the same. However, the triode grid requires a higher amplitude video signal. This higher amplitude is readily available in many receivers at the output of the video amplifiers. In the case of triodes, a peak-to-peak video signal in the order of 70 V to 80 V is used.

14.5 DELAYED AGC AND DIODE CLAMPING IN AGC CIRCUITS

Before other keyed-AGC systems are considered, there is a method used in vacuum tube receivers by which the AGC voltage fed to the RF amplifier is varied differently from the AGC voltage applied to the video-IF stage(s). This difference arises because the RF amplifier must be operated, under weak signal conditions, at maximum gain until the input signal reaches a value of about 500 μV. It is desirable to present as much signal to the mixer as possible, since the mixer is one of the greatest sources of noise in the receiver. With a large input signal, the mixer output signal-to-noise ratio

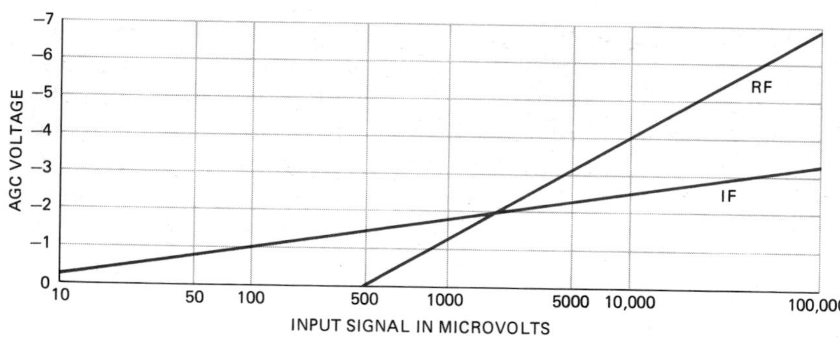

Figure 14.7 The desired variation of AGC voltage versus the input signal for the controlled vacuum tube RF and IF stages.

will be more favorable than it would be if the gain of the RF amplifier were less than maximum for weak signals.

Actually, what we are seeking to do is to delay the application of an AGC voltage to the RF amplifier until the input signal attains a level of about 500 μV. Thereafter, the AGC voltage at this stage should rise fairly rapidly to avoid overloading the mixer. Figure 14.7 shows the desired variation of the AGC voltage versus signal strength for the RF and IF stages. Note that IF control is initiated immediately. The RF control is first delayed and then made to rise quite sharply.

DELAYED AGC CIRCUIT

One method of obtaining a delay in AGC for RF is shown in Figure 14.8. Tube V_1 is the keyed AGC tube which operates in the same fashion as the tube of Figure 14.6. The AGC voltage which V_1 develops appears across C_2. From this point it is distributed to the video IF and RF stages through R_2, R_3, R_4, and R_5. Of these components, R_3 and C_2 are readily recognizable as an additional filter for the AGC voltage fed to the first video-IF stage. Resistor R_2 is an isolating resistor designed to keep the AGC variations at the RF tube distinct from the AGC variations occurring at video-IF tubes.

Diode V_2 and resistors R_4 and R_5 all connect point A to the line that goes to the RF amplifier. If we first concentrate only on R_4 and R_5 and disregard V_2, then the +310 volts applied to one end of R_5 divides between R_5 and R_4. However, since R_5 is so much larger than R_4, 2 volts appears across R_4 making point A 2 volts positive with respect to the ground. If diode V_2 were not present, this +2 volts would be applied to the control grid of the RF amplifier. The tube conducts by connecting the plate of V_2 to point A and the cathode of V_2 to ground. The plate resistance of V_2 under these conditions is low enough so that resistor R_4 is shunted by what amounts to nearly a short circuit. The voltage at point A drops to almost zero. Furthermore, point A remains fairly close to zero as long as the diode is conducting.

While all this is happening at point A, point B remains at whatever negative voltage V_1 develops across C_2. It is the purpose of R_2 to provide some isolation between points A and B. However, as the signal level rises and point B becomes increasingly negative, point A becomes less and less positive until the voltage across C_1 is high enough to make point A slightly negative. At this moment, V_2 stops conducting and the voltage at point A becomes more and more negative as the generated AGC voltage increases. By properly proportioning the resistive divider

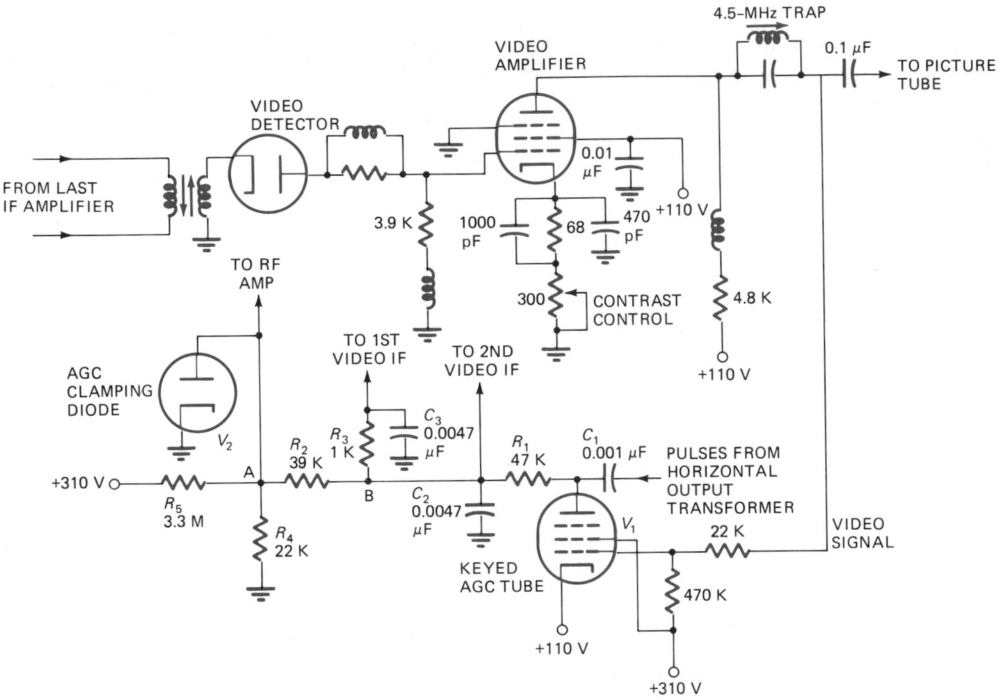

Figure 14.8 A vacuum tube keyed-AGC system employing delayed AGC for the RF amplifier. The clamping diode V_2 is used to provide the delay, as shown in Figure 14.7.

in the RF branch, the AGC voltage at point A can be made to rise faster than the AGC voltage reaching the video-IF stages. This results in the desirable situation illustrated graphically in Figure 14.7.

An important thing to remember is that when the incoming signal is weak, the AGC negative bias is small. Therefore, the potential at point A is positive and near zero because of the presence of the clamping diode. The controlled video-IF amplifiers have a bias close to −1 volt under the same signal condition. But as the signal strength increases, so does the negative AGC bias. Part of it overcomes the slight positive potential at point A, driving the clamping diode into nonconduction and raising the negative grid bias of the RF amplifier.

14.6 AGC SYSTEMS IN SOLID-STATE RECEIVERS

There are many similarities between the AGC systems used in tube receivers and those used in solid-state receivers. For example, the AGC circuitry in a solid-state receiver may or may not be keyed. However, the keyed-AGC system is strongly preferred. Also, the AGC voltage to the RF amplifier is generally "delayed" so that the gain is not reduced for weak signals. As in the case of tube receivers, the AGC voltage in a transistor receiver is delivered to the RF and IF stages. (Normally, only the first two IF amplifiers are controlled by the AGC voltage.) Also the AGC voltage in solid-state receivers may be amplified to obtain a greater change in gain for a small change in signal strength.

While there are similarities between the AGC systems in tube and transistor receivers, there are also some very important differences. Tube circuits differ from transistor circuits in that different characteristics of the same type of transistor may strongly affect the operation of the AGC circuit. For this reason, transistors that are used for replacements in tuner and IF stages should closely match the characteristics of the transistors being replaced. The transistors must not only have similar characteristics, but they must also be similar types. Transistors of different types should not be interchanged.

TRANSISTOR AGC BIAS

Transistor Current Gain The term "beta" (B) is applied only to common-emitter (CE) transistor amplifiers. It is a measure of the **current gain** of this type of circuit. Beta may be defined as the ratio of the change of collector current to the change of base current that produced the change of collector current. It is expressed by

$$B = \frac{I_c}{I_b}$$

For example, if a change of collector current of 100 μA is caused by a change of base current of 1 μA, then beta is,

$$B = \frac{100}{I} = 100.$$

The magnitude of beta for a given transistor is a function of the base-to-emitter forward bias. The beta may be increased or decreased by changing the base-emitter bias. This is the means of obtaining AGC action.

Amount of Bias Change In the case of a silicon transistor, the amount of AGC bias variation is much less than for a vacuum tube. To change from cutoff to saturation requires a change of about 0.5 V. In the case of a germanium transistor, the required bias change is only about 0.25 V.

As will be explained presently, a specific value of forward bias on a transistor will result in maximum current gain. Any change of this bias, either an increase or a decrease, will result in a *reduction* of amplifier current gain. This characteristic leads to two methods of transistor AGC control. These are called "forward AGC" and "reverse AGC."

Forward AGC With "forward AGC," the gain of a transistor is reduced by *increasing* the base-to-emitter bias, Figure 14.9. As the bias is increased toward saturation, the gain of the transistor decreases. For an NPN transistor, with AGC voltage applied to the base circuit, a *positive* voltage increase will *reduce* amplifier gain. For a PNP transistor, with AGC voltage applied to the base circuit, a *negative* voltage increase will *reduce* amplifier gain.

Note: In practice, an AGC-controlled amplifier is not operated at the peak of the curve shown in Figure 14.9. Rather, it is operated on one side or the other of the peak. The side depends upon whether forward or reverse bias is used. This insures stability of

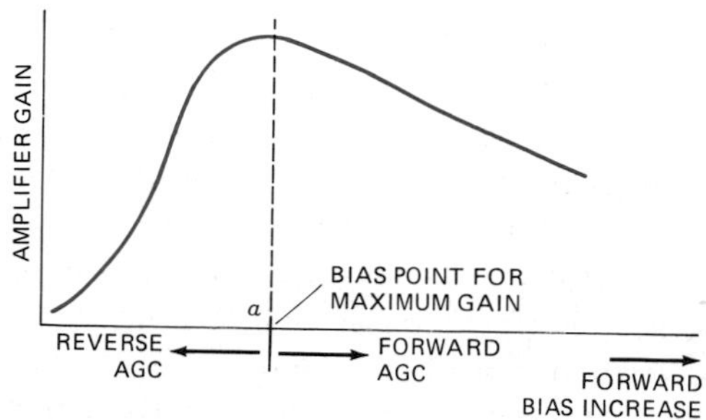

Figure 14.9 This curve shows how the AGC voltage can reduce the transistor's gain by either increasing or decreasing the forward bias.

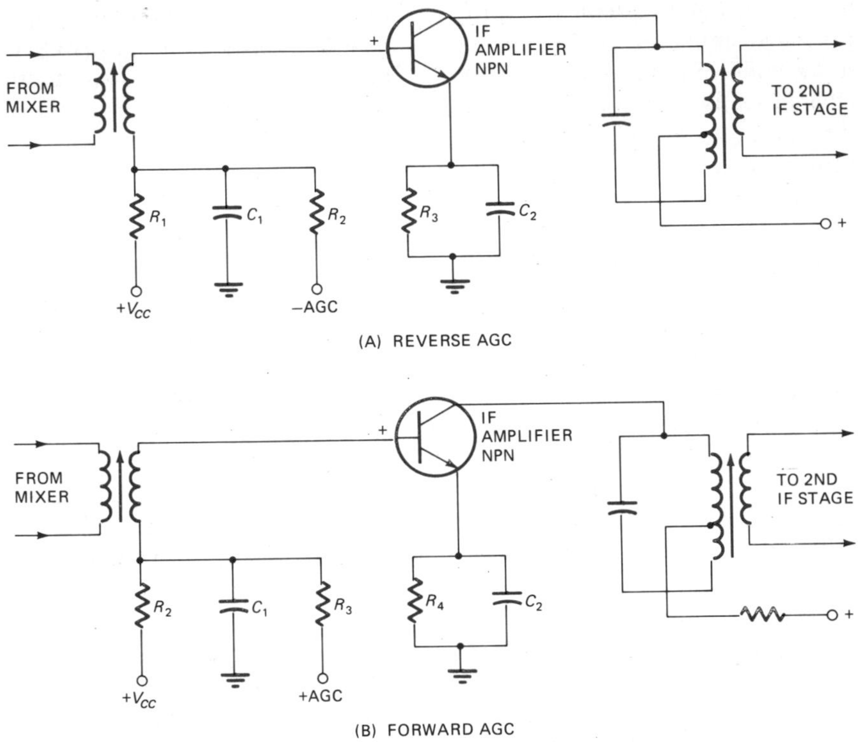

Figure 14.10 Comparison of NPN transistor amplifiers controlled by AGC. **A.** Reverse AGC requires a negative AGC voltage source. **B.** Forward AGC requires a positive AGC voltage source.

operation since the operating points will not pass over the peak from one slope to the other.

Reverse AGC With "reverse AGC," the gain of a transistor is *reduced* by *decreasing* the base-to-emitter bias. This is also shown in Figure 14.9. As the bias is decreased toward cutoff, the transistor gain decreases. For an NPN transistor, with AGC voltage applied to the base circuit, *reducing* the *positive* voltage at the base will reduce amplifier gain. For a PNP transistor, a *negative* base-voltage *decrease* will *reduce* amplifier gain.

ADVANTAGE OF REVERSE AGC OVER FORWARD AGC

To compare the advantages of forward AGC and reverse AGC, we will first look at a typical amplifier system for each type. Figure 14.10 shows two amplifiers with AGC bias. In the circuit of Figure 14.10A, *reverse* AGC is utilized. This means that the gain of the stage is reduced when the AGC voltage makes the base-emitter bias of the NPN transistor less positive. (This, of course, is only true for the NPN-type transistor shown. If a PNP-type transistor is used, the voltages will be the reverse polarity of those pictured.) Note that the collector circuit is connected directly to the B+ line through the output transformer primary.

In Figure 14.10B the amplifier employs *forward* AGC bias. In this case, an increase in the positive DC voltage from the AGC line reduces the base voltage and the gain of the stage. Note that a more positive base voltage increases the collector current, causing an increase in the voltage drop across the resistor R_1. The result of an increase in the collector current is to reduce the collector voltage, thus decreasing the gain of the transistor by operating closer to saturation.

The base-to-collector junction of a transistor is normally reverse biased. You will remember from your studies of the voltage-variable capacitor that the amount of reverse voltage across a P-N junction affects the junction capacitance in an inverse ratio. In other words, the greater the amount of reverse voltage, the smaller the amount of junction capacitance. In comparing the circuits of Figure 14.10A and Figure 14.10B, it can be seen that for the reverse AGC a larger reverse voltage will appear across the base-collector junction. Therefore, the junction capacitance is smaller than in the circuit of Figure 14.10B because the collector voltage of the latter is lower, due to the drop across R_1. The lower value of junction capacitance is desirable in RF amplifier stages because it permits a higher gain. A change in the AGC voltage in the circuit of Figure 14.10A cannot produce a large change in the base-collector junction capacitance. However, in the circuit of Figure 14.10B, a small increase in the base-emitter bias causes an increase in the collector current. As a result, the collector-base junction voltage is reduced. This, in turn, increases the base-collector junction capacitance. From the foregoing, it can be concluded that for the forward AGC circuit, the base-collector junction capacitance will be larger and will vary more. This is a disadvantage of the circuit when compared to reverse AGC. An advantage of the reverse AGC system is that the input and output impedances vary less for changes in the reverse AGC voltage.

ADVANTAGE OF FORWARD AGC OVER REVERSE AGC

One of the advantages of forward AGC can be understood from the curve of Figure 14.9. Note that the region of the curve employed for the reverse AGC is nonlinear. Therefore, any change in gain will not be in direct proportion to a change in AGC voltage. Compare this with the linear slope on the portion of the curve employed for the forward AGC. Note that the change in gain is directly proportional to the change in bias voltage.

It is also apparent from the curve of Figure 14.9 that with reverse AGC, a very strong signal can produce a sufficient decrease in AGC bias to drive the amplifier into the cutoff region. If this happens, severe distortion takes place. To avoid this distortion, special circuitry is used to limit the amount of voltage applied to the stage. On the other hand, when using the forward AGC there is very little danger of driving the amplifier into the nonlinear region. As the forward bias is increased, the transistor is driven toward saturation. The collector voltage approaches, but never reaches zero volts with respect to the base.

We have briefly covered some of the advantages and disadvantages of forward and reverse AGC. It must be remembered that the choice of forward or reverse AGC is also influenced by the type of transistor employed for an amplifier. For example, transistors of the "mesa" type operate better with reverse AGC. Epitaxial planar transistors operate better with forward AGC.

Some television receivers may use only

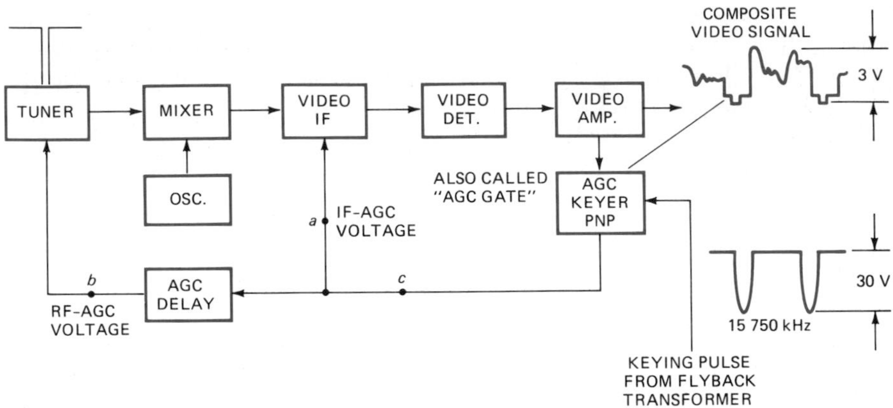

Figure 14.11 Block diagram of a keyed-AGC system. The video and flyback pulses are negative since the AGC keyer is a PNP transistor. Pulse voltages will vary depending upon the receiver design.

forward AGC or only reverse AGC. Other receivers may use combinations of forward AGC and reverse AGC for different controlled RF and IF stages. In some receivers, forward AGC is used for one IF amplifier and reverse AGC is used for the next amplifier. The reason for using both forward and reverse bias in the IF-amplifier stages is that the stability of the system is increased. The effect of changing bias on the bandwidth is also reduced.

HOW AGC VOLTAGES ARE OBTAINED IN SOLID-STATE RECEIVERS

Figure 14.11 shows a block diagram of a typical keyed-AGC system used in solid-state receivers. This is a popular type of AGC system used in both tube and transistor receivers.

One input signal to the AGC keyer comes from a video amplifier. Remember that it is only the sync pulses which are of interest as far as the keyed-AGC system is concerned.

An input pulse from the flyback transformer, or from another point in the horizontal output stage, is also fed to the AGC keyer. The function of the keyer, or the **AGC gate,** is to conduct only during that portion of the signal corresponding to the sync pulses. Since the amplitude of the sync pulses is the best indication of the amplitude of the input signal, the AGC keyer is turned on only during the sync-pulse duration. The amount of keyer conduction and of AGC voltage production is a function of the sync-pulse amplitude, which depends upon the received signal strength.

As previously mentioned, both the sync pulses and flyback pulses must be present simultaneously to turn on the keyer (or gate). This is true for both transistor and tube keyers. The pulses shown in Figure 14.11 are both of *negative* polarity. This is correct for a PNP keyer used in some AGC systems. The negative-sync pulses are fed to the transistor base. The negative flyback pulses are fed to the collector.

AGC to IF Stages

The AGC voltage from the AGC keyer is filtered and it may or may not be amplified. It is then fed (Point C in Figure 14.11) to the video-IF stages and to the tuner-RF stage. Usually, only the first video-IF amplifier, or the first *and* second video-IF amplifiers, is controlled by the AGC voltage (Point A in Figure 14.11). It is common practice to employ either direct coupling or stacked amplifiers in the video-IF section. This means that an AGC voltage to one amplifier will affect the conduction to all other amplifiers which are directly coupled to it. This is important to remember if you are troubleshooting in a video-IF stage and cannot obtain the required amount of conduction through one of the transistors in that section.

AGC to RF Amplifier

The AGC voltage from the keyer is also fed to the tuner RF amplifiers, usually through some type of AGC-delay circuit. As previously mentioned, the purpose of the delay is to permit the RF amplifier to conduct at full amplification on weak signals. If there is an amplifier in the AGC circuit, it is usually connected at point C so that it controls the DC bias to both the RF amplifier and the video-IF amplifier section. The usual practice is to make some provision for varying the gain of the AGC amplifier. This regulates the amount of AGC voltage, and hence, regulates the gain of the RF and IF amplifiers. This is done with an AGC control. The AGC control is adjusted when the receiver is tuned to a very strong or a very weak station, depending upon the manufacturer's recommendation. The object is to adjust the AGC-bias voltage so that the receiver will not be overloaded by a strong signal or produce excessive noise on a weak signal.

14.7 NOISE CANCELLATION CIRCUITS

The use of keyed-AGC systems provides improved noise immunity. However, the noise immunity of AGC (and sync) systems can be improved further by the use of **noise cancellation circuits.** These circuits reduce the effects of impulse noise, such as that generated by automobile ignition systems, sparking-motor commutators, and momentary switch-contact interruptions. Noise impulses in keyed-AGC systems, which appear within the duration of the horizontal-keying pulse, can affect the AGC voltage level. Such impulses must be approximately equal to, or greater than, the sync-pulse amplitude. The reason for this is that the horizontal-keying pulses can be up to twice the duration of the sync pulses, which are 5.08 μs wide. Noise pulses which occur on either side of sync pulses, but are *within* the duration of the horizontal-keying pulses, can key on the AGC gate. This causes the AGC voltage to change.

DIODE NOISE GATE

One of the simpler noise cancellation circuits is the diode-noise gate shown in Figure 14.12. Figure 14.12A is a simplified block diagram and Figure 14.12B is a simplified schematic diagram of this scheme.

The output of the video detector is the composite-video signal with sync pulses in the negative-going direction. As shown by the video signals in Figure 14.12B, bias at the anode of the diode-noise gate (D_1) is set by R_2 to a positive value. This value is such that the sync pulses remain slightly above zero volts. However, the positive bias causes full conduction of D_1. Provided no large amplitude noise pulses are present, D_1 now looks like a short and passes the composite video signal without change.

If a noise pulse (or pulses) appears on the video signal and has an amplitude greater than the sync pulses, it will drive into the negative voltage region. The negative portion will *cutoff* D_1 for the duration of

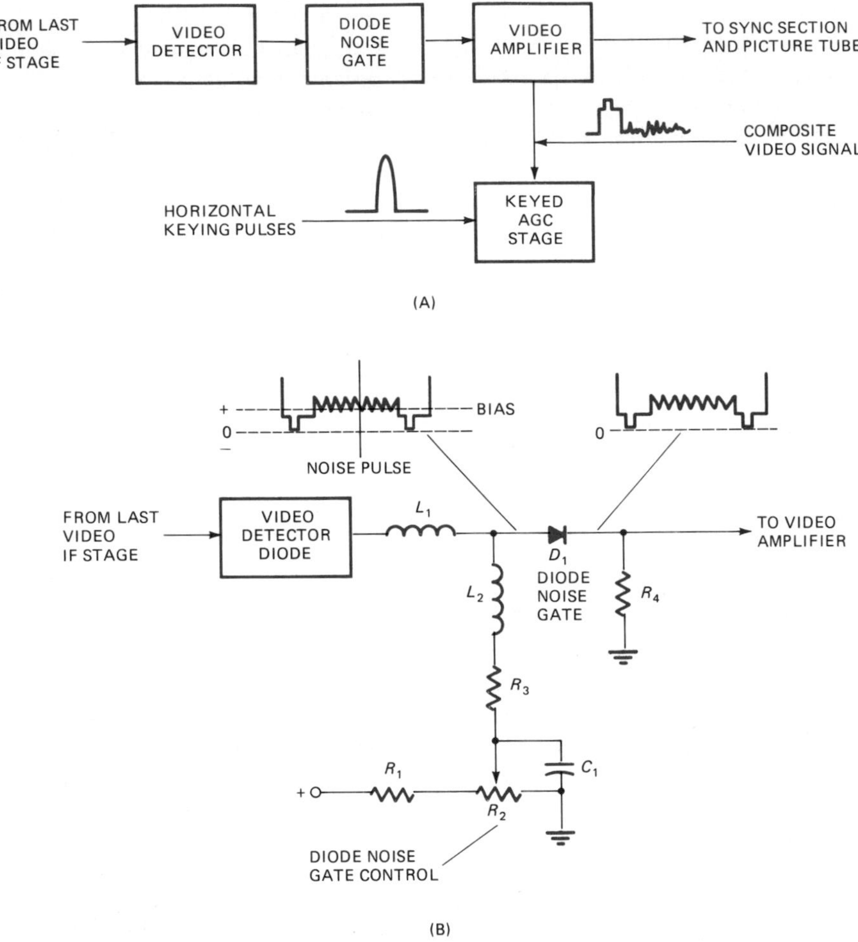

Figure 14.12A-B. A simple diode noise-cancellation circuit. **A.** Simplified block diagram of the system. **B.** Simplified schematic diagram of the system. Note that the noise pulse is removed following the diode-noise gate.

the noise pulse. Consequently, D_1 now looks like an open circuit and will not pass the noise pulse to the video amplifier.

NOISE-INVERTER CIRCUIT

Another type of noise cancellation scheme is known as the **noise-inverter** circuit. A block diagram of this scheme is shown in Figure 14.13. The composite video signal with positive going sync and noise pulses appears at the input of the keyed-AGC stage. A composite-video signal is taken from the video detector output and fed to the noise gate. The noise gate is a noninverting (common base) amplifier. It is biased so that only noise pulses greater in amplitude than sync pulses will cause it to conduct. The output of the noise gate will consist only of noise pulses having negative polarity. These are amplified to the same amplitude as the positive polarity noise pulses coming through on the composite

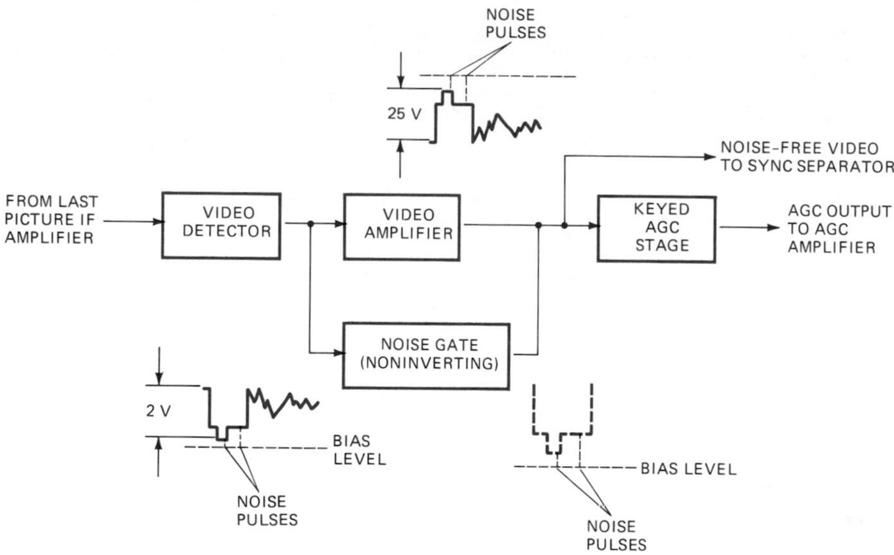

Figure 14.13 Simplified block diagram of a noise cancellation scheme utilizing noise inversion. The noise pulses at the keyer input are equal in amplitude but opposite in polarity. Thus, they cancel at that point. (See text.)

video signal from the video amplifier. The two sets of noise pulses combine and cancel at the base of the keyed-AGC stage. Thus, the keyed-AGC stage is provided with virtually noise-free sync pulses.

Noise-Free Video We have been concerned with noise reduction for the AGC system. However, this same circuit is also used to provide a virtually noise-free, composite-video signal for the sync separators. This improves the stability of horizontal and vertical synchronization.

Figure 14.13 shows noise pulses only in the vicinity of the sync pulses. Noise pulses may also appear superimposed on the video signal *between* sync pulses. Any of the noise pulses exceeding sync-pulse amplitude will be cancelled, as previously described.

Noise Inverter Schematic Diagram A simplified schematic diagram of a noise inverter scheme is shown in Figure 14.14. This is the circuit upon which the block diagram of Figure 14.13 is based.

The noise gate is NPN transistor Q_1. Q_1 is a common base (noninverting) amplifier. The fixed bias for Q_1 is set by the adjustment of R_3, the noise-bias control. This control normally sets the DC base voltage at approximately 1 V. With the signal, the emitter voltage of Q_1 is about 3.0 V. Thus, Q_1 is cut off for the composite-video signal fed to its emitter. However, *negative* noise pulses that exceed sync-pulse amplitude will cause Q_1 to conduct. The negative-noise pulses (only) are amplified and appear as negative-noise pulses at the base of Q_2, the keyed-AGC stage.

Simultaneously, the composite video signal from the video amplifier, with positive-going sync and corresponding positive-noise pulses, is also fed to the base of Q_2. As mentioned previously, the positive and negative noise pulses cancel at the base of Q_2.

This noise-cancelling circuit is also used to provide noise free, composite video for the sync-separator stages. The video from the video amplifier is coupled through C_1

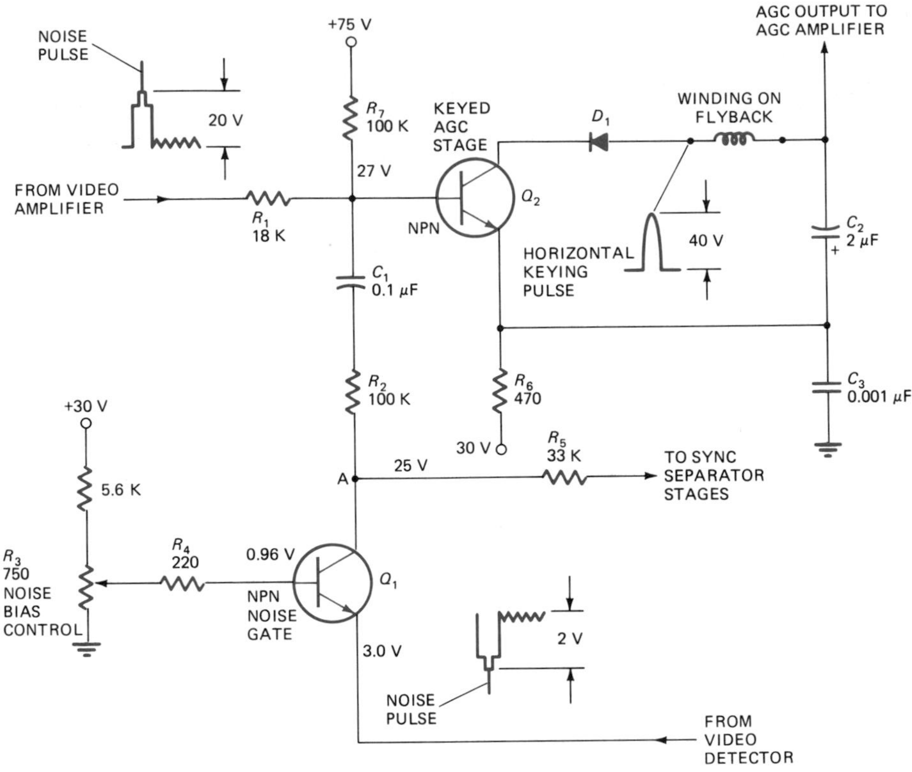

Figure 14.14 Simplified schematic diagram of a noise inverter scheme. This schematic is the basis for the block diagram of Figure 14.13.

and R_2 to the junction of the Q_1 collector and R_5. At this point, the noise pulses are reduced and the essentially noise free video is sent through R_5 to the sync-separator stages.

14.8 SOLID-STATE AGC SYSTEM

A solid-state AGC system is pictured in the simplified schematic diagram of Figure 14.15. This system is used in the RCA XL-100 color television receiver series. Basically, it consists of the following sections: (1) an AGC-keyer stage (Q_6), (2) an R-F delay AGC stage (Q_7) and, (3) a noise-inverter stage (Q_5). For the circuit diagram involving Q_5, see Figures 14.16 and 14.18. The noise-inverter stage will be discussed first. The complete schematic diagram (Figure 14.18) will be shown later.

NOISE-INVERTER STAGE

A simplified schematic diagram of the noise-inverter stage is shown in Figure 14.16. Composite video information from the video buffer, including any noise pulses present, is applied to the anode of diode CR_2. This video is also applied, through R_{53} and R_{35}, to the base of the AGC keyer (Q_6) in Figure 14.15. In addition, the same video is fed via R_{44} to the sync separators.

Action of CR_2 Consider the effect of the

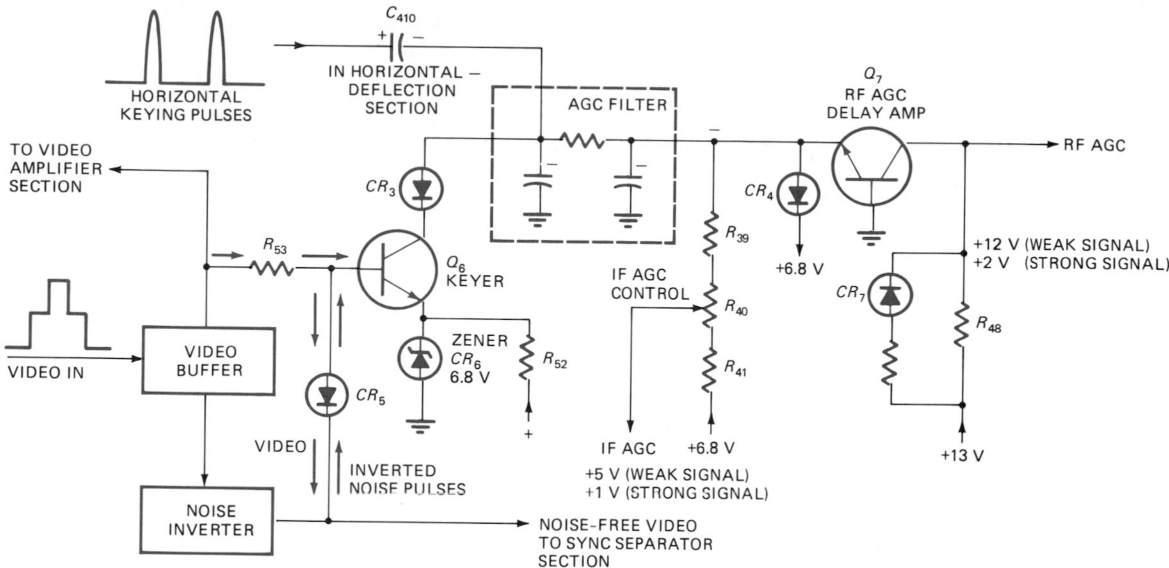

Figure 14.15 Simplified schematic diagram of the AGC system used in the RCA XL-100 Series. (See text.) *(Courtesy of RCA Consumer Electronics.)*

video signal only as it is applied to the anode of diode CR_2. CR_2 is reverse biased by the application of +6.8 V to its cathode. In addition, it has a barrier-threshold voltage of about 0.5 V. Thus, CR_2 will not conduct unless the input to its anode exceeds 6.8 V + 0.5 V = 7.3 V.

Action of Q_5 Note in Figure 14.16 that the peak sync-pulse amplitude is 7.5 V. That means that CR_2 will pass clipped sync pulses with an amplitude of 7.5 − 7.3 = 0.2 V. However, this sync-pulse amplitude is insufficient to cause conduction in the noise inverter, Q_5. This is because Q_5 is zero biased and the 0.2-V amplitude will not overcome the base-to-emitter, barrier-threshold voltage. However, noise pulses which exceed approximately 0.7 V will cause Q_5 to conduct and an inverted, amplified replica of the noise pulses will appear at its collector.

As described in a prior section, the inverted noise pulses from the collector of Q_5 will cancel the noninverted noise pulses on the composite video signal. This cancellation occurs at point A of Figure 14.16 for the AGC keyer input. It occurs at point B of Figure 14.16, for the sync-separator input.

AGC OPERATION

The operation of the AGC circuitry is explained with the aid of the simplified schematic diagram of Figure 14.15.

Video Buffer and Noise Inverter The video buffer, an emitter follower, supplies composite video with positive-sync polarity to the base of the keyer, and, through diode CR_2, to the base of the noise inverter. The same video is fed through diode CR_5 to the sync-separator section and to the video-amplifier section. Diode CR_5 provides isolation of the keyer and sync circuits. At the same time, it permits the *inverted* noise pulses to be fed to the base of the keyer to cancel any noise on the composite video signal. Noise

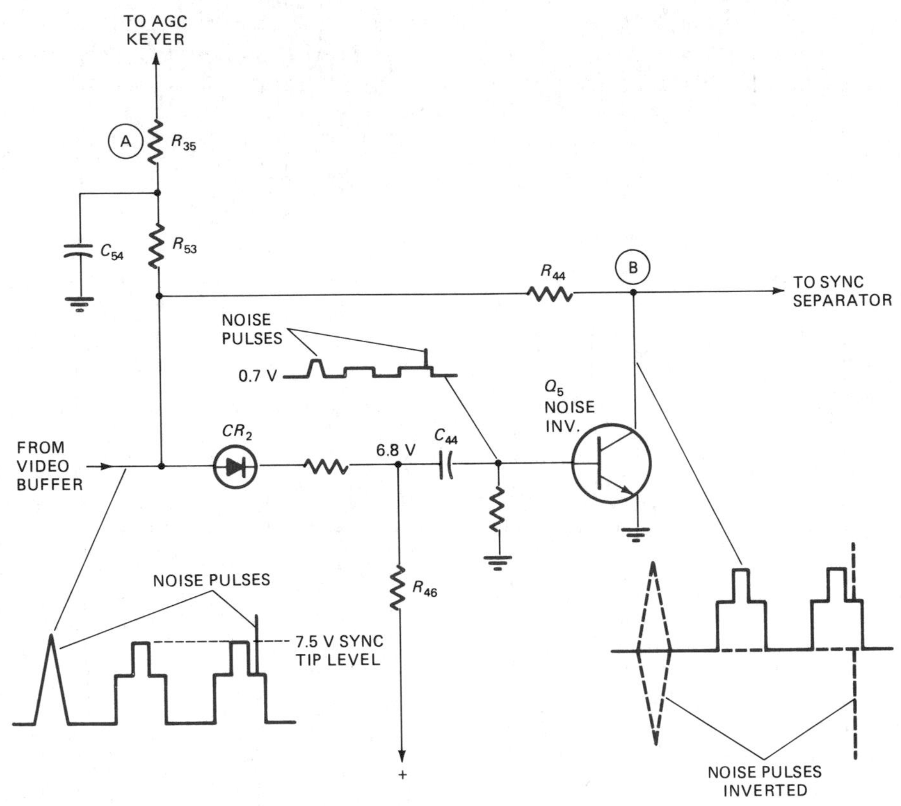

Figure 14.16 The operation of a noise-inverter circuit. This circuit reduces noise-pulse interference in AGC and sync systems. *(Courtesy of RCA Consumer Electronics.)*

is also cancelled on the video signals fed to the video-amplifier section and sync-separator section. This system of noise cancellation was explained previously.

The Keyer Stage As mentioned above, composite, noise-free video is fed to the base of the keyer (Q_6). Horizontal keying pulses are applied to the collector of Q_6. Q_6 is reverse biased so that only the amplitude of sync pulses will cause it to conduct. Q_6 will conduct in accordance with the sync-pulse amplitude, which is a function of the incoming signal strength. Thus, capacitor C_{410} will charge negatively, through diode CR_3, from Q_6 conduction. Between pulses, C_{410} discharges into the AGC filter which provides a relatively pure DC at its output. Diode CR_3 provides isolation of the negative voltage at C_{410} from the collector of the keyer.

The IF-AGC Voltage The negative voltage developed by the AGC keyer is applied through R_{39} to one end of R_{40}, the IF-AGC control. A positive 6.8 V is applied through R_{41} to the other end of the control. The negative-AGC voltage modifies the positive voltage to produce the IF AGC. This is about +5 V for a weak signal and +1 V for a strong signal. The IF AGC is applied to gate G_2 of the first and second IF-amplifier transistors. These transistors are N-channel, dual-gate MOSFETs. The effect of the response of the AGC to varying signal strengths is to adjust the IF gain accordingly. As a result, a constant video-signal amplitude is maintained. This prevents the picture contrast from changing with different signal strengths.

The RF AGC As previously discussed, the maximum signal-to-noise ratio in a TV receiver can be maintained by *delaying* the application of AGC to the RF stages. This means that AGC will not be applied until the input signal exceeds a predetermined strength. Maximum RF amplifier gain is maintained until the signal level is high enough to cause objectionable mixer-harmonic generation. In this system, the level was selected at approximately 1 000μV (75 ohms) at the antenna input. Figure 14.17 is a graph of AGC, RF, and IF gain versus input signal strength. Note on this graph that for input signals stronger than 1 000μV

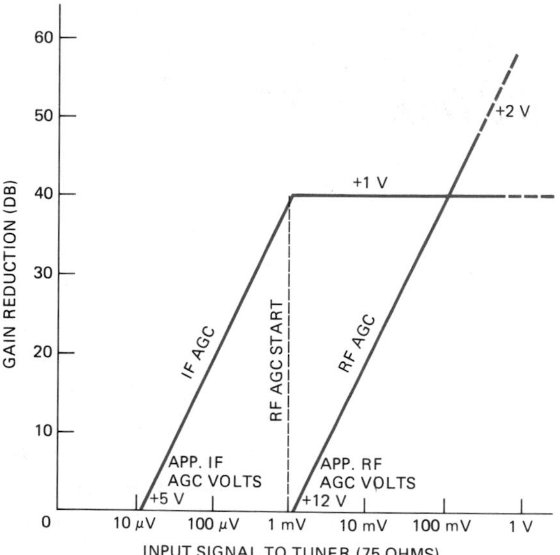

Figure 14.17 Graph of RF and IF gain versus input-signal strength for an RCA XL-100 Series color TV receiver. *(Courtesy of RCA Consumer Electronics.)*

(1 mV), the IF gain is held constant and the receiver gain is controlled by the RF amplifier.

Delay Operation The RF-AGC delay is provided by the operation of Q_7, the RF AGC-delay amplifier. Q_7 is a grounded-base amplifier, with no applied bias. The emitter-to-base junction, barrier-threshold voltage is approximately −0.7 V. Therefore, Q_7 will not conduct until the negative AGC voltage from the AGC filter exceeds 0.7 V. This constitutes the delay. This value of 0.7 V is exceeded when the input signal exceeds 1 000μV. At the signal level of 1 000V, the RF-AGC voltage is about 12 V, producing maximum RF-amplifier gain. However, as the input signal strength increases about 1 000μV, the negative AGC voltage input to the base of Q_7 increases. This increases Q_7 conduction. For strong signals above 100 000μV, the RF AGC voltage at the output of Q_7 decreases to about +2 V. Referring to Figure 14.17, a strong signal will produce a gain reduction greater than 40 dB.

Q_7 Protection Diode CR_4 provides transient-pulse protection for transistor Q_7. Any keying-pulse transients greater than +6.8 V are shunted through CR_4. Diode CR_7 functions only during the reception of exceptionally high signal levels. It acts to reduce the RF AGC-loop gain and to prevent instability in this loop.

IF AGC Control This control, R_{40} in Figure 14.15, is adjusted with a 1 000-V RF input. It is used to set the IF-amplifier gain to the point where Q_7 just begins to conduct and start RF-amplifier gain reduction. In effect, we may call R_{40} an "RF delay" control.

The complete schematic diagram for this AGC system is given in Figure 14.18. Several pertinent waveforms and their amplitudes are included.

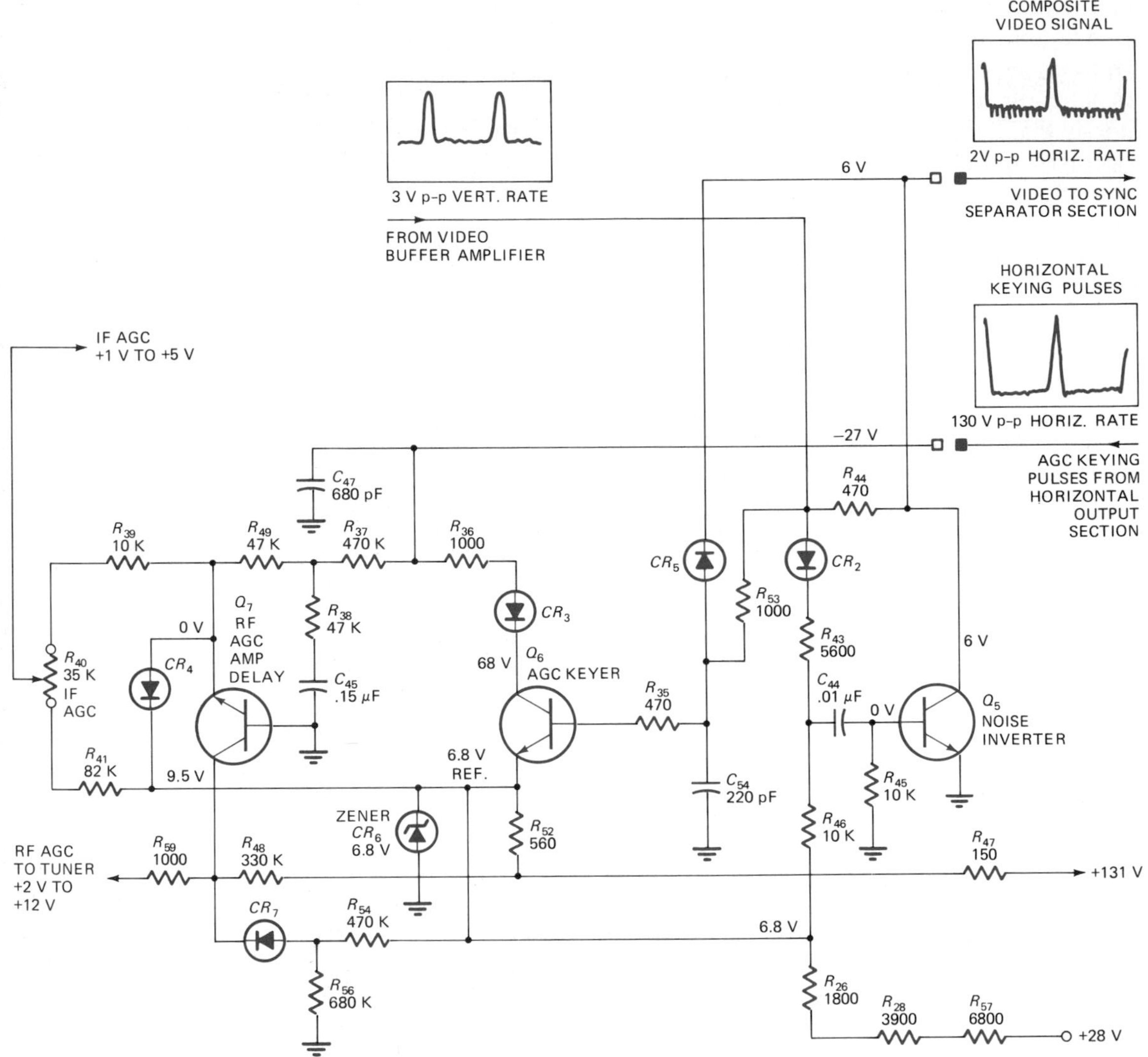

Figure 14.18 A schematic diagram of a solid-state AGC system. The circuit includes RF-AGC delay (Q_7) and a noise inverter (Q_5).

14.9 AGC SYSTEM IN AN RCA IC

When an AGC system is incorporated into an IC, it is usually only one of several circuits on a monolithic chip. The AGC system to be described in this section is part of an RCA 16-pin, dual-in-line, monolithic integrated circuit.

SIMPLIFIED IC BLOCK DIAGRAM

A simplified block diagram of the IC is given in Figure 14.19. The IC is composed of four major sections, as follows:

1. AGC processor
2. Noise processor (inverter) circuit

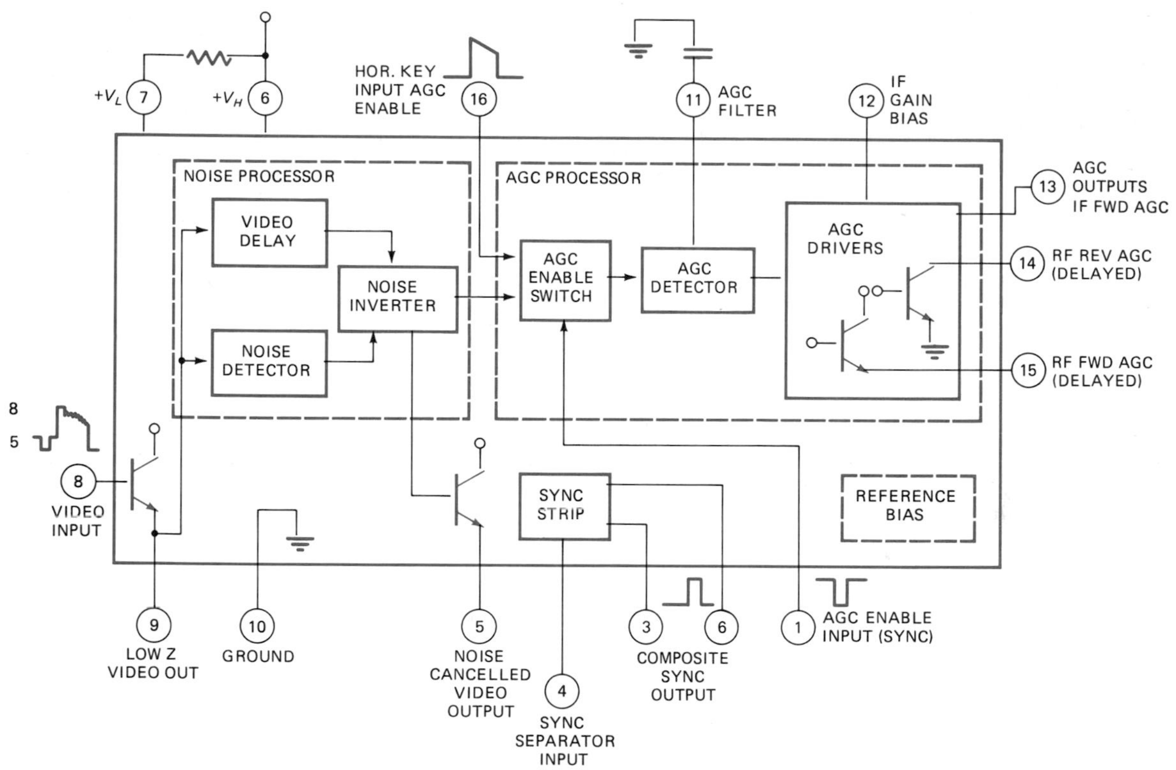

Figure 14.19 Simplified block diagram of a monolithic silicon IC containing an AGC system and other circuits. *(Courtesy of RCA Solid-State Division.)*

3. Sync-separator circuit
4. Internal DC-reference supplies

Only those sections relating to AGC operation will be discussed here in any detail.

The AGC control signals provided by this IC are

1. Undelayed-forward AGC for the IF
2. Delayed-forward AGC for tuners using bipolar transistors
3. Delayed-reverse AGC for tuners using FETs

The noise inversion principle is employed in this IC only for signals fed to the sync separator (from Terminal 5 to Terminal 4 in Figure 14.19). However, a different method of noise immunity is employed for the AGC system, as will be described here. The noise inversion scheme involving sync separators will be described in a later chapter. It should be noted that the IC terminal numbers correspond in Figures 14.19, 14.20, and 14.21.

14.10 EXPANDED BLOCK DIAGRAM OF RCA IC NOISE PROCESSOR

Figure 14.20 is an expanded block diagram of the noise processor shown as part of Figure 14.19. Figure 14.20A is the block diagram and Figure 14.20B shows several pertinent waveforms.

The composite-video signal, containing noise pulses, is fed to Terminal 8 of the IC,

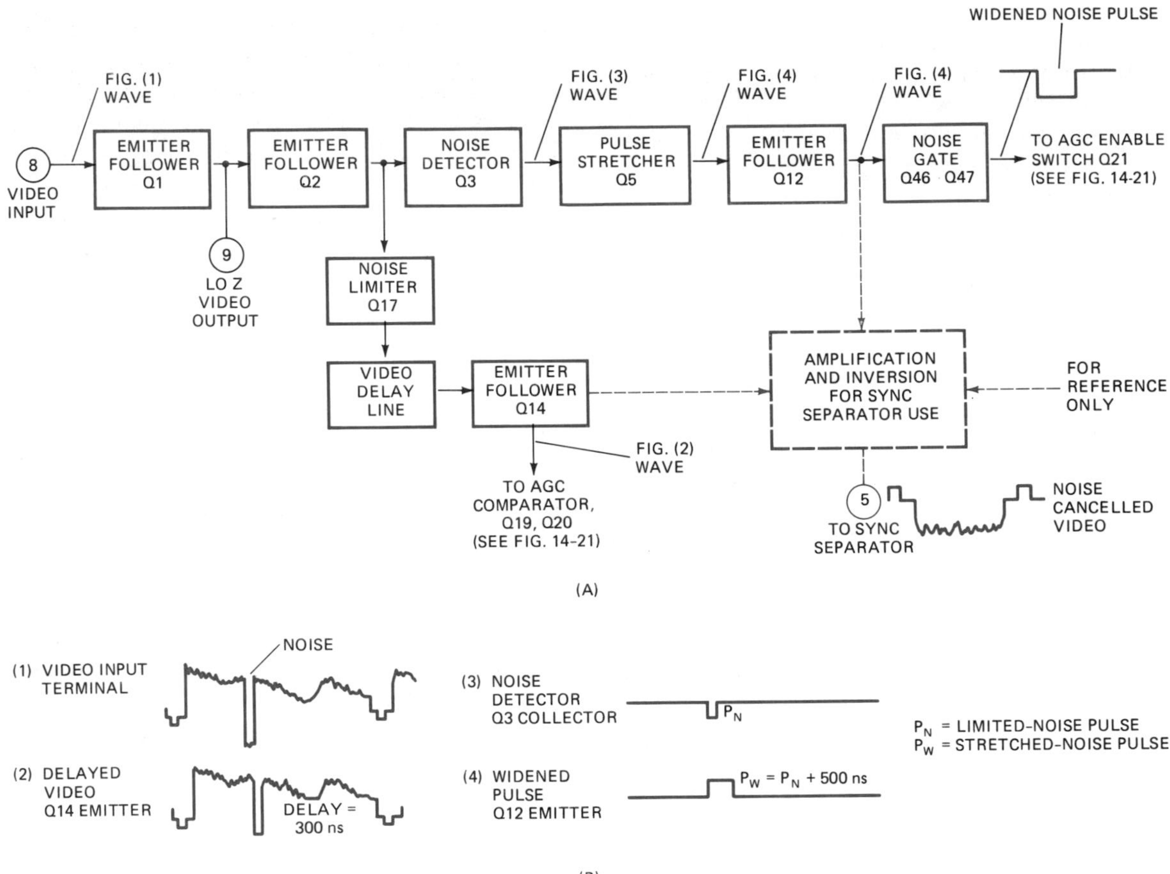

Figure 14.20A-B. The noise processor of Figure 14.19. A. Expanded block diagram. B. Waveforms for part A. The waveform numbers correspond to the numbers on the block diagram. Terminal numbers, such as eight, correspond to those on Figure 14.19.

Figure 14.20B (1). An unprocessed composite-video signal is available at Terminal 9, one output of Q_1. It can be amplified, if desired. The Q_1 output is also fed to emitter-follower Q_2.

NO NOISE PULSES PRESENT

If no negative-going noise pulses are present, Q_3, Q_5, and Q_{12} are cutoff. However, the output of Q_2 is fed through impulse-noise limiter Q_{17}, through the video-delay line (300 ns delay), and through emitter follower Q_{14}. The output wave of the emitter follower, Q_{14}, is shown in Figure 14.20B (2). This wave is fed to the AGC comparator (Q_{19}, Q_{20}) on the AGC processor (to be covered presently). The reason for the delay will not be dwelt upon here since it has significance only with regard to the composite-video signal fed to the sync separator (Terminal 4 of Figure 14.19). At this point, it is sufficient to say that the delay is part of the noise-cancelling scheme for sync. This operation will be discussed in a later chapter covering sync-separators.

Note: For simplicity, the AGC noise gate is not shown in Figure 14.19. Also, the noise-inverter block shown coupled to the AGC enable switch in Figure 14.19 is used only in connection with the video sent to the sync separator.

NOISE PULSES PRESENT

If negative-going impulse noise is present in sufficient amplitude, noise detector Q_3 conducts. In turn, Q_3 turns on transistors Q_5 and Q_{12}. Q_5 functions as a noise-pulse stretcher. Its output is a pulse about 500 nanoseconds wider than the original detected noise pulse. See Figure 14.20B (4). This widened noise pulse is fed through emitter follower Q_{12} to the noise gate (Q_{46}, Q_{47}).

The output of the noise gate is a negative going, widened-noise pulse. This pulse is fed to the AGC enable switch Q_{21}. Q_{21} is part of the AGC processor, which will be discussed presently. Under adequate noise conditions, the noise gate disables the AGC "generator" for the duration of the noise-gate pulse. This means that the AGC voltage will *not* respond to the noise pulses.

14.11 EXPANDED BLOCK DIAGRAM OF RCA IC AGC PROCESSOR

Figure 14.21 is an expanded block diagram of the AGC processor shown as part of Figure 14.19. The AGC system provides control voltages to the RF and IF amplifiers, according to the amplitude of the detected video signal. Consequently, a substantially constant amplitude video signal is maintained by means of AGC.

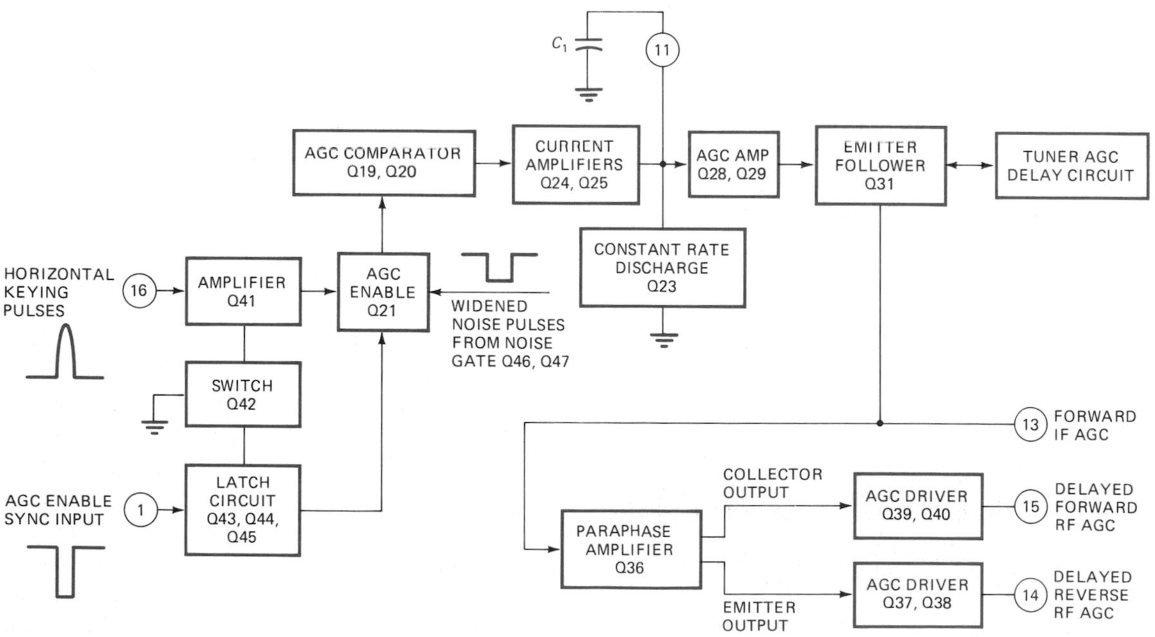

Figure 14.21. AGC processor of Figure 14.19, expanded block diagram. Terminal numbers, such as 8, correspond to those on Figure 14.19.

In this particular system, the sync-tip voltage is compared to an internal reference voltage during the horizontal-keying pulse interval. In addition, there is a **coincidence-gate circuit** which compares the *timing* of the sync pulses with that of the horizontal keying pulses. This *insures* that AGC voltage will only be developed when these pulses coincide, as explained later.

AGC ENABLE AND COMPARATOR

The AGC enable circuit (Q_{21}) receives inputs as follows: (1) horizontal-keying pulses (through Q_{41}), (2) noise-gate pulses, and (3) an input from the latch circuit (Q_{43}, Q_{44}, Q_{45}). The composite-video signal from Q_{14} and the AGC enable circuit connect to the AGC comparator (Q_{19}, Q_{20}).

In normal operation, Q_{21} is turned on during the period of the horizontal-keying pulse. In turn, Q_{21} "enables" (energizes) the AGC comparator. The DC level of the sync pulses (on the composite-video signal from Q_{14}) is now compared to the comparator-reference voltage. The comparator detects the *difference* between these two voltages.

This difference, a DC voltage, is fed to the current amplifiers (Q_{24}, Q_{25}). The amplified DC output from Q_{25} is used to charge the externally mounted (through Terminal 11) AGC-filter capacitor (C_1) negatively. The constant-rate discharge transistor Q_{23} is also gated on during the horizontal-keying period. During this time, Q_{23} provides a constant rate of discharge for C_1. For a given comparator difference output, the net charge on C_1 is the difference between the charge and discharge currents of C_1.

If the input signal strength decreases to a lower level, the C_1 charging current will decrease. There will be a lower net charge on C_1. As a result, the TV receiver gain is increased. The increased gain causes the sync-pulse amplitude to increase to its former value and a condition of equilibrium is reached. At this point, the TV receiver gain remains constant. The charge and discharge currents of C_1 are equal. The converse of this action is also true when the input signal strength increases.

The negative voltage of capacitor C_1 is applied to the AGC amplifier (Q_{28}, Q_{29}) and to the emitter follower (Q_{31}). Outputs from Q_{31} go to Terminal 13 as the undelayed IF-forward AGC, and also to amplifier Q_{36}. The tuner-AGC delay circuit also connects to Q_{31}. Tuner-delay bias is applied to the emitter of Q_{31}.

THE LATCH CIRCUIT

There are times when the TV receiver may be out of horizontal sync, such as when changing channels. At such times, video information may be sampled by the AGC system, instead of sync pulses. If this happens, horizontal synchronization will be affected.

To prevent this situation (refer to Figure 14.21), negative-sync pulses are fed into the latch circuit (Q_{43}, Q_{44}, Q_{45}). During the sync-pulse period, a current flows in the emitter circuit of Q_{45}. However, if a horizontal-keying pulse is not present simultaneously, the current is bypassed to ground. The AGC enable (Q_{21}) is not activated and the AGC comparator remains *off*. On the other hand, if the sync pulses are *coincident* with the horizontal-keying pulses, the latch circuit activates Q_{21}. This turns on the AGC comparator for normal operation.

AGC OUTPUT STAGES

As mentioned above, an output from the emitter of Q_{31} goes to paraphase amplifier Q_{36}. The collector output of Q_{36} is fed to the AGC driver (Q_{39}, Q_{40}). The output of this stage is the delayed forward-RF AGC

(Terminal 15) for tuners using bipolar transistors.

The emitter output of Q_{36} is fed to the AGC driver (Q_{37}, Q_{38}). The output of this stage is the delayed reverse-RF AGC (Terminal 14) for tuners using FETs.

14.12 TROUBLES IN AGC CIRCUITS

Chapter 27 is devoted entirely to the methods of troubleshooting TV receivers. Therefore, troubleshooting methods will not be discussed in this section. What follows is a presentation of some of the most common troubles caused by AGC malfunctions and their effect on the picture.

SYMPTOM: NEGATIVE PICTURE

In a negative picture, the black and white values are reversed, similar to the appearance of a photographic negative. Basically, this trouble is caused by insufficient AGC voltage supplied to either the RF or IF stages. In order to produce a negative picture, it is necessary to highly overdrive an IF stage. Because of the high-signal level present, the last IF stage is frequently the one which may be overdriven. Possible causes of a negative picture are (1) improperly adjusted AGC control or switch, (2) defective transistors (or tubes) in the AGC circuit, (3) defective video-IF or video-amplifier transistors or tubes, or (4) positive voltage on the AGC line.

SYMPTOM: NO PICTURE AND NO SOUND (RASTER PRESENT)

This symptom occurs when there is a complete lack of signal reaching the picture tube. This is generally caused by an excessive AGC voltage applied to the RF- or IF-amplifier stages, or to both. The excessive AGC voltage causes one or more of these stages to be cut off. This type of trouble can appear in tube, hybrid, or solid-state receivers. The possible cause for this type of trouble is usually to be found in defective parts in the AGC circuit.

Note: This same symptom can also be caused by defects in the RF, mixer, IF, video-detector or sections.

SYMPTOM: OVERLOADED PICTURE WITH INTERCARRIER BUZZ

An overloaded picture is easily recognized by the dark overall appearance of the picture (Figure 14.1). If the overloading causes sync compression in the IF stages, there may be horizontal picture pulling, or even vertical rolling. In addition, an intercarrier buzz may be present, particularly on strong signals. This is due to the modulation of the sound IF signal with the 60-Hz sync pulses.

This trouble is caused by insufficient AGC voltage and may be the result of something as simple as an incorrect AGC control adjustment. If the control is set properly, transistor, tube, or parts defects will probably be present. Note that this same symptom may be caused by defective IF stages.

Note: In general, it is frequently necessary to isolate the cause of AGC-type troubles between the actual AGC circuits and the RF, IF, and video-amplifier sections. This is necessary because many symptoms which appear to be caused by a defective AGC section may actually be caused by defects in the other circuits mentioned previously. The manner of such isolation will be described in Chapter 27.

SUMMARY OF CHAPTER HIGHLIGHTS

1. Automatic gain control (AGC) functions to keep the picture contrast at a fixed level even though the input signal strength may be varying.
2. AGC also helps to obtain more stable picture synchronization, to prevent picture overload and to reduce noise (snow) in a weak-signal picture (Figure 14.1).
3. The composite-video signal reference used to establish the AGC voltage is the amplitude of the sync pulses (Figure 14.2).
4. The AGC voltage varies in accordance with the strength of the input signal. In turn, it controls the gain of the RF and IF amplifiers. As a result, the composite-video signal amplitude remains constant (within practical limits).
5. The gain of a vacuum tube, FET, or bipolar transistor can be varied by changing its bias. This is the basis of AGC.
6. With vacuum tubes or FETs, negative AGC bias is employed. The greater the negative bias, the lower the stage gain.
7. With bipolar transistors, the applied AGC bias may be either a negative or a positive voltage. The polarity depends upon whether a PNP or an NPN transistor is used.
8. Two major types of AGC systems are the peak-AGC and the keyed-AGC systems.
9. Delayed AGC is used for the RF amplifier to maintain maximum RF gain for weak signals.
10. The peak-AGC system is the simplest. However, it has important disadvantages, including the inability to follow "flutter" (Figure 14.3).
11. Keyed-AGC systems overcome the disadvantages of peak-AGC systems. They eliminate "flutter" effects and are less susceptible to noise interference (Figure 14.4A).
12. The keyed-AGC system develops AGC voltage only when the horizontal flyback pulses and the sync pulses are both present and coincident (Figure 14.4B).
13. In Question 12, negative AGC voltage is developed when the keyer conducts and charges a capacitor.
14. In Question 12, an AGC filter smooths the AGC-voltage ripple. For a keyed-AGC system, the filter need be effective only for the 15 750-Hz horizontal frequency.
15. Vacuum tube AGC keyers may use either triodes or pentodes. The basic operation is the same in either case.
16. With delayed AGC, the AGC voltage applied to the RF amplifier is delayed until the input signal exceeds 500 to 1 000μV. (Figures 14.7 and 14.8)
17. Transistor and vacuum tube AGC systems operate in a similar manner. However, transistor circuits have different characteristics.
18. For a silicon transistor, the AGC bias variation is about 0.5 V. For a germanium transistor, the variation is only about 0.25 V.
19. In "forward AGC," the gain of a transistor is reduced by *increasing* the base-to-emitter bias (Figures 14.9 and 14.10).
20. In "reverse AGC," the gain of a transistor is reduced by *decreasing* the base-to-emitter bias (Figures 14.9 and 14.10).
21. Some TV receivers use only forward *or* reverse AGC. Others use combinations of forward and reverse AGC (Figures 14.9 and 14.10).
22. The amount of AGC voltage developed

is a direct function of the amplitude of the sync pulses. This is true of both vacuum tube and transistor AGC systems.

23. The noise immunity of keyed-AGC systems can be further improved by the use of noise-cancellation circuits. These circuits reduce the effects of impulse noise on the AGC voltage (Figures 14.12, 14.13, and 14.14).
24. Two types of noise-cancellation circuits are the diode-noise gate and the noise inverter circuit (Figures 14.12, 14.13, and 14.14).
25. An IC AGC system is only one of several circuits on a monolithic chip. The chip shown in Figure 14.19 contains (1) an AGC processor, (2) a noise processor, (3) a sync separator, and (4) DC-reference supplies.
26. The AGC signals provided by the IC of 25 above are (1) undelayed-forward IF AGC, (2) delayed-forward AGC, and (3) delayed-reverse AGC (Figure 14.19).
27. See Figure 14.20. If negative-going noise of sufficient amplitude is present, noise detector Q_3 conducts. In turn, Q_3 turns on Q_5 which acts as a noise-pulse stretcher.
28. See Figure 14.20. The widened noise pulse is fed to the noise gate. Under adequate noise conditions, the noise gate prevents the AGC "generator" from operating for the noise-pulse duration.
29. See Figure 14.21. During normal operation, AGC enabler Q_{21} energizes the AGC comparator. The comparator now detects the *difference* between the DC level of the sync pulses and the reference voltage.
30. See Figure 14.21. After amplification, the *difference* voltage (Question 29) charges the AGC-filter capacitor, C_1, negatively. This is the basic AGC voltage that varies with signal strength.
31. See Figure 14.21. The "latch" circuit prevents AGC voltage from being developed if the sync pulses are not coincident with the horizontal keying pulses.
32. A "negative" picture may be caused by insufficient AGC voltage on an RF or If stage(s).
33. Excessive AGC voltage may result in a condition of "no picture, no sound," but with a raster present.
34. An overloaded picture (Figure 14.1), with intercarrier buzz, is caused by *insufficient* AGC voltage.

EXAMINATION QUESTIONS

(Answers are provided at the back of the text.)

Part A Supply the missing word(s) or number(s).

1. Preventing picture ________ is an important AGC function.
2. The ________ is the basic reference for all AGC systems.
3. The ________ AGC system rectifies the sync pulses.
4. The AGC keyer requires ________ inputs, which must be ________.
5. The peak-AGC system has a (an) ________ time constant filter.
6. Rapid fluctuations of picture intensity may be caused by ________.
7. The ________ keying pulse is one of the inputs to an AGC keyer.
8. A ________ -Hz ripple is not developed by keyed-AGC systems.

9. It is necessary to ________ the sync pulse level at the input to an AGC keyer.
10. The basic AGC voltage is obtained by _____
11. The RF AGC is ________ to maintain maximum weak-signal gain.
12. AGC in a solid-state receiver may be either ________ or ________ AGC, or a combination of both.
13. A voltage change of about ________ V can change a silicon transistor from cutoff to saturation.
14. A lower value of base-to-collector junction capacitance is present in ________ AGC controlled amplifier.
15. An AGC keyer is turned on only during the duration of the ________.
16. The AGC control may be adjusted on a (an) ________ signal, or a (an) ________ signal, depending upon the manufacturer's recommendation.
17. Noise impulses which occur during the duration of the ________ keying pulse can affect the AGC voltage.
18. ________ and ________ noise pulses cancel each other in a noise circuit.
19. ________ V to ________ V is a common level for RF AGC delay.
20. A (an) ________ is sometimes employed in an AGC system to insure no AGC voltage is developed when the input pulses do not coincide.
21. Re 20 above, the circuit which monitors the operation is also known as a (an) ________ circuit.
22. A (an) ________ picture may appear if there is insufficient AGC voltage.

Part B Answer true (T) or false (F).

1. The blanking-pulse level is commonly used as the AGC system reference.
2. Only the video signal is stabilized against input signal variations.
3. Two general types of AGC systems are the FET system and the keyed system.
4. The portion of a tube characteristic used for AGC control is that approaching cutoff.
5. AGC action is such that a stronger signal will develop less AGC voltage.
6. With transistor amplifiers, either positive or negative AGC control voltages may be used.
7. In forward AGC, increasing the bias decreases the amplifier gain.
8. In a keyed-AGC system, the vertical-blanking interval pulses do not affect the AGC voltage improperly.
9. The RF-AGC delay is approximately 500 μs.
10. A peak-AGC system operates on the peak-voltage value of the horizontal-keying pulses.
11. In a keyed-AGC system, the system removes the 60-Hz and 15 750-Hz components.
12. The peak-AGC system develops AGC voltage even for weak signals.
13. In a simple, keyed-AGC system, noise pulses are ineffective if they occur during the horizontal retrace periods.
14. An important advantage of keyed AGC is the elimination of "airplane flutter."
15. The two inputs to an AGC keyer are the sync pulses and the horizontal-keying pulses.
16. An AGC keyer will not "fire" unless its two input signals are coincident.
17. An AGC keyer is turned on for a maximum period of approximately 10μs.
18. The AGC-keying pulse is one output of the horizontal oscillator.
19. For vacuum tube and transistor amplifiers, the applied AGC voltage always has a negative polarity.
20. An advantage of the peak-AGC system is that it does not develop a 60-Hz ripple.
21. Sync-pulse level alignment is required for the proper operation of an AGC system.
22. The gain of a transistor may be varied by changing its base-to-emitter bias.
23. With reverse AGC, the gain of a transistor is reduced by decreasing the base-to-emitter bias.
24. All solid-state television receivers use a combination of forward and reverse bias.
25. The diode-noise gate is a type of noise-inverter circuit.
26. A "coincidence gate" insures that AGC voltage will be developed only when the timing of the sync and keying pulses are correct.

REVIEW ESSAY QUESTIONS

1. Briefly discuss the need for AGC in a television receiver.
2. Draw a simple schematic diagram of a peak-AGC system.
3. For the diagram drawn in Question 2, explain briefly how the system operates.
4. List the disadvantages of a peak-AGC system.
5. Draw a simple block diagram of a basic keyed-AGC system.
6. For the diagram drawn in Question 5, explain briefly the operation of the system.
7. In Figure 14.4A, explain briefly how a negative voltage is developed in C_1.
8. Explain the significance of Figure 14.7.
9. In Figure 14.8, describe the function of C_1, R_3C_2, V_2, R_4R_5.
10. Draw a simple schematic diagram showing how forward AGC is applied to an NPN-transistor amplifier.
11. For the diagram drawn in Question 10, refer to Figure 14.9 and explain the operation of forward AGC.
12. Explain briefly how it is possible to use either positive or negative AGC-control voltage with transistor amplifiers.
13. Refer to Figures 14.15 and 14.17. In your own words, briefly explain the operation of the RF-AGC delay circuit.
14. Define the following: peak AGC, keyed AGC, coincidence circuit, delayed AGC, and noise cancelling.
15. In your own words, explain how a keyed-AGC system overcomes the disadvantages of the peak-AGC system.
16. Refer to Figure 14.19. In your own words, briefly explain the function of the noise detector and the noise gate.
17. Refer to Figure 14.21. In your own words, briefly explain the function of the AGC enable and the AGC comparator.
18. In your own words, briefly explain the operation of a noise cancellation scheme. Refer to Figure 14.13.
19. Refer to Figure 14.14. Briefly explain the operation of the noise gate.
20. What is the function of CR_6 in Figure 14.18?
21. In your own words, explain how a 60-Hz ripple is developed in a peak-AGC system.
22. Referring to No. 21, what picture problem may the 60-Hz ripple cause?
23. Refer to Figure 14.16. In your own words, briefly explain the operation of the noise inverter.
24. Explain how 60-Hz intercarrier buzz may occur.

EXAMINATION PROBLEMS

(Selected problem answers at the back of the text.)

1. Make up a simple table showing the relative advantages and disadvantages of peak AGC versus keyed AGC.
2. What are the approximate durations of the horizontal blanking pulse, the vertical-sync pulse interval, the horizontal-sync pulse, and the horizontal keying pulse?
3. In Figure 14.6, identify the AGC filter.
4. Refer to Figure 14.7. What are the approximate RF and IF-AGC voltages for signal strengths of 100μV, 400μV, 2 500μV, and 80 000μV?
5. In Figure 14.8, calculate the time constants for (a) R_1C_2, (b) R_1C_3, (c) $C_1R_1R_2R_4$.
6. What is the dB gain, if the antenna input signal is 10 000μV and the first video-amplifier output signal is 2 V?

Chapter 15

15.1 Video signal requirements of picture tubes
15.2 Video signal amplitude
15.3 Eye resolving power
15.4 Effects of loss of low- and high-video frequencies
15.5 The DC component of a video signal
15.6 Phase distortion
15.7 Square-wave response of video amplifiers
15.8 Video requirements for color TV
15.9 Types of video amplifiers for monochrome and color sets
15.10 Comparison of tube and solid-state video amplifiers
15.11 Contrast controls in video amplifiers
15.12 Automatic contrast control
15.13 Automatic brightness and contrast control
15.14 Video peaking
15.15 Improved video peaking system
15.16 The comb filter
15.17 Elementary comb filter
15.18 Comb filter block diagram
Summary

VIDEO AMPLIFIERS

INTRODUCTION

To this point, the television signal has been received and amplified by an RF stage. It was then converted to another frequency by means of a mixer and further amplified by the IF stages. Finally, it was rectified by the diode detector. The signal amplitude at the output of the video second detector is not strong enough to drive the picture tube directly. Hence, further amplification is necessary. This is provided by the video amplifiers. Typically, the output of a video detector is about 2 V, peak-to-peak. The picture tube requires a peak-to-peak signal of 80 V to 150 V (or greater) for maximum picture contrast.

Many luminance-video amplifiers for current television receivers amplify the video band from about 30 Hz to about 3.2 MHz. The high-frequency end must be restricted to avoid interference caused by the 3.58-MHz color subcarrier. However, a more recent development for television receivers is the **comb filter.** This system makes it possible to amplify the entire range of broadcast luminance information up to 4.2 MHz, without interference appearing on the screen. This system will be discussed in this chapter. In monochrome receivers, the amplifier stages following the video detector are called **video amplifiers.** In color receivers, however, the comparable stages are called **luminance amplifiers,** the **"Y" channel,** the **luminance section,** or the **luminance channel.**

By the time you have completed the reading and work assignments for Chapter 15, you should be able to:

- Explain the meaning of the resolving power of the human eye.
- Draw a simple block diagram of a color TV video amplifier from the last video IF to the picture tube. Indicate all inputs and outputs. Label all blocks correctly. Explain the operation.
- Draw the luminance amplifier response curve of a conventional (non-comb filter) luminance section. Show

all the pertinent frequency points.

- Discuss briefly the operation and advantages of a comb filter.
- Discuss briefly the function of the luminance section delay line.
- Draw a simple circuit for emitter contrast control and explain its operation.
- Explain how picture contrast may be varied automatically.
- Define the following: H, $\frac{H}{2}$, blanking level, the DC component of the video signal, video peaking, picture sharpness, LDR, barberpole pattern, "Y" signal, and LED.
- Draw a simple schematic diagram of an automatic contrast control circuit and explain its operation.
- Explain how the transversal filter peaking system works.
- Understand how to check a video amplifier using square waves.
- Discuss the effect on the image of insufficient low- and high-frequency response.
- Explain how the bandpass of a video amplifier is improved by compensating networks.
- Explain what the chroma signal sidebands are.

15.1 VIDEO SIGNAL REQUIREMENTS OF PICTURE TUBES

To determine the characteristics of a video amplifier, let us first look ahead to the requirements of the picture tube. We can then better determine how the video amplifiers should meet these needs.

A video signal must possess certain attributes to produce a suitable image on the screen of the picture tube. First, the video information must be in the same form that it was when it was originally developed at the studio. Any change in waveshape or any loss of frequency will alter the image produced on the picture tube screen. Second, the signal must possess the proper polarity. Otherwise, it will produce a negative picture at the picture tube. Finally, the video signal must be strong enough to vary the intensity of the picture tube scanning beam. This produces a range from bright to dark values on the screen. Unless the amplitude of the video signal is large enough to fully drive the picture tube, the image on the screen will appear washed-out because it lacks sufficient contrast to provide a satisfactory image. Typical video signals are shown in Figure 15.1. These three requirements are basic to every picture tube. However, the extent to which they are met will frequently vary from receiver to receiver. To understand the relative importance of each characteristic, let us examine each in more detail.

VIDEO-SIGNAL BANDWIDTH

The video signal is developed by the second detector. It is then passed through one or more video amplifiers before being applied to the picture tube. The signal contains the blanking, synchronizing, and video information. The actual video bandwidth that must be passed by the video amplifiers depends upon the particular TV receiver involved. For inexpensive monochrome sets, a bandwidth extending from about 30 Hz to 2.5 MHz is adequate, particularly if a small size picture tube is used. The larger screen monochrome and most color sets may extend the high-frequency bandpass to about 3.2 MHz. With a comb filter, the response goes to 4.2 MHz.

Uniform Amplifier Response The video information deals directly with the detail which forms the picture. The low frequencies in the video signal produce the larger objects in the scene and the high frequencies

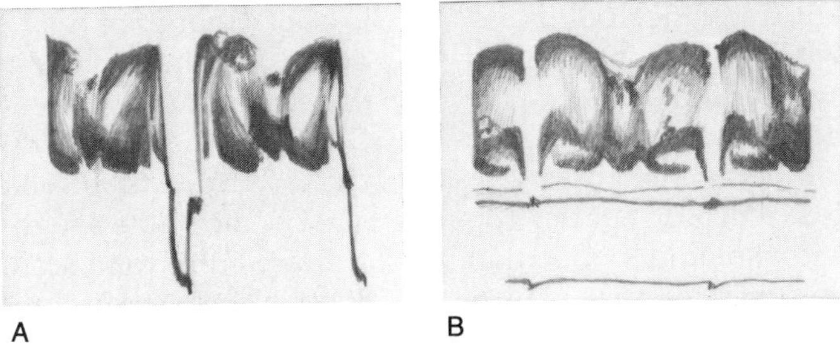

Figure 15.1A-B. Typical video waveforms at the grid of a picture tube. Peak-to-peak amplitude is approximately 100 V. A. Two horizontal lines of video signals. B. Two vertical fields of video signals.

in the video signal produce the fine-picture detail. The high frequencies may also contain color-video information. It is important for a video amplifier to have a uniform response over the entire range. Otherwise, either or both ends of the video spectrum will suffer. We shall learn in our subsequent study of video amplifier circuits what precautions are taken to insure that the response does not fall off too soon at either the high or the low ends. We shall also see why certain manufacturers of monochrome receivers purposely restrict the video bandwidth at the high end because of economy or because of the small size of the picture tube screen.

VIDEO-SIGNAL POLARITY

It was noted that the video signal must possess a certain polarity when applied to the picture tube. Otherwise a reverse, or negative, image will be produced on the screen. Two lines of a typical video signal are shown in Figure 15.2. The signal is drawn with the black level being most negative. Whether it possesses this particular polarity at the output of the video second detector depends on the detector circuit. This we have already seen. When the video signal reaches the cathode ray tube with this polarity, it must be applied to the control grid. The black level will serve to cut off the electron beam. Since the video variations are relatively more positive, they permit electrons to pass the control grid and reach the screen. The brightest portion of the video signal will be produced by the most positive voltages in this signal. These positive voltages represent the highlights in the image.

The video signal can also be applied to the cathode of the picture tube. In this case, it is necessary that the video-signal polarity

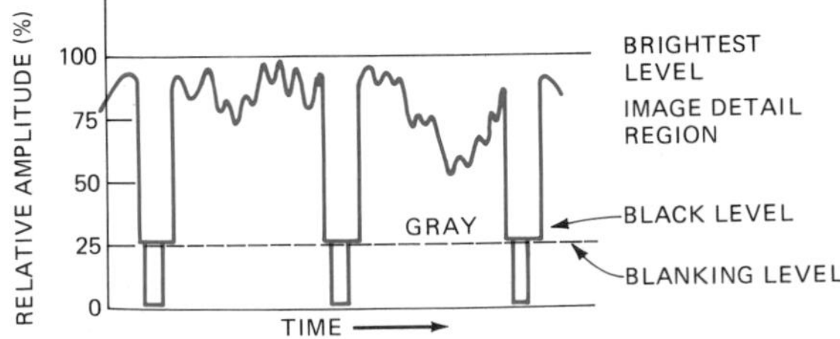

Figure 15.2 Two horizontal lines of a typical video signal.

be reversed; that is, the sync pulses will be more positive than the video-signal variations.

15.2 VIDEO SIGNAL AMPLITUDE

The amplitude of the signal that is applied to the picture tube governs the contrast of the image that appears on the screen. To understand this dependence better, consider the typical monochrome transfer characteristic curve shown in Figure 15.3. This curve shows the relationship between the control-grid voltage and the intensity of any spot produced on the screen by the electron beam. For example, if the control-grid voltage is 75 volts (with respect to the cathode), the beam is completely cut off and nothing is seen on the screen. This is the condition when the screen is black. If we lower the grid bias voltage to −30 volts, the screen illumination produced can be determined in the following manner. Start at the −30 volt point on the horizontal axis and draw a straight line vertically until the curve is reached. Then draw a line to the right. A line so drawn will fall within the area marked "gray." If we continue to reduce the control-grid bias voltage (perhaps to −15 volts), then, according to the previous procedure, we shall see that the screen will become brighter.

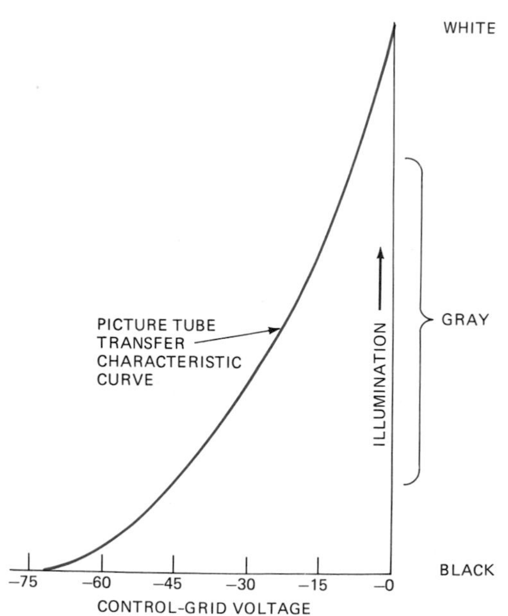

Figure 15.3 A typical transfer characteristic curve of a picture tube.

LARGE VIDEO SIGNAL

It is the purpose of the incoming video signal to vary the control-grid bias of the picture tube. Then the desired variation of screen brightness is produced as the electron beam is moved back and forth across the screen. The first step is to establish the proper operating bias for the picture tube. Let us say that without any incoming signal this bias is adjusted to 37.5 volts, Figure 15.4. The video signal is now applied. It will distribute itself about the operating point so that as much signal area appears on one side of the point as on the other. This distribution is also indicated in Figure 15.4. Actually, the operating bias is adjusted so that the blanking voltage just reaches the cutoff level of the beam. If we examine the video-signal variations and the brightness which they produce on the screen, we see that the maximum white is produced on the screen for the video signal that extends the farthest to the right. For that portion of the video signal that does not extend quite as far, less screen illumination is produced. The signal then falls, perhaps within the gray areas. Finally, whenever the blanking pulses appear, the voltage on the tube reaches the −75 volt cutoff point and the screen goes black.

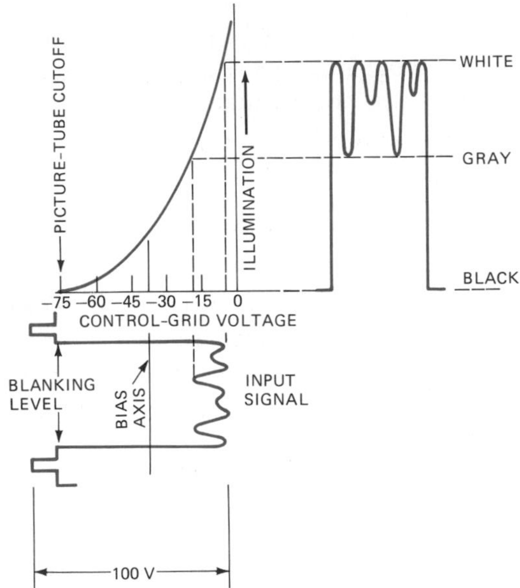

Figure 15.4 The effect of a normal video signal at the control grid of a picture tube.

SMALL VIDEO SIGNAL

The situation shown in Figure 15.5 prevails if a smaller video signal is received (60 V). First, the DC-grid bias on the tube is adjusted as before, until the blanking level of the incoming signal plus the DC-bias voltage reaches the cutoff point. The signal variation to the right of the cutoff point will then produce varying levels of illumination on the screen. Since the video signal in this case has a smaller peak-to-peak variation than the preceding signal, it will not produce the same range of screen illumination. The present signal extends from the cutoff to a point less than halfway between the maximum white and gray. There is therefore less contrast in the picture between the dark and light areas. If the video-signal amplitude is further reduced, we shall eventually have a washed out picture. This means that there will be little difference between the dark and light areas of the picture.

SCREEN SIZE AND VIDEO-SIGNAL DETAIL

The size of the screen on which an image is placed will also govern how much fine detail the image should possess. In a 525-line television system, there are only about 483 *active lines*. This means that only about 483 lines actually contain picture information. The rest occur during the vertical retrace. The 483 active lines of video information can be placed on a 5-in (12.7-cm) screen as well as on a 23-in (5.8-cm) screen. The amount of detail necessary for the smaller tube is not as great as for the larger screen. The reason stems from the resolving power of the human eye.

15.3 EYE RESOLVING POWER

The **resolving power of the eye** is the ability of the eye to distinguish between ob-

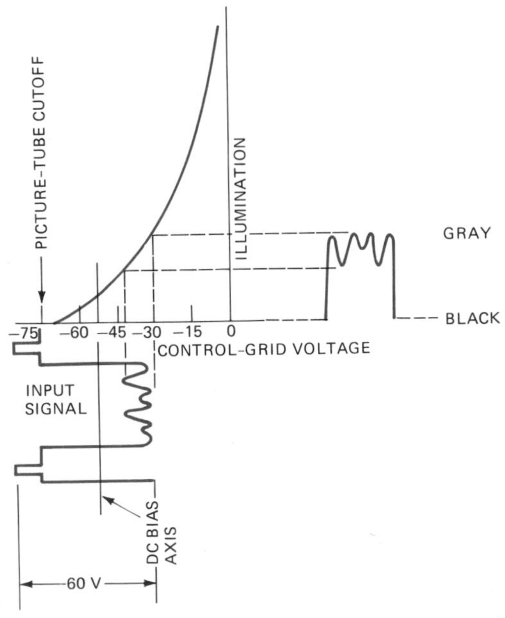

Figure 15.5 The effect of a small video signal at the control grid of a picture tube.

jects that are placed close together. As an example, consider the card shown in Figure 15.6 with two narrow lines located side by side. While the card is held fairly close to the eye, it is possible for an observer to see each line separately. As the card is slowly moved away from the eye, it becomes increasingly difficult to see each line distinctly. Eventually, a point is reached where the eye is just capable of distinguishing between them. This point is the limit of the resolving power of the eye for these two lines.

The farther apart the lines are, the more easily they can be distinguished by the eye at any given viewing distance. For the average person, it is claimed that if the two objects subtend an angle of one minute or more at the observer's eye, they can be seen as distinct units. This angle is known as the **minimum resolving angle of the eye.** It is illustrated in Figure 15.6.

The **critical resolving distance** is the distance that the observer must be from the objects in order to have the one-minute angle subtended at the eye. If the observer is farther away than this distance, the two objects merge into one. With television, it is necessary for the observer to remain outside the critical resolving distance. Coming closer only reveals the separate scanning lines. This hampers the illusion of continuity.

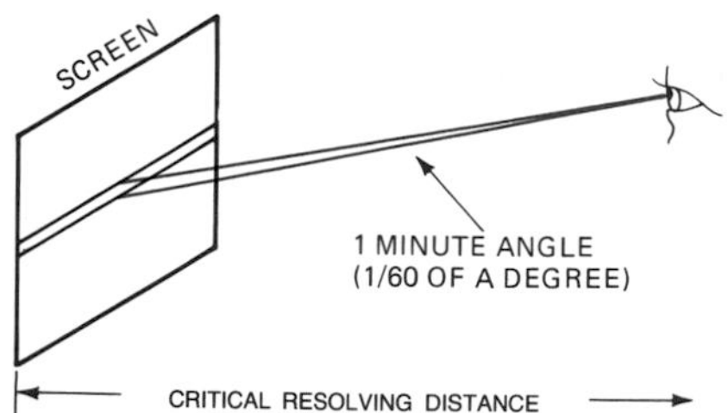

Figure 15.6 The power of the human eye to resolve or separate two objects that are closely spaced depends upon the distance from the eye to the objects. If the objects subtend a 1-min angle at the eye, they may be seen separately. In television viewing, the eye should be further from the screen.

TV VIEWING DISTANCE

From the foregoing line of reasoning, it would seem possible to calculate the exact viewing distance for an object of any size. With television images, an observer can actually approach the screen closer than the calculated figure and still be unable to distinguish one line from another. This is possible because the resolution of two lines depends not only on their separation, but also on the amount of light of the lines and their relative motion. The stronger the light, the more clearly they stand out. Under these conditions, the critical resolving distance increases.

On the other hand, the introduction of motion tends to make the line of demarcation less clear-cut. The objects blend into each other at much smaller distances than they would if they were stationary. The latter condition prevails for television images. Hence, the observer may view the screen from closer distances than is possible if the motion was absent. In addition, the positions of the lines of the picture tube tend to change slightly during each scanning run because it is impossible to obtain perfect synchronizing action. This further obscures any clear division between the lines.

Placing the same 483 active lines of picture information on a 19-in (48.3-cm) screen as on a 7-in (17.8-cm) screen means that the proper viewing distance for the larger screen is greater than that for the smaller screen. With the smaller screen, the ideal viewing distance is generally so short that the observer ordinarily never comes this close to the screen. Therefore, many of the finer details of the picture are not seen, even though they are present on the screen. Manufacturers of monochrome sets take advantage of this fact to design video amplifiers for small-screen receivers with bandwidths less than 3.2 MHz. As the picture tube screens become larger, it is important to make the bandwidths of the video amplifiers wider.

15.4 EFFECTS OF LOSS OF LOW- AND HIGH-VIDEO FREQUENCIES

The response of uncompensated RC amplifiers and their low- and high-frequency compensation will be discussed later. At this point, we wish to discuss the effects on the television picture of insufficient low- and high-frequency response of video amplifiers.

FREQUENCY RESPONSE CURVE

An ideal frequency response curve for a video amplifier is given in Figure 15.7. Although this curve is essentially flat from 30 Hz to 4.2 MHz, many current receivers cut off the high-frequency response at about 3.2 MHz. This is to reduce beat interference which may be caused by the 3.58-MHz color subcarrier. However, a more recent TV receiver development, the **comb filter,** makes it possible to extend the video (luminance) amplifier response to 4.2 MHz, without interference from the 3.58-MHz subcarrier.

Frequency Distortion

Frequency distortion is introduced when there is unequal amplification of all desired frequencies over the passband of an amplifier. In Figure 15.7, the ideal response curve is flat from 30 Hz to 4.2 MHz. No frequency distortion will occur for TV video signals within this passband. The midband response of a video amplifier can generally be considered flat. However, the high- or low-frequency ends of the desired passband may fall off prematurely. If this happens, the resultant picture will be distorted.

To provide the required video-amplifier bandpass, both low-frequency and high-frequency compensating networks must be used. (The performance of an uncompensated video amplifier and the design of compensation networks will be covered in Chapter 16.) If compensation is not employed in video amplifiers, the response at the low-frequency end may fall off below about 100 Hz. Also, the high-frequency end may fall off after about 500 kHz.

Insufficient Low-Frequency Response

Instead of the video-amplifier frequency response being essentially flat down to about 30 Hz (Figure 15.7), it may fall off sooner. For example, it may fall off at 500 Hz. This condition can result from a defect of a low-frequency compensation network (to be described later). The indication of poor low-frequency response on a test pattern is shown in Figure 15.8. Note the dark background shading and that large objects in the pattern appear smeared. Low frequencies are in the range of 30 Hz to 100 kHz. These frequencies are responsible for the reproduction of large picture areas.

Insufficient High-Frequency Response

The ideal high-frequency response shown in Figure 15.7 is flat to 4.2 MHz. However, unless a comb filter is used, it is generally flat to a maximum value of 3.2 MHz. This response can fall off much sooner due to a defect in a high-frequency compensation network. In this case, fine-picture detail will be lost. Sudden changes of shading cannot be reproduced and the picture lacks "sharpness." Since vertical detail is for larger areas, the loss of small-detail information occurs only in the horizontal direction. Therefore, lower frequencies are involved.

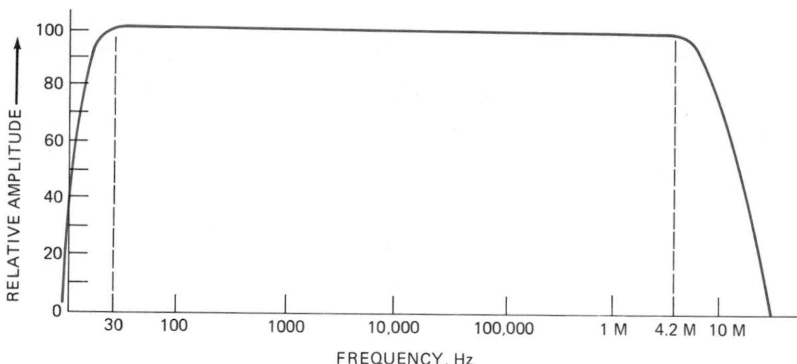

Figure 15.7 Ideal response curve for a video amplifier. It is essentially flat from 30 Hz to 4.2 MHz. (See text.) Unless a "comb filter" is used, most receivers cut off the high-frequency response at about 3.2 MHz or less.

15.5 THE DC COMPONENT OF A VIDEO SIGNAL

To know what a loss of response at the low frequencies means, let us examine the camera signal. It contains the picture information. A section of the signal obtained from the scanning of one line is shown in Figure 15.9. On either end of the line, we find the blanking and synchronizing pulses. These pulses always reach the same voltage (or current) values whenever they are inserted into the signal. The elements of the image itself are represented by the varying voltages between the pulses. Naturally, they differ from one line to the next.

Figure 15.8 A visual indication of poor low-frequency response in a video system. *(Courtesy of RCA.)*

In addition to the AC variations that make up the video signal, such as synchronizing pulses, blanking pedestals, and video information, there is another important part of the signal called the **DC component.** Examine the two video signals placed side by side in Figure 15.9. The blanking levels of both are of the same height and the AC variations of each signal are identical. The only difference is in the average level of the AC variations of Figure 15.9A as compared with the average level of the AC portion of the signal of Figure 15.9B. It is clearly shown that the average level of the signal in Part B is higher. This average level represents the background illumination of the scene at that line and is the DC component of the video signal. The background illumination may vary from line to line. But generally, it changes slowly over the entire scene, and adjacent lines will have almost equal DC components.

When the value of the DC component is high, as in Figure 15.9B, the people and objects in the scene being televised will appear against a dark background. This is true because with negative transmission every value is reversed. The darker the scene (or element), the greater the amplitude. As the scene becomes brighter, there is correspondingly less amplitude, and the AC variations of the video signal move close to the zero axis. Hence, as the DC value is less in Figure 15.9A than in Figure 15.9B, the background illumination of Figure 15.9A will be brighter. However, neither the people nor the objects will have changed. A lighted background will convey to the viewer the impression of daylight, sunshine, and clear weather. A darker background, on the other hand, will give the viewer the impression of night.

At the transmitter, the DC component of the video signal can be inserted manually by an operator viewing the scene from a monitor or automatically by using the average signal level derived from the viewing tube. If the latter cannot be accomplished, the light from the scene is allowed to fall onto a photoelectric tube at the camera position. The DC component is derived in this manner. Once obtained, it is inserted into the video signal, raising the AC component to the desired level.

From our discussion of the DC component, we can see that the average illumination of a scene may change with each frame, or 30 times a second. Of course, if the exact scene is televised without any variations, the average illumination remains constant. Actually, each frame scanned at the camera has a somewhat different average value. In order to obtain the correct shading of the image background at the receiver, it is necessary that all transmitting and receiving circuits be capable of passing 30 Hz without too great an attenuation. Any poor response will result in incorrect values for the background illumination and the left-to-right stretching or smearing of large objects.

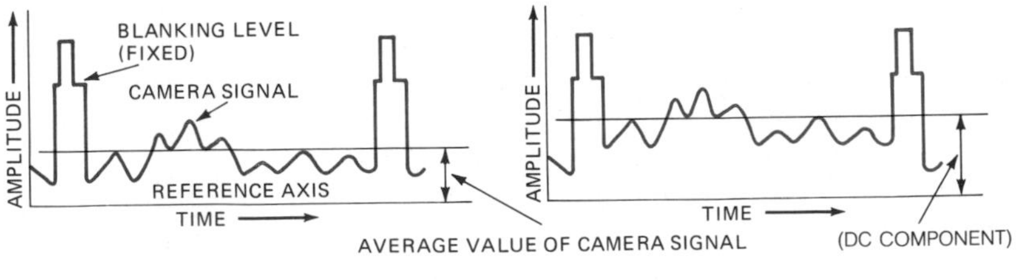

Figure 15.9A-B. The height of the camera-signal variations above the reference axis represents the amount of background illumination that the line (or scene) will possess. The average value is known as the DC component of the video signal.

15.6 PHASE DISTORTION

Phase distortion is capable of distorting the image on the picture tube screen. Although phase distortion can be tolerated in an audio amplifier, it must be given careful attention when a video amplifier is designed. Since phase distortion is such an important factor, a brief discussion at this point is required. **Phase distortion** (also called **phase/frequency distortion)** occurs in an amplifier when the phase shift is not proportional to frequency over the desired passband.

RESULTS OF PHASE DISTORTION

To correlate phase distortion and its effect on the television picture, let us study the dependence of phase distortion upon time delay. At the low frequencies, the phase angle between input and output voltages increases to a maximum of 90° as the frequency decreases. Suppose that a video signal is sent through an RC network containing (among others) two frequencies, 40 Hz and 90 Hz. We know that the 40 Hz wave will receive a greater phase delay than the 90-Hz wave. Let us assume that the 40-Hz wave is shifted 45 degrees and the 90 Hz wave, 10 degrees. Obviously the two waves will no longer have the same relationship at the output that they had at the input. By simple mathematics, it is possible to compute their difference.

A 40 Hz wave takes $\frac{1}{40}$ second to complete one full cycle, or 360 degrees. With $\frac{1}{40}$ second for 360 degrees, it will take $\frac{1}{320}$ second for the wave to change 45 degrees; $\frac{1}{320}$ second is approximately 0.003 second. Thus, there will be a time difference of 0.003 second between a maximum occurring at the input to the next stage and that occurring at the output of the preceding stage. The appearance of the maximum at the next tube will lag behind the other by 0.003 second.

The 90-Hz wave has a 10° phase angle introduced into it. One cycle, or 360°, of a 90-Hz wave occurs in $\frac{1}{90}$ second. Ten degrees will require only $\frac{1}{3240}$ second, or approximately 0.000 3 second. Thus, the input variations will differ by 0.0003 second for the 90-Hz wave.

At the picture tube screen, the electron beam moves across a 12-in (30.5-cm) screen a distance of one inch from left to right in about 0.000007 second. The time interval is extremely short, and if waves containing the 40- and 90-Hz frequencies receive the time displacement computed previously, the end result is a displacement of the picture elements that they represent.

Low-Frequency Time Delay At low frequencies a slight time delay causes certain parts of the object to be displaced from the correct position. The visible consequence of this displacement is smearing. Since the beam moves from left to right, the extended stretching of large objects will always be toward the right, or in the direction that the beam is moving. Only large objects are affected because they are the only ones represented by the lower frequencies.

High-Frequency Phase Distortion At the high-frequency end of the video signal, phase distortion results in the blurring of the fine detail of the picture. The larger the size of the picture tube screen, the more evident is this defect. This is another reason why the larger sets require more careful design and construction.

ELIMINATING PHASE DISTORTION

Phase distortion can be eliminated if the phase difference between the input and output voltages is zero, or if a proportional amount of delay is introduced for each frequency. Thus, a phase delay of 45° at 60 Hz is equivalent to a 90° delay at 120 Hz, etc. The first introduces a delay of approximately 0.002 second, similar to 90° at 120 Hz. The net result is that all the picture elements are shifted the same amount and correction is attained by positioning the picture.

Phase shifts introduced by the electrical constants of one stage are added to those of any other stage. The total phase delay of a system is equal to the sum of all the individual phase delays.

IMPORTANCE OF LOW-FREQUENCY PHASE DISTORTION

Phase distortion is more critical at the low-video frequencies. This follows because at these frequencies, a minor degree of phase distortion results in a relatively large time delay. This causes stretching and smearing of large objects. On the other hand, the actual time delay introduced by high-video frequency phase distortion is relatively small. Thus, the blurring effect of fine picture detail will not be as noticeable as smearing caused by low-frequency phase distortion.

15.7 SQUARE-WAVE RESPONSE OF VIDEO AMPLIFIERS

There are a number of ways to check the response of a video amplifier. One method is to feed a wide range of frequencies into the amplifier, one at a time, and measure the output signal amplitude. A graph is then plotted showing the output signal amplitude versus the frequency. When this method is used, it is necessary to maintain the input signal voltage to the amplifier at a very precise value. Also, the instrument used to measure the output signal amplitude must respond equally to all frequencies used for testing. This method is very time consuming and is used only for very precise measurements.

SWEEP GENERATOR METHOD

Another method of checking video amplifier response is to use a sweep generator. The generator sweeps back and forth throughout the range of video frequencies. An oscilloscope sweeps in step with the sweep generator and displays the amplifier response. This technique is the same as that used for aligning the IF stages.

SQUARE WAVE TESTING

It has been shown that a square wave consists of a fundamental frequency and a large number of odd harmonic frequencies. The greater the number of odd harmonics, the more nearly the wave approaches a perfect square or rectangle. If a square wave is passed through a video amplifier, it should appear at the output undistorted. If there is any frequency or phase distortion present, the square wave will be modified. The block diagram of Figure 15.10 shows how a square-wave test may be performed on a video-amplifier stage.

Interpreting the Waveforms. Interpreting the waveforms correctly requires a little practice but it is a quick method of determining video-amplifier response. It might be useful to remember that the horizontal portion of the square wave represents a period when the amplifier maintains a steady voltage. This steady voltage is like a DC voltage for the duration of the horizontal portion of each half cycle. Now, the lowest possible frequency is 0 Hz, which is a DC voltage.

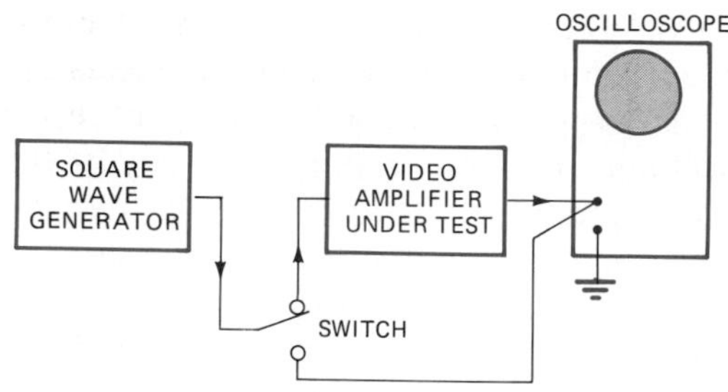

Figure 15.10 Test setup for performing a square-wave test on a video amplifier. The switch makes it possible to recheck the amplitude of the square wave whenever the generator frequency is changed.

Any change in the horizontal portion of the square wave represents a poor low-frequency response. The sides of the square wave represent a condition where the voltage changes almost instantly from one point to another. This rapid change corresponds to a high-frequency signal. The voltage of a high-frequency signal changes very rapidly from one point to another. This means that any distortion of the steep sides of a square wave represents problems with the high-frequency response.

INPUT TO VIDEO AMPLIFIER	SHAPE OF OUTPUT WAVE SEEN ON CRT SCREEN	INTERPRETATION
 60 HERTZ SQUARE WAVE		GOOD LOW–FREQUENCY RESPONSE AND NEGLIGIBLE PHASE SHIFT
		LEADING LOW–FREQUENCY PHASE SHIFT AND LOW–FREQUENCY ATTENUATION
		LAGGING LOW–FREQUENCY PHASE SHIFT
 25 kHz SQUARE WAVE		GOOD HIGH–FREQUENCY AND TRANSIENT RESPONSE
		POOR HIGH–FREQUENCY RESPONSE
		EXCESSIVE HIGH–FREQUENCY RESPONSE (OSCILLATION) AND PHASE SHIFT DISTORTION
	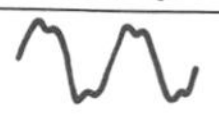	EXCESSIVE OR INSUFFICIENT MID–FREQUENCY RESPONSE AND PHASE SHIFT DISTORTION

Figure 15.11 Common symptoms of video amplifier troubles as indicated by the square-wave test.

A low-frequency square wave is fed to the amplifier to test the amplifier's low-frequency response. A high-frequency square wave is used to test the high-frequency response. Figure 15.11 shows the various symptoms that may be expected with video amplifiers. One important precaution should be observed when making any of the tests mentioned. The input signal to the video amplifier must not drive the amplifier into the nonlinear portion of its characteristic curve. This might cause the top and bottom of the wave to be cut off. It would then be impossible to tell if the amplifier is introducing distortion. Since transistor amplifiers can be damaged by overloads, it is especially important in solid-state receivers to hold the amplitude of the test signal within the prescribed values.

15.8 VIDEO REQUIREMENTS FOR COLOR TV

Although the subject of color television was covered more completely in previous chapters, a comparison of monochrome and color television video amplifiers is useful at this point. In color television receivers, the video-amplifier section accomplishes the same basic purpose as in monochrome receivers. In both, it raises the amplitude of the video signal enough to drive the picture tube to full contrast. In monochrome receivers, the video signal is applied to the single gun of the picture tube to control the brightness of the various portions of the screen as the beam is scanned over the face of the tube.

In the case of a color set using a three-gun color tube and producing color pictures, two types of video signals are simultaneously applied to the three guns. One is the "Y" signal which contains all the information normally found in a monochrome-video signal. The other video signal consists

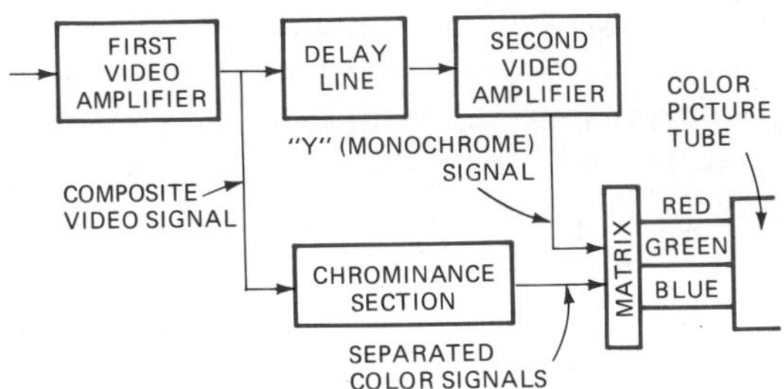

Figure 15.12 A simplified block diagram of a video-amplifier section for a color TV receiver.

of three color video signals. These are individually applied to the three guns of the color tube. It is the combination of these two types of video signals that causes the picture tube to produce color pictures, as shown in Figure 15.12.

VIDEO-AMPLIFIER BANDWIDTH

As mentioned previously, the overall video-amplifier bandwidth for a color TV receiver is about 3.2 MHz (except where the comb filter is used). However, in order for the chroma subcarrier (3.58 MHz) and its sidebands to reach the chroma (color) amplifier, a uniform bandpass to at least 3.6 MHz is required. To provide this response and also an overall high-video frequency response not exceeding 3.2 MHz for the monochrome (Y) signal, the video signal for the chroma circuits is generally taken from the output of the first video amplifier (video preamplifier), Figure 15.12.

Since the response of the first video amplifier extends to at least 3.6 MHz, the chroma signals can reach its output. However, the response of the following luminance video amplifier(s) is limited to 3.2 MHz. In this way interference from the 3.58-MHz color subcarrier is reduced.

The upper chroma sideband actually extends to about 4.1 MHz, while the response (for economy) of the first video amplifier may be flat only to about 3.6 MHz. The effect of this high-frequency dropoff is compensated by a corresponding rising frequency response characteristic of the chroma amplifier receiving the output of the first video amplifier. The combined response characteristics provide a flat frequency and phase response extending to 4.1 MHz. A chroma bandwidth less than 4.1 MHz will result in serious color deterioration.

THE "Y" SIGNAL AND THE DELAY LINE

When a color receiver is reproducing a monochrome picture, the color circuits are automatically disabled. Only the "Y" signal is applied to the three guns. This signal is proportioned to each gun so that the combination of the red, green, and blue phosphors, when illuminated, will produce white light. A monochrome picture is now produced by the same general process used to create a monochrome picture in a strictly monochrome receiver. Figure 15.13 shows the video signal being applied simultaneously to the three guns of a color tube to produce a monochrome picture. This video signal is the "Y" signal, and not the color-video signals.

Delay Line In Figure 15.12 the composite

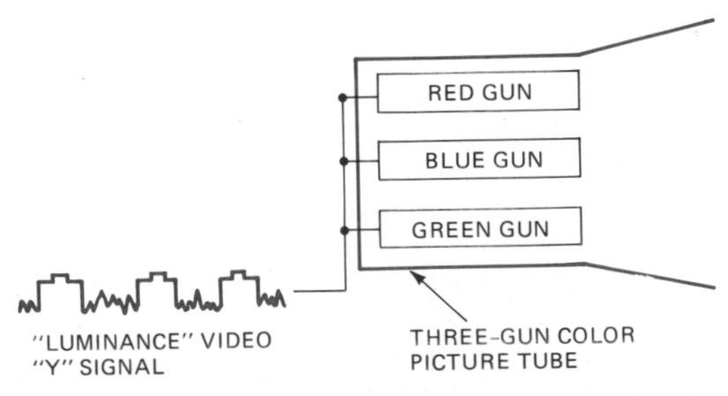

Figure 15.13 When a three-gun color picture tube is used to produce a black-and-white picture, the video ("Y") signal is applied simultaneously to the three guns in the proper proportions.

video signal that contains the color information is fed to the chroma section where individual color-video voltages are generated. It is then fed to the correct gun of the color tube. In passing through the chroma bandpass filters (or amplifiers), a delay is introduced into the color-video signals. This causes them to be delayed with respect to the "Y" signal. To produce a correct color picture, it is necessary to introduce a compensating delay into the "Y" signal channel so that the "Y" signal arrives at the three guns exactly in step with its related color-video signals. This delay is provided by the delay line, shown in Figure 15.12.

Luminance Channel Another name given to the "Y" signal is the "luminance signal." The channel carrying this signal is sometimes called the "luminance channel." This channel is part of the video amplifier in a color set.

Having studied the general requirements of the video section in a color set, we shall now look at some block diagrams of video amplifiers. This will give us a better understanding of the various signal paths and operational requirements of the video section.

15.9 TYPES OF VIDEO AMPLIFIERS FOR MONOCHROME AND COLOR SETS

MONOCHROME VIDEO AMPLIFIERS

The video section of a television receiver comprises all of the stages between the video detector and the picture tube. We will first discuss a block diagram of a typical video amplifier in a monochrome receiver and then a block diagram of a typical luminance channel in a color television receiver.

Block Diagram

Figure 15.14 shows a block diagram of a video-amplifier section in a monochrome television receiver. The video section is enclosed within the dotted lines. The complexity and design of this section varies from manufacturer to manufacturer. However, the block diagram of Figure 15.14 shows a typical video-amplifier section. In this discussion, the video-amplifier section is assumed to include all of the stages that handle the video signal, from the video detector to the picture tube. Although the DC restorer is included in the video section of the block diagram, it is important enough to be treated in a separate chapter in this text.

The output of the video detector is delivered to the sound takeoff point, and to the first video amplifier, through a 4.5-MHz trap. In some monochrome receivers the synchronizing pulses are also taken from the output of the video detector. Also, in some monochrome receivers the sound takeoff occurs after one or more stages of video amplification.

The first video amplifier accepts the composite video signal from the detector and amplifies it. In some cases, a single stage of video amplification is sufficient to develop enough signal strength to drive the picture tube (about 100 V, peak-to-peak). If more than one stage of video amplification is used, careful attention must be paid to the coupling circuits between amplifiers, and between the last video amplifier and the picture tube. This insures that the full desired range of video frequencies is passed from stage to stage (See Chapter 16).

Meaning of Gamma

Contrast control may be accomplished by controlling the gain of one of the video amplifiers. Changing the gain of the amplifier causes a change in the amplitude of the signal delivered to the picture tube. The larger

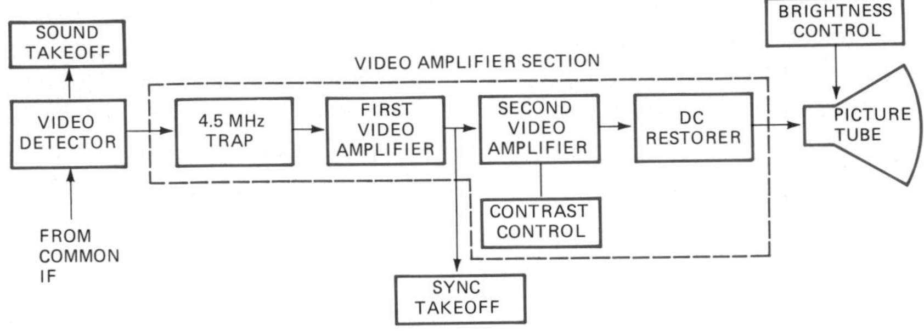

Figure 15.14 Block diagram of the video section in a monochrome receiver.

the difference in the amplitude between the maximum and minimum voltages, the greater the contrast of the picture. The ideal picture on the receiver picture tube occurs when the light and dark areas in the reproduced picture are in the same ratio as the light and dark areas in the original scene. Television engineers have a measurement that compares the contrast in the original and reproduced pictures. They call this measurement the *gamma* (γ).

$$y = \frac{\text{Ratio of bright to dark areas in the reproduced picture}}{\text{Ratio of bright to dark areas in the original picture}} \qquad (15.1)$$

There are no units of measurement for gamma. If the contrast of the original picture is identical to the reproduced picture contrast, gamma is equal to one ($\gamma = 1$). Excessive contrast means a gamma greater than one ($\gamma > 1$). A washed-out picture has a gamma value that is less than one ($\gamma < 1$). When a monochrome picture is being reproduced, the presence of colors in the original picture tends to reduce the contrast. Therefore, engineers recommend that the gamma be adjusted to a value greater than one. In such cases, a value of 1.4 ($\gamma = 1.4$) is considered to be ideal.

The DC Restorer

In the discussion on waveforms (Figure 15.9) it was shown that the DC level of the video signal changes from scene to scene. A picture with large white areas has a different DC level than a picture with large dark areas. If this DC level is lost in the coupling

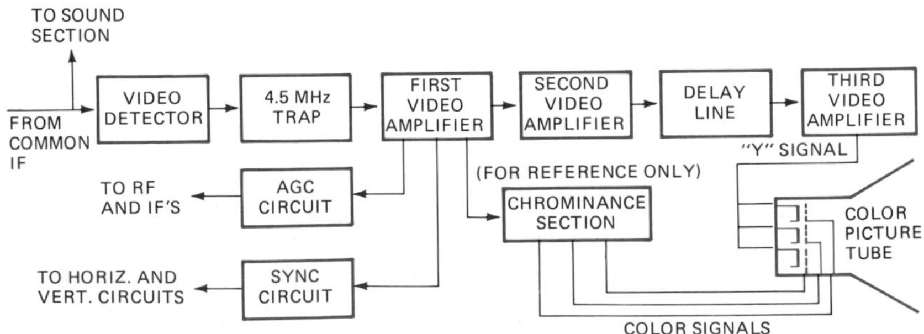

Figure 15.15 Simplified block diagram of the luminance channel of a color TV receiver.

circuits in the video section, then it must be restored. That is the purpose of the **DC restorer** illustrated in the block diagram of Figure 15.14. If direct coupling is used from the video detector to the picture tube, the DC level will be preserved and DC restoration is not necessary.

COLOR VIDEO AMPLIFIERS

Figure 15.15 shows a block diagram of a color set luminance channel. At this time we are interested primarily in the circuits that produce the "Y" signal, since this is the equivalent of the monochrome-video signal. However, a brief description of the related sections will also be given. As shown in Figure 15.15, the sound takeoff precedes the video detector in color sets. At the output of the video detector, the composite video signal is delivered to the first video amplifier through a 4.5-MHz trap. The color-burst signal centers around a 3.58-MHz color subcarrier frequency and is present in the video-detector output. Although the color subcarrier is not actually transmitted, its sidebands are present. These sidebands and the 3.58-MHz burst may heterodyne with the 4.5-MHz sound signal to produce frequencies that are within the video range. The heterodyne frequencies (if present) will appear on the screen of the color tube as interference. The 4.5-MHz trap reduces the 4.5-MHz level and eliminates such interference.

The first video amplifier is similar to the first video amplifier in a monochrome receiver. In older color receivers, the video amplifier(s) was required to pass color information to at least 4.2 MHz. However, in current color receivers this requirement for the first video amplifier is only about 3.6 MHz. This is because compensation for the higher color-video sidebands is provided in the chrominance section of the receiver, as described previously. Some color receivers have their color-video signals tapped off directly after the video detector. This makes it possible to restrict the first video-amplifier bandwidth to less than 3.6 MHz. In the case of other color receivers, the color-video sidebands are frequently taken from the output of the first video amplifier. In this way the bandwidth of the subsequent amplifier(s) can be reduced to 3.2 MHz.

The output of the third video amplifier is the final "Y" (or monochrome) signal. It is applied in many TV sets to the three cathodes of the color tube. As shown in Figure 15.15, the three color-video signals are applied to the appropriate color guns. In this case, the mixing (or matrixing) of the two signals actually takes place in the guns, producing the required color-video signals for red, green, and blue.

Some color TV sets use a two-stage video amplifier instead of the three-stage type described previously. In this case, the video outputs for the color section and the delay line are taken from the first video amplifier. The output of the second-video amplifier will then go to the color-tube cathodes.

15.10 COMPARISON OF TUBE AND SOLID-STATE VIDEO AMPLIFIERS

There are some advantages of transistors over vacuum tube amplifiers. There are also a few disadvantages. Some of the obvious advantages include very small physical size, low power supply voltage requirement, low heat dissipation, and low cost. Greater reliability is another factor in the popularity of the transistor. On the other hand, it is more difficult to design transistors to accommodate large signal amplitudes. The required video-signal amplitude to a picture tube may be as high as 150 V. Tube circuits can readily produce this much signal voltage. However, it is more difficult to obtain linear amplification from transistors for high-signal amplitudes.

It is desirable to have the video detector connected to an amplifier with a high input impedance. While tube amplifiers have a characteristically high input impedance, conventional transistor amplifiers do not. FETs have the high input impedance advantage of vacuum tubes and the solid-state advantage of transistors. Of the four general methods of coupling tube amplifiers, direct coupling provides the best overall frequency response. However, this also has disadvantages. The power supply requirement for tubes that are direct coupled is quite rigid. Transistors are more easily direct coupled for two reasons. First, they operate at lower voltages, and the increased voltage requirement arising from cascading is not as difficult to obtain. Second, transistors can be put into special direct coupled configurations that are not possible with tubes. The tube must always be connected into the circuit, so that the plate is positive with respect to the cathode. Transistors, on the other hand, can be connected so that the collector is positive (in the case of NPN transistors), or they may be connected so that the collector is negative (in the case of PNP transistors). The combination of one NPN and one PNP transistor in a direct coupled configuration is sometimes referred to as a complementary amplifier. Figure 15.16 shows a complementary amplifier. Not *all* transistorized video amplifiers are complementary amplifiers. As a matter of fact, the first stage following the video detector is more often an emitter follower because of the high input impedance of this configuration. The emitter follower delivers its output to a grounded base or grounded emitter amplifier. However, designers have a greater range of design choices when transistor amplifiers are used.

In some of the early transistorized television receivers the ability of tubes to handle high-signal voltages resulted in hybrid designs. A hybrid amplifier is one that uses both tubes and transistors. In the video-amplifier stages of these receivers, the output of the video detector is fed to a transistor amplifier. This, in turn, delivers its output signal to a vacuum tube video-output amplifier.

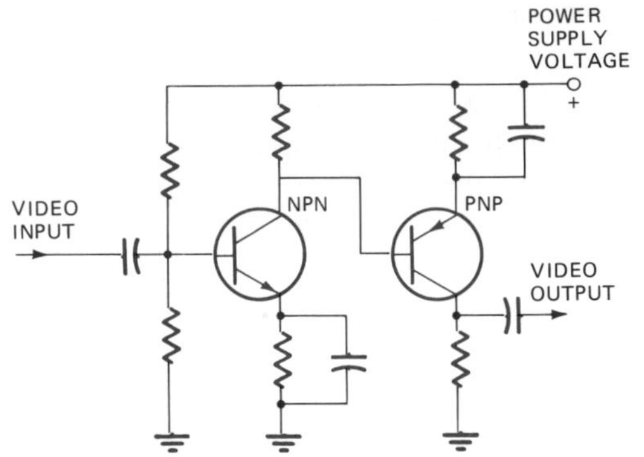

Figure 15.16 Complementary amplifiers.

15.11 CONTRAST CONTROLS IN VIDEO AMPLIFIERS

The **contrast control,** also known as the **picture control** or **pix control,** regulates the amount of video signal reaching the picture tube. It is manually operated and is adjusted by the viewer until the relationship between the light and dark areas of the picture meets the particular requirements. If the room is light, it may be necessary to increase the contrast. If the room is dark, the contrast may be decreased. In either case, when the contrast control is adjusted, there is a regulation of the intensity, or the amplitude, of the video or luminance signal that reaches the picture tube.

CONTROL METHODS

Picture intensity may be regulated in several ways. The bias of one or more video-

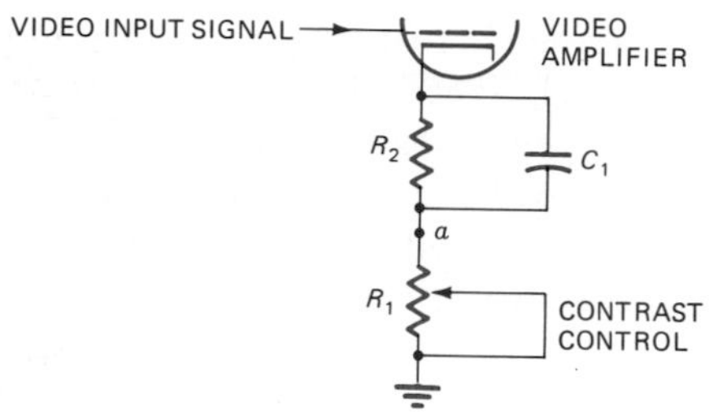

Figure 15.17 Cathode degeneration control of video amplitude.

IF amplifiers will vary the gain of these stages. This variation, in turn, controls the signal amplitude of the picture. The automatic-gain control voltage adjusts the gain of the video-IF stages. For a manual contrast control, it is a more common practice to control the picture intensity in a video-amplifier stage. This is normally accomplished in one of two ways: (1) control the gain of one of the video amplifier stages, or (2) control the amplitude of the signal delivered from one stage to the next.

CONTROLLING CONTRAST BY CONTROLLING GAIN

Figure 15.17 shows a circuit that varies the gain of a vacuum tube video amplifier by controlling the amount of degeneration introduced in the cathode circuit. When the arm of variable resistor R_1 is moved to Point A, maximum gain (and contrast) will result because there is now no degeneration in the cathode circuit. When the arm of R_1 is at the ground position, maximum degeneration and minimum gain (and contrast) occur. Resistor R_2 maintains a minimum value of operating bias at the maximum contrast position of R_1. Capacitor C_1 prevents degeneration across R_2.

Transistor Circuit Figure 15.18 shows the equivalent transistor-video amplifier with

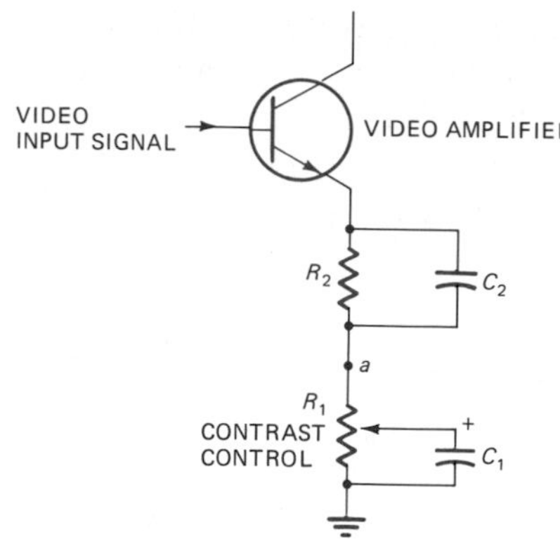

Figure 15.18 Emitter contrast control of a transistor video amplifier.

the contrast control in the emitter leg. The position of the arm of the control determines the gain of the stage by controlling the amount of degeneration introduced. When the arm is at Point A, resistor R_1 is bypassed by C_1 and there is no degeneration. At this point the gain will be maximum. When the arm is moved toward the ground, the amount of degeneration is increased and the gain is decreased. C_1 is an electrolytic capacitor with a large value that is capable of preventing degeneration even at the lowest video frequencies.

Emitter-Driver A commonly encountered transistor video system employs an emitter-

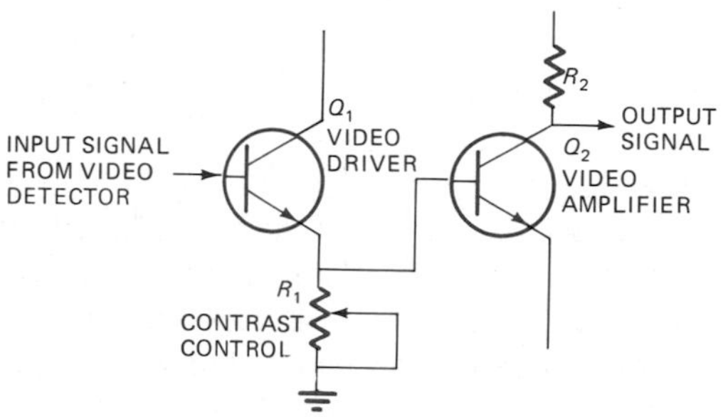

Figure 15.19 Simplified diagram of contrast control in a video-driver stage. Stage Q_1 is frequently called an emitter-driver.

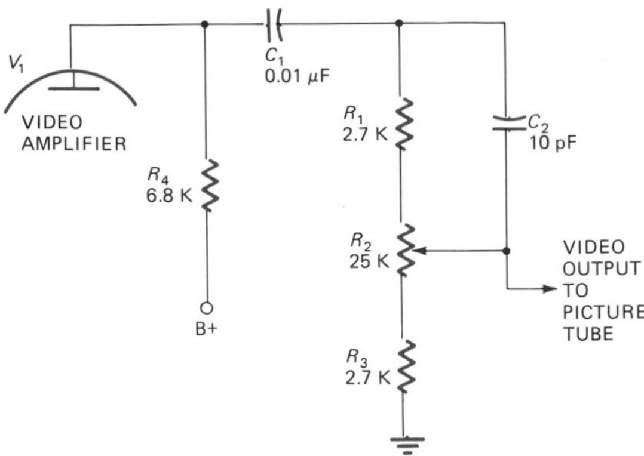

Figure 15.20 A contrast control in the plate circuit of a vacuum tube video amplifier.

follower in the stage immediately after the detector. The emitter-follower stage is often called the emitter-driver. The emitter-driver may be either direct coupled or capacitively coupled to the next stage. Figure 15.19 shows a contrast control in the emitter-driver system. The adjustment of R_1 will affect the amount of degeneration of Q_2, and the gain of that stage. Thus, changing the resistance of R_1 will alter the amount of signal delivered to Q_2.

CONTROLLING CONTRAST BY CONTROLLING SIGNAL AMPLITUDE

A disadvantage of the circuits shown in Figures 15.17 through 15.19 is that the bias changes when the contrast control is varied. As a result, the operating point of the amplifier is changed. If it is not carefully designed, amplitude distortion of the video signal may result. To overcome this disadvantage, circuits may be used that control the amplitude of the video signal delivered from one stage to another. In this way the bias can remain fixed.

Vacuum Tube Circuit Figure 15.20 shows a simplified schematic diagram of a vacuum tube video amplifier circuit. In this circuit, contrast control R_2 acts as a signal-voltage divider. The desired amount of video signal is selected by the position of the center arm of R_2. Resistor R_3 insures that a selected minimum amplitude of video signal will be available at the minimum contrast position of R_2. R_2 is in a high-impedance circuit because of the input impedance of the following stage. Therefore, its high-frequency response is subject to change as the center arm position is varied. However, if the correct value of capacitor C_2 is chosen for the particular circuit, the high-frequency response characteristics of the contrast control circuit will not change.

Video Driver Circuit Figure 15.21 illustrates how a transistor video driver (emitter-follower) is connected to a potentiometer-type contrast control. Since the contrast control is in the low-impedance emitter circuit, it has a low-resistance value (300 ohms). This is much lower than the value of R_2 in Figure 15.20 (25 kΩ). As a result, the position of the contrast control in Figure 15.21 is far less critical with regard to changes of frequency response. Thus, high-frequency compensation, as used in Figure 15.20, is not required here.

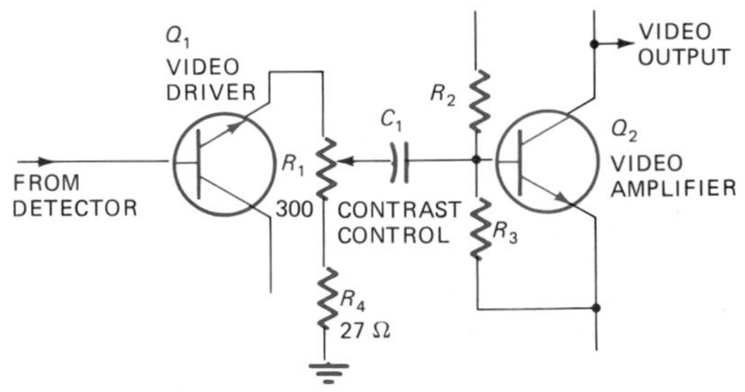

Figure 15.21 Controlling the contrast by controlling the signal amplitude output of a transistor video-driver stage.

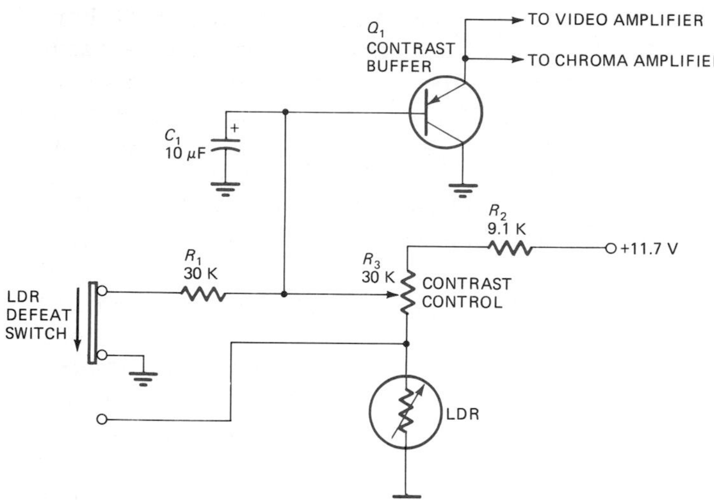

Figure 15.22 Simplified diagram of an automatic contrast control circuit. The LDR (light-dependent resistor) provides the automatic control feature. *(Courtesy of RCA Consumer Electronics.)*

15.12 AUTOMATIC CONTRAST CONTROL

Some television receivers contain a simple circuit which automatically adjusts contrast in response to changes of ambient lighting conditions. A simplified schematic diagram is shown in Figure 15.22. This scheme is used in the RCA XL-100 TV receiver series.

THE LDR

The operation of this circuit depends upon the characteristics of an LDR (light-dependent resistor). This device is a photoconductive cell. In other words, it is a photocell whose resistance varies inversely with the intensity of the light striking its active material. The LDR is mounted on the front panel. Figure 15.23 shows a typical LDR employing a cadmium-sulfide wafer. The electrical resistance of an LDR is high with low light intensity and low with high light intensity.

Circuit Operation In Figure 15.22, the resistance element of the LDR is connected in series with the lower end of the contrast control and ground. If the ambient-light intensity is low, the LDR resistance becomes high. This increases the DC voltage at the center arm of the contrast control. This increased voltage is applied through the contrast buffer Q_1 to the video amplifier. The gain of the video amplifier is now decreased. This reduces the contrast to an appropriate level.

Conversely, the LDR resistance is lowered with high-ambient-light intensity. Now,

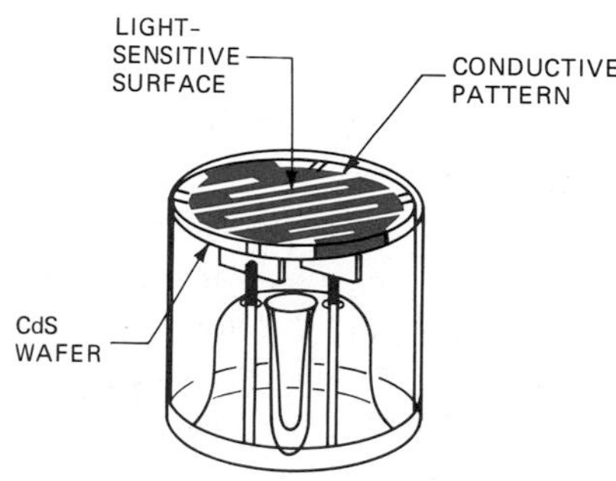

Figure 15.23 Photoconductive cell (LDR) using a cadmium sulfide wafer.

the DC voltage at the center arm of the contrast control is reduced. The reduced voltage causes an increase in the gain of the video amplifier. This results in an increase in the contrast, as required, for the high ambient light.

LDR Defeat Switch In Figure 15.22, an LDR defeat switch is included. This is a customer-operated switch which gives the option of turning off the automatic contrast control feature. In this case, the manual contrast control functions normally. In the position shown in Figure 15.22, the LDR is functioning. When the switch bar is moved to the bottom position, the LDR is shorted and the bottom end of the contrast control connects to ground.

Color Tracking Although the details are not shown in Figure 15.22, another output from the contrast buffer Q_1 is fed to the chrominance circuit. This provides automatic color saturation control in addition to automatic contrast control. After the customer has set the *color* control to the desired conditions, color saturation requirements will *track* as the contrast control is varied or as the LDR senses changes in ambient lighting conditions.

15.13 AUTOMATIC BRIGHTNESS AND CONTRAST CONTROL

A single LDR may simultaneously provide automatic control of both brightness and contrast. A typical arrangement is shown in the simplified schematic diagram of Figure 15.24.

AUTOMATIC BRIGHTNESS CONTROL

Note that the collector of the video-output amplifier Q_1 is DC coupled to the cathode of the picture tube. Therefore, any change in the DC collector voltage results in a change in the picture tube bias and brightness.

First, assume that the ambient light level is relatively low. As a result, the resis-

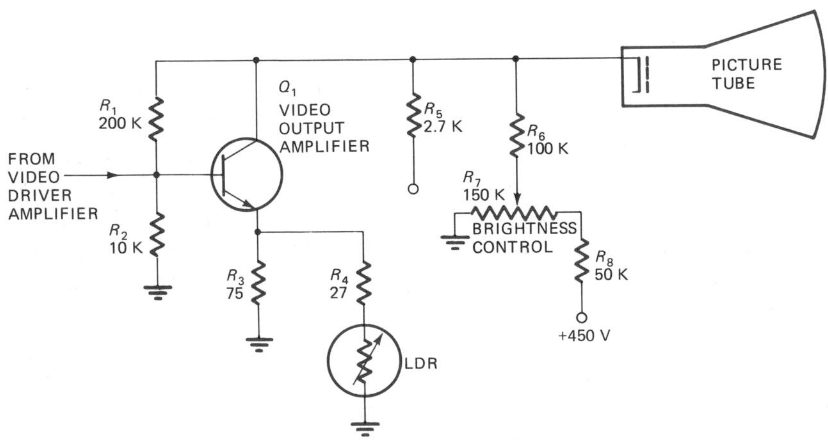

Figure 15.24 Simplified diagram showing how both brightness and contrast may be automatically controlled by a single LDR.

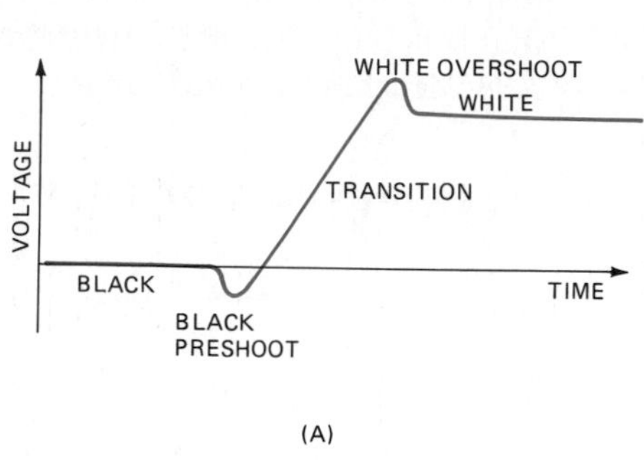

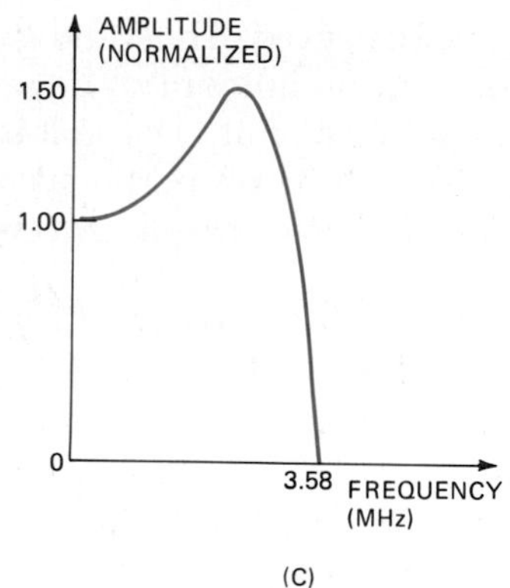

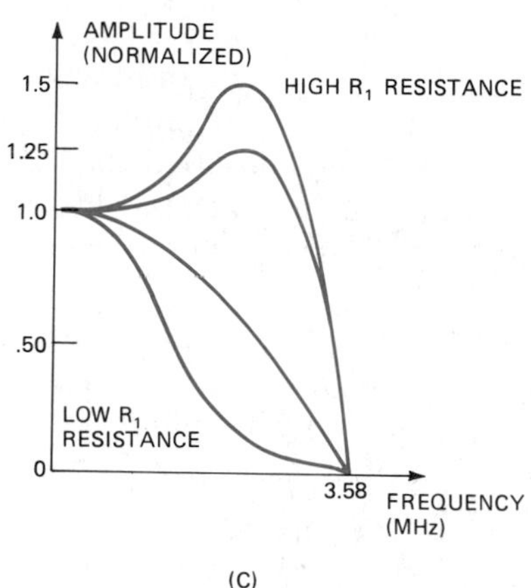

Figure 15.25A-C. Video peaking (sharpness) response characteristics. **A.** The desired video peaking circuit transient response. Note the symmetrical preshoot and overshoot characteristics. **B.** The ideal video peaking amplitude response. **C.** The effect on the response of the system in Figure 15.26 of varying the customer-operated peaking (sharpness) control. *(Courtesy of RCA Consumer Electronics.)*

tance of the LDR is high. Since the LDR resistance parallels (with R_4) the emitter resistor, R_3, the total emitter resistance will be a maximum value. This resistance will reduce the effective forward base-to-emitter bias of Q_1. In turn, the collector current of Q_1 will be reduced, with a consequent increase of collector voltage. As shown in Figure 15.24, the collector is DC-coupled to the cathode of the picture tube. A more positive voltage now appears on the picture tube cathode. This results in decreased brightness, as now required.

Conversely, if the ambient light level is relatively high, the LDR resistance will be low. This will result in a low total emitter resistance and increased forward bias for Q_1. The collector voltage and the picture tube cathode voltage will become less positive and the picture tube brightness will increase to counter the higher ambient light level.

AUTOMATIC CONTRAST CONTROL

The LDR of Figure 15.24 also provides automatic contrast control. The emitter resistance of Q_1 is unbypassed. Thus, the gain of Q_1 is a function of the amount of degeneration from its emitter circuit. The amount of degeneration, in turn, is proportional to the amount of effective resistance in the emitter circuit.

If the ambient light level is high, the LDR resistance (and the effective emitter resistance) is low. Emitter degeneration is reduced and the gain of Q_1 is *increased.* This increases the contrast, as required by the higher ambient light level.

With a low ambient light level, the high value of LDR resistance causes an increase in emitter degeneration. The subsequent lowered gain of Q_1 reduces the picture tube contrast in accordance with the requirement of a low ambient light level.

A 27-ohm resistor, R_4, is placed in series with the LDR. With this resistor in the circuit, the shunting resistance for R_3 can never be less than 27 ohms. This means that excessive brightness and contrast are prevented at high ambient light levels.

15.14 VIDEO PEAKING

As explained previously, the overall response of the video amplifier is limited to about 3.2 MHz unless a comb filter is used. Unfortunately, the use of a comb filter results in some loss of picture definition. To obtain sharp images, the edge (or transient) characteristics of the video system from white to black (or from black to white in the image) should provide an ideal transient response (see Figure 15.25A). Note the symmetrical preshoot and overshoot and the absence of *"ringing"*. If "ringing" were present, image quality would be degraded. The preshoot and overshoot sharpen the edges of an image and thus provide greater (apparent) picture detail.

To obtain the characteristic of Figure 15.25A, the ideal video-amplifier amplitude response of Figure 15.25B is required. Note that this response still maintains a null at the 3.58-MHz chroma subcarrier.

SHARPNESS CONTROL

Because ghosts, noise, or program-content errors may exist in a received TV signal, it is an advantage if the customer can control the degree of video peaking being introduced. This is done by the use of a *"sharpness"* control. A simple method of providing such control is with the prepeaked response of Figure 15.25B in conjunction with a simple, variable "rolloff" circuit shown in Figure 15.26. In this simplified diagram, the rolloff circuit consists of C_1 and R_1. Note the similarity of this circuit to a simple audio-tone control.

The effect of varying R_1 is shown in Figure 15.25C. As the resistance of R_1 (the sharpness control) is decreased, the effect of the high-frequency bypassing of C_1 is increased. At the maximum resistance condition of R_1, C_1 has virtually no effect on the high frequencies. Thus, the customer has the choice of varying the image from a condition of maximum sharpness to a "soft" picture. However, at the maximum sharpness condition, the effect of any noise (or other high-frequency interference) will be most

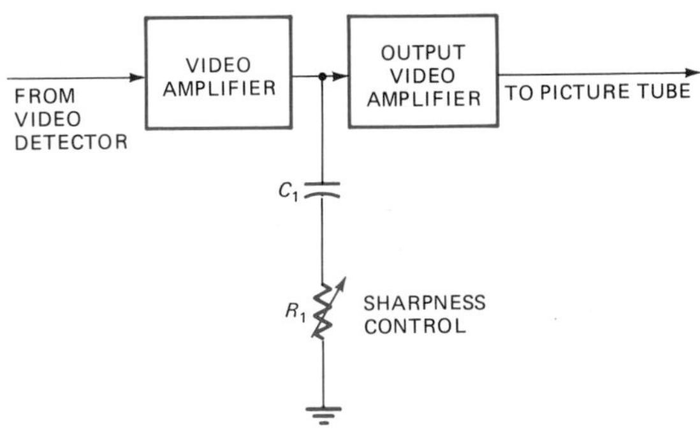

Figure 15.26 Simplified diagram showing the connection of a sharpness (video peaking) control.

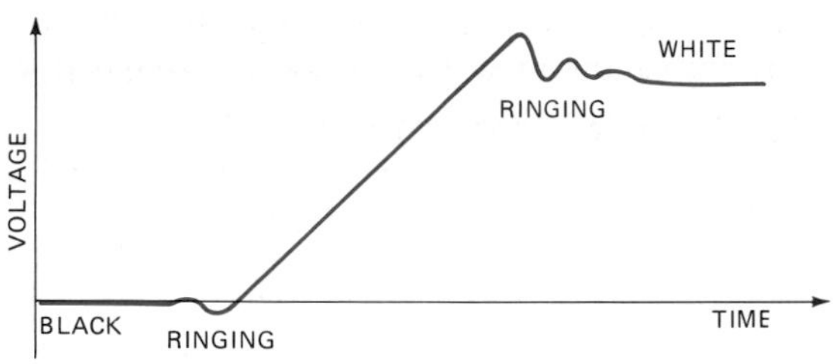

Figure 15.27 Transient response of peaking circuit with nonlinear phase characteristics.

noticeable. In such conditions, the customer can reduce the interference by decreasing the video peaking effect with the sharpness control. The **sharpness control** may also be labelled the **peaking control** or the **fidelity control** on some TV receivers.

15.15 IMPROVED VIDEO PEAKING SYSTEM

An important disadvantage of the video-peaking scheme described above, is its inability to maintain linear-phase response over the range of the sharpness control. If the phase response is not linear, ringing can result, as shown in Figure 15.27. This can degrade the image quality. In order to overcome this disadvantage, an improved circuit is used in the RCA XL-100 TV receiver series. This circuit is known as the **transversal filter peaking system.**

TRANSVERSAL FILTER PEAKING SYSTEM

A simplified diagram of the transversal filter peaking system is shown in Figure 15.28. This system consists of a tapped-delay line, a subtractor, and an adder. The system operates as follows: the video signals—e_a, e_b, and e_c, are delayed linearly by the tapped-delay line. They are then added or subtracted together in appropriate amounts. The results are the separate high- and low-frequency components of the input-video signal.

The Subtractor Signals The original unpeaked video signal, e_b, is applied to the subtractor. Also applied to the subtractor are the outputs of the other two taps of the delay line, e_a and e_c. These outputs are applied as $\frac{1}{2}(e_a + e_c)$. The signal $\frac{1}{2}(e_a + e_c)$ is subtracted from signal *eb*. This subtraction produces only the high-frequency component of the video signal, *eh*.

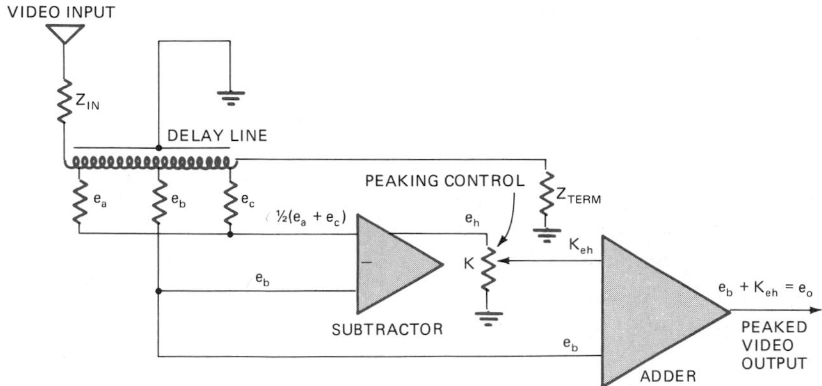

Figure 15.28 A simple transversal filter peaking system. *(Courtesy of RCA Consumer Electronics.)*

The Adder Signals The original unpeaked video signal, e_b, is also fed to the adder together with a selected fraction of the high-frequency component, *eh.* This fraction of the high-frequency component is determined by the customer operating the peaking control "*K*". This fraction, therefore, is designated as K_{eh}. As a result, the output of the adder consists of the original unpeaked video signal, e_b, plus a selected amount of the video signal high-frequency component K_{eh}, or $e_o = e_b + K_{eh}$.

With this system, the customer can increase or decrease the amount of video peaking by manipulating the peaking control. This eliminates the disadvantage of nonlinear phase response.

15.16 THE COMB FILTER

The **comb filter** is a recent innovation for color television receivers. It makes it possible to extend the luminance video amplifier response flat to 4.2 MHz. This is a full 1 MHz more than is possible with more conventional circuitry. As we have seen with conventional circuitry, the video response to the picture tube must be limited to about 3.2 MHz in order to reduce interference patterns in the picture. By extending the response to 4.2 MHz, it is possible to reproduce the full picture detail as it is originally transmitted. No video peaking techniques are used with comb filter circuitry.

WHAT DOES A COMB FILTER DO?

In the more conventional color TV receiver systems, a bandpass filter is used to separate the video and chroma signals. As part of the system function, the luminance (Y) signal response must be limited to about 3.2 MHz. The comb filter system eliminates the requirement for a 3.58 MHz trap in the luminance circuit. This permits the full 4.2-MHz luminance information to be displayed on the picture tube. The use of the comb-filter system provides a 25% increase of horizontal (fine detail) resolution. It also reduces interfering dot patterns and *barber pole,* vertical-moving color effects.

What is a Comb Filter? The name comb filter comes from the way it functions. It selects some frequencies of a given band to be passed, while rejecting other frequencies within the *same* band. It "combs out" of the band the desired (chroma) frequencies. The luminance signals are processed separately, as explained later.

Factors Necessary for the Operation of a Comb Filter Two basic factors which are normal in the NTSC color television system are utilized for the comb-filter scheme.

1. The luminance information has the same phase from scanning line to scanning line. From one line to the next, the luminance information is basically unchanged. This information occurs in energy groups (See Chapter 8) centered around frequencies that are multiples of the horizontal rate (15 734.26 Hz).
2. The chroma information has its *phase reversed* from one scanning line to the next. The chroma information is basically unchanged from one line to the next. This information occurs in energy groups (See Chapter 8) that are odd multiples of one-half the horizontal scanning rate. The comb filter will reject those frequencies (luminance) which have the same phase from line-to-line (in a field). The comb filter will pass with double amplitude those frequencies (chroma), which have the *opposite* phases from line-to-line (in a field).

Frequencies Rejected and Passed

The luminance frequencies that are *rejected* by the comb filter are integer multiples of the line rate. Examples of the rejected frequencies are the base frequency of 15 734.26 Hz, the second multiple frequency of 15 734.26 × 2, or 31 468.52 Hz, the sixth multiple frequency of 15 734.26 × 6, or 94 405.56 Hz, and so on.

Chroma frequencies passed (with double amplitude) by a comb filter are odd multiples of one-half the line rate. Examples of these frequencies are the third multiple frequency of $\frac{15\,734.26}{2} \times 3$, or 23 601.39 Hz, the seventh multiple frequency of $\frac{15\,734.26}{2} \times 7$, or 55 069.91 Hz, the 35th multiple frequency of $\frac{15\,734.26}{2} \times 35$, or 275 349.55 Hz, and so on.

As explained in Chapter 8 and again below, the chroma frequencies are spaced exactly between the gaps of the luminance frequencies.

By the use of a comb filter system, the color signals are effectively "combed out" of the composite-video signals and sent to the chrominance-receiver section. At the same time, the *full bandwidth* luminance signal is sent to the luminance-amplifier section.

Interleaved Chroma and Monochrome Signals

Figure 15.29 shows in a somewhat different manner how the chroma and mono-chrome signals are *interleaved* in the composite video signal. Note in Figure 15.29A that luminance energy bursts recur at the color horizontal-line frequency (15 734.26 Hz), or every 63.5 μs. Also note that the chrominance energy bursts occur halfway between the luminance energy bursts. A simplified diagram in Figure 15.29B shows how the chrominance energy is interleaved with the luminance energy over the entire video spectrum.

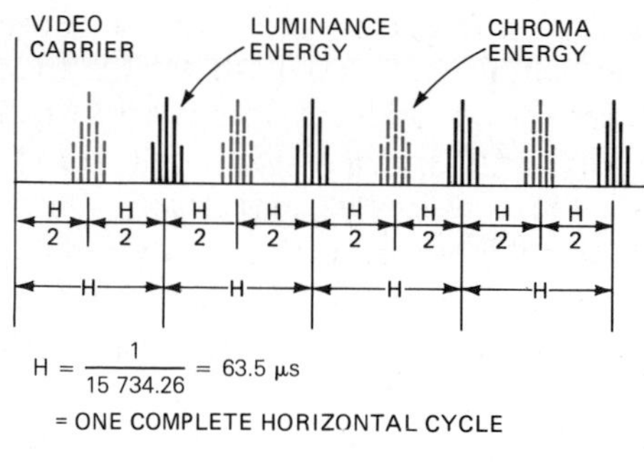

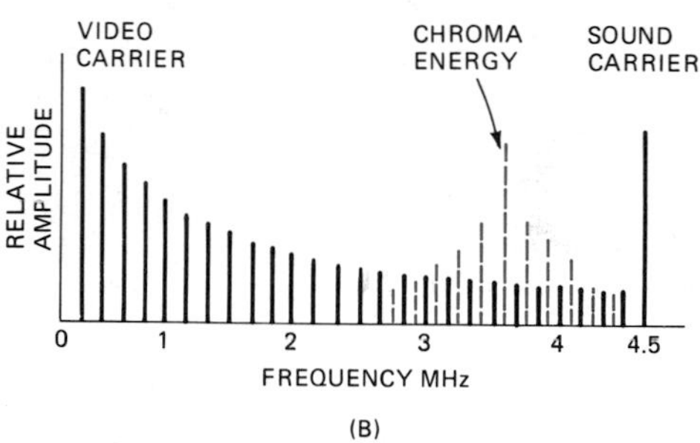

Figure 15.29A-B. Illustrating the interleaving of chroma and luminance signals. **A.** The spacing of luminance and chroma energy bursts with respect to the horizontal period. **B.** Interleaving of chroma and luminance energy bursts in the video spectrum, shown in simplified form.

15.17 ELEMENTARY COMB FILTER

The simplified block diagram of Figure 15.30 explains the basic principle of a comb filter. The composite video signal (V) is fed to the input of a 63.5-μs delay line. Time (H) is the period of one complete horizontal cycle. If the input wave is a 15 734.26-Hz sine wave (color, horizontal-line frequency), the delay line retards the input wave by a full 360°. Now the phases of the input and output signals of the delay line are identical. For any frequency that is an exact multiple of the horizontal-line rate, the output waves will also be in phase with the input waves.

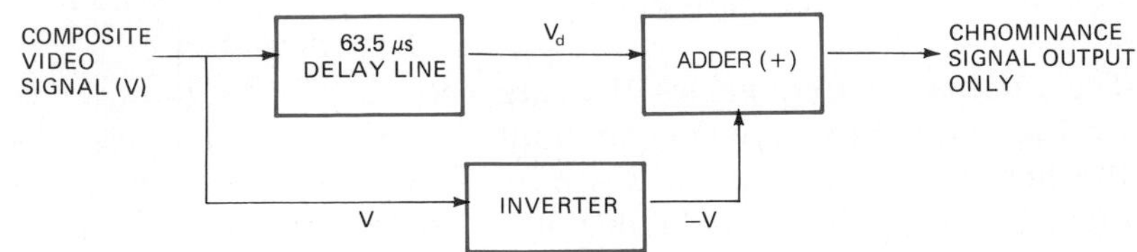

Figure 15.30 A simplified block diagram of an elementary comb filter. (V_d = V delayed)

THE ADDER

In Figure 15.30, note that the undelayed, composite-video signal (V) goes through a polarity inverter and becomes −V. When V and −V are fed to the adder, they tend to cancel.

If the delayed output V_d is at the horizontal-line frequency or a multiple thereof, both the input and output signals of the delay line will be in phase and will subtract completely. Thus, for luminance signals there will be no output from the elementary comb filter (Review Figure 15.29).

As the input frequency to the comb filter deviates from a whole multiple of the horizontal-line rate, cancellation is no longer complete and an output signal (chrominance) appears at the output of the comb filter. The output signal will be a maximum when the input signal is an odd multiple of $\frac{H}{2}$. Refer to Figure 15.29A. This means an odd multiple of one-half of the horizontal line frequency, which is $\frac{15\,734.26}{2}$, or 7 867.13 Hz. As shown in Figure 15.29A, multiples of 7 867.13 represent chrominance information, which now is the only information appearing at the output of the elementary comb filter.

Double Chrominance Amplitude At the output of the filter, the chrominance information actually doubles in amplitude. The reason for this is that chrominance signals are delayed by 63.5 μs in passing through the delay line. This time is the period of one horizontal cycle. Thus, by the time the delayed signal (V_d) arrives at the adder, the non-delayed but inverted chrominance signal (−V) from the *next* line also appears at the adder. Since the chrominance signals at the adder are from adjacent lines, they would normally be out of phase. However, because of the inverter they are actually in phase at the adder and are summed to twice the original amplitude. (The manner of handling the luminance information will be explained later.)

15.18 COMB FILTER BLOCK DIAGRAM

The complete block diagram of a comb filter used in some Magnavox TV receivers is shown in Figure 15.31. The composite video signal (V) is fed through an amplifier to the 63.5-μs delay line which is basically the same as the one shown in Figure 15.30. The amplifier is employed to overcome the insertion loss of the delay line (6 dB). Note that the bandwidth of the delay line extends only from 3- to 4-MHz. This range covers all of the desired chroma sidebands, and also includes the 3- to 4-MHz luminance information.

LUMINANCE SIGNAL CANCELLATION

The delayed video signal, V_d, is passed through a polarity inverter and becomes $-V_d$. The $-V_d$ signal and the original, but attenuated, V signal are applied to an adder (+). When the $-V_d$ and V signals are added, the luminance signals in the range of about 3 to 4 MHz cancel, as explained in connection with Figure 15.30. Note how-

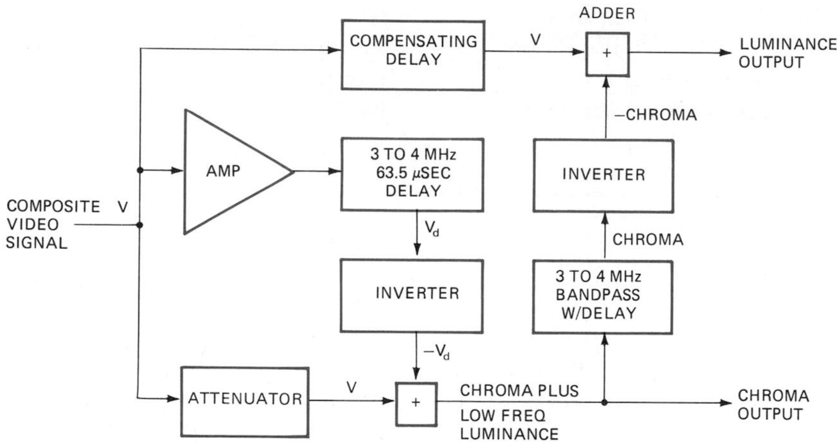

Figure 15.31 Complete block diagram of a comb filter circuit. *(Courtesy of Magnavox.)*

ever, that at this point, the lower frequency luminance signals (below 3 MHz) from the input (V) will still appear at the output of the adder, reaching this point by way of the attenuator. This is true because the 63.5 μs delay line passes video information from about 3 to 4 MHz and only this band of luminance information is cancelled here.

ADDER OUTPUT

The output of the adder consists of the *combed out* chroma information, plus the luminance information below about 3 MHz. The lower frequency luminance information is fed to the chrominance receiver section together with the combed out chrominance signals. However, these luminance signals have no effect in the chrominance section. The reason is that frequencies below about 3 MHz are filtered out in the chrominance section.

LUMINANCE PROCESSING

In Figure 15.31, we see that the chroma signals plus luminance signals below about 3 MHz are applied to a 3- to 4-MHz bandpass filter, which has an inherent delay. The filter passes the chroma information but rejects the low-frequency luminance information. The chroma information (alone) is applied to an inverter which reverses its polarity. Thus the output of the inverter is only "−chroma". The "−chroma" is applied to an adder (+) together with the composite-video signal (V). Because of the **compensating delay,** the chroma signal in (V) and the "−chroma" signal from the chroma inverter arrive at the adder at the same time, but with opposite polarities. Thus, the chroma signal is cancelled in the adder. As a result, the full bandwidth (4.2-MHz) luminance signal is the only output of the adder and is sent to the luminance section of the receiver.

SUMMARY OF COMB FILTER OPERATION

In summary, we see that the chroma signal has been combed out of the composite-video signal and sent to the TV receiver chroma section. In addition, *all chroma information* has been removed from the *full bandwidth* (4.2-MHz) luminance signal. This is then sent to the TV receiver luminance section. The 3.58-MHz trap is no longer required in the luminance section. In addition, annoying dot pattern interference and *barber pole* color effects have been eliminated by the comb-filter system. And as previously mentioned, the horizontal detail is increased by 25%.

SUMMARY OF CHAPTER HIGHLIGHTS

1. The basic function of the video (or luminance) amplifier section is to increase the amplitude of the video signal to a level sufficient to drive the picture tube to full contrast (Figures 15.14 and 15.15).
2. The output (typically) of a video detector is about 2 V peak-to-peak. The picture tube requires a peak-to-peak video input signal amplitude of 100 V to 150 V, or greater, depending upon the particular tube. (Figure 15.4)
3. The full, transmitted video-signal bandwidth is about 4.2 MHz. However, except where a "comb filter" is used, the video section response is limited to about 3.2 MHz. This is done to reduce interference patterns on the picture tube. (Figure 15.7)
4. In small screen monochrome receivers, the video response may be limited to 2.5 MHz. However, in color TV receivers using comb filters, the luminance video response extends to the full 4.2 MHz. (Figure 15.7)
5. The amplitude of the video signal applied to the picture tube determines the contrast of the image on the screen. (Figures 15.3 and 15.4)
6. Small screen TV receivers do not require as much fine picture detail to produce a presentable picture. This is because of the *resolving power of the eye.* (Figure 15.6)
7. The ideal video (or luminance) amplifier frequency response is from about 30 Hz to 4.2 MHz. (Figure 15.7 and Item 3 in the Summary).
8. Unequal amplification of all desired frequencies in an amplifier passband is called "frequency distortion".
9. To provide the required low- and high-frequency video amplifier response, compensating networks are used. (Described in Chapter 16.)
10. If the low-frequency response is inadequate, the picture background will be too dark and large objects will appear to be smeared. (Figure 15.8)
11. If the high-frequency response is not adequate, the picture will lack *sharpness.* This is caused by a reduction of fine-picture detail.
12. The DC component of the video signal determines the background brightness. (Figure 15.9)
13. If the phase shift of an amplifier is not proportional to frequency, the result is *phase distortion.*
14. Phase distortion causes picture smearing (low frequency) or blurring (high frequency).
15. A convenient way to check the response of a video amplifier is to check it with low and high frequency square waves. (Figure 15.11)
16. Contrast control may be accomplished by changing the gain of one of the video amplifiers. It may also be accomplished by regulating the amount of signal fed to a video amplifier. (Figures 15.17 through 15.21)

17. In a color TV receiver video (luminance) section, a delay line is inserted in series with the signal path. It is necessary to delay the luminance signal so that it will coincide with the color signal fed to the picture tube.
18. The gain of a video amplifier (and thus contrast) may be varied by changing the amount of amplifier degeneration. (Figures 15.18 and 15.19)
19. Automatic contrast control is provided by the use of a light-dependent resistor (LDR). (Figures 15.22 and 15.23)
20. Both automatic brightness and automatic contrast control may be simultaneously provided by the use of a single LDR. (Figure 15.24)
21. In many TV receivers, a video peaking circuit and (sharpness) control is provided. This gives the user the option of displaying a "sharp" or a "soft" picture. With a "soft" picture, interference is reduced on the picture tube. (Figures 15.25 and 15.26)
22. An improved video peaking system employs a "transversal" filter. This system provides linear phase response over the range of the sharpness control. (Figure 15.28)
23. The *comb filter* system used with some color TV receivers enables the full 4.2-MHz video band to be displayed. This provides a 25% increase of horizontal detail and the elimination of dot patterns and "barber pole" effect.
24. A *comb filter* derives its name from the fact that it "combs out" the chroma signal from the composite video signal. This is possible because of the interleaving of the chrominance and luminance video signals. (Figures 15.29 and 15.30)
25. A comb filter will reject frequencies that are integer multiples of the horizontal line frequency (15 734.26 Hz). It will pass frequencies that are odd integer multiples of one-half the horizontal line frequency (7 867.13 Hz). (Figure 15.29A)
26. In the elementary comb filter of Figure 15.30, the only output is the chrominance signals.
27. In a comb filter system, there is no 3.58-MHz trap. (Figure 15.31)
28. In the block diagram of Figure 15.31, the 3- to 4-MHz bandpass circuit rejects the low-frequency (30-Hz to 3-MHz) video signals but passes the chroma signals. These are inverted and cancel the chrominance signals applied to the luminance adder. (Figure 15.31)

EXAMINATION QUESTIONS

(Answers are provided at the back of the text.)

Part A Supply the missing word(s) or number(s).

1. ________ volts peak-to-peak is a typical output from a video detector.
2. A (An) ________-MHz luminance-amplifier bandwidth is typical for a color TV set, not using a comb filter.
3. When a comb filter is used, the luminance bandwidth may extend to ________ MHz.
4. The desired (flat) low-frequency response for a video amplifier is ________ Hz.
5. A peak-to-peak video signal of ________ volts (or greater) is typically required for full contrast of a picture tube.

6. _______ of large objects results from insufficient _______-frequency response of a video amplifier.
7. The bandpass of a video amplifier is improved by the use of _______ networks.
8. A lack of sufficient _______ frequency response in a video amplifier produces a picture that lacks _______.
9. Phase distortion occurs when the _______ is not proportional to frequency over the amplifier passband.
10. Ringing of a square wave used to check video-amplifier response indicates _______ frequency response.
11. A (An) _______ is used to bring the luminance signals into phase with the chrominance signals.
12. A (An) _______ is used to achieve automatic contrast control.
13. A scheme of _______ is used to improve the sharpness of the TV picture.
14. A comb filter system can operate because the luminance and chrominance signals are _______.
15. In a color set, the first video amplifier must pass a bandwidth of at least _______ MHz to include the chroma information.
16. Contrast control may be achieved by changing the _______ of a video amplifier.
17. Compensation for the amplitude response at the high chroma sidebands, is achieved in the _______.
18. Varying the emitter _______ of a video amplifier, is one method of providing contrast control.
19. The electrical resistance of an LDR varies _______ with light.
20. A transversal filter is part of an improved _______ system.
21. The 63.5-μs _______ is a vital part of a filter system.
22. The phase of the chroma signal _______ on each scanning line in sequence.
23. Vertical moving color interference patterns, are called _______ patterns.

Part B Answer true (T) or false (F).

1. The use of a comb filter extends the video response to 3.58 MHz.
2. The desired flat low-frequency response of a luminance amplifier extends to 60 Hz.
3. The video signal applied to the grid of a picture tube must have sync pulses in the negative direction.
4. The peak-to-peak video signal amplitude determines the picture contrast.
5. A small screen TV receiver must have a video frequency response no higher than 2.5 MHz.
6. If two objects subtend an angle of one minute or more, they can be seen as separate units.
7. If a video amplifier has insufficient high-frequency response, a monochrome picture will show "ringing" effects.
8. Phase distortion and frequency distortion produce the same picture defects.
9. High- and low-frequency compensation of the luminance amplifier section is required only if a comb filter is used.
10. The *sharpness* of a picture is a function of the video section midband response.
11. The DC component of a composite video signal is used as a reference to provide automatic brightness control.
12. With phase distortion at low frequencies, large objects will be stretched toward the right.
13. If the high-frequency square wave shows ringing, excessive high-frequency amplifier response is indicated.
14. The top and bottom of a square wave can indicate the high-frequency response of an amplifier.
15. The luminance signal is also known as the "Y" signal.
16. The upper chroma sideband extends to about 4.1 MHz.
17. Contrast control is frequently performed by changing the gain of the last common-IF stage.
18. Contrast controls in high-impedance circuits must be frequency compensated.
19. All automatic brightness control circuits also control contrast and hue.
20. Automatic contrast circuits operate by virtue of the characteristics of an LED.

21. Video peaking is regulated with the *sharpness* control.
22. A *transversal* delay line is an integral part of the comb filter.
23. The comb filter is also an impulse-noise limiter.
24. The use of a comb filter makes possible a 25% increase in horizontal resolution.
25. An important factor in the composite video signal, essential for comb filter operation, is the chroma signal phase inversion from line-to-line.
26. A comb filter will reject luminance signals, but will pass chroma signals at 50% of their original amplitude.

REVIEW ESSAY QUESTIONS

1. Draw a simple block diagram of a color TV video amplifier section from the last video IF to the picture tube. Indicate all inputs and outputs. Label all blocks correctly.
2. For the diagram drawn for Question 1, explain briefly the operation of the section.
3. Explain briefly, why a 3.58 MHz trap is used in a conventional video amplifier section.
4. Draw the luminance amplifier response curve of a conventional (noncomb filter) luminance section. Show all pertinent frequency points.
5. Discuss briefly the function of the luminance section delay line.
6. With the aid of a simple diagram, explain how the peak-to-peak video signal amplitude affects picture contrast.
7. How does the horizontal blanking pulse cut off the picture tube beam(s)?
8. Draw a simple circuit for emitter contrast control.
9. Explain the operation of the circuit drawn for Question 8.
10. Is phase distortion more critical at low- or high-video frequencies? Why?
11. Define the following: (a) H, (b) $\frac{H}{2}$, (c) blanking level, (d) DC component of the video signal.
12. In a color TV transmission, what is the horizontal scanning frequency? The vertical scanning frequency?
13. If the video-amplifier response cuts off at 2.0 MHz, what is the effect on the picture?
14. If the low-frequency response of the video amplifier does not extend below 200 Hz, how will this affect the picture?
15. In your own words, explain the meaning of the *resolving power* of the human eye.
16. Draw simple sketches of the square wave responses indicating (a) leading low-frequency phase shift, (b) good high-frequency and transient response, and (c) ringing.
17. Draw a simple schematic diagram of an automatic contrast control circuit. In your own words briefly explain the operation of this circuit.
18. What is the difference between frequency distortion and phase distortion?
19. Draw a simple diagram showing the connection of a sharpness control. Explain its operation.
20. In your own words, briefly explain how the transversal filter peaking system works. (Figure 15.28)
21. How does the luminance signal phase vary from line-to-line? The chrominance signal phase?
22. Draw an elementary comb filter diagram. Briefly explain how it operates, in your own words.
23. In Figure 15.17, what are the functions of R_2 and C_1?
24. In Figure 15.22, briefly explain the functions of (a) C_1, (b) R_2, and (c) LDR.
25. In Figure 15.28, briefly explain the operation of the peaking control in the overall circuit function.
26. Refer to Figure 15.31. In your own words, explain how the chroma output consists of the chroma plus the low-frequency luminance signals.
27. In Question 26, how is the low-frequency eliminated in the chroma section?

EXAMINATION PROBLEMS

(Selected problem answers are given at the back of the text.)

1. Calculate the following frequencies: (a) 25th multiple of the color horizontal-line frequency and (b) 19th multiple of one-half the color horizontal-line frequency.
2. For a phase angle of 45°, calculate the time (phase) delay (μs) for frequencies of (a) 3.5 MHz, (b) 500 kHz, and (c) 100 Hz.
3. For Figure 15.30, add and label blocks required to produce the 4.2 MHz luminance frequency band. (Hint: refer to Figure 15.31)
4. In Figure 15.3, what is the range of control-grid voltage within the indicated gray region.
5. In Figure 15.29A, if the chroma energy burst at the extreme left centers around 7 367.13 Hz, what is the center frequency of the last chroma energy burst to the right?

Chapter

16

16.1 The gain of a pentode video amplifier
16.2 The gain of a transistor video amplifier
16.3 High-frequency behavior of video amplifiers
16.4 Shunt peaking
16.5 Series peaking
16.6 Series-shunt peaking
16.7 Degenerative high-frequency compensation
16.8 Low-frequency compensation
16.9 The selection of tubes and transistors for video amplifiers
16.10 Typical transistor video amplifier circuits
16.11 Integrated circuits
16.12 Troubles in video amplifier circuits
Summary

VIDEO AMPLIFIER DESIGN

INTRODUCTION

In Chapter 15, the requirements necessary for the reproduction of television images are discussed. The methods whereby these requirements are met in practice, represent an important consideration in television today.

Almost without exception, the type of video amplifier that can be used to provide the necessary bandwidth is restricted to direct-coupled or resistance-capacitance (R-C)-coupled amplifiers. **R-C amplifiers** have the advantages of compactness, simplicity, and economy. Direct-coupled amplifiers have a perfect low-frequency response, due to the absence of reactive components in the coupling circuit.

A flat response is obtained in the middle range of frequencies (from 200 Hz to approximately 4 000 Hz) with the conventional R-C amplifier. The frequency and phase characteristics of the amplifier, throughout the middle range, are suitable for use in video amplifiers. This section of the curve requires no further improvement. However, the responses at either end of the curve are far from satisfactory and corrective measures must be taken. Fortunately, any changes made in the circuit to improve the high- or the low-frequency response of the amplifier will generally not react on each other (with one limitation, to be explained later). Each end can be analyzed separately and independently. In monochrome TV receivers, as few as one video amplifier stage may be present. However, in color-TV receivers, there are between two and five stages. The number of stages depends upon design and whether or not solid-state components are used.

In this chapter, we will analyze the low- and high-frequency behavior of R-C coupled and direct-coupled video amplifiers. We will study methods of improving the frequency and phase responses. Some typical video amplifier circuits will be discussed, including an IC type. We will also cover some typical troubles that may occur in video amplifiers.

By the time you have completed the reading and work assignments for Chapter 16, you should be able to:

- Define the following symbols: R_L, C_T, R_F, C_F, C_c, C_m, and R_{damp}.
- Explain the "alpha" and "beta" cutoff frequencies.
- Discuss the high-frequency and the low-frequency limiting factors of a video amplifier.
- Name four types of high-frequency compensation and two types of low-frequency compensation.
- Explain basically how high- and low-frequency compensation functions.
- Briefly explain the "Miller effect". Why is this effect important in video amplifiers?
- List the desirable characteristics of transistors and tubes used as video amplifiers.
- Draw a simple schematic diagram of a transistor video amplifier.
- Explain briefly why large values of coupling capacitors are required in transistor video amplifiers. Discuss a possible disadvantage of using large values.
- Compare three types of high frequency-peaking-coil compensation schemes with regards to frequency response, phase response, and gain.
- Compare the characteristics of common-emitter and common-base amplifiers.
- Explain briefly why an emitter-follower may be used as a video driver.
- Explain why a stacked amplifier may be used to operate a picture tube.
- Calculate the gain and frequency response of a video amplifier.
- Calculate the values of high- and low-frequency compensating components for a video amplifier that are needed to achieve a given bandwidth.

16.1 THE GAIN OF A PENTODE VIDEO AMPLIFIER

The gain equation of a vacuum-tube pentode amplifier is given by

$$A_e = g_m \times Z_L \qquad (16.1)$$

where

A_e is the voltage gain
g_m is the tube mutual conductance
Z_L is the total plate circuit impedance

This equation is convenient to work with and can be used when the internal plate resistance of the tube is considerably greater than the load resistor.

Thus, as a measure of the amplification of an amplifier, it is necessary simply to multiply the mutual conductance of the tube in question, by the impedance of the plate load. This does not always give an accurate value because of other impedances which affect the plate circuit and which frequently have a significant effect on the total value of impedance that the tube sees as a load. However, the foregoing procedure can provide a rough indication. In the discussion that ensues, we shall consider the impedance of the plate as being purely resistive and use R_L in place of Z_L. This is permissible for video amplifiers in the mid-frequency range only.

THE COUPLING NETWORK

The g_m, or mutual conductance, of a tube is governed by the particular tube used and the plate current flowing through the tube. The latter is dependent upon the B+ applied to the plate. The second part of Equation 16.1 is the load resistance into which the tube works. To see what fully constitutes this load, consider Figure 16.1. Here we have the coupling network between the output of one amplifier and the input of the following stage. In addition to the plate-

Table 16.1 **TRANSISTOR CONFIGURATIONS—COMPARISON OF CHARACTERISTICS**

CHARACTERISTIC	COMMON EMITTER (CONVENTIONAL)	COMMON COLLECTOR (EMITTER FOLLOWER)	COMMON BASE
Input Impedance	Moderate (500–1 500 Ω)	High (25k–500k Ω	Low (30–150 Ω)
Output Impedance	Moderate (30k–50k Ω)	Low (50–1 000 Ω)	Highest (300k–1m Ω)
Voltage gain	High (250–1 000)	Less than 1	High—may be somewhat higher than for the common emitter (500–1 750)
Current gain	High (25–55)	High (Approximately the same as a common emitter)	Less than 1
Power gain	Highest of the three configurations (25–40 dB)	Lowest of the three configurations (10–20 dB)	Slightly lower than for the common emitter (20–30 dB)
Phase inversion between input and output signals	Yes	No	No

THE COMPLEX Z_L

Thus, what initially appears to be a fairly simple circuit, consisting of two resistors and a coupling capacitance, actually turns out to be a fairly complex network containing three additional capacitances which are ordinarily not visible. These capacitances, combined with the resistances, form a complex quantity, Z_L. However, not all of these components need to be considered when dealing with any specific section of the overall video response. This fact will become evident as we consider the operation of these amplifiers, first at the high-frequency end, then over the mid-frequency section, and finally at the low-frequency end.

16.2 THE GAIN OF A TRANSISTOR VIDEO AMPLIFIER

The coupling network for transistor video amplifiers is very similar to that for vacuum-tube video amplifiers. This is shown in Figure 16.2. The value of R_L in both cases is within the same range, about 2.2 to 8.2 kΩ. The low value of R_L in both cases is within the same range, about 2.2 to 8.2 kΩ. The low value of R_L is required to provide the necessary high-frequency response, as will be explained shortly.

The input impedance of a transistor is much lower than that of a vacuum tube. Thus, to achieve the desired low-frequency response of the RC-coupling network, large values of C_c are required. This is on the order of 10 to 25 μF and electrolytic capacitors are commonly used. These values compare with the 0.01 to 0.22 μF capacitors used as C_c for vacuum-tube circuits. These lower values are also used, however, for transistor amplifiers coupled to the high-input impedance of the picture tube.

COMMON-EMITTER AMPLIFIERS

Common-emitter amplifiers are also known as **grounded-emitter amplifiers.**

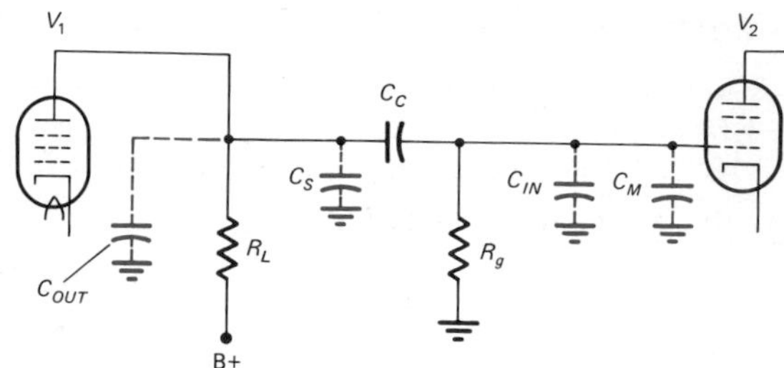

Figure 16.1 The complete coupling network between two vacuum tube video amplifier stages.

load resistor, R_L, we also see C_c, the coupling capacitor, and R_G, the grid resistor of the following tube. These three components are ordinarily wired into the circuit. Also present, but not physically wired into the circuit by the designer, is the output capacitance of the first tube, C_{out}, and the input capacitance of the following tube, C_{in}, plus two additional shunt capacities, C_s and C_M. C_s is the stray capacitance which exists across the circuit because of the wiring between the stages, the capacitance that R_L or R_G may have with respect to the chassis, and any capacitance that C_c itself may develop with respect to the chassis. This stray capacitance, while it is seldom greater than 5 or 6 picofarads (pF), must be taken into account when dealing with the high frequencies which pass through a video amplifier.

THE MILLER EFFECT

C_M is a capacitance which is reflected from the plate of V_2 to its grid circuit. C_M is an effective capacitance at the grid of the tube, caused by the so-called **Miller effect.** C_M is equal to the grid-to-plate capacitance (C_{gp}), multiplied by $(1 + A_e)$, which is the voltage gain of the stage. Because of the Miller effect, the realized grid-input capacitance may be many times greater than the actual, physical capacitance.

Emitter-Follower Amplifier The **emitter follower** (also known as the **common collector**) transistor configuration has characteristics that are similar to the cathode follower. For example, the voltage gain of an emitter-follower is slightly less than one, but it may be considered to be equal to one in most cases. Also, it has a high-input impedance (25k–500k Ω) and a low-output impedance (50–1 000Ω). We have already encountered the emitter-follower as a driver for a video amplifier in Chapter 15. Although an emitter-follower may have a high-current gain (25–55), its power gain (10–20dB) is the least of any of the three basic transistor circuits.

Grounded-Base Amplifier The **grounded-base configuration** is also called a **common-base circuit.** It has approximately the same gain as a conventional transistor amplifier, and it is similar to its counterpart—the grounded-grid amplifier. The input and output impedances of the grounded-base transistor circuit are the reverse of those given for the emitter-follower circuits. That is, the input impedance is low (30 to 150 ohms) and the output impedance is high (300k to 1M ohm). In applications where the ambient temperature may change over a wide range of values, the common-base circuit is more stable than the common-emitter circuit.

The grounded-base circuit is used extensively in high-frequency circuits. The high-frequency limit is set by the input capacitance of the transistor and the *transit time* of the charge carriers. The transit time is the time it takes an electron or hole to travel from emitter to collector through the semiconductor material comprising the base of the transistor. Some transistors, such as Drift and Mesa types have been designed specifically to reduce input capacitance and transit time.

The characteristics of the common-emitter, the common-base, and the emitter-follower circuits are summarized in Table 16.1.

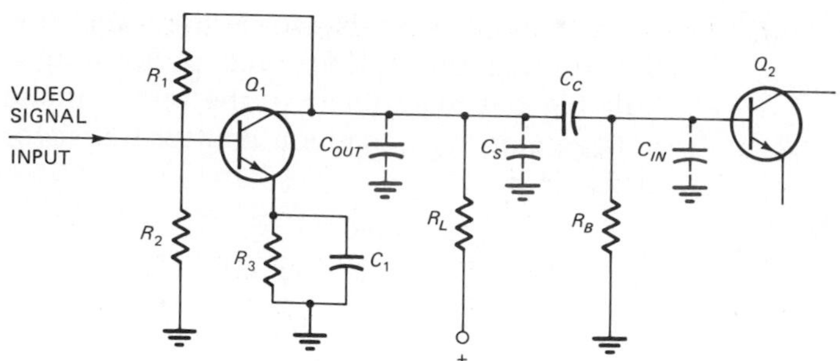

Figure 16.2 The complete coupling network between two transistor video amplifier stages.

Transistor amplifiers cannot be analyzed as simply as vacuum-tube amplifiers. However, an approximate voltage gain can be determined readily. The method is given in this section. The circuit in its basic form, shown in Figure 16.3 is used extensively as a voltage amplifier. It has a medium value of input impedance (500–1 500 Ω) and a medium value of output impedance (30 k–50 k Ω). The voltage gain (250–1 000) and the power gain (25–40 dB) are both high in a common-emitter amplifier.

The voltage gain (of any amplifier) is the output-signal voltage (V_o) divided by the input-signal voltage (V_i). The output voltage in Figure 16.3 is the signal current in the collector circuit multiplied by the load resistance R_L.

Thus

$$V_o = i_e R_L \tag{16.2}$$

where

V_o is the output signal voltage,

i_e is the collector signal current,

R_L is the load resistance.

The input voltage is equal to the base signal current (i_b) times the input resistance (R_i) of the transistor.

Thus

$$V_i = i_b R_i \tag{16.3}$$

where

V_i is the signal voltage applied to the input terminals,

i_b is the signal current flowing in the base, and

R_i is the input resistance of the transistor.

The approximate voltage gain equation can now be derived:

$$\begin{aligned} A_e &= \frac{V_o}{V_i} \\ &= \frac{i_e R_L}{i_b R_i} \\ &= \frac{i_e}{i_b} \cdot \frac{R_L}{R_i} \end{aligned} \tag{16.4}$$

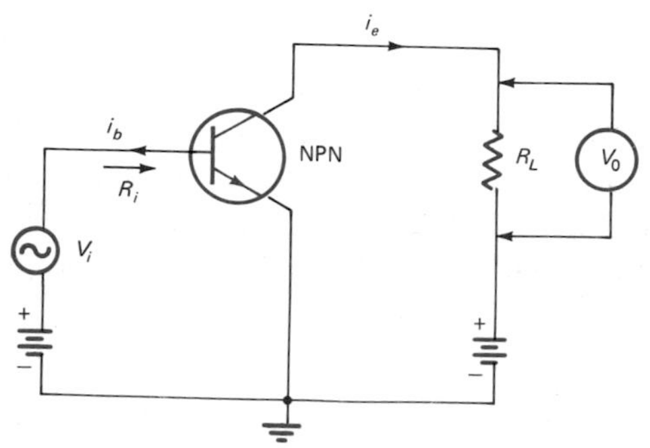

Figure 16.3 Simplified circuit for a grounded-emitter amplifier.

BETA AND ALPHA

It is now appropriate to introduce two new terms associated with transistor amplifiers. The terms are *beta* and *alpha*. **Beta** (β) is applied only to common-emitter circuits and is a measure of the "current gain" of such circuits. Beta is defined as the ratio of the change of collector current (ΔI_c) to the change of base current (ΔI_b) which produced it. Common values lie between 25 and 100. Beta is expressed by

$$\beta = \frac{\Delta I_c}{\Delta I_b} \qquad (16.5)$$

Alpha (α) is applied only to common-base amplifiers and is a measure of the "current gain" of such circuits. Alpha is defined as the ratio of a change of collector current (ΔI_c) to the change of emitter current (ΔI_e) which produced it. The current gain is always less than one and is commonly 0.91 to 0.99. Alpha is expressed by

$$\alpha = \frac{\Delta I_c}{\Delta I_e} \qquad (16.6)$$

If the AC base current is held to a relatively small value, then the relationship i_e/i_b (from equation 16.4) is approximately equal to the transistor *beta* (β). In a grounded-emitter circuit, beta is called the **common-emitter forward current transfer ratio** represented by h_{fe}. The **transistor alpha (α),** often represented by h_{fb}, is the grounded-base forward current gain of a transistor, and it is related to β by the equation

$$\beta = \frac{\alpha}{1 - \alpha} \qquad (16.7)$$

Subsituting $\alpha/(1 - \alpha)$ for $\frac{i_e}{i_b}$ in equation 16.4

$$A_e = \frac{i_e}{i_b} \cdot \frac{R_L}{R_i} = \frac{\alpha}{1 - \alpha} \cdot \frac{R_L}{R_i} \qquad (16.8)$$

The Input Resistance

The approximate input resistance (R_i) of a junction transistor is given by

$$R_i = \frac{r_e + r_b(1 - \alpha)}{(1 - \alpha)} \qquad (16.9)$$

where

r_e is the emitter resistance of the transistor, and

r_b is the base resistance of the transistor. Substituting this fraction and performing a little algebra manipulation, the approximate voltage gain of a common emitter-transistor circuit is

$$A_e = \frac{\alpha}{(1 - \alpha)} \cdot \frac{R_L}{\dfrac{r_e + r_b(1 - \alpha)}{(1 - \alpha)}}$$

$$A_e = \frac{\alpha}{(1 - \alpha)} \cdot \frac{R_L(1 - \alpha)}{r_e + r_b(1 - \alpha)} \qquad (16.10)$$

$$A_e = \frac{\alpha R_L}{r_e + r_b(1 - \alpha)} \approx \frac{R_L}{r_e}$$

This equation shows that increasing the value of R_L will increase the voltage gain. It also shows that the gain is greater if transistors with larger values of alpha are used.

High Frequency Response

High-frequency circuits have a reduction in gain because of the input capacitance of the amplifying device. The alpha rating of a transistor varies with the frequency. It follows, then, that the beta rating also varies with the frequency. The relationship between the frequency and the alpha and the beta of a transistor is shown in Fig. 16.4. The frequency where the alpha falls off to 0.707 (−3 dB) of its 1-kHz value is called the **alpha**

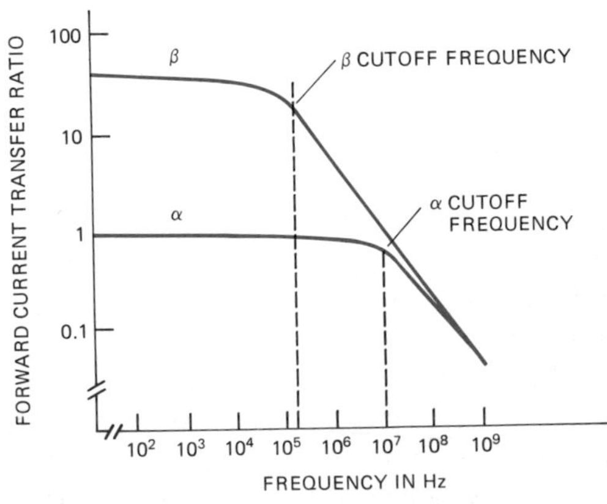

Figure 16.4 A comparison of α and β cutoff points.

cutoff frequency. The point where beta falls off to 0.707 (−3 dB) of its 1 kHz value is called the **beta cutoff frequency.** These curves show that the β cutoff, which is associated with a grounded-emitter configuration, is lower than the alpha cutoff which is associated with a grounded-base configuration. In other words, the grounded-base circuit has a higher cutoff point than the grounded-emitter circuit. However, the gain bandwidth product is approximately the same for both configurations.

16.3 HIGH-FREQUENCY BEHAVIOR OF VIDEO AMPLIFIERS

The high-frequency response of a video amplifier depends upon two things: the nature of the amplifying device and the type of impedance between amplifier stages. We have seen that the mutual conductance (g_m) of the tube or the α, β of a transistor will affect the gain of the amplifier. We have also seen that the input and the capacitance values are the determining factors of the highest frequency that can be amplified. We shall now direct our attention to the inter-stage impedances, and how they relate to bandwidth.

VACUUM-TUBE AMPLIFIERS

In considering the high-frequency operation of the circuit of Figure 16.1 we did not have to include the coupling capacitor C_c. The reason for this is that C_c will generally have a value of approximately 0.1 μF, and the high-frequency end of a video-amplifier-response curve generally falls off at about 3 to 4 MHz. At these frequencies, C_c has a negligible reactance. All we need include are the two resistances, R_L and R_g, plus the shunt capacitances which are present in the circuit. The high-frequency version of the network between V_1 and V_2 is now as illustrated in Figure 16.5. We could simplify this circuit even more by showing only R_L and omitting R_g. This simplification can be made because R_g is considerably higher in value than R_L. The two resistances in parallel will provide a total resistance very close to the value of R_L. However, for the sake of those instances when R_g may not be negligible in its effects on the parallel circuit, we shall retain R_g.

Value of CT

To determine the high-frequency gain of an amplifier with the interstage network shown in Figure 16.5 we must take not only R_L into consideration, but also the four shunt capacitances. See Figure 16.1. These capacitances are in parallel and their total value is equal to the sum of the separate units. That is

$$C_T = C_{out} + C_s + C_{in} + C_M \quad (16.11)$$

where

C_T is the total capacitance of the circuit,
C_{out} is the plate capacitance of the first tube,
C_s is the stray wiring capacitance of the circuit,
C_{in} is the input capacitance of the second tube,
C_M is the capacitance due to the Miller effect.

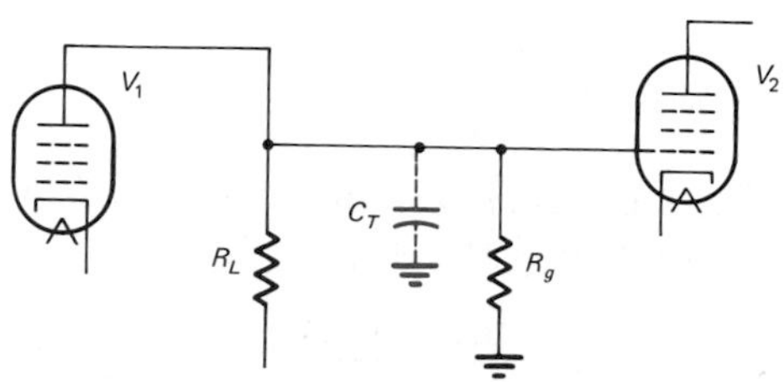

Figure 16.5 The plate-load resistor R_L and C_T are in parallel with each other. The coupling capacitor C_C has been omitted, because it does not enter into the high-frequency gain calculations.

To evaluate C_T numerically, we must know the exact values of each of its four components. The values of C_{out} and C_{in} can be obtained from a tube manual. A typical value of C_{out} is 3 pF, and for C_{in} it is 7 pF. C_s will vary with the circuit, but generally it falls at about 6 pF. Still remaining is the determination of C_M.

The input capacitance due to the Miller effect is equal to

$$C_M = C_{gp}\,(1 + A_e) \qquad (16.12)$$

where

A_e is the voltage gain of the stage.

If we assume a value for C_{gp} of 0.05 pF and a stage gain of 20, then the Miller capacitance is equal to

$$\begin{aligned} C_M &= 0.05\,(1 + 20) \\ &= 1.05 \text{ pF} \end{aligned}$$

We can now compute the value of C_T.

$$\begin{aligned} C_T &= C_{out} + C_s + C_{in} + C_M \\ &= 3 + 6 + 7 + 1.05 \\ &= 17.05 \text{ pF} \end{aligned}$$

When $R_L = X_{CT}$

This capacitance is obviously not negligible when high frequencies are involved. It will certainly affect the total plate impedance at the high-frequency end of the video-response curve. The plate-load resistor, R_L, and the total capacitance C_T are in parallel with each other as shown in Figure 16.5. It is useful to determine at what frequency the reactance of C_T (we shall call it X_{CT}) equals R_L. When this point is reached,

$$R_L = X_{CT}$$

Substituting $1/2\pi fC_T$ for X_{CT}:

$$R_L = \frac{1}{2\pi fC_T} \qquad (16.13)$$

Solving this equation for f:

$$f = \frac{1}{2\pi C_T R_L} \qquad (16.14)$$

At this frequency (f), the total impedance in the plate circuit is $1/\sqrt{2}$ of the value it has at lower frequences.[1] At lower frequencies X_{CT} is so large it can be disregarded. The expression $1/\sqrt{2}$ is equal numerically to 0.707. In terms of decibel loss, 0.707 represents a decrease of 3 dB. Hence, at the frequency f, the amplifier gain is 3 dB down from its gain at lower frequencies.

To Increase f

It is apparent from equation 16.14 that in order to raise frequency f, either C_T or R_L must be decreased in value. The curves in Figure 16.6 demonstrate how the bandwidth of a video amplifier is broadened by lowering the value of the plate-load resistor. The relationship between R_L and the frequency response was mentioned briefly in Chapter 15.

Now we have given the mathematical explanation of why low-valued load resistors are used in video amplifiers. Unfortunately, Figure 16.6 reveals that lower value load resistors also provide less stage gain. Hence, while this method of increasing the response of a stage with regard to frequency is useful, it cannot be carried too far if any useful gain from the stage is to be achieved.

Note that although this discussion is based on vacuum-tube circuits, the same principles also apply to transistor circuits.

---

[1]This can be seen readily. Let us assume that R_L and X_{CT} each have a value of 1 ohm. Then, since both are in parallel, and since one is a resistor and the other is a reactance,

$$\begin{aligned} Z &= \frac{R \cdot X_{CT}}{\sqrt{(R)^2 + (X_{CT})^2}} \\ &= \frac{1 \cdot 1}{\sqrt{(1)^2 + (1)^2}} \\ &= \frac{1}{\sqrt{2}} \end{aligned}$$

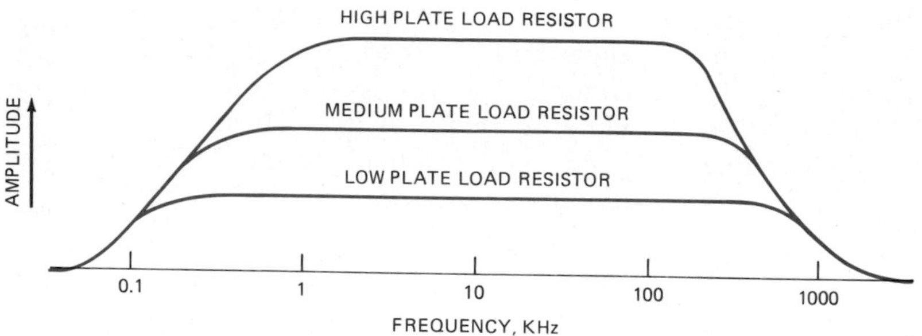

Figure 16.6 By lowering the plate-load resistor value, it is possible to increase the extent of the flat portion of the response curve.

TRANSISTOR AMPLIFIERS

The coupling circuit between two R-C coupled transistor amplifiers is very similar to the vacuum-tube circuit just discussed. However, the transistor input and output impedances are different due to the nature of its construction. It will be helpful to look at the characteristics that determine the transistor impedances, and compare them with the vacuum-tube impedances.

Transistor Input Resistance The grid of a vacuum tube presents an open circuit as far as DC is concerned, provided that the grid is maintained negative with respect to the cathode. Such is not the case with the transistors. The input resistance of a transistor is seen to be a combination of three resistors. Figure 16.7 pictures the combination that is R_e, R_b, and R_c. These are the internal emitter, the base, and the collector resistances respectively. In this illustration the input capacitances are not shown. They will be considered presently. The resistors R_1 and R_2 form a voltage divider network to provide the correct emitter-to-base transistor bias. Now the collector-base junction of a transistor is normally reverse biased. Therefore, the collector resistance is high and can be disregarded in our discussion. The base and the emitter resistances are in series. The combination is in parallel with R_2, the input resistor across which the signal is developed. The resistor R_1 is AC grounded by a filter capacitor C at the power supply, so it is effectively in parallel with R_2 and the input resistance $(R_b + R_e)$ of the transistor.

The equivalent input resistance of the transistor amplifier as seen by the input signal is shown in Figure 16.8. It is obvious from this illustration that the effect of the internal transistor resistances is to **lower** the total input resistance of the amplifier. This is

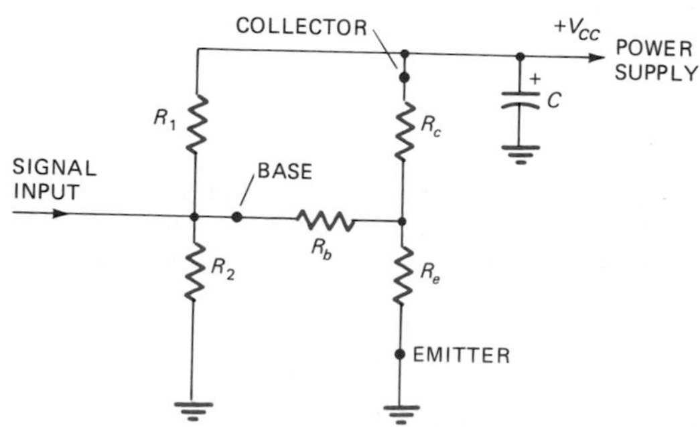

Figure 16.7 Resistances seen when looking into the base of a common-emitter transistor amplifier.

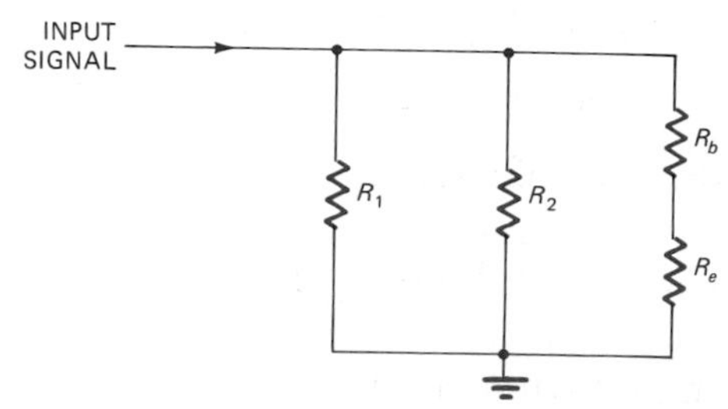

Figure 16.8 A simplified circuit of resistance as seen by the input signal.

an important point, because the input resistance of a transistor amplifier will be an influence on the high-frequency response.

The circuit of Figure 16.8 is a simplification given for the purpose of understanding the transistor amplifier. The actual input resistance is more complicated because the internal transistor resistances vary with the temperature and with the amount of junction current flowing.

The junction capacitance in a transistor introduces the same effect as the interelectrode capacitance in the tubes.

Semiconductor Capacitance

The collector-base junction of a transistor is normally reverse biased, while the emitter-base junction is forward biased. When a semiconductor junction is forward biased, it presents a greater capacitance than if it were reverse biased. The reason is that the forward-biased junction has a preponderance of charge carriers present at the junction, while the reverse-biased junction charge carriers are forced apart at the junction, introducing a wider, effective dielectric at the junction. (See also the discussion of the varactor in Chapter 10.) From the foregoing discussion, we can see that the emitter-base junction of a transistor will normally present a somewhat greater capacitance to the input signal than the collector-base junction.

It should be emphasized that the amplitude of the input signal to a transistor is very much dependent upon the emitter-base junction. There are other factors besides the junction bias that determine the amount of junction capacitance. The physical size of the junction is an important consideration. The capacitance between the transistor leads must also be considered. These factors are controlled by the manufacturer in producing high-frequency transistors.

It is an interesting fact that the junction capacitance is made use of in certain applications. The **voltage-variable capacitors** operate on this principle. These capacitors are used in automatic-tuning circuits. The amount of capacitance in such circuits is varied by controlling the amount of reverse bias present.

Transistor Input Capacitance

The input capacitance of a transistor amplifier is higher than would be expected from the presence of the junction capacitance alone. This increase in capacitance is due to the Miller effect, a problem that we have already encountered in vacuum-tube amplifiers. The Miller effect refers to the increase of input capacitance with amplification. As in the case of the tube amplifiers, it becomes larger as the amplifier gain is increased.

As with vacuum tubes there are also wiring and stray capacitances that affect the high-frequency response of transistor amplifiers. These capacitances are identical to those discussed for the tube circuit. The circuit of Figure 16.9 can be used to represent two R-C coupled transistor amplifiers. All of the capacitance values are combined to make C_T. The input transistor is represented as a voltage generator (e) and a series resistor (r). The voltage at the base of the following transistor is, V_o. The upper frequency response

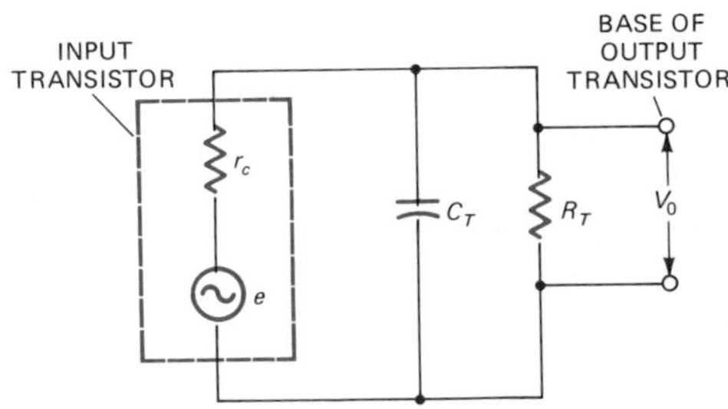

Figure 16.9 A simplified circuit of a transistor amplifier at high frequencies.

of this circuit is similar in shape to the response curve for a tube amplifier. In other words, as the frequency rises, the output voltage, V_o, decreases due to the shunting effect of C_T.

16.4 SHUNT PEAKING

A method which is useful in extending the high-frequency response of an amplifier is the addition of a small inductance (L_s) in parallel with the total capacitance (C_T) of the stage. This is usually accomplished by placing the inductance in series with the load resistor of the input stage. The inductance is designed to neutralize the effect of the shunting capacitances, at least to the extent that we can improve the amplifier response at the upper frequencies. This method is known as **shunt peaking.**

A circuit diagram using this type of compensating inductance is shown in Figure 16.10. This method, as well as typical values for R_L and L hold true for both transistor and vacuum-tube video amplifiers.

SHUNT PEAKING DESIGN PROCEDURE

The procedure for finding actual values for R_L and L_s is as follows. First, the highest frequency at which it was desired to have the response remain flat would be specified. In a video amplifier, this would usually be between 3 and 4 MHz. Once the type of tube or transistor to be used has been decided upon, the capacitance values that make up C_T can be obtained as previously described. The wiring capacitance, which is also needed for finding C_T, could be measured from a circuit layout, or it could be estimated. With both f and C_T known, the value of R_L could be obtained from the equation:

$$R_L = \frac{1}{2\pi f C_T} \qquad (16.15)$$

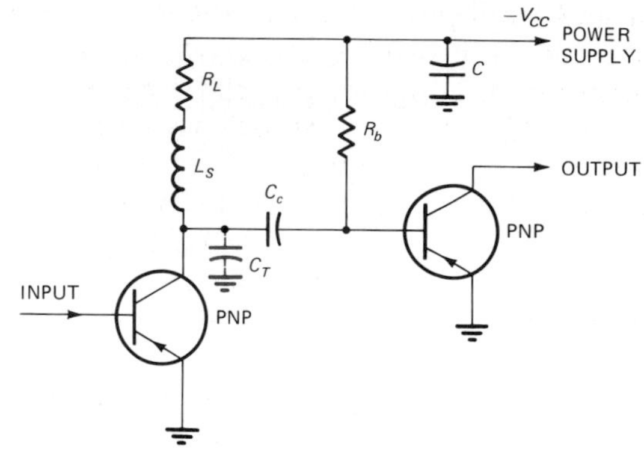

Figure 16.10 Shunt-peaking compensation in a transistor-video amplifier.

Also, since X_L should have a value equal to one-half R_L,

$$X_L = 0.5R_L = \frac{0.5}{2\pi f C_T} \qquad (16.16)$$

By substituting $2\pi fL$ for X_L

$$2\pi f L_S = \frac{0.5}{2\pi f C_T}$$

or

$$L_S = \frac{0.5}{4\pi^2 f^2 C_T} \qquad (16.17)$$

and

$$L_S = 0.5 C_T R_L^2$$

Typical values of R_L range from 2 200 ohms to 8 200 ohms. Figure 16.11 demonstrates the effect when L_s is too high an inductance value (Curve 4) and when L_s is too low a value (Curve 1). A small amount of overpeaking may sometimes be employed to sharpen the fine detail in the picture. Too much peaking, however, will lead to ringing. This is a condition where multiple lines follow the edge of an object. This condition is sometimes mistaken for ghosts.

Figure 16.12 reveals another way of connecting a shunt-peaking coil. In this case, the coil is in the base leg of the second-transistor stage, but it is still in parallel with C_T.

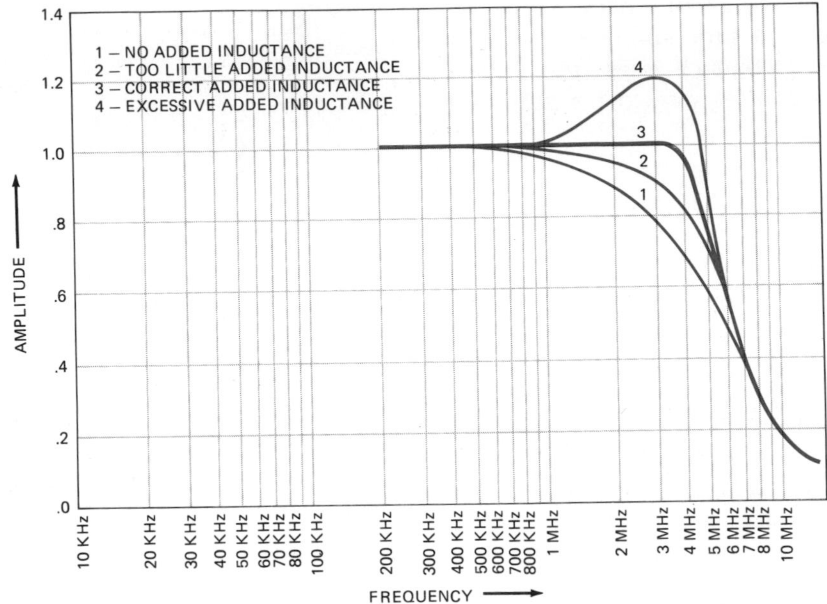

Figure 16.11 The effect on amplifier response of the insertion of various amounts of peaking inductance in the plate load.

An advantage of this circuit is that the amount of current through the coil is considerably less than when it is connected in series with the load. This avoids core saturation, thus maintaining the correct inductance value, regardless of the collector current. However, there is no significant change in the shape of the response curve with the circuit of Figure 16.12 as compared with the response for Figure 16.10.

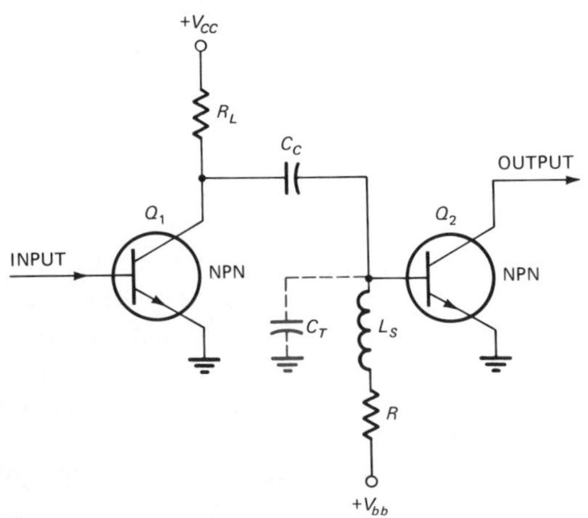

Figure 16.12 Another approach for achieving shunt-peaking compensation.

16.5 SERIES PEAKING

Another method of improving high-frequency response is to insert a small coil (L_c) in a series with the coupling capacitor, as illustrated in Figure 16.13. This method provides higher gain and better phase response than shunt peaking. The improved gain of this type of coupling is due to the fact that the components of C_T are no longer lumped together in one unit, but are separated. On the left-hand side of the series inductance is the output capacitance of the preceding amplifier, and on the other side is the input capacitance of the next amplifier. With this separation, the load resistor R_L may be

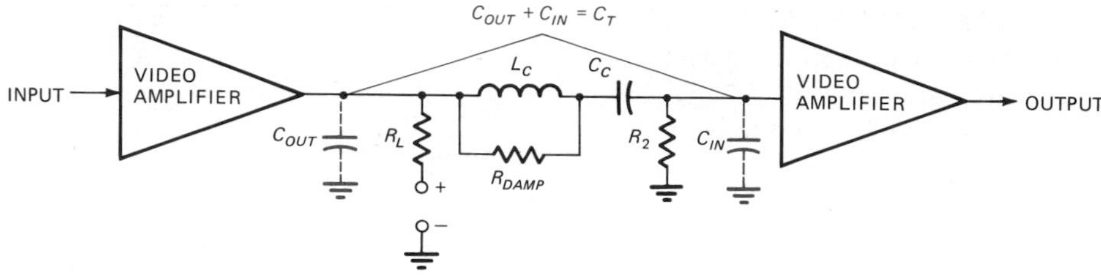

Figure 16.13 Illustrating the series peaking method of high-frequency compensation. As with shunt peaking, this method is applicable to either transistors or vacuum tubes. R_{DAMP} prevents ringing in L_C.

higher in value, because only C_o is directly across it and not the larger C_T. As C_o is smaller than C_T, its capacitive reactance is greater, and it will have less of a shunting effect on R_L. A larger value of R_L is then possible, actually 50% larger. Thus,

$$R_L = \frac{1.5}{2\pi f C_T} \quad (16.17)$$

In this case C_T is the *total* capacitance shunting R_L, that is, $C_o + C_i$ in Figure 16.13.

It has been found that the best results are obtained when the ratio of C_i to C_o is approximately 2. This is accomplished (or closely approached) by the proper choice of the tubes used for the amplifiers. The value of the series coil, L_c, is given as

$$L_c = 0.67 C_T R_L^2 \quad (16.18)$$

Figure 16.14 shows a transistor video amplifier with series peaking compensation provided by L_c. The resistor R_2, in parallel with L_c, is called a **swamping** or **damping resistor.** The distributed capacitance of any coil in conjunction with the circuit capacitance will act with its inductance to produce resonance. The overall effect is to cause the output circuit of the video amplifier to resonate, or overcompensate, over a narrow band of frequencies. This is undesirable in view of the fact that a flat response is needed. The swamping resistor reduces the Q of the tuned circuit formed by the peaking coil, and the distributed and circuit capacitances. The standard practice is to wind the peaking coil on the resistor, using the resistor as a coil form. The assembly is then covered with a compound to seal it against moisture.

16.6 SERIES-SHUNT PEAKING

It is possible to combine shunt and series peaking and obtain the advantages of both. Such combinations appear in both tube and transistor circuits. The shunt coil is designed to neutralize the output capacitance

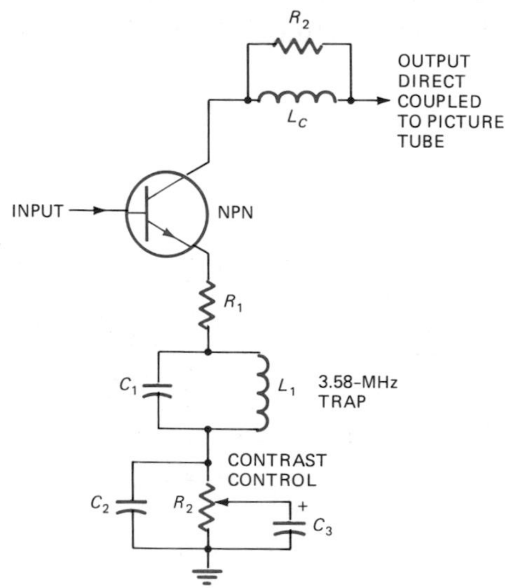

Figure 16.14 Another series peaking compensation circuit.

of the preceding amplifier while the series coil combines with the input capacitance (and stray-wiring capacitance) of the next amplifier. With this double combination, it is possible to extend the bandwidth to 4 MHz, which is more than can be derived through the use of the shunt peaking alone. Furthermore, the phase distortion of the combined compensation in the coupling network is lower than either of the two preceding types.

An amplifier using both shunt and series peaking combined is shown in Figure 16.15. A swamping resistor is shunted across the series coil to minimize any sharp increase in the circuit response due to the combination of the series coil inductance and its natural or inherent capacitance. The coil is specifically designed to have a natural frequency considerably above the highest video frequency. In production, however, a certain number of coils will be produced with natural resonant frequencies within the range covered by the amplifier. The value of the swamping resistor is generally 4 or 5 times the impedance of the series coil at the highest video frequency.

For the combination circuit, the values of the shunt-peaking inductance (L_s), the series-peaking inductance (L_c), and R_L are obtained from the following relationships:

$$R_L = \frac{1.8}{2\pi f C_T} \qquad (16.19)$$

where

$C_T = C_i + C_o$

$L_s = .12 C_T R_L^2$ (shunt coil)

$L_c = .52 C_T R_L^2$ (series coil), and

f is the highest frequency at which the response remains uniform.

These equations hold true only if the input capacitance of the second amplifier (C_i) and output capacitance of the first amplifier (C_o) are related by the ideal ratio:

$$\frac{C_i}{C_o} = 2$$

16.7 DEGENERATIVE HIGH-FREQUENCY COMPENSATION

This scheme can be used either with transistors or vacuum tubes and is illustrated in Figure 16.16. Emitter (or cathode) resistor R_1 is bypassed by a low value of C_1, such as 470 pF. At frequencies up to perhaps 1.5 MHz, the gain of video amplifier No. 1 is lower than normal because of degeneration caused by the high reactance of C_1 in parallel with R_1.

Above about 1.5 MHz, the lowering re-

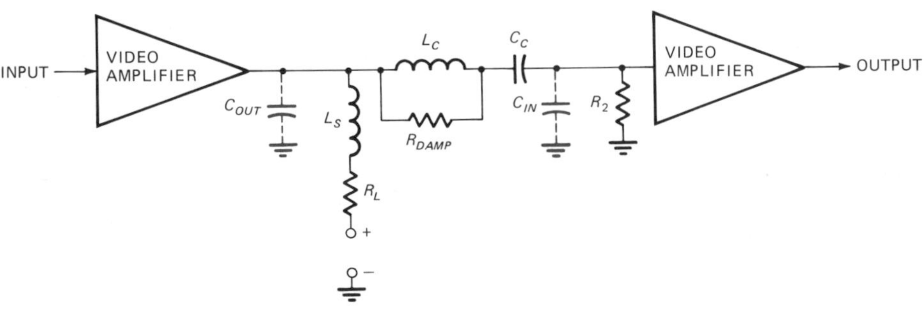

Figure 16.15 Illustrating the series shunt peaking method of high-frequency compensation. This method is applicable to either transistors or vacuum tubes.

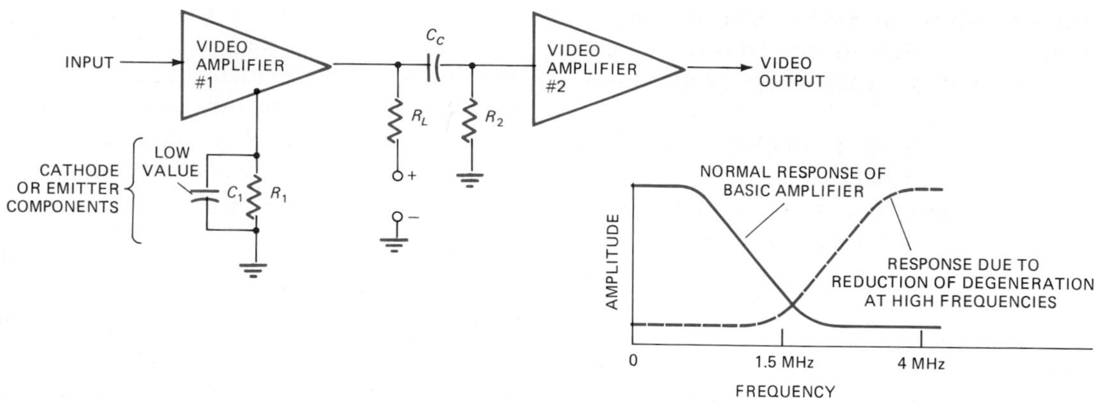

Figure 16.16 Illustrating the degenerative method of high-frequency compensation. Inset shows the effect of the compensation. The two curves of the inset combine to provide flat response to 4 MHz (approx.). (See text.)

actance of C_1 begins to provide improved bypassing for R_1. Thus, the amount of degeneration begins to decrease and the gain of the amplifier increases. This is shown in the inset for Figure 16.16. The high-frequency response of the amplifier may be extended flat by this method, to 4 MHz or somewhat higher. The overall gain of an amplifier using this type of high-frequency compensation, will always be less than if the preceding methods of high-frequency compensation are used.

16.8 LOW-FREQUENCY COMPENSATION

Let us determine now what changes can be made to the circuit to improve its low-frequency response. At this end of the band, it is possible to disregard the shunting capacitances since their reactance, given by

$$X_c = \frac{1}{2\pi fC}$$

is very high, and they do not affect the low-frequency signal voltages to any appreciable extent. Now, however, it becomes necessary to include the coupling capacitor. Figure 16.17 shows the R-C coupled amplifier. The operation of the circuit, as explained in conjunction with Figure 16.17 is such that the lower the frequency, the greater is the effect of the coupling capacitor. The response gradually falls off at low frequencies because the reactance of C_c becomes dominant, and an increasing portion of the output voltage of video amplifier No. 1 is lost across C_c. The

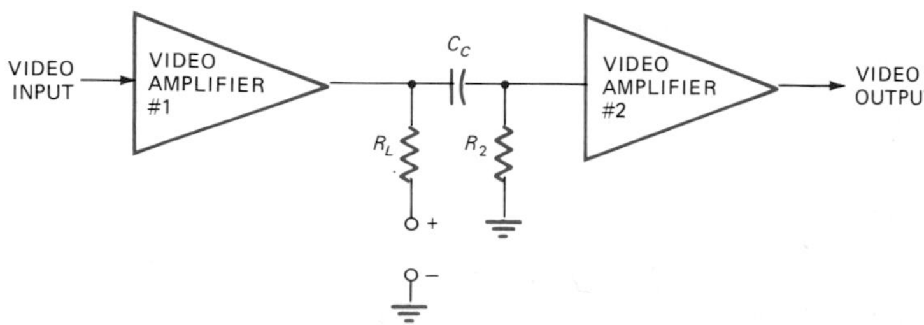

Figure 16.17 Simplified diagram of a video amplifier for discussion of low-frequency response. (See text.)

phase delay of the signal begins to change, eventually approaching 90°. As a result, the background illumination of the reproduced image is affected.

In order to make the response more linear at the low frequencies, either C_c may be made larger (so that it will have less reactance) or R_2 may be made larger or both may be made larger. There are however practical limits to their values.

Compensating Circuit By inserting a resistor R_F and a capacitor C_F in the output circuit of the amplifier as indicated in Fig. 16.18, it is possible to improve the low-frequency response without making either R_2 or C_c too large. (If these components are too large, this can lead to circuit operation problems.) R_f and C_f are the two added components. They form the **low-frequency compensation** circuit. The gain of an amplifier is directly related to the size of the plate-load resistor (or collector-load resistor). At low frequencies, the output load in the circuit of Figure 16.18 is R_L and R_f in series. The reactance of C_f is high at low frequencies, and it appears as an open circuit. At the higher frequencies, the reactance of C_f is low enough to present a short circuit to ground for the signal. This means that the plate load is comprised only of R_L at the higher signal frequencies and the gain is reduced. By amplifying the low frequencies with a higher gain, the loss of the signal for low frequencies at the coupling capacitor is overcome. An additional advantage of the low-frequency compensation is the reduction in phase-shift distortion in both tube and transistor circuits.

The value of C_f in Fig. 16.18 is obtained from the expression

$$R_LC_f = C_cR_2 \qquad (16.20)$$

where

R_L, C_c, and R_2 have previously been assigned values, and

C_f is the value of the low-frequency compensating capacitor.

Load resistor R_L will be determined by the highest frequency to be passed by the amplifier, and C_c and R_2 will be as large as possible. Finally, R_f should have a resistance which is at least 20 times larger than the reactance of C_f at the lowest frequency to be passed.

Emitter (or Cathode) Circuit The network comprised of C_f and R_f provides the greatest amount of low-frequency compensation, but there are two more components which influence the low-frequency response. These components are the emitter or cathode resistor, R_k, and the emitter or cathode-bypass

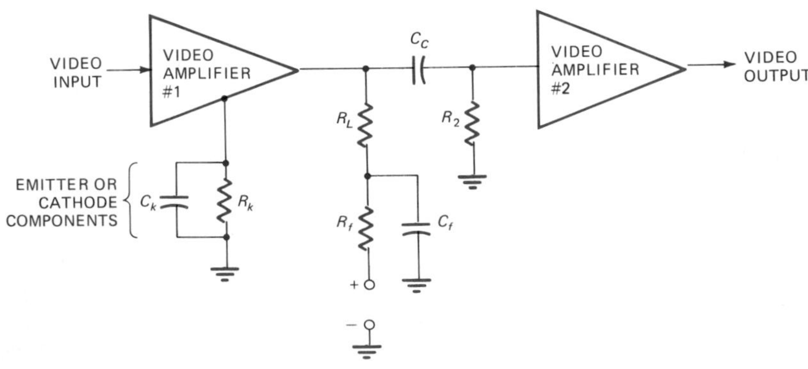

Figure 16.18 Illustrating a common method of low-frequency compensation. R_f and C_f have been added to Figure 16.17 to provide this compensation.

capacitor, C_k. Both should be chosen to satisfy the following expression:

$R_kC_k = R_fC_f$, where

R_k, C_k, R_f and C_f are as shown in Figure 16.18.

By leaving the emitter or cathode resistance unbypassed, or by dividing it into two resistances with one unbypassed, the gain of the amplifier is decreased at all frequencies. However, the low-frequency response is *improved* by the addition of degeneration. This occurs because the bypass capacitor across the cathode resistor has a reactance related to frequency. High frequencies are effectively bypassed, but low frequencies are only partially bypassed.

Design Procedure In the design procedure of video amplifiers, the values of the high-frequency compensating components are selected first. These include R_L, L_s, and L_c. Next, the low-frequency compensating components, C_f and R_f, are computed, and, finally, R_k and C_k. The values of each of the latter two resistors are determined by the voltage needed for proper operation as recommended by the manufacturer. This requirement imposes a limitation. However, since we are concerned with a time constant in each instance (as $C_f \times R_f$, and $R_k \times C_k$) rather than the individual value of each part, we can usually satisfy all the required conditions.

When the high- and low-frequency compensating circuits are simultaneously applied to a video amplifier, the circuit appears as shown in Figure 16.19. The frequency-and-phase response of this amplifier is plotted in Figure 16.20.

PASSING THE DC COMPONENT

As previously mentioned, the average value, or DC component of the video signal represents the average picture brightness. Thus, it is essential that this DC component is passed onto the grid (or cathode) of the picture tube. In the video amplifier shown in Figure 16.18 the DC component is lost because it will not pass the coupling capacitor, C_c. This problem and its solutions are discussed more fully in Chapter 17.

The DC Shunting Resistor One method of passing the DC component using a circuit similar to Fig. 16.18 is shown in Figure 16.21. This method consists simply of shunting C_c with a high value resistance, R_c (about 0.5MΩ). The DC component now passes

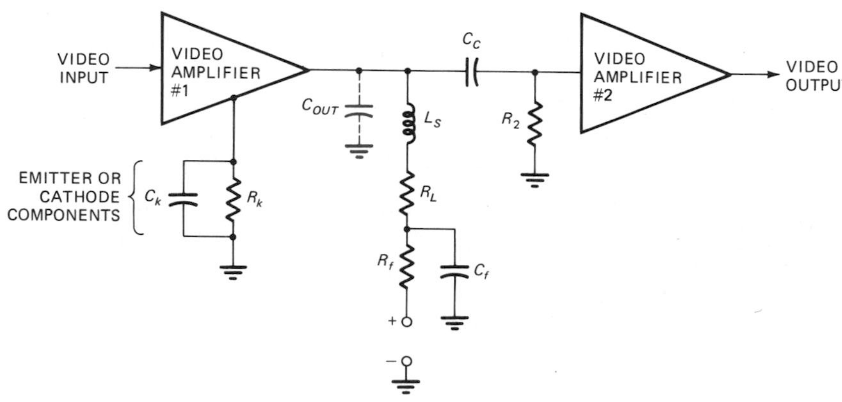

Figure 16.19 A video amplifier which incorporates shunt high-frequency compensation as well as low-frequency compensation.

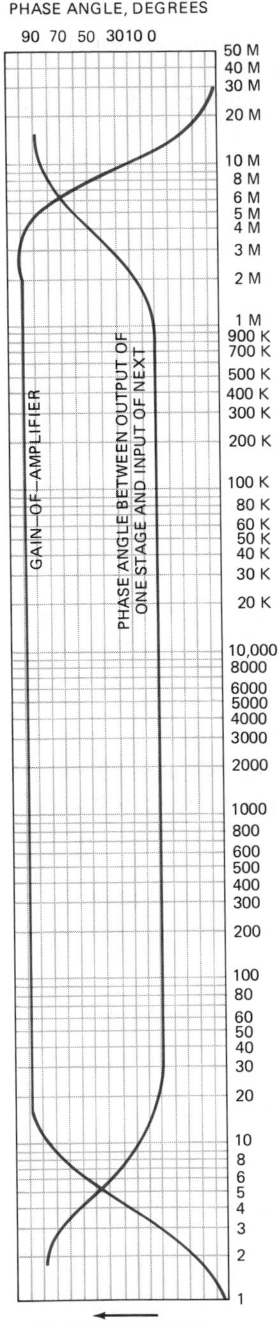

Figure 16.20 The frequency and phase response of a fully compensated resistance-coupled amplifier.

through R_c to the input of the following video amplifier. Now, the low frequency response of the amplifier is flat down to DC.

DIRECT COUPLING

If the output of video amplifier No. 2 in Figure 16.21 is direct coupled to the picture tube, the DC component is maintained. (However, if AC coupling is employed, a scheme called "DC reinsertion" is employed, as explained in Chapter 17.)

If direct coupling is used between the output of the video detector and the input to the picture tube, the DC component is maintained and in addition no low-frequency compensation is required. A direct-coupled amplifier is flat to zero Hz. However, such amplifiers must be operated with stable power supplies. Otherwise, small DC voltage changes in low-level stages will be amplified to cause larger changes of operating potentials in the output stage.

Direct-Coupled Vacuum-Tube Circuit A two-stage, direct-coupled, vacuum tube video-amplifier system is shown in Figure 16.22. Note how the voltages start at -130 volts at the grid of V_1 and work up to $+225$ volts at the plate of V_2. This increase in the operating voltages from stage to stage is a characteristic of the direct-coupled amplifiers and is sometimes called **level shifting.** If we add another stage after V_2, the grid of this third stage will be positive because of its connection to the plate of V_2. This, in turn, means that the cathode and plate of this third tube will have to be even more positive. By adding enough stages, the B+ voltage required soon rises to extremely high values. Fortunately, tube-type television receivers seldom require more than two stages of video amplification. Hence, direct coupling can be successfully employed.

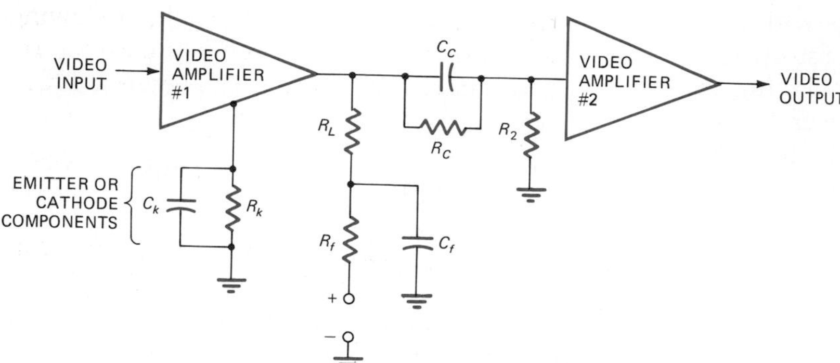

Figure 16.21 A low-frequency compensated video amplifier which passes the DC component (through R_c) of the video signal. Low-frequency compensation is achieved as follows:

$$\frac{R_L}{R_2} = \frac{R_f}{R_c} = \frac{C_c}{C_f}$$

In the circuit of Figure 16.22, the balance between the amplification accorded the *medium and high frequency* and the *low frequency and DC* components is carefully maintained. C_1 and R_1 form one equalizing network wherein the high-frequency component is left alone while the low-frequency component is reduced. C_2, C_3, R_2, R_3, and R_4 form another such network. Finally, L_1, C_4, and R_5 constitute a 4.5 MHz trap to remove any 4.5 MHz signal that may be present in the circuit.

The complete schematic diagram of a direct-coupled, transistor video-amplifier section and its explanation, is given in Chapter 17.

16.9 THE SELECTION OF TUBES AND TRANSISTORS FOR VIDEO AMPLIFIERS

Tubes The ability to provide high gain and to handle signals up to 4 MHz are the primary considerations in the selection of tubes for use as video amplifiers. To achieve high gain, pentodes and beam-power tetrodes are the favored types, although occasionally a triode or a dual triode is utilized. The higher mutual conductance values of pentodes compared to triodes makes the pentodes more useful in video-amplifier circuits. In addition, pentodes are capable of handling

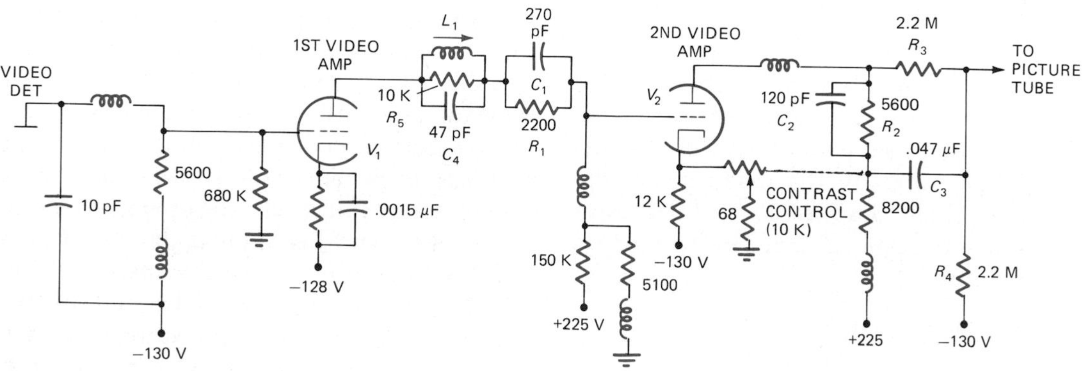

Figure 16.22 A two-stage, direct-coupled vacuum tube video-amplifier system. *(Courtesy of RCA.)*

relatively large signal swings without introducing excessive amplitude distortion.

The requirement of amplifying signals up to 4 MHz is met by low interelectrode capacitance, particularly between the plate and grid. The figure of merit of a high-frequency tube is given by the expression

$$\frac{g_m}{C_{in} + C_{out}}$$

The larger the value of this expression, the better the tube is for use as a video amplifier.

Transistors Transistors used for video amplifiers should have a high gain-bandwidth product. The beta of the transistor should be high at both DC and high frequencies, and of course, the input capacitance should be small. The high-signal voltages that are needed for driving the picture tube means that output transistors with a high-breakdown voltage rating must be chosen. It is difficult to make transistors with a high β, low-input capacitance, high-power gain, and high-breakdown rating all in the same device, so design engineers are faced with compromises between these characteristics.

16.10 TYPICAL TRANSISTOR VIDEO AMPLIFIER CIRCUITS

It is not possible to include every variation of video amplifiers that has been used by the manufacturers. However, a few examples of the actual circuits will be instructive. We have already discussed some representative video amplifiers, and we have also discussed some circuits in relation to contrast controls.

Emitter Follower Figure 16.23 shows an emitter-follower connected as a video driver. The input signal from the video detector is delivered to the base of the driver via capacitor C_1. This is an electrolytic capacitor with a relatively large value of 10 microFarads or more. The large value serves to extend the frequency response by reducing the reactance of the coupling network at low frequencies. The resistors R_1 and R_2 form a voltage divider network for obtaining the required DC base-bias voltage. It will be noted that the collector of transistor Q_1 is connected directly to ground, and the output signal is taken from the emitter across resistor R_3. The emitter circuit, and the voltage-divider network in the base, both obtain their operating voltage from the same source. This makes it necessary to supply a decoupling filter comprised of R_4, C_5. Decoupling filters are used whenever more than one signal source is connected to the same point on a power source. This prevents low-frequency oscillation (motorboating). The decoupling circuit has the same configuration as the low-frequency compensating network discussed in Section 16.8. In fact, such a circuit may serve the dual role of frequency compensation and decoupling. If the purpose is decoupling only, then the resistance of R_4 will be quite low, usually less than 100 ohms. The important thing to remember is that you cannot interpret such a circuit by configuration alone, but must take the value of the components into consideration.

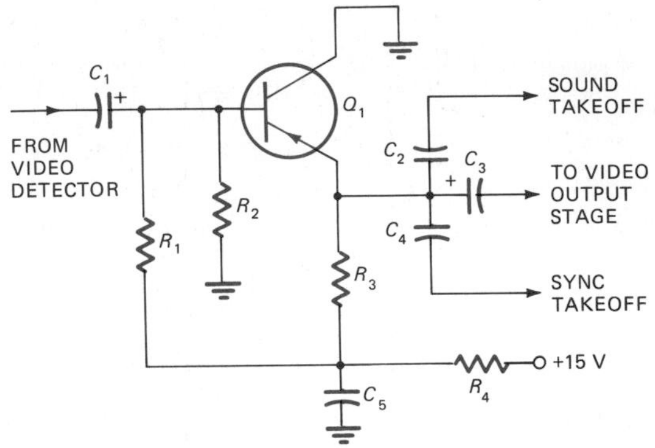

Figure 16.23 An emitter-follower used as a video driver.

The output signal of the emitter follower is delivered to the sound takeoff and the sync circuits through capacitors C_2 and C_4, respectively. Electrolytic capacitor C_3 delivers the video signal to the video-output stage. Note that no peaking compensation is used in the circuit which attests to the very wide frequency response of emitter-follower configurations. In actual practice it is possible to obtain the necessary 4 MHz bandwidth with such a circuit, without compensating networks.

Stacked Amplifier Figure 16.24 shows a transistor video-amplifier output stage that uses two transistors connected in series. This type of circuit is called a **stacked amplifier.** (It is also called a **cascode amplifier.**) It has the obvious advantage that the relatively large signal voltage needed to operate the picture tube is divided between the two transistors, thus reducing the drop across each transistor and decreasing the likelihood of a voltage breakdown. The input signal from the video driver is delivered through electrolytic capacitor C_1 to the base of Q_1. The voltage divider R_3 and R_4 establishes the base-bias voltage for the stage. Three emitter resistors (R_5, R_6, and R_7) are connected in series in order to obtain the desired frequency compensation. Resistor R_5 is unbypassed and provides the current feedback for the amplifier at all frequencies. Current feedback, or degenerative feedback as it is sometimes called, serves to broaden the response of the amplifier and stabilizes its temperature performance. Resistor R_6 is bypassed for high frequencies, but not for low frequencies. This means that R_5 and R_6 are both in the emitter circuit as far as the low frequencies are concerned. However, only R_5 is in the circuit for high frequencies. C_2 and C_3 effectively bypass the higher frequencies around R_6 and R_7. The overall result is that a different amount of degeneration occurs for low and high frequencies. The emitter circuit for Q_1 may be considered to be a form of high-frequency compensation. The series peaking coil L_1 and its swamping resistor R_8 are used to further improve the high-frequency response of the amplifier.

The necessary base bias for transistor Q_2 is obtained with the voltage divider R_1 and R_2. This voltage divider is different in principle than the one formed by R_3 and R_4. You will note that resistor R_1 is tied to the collector side of load resistor R_9. The signal voltage developed across R_9 is fed back to the

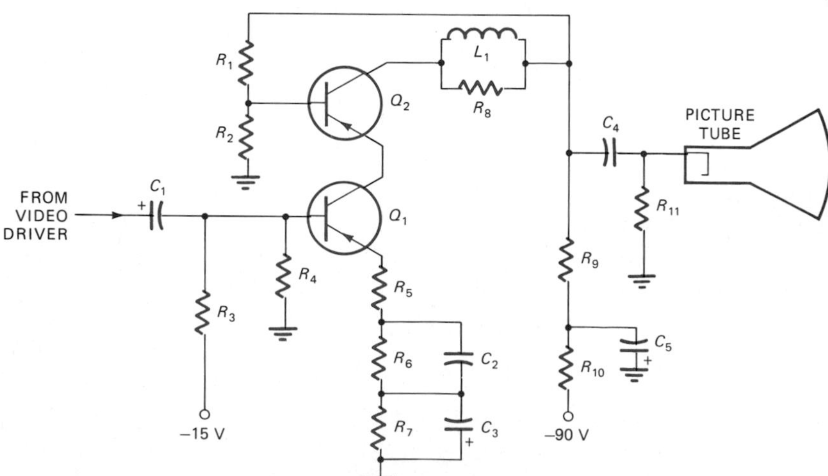

Figure 16.24 Video-output stage using stacked amplifiers.

base of Q_2 through R_1. This feedback is *degenerative* because the collector voltage is 180° out of phase with the voltage on the base. Earlier in this chapter we made a general statement that *the higher the gain at mid-frequencies, the narrower the bandwidth*. By introducing degenerative feedback, we will at the same time increase the frequency response of the amplifier. Compare this voltage divider with the one used to establish the base bias for Q_1. The high-voltage end of resistor R_3 is connected directly to the -15 volt source which is an unvarying voltage, and so no feedback is obtained in this case. Essentially, then, there are two feedback networks employed in the circuit of Figure 16.24. The degenerative *current* feedback is due to the unbypassed emitter resistor R_5 (and also R_6 at lower frequencies). The degenerative *voltage* feedback is due to the connection of R_1 to the emitter side of the load resistor.

A decoupling filter comprised of R_{10} and C_5 is needed, because there are other amplifiers (not shown) connected to the -90 volt source. The input signal to the cathode of the picture tube is delivered by C_4, and the signal voltage is developed across R_{11}.

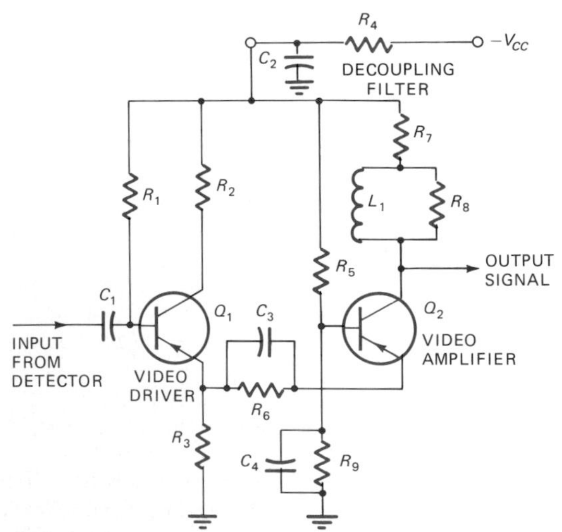

Figure 16.25 A grounded-base video amplifier.

Grounded-base Video Amplifier You will recall that the grounded-base circuit has a low-input impedance and a high-output impedance. This characteristic, together with the fact that the grounded-base circuit has a higher-frequency cutoff, makes it useful in the circuit of Figure 16.25. The low-output impedance of the emitter follower (Q_1) approximately matches the low-input impedance of amplifier Q_2, thus providing an optimum power transfer. The input signal for Q_2 is delivered to the emitter, and the output circuit is taken from the collector. The base is AC grounded by capacitor C_4.

16.11 INTEGRATED CIRCUITS

Integrated circuits are ideally suited for broadband and high-frequency applications, but they were slow to appear in video-amplifier circuitry. Among the reasons is the relatively high output signal voltage requirement imposed by the picture tube. One of the first applications of integrated circuits in video amplifiers was in the video-driver stage where the signal amplitude is relatively low.

Many integrated circuits are comprised of monolithic circuits on a small single chip and include transistors, diodes, resistors, and capacitors. A description of this type of integrated circuit is given in Chapter 5 and again in Chapter 20, in greater detail. Some integrated circuits, however, contain only transistors (active elements) with the resistors and capacitors (passive elements), all being externally connected. The latter type of integrated circuit is shown schematically in Figure 16.26, and the integrated circuit connected to its external passive elements to form a video-driver amplifier, is shown in Figure 16.27(A). The external connection numbers for the integrated circuit are seen correspondingly on both of the aforementioned diagrams. The integrated circuit of Figure 16.26 is that of the RCA, CA3018

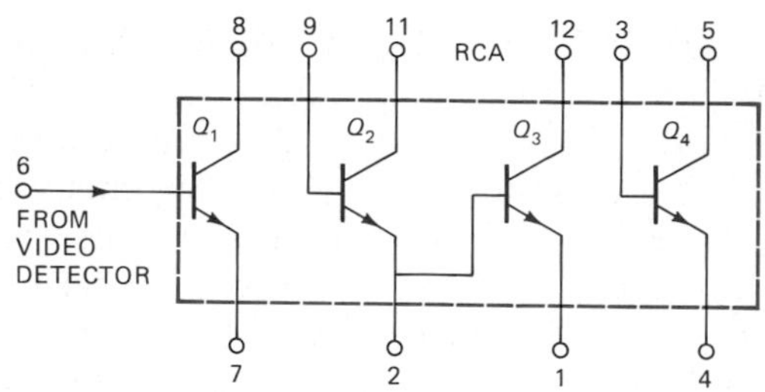

Figure 16.26 An integrated circuit that can be used as a video-driver amplifier.

unit. This consists of four silicon epitaxial transistors on a single chip, mounted in a 12 lead, TO-5 style (round) can. Two of the transistors are isolated, and the remaining two are emitter-based coupled. In addition to being adaptable for use as a video amplifier, this same IC may also be used to amplify RF, IF (through 100 MHz), and audio signals. Of course, the passive (discrete) components would be different for each type of use. Such passive circuits as resistors, capacitors, and inductors would be externally connected to the terminals of an integrated circuit. The circuit is designed to operate in a range of temperatures from −55 °C to +125 °C.

Video Driver Figure 16.27 A shows an RCA Integrated Circuit CA3018 connected as a video driver. The discrete components are located outside the dotted lines. The input signal from the detector appears as a voltage

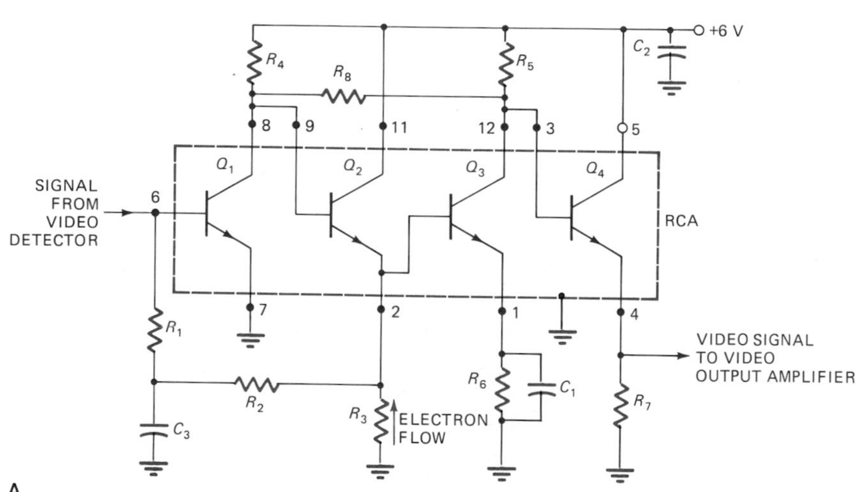

Figure 16.27A. The integrated circuit of Figure 16.26 connected as a video driver. *(Courtesy of RCA.)* **B.** One TO-5 type (round can) and two "flat-pack" integrated circuit packages. *(Courtesy of Texas Instruments, Inc.)*

across R_1. This signal is applied to the base of Q_1. The DC operating voltage on the base of this transistor must be positive, since it is an NPN transistor. The required positive bias voltage is acquired from the emitter resistor of Q_2 as a result of an electron flow from the ground to the emitter. The positive signal voltage from the emitter is filtered by R_2 and C_3.

Transistor Q_1 is connected as a conventional amplifier with the output signal developed across R_4. This signal is direct coupled to the base of Q_2. Transistor Q_2 is connected internally in the integrated circuit as an emitter follower, the output being fed to the base of the conventional amplifier Q_3. The emitter resistor of Q_3 is bypassed for high frequencies, but not for low frequencies. This means that the high frequencies will receive a greater amount of amplification in this stage. The output of Q_3 is delivered to emitter follower Q_4, and the output of this stage goes to the video-output stage.

You will notice that all of the stages in the circuit of Figure 16.27A are direct coupled. Thus, there is no need for low-frequency compensating networks. In addition, there is no phase distortion at low frequencies.

Three different kinds of integrated circuit packages can be seen in Fig. 16.27 B. The round can is similar to the one used in the video amplifier just described. The others illustrate two configurations of "flat-pack" circuits. The flat pack at the left is built to be wired into a printed-circuit board, while the one on the right is a plug-in unit.

16.12 TROUBLES IN VIDEO AMPLIFIER CIRCUITS

This section is not intended to instruct the student in the methods of troubleshooting video-amplifier circuits. These methods are discussed in Chapter 27. In this section,

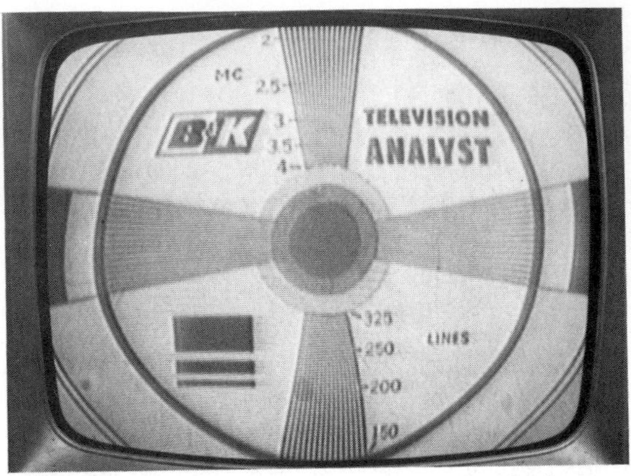

Figure 16.28 A normal test pattern as displayed on the screen of a typical monochrome TV set.

we will become acquainted with the nature of the most common troubles, as they are seen on the picture tube. Only video troubles common to monochrome sets, or to the monochrome portion of color sets are presented here. Video troubles relating specifically to the color portion of color sets are discussed in Chapter 27.

Normal Test Pattern A normal test pattern, as seen on the picture-tube screen of a typical monochrome-TV set is pictured in Figure 16.28. Note that the vertical lines in the top half of the pattern fade out a little above 3 MHz, which is the overall high-frequency response (video) of the receiver. This response is typical of many monochrome sets. Note the good contrast throughout the pattern and the absence of smearing, following the letters and the large areas, which indicates correct low-frequency response.

Symptom: No Video on Screen A completely inoperative video-amplifier section, with the remainder of the set operating normally, will result in a situation of a normal raster, but no picture. The appearance of this condition is shown in the photo of Figure 16.29. Since there is no video at all, this

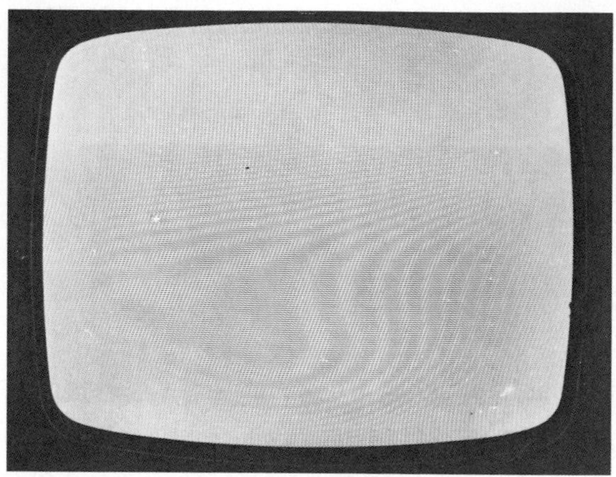

Figure 16.29 The appearance of the TV screen when the video amplifier is completely inoperative.

condition is generally caused by a major malfunction of the video-amplifier section. This trouble also may be caused by a lack of operating voltages to one or more of the video-amplifier stages. It may also be caused by a defective tube or transistor. Other common causes are: a defective video-detector diode, an open-coupling capacitor, an open-peaking coil, or a break in the printed-circuit wiring.

Symptom: Poor High-Frequency Response Referring to Figure 16.28, we note that the vertical lines in the upper half of the test pattern cease to be visible after about 3MHz. If the high-frequency response of the video amplifier decreases substantially because of a defect in the video amplifier (or video IF) section, this would be apparent in the test pattern by the disappearance of the upper section vertical lines at a frequency appreciably lower than 3 MHz. For example the lines might disappear at 2 MHz. (The lower vertical lines would be similarly affected and would disappear closer to the lower end of the pattern.) Since the high-video frequencies are responsible for reproducing the smallest picture details, a loss of the high frequencies would manifest itself by a lack of fine picture detail in a TV picture. There would be an appearance of fuzziness, somewhat similar to the appearance of a photo which has been excessively enlarged. The viewer would note that the edges between the adjacent light and dark areas would not be sharply delineated. Also, small printed letters could not be easily read. A smeary picture may or may not also be present. This would depend upon the degree of the high-frequency phase shift that accompanies the loss of the high-frequency response. The greater the phase shift, the more the smeariness that would result. The picture may give the impression of being out of focus. However, in this case the scanning lines would be in focus, showing that the picture-tube focus is not at fault. Some possible causes for the poor high-frequency responses are: a defective peaking coil, a defective bypass capacitor, an incorrect type of replacement transistor, increased value of the load resistor, or a defective shunt capacitor in the video detector circuit.

Figure 16.30 Excessive low-frequency response and phase distortion. Note the excessive contrast, the smearing of large areas, and the dark horizontal wedges.

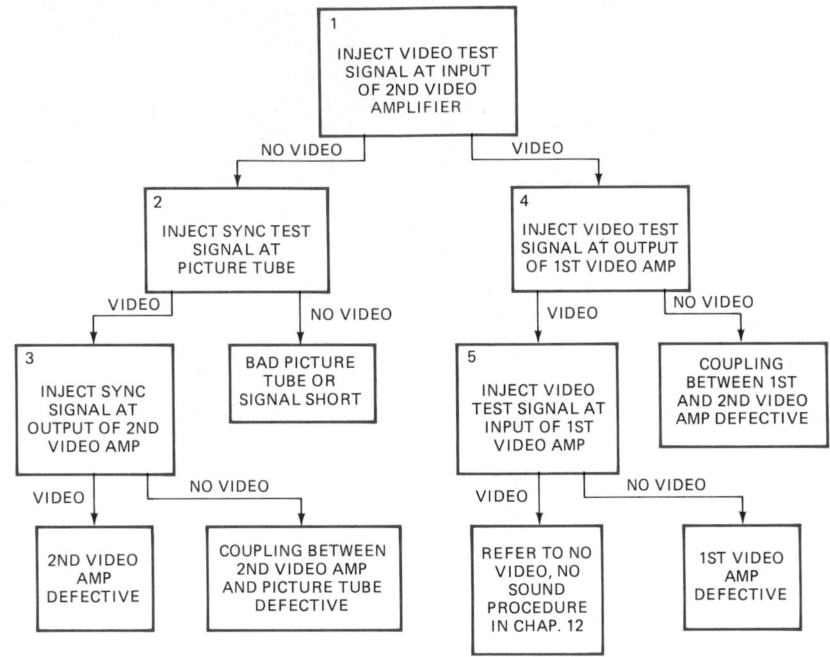

Chart 16.1 Illustrating a practical troubleshooting procedure for the condition of no video, sound normal *(Courtesy of B & K-Precision, Dynascan Corp.).*

Symptom: Poor Low-Frequency Response The lower video frequencies (below about 100 kHz) are responsible for reproducing the larger areas of the picture properly and also provide the correct overall contrast. If there is an insufficient low-frequency response (loss of the lower frequencies), the overall picture will give a weak appearance, with poor contrast. However, smearing will probably not take place and the picture, although weak in appearance will still appear sharp.

An excessive low-frequency response (increased amplitudes at the lower frequencies), will produce excessive contrast and smearing caused by the resulting phase distortion. This condition is shown in Figure 16.30 (compare with Figure 16.28). Although not shown too clearly in the figure, the high-frequency response is not affected. However, the pattern shows excessive smearing to the right of all the large areas, and the overall picture is darker than normal. Note that the shaded, circular areas in the middle of the pattern tend to blend into each other. In addition, the horizontal wedges are darker than the vertical wedges, showing that there is an excessive low-frequency response.

Poor low-frequency response may be caused by a defective interstage coupling network, particularly a defective coupling capacitor. On the other hand, excessive low-frequency response is most likely caused by a defective low-frequency compensation network. In this case, the low-frequency compensation capacitors would be suspect.

Symptom: No Video, Sound Normal Chart 16.1 illustrates a practical troubleshooting procedure when there is no video, but normal sound.

Symptom: Hum Bars on the Screen Figure 16.31 shows a typical case of a 60-Hz hum bar on the screen. This results from the introduction of a 60-Hz voltage into the video-amplifier section. In a tube set, this can be caused by heater-to-cathode leakage of a video-amplifier tube. Since transistor video amplifiers have no heaters, 60-Hz hum bars will not be produced in such amplifiers. However, it is possible to have 120-Hz hum bars in both transistor and tube sets, if the DC power supply voltages to the video-amplifier stages are poorly filtered. In this case, the appearance on the screen would be similar to that shown in Figure 16.31, except that there would be two (rather than one) horizontal hum bars visible. Note that 60- or 120-Hz hum voltages may be produced by modulation in any of the RF or IF stages. This would be rectified by the video detector and appear on the screen as can be seen in Figure 16.31.

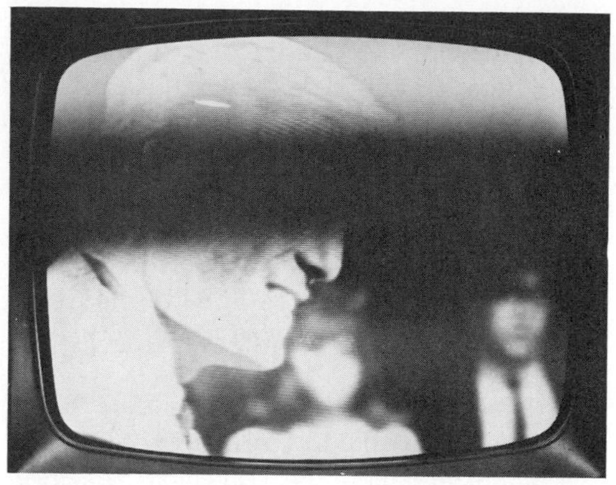

Figure 16.31 A 60-Hz hum bar in the video signal. If this was caused by a 120-Hz power-supply hum, there would be two dark bars. Note that there is no bending of the picture, indicating that the hum voltage does not affect the sync circuits.

SUMMARY OF CHAPTER HIGHLIGHTS

1. The general types of video amplifiers are resistance-capacitance (R-C) coupled amplifiers and direct-coupled amplifiers. (Figures 16.2 and 16.22)
2. The response of a basic R-C coupled amplifier is unsatisfactory at both the high and low-frequency ends for video amplifier use.
3. The basic R-C coupled amplifier must be compensated both at the low and the high frequencies in order to be suitable as a video amplifier. (Fig. 16.19)
4. The gain of a pentode-video amplifier is expressed by

   where $A_e = g_m \times Z_L$

   A_e is the voltage gain

   g_m is the tube mutual conductance

   Z_L is the total plate-circuit impedance
5. The high frequency response of a video amplifier is limited by the various shunt capacitances, designated as C_I.
6. The value of a video amplifier collector (or plate) load resistance is limited by the required high-frequency response. Typically, the resistance ranges from 2.2 to 8.2 kΩ.
7. A monochrome-TV receiver may have as few as one video amplifier. Color-TV receivers generally have between two and five (solid-state) stages.
8. The low-frequency response of an R-C coupled amplifier is limited by the time constant of the coupling capacitor (C_c) and the following input resistor. (Figure 16.2)
9. For transistor-video amplifiers, C_c may be in the order of 10 to 25μF. For vac-

uum-tube video amplifiers, common values lie in the range of 0.01 to 0.22μF.

10. "Beta" is a measure of the "current gain" of a common-emitter amplifier. "Alpha" is a measure of the "current gain" of a common-base amplifier. (Figure 16.4)
11. The high frequency where the "alpha" falls off to 0.707 (−3dB) of its 1kHz value is called the *alpha-cutoff frequency*. (Figure 16.4)
12. The high frequency where the "beta" falls off to 0.707 (−3dB) of its 1 kHz value is called the *beta-cutoff frequency*. (Figure 16.4)
13. The emitter-follower circuit has a high-input impedance (25k to 500kΩ) and a low-output impedance (50 to 1 000 Ω). (Figure 16.23).
14. The grounded-base circuit has a low-input impedance (30 to 150 Ω) and a high-output impedance (300k to 1MΩ). (Figure 16.25).
15. The characteristics of the common-emitter, common-base, and emitter-follower circuits, are summarized in Table 16.1.
16. The high-frequency gain of a video amplifier is affected by four shunt capacitances (1) C_{out}, (2) C_{stray}, (3) C_{in} and (4) C_M. C_M is due to the "Miller" effect. (Figure 16.1)
17. The frequency at which the reactance of C_T (total shunt capacitance) is equal to R_L (amplifier load resistor) is expressed by

$$f = \frac{1}{2\pi\, C_T\, R_L}$$

18. One method of improving a video amplifier's high-frequency response, is by shunt peaking. The peaking coil (L_s) is placed in series with the load resistor (R_L). (Figure 16.10)
19. Typical values of R_L in a video amplifier, range from 2 200 to 8 200 Ω. Values of peaking coils used for shunt, series, or series-shunt compensation, may range from 30 to 300μH.
20. In the series-peaking method, the peaking coil (L_c) is inserted in series with the coupling capacitor. This method provides higher gain (50%) and better phase response than shunt peaking. (Figure 16.13)
21. In the series-shunt peaking method, both peaking coils described in Questions 18 and 20 above are utilized. This method provides a gain 80% greater than with shunt peaking alone. (Figure 16.15)
22. The high-frequency response of a video amplifier can be extended by the use of degeneration in the emitter (or cathode) circuit. With this scheme, the amplifier gain is always less than in the preceding three methods. (Figure 16.16.)
23. The low-frequency response of a video amplifier may be improved by the addition of an R-C network in series with the amplifier-load resistor. (Figure 16.18)
24. In Question 23, the amplifier gain increases at low frequencies because the reactance of C_F increases. This effectively increases the output load resistance and thus, the amplifier gain.
25. One method of passing the DC signal component through an amplifier stage is to shunt the coupling capacitor (C_c) with a high value of resistance (R_c). (Figure 16.21)
26. If direct coupling is utilized from the video-detector output to the picture-tube input, the DC signal component is maintained. Furthermore, the low-frequency response is maintained down to zero Hz. (Figure 16.22)
27. An emitter follower may be used as a video driver. Because of its inherent wide-band response, no high-fre-

quency compensation is required. (Figure 16.23)

28. A *stacked* (or *cascode*) *video amplifier* utilizes two transistors connected in series. The large amplitude signal needed to operate the picture tube is divided between two transistors, reducing the possibility of transistor-voltage breakdown. (Figure 16.24)
29. IC's are used for video-driver stages, where the signal amplitude is relatively low. (Figure 16.26, 16.27)
30. The symptoms of normal sound, normal raster, but no video, can be caused by a defect situated between the video detector and the picture tube. (Figure 16.28)
31. Poor video high-frequency response may be caused by a defective peaking coil or an incorrect type of replacement transistor.
32. Poor low-frequency response may be caused by a defective coupling capacitor. Excessive low-frequency response may be caused by a defective low-frequency compensation network. (Figure 16.30)

EXAMINATION QUESTIONS

(Answers are provided at the back of the text.)

Part A Supply the missing word(s) or number(s).

1. Video amplifiers are of the _______ coupled type or the _______ coupled type.
2. The gain of a pentode amplifier varies directly with its _______ conductance and plate circuit _______.
3. The number of video-amplifier stages varies from _______ to _______, depending upon receiver design.
4. With transistor-video amplifiers, _______ values of coupling capacitors are required.
5. The expression $\frac{I_c}{I_b}$ refers to the _______ of a transistor.
6. The frequency where Alpha falls off to 0.707 of its 1 kHz value is called the _______.
7. The voltage gain of an emitter follower is _______ than one.
8. When X_{CT} = R_L, their impedance is _______ of its value at low frequencies.
9. In shunt peaking, the added inductance is placed in series with the _______.
10. With series peaking, R_L can be _______% larger than for shunt peaking.
11. To prevent "ringing" in a series peaking coil, a _______ _______ is placed across it.
12. When a small value of capacitance is placed across an emitter (or cathode) resistor, we have _______ _______.
13. The presence of C_T limits the _______ _______ response of a video amplifier.
14. The coupling time constant determines the _______ response of an uncompensated video amplifier.
15. The DC component of the video signal represents the _______ of the picture.
16. If an amplifier is _______, it does not require low-frequency compensation.
17. A high _______ is required of transistors used in video amplifiers.
18. A transistor video driver is usually connected as an _______.
19. A defective _______ can cause poor high frequency response in a video amplifier.
20. Excessive low-frequency response can result from a defective low-frequency _______.

Part B Answer true (T) or false (F).

1. In a TV receiver without a comb filter, the luminance video amplifier high-frequency response is limited to 3.6 MHz.

2. The "Miller effect" refers to the increase of amplifier gain with increasing frequency.
3. The maximum number of luminance video-amplifier stages in any color-TV receiver, is two.
4. The values of output load resistors for both transistor and vacuum-tube video amplifiers are similar.
5. Electrolytic capacitors are used for interstage coupling in vacuum-tube video amplifiers.
6. Low-frequency response of a video amplifier is inversely proportional to the value of the amplifier-load resistance.
7. A common-emitter amplifier has medium values of input impedance and output impedance.
8. In Question 7, this type of amplifier has low values of voltage gain and power gain.
9. "Alpha" is a measure of the current gain of a common-base amplifier.
10. The frequency where "beta" falls off to 0.707 of its 1 kHz value is called the *beta* cutoff frequency.
11. An emitter follower has high-input impedance and low-output impedance.
12. The total-shunt capacity (C_T) of a video amplifier, may be expressed by $C_T = C_{out} + C_s + C_{in} + C_M$.
13. When the shunt-capacitive reactance is equal to the load resistance of a video amplifier, the gain decreases by 6dB.
14. Shunt peaking provides the least gain of the three peaking-coil configurations.
15. Series peaking has higher gain but poorer phase response than shunt peaking.
16. Series-shunt peaking can provide 80% greater gain than shunt peaking.
17. In degenerative high-frequency compensation, a low value of emitter (cathode) resistance is used to provide a rising high-frequency response.
18. In a low-frequency compensation network, the value of C_f is $C_f = \dfrac{C_c R_2}{R_L}$
19. The DC component of a video signal can be passed through a coupling network by bypassing the coupling capacitor with a high-value resistor.
20. All direct-coupled video amplifiers do not require high-frequency compensation.
21. Direct-coupled video amplifiers should be operated from a stable DC power source to prevent *level* shifting.
22. The figure of merit of a high-frequency tube is given by

$$\frac{g_m}{C_{in} + C_{out}}$$

23. Transistors used as video amplifiers should have high "beta" and low input capacitance.
24. In emitter followers, a 4 MHz bandwidth can be achieved without high-frequency compensation.

REVIEW ESSAY QUESTIONS

1. Briefly discuss the high-frequency limiting factor(s) of a video amplifier.
2. Briefly discuss the low-frequency limiting factor(s) of a video amplifier.
3. Draw a simple schematic diagram of a transistor video amplifier employing shunt peaking and a low frequency compensation network. Show all shunt capacitances.
4. In Question 3, and in your own words, explain how shunt peaking works.
5. In Question 3, and in your own words, explain the operation of the low-frequency compensation network.
6. Name four types of high-frequency compensation.
7. Name two types of low-frequency compensation.
8. Define the following symbols; R_L, C_T, R_F, C_F, C_c and C_M.
9. In Figure 16.4, explain briefly the significance of the "alpha" and "beta" cutoff frequencies.

10. In your own words, briefly explain the "Miller effect".
11. In Figure 16.11, briefly explain the rise in curve No. 4.
12. Explain how series peaking can provide 50% more gain than shunt peaking. (Figure 16.13).
13. Draw a simple schematic diagram of a transistor-video amplifier employing degenerative high-frequency peaking.
14. In Question 13, briefly explain how high-frequency compensation is achieved.
15. Refer to Figure 16.30. Briefly explain why there is smearing of large areas.
16. In Question 15, what circuit condition might create this problem.
17. Name the desirable characteristics of transistors and tubes used as video amplifiers.
18. Explain briefly why large values of coupling capacitors are required in transistor video amplifiers. What might be a disadvantage of using such large values?
19. If a transistor amplifier is R-C coupled to a picture tube grid (or cathode), is a large value of coupling capacitance used? Why?
20. Refer to Table 16.1. Which transistor configuration has the highest input impedance? The lowest input impedance? The lowest current gain? The highest power gain? The lowest voltage gain?
21. Describe in your own words, the physical meaning of "beta" and of "alpha".
22. Compare the three types of peaking-coil compensation schemes with regard to frequency response, phase response, and gain.
23. In Figure 16.21, what is the function of R_c? How does it perform its function?
24. In Question 23, what additional circuitry would be required if R_c were not present?
25. Explain briefly why an emitter follower may be used as a video driver.
26. Why may a stacked amplifier be used to operate a picture tube. (Use your own words.)
27. In Figure 16.26, explain in your own words how the required wide bandwidth (low and high frequencies included) is achieved.
28. In Figure 16.13, state the function of the following components, R_L, R_{damp}, L_c, and R_2.
29. In Figure 16.14, explain the operation of R_2 briefly.
30. In Figure 16.24, why is R_5 unbypassed?

EXAMINATION PROBLEMS

(Answers to selected problems appear at the back of the text.)

1. Refer to Figure 16.22. Identify (1) shunt-peaking coil for V_1, (2) shunt-peaking coil for V_2, (3) D-C coupling resistor for V_2.
2. A pentode-video amplifier has a g_m of 12 000 μ and an R_L of 2 700 Ω. Calculate the mid-frequency gain.
3. In Question 2, the grid to plate capacitance is 2 pF. Calculate the Miller effect capacitance (C_M).
4. In Figure 16.4, to a close approximation (a) What is the "beta"-cutoff frequency? (b) The "alpha"-cutoff frequency? (c) In both of these cases, how does the cutoff gain compare with the gain at 1 kHz?
5. Compare the relative output impedance, current gain, and power gain of common-emitter and common-base amplifiers. Draw a simple table to show this. Use words and typical values.
6. Given the following:

$$C_{out} = 2\text{pF}$$
$$C_s = 3\text{pF}$$
$$C_{in} = 4\text{pF}$$
$$C_M = 10\text{pF}$$
$$R_L = 2\,700 \text{ ohms}$$

Calculate F_2, the frequency at which the reactance of C_T equals R_L. Hint:

$$F_2 = \frac{1}{2\pi R_L C_T}$$

7. In Question 6, calculate the value of a shunt-peaking coil to provide flat-frequency response to 3.2 MHz.

8. In Question 6, calculate the values of the shunt and series-peaking coils to provide flat-frequency response to 4.2 MHz.
9. Given the following:
   R_L = 2 700
   C_c = .01 μF
   R_2 = 500k Ω (input resistor)

   Calculate the values of C_F and R_F required to provide low-frequency compensation.
10. In Question 9, assume a value of 200 Ω for R_K (or R_E) and calculate the value of C_K (or C_E).

Chapter 17

17.1 The DC component of video signals
17.2 Reinserting the DC component
17.3 DC reinsertion with a diode
17.4 Television receivers that do not employ DC restoration
17.5 DC restorers in color receivers
17.6 Troubles in DC restorer circuits
Summary

DC REINSERTION[1]

INTRODUCTION

The composite-video signal contains several distinct components. Each serves a definite purpose. First, there is the AC component which represents the detail in the image. Second, there is the DC component which governs the overall background shading of the picture. Both components are separate and each may be varied independently of the other. Finally, there are the blanking and synchronizing pulses.

The two preceding chapters dealt with video amplifiers. They are primarily concerned with the AC component of the video signal. Here we are concerned with the DC component, its function within the video signal, what happens when it is removed from the signal, and how it is reinserted.

When you have completed the reading and work assignments for Chapter 17 you should be able to:

- Define the following: blanking level, DC signal component, AC signal component, DC restorer, clamper, and average scene brightness.
- Explain the operation of a DC restorer, with the aid of a simple schematic diagram.
- Draw simple sketches of video signals indicating the loss of the DC component and the reinsertion of the DC component.
- Explain briefly why the blanking-pulse level should be clamped at the cutoff point of the picture tube.
- Describe how the DC signal component may be lost.
- Describe briefly two possible effects of the loss of the DC signal component in a TV receiver.
- Know what a practical DC reinsertion time constant should be.

[1]The name **DC reinsertion** circuit is common throughout the television field. However, the terms **clamping circuit** or **DC restorer** are also used. They refer to the same thing and may be used interchangeably.

- Explain briefly why DC reinsertion is not needed between direct-coupled stages.
- Explain briefly why an artificial reference voltage is established for DC restorers used in color-TV receivers. Also, why three DC restorers are required for color-TV receivers.

17.1 THE DC COMPONENT OF VIDEO SIGNALS

Several lines of a typical video signal are shown in Figure 17.1. Between every two successive synchronizing and blanking pulses, we have the camera-signal variations, ranging from white (at the most positive value) to black at the level of the blanking pulse. The signals are shown in the positive picture-phase form. When applied to the control grid of a picture tube, each different value of video voltage produces a different spot intensity on the screen of the tube. From all these light gradations we obtain the image.

SHIFTING THE DC LEVEL

Suppose now that we take a video signal. While maintaining the same camera-signal variations, we first move these variations closer to the blanking-pulse level. This situation is shown in Figure 17.2A. Next, we shift the same variations as far away as possible from the blanking pulses as shown in Figure 17.2B. What will be the visual result in each instance? Since the blanking level represents the point at which the picture tube beam is supposed to be cut off, moving the video signal closer to this level means that the overall background of the image will become *darker*. On the other hand, when the video-signal variations are farther away from the blanking level, the background of the image becomes *brighter*. Note, however, that because the video-signal variations are identical in each instance, the same scene is obtained. The only thing we have altered by shifting the relative position of the video signal is the *background brightness*. In the first instance, it is dark; in the second, it becomes bright. We can simulate the same condition in a room by increasing or decreasing the intensity of the electric lights. This change does not affect the objects in the room; it merely affects the overall brightness of the scene.

To distinguish between the camera-signal variations and the average level of these variations (or the average distance of these variations from the blanking level), it has become standard practice to call the latter the DC component and the former the AC component of the video signal. The average level of the signal can be altered by the insertion of a DC voltage. This raises or lowers it and changes the background brightness of the image.

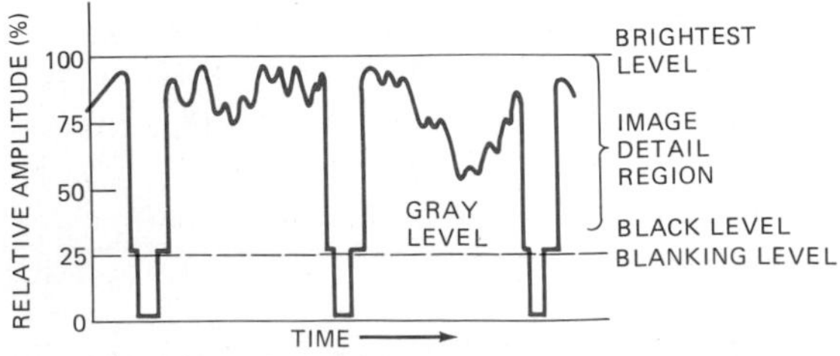

Figure 17.1 Several lines of a typical video signal.

Level of the Blanking Pulses At the transmitter, the level of the blanking pulses is established as the blacker-than-black level. At this point the electron beam in the receiver picture tube is cut off, and the screen for that point becomes dark. When the AC video-signal variations obtained from the camera tube are combined with this blanking voltage and the sync pulses, we have a complete video signal. At any point along the

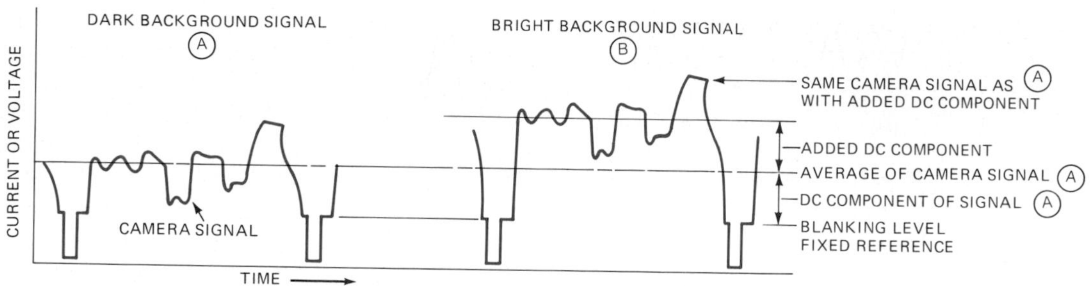

Figure 17.2 Two video signals containing the same detail (AC component) but different background brightness (DC component).

program line, the distance between the average level of the AC video signal and the blanking level may be varied (through the insertion of a DC voltage). This produces the desired shading or background brightness as dictated either by the program director or by the scene itself. Note that, since the DC voltage moves the video-signal variations closer to (or farther away from) the *blanking level,* we are using this level as a reference. Therefore, the level of the blanking pulses must always remain fixed. The signal is transmitted with this relationship maintained.

Loss of the DC Component The second-detector output in the receiver contains the full video signal, as shown in Figure 17.1. The blanking pulse of each line is aligned to the same level. However, when the signal is passed through *R-C* coupled video-frequency amplifiers, the blanking pulses of the various lines are no longer aligned. This is because the coupling capacitors cause the video signal to possess equal positive and negative areas about the zero axis.

This situation is encountered in many circuits although in slightly different form. Suppose we take three 60 Hz AC voltages and three DC voltages and combine them to form the signals indicated in Figure 17.3A. Voltages of this type are frequently found in power supplies where the AC wave represents the ripple. For the sake of this discussion, we have provided enough DC voltage for each AC voltage, so that the positive peaks of all three waves reach the same level. Now, let us pass these voltages through a capacitor. The result is shown in Figure 17.3B. By removing the DC voltages, each wave has as much area above the axis as below it. Because of this, the positive peaks of the waves are no longer at the same level.

Let us look at the equivalent situation in a television system. In Figure 17.3C there are shown three video signals taken at different moments from a television broadcast, which represent three lines. One line is almost white, one is gray, and one is dark or black. As they come out of the video-second detector, all the blanking voltages are aligned to the same level. After passing these three signals through a coupling capacitor, the signals possess the form indicated in Figure 17.3D. For each signal, the *area above* the axis is equal to the *area below* the axis. Because of this distribution, the blanking voltages of the signals are no longer at the same level, and we say that the DC component of the video signal is "missing." The question now is: What effect will this variation in the blanking level have on the image produced on the screen?

Effect of Loss of DC Component on the TV Picture The top of each blanking pulse rep-

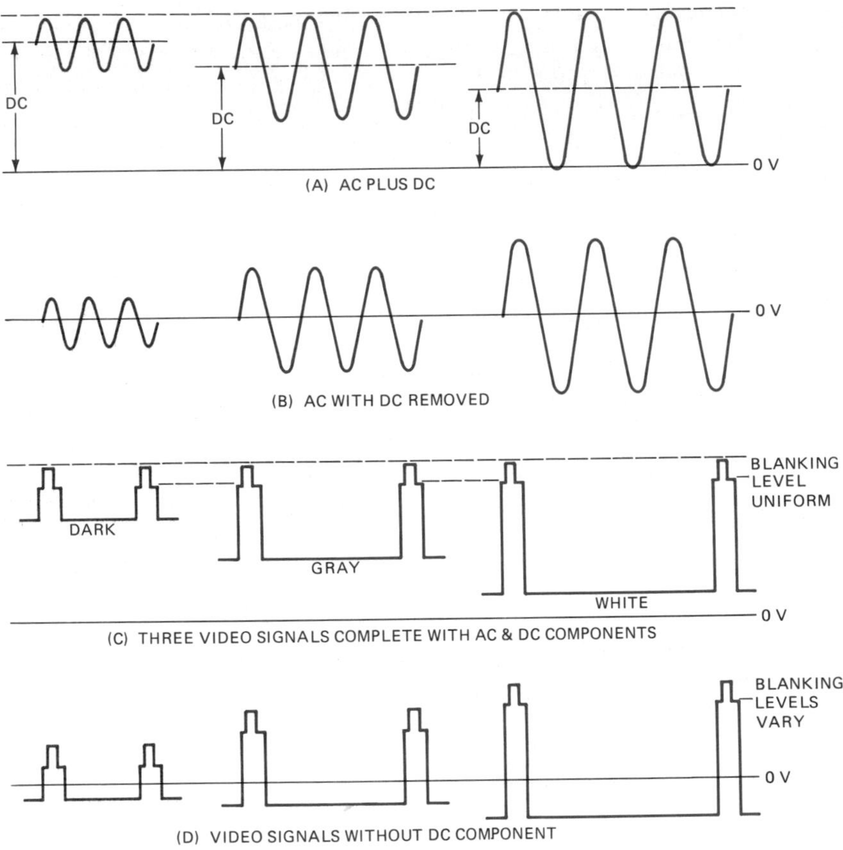

Figure 17.3 An illustration of the effect of removing the DC component from the video signal.

resents the darkest possible level of each line. Since all lines in an image should have the same reference (or black) level, the tops of all blanking pulses should have the same voltage value. This was true in the receiver just before we passed the detected-video signal through a coupling capacitor in the video-frequency amplifier system. After passage through this capacitor, the blanking-pulse levels were no longer aligned to the same level. Now if we apply the three signals to a picture-tube grid (reverse polarity of Figure 17.3D) what happens?

First of all, when the signal corresponding to a white line reaches the picture-tube grid, we manually adjust the brightness control (which controls the bias for the picture tube) to the point where the blanking-pulse level just drives the tube into cutoff. Thus, as long as the signal remains at the level it had when we adjusted the brightness control, the negative voltage of the blanking pulse—added to the negative bias set by the brightness control—will darken the screen at the blanking-pulse level.

Now, if the gray-video signal comes to the picture tube, we see that its blanking-pulse level is less negative than the blanking pulse level of the previous video signal. In this case, the beam will not be cutoff by the blanking pulse. Then the beam retrace will be visible. We could produce the proper cutoff conditions by increasing the negative bias on the image tube. However, this procedure would be impractical for the continually-changing brightness levels. For scenes that are continually changing, the background shading varies too rapidly to be adjusted manually by the viewer. As a result, if the brightness control is set for a very bright picture, we will then see the retrace lines when a darker picture arrives. Conversely, if the brightness control is set for a darker image part of the detail will be lost when a lighter image is viewed, because of the greater picture tube-grid bias. The situation is aggravated even more when a dark video signal arrives. Now, we require an even greater negative bias and when the brightness control is set correctly for this signal, it is much too negative for any of the two previous signals. If either of these two other signals is viewed with the bias set for this last signal, the image will appear too dark. On the other hand, when it is correctly set for a white picture, a black picture will appear *too light,* with the retrace lines visible. The only solution to this state of affairs is to always return all blanking voltages to the *same level* just as we found them in the incoming signal. *This, then, is the function of the DC restorer in the receiver.*

Function of DC Reinsertion Every picture tube has a definite characteristic curve. For a given amount of grid-bias voltage, a definite amount of light appears on the screen. All blanking pulses are purposely placed on the same level in order that the cathode-ray tube will react to them in the same manner throughout the entire reception of the signal. The same is true of white, gray, black, or any other shade that is transmitted to the scene. Any one shade must produce the same illumination on the cathode-ray tube screen each time its corresponding voltage is present on the control grid of the picture tube. However, this cannot occur unless all video signals have *the same reference level.* It is here that the usefulness of the DC component becomes apparent. Through the use of this inserted voltage, all blanking and synchronizing pulses are leveled off. The image detail attached to these pulses is likewise correctly oriented.

To operate the television receiver properly, the AC-video signals must be brought to the same relative level that they had before the removal of the DC component in the AC-coupling circuits of the video amplifiers. Of course, if direct coupling is used between detector output and the picture tube, the DC level will not be lost. Consequently, DC insertion is not needed. However, for receivers that employ *R-C* coupling in the video stages, the problem resolves itself into one of reinserting a DC voltage that will take the place of the one removed. Consequently, there is a need here for special DC-reinsertion networks.

17.2 REINSERTING THE DC COMPONENT

To understand how DC reinsertion is possible, it is necessary to know that removing the DC component from a video signal does not change its shape. It merely changes its reference level. This is evident when Figure 17.2A and 17.2B are compared. The same variations in the AC components still occur. The relationship of the AC signal to the blanking and synchronizing pulses remains the same, with or without the DC component. It is also seen that brighter picture signals result in greater separation between the picture-information variations and the above-mentioned pulses. As the scene becomes darker, these two components move closer together.

It is from these relationships that we are able to reinsert the DC component. Our purpose is achieved if we can develop a variable

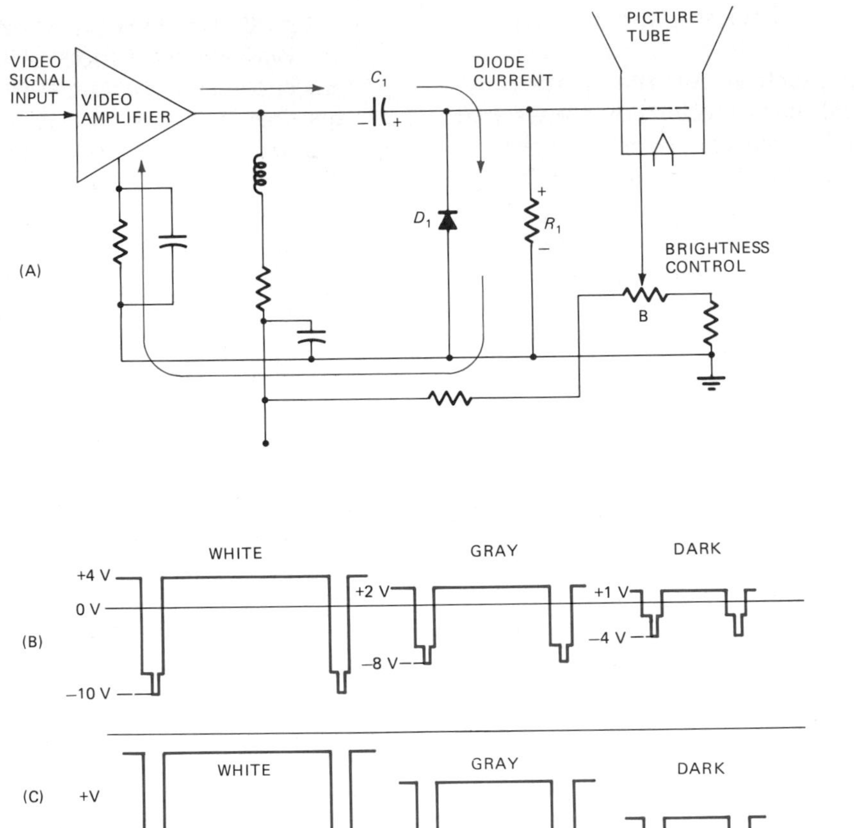

Figure 17.4 Illustrating DC reinsertion with a diode circuit. **A.** Simplified diagram of a video amplifier circuit incorporating diode reinsertion. **B.** Changes of blanking level without DC reinsertion. **C.** Stabilized blanking level with diode DC reinsertion.

bias that will affect each change in the blanking and synchronizing pulse voltage and bring all pulses to one common level. If a video signal in its AC form is applied to the input of a diode where the process of DC restoration occurs, the pulses will return to the same level. Then, with all the pulses lined up again, the video signals can be applied to the picture tube.

17.3 DC REINSERTION WITH A DIODE

The most common method of DC reinsertion is using a simple diode circuit. This is illustrated in Figure 17.4. The signal here is in its AC form until it reaches the input to the DC restorer circuit, composed of capacitor C_1, resistor R_1, and diode D_1. At the DC restorer, the signal regains its DC component, as explained presently. The only addition to the normal circuitry of Figure 17.4A is the diode, D_1.

Signal Prior to DC Reinsertion

Figure 17.4B illustrates how the blanking levels of the three signals shown may shift at the grid of the picture tube. This condition would occur if diode D_1 was not present. It would happen because of the inability of the coupling capacitor C_1, to pass the signal DC component, as previously mentioned.

Signal with DC Reinsertion

When diode, D_1, is added to the circuit, as shown in Figure 17.4A, its action causes the blanking levels of the three signals to line up at the same voltage value, as shown in Figure 17.4C.

This circuit functions to align the blanking (and sync) pulses at a common level because the diode charges C_1 to approximately the peak value of each sync pulse. The peak value here is the magnitude of voltage from zero (average) to the negative peak of the sync pulses. It is this (negative) portion of the signal which causes D_1 to conduct.

Note in Figure 17.4B that the respective negative-peak voltages are -10V, -8V, and -4V, for the white, grey, and dark signals. When the white signal appears, C_1 charges to approximately $+10$ volts on the diode side. This voltage appears in series with the signal and has the effect of "lifting" the signal 10 volts in the positive direction. This places the sync pulse at the zero-volt level. (See Figure 17.4C). Similarly, the grey signal causes a $+8$ volt charge to appear at C_1, "lifting" the signal 8 volts in the positive direction. Again, the sync pulse is placed at the zero-volt level. Lastly, the dark signal charges C_1 to $+4$ volts and its sync pulse is placed on the zero-volt level.

The Practical Case

The above example is simplified for the purpose of the explanation. If the brightness level of a picture actually changes from line-to-line, it does so much more gradually. However, the normal operation is the same as explained above, with negative-peak voltages of the sync pulses being lined up at the zero-volt level. This process is also frequently referred to as "clamping".

In this example, the signal is fed to the picture-tube grid. The picture-tube bias is set by the brightness control in its cathode circuit. When this control is properly adjusted, the blanking pulses will drive the picture tube to cutoff.

DC reinsertion is not completely effective in maintaining the proper signal-blanking level. Also, the setting of the brightness control affects the blanking level. Therefore, in many television receivers, additional horizontal and vertical retrace-blanking circuits are provided to insure proper retrace blanking.

The DC Reinsertion Time Constant

As mentioned above, the average scene brightness changes little (if any) from line to line. Depending upon the scene, it may or may not change from frame to frame. Thus, the reinsertion-time constant of C_1R_1 should be long with respect to the period of a horizontal line, but comparable to that of a frame. A value such as .03s is typical.

DC Reinsertion For Positive-Sync Polarity

In Figure 17.4, the video signal was fed to the grid of the picture tube. Thus, the polarity of the sync and blanking pulses had to be in the negative direction. If the video signal was fed instead to the cathode of the picture tube, the video signal would have to be inverted. Then, the sync and blanking pulses would be going in the positive direction. In such a case, the only change required to maintain DC reinsertion would be to reverse the diode connections. The tips of the sync pulses would now be clamped to zero, with the remainder of the video signal appearing in the negative direction.

17.4 TELEVISION RECEIVERS THAT DO NOT EMPLOY DC RESTORATION

Cost is a strong determining factor in the design of commercial television receivers. If it is possible to reduce the cost of a set without compromising picture quality too much, this sacrifice is frequently made. A number of receivers have been manufactured that do not employ DC restoration, nor do they possess a DC path between the video second detector and the picture tube. In other words, the DC component is removed from the signal and never reinserted.

Loss of the DC component will tend to make the overall picture darker. To counteract this, the viewer generally turns up his brightness control. This, in turn, frequently causes the vertical retrace lines to become visible. The continued presence of these lines during normal broadcasts will prove to be annoying. To rid the screen of these retrace lines, it has become standard practice to apply an additional negative pulse to the grid of the picture tube during the vertical retrace interval. (A positive pulse fed to the picture-tube cathode will achieve the same results. Generally the pulse is applied to the element not receiving the video signal.) The pulse biases the tube to cutoff, prevents electrons from passing through the tube, and effectively removes the vertical retrace lines for any normal position of the brightness and contrast controls.

It is true that removal of the DC component will reduce the contrast range of the image. However, this has been partly offset by the development of screen phosphors possessing wider contrast ranges. It is doubtful whether any viewer can tell the difference when the DC component is missing.

DIRECT COUPLING

When direct coupling is employed between all stages in the video amplifier section, including the coupling stage between the video detector and first video amplifier and between the last video amplifier and the picture tube, then DC restoration is not required. The DC component is preserved.

The disadvantage of direct coupling more than one tube-type video stage is that the power-supply voltage must be higher than is required for a single amplifier stage. Also, any small amount of change in the DC level will be amplified by the first video amplifier and reamplified by the second video amplifier. The demands on the power supply include the requirements of high voltage, good regulation, and exceptionally-good filtering. These result in a preference for DC coupling only for the video-output stage.

It is an easier matter to direct-couple transistor amplifiers. The emitter-collector voltage is not very large to begin with, so three or four stages can be direct coupled without a voltage in excess of 100 V. The low impedance of transistor circuits makes it easier to obtain a ripple-free DC voltage. Figure 17.5 shows a direct-coupled transistor video-amplifier circuit. The detector (D_1) is direct coupled to the first video amplifier (Q_1) through a peaking coil L_{212}. This is an emitter-follower stage which develops the signal across R_{248}. The base of the second video amplifier (Q_2) receives its signal directly from the emitter of Q_1. The second amplifier is a conventional (common-emitter) amplifier stage direct coupled to the picture tube through the contrast control, diode D_2, and the peaking coil L_{218}. (Diode D_2 is used to compensate for nonlinearity in the transistor amplifier (Q_2) on high-amplitude signals.)

When direct coupling is used, the brightness control may be part of a video-amplifier circuit rather than part of the picture-tube bias circuit. Also retrace blanking

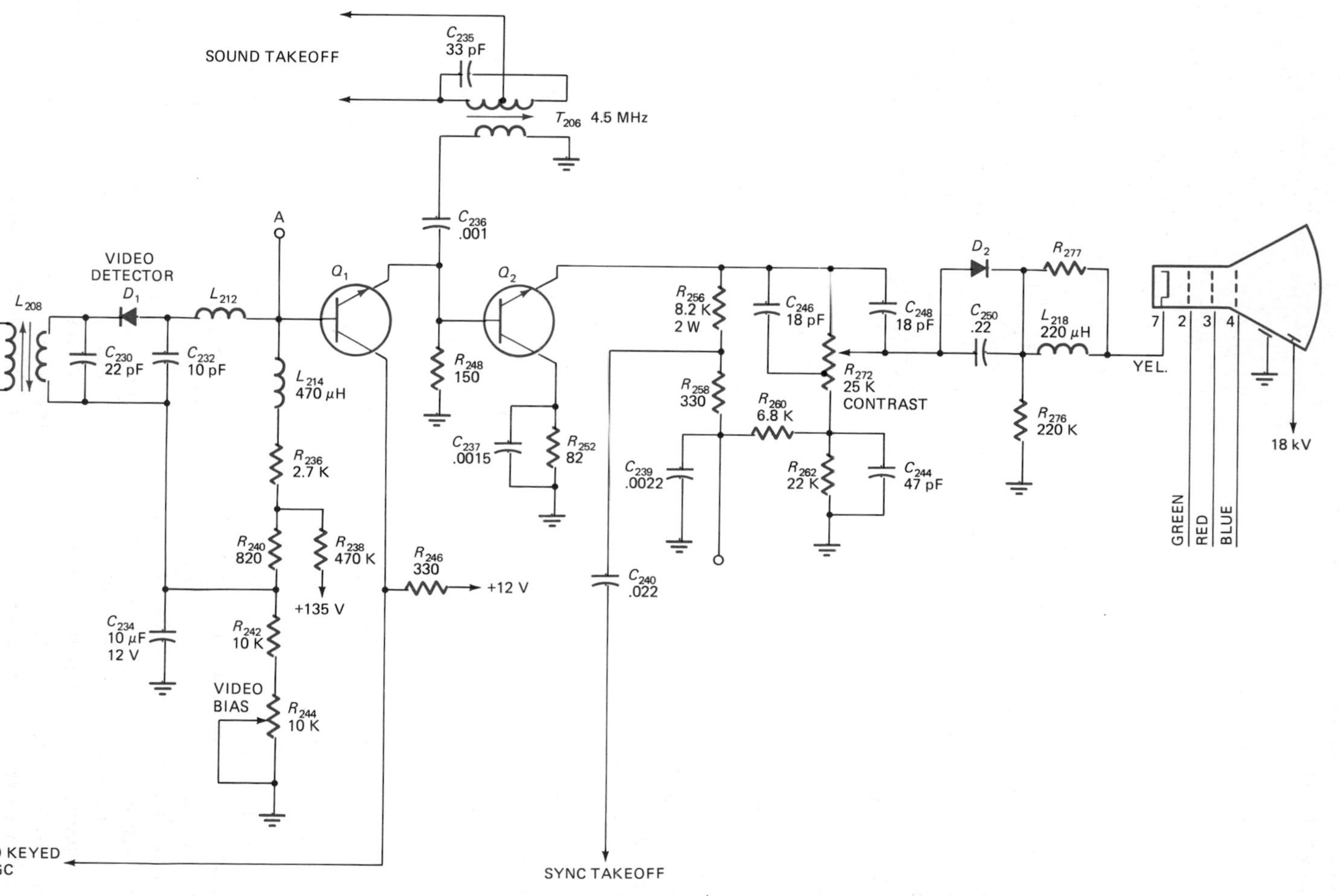

Figure 17.5 Video detector and video amplifier stages, direct-coupled to the picture tube of a color TV receiver.

signals may be fed to a video-amplifier stage rather than to the cathode or grid of the picture tube.

17.5 DC RESTORERS IN COLOR RECEIVERS

Although the subject of color receivers has been covered in detail previously, the need for DC restorers in color-TV circuits is discussed here at this time.

As a general rule, whenever a video signal is delivered to the picture tube of a television receiver, whether it is a color or a black-and-white receiver, then that signal must have a DC voltage reference if the maximum picture effectiveness is to be achieved. In this chapter it has been shown that the DC reference is lost whenever the signal passes through a capacitive coupling network. In a color receiver, the video signal (which is usually called the *luminance signal* in color receivers) may, in some designs, be delivered to the picture tube through capacitive-coupling circuits. In this event the same problem of DC restoration occurs as with the video signal in black-and-white receivers. The block diagram in Figure 17.6 shows that the luminance signal passes through a video-amplifier stage, a delay line, and then through additional video-amplifier stages. The purpose of the delay line is to keep the luminance signal in phase with the color signals which must pass through different amplifier sections. If capacitive coupling is used with any of the video-amplifier stages, then DC restoration is needed. The circuitry is the same as that used in monochrome receivers.

In the block diagram of Figure 17.6 the color signal is taken from the first video amplifier and fed through bandpass amplifier stages, then into color processing circuits. It should be pointed out that color receivers vary in their circuitry according to the manufacturer's preference, and this is only a typical color receiver. However, most receivers have a section called the *bandpass amplifier*. These are simply tuned amplifier stages which pass only the color signals that are grouped around the 3.58 MHz color subcarrier.

CHROMA SECTION BLOCK DIAGRAM

Figure 17.7 shows the block diagram of a color-receiver chroma section. The color-killer control voltage prevents the color-bandpass amplifier stage from producing an output signal when there is no color signal being received. It is like a switch that automatically turns the amplifier on when a color transmission is being received. The color bandpass amplifiers deliver the color signals simultaneously to three demodulators—one for each of the primary colors. At the same time, the 3.58 MHz reference signal is also

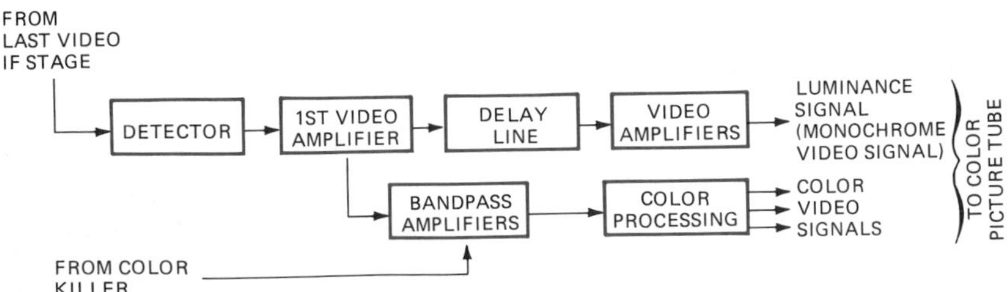

Figure 17.6 Simplified block diagram of luminance and color sections.

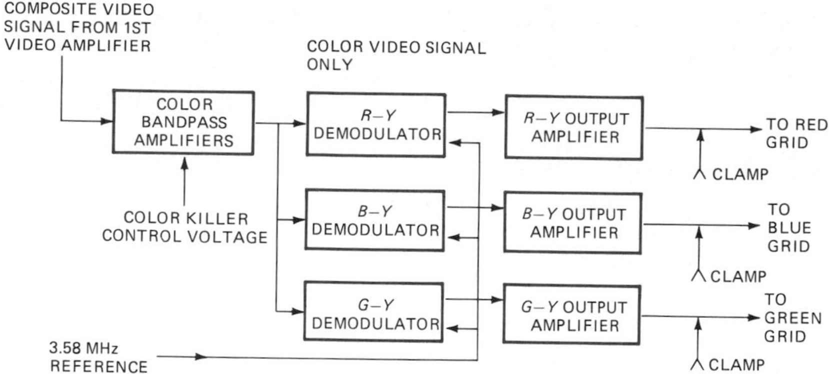

Figure 17.7 Simplified block diagram of a color section showing clamping (DC reinsertion) voltage delivered to each amplifier output.

fed to the demodulators. The output of each demodulator is then amplified before it is fed to the respective color grid. The amplifiers that follow the demodulators are capacitively coupled, and DC restoration must be used. Note that in the block diagram of Figure 17.7 each color signal output is *clamped.* This is another way of saying that the DC level has been restored.

Since there are no sync or blanking signals present in the chroma-video signals to act as a reference, an artificial reference is established by the clamping amplifier circuit shown in Figure 17.8. Note that while each color grid is clamped separately by the action of its own particular signal, the DC reference established by the clamping amplifier is used for all three color grids.

A negative pulse from the flyback transformer is fed to the emitter of clamping amplifier, Q_{101}, and causes conduction of Q_{101} only during the negative pulse time. Diode D_{101} removes any positive portion of the flyback pulse. Inductor L_{101} suppresses radiations from this circuitry. The negative, amplified pulse appearing across the picture tube bias control and resistor R_{104} is clamped to a maximum value of 180 volts by the zener diode D_{102}. A portion of the negative pulse voltage, as determined by the setting of the picture tube bias control, is applied to each of the three clamping (DC restoring) diodes, D_{103}, D_{104}, and D_{105}. The action of the clamping diodes causes the three grids of the color tube to be initially clamped to the bias pulse voltage, or about minus 80 V. In Figure 17.8, only one color amplifier is shown in detail, to simplify the diagram. However, the other two color amplifiers are connected to the clamping circuit in the identical manner.

When the chroma signals are passed through the coupling capacitors in the output of each chroma amplifier, such as C_{103} in the collector circuit of Q_{102}, the DC level is lost. It is necessary to provide DC restoration individually for each color grid, so that the proper color brightness values will be reproduced. This is accomplished by the same basic method as described previously for the composite video signal in a monochrome set. Each chroma signal causes conduction of its respective clamping diode, charging and discharging its respective coupling capacitor (for example, C_{103}, for the *R-Y* amplifier). This is in accordance with the average DC level of the chroma signal from each color amplifier. In turn, this causes the DC voltage applied to each color-tube grid to vary independently in accordance with the average

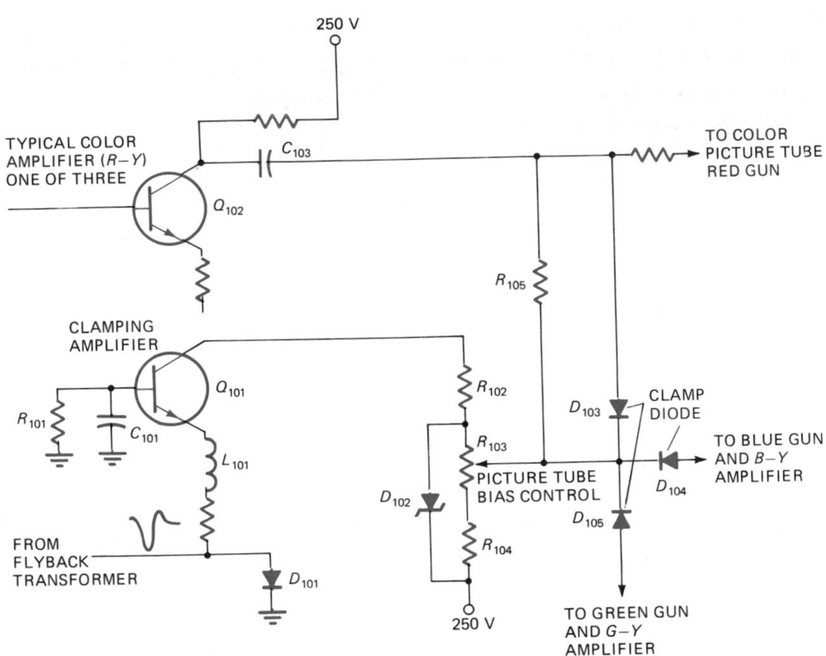

Figure 17.8 Circuit for obtaining the clamping voltage. Only one color-amplifier circuit (R-Y) is shown in detail. The B-Y and G-Y amplifier circuits are connected in the identical manner as the R-Y amplifier circuit.

color brightness value of each color signal. In this manner, the color brightness value is caused to track with the values originally seen at the color-camera tubes.

17.6 TROUBLES IN DC RESTORER CIRCUITS

Since DC restoration affects the bias voltage on the picture tube, it is reasonable to expect that a faulty DC restorer can result in a complete loss of picture and/or brightness. If the DC restorer circuit is faulty because of a change in the value or a defect in a component, the DC level may be lost. Also, since the bias voltage on the picture tube does not vary properly with changes in the average DC value of the video signal when the restorer is not functioning properly, it is possible for the picture background to become darker. Loss of contrast may be a symptom of DC restorer trouble. If the brightness control is then turned up, vertical retrace lines may become visible.

In color receivers it is possible to have a defective DC restorer in the luminance circuit as well as in the chroma circuitry. If the DC restorer in the luminance channel only becomes defective, the result may be poor picture contrast which will be noticeable on monochrome pictures. There may also be some loss of monochrome portions of the color picture. The result will be that poor color displays may be seen when a color picture is being viewed, and poor contrast may be seen when a monochrome picture is displayed.

If the clamping circuit in one of the color sections is faulty, it can result in a complete loss of one color, or in a washed-out color. DC voltage checks on the grid and the cathode circuits of the picture tube should be made if a DC restorer circuit is suspected. Remember that a DC restorer fault will change the DC bias on the affected gun of the color tube.

SUMMARY OF CHAPTER HIGHLIGHTS

1. The composite-video signal, as transmitted, contains a DC component which represents the average background illumination. (Figures 17.2, 17.3)
2. The DC component is the average value of the video-signal variations. (Figures 17.2, 17.3)
3. At the TV transmitter, the level of the blanking pulses is established as the "blacker-than black level". This is the level at which the picture tube should be cut off. (Figures 17.2, 17.3)
4. At the TV transmitter, the DC (or average) level of the video signals may be varied to produce the desired background illumination.
5. The blanking-pulse level is used as the video signal amplitude reference. This level must always remain fixed in transmission. (Figures 17.1, 17.2)
6. If the video signal is passed through R-C coupled video amplifiers, the DC component and the correct alignment of the blanking pulses is lost. (Figures 17.3, 17.4)
7. If the signal-DC component is lost (1) retrace lines may not be blanked, (2) sync may become unstable, and (3) dark and light scenes will have incorrect background illumination. (Figures 17.3D, 17.4B)
8. The function of a DC restorer (clamper, DC reinsertion circuit) is to reinsert the DC component of the signal. This means that the blanking-pulse levels will all be at the same point. (Figure 17.4)
9. When the video signal DC component is lost, the shape of the video signal is *not* lost. The reference level of the signal is the only thing lost. (Figure 17.4)
10. The most common method of reinserting the DC component employs a diode. (Figure 17.4)
11. The DC restorer diode causes the blanking-pulse levels to be realigned. In this manner, it restores the DC video signal component. (Figure 17.4)
12. In Question 11 the diode charges the coupling capacitor approximately to the peak voltage of the sync pulses. This causes the alignment of the sync (and blanking) pulses. (Figure 17.4)
13. A DC restorer diode can be connected to restore the DC component to a video signal having either positive or negative-going sync polarity. (Figure 17.4 and Section 17.3)
14. In Question 13, the diode connections need merely to be reversed to accommodate either sync-pulse polarity. (Section 17.3)
15. When the TV receiver-brightness control is properly adjusted, the blanking pulses will drive the picture tube to cut off.
16. In Question 15, this procedure may not be practical for the average viewer. Therefore, in most TV receivers addi-

tional horizontal and blanking pulses are supplied to insure blanking of the retrace lines.

17. The average scene brightness changes little if any, from line to line. However, it may change from frame to frame.
18. In Question 17, the DC reinsertion time constant (R_1C_1 of Figure 17.4) should be long with respect to the period of a horizontal line, but comparable to that of a frame. About .03 s is typical. (Section 17.3)
19. When direct coupling is employed from the output of the video detector to the input of the picture tube, DC reinsertion is not required. This is because the DC component is not lost. (Figure 17.5)
20. DC restorers used in color-TV receivers operate on the same principle as in monochrome receivers. However, since sync and blanking pulses are not available in the chroma section, an artificial reference voltage is established. (Figures 17.7, 17.8)

EXAMINATION QUESTIONS

(Answers are provided at the back of the text.)

Part A Supply the missing word(s) or number(s).

1. The DC component of the video signal represents the ______ ______.
2. The ______ is another way of describing the DC component of a video signal.
3. The overall image background will become ______ if the video-signal variations are moved closer to the blanking level.
4. The video signal variations are called the ______ of the signal.
5. The ______ level is used as the level to set the picture background illumination.
6. ______ is unnecessary if the video signal is DC coupled from the video detector to the picture tube.
7. If the DC signal component is lost ______ ______ may become visible.
8. A line drawn so that the signal areas above and below it, represents the ______ of the signal.
9. When a video signal is passed through a capacitor, its ______ is lost.
10. The ______ will be incorrect if the video signal DC component is lost.
11. The circuit that reinserts the DC signal component is called a ______ or a ______.
12. The ______ of a video signal is not changed by the loss of its DC component.
13. The addition of a simple ______ is a common method of DC reinsertion.
14. The clamping diode causes the ______ and ______ pulses to be correctly aligned.
15. Because of the clamping diode, the ______ charges to the peak value of the sync pulse amplitudes.
16. The DC reinsertion time constant should be approximately equal to the period of one ______.

Part B Answer true (T) or false (F).

1. A video signal with a dark background will have its variations far from the blanking level.
2. The average value of a signal is the same as its DC component.
3. Retrace lines may appear if the DC signal component is missing.
4. A DC restorer is also known as a "clipper" circuit.
5. DC reinsertion is required for all types of vacuum-tube circuits.
6. DC restoration is accomplished by adding

the voltage across the diode, to the signal voltage.

7. The DC restorer diode also provides automatic-brightness control (ABC).
8. A diode-DC restorer is effective only if the video signal has negative-going sync pulses.
9. Some television receivers do not employ DC reinsertion even when R-C coupling is used.
10. In TV receivers employing additional vertical and horizontal-blanking pulses, DC reinsertion is unnecessary.
11. For DC reinsertion in chroma amplifiers, an artificial DC reference must be provided.
12. A faulty DC restorer can cause a change in the picture-background illumination.
13. If only one DC restorer in a color receiver is defective this can cause a loss of one color.

REVIEW ESSAY QUESTIONS

1. Draw simple sketches of video signals indicating (1) loss of the DC component and (2) reinsertion of the DC component.
2. Draw a simple schematic diagram of a transistor, R-C coupled video amplifier, including DC reinsertion.
3. In Question 2, briefly explain the operation of DC reinsertion, in your own words.
4. Explain briefly why the blanking-pulse level should be clamped at the cutoff point of the picture tube.
5. In Question 4, draw a simple sketch of a picture tube characteristic curve (brightness vs. bias), showing the correct position of a video signal in relation to the curve.
6. A TV receiver uses direct coupling from the video detector to the picture tube. What kind of DC reinsertion is required for positive-going sync? Why?
7. In Figure 17.4A give the function of (1) C_1, D_1, R_1.
8. Define the following: (1) blanking level, (2) DC signal component, (3) AC signal component, (4) DC restorer, and (5) clamper.
9. Refer to Figure 17.5. What is the function of (1) C_{232}, (2) C_{234}, (3) L_{214}, (4) D_2, and (5) L_{218}?
10. In Figure 17.7, explain briefly why three clamping diodes are required.
11. In Figure 17.8, identify the clamping diode for the red gun. What is the function of R_{103}?
12. In Figure 17.8, how is the clamping-reference voltage established? Why is this scheme used?
13. Describe briefly two possible effects of a loss of the DC signal component in a TV receiver. (In your own words.)

EXAMINATION PROBLEMS

(Selected answers are provided at the back of the text.)

1. A DC reinsertion network consisting of a .01μF capacitor and a resistor, are to have a time constant of .05 s. Calculate the required value of resistance.
2. In Figure 17.9, draw a horizontal line indicating the approximate average value of the waveform. What are the approximate positive and negative voltage values of the AC waveform?
3. Draw simple illustrations of (1) a video waveform with high-background illumination, (2)

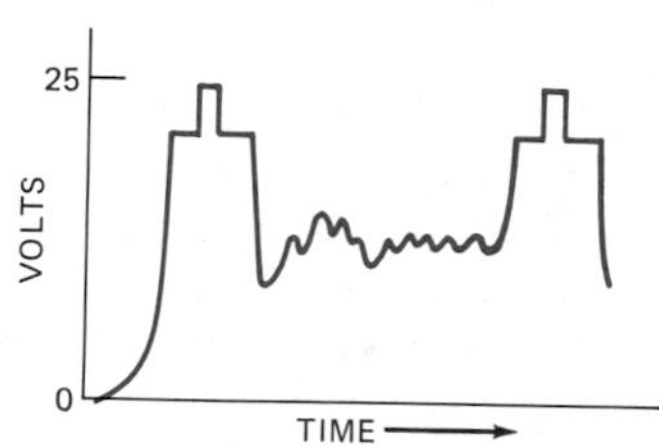

Figure 17.9 Diagram for Problem 2.

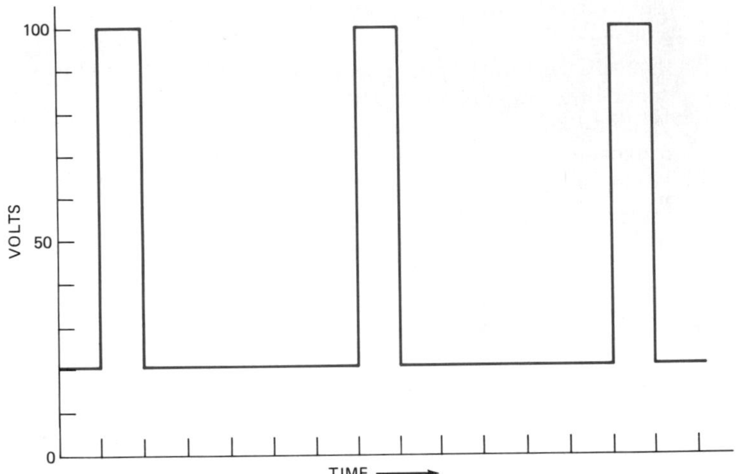

Figure 17.10 Diagram for Problem 4.

a video waveform with moderate-background illumination, and (3) a video waveform with dark-background illumination.

4. Refer to Figure 17.10. This is a highly simplified representation of a video signal. What is the average value (DC component voltage)? (Hint: The average value separates equal areas above and below the average.)
5. In Question 4, what are the positive and negative peak voltages with respect to the average value?
6. Refer to Figure 17.11. Assuming the wave of Figure 17.10 to be the output of the video amplifier, what value of DC voltage will appear at the grid of the picture tube?

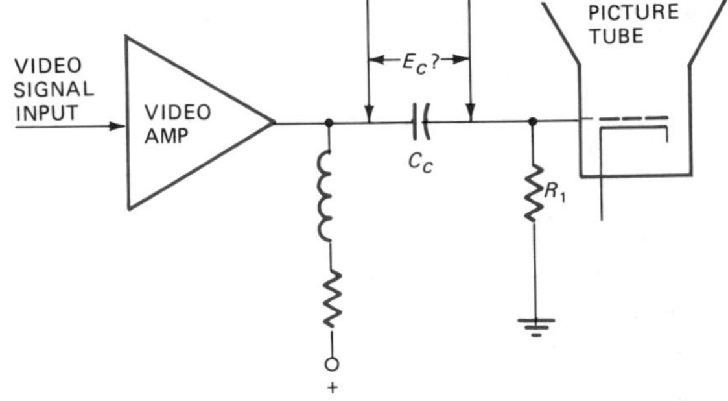

Figure 17.11 Diagram for Problem 6.

Chapter 18

18.1 Picture-tube specifications
18.2 The electron gun
18.3 Electron beam deflection
18.4 Electromagnetic deflection with electrostatic focus
18.5 Ion spots
18.6 Problems in obtaining brightness and contrast
18.7 Rectangular screens and tube-safety shields
18.8 Color-picture tube with a delta gun and color-dot triads
18.9 RCA in-line gun, color-picture tube
18.10 Trinitron picture tube
18.11 Tri-potential color tube
18.12 Picture-tube protection considerations
18.13 Troubles in monochrome and color-picture tubes
Summary

TV PICTURE TUBES

INTRODUCTION

Monochrome and color-TV picture tubes are specialized forms of cathode-ray tubes. A monochrome-picture tube has a single electron gun. This gun directs a finely focused beam of electrons toward a fluorescent material (or phosphor) coated on the inside of the screen. This material glows white when struck by the electron beam. Between the gun and the screen are deflection coils which deflect the beam horizontally and vertically to form a raster. The brightness of the screen at any point depends upon the number (and velocity) of electrons striking that point. Therefore, the brightness of the picture may be controlled by varying the grid voltage with respect to the cathode voltage of the picture tube.

A color-picture tube has three separate electron beams. The screen is made either of many dot triangles (triads) or thin-vertical stripe groups of red, green, and blue phosphors. Each of the three beams activates only a red, green or blue phosphor individually. Color phosphor dots (or stripes) are positioned so close together on the picture tube screen that the eye cannot distinguish them. When all three adjacent color phosphors are simultaneously illuminated in the correct proportions by electron beams, the screen radiates a white light. Various mixture colors are obtained by controlling the strength of the individual electron beams striking each color phosphor. The brightness (or saturation) of each color is controlled by varying the grid or cathode voltage of its respective electron beam element. In order to insure that each electron beam strikes only one particular color phosphor dot or stripe, a "shadow mask," or "aperture mask," is located near the screen. It is very important that each electron beam be carefully directed to pass through the desired hole in the mask and strike only the desired color phosphor.

This chapter covers the theory and operation of the picture tubes used in monochrome and color receivers. Also discussed are some of the more recent developments in color-picture tubes.

When you have completed the reading and work assignments for Chapter 18, you should be able to:

- Define the following terms: delta gun, in-line gun, preconvergence, dynamic convergence, static convergence, P_{22}, P_4, color triads, color stripes, aperture mask, aquadag, electromagnetic deflection, yoke, ion trap, tri-potential focus ions, ABL, crossover point black surround.
- Define the term deflection angle and explain how it is measured.
- Understand the picture tube identification code.
- List some functions of the aluminized backing of the phosphor screen.
- Understand the structure and operation of color picture tube electron guns.
- Explain two methods of reducing the implosion hazard in picture tubes.
- Discuss the different types of yokes.
- Draw a simple schematic diagram of a "tri-potential" color tube. Label all parts. Discuss the advantages.
- Discuss three basic types of color picture tubes.
- Explain the advantages of the in-line gun over the delta gun.
- Understand the precautions taken to prevent excessive exposure to x-rays.
- Discuss the operation of an automatic-brightness limiter (ABL).
- Explain the precautions that should be taken in handling a picture tube.
- Discuss the components on the color picture tube neck.
- Discuss the importance of a narrow picture-tube neck diameter.
- Explain why only electromagnetic deflection is used with TV picture tubes.
- Discuss the operation of three types of shadow masks.
- Discuss the two types of convergence.
- Know why the combination of horizontal, in-line guns and vertical-phosphor stripes effectively eliminates the need for vertical convergence.
- Discuss some common troubles in picture tubes and how to recognize them.

18.1 PICTURE-TUBE SPECIFICATIONS

In this section we will consider some of the important picture-tube specifications.

PICTURE-TUBE NUMBERING

All picture tubes, whether monochrome or color, are identified by a specific-code scheme consisting of numbers and letters. This code is different for each tube having a differing characteristic from all others.

For example, a certain color-picture tube may have the designation, 25VBHP22. The number "25" is the approximate diagonal measurement in inches of the external dimension of the picture tube. The "V" following the "25" indicates that the actual *viewing* diagonal is close to 25 in (63.5 cm). For this tube, the *viewing* diagonal is actually 24.658 in (62.6 cm) inches. The external diagonal is slightly larger than the viewing diagonal.

The use of the "V" designator is a recent addition. For example, take the case of a tube designated, 25BHP22, which does not include the "V". With tubes not using the "V" designator, the actual viewing diagonal is appreciably less than the external-tube diagonal. In this particular case, the viewing diagonal is 22.995 in (58.4 cm) for the 25BHP22 tube.

The Center Letters The letters following the screen size (here "BH") are assigned alphabetically for each screen size, by the manufacturer. Different letters indicate different tube characteristics. These characteristics are described in the manufacturers' data sheets.

The Phosphor Numbers Color-picture tube red, green, and blue phosphors are designated as "P22". Sometimes a letter follows, as in "P22A". This indicates a change in the tube. Manufacturers data must be consulted for details. All monochrome-tube phosphors are designated as "P4".

Some manufacturers add their own coding prefix, consisting generally of two or three letters to the tube identification code. Thus, a certain Sylvania color-picture tube is designated as MV25VEMP22. Here, the manufacturers prefix is "MV". The addition of the prefix does not indicate any change in the electrical characteristics of the tube, or its possible interchangeability with an equivalent type.

HEATER VOLTS/MA

Color-picture tubes have three heaters. However, these are paralleled to one pair of tube-plug pins. The heater voltage is 6.3 V AC. A common heater-current drain for color tubes is 900 mA, but some color tubes draw as much as 1 900 mA.

Most monochrome-picture tubes use a heater voltage of 6.3 V AC at 450 or 600 mA. However, monochrome picture tubes for portable operation may use heater voltages of 2.35, 2.68, 4.2, 9.45 (300 mA), 11 (77 or 140 mA), or 12 (about 80 mA).

ANODE VOLTAGES

The anode voltage (or "high voltage") is responsible for most of the electron-beam(s) acceleration in picture tubes. Anode voltages for color-picture tubes range from approximately 22 to 32 kV. The average total-anode current for a color-picture tube with 30 kV applied to the anode, may be approximately 2 000 μA.

For monochrome-picture tubes, the anode voltage ranges from approximately 9 kV (8-in or 20.32 cm tube) to 25 kV (23-in or 58.4 cm tube). However, it must not be inferred that all 23-in (58.4 cm) monochrome-picture tubes operate at 25 kV. Some operate at 18 kV.

AQUADAG COATINGS

All picture tubes have graphite coatings on the internal and external surfaces of the glass bell. These conductive-graphite coatings are commonly known as **aquadag coatings,** since they are made of water-based graphite solutions. The coatings are indicated in Figures 18.12 and 18.14 and can be clearly seen as a black area in Figure 18.19.

Internal Coating The internal coating extends into the neck of the tube, as shown in Figures 18.12 and 18.14. In the neck, it connects by spring contacts to a portion of the electron gun. The internal coating is the anode of the picture tube. The receiver-"high voltage" (up to about 32 000 volts) is applied to the internal coating via a small recessed socket. The anode provides a uniform electric field to accelerate the electron beam(s).

External Coating The external coating is insulated from the internal coating (anode) by the glass bell. However, the external coating is grounded by a spring arrangement. The capacitance (approximately 1 500 pF) is formed by the two coatings and the glass is used as a high-voltage filter capacitor.

Caution! This high-voltage filter capacitor has very low leakage. Before removing a picture tube, be certain to discharge this capacitor. An insulated probe connected to ground (first) and touched to the recessed anode connection will safely discharge the capacitor. If the tube has been operated outside of a receiver, simply short the external and internal coatings together with a well insulated lead.

BEAM DEFLECTION ANGLE

The beam-deflection angle is the *total angle* through which the electron beam can be deflected by the yoke-magnetic field. The angle is restricted to that which causes the electron beam to reach the extreme ends of the viewing screen. The deflection angle is defined as the total angle. Thus, a deflection angle of 114° means a deflection of plus and minus 57° from beam center.

The deflection angle is usually defined as the **diagonal deflection angle,** since this is the greatest of the three possible deflection angles. The other two are the horizontal and the vertical deflection angles. Typical deflection angles for a 25-in color picture tube are:

Diagonal – 100°
Horizontal – 86°
Vertical – 67°

TUBE LENGTH AND NECK DIAMETER

For a given screen size, the greater the deflection angle, the shorter the overall length of the tube. A shorter tube means less depth required of the cabinet. For example, one 90° deflection angle color tube, has an overall length of 18.42 in (46.79 cm). However, another color-tube with a 110° deflection angle, has an overall length of only 14.25 in (36.20 cm). This is a reduction in overall length of 4.17 in (10.6 cm).

Increased deflection angles require that the yoke magnetic-field strength be increased. In turn, the tube-neck diameter must be decreased to place the magnetic field closer to the electron beam. This can be seen from the following examples:

Type	Deflection Angle	Neck Diameter
Color	70°	2 in (5.08 cm)
Color	90°	$1\frac{7}{16}$ in (3.63 cm)
Color	100–114°	$1\frac{1}{8}$ in (2.9 cm)

Monochrome-picture tubes have neck diameters similar to those of the color tubes listed above.

TUBE BASES

There are in excess of 24 different bases for monochrome and color-picture tubes. Therefore, the manufacturers diagram should be consulted in each case. Monochrome picture tubes generally have either 8- or 12-pin bases. Those tubes using 70° or 90° deflection generally have 12-pin bases. Those with 100° to 114° deflection generally have 8-pin bases.

Color-Tube Bases Color-picture tubes generally have either 13- or 14-pin bases. An example of a tri-potential color tube (MV25VEMP22) 13-pin base and its legend, is shown in Figure 18.1A. This tube is described later in this chapter.

Monochrome-picture tubes (as previously mentioned) generally have 8- or 12-pin bases. Figure 18.1B is an example of a 23TP4 monochrome-picture tube 12-pin base and its legend. The designation "13A" for the color tube and "12L" for the monochrome tube are industry standards.

PICTURE-TUBE INTERCHANGEABILITY

Picture tubes of one manufacturer may often be replaced by those of another manufacturer. However, a replacement tube may not always have the same number as the original tube. For example, a type 19VBRP22 may be directly replaced by an RCA type H-

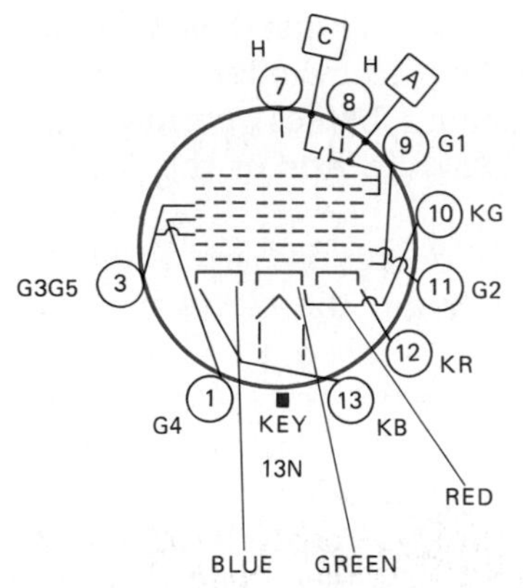

TERMINAL CONNECTIONS
PIN 1 GRID NO. 4
PIN 2 GRIDS NO. 3 & 5
PIN 7 HEATER
PIN 8 HEATER
PIN 9 GRID NO. 1
PIN 10 CATHODE (GREEN)
PIN 11 GRID NO. 2
PIN 12 CATHODE (RED)
PIN 13 CATHODE (BLUE)
A – ANODE, GRID NO. 6 (J1-21 BULB CONTACT)
C – EXTERNAL CONDUCTIVE COATING

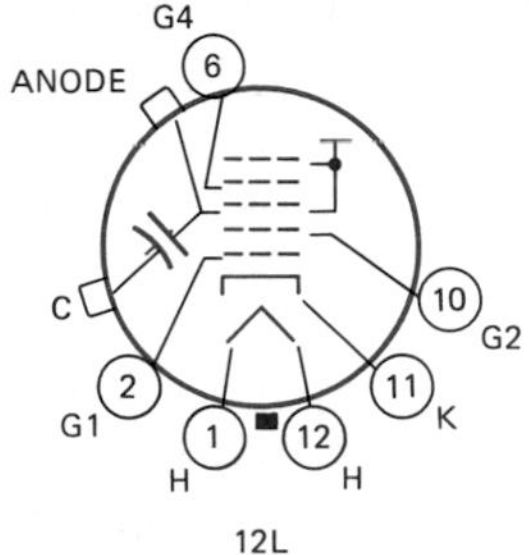

TERMINAL CONNECTIONS
PIN 1 HEATER
PIN 2 CONTROL GRID
PIN 6 FOCUS ELECTRODE
PIN 10 ACCELERATING GRID
PIN 11 CATHODE
PIN 12 HEATER
C – EXTERNAL CONDUCTIVE COATING
CL – INTERNAL CONDUCTIVE COATING
G3 + G5 + CL CONNECTED INTERNALLY

Figure 18.1A. The 13-pin base for a 25-inch color tube (tri-potential type MV25VEQP22). *(Courtesy of Sylvania Electronic Components.)* **B.** The 12-pin base for a 24-inch monochrome picture tube (23TP4). *(Courtesy of RCA Picture Tube Division.)*

19VEDP22, H-19VHBP22, etc. Thus, when it is necessary to interchange picture tubes, an interchangeability list must be consulted. These are furnished by picture-tube manufacturers and by technical publishers.

18.2 THE ELECTRON GUN

We will discuss in this section the basic electron gun as used in monochrome-picture tubes. The operating principle of the electron guns used in color-picture tubes is essentially the same and will be discussed in detail later in this chapter.

An electron gun includes (1) a heater, (2) a cathode, (3) a control grid, and (4) accelerator and focusing grids. The grids are unlike those used in vacuum tubes, in that they consist of metal cylinders having a small opening in their center.

BEAM FORMATION

The formation of the electron beam starts at the cathode. The emitting surface is composed of thoriated tungsten or barium and strontium oxides. It is restricted to a small area so that the emitted electrons progress only toward the fluorescent screen. They would serve no useful purpose in any other direction. The emitting material is thus deposited on the end of the nickel cathode cap that encloses the heater in the manner shown in Figure 18.2. After emission, the electrons are drawn by the positive-anode voltages into electric-lens systems. These

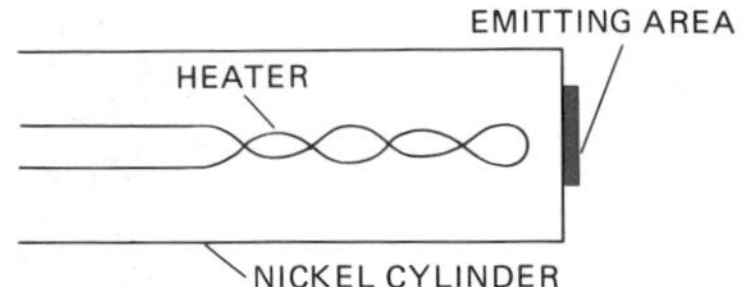

Figure 18.2 Cathode and heater construction for a picture tube.

form and focus the electrons into a sharp, narrow beam that finally impinges on the fluorescent screen in a small round point.

The use of the word "lens" may puzzle the reader who thinks of this term only in connection with light rays, not electron beams. The purpose of a glass lens is to cause light rays either to diverge from or to converge to a point. The same results can be achieved electronically; hence the reason for the carry-over of the name.

THE FIRST-LENS SYSTEM

In the first lens, we find the cathode, the control grid, and the first anode arranged in the manner shown in Figure 18.3. The grid, it is noticed, is not the familiar mesh-wire arrangement found in ordinary vacuum tubes. For the present purpose, it is a small hollow cylinder with only a small pinhole through which the electrons may pass. This restricts the area of the cathode that is effective in providing electrons for the beam and aids in giving the beam sharpness. Following the grid cylinder is the first anode. Here, again, baffles permit only those electrons near the axis of the tube to pass through.

Because of the energy imparted to them by the heated cathode, the electrons leave the cathode surface with a small velocity. With no positive electric force (or field) to urge them forward, the electrons tend to congregate in the vacuum space just beyond the cathode and form a space charge. Eventually, just as many electrons will leave the heated cathode surface as are repelled by the negative space charge. A state of equilibrium will exist. This condition can be overcome and a flow of electrons allowed to take place down the tube, if a high positive voltage is placed on the first anode.

The first anode, which is a hollow cylinder, does not have its electric field contained merely within itself. It also reaches into the surrounding regions. To be sure, the farther away from the anode, the weaker the strength of the field. With zero and low-negative potentials on the control grid, the influence of the positive-anode field extends through the baffle of the control grid right to the cathode surface. Electrons leaving this surface are urged on by the positive electric field and accelerated down the tube, with the baffle restricting their direction to very small angles with the axis of the tube.

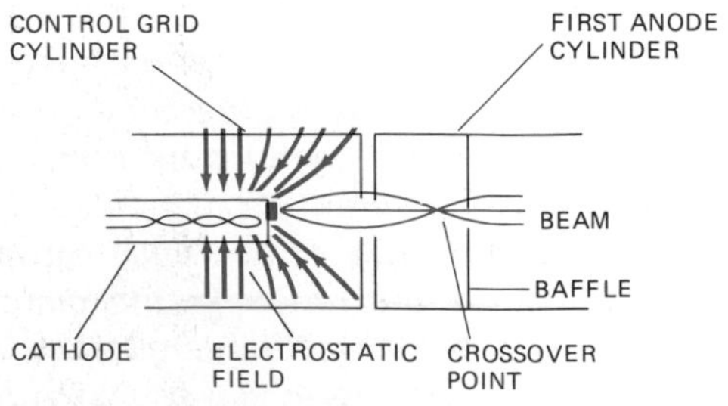

Figure 18.3 The first lens system of a picture tube.

The Crossover Point Figure 18.3 illustrates the distribution of the electrostatic lines between the cathode and control grid. It is interesting to note that these lines are not straight. They tend to curve, the amount of curvature being influenced by the distance from the first anode and the control grid and by the voltages on these elements. Picture tube design engineers use such field distribution diagrams to determine the effect of each electrode on the electrons at the cathode and in the beam.

As a result of the bending of the electric field at the cathode, it can be proved by means of vectors that all electrons passing through the small hole in the control-grid baffle will come to a focus or converge toward a small area located just inside the first anode. This region is on the axis of the tube and is known as the **crossover point.** The effect of the electric field is such that electrons near the outer edges of the control-grid opening travel at an angle in order to get to the crossover point, whereas electrons on the axis of the lens move straight forward to this point. The direction of some of the electrons is shown in Figure 18.3.

It is well to keep in mind that the shape of the electric field is determined by the placement of the electrodes and the voltages applied to them. The electrons are forced to converge toward the crossover point, because this point can more readily serve as the supply source of the beam electrons than the cathode from which they initially came. The area of the crossover point is more clearly defined than the relatively larger cathode surface, and it has been found that the electron beam is easier to focus if the crossover area is considered as the *starting point* rather than the cathode itself. The electrons that compose the final beam are then drawn from the crossover point while other electrons come from the cathode to take their place. The greater the number of electrons drawn from this point, the brighter the final image on the fluorescent screen.

For ordinary purposes, a negative bias is placed on the grid. In the larger picture tubes, the bias may be as high as −60 volts or more. With a negative voltage on the control grid, the extent of the positive electric field is modified, and it no longer affects as large an area at the cathode surface as it did previously with zero grid volts. Now, only electrons located near the center of the cathode are subjected to the positive urging force, and the number of electrons arriving at the crossover point is correspondingly less. The intensity of the final electron beam likewise decreases. In the television receiver, the video signal is applied to the control grid, and the resulting variations in potential cause similar changes in electron-beam intensity.

For the beam arriving at the screen to remain in focus once the controls have been set, the position of the crossover point must remain fixed. With the normal variations of control-grid voltage, this condition is obtained. With large variations, however, the position of the crossover point tends to change, moving closer to the cathode as the grid becomes more negative. Thus, a certain amount of defocusing will take place. Proper design generally keeps this at a minimum, and for most of the voltage variations encountered in television work, defocusing is scarcely noticeable.

To summarize the purpose of the first-lens system: we see that electrons leaving the cathode surface are forced to converge to a small area near the anode. This offers a better point for the formation of the beam and its subsequent focusing.

THE SECOND-LENS SYSTEM (ELECTROSTATIC FOCUSING)

The second-lens system draws electrons from the crossover point and brings them to a focus at the viewing screen. The system consists of the first and second anodes, as shown in Figure 18.4. The second anode is operated at a higher potential than the first anode, is larger in diameter, and frequently overlaps the first anode to some extent. It is at the point of overlap of the two anodes that the second lens is effective. It is also here that the focusing action of the electron beam takes place. Electrons, when drawn from the crossover point established by the first lens system, are not all parallel to the

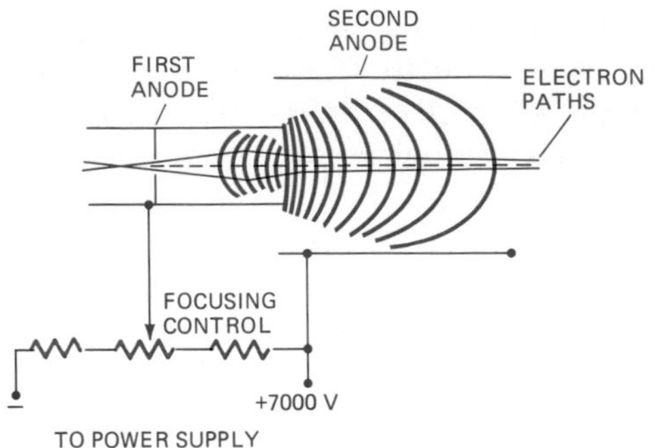

Figure 18.4 The second lens system. The focusing of the electron beam at the viewing screen is accomplished by varying the voltage at the first anode.

axis of the tube. Some leave at various small angles. The beam thus tends to diverge, and it is due to the second lens that these diverging electrons alter their path and meet at another point on the axis. This second point is at the screen. Those electrons moving straight along the axis of the tube are not affected, in direction, by the focusing action of the second lens.

Second-Lens Operation The operation of the second lens depends upon the different potentials that are applied to the first and second anodes and the distribution of the resulting electric field. The equipotential lines for this lens are drawn in Figure 18.4. It is to be noted that the curvature of these lines changes at the intersection of the two anodes. On the left-hand side, the electric-field lines are convex to the approaching electron beam, while to the right of the intersection the lines are concave. The effect of these oppositely shaped electric-field lines on the beam is likewise opposite. Since we have seen that some of the electrons tend to diverge after they leave the crossover point, the field distribution must be designed to overcome such a tendency. In action, the convex equipotential lines force the electrons to converge to a greater extent than the concave lines cause the electrons to diverge. Inasmuch as the convergence exceeds the divergence, the net result is a focusing of the electrons at the screen.

The ratio of the voltages, the size of the anode cylinders, and their relation to each other will determine the distribution and curvature of the electric lines of force; the latter, in turn, will determine the amount and the point at which the focusing takes place. In picture tubes, the ratio of the first to the second-anode voltages ranges from 3 to 1 to more than 6 to 1.

In order that the electron beam leaving the crossover point shall not diverge too much, a baffle is placed just beyond this point, similar in construction to the baffle previously described for the control grid. The baffle again limits the width of the electron beam to the desired size. Practically, focusing control can be accomplished by varying the voltage on the first anode by an arrangement shown in Figure 18.4. This is one way of altering the voltage ratio between the first and second anodes and, with it, the distribution of the electric lines of force of the lens system. An approximate optical analogy of the lens system is shown in Figure 18.5 and may prove helpful in indicating the operation of the electric system.

Focusing Methods Older picture tubes (prior to 1959) used electromagnetic focusing. This required the use of a focus coil positioned around the neck of the picture tube. Focus was obtained by varying the current

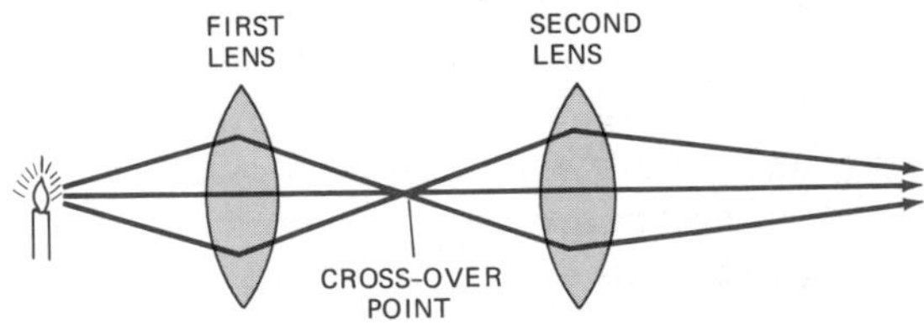

Figure 18.5 The glass lenses used in focusing light rays illustrate the similarity between light-wave and electron-beam focusing.

through the focus coil. However, all monochrome and color-picture tubes manufactured after 1959 use a form of electrostatic focusing, similar to that described above.

18.3 ELECTRON BEAM DEFLECTION

Two methods of beam deflection are in use for cathode-ray tubes, but only one method is useful for television-picture tubes. The two are: (1) electrostatic deflection, and (2) electromagnetic deflection.

With electrostatic deflection, two pairs of deflection plates are situated at the end of the electron gun nearest the screen. One pair of plates is positioned horizontally (the vertical-beam deflection plates) and the other pair is positioned vertically (the horizontal-beam deflection plates.) The electron beam passes between the two sets of deflection plates on its way to the screen. Beam deflection is accomplished by the application of suitable deflection voltages to the plates.

ELECTROSTATIC DEFLECTION

Electrostatic deflection is widely used for cathode-ray tubes in oscilloscopes. However, only electromagnetic deflection is used with TV-picture tubes. The reasons for this can be readily seen from the following:

1. TV-picture tubes utilize appreciably greater deflection angles than cathode-ray tubes used in oscilloscopes. (Up to 114° vs about 40° to 50°.)
2. TV-picture tubes utilize much higher acceleration voltages (30 000 V versus about 5 000 V).

With electrostatic deflection, an increase of deflection angle, or an increase of acceleration voltage will require appreciably higher-deflection voltages to be applied to the deflection plates. For electrostatic deflection to be used in TV-picture tubes, excessively high-deflection voltages would have to be generated.

ELECTROMAGNETIC DEFLECTION

On the other hand, electromagnetic deflection, which employs magnetic-deflection coils (the yoke) offers a more practical solution for TV-picture tubes. In this system, the required deflection currents can be generated at reasonable voltage levels. There are also other advantages to using electromagnetic deflection and these will be discussed here and in Chapters 22 and 23.

Magnetic Theory To better appreciate the manner in which magnetic deflection can take place, it will be well to briefly review some basic magnetic theory.

It is well known that a wire carrying a current has a circular magnetic field set up around it, as shown in Figure 18.6A. Suppose the wire is placed in a magnetic field parallel to the magnetic lines of force (see Figure 18.6B). There will be no interaction between the magnetic lines of the field and those set up by the wire. Why? Because the two *fields* are at right angles to each other.

For the opposite case, illustrated in Figure 18.6C, the current-carrying wire is placed at right angles to the field lines of magnetic force. Above the wire the lines of both fields add; underneath the wire they oppose and tend to cancel each other. Experiment indicates that a resulting force will act on the wire in such a way that it moves from the stronger part of the magnetic field to the weaker part. This is indicated in the figure. The illustration represents the two extreme angles that the wire and the field can make with each other. Intermediate positions

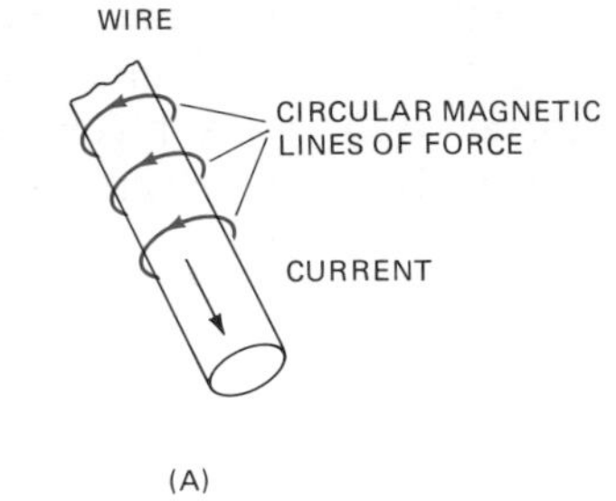

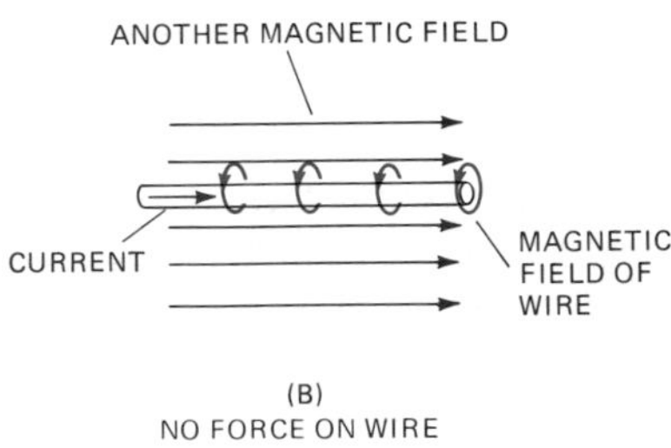

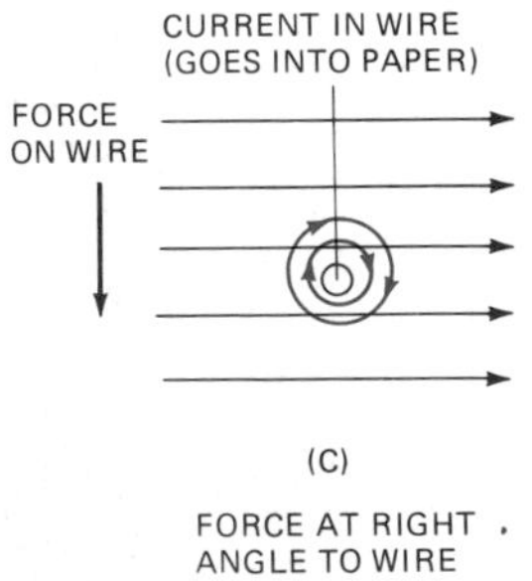

Figure 18.6A. The magnetic field of a wire carrying a current. **B.** When the wire is placed parallel to a magnetic field, no force is exerted on the wire. **C.** When the wire is perpendicular to a magnetic field, the wire is subjected to a force.

(those between 0° and 90°) will cause intermediate values of force to act on the wire.

Force On An Electron Beam The transition from a wire carrying electrons to the electrons themselves without the wire is quite simply made. With only electrons moving through space, the same circular magnetic field is set up about their path. From the preceding discussion we know that electrons traveling parallel to the lines of force of an additional magnetic field experience no reaction from this field. On the other hand, if they enter the magnetic field at an angle to the flux lines, a force will be brought to bear on them and their path will be altered.

It is well to reiterate that for an electron to react with a magnetic field: (1) The electron must be moving, otherwise it does not generate a magnetic field; and (2) the moving electron must make an angle with the magnetic field in which it is traveling.

SUMMARY OF ELECTROMAGNETIC DEFLECTION

Little additional information needs to be added at this point to understand the action of the deflection coils on the electron beam. Two sets of coils are placed at right angles to each other and mounted on the section of the tube neck where the electron beam leaves the focusing electrode and travels toward the screen. There are four coils in all (two in each set), with opposite coils comprising one set. These are connected in series in order to obtain the proper polarity (see Figure 18.7).

At the deflection coils, the magnetic fields are at *right angles* to the path of the beam. The beam, in moving through these fields, has a force applied which is at right angles to the forward motion of the electrons and the direction of the magnetic lines of force. The influence of the field ends when the electrons pass the yoke. Any sideward or up-and-down motion imparted to the electrons while in the field is retained. By varying the direction of the flow of current through the vertical and horizontal-deflection coils, it is possible to reach all points on the screen. This type of deflection is used with all present-day television-picture tubes.

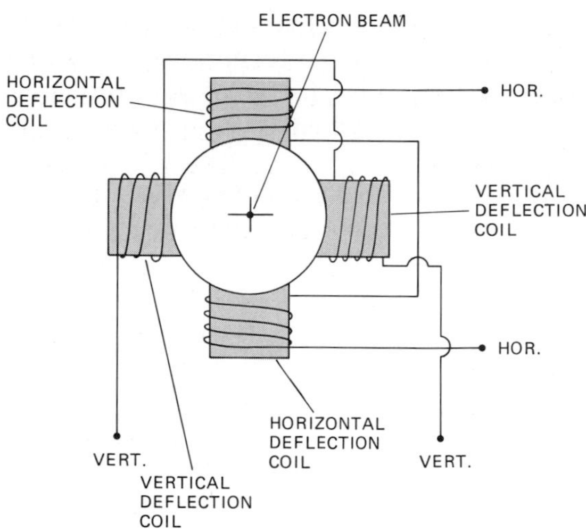

Figure 18.7 Arrangement of coils for electromagnetic deflection.

Physical Placement Figure 18.8 shows the actual physical placement of the deflection coils. For horizontal deflection, the coils are vertically placed, whereas, for vertical deflection, the coils are horizontally mounted. This reverse placement of the coils is due to the fact that the force on traveling electrons

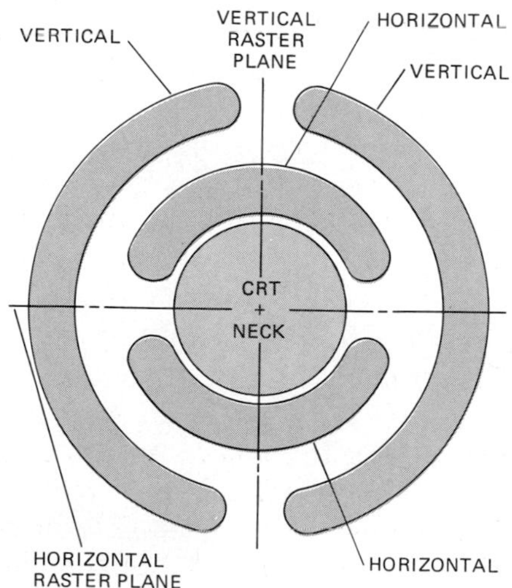

Figure 18.8 The actual physical placement of the deflection coils about the neck of the picture tube.

in a magnetic field is at right angles to both the direction of the motion and the lines of the field. After the coils have been oriented, sawtooth-shaped current variations are sent through them. The magnetic-field flux follows these current changes and causes the electron beam to move back and forth (or up and down) across the screen, sweeping out the desired pattern.

Yoke Assembly The entire assembly of deflection coils is known as a "deflection yoke." Two typical commercial units are shown in Figure 18.9. Note how the forward windings lap over the front edge of the yoke housing. The yoke is thus positioned right up against the flare of the tube in order to achieve the complete coverage of the full

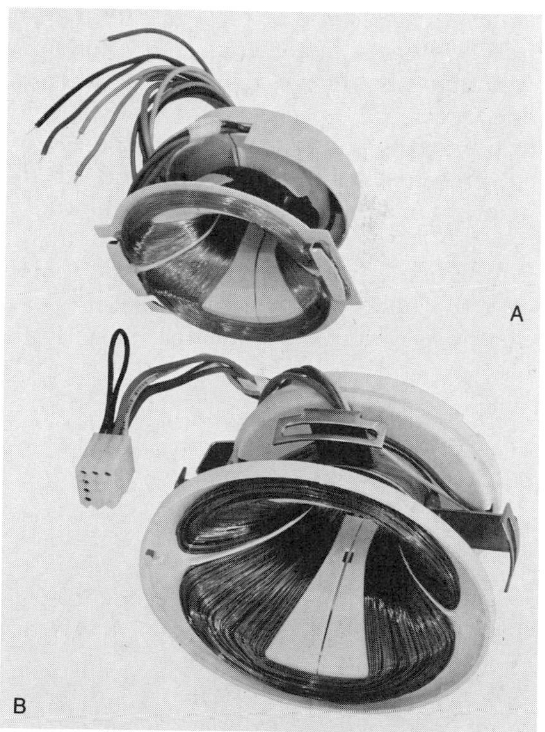

Figure 18.9 Two typical deflection yokes. **A.** A 90° deflection yoke. **B.** A 110° deflection yoke. Wider angles of deflection make possible shorter picture tubes. *(Courtesy of Triad-Utrad.)*

screen area. This is particularly important for wide-angle tubes (110° or more).

The deflection windings in Figure 18.8 are shown wound uniformly; that is, there is no variation in the winding thickness from end to end. This type of winding was characteristic of the yokes employed when narrow-angle picture tubes were prevalent. As the deflection angle increased, it was found that the magnetic field produced was not uniform, particularly when the beam was deflected toward the edges of the raster. Visually, an elongated spot was produced, tending to develop an out-of-focus condition.

A more uniform field is developed when a cosine-type winding is employed as in Figure 18.9. In this arrangement (Figure 18.10) the thickness of a deflection winding varies as the cosine of the angle from a central reference line varies. For horizontal windings, the reference line is the horizontal line through the center of the yoke. For vertical windings, it is the vertical line. Nearly all present-day yokes are wound in this manner, or in a cosine-squared fashion.

Yoke Positioning When a yoke is inserted over the neck of the picture tube, it is very easy to position it so that the image is not properly oriented. This is indicated in Figure 18.11. In this case, correction may be accomplished by rotating the yoke until the image is properly positioned.

Yoke Parameters Some important parameters of the two yokes shown in Figure 18.9 are as follows:

A. For the 90° yoke:

	Inductance	DC Resistance
Horizontal	1.40 mH	1.15 ohms
Vertical	5.65 mH	3.30 ohms

B. For the 110°–114° yoke:

	Inductance	DC Resistance
Horizontal	24 mH	42 ohms
Vertical	67 mH	28 ohms

Note the higher inductance values for the 110°–114° yoke. This is required to develop the stronger magnetic fields for wide-angle deflection.

18.4 ELECTROMAGNETIC DEFLECTION WITH ELECTROSTATIC FOCUS

This combination is commonly used for both monochrome and color-picture tubes.

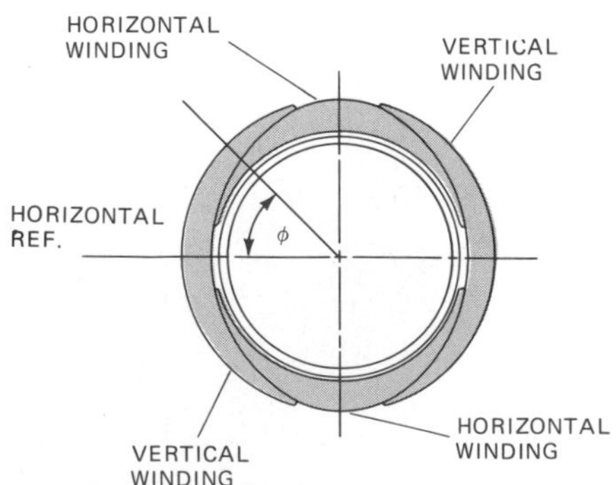

Figure 18.10 The appearance of cosine windings in a deflection yoke. Note the uneven thickness of each winding.

Figure 18.11 Image incorrectly positioned because of the improper placement of the deflection yoke.

Electrostatic focusing is particularly suited to wide-angle tubes, because the beam produced by the gun is smaller in diameter and there is less defocusing when the beam is swung to the edges of the raster. Also, this form of focusing is useful to short guns by eliminating the external focus magnet.

Electrostatic-Focus Gun A cross-sectional view of an electrostatic focus gun is shown in Figure 18.12. The labeling of the various elements in the gun assembly as consecutive grids is done frequently to simplify identification. Thus, the control grid is Grid Number 1. The first anode following this is Grid Number 2. The second anode is Grid Number 3, etc. Note that this change in name does not in any way alter the function or construction of the electrodes. Grid Number 3 serves as the accelerating anode. It contacts and operates at the high potential of the aquadag coating.

The elements, Grids Number 4 and Number 5, provide the focusing field which directs the electron beam. The voltage applied to Grid Number 4 is lower than that which Grids Numbers 3 or 5 receive. It is frequently made variable to permit the adjustment of the focus voltage to the proper value. Grid Number 5 (which structurally surrounds Grid Number 4) is connected internally to Grid Number 3 and operates at the same potential as Number 3.

Focus Voltage The voltage applied to Grid Number 4 depends upon the manner in which Grids Numbers 3, 4, and 5 are constructed. The first monochrome electrostatic-focus tubes manufactured required that the potential of Grid Number 4 be on the order of 20% of the accelerating (or anode) voltage. (This is still required in color tubes.) This meant that voltages between 2 000 and 3 000 V had to be made available. A special potentiometer, inserted in this circuit, permitted the adjustment of this voltage for the sharpest picture focus.

In subsequent designs of monochrome tubes, it was found that by constructing

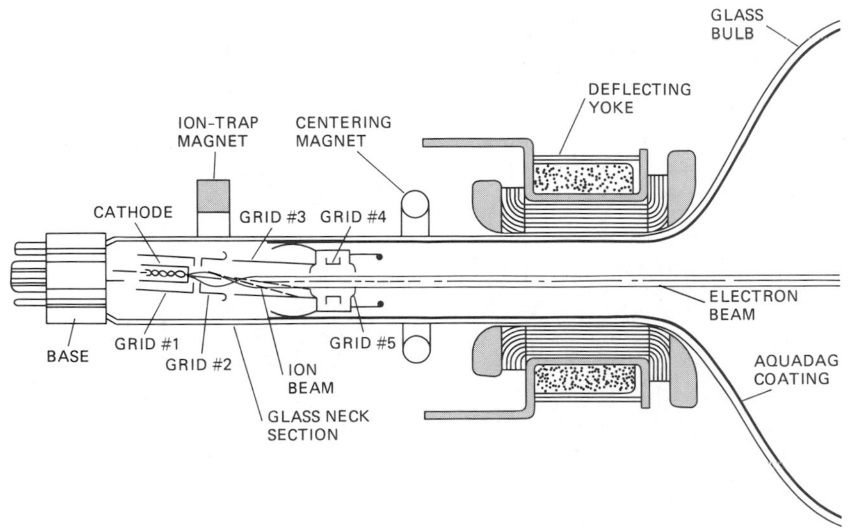

Figure 18.12 Internal structure of an electrostatic focus tube. The deflection is accomplished magnetically.

Grids Numbers 3, 4, and 5 to closer tolerances, the necessary focusing action could be obtained with voltages on the order of 150 to 300 V. These latter values can be obtained directly from the low-voltage or boost power supply. The special circuit required when several thousand volts are needed is thereby avoided.

With low-voltage focusing, the exact voltage is not critical. Therefore, no focus control is provided. However, the desired focus voltage may be preset at the factory.

Zero-Voltage Focus By modifying the structure of Grids Numbers 3, 4, and 5, for monochrome picture tubes, to the configuration shown in Figure 18.13, it is possible to obtain focus with zero voltage on Grid Number 4. With this design no focus voltage is required and the focus grid may be simply grounded. This scheme has the obvious advantage of simplicity but may not provide the sharpest focus in all tubes.

Centering Magnets In order to center the picture on an electrostatic focus, monochrome-picture tube, two permanent-magnet rings are utilized. These are placed around the neck of the picture tube about $\frac{3}{8}$ in (0.93 cm) behind the deflection yoke (See Figure 18.12B). As shown in the figure, tabs are provided for the adjustment (rotation) of the magnet rings. By spreading the tabs apart and rotating both rings, the picture may be correctly centered.

Centering magnets are not used with color-picture tubes. When this type of tube is correctly adjusted, centering is automatically achieved.

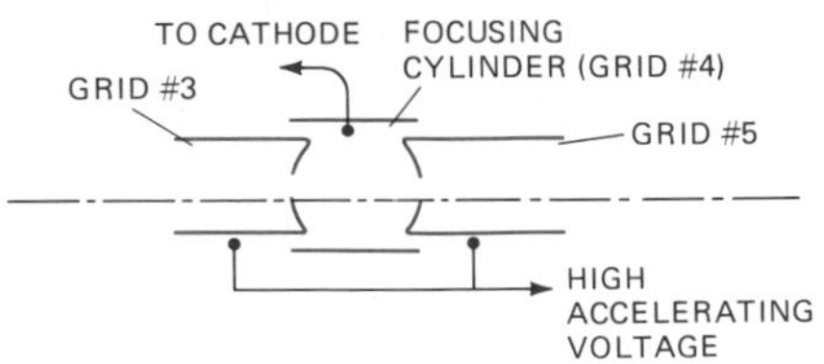

Figure 18.13 The modified arrangements of grids numbers 3, 4, and 5 in the zero-voltage focus tube.

18.5 ION SPOTS

The requirement for the ion-trap magnet shown in Figure 18.14, will now be briefly discussed. Note that these magnets have been used only with older-picture tubes. More recent picture tubes have an aluminum coating over the screen phosphor, which eliminates the requirement for the magnet and special electron-gun combination.

Effect of Ions Another important matter is the elimination of the ion spot in tubes using electromagnetic deflection. No matter how carefully a tube is degassed or how well a cathode-coating is applied, ions will be found in the electron beam. These ions are either gas molecules which have acquired an electron or else molecules of the outside coating material of the cathode. These ions possess the same charge as the electrons and are sensitive to the same accelerating voltages. In tubes employing electrostatic deflection, the ions and the electrons are similarly deflected. For all practical purposes, they may be considered as one. However, when

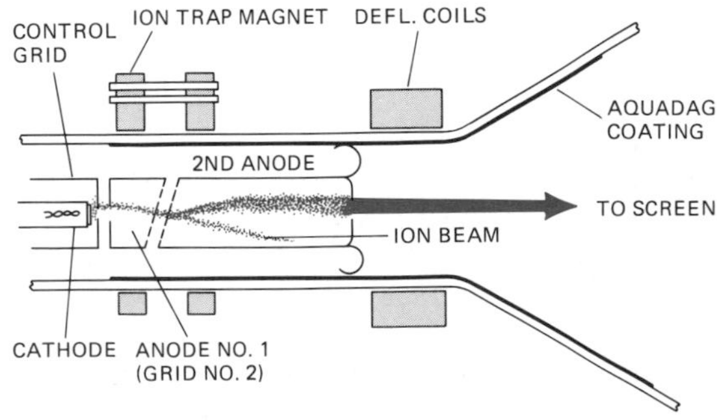

Figure 18.14 A simplified drawing illustrating the principle of the diagonal-cut (or slash-field) ion trap.

electromagnetic deflection is employed, these heavier ions are hardly deflected. As a result, they strike the center of the screen in a steady stream. In time, they deactivate the fluorescent material in this area. When the electrons in the scanning beam subsequently pass over this section of the screen, no light is emitted. To the observer this section appears as a dark patch.

Diagonal-Cut Ion Trap One approach to the prevention of ion spots is the diagonal-cut ion trap (See Figure 18.14). The electrons and ions are emitted by the electron gun and are accelerated forward. The first and second anodes are so designed that the gap between them is oblique. The first anode has a low positive voltage; the second anode has a high positive voltage. The electrons, as they leave the cathode, are attracted forward by the first anode. However, the oblique gap between the first and second anode causes the electric field here to become warped, and the electrons and ions crossing the gap are bent in toward the second anode. With no other forces applied, the electrons and ions will strike the second anode and will be prevented from reaching the screen.

However, if a magnetic field is introduced at right angles to the electrode, the electrons receive a counterforce deflecting them upward. This permits them to continue through the gun. The ions, because of their greater mass and because the magnetic field scarcely deflects them, strike the second anode and are removed from the beam path. This magnetic field is furnished by the ion-trap magnet shown in Figure 18.14. When the magnet is correctly adjusted, maximum brightness with a full raster is present. If incorrectly adjusted, in the worst case, no brightness will appear.

Aluminized Screens As mentioned briefly above (and in greater detail subsequently) in all fairly recent tubes, a thin layer of aluminum is deposited upon the phosphor. This is used for both monochrome and color tubes.

The depth of penetration of any particle is governed by the relationship

$$\text{where depth of penetration} = \frac{K\,V_{e},}{m}$$

- K = Constant
- V_e = Energy of particle
- m = Mass of particle

Since an ion has considerably more mass than an electron, its depth of penetration is less. By properly proportioning the thickness of the metallic screen, the ions are excluded. However, the electrons in the beam are able to pass through the aluminum virtually unimpeded.

18.6 PROBLEMS IN OBTAINING BRIGHTNESS AND CONTRAST

The principal objective in the design of a picture tube is the production of an image having good brightness and high contrast. When the electron beam strikes the back side of the fluorescent screen, the light which is emitted distributes itself in the following approximate manner:

50% of the light travels back into the tube.
20% of the light is lost in the glass of the tube by internal reflection.
30% of the light reaches the observer.

Thus, of all the light produced by the electron beam (and this, itself is a highly inefficient process), only 30% reaches the observer.

Image contrast is impaired because of the interference caused by light which is returned to the screen after it has been reflected from some other points. Some of these sources of interference are given here in the order of their importance:

1. Halation
2. Reflections due to the curvature of the screen

3. Reflections at the surface of the screen face
4. Reflections from inside the tube

Halation If we take a cathode-ray tube and minutely examine the light pattern produced by a stationary electron beam, we find that the visible spot is surrounded by rings of light. These rings of light are due to a phenomenon known as **halation** (See Figure 18.15). The light rays which leave the fluorescent crystals at the inner surface of the tube face travel into the glass and are refracted. Those rays which make an angle greater than θ do not leave the glass when they reach the outer surface, but instead they are totally reflected back into the glass. At each point where these reflected rays strike the fluorescent crystals they scatter. It is this scattering of the rays that produces visible rings on the screen. These rings cause a hazy glow in the region surrounding the beam spot and reduce the maximum possible detail contrast. Contrast, it will be recalled, is the ratio of the brightness of two points, one of which is being bombarded by the electron beam, the other of which is under cutoff conditions. It is desirable to have this ratio as high as possible in order to achieve "rich-looking" or high-quality images. Due to the scattering of the light, however, areas which should be in total darkness receive some light, and the result is a reduction in the contrast ratio. A distinction is usually made between the detail-contrast ratio and the overall field contrast. The field-contrast ratio compares two sections of the screen which are widely removed from each other. Halation affects only the detail contrast.

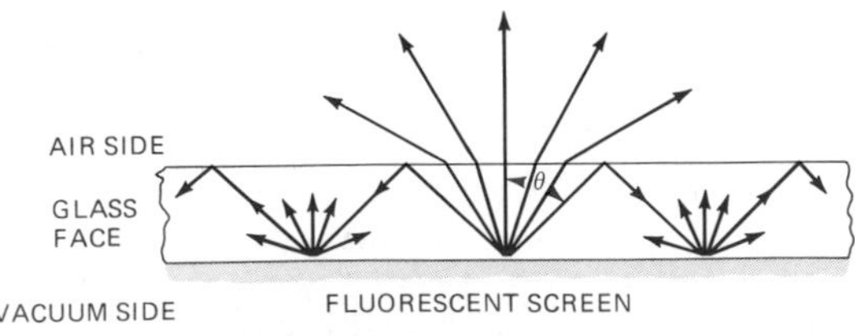

Figure 18.15 Reflections between the two surfaces of the glass can cause halation.

Reflections due to the Curvature of the Screen Reflection arising from the curvature of the screen, as shown in Figure 18.16, causes loss of contrast. The remedy is the use of a flat screen. Much progress has been made in this direction, since the screen curvature greatly restricts the useful image area.

Reflections at the Surface of the Screen Face Light rays, when traveling from one medium to another, always lose a certain amount of energy at the intersection of the two media. At the cathode-ray tube screen, some light is reflected when it reaches the dividing surface between the air and the glass of the tube. The reflected light travels back to the inner surface and then back to the outer surface again. At each dividing surface, some of the light continues onward and some is reflected back into the glass. Absorption and dispersion quickly reduce the strength of these rebounding rays.

Reflections from inside the Tube In Figure 18.17 we see how reflections from the inside surfaces of the tube can act to decrease the field contrast of the image. The loss of contrast from this source of interference can be made quite low by a special shaping of the walls of the bulb, as seen in Figure 18.17, and the use of the black aquadag coating.

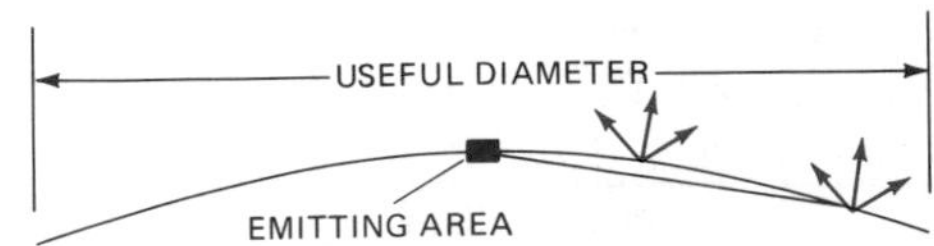

Figure 18.16 Diffusion effects in non-flat screen.

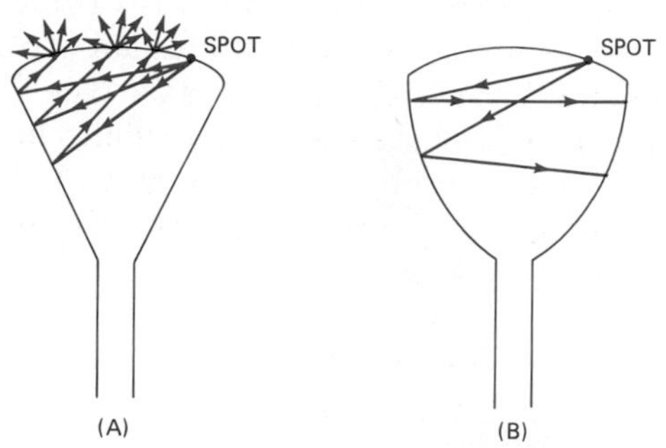

Figure 18.17 The shape of the tube can reduce the internal reflections.

The aquadag coating is also useful for electrical purposes. It acts as a shield and a path for the return of the secondary electrons emitted from the fluorescent screen. Secondary electrons must be emitted by the screen. If not, the negative charge accumulation on the screen would soon become great enough to prevent the electron beam from reaching it.

One step taken toward improving screen brightness and contrast has been the addition of an extremely thin film of aluminum on the back of the fluorescent screen. The film is sufficiently thin to permit the electrons in the scanning beam to reach the fluorescent crystals. It will prevent, however, any of the light which is generated by the screen crystals from traveling back into the tube. This can be seen in Figure 18.18. The light which previously went back into the tube is now reflected toward the observer. This is one improvement. In addition, the overall field contrast is improved as much as ten times. However, the detail contrast is not noticeably affected, since it is governed primarily by halation, and the addition of the aluminum layer does not affect this condition. An additional purpose which the aluminum film serves is to prevent undesirable effects due to poor secondary emission from the screen. It also greatly increases the range of substances which can be used as screen phosphors.

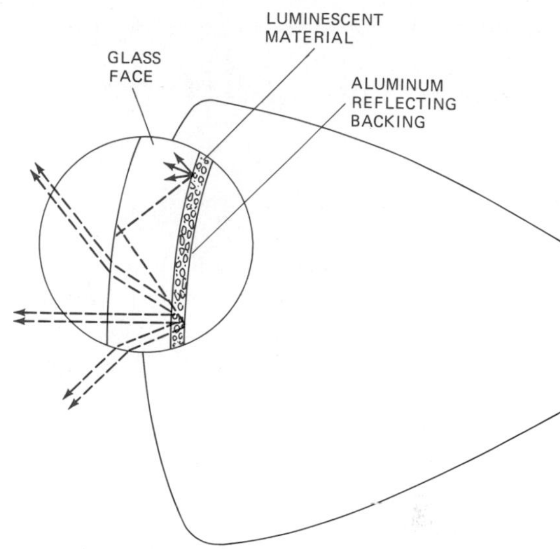

Figure 18.18 An aluminum backing over the fluorescent screen prevents light from traveling back into the tube.

18.7 RECTANGULAR SCREENS AND TUBE-SAFETY SHIELDS

Rectangular Screens It had been recognized for many years that a rectangular image on a circular screen is wasteful of screen area and cabinet space. The sensible solution was a rectangular screen dimensioned in the standard 4:3 ratio of the transmitted image. At first, manufacturing difficulties and obstacles presented by the rectangular shape of the tube prevented mass production. In time, however, these were overcome and rectangular tubes are now used exclusively.

A rectangular tube with a 21-in (53.3 cm) screen is illustrated in Figure 18.19A. This has a fairly long neck and a conventional socket base. A 110° picture tube possessing a shortened neck and a modified plastic base is pictured in Figure 18.19B. With the deflection yoke in place, as shown, adequate room remains for the centering rings.

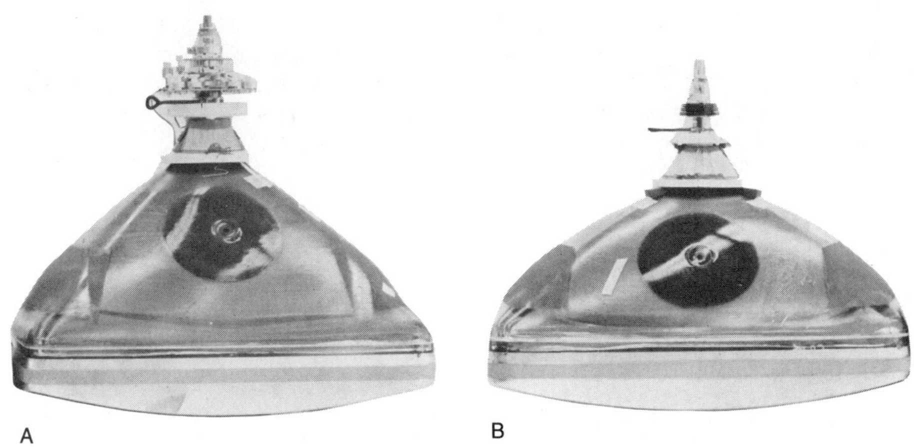

Figure 18.19A. Rectangular 25-in picture tube. The deflection angle is 90°. **B.** A 110° picture tube. Note the shortened neck, and a modified plastic base. *(Courtesy of GTE Sylvania.)*

Tube Shields In older TV receivers, a plate-glass shield is mounted close to and in front of the picture tube as a safeguard to the viewers against possible serious injury should the tube implode. This shield, however, is not always easily removable. This makes it difficult to wipe dust or fog from either the tube face or the inner shield surface. There is also a loss of contrast stemming from light reflections between the shield and the tube.

The separate safety shield has long been replaced by a flat sheet of glass possessing the same contour as the external surface of the tube screen and permanently bonded to this screen by an epoxy resin (See Figure 18.20). The result is a virtually implosion-proof tube that is easy to keep clean, pro-

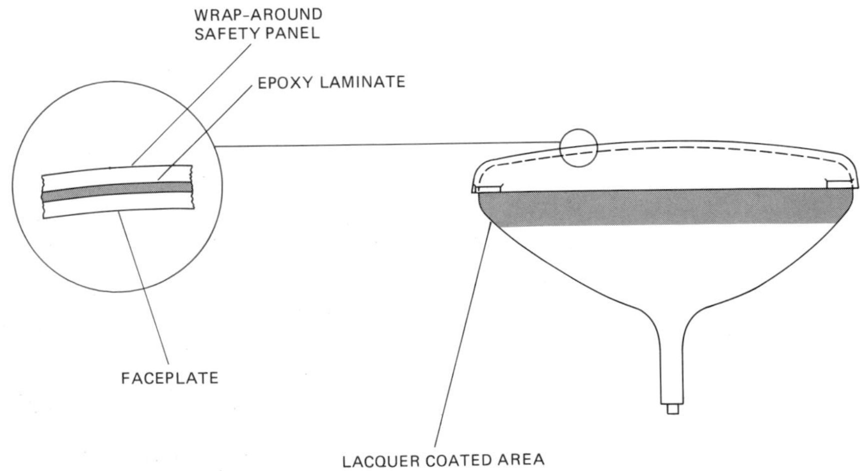

Figure 18.20 Cross section of tube possessing a bonded shield. Lacquer band protects critical area of the bulb behind the wrap-around panel.

duces a brighter, sharper image, and permits a picture shaped more nearly rectangular. Finally, the use of this technique has made it possible to design a bulb having a flatter face. This reduces distortion at wide-viewing angles.

18.8 COLOR-PICTURE TUBE WITH A DELTA GUN AND COLOR-DOT TRIADS

A color-picture tube using a delta-electron gun and color-dot phosphor triads on the screen are described in this section. There are also two other basic designs of color-picture tubes. These are (1) the three-gun, in-line type, using vertical color stripes and (2) the one gun (three beam), in-line type, which also uses vertical-color stripes. A simple comparison of the three basic types is shown in Figure 18.21. Details of all three systems will be given as we progress in this chapter.

BASIC DELTA-GUN TUBE OPERATION

(Refer also to Figure 18.21A) A delta-gun color-picture tube, using color-phosphor dots is shown in Figure 18.22. This illustration should be referred to in subsequent discussions.

The screen of the tube possesses three different color-emitting phosphors. This, of

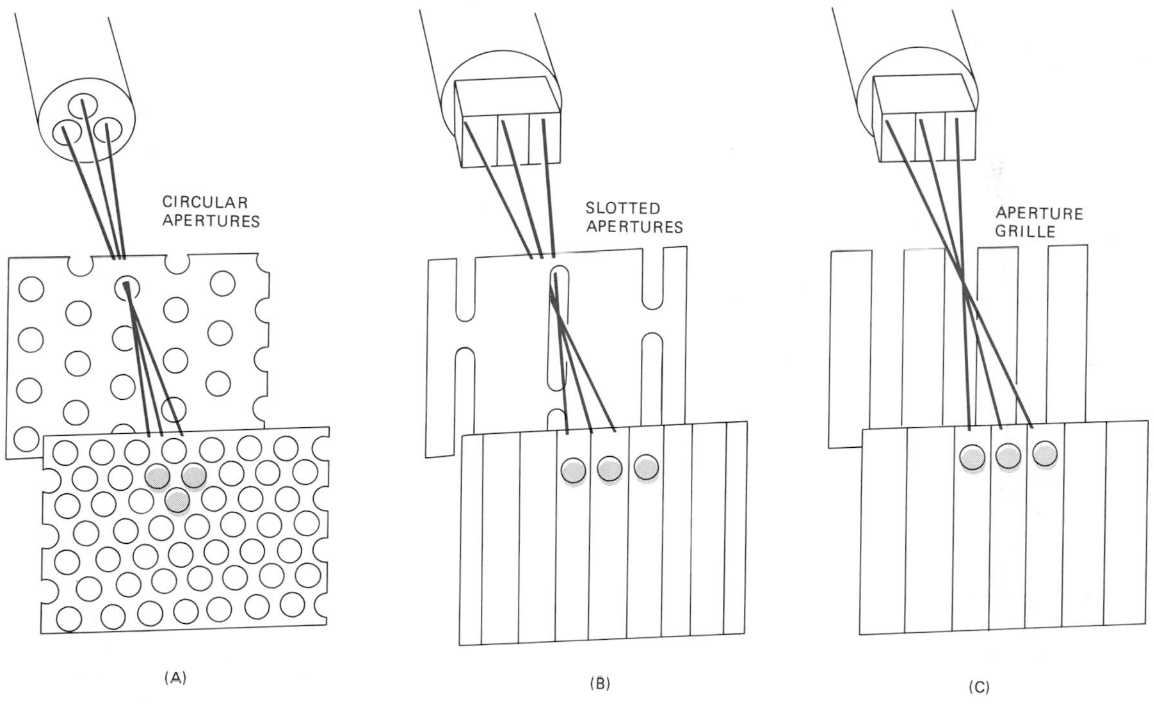

Figure 18.21A-C. The three basic types of color picture tubes. **A.** Delta-electron gun used with circular-aperture mask and color phosphor-dot trios. **B.** The RCA in-line gun, used with slotted-aperture mask and vertical color-stripe trios. **C.** The Sony Trinitron in-line gun, used with aperture-grille mask and vertical color-stripe trios.

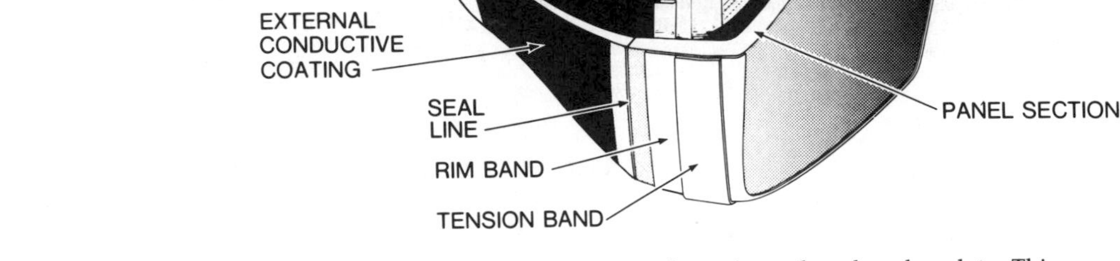

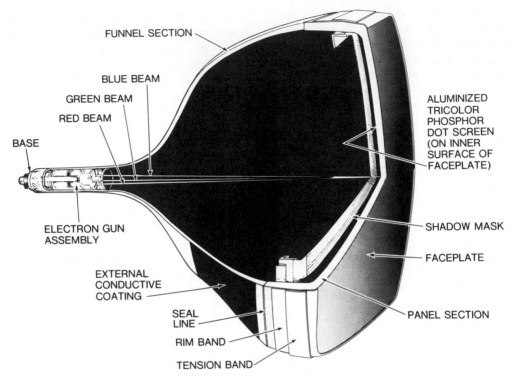

Figure 18.22 A cutaway view of a delta-gun color picture tube, using color phosphor dots. This is a type 25VEHP22. *(Courtesy of Zenith Radio Corp.)*

course, is basic to the entire color-television system which employs the three primary colors—red, green, and blue—to synthesize the wide range of hues and tints required for the satisfactory presentation of a color picture.

Each gun is concerned with one type of phosphor. Thus, one of the electron guns develops an electron beam which strikes, say, the phosphor which emits red light. This gun is labeled the "red gun." A second electron gun directs its beam only at the green phosphor dots and it is the "green gun." The third gun is concerned in similar manner only with the blue dots. In each case, it is not the color of the phosphor referred to, but the light which this phosphor gives off when struck by an electron beam. The actual color of the substance and the color of the phosphorescent light it emits do not necessarily bear any relationship to each other.

Producing Colors The overall color that is seen on the screen is determined by two general factors (1) the phosphors which are being bombarded by the three guns, and (2) the number of electrons which are contained in each beam. Thus, suppose you turn one beam off completely—say, the beam from the red gun. Then, only the blue and green dots will be emitting light, and what you see is a mixture of blue and green light which can range from a greenish blue to a bluish green. This color is called "cyan." The exact color is determined by which of the two beams is the stronger.

In the same way, we could cut off the

green gun, leaving only the red and blue guns in operation. Now the screen color would fall somewhere in the purplish range and this color is known as "magenta." If the blue gun were stronger than the red gun, the color would appear closer to blue, say bluish purple. On the other hand, if the red beam were made more intense, the resultant color would be nearer a purplish red.

It is, of course, not necessary to turn any gun off. All three may be operating simultaneously. When they do, you generally see the lighter or pastel shades on the screen. This is because red, green, and blue combine in some measure to form white. Although white may not be predominant, it will mix with whatever colors are present and serve to lighten, or *desaturate* them.

Dot Triads The phosphorescent dots which produce the colored light are arranged on the screen in any orderly array of small triangular groups, each group containing a green-emitting dot, a red-emitting dot, and a blue-emitting dot. The combination of the three dots is called a triad (See Figure 18.23). The actual number of such dots, for a 21-in screen, like the one in Figure 18.24, is somewhere in the neighborhood of 1 071 000.

With 1 071 000 dots on the screen, there are 357 000 triads. Each dot has a diameter of approximately 16 mils (.016 in). If all three dots in a group are bombarded at the same time, the combined red-, green-, and blue-light output will present one mixed color to the observer's eyes.

ELECTRON-GUN STRUCTURE

At the other end of the color-picture tube, there are three parallel, closely-spaced electron guns which produce three independent electron beams (See Figure 18.25). Each gun consists of a heater, a cathode, a control grid (Grid Number 1), an accelerating (or screen) grid (Grid Number 2), a focusing electrode (Grid Number 3), and a converging electrode (Grid Number 4). The heaters of all three guns are in parallel and require only two external connections to the tube base. Each grid has its own base pin. The focusing electrodes (Grid Number 3) of all the guns are electrically connected, because one overall voltage variation will bring all three beams to a focus at the phosphor dot screen.

The final electrode in the gun structure is Grid Number 4, the converging grid. This is a cylinder of a small diameter, which is internally connected to (and operated at the same high potential as) the aquadag coating

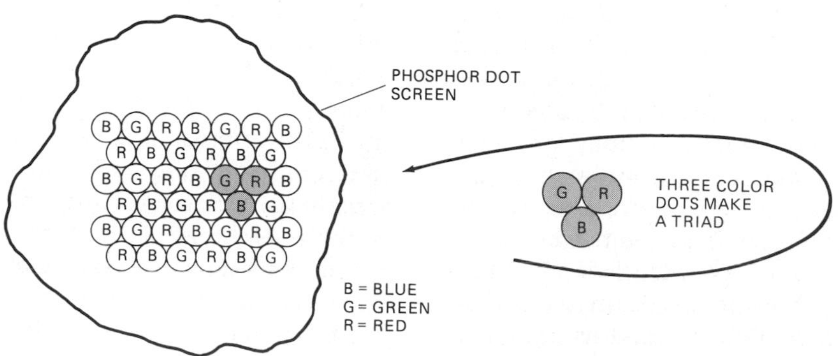

Figure 18.23 The phosphor dot screen of a delta-gun color picture tube.

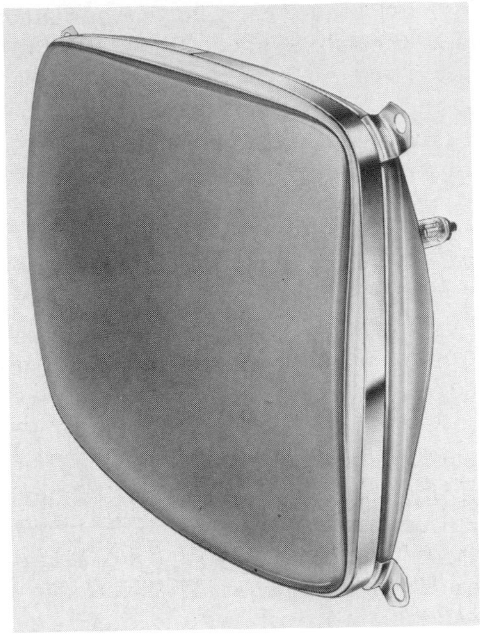

Figure 18.24 A 21-in color picture tube. *(Courtesy of RCA.)*

Figure 18.25 Electron-gun assembly of a 21-inch color picture tube. Three electron-gun structures are employed, although only two are clearly visible.

(22 000 to 32 000 volts). Also associated with each Number 4 Grid is a pair of pole pieces. These are mounted above each grid. External coils on the neck of the tube induce magnetic fields in each set of pole pieces, as shown in Figure 18.26. These fields force the three beams to converge, so that each beam will strike the proper phosphor dot of a triad at any one instant of time. That is, one beam will strike the red dot, a second beam will strike the green dot, and the third beam will hit the blue dot, all three dots being in the same triad. The three dots are bunched so close together that the light they produce combines and appears to the eye as a single color. In the absence of the proper converging action, it is possible for the beams to hit phosphor dots at sufficiently separated points so that an observer sees three individual points of light. Under these conditions, proper mixing of colors to obtain different hues is not possible.

THE SHADOW MASK

Proper beam convergence is an important aspect of delta-gun picture-tube operation. Thus, to insure that each beam strikes only one type of phosphor dot, a mask, called a **shadow** or **aperture mask,** is inserted between the electron guns and the phosphor dot screen (See Figure 18.27). The mask is positioned in front of and parallel to the screen. It contains circular holes, equal in number to the dot triads. Each hole is so aligned with respect to its group that any one of the approaching beams can "see" and therefore strike only one phosphor dot. The remaining two dots of the triad are hidden by the mask; that is, the two other dots are in the "shadow" of the mask opening—hence the name of shadow mask.

What is true for one beam is true for the other two beams. Each can also see one phosphor dot. In this way, it is possible to

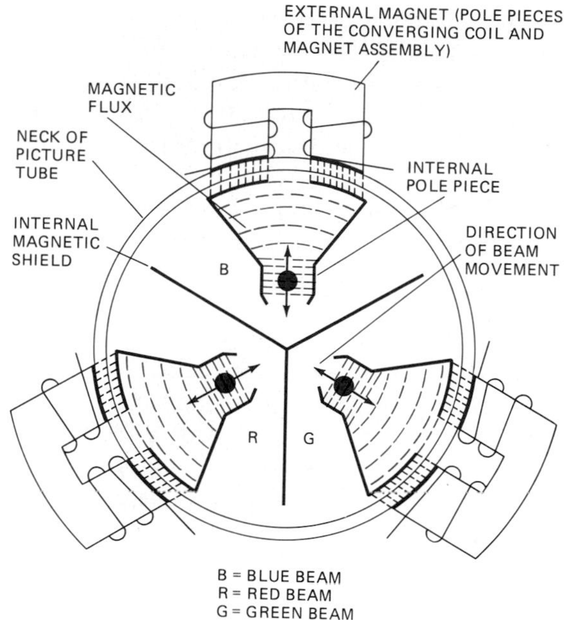

Figure 18.26 External coils mounted on the neck of the picture tube induce magnetic fields in each set of pole pieces. These fields force the three beams to converge, so that each beam will strike the proper phosphor dot in each triad.

minimize color contamination which occurs when a beam either hits the wrong dot or overlaps several dots at the same time.

STATIC AND DYNAMIC CONVERGENCE

In the foregoing discussion, beam convergence was covered in a general manner. Actually, there are two types of convergence: **static convergence** and **dynamic convergence.** In static convergence the positions of the beams are adjusted by using either fixed-DC voltages or fixed-magnetic fields. As a further aid in this action, the electron guns are tilted inward slightly. If the adjustments are made carefully, the beams will converge properly over the central area of the screen.

Dynamic Convergence To maintain this converged condition of the beams as they swing away from the center, it is also nec-

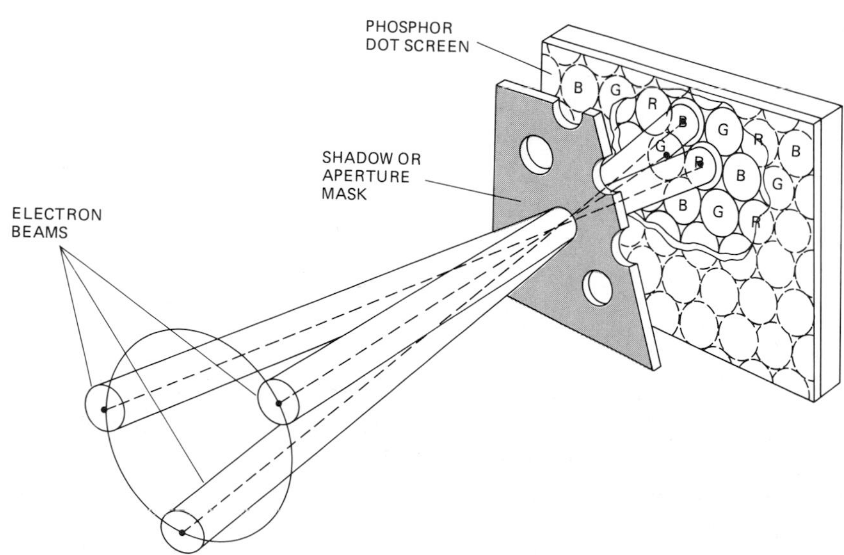

Figure 18.27 Diagram of the mask and screen, showing the convergence of the three beams at a single hole in the shadow (aperture) mask. Note that the converged beams pass through the hole and strike their respective phosphor dots.

essary to vary their relative angles slightly. This process of changing the beam angle so that it will be in step with the scanning is referred to as **dynamic convergence.** It is required because the distance traveled by the beams increases as they swing away from the center of the screen. The swing away from the center, in turn, occurs because the curvature of the screen is not perfectly spherical. Beams which are converged at the screen center will tend to converge in front of the shadow mask at points away from the center (See Figure 18.28).

A moment's reflection will reveal that the extent of convergence changes, the farther the beams are from the center of the screen. Furthermore, there is a direct relationship between the convergence needed at any one point and the instantaneous horizontal- and vertical-deflection current values. Thus it is possible to obtain whatever correction currents are needed from the vertical and horizontal deflection systems. These additional currents are known as **dynamic convergence currents** to distinguish them from the DC or static convergence which is made over the *central area*. Where dynamic-magnetic convergence means are employed, the static adjustment is usually made with permanent magnets. The dynamic convergence is then achieved by introducing varying magnetic fields via convergence coils mounted on the neck of the picture tube.

Convergence Current The basic form of the dynamic convergence correcting current is parabolic, as pictured in Figure 18.29. When the three beams are in the center of the screen, the correction current is zero. On either side of the center, however, the current varies, and the combined effect of the correction (that is, dynamic and static fields) is to keep the beams properly converged at every point of the screen.

On some receivers the convergence circuitry and controls are contained on a separate, small panel which is easily removed

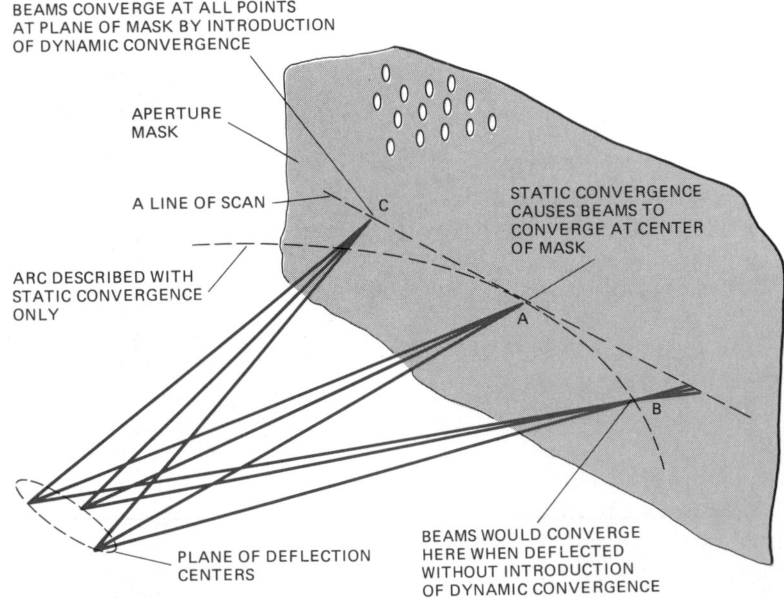

Figure 18.28 Dynamic convergence voltages are required to cause the three beams to converge at the picture edges because the screen and the aperture (or shadow) mask are not perfectly spherical.

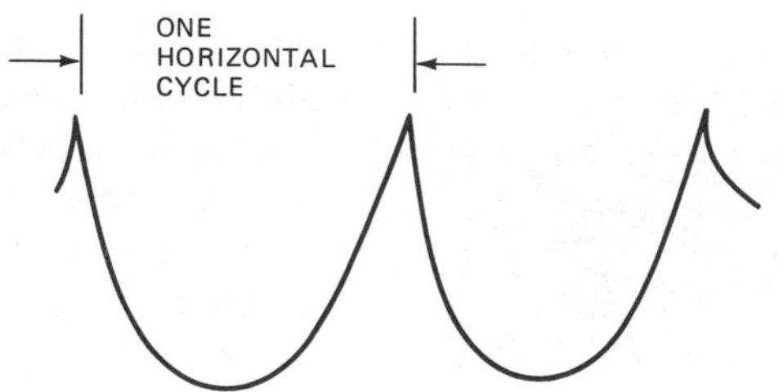

Figure 18.29 The form of the dynamic convergence current is parabolic.

and positioned for convenience in performing convergence adjustments. The connection to the main chassis is made via a cable-and-socket arrangement. Generally, there are six controls for adjusting the vertical convergence and six controls for adjusting the horizontal dynamic convergence.

EXTERNAL PICTURE TUBE COMPONENTS

The components which are mounted on the neck of the delta-gun picture tube will now be discussed (See Figure 18.30). The deflection yoke is, to a considerable extent, similar to the deflection yoke used in a black-and-white receiver. However, its design is more complex because three beams must be deflected instead of one, and it is of the utmost importance that a symmetrical and uniform magnetic field be maintained throughout the deflection area.

Purity Assembly Another component found on the neck of the color-picture tube is the purity-magnet assembly. This device adjusts the axis of each electron beam so that it approaches each hole in the shadow mask at the proper angle to strike the appropriate color phosphor dot. In other words, the purity magnet provides for the proper alignment of the three beams with respect to the phosphor-dot screen and the shadow mask. When this component is properly set, a uniform color field will be obtained for each gun. This is referred to as "purity". For example, with only the red gun in operation a uniform red raster should be observed. Any departure from pure red at any point on the screen indicates that the beam is striking phosphor dots other than red. Similarly, when only the green gun is in operation, a uniform green raster should be obtained, and when only the blue gun is active, a blue field should be visible.

Convergence Coils The larger screen color tubes utilize magnetic convergence and, toward that end, they employ three sets of convergence coils. Each is positioned directly over the pole pieces which are internally associated with each Number 4 Grid. The magnetic fields set up by the coils are coupled through the glass neck of the tube to the internal pole pieces.

The latter serve to shape and confine the fields, so as to affect only the particular electron beams to which the individual pole pieces correspond. For example, the change in the convergence angle of the red beam is a function only of the current through the external coil which couples to the internal set of pole pieces adjacent to the red beams. Similarly, the currents through the green and blue external magnets affect respectively only the green and blue beams.

Each external coil possesses two separate windings to provide for horizontal- and vertical-dynamic-convergence correction. For the static-convergence adjustment, each coil has a small permanent magnet associated with it. Its position can be varied to achieve center-screen static convergence.

A diagram of the individual dynamic convergence-control magnets was shown in Figure 18.26. The heavy dots represent the individual electron beams as they pass through the gun on their way to the screen.

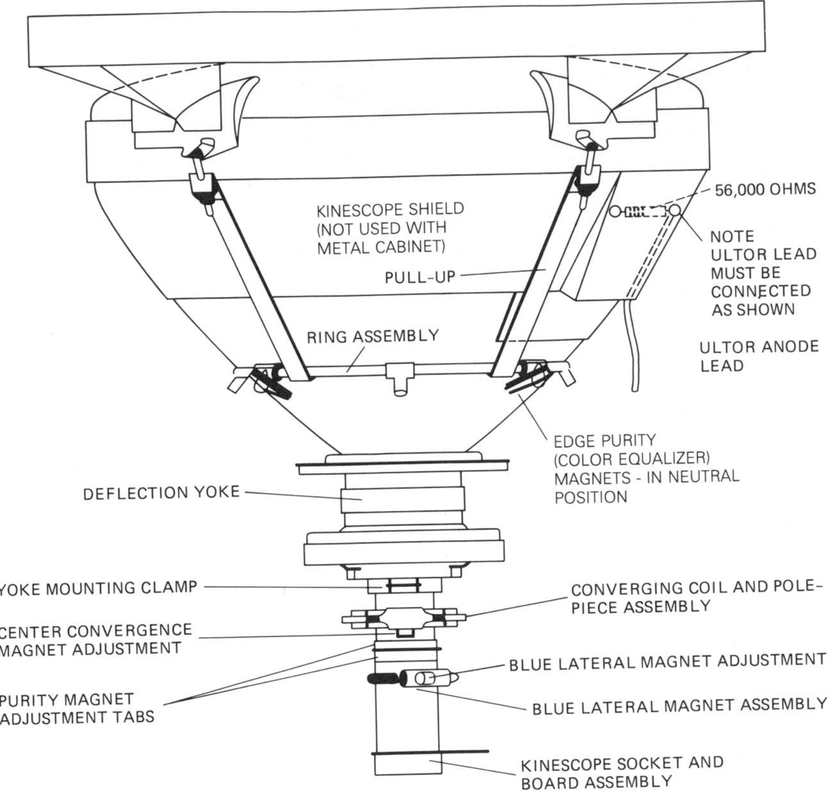

Figure 18.30 The location of the external components on the neck of the tri-color picture tube. *(Courtesy of RCA.)*

The arrows at these beams indicate their direction of movement. Note that the red and green beams are confined to paths which make an angle of 60° on either side of a perpendicular axis. The blue beam, on the other hand, can move only vertically.

Blue-Lateral Magnet Now it can readily happen that, although the color dots of the green and red beams fall within the triad of phosphors, those of the blue beam do not. This means that, while it is always possible to make the red and green beams (or color dots) converge, it may not be possible to have the blue beam meet the other two. Still another adjustment is required, that of being able to move the blue beam from side to side. To effect this, a special blue-beam lateral-positioning magnet is also found on the neck of the tube. This makes correct convergence of the three beams at the center of the screen always achievable (see Chapter 26 for the convergence procedure).

Aluminized Layer No ion traps are used with color-picture tubes, because the color screen is aluminized. The layer of aluminum presents a barrier to any oncoming ions and prevents them from reaching and damaging the screen. Electrons, having only 1/1800th of the mass of an ion, encounter little difficulty in passing through this aluminum layer.

SOME IMPROVEMENTS IN DELTA-GUN TUBES

The size of the holes in the shadow mask is extremely small. As an example, in a typical 25-in color tube the separation center-to-center between the holes is only 0.028 in (0.71 mm). The holes have a diameter of only 0.01 inch (0.254 mm). The mask, which contains over 400,000 such holes, is manufactured with a photo etching process. The holes at the center of the mask are larger than those at the edge which gives the picture an overall brighter appearance. Once it is completed, it is exactly positioned into the tube to an accuracy within 0.00025 inch (.0064 mm).

Two important improvements have been made in the design of the shadow mask. One of these simplifies the problem of dynamic convergence and the other improves the quality of the color picture.

Shadow-Mask Improvements In the early types of three-gun color tubes, the problem of dynamic convergence was aggravated by the fact that the shadow mask was flat and the screen of the tube was curved. One of the improvements in the design of shadow masks is a curved mask like the one which can be seen in Figure 18.31. Use of the curved mask simplifies the problem of dynamic convergence.

The second improvement in shadow-mask designs is illustrated in Figure 18.32. When the electron beam strikes the holes of a shadow mask at an angle as shown in Figure 18.32A, some of the electrons are reflected from the sides of the hole. These electrons strike phosphor dots in triads other than the ones they are supposed to strike. To get around this problem, the edge of the hole is shaped as shown in Figure 18.32B. Note that the edge of the hole is now very small, thus eliminating undesirable reflections.

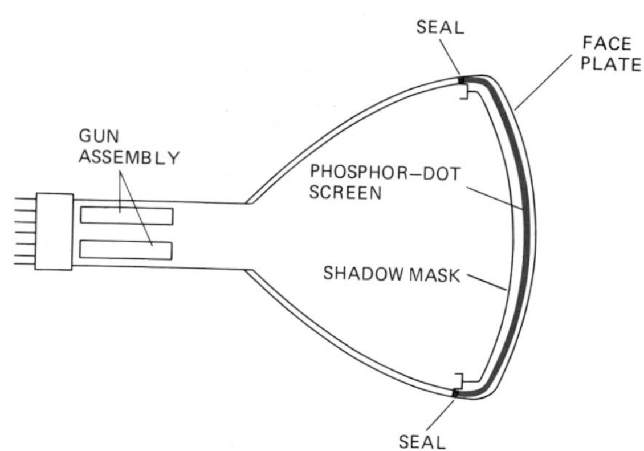

Figure 18.31 The use of a curved shadow mask simplifies the problem of dynamic convergence.

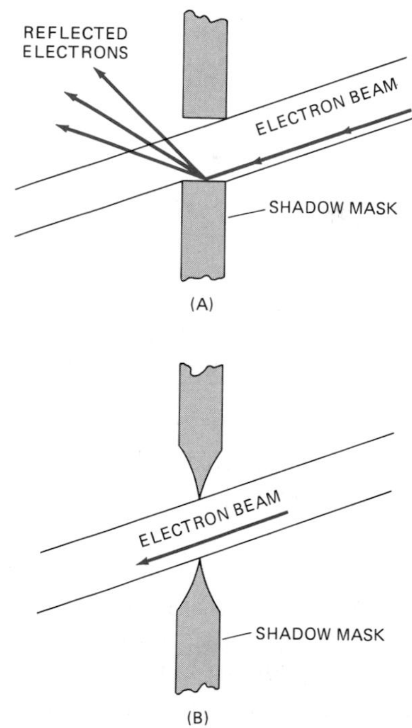

Figure 18.32A-B. The hole in the shadow mask affects the quality of the picture. **A.** Electrons are reflected from the edge of the hole in the shadow mask. **B.** By shaping the holes in the shadow mask as shown here, undesired electron reflections are practically eliminated.

Black Surround The triads on the inside surface of the picture-tube screen emit light when struck by an electron beam. When there is a bright light striking the face of the tube, it is reflected. The reflected light combines with the light from the triads, and the overall effect is to give the television picture a washed-out appearance. This problem has been virtually eliminated by the use of the **black surround** which is simply an opaque, jet-black material between the dots in the triads. Since ambient light is not reflected from the front of the screen, the tube manufacturer does not need to use a filter-type glass screen that dims the picture. To the viewer the result is a much brighter color picture.

Rare Earth Phosphors Brighter color pictures have also been achieved by the use of rare-earth phosphors instead of sulphide phosphors. Not only do the rare-earth phosphors radiate light much more efficiently, they also have the advantage that they reflect white light (since the phosphors are white) rather than the yellow light reflected by sulphide phosphors. While the reflected white light still dilutes the red light being emitted by the red rare-earth phosphor, this is still preferable to the red-yellow mixing that occurs with the sulphide phosphor.

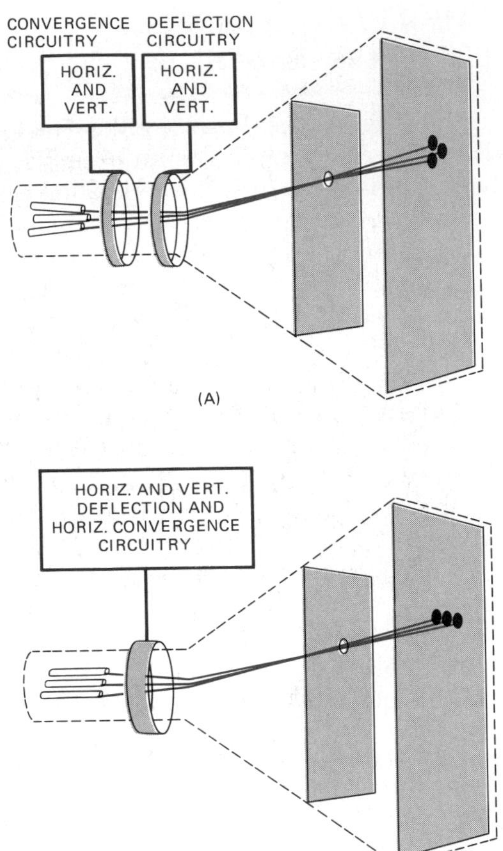

Figure 18.33 Comparison of the delta and in-line cathode systems in color picture tubes. **A.** A conventional tube with the cathodes in a delta formation. **B.** The in-line tube. Note that the cathodes and the three color dots are in line horizontally. Many in-line tubes use vertical color stripes instead of color dots (see Figure 18.21.)

18.9 RCA IN-LINE GUN COLOR-PICTURE TUBE

The problem of convergence has been simplified by the use of the color-picture tubes having in-line cathodes. A comparison of the conventional delta arrangement and the in-line arrangement for the cathodes is pictured in Figure 18.33. As shown in Figure 18.33A, the delta arrangement results in the use of two separate beam-control systems: one for converging the beam and the other for deflecting the beam. When the cathode and the triads are arranged horizontally as shown in Figure 18.33B, the vertical convergence remains constant with vertical deflection and thus, no vertical convergence correction is required. Horizontal convergence correction is still needed. General Electric first used color-picture tubes with in-line cathodes for their portable color-television receivers. (See also Figure 18.21.)

A comparison between a delta-gun tube

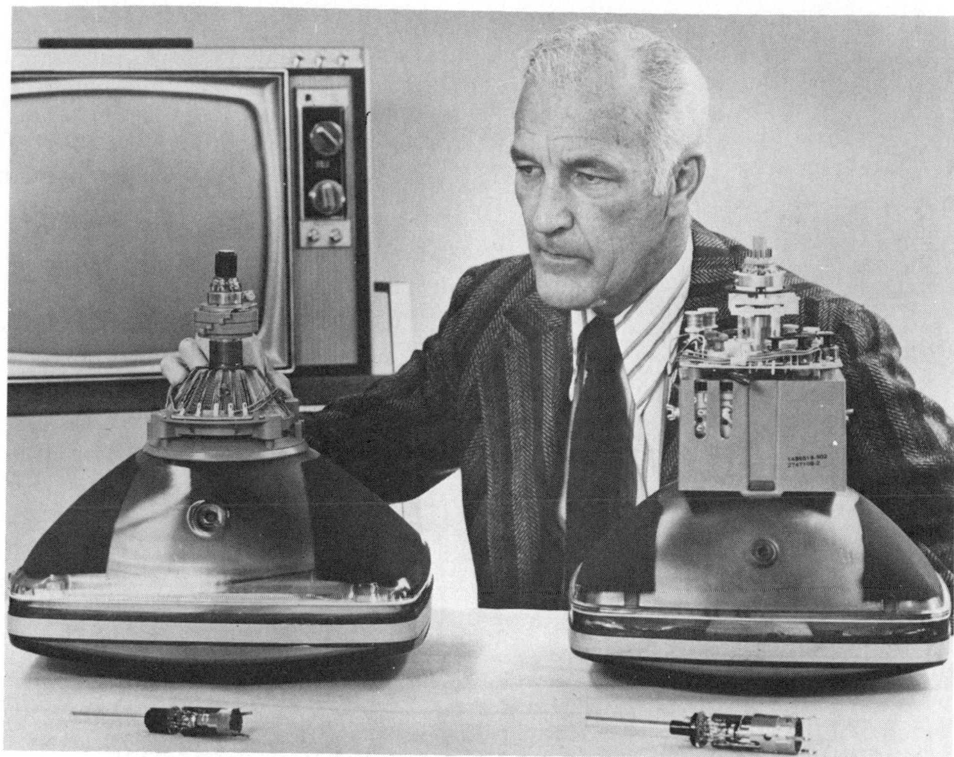

Figure 18.34 Comparison between delta-gun color tube (right) with its convergence yoke and circuitry, and in-line color tube (left) with its convergence magnets. The in-line tube is shorter than the delta-gun tube. *(Courtesy of RCA Picture Tube Division.)*

and an in-line tube with vertical-phosphor stripes, is shown in Figure 18.34. Note the relative simplicity of the neck-mounted components on the in-line tube. With the in-line tube, many convergence-circuit components and adjustments needed to align the three-electron beams, are eliminated. The delta-gun tube in Figure 18.34 uses a 12-adjustment convergence yoke. With this scheme, the color tube generally must be converged in the user's home. Further, reconvergence is sometimes required because of the aging of receiver components.

Advantages of In-Line Gun Tube A basic advantage of the RCA in-line gun tube is the elimination of all convergence adjustments after the tube leaves the factory. A precision-wound, toroid-deflection yoke is installed and cemented into place at the factory. The design and placement of this yoke is the key to the tube's "self-convergence" capability. In addition, a magnet assembly for purity adjustments and static convergence is also cemented to the tube neck. The positions of the yoke and magnets is critical, thus the cementing. However, once set, these positions hold for the life of the tube. No dynamic convergence is required for this tube. The yoke and magnet assembly is shown in Figure 18.35.

Because the components are perma-

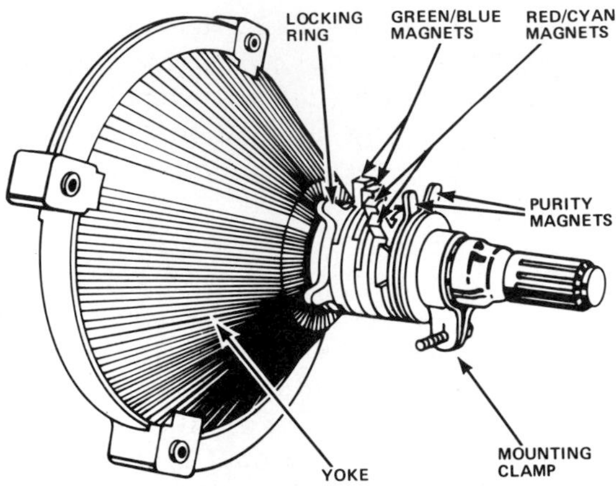

Figure 18.35 Details of the yoke and magnet assembly for the RCA in-line gun tube. *(Courtesy of RCA Picture Tube Division.)*

nently cemented to the tube neck, the entire assembly (preconverged) is changed if replacement is required.

Tube Screen Referring to Figure 18.21B, note the slotted-aperture mask and the vertical-color stripes. The vertical phosphor stripes are an important feature of the preconverged-tube design. This is true because with vertical-phosphor stripes, there cannot be vertical misregister of the electron beams. Also, as mentioned previously, with the three electron beams in the horizontal plane, no vertical-convergence correction is needed.

Gun Construction The in-line gun has three separate cathodes providing three electron beams. However, there is only one electron gun which is common to and controls all three beams. This basic scheme is somewhat similar to that of the Sony Trinitron color tube described in the next section.

Typical Characteristics Some characteristics of an RCA in-line 19VGQP22 color tube are as follows:

Deflection angle	90°
Maximum-anode voltage	32 kV
Focus volts	17-20% anode voltage
Screen diagonal	18.897 in (48 cm)
Tube length	16.667 in (37.3 cm)

Note: Not all in-line tubes are preconverged. Some types require horizontal dynamic and static convergence. This is true, for example, of the "tri-potential" tube, which is described subsequently.

18.10 TRINITRON PICTURE TUBE

The Trinitron color-picture tube employs three cathodes, but only a single electron-gun focusing lens. The color phosphors on the screen of a Trinitron are in alternating vertical color stripes, and an aperture grill is placed in front of the screen.

Figure 18.36 shows the arrangement of the electron gun, the aperture grid, and the vertically-striped color phosphors. (See also Figure 18.21C.) Note that the three electron beams leave the cathodes and are focused in a single electron lens. Convergence is accomplished by four convergence-control plates. (These are shown more clearly in the top view of Figure 18.37.)

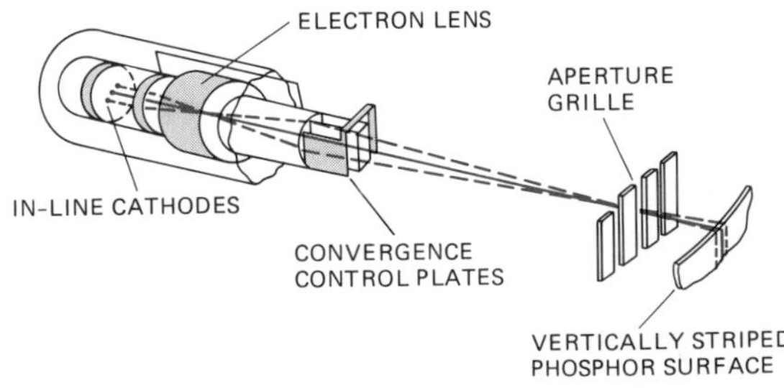

Figure 18.36 The basic components of the Trinitron tube. (See also Figure 18.21C.)

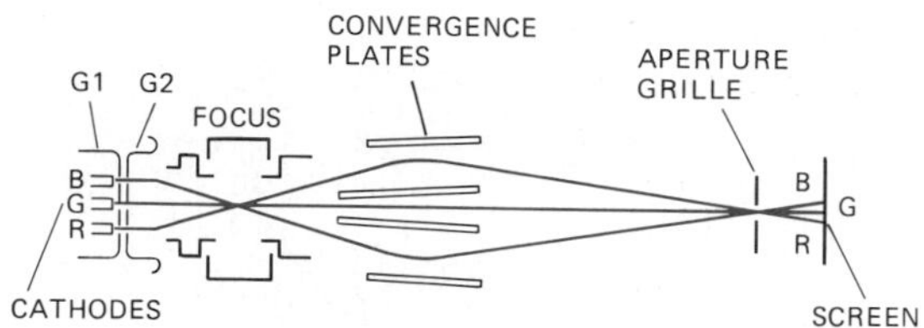

Figure 18.37 Top view of the beam passing through the various sections of the Trinitron tube.

The Electron Gun Figure 18.37 shows an illustration of a Trinitron gun, the aperture grill, and the screen. In this illustration, G_1 is the control grid, and G_2 is the accelerating grid which moves the electrons toward the screen. Note the single electrostatic-focus lens. As shown in Figure 18.37, the Trinitron tube illustrated has a single control grid, accelerating grid and focus lens for all three-electron beams. Consequently the video signals must be applied to the three individual cathodes in this particular tube. The common-control grid configuration is used only in the smaller size Trinitrons. (Up to 12 in diagonal screen size.) For larger Trinitrons, individual control grids are used for each electron beam. However, the accelerating grid and focus lens are still common to all three beams. With individual control grids, the video signals may be applied either to the control grids or to the cathodes. By utilizing a common focus lens, the Trinitron can achieve small electron-beam size, as well as uniform focus and beam shape for all three beams.

TRINITRON CONVERGENCE

Because of the horizontal, in-line gun and the vertical color-phosphor stripes (as with the RCA in-line tube), no vertical-convergence circuits or adjustments are required. However, horizontal electrostatic, static and dynamic convergence are required. This is provided by a combination of DC and (for some tube sizes), AC voltages applied to the convergence plates shown in Figure 18.37.

No Green Convergence The two middle-convergence plates are connected to the anode, inside of the tube. The green electron beam passes between these plates and is not affected by them since they are both at the same potential. These middle plates also act to shield the green electron beam from the DC (and AC) convergence electrostatic fields developed at the two outer-convergence plates.

Red and Blue Static-Convergence Adjustments We see now that there is no convergence adjustment for the green electron beam. The red and blue electron beams, however, may be moved horizontally by the potential difference between the outer plates and the middle plates. As mentioned previously, the middle plates are both at anode potential. However, the outer plates each have an individually adjustable DC voltage applied, which averages about two percent lower than the anode voltage. By adjustment of these two DC voltages, horizontal static convergence of the three beams is achieved.

Dynamic Convergence This static convergence is all that is required for Trinitrons larger than 12 in to maintain correct convergence over the entire screen. However, in 12-in or smaller tubes, a parabolic wave (described previously) is added to the DC convergence voltages to achieve full-screen convergence.

APERTURE GRILLE AND VERTICAL-COLOR STRIPES

Figure 18.38 illustrates the manner in which the vertical-color stripes are related to the aperture grille. The aperture grille has one vertical slot corresponding to each set of red, green, and blue vertical-phosphor stripes. Each green phosphor stripe is centered in a vertical slot. As previously mentioned, the green electron beam is in the center of the three beams as they sweep horizontally. Thus, each time the three beams pass through a vertical slot, they are capable of illuminating small portions of the red, green, and blue vertical-phosphor stripes. Of course, the actual degree of illumination which occurs at any point during scanning depends upon the strength of the applied video signals.

18.11 TRI-POTENTIAL COLOR TUBE

This type of tube features a "tri-potential" focus-lens system, small neck diameter (1.146 in, 2.91 cm), 100° deflection angle, in-line guns and vertical color stripes separated by opaque black lines. A slotted aperture mask is used. The mask and screen are illustrated in Figure 18.39.

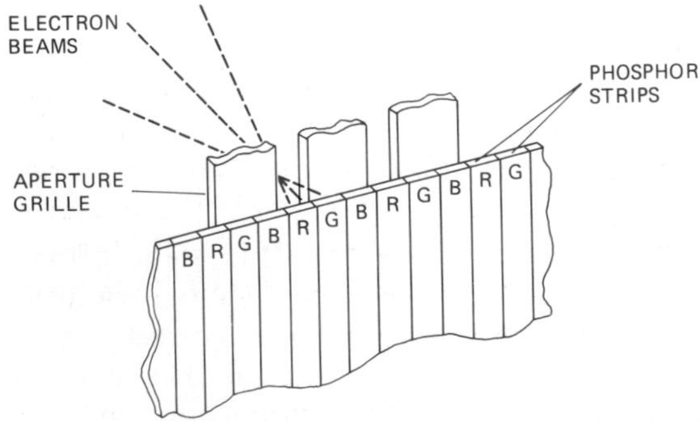

Figure 18.38 The phosphors of the Trinitron are in vertical strips. Note the relationship of the aperture grille behind the color strips.

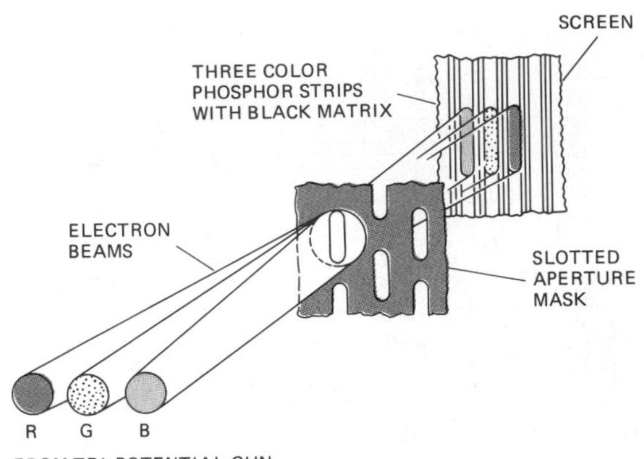

Figure 18.39 The three electron beams from a tri-potential electron gun pass through a slotted aperture mask to the vertical color-phosphor strip screen. Each color strip is separated from the next by an opaque black line. *(Courtesy of Quasar Electronics Company.)*

In-line guns and vertical phosphor stripes eliminate the need for vertical convergence. Horizontal convergence is achieved through the use of a tiltable, convergence-correction, deflection yoke. The yoke axis may be tilted slightly in the horizontal and vertical planes to converge the three electron beams.

Figure 18.40 shows a comparison between a bi-potential 25-in (63.50 cm) color tube and a 25-in (63.50 cm) tri-potential color tube. The tri-potential tube is shorter and requires only a static convergence and purity-magnet assembly. Dynamic convergence components are not required for this tube. [Compare the tube and yoke assembly of Figure 18.40 with those of Figure 18.34 (left).]

Uncontrolled thermal expansion of the slotted-aperture mask can cause loss of purity. This is prevented by a temperature-compensated slotted-aperture mask (TCM).

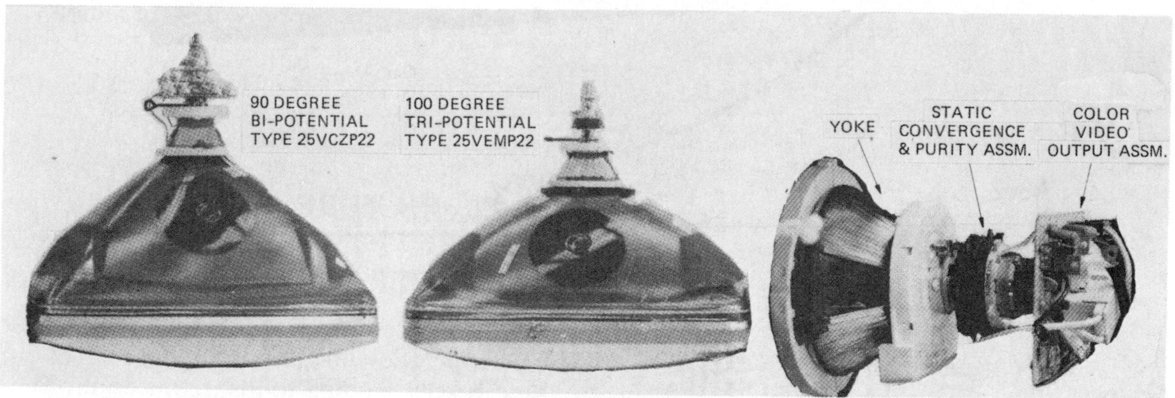

Figure 18.40 Comparison of bi-potential and tri-potential color tube and neck components. The tri-potential yoke and magnet assembly with attached color video output amplifier assembly is shown at the right. Compare it with Figure 18.34. *(Courtesy of Quasar Electronics Company.)*

TCM maintains optimum field purity during initial warm-up and continued operation to improve contrast.

TRI-POTENTIAL FOCUS

The tri-potential focus lens design provides reduced spot size which now makes practical the use of in-line guns with tubes larger than 19 in (48.26 cm). In-line gun tubes as large as 25 in (63.50 cm) are now made with this system.

A cutaway drawing of the tri-potential electron gun is shown in Figure 18.41. The tube is shown schematically in Figure 18.42, which also shows the base diagram and the legend of the base-pin connections.

The Lens System The most significant improvement in this tube is in its focus system. *Two* focus voltages (7 kV and 12 kV) are employed (instead of one). The additional 12 kV focus voltage is applied to Grids 3 and 5. As shown in Figure 18.42 these grids are situated before and after the original focus Grid, G4. The effect is to extend the field in the focus lens. This reduces beam diameter and also beam deviation from the desired paths. A separate 12 kV potential is developed at the horizontal-output transformer (H.O.T.) assembly to be applied to Grids 3 and 5. The name "tri-potential" arises from the fact that there are three different potentials on Grids 2, 3 and 5, and 4.

The Phosphors The "MV" at the beginning of Sylvania type MV25VEMP22 signifies a specific phosphor system. This system provides bright red, green, and blue fields, as well as white-field brighteners. Bright reds result from the improved europium-activated, yttrium-oxide phosphor. Bright blues and greens result from an alteration in the chemical composition of the sulphide phosphors. In addition, the phosphors are deposited on the screen in a dry state. This enables the achievement of optimum brightness and uniformity.

Tube Replacement Unlike the preconverged RCA in-line tube described previously, the picture tube and/or the deflection yoke may be replaced separately.

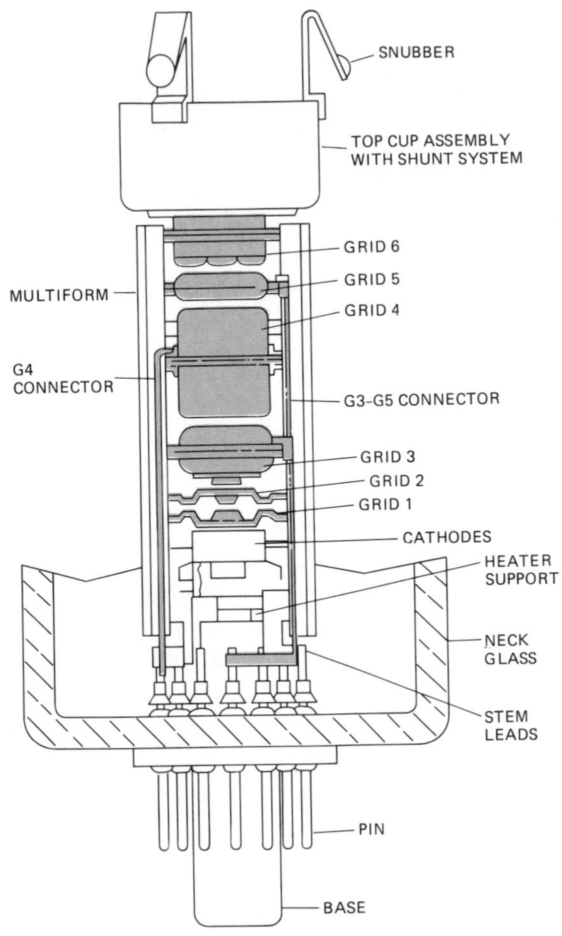

Figure 18.41 The tri-potential electron gun. Additional focus elements G_3 and G_5 are connected together and appear on both sides of focus element G_4. See Figure 18.42 for the schematic. *(Courtesy of Sylvania Electronic Components.)*

TRI-POTENTIAL TUBE CHARACTERISTICS

Some important characteristics of a Sylvania MV25VEMP22 are as follows:

Deflection angle	100°
Maximum-anode voltage	33 kV
G_3, G_5 voltage	approx. 40% of anode voltage
G_4 voltage	approx. 22% of anode voltage
Screen diagonal	24.658 in (62.6 cm)
Tube length	18.736 in (47.6 cm)

18.12 PICTURE TUBE PROTECTION CONSIDERATIONS

The following important topics should be carefully studied to insure maximum reliability from monochrome and color-picture tubes and safety for technicians and set owners.

SHOCK HAZARD

The high voltages at which picture tubes are operated (up to 33 000 V) may be very dangerous. The method for safely discharging the capacitor formed by the internal and external picture tube conductive coatings have been given in Section 18.1 and should be reviewed at this time. Extreme care must be taken in the servicing or adjustment of any high-voltage circuit. Note that discharging the high voltage to *isolated* metal parts, such as a cabinet or control brackets, may constitute a shock hazard. Be certain the *ground* you short to, is common to most receiver circuits.

IMPLOSION PROTECTION

A basic method of bonding safety glass to the tube face was described in Section 18.6. This protects the user from possible tube implosion. Other methods of implosion protection include the following:

1. A rimband affixed around the skirt of the face-plate and secured by tension

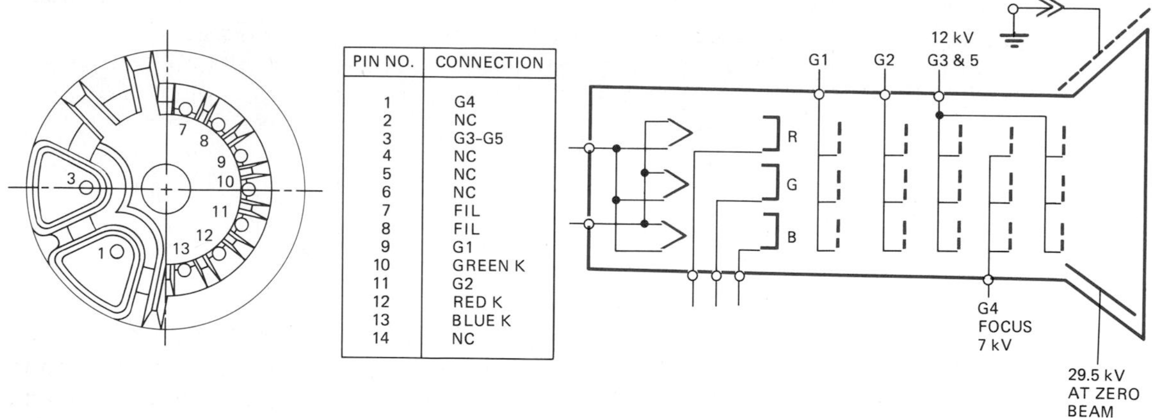

PIN NO.	CONNECTION
1	G4
2	NC
3	G3-G5
4	NC
5	NC
6	NC
7	FIL
8	FIL
9	G1
10	GREEN K
11	G2
12	RED K
13	BLUE K
14	NC

Figure 18.42 Schematic diagram and base connections for the 25VEMP22 tri-potential color picture tube. Note the new focus grids, G_3 and G_5, which are connected to a potential of 12 kV. *(Courtesy of Quasar Electronics Company.)*

bands over the rimband. (See Figures 18.22 and 18.43.)

2. A metal band (T-band) around the skirt of the face-plate and secured by a clip.
3. A resin-filled steel shell (shellbond) affixed around the front end of the skirt of the face-plate.

Possible Tube Hazard It should be carefully noted, that a picture-tube implosion may cause serious injury from flying glass. A 25-in (63.5 cm) picture tube may have a total external force on the glass surface of possibly 18 000 lb (8165 Kg). Consequently, picture tubes must be handled with great care and disposed of properly. The following precautions should always be adhered to.

Picture tubes should be kept in the shipping box or similar protective container until just prior to installation. To prevent possible injury from flying glass in the event a tube breaks, wear heavy protective clothing (including gloves and safety goggles with side shields) in areas containing unpacked and unprotected tubes. Handle the picture tube with extreme care. Do not strike, scratch, or subject the tube to more than moderate pressure. Particular care should be taken to prevent damage to the seal area.

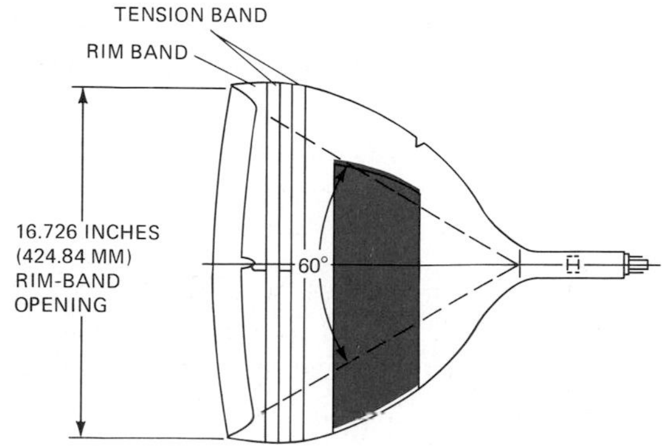

Figure 18.43 The implosion protection for a color picture tube, type 25VEHP22. Note the rim band secured by two tension bands. This is a side view. *(Courtesy of RCA Picture Tube Division.)*

X-RAYS

The level of x-ray radiation has been limited by Federal standards to a level of 0.5 milliroentgen per hour (mR/h). This level is to be measured at a point not greater than 2 in (5.08 cm) from any surface of a television

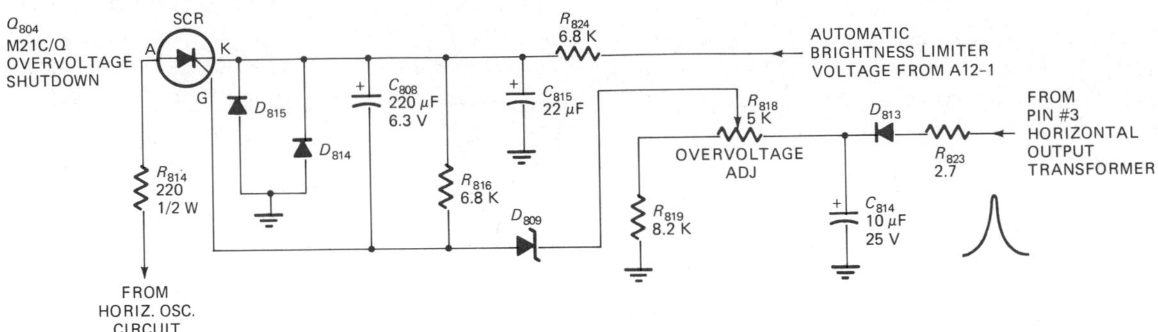

Figure 18.44 Schematic diagram of an overvoltage or excessive beam current shutdown circuit. *(Courtesy of Quasar Electronics Company.)*

receiver in its cabinet. X-ray radiation is limited from the tube face by the use of face glass with a lead or strontium content. The receiver manufacturer is responsible for limiting x-ray radiation from other portions of the tube or high-voltage circuits, by adequate shielding.

Generally, the x-ray level will be below the 0.5 mR/hr limit as long as the anode-high voltage is maintained within its recommended limits.

OVERVOLTAGE OR EXCESSIVE-BEAM CURRENT SHUTDOWN

If the anode-high voltage exceeds the recommended limit, many TV receivers provide a circuit which automatically shuts down the high voltage. The circuit to be described here, also shuts down the high voltage to protect the picture tube, if its electron-beam current becomes excessive. The circuit is shown in Figure 18.44.

The key component in this circuit is the silicon-controlled rectifier (SCR), Q804. If Q804 is caused to conduct, the B+ to the horizontal oscillator is reduced and the oscillator shuts off. If this happens, anode-high voltage is no longer developed.

Overvoltage Shutdown Positive pulses from the horizontal-output transformer are developed during horizontal retrace. These charge capacitor C814 through diode D813. The amount of charge (voltage) is a function of the pulse amplitude which in turn is directly related to the high voltage. If the high voltage exceeds a predetermined amount (set by R818), zener diode D809 will conduct. When the zener conducts, the positive voltage now on the gate of SCR Q804, causes Q804 to conduct. As previously mentioned, this shuts off the high voltage. It will not come on again until readjusted to the proper amount.

Excess Beam-Current Shutdown Automatic-Brightness Limiter (ABL) voltage is fed to the shutdown circuit through R824 and charges C815. With normal beam current, this voltage is positive and is applied to the anode of zener D809. Positive voltage at the anode of the zener will prevent it from conducting. Diode D814 clamps this positive voltage to a maximum of +0.6 V.

As the beam current increases, the ABL voltage decreases and will become *negative* with excessive beam current. The negative voltage is limited to −0.6 V by diode D815. Negative voltage from C815 is applied to the anode of zener D809 causing it to conduct. Again, SCR Q804 conducts and shuts off the horizontal oscillator. It will not come on again until the beam current is reduced to a safe level.

AUTOMATIC BRIGHTNESS LIMITER (ABL)

In addition to excess-beam current shutdown circuits, many color receivers contain an automatic brightness limiter (ABL) circuit. An ABL circuit functions to prevent "blooming" of the pictures. Blooming refers to an increase in the size of the picture, from a decrease of anode voltage. This may be caused by high-beam current drain resulting from a high-level video signal having a bright background. While the anode voltage is generally regulated, the high-voltage power supply can go out of regulation if the beam current exceeds the regulator limit of operation. Blooming occurs because the amount of deflection increases with a decrease of anode voltage (less beam acceleration). This condition is usually accompanied by some defocusing of the picture. An ABL circuit prevents blooming by preventing the total beam-current drain from reaching a value capable of causing blooming.

ABL CIRCUIT OPERATION

A simplified schematic diagram of an ABL circuit is shown in Figure 18.45. In this circuit, beam current is limited by reducing the video signal amplitude and brightness. In addition, peak beam-current excursions are limited at the picture tube control grids.

Limiting Video-Signal Amplitude Beam current flows through R_{435} and R_{425} to +27 V. A portion also flows via R_{15} through the base-emitter junction of beam limiter Q_2, to

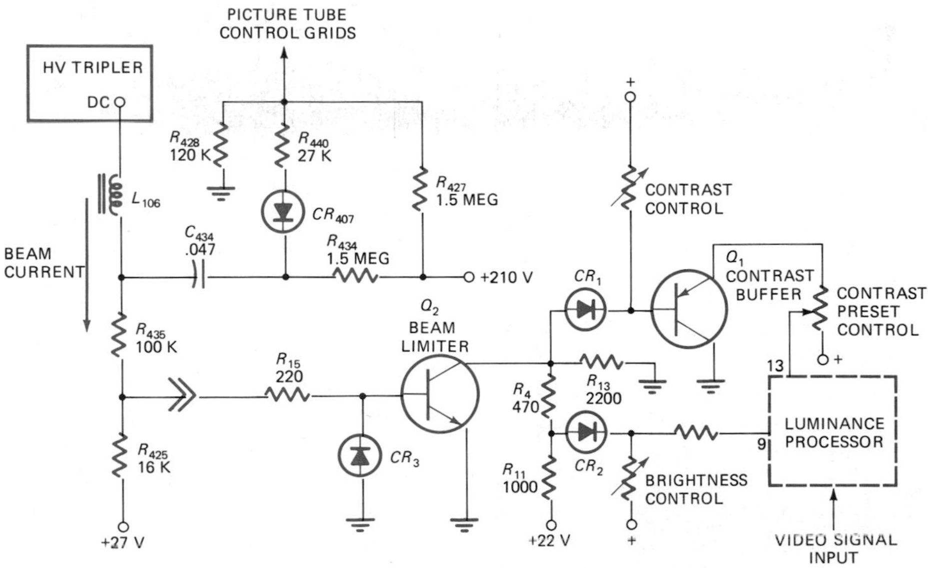

Figure 18.45 Simplified schematic diagram of an automatic brightness limiter (ABL) circuit. *(Courtesy of RCA Consumer Electronics.)*

ground. As the beam current increases, the voltage at the base of Q_2 becomes less positive, causing its collector voltage to become more positive. A sufficient rise of Q_2 collector voltage causes CR_1 to conduct and the positive voltage on the collector of Q_2 drives contrast buffer Q_1 toward cutoff. A more positive voltage is now fed via the contrast-preset control to Pin 13 of the luminance-processor module. The effect here is to reduce the video signal amplitude.

Limiting Brightness A rise of Q_2 collector voltage somewhat greater than that which caused CR_1 to conduct, will cause conduction of CR_2. If this occurs, the increased positive collector voltage of Q_2 causes the voltage at Pin 9 of the luminance processor module to become more positive. This, in turn, reduces the brightness (beam current) of the picture tube.

Peak Beam-Current Excursions A third action of this ABL circuit affects the picture-tube control grids. Peak-beam current variations are differentiated (peaked) by C_{434}, CR_{407}, R_{440} and R_{428}. As a result of the differentiation, peak-beam currents cause negative pulses to appear at the cathode of diode CR_{407}. Those negative pulses which are of sufficient amplitude to overcome the diode-back bias (0.6 V) are coupled through CR_{407} to the picture tube-control grids. These negative pulses reduce tube condition and limit beam current for the duration of peak-beam current excursions.

SPARK GAPS

Picture tubes operate with very high voltages, up to about 33 kV. Because of this and the close spacing of elements of the electron gun, there is a potential for internal arcing. **Arcs** (or **flashovers**) can cause picture-tube damage or transistor-circuit failures. To prevent such problems, spark gaps are provided on the picture tube base socket. Spark gaps are provided for all tube elements except the heaters. (On older receivers, the spark gaps may be external to the picture tube.)

The spark gaps are designed for breakdown voltages slightly above the maximum expected operating potentials. Typical spark gap breakdowns range from approximately 1 500 V to 5 000 V. In some instances, the spark-gap breakdown voltage must be higher than a component maximum-voltage rating. In such cases, current-limiting resistors (about 1 000 ohms) may be inserted in series with each applicable picture-tube base lead.

In addition to being molded into the picture tube base socket, spark gaps may be found in other forms. One type is an integral part of a ceramic disc capacitor. Another type consists of two electrodes mounted in a ceramic molding.

DC-HEATER BIASING

In order to reduce the possibility of heater-to-cathode shorts in color-picture tubes, a DC bias is placed on the heaters. The picture-tube heaters are supplied from a separate filament transformer (6.3 VAC), which is isolated from ground. B+ is connected to the heaters through a resistance of between 100 k and 1 MΩ. The bias voltage brings the heater voltage close to the cathode voltage, thus reducing the possibility of arcing and shorting between these elements. Generally, the heaters are required to be maintained somewhat negative with respect to the cathodes, for the best protection.

Note: Details involving special waveforms for color-picture tubes, as well as the setup and adjustment procedures for color-picture tubes will be found in Chapter 26.

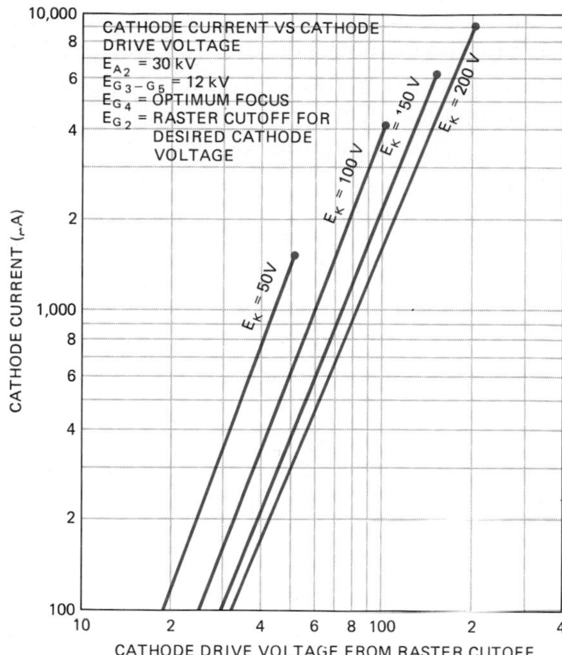

Figure 18.46 The cathode-drive characteristics of a tri-potential color tube, type MV25VEMP22. *(Courtesy of Sylvania Electronic Components.)*

This includes such topics as (1) convergence and convergence wave shapes, (2) purity, (3) pincushion correction, (4) degaussing, and (5) necessary test equipment and their uses.

18.13 TROUBLES IN MONOCHROME AND COLOR-PICTURE TUBES[1]

In this section some of the symptoms of defective monochrome and color-picture tubes are discussed.

Symptom: No Brightness There are a number of things in a television receiver that can cause a complete loss of brightness. For example, the high-voltage section can be defective, or the picture-tube grid or cathode circuit can be defective. In the event that there is no brightness, these things should be checked in the normal troubleshooting procedure. One cause of loss of brightness is a burnout of the filament in the picture tube. As within the other vacuum tubes, failure of the filament is catastrophic. Fortunately, in modern television picture tubes, both monochrome and color, it is rare for the filament to burn out.

Symptom: Low Brightness, Cannot Set Contrast Control Properly If the picture is very slow to appear after the set is turned on, and if the picture is weak with very poor contrast (often with a silvery appearance), then the picture tube may be gassy, or its cathode emission may be low. If the problem is low emission, some technicians prefer to install a *picture-tube brightener* which is nothing more than a step-up voltage transformer for increasing the filament voltage. This makes the filament hotter and increases the

[1]Additional information on this subject can be found in Chapter 26. This includes a different type of tube brightener and the use of a picture tube test jig.

emission from the cathode which, in turn, makes the tube useable for an additional amount of time.

Symptom: Excessive Brightness When the brightness control has no control over the amount of light coming from the picture tube, and the tube is at full brightness at all times, the problem may indicate a shorted control grid in the tube.

Symptom: Broad Dark Horizontal Bars in the Picture This type of problem can occur with both monochrome and color-picture tubes. It indicates a short between the picture-tube heater and cathode. It does not necessarily mean that the picture tube must be considered to be useless. Some technicians prefer to use an isolation transformer in the heater circuit. The transformer has a "floating secondary" which means that the heater and cathode voltage will be the same. This isolation transformer is especially useful for color picture tubes which are very expensive to replace.

Symptom: Large Blotches of Color that Cannot be Adjusted This is usually an indication that picture tubes or related mounting components have become magnetized. The magnetic field causes deflection on the beam in the color-picture tube which disturbs the proper beam landing needed for convergence. Hence, it is necessary to *degauss*—that is, demagnetize—the picture tube and mountings. A large coil connected to the AC power line is used for this purpose.

In many receivers the degaussing is accomplished automatically when the set is turned on. (See also Chapter 24.)

Symptom: Only Two Colors are Available in a Three-Gun Picture Tube This type of problem is normally due to one defective gun in the color-picture tube, or to defective circuitry or voltages associated with the gun.

Symptom: Only One Color is Weak in a Color Tube This is generally caused by weak emission from only one cathode. (May also be caused by a circuit problem.) A type of picture-tube "brightener" described in Chapter 26 may help or cure this problem for a time.

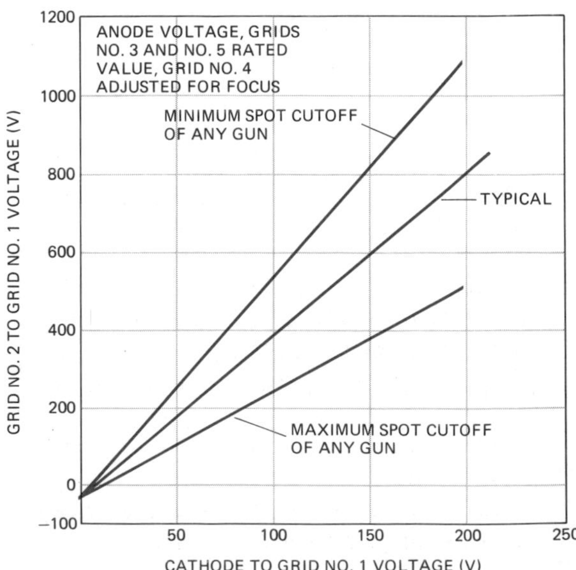

Figure 18.47 Spot cutoff characteristics of a tri-potential color tube, type MV25VEMP22. *(Courtesy of Sylvania Electronic Components.)*

SUMMARY OF CHAPTER HIGHLIGHTS

1. A monochrome-picture tube has a single electron gun. The gun directs a fine beam of electrons onto a phosphor (P4)-coated screen. The screen is illuminated by the impact of the electron beam. (Figure 18.12)
2. A color-picture tube may have one or three electron guns. In either case, three-electron beams are generated, representing red, blue, and green colors. (Figure 18.21)
3. A color-picture tube screen is composed of red, green, and blue phosphor dot-triads or of adjacent groups of red, green, and blue vertical-phosphor stripes. (Figure 18.21)
4. Depending upon the illumination of the red, green, and blue phosphors, any desired color may be produced. Also, white may be produced by the proper percentages of red, green, and blue illumination.
5. All color-picture tubes employ some form of aperture mask near the screen. This mask determines that the red, green, and blue electron beams strike only their respective color phosphors. (Figures 18.21, 18.22, 18.27, 18.31 and 18.32)
6. As part of a picture-tube identification code, the number "25" indicates the approximate diagonal measurement in inches of the external dimension of the picture tube. If a "V" follows the "25", this indicates the actual *viewing* diagonal is very close to 25 in (63.5 cm).
7. The screen phosphor for monochrome-picture tubes is designated as P4. For most color-picture tubes, the designation is P22.
8. Anode voltages for picture tubes range from approximately 9 to 33 kV.
9. The inner and outer "aquadag" coatings of a picture tube, form a high-voltage capacitor. This should be carefully discharged before handling the tube.
10. Electron-beam deflection angles range from about 70° to 114°. These numbers are the total (included) angles.
11. A basic electron gun includes (1) a heater, (2) a cathode, (3) a control grid, and (4) accelerator and focusing grids. The electron gun forms and focuses the electron beam and provides initial acceleration. (Figure 18.13)
12. All picture tubes manufactured after 1959 use electrostatic focusing. This eliminates the need for a focus coil on the neck of the tube. (Figure 18.12)
13. All TV-picture tubes use electromagnetic deflection. Electrostatic deflection for TV-picture tubes is impractical because of the high-deflection voltages that would be required. (Figure 18.12)
14. A picture-tube yoke includes the horizontal and vertical-deflection coils. Cosine, cosine squared, or toroid windings are widely used. (Figure 18.9)
15. Some monochrome picture tubes are focused with zero volts on the focusing element of the electron gun. (Figure 18.13)
16. Generally, centering in monochrome-picture tubes is accomplished with the aid of two permanent-magnet rings on the neck of the picture tube. These are not required on color-picture tubes. (Figure 18.12)
17. An aluminum coating over the picture-tube phosphor(s) prevents ion spots and thus eliminates the need for ion-trap magnets and special electron-gun configurations. (Figures 18.18 and 18.22)
18. One type of picture-tube safety shield is a sheet of glass which is bonded by

an epoxy resin to the face of the tube. (Figure 18.20)

19. Three basic types of color-picture tubes are (1) the delta gun with color-dot triads, (2) the three-gun, in-line type, using vertical-color stripes and (3) the one gun, three beam, in-line type, using vertical-color stripes. (Figure 18.21)
20. All color tubes use some form of shadow mask immediately preceding the phosphor screen. The purpose of these masks is to insure that each electron beam will strike only the correct color phosphor. (Figures 18.21, 18.22 and 18.27)
21. Whether phosphor-dot triads or vertical-stripe phosphor triads are used for the screen of a color-picture tube, they are arranged in groups of repetitive red, green, and blue-emitting phosphors. (Figure 18.21)
22. With delta-gun color tubes, a separate convergence yoke and static-convergence magnets are used. There may be up to 12 convergence adjustments required. These may have to be repeated from time to time. (Figures 18.26–18.31)
23. The term "purity" refers to having a full raster of separate red, green, and blue colors on the picture tube.
24. The two types of convergence are (1) dynamic and (2) static. With dynamic convergence, a parabolic waveform is passed through a convergence yoke or to convergence plates. Static convergence is accomplished by permanent magnets, or DC voltages. (Figures 18.26 – 18.30)
25. Color-picture tubes employing in-line guns have simpler convergence procedures. No vertical-convergence correction is required. (Figures 18.21, 18.33, 18.34, and 18.35)
26. The RCA in-line gun color-picture tube is preconverged at the factory. No convergence is required (or can be performed) after tube installation. (Figures 18.34 and 18.35)
27. In Question 26, if replacement is required, the entire assembly including picture tube, deflection yoke, and magnet assembly is changed as a unit. (Figures 18.34 and 18.35)
28. The Trinitron-color tube employs three cathodes and a single-focusing lens. It is an in-line beam gun tube, with an aperture grille and vertical color-phosphor stripes. (Figures 18.21, 18.36, 18.37, and 18.38)
29. The Trinitron-color tube requires no vertical-convergence circuits or adjustments. However, horizontal electrostatic, static, and dynamic convergence are required. (Figure 18.37)
30. In the Trinitron, the "green" electron beam has no convergence adjustment. The "red" and "blue" electron beams are caused to converge with the "green" electron beam by means of DC and AC convergence voltages applied to the convergence plates. (Figure 18.37)
31. The "tri-potential" color tube features a tri-potential focus-lens system. This employs two focus voltages of 7 kV and 12 kV and results in sharper focus, not previously possible for in-line gun tubes. (Figures 18.41 and 18.42)
32. The tri-potential focus system has made it possible to manufacture in-line gun color tubes larger than 19 in (48.26 cm).
33. Tri-potential color tubes employ in-line guns, slotted aperture masks and vertical, phosphor-color stripes, separated by opaque-black strips.
34. In tri-potential tubes no vertical convergence is required. Horizontal convergence is achieved through the use of a tiltable, convergence-correction deflection yoke. (Figure 18.40)
35. Picture tubes present possible hazards from high-voltage shock and implo-

sion. Extreme care must be used in handling these tubes (Sections 18.1, 18.6, and 18.12)

36. Several methods are used for picture-tube implosion protection. Two are shown in Figures 18.20 and 18.43 (See also Section 18.12)
37. The level of x-ray radiation is limited by Federal standards to 0.5 milliroentgen per hour (mR/h). This level (or less) is met by insuring that the anode-high voltage does not exceed a specified maximum. (Section 18.12)
38. In Question 37, other precautions are the use of a face plate with a lead or strontium content and by adequate shielding.
39. Many color-TV receivers employ a circuit which shuts off the anode voltage if it exceeds the recommended maximum value. Some of these circuits also protect the picture tube against damage from excessive beam currents. (Figure 18.44)
40. Many color-TV receivers include an automatic-brightness limiter (ABL) circuit. An ABL circuit functions to prevent "blooming" of the picture (Section 18.12)
41. To prevent picture tube or transistor damage from high-voltage arcs, spark gaps are placed in the base of the picture tube (Section 18.12)
42. Spark gaps are provided for all picture-tube elements except the heaters. Typical spark-gap breakdowns range from 1 500 to 5 000 V (Section 18.12)
43. To reduce the possibility of picture tube, heater-to-cathode shorts, a DC bias is placed on the heaters (Section 18.12)
44. A color-picture tube which has the symptom of low brightness in one gun probably has low emission from the cathode of that gun.
45. If only two colors appear on a color tube, the problem is probably due to one defective gun, or to defective receiver circuitry.

EXAMINATION QUESTIONS

(Answers are provided at the back of the text.)

Part A Supply the missing word(s) or number(s).

1. A ______ coating on the screen of picture tubes, glows when struck by an electron beam(s).
2. The screen of a color tube may consist of ______, or ______ phosphors.
3. The ______ placed in front (gun side) of the screen insures that the three beams of a color tube will strike only the correct color phosphors.
4. The monochrome picture tube phosphor is designated ______ .
5. Most color-picture tubes use a heater voltage of ______ VAC.
6. The ______ of a picture tube consists of an aquadag coating.
7. A smaller tube neck enables greater ______ to affect the electron beam(s).
8. Generally ______ or ______ pin bases are used for color tubes.
9. An electric ______ system is used to focus an electron beam in picture tubes.
10. Only ______ deflection is used with picture tubes.
11. Currently ______ degrees is the widest deflection angle.

12. A deflection yoke contains the ________ deflection coils.
13. Most picture tubes employ ________ focus with ________ deflection.
14. ________ are used for picture centering on monochrome-picture tubes.
15. An aluminized screen eliminates the need for ________ magnets.
16. The possibility of tube ________ is reduced by bonding a glass shield to the tube face.
17. In-line picture tubes generally have the color phosphors in the form of ________ stripes.
18. The ________ color tube has three separate electron guns.
19. A separate convergence yoke is used with a ________ color tube.
20. ________ is the term for uniform color fields.
21. Reflected light from a color-tube face is greatly reduced by the use of ________ opaque material between color phosphor dots or stripes.
22. The RCA ________ color-picture tube is pre-converged at the factory.
23. In Question 22, this tube uses a ________ mask and ________ stripes.
24. In Question 22, this tube has ________ electron gun(s).
25. The Trinitron tube has three separate cathodes and ________ electron gun.
26. The Trinitron achieves sharp focus by the use of a common ________.
27. AC and DC voltages applied to ________ in the Trinitron are used to perform horizontal convergence.
28. The tri-potential focus lens makes possible the use of ________ guns in color tubes larger than 19 in (48.26 cm).
29. ________ phosphors and a slotted ________ are used with the tri-potential color tubes.
30. Two focus voltages of ________ kV and ________ kV are used with the tri-potential color tube.
31. Picture-tube ________ can cause serious injury from flying glass.
32. If the ________ voltage is within specification, x-ray radiation is usually within recommended limits.
33. An ABL circuit functions to prevent picture ________.
34. ________ are used in the base of color tubes to protect the tubes and/or associated circuits.
35. Picture tube emission can often be improved through the use of a ________.

Part B Answer true (T) or false (F).

1. A number such as "25" in a picture tube identification code, indicates the horizontal screen measurement in inches.
2. Monochrome-picture tube phosphors are designated either as "P4" or "P22".
3. Anode voltages for color-picture tubes range from approximately 22 to 33 kV.
4. The deflection angle is measured diagonally for picture tubes.
5. Picture-tube deflection angles from 110° to 140° are common.
6. Monochrome and color picture tubes have similar neck diameters.
7. Monochrome picture tubes with 114° deflection, generally have 12-pin bases.
8. Replacement picture tubes must always have the exact code number as the original tube.
9. An electron gun includes a cathode, control grid, and suppressor grid.
10. The "crossover point" of an electron gun occurs just before the shadow (aperture) mask.
11. All color-picture tubes employ magnetic deflection and focus.
12. Popular types of deflection-yoke windings include (1) cosine, (2) cosine squared, and (3) toroid.
13. Typical inductance values for a 110° yoke are; 240 mH horizontal and 670 mH vertical.
14. For some color-picture tubes, a 0 V focus lens is used.
15. Centering magnets are used to center the picture on monochrome-picture tubes.
16. Ion-trap magnets are not required for picture tubes with aluminized screens.
17. In-line gun, color-picture tubes do not re-

quire vertical-convergence circuits.

18. "Black surround", refers to the aquadag coating on the outside of picture tubes.
19. Red, green, and blue light combined in the proper proportions, will produce white light.
20. Static convergence is performed by permanent magnets.
21. The RCA in-line gun color tube has all its neck components permanently cemented in place.
22. The Trinitron tube has a single focus lens.
23. The tri-potential color tube has three additional focus elements, each having a different potential applied to it.
24. The high-voltage capacitor, a part of all picture tubes, must be carefully discharged before handling.
25. The Federal limit of x-ray radiation from TV sets is 0.5 milliroentgen per hour (mR/h).
26. An overvoltage shutdown circuit turns off the anode voltage if it exceeds the specified limit.

REVIEW ESSAY QUESTIONS

1. Refer to Figure 18.12A. Briefly describe the function of each part of the electron gun in your own words.
2. In Figure 18.12B, describe the operation of the centering magnets in your own words.
3. In Figure 18.14, explain briefly the operation of the diagonal-cut ion trap and its magnet.
4. What is the meaning of the term "deflection angle"? How is it measured?
5. What are some functions of the aluminized backing of the phosphor screen? (See Figures 18.18 and 18.22.)
6. Explain briefly two methods of reducing the implosion hazard in picture tubes.
7. Define the following terms briefly: (1) delta gun, (2) in-line gun, (3) preconvergence, (4) dynamic convergence, and (5) static convergence.
8. Draw a simple, rough sketch showing and labeling the components mounted on the neck of an RCA in-line gun, preconverged-color tube.
9. Draw a simple sketch showing the basic operation of a Trinitron. Include convergence plates and some phosphor strips. (Two separate sketches are acceptable if desired.)
10. Make a simple schematic drawing of a "tri-potential" color tube. Label all parts.
11. In Figure 18.44, what is the function of D813 and C814?
12. In Question 11, what may cause SCR Q804 to be turned on? What happens if Q804 is turned on?
13. In Figure 18.45, explain briefly in your own words how the brightness is limited by this circuit.
14. Explain briefly the advantages of the in-line gun over the delta gun.
15. How is a "tri-potential" tube superior to other in-line gun tubes? Why?
16. What type of shadow (aperture) mask is used with (1) a delta-gun, dot-triad tube, (2) an RCA in-line gun tube and (3) a Trinitron?
17. Explain briefly the meaning of the various portions of the color-tube identification number, 23VABP22.
18. What is an alternate function of an overvoltage-shutoff circuit? Briefly describe the operation.
19. Explain in your own words the precautions you should take in handling any picture tube.
20. What is the importance of a narrow picture-tube neck diameter? Explain.
21. In your own words, explain briefly why only electromagnetic deflection is used with TV picture tubes.
22. Describe briefly how a shadow (aperture) mask performs its function.
23. Why does the combination of horizontal, in-line guns and vertical-phosphor stripes effectively eliminate the need for vertical convergence?
24. In Question 23, why is horizontal convergence still needed?
25. In your own words, briefly describe how static and dynamic convergence is accomplished in the Trinitron.

Chapter 19

19.1 Types of power supplies
19.2 Rectifiers
19.3 Filters
19.4 Transformer and transformerless power supplies
19.5 Voltage multipliers
19.6 Voltage regulators
19.7 The zener diode regulator
19.8 The series-pass transistor regulator
19.9 Three-terminal voltage regulators
19.10 Switching regulators
19.11 Solid-state monochrome TV power supplies
19.12 A hybrid color television power supply
19.13 Horizontal frequency power supply
19.14 Troubles in low-voltage TV power supplies
Summary

LOW-VOLTAGE TELEVISION POWER SUPPLIES

INTRODUCTION

This chapter discusses various types of TV, low-voltage power supplies. This will also include the horizontal frequency type of low-voltage power supply. The high-voltage power supplies required for picture tubes are discussed in Chapter 22.

A low-voltage power supply provides B+ (or B−) voltages for transistors or vacuum tubes. It also provides AC heater voltages for the picture tube and any other vacuum tubes that may be used.

Older color-TV receivers drew about 250 W from an AC line. However, because of the recent emphasis on energy conservation, this drain has been appreciably reduced. For example, a recent solid-state 19-in (48-cm) Sony color-TV receiver draws an average of only 75 W. Modern, small-screen, portable monochrome sets draw only 35 to 45 W, whereas older similar receivers drew 75 to 100 W.

Solid-state TV receivers require B+ voltages ranging from approximately 5 to 250 V. Vacuum tube receivers generally use B+ voltages ranging from approximately 100 to 400 V. Some vacuum tube receivers require higher B+ voltages, as high as perhaps 750 V. Such voltages are obtained from a "B+ boost" scheme, described in Chapter 22.

When you have completed the reading and work assignments for Chapter 19, you should be able to:

- Define the following terms: pi filter, *LC* filter, *RC* filter, voltage multiplier, voltage regulator, zener diode, derived voltage, switching regulator, series-pass regulator, protection circuit, three-terminal regulator, start-up voltage, run voltage.

- Describe the characteristics of the three general types of low-voltage power supplies for television receivers.
- Explain how derived voltages are produced.
- Discuss the differences in B+ and B− voltages required for vacuum tube and solid-state TV receivers.
- List the different types of rectifier circuits.
- Explain how power supply filters work.
- Explain how a power supply protection circuit works.
- Understand the operation of voltage multipliers.
- Explain the operation of a series-pass regulator.
- Explain the operation of switching regulators.
- Recognize symptoms of a defective low-voltage power supply.
- Service the low-voltage power supply of a television receiver.

19.1 TYPES OF POWER SUPPLIES

TV receiver low-voltage power supplies can be classified generally as follows:

1. Power-transformer-operated power supplies. These provide isolation from the AC power line.
2. Transformerless power supplies. These operate directly from the AC line. Some have voltage multipliers (generally doublers). These supplies are *not* isolated from the AC line. An isolation transformer should be used for safety when servicing TV receivers with such supplies.
3. "Derived" power supplies. Most of the power is supplied by rectifying several outputs of a horizontal output transformer. In some cases, a power supply that is not isolated from the AC line may be used to power the horizontal output transistor. An isolation transformer should be used when servicing this type of power supply.

Power supplies may also be classified by the type of rectifier system used:

1. Half-wave rectifier
2. Full-wave rectifier
3. Bridge rectifier
4. Voltage-multiplier rectifier, generally doublers (or triplers for the picture tube anode)

Power supplies may be unregulated or regulated. Most monochrome receivers (particularly those using vacuum tubes) use unregulated power supplies. In these sets, changes in line voltage or in receiver load current may produce changes in picture size without noticeably affecting picture quality. However, the voltage requirements of solid-state TV receivers, and particularly solid-state color receivers, are more critical. These receivers frequently have voltage regulators for both the low-voltage supplies and the anode high-voltage supply.

19.2 RECTIFIERS

Rectifiers, which are devices for changing AC to DC, come in many sizes, shapes, and prices. Until the early 1950s, nearly all TV power supplies used tubes for rectifiers. As selenium, germanium, and silicon rectifiers became more plentiful and cheaper, they gradually replaced tubes. Today, every manufacturer uses silicon rectifiers, and this type will probably be the mainstay for many years to come.

SILICON RECTIFIERS

Nearly all new transistors, diodes, and ICs are made with silicon. It is silicon that has enabled solid-state electronics to take over most of the functions of vacuum tubes. Silicon devices can withstand high temperatures. They are stable for long periods of time because of their resistance to the deleterious effects of air. Their leakage current in the reverse direction is low, and they have a low internal voltage drop (less than 1 V) in the forward direction. Also, they may have current ratings in excess of 10 A.

In the **silicon rectifier,** a small wafer of silicon is securely fastened, generally by welding, to a copper plate, which is then attached to the rectifier housing (Figure 19.1). This terminal is the cathode, and the copper makes an electrical connection to the silicon. By connecting the copper to the rectifier case and then mounting the case to a chassis, an efficient heat dissipator is utilized.

The other terminal of the rectifier is formed by alloying a small dot of gold antimony into the opposite face of the silicon wafer. A lead is then connected to this dot, and this lead, brought out of the case but insulated from it, becomes the anode of the rectifier.

Precaution One precaution must be observed when using silicon rectifiers. Their forward resistance is so low that a high current may surge through them to the filter capacitors when a set is turned on. In Figure 19.2, a 10-ohm resistor (R_1) limits the surge current to a safe value. Let us determine what this current is with and without the 10-ohm resistor. The IN2071 diode has a forward resistance of approximately 1 ohm at high currents. At the instant the set is turned on, the 200-μF input capacitor C_1 looks like a short circuit. However, the peak value of the first half cycle of the voltage is equal to $120(1.414) = 170$ V. Thus, a maximum current of

$$I = \frac{V}{R} = \frac{170}{1} = 170 \text{ A}$$

will flow during the first half cycle. This diode is capable of withstanding only a 25-A surge. The 10-ohm resistor shown in Figure 19.2 keeps this initial surge down to

$$I = \frac{170}{(10 + 1)} = 15.5 \text{ A}$$

19.3 FILTERS

The DC output from rectifiers has a large amount of ripple. Half-wave rectifiers

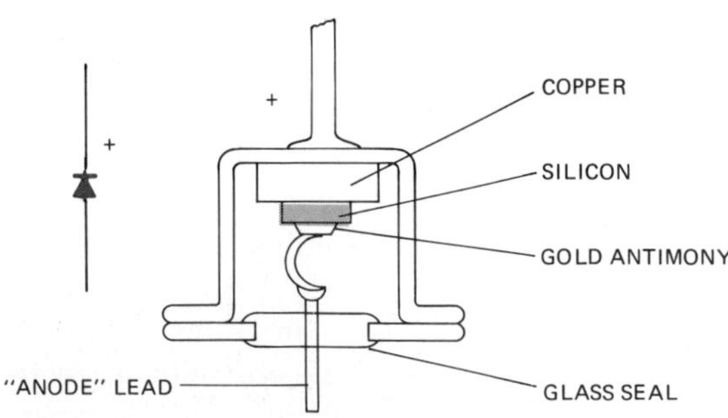

Figure 19.1 Internal construction of a silicon rectifier diode.

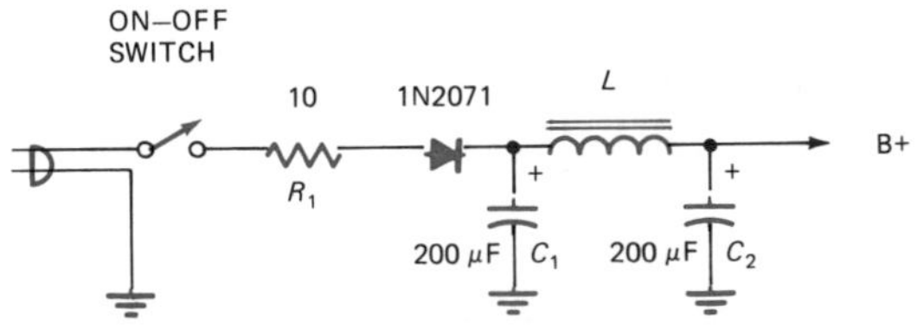

Figure 19.2 Simplified circuit diagram of a silicon diode half-wave rectifier circuit. Resistor R_1 limits diode surge currents to a safe value.

have a 60-Hz ripple, and full-wave rectifiers have a 120-Hz ripple. Some electronic devices can use this form of DC with no further processing. For television receivers, however, most of the ripple must be removed. In this section, we shall examine a few of the many ways ripple is removed from rectified AC power. Only passive filters will be considered. These contain resistors, capacitors, and inductors. We shall close the section by briefly discussing line filters and heater filters.

RC FILTERS

A common *RC* filter is the **pi filter** shown in Figure 19.3A. For low-voltage power supplies, this type of filter is ideal. There are two reasons for this. First, the resistor is much smaller and lighter than the choke used in higher-voltage power supplies. Second, high-capacitance, low-voltage capacitors are quite small. Some TV receivers use up to several thousand microfarads for these capacitors. The exact *R* and *C* values depends upon the ripple, current, and regulation requirements. For a given capacitor, a larger *R* will produce less ripple. However, this provides a lower output voltage and poorer regulation.

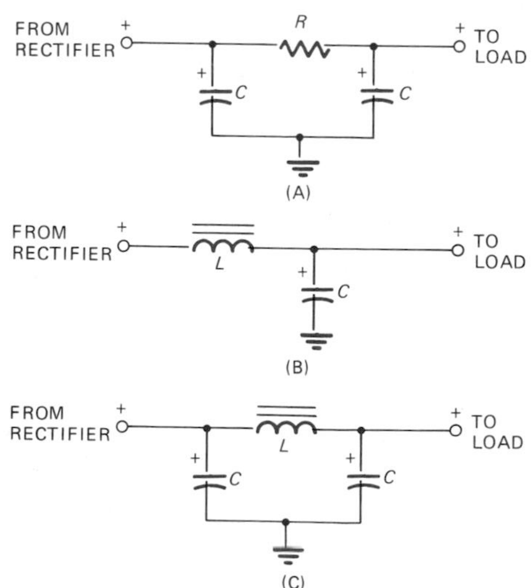

Figure 19.3A-C. Some of the filter circuits found in TV receivers: **A.** RC π circuit; **B.** choke-input LC circuit; **C.** capacitor-input LC circuit.

CHOKE INPUT LC FILTERS

Inductance and capacitance filters also come in several forms. Again, the exact circuit values depend on whether ripple or regulation is more important.

The **choke input LC filter** is shown in Figure 19.3B. This is the most commonly used *LC* filter. The choke is chosen so that it presents to the ripple, an impedance that is high relative to the reactance of the filter capacitors. The output voltage (if used with a full-wave rectifier) is usually between 60 and 70% of the peak sine-wave voltage. The choke input *LC* filter has better voltage regulation than any other filter that uses only passive components.

CAPACITOR INPUT LC FILTER

The pi circuit *LC* filter is usually called the capacitor input *LC* filter. It is shown in Figure 19.3C. The rectified current is fed directly into a capacitor before going to the choke. This results in an output voltage greater than 90% of the peak sine-wave voltage. However, regulation is not as good as with the choke input filter.

TYPICAL COMPONENT VALUES

The values of filter components depend somewhat upon whether the filter is used in a vacuum tube or solid-state receiver. In tube receivers, typical capacitor values range from 25 to 200 μF. Filter choke inductances range from 2 to 15 H. Part of the filter in a vacuum tube receiver may be an *RC* circuit, as shown in Figure 19.3A. These filters are practical when feeding circuits having relatively low current drain. This prevents an excessive voltage drop across the series filter resistor. In high-current-demand circuits, a filter choke is preferred.

In transistor TV receivers, the filter "sees" a considerably lower impedance than in tube receivers. Most transistors operate at low voltages, such as 20 to 25 V. Consequently, large filter capacitors (2 000 μF or more) are used in such filters. Large capacitors are needed because, as in all filters, the capacitors must offer a low impedance relative to the impedance in the receiver circuits.

19.4 TRANSFORMER AND TRANSFORMERLESS POWER SUPPLIES

Many TV receivers use power transformers. A power transformer is more expensive than some of the components that replace it in cheaper sets. Many advantages are possible by the proper use of a power transformer. They are:

1. Since the AC line is isolated from the B+ and filaments, there is no shock hazard to persons touching the chassis or metal cabinet.
2. B+ voltages are generated more efficiently. This results in longer receiver life and better voltage regulation.
3. The filament voltages for tubes that require a high DC bias are easily available on a separate winding.
4. The center-tapped secondaries allow full-wave rectifier circuits to use only two diodes instead of four.
5. Since some tubes are connected in parallel, they need not all have the same filament current requirements.
6. Some better-quality color-TV sets use a constant-voltage transformer. Its output voltage is not affected by input voltage changes between 105 and 135 V AC.
7. The set is easily adjusted for 110, 120, or 240 V operation if a tapped transformer primary is used.

The input sections of two tranformer power supplies are shown in Figure 19.4. The supply in Figure 19.4A totally isolates the AC line from all circuits. The circuit in Figure 19.4B is used with a polarized plug, though it can be used on a floating AC line. The 2.2-ohm resistor acts as a relatively short-time constant filter, together with the 0.05-μF capacitor. This filter reduces AC line transients. Its impedance at 60 Hz is high enough to effectively isolate the chassis ground from the AC line.

TRANSFORMER POWER SUPPLIES FOR TUBE RECEIVERS

Some TV vacuum tube receivers use a transformer only for the generation of B+ voltages and the picture tube heater voltage. This is an inexpensive approach because it is quite easy to connect all of the filaments in series across the AC line. However, many receivers require a DC bias of 100 to 200 V on the picture tube and damper tube heaters. This is needed so that the voltage difference between the heater and cathode does not become too large. If it did, there might be a heater-cathode short. A separate filament winding may be provided for this purpose.

Figure 19.5 shows such a typical TV-receiver power supply. A filament winding for the picture tube and damper tube filaments is also illustrated. This winding is biased to approximately two-thirds of the B+ voltage. Note that one side of the AC line is

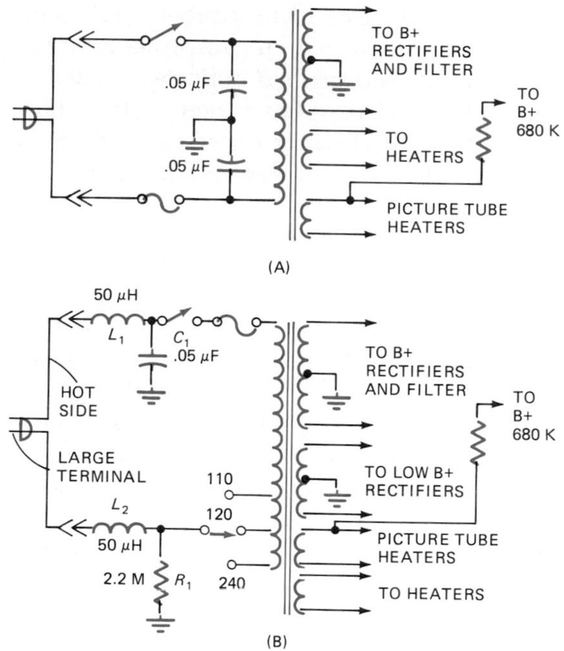

Figure 19.4A-B. The input portions of transformer power supplies: A. The conventional transformer; B. with tapped primary, two B+ windings, and polarized plug.

grounded; similarly, one side of the filament string must be grounded.

TRANSFORMERLESS TV POWER SUPPLY

A simple transformerless TV power supply is shown in Figure 19.6. This is a **half-wave rectifier** with a capacitor input filter. This particular power supply is for a 14-in (3.6-cm) portable TV receiver.

Series Heaters The heaters for all tubes in Figure 19.6 are connected in series and to the AC line through a 38-ohm dropping resistor. Note that the sum of all heater voltages equals 101 V. The balance of the line voltage (assuming 115 to 120 V AC) appears across the 38-ohm resistor. Since series tubes normally draw 0.45 A, the voltage drop across the resistor is

$$V = IR = 0.45 \times 38 = 17 \text{ V}$$

Thus, we have total drop of

$$101 \text{ V} + 17 \text{ V} = 118 \text{ V AC}$$

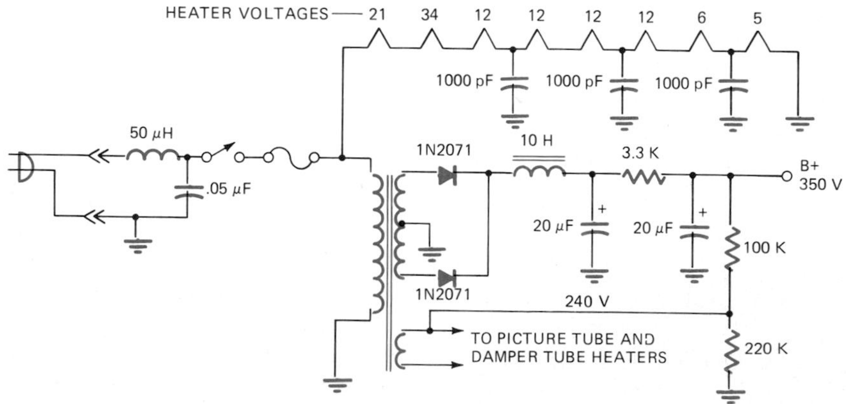

Figure 19.5 A vacuum tube TV receiver power supply, which uses a power transformer and a full-wave rectifier to supply B+. The picture tube and damper tube heater voltages are supplied by a separate isolated winding. Series filaments are used for all other tubes connected across the AC line.

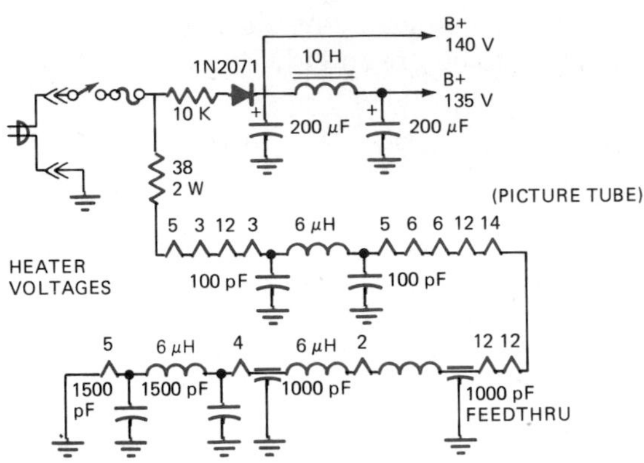

Figure 19.6 A simple half-wave transformerless power supply for a 14-in portable TV receiver.

Rectifier Circuit In Figure 19.6, a simple half-wave rectifier and filter furnish the B+ power. Note that the B+ values are higher than the 118 V AC average value. This is because the input filter charges (with no load) to the peak of the AC line voltage, or to approximately 168 V peak. However, the load currents through the 10-ohm surge resistor and filter choke drop the voltages to +140 V and +135 V, as shown in Figure 19.6. The 140-V output line has a large ripple and is used only for an audio output stage. Either push-pull or negative feedback must be used in this audio amplifier to prevent hum.

Note: This type of power supply is *not isolated* from the AC line. To avoid severe electric shock, always service TV receivers with such supplies with an isolation transformer.

19.5 VOLTAGE MULTIPLIERS

Large-screen vacuum tube receivers require B+ values of about 250 V. To achieve such values in a transformerless receiver, a voltage-doubler rectifier circuit must be used. In addition, voltage triplers are often used to obtain the desired high voltage for the anode of color picture tubes. (A circuit that uses a voltage tripler to provide anode high voltage is described in Section 19.9)

There are two basic types of voltage doublers: the half-wave (or cascade) doubler and the full-wave doubler.

THE HALF-WAVE DOUBLER

The more common of the two types of doublers is the **half-wave doubler,** shown schematically in Figure 19.7A. It is frequently used as the power supply for transformerless receivers. The half-wave doubler has a ripple frequency of 60 Hz. Thus, it is more difficult to filter than the full-wave doubler, which has a ripple frequency of 120 Hz. In addition, the regulation of the half-wave doubler is poorer than that of the full-wave type.

The half-wave doubler has one important advantage when used for TV receivers. This advantage can be seen with the aid of Figure 19.7A. Note that the grounded side of the AC line is also the DC ground. This connection is important because it reduces un-

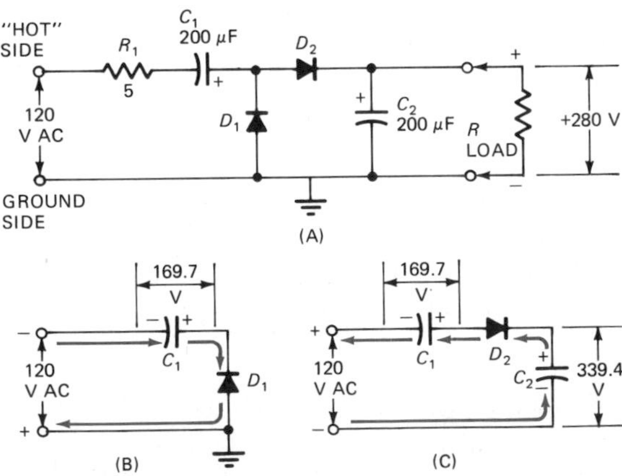

Figure 19.7A-C. A half-wave voltage doubler. A. The circuit diagram. B. Simplified diagram showing charging of C_1 to 169.7 V (no load). C. Simplified diagram showing charging of C_2 to double voltage, or 339.4 V, with no load.

desired 60-Hz hum pickup in the TV receiver circuits.

Note: This supply (as well as the full-wave type) is a *nonisolated* power supply. To avoid the possibility of severe electric shock, always service receivers using such supplies with an *isolation transformer*.

The half-wave voltage-doubler circuit of Figure 19.7A, is typical of those used in a number of TV receivers. The values shown are typical. Resistor R_1, as described for Figure 19.2, provides surge protection for diodes D_1 and D_2. The theoretical output of the doubler (without a load) should be

$$120 \text{ V AC} \times 1.4 \times 2 = 339.4 \text{ V DC}$$

However, in most receivers, the current load will reduce this to approximately 280 V DC. The operation of this doubler will now be described with the aid of the simplified diagrams.

Charging C_1 When the AC voltage has the polarity indicated in Figure 19.7B, capacitor C_1 charges through diode D_1, as shown. Capacitor C_1 charges to the peak value of the AC input, which is 120 V AC × 1.4 = 169.7 V. (For the sake of simplicity, a no-load condition is assumed.) At this time, diode D_2 is reverse-biased and is inoperative.

Charging C_2 When the AC voltage has the reverse polarity, as shown in Figure 19.7C, capacitor C_2 will charge to *twice* the peak AC value, or 339.4 V (no load). This happens as follows:

1. From the previous AC half cycle, capacitor C_1 is charged to 169.7 V. Note that the polarities of the AC line voltage and the charge across C_1 are now *series adding*. (This is similar to two batteries in series.)
2. The sum of the two voltages, 169.7 V AC (peak) + 169.7 V(C_1), now charges capacitor C_2 to 339.4 V, which is twice the peak AC line voltage, through diode D_2. (Diode D_1 is now reverse-biased.)

Although 339.4 V DC is shown as the output voltage, this is done only for simplicity. In an actual set the output voltage is reduced by the drop across R_1 and by the TV-receiver current (i.e., load) requirements.

Note that capacitor C_2 charges only during one half cycle of the input AC wave. Therefore, this is effectively a half-wave circuit, and it has a 60 Hz ripple frequency.

THE FULL-WAVE DOUBLER

The **full-wave doubler** offers two advantages over the half-wave doubler: higher ripple frequency (120 Hz) and better regulation. However, as seen in Figure 19.8A, the full-wave doubler does not have a common ground connection for one side of the AC line and B+, which is the case for a half-wave doubler. As a result, this circuit is used only with a transformer.

Capacitors C_1 and C_2 in Figure 19.8A are connected in series. The doubled output voltage is taken across *both* capacitors.

For simplicity in explaining how the full-wave doubler works, a no-load condition is assumed at this time. In Figure 19.8A, however, a loaded condition is shown. This load will reduce the maximum voltage output of the power supply.

Charging C_1 When the AC voltage has the polarity indicated in Figure 19.8B, capacitor C_1 charges through diode D_1 as shown. With no load, C_1 charges to the peak value of the AC voltage, which is 169.7 V.

Charging C_2 When the AC voltage has the reverse polarity, as indicated in Figure

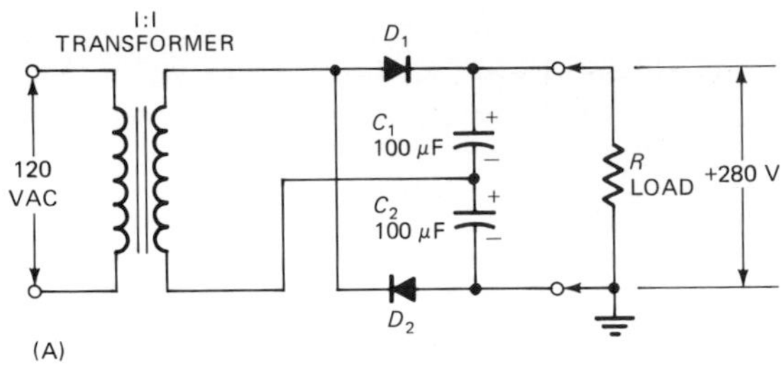

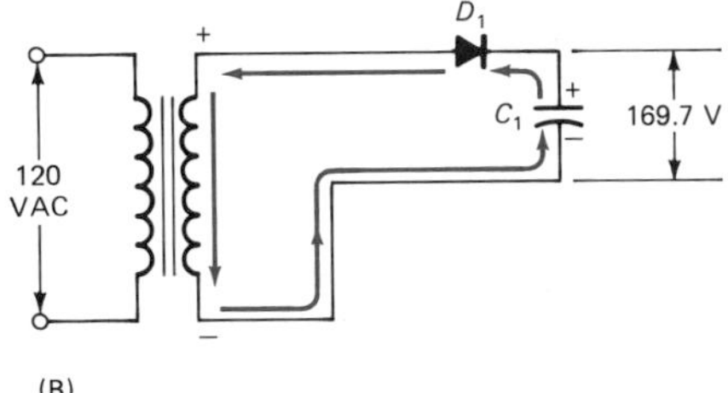

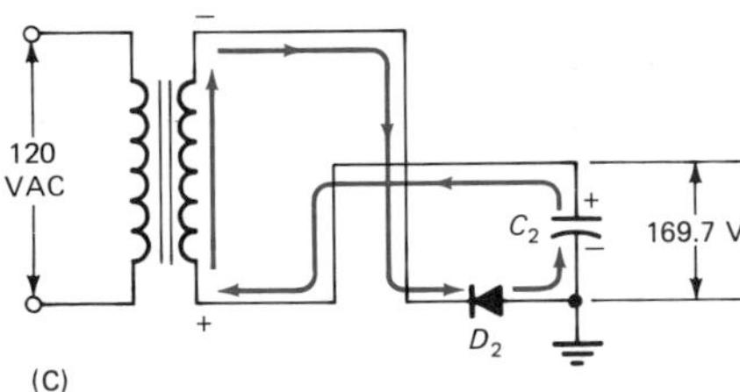

Figure 19.8A-C. The full-wave voltage doubler. **A.** The circuit diagram. **B.** Simplified diagram showing C_1 charging to +169.7 volts (no load). **C.** Simplified diagram showing C_2 charging to +169.7 V (no load). The no-load output voltage of +339.4 V is taken across the two capacitors in series. The loaded voltage, as shown in Part A, however, is +280 V.

19.8C, capacitor C_2 charges through diode D_2. With no load, C_2 also charges to the peak AC value.

Series Capacitors Note in Figure 19.8A that C_1 and C_2 are connected so their voltages are *series-aiding*. This is similar to two batteries in series, as previously mentioned. The no-load voltage across the two capacitors is 339.4 V. However, this value drops when a load is applied. Here, the loaded output voltage is assumed to be 280 V, which is a practical value.

Since either C_1 or C_2 is being charged on both half cycles of the AC wave, the ripple frequency is 120 Hz. Thus, this circuit is a full-wave voltage doubler.

HALF-WAVE VOLTAGE TRIPLER

The **half-wave voltage tripler** is made up of a half-wave voltage doubler (see Figure 19.7), plus an additional half-wave rectifier (D_3) and filter capacitor (C_3). A schematic diagram of such a unit is shown in Figure 19.9. (Note that a half-wave voltage **quadrupler** would require a fourth half-wave rectifier and filter capacitor added to a tripler.)

Voltage Tripler Operation

As explained in the discussion of half-wave voltage doublers, the doubled voltage appears across capacitor C_2 (Figure 19.9). Now consider that the AC input polarity is such that the bottom of the transformer secondary is positive. This voltage, in series with the doubled voltage across capacitor C_2, charges through diode D_3. Current flow is shown by the arrows in Figure 19.9. The peak AC voltage in series with the voltage across C_2 charges C_3 to *triple* the input voltage.

With no load, the voltage across C_3 would be $120 \times 1.4 \times 3 = 509$ V. However, in Figure 19.9, this voltage is shown as 420 V to indicate the effect of a practical load. It should be pointed out that the greater the amount of voltage multiplication, the poorer the voltage regulation. However, this is generally not a problem for devices using voltage multipliers. For example, voltage triplers are commonly used to increase the high voltage for a picture tube anode. In many cases, this increase for color picture tubes is from about 9 kV to 27 kV.

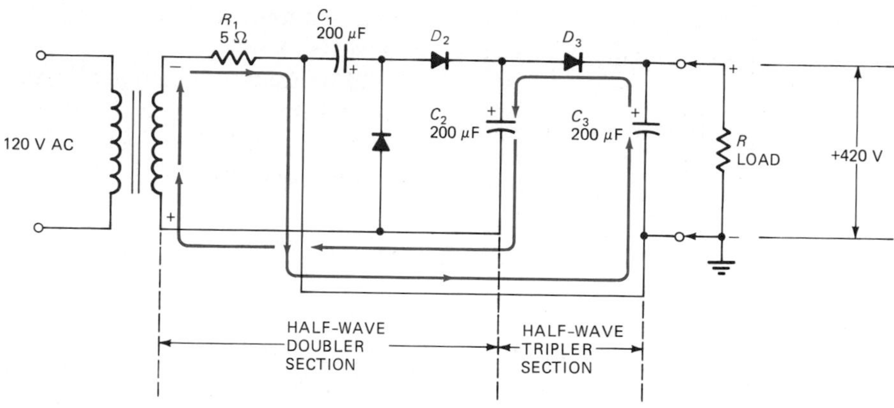

Figure 19.9 Schematic diagram of a half-wave voltage tripler. Note that it consists of a half-wave doubler, plus an additional rectifier section. Compare it with Figure 19.7. Arrows indicate charging of C_3 to triple voltage (see text).

19.6 VOLTAGE REGULATORS

Power supply voltage regulation refers to the ability of the circuit to maintain the output DC voltage relatively constant regardless of changes in AC line voltage or in load current. The low-voltage, B+ power supplies in vacuum tube monochrome TV receivers are not regulated because B+ values are not critical in these sets. However, in vacuum tube color-TV sets, the high voltage to the picture tube anode is regulated. This is done to maintain proper beam convergence and focus beam current (brightness).

Regulated Supplies In solid-state TV receivers and particularly in color receivers, it is essential that the B+ (or B−) supplies be regulated. Most transistors operate at fairly low voltages (5 to 25 V). An appreciable change in the operating voltage of these transistors can cause important changes in their operating characteristics. This in turn can seriously affect the performance of a solid-state color or monochrome TV receiver.

This is also true of transistors operating at higher voltages. For example, the horizontal output transistor may operate at 110 V. Appreciable changes of this voltage will cause corresponding changes in raster width. Accompanying changes will take place in the anode high voltage of the color picture tube as well as in the focus and screen potentials.

In addition, many solid-state color TV receivers derive most of their B+ (or B−) voltages from windings on the horizontal output transformer. (This type of power supply is described later in the chapter.) Any appreciable change in the B+ voltage to the horizontal output transistor changes these *derived* voltages. By regulating all of the B+ (or B−) potentials, operation of a solid-state color-TV receiver may be achieved over a practical range of AC line voltage (105 to 135 V) receiver-load current variations.

Types of Regulators There are several types of voltage regulators, categorized generally as follows:

1. Zener diode regulators
2. Series-pass transistor regulators
3. Three-terminal regulators
4. Switching regulators

19.7 THE ZENER DIODE REGULATOR

Zener (or avalanche) **diodes** operate on the basis of a reverse-current breakdown. This breakdown (or avalanche, or zener) voltage varies from 2 to 200 V for different zener diodes. As the applied reverse voltage increases to a specific value for a particular diode, the diode breaks down. At this point, the reverse current increases rapidly. This is the useful operating condition. In the operating region, the voltage *across* the zener diode is nearly constant for a wide range of zener current. This voltage is equal to the design value of the particular zener diode.

Zener diodes are used to provide fixed reference voltages for transistor voltage regulators. They are also used as simple shunt voltage-regulating devices. In the latter case the output voltage will be held constant at the zener voltage.

Zener Current The maximum operating zener current is limited by the permissible power dissipation of the device. Maximum zener currents range from a few milliamperes to tens of amperes. Zener diodes with a breakdown voltage of less than 6 V have a negative temperature coefficient, that is, the breakdown voltage decreases with increasing temperature. For breakdown voltages above 6 V, the zener diodes have a positive temperature coefficient i.e., the breakdown voltage increases with increasing temperature.

Zener Diode Regulator Circuit A simple voltage-regulating circuit using a zener diode is shown in Figure 19.10. Values shown are for an unregulated input DC voltage of 11 V and a regulated output voltage (zener voltage) of 7.5 V. The regulated voltage remains essentially constant for un-regulated input voltages from approximately 9.5 V to 14 V and for load currents of *less* than 4.5 mA.

The operation of this regulator circuit is as follows. If the input voltage is greater than 9.5 V (as it is in Figure 19.10), the zener will break down and 7.5 V will appear across its terminals. The difference between the input and zener voltages, 11 − 7.5 = 3.5 V, appears across the series resistor R_1. Thus, a current of I = 3.5/500 = 7 mA is in R_1. This current is divided between the load (2.2 kΩ) and the zener diode. The load current is I = 7.5/2 200 = 3.41 mA. The zener current is therefore 7.0 − 3.41 = 3.59 mA.

Input-Voltage Changes If the *input voltage increases*, there will be a higher zener diode current and a greater voltage drop across R_1. The output voltage remains constant at 7.5 V.

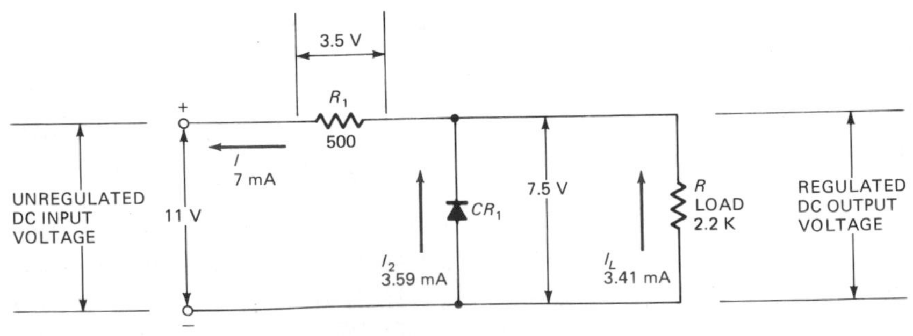

Figure 19.10 Schematic diagram of a simple voltage regulator using a zener diode.

A *decrease* of *input voltage* results in less current through the zener diode and a smaller voltage drop across R_1. Again, the output voltage remains at 7.5 V.

Load Changes If the *load current increases* (input voltage constant), the voltage drop across R_1 increases. This results in less zener current but a constant voltage across the zener diode of 7.5 V. Similarly, if the *load current decreases* (input voltage constant), the voltage drop across R_1 decreases, resulting in higher zener current. As before, the voltage across the zener diode remains constant at 7.5 V (the output voltage).

In Summary It is seen from the above, that input voltage or load changes are compensated by corresponding changes of zener diode current and the voltage drop across R_1. However, within specifications, the voltage across the zener diode (the output voltage) remains constant at its design value.

19.8 THE SERIES-PASS TRANSISTOR REGULATOR

The zener diode regulator is a simple circuit, but it has several disadvantages:

1. Appreciable power dissipated in the series resistor (R_1) and in the zener diode is wasted.
2. The voltage across the zener diode changes by a few percent as the current through it changes.
3. The output current is limited.
4. The zener circuit does not provide any appreciable filtering of a large ripple voltage.

These disadvantages are overcome by a series-pass transistor regulator circuit. A block diagram of this circuit is shown in Figure 19.11A and the schematic diagram in Figure 19.11B.

Circuit Operation The series-pass transistor regulator in Figure 19.11 includes protection for series-pass transistor Q_1 against excessive load currents. It also protects against damage to the load in the event of a shorted Q_1.

Transistor Q_1 is the series-pass transistor, Q_2 is the error amplifier, and zener diode CR_1 provides the fixed reference voltage required. The voltage-monitoring network consists of the effective resistance of R_3 and R_4. The desired output voltage is obtained by adjusting the potentiometer portion of R_3 and R_4. The element that actually controls (and stabilizes) the DC output voltage of the regulator is series-pass transistor Q_1, this by changes in the voltage drop from collector to emitter (V_{CE_1}). These changes add to or subtract from the input voltage (V_I) as required to maintain a constant output voltage (V_O). Changes of V_{CE_1} are caused by changes of the base-to-emitter voltage of Q_1 (V_{BE_1}).

The output voltage will tend to rise because of an increase of the input DC voltage (V_I) or because of a decrease in the load (increase of R_{load}). In either case, V_4 will rise. Since V_4 is the base-to-emitter voltage (V_{BE_2}) for Q_2, the collector current of Q_2 will increase through R_1. This causes a drop in the base-to-emitter voltage (V_{BE_1}) of Q_1, which in turn decreases the emitter-to-collector current of Q_1. This increases the voltage drop across Q_1 (V_{CE_1}), and the output voltage (V_O) is reduced to its nominal value.

The output voltage will tend to fall either because of a decrease of the input voltage V_I or because of an increase in the load (decrease of R_{load}). Now V_4 (V_{BE_2}) will drop. The reduced current through Q_2 and R_1 causes an increase of V_{BE_2}. In turn, the current will increase through series-pass transistor Q_1. This decreases the voltage drop

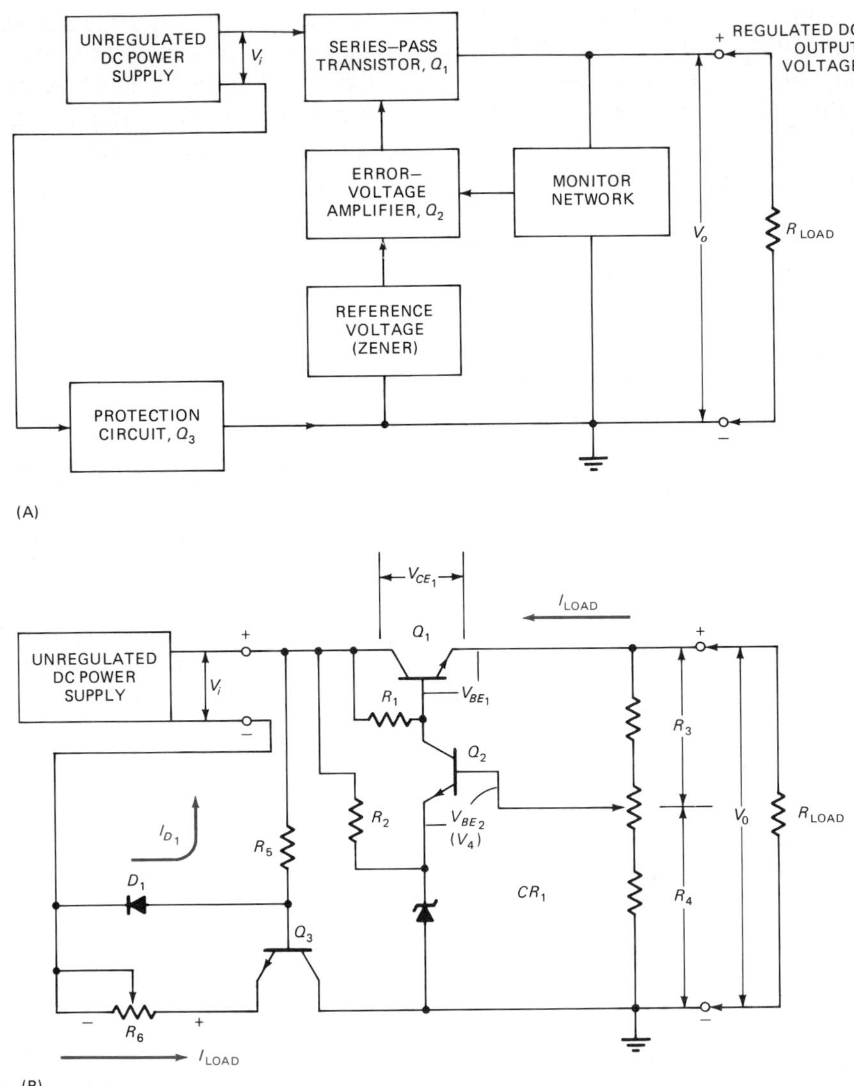

Figure 19.11A-B. The series-pass regulator circuit, with protection for Q_1. **A.** The block diagram. **B.** The schematic diagram (see text).

across Q_1 (V_{CE_1}), and the output voltage (V_O) is increased to its nominal value.

Since the load current passes through Q_1, this transistor may have to dissipate appreciable amounts of power. Therefore, a power transistor is used for Q_1, which is mounted on a heat sink.

The Protection Circuit Two kinds of protection are provided by the protection circuit. (1) It protects against damage to the load in the event of a series-pass transistor short circuit. (2) It prevents damage to the series-pass transistor Q_1 in the event excessive load current is drawn. (Many protection circuits protect only for item 2 above.)

The protection circuit in Figure 19.11A is in *series* with the negative leg of the power supply. (The same power supply current flows through the negative and positive legs.) Basically, here is how the protection circuit works. In the event of either (1) or (2) the tendency is for the output voltage and current to rise. When this happens, the protection circuit increases the effective series resistance of Q_3. This reduces the power supply output voltage and current to a safe value. Note that Q_3 operates only if (1) or (2) occur.

During normal operation, protection transistor Q_3 in Figure 19.11B is saturated because of the selected value of R_5. Thus, its emitter-to-collector resistance is practically zero, and the protection circuit does not affect the operation of the regulator. If there should be an emitter-to-collector short circuit in the series-pass transistor Q_1, the output voltage V_O and current I_{load} will rise. Thus, the current through R_6 and the voltage across it will increase.

Normally, diode D_1 does not conduct because of its inherent reverse barrier voltage (about 0.7 V). However, increased current through R_6, caused by the short-circuited Q_1, increases the voltage across R_6 enough to overcome the barrier voltage. Now D_1 conducts through R_5, reducing the base voltage (and current) of Q_3, and Q_3 comes out of saturation. As a result, the emitter-to-collector resistance of Q_3 is increased and the power supply current (and voltage) are reduced to a safe value that will protect the load from damage.

Resistor R_6 is preadjusted so that, if Q_1 short-circuits (or if there is an excessive current demand from the load), D_1 will conduct and cause the emitter-to-collector resistance of Q_3 to increase. This, of course, will limit the output current (and voltage) to a safe value.

The protection circuit operates to protect Q_1 from excessive load current in the same way. If the load across the regulator short-circuits or if the load current simply increased beyond a safe value, the current through R_6 would cause D_1 to conduct and Q_3 to increase its emitter-to-collector resistance. This would limit the output current to a safe value.

Ripple Reduction Because there are two amplifying elements in Figure 19.11B, namely, Q_1 and Q_2, this circuit is considerably more sensitive to input voltage changes than the zener regulator. Ripple voltage from the unregulated DC power supply is "seen" by the series-pass regulator as rapid variations (60 or 120 Hz) of input voltage. The regulator circuit responds to the ripple voltage as it would to any other variation of regulator input voltage. That is, it tries to cancel any such variations. Although cancellation is not complete, the ripple is appreciably reduced by the regulator circuit. This action allows a reduction in the requirement for passive (choke and capacitor) filtering at the regulator output.

19.9 THREE-TERMINAL VOLTAGE REGULATOR

A **three-terminal voltage regulator** is constructed as a monolithic (one-piece) conductor IC. That is, all circuit components are manufactured on a single slice of silicon substrate (a "chip"). The name *three-terminal* derives from the fact that only three connections are made to the IC. As shown in Figure 19.12A, these connections are (1) input, (2) output, and (3) ground.

Characteristics The IC units contain all the necessary components to provide voltage regulation and automatic excess-current pro-

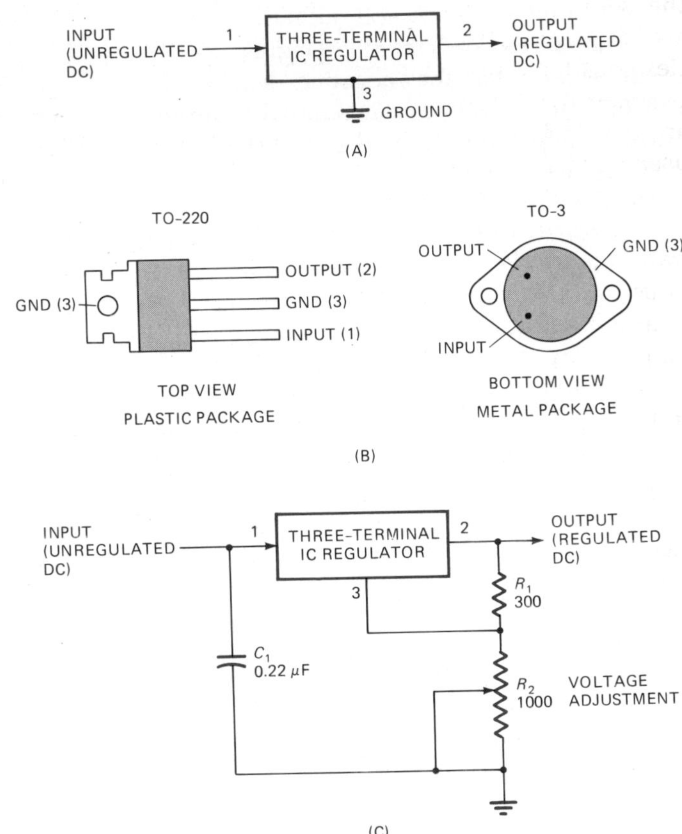

Figure 19.12A-C. The three-terminal voltage regulator. **A.** The basic block diagram. **B.** Two types of IC packages. **C.** Modification to permit a variable voltage output.

tection for the chip. Included in the chip are a high-gain error amplifier, sensing resistors and transistors for current limiting, a voltage reference, and series-pass transistors. Typical output voltage ratings are 5, 6, 8, 12, 15, 18 and 24 V. These are preset for individual ICs. Load current ratings are available up to 3 A.

The ground connection of the IC is usually connected to the chassis. It is therefore not necessary to insulate the IC from ground. Also, the chassis will then act as a heat sink for the IC.

The current capabilities of the IC can be extended by using it to control the operation of a higher-current series-pass transistor circuit (previously described). In this case, the current-limiting properties of the IC are effective for the composite circuit.

ICs come in various package styles. Two are shown in Figure 19.12B: plastic package TO-220 and metal package TO-3.

Adjustable Voltage Output Simple external circuitry can be added to the IC, if desired, to provide an adjustable regulated voltage output. This is shown in Figure 19.12C.

19.10 SWITCHING REGULATORS

The series-pass transistor regulator described is called a linear regulator, because the series-pass transistor is always ON. A **switching regulator** differs because its series-pass transistor is switched ON and OFF periodically. Typical switching frequencies are 5 to 100 kHz. Switching regulators are used where small size and high efficiency are required.

In a switching regulator, the series-pass transistor is always fully ON or fully OFF. Thus, there is much less heat produced than in a linear regulator. This is true for a wide range of input-output voltages and currents. As an example, a given linear regulator heat sink must dissipate 230 W. The equivalent switching regulator heat sink must dissipate only 30 W.

Switching regulators may be made with discrete components. However, as in the case of the three-terminal regulator, they are usually found in monolithic IC form.

At times it is desirable to have a higher current than can be handled by a monolithic switching regulator. This may be achieved by adding an external series-pass transistor. However, all functions are still controlled by the monolithic device. Note that, as previously mentioned, a similar solution may be used with the three-terminal regulator.

A switching regulator is used to obtain the +110 V regulated supply for the RCA *derived-power supply* to be described. This voltage powers the horizontal output stage. This particular circuit uses a silicon-controlled rectifier (SCR) as the switched series-pass device.

Basic Circuit The basic circuit of a switching regulator is shown in Figure 19.13. Compare with Figure 19.11A. The series-pass transistor is again Q_1. The A_1 takes the place of the error-voltage amplifier Q_2 in Figure 19.11A. The reference voltage can be supplied by a discrete zener diode, or the zener may be part of a monolithic assembly. In fact, all the parts of Figure 19.13 can be contained in a chip except for the unregulated DC supply and filter choke L_1.

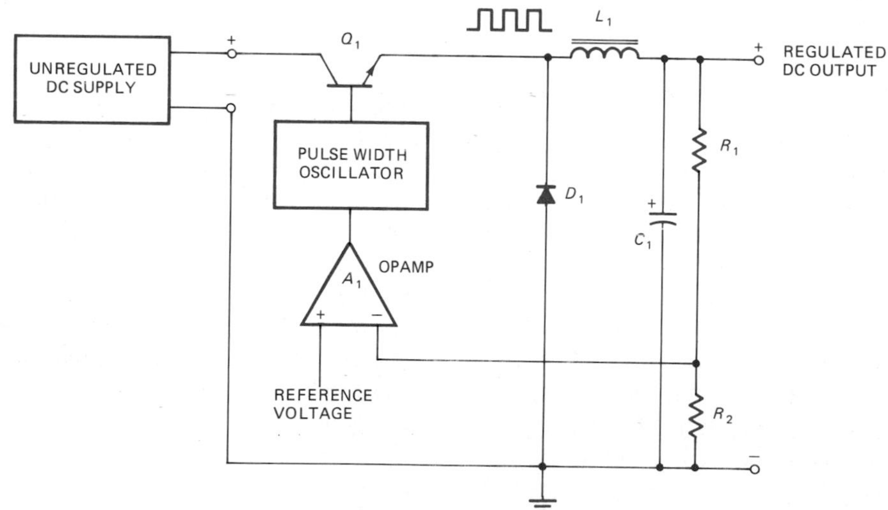

Figure 19.13 Simplified schematic diagram of a switching voltage regulator. For a monolithic type, all parts, except the unregulated DC supply and L_1, would be contained in the "chip", including the source of the reference voltage.

Pulse Width Oscillator The major difference between the circuit in Figure 19.13 and that in Figure 19.11A is the **pulse width oscillator**, which is the heart of switching regulator operation. The oscillator wave form switches Q_1 ON and OFF periodically (for example, at 20 kHz). The output of Q_1 is (as shown) a rectangular wave that is applied to the filter consisting of L_1 and C_1. Because of the high ripple frequency (say, 20 kHz), L_1 and C_1 can have relatively small values. The output of the filter is the regulated DC.

Regulation and Protection Regulation by the switching circuit of Figure 19.13 is performed in basically the same manner as described for the circuit of Figure 19.11B. (For another type of switching regulator and its operation in an RCA color TV receiver, see Section 19.13, "Horizontal-Frequency Power Supply".) Protection of the regulator parts and of the load may also be provided for a switching regulator, as discussed in connection with Figure 19.11B.

19.11 SOLID-STATE MONOCHROME TV POWER SUPPLIES

Solid-state TV receivers use lower DC voltages than vacuum tube sets. The current requirements are higher because transistors are low-impedance devices. Some portable solid-state TV power supplies generate only 6 or 12 volts DC which makes them easy to switch over to battery operation. Figure 19.14 shows a typical low-voltage power supply which may be operated from the AC line or from a battery.

The ON-OFF switch has a third position for charging the batteries without operating the receiver. The batteries are also charged during receiver operation; in addition, they serve as voltage regulators during normal operation. They perform in a manner similar to the parallel combination of a zener diode and a filter capacitor. The 100-ohm resistor limits the battery charging current to a recommended value. At the same time, this resistor is part of the *RC* filter network. The 10-ohm resistor limits the initial capacitor charging current to a safe value for the 1N2070 diodes.

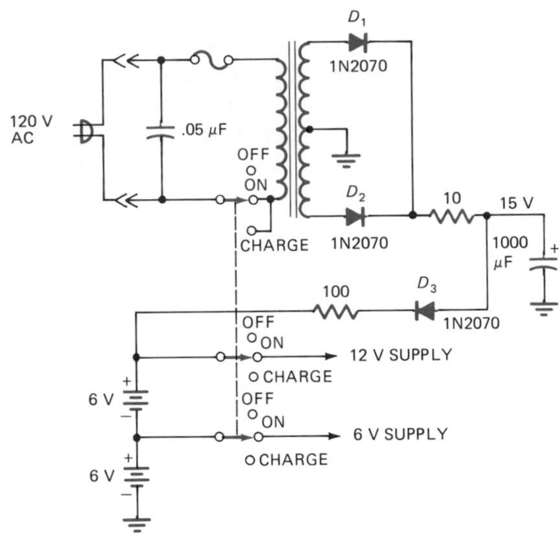

Figure 19.14 The power supply for an AC or battery operated monochrome transistor TV receiver.

High and Low DC Voltages Some solid-state TV receivers utilize both low-voltage and high-voltage transistors. This requires an assortment of voltages. However, to reduce cross coupling between receiver circuits, low-impedance supply sources are required. Some receivers have 3 or 4 rectifier-filter circuits for this reason. An assortment of bleeder resistors would not provide this low impedance. A power supply of this type is shown in Figure 19.15. The +24 and −20 output voltages drive most of the low-level stages in the set. A zener regulator is used for the +24 supply, and a series-pass regulator is required for the +20-V supply.

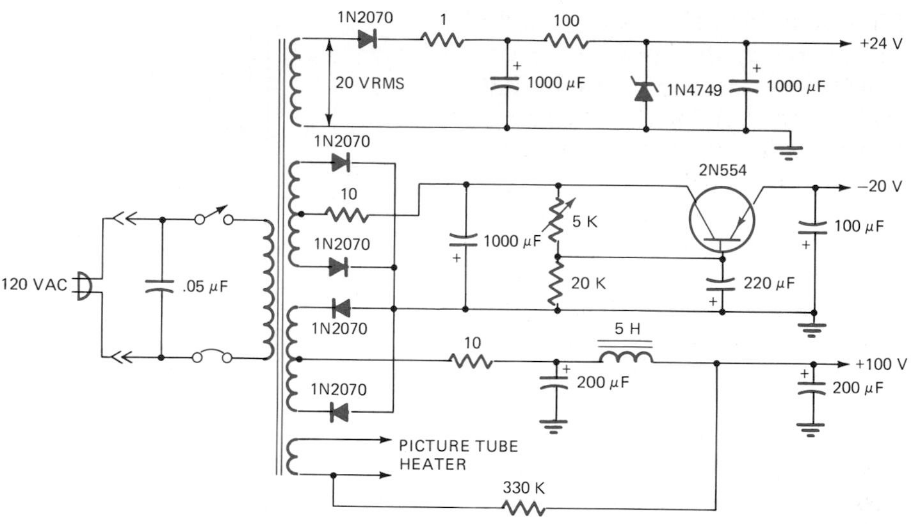

Figure 19.15 Power supply for a solid-state TV receiver that requires several low-impedance output voltages.

The +100-V output voltage powers the video, vertical, horizontal, and audio output stages. Note the wiring arrangement of the two full-wave rectifier circuits. Some receivers do this so that part of the diode can be physically connected to ground. The ground is then used as a heat sink, and no insulator is required between ground and the rectifier body.

19.12 A HYBRID COLOR TELEVISION POWER SUPPLY

In general, color TV power supplies are larger and more complex than those in monochrome sets because color TV receivers have more circuits.

Figure 19.16 shows a power supply for a hybrid, large-screen color receiver. This set has several vacuum tubes. Most of the transistors receive their power from the +20 V developed by the series-pass regulator. The tuner also uses +20 V with extra filtering. An adjustment (R_1) is available for setting the exact voltage of the +20-V supply. The other side of the transformer winding of the +20-V supply is used for the tube heaters. The +290-V B+ circuit has a degaussing circuit in series with it. A thermistor and varistor (voltage-variable resistor) circuit control the current in the degaussing coil. As the circuit warms up, the resistance of the thermistor decreases. At the same time, the varistor resistance increases. The net result is a surge of current through the degaussing coil for a few seconds before the set warms up. Very little current flows in the degaussing coil during normal receiver operation.

Several other things are worth noting in Figure 19.16. The picture tube heater is biased to +140 V to prevent cathode-to-heater shorts. The 0.01μF capacitors across the diodes are for transient protection. Solid-state diodes in the rectifier circuits have a much longer life with these capacitors present. In some receivers, the diode will soon burn out if the capacitor is removed.

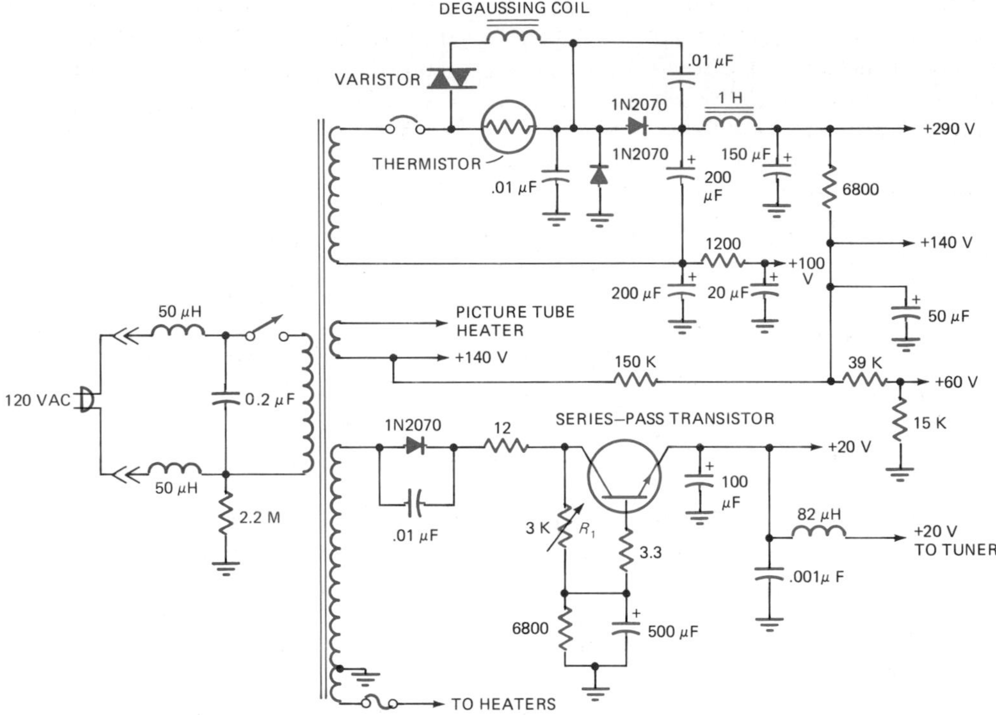

Figure 19.16 Power supply for a transistorized color TV receiver, which also has several tubes.

19.13 HORIZONTAL FREQUENCY POWER SUPPLY

The advantages of transformer power supplies are described in Section 19.4. However, a transformer adds appreciable bulk and weight to a TV receiver. In addition, the low ripple frequency (60 or 120 Hz) requires high-capacity filter capacitors. Finally, some B+ voltages may require additional filtering in the form of a filter choke.

As mentioned at the beginning of the chapter, B+ (and B−) voltages can be obtained from the horizontal output transformer. Such voltages are called *derived* voltages. When most of the B+ (and B−) supplies are derived from the horizontal output ("flyback") transformer, the following advantages are realized:

1. Bulk and weight of the TV receiver are reduced.
2. The powerline is isolated from the receiver.
3. The ripple frequency is higher (for color-TV receivers it is 15 734.26 Hz).

The high ripple frequency simplifies filtering. This permits the use of smaller filter capacitors and eliminates the need for filter chokes.

The derived-voltage system described below is from an RCA XL-100 series receiver.

"START" AND "RUN" VOLTAGES

Overview In the TV receiver whose power supply is described here, most of the circuits, including the horizontal oscillator circuit, are powered from derived voltages. (The exception is the horizontal output transistor, which is powered from a transformerless, line-operated B+ supply, to be described. In order to produce the derived voltages, however, the horizontal oscillator must first be *started*. This is essential, so that the horizontal output transistor receives its drive signal.

The *start-up* voltages for the horizontal oscillator (buffer and driver) are *momentary* voltages developed when the TV receiver is first turned on. However, these start-up voltages are sufficient to begin horizontal oscillator-buffer and driver operation. Once these circuits are operating, the derived voltages provide the normal operating, or "run," voltages for the oscillator buffer and driver. As previously mentioned, the horizontal output transistor is independently powered from a line-operated supply.

HORIZONAL OUTPUT STAGE POWER SUPPLY

A simplified schematic diagram showing how the start-up B+ voltages and the horizontal output stage B+ voltage are developed is shown in Figure 19.17. The horizontal output stage B+ supply will be considered first.

CAUTION: The derived voltages are isolated from the AC line, but the horizontal output stage power supply is *not*. Thus, for safety, it is essential, when servicing this type of receiver, to always connect it to the AC line through an isolation transformer.

Input Circuit The 120 V AC is applied to the input of a bridge rectifier (CR_{201}, CR_{202}, CR_{203}, CR_{204}), through line choke L_{201}. This choke, in conjunction with varistor RV_{201}, protects the bridge rectifiers from line surges or transients. Resistor RF_{201} is a *fusible* type. It provides additional rectifier protection by limiting the peak currents. In addition, it provides overload protection against rectifier damage from a possible B+ overload.

Rectifier Output The output voltage of the bridge rectifier, nominally +150 V, is applied through one winding of the start-up transformer T_{201} to filter choke L_{403}. Additional filtering is provided by filter capacitor C_{304}, and the filtered +150 V is fed to the input of a switching regulator. The output of the regulator (+110 V) is fed to the collector circuit of the horizontal output transistor.

Power Supply Regulation In addition to the +110-V regulator mentioned above, all voltage-critical circuits in this RCA color-TV receiver are powered by regulated voltages. For example, a +22-V regulator located in the video module powers this module. It also powers the horizontal oscillator, the tuners, and the IF/AFT module. This same supply also feeds the chroma module. Here it is additionally regulated at +11.2 V for use in this module.

With the (solid-state) regulator system described above, the receiver can perform consistently over a range of AC line voltages from 105 to 135 V. This is particularly important because "brownouts" (small voltage cutbacks by the power company during peak usage periods) occur in many parts of the country.

START-UP POWER SUPPLY CIRCUITS

The operation of the *start-up* circuits will be explained with the aid of Figure 19.17. During normal receiver operation, the output of the bridge rectifier (+150 V) is fed through the primary of start-up transformer T_{201} and filter choke L_{403} to filter capacitor

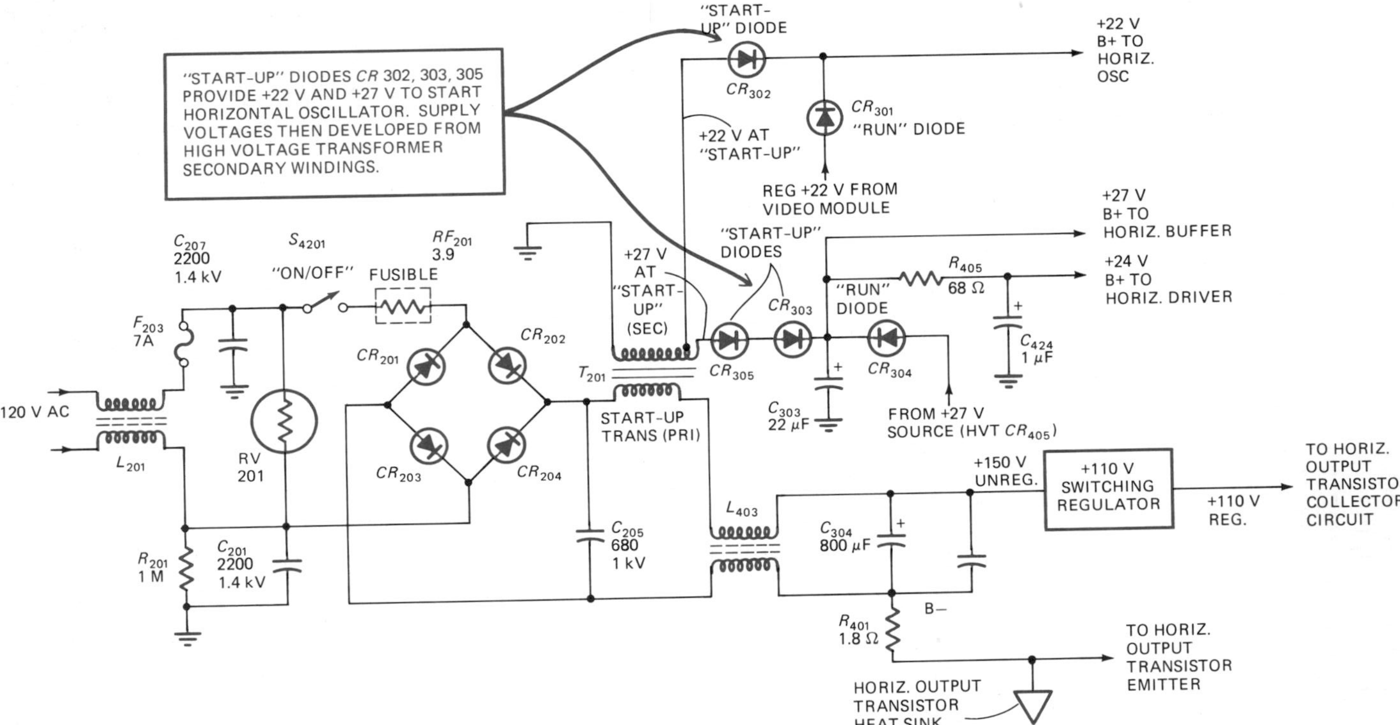

Figure 19.17 Simplified diagram of a derived-voltage power supply. Unless otherwise stated, all capacitor values are in pF. *(Courtesy of RCA Consumer Electronics.)*

C_{304}. From there it is applied to the +110 V horizontal regulator. However, our immediate concern here is with conditions prior to and at the instant of closing ON/OFF switch S_{4201}.

Charging Of C_{304} When ON/OFF S_{4201} is open, there is no current in start-up transformer T_{201} or filter choke L_{403}. Filter capacitor C_{304} is completely discharged. At the instant S_{4201} is closed, C_{304} appears to be a short circuit and maximum current flows into the capacitor through the T_{201} primary and L_{403}. This transient current through the T_{201} primary induces a voltage into its secondary winding. The polarity is such that a positive voltage is induced at the right side of the T_{201} secondary.

Horizontal Oscillator Start-Up Voltage A transient start-up voltage of +22 V is now applied through start-up diode CR_{302} to the horizontal oscillator. This voltage is sufficient to start oscillations. At this time, *run* diode CR_{301} is reverse-biased since power is not yet available from the +22-V regulator operating from the +27-V *derived* supply.

Horizontal Buffer And Driver Start-Up Simultaneous with the above, start-up voltages of +27 V (buffer) and +24 V (driver) are passed through start-up diodes CR_{303} and CR_{305}. As above, run diode CR_{304} is reverse-biased by the start-up voltage because the derived supply is not yet operating. These two start-up voltages are sufficient to *initially* power the horizontal buffer and driver stages.

Since run diodes CR_{301} and CR_{304} are reverse-biased during the start-up period, which is very short, they provide isolation between the start-up and run circuits. When the derived (run) supplies are active, start-up diodes CR_{302}, CR_{303}, and CR_{305} become reverse-biased, isolating the start-up and run supply circuits.

RUN POWER SUPPLY CIRCUITS (DERIVED POWER)

After the horizontal oscillator, buffer, and driver are powered by the start-up circuits, the horizontal output deflection circuits become active. At that time, the run (derived) power supply circuits become active. The start-up circuits now become inoperative until the next time the TV receiver is turned off and then on again. As previously mentioned, all receiver circuits except for the horizontal output transistor are powered from the derived power supplies. These receive their energy from the horizontal output transformer (flyback).

+110-V REGULATOR

The +110-V regulator is a switching regulator, the principles of which were explained in Section 19.6. However, there are some important differences in this regulator, and these will be described with the aid of the simplified block diagram of Figure 19.18. Note that the use of a switching regulator results in a significant reduction in the power consumption of a TV receiver.

Basic Operation As noted previously, a switching regulator may be thought of as an electronic switch. This "switch" turns the B+ to the load ON and OFF. The load is the horizontal output stage Q_{401}. Filter capacitor C_{401} stores the energy of the pulses and supplies it to the load at a constant rate.

In Figure 19.18, the switch is actually an SCR (silicon-controlled rectifier). For simplicity, though, it is shown as a switch (S). When this switch is closed, B+ is applied to the load. Note that load current must flow through inductor L_{402}. A parallel resonant circuit is formed by L_{402} and C_{401}. This circuit presents a series impedance at the output of

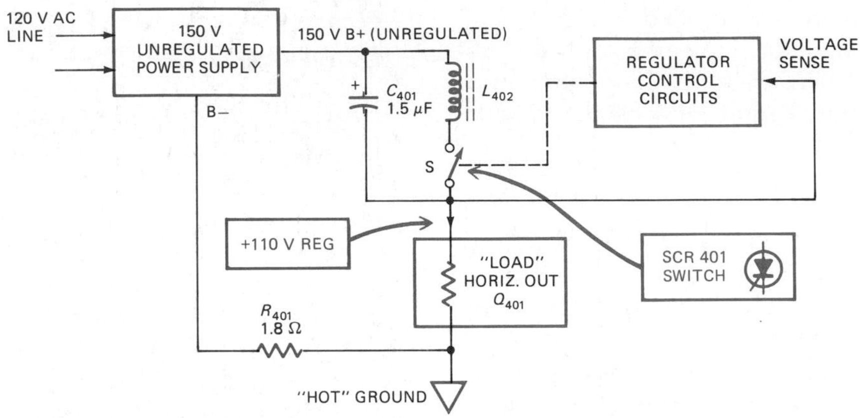

Figure 19.18 Simplified block diagram of the +110-V switching regulator. *(Courtesy of RCA Consumer Electronics.)*

the +150-V unregulated supply. As a result, the current through filter capacitor C_{401} (and the voltage across it) may increase at a rate determined by L_{402}.

This regulator is a *closed-loop* system. Note the connection from switch S to the regulator-control circuits. This is labeled "voltage sense." When the load voltage reaches +110 V, the regulator control circuits sense this and open the switch. The switch remains open until the control circuits sense that the load voltage has dropped. When that happens, the control circuits cause "switch" S to close and C_{401} is recharged to the desired +110 V. In this way, the B+ is maintained at a regulated +110 V.

SCR Operation The operation of S (actually an SCR) is explained with the aid of the simplified diagram of Figure 19.19. Note that a winding of the horizontal flyback transformer is in series with L_{402} and the *anode* of SCR_{401}.

The SCR is turned ON by a gate pulse from the regulator control circuits (Figure 19.18) via driver transformer T_1. Once an SCR is turned ON, it can be switched OFF only by momentarily forcing its anode current to zero. This is done during the horizontal retrace period by applying a negative 300-V peak pulse to the anode, as shown in Figure 19.19. Thus, the ON/OFF function of the switch S of Figure 19.18 is accomplished.

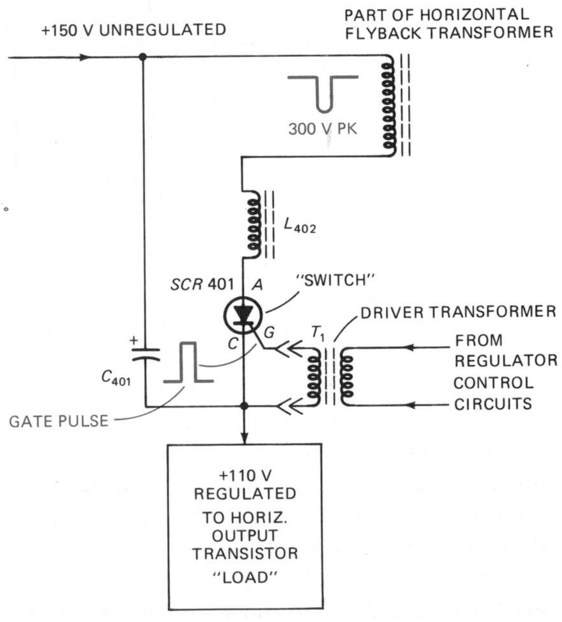

Figure 19.19 Simplified diagram to illustrate the operation of SCR_{401}, the "switch" of Figure 19.18. *(Courtesy of RCA Consumer Electronics.)*

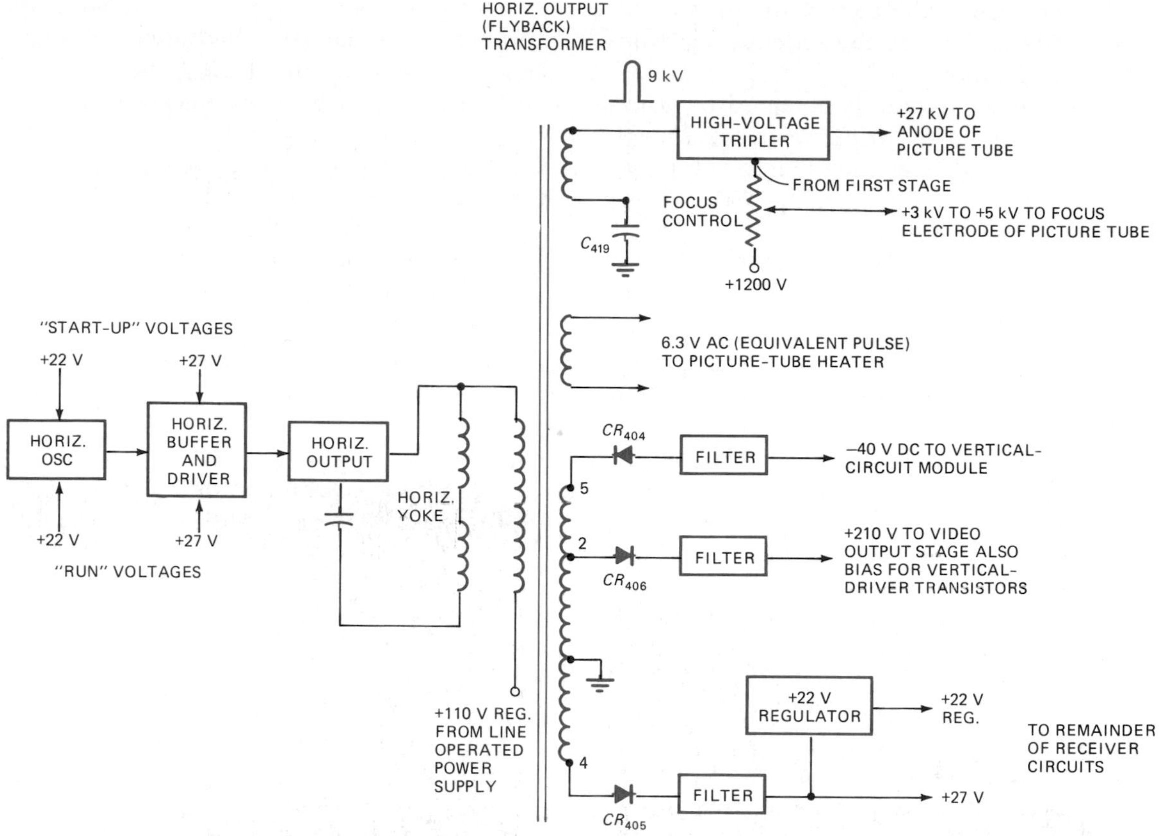

Figure 19.20 Simplified diagram showing "derived" voltages and horizontal stages.

Regulation by the SCR Duty Cycle The longer SCR_{401} is on, the higher the voltage will be at the regulator output. If the SCR conducts for a shorter time, the regulator output voltage will decrease. Thus the regulator output voltage can be controlled by varying the ON time of the SCR. This is accomplished by the regulator control circuits.

It should be noted that SCR_{401} is turned on just after the start of each horizontal trace and is turned off by the horizontal flyback pulse during retrace. In addition, the charging time constant of capacitor C_{401} is considerably longer than the time of one horizontal cycle. Consequently, a number of horizontal cycles must occur for each recharging of C_{401}.

DERIVED VOLTAGES

Figure 19.20 is a simplified diagram showing the various derived voltages, which are:

1. Picture tube anode voltage, +9 kV, tripled to +27 kV
2. Picture tube variable-focus voltage, +3 kV to +5 kV
3. Picture tube screen voltage, 1 200 V.
4. Picture tube heater voltage, horizontal pulse equivalent to 6.3 V ac
5. −40 V, +210 V, and +27 V

Functions of Derived Voltages The **anode voltage** is approximately 27 kV (with tripler).

It is connected to the picture tube high-voltage anode to provide the required electron-beam acceleration.

The **focus voltage** is obtained by using the first stage of the tripler as a focus rectifier. This voltage is applied to one end of the focus control. The other end is connected to a point of 1 200 V (Figure 19.20). The focus control makes available a voltage, variable from 3 kV to 5 kV, for the focus electrode of the color picture tube.

The **screen voltage** is a voltage of approximately 1 200 V and is normally present across capacitor C_{419}. This is the high-voltage, secondary, return capacitor. This voltage is applied to the screen element of the color picture tube.

The **heater voltage** is furnished by a horizontal pulse from a separate secondary of the horizontal output transformer. The pulse furnishes a current equivalent to that of an applied 6.3 V AC potential to the color picture tube heater.

Note: An AC meter will *not* measure 6.3 V AC because this is a pulse voltage.

As shown in Figure 19.20, the −40 V DC is derived from Terminal 5 of the large secondary and is rectified by diode CR_{404}. This voltage powers the vertical circuit module, which includes the vertical oscillator and output transistors. Although this voltage is taken from the top of the secondary, only −40 V DC is developed because the negative overshoot of the horizontal pulse is rectified. This pulse is appreciably smaller than the positive pulse.

A +210-V DC voltage is derived from Terminal 2 of the large secondary winding and is rectified by diode CR_{406}. This voltage powers the video output stage. It also is used as a base-bias supply for the vertical driver transistors in the vertical circuit module.

A +27-V DC supply is derived from Terminal 4 of the large secondary winding and is rectified by diode CR_{405}. Most of the remaining circuitry in the receiver is powered either from this source or from a regulated +22-V DC supply fed from the +27-V DC source.

Horizontal Stages The horizontal oscillator, buffer, driver, and output stages are also shown (in simplified form) in Figure 19.20. The power sources for these stages have already been discussed.

19.14 TROUBLES IN LOW-VOLTAGE TV POWER SUPPLIES

The DC voltages from a power supply will generally have only one of three types of trouble. These are (1) total loss of one or all voltages, (2) reduction in one or more voltages, and (3) an AC ripple superimposed on one or more DC voltages. The AC voltage used for vacuum tube heaters will most likely be either operative or nonoperative.

NO DC VOLTAGES

If all the DC voltages fail simultaneously, there will be no picture and no sound on the TV. Defects that can cause this condition include:

1. Open B+ fuse or circuit breaker
2. Shorted filter capacitor
3. Defective rectifier(s)
4. Open filter choke or filter resistor
5. Open surge current resistor and/or fusible resistor

Chart 19.1 will aid in localizing this problem.

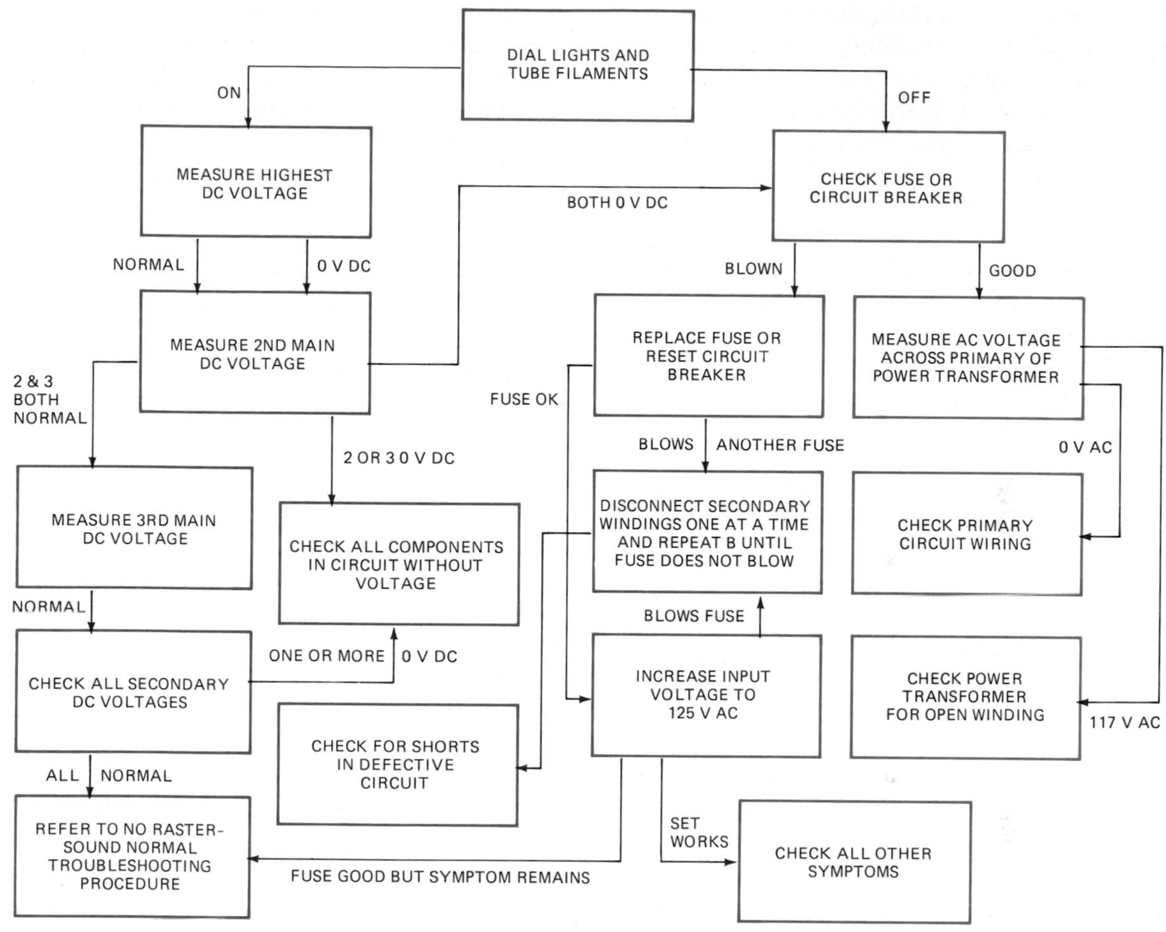

Chart 19.1 Illustrating a troubleshooting procedure to help localize the problem of no raster, no sound. *(Courtesy of B & K-Precision, Dynascan Corp.)*

LOSS OF ONLY ONE DC VOLTAGE

The effect on the receiver will depend upon which stage(s) is being powered by the failed voltage. Some defects that can cause this condition are:

1. Open filter resistor
2. Open decoupling resistor
3. Shorted bypass capacitor

REDUCTION OF DC VOLTAGES

A reduction of all DC voltages can cause reduced picture size (vertically and horizontally). It may also cause either a complete loss of raster or a "blooming" raster with reduced brightness. In this case, the focus is generally poor. Some conditions that may cause these problems are:

1. Brownout, where the AC line voltage

may drop as low as 105 V (with a properly regulated power supply, brownout will have little or no effect)
2. Defective power supply regulator.
3. Open input power supply filter capacitor or reduced capacity
4. Defective rectifier(s)
5. Excessive loading, caused by one or more defective components in the receiver.

POWER SUPPLY RIPPLE

Power supply ripple refers to *excessive* AC ripple on the DC supply voltages (some amount of ripple is normally present). The basic cause is inadequate filtering. If caused by a defective (open) filter capacitor, the increased power supply impedance may cause *cross-coupling* between some stages. "Motorboating" (low frequency oscillation) may result.

An open diode in a bridge rectifier (full-wave) can result in half-wave rectification. The ripple frequency will change from 120 Hz to 60 Hz, and inadequte filtering may result. In addition, an electronic regulator provides some filtering. Thus, a defective regulator may cause increased ripple.

Wide, horizontal, alternating black and white bars are caused by 60-Hz or 120-Hz hum. For 60 Hz, there will be one white and one black bar. For 120 Hz, there will be two of each.

If the ripple affects the horizontal scanning circuits, the raster will show "pulling" horizontally. Each vertical side of the raster will be fashioned into one sine wave for 60-Hz ripple or two sine waves for 120-Hz ripple. If the ripple affects the horizontal synchronizing circuits, there will be horizontal pulling but only in the picture, not in the raster.

Ripple in the vertical synchronizing circuits may result in poor vertical synchronization.

A power supply ripple can also cause hum in the sound and a pulsating picture (motorboating).

TROUBLES IN A HORIZONTAL FREQUENCY POWER SUPPLY

There are some problems that can occur only in horizontal frequency power supplies. (Refer to Figures 19.17 and 19.20).

A no picture, no sound condition could be caused by a defect in (1) the +150-V bridge rectifier circuit, (2) the start-up circuits, (3) the horizontal +110-V regulator, or (4) the horizontal circuits or flyback transformer. If any one of these circuits is not working correctly, the derived B+ voltages will not develop and the set will be completely inoperative.

Although not previously discussed, there is a +27-V *shutdown* circuit in the horizontal frequency power supply system. If the derived B+ voltages rise beyond permissible limits, an *overvoltage protection* circuit shuts off the horizontal oscillator. Thus, all derived voltages will cease and the set will be "dead". The AC power supply must be turned OFF and then ON to reactivate the set. However, if the problem persists, the set will shut off again and servicing is required. The usual cause for this problem is excessive voltage output from the +110-V regulator. This may be caused by a shorted SCR or filter capacitor.

As seen in Figure 19.20, there are several derived voltages obtained from the secondaries of the horizontal flyback transformer. Failure of any one of these voltages will affect the operation of the circuits they power.

SUMMARY OF CHAPTER HIGHLIGHTS

1. Three types of low-voltage TV receiver power supplies are (1) transformerless, or line-operated, types, (2) transformer types, and (3) horizontal frequency (derived) types.
2. A low-voltage power supply provides B+ or B− voltages for transistors or vacuum tubes. (Figure 19.16)
3. Solid-state TV receivers require DC voltages of approximately 5 to 250 V.
4. A power transformer power supply provides isolation from the AC line. A transformerless supply does not, and so an isolation transformer should always be used in servicing it. (Figures 19.4 and 19.6)
5. A *derived* power supply gets its voltages from secondary windings of the horizontal output transformer. (Figure 19.17)
6. Power supplies may be classified as (1) half-wave, (2) full-wave, (3) bridge (full-wave), or (4) voltage-multiplier. (Figures 19.5 to 19.9 and 19.17)
7. Power supplies may be unregulated or regulated. Most solid-state TV receiver supplies are regulated, particularly ones for color sets. (Figures 19.5 and 19.17).
8. Power supplies use silicon diode rectifiers. These are reliable, compact devices that can withstand high temperatures. (Figure 19.1)
9. A power supply filter greatly reduces the ripple on the rectified DC wave. (Figure 19.3)
10. Filters are classified as *RC* and *LC* filters. Transistor voltage regulators also reduce ripple. (Figures 19.3 and 19.11)
11. Power supply voltages may be increased by the use of voltage multipliers. The most common ones are voltage doublers and triplers. (Figures 19.7 to 19.9)
12. The most common voltage doubler is the half-wave type. It has the advantage of a common ground connection. (Figure 19.7)
13. Voltage triplers are frequently used to increase the high voltage applied to the anode of a picture tube. (Figure 19.20)
14. Voltage regulation refers to the ability of a power supply to maintain its DC voltage constant (within specifications) regardless of changes of AC line voltage or load current.
15. Voltage regulators may be classified as (1) zener diode regulators, (2) series-pass regulators, (3) three-terminal regulators, and (4) switching regulators. (Figures 19.10 to 19.13 and 19.18)
16. The zener diode regulator is simple, but not as effective as the other types. It does not appreciably reduce ripple and provides limited output current. (Figure 19.10)
17. A series-pass transistor regulator provides higher output currents and better regulation than a zener-diode regulator. It may also include protection circuits against damage to the load or to the series-pass transistor. It also reduces ripple appreciably. (Figure 19.11)
18. A three-terminal regulator is usually in monolithic form. Only three connections are made to the IC: These are (1) input, (2) output, and (3) ground. (Figure 19.12)
19. Typical three-terminal regulator output voltages range from 5 to 24 V. By the addition of simple external circuitry,

the output voltage can be made adjustable. (Figure 19.12)

20. In a *switching* regulator, the series-pass transistor is fully on and fully off periodically. Because of this, considerably less heat is produced. (Figure 19.13)
21. Typical *switching* regulator frequencies range from 5 to 100 kHz. These regulators are frequently monolithic IC units.
22. Switching regulators are used where high efficiency and small size are required. (Figures 19.13 and 19.18)
23. The pulse width oscillator is the heart of a switching regulator. (Figure 19.13)
24. The horizontal frequency power supply used in many television receivers provides reduced size and weight as well as increased efficiency. (Figures 19.17 to 19.20)
25. In the horizontal frequency power supply, ripple can be more easily filtered because it occurs at a high frequency (15 734.26 Hz).
26. In the horizontal frequency power supply derived DC voltages are obtained by rectifying outputs from a secondary winding of the horizontal output (flyback) transformer. (Figure 19.17)
27. Derived voltages are developed only when the horizontal output transistor drives the horizontal output transformer in the normal manner. (Figure 19.20)
28. *Start-up* voltages are momentary. They are required to begin operation of the horizontal stages. (Figure 19.17)
29. *Run* voltages are the derived voltages that power most of the TV receiver. (Figure 19.20)
30. In the RCA XL-100 series color-TV receiver, the horizontal output transistor is powered by 110 V. (Figures 19.17 and 19.18)
31. The voltage that powers the RCA XL-100 series receiver is developed by a bridge rectifier operated from the AC line and a switching regulator that uses an SCR as the switch.
32. Because of the regulator system in the RCA XL-100 series receiver, consistent performance is achieved over a range of AC line voltages from 105 to 135 V AC.
33. The XL-100 series receiver regulator system also stabilizes operating voltages that result from load changes.
34. The regulator output voltage of the XL-100 series receiver is controlled by varying the ON time of the SCR.
35. Some of the derived voltages in the XL-100 series receiver are (1) picture tube anode and focus voltages, (2) picture tube screen and heater voltages, and (3) B+ and B− voltages for the various circuits (except the horizontal output transistor, which receives +110 V from the switching regulator).
36. Some causes of a no picture, no sound condition are (1) open fuse or circuit breaker, (2) shorted filter capacitor, and (3) defective rectifier(s).
37. Some causes of reduced DC voltages are (1) defective power supply regulator, (2) defective rectifier(s), and (3) excessive loading.
38. Excessive power supply ripple may be caused by (1) a defective regulator or (2) an open filter capacitor.
39. Excessive ripple in the horizontal scanning or synchronizing circuits may cause horizontal picture "pulling."
40. In a horizontal frequency power supply television receiver, some causes of a no picture, no sound condition may be (1) defective bridge rectifier, (2) defective start-up circuit, and (3) defective horizontal circuit(s).

EXAMINATION QUESTIONS

(Answers are provided at the back of the text.)

Part A Supply the missing word(s) or number(s).

1. Features of a horizontal frequency power supply are reduced ________ and increased ________.
2. When testing a transformerless TV receiver, an ________ ________ should be used.
3. Solid-state TV receivers generally have ________ power supplies.
4. ________ type rectifiers are used in TV receivers.
5. ________ and ________ filters are commonly used in TV receivers.
6. Power supplies should present ________ impedance to the load.
7. Transformer power supplies are ________ from the AC line.
8. The ripple frequency of a half-wave rectifier is ________ Hz.
9. ________ volts is the peak voltage output from an unloaded half-wave rectifier with an AC line voltage of 120 V.
10. An unloaded half-wave voltage tripler will have an output voltage of ________ V if the AC line voltage is 120 V.
11. The power supplies in most ________ TV receivers are regulated.
12. ________ voltages are DC voltages obtained from the horizontal output transformer.
13. Many color-TV receivers have regulated supplies that can accommodate AC line voltage changes from ________ to ________ V AC.
14. The ________ diode operates by reverse-current breakdown.
15. The series-pass regulator reduces ________ on the DC output.
16. Protection against excessive load currents in a series-pass regulator can be provided by a ________ circuit.
17. Three-terminal voltage regulators are constructed as a ________ IC.
18. The three connections of a three-terminal regulator are ________, ________, and ________.
19. A ________ regulator is one in which the series-pass element is either fully ON or fully OFF.
20. Considerably reduced heat production is a major advantage of a ________ regulator.
21. ________ Hz is the ripple frequency of a derived power supply receiving a color program.
22. The voltages required to begin operation of a color TV receiver using derived voltages are called ________ voltages.
23. In an RCA XL-100 series color-TV receiver, a ________ is used to increase the picture tube anode voltage.
24. The series-pass device of the +110-V supply in an XL-100 series receiver is an ________.
25. A reduction of all DC voltages may cause the ________ dimensions to decrease.

Part B Answer true or false.

1. A derived voltage power supply can power only the horizontal output transistor.
2. The highest B+ used in any solid-state TV receiver is 80 V.
3. When transformerless TV receivers are being serviced, an isolation transformer should be used.
4. Derived B+ voltages are obtained by special primary windings on the horizontal output transformer.
5. The ripple frequency from a bridge rectifier is 120 Hz.
6. Solid-state TV receiver power supplies are generally regulated.
7. Most current TV manufacturers use selenium rectifiers.
8. A series-pass transistor regulator also acts as a ripple filter.
9. A high output impedance is desirable for a TV receiver power supply.

10. Transformer-type power supplies are isolated from the AC line.
11. The input filter capacitor of a half-wave rectifier power supply (no load) charges to the peak value of the applied AC line voltage.
12. Half-wave, full-wave, and bridge-type voltage multipliers are found in TV receivers.
13. The voltage multiplier for B+ voltages most commonly found in TV receivers is the half-wave doubler.
14. The simplest TV receiver voltage regulator is the zener diode type.
15. It is not necessary to ever regulate derived B+ voltage.
16. The protection circuit in a voltage regulator can protect only the regulator circuits.
17. A zener diode regulator is normally capable of supplying very high load currents, providing the correct zener is chosen.
18. A series-pass transistor regulator can regulate voltages to closer tolerances than a zener diode regulator.
19. A three-terminal IC voltage regulator includes series-pass transistors and current-limiting resistors.
20. Typical output voltage ratings for a three-terminal regulator range from 5 to 24 V.
21. Typical switching frequencies for switching regulators range from 5 to 100 kHz.
22. The heart of a switching regulator is the pulse-width oscillator.
23. In a derived voltage power supply, no filtering is required because of the very high ripple frequency.
24. In a derived voltage power supply, regulation is provided for line voltage and load variations.
25. *Start-up* voltages for a derived power supply occur only once each time the TV receiver is turned on.
26. *Run* voltages for a derived voltage power supply originate at the secondary of the horizontal output transformer.
27. A condition of no picture and no sound can be caused by an open power supply filter capacitor.
28. In a brownout, the line voltage may drop as low as 105 V ac.
29. In a derived voltage power supply, a failure of the start-up circuits will result in a "dead" TV receiver.

REVIEW ESSAY QUESTIONS

1. Explain the operation of the circuit shown in Figure 19.2.
2. In Question 1, explain how resistor R_1 performs its function.
3. Define the following: (1) transformerless power supply and (2) derived voltage power supply.
4. Explain why in some cases an isolation transformer is required, in your own words.
5. List the various types of power supplies discussed in this chapter.
6. What is the most commonly used rectifier? Why?
7. In Figure 19.4B, explain (1) the function of L_1, L_2, C_1, and R_1, and (2) why B+ is applied to the picture tube heaters.
8. Explain the operation of an unloaded half-wave voltage doubler. Use simple diagrams to illustrate your answer.
9. Explain briefly the operation of D_3 and C_3 in Figure 19.9.
10. Discuss the requirements for regulating power supplies in solid-state TV receivers.
11. Explain what is meant by a series-pass transistor regulator.
12. What is a protection circuit and what does it protect in a series-pass transistor regulator?
13. What is the function of (1) R_2, (2) CR_1, (3) Q_1, and (4) R_3–R_4 in Figure 19.11B?
14. Explain briefly the basic operation of the circuit shown in Figure 19.11B, neglecting the protection circuit (Q_3).
15. Explain how the protection circuit in Figure 19.11B works when there is excessive load current.
16. Explain the operation of the switching regulator in Figure 19.13.
17. In the circuit shown in Figure 19.13, what

are the advantages of a switching regulator over a series-pass regulator?

18. What is the function of the operational amplifier A_1 in Figure 19.13?
19. Define (1) start-up voltage, (2) run voltage, (3) derived voltage, (4) ON time, and (5) OFF time.
20. Draw a simple block diagram equivalent to Figure 19.17. Label all blocks and voltages.
21. What is the function of (1) T_{201}, (2) CR_{302}, and (3) CR_{304} and L_{403} in Figure 19.17?
22. In Figure 19.18, describe the function and operation of SCR_{401}.
23. In Figure 19.19, what is the function of the 300-V peak pulse?
24. Using Figure 19.20, describe briefly how derived voltages are produced.
25. In Figure 19.20, does regulating the +110-V supply also have a regulating effect on the derived voltages? Explain your answer.

EXAMINATION PROBLEMS

(Selected problem answers are provided at the back of the text book.)

1. Make a simple table showing all the types of power supplies discussed in this chapter and listing the following: (1) name of power supply (including multipliers), (2) number of diodes, (3) ripple frequency, and (4) unloaded output voltage for 120-V AC input.
2. Calculate the unloaded DC output voltage for a voltage tripler with an input of 113 V AC.
3. In Figure 19.17, identify (1) the bridge rectifier, (2) the varistor, (3) the 27 V start-up diode(s), and (4) the 150-V filter.
4. In Figure 19.11, identify the following transistors and diodes: (1) series pass, (2) error-voltage amplifier, (3) protection diode, and (4) zener diode.
5. If, in Figure 19.21, I_2 = 125 mA, and I_L = 75 mA, what is (1) the voltage across R_1, (2) the current in R_1, (3) the value of R_1.

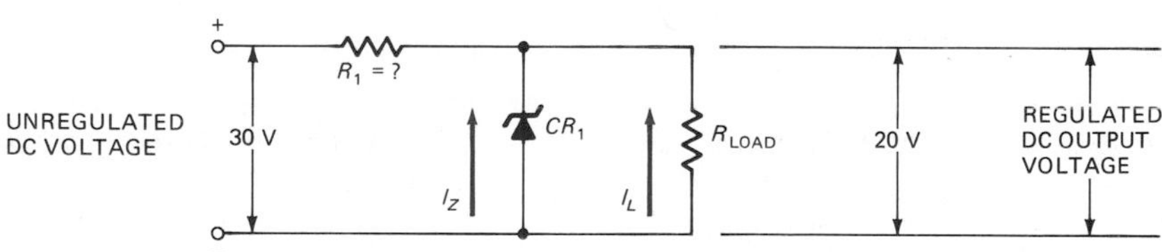

Figure 19.21 Figure for Examination Problem 5.

Chapter 20

20.1 Effects of loss of synchronization
20.2 Synchronizing pulse and video signal separation
20.3 Synchronizing pulse separator circuits
20.4 Reduction of noise to improve synchronization
20.5 Transistor noise-cancellation circuits
20.6 Synchronizing pulse separator in an RCA IC
20.7 Vertical and horizontal pulse separation
20.8 Function of the equalizing pulses
20.9 Vertical synchronizing and equalizing pulse display (hammerhead)
20.10 Synchronizing troubles.
Summary

SYNCHRONIZING CIRCUITS

INTRODUCTION

In Chapter 3 we studied the requirements for horizontal and vertical synchronization. In addition, we studied the form of the synchronizing pulses.* We saw that these are a part of the composite video signal. It was mentioned that the synchronizing pulses must be separated from the composite video signal before they can be used. These pulses are the **horizontal, vertical,** and **equalizing pulses.**

Let us very briefly review the functions of the synchronizing pulses. Each horizontal line has associated with it a horizontal synchronizing pulse. This pulse synchronizes the TV receiver's horizontal deflection oscillator with that of the TV transmitter. As a result, the TV receiver's picture elements are correctly reproduced horizontally in step with those of the TV transmitter. The vertical synchronizing pulses lock the picture fields (and frames) to prevent them from moving vertically. They do this by synchronizing the TV receiver's vertical deflection oscillator with that of the TV transmitter. As described in Chapter 3, equalizing pulses ensure proper interlacing of odd and even fields.

The pulse repetition rate (PRR) of the synchronizing pulses is the same for monochrome and color-TV broadcasts.

This has been described previously and is mentioned here only as a review. For these broadcasts, the horizontal synchronizing pulse (and horizontal line) frequency is 15 734.26 Hz. The vertical synchronizing (and field) frequency is 59.94 Hz.

Many color-TV receivers do not have vertical or horizontal deflection oscillators. In these receivers, digital "countdown" circuits provide the correct horizontal and vertical scanning rates from a stabilized, voltage-controlled oscillator operating at 31 468.52 Hz. The oscillator is stabilized with

*It is recommended that at this point, the reader carefully review the material in Chapter 3.

horizontal synchronizing pulses as a reference. (More about this in Chapter 21.) The horizontal scanning frequency is 31 468.52/2 = 15 734.26 Hz. Similarly, the vertical scanning frequency is 31 468.52/525 = 59.94 Hz. The digital countdown system is more stable and more independent of noise pulses than deflection oscillators. Countdown circuits are packaged as ICs.

In this chapter we are particularly interested in the circuits that separate the horizontal and vertical synchronizing pulses from the composite video signal and from each other. We shall also study noise cancellation in synchronizing circuits. (Horizontal AFC circuits, which are used in all television receivers except for those that have countdown circuits, are covered in Chapter 22.)

When you have completed the reading and work assignments for Chapter 20, you should be able to:

- Define the following terms: hammerhead pattern, synchronizing pulse clipper, synchronizing pulse separator, synchronizing pulse amplifier, impulse noise gate, *RC* integrator, *RC* differentiator, equalizing pulse, countdown circuit, synchronizing pulse compression, and noise clipper.
- Discuss the three possible variations of synchronization loss and their effect on the TV picture.
- Discuss all the components of the composite video signal.
- Explain how the synchronizing pulses are separated from the composite-video signal (three methods).
- Understand how horizontal and vertical synchronizing pulses are separated.
- Explain the operation of a synchronizing pulse amplifier.
- Discuss the need to reduce noise on synchronizing pulses.
- Explain how a noise gate works in a synchronization circuit.
- Discuss other noise cancellation circuits.
- Describe the function of the equalizing pulses.
- Understand the common types of troubles which might occur to produce improper horizontal and vertical synchronization and their causes.

20.1 EFFECTS OF LOSS OF SYNCHRONIZATION

The effect(s) on the picture of loss of synchronization will depend upon one of the following factors: (see Figure 20.1).

1. Loss of vertical synchronization
2. Loss of horizontal synchronization
3. Loss of both vertical and horizontal synchronization.

LOSS OF VERTICAL SYNCHRONIZATION

If only vertical synchronization is lost, the picture will not be broken up. However, it will roll upward or downward at a speed that depends upon the setting of the vertical hold control.[1] This is shown in Figure 20.1A. The picture can be momentarily stopped by careful adjustment of the vertical hold control, but it will not remain locked.

In Figure 20.1A, the black horizontal bar and pattern are caused by the vertical synchronizing, equalizing, and blanking pulses. These are not normally visible for a locked-in picture. They belong below the picture and off the screen. They can be seen here

[1]Receivers with countdown circuits, which are covered in Chapters 21 and 22, do not have a vertical or horizontal hold control. The discussions in this chapter do not refer to such receivers.

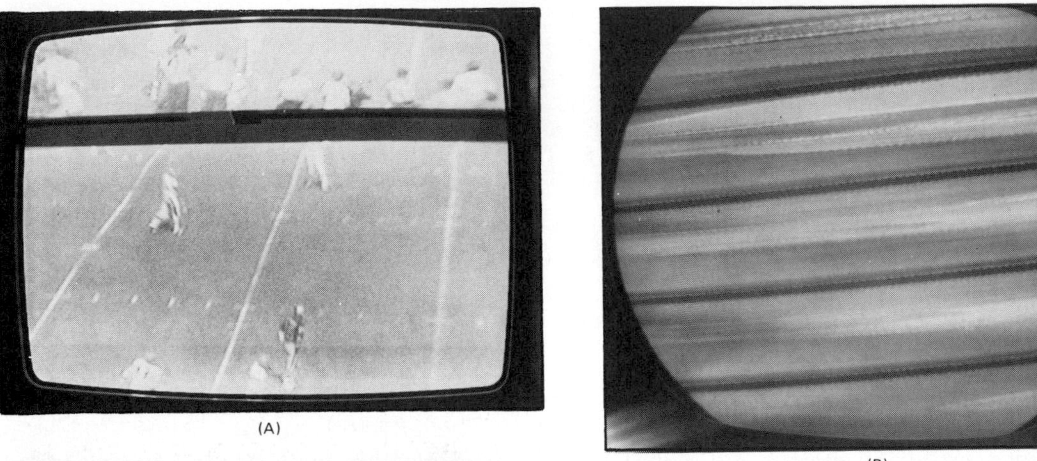

(A)

(B)

(C)

Figure 20.1A-C. The effects on the TV picture of malfunctioning sync. **A.** Picture rolls vertically with no vertical sync. **B.** Loss of horizontal sync causes picture to tear in diagonal strips. Number of strips may change continuously. **C.** Loss of both horizontal and vertical sync. Pattern will be moving vertically and changing diagonally.

because the picture is not synchronized and is slowly rolling.

The blackest part of the pattern is called a **hammerhead** which it resembles.

LOSS OF HORIZONTAL SYNCHRONIZATION ONLY

The correct color horizontal scanning frequency is 15 734.26 Hz. When the TV receiver horizontal oscillator is synchronized in frequency and phase with the TV transmitter horizontal oscillator, the picture is whole and properly centered. However, if the TV receiver horizontal oscillator frequency deviates by 59.94 Hz (or multiples of this value), the picture "tears" into diagonal segments, as shown in Figure 20.1B. Each diagonal segment represents a deviation of 59.94 Hz from 15 734.26 Hz. Since there are five diagonal segments in this illustration, the horizontal scanning frequency must have deviated by 5 × 59.94 Hz = 299.7 Hz.

It is possible to know whether the horizontal frequency is above or below the correct value by the direction of slant of the diagonal bars. If the bars slant downward from right to left, the frequency is below the correct value. If they slant downward from left to right, the frequency is above the correct value.

Note in Figure 20.1B that a black bar separates each diagonal segment. These bars are formed from portions of the horizontal blanking interval. This blanking interval is not ordinarily visible, since it is off screen on both sides of the picture.

If the condition pictured in Figure 20.1B is due to a complete loss of horizontal synchronization, the number of diagonal bars will generally vary with time. However, the pictured condition, if stationary, can be caused by misadjustment of the TV receiver's horizontal hold control or by normal aging of components. In these cases, the condition can usually be remedied by readjusting the horizontal-hold control.

LOSS OF BOTH VERTICAL AND HORIZONTAL SYNCHRONIZATION

This condition is pictured in Figure 20.1C. Here, the pattern rolls vertically, and the number of diagonal bars varies with time. There are several possible causes for this condition: (1) failure of synchronizing pulse clipper (separator) (see Figure 20.2), (2) synchronizing pulse compression in a common IF stage, or (3) AGC malfunction.

20.2 SYNCHRONIZING PULSE AND VIDEO SIGNAL SEPARATION

Before the synchronizing pulses of a composite video wave can be used, they must be separated from the other portions of the signal. The separation may be made anywhere from the video detector to the last video stage before the cathode-ray tube. In practice, the input for the synchronizing stages is generally taken from a point beyond the video detector, usually at the output of one of the video amplifiers. At such points, the signal has sufficient amplitude and is in proper form for controlling horizontal-and vertical-deflection oscillators with a minimum of additional stages. For example, circuit designers often do not apply the video signal to the synchronizing pulse separator until it has passed through the first video amplifier. In this way, an extra pulse amplifier is not needed.

SIGNAL IN DC FORM

In order to separate the synchronizing pulses from the incoming wave, the signal must be in its DC form. This should be evident from the illustrations in Chapter 17, which show the AC and DC forms of a video signal. Although the signal is always in its DC form at the output of the detector, it may not be so at the output of a following amplifier. In this case, DC restoration is necessary.

BLOCK DIAGRAM OF SYNCHRONIZER CIRCUITS

The simplified block diagram Figure 20.2 illustrates the general path of the synchronizing pulses in a television receiver. This chapter is concerned with the first three blocks in the figure: (1) the synchronizing pulse clipper (or separator), where the synchronizing pulses are removed from the video signal, (2) the high-pass filter (or differentiator), which separates the horizontal synchronizing pulses from the combined

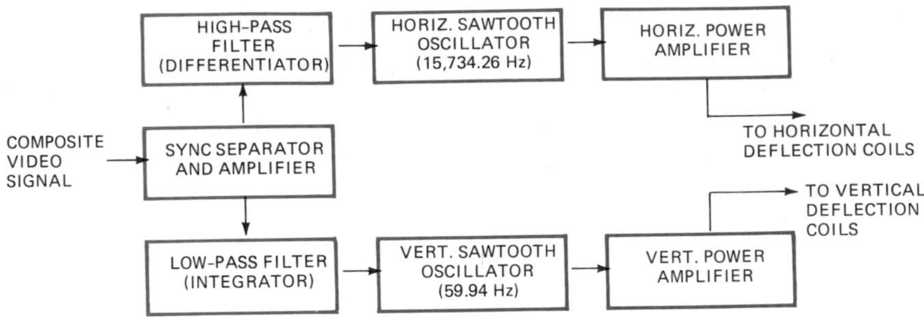

Figure 20.2 Simplified block diagram of the synchronizing and deflection sections of a television receiver.

synchronizing signal, and (3) the low-pass filter (or integrator), which separates the vertical synchronizing pulses from the combined synchronizing signal.

Types of Circuits The circuit that separates the synchronizing pulses from the rest of the video signal is called the **synchronizing pulse clipper** or **synchronizing pulse separator.** Both horizontal and vertical synchronizing pulses are clipped by this circuit. These two pulses are then further separated in a later stage. Practically any type of circuit may be used for synchronizing separation. The only thing necessary is that the circuit be biased so that only the pulse portions of the video wave cause current to flow.

OSCILLOSCOPE PATTERNS

Figure 20.3 shows oscilloscope patterns for the composite video signal and for the separated synchronizing pulses at the output of a synchronizing pulse separator. This particular type of synchronizing pulse separator (transistor or tube) produces polarity inversion of the synchronizing pulses; diode separators (to be discussed shortly) do not produce polarity inversion.

Horizontal Synchronizing Separation Pulse Figure 20.3A shows the composite video signal *input* to a synchronizing pulse separator. Here the oscilloscope sweep rate is 15 750/2 = 7 875 Hz. As a result, the screen shows two horizontal lines of composite video containing two horizontal synchronizing pulses

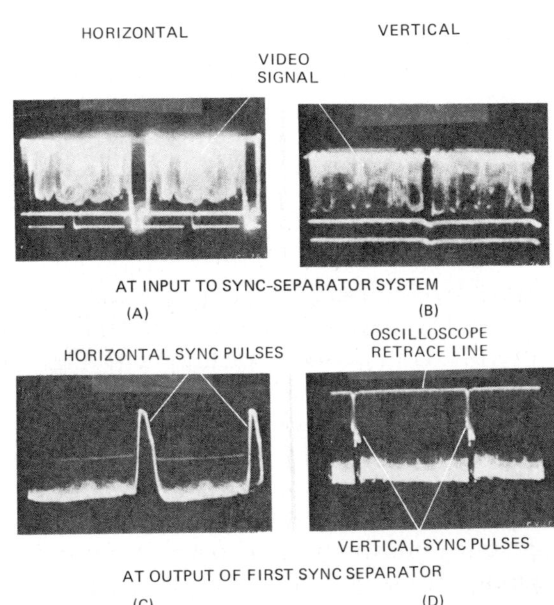

Figure 20.3 Showing sync-pulse separation from the composite-video signal. **A.** The composite-video signal on an oscilloscope at horizontal-frequency sweep rate $\frac{(15\ 750\ \text{Hz})}{2}$. **B.** The composite-video signal on an oscilloscope at vertical-frequency sweep rate $\frac{(60\ \text{Hz})}{2}$. **C.** Horizontal-sync pulses after separation (inverted polarity). **D.** Vertical-sync pulses after separation (inverted polarity).

(negative-going). Compare with Figure 20.3C which shows the *output* of the separator with the oscilloscope sweep rate set as above. Here we see two separated horizontal synchronizing pulses, inverted in polarity. (The small residual video signal showing at the bottom of the wave form would normally be eliminated by an additional separator stage.)

Vertical Synchronizing Pulse Separation Figure 20.3B shows the same composite video signal *input* to a synchronizing pulse separator. However, now the oscilloscope sweep rate is 60/2 = 30 Hz. Actually, two fields are shown here. However, because of the way the oscilloscope is synchronized to the signal, one of the vertical synchronizing pulses is at the extreme right and not clearly visible. The *output* of the separator, viewed with a 30-Hz oscilloscope sweep rate, is shown in Figure 20.3D. Here, two vertical synchronizing pulses, inverted in polarity, are clearly visible.

Observing Synchronizing Pulses When observing synchronizing pulses on an oscilloscope screen, you will find that the horizontal pulses stand out more distinctly than the vertical pulses. This is because the structure of the horizontal pulse is simpler than that of the serrated vertical pulse. Also, the horizontal pulse takes up a greater percentage of a line than the vertical pulse does of a field. Hence, there is more of the pulse to be observed when the scope-scanning rate is set to the proper value. These facts are borne out by the oscillograms of vertical- and horizontal-sync pulses shown in Figure 20.3.

20.3 SYNCHRONIZING PULSE SEPARATOR CIRCUITS

There are many types of circuits used to separate synchronizing pulses from the video signal. In this section, we begin with a simple diode synchronizing pulse separator and then advance to the more complicated tube and transistor circuits. Emphasis will be on transistor synchronizing pulse separators.

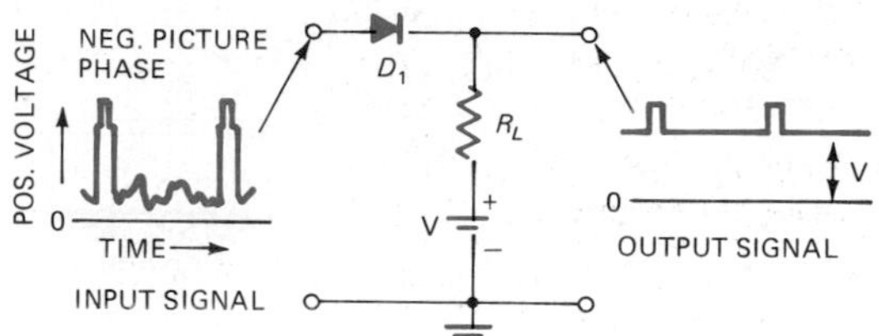

Figure 20.4 A diode clipper operating with input video signals having a negative picture phase.

DIODE SEPARATORS

A diode synchronizing pulse separator circuit is shown in Figure 20.4. The video signal is applied between the anode and ground. The output voltage is developed across the battery V and the load resistor R_L. The small battery is inserted with its negative terminal toward the anode. This prevents current from flowing until the video signal acting on the diode becomes positive enough to counteract the positive biasing voltage. Current then flows. If the circuit constants are chosen properly, current will flow only at those synchronizing pulses. The output will consist only of these short pulses of current. The picture phase at the input of this diode must be negative, as in Figure 20.4.

Inverted Diode Inverting the diode, as in Figure 20.5, it is possible to apply a positive picture phase to the circuit and again obtain only the pulse tips across R_L. Note that the battery must also be inverted for this circuit. The DC biasing voltages necessary for these diodes may be taken from a low-voltage power supply.

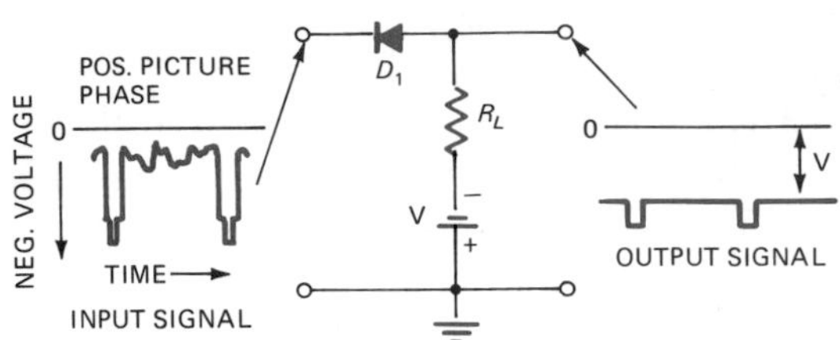

Figure 20.5 An inverted diode clipper, suitable for input signals having a positive picture phase.

Automatic Bias It is generally not practical to use a bias battery or power supply DC voltage for the diode clipper. We require an arrangement that is completely automatic in its operation, altering its operating point as the amplitude of the received carrier varies. A simple yet effective circuit is shown in Figure 20.6. The diode clipper uses the time constant of R and C to bias the circuit so that all but the synchronizing pulses are eliminated. Capacitor C and resistor R form a low-pass filter with a comparatively long-time constant, equal to approximately three horizontal lines. Therefore, the voltage developed across R (and C) is determined by the highest signal voltage applied across the input terminals. This, of course, is the synchronizing pulses. Throughout the remainder of the line, the anode is never driven positive enough to overcome the positive cathode bias, even though the video voltage is active.

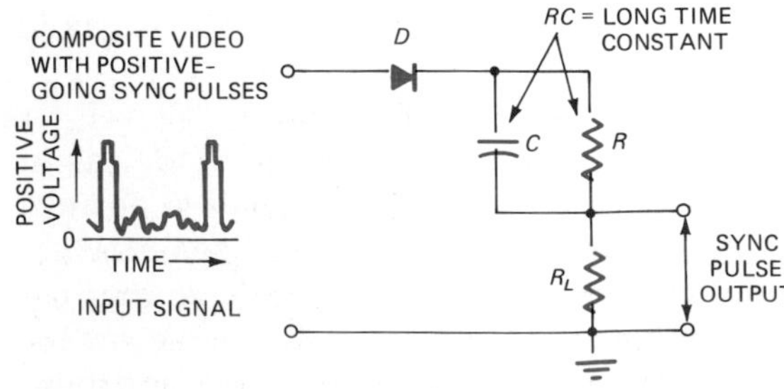

Figure 20.6 A practical diode clipper with the battery removed.

TRIODE SYNCHRONIZING PULSE SEPARATORS

Since a triode or a pentode can do anything a diode can do, and amplify as well, it is natural to use these tubes as synchronizing pulse separators. A triode synchronizing pulse separator is shown in Figure 20.7. The triode V_1 is biased by grid-leak bias developed across C_1 and R_1. The pulses in the video signal fed to V_1 are the most positive components of the signal. Electrons will flow in the grid circuit, charging C_1. Because of the high value of R_1, the charge on C_1 will leak off slowly. This develops a fairly steady bias voltage across the grid resistor. This bias voltage prevents plate current from flowing except for the most positive values of the incoming signal, which are the synchronizing pulses. A fairly low plate voltage causes V_1 to saturate readily. This squares off the synchronizing pulses and limits any noise pulses in the signal.

Minimizing the Effect of Noise Pulses Resistor R_2 isolates (decouples) the synchronizing pulse separator tube from the video amplifier where the signal is obtained. In this way, the input circuit of V_1 and its capacity do not overload the video amplifier plate circuit. If this circuit were overloaded, picture quality would deteriorate. Resistor R_3 and capacitor C_2 minimize the effect of any noise pulses in the received signal. If a strong noise pulse (extending in the positive direction) should come along and R_3 and C_2 were not present, the resulting electron flow in the grid circuit of V_1 would charge C_1 to a fairly high negative voltage. Because R_1 is so large, it would take a long time for C_1 to discharge through R_1. Then synchronization would be lost before the regular signal again produced a current flow through V_1. To avoid this, R_3 and C_2 are placed in the grid circuit. When a noise pulse comes along, C_2 absorbs the additional current flow the noise

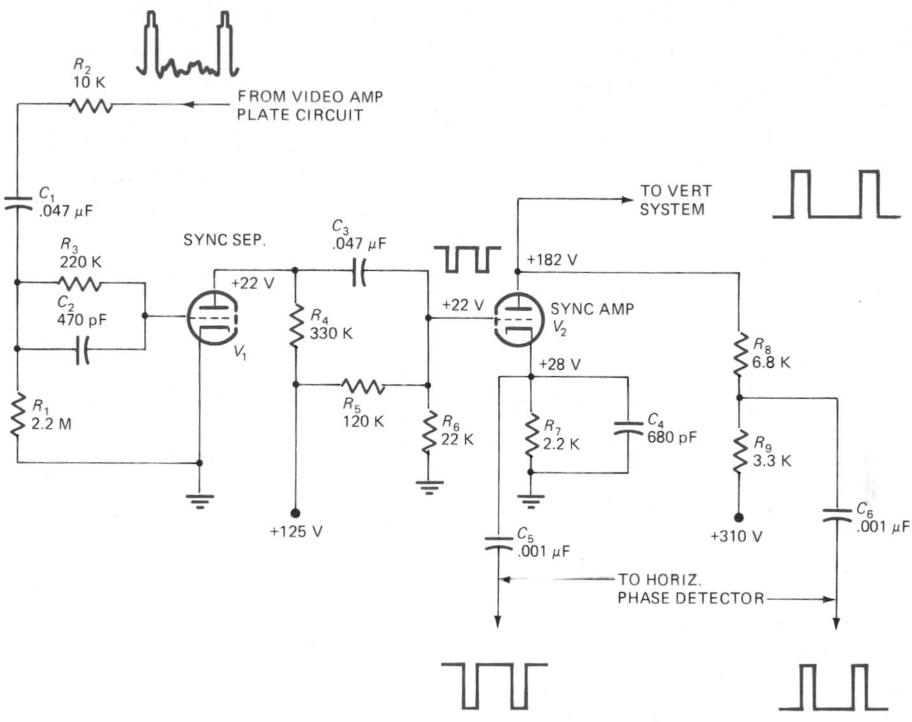

Figure 20.7 A triode (V_1) sync separator.

pulse produces. Then C_2 discharges fairly quickly through R_3, which is one-tenth the value of R_1.

V_1 Conduction The video signal arriving at the grid of V_1 has a peak-to-peak amplitude of 60 V (Fig. 20.8). However, the grid-leak bias developed by C_1 and R_1 permits V_1 to conduct only when the synchronizing pulses are active. For the rest of the video signal (below the sync pulses) the tube is cut off by the bias. The low plate voltage of the tube also helps in this process.

Note that not all of the synchronizing pulse produces a corresponding change in the plate circuit. Near the pulse peak, the grid is almost at zero bias and the tube is passing as much current as it can with the low level of plate voltage. In other words, it is operating at saturation, and a further rise in input voltage produces little additional output voltage. This is why the very tip of the synchronizing pulse in Figure 20.8 is shown unshaded. This small upper section does not develop much additional output voltage. In essence, this squares off the top of the output pulse.

Synchronizing Pulse Amplifier From V_1, the synchronizing pulses, which are now negative-going, travel to V_2, the **synchronizing pulse amplifier.** The grid of this tube is made 22 V positive by R_5 and R_6. This causes a considerable amount of current to flow in both the grid circuit and the plate circuit. Both currents return to the cathode by way of R_7. The result is 28 V at the cathode. This makes the grid negative with respect to the cathode by 6 V. This voltage, together with the negative portion of the synchronizing

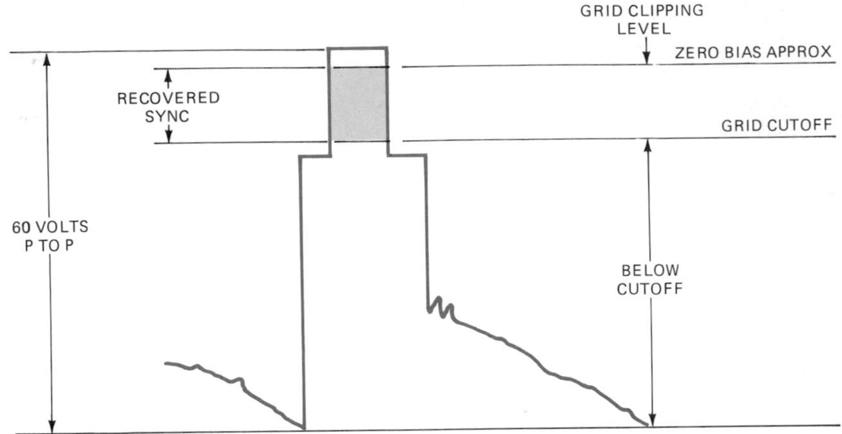

Figure 20.8 The video signal at the grid of V_1 in Figure 20.7. Only the shaded portion is allowed to pass through V_1.

pulse from V_1, quickly drives V_2 into cutoff and helps to square off this end of the applied pulses. By the same token, since the grid already has a positive voltage (from B+), the positive portion of the pulses from V_1 cannot drive the grid far before plate current saturation is reached. Thus, the positive portion of the signal is also clipped.

The synchronizing pulse amplifier clips both the positive and negative extremes of the synchronizing pulse fed to it from V_1. This amplifier section operates at considerably higher plate voltage, and the output is consequently greater than that of the synchronizing pulse separator stage alone.

The synchronizing pulses at the plate of V_2 are in the positive direction and are fed to the vertical integrator network. At the same time, positive pulses from R_9 and negative pulses from the cathode are fed to the horizontal phase detector.

TRANSISTOR SYNCHRONIZING PULSE SEPARATORS

The simple diode synchronizing pulse separator is easily converted to a transistor separator. This is possible because the transistor input terminals (the base-to-emitter terminals) are actually a diode. The transistor merely adds gain to the basic diode separator. Transistors are ideally suited for synchronizing pulse separators because their input-output characteristics resemble those of a sharp-cutoff pentode. Transistor sync separators will be examined in detail, so that the reader may appreciate why transistors fit this application so well.

Transistor Characteristics TV receivers use silicon transistors because of their superior temperature characteristics and reliability. A silicon transistor requires approximately 0.5 V across its base-to-emitter terminals before it begins to conduct. As the base-to-emitter voltage varies from 0.5 to 0.7 V, the transistor collector voltage also varies. However, when the base-to-emitter voltage rises above 0.7 V, the collector saturates at approximately 0.3 V. These three situations are summarized in Table 20.1.

Circuit Operation The characteristics listed in Table 20.1 make it possible to perform synchronizing pulse separation, shaping, and gain in one transistor stage. Figure 20.9A shows the major part of the circuit required to perform these functions. By adding only one resistor (R) to the circuit shown

Table 20.1 **COLLECTOR VOLTAGE VS BASE-TO-EMITTER VOLTAGE**

INPUT (BASE-TO-EMITTER VOLTAGE)	OUTPUT AT COLLECTOR (ASSUME A 20-V COLLECTOR SUPPLY)
Below 0.5 V	20 V (cutoff)
Varies from 0.5 to 0.7 V	20 to 0.3 V
Above 0.7 V	0.3 V (saturation)

in Figure 20.9A, the complete separator circuits shown in Figure 20.9B, 20.9C, and 20.9D can be made.

A video signal with negative picture phase is applied to the input of Figure 20.9A. Assume the peak-to-peak video voltage is 1 V. This signal passes through C and is applied to the base-to-emitter terminals of Q_1. Before this signal rises to 0.5 V, the collector voltage remains at +20 V (cutoff). However, as the base voltage rises above 0.5 V, it begins to draw a current from C. At the same time, the collector voltage swings from +20 V toward 0.3 V. When the input video voltage is +1 V (peak), most of this voltage appears across the base-to-emitter terminals. However, when the video signal again drops to zero, a 0.5 V charge is left on the capacitor C. This charge remains because it has no place to go after the transistor base stops drawing current (at 0.5 V). The polarity of this 0.5 V charge is shown in Figure 20.9A.

When the other horizontal synchronizing pulses arrive, they find a new situation. They must overcome the 0.5-V base-to-emitter cutoff voltage in series with the 0.5-V charge across the capacitor C. If these pulses are at the same voltage as the first pulse (+1 volt peak), they will not cause the base to come out of cutoff. Now, how does resistor R, added as shown in Figure 20.9B, make a complete synchronizing pulse separator? If the synchronizing pulse portion of the video wave form is 0.25 V in amplitude, then we should lose exactly that much voltage from C during each horizontal scan period (63.5 μs). This will allow only the synchronizing pulse to turn the transistor on. The rest of the video wave form will be blocked by the remaining 0.25 V on C and the 0.5-V base-to-emitter cutoff voltage. Since 0.25 + 0.5 = 0.75, the bottom, 0.75-V portion of the video signal will not pass through the separator.

The circuit shown in Figure 20.9B partially discharges C to ground through a 2 700-ohm resistor. The circuit in Figure 20.9C partially discharges C from one of its plates to the other through a 3 300-ohm resistor. These two circuits have the same time constant because R in Figure 20.9B is not the total resistance around C. Resistor R must be added in series to the value of 600 ohms from the previous stage to make 3 300 ohms. Thus both circuits have a time constant of 3 300 ohms × 0.03 μF = 100 μs. This is approximately the time of 1.5 horizontal scan periods (1.5 × 64 μs = 96 μs). In one horizontal scan period, C will discharge for two-thirds of a time constant. This is long enough to remove 0.25 V from C.

The circuit of Figure 20.9D discharges C in a different way. A 220-kilohm resistor feeds electrons from the negative side of C to the positive supply voltage. This is a much faster way to discharge C, and so R must have a large resistance, typically several hundred thousand ohms.

Resistor R in Figure 20.9B, 20.9C, and 20.9D performs a critical function. It discharges capacitor C by approximately the height of the synchronizing pulse during each horizontal scan period. If R is too large,

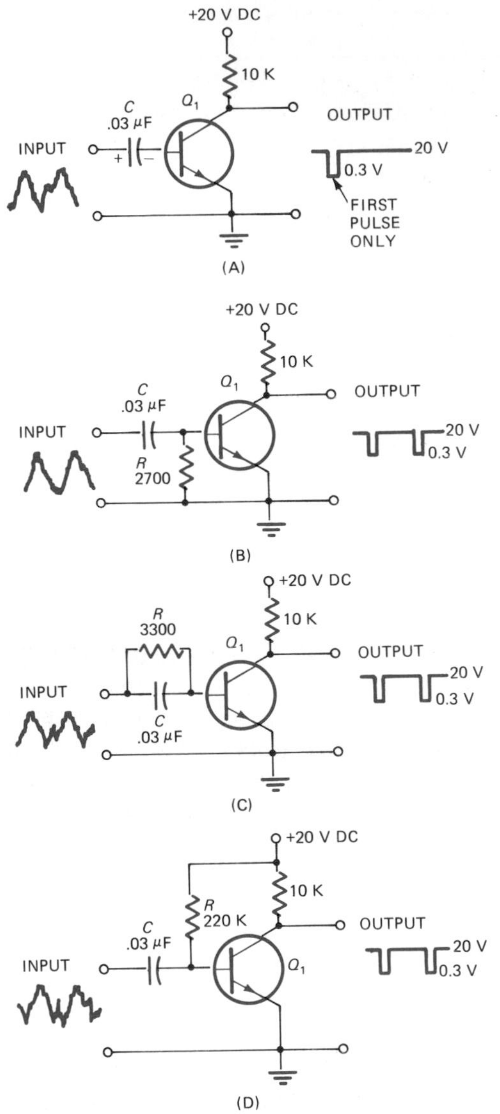

Figure 20.9A-D. The evolution of three types of transistor sync separators: A. basic circuit; B. discharge of C directly to ground; C. discharge of C around itself; and D. discharge of C to the power supply.

C will not be sufficiently discharged. This will keep the separator from using the full voltage of the pulse. If R is too small, C will discharge too much. As a result, some of the blanking voltage or video signal will pass through the synchronizing pulse separator.

20.4 REDUCTION OF NOISE TO IMPROVE SYNCHRONIZATION

Noise in the synchronizing pulses is much more serious than noise in the video or audio signals. The human eye and ear tend to ignore a certain amount of video and audio noise. However, just one noise pulse at the right time can cause a receiver to momentarily lose vertical or horizontal synchronization. For this reason, almost all TV receivers have circuitry to minimize the effect of noise associated with the synchronizing pulses. These circuits are known variously as (1) noise gates, (2) noise clippers, (3) noise inverters, or (4) noise cancelers.

IMPULSE NOISE

Impulse noise can be man made, or a result of atmospheric electrical disturbances. Such noise can be picked up by the receiver antenna. It can also be conducted to the set through the power line. Some sort of filtering is invariably found at the AC line input to most TV receivers to minimize power line pickup. (The triode separator shown in Figures 20.7 and 20.8 also functions as a noise clipper, as previously mentioned.)

Note: Some noise-cancellation circuits were described in Sections 14.6 to 14.9. This information should be reviewed at this time.

20.5 TRANSISTOR NOISE-CANCELLATION CIRCUITS

TRANSISTOR NOISE GATE

The series method of noise cancellation uses a stage called the **noise gate.** A transistor version of this device is shown in Figure 20.10. The noise gate Q_2 is in series with the

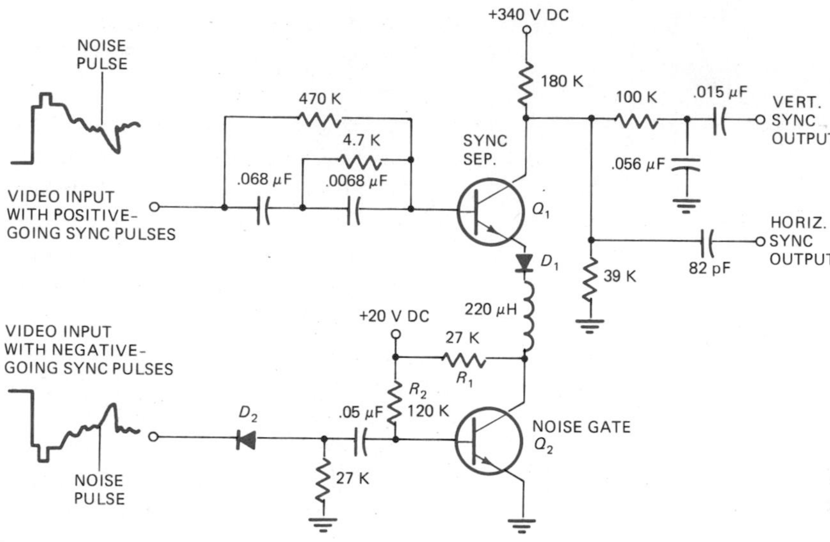

Figure 20.10 A transistorized noise gate and sync separator. *(Courtesy of Sylvania.)*

emitter of the synchronizing pulse separator Q_1. If the noise gate is fully on, as it normally is, the voltage between its collector and emitter will be only 0.3 V. The separator operates normally when the noise gate is on. If the noise gate is momentarily turned off, the separator emitter voltage rises to +20 V. This voltage is developed by the 20 V supply connected to R_1. With a positive bias on the separator emitter, no synchronizing pulse separation can take place.

Noise Gate Turn Off What can cause the noise gate to turn off? Note that it is driven with video which has negative-going sync pulses. The separator is driven with a video of the opposite phase. The normal video signal applied to the base of the noise gate does not have sufficient amplitude to overcome the bias of R_2. However, if a noise spike with a negative amplitude beyond the synchronizing pulse tip appears, it will turn the noise gate off. Noise pulses that are not larger than the synchronizing pulse tip will not cause the noise gate to turn off.

Horizontal and vertical synchronizing pulses are separated by the filter circuits at the collector of Q_1. These circuits will be discussed soon. For now, it suffices to know that the vertical synchronizing pulses are separated by an ***RC* integrator.** The horizontal synchronizing pulses are separated by an ***RC* differentiator.**

TRANSISTOR NOISE CLIPPER AND NOISE GATE CIRCUIT

Figure 20.11 is another schematic diagram of a transistor synchronizing pulse separator and noise-cancellation circuit. The circuit consists of synchronizing pulse amplifier Q_{300}, separator Q_{301}, noise clipper Q_{302}, and noise gate Q_{303}. Oscilloscope wave forms are shown at the input and output of Q_{300} and at the output of Q_{301}.

Separator Operation The peak-to-peak voltage of the composite video signal at the input to synchronizing pulse amplifier Q_{300} is approximately 2.8 V. The synchronizing

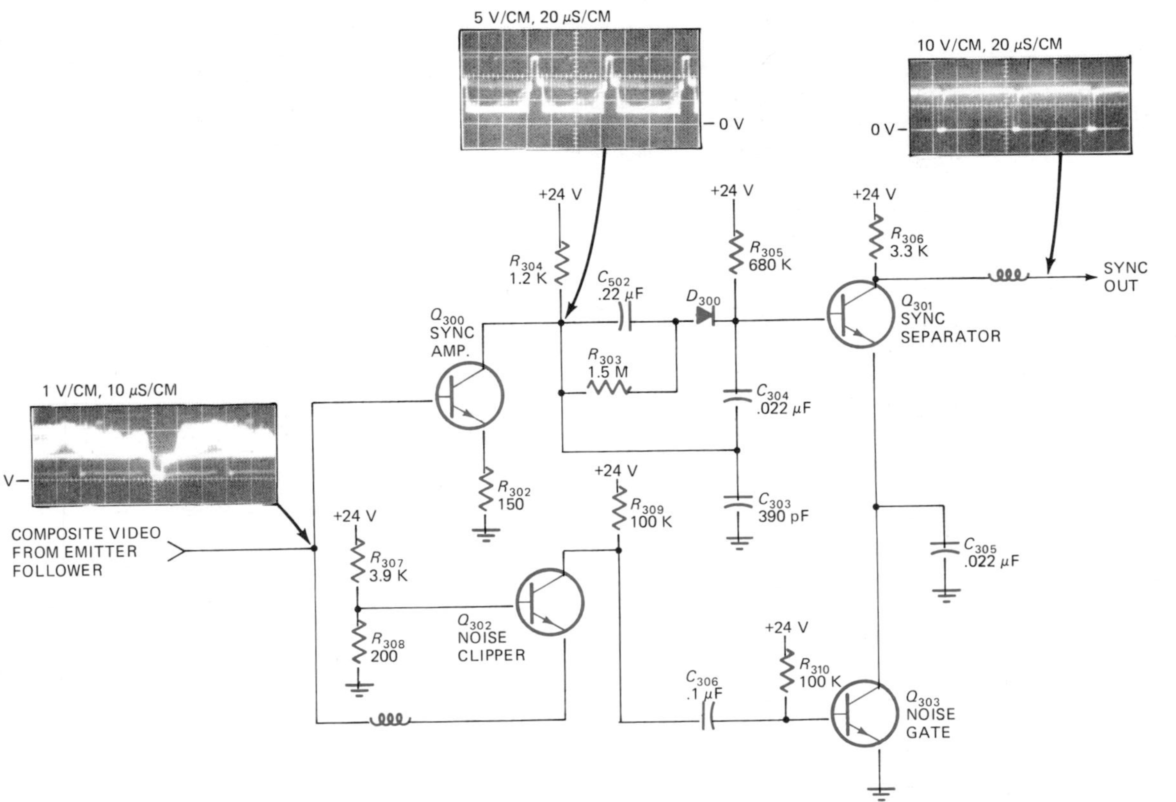

Figure 20.11 A transistor sync separator and noise-cancellation circuit. The first number above each oscillogram indicates the vertical deflection sensitivity in volts/centimeters. (A centimeter is one box.) The second number is the oscilloscope sweep speed in μs/cm. *(Courtesy RCA Solid State Division.)*

pulse amplitude is about 1 V. This composite video is applied to the base of amplifier Q_{300}. It comes from an emitter follower not shown in Figure 20.11. It is important to note, however, that Q_{300} is DC base-biased directly from the emitter voltage of the prior emitter follower. This base bias is about +2.3 V. The synchronizing pulse amplifier amplifies and inverts the input signal. As shown in Figure 20.11, this amplified composite video signal has a peak-to-peak value of about 10 V. The synchronizing pulses are now in the positive direction and have an amplitude of about 5 V.

The amplified composite video signal is applied to the base of synchronizing pulse separator Q_{301}. The base of Q_{301} is biased by the emitter-to-base junction voltage (about 0.7 V) and through resistor R_{305}. Because of the base bias, Q_{301} will conduct only on the positive synchronizing pulses. The pulses are inverted and amplified by Q_{301} to a value of about 18 V. (see Figure 20.11.)

Noise Cancellation In Figure 20.11, the input composite video signal is also fed to the emitter of noise clipper Q_{302}, which is base-biased to 1.17 V from the 24-V supply by the voltage divider R_{307} and R_{308}. The base-bias voltage is such that Q_{302} cannot conduct except on *negative* noise peaks that *exceed* the amplitude of the negative-going synchronizing pulses. The emitter of Q_{302} is biased to about 2.3 V.

The only signals appearing at the collector of Q_{302} are negative-noise peaks which exceed the amplitude of the sync pulses. (Note no phase inversion in Q_{302}.) These negative noise peaks are applied to the base of noise gate Q_{303} via C_{306}. Gate Q_{303} is base-biased by the emitter-to-base junction voltage (0.7 V) and through R_{310} to the 24-V supply. It is *normally conducting*. As in Figure 20.10, synchronizer separator Q_{301} and noise gate Q_{303} are connected in series. Thus, in order for Q_{301} to conduct, Q_{303} must also conduct.

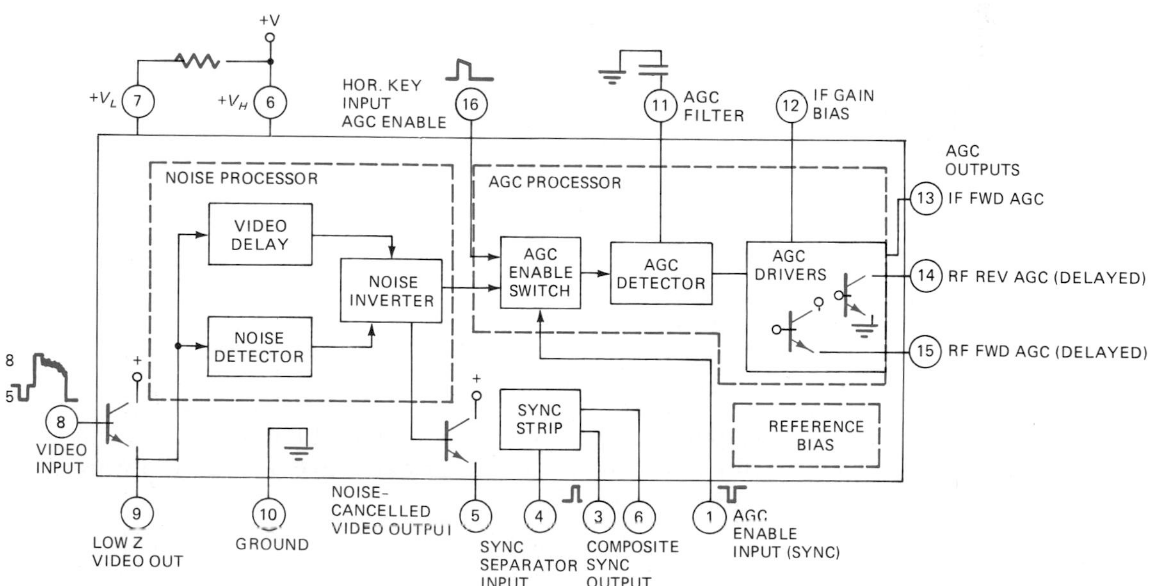

Figure 20.12 Simplified diagram of a monolithic-silicon IC containing a sync separator, noise inverter, and other circuits. *(Courtesy of RCA Solid State Division.)*

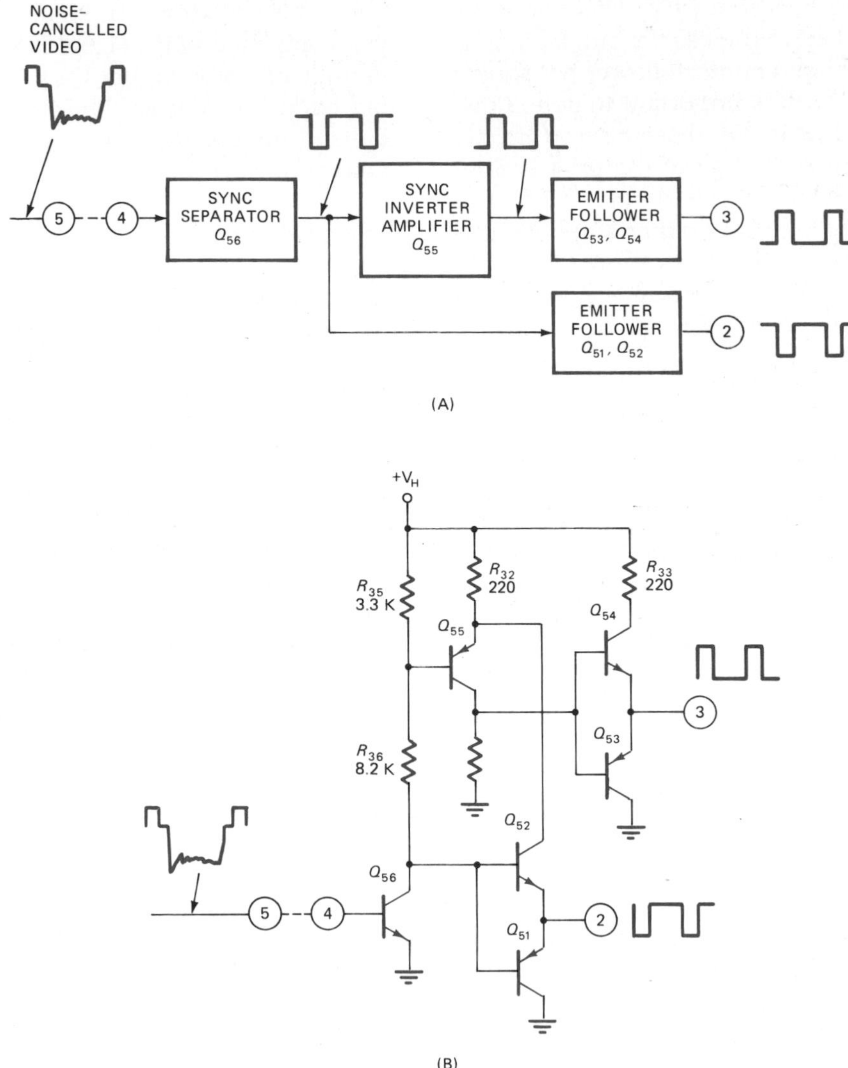

Figure 20.13A-B. The sync-separator section of the RCA IC. **A.** The block diagram. **B.** The schematic diagram. Terminal numbers (except No. 2) correspond with those on Figure 20.12. *(Courtesy of RCA Solid State Division.)*

Under noise conditions that exceed the synchronizing pulse amplitude, the negative noise pulses at the base of Q_{303} turn off this transistor. Consequently, Q_{301} is simultaneously turned off and there is no synchronizing pulse output for the duration of the noise peaks. The vertical and horizontal sweep circuits become free-running for the short duration of the noise peaks. This is not ordinarily noticeable in the TV picture.

20.6 SYNCHRONIZING PULSE SEPARATOR IN AN RCA IC

The synchronizing pulse separator discussed in this section is part of the IC described in Section 14.8. Refer also to Figure 14.20 for a block diagram of the noise processor (Figure 20.12). The IC block diagram was originally introduced in Chapter 14 as Figure 14.19. It is again included here for convenience. For an explanation of the noise processor, refer to Section 14.8. Note that in Figures 14.19 and 14.20 the output at Terminal 5 is a noise-cancelled composite video signal. This signal is the input to the synchronizing pulse separator stage.

In Figure 20.12, the separator is labeled "sync strip." The composite video input is fed into Terminal 4 of the IC. The separated synchronizing pulse output is taken from Terminal 3. Terminal 6 accepts the DC voltage that powers the synchronizing strip. Although not shown in Figure 20.12 (for simplicity), there is a Terminal 2 synchronizing pulse output. This is an inverted-polarity replica of the Terminal 3 output. It is shown in the sync separator block diagram of Figure 20.13A and its schematic shown in Figure 20.13B. Terminal numbers (except 2) correspond with those of Figure 20.12.

Synchronizing pulse separation from the composite video input to Terminal 4 is accomplished by separator Q_{56}. This separator Q_{56} is normally off because of the inherent 0.7 V emitter-to-base junction voltage. It conducts only for the synchronizing pulse portion of the composite video signal. The output of Q_{56} consists of negative synchronizing pulses which are fed both to synchronizing pulse inverter amplifier Q_{55} and to emitter follower Q_{51}, Q_{52}. Inverter Q_{55} serves only to reverse the polarity of the synchronizing pulse fed to emitter follower Q_{53}, Q_{54}. The low-impedance output of Q_{53}, Q_{54} is positive synchronizing pulses at Terminal 3 (see Figure 20.13).

The output of emitter follower Q_{51}, Q_{52}, which is not in the path of inverter amplifier Q_{55}, consists of negative synchronizing pulses at Terminal 2.

Both positive and negative synchronizing pulse outputs are provided from the IC to accommodate specific TV receiver designs.

20.7 VERTICAL AND HORIZONTAL PULSE SEPARATION

The separation of the vertical and horizontal pulses from each other is based on frequency (or wave form) difference and not on amplitude (which is the same for both). The two types of pulses are shown in simplified form in Figure 20.14. Note that the horizontal pulse lasts for a much shorter time (5 μs) than the vertical pulse. Therefore a low-pass filter (integrator) develops the vertical pulse voltage at its output, and a high-pass filter (differentiator) has only the horizontal pulse voltage at its output. These two distinct pulses can then be fed to their respective oscillator circuits to provide frequency control.

HORIZONTAL SYNCHRONIZING PULSE FILTER

Let us consider a high-pass filter (differentiator) and its effect on the horizontal synchronizing pulses (Figure 20.15). The filter in the diagram has a time constant of

$$\begin{aligned} T &= RC \\ &= 2000 \text{ ohms} \times 50 \text{ pF (picofarads)} \\ &= 2000 \text{ ohms} \times 0.00005\ \mu\text{F} \\ &= 0.1\ \mu\text{s} \end{aligned}$$

A time of 0.1 μs is short relative to the 5-μs duration of the horizontal synchronizing pulse.[1] When the first edge of the horizontal

[1] Any time constant one-fifth the duration of an applied pulse is said to be short with respect to that pulse. By the same token, any time constant five times longer than the duration of an applied pulse is said to be long.

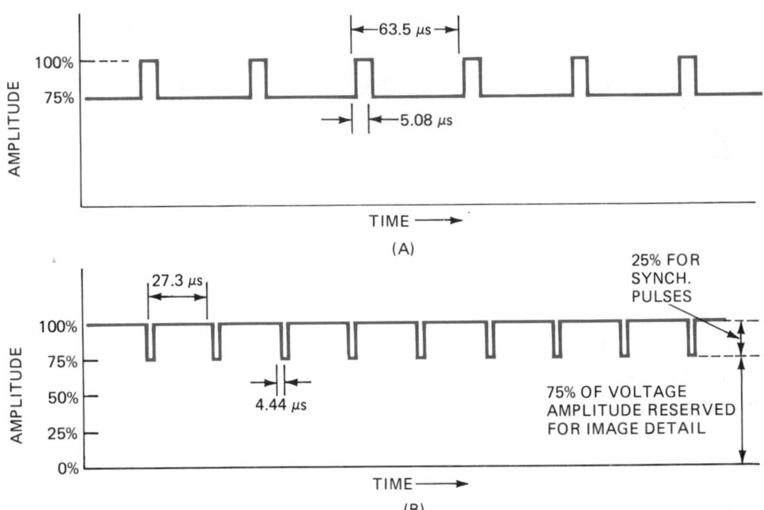

Figure 20.14A-B. The differences in wave forms of A. horizontal and B. vertical pulses. (Simplified form to show essential pulse shape differences.)

synchronizing pulse in Figure 20.15, known as the leading edge, is applied, current flows for a moment through the resistor (producing a sharp positive pulse). This charges the capacitor to the full pulse voltage. Once the capacitor is fully charged, nothing else happens along the flat portion of the pulse because a capacitor (and hence, a capacitor and a resistor in series) reacts only to changing (AC) voltages, not to steady (DC) voltages. At the lagging edge of the pulse, where the voltage drops suddenly, there is another short flow of current, this time in the opposite direction. This discharges the capacitor and produces a sharp negative pulse. The result of the application of the square-wave synchronizing pulse to the input of the high-pass (short-time-constant) filter is the output wave shown in Figure 20.15. This is a *differentiated* pulse.

DIFFERENTIATOR OUTPUT

Each incoming synchronizing pulse gives rise to two sharp pulses at the output of the filter. One of these pulses is above the reference line, and the other is below. This happens because one is obtained when the front edge of the incoming pulse acts on the filter and the other is obtained when the lagging edge arrives. Only one (the positive one) of these two output pulses is needed to control the horizontal sweep oscillator.

HORIZONTAL SYNCHRONIZATION DURING VERTICAL SYNCHRONIZING PULSE INTERVAL

The action of a high-pass filter indicates how the serrations of the vertical synchronizing pulse permit control of the horizontal synchronizing oscillator when the vertical synchronizing pulse is being applied. In Figure 20.16 the input wave and the output pulses of a high-pass filter are shown. Of all those present, only the positive pulses that occur at the proper time (1/15 750 s) affect the horizontal oscillator. These active pulses are indicated by *A* in the figure. Note that all active pulses are evenly spaced and differ by 1/15 750 s. The conditions illustrated in Figure 20.16A occur only when the vertical

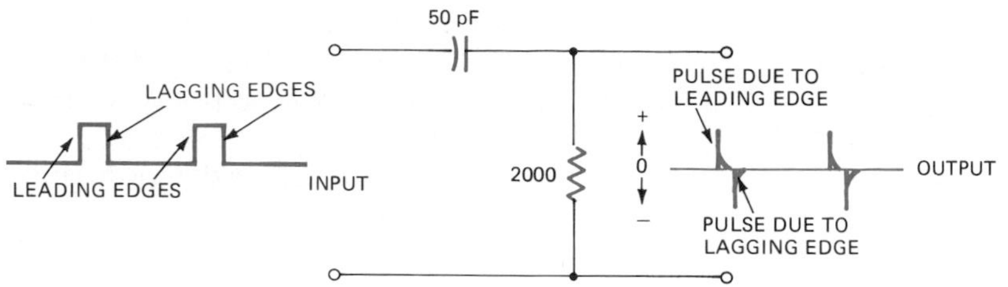

Figure 20.15 A high-pass filter and its effect on the horizontal synchronizing pulses.

pulses are inserted at the end of a full scanning line (even fields). Figure 20.16B shows the situation when the field ends on a half line (odd fields). When this happens, different equalizing and serrated vertical synchronizing pulse[1] pips control the horizontal oscillator. Because of the difference in field ending, control shifts to those pips that were inactive in Figure 20.16A. However, the shift in no way interferes with the timing of the horizontal oscillator. This shift from field to field illustrates why all the equalizing and vertical pulses are designed to produce synchronizing pulses *twice* in each horizontal line interval.

Effect of Vertical Synchronizing Pulse The serrated vertical synchronizing pulse lasts for 190.5 μs (3 H). This is shown in Figure 20.16B. This pulse is applied to a differentiator as a series of short pulses, as shown in the figure. These pulses (properly spaced at 1 H) synchronize the horizontal oscillator, but they are not useful for synchronizing the vertical oscillator. This is because the vertical oscillator requires one synchronizing pulse each 1/60 of a second (monochrome rate). In order to obtain a suitable series of vertical synchronizing pulses, the string of pulses must be passed through a low-pass filter or *integrator*.

[1]The vertical synchronizing pulse is divided into six individual pulses. This is to maintain horizontal synchronization during the vertical synchronizing pulse interval.

VERTICAL SYNCHRONIZING PULSE FILTERS

For vertical pulse separation, a low-pass or long-time constant filter of the type shown in Figure 20.17 is used. This filter is identical to the high-pass filter except that the positions of the capacitor and resistor have been interchanged. The output is obtained from the voltage across the capacitor. Also, the time constant of the capacitor and

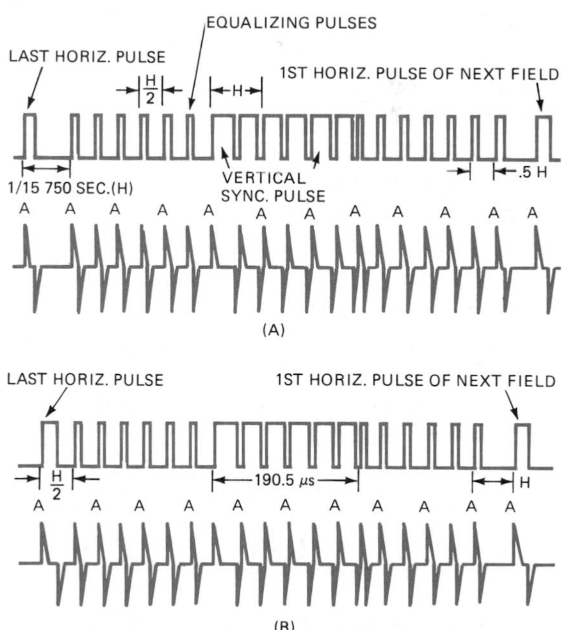

Figure 20.16A-B. The conditions during vertical pulses in **A.** The first field and **B.** The second field. The pips labeled A control the horizontal-sweep oscillator.

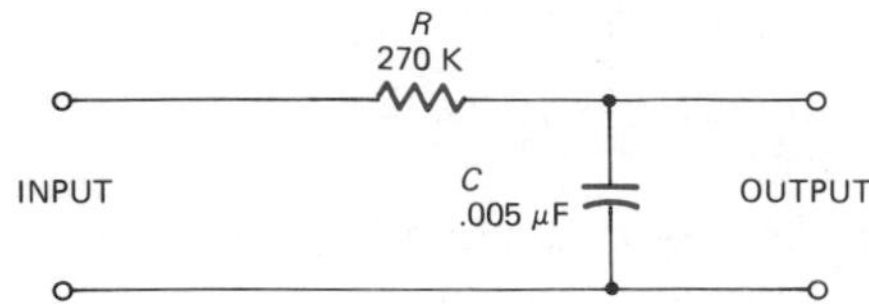

Figure 20.17 A low-pass filter, or integrator. The capacitor combines (or integrates) all the serrated vertical pulses until the output voltage rises to the level necessary for the vertical-sync oscillator to react.

resistor is *much greater* than in the previous filter. A long-time constant means that the capacitor charges and discharges slowly and does not respond as quickly as a filter with a shorter time constant to rapid changes in voltages. Hence, when a horizontal pulse arrives at the input of this filter, its leading edge starts a lesser current flow through the resistor and the capacitor begins to charge. This charging process is slow, however. Therefore, the lagging edge of the wave reaches the filter shortly afterwards and reverses the current flow. This brings the capacitor back to its previous value. There is very little change during this short time interval. The vertical-deflection oscillator is designed so that it does not respond to these small fluctuations.

What is true about how the horizontal synchronizing pulses affect the vertical integrator is even more true about the equalizing pulses. These last for a much shorter time than the horizontal synchronizing pulses. Essentially, then, there is no chance that these pulses will affect the vertical oscillator. Figure 20.18 indicates the output voltage of the integrator from the application of these pulses. The voltage level is below the dotted line, which is the point the voltages must reach in order to affect the vertical oscillator.

The voltage across the capacitor (output) begins to build up when the serrated vertical synchronizing pulse is reached. Note in Figure 20.14 that each vertical synchronizing pulse section is 27.3 μs wide and separated by 4.44 μs. Compare this with the horizontal pulse, which is 5.08 μs wide and separated by 63.5 − 5.08 = 58.42 μs. As previ-

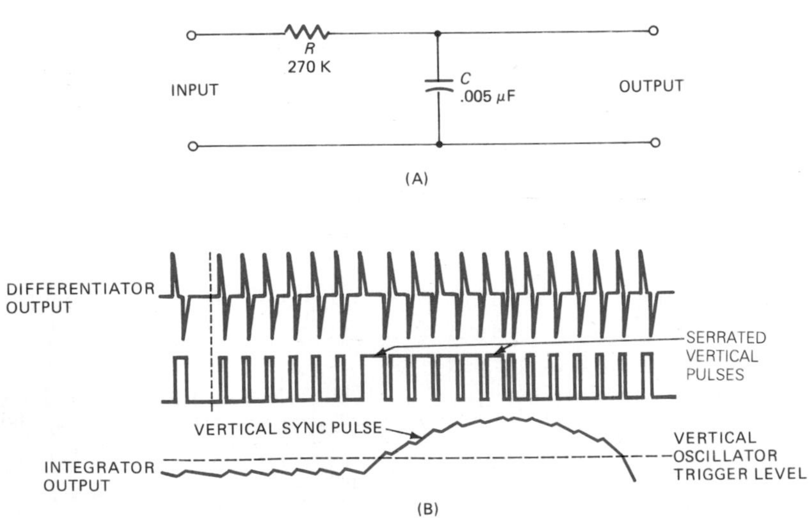

Figure 20.18A. Low-pass filter (integrator) for separating the vertical and horizontal pulses. **B.** A wave form of the rise in voltage across the capacitor due to vertical pulses.

ously mentioned, this characteristic keeps a charge from accumulating in the integrator capacitor.

VERTICAL SYNCHRONIZING PULSE INTEGRATION

Consider the effect of the serrated vertical pulses on the integrator. During the first of the six pulse sections (27.3 μs), the integrator capacitor charges to roughly 30% of the applied peak synchronizing voltage. However, during the space between the six pulse sections (4.44 μs), there is relatively little time for the capacitor to discharge and so it retains most of its initial charge. For each successive pulse section in the train of six, the capacitor charge increases until the end of the sixth pulse section is reached. After this sixth pulse section, the capacitor discharges and responds (as explained above) to the horizontal synchronizing pulses. The net result is the formation of a single vertical synchronizing pulse from the six pulse sections. This is clearly shown by the bottom wave of Figure 20.18B.

20.8 FUNCTION OF THE EQUALIZING PULSES

In Figure 20.19, the buildup of a vertical synchronizing pulse across the output of the vertical filter is shown. Figure 20.19A shows the buildup for the vertical pulse that comes at the end of a line. Figure 20.19B shows the buildup for the pulse that comes in the middle of a line. In Figure 20.19A, each horizontal pulse causes a slight rise in voltage across the output of the vertical filter, but this is reduced to zero by the time the next pulse arrives. Hence, there is no residual voltage across the vertical filter due to the horizontal pulses. Only when the long, serrated vertical pulse arrives is the desired voltage increase obtained.

The situation is slightly different in Figure 20.19B, however. Here, the last horizontal pulse is separated from the first vertical pulse by only half a line. Therefore any horizontal voltage developed in the vertical filter does not have as much time to reach zero before the first vertical synchronizing pulse section arrives. This means that the vertical

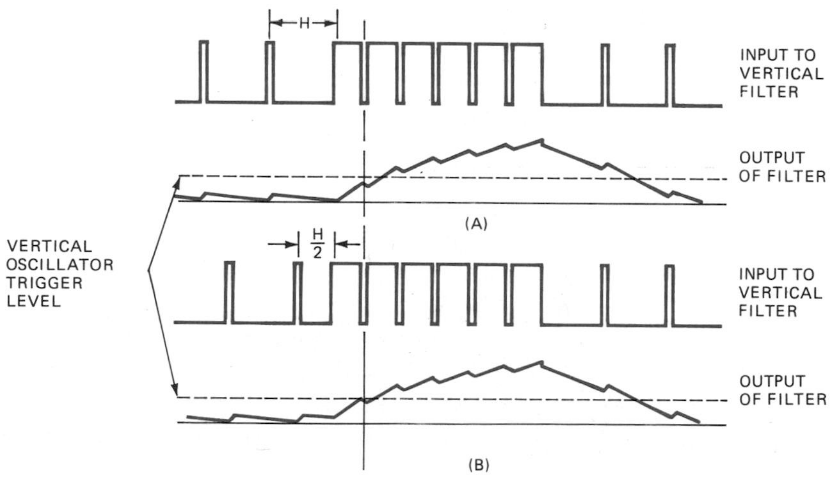

Figure 20.19A-B. The difference in voltage conditions of a vertical sync pulse when no equalizing pulses are used: **A.** First field and **B.** second field. Note that the trigger level at *B* is reached sooner than that at *A*. This would affect interlace.

buildup starts not from zero, as in Figure 20.19A, but from some low voltage. As a result, the dotted line (trigger level) is reached sooner. Since the dotted line represents the firing point of the vertical oscillator, the oscillator is triggered a fraction of a second too soon. The actual time involved is quite short, but it is sufficient to affect the interlacing of the odd and even fields.

ADDING EQUALIZING PULSES

When equalizing pulses are inserted before every vertical synchronizing pulse, the voltage level established before the start of each serrated vertical pulse is essentially the same for odd and even fields. Therefore the vertical oscillator is triggered at the proper moment in each instance.

In Figure 20.16, the spacing between the last equalizing pulse and the first serrated vertical pulse is constant at H/2 for odd and even fields. Now the vertical synchronizing pulse rises identically for odd and even fields. This means that the vertical oscillator is triggered (synchronized) at identical times and correct interlace is maintained.

THE VERTICAL PULSE

From a comparison of the vertical and horizontal pulses in Figure 20.18B, we might get the impression that the vertical pulse is not very sharp. This is because it is extended over six serrated pulses. If it were drawn on a larger scale, this pulse would also be sharp (Figure 20.20). As far as the vertical oscillator is concerned, this pulse occurs rapidly and represents a sudden change in voltage.

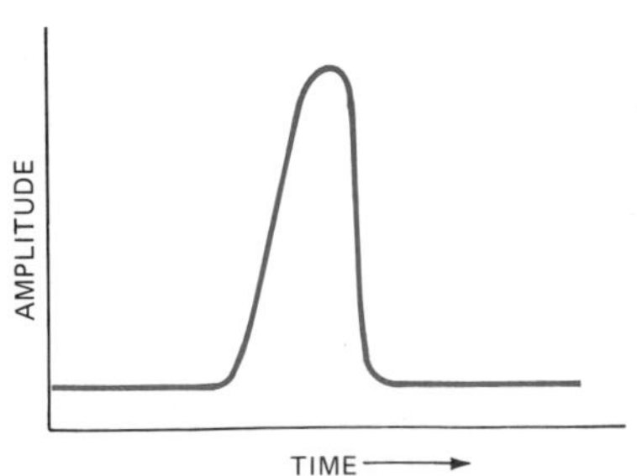

Figure 20.20 The appearance of the vertical-sync pulse at the output of an integrator.

20.9 VERTICAL SYNCHRONIZING AND EQUALIZING PULSE DISPLAY (HAMMERHEAD)

It is possible to see the vertical synchronizing and equalizing pulses on the picture tube screen. The appearance of these pulses is shown in Figure 20.21. This display is frequently called a "hammerhead" pattern because of its similarity to a hammer. This hammerhead pattern appears on the screen because the entire composite video signal, including all synchronizing and blanking pulses, is being fed to the picture tube.

It is sometimes useful to examine the hammerhead pattern because the contrast of different portions of it can indicate the relationship between the synchronizing pulse and blanking pulse amplitudes.

HOW TO VIEW THE HAMMERHEAD PATTERN

Vertical synchronizing, equalizing, and blanking pulses are normally out of sight beyond the bottom and top of the screen. In order to view the hammerhead pattern, it is necessary to carefully adjust the vertical-hold control to move the pattern vertically to the approximate center of the screen. Next, adjust the brightness control so that the vertical blanking interval (Figure 20.21) is gray and slightly darker than the rest of the screen. In a normally operating TV receiver, the vertical synchronizing and equalizing pulses will be black, as seen in Figure 20.21.

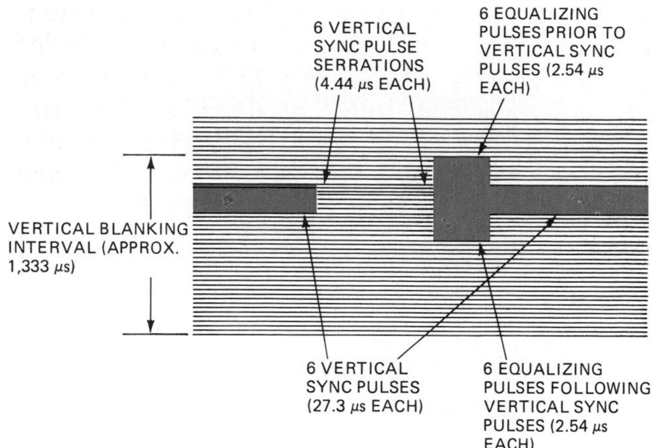

Figure 20.21 The appearance of the "hammerhead" pattern on the picture tube screen. This shows equalizing and serrated vertical-sync pulses, as well as the vertical-blanking interval. *(Courtesy of RCA Consumer Products.)*

This happens because the vertical blanking pulses normally drive the picture tube to cutoff. However, now that the blanking is gray, the synchronizing pulses which are 25% higher in amplitude, drive the picture tube to cutoff and consequently appear black.

By observing the appearance of the hammerhead in a normally operating set and comparing it with the set being serviced, it may be possible to determine whether or not synchronizing pulse compression is occurring. If it is, the hammerhead will not be black enough. Synchronizing pulse compression can be caused by a defect in the AGC, common IF, or video amplifier sections.

20.10 SYNCHRONIZING TROUBLES

Synchronizing troubles may take one of three possible forms:

1. No horizontal or vertical synchronizing (Figure 20.1C).
2. No vertical synchronizing (Figure 20.1A).
3. No horizontal synchronizing (Figure 20.1B).

Each of these will be discussed in turn.

NO HORIZONTAL OR VERTICAL SYNCHRONIZING

This trouble may be caused by a defective synchronizing pulse separator or amplifier. Note that these stages are common to both the horizontal and vertical synchronizing pulses (Figure 20.2).

The synchronizing pulses may become compressed, as shown in Figure 20.2. This can be caused by a defective common IF stage or an incorrect setting or defect in the AGC system. In the latter case, there will be insufficient AGC voltage and subsequent video signal overloading in one or more of the common IF amplifier stages. When the synchronizing pulse is compressed, separation will be incomplete. Possibly, some video signal may also be separated. As a result, the synchronizing signals will not be clearly defined. Unstable horizontal and vertical synchronization can result. In extreme cases, the picture will "flag wave," that is, it will appear somewhat as a flag waving in the wind.

NO VERTICAL SYNCHRONIZATION (NON COUNT DOWN CIRCUITS)

When the symptom is a vertically rolling picture, the vertical hold control should be adjusted first. It may be incorrectly set.

This trouble may be caused by a defect in the vertical oscillator or a defect between the output of the separator and the synchronizing input to the vertical oscillator. In the latter case, the problem most likely is either a defective integrator capacitor or resistor, or an open coupling capacitor between the synchronizing pulse separator and the vertical oscillator. (Since the horizontal synchronization is working, it must be assumed that the synchronizing pulse separator, and amplifier if present, are normal.

Vertical Oscillator It is a simple matter to determine whether the vertical oscillator or the synchronizing signal is at fault. (For TV receivers that use countdown circuits, this technique does not apply. See Chapter 21.) Operate the vertical hold control while looking at the picture. If the picture can be made to roll *both* upward and downward, the vertical oscillator is working properly. If the picture will not roll both ways, the trouble is in the vertical oscillator. In this case, the *RC* frequency-determining parts are usually at fault.

SEPARATE SYNCHRONIZING AMPLIFIERS

Some TV receivers use separate synchronizing pulse amplifiers for the vertical and horizontal oscillators. In this case, vertical synchronization problems can be caused by a defect in the vertical synchronizer pulse amplifier stage.

VERTICAL JITTER

In the condition known as vertical jitter, the picture bounces rapidly up and down, but does not roll. The interlace may also be affected. Vertical jitter can be caused by:

1. Defective synchronizing pulse separator or amplifier stage
2. Defective vertical integrator network
3. Horizontal flyback pulses introduced into the AGC line (this may cause variations in vertical pulse amplitude, which can cause the jitter. AGC decoupling filter(s) may be at fault).

NO HORIZONTAL SYNCHRONIZATION

If the picture is not stable in the horizontal direction, try to adjust the horizontal hold control first. It may be incorrectly set.

This trouble may be caused by a defect in the horizontal oscillator or in its associated AFC circuit (Chapter 21). Also, there may be a defect between the output of the synchronizing pulse separator and the input to the horizontal oscillator. In the latter case, the most likely source of the problem is a defective differentiator or coupling capacitor. (It is assumed that the output from the synchronizing pulse separator is normal.)

Horizontal Oscillator The horizontal oscillator operation can be checked easily. Vary the setting of the horizontal hold control. If a complete picture can be achieved, even momentarily, the horizontal oscillator is working properly. The trouble then is in the horizontal AFC circuit or inadequate synchronizing pulses.

As in the case of the vertical oscillator, there may be a separate amplifier for the horizontal synchronizing pulse. This should be kept in mind if the horizontal synchronizing pulses at the horizontal oscillator are not normal.

Horizontal Picture Bending If 60-Hz or 120-Hz hum gets *only* into the synchronizing circuits, it will modulate the synchronizing pulses. The result in the case of horizontal synchronization will be horizontal-picture bending with either 1 (60 Hz) or 2 (120 Hz) roughly sinusoidal-bending patterns occurring in the picture. However, note that in this situation the sides of the raster will be straight because only synchronizing pulses are being affected.

SUMMARY OF CHAPTER HIGHLIGHTS

1. The horizontal and vertical synchronizing pulses are fed to their respective horizontal and vertical oscillators. (Figure 20.2)
2. The function of these pulses is to synchronize the TV receiver's deflection oscillators with those at the TV transmitter.
3. The synchronizing pulse rates for monochrome and color transmissions are 15 734.26 and 59.94 Hz.
4. Some TV receivers use countdown circuits rather than oscillators in the vertical and horizontal deflection circuits.
5. If only vertical synchronization is lost, the picture will roll vertically. If only horizontal synchronization is lost, the picture will break up into diagonal strips. If all synchronization is lost, the picture will assume a combination of the two effects mentioned above. (Figure 20.1)
6. By rolling up the picture, a *hammerhead* vertical synchronization pattern can be observed. This pattern is useful in checking synchronizing pulse amplitude. (Figure 20.21)
7. Synchronizing pulse compression may affect both horizontal and vertical synchronization. This condition can be caused by a defect in a common IF stage or in the AGC system. (Figure 20.22)
8. The horizontal synchronizing pulses are the output of a differentiator, and the vertical synchronizing pulses are the output of an integrator. (Figure 20.3)
9. A synchronizing pulse separator stage clips the synchronizing pulses from the composite video signal. (Figure 20.3)
10. A diode synchronizing pulse separator with automatic self-bias can supply either positive or negative synchronizing pulses. (Figures 20.4 to 20.6)
11. Transistors or vacuum tubes can be used as synchronizing pulse separators. (Figures 20.7 and 20.9)
12. It is desirable to reduce the effect of noise pulses that may accompany synchronizing pulses. This is because some noise pulses may cause a temporary loss of synchronization. (Figure 20.10)
13. A noise gate may be connected in series with a synchronizing pulse separator. When the noise gate is turned off by a noise pulse, the separator is also momentarily OFF. (Figure 20.11)
14. A differentiator produces a positive and negative *spike* for each synchronizing

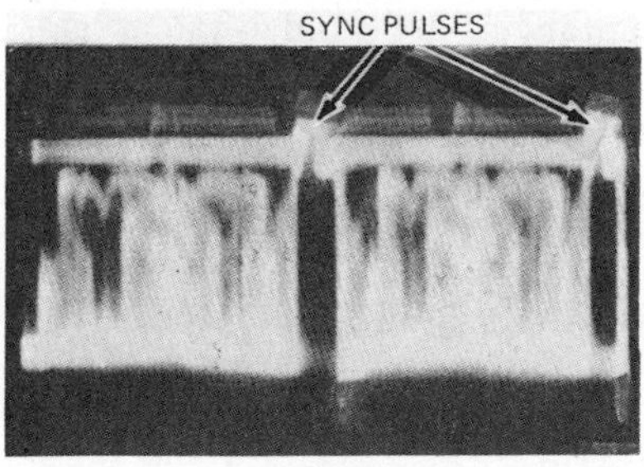

Figure 20.22 A video signal in which the sync pulses have been almost compressed. Note that the video signal extends almost to the level of the pulses.

pulse. Generally, only the positive spikes are used for horizontal synchronization. (Figure 20.15)

15. In Number 14 these spikes are also produced for selected equalizing and vertical synchronizing pulse sections. This ensures continuous horizontal synchronization. (Figure 20.16)
16. An integrator produces one vertical synchronizing pulse from each group of six serrated pulses. The vertical synchronizing pulse rate is 59.94 Hz. (Figures 20.17 and 20.18)
17. Equalizing pulses that occur before vertical synchronizing pulses maintain correct interlace. (Figure 20.19)
18. Equalizing pulses maintain correct interlace by ensuring that the integrator capacitor starts its charge from the same level on odd and even fields. (Figure 20.19)
19. Synchronizing trouble may take one of three possible forms (1) no horizontal or vertical synchronization, (Figure 20.1C) (2) no vertical synchronization (Figure 20.1A) or (3) no horizontal synchronization. (Figure 20.1B)
20. A condition of no horizontal and no vertical synchronization can be caused by a defective synchronizing pulse separator or amplifier.
21. If either horizontal *or* vertical synchronization is not working, the oscillator operation can be checked by adjusting the appropriate hold control. If a whole picture can be at least momentarily achieved, the oscillator is probably working properly.
22. A defective horizontal AFC circuit can cause loss of horizontal synchronization.

EXAMINATION QUESTIONS

(Answers are provided at the back of the text.)

Part A Supply the missing word(s) or number(s).

1. The synchronizing pulses are part of the ________ signal.
2. ________ pulses help to maintain correct interlace.
3. The master oscillator frequency in a countdown deflection system is ________ Hz.
4. A rolling picture indicates a loss of ________ synchronization.
5. The horizontal frequency is ________ the correct value if the diagonal segments slope downward from right to left.
6. A defective ________ can cause loss of both horizontal and vertical synchronization.
7. A ________ -pass filter provides the vertical synchronizing pulse. It is also called a(n) ________.
8. The ________ stage removes the synchronizing pulses from the composite video signal.
9. ________ pulses of sufficient amplitude can cause a momentary loss of synchronization.
10. A ________ reduces noise effects on synchronization by cutting off the ________ momentarily.
11. A ________ constant circuit is used as a differentiator.
12. Every other vertical synchronizing pulse section or ________ pulse is used for horizontal synchronization.
13. Each vertical synchronizing pulse section has a total duration of ________ μs.
14. There are ________ equalizing pulses before the vertical synchronizing pulse section and ________ equalizing pulses following.

Part B Answer true (T) or false (F).

1. Horizontal synchronization is very briefly interrupted during vertical blanking.
2. Equalizing pulses equalize the number of lines in the odd and even fields.
3. For color and monochrome TV, the horizontal scanning rate is 15 724.26 Hz.
4. *Countdown* circuits are generally constructed of discrete parts.
5. Failure of an integrator will result in a loss of vertical synchronization only.
6. If the picture tears into three diagonal segments, the horizontal frequency is off by about 180 Hz.
7. A countdown circuit clips the synchronizing pulses from the composite video signal.
8. When two complete horizontal lines appear on an oscilloscope screen, the oscilloscope sweep frequency is set to 7 875 Hz.
9. Synchronizing pulse separation can be accomplished with diodes, triodes, tetrodes, pentodes, bipolar transistors, and FETs.
10. Noise pulses that exceed blanking pulse level cannot cause synchronization problems.
11. A noise gate is turned ON by high-amplitude noise pulses.
12. A differentiator is a high-pass filter.
13. Each vertical synchronizing pulse section is 27.3 μs wide.
14. Each equalizing pulse is 4.44 μs wide.
15. The hammerhead pattern is made up of serrated vertical synchronizing pulses and equalizing pulses.
16. Synchronizing pulse compression may occur as a result of a defective common IF stage or AGC circuit.

REVIEW ESSAY QUESTIONS

1. Discuss the requirement for horizontal synchronizing pulses, vertical synchronizing pulses, and equalizing pulses.
2. Why are the vertical synchronizing pulses serrated?
3. Draw a simple block diagram showing the following: synchronizing pulse separator, synchronizing pulse amplifier, noise gate, integrator, differentiator, vertical oscillator, and horizontal oscillator. Include all applicable wave forms.
4. Draw a simple sketch of a composite video signal showing the video signal and all types of synchronizing and blanking pulses. Label all parts.
5. List the following pulse widths and/or spacings: (1) horizontal synchronizing pulse, (2) equalizing pulses, (3) vertical synchronizing pulse section, (4) spacing between vertical synchronizing sections pulse, (5) spacing between horizontal synchronizing pulses, and (6) spacing between vertical synchronizing pulses.
6. List the pulse repetition frequencies of the pulses listed in Question 5.
7. Explain what is meant by a *countdown* circuit. What, if any, are its advantages?
8. What is the meaning of synchronizing pulse compression?
9. In Question 8, what are some possible causes?
10. Discuss the operation of the noise gate in a synchronizing circuit.
11. What would happen to the TV picture if there were no equalizing pulses? Why?
12. In Figure 20.21, explain in your own words how a *hammerhead* pattern is constructed.
13. In Question 12, why is this pattern useful?
14. List all the causes you know for a loss of horizontal synchronization. How does each affect the picture?
15. In Figure 20.9B–D, explain the function and operation of resistor R.
16. Explain briefly in your own words, how the circuit in Figure 20.7 functions as a noise clipper.
17. Discuss the possible problems that may cause a condition of no horizontal or no vertical synchronization.
18. How would you determine from the receiver screen that hum was only in the synchronizing circuits?

EXAMINATION PROBLEMS

(Selected problem answers are provided at the back of the text.)

1. What is the time occupied by the two sets of equalizing pulses and the synchronizing vertical pulses for one field?
2. The color field frequency is 59.94 Hz, and there are 525 lines to a frame. Show the calculations to arrive at a master countdown oscillator frequency that is twice the normal color horizontal oscillator frequency.
3. Give the effective frequencies of: (1) equalizing pulses and (2) vertical synchronizing pulse sections. (Hint: $f = 1/2\,t$, where t = the time of one complete cycle in seconds.)
4. A circuit to be used as a differentiator has the following values: R = 200 k, C = 50 pF. (1) Would this circuit work as a differentiator for horizontal sync pulses? (2) Why?
5. Draw a simple integrator circuit. Indicate values that would provide the minimum required time constant. (Hint: The time constant should be at least five times greater than the width of one vertical synchronizing pulse section.)
6. In Figure 20.11, identify the base-bias network for Q_{302}.
7. A color picture is out of horizontal synchronization and shows nine diagonal segments sloping downward from left to right. Calculate the horizontal frequency.

Chapter 21

21.1 Requirement for sawtooth current
21.2 Need for trapezoidal wave forms
21.3 The vacuum tube blocking oscillator
21.4 Vacuum tube blocking oscillator, sawtooth generator
21.5 Transistor blocking oscillators
21.6 A transistor blocking oscillator vertical-deflection system
21.7 Vacuum tube plate-coupled multivibrators
21.8 Plate-coupled, multivibrator sawtooth generator
21.9 Combination vacuum tube multivibrator and output stage
21.10 Cathode-coupled multivibrators
21.11 Cathode-coupled, multivibrator sawtooth generator
21.12 Synchronizing the multivibrator
21.13 Transistor multivibrator oscillators
21.14 Transistor, combination multivibrator, and output stage
21.15 Miller integrator, vertical-deflection circuit
21.16 Digital vertical countdown circuits
21.17 Troubles in vertical-deflection oscillator systems
Summary

VERTICAL OSCILLATORS AND DIGITAL COUNTDOWN CIRCUITS

INTRODUCTION

In order to form a picture on the screen of the TV receiver that is synchronized with the one generated at the transmitter, it is necessary to first produce a synchronized raster. Having produced such a receiver raster, the video information automatically "paints" a copy of the transmitted picture.

While the actual movements of the electron beam (or beams) in the picture tube are controlled by magnetic fields produced in the horizontal and vertical yoke coils, the proper synchronized horizontal and vertical scanning rates must first be created by synchronized oscillators.* This, then, is the function of the horizontal and vertical deflection oscillators. These oscillators and their closely associated circuits must create suitable driving wave forms at the correct synchronized frequencies. For vertical deflection, the frequency is 59.94 Hz; for horizontal deflection, it is 15 734.26 Hz. (The principles of scanning and synchronization are explained in Chapter 3.)

The driving wave forms created by the vertical and horizontal deflection oscillators are applied to power amplifiers, which are required to provide sufficient current to drive the yokes. As we shall see, these driving wave forms may be either pure or modified sawtooth waves, depending upon the deflection frequency and upon whether we are using vacuum tube or transistor circuits. The basic types of deflection oscillators are the same for vacuum tube and transistor circuits, as well as for monochrome and color

*Synchronized horizontal and vertical scanning rates are produced in some television receivers by digital (countdown) circuits. Vertical digital countdown circuits are described in this chapter.

circuits. However, as will be shown in this chapter, some color sets may use circuits not normally found in monochrome sets.

In general, monochrome TV receivers use blocking oscillators or a variation of the multivibrator oscillator, in which the vertical output tube serves as part of the oscillator. Color-TV receivers may use the same type of circuits for vertical deflection. In addition, variations such as the RCA Miller Integrator circuit will be found. All of these vertical deflection oscillators are explained in this chapter.

Note: Although the deflection oscillator circuits for monochrome and color sets can be quite similar, there are important differences in the output amplifier circuits. These are explained in Chapters 23 and 24.

When you have completed the reading and work assignments for Chapter 21, you should be able to:

- Define the following terms: sawtooth wave, deflection oscillator, sawtooth generator multivibrator, blocking oscillator, Miller Integrator, lin-clamp circuit, vertical digital countdown system, time constant, trapezoidal wave driver, VCO, thermal runaway, set pulse, and reset pulse.
- Explain the operation of vertical blocking oscillators and vertical multivibrators.
- Explain how sawtooth generators work.
- Explain how the Miller Integrator works.
- Explain how a vertical digital countdown system works.
- Compare the operation of two basic types of vertical deflection multivibrators.
- Explain what a sawtooth wave is and how it is generated.
- Describe the operation of a transistor trapezoidal wave generator.
- Discuss the synchronization of vertical oscillators.
- Understand the operation of vertical deflection controls.
- Understand some common troubles in the vertical oscillators and countdown systems and how to isolate them.

21.1 REQUIREMENT FOR SAWTOOTH CURRENT

The electron beam must be scanned across the screen at a horizontal rate of 15 734.26 Hz. At the same time, it is moved slowly down the screen at a vertical rate of 59.96 Hz. Its path, as explained in Chapter 3, is not straight across the screen, but tilted slightly downward. At the end of the line, it is brought quickly back to the left-hand side of the screen. The type of current in the horizontal and vertical deflection windings that will accomplish this motion is the **sawtooth wave** shown in Figure 21.1. This wave gradually rises linearly and then, when it reaches a certain height, returns rapidly to its starting value. The process then repeats itself, 15 734.26 times per second for the horizontal oscillator and 59.94 times per second for the vertical oscillator. Before sawtooth waves of current can be produced, sawtooth waves of voltage must be produced. The voltage wave is then used to drive a power amplifier, which supplies the sawtooth current.

DEVELOPING A SAWTOOTH VOLTAGE

The simplest way to get the desired gradual rise followed by sudden drop is to charge and discharge a capacitor. If a capacitor is placed in series with a resistor and a voltage source, the flow of current through the circuit will cause the voltage across the

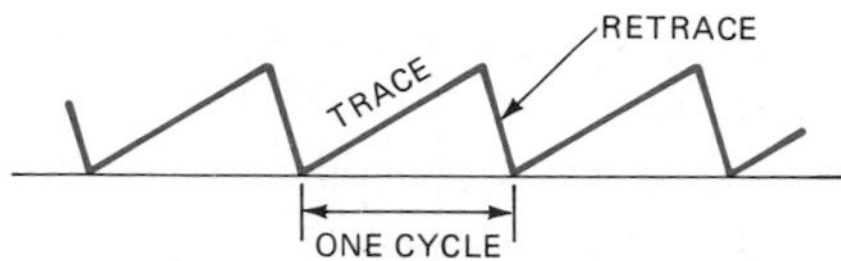

Figure 21.1 A sawtooth-wave current. This type of current is required for the horizontal and vertical deflection coils.

capacitor to rise in the manner indicated by Figure 21.2. This is a universal capacitor charging curve for any value of capacitance *(C)* and any value of resistance *(R)*. The product *RC* is designated by *T* and is called a *time constant*, where *R* is in ohms, *C* in farads, and *T* in seconds. The time constant of Figure 21.3 is

$$T = RC = (1 \times 10^6) \times (1 \times 10^{-6}) = 1 \text{ s}$$

When connected to a voltage source *(V)* as shown in Figure 21.3A, the capacitor will charge exponentially, as shown in Figure 21.2. In five time constants ($5 \times RC$), the capacitor is considered (for practical purposes) to be fully charged. Although the overall charging curve is exponential, the first 10 to 20% is fairly linear and can be used for the trace portion of a sawtooth wave.

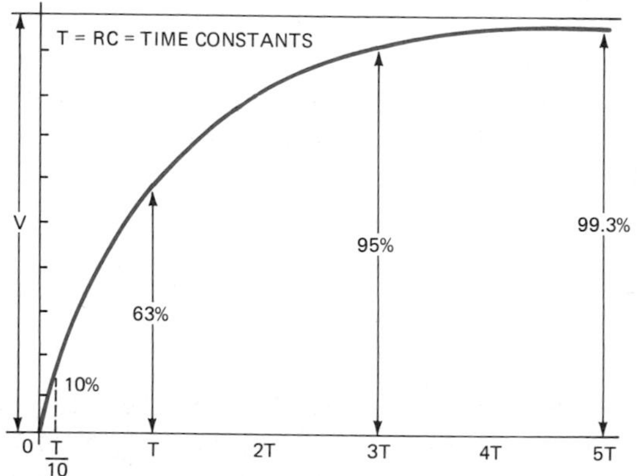

Figure 21.2 The manner in which the voltage across a capacitor increases when a potential is applied through a series resistor.

SIMPLE SAWTOOTH GENERATOR

Figure 21.3 shows the operation of a simple **sawtooth generator.** Initially, both switches are open. When SW_1 is closed (Figure 21.3A), capacitor *C* is permitted to charge to 10% of the charging voltage *V*. This produces the trace portion of the sawtooth wave (Figure 21.3C). At this time, SW_1 is opened and SW_2 is closed. The capacitor now discharges rapidly. This produces the retrace portion of the sawtooth wave.

In actual circuits, the charging circuit remains intact and the capacitor is discharged periodically by a transistor or vacuum tube.

21.2 NEED FOR TRAPEZOIDAL WAVE FORMS

In electromagnetic deflection systems, the driving force in the picture tube is a magnetic field. To develop such fields, current is required, a **sawtooth deflection** current. However, in order to achieve a sawtooth current flow through the deflection coils, we frequently must apply to these coils a voltage that has a modified sawtooth form.

The form of the voltage wave to be applied to the deflection coils is derived by analyzing the components of the coils and their action when subjected to voltages of various shapes. Each coil contains inductance *plus* a certain amount of resistance. So far as the resistance is concerned, a sawtooth voltage will result in a sawtooth current (Figure 21.4A).

For the inductance, considering a pure inductance, a voltage having the form shown in Figure 21.4B is needed for sawtooth current flow. Combining both voltage waves, the result obtained varies in the manner shown in Figure 21.4C. A voltage of this type, when placed across the deflection coils, will give a sawtooth current. The magnetic flux, varying in like manner, will force

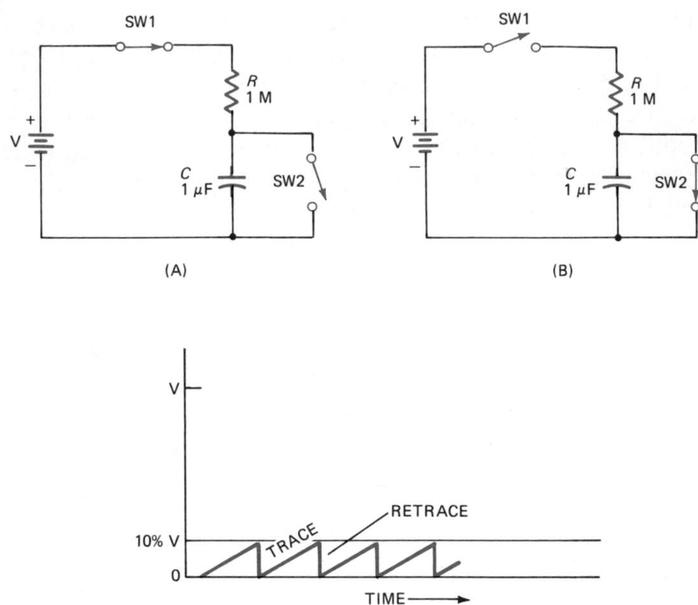

Figure 21.3A-C. Simple method of producing a sawtooth wave.

the electron beam to sweep across the screen properly. Note that the resultant wave is not obtained by combining the two voltage waves in equal measure. If the deflection circuit contains more inductance than resistance, the resultant wave will be closer in form to Figure 21.4B. On the other hand, if the resistance predominates, then the resultant wave will more closely resemble Figure 21.4A. Hence, one may expect to find variations of this deflection wave ranging from almost a pure sawtooth wave to that shown in Figure 21.4C.

TUBE DEFLECTION CIRCUITS

In vertical deflection circuits that are driven by vacuum tubes, the ratio of the yoke inductive reactance to its resistance, is

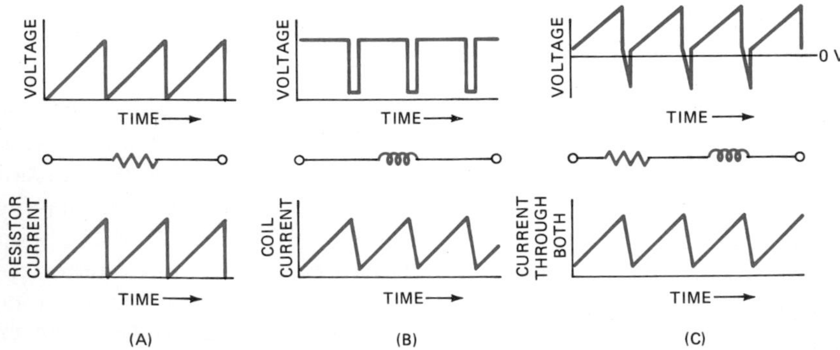

Figure 21.4A-C. When the voltage shown above the electrical component is applied, the sawtooth current wave shown below the component is obtained.

such that the *resistance* is an appreciable part of the ratio. In this case, it is generally found that a trapezoidal wave of voltage across the vertical yoke is required to produce a sawtooth of current through it. In vacuum tube horizontal deflection circuits, however, the situation is quite different. In this case, as a result of the much higher deflection frequency, the *reactance* of the horizontal yoke is very much higher than its resistance. Consequently, the horizontal yoke looks almost like a pure inductance, and a square wave of voltage across the yoke will produce a sawtooth current through it. Note that the square wave of voltage across the yoke is produced by applying a sawtooth driving voltage to the input of the horizontal output amplifier, which drives a sawtooth current through the yoke.

TRANSISTOR DEFLECTION CIRCUITS

When transistors are used for the vertical and horizontal output amplifiers, the input wave to the output transistor is generally not trapezoidal but close to a pure sawtooth. The reason for this is that the transistor represents a very low driving impedance for the yokes, which now appear to be mainly inductive. When tubes are used for the output amplifiers, the series plate resistance must be considered part of the yoke resistance. This series plate resistance is much higher than the equivalent yoke driving resistance of a power transistor.

GENERATING THE TRAPEZOIDAL VOLTAGE

The next problem is to generate this trapezoidal voltage. This can be accomplished by obtaining the output from the charging capacitor and a series resistor in place of the capacitor alone. The circuit is shown in Figure 21.5. This modified wave form is generated in the output circuit of the deflection oscillator. Resistors R_1 and R_2 and capacitor C_1 form the normal sawtooth. A lower-peaking resistor R_3 connected to the side of C_1 which is normally grounded provides the additional modification that results in a trapezoidal shape.

With V_1 cut off, C_1 charges through R_1, R_2, and R_3. Resistor R_3 is small compared to R_1 and R_2 and during this interval produces the positive portion of the rectangular wave across it. During the retrace period, with V_1 conducting, C_1 discharges through V_1 and R_3. The discharge current that now flows through R_3 and V_1 develops the negative segment of the rectangular wave across R_3. This rectangular wave and the sawtooth add to form the trapezoidal voltage wave form required by a vacuum tube vertical output amplifier.

21.3 THE VACUUM TUBE BLOCKING OSCILLATOR

The blocking oscillator is one of the most popular deflection oscillators used in television receivers. In this oscillator, feedback of energy from the plate to the grid

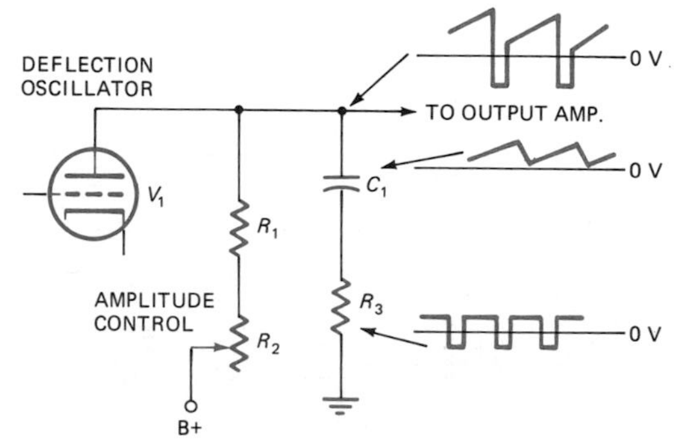

Figure 21.5 Method of generating deflection voltages required to produce sawtooth currents in deflection coils.

must occur and a transformer is used for this purpose. Any change of current in the plate circuit will induce a voltage in the grid circuit that will aid this change.

GRID GOES POSITIVE

To examine the situation in detail, consider the operation of the oscillator when a disturbance in the circuit increases the plate current. To aid this increase, a positive voltage is induced in the grid through transformer T (Figure 21.6A). With the grid more positive than before, more plate current will flow, resulting in the grid becoming rapidly positive. A positive grid means that electrons will flow in the circuit, charging capacitor C_2.

The plate current increases until saturation occurs. At that time, there is no further induced positive grid voltage. The coil voltage now rapidly reverses, cutting off the tube. Now the charge in capacitor C_2 decreases exponentially until the bias passes through cutoff. At this time, tube conduction resumes and the cycle begins again (Figure 21.6A).

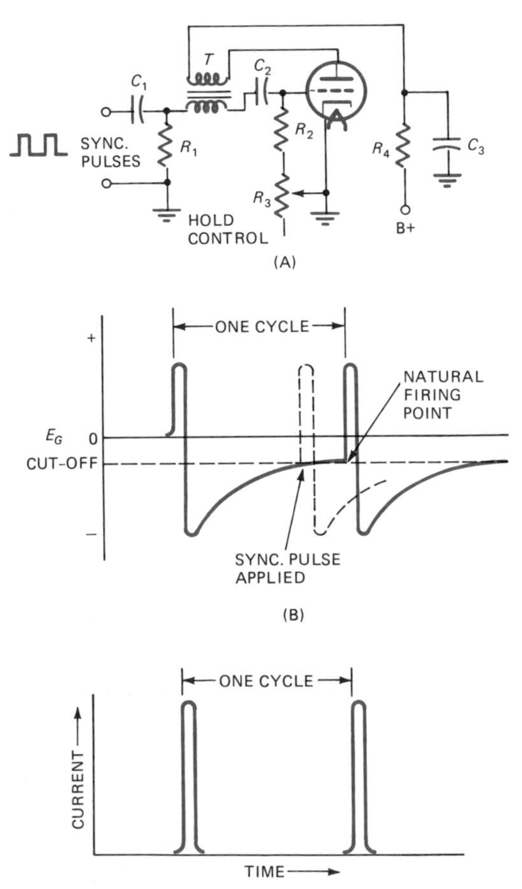

Figure 21.6A-C. **A** The blocking oscillator. **B.** The grid voltage variation. **C.** The form of the plate current.

DISCHARGE OF C_2

Because of the slow discharge of C_2, electrons that have accumulated remain in sufficient numbers to give the grid a large negative bias, sufficient to block or stop the plate current flow. Gradually the electrons accumulated on C_2 pass through R_2, R_3, and R_1 back to the other plate of C_2 and the negative bias on the grid slowly becomes less. When the discharge is such that the grid bias becomes less than cutoff, the plate current starts up, quickly reaches its high value, drives the grid positive, and the process repeats itself. Thus, during every cycle there is a short, sharp pulse of plate current followed by a period during which the tube is blocked until the accumulated negative charge on C_2 leaks off again. The frequency of these pulses is determined mainly by C_2, R_1, R_2, and R_3.

SYNCHRONIZING PULSE AT THE GRID

The form of the voltage drop across R_2 and R_3 can be seen in Figure 21.6B. In Figure 21.6C, the plate current pulse occurs once in every cycle. It is possible to control the frequency of this oscillator if a positive pulse is injected into the grid circuit at the time indicated in Figure 21.6B. To be effective, the frequency of the controlling pulse must be near and slightly higher than the free running frequency of the oscillator. By *free running frequency* is meant the natural frequency at which it will oscillate if permitted to function alone. This frequency is controlled by C_2, R_1, R_2, and R_3.

The point at which the synchronizing pulse should be applied to the grid of the oscillator is illustrated on the curve of Figure 21.6B. A positive pulse applied to the oscillator grid when it is at this point of its cycle will bring the tube sharply out of cutoff and cause a sharp pulse of plate current to flow. Then, at the application of the negative differentiated pulse of the horizontal synchronizing voltage which follows immediately, the oscillator is no longer in any position to respond. The grid has now become so negative that it is unaffected by the second negative synchronizing pulse. It is only when the grid capacitor C_2 is almost completely discharged that any pulse will effectively control the frequency of the oscillator. This accounts for the firm control of the synchronizing pulses. Equalizing pulses at the halfway point in the oscillator cycle are not strong enough to bring the tube out of cutoff. This also explains why a positive synchronizing pulse is required.

In short, then, the synchronizing pulse controls the start of the oscillator cycle. If left alone, the oscillator would function at its natural period, which, more often than not, would not coincide with the incoming signal. Through the intervening action of the synchronizing pulse, both oscillator and signal are brought together, in step. Naturally, for effective control, the synchronizing pulse and oscillator frequency must be close enough together to permit the locking in of the oscillator. However, the synchronizing pulse frequency must always be higher than the natural oscillator frequency.

THE HOLD CONTROL

Resistor R_3 is made variable in order to provide adjustment of the oscillator frequency. It is commonly known as the hold control, because it can be varied until the frequency of the blocking oscillator is held in synchronization with the incoming pulses.

21.4 VACUUM TUBE BLOCKING OSCILLATOR SAWTOOTH GENERATOR

A simple and inexpensive method of charging and discharging a capacitor to produce sawtooth waves is given in Figure 21.7. The triode V_1 is connected as a blocking oscillator, and sawtooth capacitor C_3 is placed in the plate circuit. From the preceding discussion of the operation of these oscillators, we know that a short, sharp pulse of plate current flows once in every cycle. During the remainder of the time, the grid is negatively biased beyond cutoff and no current flows in the plate circuit.

CHARGING C_3

During the time no plate current is flowing, C_3 is charging because one side of this capacitor connects to the positive terminal of the power supply through resistors R_1 and R_4 and the opposite side is attached to ground. The charge that the capacitor absorbs assumes the polarity shown in Figure 21.7.

DISCHARGING C_3

When plate current starts to flow, it is for only a very short period, and during this time the resistance of the tube becomes very low. Capacitor C_3 then quickly discharges through

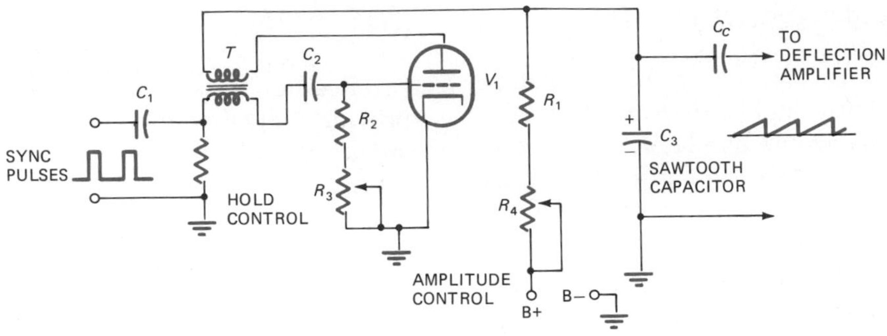

Figure 21.7 A simple method of obtaining a sawtooth voltage from a capacitor, using a blocking oscillator.

this low-resistance path. At the end of the short pulse of plate current, the grid has been driven very negative by the accumulation of electrons in C_2 and the tube becomes nonconducting again. Capacitor C_3 no longer has this easy path for discharging and starts to charge, as previously explained. The sawtooth variation in voltage across C_3 is transmitted to the next tube, an amplifier, through coupling capacitor C_c. The process repeats itself, either at the horizontal scanning frequency or at the vertical frequency, depending upon the oscillator constants.

SYNCHRONIZING PULSE ACTION

It will be noted from the foregoing that the instant the synchronizing pulse arrives at the oscillator, it triggers the oscillator. The tube then becomes conducting, and the capacitor developing the sawtooth voltage discharges. Hence, whenever a pulse arrives at the grid of the blocking oscillator, the capacitor discharges and the electron beam is brought back from the right-hand side of the screen to the left-hand side. This action is true of all such synchronizing oscillators.

HEIGHT CONTROL

Resistor R_4 may be made variable to permit adjustment of the amplitude of the sawtooth. As more of its resistance is placed in the circuit, the amount of charging current reaching C_3 is lessened, with a subsequent decrease in the voltage developed across C_3 during its period of charging. A small charging voltage at C_3 means, in turn, a small sawtooth voltage applied to the deflection amplifier. The length of the motion of the electron beam is consequently shortened. In a vertical-deflection oscillator circuit, this control will affect (and adjust) the height of the picture. It is labeled height control. Horizontal-deflection oscillators generally do not contain this control, as the width is usually controlled in another manner, to be described later.

HORIZONTAL-DEFLECTION OSCILLATOR

The blocking oscillator circuit illustrated in Figure 21.7 is used primarily for vertical deflection. Blocking oscillators are also used to generate synchronized deflection sawtooth waves at 15 734.26 Hz for the horizontal system. The basic principle of these blocking oscillators is the same as for the ones described previously. However, for horizontal deflection, the oscillator is invariably controlled by an AFC system. Horizontal-deflection oscillators and AFC systems are described in Chapter 22. (Horizontal- and vertical-deflection output amplifier systems are described in Chapters 23 and 24.)

21.5 TRANSISTOR BLOCKING OSCILLATORS

Transistor blocking oscillators are used for both vertical- and horizontal-deflection systems. While the principle of operation applies equally to both horizontal and vertical oscillators, our discussion at this point is directed mainly toward vertical oscillators. Horizontal deflection transistor oscillators and their associated AFC systems are discussed in Chapter 22.

The basic operation of tube and transistor blocking oscillators is the same, although there is a difference in the method used to turn the transistor oscillator on and off. This frequently involves the action of the sawtooth wave generated by the transistor circuit itself, rather than by a grid-leak bias action, as used in tube circuits. In transistor deflection oscillator circuits, we frequently find one or more driver stages between the oscillator and the power amplifier because in those instances the oscillator output is not adequate to fully drive the output stage. However, if a small picture tube screen is used and the oscillator transistor is powerful enough, a direct connection between the oscillator and the output amplifier will produce the necessary power to adequately drive the deflection coils.

A TYPICAL VERTICAL TRANSISTOR BLOCKING OSCILLATOR AND SAWTOOTH GENERATOR

A typical example of a transistor vertical blocking oscillator and sawtooth generator is shown in Figure 21.8. The tightly coupled windings on transformer T_1 provide positive feedback from collector to base of transistor Q_1. Transistor Q_1 is operated Class C and is turned on (conducts) for brief intervals.

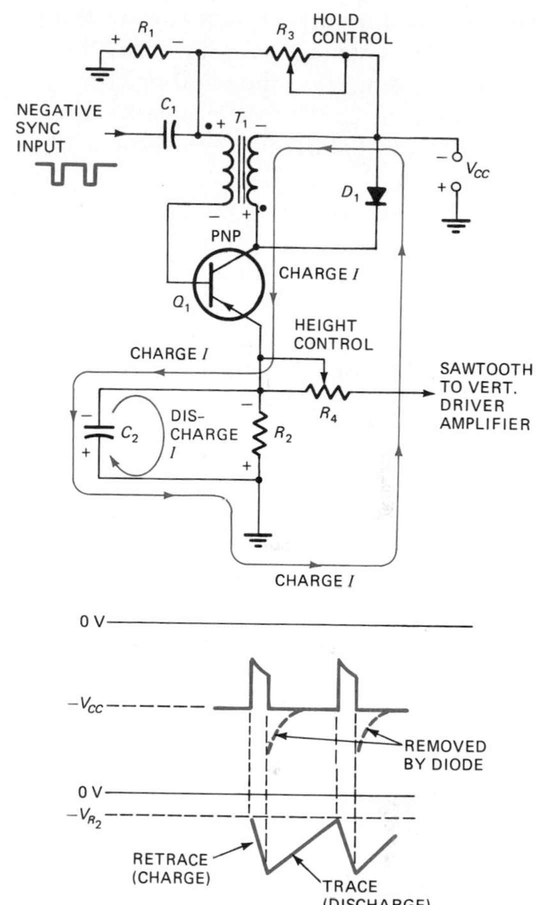

Figure 21.8 A vertical-deflection sawtooth generator that uses a transistor blocking oscillator.

This circuit operates similarly to its tube counterpart. An interesting feature to note is that one RC network, C_2–R_2, performs the dual role of determining the oscillator frequency and generating the sawtooth wave.

When the supply voltage is applied, the circuit operates as follows. A negative forward bias voltage appears at the base of Q_1 (PNP transistor) due to the voltage divider R_1–R_3. Capacitor C_2 is not charged.

Q_1 Conducts Transistor Q_1 conducts, and C_2 starts to charge in a negative direction.

As the collector current increases, in the primary of T_1, a voltage is induced in the secondary winding (i.e., the winding connected to the base of Q_1) with the polarity shown. Note that the polarizing dots show a phase reversal in T_1.

The base voltage now consists of two voltages in series to ground: the voltage across R_1 and the T_1 secondary voltage. Both voltages are series-aiding and force the base to a higher negative voltage (more forward bias). Transistor Q_1 thus conducts even more strongly, and C_2 charges rapidly (and linearly) to produce the *retrace portion* of the sawtooth wave.

Q_1 Cuts Off The regenerative cycle in which C_2 continues to charge to a higher negative voltage goes on until Q_1 saturates. Saturation causes the collector current in Q_1 to level off at the saturation level. T_1 now has a strong but steady magnetic field built up in the primary winding. No voltage is induced in the T_1 secondary winding at the moment the collector current saturates. (Voltage is generated in a transformer winding only so long as current is changing.) The base voltage on Q_1 drops to the voltage developed across R_1. At this time, the C_2 voltage (reverse bias) exceeds the base voltage (forward bias) and transistor Q_1 cuts off.

The steady magnetic field in T_1 now collapses and induces a voltage pulse in both windings opposite the polarity shown in Figure 21.8. The negative collector pulse is clipped because of diode D_1, becoming forward-biased.

Producing the Trace With Q_1 now cut off, C_2 begins to discharge through R_2, producing the trace portion of the wave form in the positive-going direction. Transistor Q_1 remains cut off until the voltage across C_2–R_2 decreases to the same value as the forward-bias voltage across R_1. When this happens, the base-to-emitter junction again becomes forward-biased, transistor Q_1 conducts, and the charging cycle begins again. The relatively slow discharge of C_2 through R_2 develops the trace portion of the sawtooth and establishes the trace timing interval.

Synchronization Lock in, or synchronization, is accomplished by applying negative-going synchronizing pulses to the base. As in tube blocking oscillators, the natural repetition frequency of the oscillator must be *slightly lower* than the synchronizing pulse frequency. Synchronizing pulses force the transistor into conduction before C_2 completely discharges and causes transistor conduction.

Free-Running Frequency The oscillator free-running frequency is controlled primarily by the R_2–C_2 network in the emitter circuit. Notice that during discharge, transistor Q_1 is cut off and the R_2–C_2 network is *isolated* from the remainder of the circuitry. This helps to provide excellent frequency stability. The natural frequency may be shifted over a narrow range by changing the DC forward bias (across R_1) applied to the base through the T_1 secondary winding. Variable resistor R_3 is part of the bias network and performs this hold control function. An increase in negative voltage across R_1 (an increase in forward bias) allows transistor Q_1 to break into conduction sooner. The natural frequency is then raised.

Removing Transients At the time Q_1 cuts off, the voltage transient pulses appearing across the T_1 secondary and primary might be troublesome because they may exceed the collector-to-base breakdown voltage rating of Q_1. Such transients are removed by connecting a diode across the primary winding, as indicated in Figure 21.8. The diode conducts when the T_1 voltage polarities are opposite those shown and the magnetic field energy is then dissipated harmlessly through the diode. The sawtooth output is applied to the vertical driver stage.

ANOTHER TRANSISTOR BLOCKING OSCILLATOR AND SAWTOOTH GENERATOR

The circuit of Figure 21.9 is similar in operation to a vacuum tube blocking oscillator in that the sawtooth-forming capacitor, C_1, charges during the trace period and discharges during the retrace. Circuit action to form the required sweep sawtooth is as follows.

When the collector supply voltage ($-V_{cc}$) is first applied, the charge on sawtooth-forming capacitor C_1 is zero. Transistor Q_1 is cut off by the reverse-bias voltage across R_7, which is applied to the emitter. Since the base is essentially at ground potential, the trace starts at time t_1 as C_1 charges through R_4 toward $-V_{cc}$. As the voltage across C_1 increases negatively, a portion of it is fed to the base of Q_1 through the T_1 secondary, as a forward bias from voltage divider R_1 and R_2. Time t_2 represents the end of the sweep as the forward bias across R_2 just exceeds the reverse bias across R_7. Transistor Q_1 conducts at this instant, and the collector current through transformer T_1 causes the base voltage to swing highly negative. The voltage across R_2 and the secondary voltage of T_1 are now series-aiding to drive transistor Q_1 into saturation in order to begin the retrace portion of the sawtooth.

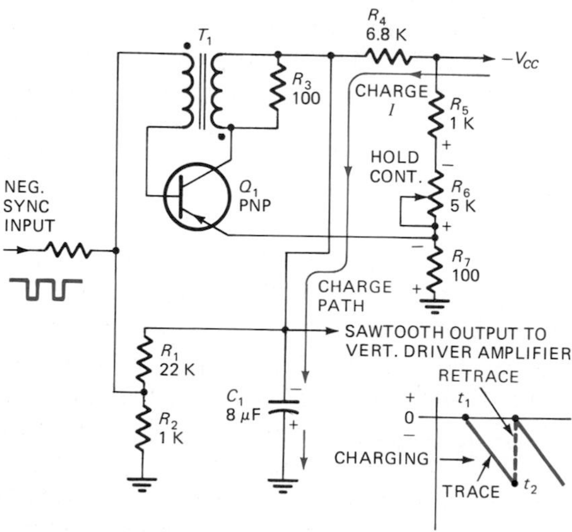

Figure 21.9 Another vertical-deflection sawtooth generator that uses a transistor blocking oscillator.

The transistor, the primary of T_1 and R_7 form a low-resistance discharge path for C_1. At the end of this rapid discharge, C_1 has discharged to nearly zero and Q_1 is cut off again by the reverse bias across R_7. This completes the retrace.

Frequency control is accomplished by varying the amount of reverse bias to the Q_1 emitter. An increase in negative voltage by R_6 adjustment increases the time Q_1 is held in at a cutoff. This increases the time C_1 has to charge and effectively decreases the frequency of the oscillator cycle.

Resistor R_3 (100 ohms) performs the function of a damping resistor across the T_1 primary to reduce the retrace pulse transient to safe limits. The Q_1 base-to-collector junction requires this, or else voltage breakdown might occur and Q_1 would be destroyed.

DRIVER AMPLIFIERS

Sawtooth wave forms are generated in tube circuits by allowing a capacitor to charge to a fraction of the applied voltage. The voltage across the capacitor is then fed to the grid of a vacuum tube (output amplifier). The very high input impedance of the tube grid circuit does not load the capacitor.

However, in transistor circuits, the sawtooth is fed to the base circuit of the following stage and the input impedance of this stage, representing a load on the sawtooth capacitor, may be only a few hundred ohms. This low resistance across the sawtooth capacitor draws appreciable current from the capacitor, lowers the sawtooth voltage amplitude, and destroys linearity.

A solution to the problem of loading the sawtooth-forming capacitor (and increasing the sawtooth current gain to drive the output amplifier) is to feed the sawtooth into a special driver amplifier. This amplifier must present a high input impedance to the sawtooth capacitor so as not to load it and a low output impedance so as to match the low input impedance of the following amplifier. A common collector amplifier (emitter follower) does not severely load sawtooth capacitor C_1 and yet has a low output impedance to match the output amplifier stage. Figure 21.10 shows this circuitry in a typical arrangement. Direct coupling is used to the output amplifier, and the requirement for an expensive high-capacity coupling capacitor is eliminated.

21.6 A TRANSISTOR BLOCKING OSCILLATOR VERTICAL-DEFLECTION SYSTEM

The schematic diagram of a complete GE vertical-deflection system is given in Figure 21.11. This consists of a blocking oscillator (Q_{601}), a driver stage (Q_{602}), and a vertical output stage (Q_{603}). Explanations of additional vertical-deflection systems are given in Chapter 24. This circuit is presented at this time mainly to show an example of a complete vertical-deflection system. Note that the blocking oscillator, the sawtooth generator, and Q_{601} are of the same basic type as shown in Figure 21.8, with the sawtooth capacitor (C_{604}) and its shunting resistor (R_{606}) located in the emitter circuit. However, note that Q_{601} is an NPN transistor and so all voltages are reversed compared with the PNP circuit of Figure 21.8. Also note that blocking oscillator transformer T_{601} has a third winding. This winding provides a means of coupling the vertical synchronizing pulses into the base circuit of Q_{601}.

THE VERTICAL DRIVER

Driver Q_{602} is of the common emitter type. The sawtooth output of the driver is capacity-coupled to the vertical output stage by capacitor C_{608}. The output of the power stage, Q_{603}, is choke-coupled to the vertical yoke by the primary of vertical choke transformer T_{602}. This provides the proper impedance match for the output transistor and the yoke.

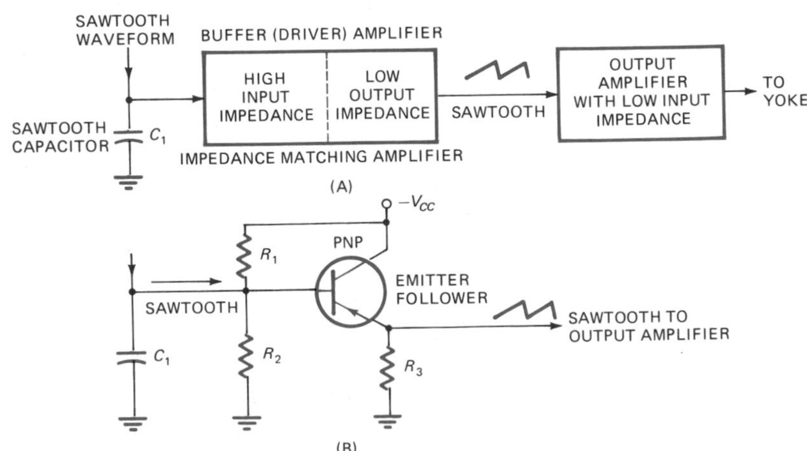

Figure 21.10A-B. A typical transistor-driver amplifier connected as an emitter follower.

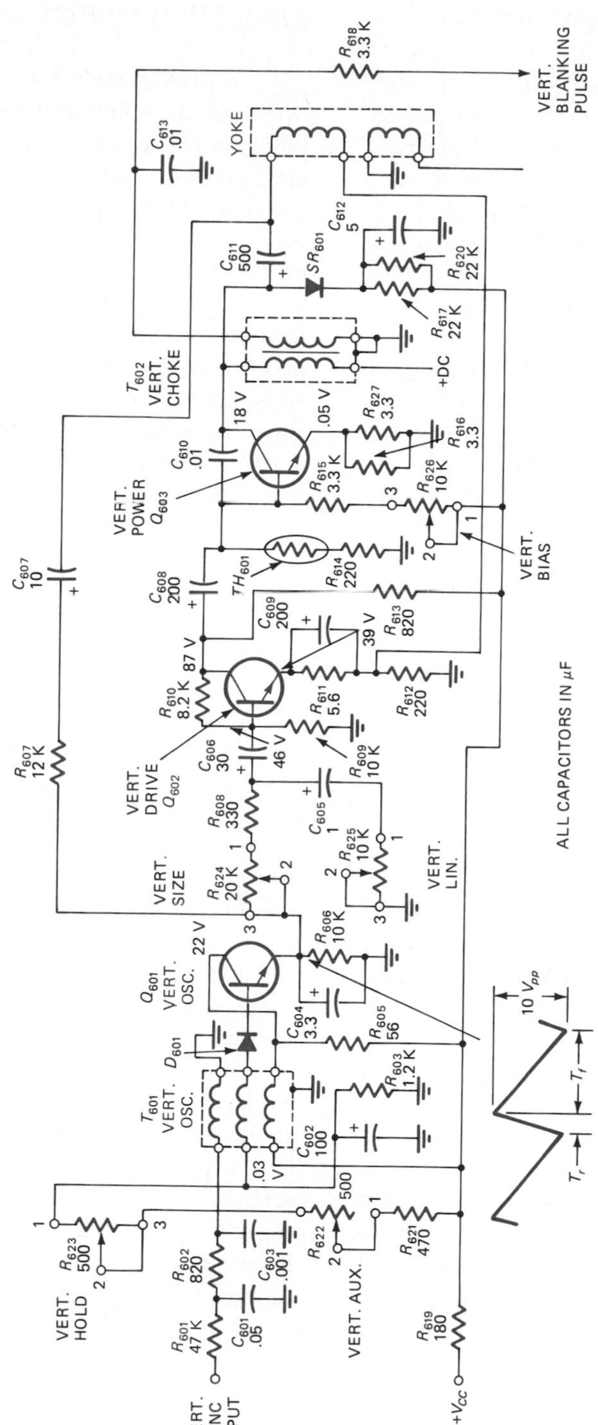

Figure 21.11 Vertical-deflection system with an oscillator-driver (buffer) and output transistors. *(Courtesy of General Electric Company.)*

ADJUSTING FREQUENCY

The vertical oscillator frequency is varied by changing the value of the bias applied through D_{601} to the base of the vertical oscillator (Q_{601}). Two controls, vertical hold R_{623} and vertical auxiliary R_{622}, are provided for this purpose. Resistor R_{622} should be adjusted so that a vertical lock-in occurs when R_{623} is set at the mechanical center of its range. The output of the blocking oscillator circuit is a sawtooth wave that is amplified by vertical driver Q_{602}.

OUTPUT CIRCUITS

The output of Q_{602} is further amplified by vertical output transistor Q_{603} and is then applied to the vertical coils of the deflection yoke. The function of C_{611} is to block out the DC component of the vertical output signal and to allow only the AC component to be applied to the vertical deflection coils.

LINEARITY AND SIZE

The vertical linearity control R_{625} and C_{605} combine to form an integrating circuit that is used to regulate the shape of the sawtooth wave generated by the vertical blocking oscillator. By varying the resistance of R_{625}, the linearity at the start of the vertical scan (upper portion of the picture) can be controlled. Resistors R_{612} and R_{607} and capacitor C_{607} provide feedback which tends to correct distortion introduced into the vertical circuit by Q_{602} and Q_{603}. Resistor R_{607} and capacitor C_{607} affect the linearity at the end (lower portion of the picture) of the vertical scan. The vertical size control (R_{624}), located in the emitter circuit of the vertical oscillator (Q_{601}), regulates the overall vertical size of the picture and also controls the linearity of the lower portion of the picture.

LINEAR OUTPUT AMPLIFIER

Vertical output transistor Q_{603} must be operated as a linear amplifier. The vertical bias control (R_{626}) is used to regulate the linear operation of this stage. Misadjustment of this control can produce a nonlinear condition in either the upper or the lower portion of the picture.

Q_{603} Protection A thermistor (TH_{601}) has been installed close to the vertical output transistor (Q_{603}) to prevent excessive current flow through its collector circuit, which could result from an increase in the ambient temperature. Thermal runaway, resulting in excessive collector current drain, could quickly destroy Q_{603}. Should the temperature of Q_{603} rise, the resistance of TH_{601} would decrease, lowering the drive to Q_{603} and hence counteracting a collector current rise due to the temperature. Transistor Q_{603} is further protected by a network consisting of the vertical pulse limiter diode (SR_{601}), C_{612}, R_{617}, and R_{620}.

21.7 VACUUM TUBE PLATE-COUPLED MULTIVIBRATORS

Essentially, the **multivibrator** is a two-stage resistance-coupled amplifier with the output of the second tube fed back to the input of the first stage. Oscillations are possible in a circuit of this type because a voltage at the grid of the first tube causes an amplified voltage to appear at the output of the second tube, which has the same phase as the voltage at the grid of the first tube. This is always the case with an even number of resistance-coupled amplifiers, but never with an odd number. The output of an odd number of such stages is always 180° out of phase with the voltage applied at the input of the first tube. The two voltages would thus oppose, rather than aid, each other.

Two types of resistance-coupled multivibrators are used in television receivers—plate-coupled and cathode-coupled. A third type uses the vertical output amplifier as one stage of the multivibrator. All three types are discussed in this chapter.

CIRCUIT OPERATION

The circuit diagram of a basic free-running, balanced, plate-coupled multivibrator is shown in Figure 21.12. There are four different conditions during one complete cycle of operation:

1. An extremely rapid change from V_1 conducting to V_2 conducting
2. A long period during which V_1 is cut off and the circuit is relaxed
3. A second rapid change as V_1 conducts and drives V_2 beyond cutoff
4. A long period during which V_2 is cut off and the circuit is relaxed

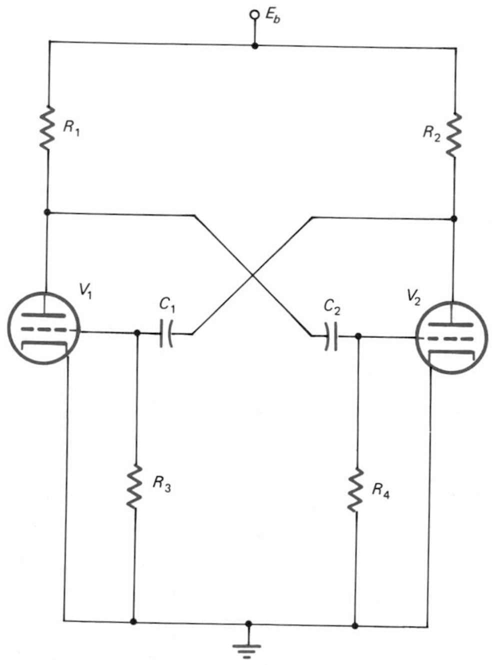

Figure 21.12 The basic free-running multivibrator circuit.

The cycle repeats as Condition 1 follows Condition 4. The rapid changes just described are indicated by the vertical parts of the wave shapes at the left of Figure 21.13C. These changes ordinarily occur in a fraction of a microsecond. The long relaxation periods are represented by the nearly constant voltages shown in Figure 21.13C, between switchover times.

First Switching Action When the filament voltages are on and the plate voltage is applied, current flows in both tubes. If the circuit components are the same in both tube circuits and if both tubes are of the same type, the circuit is said to be symmetrical and both currents should be equal. It is impossible to obtain a perfect balance in the circuit, however, because of the nonuniformity of available parts and because of the irregular way emission from the cathode occurs. Any slight current change in one tube that is not accompanied by an equal change in the other tube will start oscillation (Figure 21.13A and the wave shapes of Figure 21.13C.)

Assuming that a slight current increase has occurred in tube V_1, the following events then occur in rapid order. The increase of current in V_1 increases the voltage drop across R_1, which lowers the plate-to-ground voltage at V_1. When the plate voltage of V_1 drops, the grid voltage of V_2 also instantaneously drops by the same amount. This happens because the voltage *across* C_2 cannot change instantaneously. The grid voltage drop cuts off V_2, and its plate voltage rises. Capacitor C_1 is between this plate and ground and therefore must charge to the higher voltage. The charging current flows through R_3 toward the plate supply voltage, as shown by solid arrows in Figure

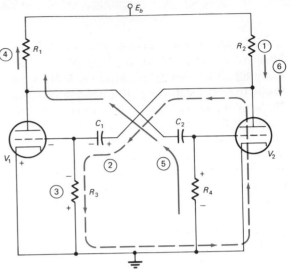

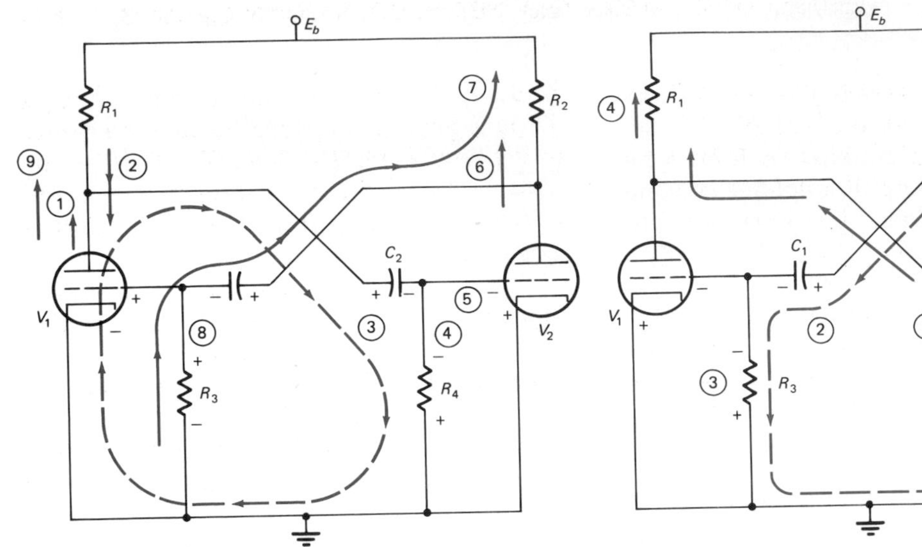

(1) CURRENT INCREASES IN V_1.

(2) VOLTAGE DROPS AT PLATE OF V_1 DUE TO INCREASED VOLTAGE ACROSS R_1.

(3) CAPACITOR C_2, NOW CHARGED HIGHER THAN APPLIED VOLTAGE, DISCHARGES TOWARD NEW LOWER VOLTAGE; DOTTED LINES SHOW CLOSED DISCHARGE CIRCUIT.

(4) DISCHARGE CURRENT MAKES TOP OF R_4 NEGATIVE WITH RESPECT TO BOTTOM.

(5) SINCE GRID AND CATHODE (V_2) ARE CONNECTED TO R_4, GRID BECOMES NEGATIVE WITH RESPECT TO CATHODE.

(6) MORE NEGATIVE GRID AT V_2 DECREASES CURRENT, RAISING VOLTAGE AT PLATE OF V_2.

(7) A CURRENT FLOWS THROUGH PATH OF SOLID ARROWS TO CHARGE C CAPACITOR C_1 TO NEW HIGHER VOLTAGE.

(8) THE CHARGING CURRENT MAKES THE TOP OF R_3 POSITIVE WITH RESPECT TO BOTTOM AND CONSEQUENTLY, THE GRID OF V_1 IS POSITIVE.

(9) THIS CAUSES A GREAT INCREASE IN CURRENT IN V_1, WHERE THE ORIGINAL SMALL INCREASES OCCURRED.

(A)

(1) AS C_2 DISCHARGES TO LESS THAN CUTOFF VOLTAGE, CURRENT FLOWS, LOWERING VOLTAGE AT V_2.

(2) CAPACITOR C_1 DISCHARGES.

(3) THE (DOTTED LINE) CURRENT MAKES THE TOP OF RESISTOR AND GRID NEGATIVE.

(4) THE NEGATIVE GRID VOLTAGE LOWERS PLATE CURRENT SO PLATE VOLTAGE GOES UP.

(5) CAPACITOR C_2 NOW RECHARGES AND THE CURRENT (SOLID ARROWS) MAKES GRID POSITIVE.

(6) THIS INCREASES THE CURRENT MORE.

(B)

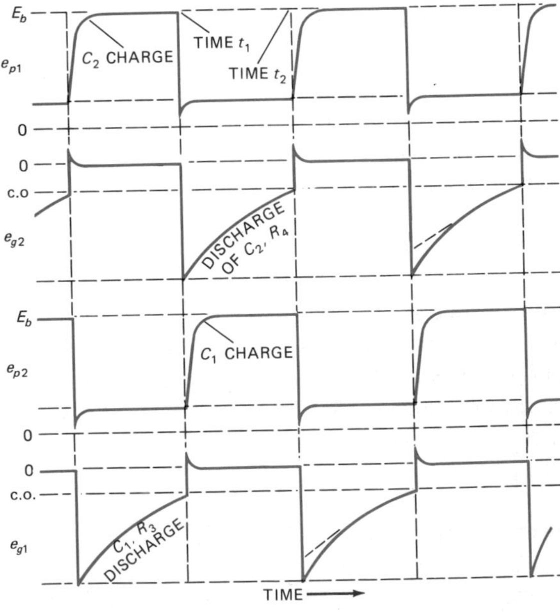

(C)

Figure 21.13A-C. The operation of a balanced free-running multivibrator. **A.** The first switching action. **B.** The second switching action. **C.** The waveforms of the multivibrator.

21.13A. This current causes the top of R_3 to be positive, and therefore the grid of V_1 is also positive. The positive grid causes the original slight current increase to become a tremendous current increase. As the changes go around the circuit, a large voltage drop occurs at the plate of V_1, the grid of V_2 is made more negative, current in V_2 is decreased more and plate voltage at V_2 goes up, making the grid of V_1 more positive and further increasing the V_1 plate current.

V_2 Cuts Off and V_1 Saturates This is a cumulative process what continues until the current in V_2 is decreased to zero by a grid voltage far below cutoff. With zero current through V_2, the plate voltage becomes equal to the supply voltage. Capacitor C_1 quickly charges to this value by grid current flow in V_1. After that, current through R_3 is zero and so the grid-to-cathode voltage is zero. The plate voltage of V_1 remains at a constant low value while the grid voltage is zero. Therefore capacitor C_2 continues to discharge through V_1. While discharging, the current (dotted lines in Figure 21.13A) holds V_2 at cutoff. Everything is at a standstill in the circuit except for the slow discharge of C_2. This corresponds to Condition 4 listed above.

The slow discharge of C_2 continues, with the voltage across C_2 becoming lower and lower. The voltage across R_4 continuously equals the C_2 voltage in accordance with Kirchhoff's voltage law, and so the R_4 voltage decreases. When a voltage just below the cutoff voltage for V_2 is reached, a slight current starts flowing through V_2. This ends the period of inactivity in the circuit as rapid changes follow this slight current (Figure 21.13B and the wave shapes of Figure 21.13C).

Second Switching Action The current lowers the V_2 plate voltage. Capacitor C_1 cannot change its voltage instantaneously, and this instantly makes the grid of V_1 negative. The resultant decrease in plate current raises the V_1 plate voltage. Capacitor C_2 now recharges to the new V_1 plate voltage. The C_2 charging current makes the grid of V_2 positive, and the original small current is increased tremendously.

This regenerative process continues until V_1 is cut off by the discharge of C_1. At this point, V_1 plate voltage rises quickly to the plate supply voltage, C_2 charges swiftly to supply voltage through V_2 grid current, and the plate voltage of V_2 stabilizes at a low value. This ends the rapid changes during Condition 1.

V_1 Cuts Off and V_2 Saturates The second inactive period occurs as V_1 is held nonconducting by the slow discharge of C_1. The discharge continues until the C_1 voltage is just below cutoff for V_1. This period is as described for Condition 2.

The Cycle Repeats The slight current in V_1 then repeats the first rapid change described above. In the wave shapes in Figure 21.13C, the long period between time t_1 and t_2 is the time that V_2 is cut off. The exponential grid voltage decreases of e_{g2} shows how long capacitor C_2 can hold V_2 nonconducting. Time t_2 is the rapid change just described, and the half cycle following that is the time V_1 is cut off.

Summary The operation of this circuit consists of long periods when one tube conducts a high current while the other tube is cut off, followed by an extremely rapid change to the other tube conducting and the first tube cut off.

21.8 PLATE-COUPLED MULTIVIBRATOR SAWTOOTH GENERATOR

Figure 21.14 illustrates how the multivibrator can control the charge and discharge of a capacitor, thereby developing the required sawtooth voltages. The same multivibrator is used, with the addition of sawtooth capacitor C_3. When tube V_2 is not conducting, the power supply will slowly charge C_3 through resistor R_2. The moment that the grid voltage of V_2 reaches the cutoff point of the tube, the tube starts to conduct and its internal resistance decreases. Capacitor C_3 then discharges rapidly through the tube. During the next cycle, V_2 is again nonconductive and again C_3 slowly charges. Capacitor C_4 transmits the voltage variations appearing across C_3 to the next amplifying tube. Resistor R_3 is made variable to permit adjustment of the multivibrator so that it can be locked in with the synchronizing pulses. Hence R_3 is the hold control.

The desired form of the sawtooth output wave form is a slow rise in voltage followed by a rapid decrease. Toward that end, C_1 and R_3 of Figure 21.14 are designed to have a considerably longer time constant than C_2 and R_4. Capacitor C_1 and resistor R_3 will discharge slowly, maintaining V_2 in cutoff while C_3 slowly charges. During this interval, V_1 is conducting. Upon the application of a negative synchronizing pulse to the grid of V_1, this tube is forced into cutoff while V_2 rises sharply out of cutoff and into conduction. Capacitor C_3 now discharges rapidly. Because C_2 and R_4 have a small time constant, V_1 does not remain cut off very long. As soon as C_3 has discharged, V_1 begins to conduct, again cutting off the plate current of V_2. The ratio of the time constants of C_1, R_3 and C_2, R_4 is approximately 11:1.

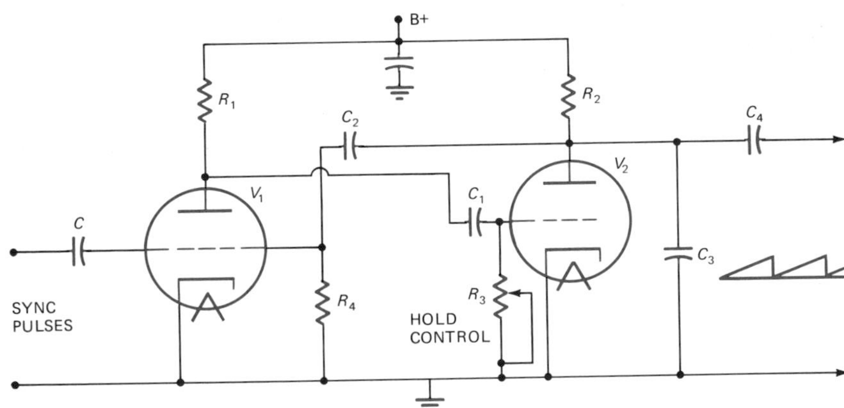

TYPICAL VALUES OF COMPONENTS

60 HERTZ	15,750 HERTZ
R_3 – 2.2 MEGOHMS	R_3 – 220 000 OHMS
R_4 – 1.0 MEGOHMS	R_4 – 100 000 OHMS
R_1 – 100,000 OHMS	R_1 – 47 000 OHMS
R_2 – 2.2 MEGOHMS	R_2 – 500 000 OHMS
C_3 – 0.1 μF	C_3 – 500 pF
C_2 – .01 μF	C_2 – .001 μF
C_1 – .05 μF	C_1 – .005 μF

Figure 21.14 How a multivibrator may be connected to control the charging and discharging of a capacitor to derive sawtooth waves.

21.9 COMBINATION VACUUM TUBE MULTIVIBRATOR AND OUTPUT STAGE

Another vertical-deflection multivibrator system is shown in Figure 21.15. Here two tubes serve as both the multivibrator and the output amplifier, an arrangement made possible by using the output tube to complete the multivibrator circuit. There is a feedback path from the plate of V_2 to the grid of V_1. This path is formed by R_1 and C_1. By transferring back energy that arrives in phase at the grid of V_1, the oscillating action of a multivibrator is achieved. At the same time, a transformer in the plate circuit of V_2 feeds output signals to the vertical-deflection coil. Thus, by using the output tube as the second multivibrator tube, both functions are accomplished with only two tubes. This arrangement is economical and for this reason has found widespread use.

The rest of the circuit is fairly conventional and follows the previous circuit quite closely. A potentiometer R_2 in the grid circuit of V_1 serves as the vertical hold control. A second potentiometer (R_3) in the plate circuit of V_1 varies the amount of voltage fed to capacitor C_2 and resistor R_4. Resistor R_3 is thus the height control. The trapezoidal wave form developed across C_2 and R_4 is then transferred by C_3 to the grid of the output tube, where it is amplified and applied to the vertical-deflection coils. Part of this wave is fed back through R_1 and C_1 to the grid of V_1 to keep the multivibrator oscillating.

LINEARITY

Vertical linearity is accomplished by a potentiometer in the cathode leg of V_2. Varying this control changes the bias of V_2. In turns, this changes the operating characteristic of V_2 and thus affects the linearity of the output wave.

21.10 CATHODE-COUPLED MULTIVIBRATOR

Another vacuum tube deflection oscillator commonly found in television receivers is

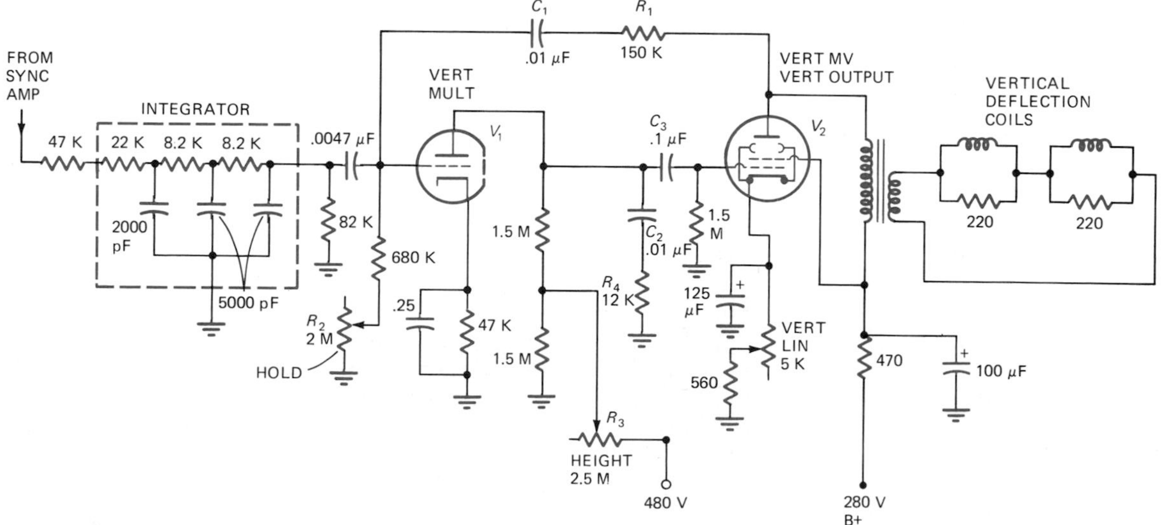

Figure 21.15 In this system, the output tube is also part of the multivibrator.

the cathode-coupled multivibrator. The basic schematic diagram is shown in Figure 21.16A. The feedback between the tubes is accomplished in two ways: (1) through coupling capacitor C_2 and (2) through the unbypassed cathode resistor R_K.

The multivibrator in Figure 21.16A is said to be cathode-coupled because the coupling from V_2 to V_1 takes place through the common cathode resistor R_K. In this circuit, V_1 is cut off by the voltage produced across R_K by the plate current of V_2 flowing through this resistor. Then, when V_2 is cut off, in the manner described below, V_1 is conducting strongly. Thus, the conduction between tubes switches back and forth.

For the following explanation, refer to the schematic diagram of Figure 21.16A and the wave forms of Figure 21.16B.

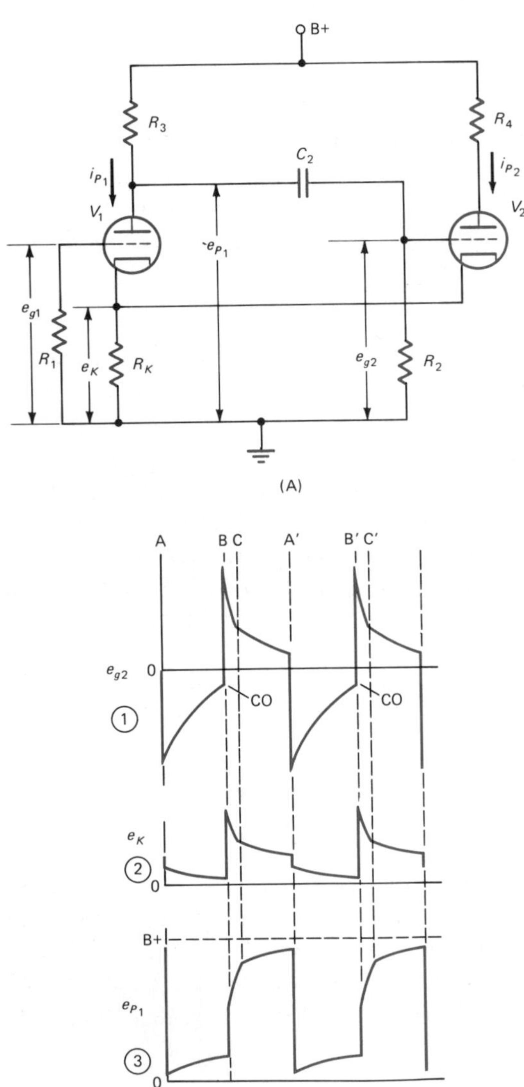

Figure 21.16A-B. The basic cathode-coupled multivibrator. A. The schematic diagram. B. The waveforms.

V_2 CUTS OFF, V_1 SATURATES

If there is no plate potential applied to the circuit of Figure 21.16A, there is no charge on C_2 and the grids of both V_1 and V_2 are at ground potential. When the plate potential is suddenly applied, both tubes start to conduct. Since the plate current of both tubes flows through R_K, the cathode potential of both V_1 and V_2 increases above ground, producing a bias voltage that limits the magnitude of the current flowing in the tubes. The flow of the plate current in V_1 (i_{p1}) through R_3 reduces the voltage at the plate of V_1. Since the voltage across C_2 cannot change instantaneously, the drop in voltage that takes place at the plate of V_1 is coupled to the grid of V_2, further reducing the current through this tube. This reduction of current decreases the voltage developed across R_K, and so i_{p1} increases. This increased current causes e_{p1} to drop further, and the conduction of V_2 is decreased even more. This action is cumulative, ending with the current in V_2 reduced to zero and the current in V_1 at a maximum. Although this step-by-step description of multivibrator action might give the impression that a considerable length of time is involved, V_2 is driven beyond cutoff almost instantaneously, as at *A* in Figure 21.16B.

V_1 CUTS OFF, V_2 SATURATES

Tube V_2 is held beyond cutoff during the time that C_2 discharges through R_2, R_p of V_1, and R_K. The discharge current through R_2 produces a voltage at the grid of V_2 that is negative with respect to ground and decreases exponentially as the capacitor discharges. At time *B* in Figure 21.16B, the voltage at the grid of V_2 reaches cutoff and V_2 begins to conduct. The current drawn through V_2 also flows through R_K, producing an increased voltage across this resistor. This increased voltage makes the bias on V_1 more negative so that less plate current flows through this tube. As a result of the decrease in i_{p1}, the voltage at the plate of V_1 increases. Since the voltage across C_2 cannot change instantaneously, the grid of V_2 is driven positive, which further increases the plate current of V_2. This action is cumulative and results in the current through V_1 being reduced to zero almost instantaneously and the current through V_2 being increased to a maximum. Tube V_1 is cut off by the voltage developed across R_K when the high plate current of V_2 flows through it as a result of driving the grid of V_2 positive (Figure 21.16B, time *B*).

At time *B* the grid of V_2 is driven highly positive, causing i_{p2} to be large, so that e_K rises quickly. Because grid current is drawn while e_{g2} is more positive than e_K, capacitor C_2 charges relatively quickly through R_K, R_{GK} of V_2, and R_3. The capacitor has charged sufficiently by time *C* that e_{g2} is reduced to cathode potential at this time. Since the grid of V_2 is still positive with respect to ground, the charging of C_2 continues, but at a lower rate. The charging rate is lower because the resistance of the charging path is through R_3 and R_2 when grid current is not drawn. This results in a longer time constant than that of the path through V_2. As C_2 charges, the bias on V_2 decreases, and this causes i_{p2} to decrease, which in turn decreases e_K. The grid of V_1 is held constant at ground potential, so that this tube remains cut off as long as e_K is positive relative to ground by more than cutoff voltage. When e_K drops to the cutoff voltage, however, V_1 conducts and rapidly cuts off V_2 by feeding a large negative-going voltage through C_2 to the grid of V_2.

FREE RUNNING

Since the conduction of V_2 is unable to maintain V_1 nonconducting, the multivibrator of Figure 21.16A is free-running. No trigger pulses are required to operate this circuit because it has no stable equilibrium condition.

21.11 CATHODE-COUPLED MULTIVIBRATOR SAWTOOTH GENERATOR

A cathode-coupled multivibrator sawtooth generator suitable for use as a vertical- or horizontal-deflection oscillator is shown in Figure 21.17. The principle of operation is the same as for Figure 21.16. In Figure 21.17, however, note the addition of the hold (frequency) control R_4, the height control R_2 with series resistor R_1, and the sawtooth capacitor C_2.

HOLD CONTROL

The hold control R_4 can vary the time constant in the grid circuit of V_2. This time constant is composed of C_1, R_S, and R_4. It determines the length of time V_2 is cut off during a cycle and thus the free-running frequency.

HEIGHT CONTROL

The height control R_2 can vary the charging time constant for C_2. A shorter

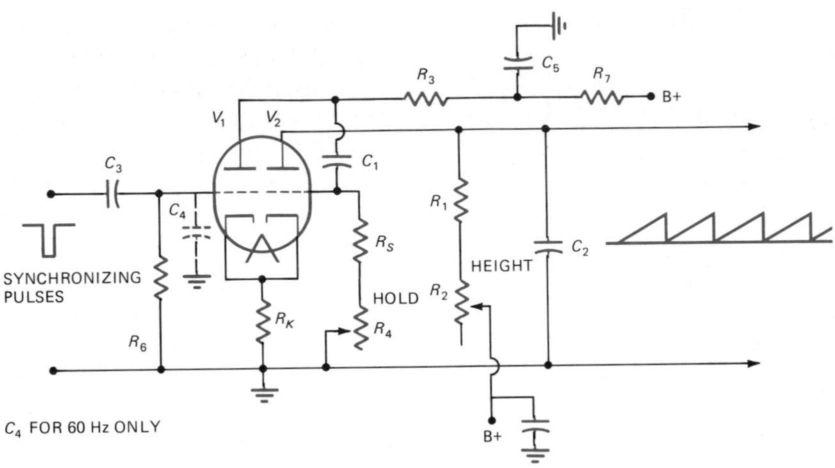

TYPICAL VALUES OF COMPONENTS

60 HERTZ	15 750 HERTZ
R_1 – 1.0 MEGOHM	R_1 – 470 000 OHMS
R_2 – 2.0 MEGOHMS	R_2 – 500 000 OHMS
R_3 – 100,000 OHMS	R_3 – 47 000 OHMS
R_4 – 1.2 MEGOHMS	R_4 – 50 000 OHMS
R_5 – 1.2 MEGOHMS	R_5 – 33 000 OHMS
R_6 – 2.2 MEGOHMS	R_6 – 2,000 OHMS
R_7 – 100 000 OHMS	R_7 – 100 000 OHMS
C_1 – .01 μF	C_1 – .001 μF
C_2 – .1 μF	C_2 – 500 pF
C_3 – .01 μF	C_3 – 50 pF
C_4 – .001 μF	C_4 – NOT NECESSARY
C_5 – .1 μF	C_5 – .006 μF
R_K – 470 OHMS	R_K – 470 OHMS

Figure 21.17 A cathode-coupled sawtooth generator. This can be used as either a vertical- or a horizontal-deflection oscillator. Compare with Figure 21.15. (Control labels are for a vertical oscillator.)

time constant results in a larger sawtooth amplitude and thus greater picture height. The converse is also true.

CONTROL LABELS

The control labels in Figure 21.16 are for a vertical oscillator. If this circuit is used as a horizontal oscillator, the height control would become the width control.

SYNCHRONIZATION

Note that oscillator synchronization is accomplished by *negative* synchronizing pulses applied to the grid of V_1. This is common practice with cathode-coupled multivibrators and will be discussed in the next section. Note that with plate-coupled (or collector-coupled) multivibrators, either positive or negative synchronizing pulses may be used.

21.12 SYNCHRONIZING THE MULTIVIBRATOR

The phrase *synchronizing an oscillator* is frequently used in describing the operation of television circuits. There are, however, many students who are not completely clear as to the exact mechanism of this synchronization. To clarify this point the following explanation is offered.

In a television receiver, the synchronizing pulses of the incoming signal take control of the free-running sweep oscillators and lock them into synchronization with the pulse frequencies. We are referring, of course, to the horizontal and vertical synchronizing pulses. It is highly improbable that the first pulse, when it reaches the oscillator, arrives at such a time as to force the free-running oscillator exactly into line. Generally, this does not occur until several pulses of the incoming signal have reached the sweep oscillator. Let us examine the means whereby the deflection oscillator is gradually forced into synchronization with the incoming synchronizing pulses.

SYNCHRONIZING THE MULTIVIBRATOR

In order to synchronize an oscillator, the pulses must be applied to the oscillator input. In Figure 21.18 we have the grid-voltage wave forms of a multivibrator and, beneath

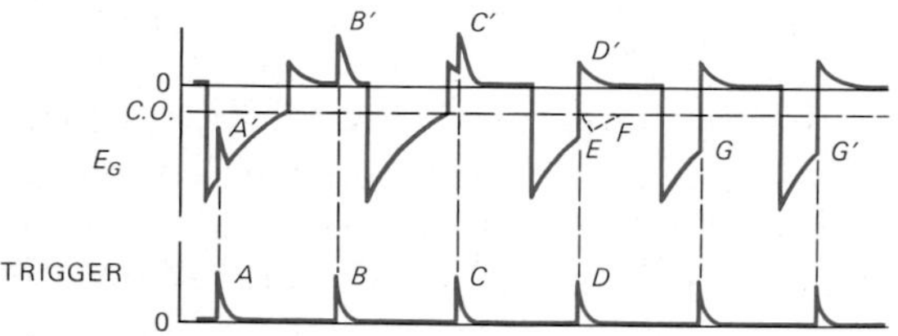

Figure 21.18 How the sync pulses lock in the multivibrator oscillator.

them, the triggering pulses as they are received from the preceding pulse separator networks. Suppose the first pulse, *A*, arrives at a time when the grid is quite negative, and thus this pulse is unable to bring the tube out of cutoff. The second pulse, *B*, arrives when the tube is conducting. Thus, it drives the grid more positive and has very little effect on its operation. The conditions for the third pulse are similar to those for the second pulse. The fourth pulse, *D*, arrives at a time when the grid of the tube is negative. However, this pulse is able to drive the grid positive, thereby initiating a new cycle. Thereafter, each succeeding pulse arrives at a time when it will bring the tube out of cutoff, and the sweep oscillator is securely locked in as long as the pulses are active. It is important that the pulses reach the grid of the oscillator when it can raise the tube *above cutoff.* Unless it can do this, it will be without the ability to lock in the oscillator.

NEGATIVE SYNCHRONIZING PULSES

Negative synchronizing pulses may be used to synchronize a multivibrator. As mentioned above, they are commonly used for cathode-coupled (or emitter-coupled) multivibrators. The synchronization is accomplished as illustrated in Figure 21.18.

The negative synchronizing pulses are applied to the grid of V_1 in Figure 21.17 when V_1 is conducting. These pulses are inverted and amplified by V_1 and applied as *positive* pulses to the grid of V_2. Here, they initiate the switching action to synchronize the multivibrator.

EFFECT OF NOISE

One final word about the foregoing oscillators. As the grid voltage approaches the cutoff value, it becomes increasingly sensitive to noise pulses that may have become part of the signal. A sufficiently strong interference pulse, arriving slightly before the synchronizing pulse, could readily trigger the oscillator prematurely. When this occurs in the horizontal oscillator, the electron beam is returned to the left-hand side of the screen before it should be and the right-hand edge becomes uneven. Severe interference may cause sections of the image to become "torn." To prevent this form of image distortion television receiver manufacturers use synchronizing circuits that respond only to long-period changes in the pulse frequency. Since interference flashes seldom have regular patterns, they cannot affect these special circuits. Several such horizontal-deflection systems are analyzed later in this chapter. Vertical-deflection systems are relatively immune to impulse noise because most of it is filtered out in the vertical integrator circuit preceding the vertical-deflection oscillator.

Note: The principles of synchronization described above apply basically to either tube or transistor circuits.

21.13 TRANSISTOR MULTIVIBRATOR OSCILLATORS

Figure 21.19 is a diagram of a transistor collector-coupled multivibrator whose operation is similar to the tube circuit of Figure 12.12. These oscillators are used for both horizontal- and vertical-deflection systems. The circuit oscillates because of coupling from the collector of Q_1 to the base of Q_2 and feedback coupling from the collector of Q_2 to

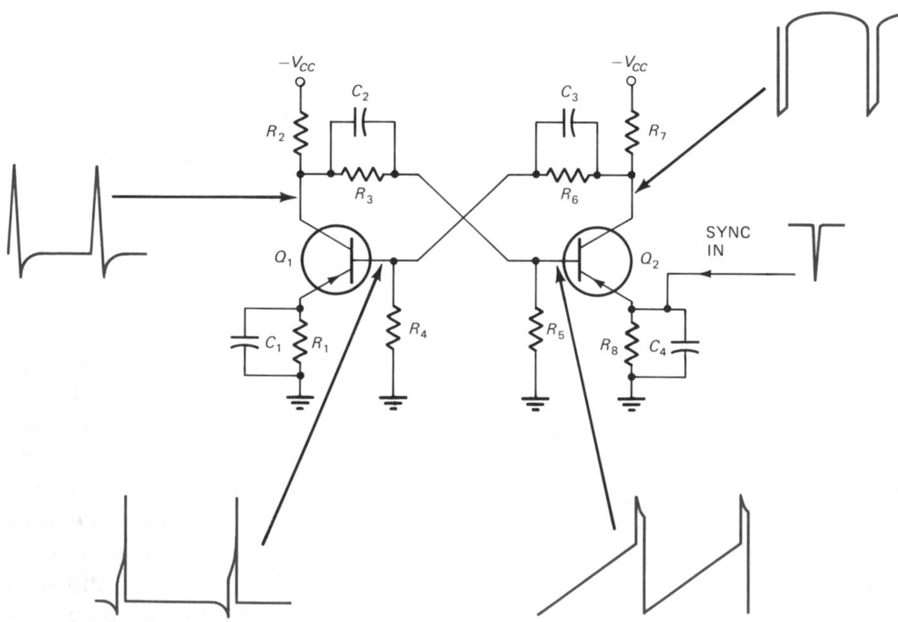

Figure 21.19 A free-running, collector-coupled, transistor multivibrator. Resistors R_3 and R_6 are added for proper base bias.

the base of Q_1. Both transistors shown are PNP.

PNP transistors are cut off when the base is driven positive with respect to the emitter. The discharge of each coupling capacitor (C_2 or C_3) through the base bias resistors (R_4 or R_5) produces this cutoff voltage. The period of cutoff for Q_1 and Q_2 is determined by the RC time constant of their respective coupling capacitor and resistor (C_3, R_4 or C_2, R_5).

The following properties of a transistor amplifier circuit should be considered in the analysis of multivibrator oscillator circuits. These properties are applied to the common-emitter configuration used here because a 180° phase shift occurs (as in an electron tube amplifier) between the output and input signals.

1. An increase in base current causes an increase in collector current through the transistor. Conversely, a decrease in base current causes a decrease in collector current.
2. An increase in collector current causes the collector voltage to decrease. A decrease in collector current causes the collector voltage to increase toward the value of the source voltage. With PNP transistors, an increase in collector current causes the collector voltage to become less negative. With NPN transistors, the collector voltage becomes less positive.
3. For normal functioning of a transistor amplifier, the base-emitter diode is forward-biased and the collector-base diode is reversed-biased. The polarity is determined by the type of transistor used (PNP or NPN).
4. A transistor is saturated when a further increase in base current causes no further increase in collector current.
5. A transistor is cut off when the base-emitter junction is reverse-biased.

OPERATION

The free-running multivibrator is essentially a nonsinusoidal, two-stage oscillator in which one stage conducts while the other is cut off, until a point is reached at which the stages reverse their conditions. That is, the stage that has been conducting cuts off and the stage that has been cut off conducts. This oscillating process is normally used to produce a square-wave output. Most transistor multivibrator circuits are counterparts of those using electron tubes, whose operation has been described. For example, the collector-coupled transistor multivibrator operates on the same principle as the plate-coupled tube multivibrator of Figure 21.12. In addition, the emitter-coupled multivibrator of Figure 21.20 operates in the same manner as the cathode-coupled multivibrator of Figure 21.16. Because of this similarity of operation, no individual discussion of the operation of the basic transistor multivibrators will be given here. The student should refer to the appropriate vacuum tube multivibrator discussion for an explanation of the equivalent transistor multivibrator operation.

21.14 TRANSISTOR, COMBINATION MULTIVIBRATOR, AND OUTPUT STAGE

The circuit of Figure 21.21 is a multivibrator circuit for vertical deflection in which the output stage forms a portion of the multivibrator. In the transistor circuit, however, note that a driver stage Q_2 is located between the vertical discharge transistor and the vertical output transistor. The driver serves two important functions. The first is to isolate sawtooth-forming capacitors C_2 and C_3 from the low-input impedance of the vertical output transistor (as previously dis-

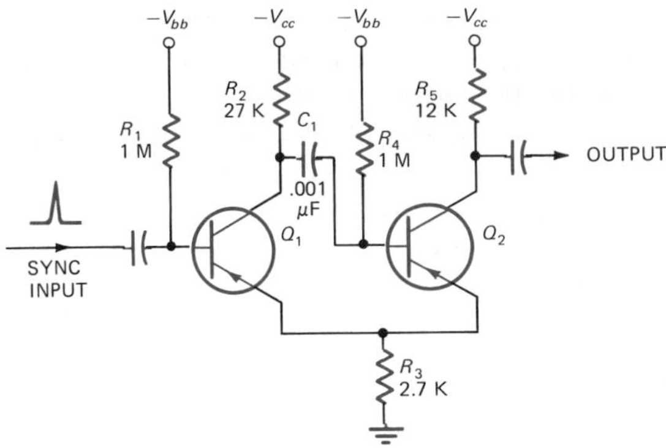

Figure 21.20 An emitter-coupled multivibrator.

cussed). The second function of the driver transistor is to increase the sawtooth driving-current input to the output transistor to the required value. Note, however, that the driver is not an inherent part of the multivibrator action of the circuit. The driver is an emitter follower, which matches the relatively high impedance of the collector circuit of Q_1 to the low impedance of the base circuit of Q_3.

C_2, C_3 CHARGE

At the beginning of the vertical trace (top of picture), the sawtooth capacitors C_2 and C_3 are completely discharged and the collector, base, and emitter of Q_1 are all at the same potential, preventing Q_1 from conducting. To begin the trace, the two capacitors start to charge through R_5 and R_6 toward $+V_{cc}$. As the capacitors begin to accumulate a charge, electrons flow into the top plates, making the polarity of these plates negative with respect to ground. This negative potential is DC-coupled to the bases of both Q_2 and Q_3, representing forward bias for these two transistors. The increasing current through Q_3 flows through the vertical yoke coils, causing the picture tube electron beam(s) to begin to be deflected downward.

At the same time, the voltage across Q_1 is increasing because of the increasing charge across C_2 and C_3. This causes a negative voltage to appear at the collector of Q_1 and also at its base, through resistors R_2 and R_3. However, Q_1 does not conduct to any

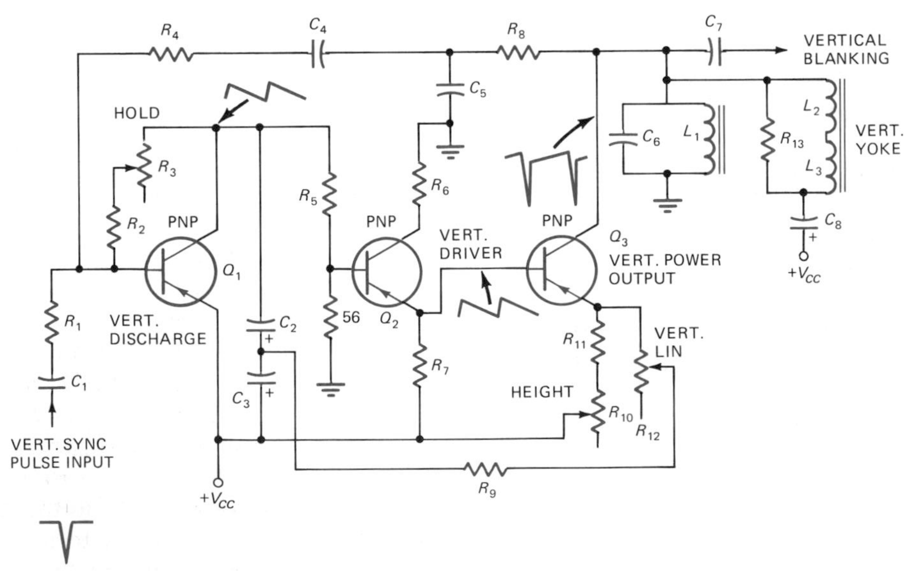

Figure 21.21 A combination vertical multivibrator-output circuit, employing an intermediate driver stage. *(Courtesy of Quasar Electronics.)*

appreciable extent until the collector and base voltage have risen to a critical value. During this time, the sawtooth capacitors continue to charge and the electron beam(s) in the picture tube continues to be deflected toward the bottom of the tube. In practice, the sawtooth wave across the capacitors will reach a peak-to-peak amplitude of about 3 V before Q_1 conducts heavily to discharge them. Remember that during this trace-forming time transistor Q_1 is essentially non-conducting, and Q_2 is conducting more and more heavily as the trace approaches the bottom of the picture tube.

C_2, C_3 DISCHARGE

When the trace reaches the bottom of the picture tube, Q_1 begins to conduct heavily and sawtooth capacitors C_2 and C_3 begin to discharge rapidly through the collector to the emitter path of Q_1. This causes a positive-going (although still negative) voltage to be applied to the bases of Q_2 and Q_3. The positive-going voltage causes a negative pulse to appear at the collector of Q_3, as shown in Figure 21.21. This negative pulse is coupled to the base of Q_1 through R_8, C_4, and R_4 and reinforces the current conduction of Q_1, which had initiated this part of the cycle. As a result of this regenerative feedback, Q_1 is driven into saturation and Q_3 is cut off. During the Q_1 saturation, the sawtooth capacitors are quickly discharged and return to approximately zero voltage to complete the retrace portion of the sawtooth wave. At this point, Q_1 again becomes cut off because of the lack of voltage on its collector and base. The cycle then continues as previously described.

HEIGHT AND LINEARITY CONTROLS

The height control R_{10} provides an adjustment of the amount of degeneration introduced into Q_3 and thus controls its output amplitude. A portion of the sawtooth wave in the emitter circuit of Q_3 is fed through the vertical linearity control R_{12} and resistor R_9 to capacitor C_3, where it is integrated to form a parabolic wave. This wave is used to modify the shape of the sawtooth wave fed to the base of Q_2 and Q_3 to adjust the linearity of the vertical sweep.

HOLD CONTROL

The hold control R_3 adjusts the amount of forward-base bias and thus determines the point at which Q_1 begins conducting on each cycle. In this manner, the hold control adjusts the free-running frequency of the system. Synchronization is accomplished by feeding negative vertical synchronizing pulses to the base of Q_1. Each pulse causes Q_1 to turn on before the free-running time and thus locks the oscillator to the synchronizing pulse rate.

21.15 MILLER INTEGRATOR VERTICAL-DEFLECTION CIRCUIT

Some RCA color television chassis utilize a novel type of vertical oscillator deflection circuit known as the **Miller integrator** or **Miller rundown circuit.** This important circuit differs from any of the vertical-deflection circuits discussed thus far.

A simplified block diagram of this system is shown in Figure 21.22. The basic operation of this system follows. A linear sawtooth wave is developed across the integrator capacitor C_1 by a rather unique method. To understand how this is accomplished, it is first necessary to briefly review some basic theory regarding the charging of capacitors. In order to develop a truly linear sawtooth voltage wave across a capacitor, a

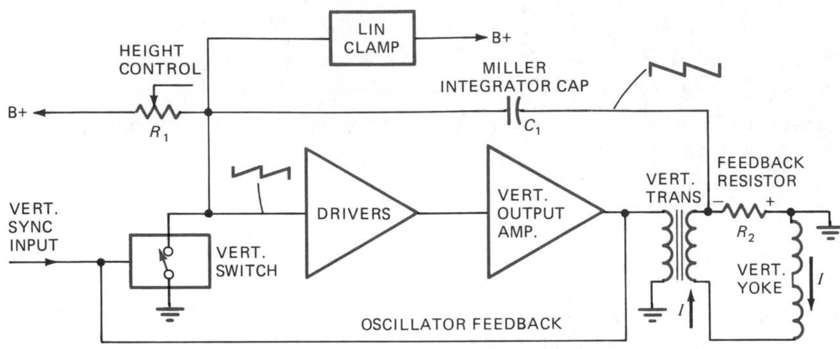

Figure 21.22 Simplified diagram of the RCA Miller integrator vertical-sweep system. *(Courtesy of RCA.)*

constant current must be used to charge the capacitor. In the usual sawtooth circuit, the wave is developed across the capacitor by the simple expedient of charging it from a power supply, through a resistor. If the charging is terminated when the capacitor has charged only to a relatively small percentage of the power supply voltage, the sawtooth will be fairly linear, although it will possess some curvature toward its peak. Because the sawtooth is nonlinear, some factor must be incorporated into the vertical circuit to compensate both for the nonlinearity and for other circuit factors that may affect the vertical sweep linearity. In many circuits, this is accomplished by the adjustment of a vertical linearity control. Because the Miller integrator circuit produces a linear vertical sweep by virtue of its design, no vertical linearity control is required. In the discussion that follows, refer to Figure 21.23.

CONSTANT CURRENT CHARGE

We said that to develop a linear sawtooth voltage wave, it is necessary to pass a constant current through the sawtooth-forming capacitor, in this case, C_1. One method of accomplishing this is to cause the usually grounded side of the sawtooth capacitor to vary its voltage in the opposite direction and in the same phase and amplitude, as it is varying its voltage on its charging side. That is to say, while the voltage of the left side of C_1 in Figure 21.22 is rising in the positive direction, the voltage of the right side of this capacitor is decreasing at the same rate simultaneously. The sawtooth voltages on the left and right sides of C_1 have opposite polarities and thus are additive across the capacitor. In effect, the negative sawtooth at the right side of C_1 adds to the power supply voltage and provides C_1 with a charging voltage that is increasing at the exact same rate that the sawtooth is being formed. The net effect is to produce a constant current through C_1 rather than the exponentially decreasing capacitor current found in the usual capacitor charging circuit. Because of the constant current passing through C_1 during its normal charging time, a linear sawtooth wave is formed across C_1, and this wave is used to drive the vertical yoke (after passing through several amplifier stages) and the vertical output transformer.

NEGATIVE SAWTOOTH

The negative sawtooth wave at the right side of C_1 is developed as follows. At the be-

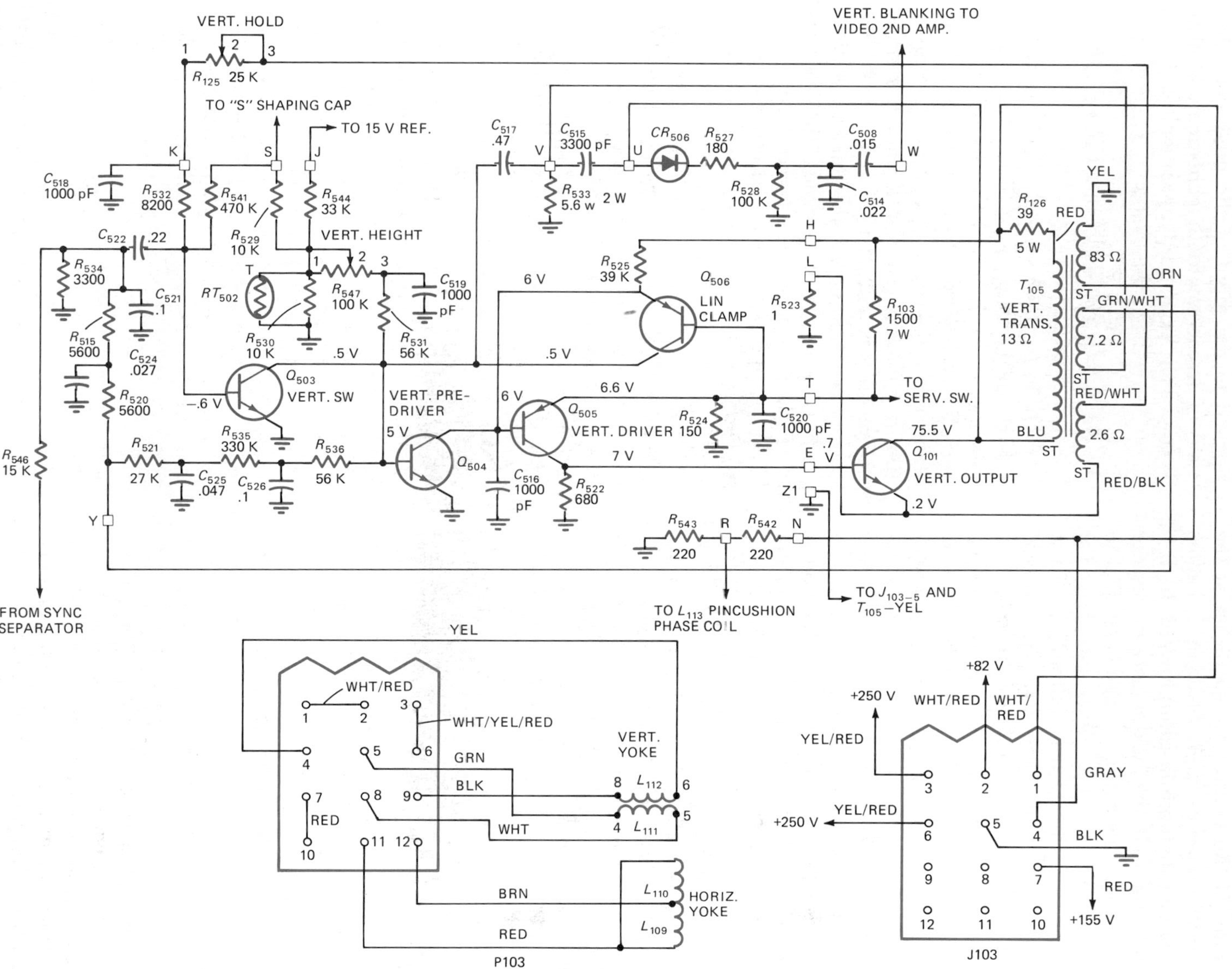

Figure 21.23 Schematic of the RCA Miller integrator vertical-sweep system. (*Courtesy of RCA Consumer Products.*)

ginning of the vertical sweep (beam at the top of the screen), the left side of C_1 begins to charge (sawtooth fashion) toward the B+ supply through the height control resistor R_1. This positive-going voltage is applied to the predriver and the driver and arrives at the base of the vertical output amplifier, again as a positive-going wave. The vertical output stage is an NPN transistor, and so the positive signal on its base causes the collector current to increase in proportion to the signal on the base. This increasing collector current (beginning of the sawtooth) is coupled through the vertical output transformer and flows through the vertical-deflection yoke and through the series feedback resistor R_2 (R_{533} in Figure 21.23). The direction of the yoke current is such that the voltage at the left side of R_2 (Figure 21.22) is going in the negative direction at the same time, and at the same rate, as the voltage at the left side of capacitor C_1 is going positive. The value of R_2 is carefully chosen at the value of 5.6 ohms to ensure that the amplitude of the negative sawtooth wave is equal in amplitude to that of the positive sawtooth wave applied to the input of the predriver. This type of linear sawtooth generation is also known as a "bootstrap" operation.

VERTICAL SWITCH OPERATION

The free-running oscillator operation of this circuit is made possible by means of a vertical switch transistor that uses feedback signals originating at the vertical output transformer. Vertical synchronizing pulses (positive polarity) are also applied to the input of the switch transistor and lock the receiver vertical sweep to the transmitted vertical sweep. Simply stated, the vertical switch transistor discharges the sawtooth capacitor C_1 quickly at the end of the vertical trace. This initiates the vertical retrace, after which the next vertical trace begins.

The action of the vertical switch transistor can be made clearer with the aid of the simplified diagram of Figure 21.24. Note that there are two feedback paths, at the top and bottom of the diagram, from the vertical output transformer to the base of the switch

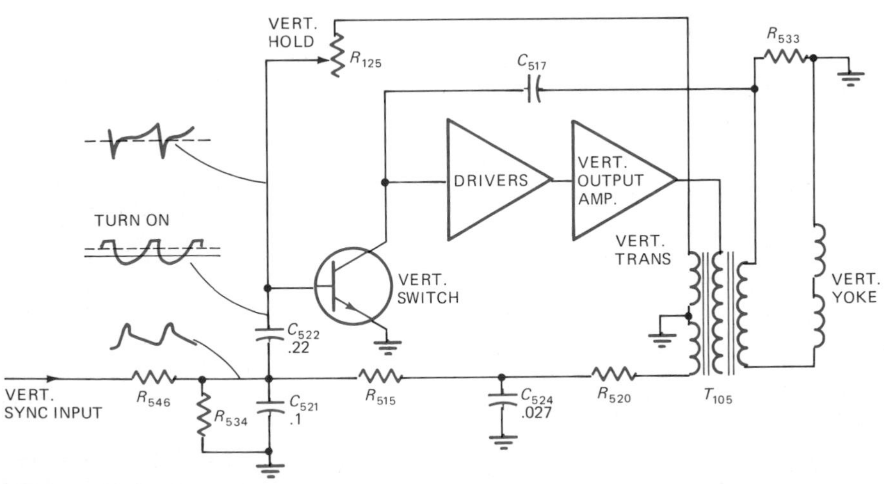

Figure 21.24 Simplified schematic diagram of the RCA Miller integrator vertical-sweep system. *(Courtesy of RCA Consumer Products.)*

transistor. The path that makes it possible to sustain free-running oscillations is the one that includes R_{520}, R_{515}, and C_{522}. The wave fed back through this path becomes integrated by its network and appears at the base of the switch transistor as a modified parabola. When the positive peak of this wave appears above a threshold voltage (represented by the dotted line), it forward-biases the switch transistor causing it to conduct heavily and to discharge the sawtooth capacitor C_{517} (C_1 in Figure 21.22). (The components in Figure 21.24 are numbered to correspond to Figure 21.23.) Following the turn-on period shown in Figure 21.24, the switch transistor is turned off sharply by the negative-going portion of the parabola applied to the base. This initiates the start of the next vertical trace as C_{517} once again begins to charge from B+.

Second Feedback Path The second feedback path from the vertical output transformer is through the vertical hold control, R_{125} to the base of the switch transistor. The portion of this wave form (Figure 21.24) above the dotted lines adds to the parabola turn-on voltage. In the free-running state, it is the combination of the two previously mentioned wave forms that is responsible for the rapid and stable turn-on action of the switch transistor. By using two wave forms, the turn-on action is made relatively immune from noise impulses that may be present on the synchronizing input line. The vertical hold control may be varied to modify the shape of the second feedback wave. As a result, the total switch transistor base turn-on voltage is changed and this in turn changes the turn-on time and thus the free-running frequency of the system.

Vertical Synchronizing Pulses The third signal applied to the base of the switch transistor base is the vertical synchronizing pulses, which are integrated by R_{546}, R_{534}, and C_{521}. Integration helps to remove noise pulses from the synchronizing signal. The vertical synchronizing pulses are then added to the other two wave forms. This added synchronizing pulse voltage actually determines the turn-on time of the switch and thus locks in the frequency of this vertical-deflection system to the incoming signal.

LIN-CLAMP CIRCUIT

In order to preserve the linearity of the sawtooth wave at the beginning of the sweep, it is necessary to momentarily connect the left side of C_{517} to a charging voltage higher than the +15 V obtained from the height control. A simplified schematic indicating the operation of the lin-clamp circuit is shown in Figure 21.25. At the end of the vertical sweep (beam at bottom), the vertical switch transistor is turned on, discharging sawtooth capacitor C_{517}. The negative-going sawtooth wave at the input to the predriver produces a positive-going wave at its collec-

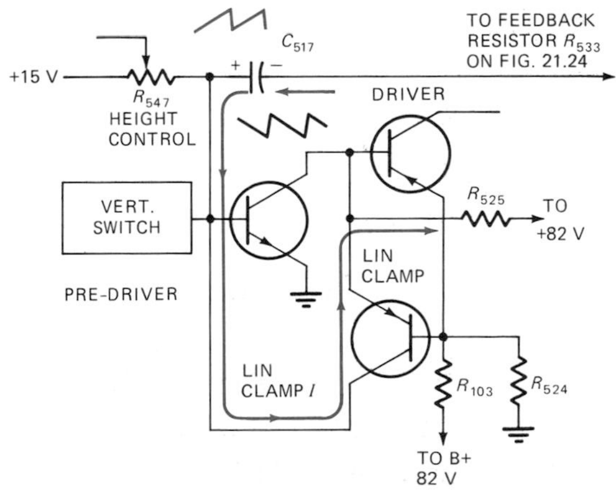

Figure 21.25 Simplified diagram to illustrate RCA Miller integrator's vertical-linearity clamp operation. Arrows indicate momentary linearity-clamp charging current to C_{517}. *(Courtesy of RCA Consumer Products.)*

tor circuit. Since this circuit is also connected to the emitter of the lin-clamp transistor (PNP), it provides forward bias for this transistor at the peak of the positive-going voltage. Thus, at the end of the sweep and at the beginning of the next vertical sweep, the lin-clamp transistor is turned on.

Overcoming Yoke Inductance Note in Figure 21.25 that there is now a charging current path for C_{517}. This path passes through the lin-clamp transistor and R_{525}, to a source of +82 V. This source voltage is considerably higher than the normal +15 V source at the height control input. The effect of connecting C_{517} to this higher source (momentarily) is to cause it to start charging more rapidly than if it were only connected to the +15-V source at all times. The momentary higher charging voltage applied through the lin clamp is necessary because, as shown in Figure 21.22, sawtooth capacitor C_{517} must charge effectively through the inductance of the yoke and thus requires a higher initial charging voltage to overcome the current opposition of the yoke inductance.

C_{517} Returned to Normal Charging Source Once the yoke current has started to flow, C_{517} is returned to its normal charging source of +15 V through the height control. This happens as follows. When C_{517} begins to charge through the lin-clamp circuit, its left plate starts rising in the positive direction. This positive-going voltage is applied to the base of the predriver and produces a negative-going sawtooth voltage at the collector. Since the predriver collector is connected to the emitter of the lin-clamp transistor (PNP), the negative-going voltage cuts

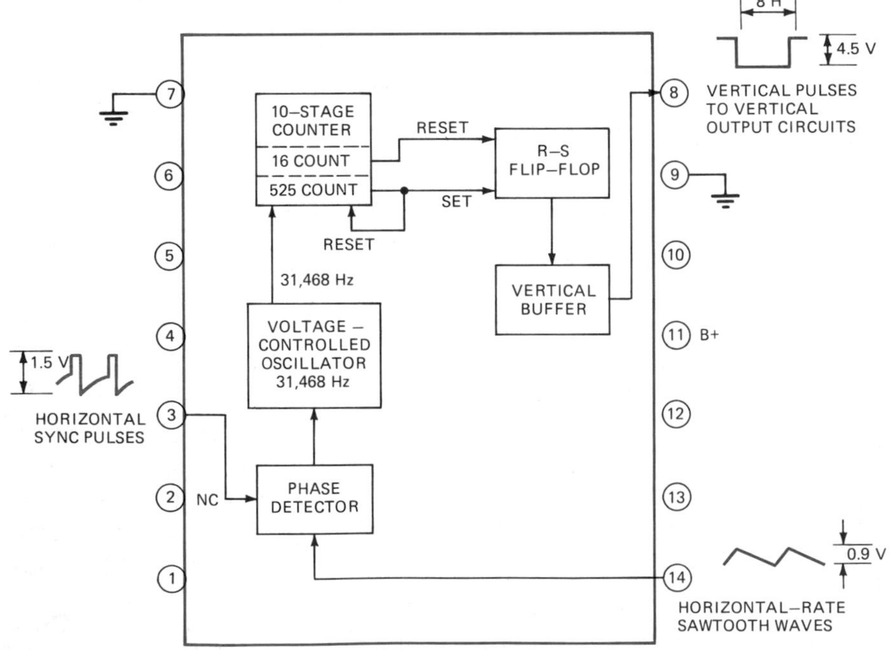

Figure 21.26 Simplified block diagram of RCA 14-pin DIP IC, showing major stages producing vertical countdown (see text.) Output at pin 8 consists of pulses at 59.94 Hz.

off the lin-clamp transistor and C_{517} can now charge only through the height control toward +15 V, as previously explained.

21.16 DIGITAL VERTICAL COUNTDOWN CIRCUITS

Some TV receivers have no vertical- or horizontal-deflection oscillators. The functions of these circuits are performed by digital countdown circuits. The circuits described here are part of an RCA 14-pin DIP IC package. This IC contains the circuitry to produce both vertical and horizontal sweep frequencies. Only the vertical circuitry is described in this chapter. The horizontal circuitry is described in Chapter 22.

SIMPLIFIED VERTICAL RATE BLOCK DIAGRAM

A block diagram of the IC, showing how the vertical rate pulses are developed, is given in Figure 21.26. (Certain functions and stages have been omitted for the sake of simplicity.)

Deriving the Vertical Pulse Rate The vertical pulse rate of 59.94 Hz is derived by dividing the voltage-controlled oscillator (VCO) frequency of 31 468 Hz by 525. This is accomplished by a section of the ten-stage counter shown in Figure 21.26. Since the vertical rate is always correct, no vertical hold control is needed.

Why 31 468 Hz? The value 31 468 Hz is derived as follows. Each field is composed of 262.5 scanning lines. One way to arrive at the required frequency of 59.94 Hz is to divide the horizontal scan frequency of 15 734 Hz by 262.5. However, this is not practical because today's digital dividers can divide only by *whole* numbers. To overcome this problem, the VCO (clock) is operated at twice the horizontal frequency (31 468 Hz) and then divided by 525.

VCO STABILIZATION

The VCO is frequency-stabilized by the DC voltage output of a phase detector (similar to phase-locked loop operation). The phase detector compares the frequency and phase of the received horizontal synchronizing pulses with that of sawtooth waves derived from horizontal output transformer pulses. Any difference results in the output of a DC control voltage which is used to correct the VCO frequency. Note that in this case the VCO frequency is twice that of the waves fed to the phase detector.

COUNTER OPERATION

The vertical pulses that drive the vertical output circuits are actually developed by the R-S flip-flop and fed through the vertical buffer (Figure 21.26). However, the R-S flip-flop requires a *set* pulse to initiate its output and a *reset* pulse to cut it off. These control pulses are provided by the ten-stage counter as follows:

1. The VCO oscillates at *twice* the horizontal frequency. Thus, count 525 corresponds to 262.5 (lines). This is the *beginning* of the *next field* and the count is now reset to zero.
2. New count zero causes a set pulse to appear at the R-S flip-flop. This *initiates* the vertical output pulse. Count zero always occurs after the reset of the ten-stage counter, which happens after count 525 (262.5 lines).
3. The R-S flip-flop remains activated

until it receives a reset pulse from the 16-count line of the 10-stage counter.

4. At the end of 16 counts (8 lines), the R-S flip-flop is reset and ends the vertical output pulse. Thus, the width of this pulse is 8 H(8 × 63.5 μs).
5. The count continues to 525, at which time the cycle repeats. This process is continuous as long as a TV station is being received.

The vertical output pulses are fed to the vertical output circuits and the vertical deflection yoke at the rate of 59.94 Hz.

Details regarding the horizontal countdown circuits are given in Chapter 22.

21.17 TROUBLES IN VERTICAL-DEFLECTION OSCILLATOR SYSTEMS

Trouble in the vertical oscillator system can be grouped into four general categories: (1) no oscillation, (2) incorrect frequency or failure to synchronize, (3) height and/or linearity problems, and (4) interlace problems.

NO OSCILLATION

If the vertical oscillator does not operate or if any stage following the oscillator is completely inoperative, there will be no vertical sweep, although there may be a normal horizontal sweep. The symptom of this problem is a single bright horizontal line on the picture tube, indicating that only the horizontal-deflection circuits are operative. There are many possible causes for this symptom, including defective tubes or transistors, a defective vertical blocking oscillator transformer, a loss of power supply voltage, and, of course, defective parts. For more detailed troubleshooting procedures for this and other symptoms caused by problems in both vertical- and horizontal-oscillator systems, see Chapter 27.

INCORRECT FREQUENCY OR NO SYNCHRONIZATION

Incorrect vertical sweep frequencies may be caused by an improper setting of the vertical hold control or by defective parts in the vertical-oscillator circuit. The appearance of this symptom is that a number of pictures appear vertically, instead of the desired single picture. You should remember that when a combination oscillator vertical output system is employed, using two tubes or transistors (the transistor circuit may also include a driver stage), the components in both stages may affect the oscillation frequency.

Loss of vertical synchronization may be caused by a lack of the vertical synchronizing pulses being fed into the vertical oscillator. This should be checked first, since it indicates that the problem precedes and is not

Figure 21.27 This picture shows lack of vertical synchronization.

in the vertical oscillator system. If synchronizing pulses are present, they should be of the proper wave form and amplitude. The vertical integrator circuit, at the input to the vertical oscillator, sometimes becomes defective and may cause this problem. Vertical oscillator operation should also be checked to see that it is capable of being adjusted to the correct frequency. This may be determined by adjusting the hold control and trying to obtain a momentarily stationary, single picture on the screen. The effect, on the screen, of a lack of vertical synchronization, is shown in Figure 21.27. In such a case, the picture may drift slowly up or down, depending on the setting of the vertical hold control.

HEIGHT/LINEARITY TROUBLES

Height and linearity troubles are frequently related but sometimes will occur separately. Height problems, caused by the vertical oscillator system, are generally the result of a lower-than-normal amplitude of the generated vertical-deflection sawtooth. This may also affect the vertical linearity. The settings of the height and linearity controls should be checked before looking elsewhere for the source of trouble, as these may simply be misadjusted. Note that these controls interact and that both of them must be adjusted each time adjustment is required. These same vertical problems may also be caused by a defective tube(s) or transistor(s), by defective controls or other parts, or by improper power supply voltages. Both height and linearity problems may also be caused by the vertical output circuit (Chapter 24), and thus the problem must first be isolated between the vertical oscillator system and the vertical output system. In some sets, the two functions are combined in a combination circuit, and this must be taken into account when troubleshooting these problems.

INTERLACE TROUBLES

As described in Chapter 3, a television picture is composed of two interlaced fields that compose one frame, or one complete picture. If the two fields do not interlace properly, the picture will lose detail. There may also be some slight vertical movements of the scanning lines if the problem is such as to allow variations of the interlace spacing between scanning lines. If alternate scanning lines "pair," visible spaces may actually appear between scanning lines. Poor interlace can be caused by a defect in the vertical integrator circuit or by the introduction of horizontal pulses into the vertical-deflection system. The horizontal pulses may be introduced either by a direct conducting path or by radiation.

VERTICAL DIGITAL COUNTDOWN CIRCUIT TROUBLES

It will be remembered that the digital circuits are all on one monolithic IC. Thus, no repairs are possible on this unit. If it is determined that the trouble lies with this IC (Figure 21.26) and not with the vertical output circuits, it will be necessary to replace the IC.

It is a relatively simple matter to isolate the trouble between the countdown IC and the vertical output circuits. Referring to Figure 21.26, the IC can be disconnected at Terminal 8 and a 60-Hz square wave from a function generator (at 4.5 VPP) fed into the output circuits. If there was no vertical deflection and now there is, the trouble is in the countdown IC. If no vertical deflection now occurs, the trouble is in the output circuits.

SUMMARY OF CHAPTER HIGHLIGHTS

1. The countdown circuit, or vertical-deflection oscillator, develops the correct vertical scanning rate. This is 59.94 Hz.
2. Vertical scanning rates are developed by (1) blocking oscillators, (2) multivibrators, (3) Miller integrator circuits, (4) digital countdown circuits. (Figures 21.6, 21.8, 21.12, 21.16, 21.19, 21.23, and 21.26)
3. A common method of developing a sawtooth wave form is by the charge and discharge of a capacitor. (Figure 21.3)
4. In five time constants ($T = RC$), a capacitor is considered to be fully charged. About the first 10 to 20% of the charging curve is fairly linear. (Figure 21.2)
5. A trapezoidal wave form is required to drive a sawtooth current through the vertical yoke coils. (Figure 21.4)
6. A blocking oscillator blocks because of the long time constant in the grid (or base) circuit. (Figures 21.6 and 21.8)
7. A blocking oscillator can become a sawtooth generator by the addition of a sawtooth capacitor and charging resistor. (Figures 21.7 and 21.8)
8. Blocking oscillators may be used to generate either vertical or horizontal sawtooth wave forms.
9. Transistor blocking oscillators operate on the same operating principle as tube blocking oscillators. However, the action of a self-generated sawtooth in the transistor circuit is used to turn the transistor on and off. (Figures 21.7 and 21.8)
10. With a transistor blocking oscillator and sawtooth generator, it is desirable to feed the sawtooth into a driver prior to the output stage. This prevents loading of the sawtooth generator (Figure 21.10)
11. A multivibrator is a nonsinusoidal oscillator that is used as a deflection oscillator.
12. A multivibrator circuit is basically a two-stage resistance-coupled amplifier, with the output of the second stage fed back (in phase) to the first stage. (Figures 21.12 and 21.13)
13. Two common types of multivibrators are the (1) plate-coupled (or collector-coupled) type and, (2) the cathode-coupled (or emitter-coupled) type. (Figures 21.12 and 21.16)
14. Multivibrators are commonly used in television receivers as sawtooth generators. (Figures 21.14, 21.15, 21.17, and 21.21)
15. In synchronizing a deflection oscillator, the free-running oscillator frequency is set slightly lower than the synchronizing pulse frequency. (Figures 21.6 and 21.18)
16. A two-stage circuit may be designed as a combination vertical-deflection oscillator and output stage. (Figures 21.15 (tube) and 21.21 (transistor))
17. In a cathode-coupled (or emitter-coupled) multivibrator, one path of feedback is through the unbypassed cathode (or emitter) resistor. (Figure 21.16)
18. Positive synchronizing pulses are required for a tube-type blocking oscillator. An NPN type requires positive pulses, and a PNP type requires negative pulses. Mulivibrators may be synchronized by either positive or negative pulses. (Figures 21.6 and 21.18)
19. The Miller integrator vertical-deflection circuit is different from a blocking oscillator or multivibrator. It is a unique oscillating circuit. (Figures 21.23, 21.24, and 21.25)
20. The digital vertical countdown circuit is

not an oscillator. It operates by counting pulses from a 31 468 Hz VCO. (Figure 21.26)

21. In Number 20, the vertical-pulse rate of 59.94 Hz in the digital circuit is derived by dividing the VCO frequency of 31 468 Hz by 525. (Figure 21.26)
22. In Number 20, the VCO is operated at 31 468 rather than 15 734 because the counter can divide only by whole numbers (525), not fractions (262.5). (Figure 21.26)
23. In Number 20 the vertical output pulse is initiated by count zero and terminated at count 16 (8 H wide) of each group of 525 counts. (Figure 21.26)

EXAMINATION QUESTIONS

(Answers are provided at the back of the text.)

Part A Supply the missing word(s) or number(s).

1. ______ Hz is the vertical scan frequency for TV.
2. The vertical scan frequency not produced by an oscillator may originate in a ______ countdown circuit.
3. The first ______ to ______ % of a capacitor-charging curve is fairly linear.
4. A ______ wave form is required to drive a sawtooth current through the vertical yoke coils.
5. The ______ frequency of a tube blocking oscillator is determined by the grid time constant.
6. The synchronizing pulse frequency for a deflection oscillator must be ______ than the free-running frequency.
7. A ______ stage will be found between a transistor deflection oscillator and the output stage.
8. The frequency of a transistor-blocking oscillator can be varied by changing the ______.
9. A plate-coupled (or collector-coupled) multivibrator may be synchronized by a ______ or ______ pulse.
10. A multivibrator is a non- ______ oscillator.
11. A multivibrator can function as a sawtooth generator with the addition of a ______ and a series-charging ______.
12. One path of feedback is through the ______ (or ______) resistor in a cathode-coupled (or emitter-coupled) multivibrator.
13. The ______ control of a deflection oscillator can vary its frequency.
14. The Miller ______ is an unconventional vertical oscillator deflection circuit.
15. In the digital vertical countdown system, the VCO frequency is ______ Hz.
16. In Question 15 the circuit that actually produces the vertical output pulses is the ______.
17. In Question 15, the ______ corrects the VCO frequency.

Part B Answer true (T) or false (F).

1. A capacitor will charge fairly linearly to about 50% of full charge.
2. The time constant of a capacitor charging circuit is equal to $R \times C$.
3. A trapezoidal wave consists of a rectangular pulse added to a sawtooth.
4. The horizontal as well as the vertical yoke coils require a trapezoidal driving voltage.
5. The hold control can vary the synchronizing pulse timing.
6. A vertical-deflection oscillator may be synchronized accurately by a 59.94-Hz sinusoidal wave.
7. Transistor blocking oscillators operate on the same principles as tube blocking oscillators.
8. Thermal runaway of a transistor may occur

as a result of a substantial increase of the ambient temperature.

9. All multivibrators are free-running.
10. At least three stages are required to make up a multivibrator deflection oscillator and an output stage.
11. In a cathode-coupled (or emitter-coupled) multivibrator, there is no RC coupling from the plate (collector) of one stage to the grid (or base) of the other.
12. In the digital vertical countdown system, the width of the vertical output pulse is 8 counts.
13. In the digital system, the zero count causes a set pulse to appear at the R-S flip-flop.
14. In the digital system, count zero always occurs after the reset of the ten-stage counter.
15. This reset occurs after count 262.5.

REVIEW ESSAY QUESTIONS

1. Discuss the requirement for deflection oscillators and digital countdown circuits.
2. Name the two basic types of deflection multivibrators. What is the essential difference between the two types?
3. How is the trace portion in Figure 21.1 formed? The retrace portion?
4. Draw a simple diagram of a transistor trapezoidal wave generator. Show the important waveforms. (Hint: refer to Figure 21.5.)
5. Draw a simple schematic diagram of a transistor blocking oscillator.
6. In Question 5, explain the operation of the circuit.
7. In Question 5, explain briefly (with simple wave forms) how the oscillator is synchronized.
8. Define *time constant*. How may this be related to the production of a fairly linear sawtooth wave?
9. Is a multivibrator synchronized with positive or negative pulses? Explain your answer.
10. Explain the operation of the hold control in Figure 21.6.
11. Describe how the sawtooth is generated in Figure 21.8 in your own words.
12. Briefly describe the function of the "driver" in Figure 21.10. What type of circuit is used?
13. In Figure 21.11, explain briefly the operation of the vertical size control, R_{624}.
14. Explain why a trapezoidal voltage wave form is *not* required to drive horizontal yoke coils.
15. When synchronizing a deflection oscillator, should its free-running frequency be higher or lower than the synchronizing frequency? Why?
16. Name all the types of deflection oscillators discussed in this chapter.
17. The digital vertical countdown circuit does not require a hold control. Name at least one other advantage of this system.
18. Refer to Figure 21.26 and explain why the VCO is operated at 31 468 Hz.
19. Describe the operation of the R-S flip-flop in Figure 21.26.
20. In Figure 21.26, at what point in its operation is the vertical output pulse initiated?

EXAMINATION PROBLEMS

(Selected problem answers are provided at the back of the text.)

1. A capacitor is charged through a resistor from a source of 300 V. From Figure 21.2, what is the voltage across the capacitor in (1) $\frac{T}{10}$ (2) $\frac{T}{5}$, (3) T, (4) $3T$, and (5) $5T$?
2. Identify these numbers: (1) 59.94 Hz, (2) 15 734.26 Hz, and (3) 31 468.52 Hz.
3. It is desired to charge a capacitor of 1 μF through a resistor of 1 megohm from a source voltage of 100 V to a voltage of 10 V. How

long (in seconds) will this take? (Hint: $T = RC$; refer to Figure 21.2.)

4. What is the time in microseconds of a VCO operating at 31 468 Hz, for (1) 16 counts, (2) 262 counts, (3) 525 counts? (Hint: time for one cycle is $T = \frac{1}{F}$).
5. In Figure 21.7, identify the circuit constants that determine the free-running frequency. Do the same for Figures 21.8 and 21.9.
6. In Figure 21.11, identify the vertical pulse integrator.
7. If Figure 21.14 is to be used as a 59.94-Hz circuit, what is the time constant of the sawtooth capacitor circuit?

Chapter 22

22.1 Fundamentals of horizontal AFC
22.2 Vacuum tube AFC and multivibrator circuits
22.3 Multivibrator stabilization
22.4 Solid-state horizontal AFC systems
22.5 Solid-state AFC
22.6 Reactance-controlled sinusoidal oscillators
22.7 Digital horizontal countdown
22.8 Troubles in horizontal oscillator and AFC systems
Summary

HORIZONTAL OSCILLATORS AND HORIZONTAL AFC

INTRODUCTION

In general, the same *basic* types of deflection oscillators are used in the horizontal deflection system as in the vertical deflection system. These are the blocking oscillator and the multivibrator types, and they are found in both the tube and solid-state versions. In addition, some horizontal-deflection systems use sinusoidal oscillators.

The same general types of circuits are used for both monochrome and color sets, although, as with the vertical systems, some unique designs in some color sets may also be found. Also, as with the vertical-deflection systems, digital countdown circuits are found in horizontal-deflection systems. (Although not previously mentioned, most TV stations now use the deflection frequencies of 59.94 Hz and 15 734.26 Hz for both monochrome and color-TV transmissions.)

Since the horizontal oscillators operate at more than 200 times the frequency of the vertical oscillators, it must be expected that stability problems for horizontal oscillators are considerably greater than for vertical oscillators, and this is indeed the case. In addition, the problem of false triggering of the horizontal oscillators by noise pulses, rather than by the horizontal synchronizing pulses, is far more serious than with vertical oscillators. A major reason for this is the manner in which the vertical and horizontal synchronizing pulses are separated before being fed to their respective deflection oscillators. In the case of the vertical-deflection oscillators, the vertical synchronizing pulses are selected by means of integrators, which are low-pass filters. Because the vertical synchronizing pulses are passed through low-pass filters, most of the noise pulses accompanying the composite synchronizing pulses are rejected by the integrator and never reach the vertical oscillator.

For horizontal synchronizing pulses, the situation is quite different. Horizontal synchronizing pulses are selected by differentiation, and differentiators are high-pass filters. Thus, most noise pulses tend to be passed, rather than rejected, by the differentiators. This problem is particularly serious because of the much higher horizontal oscillator frequency and its attendant stability problems. If the horizontal-deflection oscillator were synchronized directly by the horizontal synchronizing pulses (a situation that was true in some of the very early TV receivers), poor stability would result because of the triggering by noise impulses. This would produce horizontal "tearing" of the picture.

In order to ensure that the horizontal oscillator operates only on the correct frequency and is basically immune to noise pulses, all horizontal-deflection oscillators in today's TV receivers are controlled by an automatic frequency control (AFC) circuit, which in turn is controlled by the horizontal synchronizing pulses as well as by the horizontal-deflection wave forms. This indirect method of frequency control results in an excellent horizontal oscillator stability and almost complete immunity from the effects of noise impulses.

When you have completed the reading and work assignments for Chapter 22, you should be able to:

- Define the following terms: horizontal AFC, Hartley oscillator, digital countdown system, variable-resistance transistor, reactance transistor, differentiator, and phase detector.
- Explain why AFC is needed for horizontal oscillators.
- Discuss the basic types of horizontal-deflection oscillators.
- Explain the operation of a reactance transistor.
- Explain the operation of a horizontal AFC oscillator system.
- Explain why resonant-stabilizing circuits are used in multivibrators and blocking oscillators.
- Explain the operation of a horizontal digital countdown system.
- Understand the cause and cure of some common problems in horizontal oscillator and AFC systems.

22.1 FUNDAMENTALS OF HORIZONTAL AFC

Figure 22.1 is a block diagram of a horizontal-deflection system. The horizontal synchronizing pulses from the synchronizing pulse separator (or amplifier or phase inverter) are fed to the AFC circuit. With these pulses and others (which will be discussed shortly), the AFC circuit determines whether or not the horizontal oscillator is on frequency. If it is not, then the AFC block develops a DC voltage that is applied to the horizontal oscillator and brings the oscillator frequency back to its correct value. The oscillator develops an appropriate deflection voltage, which is then passed on to the output amplifier or amplifiers and, from there, to the horizontal-deflection coils.

This, then, is the overall action of a typical horizontal-deflection system. We shall first examine the effect of noise pulses on the horizontal deflection oscillators.

EFFECT OF NOISE PULSES

The use of the incoming synchronizing pulses to trigger and control the vertical and horizontal sweep oscillators represents the simplest, most economical, and most direct method of controlling the motion of the electron beam in the picture tube. Unfortunately, however, this method has limitations and disadvantages that outweigh its economy and simplicity.

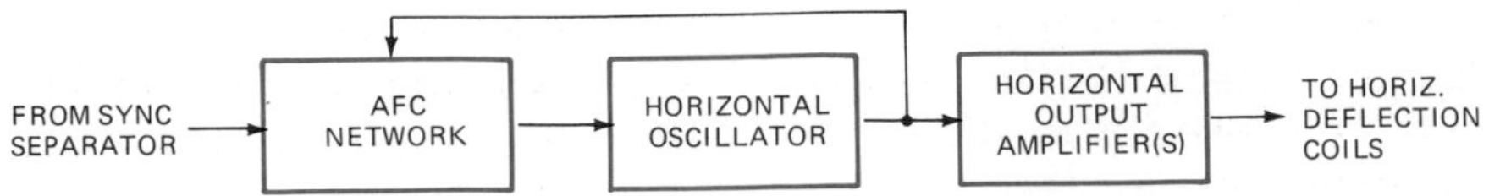

Figure 22.1 Block diagram of a typical horizontal-deflection system.

Perhaps the greatest disadvantage is the susceptibility of this method to noise disturbances arising from electrical equipment operating in the vicinity of the receiver. The noise pulses, which combine with the video signal and usually extend in the same direction as the desired synchronizing pulses, pass through the same stages as the pulses and arrive at the sweep oscillators. They do their greatest damage when they arrive during the interval between synchronizing pulses. If the amplitude of the noise pulses is sufficiently great, they will trigger the sweep oscillator, initiating a new cycle too early. When the vertical oscillator is triggered this way, the picture will move vertically either up or down until the proper synchronizing pulses in the signal can again assume control. If the horizontal oscillator is incorrectly triggered, a narrow band of lines will streak or tear across the image. When the interference is particularly heavy and persistent, the entire picture becomes jumbled.

Noise Triggers Oscillators Of the two sweep systems in a television receiver, interference is particularly destructive to the horizontal system. To understand why this is so, we must examine the nature of most interference voltages and their effect upon the vertical and horizontal sweep oscillators.

Whenever a blocking oscillator is triggered, for example by a synchronizing pulse, its grid, after a short period of conduction, becomes highly negative as a result of an accumulation of electrons on the grid capacitor. This negative voltage is sufficient to keep the tube beyond cutoff until the charge on the grid capacitor has increased to a value at which current is permitted to flow again through the tube. In most circuits now in use, the capacitor discharge occurs in the manner shown in Figure 22.2. At the start, the discharge is fairly linear. However, as the amount of charge contained in the capacitor decreases, the discharging rate decreases exponentially. In Figure 22.2 this region, usually called nonlinear, extends from Point *A* to Point *B*.

Now, when the negative charge on the grid capacitor is large, the oscillator is relatively immune to incoming positive pulses. With continued discharge, however, the immunity decreases. Experience has indicated that off-cycle triggering of the oscillator is generally concentrated in the last 15% of its discharge cycle. This is true regardless of the frequency at which the oscillator is operating. Hence, one would expect to experience equal difficulty with both deflection systems in the receiver. That this is not so is due to the nature of the noise pulses and the type of filters inserted before each sweep oscillator.

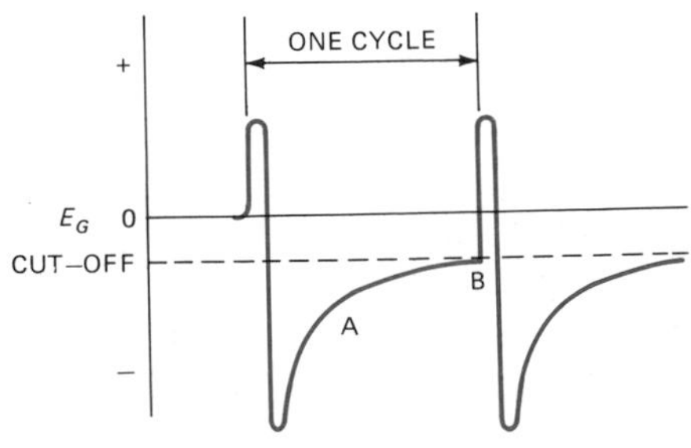

Figure 22.2 The manner in which the grid voltage of a blocking or multivibrator oscillator varies.

Effect of Filter Time Constants Noise pulses which are the most troublesome to television receivers possess a high amplitude but are narrow or of short duration. (The energy of the noise pulses is distributed over a wide range of frequencies. In order for a peak to occur, the phase relationship among the various frequencies must be such as to permit them to add, forming the high-amplitude pulse or peak.) When the pulses reach the path leading to the horizontal-sweep oscillator, they are readily passed because of the short time constant of the filter leading to the horizontal system. A filter with a short time constant is necessary because the horizontal synchronizing pulses themselves have a duration of only 5 μs. On the other hand, the filters leading to the vertical system have a long time constant and automatically act to suppress the effects of all horizontal synchronizing pulses and noise pulses of short duration. The presence of this low-pass filter (the integrating network) is largely responsible for the greater immunity to noise pulses enjoyed by the vertical system. Of course, when a wide noise pulse is received, it contains enough energy to cause off-time firing of the vertical oscillator, but the annoyance caused to the viewer from this source is seldom great. To reduce the susceptibility of the horizontal-sweep system to noise pulses of any type, several AFC (and phase control) systems have been developed.

BASIC AFC OPERATION

In each of these special control systems, an oscillator is set to operate at 15 734.26 Hz; and the output of the oscillator (through amplifiers) controls the horizontal motion of the electron beam across the screen. The next step is to synchronize the frequency of this sweep oscillator with the incoming horizontal synchronizing pulses of the signal. This step is accomplished through an intermediate stage generally known by one of the following names: control stage, AFC phase detector, or horizontal synchronizing discriminator. Whatever the name, the function of this intermediate network is to compare the frequency of the incoming horizontal synchronizing pulses with the frequency of the receiver horizontal deflection oscillator. If a difference exists, a DC voltage is developed, which, when fed back to the horizontal deflection oscillator, is used to change its frequency until it is exactly equal to that of the incoming pulses.

Note that the incoming synchronizing pulses are not applied directly to the sweep oscillator. They are merely compared (in frequency) with the output of the sweep oscillator, and, if a frequency difference exists, then a DC voltage is developed which, when fed back to the sweep oscillator, forces its frequency back into line with that of the synchronizing pulses.

THE LONG TIME-CONSTANT FILTER

By having the DC control voltage pass through a long time-constant filter before it reaches the sweep oscillator, we can eliminate the effects of any noise impulses and permit only relatively slow changes in the frequency of the synchronizing pulses (which may occur at the transmitter) to affect the sweep oscillator. Thus, a long time-constant filter somewhat similar to that present in the vertical sweep system is incorporated in the horizontal AFC system. Such a filter could not be used directly in the horizontal system because it would prevent the desired horizontal synchronizing pulses (as well as the noise pulses) from reaching the horizontal sweep oscillator. Hence the need for the indirect method outlined above.

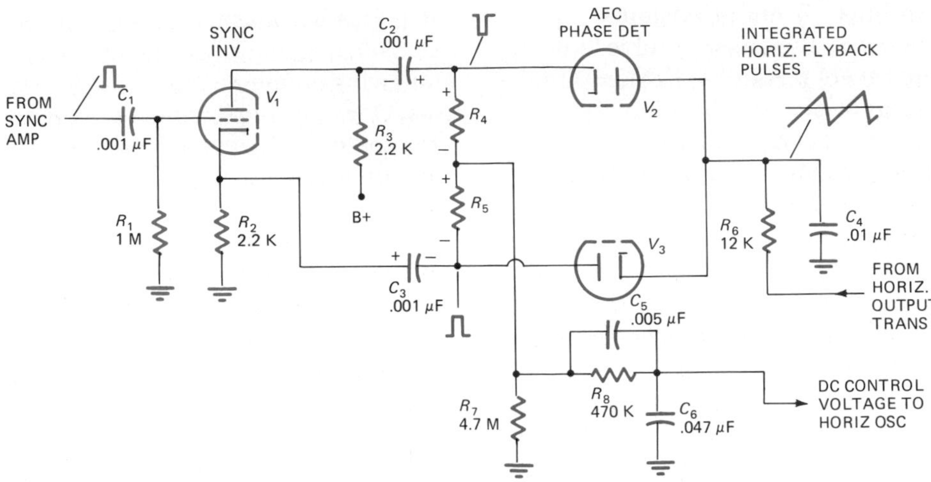

Figure 22.3 A double-diode phase detector.

22.2 VACUUM TUBE AFC AND MULTIVIBRATOR CIRCUITS

Vacuum tube phase-detector AFC systems fall into two general categories. These are diode and triode systems, according to the type of tube used. Even within each category there are several different methods of achieving the desired output control voltage, although in most instances the general operation remains the same.

DIODE PHASE DETECTOR

A diode phase-detector circuit is shown in Figure 22.3. Two diodes are connected so that they receive horizontal synchronizing pulses in phase opposition and sawtooth waves of the same polarity. From the interaction of these wave forms, a DC voltage is developed across R_7 that is governed by the relative frequency difference between the incoming synchronizing pulses and the sawtooth wave.

The Synchronizing Pulses In detail, the network functions as follows. When the synchronizing pulses are received by V_1, positive and negative pulses of equal amplitude are applied to both diode sections of the phase detector. The cathode of V_2 receives a negative pulse at the same time that the plate of V_3 receives the positive pulse. These pulses cause both diode sections to conduct. The current flowing through V_2 charges C_2 to approximately the peak value of the applied pulse, and the current following through V_3 charges C_3. The polarity of each voltage is indicated in Figure 22.3. During the interval between the pulses, each capacitor discharges, the electrons moving from C_3 up through R_5, then down through R_7 to ground, and up through R_2 to the other plate of C_3. This current thus develops a negative voltage across R_5 and R_7. Capacitor C_2 also discharges, and its electrons travel down through R_3 to ground (by way of the power supply), then up through R_7 to the junction of R_4 and R_5, and then up through R_4 to the other plate of C_2. The result of these two currents through R_7 is that the voltage produced by one current cancels the voltage produced by the other current, leaving a net potential of zero. This is desirable since these two pulses alone should produce no net control voltage.

Because of the slow discharge of C_2 and

C_3 through their respective networks, the voltages developed across R_4 and R_5 keep V_2 and V_3 from conducting until the arrival of the next pulse.

The Sawtooth Wave Fed to the phase detector is another voltage, a sawtooth wave that is developed across C_4 from pulses applied to it from the secondary of the horizontal output transformer. This sawtooth voltage has the same frequency as the horizontal oscillator, since it is the oscillator that drives the horizontal output amplifier. The sawtooth voltage is applied equally to each tube; thus, the plate of V_2 and the cathode of V_3 receive the same polarity voltage of the sawtooth wave at the same time. Hence, at the phase detector, we have both ingredients needed to check the oscillator frequency against the frequency of the incoming pulses.

Comparing Frequencies Comparison of the two frequencies is possible only at the instant the synchronizing pulses arrive because it is only at this moment that V_2 and V_3 conduct and are in a position to respond to the sawtooth voltage. Three situations are possible.

First, if the synchronizing pulses arrive at a time when the sawtooth voltage is passing through zero, current will flow through V_2 and V_3, replenishing any charge that C_2 and C_3 may have lost during the interval between pulses. No net voltage will appear across R_7, as indicated previously. This condition is the desired one because the frequency of the sweep oscillator and the synchronizing pulses are in step with each other.

The Second Situation The second situation occurs when the synchronizing pulses arrive. The sawtooth voltage is negative at this instant. (This occurs when the horizontal oscillator is running too slow.) Now V_3 will receive a positive pulse at the plate and a negative sawtooth voltage on the cathode and, hence, conduct more strongly than usual, producing a larger-than-normal voltage across R_7. At the same time, conduction through V_2 is reduced because the negative sawtooth voltage at the plate partly offsets the negative synchronizing pulse at the cathode. The reduced current flow through V_2 cannot offset the voltage that the current of V_3 develops across R_7. Hence, a resultant negative voltage is developed, which is fed to the horizontal oscillator, and its frequency is altered (in this case, speeded up).

The Third Situation In the third situation, the pulses arrive when the sawtooth voltage is positive. Now V_2 conducts more strongly than V_3, and a resultant positive voltage is developed across R_7. This voltage, fed to the controlled horizontal oscillator, slows it down or lowers its frequency to bring it in line with the frequency of the incoming pulses.

Filters C_5, C_6, and R_8 respond only to slow changes in voltage level, preventing fast-acting noise pulses from affecting the operation of the horizontal oscillator. In this way, the circuit is stabilized and the false triggering that can happen when synchronizing pulses are fed directly to the horizontal oscillator is avoided. In place of the vacuum tube diodes, solid-state diodes may be used with identical results.

DC CONTROL OF OSCILLATOR FREQUENCY

The horizontal oscillator to which the DC control voltage developed in Figure 22.3 is applied is shown in Figure 22.4. This oscillator is a cathode-coupled multivibrator containing a special resonant-stabilizing circuit in the plate circuit of the first triode. More will be said on this point later.

To understand what happens when the DC control voltage is applied directly to an

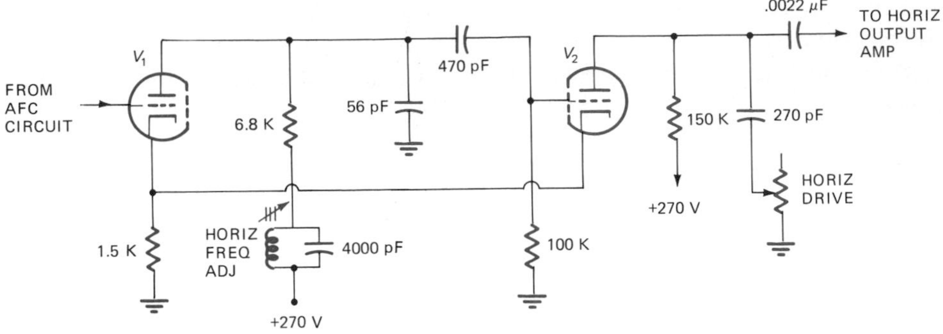

Figure 22.4 A cathode-coupled multivibrator to which the AFC circuit of Figure 22.3 would be connected.

oscillator, consider the operation of the cathode-coupled multivibrator. In this oscillator, the first triode conducts during trace time. The second triode conducts only during the retrace time. Since the cathodes are tied to the ground through a common resistor, the operating bias of the second triode is affected by the cathode voltage developed by the first triode.

The grid of the first triode is usually bypassed to ground and is not part of the feedback loop. This leaves the grid available as the controlling element of the system.

If the correcting voltage on the grid of the first triode is made positive (by the AFC network), current flow through the tube will increase and the cathode voltage will rise. This extends the cutoff time of the second triode. Since the time is lengthened before the retrace time occurs, the oscillator frequency is lowered.

Similarly, any negative voltage applied to the first triode grid lowers the cathode potential and shortens the time of the RC discharge of the second triode grid circuit. This change increases the firing rate and raises the frequency of the system.

Reversing Sawtooth Voltage Polarity It is possible, by reversing the polarity of the sawtooth voltage fed to the phase detector, to obtain control voltages of opposite polarity for the conditions of a fast or a slow oscillator. The oscillator is the controlling factor. For a cathode-coupled multivibrator, the required control voltages should have the polarity indicated. For a blocking oscillator, an opposite set of polarity voltages would be needed.

Note: Transistor multivibrators are discussed in Chapter 21. For the collector-coupled type, see Figure 21.19. The emitter-coupled type is shown in Figure 21.20.

AFC CIRCUITS USING SINGLE-POLARITY SYNCHRONIZING PULSES

In the phase detector just discussed, horizontal synchronizing pulses of both positive and negative polarity are required. A dual-diode phase detector that requires only one polarity of synchronizing pulses is shown in Figure 22.5. The cathodes of diodes D_1 and D_2 are connected, and negative-going synchronizing pulses are applied at their junction. In addition, horizontal flyback pulses from the horizontal output transformer are integrated to sawtooth form by C_3 and R_5. The sawtooth waves are also applied to the two diodes (Figures 22.6 and

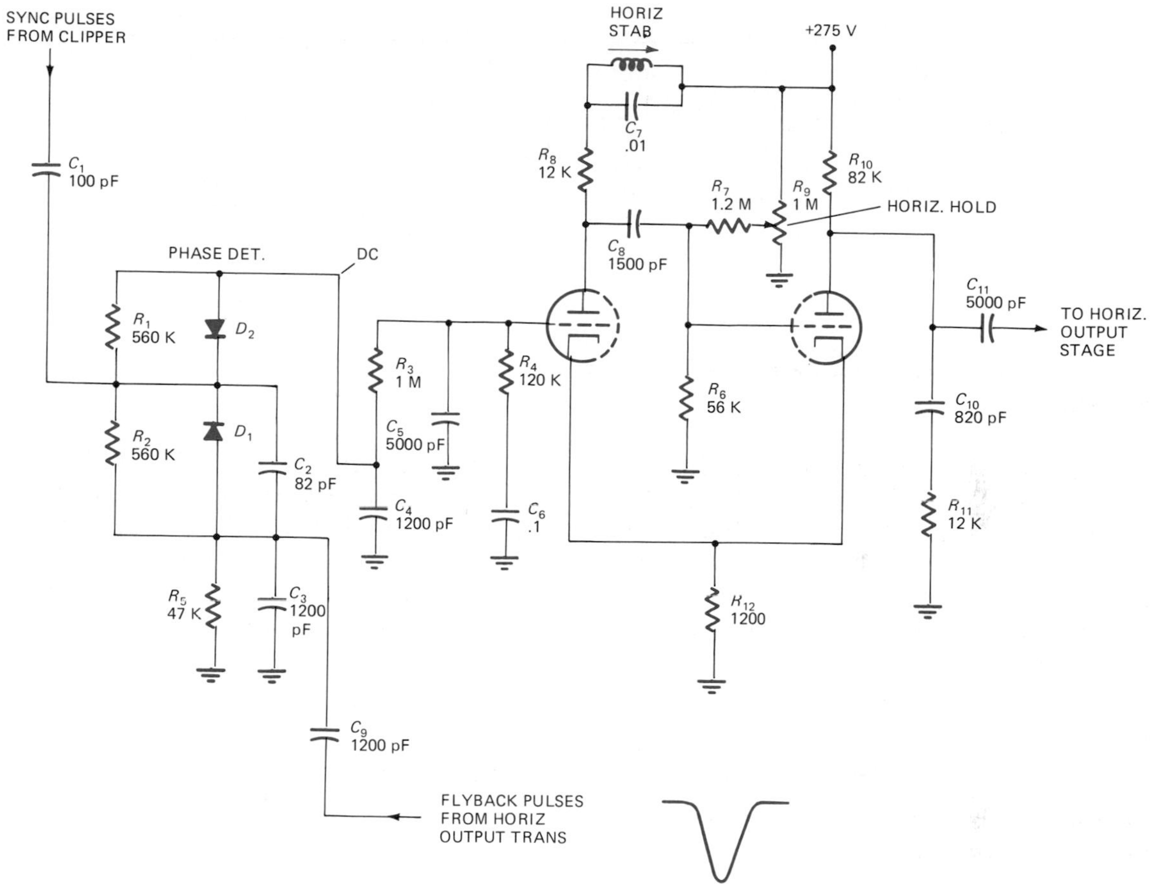

Figure 22.5 A phase detector using silicon diodes and requiring only one set of input sync pulses.

22.7). As will be described in the following paragraphs, it is the combination of the two types of waves acting on the circuit that develops the AFC correction voltage.

If the synchronizing pulses *alone* are applied to the circuit of Figure 22.5, equal and opposite voltages will be developed across resistors R_1 and R_2 and a zero output voltage will result.

If the sawtooth waves *alone* are applied to the circuit, equal and opposite voltages again will appear across resistors R_1 and R_2 and a zero output voltage will result.

A third condition that would result in a zero output control voltage is if the synchronizing pulses occur at the exact center of the sawtooth retrace (Figure 22.7A). Actually, this condition occurs only if the horizontal oscillator is correctly locked in, both in phase and in frequency.

If the oscillator is slow, the synchronizing pulse will arrive before the sawtooth retrace passes through its AC axis (Figure 22.7B). On D_2, therefore, part of the sawtooth voltage will be added to the synchronizing pulse voltage because the sawtooth voltage is on the positive half of its cycle when the synchronizing pulse occurs. Part of the sawtooth voltage on D_1 will be subtracted from the synchronizing pulse voltage

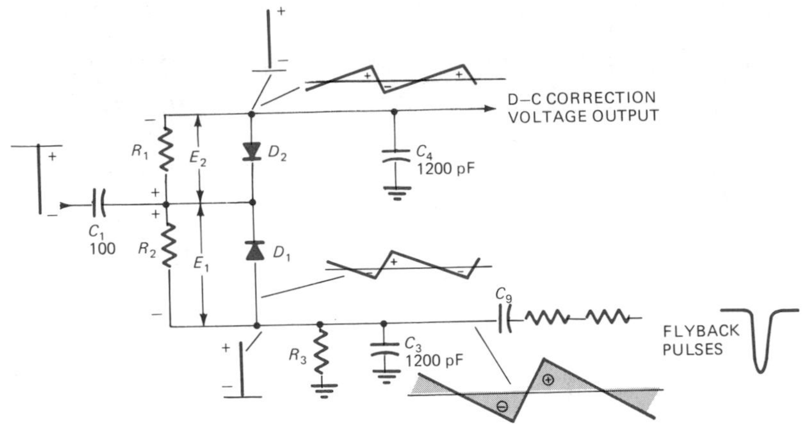

Figure 22.6 A simplified diagram of the phase detector in Figure 22.5, showing the waveforms in the circuit.

because the sawtooth retrace here is still in the negative half of its cycle. The output voltage of the phase detector in this case will be negative because the voltage drop across R_1 is greater than the drop across R_2.

With Oscillator Fast If the oscillator is fast, the sawtooth retrace will pass through its AC axis before the synchronizing pulse occurs (Figure 22.7C). On D_2, therefore, part of the sawtooth voltage will be subtracted from the synchronizing pulse. On D_1, part of the sawtooth voltage will be added to the synchronizing pulse, producing a higher voltage drop across R_2 than across R_1. This will produce a positive output voltage that slows down the horizontal oscillator.

CIRCUIT ADJUSTMENTS

In the foregoing AFC circuits, there are practically no variable controls. Hence, there is actually nothing to adjust. Variable components, however, are generally found in a multivibrator horizontal oscillator. A multivibrator is frequently used because it provides a convenient input for applying the AFC voltage and because it operates with good stability, particularly when it has a stabilizing resonant circuit in the plate circuit of the first triode.

The cathode-coupled multivibrator used with phase-detector AFC systems may have one or two adjustments. If there are two adjustments, one is horizontal hold control and the other is the movable core in the stabilizing coil. The multivibrator of Figure 22.5 is representative of this group. The hold control is accessible on the front or rear panels of the receiver and may be adjusted from time to time, as required. In a number of sets, however, the hold control is dispensed with and the only adjustment then is the movable core of the stabilizing coil. Figure 22.4 is an example of this approach. In both arrangements, the coil core should not require any attention once it is adjusted, unless some component changes value in the circuit and the stabilizing circuit is unable to keep the multivibrator synchronized. The adjustment is quite simple and requires only that the core be rotated until the picture is properly synchronized. If a hold control is present, it is set to the center of its range before the coil core is moved.

22.3 MULTIVIBRATOR STABILIZATION

Resonant-stabilizing circuits are used in multivibrators as well as in blocking oscillators. One such circuit was shown in Figure

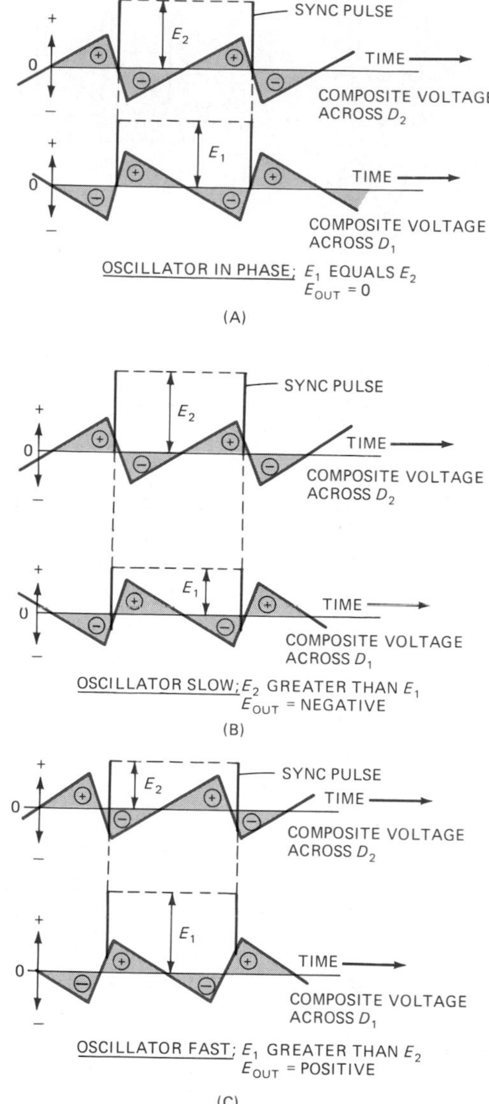

Figure 22.7A-C. The operation of the phase detector of Figure 22.6 when the sync-pulse frequency and the oscillator frequency are **A.** equal and **B.** and C. unequal.

22.4. The additional resonant coil and capacitor are placed in the plate circuit of the first triode and adjusted to 15 734.26 Hz. The presence of this circuit alters the manner in which the grid of the second triode comes out of cutoff. The wave form of Figure 22.8A appears at the plate of the first triode in the absence of the stabilizing circuit. Figure 22.8B shows the grid wave form of the second triode under the same condition. Now, when we insert the stabilizing circuit, its wave form, shown in Figure 22.8C, will add to those existing in the circuit to produce the modified wave forms shown in Figure 22.8D and E. Of particular importance is the grid wave form of the second triode. Note that it now comes out of cutoff quite sharply. A considerably stronger noise pulse will be required to trigger this tube prematurely than without the stabilizing circuit.

22.4 SOLID-STATE HORIZONTAL AFC SYSTEMS

By studying the solid-state AFC circuit in Figure 22.9, you may note that it is similar to the AFC circuit used in many vacuum tube television receivers (compare with Figure 22.5). Also included in Figure 22.9 is a set of wave forms that indicate the phase relationship between the oscillator sawtooth and synchronizing pulse wave forms, for different frequencies of the oscillator.

The phase detector diodes D_1 and D_2 are connected back to back and are enclosed in a single plastic case. R_{501} and R_{502} form the output network load across which the control voltage will be developed. Resistor R_{500} and capacitor C_{500} couple the negative horizontal synchronizing pulses into the phase detector to establish a reference frequency against which to compare the oscillator frequency and phase. The capacitor C_{500} and resistor R_{502} differentiate the horizontal and the vertical serrated synchronizing pulses, thus coupling only horizontal pulses to the AFC network. Then C_{510}, C_{503}, R_{505}, and C_{501} couple integrate positive pulses from the flyback transformer (this pulse is repeated at the oscillator frequency) into the phase-detector network for frequency comparison with the horizontal synchronizing pulses.

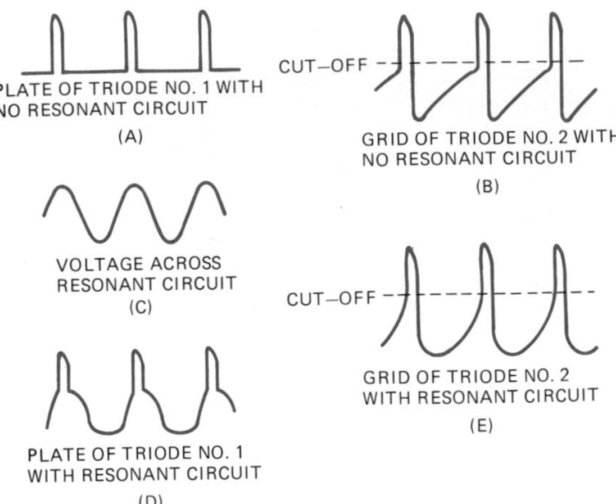

Figure 22.8A-E. The effect of a resonant stabilizing circuit on the operation of the multivibrator shown in Figure 22.4.

R_{503}, C_{502}, R_{504}, and C_{504} form the horizontal AFC antihunt network, thus preventing any AC variations from entering the horizontal oscillator circuit and causing the AFC loop to be unstable.

AFC OPERATION

To explain the operation of the AFC circuit, let us begin with the synchronizing pulse separator stage (Q_1). Prior to the arrival of the positive synchronizing pulses at its base, Q_1 is cut off and has a high collector voltage. Since C_{500} is connected to the collector of Q_1, it will be charged to nearly the supply voltage. When an incoming positive synchronizing pulse turns on Q_1, its collector voltage decreases.

Capacitor C_{500} will now discharge (dashed arrows), coupling the synchronizing pulse to the cathodes of D_1 and D_2. This discharging voltage will forward-bias both diodes (D_1 and D_2) causing them to conduct simultaneously and equally. If the negative-going sawtooth at the anode of D_1 is going through zero at this time (this is indicative of the oscillator being in phase with the synchronizing pulse), both diodes will conduct equally since both the D_1 and D_2 anodes are now at zero volts. After the pulse ends, the collector voltage of Q_1 rises. Now C_{500} will recharge (solid arrows), producing a voltage drop across the two load resistors (R_{501} and R_{502}), as indicated in Figure 22.9. Since both diodes conducted equally, equal and opposite voltages will now appear across R_{501} and R_{502} when C_{500} recharges. This being the case, the net voltage across the two resistors will be zero and no correction voltage will be developed for the oscillator across C_{501}.

If the sawtooth at the anode of D_1 is not going through zero when the synchronizing pulse arrives (indicative of the oscillator frequency being out of phase with the pulse), the conduction of D_1 will be influenced by the sawtooth wave at its anode. When the pulse is completed and C_{500} recharges, the voltages developed across the load resistors are still opposite but are now not equal. The net voltage is no longer zero. A correction voltage is now developed across C_{501} and applied to the oscillator to correct its frequency.

Wave Form Analysis We can clarify the foregoing discussion by observing Wave forms *A*, *B*, *C*, and *D* in Figure 22.9. Wave forms *A* and *B* are impressed across D_1 when the oscillator frequency is in phase

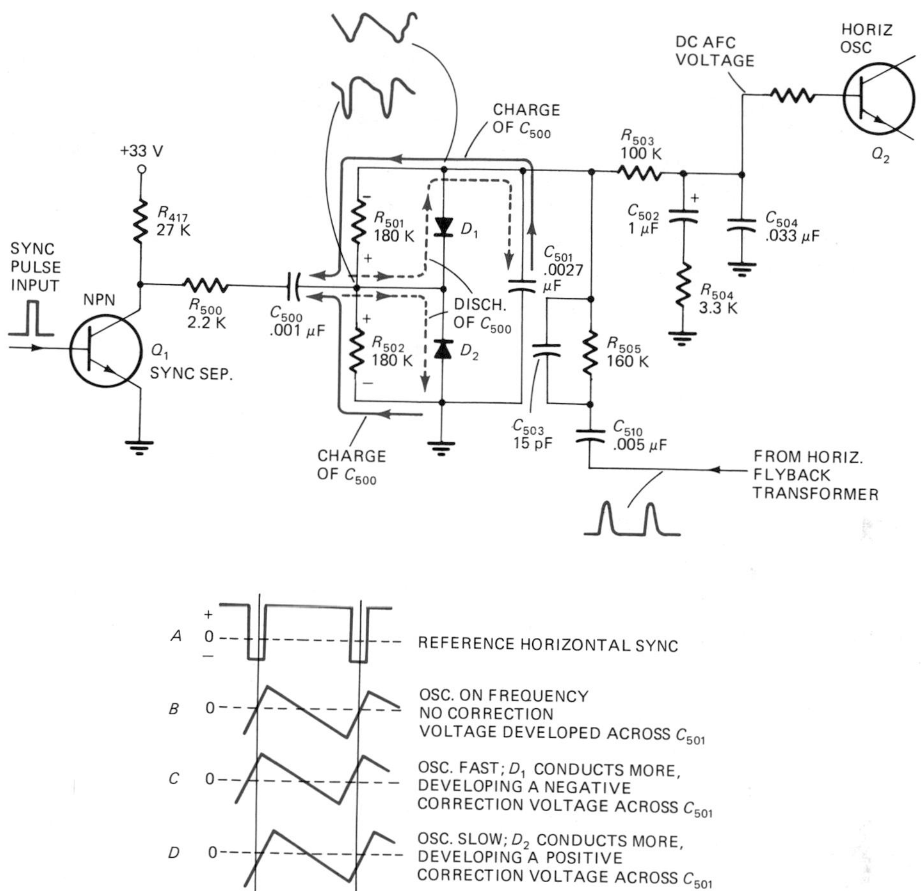

Figure 22.9A-D. Solid-state AFC circuit. *(Courtesy of Quasar.)*

with the synchronizing frequency. Because the sawtooth *(B)* is going through zero at this time, it does not alter the conduction of D_1. Hence, both diodes will conduct equally since their anodes are referenced to zero. No correction voltage is developed under these conditions across C_{501}.

Wave forms *A* and *C* are impressed across D_1 when the oscillator is leading in phase (higher frequency) with respect to the synchronizing pulse. Now when the pulse turns D_1 on, it will conduct more than D_2 since Wave form *C* caused its (D_1) anode to be positive at this time. As a result, the voltage drop across R_{501} will be larger than the voltage across R_{502} and a negative correction voltage will be developed across C_{501}. This voltage has the correct polarity to reduce the oscillator frequency and pull it back into phase lock with the synchronizing pulse frequency.

Wave forms *A* and *D* are impressed across D_1 when the oscillator is lagging in phase with respect to the synchronizing pulse. Now, when the pulse turns D_1 on, it will conduct less than D_2, since Wave form *D* caused its anode to be negative at this time. As a result, the voltage drop across R_{501} is less than the voltage drop across R_{502} and a net positive voltage is developed

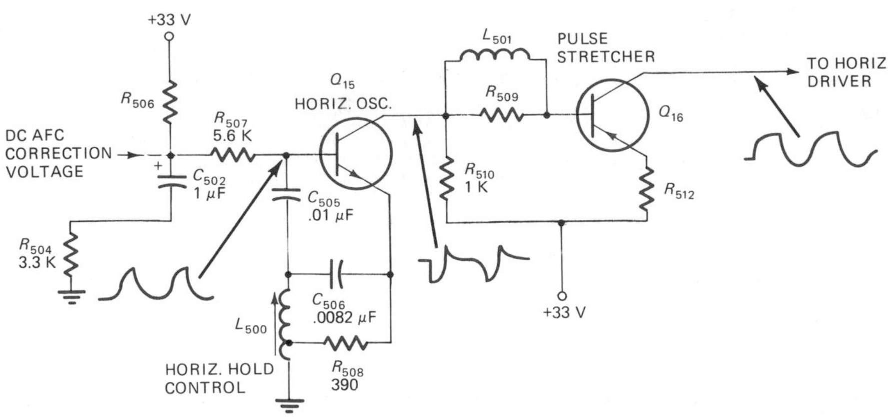

Figure 22.10 Solid-state horizontal-oscillator circuit. *(Courtesy of Quasar.)*

across C_{501}. This voltage now has the correct polarity to increase the oscillator frequency and pull it back into phase lock with the pulse frequency.

SOLID-STATE SINSUSOIDAL HORIZONTAL OSCILLATOR

The horizontal oscillator circuit that is frequency stabilized by the foregoing AFC circuit is shown in Figure 22.10. The horizontal oscillator acts somewhat like a switch, turning on and off at a predetermined rate (15 734.26 Hz). From its output, we obtain a driving signal that we can shape, amplify, and ultimately apply to the horizontal-deflection coils.

The horizontal oscillator, Q_{15}, receives its initial emitter-base forward bias for starting from R_{506} and R_{507}. The base of Q_{15} also receives a DC voltage from the AFC, which keeps the oscillator locked in frequency with the transmitter synchronizing pulses.

The horizontal oscillator is of the Hartley type and uses a tapped oscillator coil, L_{500}, in its emitter circuit to sustain oscillation. The main frequency-determining components are L_{500}, C_{506}, and R_{508}. In the output of the oscillator, a wave form is produced that resembles a square wave. This wave form is produced by the abrupt switch on of Q_{15} to saturation and the equally abrupt switch off of Q_{15}. This unsymmetric square wave is not quite suitable for switching (or driving) the horizontal output transistor and must be modified.

Inductor L_{501} and resistor R_{509} shape a more uniform square wave to be applied to the pulse shaper (Q_{16}) for further shaping and amplification. These two components form what is called a pulse stretcher. The name is very descriptive because this is exactly what they do. The inductor L_{501} opposes a change in current similar to a choke in a power supply; thus, the spikes and abrupt voltage changes on the existing square wave are filtered out by the inductor.

22.5 SOLID-STATE AFC

A solid-state RCA color TV receiver AFC circuit will now be explained. The following information was supplied by RCA and is reproduced by their permission.

HORIZONTAL AFC

The AFC circuitry used in any television receiver is designed to automatically hold the

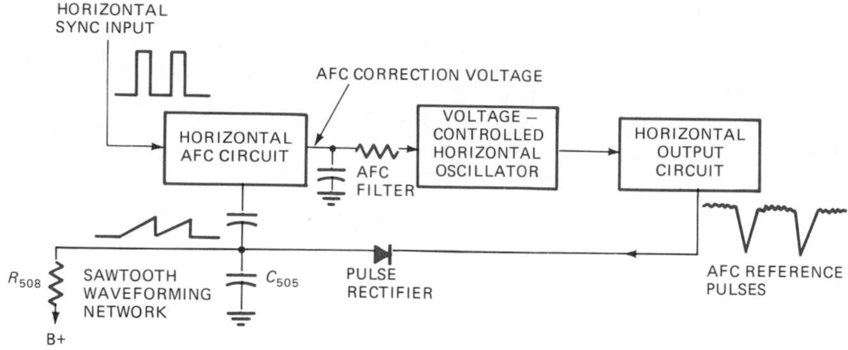

Figure 22.11 Block diagram of a solid-state horizontal AFC system. *(Courtesy of RCA Consumer Electronics.)*

horizontal oscillator at the exact frequency and phase of the received horizontal synchronizing pulses. This function is especially critical in color television circuitry because the color synchronizing pulse (burst) amplifier is keyed from pulses derived from the horizontal circuitry. As a result, any slight frequency or phase discrepancy between the occurrence of the color burst and the horizontal-keying pulses for the burst amplifier may cause an incorrect color display.

The horizontal AFC circuitry in this RCA chassis will automatically hold the horizontal oscillator at the exact frequency of the received horizontal synchronizing pulses if the oscillator free-running frequency is within ± 300 Hz of the received pulse frequency.

A block diagram of the horizontal AFC system is illustrated in Figure 22.11. The circuit develops a DC oscillator control voltage that is proportional to the difference between the frequency of the received synchronizing pulses and the operating frequency of the horizontal oscillator. The functioning of the AFC circuit is basically the same as explained for the dual-diode phase detector at the beginning of Section 22.3. Therefore, only a brief description of the AFC circuit operation will be presented here. For a more detailed discussion, refer to Section 22.3.

Sensing Oscillator Frequency The horizontal AFC circuit senses the oscillator frequency through a feedback network connected between the horizontal output transformer and the AFC circuit. The feedback wave shape at the horizontal output transformer is in the form of negative-going pulses, occurring at the operating frequency of the oscillator. A wave-shaping network (C_{505} and R_{508}) transforms these pulses into a sawtooth wave shape that is suitable for application to the AFC circuit. The DC correction voltage developed by the AFC circuit is filtered and is then applied to the horizontal oscillator. The frequency of this oscillator may be varied by the AFC voltage, causing the oscillator to operate at the exact frequency and phase of the received horizontal synchronizing pulses.

A simplified diagram of the AFC circuit is illustrated in Figure 22.12. Note the similarity to Figure 22.3. A phase-splitter circuit supplies equal but opposite-polarity synchronizing pulses to a dual-diode phase-detector network. The incoming horizontal synchronizing pulses are differentiated at the base of the phase-splitter transistor (Q_{501}). The phase-splitter output pulses are coupled to the AFC detector circuit through capacitors C_{502} and C_{503}. The sawtooth reference voltage is applied to the common diode connection, as shown.

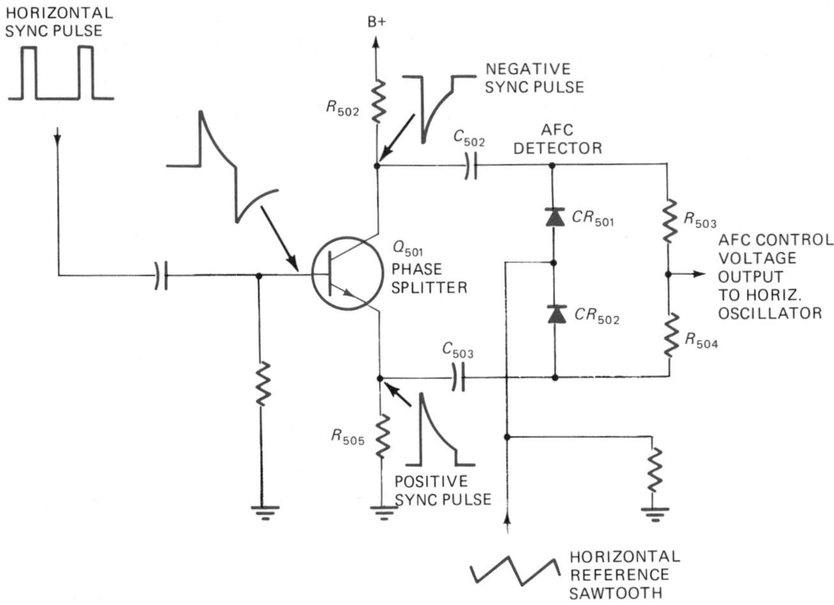

Figure 22.12 Solid-state horizontal AFC phase-splitter and detector. *(Courtesy of RCA Consumer Electronics.)*

AFC Operation Figure 22.13A illustrates the action of the diode-detector circuit when there is no frequency difference between the applied pulses and the sawtooth reference voltage. In this case, each diode is gated into conduction by the synchronizing pulses as the reference voltage passes through O. The current through each diode is therefore equal, resulting in equal but opposite charges on capacitors C_1 and C_2. The discharging action of these capacitors develops equal but opposite-polarity voltages across resistors R_1 and R_2. The voltage at the junction of these two resistors (Point *A*) is therefore OV (with respect to ground). This voltage is the AFC correction voltage. It is significant that the discharging action of the capacitors is sufficient to reverse-bias the diodes between synchronizing pulses. This action improves the noise immunity of the AFC system. Also, it should be noted that no potential will be formed at Point *A* if either the synchronizing pulses or the reference voltage is absent.

The action of the AFC detector circuit when the oscillator is operating at a frequency less than that of the applied synchronizing pulse is shown in Figure 22.13B. A decrease in the operating frequency of the oscillator is represented by a change in the relative position of the reference voltage wave form during the application of the synchronizing pulses. The pulses now gate on the diodes during the positive portion of the retrace slope, causing diode D_1 to conduct more strongly than diode D_2. The charge on C_1 therefore becomes more positive, and the charge on C_2 becomes less negative. The resulting unbalance in the current flow through R_1 and R_2 (when capacitors C_1 and C_2 discharge) causes the potential at Point *A* to become positive. This positive voltage causes an increase in the oscillator frequency, compensating for the discrepancy between the oscillator frequency and the horizontal synchronizing pulses. A similar but opposite action occurs when the oscillator frequency is higher than that of the ap-

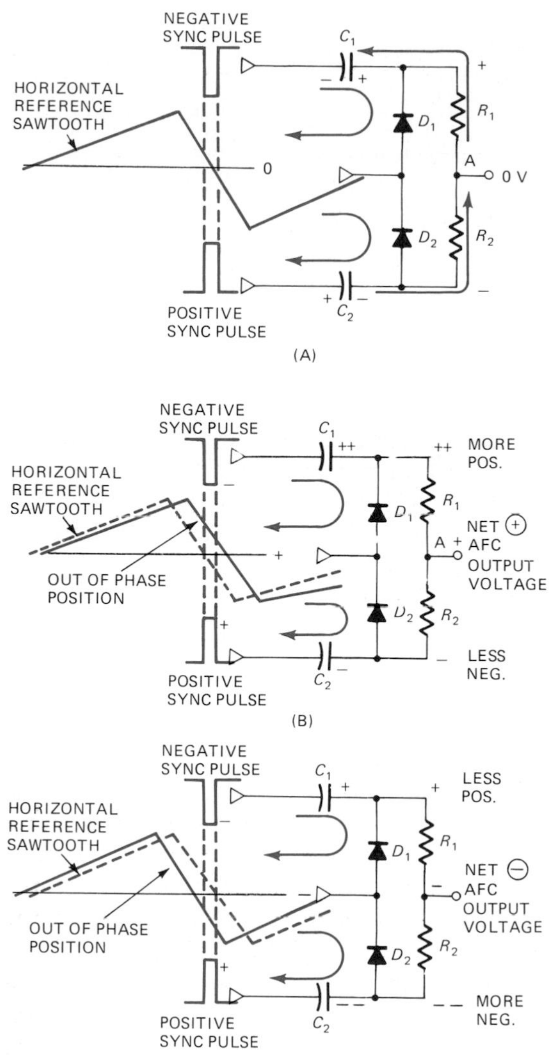

Figure 22.13A-C. The operation of the solid-state horizontal AFC system. **A.** The action of the horizontal AFC phase detector when the oscillator is in sync. **B.** The action of the horizontal AFC phase detector when the oscillator frequency is low. **C.** The action of the horizontal AFC phase detector when the oscillator frequency is high. *(Courtesy of RCA Consumer Electronics.)*

plied pulses (Figure 22.13C). In this case, the synchronizing pulses will gate on the diodes during the negative portion of the retrace slope, causing diode D_2 to conduct more than D_1. The resulting C_1–C_2 capacitor discharging action through R_1 and R_2 will cause the potential at Point A to become negative. This negative voltage will soon cause an appropriate decrease in the oscillator frequency.

AFC Limiter and Filter Circuitry The DC voltage from the output of the AFC discriminator circuit is applied to a limiting and filtering network, illustrated in Figure 22.14. The limiting network consists of two silicon diodes, CR_{508} and CR_{507}, in a front-to-back configuration. The natural silicon-diode junction barrier potential will result in diode conduction when a minimum of 0.5 V is applied across each diode. Therefore, diodes CR_{508} and CR_{507} will conduct when the AFC voltage reaches a value of + 0.5 V or − 0.5 V. This action (diode conduction) will prevent the AFC voltage applied to the oscillator from swinging more positive or negative than ± 0.5 V, thus establishing the proper range of AFC control over the oscillator frequency (hold-in range).

The AFC filter network, also illustrated in Figure 22.14, consists of R_{511}, C_{509}, C_{507}, and the distributed impedance of the detector output circuitry. The purpose of the filter is to prevent the AFC voltage from varying at undesirable frequencies (antihunt action). These undesirable frequencies include 59.94-Hz variations, which would cause horizontal bending in the picture, and 15 734.26-Hz variations, which would create raster "edge ripple."

Reference Voltage Wave Form Figure 22.15 illustrates the circuitry that forms the AFC reference voltage wave shape. A negative-going horizontal pulse is tapped from the horizontal output transformer, delayed slightly by inductor L_{501}, and applied to the pulse rectifier diode CR_{503}. An RC network consisting of capacitor C_{505} and resistor R_{508} converts the pulses to sawtooth wave shapes. During scan time, capacitor C_{505} charges through R_{508}, creating the negative

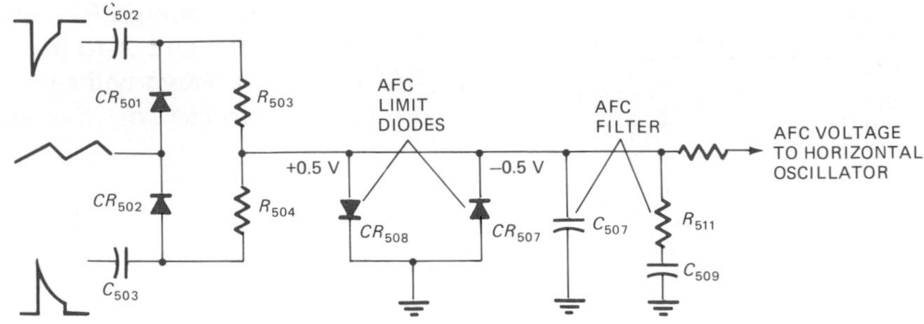

Figure 22.14 AFC limiting and filter network. *(Courtesy of RCA Consumer Electronics.)*

(trace) slope of the sawtooth. The appearance of the negative-going horizontal pulse causes CR_{503} to conduct, resulting in the quick discharging of C_{505}. This action creates the positive-going (retrace) slope of the sawtooth. Coupling capacitor C_{504} removes the DC component from the sawtooth. The resulting wave form is developed across R_{507} and is coupled to the AFC discriminator diodes CR_{501} and CR_{502}.

Horizontal Oscillator The basic schematic of the RCA solid-state horizontal oscillator circuit is illustrated in Figure 22.16. A blocking oscillator uses transformer T_{501} to maintain the regeneration and to couple the output pulses from the oscillator collector to the horizontal output circuitry.

Basically, the oscillator functions as follows. Collector pulses coupled into the base circuit by T_{501} cause the oscillator transistor to be driven into cutoff. While the transistor is cut off, C_{510} discharges through the hold control and associated circuitry to a slightly positive turn-on potential of the transistor. This action is illustrated in Figure 22.16 by the time interval from *A* to *B* of the basic oscillator wave shape. The resulting pulse of current appears in the collector circuit and is coupled back into the base circuit through T_{501}, driving the oscillator to saturation.

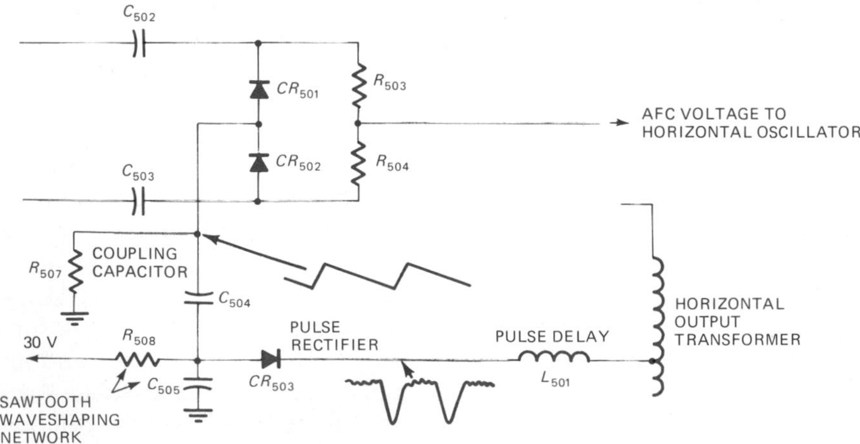

Figure 22.15 Reference sawtooth generation. *(Courtesy of RCA Consumer Electronics.)*

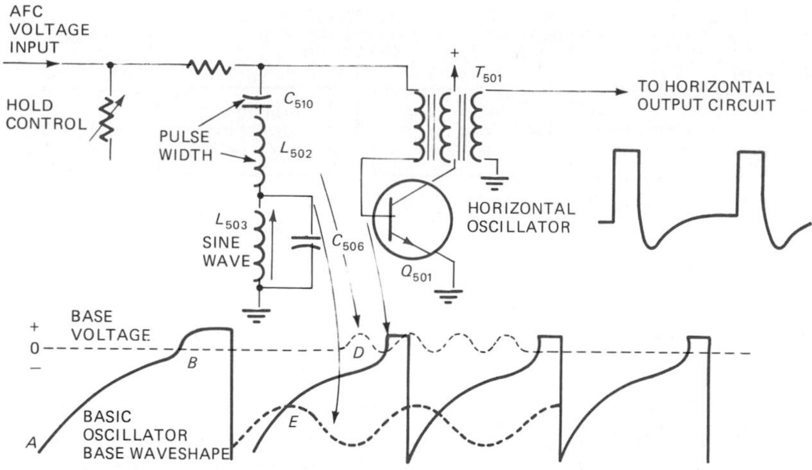

Figure 22.16 Basic solid-state horizontal oscillator. *(Courtesy of RCA Consumer Electronics.)*

The Hold Control The hold control varies the discharging time of C_{510} and thus determines the instant at which the transistor will reach its turn-on voltage (Point *B*). The hold control, therefore, varies the frequency of oscillator operation. The AFC voltage is effectively added to the charge of C_{510} and thereby alters the instantaneous oscillator-base bias developed by the discharging action of this capacitor. In this manner, the AFC voltage dictates within a relatively narrow range the exact time the C_{510} discharging voltage reaches transistor turn-on (Point *B*). As a result, the AFC voltage provides a means of controlling the oscillator frequency.

The Resonant Circuits In order to provide the necessary oscillator output pulse width and improve the stability of operation, two resonant circuits are used in the oscillator circuit. The pulse width is determined by the series-resonant circuit consisting of C_{510} and L_{502}. The oscillatory action of this circuit effectively adds the Wave form D to the basic oscillator wave shape. The period of the C_{510}–L_{502} circuit oscillation is approximately twice the desired pulse width. In this manner, the negative-going portion of one half cycle of this oscillation will cause the transistor bias to instantaneously swing below the cutoff value. The resulting blocking oscillator regenerative action will then complete the cutoff pulse.

Coil L_{503} and capacitor C_{506} constitute a parallel resonant circuit commonly called the sine wave circuit. The action of this circuit effectively adds a sine wave (Wave form E) to the basic oscillator wave shape. The frequency of the sine wave oscillation is slightly higher than the operating frequency of the horizontal oscillator. The addition causes the potential at the oscillator base to quickly pass through the transistor turn-on voltage value. In this manner, the oscillator operating frequency is made relatively immune from noise pulses and relatively independent of small changes in the supply voltage.

22.6 REACTANCE-CONTROLLED SINUSOIDAL OSCILLATORS

In Figure 22.10, the frequency of the sinusoidal horizontal oscillator was stabilized directly by the AFC correction voltage. An-

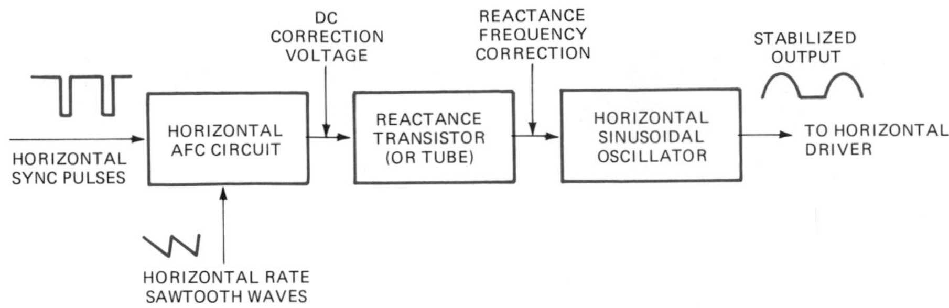

Figure 22.17 Block diagram of stabilization by a reactance transistor (or tube).

other stabilization method involves the use of a "reactance" transistor (or tube). This stage is placed between the AFC circuit and the oscillator.

Simplified Block Diagram The simplified block diagram of Figure 22.17 shows the position of the reactance transistor. Here, the correction voltage from the AFC circuit is fed to the reactance transistor. The reactance transistor circuit in turn affects the tuning and thus the stability of the sinusoidal oscillator.

Furthermore, changes in the reactance transistor's base bias (and collector current) can increase or decrease the amount of reactance. This in turn can effectively control the oscillator frequency. The changes in reactance transistor base bias are furnished by the AFC circuit to stabilize the Hartley oscillator.

The type of oscillator used (emitter-coupled Hartley) is extremely stable. Thus, base-to-emitters bias changes (as might be used) would not be adequate to stabilize its frequency. This explains the need for the reactance transistor. Because of the excellent stability of this system, no user-operated horizontal hold control is furnished.

TYPES OF REACTANCE TRANSISTORS

Two general types of reactance transistors are in use: (1) one that actually appears to be an inductive reactance, and (2) one that acts as a variable resistance in series with a capacitor and thus affects the effective capacitive reactance. Both are operated by a DC control voltage from an AFC system, and both stabilize the tuning of the horizontal oscillator.

THE INDUCTIVE-REACTANCE TRANSISTOR

A transistor (or tube) can be connected so that its collector (or plate) current *lags* its collector (or plate) voltage by 90°. The device then simulates an inductance.[1] If such a device is connected across an oscillator-tuned circuit, it can affect the oscillator frequency. Furthermore, the magnitude of the simulated inductance can be changed by varying the collector (or plate) current. Thus, the oscillator frequency can be controlled (stabilized) by controlling the base (or grid) bias. This control is provided by the horizontal AFC circuit.

A simplified schematic diagram of an inductive-reactance transistor and horizontal oscillator is shown in Figure 22.18A. A vector diagram illustrating the operation of the inductive-reactance transistor is given in Figure 22.18B.

[1]Similarly, the collector (or plate) current can be made to *lead* the collector (or plate) voltage by 90°. Now the device simulates a capacitor.

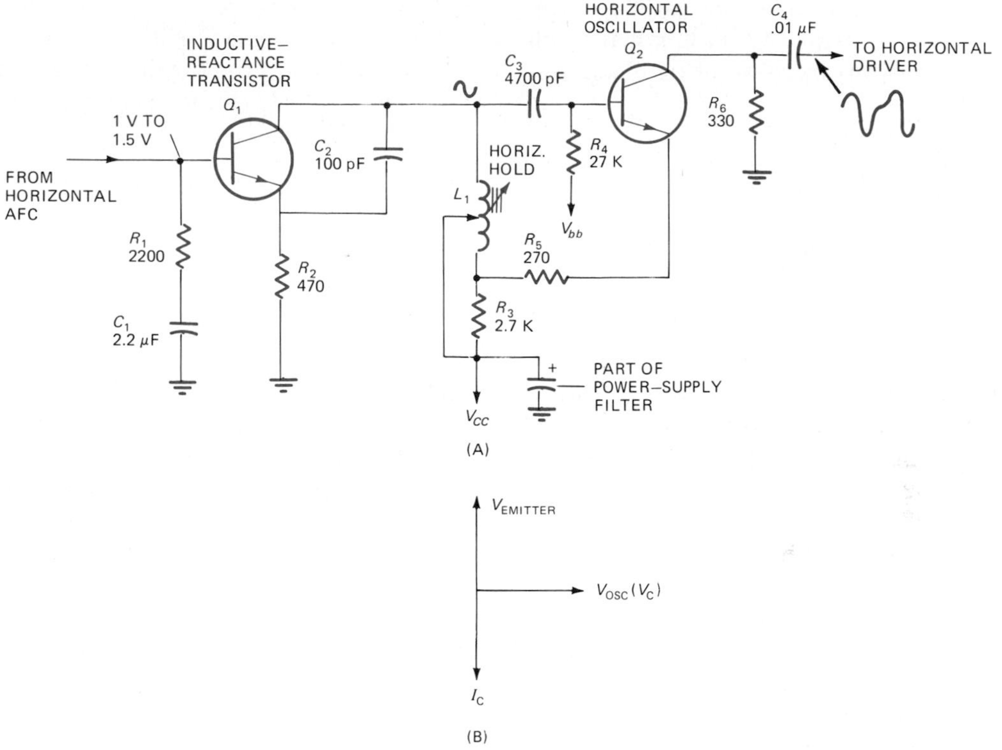

Figure 22.18A-B. Simplified diagrams showing the operation of an inductive-reactance transistor. **A.** The schematic diagram. **B.** The vector diagram (see text).

Note in Figure 22.18A that the base voltage comes from the horizontal AFC circuit. It has a control range from 1 V to 1.5 V. A phase-shift network composed of C_2 and R_2 is connected to the collector and emitter of Q_1.

Circuit Operation Because of the value of C_2, its reactance at the horizontal frequency is much greater than the resistance of R_2. Consequently, the phase-shift network is *capacitive*. (Refer also to the vector diagram of Figure 22.18B.)

The AC collector voltage (V_c) is the oscillator voltage at the horizontal frequency. This is shown as the reference vector (V_c) in Figure 22.18B. Since the phase-shift network is capacitive, the current through it leads the voltage by 90°. Thus, the voltage across R_2 ($V_{emitter}$) leads V_c by 90°. The emitter voltage and collector current (I_c) are normally 180° out of phase. In the vector diagram, we see that I_c lags V_c by 90°. Thus, Q_1 appears to be an inductor whose magnitude is controlled by the base voltage from the AFC circuit.

If the horizontal oscillator frequency should deviate from the desired value, this would be sensed by the AFC circuit. The AFC circuit would change its output DC voltage, which in turn would, through the change of inductive reactance of Q_1, correct the oscillator frequency. Note that a more positive Q_1 base voltage results in an increase of the "inductance" of Q_1 and vice versa.

The Q_1 inductance is in parallel with oscillator coil L_1. Thus, a more positive Q_1 base voltage places an *increased* inductance in parallel with L_1. The result is a higher total tuning inductance and a lower oscillator frequency. A less positive Q_1 base voltage will therefore cause an increase of oscillator frequency.

The Oscillator The horizontal oscillator in Figure 22.18A is an emitter-coupled Hartley oscillator. This oscillator is very stable and requires the use of a reactance transistor to control its frequency. In other types of oscillators, the frequency may be controlled by varying the base bias of the oscillator transistor, but this scheme is not practical for the oscillator in Figure 22.18B.

THE VARIABLE-RESISTANCE TRANSISTOR

The principle of the variable-resistance (reactance) transistor is shown in Figure 22.19A. The complete schematic diagram is shown in Figure 22.19B, which includes the horizontal oscillator and driver stages.

Variable Capacitive Reactance Referring to Figure 22.19A, we see that the horizontal oscillator tuned circuit is shunted by a capacitive circuit. This consists of capacitor C_{10} in series with the effective emitter-to-collector resistance of Q_1. This resistance varies in direct proportion to the collector current, which is affected by the AFC base bias. Since the resistance is in series with C_{10} it controls its *effective* capacitive reactance across the oscillator tuned circuit. Thus, it can change the oscillator frequency.

Controlling Oscillator Frequency The higher the transistor resistance, the less the effective capacity of C_{10} and thus the *higher* the oscillator frequency. This situation occurs if the AFC drives the base of Q_1 *less positive.*

If the base of Q_1 becomes *more positive,* this will increase collector current. Now the Q resistance *decreases,* which *increases* the effective capacitance of C_{10}. The horizontal oscillator frequency will *decrease.* In this manner, the AFC, through the reactance transistor, stabilizes the oscillator frequency.

THE COMPLETE SCHEMATIC DIAGRAM

A complete schematic diagram is shown in Figure 22.19B and wave forms in Figure 22.19C. The tuning capacitor C_{10} is shown going from the collector of Q_1 to the base of the oscillator tuned circuit (L_1–C_{13}). Its operation is described above.

The oscillator is a very stable, emitter-coupled Hartley type, similar to the one in Figure 22.10. The oscillator output is fed to the base of the driver Q_3 at an amplitude of 2.2 V peak-to-peak (Wave form 3). The driver output is increased to 20 V peak-to-peak and is basically a square wave (Wave form 4). The driver output is then transformer-coupled to the horizontal output stage.

22.7 DIGITAL HORIZONTAL COUNTDOWN

The horizontal, digital countdown system discussed here is part of the same IC described for vertical countdown (Chapter 21). In the present discussion, only those circuits peculiar to horizontal countdown will be discussed.

Simplified Block Diagram A simplified block diagram of the IC, showing only the circuits involved with the horizontal countdown, is shown in Figure 22.20. The VCO and phase-detector operation are discussed in Chapter 21. Refer to that chapter for review.

An output from the VCO at 31 468 Hz is fed to a stage called the duty cycle. This stage provides the wave form required to

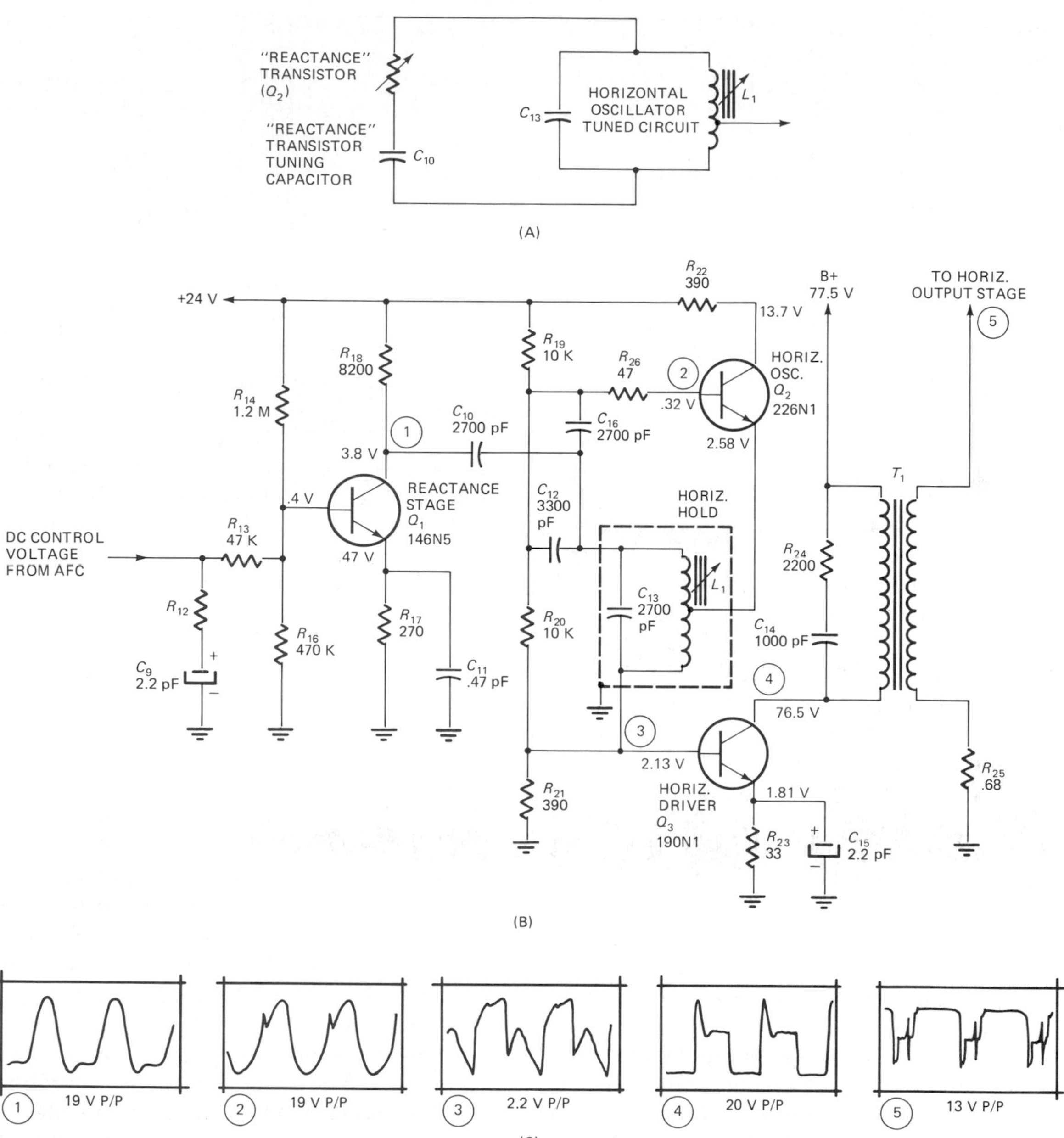

Figure 22.19A-C. The variable resistance "reactance" transistor that controls the frequency of the horizontal oscillator. **A.** Simplified diagram; **B.** complete schematic diagram; **C.** waveforms. Numbers on waveforms correspond to those on the schematic diagram (B). *(Courtesy of Magnavox Consumer Electronics.)*

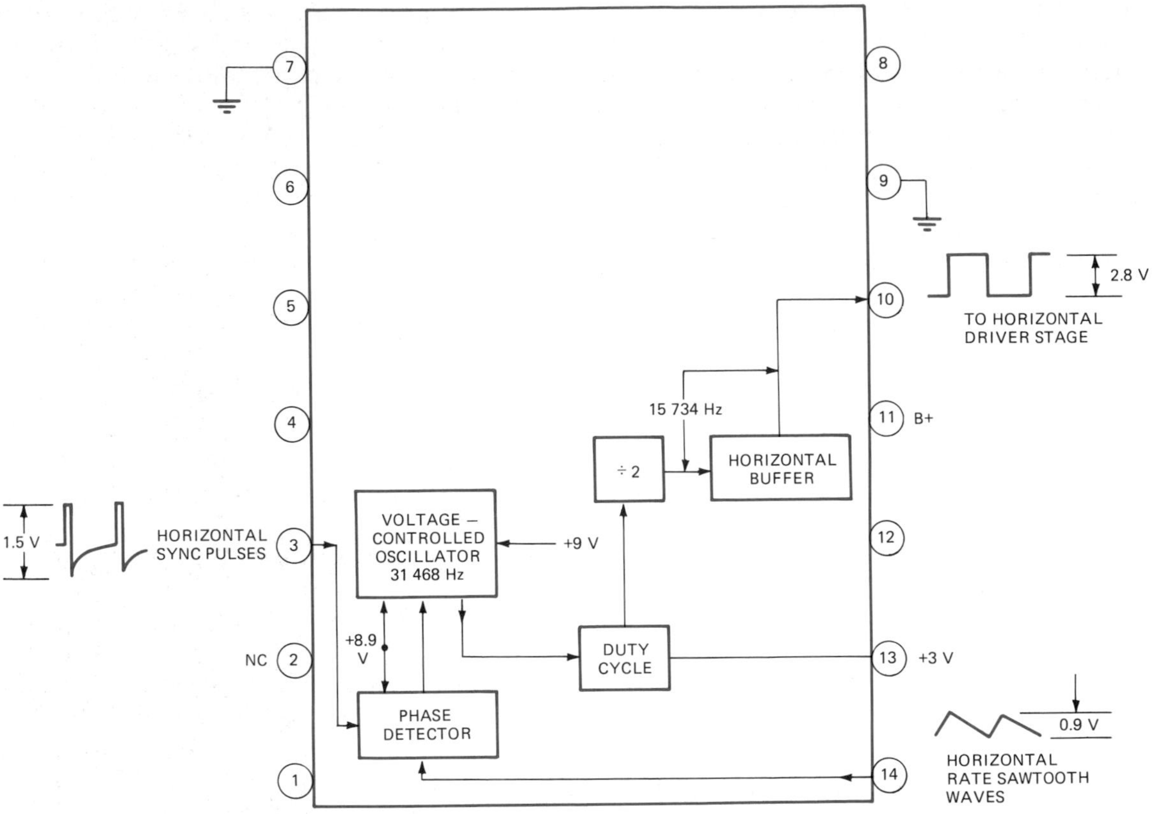

Figure 22.20 Simplified block diagram of the RCA horizontal-countdown IC. (See Figure 21.26.)

drive the following divide-by-two stage. The output of the divide-by-two stage drives the horizontal buffer amplifier at a frequency of 15 734 Hz. The output of the buffer is a square wave with a peak-to-peak amplitude of 2.8 V.

Although not shown in Figure 22.20, the output of the horizontal buffer feeds a horizontal driver, which in turn drives the horizontal output transistor.

22.8 TROUBLES IN HORIZONTAL OSCILLATOR AND AFC SYSTEMS

Trouble in horizontal oscillator and AFC systems can be generally grouped into three categories: (1) no oscillation; (2) incorrect frequency, synchronization, or phasing problems; and (3) width problems. Note that problems with this system may also affect high-voltage generation, which depends upon correct horizontal oscillator operation as well as on correct horizontal output circuit operation. You should remember that trouble in the horizontal AFC system will frequently cause changes in the oscillator frequency and may also produce improper synchronizing and phasing but will rarely cause the oscillator to stop operating.

NO OSCILLATION

If the horizontal oscillator is inoperative (or if any stage following the oscillator is completely inoperative), the screen will be blank. That is to say, no raster or illumination will appear. This results because the high-voltage supply operation depends upon the operation of the entire horizontal-deflection system. In this case, no high voltage is being produced to operate the picture tube. In this situation, it is not immediately apparent which stage of the horizontal-deflection system is at fault and isolation procedures must be instituted (Chapter 27).

FREQUENCY, SYNCHRONIZATION, AND PHASING PROBLEMS

The problem of incorrect horizontal frequency may be caused by defects in either the AFC system or the oscillator system. While the oscillator is primarily responsible for generating the horizontal frequency, this frequency is also controlled to a degree by a DC control voltage from the AFC system. However, before assuming that there is a defect in the system, you should check to see if the horizontal oscillator adjustments and controls are correctly set. Incorrect settings may be the cause of the incorrect horizontal frequency. Figure 22.21 illustrates what appears on the screen when the horizontal frequency is synchronized but incorrect. To isolate between the AFC and the oscillator circuits in a case of incorrect horizontal frequency, the AFC system should be disconnected from the oscillator and the oscillator should then be adjusted to see if a single picture can be momentarily obtained. If it can, then the fault is most likely in the AFC system.

Since the AFC system is responsible for synchronizing the oscillator to the horizontal synchronizing pulses, it is suspect whenever there is improper synchronization or phasing of the horizontal oscillator. In some horizontal systems, misadjustment or a defect

Figure 22.21 This picture is synchronized, but is displaying incorrect horizontal-sweep frequency.

in the stabilization coil circuit may cause incorrect horizontal oscillator phasing. If the phasing is incorrect, the picture will be shifted horizontally, as shown in Figure 22.22. If there is no horizontal synchronization, the symptom may be multiple pictures drifting in the horizontal direction. Remember that a loss of horizontal synchronization may also be caused by a lack of horizontal synchronizing pulse input to the AFC system, and therefore this should be checked first.

WIDTH PROBLEMS

Width problems are generally ones of insufficient width, although excessive width may also occur. Although insufficient width may be caused by defects in the horizontal oscillator system, this system is not likely to be responsible for a condition of excessive width. This most likely originates in the horizontal output system. An exception to this may be an improper setting of a horizontal *drive* control, and its setting should be checked first. Insufficient width may also be caused by an improper setting of the horizontal width control.

In the oscillator system, insufficient width may be caused by too low an amplitude of the sawtooth wave. This may be the fault of the oscillator circuit or, in the case of transistor systems, may be caused by a defective intermediate driver stage. Remember that frequently the cause of insufficient width may be a faulty horizontal output system rather than a faulty oscillator system. This must be determined before proceeding with troubleshooting. However, here again, the setting of any controls affecting the picture width should be checked before assuming that there is an actual circuit defect. Insufficient width may be caused not only by defective tubes, transistors, or other parts but also by insufficient supply voltage.

Figure 22.22 In this picture, the horizontal phasing is incorrect.

DIGITAL COUNTDOWN SYSTEM PROBLEMS

The VCO, at 31 468 Hz, as well as all associated horizontal countdown circuits, are part of a monolithic IC. (As discussed in Chapter 21, this same IC also contains the digital vertical countdown circuits.) If there is a fault in the IC (Figure 22.20), the entire unit will have to be replaced. The VCO is free-running. Even if no horizontal synchronizing pulses are present at Terminal 3 of the IC, there should be a raster and an unsynchronized picture.

If the IC is defective and no output exists at Terminal 10, there will be no high voltage and thus no raster or picture. However, the same symptoms will be present if the fault lies beyond the IC. To isolate the fault, the lead at Terminal 10 can be disconnected and a 15 734-Hz (approximately) square wave at 2.8 V peak to peak fed into the lead. If the fault is in the IC, there should now be a raster and an unsynchronized picture.

SUMMARY OF CHAPTER HIGHLIGHTS

1. The same basic types of deflection oscillators are used in the horizontal- and vertical-deflection systems.
2. As with vertical-deflection systems, digital horizontal countdown systems are also used. (Figure 22.20)
3. Unlike vertical systems, sinusoidal (Hartley) oscillators are an additional type used for horizontal systems. The other types are blocking oscillators and multivibrators. (Figures 22.4, 22.10, and 22.16)
4. Most television stations now use 59.94 Hz and 15 734.26 Hz for both monochrome and color TV transmissions.
5. The stability problems for horizontal oscillators are much greater than for vertical oscillators. (Figures 22.8, 22.9, and 22.16)
6. Horizontal oscillators are more subject to noise-caused instability than vertical oscillators.
7. All horizontal oscillators are stabilized by an AFC system. (Figures 22.1, 22.5, 22.6, 22.9, 22.11, 22.12, and 22.17)
8. The horizontal AFC ensures that the horizontal phase and frequency are correct and stable. It also ensures appreciable immunity from the effects of noise pulses.
9. Two types of wave forms are applied to horizontal AFC systems: (1) horizontal synchronizing pulses and (2) a sawtooth at the horizontal rate. (Figures 22.3, 22.5, 22.6, 22.9, and 22.11 through 22.15)
10. The horizontal AFC system output is a DC correction voltage. This voltage is then used either directly or indirectly to stabilize the horizontal oscillator frequency. (Figures 22.3, 22.6, 22.9, 22.10, and 22.13)
11. Noise pulses are greatly reduced by a vertical integrator but are freely passed by a horizontal differentiator.
12. In one type of diode phase detector, the diodes receive horizontal synchronizing pulses in phase opposition and sawtooth waves of the same polarity. (Figure 22.3)
13. In Number 12, a comparison of these two waves can result in a DC correction voltage, which will reset the oscillator frequency.

14. Another type of diode phase detector requires horizontal synchronizing pulses of only one polarity. (Figure 22.5)
15. AFC circuits generally do not have adjustable controls. However, their associated oscillators will have an adjustable frequency (hold) control. (Figures 22.3, 22.5, 22.10, 22.16, 22.18, and 22.19)
16. Resonant-stabilizing circuits are used in multivibrators and blocking oscillators. (Figures 22.4, 22.5, 22.8, and 22.16)
17. Solid-state AFC circuits are similar to their vacuum tube counterparts. (Figures 22.5 and 22.9)
18. An emitter-coupled Hartley oscillator is frequently used as the horizontal oscillator in solid-state television receivers. This oscillator is very stable. (Figures 22.10, 22.18, and 22.19)
19. A Hartley oscillator may be stabilized with a reactance transistor. (Figures 22.17 to 22.19)
20. A reactance transistor may simulate a variable inductance. It may also be used as a variable resistor in series with a capacitor.
21. In Question 19, the reactance transistor is located between the AFC system and the horizontal oscillator.
22. The horizontal digital countdown system is based upon a stabilized oscillator operating at 31 468 Hz. (Figure 22.20)
23. The oscillator frequency of a horizontal digital countdown system is divided by 2 (15 734 Hz). It is then shaped and sent to a horizontal driver as a 15 734-Hz square wave.
24. Horizontal AFC oscillator troubles can be classified as (1) no oscillation; (2) incorrect synchronizing frequency, or phasing; and (3) width problems.
25. The horizontal digital countdown system is contained in a single (monolithic) IC. If there is a fault in the IC it must be replaced.

EXAMINATION QUESTIONS

(Answers are provided at the back of the text.)

Part A Supply the missing word(s) or number(s).

1. Three types of deflection oscillators are ________, ________, and ________.
2. A monolithic ________ contains all the circuits for a digital countdown system.
3. A differentiator will pass interfering ______ pulses.
4. A DC ________ is the output of a horizontal AFC system.
5. A deflection oscillator can be triggered by ________ pulses.
6. A horizontal-deflection oscillator can be stabilized by a(n) ________ system.
7. An AFC system receives ________ waves and ________ waves.
8. In an AFC system, the sawtooth waves are developed from pulses coming from the ________.
9. The addition of a ________ coil to a deflection oscillator improves its ________ immunity.
10. A ________ oscillator is frequently used as a sinusoidal deflection oscillator.
11. A horizontal AFC system holds the oscillator to the exact ________ and ________ of the horizontal synchronizing pulses.
12. One common type of horizontal hold control varies the ________ of the horizontal oscillator tuned circuit.

13. A reactance transistor may act as a variable ________ or a variable ________.
14. A reactance transistor is positioned between the ________ system and the ________.
15. Varying the ________ can change the effective "reactance" of an inductive-reactance transistor.
16. In an inductive-reactance transistor, the ________ current lags the ________ voltage by ________ degrees.
17. In the horizontal digital countdown system, the ________ is operated at 31 468 Hz.
18. If the ________ oscillator is dead, the picture tube high voltage will be ________.

Part B Answer true (T) or false (F).

1. Basically different types of deflection oscillators are required for color and monochrome sets.
2. Stability problems for horizontal oscillators are more severe than for vertical oscillators.
3. If severe noise pulses reach the horizontal oscillator, considerable oscillator instability can result.
4. To stabilize the horizontal oscillator, the AFC output voltage must remain completely stable.
5. The horizontal differentiator time constant is effective in filtering out noise pulses.
6. A long-time constant filter is used with horizontal AFC systems.
7. Single- or dual-polarity synchronizing pulses may be used in horizontal AFC systems.
8. Operation of the horizontal AFC system depends upon the coincidence of sawtooth and horizontal synchronizing pulse waves.
9. Most horizontal hold controls are *RC* circuits.
10. In an inductive-reactance transistor, the phase-shift network is capacitive.
11. In a variable-resistance reactance transistor, the phase-shift network is inductive.
12. The emitter-coupled Hartley oscillator is very stable.
13. In the digital horizontal countdown system, the duty cycle stage provides the wave form required to drive the divide-by-two stage.
14. With incorrect horizontal oscillator phasing, there will be multiple pictures in the horizontal direction.
15. In the digital horizontal countdown system, if there are no input horizontal synchronizing pulses, there will be no raster.

REVIEW ESSAY QUESTIONS

1. Draw and completely label a simple block diagram of a reactance tube horizontal AFC oscillator system. Indicate inputs and final output.
2. In Question 1, describe briefly the operation of all stages.
3. Explain the basic difference(s) between a horizontal AFC oscillator system and a horizontal digital countdown system.
4. In Question 3, in the digital system, what corresponds to an AFC system?
5. In Figures 22.5 to 22.7, explain (in your own words) the operation of the AFC circuit.
6. In Figure 22.3, explain the purpose and operation of R_6 and C_4.
7. In Figure 22.5, if you reversed diodes D_1 and D_2, what change might you have to make to the input synchronizing pulses? Why?
8. With the aid of Figures 22.4 and 22.8, explain the purpose and operation of the resonant circuit.
9. In your own words and referring to Figure 22.16, discuss the function of L_{503} and C_{506}.
10. Briefly (in your own words), explain the operation of Q_1 in Figure 22.18.
11. In Figure 22.19, what is the purpose of capacitor C_{10}?
12. List the advantage(s) of a horizontal digital countdown system over a horizontal AFC oscillator system.

13. Briefly explain how noise pulses might affect the operation of a horizontal oscillator.
14. Why is a vertical-deflection oscillator less susceptible to noise pulses than a horizontal-deflection oscillator?
15. Name the four basic types of horizontal-deflection oscillators.
16. In Question 15, which type do you feel might be the least susceptible to noise pulses? Why?

Chapter

23

23.1 Horizontal output circuit block diagram
23.2 Vacuum tube monochrome TV circuits
23.3 Flyback high voltage
23.4 Solid-state monochrome TV circuits
23.5 Transistor horizontal output circuits
23.6 Solid-state dampers
23.7 NPN and PNP horizontal-output stages
23.8 Adjustments in monochrome TV receiver circuits
23.9 Vacuum tube color TV horizontal-output circuits
23.10 Solid-state color TV horizontal-output circuits
23.11 Pincushion (PIN) distortion
23.12 Silicon-controlled rectifier (SCR) horizontal-output circuits
23.13 Horizontal-output transformers and special assemblies
23.14 Deflection yokes
23.15 Horizontal high-voltage components
23.16 Adjustments in color TV horizontal systems
23.17 X-ray emission
23.18 X-ray protection circuits
23.19 Troubles in horizontal-deflection circuits
23.20 High-voltage troubles
23.21 Relationship between horizontal and high-voltage problems
23.22 Corona and arcing problems
Summary

HORIZONTAL OUTPUT DEFLECTION CIRCUITS AND HIGH VOLTAGE

INTRODUCTION

In Chapter 22 we discussed various types of horizontal-deflection oscillator circuits and related AFC circuits. These circuits produce an output wave form at the correct horizontal phase and frequency. However, in order to deflect the electron beam(s) of the picture tube across the screen, it is necessary to drive a sawtooth current through the horizontal yoke coils. These coils are situated around the neck of the picture tube. Not only must the current wave form have the desired shape (sawtooth), but the current amplitude must be adequate.

The deflection oscillator circuit output mentioned above does not directly provide the scanning power required. Rather, its output is used as a *driving* wave to operate the horizontal output deflection circuits.

Another important function of the horizontal output deflection circuits is to provide the high voltage(s) necessary to operate the picture tube. For color picture tubes, an anode voltage as high as 32 kV as well as a focus voltage on the order of 3 to 5 kV is required. In addition, circuits to protect color television viewers from the effects of x-rays are required by federal law.

When you have completed the reading and work assignments for Chapter 23, you should be able to:

- Define the following terms: B+ boost voltage, flyback high voltage, pincushion distortion, S capacitor, deflection yoke, saddle yoke, toroidal yoke, x-ray, corona, arcing, SCR, damper, overcurrent, overvoltage, picture foldover, trapezoidal wave, and ringing frequency.

- List five major functions of horizontal output circuits.
- Describe the operation of transistor, SCR, and vacuum tube horizontal output circuits.
- Explain how anode high voltage is developed.
- Compare saddle and toroidal deflection yokes.
- Explain how B+ boost voltage is developed.
- Describe overcurrent and overvoltage circuit operation.
- Explain (1) the reason for pincushion distortion of the raster and (2) how this problem is eliminated.
- Understand the function and operation of drive, width, and linearity control.
- Explain the operation of a linearity circuit.
- Discuss the problem of x-ray emission by a color receiver and describe the various x-ray protection circuits.
- Understand some common troubles in horizontal output circuits and how to isolate them.

23.1 HORIZONTAL OUTPUT CIRCUIT BLOCK DIAGRAM

The operation of the horizontal-deflection and high-voltage circuits can best be understood by referring to the block diagram shown in Figure 23.1. As this chapter develops, we shall see that this block diagram is generally accurate for most vacuum tube, transistor, and silicon-controlled rectifier (SCR) horizontal-deflection systems.

FREQUENCIES IN THE HORIZONTAL SYSTEM

The horizontal output amplifier must be driven by the correct voltage or current wave form in order to produce an accurate sawtooth current in the deflection coils. These wave forms are first generated in the horizontal oscillator. However, the voltage or current wave form in the horizontal oscillator may have a shape quite different from the yoke current. The reasons for this apparent paradox will become clear as we progress through the chapter.

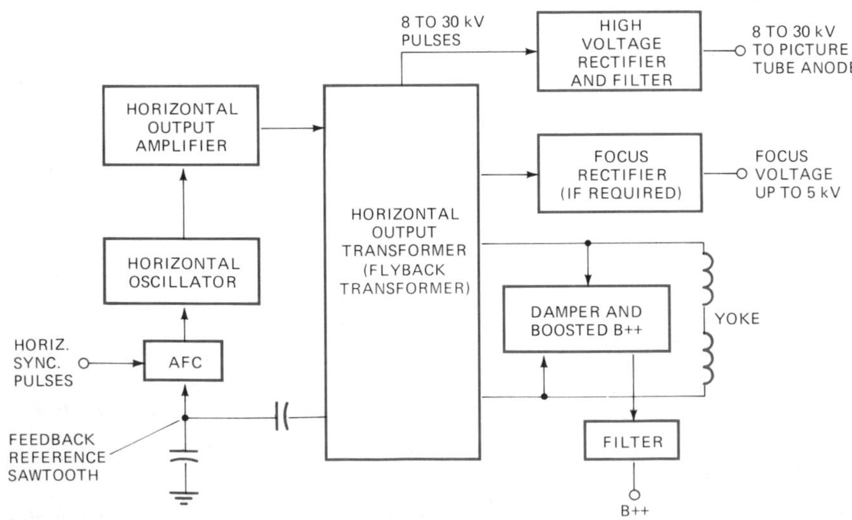

Figure 23.1 Block diagram of a typical horizontal-deflection circuit.

We are all familiar with the audio output amplifier, which accepts low-power signals, amplifies them, and then drives a loudspeaker through an impedance-matching transformer. In many ways, this is analagous to the processes carried out in the horizontal-deflection system. The horizontal oscillator provides relatively low-power signals to the grid (for a vacuum tube), base (for a transistor), or gate (for an SCR) of the horizontal output amplifier. In this output stage, the power level of the signal is amplified and fed to the horizontal output transformer. This transformer matches the impedance of the horizontal output amplifier to (1) the deflection coils, (2) the high-voltage rectifier circuit, and (3) the damper and boosted B+ circuit. Other, lower-level wave forms are also obtained from the transformer for various functions, as mentioned later.

It is difficult to draw much more of an analogy between an audio power amplifier and a horizontal output amplifier than we did in the above paragraph. The frequencies and wave forms involved in each case are quite different. The lowest frequency present in a horizontal-deflection amplifier is the 15 734-Hz scanning frequency. Since the flyback portion of the wave form is in the shape of a pulse, frequencies above 100 kHz are also present. The wave form is actually a mixture of 15 734-Hz, 31 468-Hz, 47 202-Hz, and other harmonics that extend beyond 100 kHz.

DEFLECTION YOKE AND DAMPER CIRCUITS

The horizontal deflection of the electron beam in the picture tube is the primary function of the horizontal output amplifier. This requires a large current through the deflection coils. Thus, the flyback transformer is connected as a voltage step-down, or current step-up, device for the deflection coil circuit. To preserve power, this secondary winding has an impedance close to that of the deflection coils.

As explained more fully later, the brief retrace period produces self-oscillations, and a special damper diode is used to absorb these oscillations. The energy from the first negative peak in these oscillations is passed from the damper diode to a boost filter, where it is stored and used to provide additional B+ + voltage to the horizontal output amplifier. This voltage is usually several hundred volts above the regular B+ voltage. This voltage is also used in the vertical output amplifier in some receivers. If the receiver uses solid-state horizontal and vertical output amplifiers, this additional B+ + voltage is not required.

HIGH-VOLTAGE CIRCUIT

A special high-Q secondary winding on the flyback transformer generates high-voltage, low-current pulses during the flyback period. These pulses are rectified and filtered to produce up to 32 kV DC for a color receiver* or up to 20 kV for a monochrome set. This high voltage is required for the picture tube anode. (See Chapter 18 for a detailed discussion of picture tubes.) In most receivers built before 1970, the high-voltage rectifier was a vacuum tube and a special winding on the flyback transformer was required to develop the heater voltage for this vacuum tube. In more recent sets, silicon solid-state diodes are used for the high-voltage rectifier.

OTHER FUNCTIONS OF THE DEFLECTION CIRCUITS

A number of other functions also take place in the block diagram of Figure 23.1.

---

*In many color receivers, a voltage tripler is used. In this case, the initial high-voltage pulse is about 10 kV.

Foremost among these is the sawtooth wave form that goes from the flyback transformer to the AFC circuit. This sawtooth wave form is compared with the horizontal synchronizing pulses (from the video signal) to create a frequency-correction signal for the horizontal oscillator.

One of the higher voltage taps on the flyback transformer is often used to develop a focus voltage. High-voltage pulses are rectified and filtered to provide up to 5 kV DC in color receivers.

In some television receivers, additional secondary taps on the flyback transformers are used to provide B+ ("derived") voltages to operate many circuits. This is covered in Chapter 19. Other taps on the flyback transformer are used for (1) keyed AGC (Chapter 14), (2) horizontal blanking (Chapter 20), (3) burst amplifier gating (Chapter 9), and (4) chroma amplifier gating (Chapter 9).

In the following sections we shall explore in more detail the operation of each portion of the block diagram. We shall discuss horizontal output and high-voltage circuits for both color and monochrome receivers. In each case, both vacuum tube and solid-state technology will be surveyed.

BEAM SCAN VERSUS YOKE CURRENT

The horizontal position of the beam in the picture tube depends on the magnitude of the current in the horizontal-deflection yoke. Figure 23.2 shows this one-to-one relationship between the beam location and the deflection coil current. Note that if the sawtooth current is negative, the beam is left of screen center. Likewise, a positive current places the beam to the right of screen center. The negative coil current comes from a device called a damper. The positive coil current is obtained from the horizontal output amplifier. In the following sections, we shall discuss the manner in which these two currents mix in the deflection coils (yoke) to form a linear-current sawtooth. When we say *linear,* we are referring principally to the 57.5-μs sweep time. The 6-μs retrace does not need to be linear since the screen is blanked out during that time.

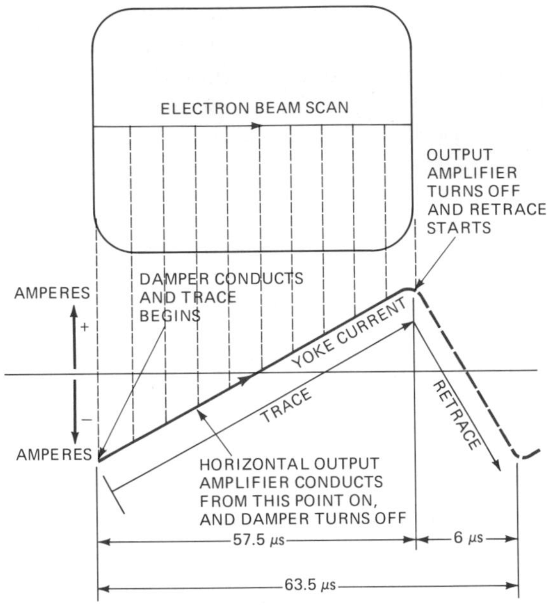

Figure 23.2 Relationship between picture tube beam position and the deflection-yoke current waveform.

23.2 VACUUM TUBE MONOCHROME TV CIRCUITS

In these circuits, a trapezoidal wave form is used to drive the horizontal output amplifier. However, note that here the requirement for a trapezoidal wave form is not the same as for driving a vertical output amplifier, which is covered in Chapter 21. In the present case, a trapezoidal wave form ensures sharp cutoff of the output amplifier, so as not to interfere with horizontal retrace and high-voltage generation. This will be clarified in subsequent sections.

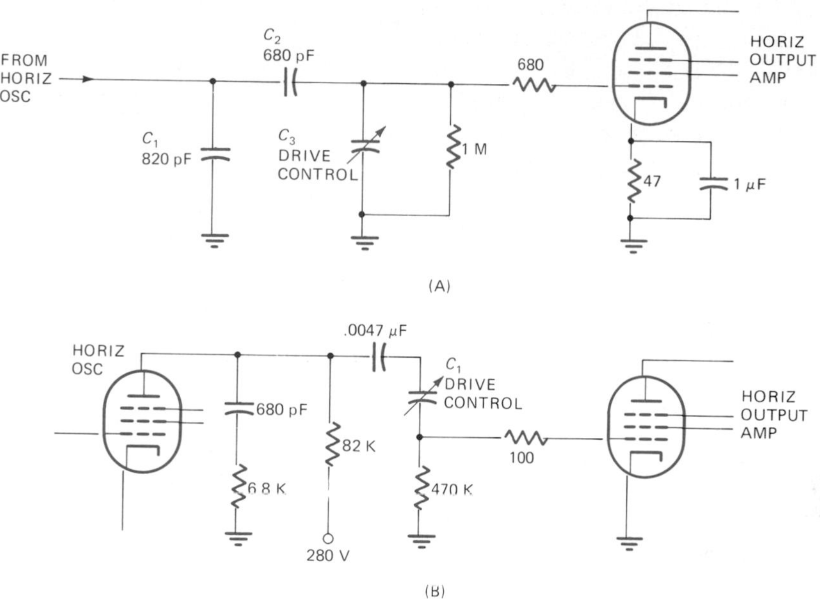

Figure 23.3A-B. Two different methods of controlling the drive voltage to the horizontal-output amplifier.

COUPLING CIRCUITS

Careful attention is paid to the transfer of the complex trapezoidal wave form from the horizontal oscillator to the output amplifier. Time constants are chosen so that no part of the trapezoidal shape is degraded while driving the grid of the horizontal output amplifier. The coupling network usually contains a variable capacitor, resistor, or inductor that controls both the amplitude and the wave shape of the trapezoidal wave form. This control is usually labeled "horizontal drive" or "horizontal wave form." It is often adjustable only with a screwdriver.

Two common coupling circuits between the horizontal oscillator and the horizontal output amplifier are shown in Figure 23.3A and B. In Figure 23.3A, C_2 and C_3 form a voltage-divider network for the voltage developed across C_1. Capacitor C_3 is the drive control, and, as its capacitance is reduced, the impedance rises and more voltage is developed across it and passed on to the output amplifier. Figure 23.3B shows a series-connected drive control. In this case, more capacitance allows a larger voltage transfer to the output stage.

THE HORIZONTAL OUTPUT AMPLIFIER

After passing through the coupling network, the trapezoidal wave form is applied to the grid of the horizontal output amplifier. A typical circuit is shown in Figure 23.4, with the wave forms developed in this circuit shown in Figure 23.5. Voltage E_1 is the voltage at the grid of V_1. The dotted line CO (Figure 23.5) is the grid cutoff voltage level. No plate current flows if E_1 is below the CO line. Current flows through V_1 from time t_2 to time t_3, from time t_4 to time t_5, etc. Thus, no plate current flows for approximately 30% of the time. The resultant plate current

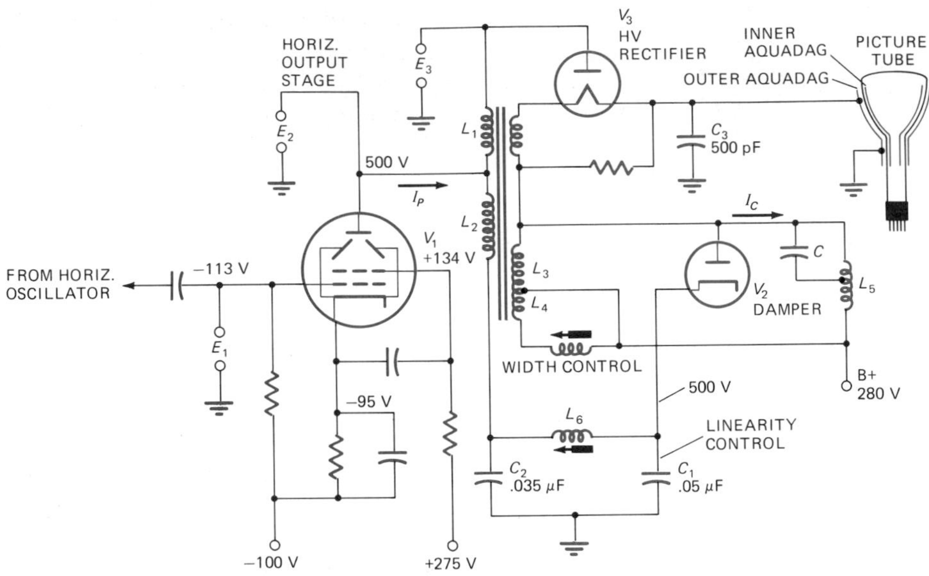

Figure 23.4 A typical vacuum tube horizontal-output amplifier and high-voltage power supply circuit.

wave form is shown in the I_p curve of Figure 23.5.

Carefully note the timing relationship between E_1 and I_p. At time t_1, V_1 is cut off and no plate current flows. Likewise, no current flows through L_2. At time t_2, voltage V_1 starts to conduct, current flows through L_2, and a field builds up around L_2. This expanding field cuts L_3, L_4, and L_1, inducing a voltage in each. At time t_3, the tube is driven sharply into cutoff, plate current drops to zero, and the field built up around L_2 quickly collapses, inducing a high voltage in L_1, L_2, L_3, and L_4. The voltage induced across L_1 and L_2 is of the order of 20 000 V or more peak to peak (monochrome receivers). This sudden collapse of the magnetic field around L_2 does not occur in zero time. It decays in a finite time determined by the time constant of L_2 and all of its loads. Some loads are directly connected, whereas others are reflected through the transformer. The collapsing field shock excites the circuit composed of L_3, L_4, L_5, and C into oscillation. The resonant frequency of this network (70 kHz) is such that the period of one-half cycle of an oscillation is about 7 μs, which is approximately equal to the flyback time of the horizontal sweep. Thus, during the first half cycle of this oscillation, the beam, which is at the extreme right-hand side of the screen, is brought back to the left-hand side. The oscillations have now served their purpose and must be stopped. For this purpose, we require a special damping circuit (see also Figure 23.2).

HORIZONTAL DAMPING

If the oscillations are not stopped after the first half cycle, they continue into the next line and interfere with the proper motion of the beam. An expedient method of damping the oscillations quickly is accomplished by means of V_2. At about time t_3, when V_1 is cut off and the field is collapsing about L_3, the top of L_3 and the plate of V_2

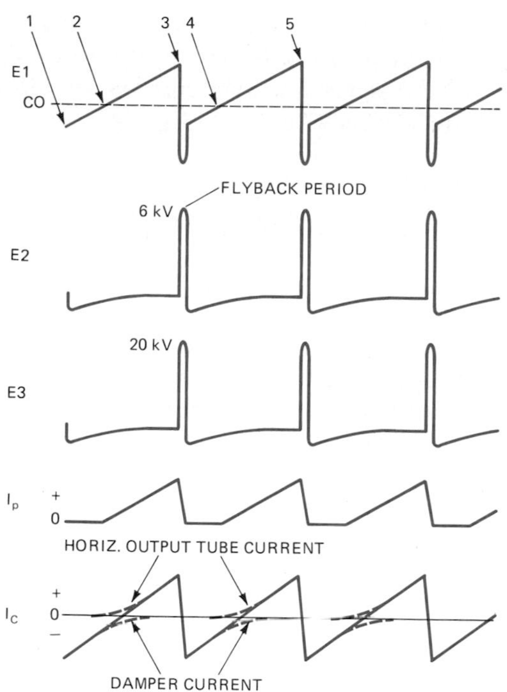

Figure 23.5 Voltage and current waveforms from the horizontal-output circuits shown in Figure 23.4. Note that the curves of the damper current and horizontal-output tube current go in opposite directions and produce a composite linear current sawtooth. (See also Figure 23.2.)

become negative, with the current flowing down through the deflection coil (L_5). When the current in the deflection coil reaches its negative peak, the voltage at the top of the coil begins to go positive because the current lags the voltage by 90° in an inductance. As soon as the voltage on the plate of V_2 becomes positive, the tube conducts and, in so doing, offers a low-resistance path for the deflection current into capacitor C_1, which thus becomes charged. The charging process is quite rapid at first, gradually slowing down as the voltage across C_1 rises. This slowing down is reflected in the gradual dying out of the current, indicated by the bottom dotted line in I_C. However, as the current approaches zero, V_1 comes out of cutoff and a magnetic field again builds up around L_2. This growing field induces a voltage in L_3, causing the current to build up to its maximum peak, time t_5 in I_C. In this way, a linear sawtooth current is generated.

V_1 and V_2 as Switches Both V_1 and V_2 in Figure 23.4 should be regarded as switches. Switch V_1 is turned on when the electron beam is about one-third of the way across the screen while the image is being traced out. It remains on until the beam reaches the far right-hand side of the screen, when the retrace starts. During all this time, V_2 is off. When the retrace starts, V_1 is also turned off and the partial oscillation in the yoke brings the beam back to the left side of the screen. Now, V_2 is turned on and the gradual dying out of energy (through absorption by C_1) brings the beam about one-third of the way across the screen. Switch V_2 now lapses into cutoff, and V_1 is turned on. Switch V_2 can conduct only when its plate voltage is greater than its cathode voltage. This means that the plate voltage must exceed not only the B+ from the power supply, but also the deflection voltage across the yoke coils. This condition occurs only during the retrace interval and for one-third of each scan period. For the rest of the cycle, the plate voltage of V_2 is lower than its cathode voltage and no conduction takes place.

Oscillations Damped As a review, a few more comments concerning the damping process are in order. This process is often misunderstood. Referring back to the E_2 and E_3 wave forms of Figure 23.5, the flyback pulse tends to be followed by several oscillations. These oscillations must not reach the yoke current, or else several bright vertical bars will appear on the left side of the picture tube screen. By placing the damper diode directly across the yoke windings, this problem is overcome. The diode clips off the negative overshoot of the first oscillation. The oscillations are thus damped beginning at this point in time.

BOOST B− (OR B+ +)

Note that the voltage on the plate of V_2 (Figure 23.4) is equal to the 280 V (B+) from the power supply plus the deflection voltage developed across L_3–L_4. The greater the deflection voltage, the greater the voltage applied to the plate of V_2 and the greater the charge received by C_1.

The charge on C_1 produces a voltage across the capacitor in which the top plate is positive and the bottom plate is negative. This polarity stems from the fact that the electrons in V_2 flow from cathode to plate, through L_5 to B+, from here to ground in the power supply, and from ground to C_1. Thus, the voltage across C_1 will be greater than the B+ voltage, and the difference can be as much as several hundred volts. The name given to this augmented voltage is boost B+, boosted B+, or simply B+ +. In Figure 23.4, the B+ + is applied to the plate of V_1 through L_6 and L_2. In addition, B+ + is frequently used to provide higher voltages for the other stages in the receiver besides the horizontal output amplifier. (Usually the vertical output amplifier uses the B+ +.)

Linearity Control Tube V_2 conducts for about 30% of the sweep. While it is conducting, the voltage across C_1 and C_2 builds up. When V_2 ceases to conduct, the voltage across C_1 and C_2 falls. This rise and fall constitutes an AC ripple on the plate of V_1. By shifting the phase of this ripple voltage, it is possible to compensate for some of the nonlinearity of the current wave form in the deflection coil. This change of phase is effected by the variable inductor L_6, which is called a *linearity* control. Capacitors C_1 and C_2 and inductor L_6 form a resonant circuit tuned to 15 734 Hz. The resonance of this circuit is indicated by a dip in the plate current of V_1. We shall discuss linearity controls in more detail in Section 23.8.

"Bootstrap" Operation The B+ + is called "boost source" in some receivers, because it is used to raise the horizontal output amplifier B+, and yet the energy to create the higher voltage comes through the horizontal output amplifier. The horizontal output amplifier appears to be "picking itself up by its own bootstraps." The ultimate boost B+ may be +500 to +700 V. The actual voltage of the B+ + can be modified in many ways. One common way is to move the damper tap on the flyback transformer up or down with respect to the yoke tap. The yoke and damper diode are then not exactly in parallel, as has been assumed in the discussion thus far. Figure 23.6A shows a circuit where the damper tap is above the yoke tap. In this instance, the B+ + is higher than if there were a direct parallel connection between the damper and the yoke.

Horizontal Output Autotransformers Before leaving the subject of dampers and boosted B+, the subject of autotransformers should be addressed. Note in Figure 23.4 that a separate winding (L_3 and L_4) on the flyback transformer is used for the damper and yoke circuit. Inductors L_3 and L_4 could also be connected in series with L_2 to form an autotransformer. Two circuits of this type are shown in Figure 23.6. With this circuit configuration, the damper diode appears to be connected upside down. An analysis of the circuit reveals that the damper and B+ + operation are not affected by this type of connection. We should carefully note that when the overshoot is positive at V_1, it is simultaneously negative at V_2. The damper works (at the start of the trace) while the horizontal output tube rests, and vice versa.

Two horizontal output circuits for 90°

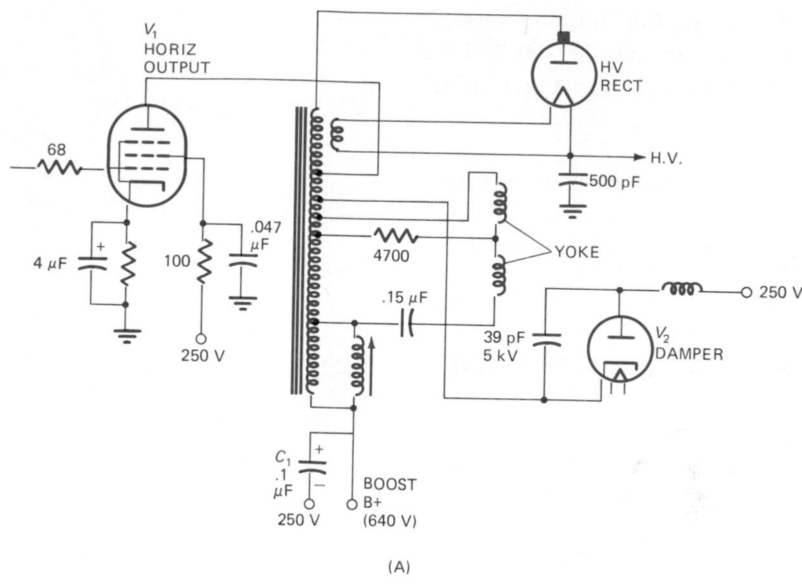

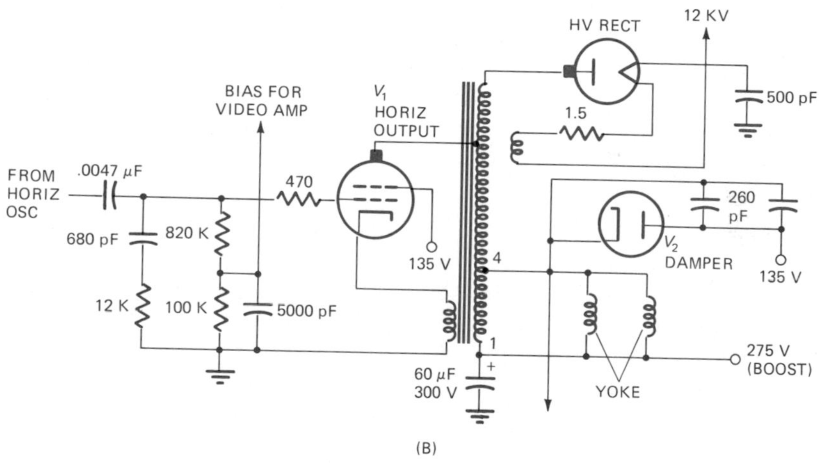

Figure 23.6A-B. The output transformer is an autotransformer in both diagrams. **A.** The damper tap is connected to produce a higher boost B+ voltage (+640 V). **B.** The damper tap is at the yoke potential. (These circuits can be used with 90- or 110-degree picture tubes.)

and 110° picture tubes are shown in Figure 23.6. In Figure 23.6A, the yoke windings are still connected in series across a portion of the horizontal output transformer. However, a connection is made between the two half sections and a suitable tap on the transformer. The 4 700-ohm resistor in this lead is designed to minimize ringing effects in the yoke and also to help balance the two yoke sections. The 0.15-μF capacitor in series with the yoke is for DC blocking.

The B++ voltage is developed across C_1 with the polarity indicated. Note that the bottom end of this capacitor connects to the plate circuit of the damper tube since both attach to the 250-V terminal in the power supply. The added voltage developed across this capacitor is equal to almost 400 V because the bottom end of C_1 has a potential of 250 V and the top end provides a boost B+ of 640 V.

In Figure 23.6B, the two horizontal yoke windings are connected in parallel. This arrangement eliminates the need for any balancing resistors.

23.3 FLYBACK HIGH VOLTAGE

Referring back to Figure 23.5, we note that a large voltage pulse (E_2) appears on the plate of the horizontal output tube during retrace. This pulse has a peak-to-peak value of 5 to 6 kV. The damper tube is not allowed to kill this positive pulse because it is useful for generating a high DC voltage for the picture tube second anode. As shown in Figure 23.4, a step-up autotransformer winding (L_1) is attached to the hot end of L_2. This winding increases the magnitude of E_2 a number of times. At the hot end of L_1, the peak-to-peak voltage may be as high as 20 kV for monochrome receivers and 30 kV for color sets.

SIMPLIFIED SCHEMATIC

Figure 23.7 is a simplified schematic showing the important circuit components in the high-voltage section. The 15- to 30-kV pulses are generated in a special high-voltage winding that is continuous with the primary, but made of finer wire. This part of the flyback transformer is nearly always an autotransformer connection. The high-voltage pulses are applied to the plate of a high-voltage, half-wave rectifier. The heater voltage for this tube, usually 1 to 3 V, is obtained by wrapping one or two turns of high-voltage wire around the flyback transformer core. One side of this heater becomes

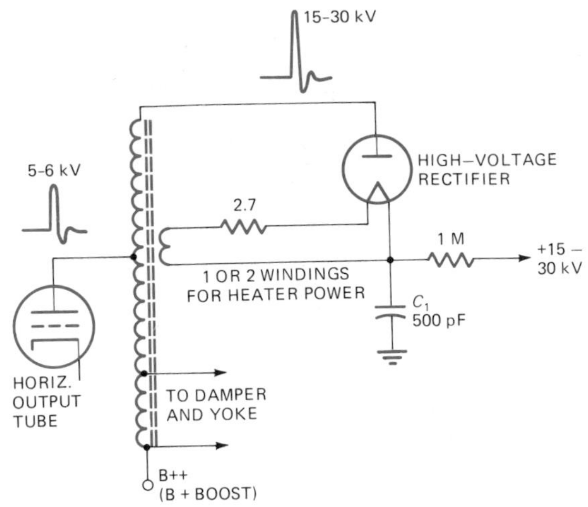

Figure 23.7 The high-voltage portion of a horizontal-output stage.

the DC output terminal for the high voltage. A high-voltage "doorknob" capacitor (C_1) filters the DC before it is sent to the picture tube second anode. Sometimes a resistor is placed in series with this lead to provide additional filtering. In some receivers, the filter capacitor is supplemented (or replaced) by the inner and outer aquadag coatings in the picture tube, which form a capacitor of fairly large capacitance (shown schematically in Figure 23.4).

Horizontal Output Transformers Two typical horizontal output transformers are shown in Figure 23.8. These are highly efficient units, using ferrite cores. Note the filament loops for the high-voltage rectifier at the bottom of the units.

23.4 SOLID-STATE MONOCHROME TV CIRCUITS

In this section we shall analyze horizontal output and high-voltage circuits that use transistors and solid-state diodes. This discussion will be restricted to monochrome-TV

A

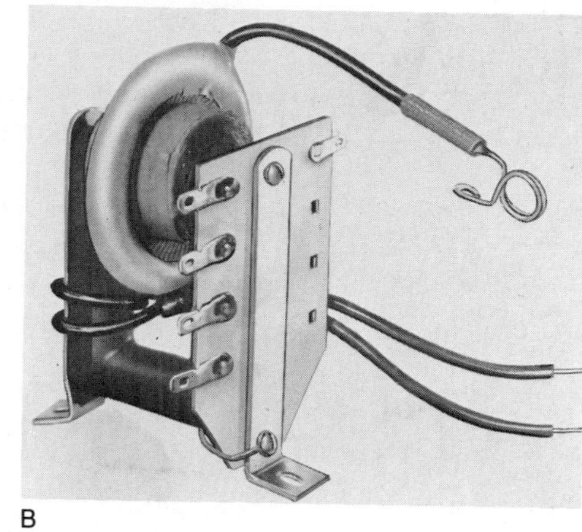

B

Figure 23.8A-B. Two typical horizontal-output transformers. The loop(s) turn at the bottom of each unit provides the filament voltage for a tube-type high-voltage rectifier.

receivers. We shall cover SCR circuits in the section on solid-state color receivers. Our first concern in this section will be the special requirements of solid-state horizontal circuits compared with their vacuum tube predecessors.

SPECIAL REQUIREMENTS FOR SOLID-STATE CIRCUITS

When transistors are used in the horizontal output amplifier, four special requirements must be met:

1. The transistors must be able to handle high currents and high voltages—both at moderately fast switching speeds.
2. The transistors must be given overcurrent and overvoltage protection.
3. The base-drive wave form must be rectangular instead of trapezoidal.
4. There must be a bias system that will protect the output transistor in the event of drive failure.

Special requirement 1 was very difficult to satisfy for large-screen receivers until the late 1960s. The transistors now used for the horizontal output stage will switch a peak power of 2 000 W or more while internally dissipating power on the order of only 1 W. Silicon devices are used extensively for this application. They offer both high-voltage breakdown capability and a fast switching speed.

Despite the capabilities of the output transistor, it must be protected against overvoltage and overcurrent. One method of reducing the overvoltage transients at the output amplifier collector is shown in Figure 23.9A. The reverse-peak voltage that appears on the collector during the retrace is reduced by adding to it a third harmonic of the sinusoidal 15 734-Hz pulse. This third harmonic is provided by a parallel-tuned circuit that is actually a part of the high-voltage system. Figure 23.9B illustrates how the fundamental pulse and its third harmonic are added to achieve a reduced pulse size.

Overcurrent Protection A method used to protect the output transistor against overcurrent is shown in Figure 23.10. Transistor Q_1 is the horizontal output transistor and Q_2 is

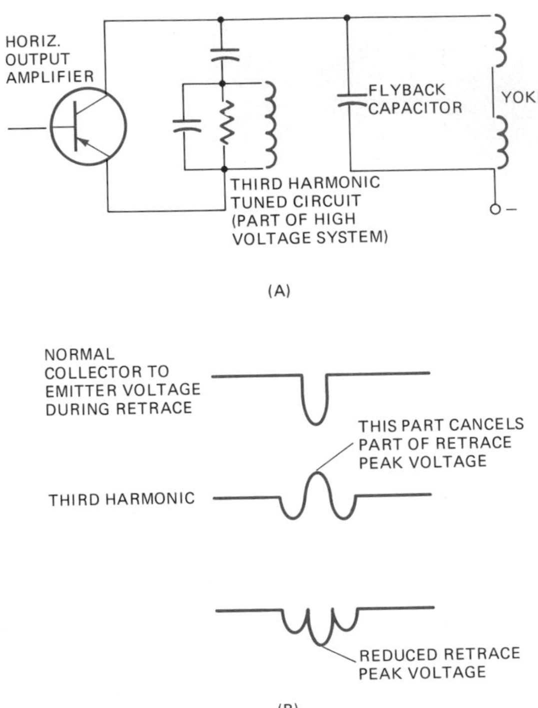

Figure 23.9A-B. Peak-pulse amplitude at the collector of the output transistor can be reduced by a tuned circuit: **A.** horizontal-output circuit; **B.** the mixing of waveforms on the collector to achieve reduced pulse voltage.

a current-limiting transistor; both are in series between B+ (+30 V) and ground. All current through Q_1 must pass through Q_2. However, the DC current (average current) through Q_2 is fixed by the bias resistor R_1. The average current through Q_1 will thus be fixed. The pulsed current through Q_1, caused by the horizontal oscillator, will be constrained to an average value determined by Q_2 and R_1.

Current limiting for the output transistor can also be accomplished by controlling its base-drive duty cycle. The base-drive voltage wave form for a PNP output stage is shown in Figure 23.11. The two wave forms in this figure illustrate the two types of base drive. Although the wave forms appear similar at first glance, they differ significantly. The ratio (duty cycle) of the turn-on drive time to the off time is (1.1/0.64) = 1.7 in Figure 23.11A. The same ratio is only 1.1 in Figure 23.11B. A circuit working normally requires only a 0.9 ratio. Thus, a circuit that is able to control this duty cycle from 0.9 to 1.7 is able to control the average current through the output transistor. This method effectively limits the average power that may be drawn from the power supply and therefore protects the output transistor.

Rectangular Base-Drive Waveform The third special requirement of transistor horizontal output amplifiers is the rectangular base-drive wave form. This is shown in Figure 23.11. It is quite different in shape when compared to the trapezoidal wave form used on the grid of a vacuum tube horizontal output amplifier. The transistor output stage drives a low-voltage, high-current transformer and yoke. The yoke thus has high inductance compared to its resistance and hence requires a rectangular (pulse) voltage drive wave form. This will be discussed in more detail shortly.

Drive-Failure Protection The last special requirement for the output transistor is a means to protect this high-power device when drive from the horizontal oscillator fails. The output transistor is biased such that no current will flow unless a pulse wave form is applied to its base. If drive fails, no pulses turn the transistor on and no power is dissipated in the device. It automatically shuts off completely when drive fails.

23.5 TRANSISTOR HORIZONTAL OUTPUT CIRCUITS

The DC series resistance of a horizontal-deflection yoke in a transistorized receiver is less than 1 ohm. The yoke resistance is usu-

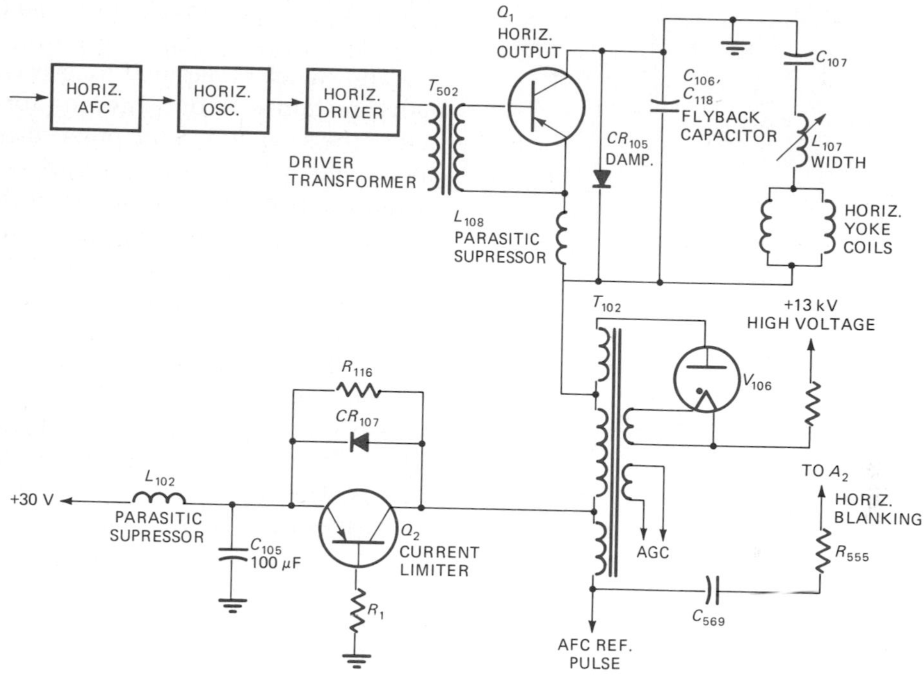

Figure 23.10 Series current-limiter protection of the horizontal-output transistor.

ally 5 to 30 ohms in a vacuum tube set. At the horizontal scanning rate of 15 734 Hz, the yoke inductive reactance in a solid-state receiver is very large relative to its series resistance. A rectangular voltage wave form is required for this type of yoke. The circuit driving the yoke must also have a very low series resistance. An ideal switch, as shown in Figure 23.12, would provide a low resistance and also a rectangular wave form. However, it would be difficult to open and close a mechanical switch 15 734 times per second. The transistor, driven between cutoff and saturation, comes closer to the characteristics of an ideal switch than does a vacuum tube. The series resistance of a high-power, high-speed transistor in the saturated state is less than 1 ohm; in the cutoff mode, the series resistance is usually above 50 000 ohms. This ON-OFF resistance range makes the transistor a potentially good horizontal output amplifier.

PRODUCING A SAWTOOTH CURRENT

Before we examine a transistor horizontal output circuit, it will be helpful to examine a sweep circuit that has an ideal switch. This is shown in Figure 23.12A. The yoke current and voltage wave forms appear in Figure 23.12B. In order to produce a linear sawtooth of current in a pure inductance, a *constant voltage* is required across the inductance. A linear sawtooth current is obtained by connecting a coil L to a constant source of voltage by means of an ideal switch. Current starts flowing in the coil by closing the switch at time t_0. If the circuit resistance is zero, the coil current will rise indefinitely in a linear fashion. In a practical circuit, the switch is held closed until one-half (the second half) of the trace is completed (time t_0 to t_1). When the switch opens, the coil current continues to flow in the same direction but

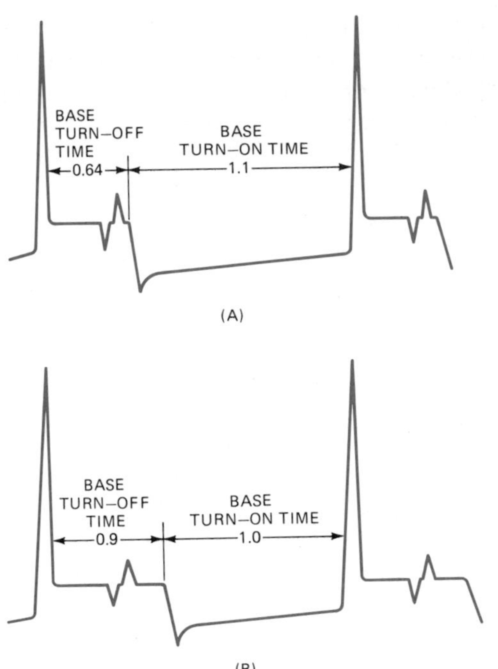

Figure 23.11A-B. Current limiting by control of the base-drive duty cycle (PNP horizontal-output transistor): **A.** duty cycle = 1.7, and **B.** duty cycle = 1.1.

flows now into capacitor C. The coil voltage variation follows a sine curve whose period is determined by $t = 2\pi\sqrt{LC}$. When the coil voltage completes one-fourth of a sine wave, the coil current is zero and the capacitor voltage is maximum. This is true because the coil and capacitor are now a resonant circuit that has completed one-fourth of an oscillation. During the next one-fourth cycle, the capacitor fully discharges and the coil current flows in exactly the opposite direction. At this instant (time t_3) the switch closes and the constant voltage causes the coil current to rise linearly. The low impedance of the battery and switch damps the oscillations. However, this time the current starts from a negative value. The linear rise continues until the switch again opens.

TRANSISTOR AS A SWITCH

The switch is required to pass current in both directions—in one direction for the first half of the scan and in the other direction for the second half. This is not possible with a vacuum tube but is easily accomplished with a saturated transistor. Figure 23.13A shows a circuit with the switch replaced by a PNP transistor. As can be seen in Figure 23.13B, if the base-to-emitter voltage is +2 V, no current flows from emitter to collector. The transistor appears to be a 50-kilohm resistor, or nearly an open circuit. During the scan period, from time t_3 to t_5 in Figure 23.12B, the base-to-emitter voltage is held at −1 V. This drives the transistor into saturation conduction and makes it appear as a resistor whose value is less than 1 ohm—essentially a short circuit. In this saturated condition, current flows easily from emitter to collector

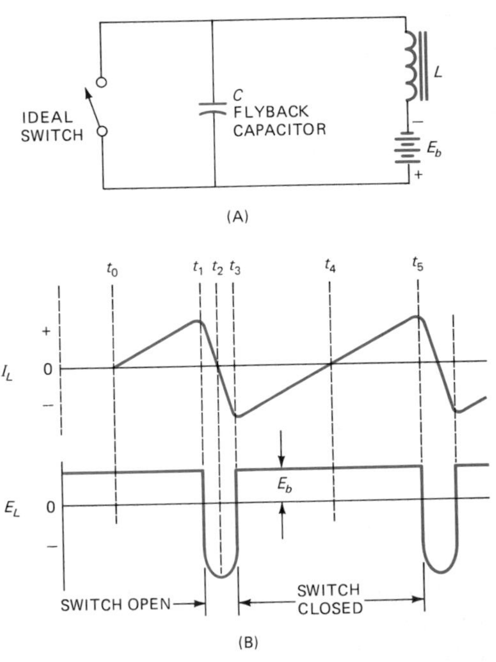

Figure 23.12A. Simple deflection circuit utilizing an ideal switch. **B.** Yoke current and voltage waveforms.

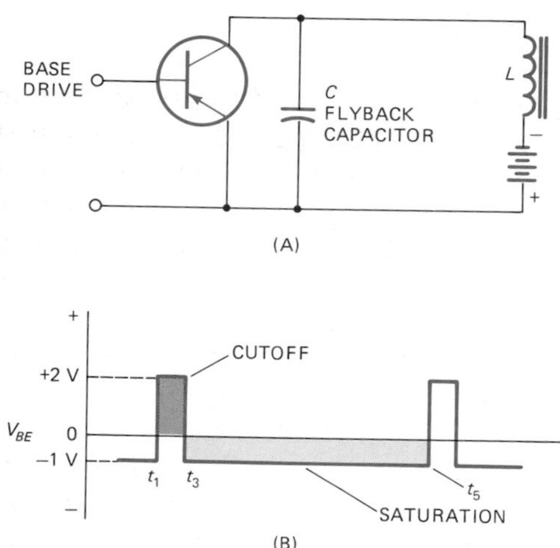

Figure 23.13A-B. Transistor functioning as an ideal switch: **A.** transistor switch circuit; **B.** base-drive voltage waveforms.

or from collector to emitter. The circuit shown in Figure 23.13A is thus capable of generating a linear-current sweep and retrace just as the ideal switch circuit did. Furthermore, the transistor circuit is capable of operating thousands of times faster than a mechanical switch.

TRACE AND RETRACE INTERVALS

The reader will recognize that the interval from t_3 to t_5 in Figure 23.12B represents the trace interval and the period from t_1 to t_3 is the retrace interval in the scanning cycle.

23.6 SOLID-STATE DAMPERS

Although a saturated transistor can carry current in either direction, it cannot reverse its direction of current fast enough at time t_3 in Figure 23.12B. At this instant, the oscillations are stopped by the low shunt resistance of the transistor and the damping diode, which are now effectively in parallel. The stored energy (current) in the yoke now begins to decay linearly toward zero, mostly through the diode. The transistor must be paralleled by a diode switch during most of this linear decay period to assure a very low resistance and linear-current decay. The diode, as seen in Figure 23.14, is connected between the emitter and the collector of the transistor. Since this diode conducts current immediately after the retrace, thereby damping self-oscillation in the circuit, it is called the damper diode. Most of the current through the yoke between times t_3 and t_4 is carried by the damper diode. The transistor will also carry part of the current but has a higher starting resistance than the damper diode and does not conduct fully until sometime between t_3 and t_4, as shown in Figure 23.15.

TRANSISTOR AND DAMPER CURRENTS

Figure 23.15 clearly illustrates the critical time relationship of the currents through the output transistor and the damper diode. The reader should study carefully each step shown at the bottom of the illustration and note the shape of each wave form at that instant.

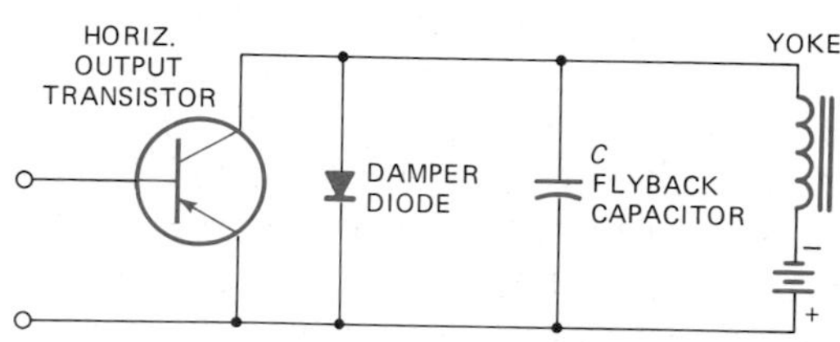

Figure 23.14 Incorporation of the solid-state damper diode (see Figure 23.15 for waveforms).

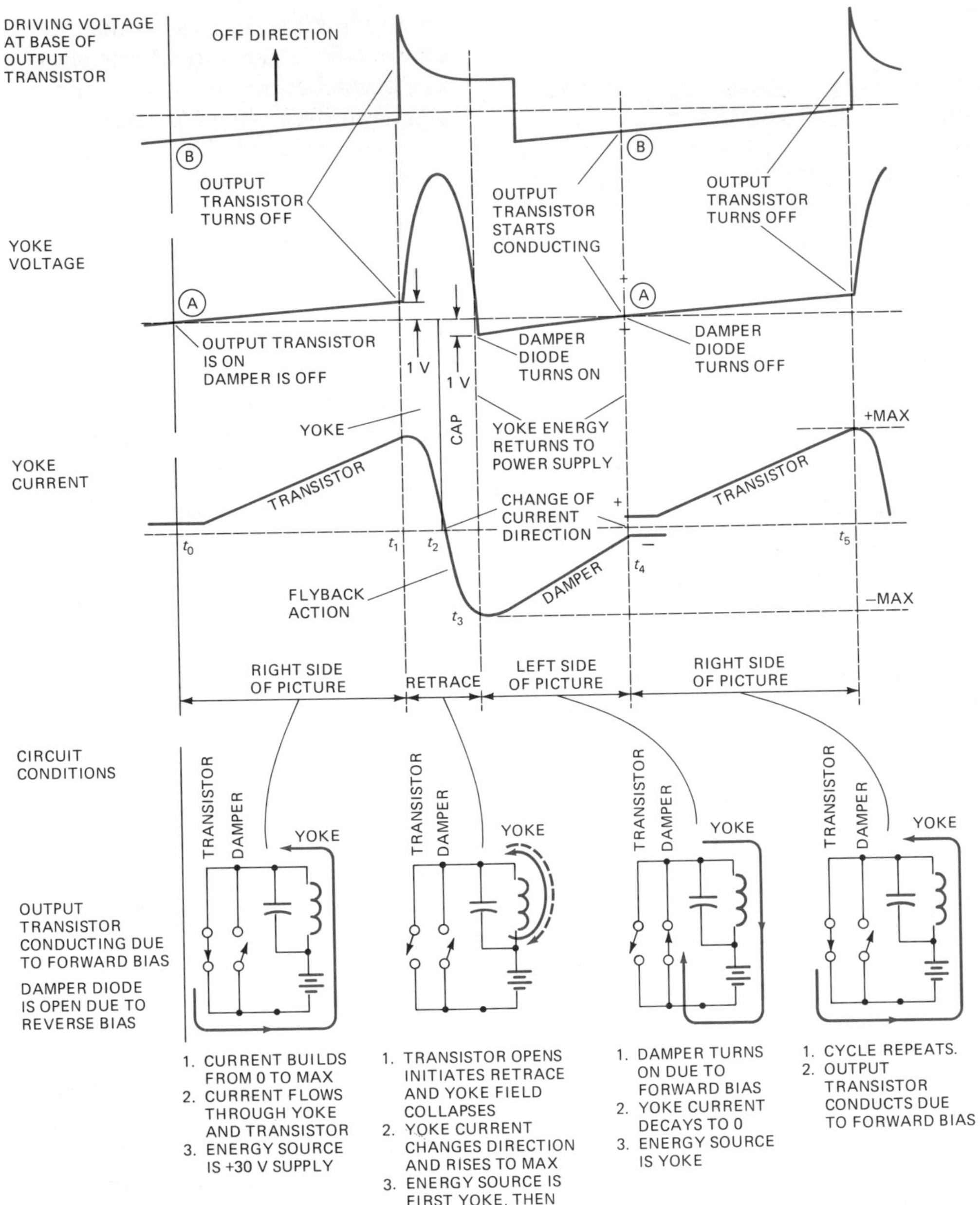

Figure 23.15 Relative time relationships of the horizontal-output operational cycle. *(Courtesy of RCA Consumer Electronics.)*

BOOSTED B+

Unlike tube circuitry, the solid-state damper diode is *not* a source of boosted B+. It is used only to carry the bulk of the yoke current when the picture tube beam is at the left side of the screen. If boosted B+ is required, a separate diode is used.

NONLINEAR SWEEP COMPENSATION

The deflection yoke in a transistor horizontal output circuit has less than 1 ohm DC resistance. However, this small resistance is enough to create a slightly nonlinear sweep. It causes a left-hand stretch and a right-hand compression on the screen. To compensate for this nonlinearity, a capacitor is placed in series with the yoke, as shown in Figure 23.16. An S-shaped voltage wave form is developed across the capacitor during scan. This compensates for the previously mentioned distortion by compressing the left-hand side and stretching the right-hand side of the picture. The capacitor is appropriately named the S capacitor.

Figure 23.16 shows how the flyback transformer is coupled to the yoke when an S capacitor is used. The capacitor also blocks DC current from flowing in the yoke. If DC current flows through the yoke, heat is generated, which in turn increases the yoke DC resistance, thus causing more nonlinearity.

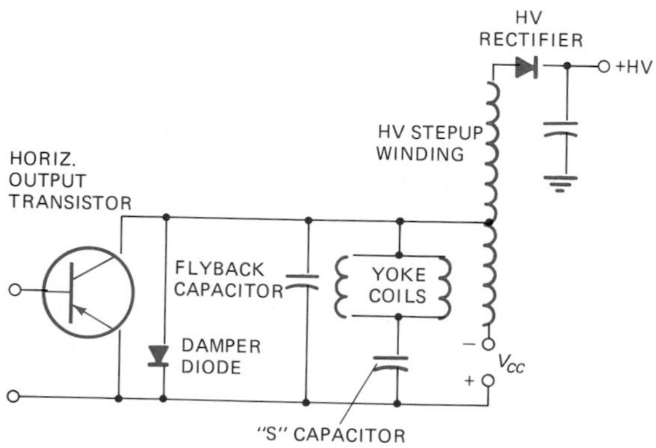

Figure 23.16 Solid-state horizontal-output stage using S capacitor.

HIGH-VOLTAGE CIRCUITS

Note the simple high-voltage circuit in Figure 23.16. The primary of the flyback transformer is placed in parallel with the damper, transistor, and yoke. The high-voltage winding provides pulses ranging from 10 to 30 kV, which are rectified and applied to the picture tube second anode. The most successful solid-state high-voltage rectifiers are made from silicon. Several examples of these devices are shown in Figure 23.17.

23.7 NPN AND PNP HORIZONTAL-OUTPUT STAGES

Figure 23.18 is an example of a typical NPN, transistor monochrome horizontal-output circuit. The high-voltage rectifier circuit is similar to that used in vacuum tube television receivers. The remainder of the circuit is solid-state.

The horizontal yoke coils are connected in parallel. They are then connected in series with the S capacitor, which is actually formed by two parallel capacitors, C_{16} and C_{17}. The yoke and the S capacitor are positioned directly across the primary of the horizontal output transformer. However, the driving wave form from the transistor must pass through a filter before it can get to the yoke and the transformer. The filter is formed by C_{13}, C_{15}, C_{24}, and L_{603}. The filter, and the damper diode, keep voltage transients from Q_1.

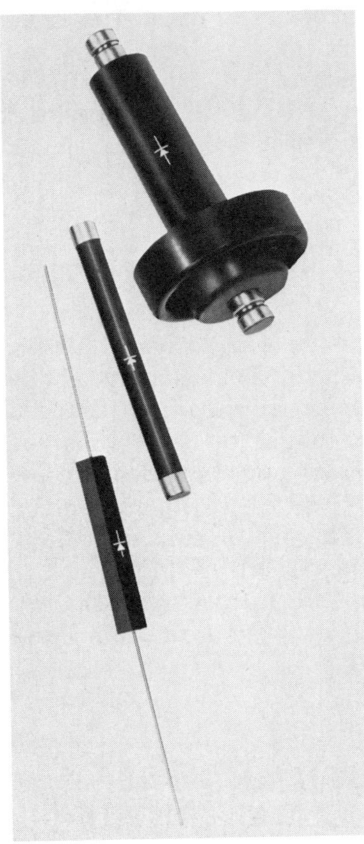

Figure 23.17 Silicon high-voltage rectifier assemblies. *(Courtesy of Varo, Inc.)*

LOSS OF DRIVE PROTECTION

Q_1 is also protected against loss of drive since it is biased below cut off. With no signal on the input transformer primary, the base-to-emitter voltage is zero. At least +0.7 V (for a silicon transistor) is needed on the base to bring the transistor out of cutoff. Thus, with no drive signal, a 0.7-V safety margin exists and no current will flow through the output transistor. When the drive signal is present, however, a −2.4-V bias builds up on the base terminal. This is caused by the charging of C_{25} during the transistor ON time and its relatively slow discharge through R_{12} during the transistor OFF time.

The pulses on the transistor collector rise to approximately 500-V amplitude during the retrace. These are rectified by diode SR_{601} and filtered with C_{23} and C_{27}. The resultant 500 V DC is used for the picture tube focus anode.

PNP HORIZONTAL OUTPUT STAGE

A solid-state horizontal output stage that uses a PNP transistor is shown in Figure 23.19. The damper diode, D_2, is connected directly between the collector and emitter of the output transistor, as usual. Another diode, D_1, provides a +80-V boost voltage since the regular B+ (a battery) is only 13 V. The base and emitter are at the same DC potential, with only the low resistance of the coupling transformer secondary between them. This assures class C operation, so that the transistor is normally off until a pulse comes through the transformer.

High-voltage pulses from the flyback transformer are rectified with the solid-state diode, D_3. The resultant 8-kV DC potential is used for the second anode of the small-screen picture tube.

23.8 ADJUSTMENTS IN MONOCHROME TV RECEIVER CIRCUITS

The horizontal-output stages of a monochrome receiver contain a number of controls and adjustments, each of which has a significant effect on the picture produced on the screen.

DRIVE CONTROLS—VACUUM TUBE SETS

Two common methods for controlling the grid drive are illustrated in Figure 23.3.

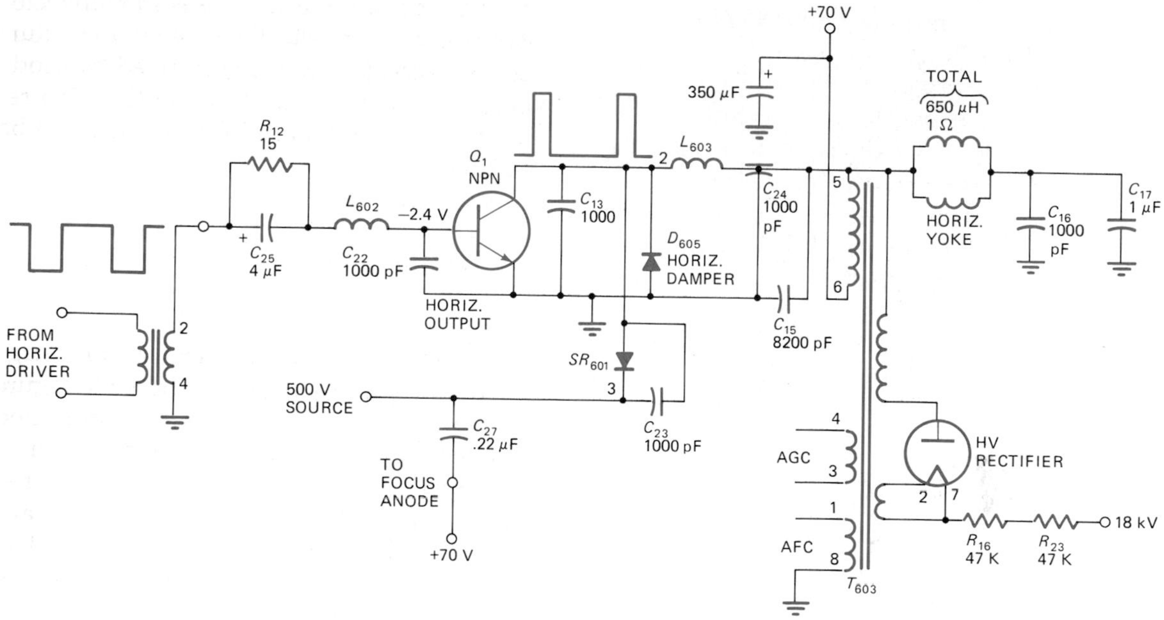

Figure 23.18 NPN horizontal-output stage. *(Courtesy of Magnavox.)*

The drive control is basically used to control the amplitude of the grid input wave form. This adjustment also has a significant effect on the linearity and width of the yoke current wave form. In many receivers, the drive control is the only width adjustment. These three controls—drive, width, and linearity—are highly *interactive*. Thus, optimum width and linearity will be achieved only after making all three adjustments *several times*.

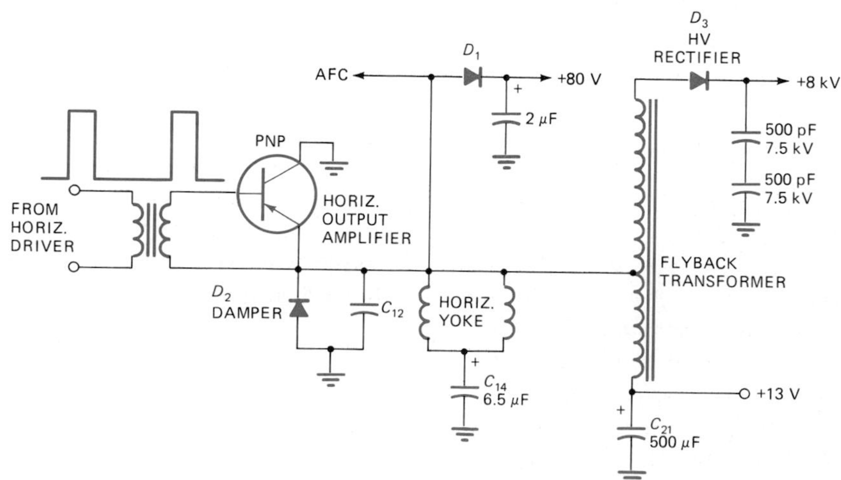

Figure 23.19 PNP horizontal-output circuit (battery operated).

WIDTH CONTROLS—VACUUM TUBE SETS

A television picture is transmitted with an aspect ratio (ratio of width to height) of 4:3. Unless this ratio is maintained in the receiver, all the objects in the televised scene will be uncomfortably distorted; figures will appear taller and thinner or shorter and broader than they should be. After the height of the picture is correctly set, the width of the picture must be adjusted. This adjustment is made by the width control or the horizontal size control. In many receivers, the width control takes the form of a slug-adjusted coil connected across part of the horizontal-output transformer secondary winding. By changing the inductance of the width coil, the amount of deflection current flowing through the deflection coils is varied and the width of the raster changes accordingly. In this fashion, picture width can be varied by 1.5 in (3.8 cm) or more. Figure 23.4 uses this type of width control.

In another method of picture width variation, a potentiometer in the screen grid circuit of the horizontal output amplifier varies the gain of the stage and, with it, the deflection voltage (and current) fed to the yoke.

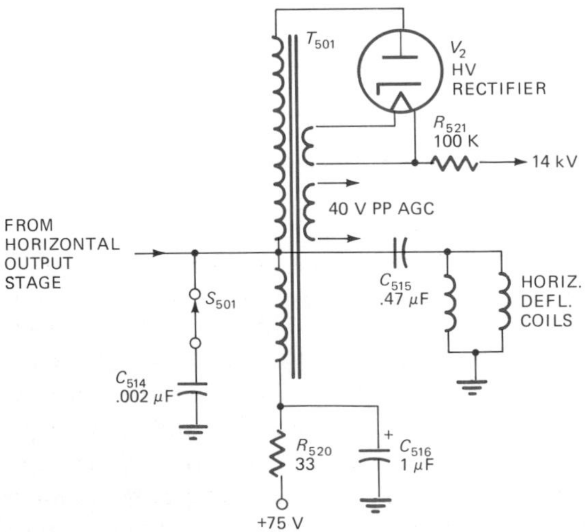

Figure 23.20 Width-control circuit which utilizes a switch-connected capacitor (see text).

A simple and practical horizontal width control can be made from a thin metal sleeve slipped between the yoke and neck of the picture tube. This forms a "losser" control (one-turn loop or short). The width is affected because the loop creates a magnetic field that reduces the sweep magnetic fields. The sweep is of minimum length when the sleeve is completely under the yoke. This method can also be used with monochrome solid-state television receivers.

WIDTH CONTROLS—SOLID-STATE SETS

A capacitive type of width adjustment is shown in the transistor circuit of Figure 23.20. A special capacitor, C_{514}, can be connected into the circuit by means of a switch (S_{501}) accessible at the rear of the chassis. When C_{514} is in circuit, the width increases; when it is out of circuit, the width decreases. Capacitor C_{514} has a loading effect when connected in the circuit, reducing the amplitude of the pulse applied to the high-voltage rectifier V_2. When V_2 conducts less, the high voltage decreases. The yoke magnetic field then has more effect on the beam and increases the picture width.

Another method, illustrated in Figure 23.21, consists of a variable inductor in series with the horizontal yoke coils. The yoke current, and therefore the width, change inversely in proportion to the magnitude of the series inductance.

LINEARITY CONTROLS

This type of control is not found in solid-state television receivers. It is not re-

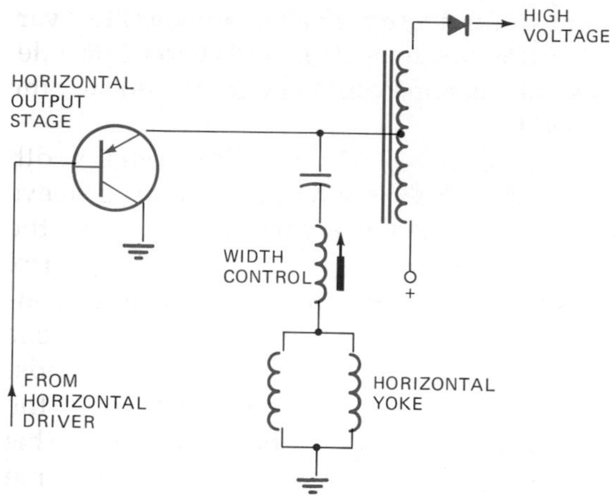

Figure 23.21 Simplified circuit showing a method of width control for a solid-state TV set.

quired in these sets because of the inherent horizontal linearity created by the circuitry. However, they are frequently found in vacuum tube television receivers.

The horizontal-output tube current sweeps the right section of the screen and the damper tube current covers the left section of the screen. A transition region is apparent in Figures 23.2 and 23.5. The final yoke current consists of a combination of currents from two tubes. They must "join up" and produce a linear wave form. A simple joining of these two currents will *not* produce a linear wave form. A wave form correction circuit is added to remove most of the nonlinearity caused by the addition or overlap of the two currents. In the circuit of Figure 23.22, the plate-load impedance of the horizontal output tube consists of two parts: the flyback transformer and a pi filter. The filter is a low-Q filter, resonant at 15 734 Hz. Resonance of this filter to the horizontal sweep rate is achieved by varying L_1. At resonance, the impedance of the filter circuit is maximum and impedes the current flowing to provide the maximum output tube efficiency. The linearity (π) filter shifts the phase of the ripple on the B+ boost voltage.

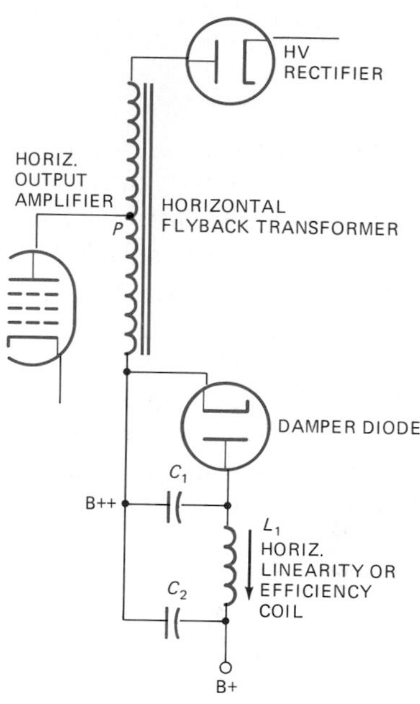

Figure 23.22 Horizontal linearity or horizontal "efficiency" control.

This modifies the output tube current wave form and permits center-screen, horizontal linearity control.

23.9 VACUUM TUBE COLOR TV HORIZONTAL-OUTPUT CIRCUITS

In general, the circuits for horizontal deflection in color receivers are quite similar to their monochrome counterparts. The principal point of departure is the power required in the deflection and high-voltage circuits. This is considerably higher in color receivers than in monochrome. The reasons for the increased power need are the larger picture tube neck diameter (which requires a larger yoke current) and the greater high voltage (30 kV). Special horizontal output tubes have been developed to meet the increased power requirements. Class C operation is used to keep the dissipation down in these tubes.

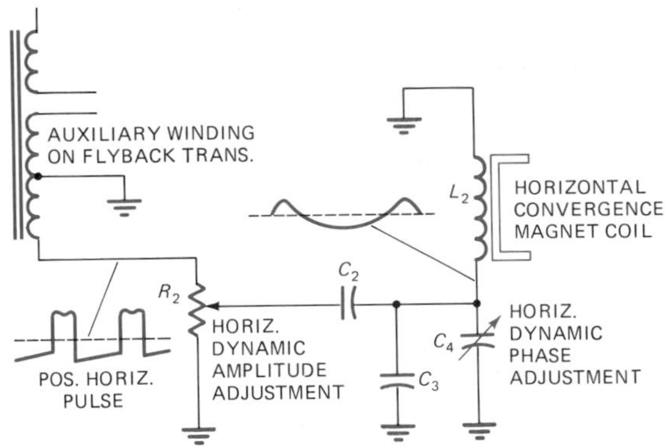

Figure 23.23 A typical horizontal dynamic convergence circuit.

This requires a grid drive sawtooth with a peak-to-peak amplitude of about 200 V.

In Chapter 18, when color picture tubes were discussed, we noted the need for static and dynamic convergence circuits. The horizontal-deflection section furnishes the basic dynamic convergence wave forms for this purpose. In order to compensate for the fact that the screen is flat and not spherical, the convergence field must be increased as the three electron beams move toward the edges of the screen. As shown in Figure 23.23, a trapezoidal signal taken from a special winding on the horizontal flyback transformer passes through a wave-shaping circuit and produces in the convergence coil a current wave form that is roughly parabolic. The amplitude of the wave form can be controlled by potentiometer R_2 and the amount of relative curvature by the "dynamic phase adjustment" capacitor C_4. Since separate convergence correction coils are needed for each of the three electron beams, the horizontal-deflection system has to supply three separately adjustable dynamic convergence signals.

A TYPICAL CIRCUIT

An example of a typical vacuum tube color TV horizontal-deflection circuit is shown in Figure 23.24. The complexity of the drawing may obscure the fact that there are six sections contained within this overall circuit:

1. Horizontal-output tube circuit.
2. Yoke-matching circuitry
3. Focus supply
4. Damper and boost circuitry
5. High-voltage supply
6. Regulator circuit

Horizontal-Output Tube Circuit Figure 23.25 shows the horizontal-output tube and its immediate circuit. Capacitor C_{125} couples in the large-amplitude sawtooth grid-drive voltage from the horizontal oscillator. The class C amplifier V_1 draws grid current on the peaks of this drive signal. During these peak intervals, C_{125} charges and then discharges through R_{169} to develop a -50-V DC bias at the control grid of the 6JE6A (V_1).

The plate circuit drives the flyback transformer, which, with its associated components, make up a tuned circuit (70 kHz). A DC milliameter meter M_1 is shown in the cathode circuit, and proper adjustment of the tuned circuit is indicated by a dip in the reading on M_1. The dip adjustment is made with the horizontal efficiency coil (L_{710} in Figure 23.24).

Yoke-Matching Circuitry Figure 23.26 presents a simplified drawing of the yoke-matching circuit. A 5-kV pulse during beam retrace develops at Point *P*. Transformer T_1 is an autotransformer, with the windings from *P* to *BB* functioning as the primary and the windings from C_2 or C_1 to *BB* as a step-down secondary. The pulse voltage is therefore stepped down with a resulting current step-up. This is necessary to match the horizontal-output tube impedance (relatively high Z) to the yoke windings (relatively low Z).

Varistor RV_{101} holds the scan size constant. A varistor changes its resistance in op-

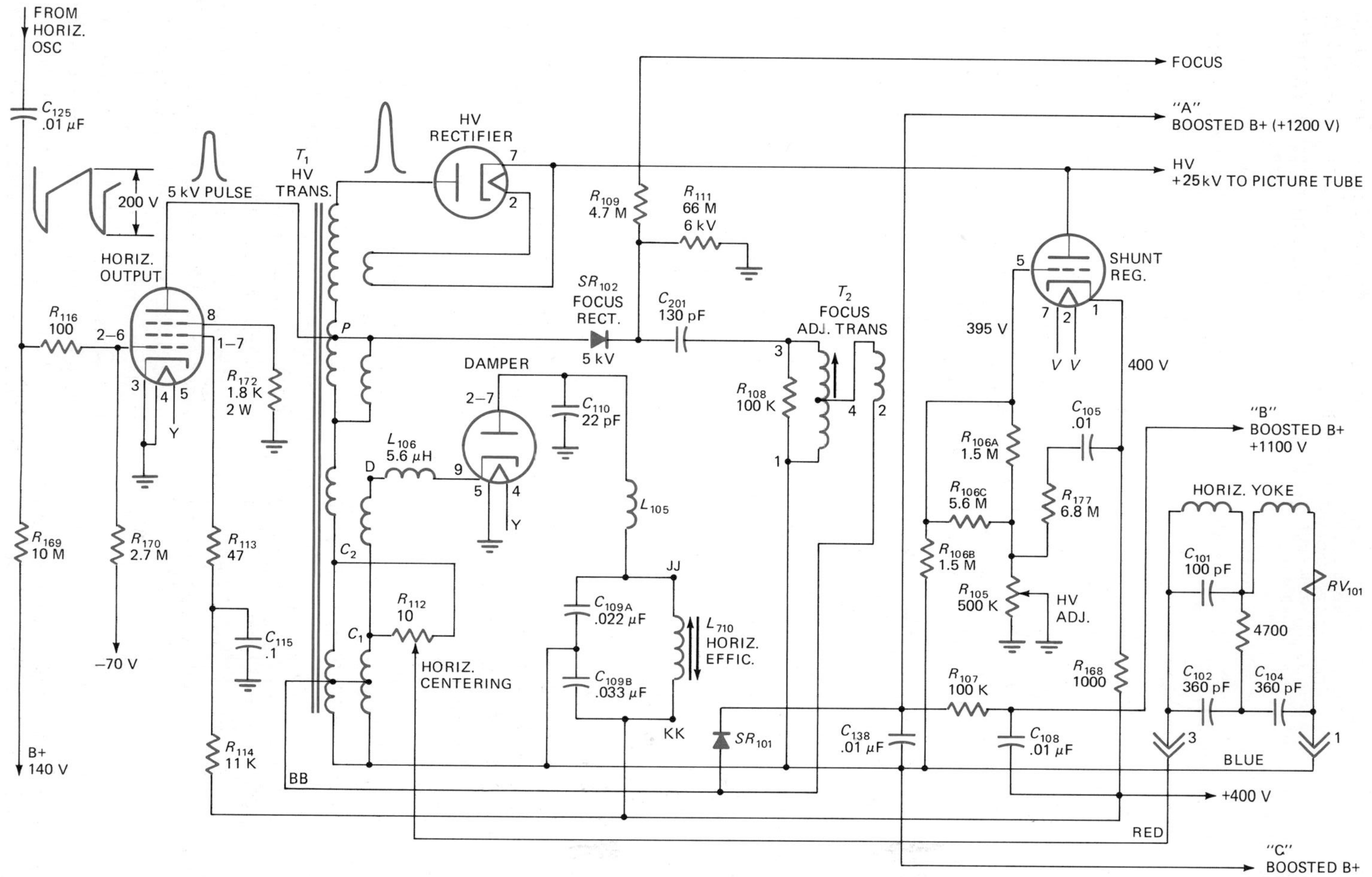

Figure 23.24 Typical vacuum-tube color TV horizontal-deflection circuit. *(Courtesy of RCA Consumer Electronics.)*

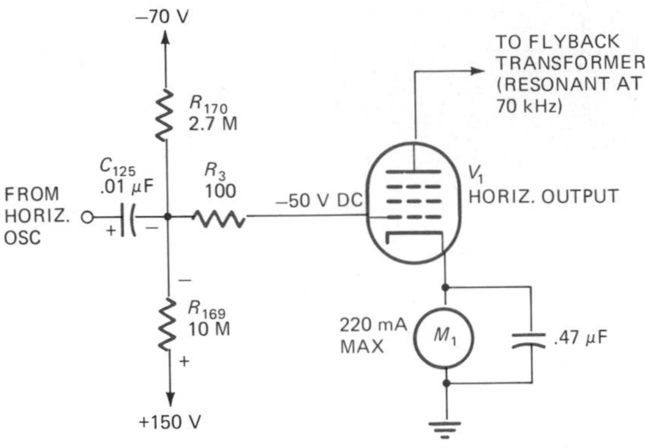

Figure 23.25 Horizontal-output tube circuitry from Figure 23.24

position to the voltage drop across it. An increase in output voltage causes RV_{101} to decrease in value. This loads the output circuit and reduces the output voltage to its normal value. A decrease in output voltage works in the opposite manner: RV_{101} increases in value and loads the output circuit less.

Focus Supply The focus supply section of Figure 23.24 is shown in Figure 23.27. If 25 kV is used on the picture tube anode, then 4

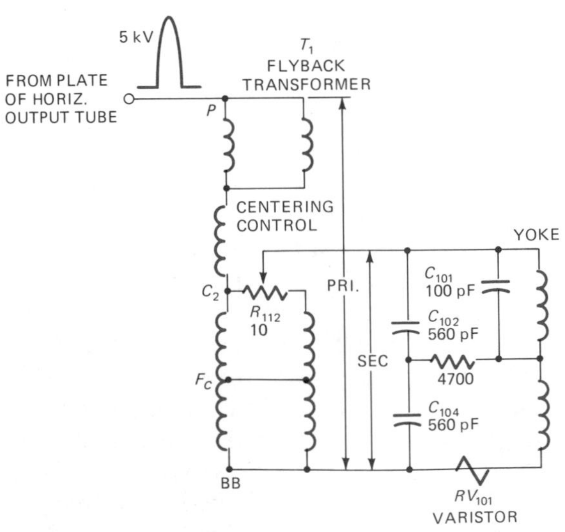

Figure 23.26 The yoke-matching circuit from Figure 23.24.

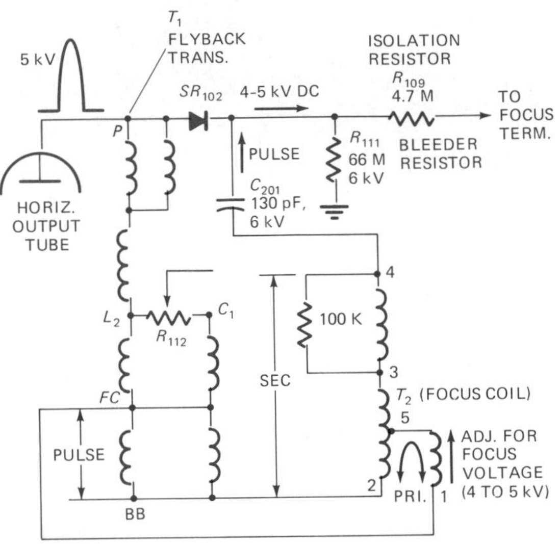

Figure 23.27 The focus supply portion of Figure 23.24.

to 5 kV is required on the focus electrode. Since a 5-kV pulse is delivered to the flyback transformer from the output tube, the focus rectifier (SR_{102}) can be attached at that point.

Damper and Boost Circuit The damper and $B++$ portion of Figure 23.24 appear in Figure 23.28. This particular receiver requires more boosted B+ than could be obtained directly from the yoke portion of the flyback transformer. Hence, an extra step-up section is added to the transformer so that an 850-V B++ may be obtained. A rugged semi-high-voltage rectifier serves as the damper diode. The rectified pulses are filtered in the π filter composed of L_{710}, C_{109A}, and C_{109B}. Coils L_{106} and L_{105} prevent parasitic oscillations. This 850 V DC is called the C-boosted B+.

Two additional B++ voltages are provided by the circuit of Figure 23.28. The diode SR_{101} has 850 V DC plus a 350-V pulse on its anode. It converts this to 1200 V DC for use in the vertical oscillator (A-boosted voltage). The same circuit also makes available 1 100 V DC for the picture tube screen grids (B-boosted voltage).

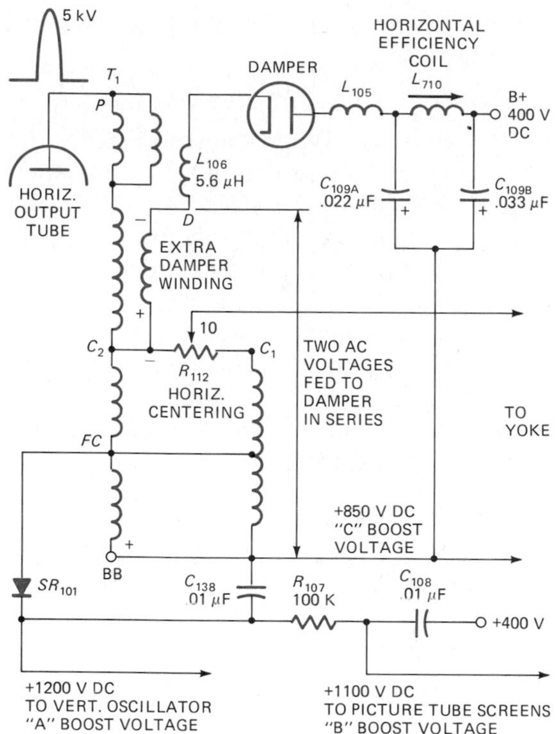

Figure 23.28 The damper and boost circuitry of Figure 23.24.

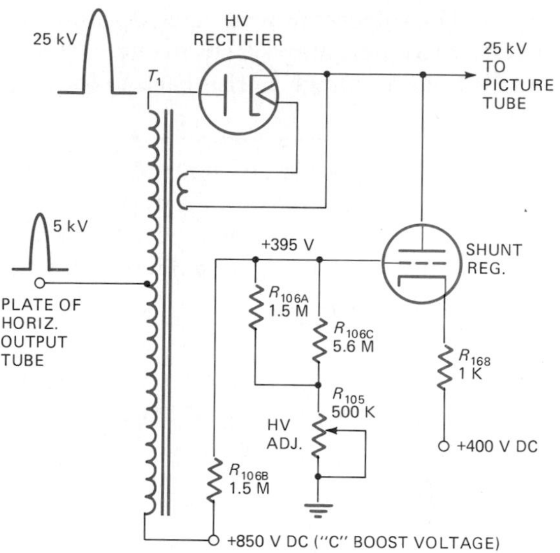

Figure 23.29 The shunt-regulator portion of Figure 23.24.

High-Voltage Supply The high-voltage transformer and rectifier tube circuit in a color receiver are quite similar to the same components in a monochrome receiver. The major difference is the larger voltage and current requirements of the color set. The high-voltage rectifier has a ruggedly built cathode to help provide this extra capability. The high-voltage rectifiers in monochrome receivers usually do not have cathodes.

Regulator Circuit Figure 23.29 is a simplified drawing of the high-voltage rectifier and shunt-regulator portion of Figure 23.24. Two basic facts regarding shunt-regulated power supplies should be emphasized. First, all power supplies have an internal impedance that causes the output voltage to vary with load changes. Second, if we desire a fixed output voltage, then the current through the internal impedance must be kept constant. This is performed with a shunt-regulator tube, which functions as part of the high-voltage load. If the picture tube draws more current, the shunt regulator draws less current, and vice versa.

It is difficult to continuously monitor the magnitude of the high voltage. Yet we must do this if we are to hold it constant. A good detection point is the boosted B+, because this varies directly with the high voltage. A sample of the boosted B+ voltage is fed to the grid of the shunt-regulator tube, as shown in Figure 23.24. Resistors R_{106B}, R_{106A}, R_{106C}, and R_{105} reduce the boosted B+ from 850 V to approximately 395 V at the grid of the tube. Resistor R_{105} is a control that establishes the bias on the shunt regulator by changing the grid voltage. Note that the cathode is held at 400 V from a separate power supply. The 400 V does not change when the high voltage and the boosted B+ change. If these latter two voltages drop, then the 395 V on the shunt-regulator grid drops. This places more reverse bias on this tube, which then draws less current and permits the high voltage to rise back to its original value.

Note: High-voltage regulators are not used with solid-state, color-TV receivers. All horizontal circuitry is powered from regulated-power supplies. This maintains a constant high voltage.

23.10 SOLID-STATE COLOR TV HORIZONTAL-OUTPUT CIRCUITS

An example of a solid-state color TV, horizontal-output circuit is shown in simplified form in Figure 23.30. This circuit is from a 19-in (48-cm) RCA XL-100 type receiver. The power supplies for this circuit are described in Chapter 19, Section 19.20, which should be reviewed at this time.

CIRCUIT OPERATION

The output transistor acts as a switch. It is either fully ON or fully OFF. (Refer also to the discussion for Figure 23.18.) The horizontal output transistor Q_{401} is powered from a regulated 110-V source (Section 19.20). The 110 V is applied to Q_{401} and the damper diode CR_{403}. An 8-V peak to peak rectangular wave is applied to the base of Q_{401}. The positive portion of this wave switches Q_{401} on for roughly the last 70% of the horizontal scan (see Q_{401} base wave form in Figure 23.31). Coil L_{401} in the Q_{401} base circuit "smooths" the base waveform.

Q_{401} Scan Portion When Q_{401} is turned ON, the electron beams scan horizontally, about the last 70% of the screen. As mentioned previously, this occurs because the current in the yoke now increases in sawtooth fashion.

Retrace Portion When the scan reaches the right side of the screen, Q_{401} is instantly turned OFF. Now the stored energy in the inductances of the yoke and flyback transformer system cause these circuits to oscillate at about 70 kHz. As a result, retrace occurs and the high-voltage pulse is simultaneously generated. (This action has already been explained in detail.) In this receiver (see Section 19.20), B+ voltages are also generated during the retrace time.

Damper Conduction At the end of the first half cycle of 70-kHz output inductance oscillation, the voltage across the inductances naturally reverses. This causes the damper diode CR_{403} to conduct, placing a low-impedance load (damping) across the inductor. Oscillation ceases at the left side of the screen. Now the stored energy causes current to flow through the yoke and damper diode CR_{403}. This provides about the first one-third of the scan, after which the current through Q_{401} provides the other two-thirds.

High-Voltage Transformer Tuned Circuit Examination of Figure 23.30 will show a fixed tuned circuit L_1–C_1 beneath and connected in series with the high-voltage transformer. This circuit is effective during retrace only and improves high-voltage regulation.

Width Control Capacitor C_{406} tunes the yoke to the retrace resonant frequency. A second capacitor, C_{407}, can be connected by a jumper to act as a width adjustment. If this capacitor is connected, it *lowers* the retrace resonant frequency. This reduces the high voltage, which increases deflection sensitivity. This increases the width of the picture.

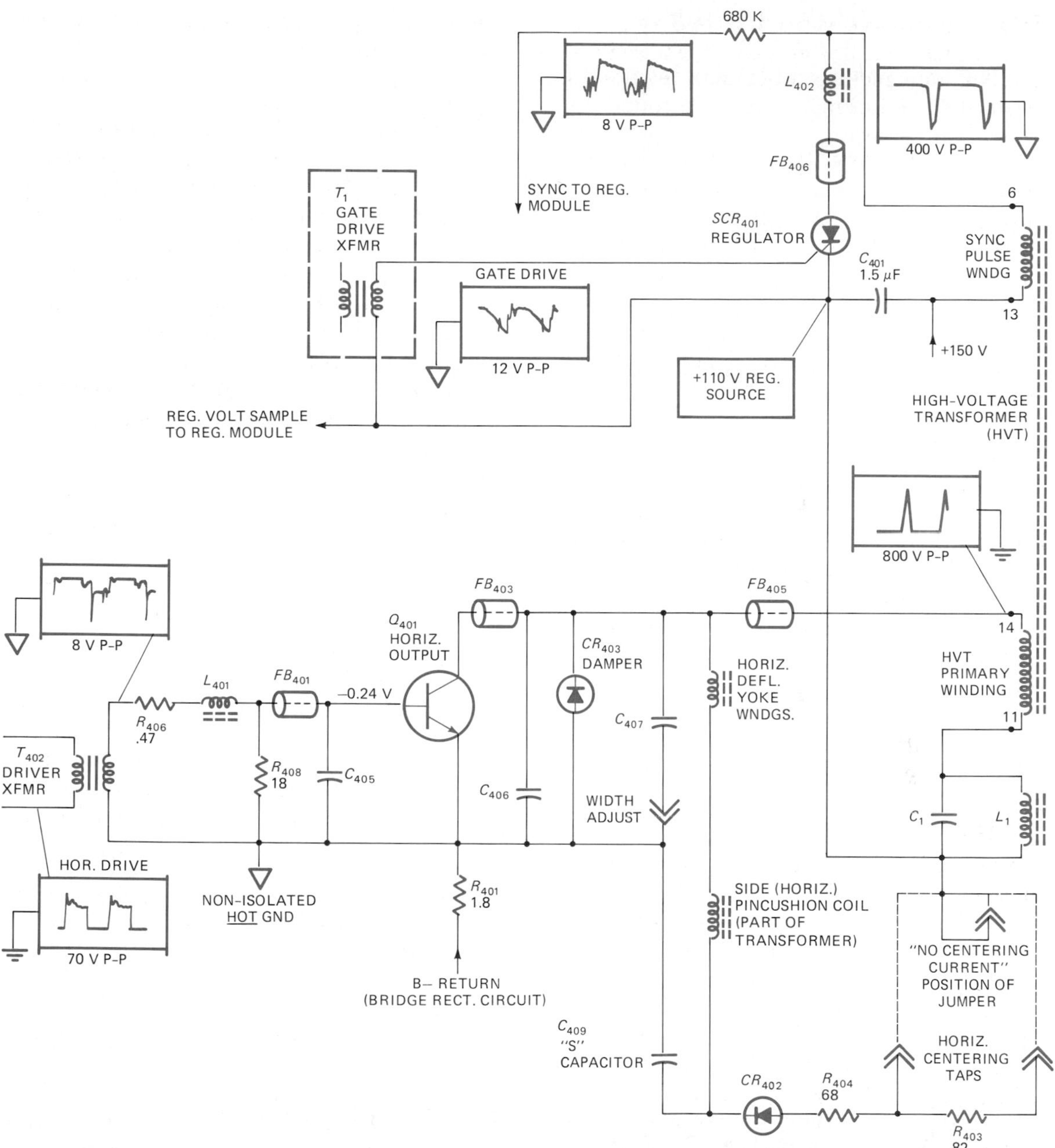

Figure 23.30 Simplified schematic diagram of a solid-state color TV horizontal-output circuit. *(Courtesy of RCA Consumer Electronics.)*

THE CENTERING CIRCUIT

Different picture tubes may require different DC centering currents through the yoke. The centering current is supplied by the 110-V regulated power supply. The centering current may be changed by selecting one of two series resistors, R_{403} or R_{404}. This is done by jumper selection. Note in Figure 23.30, that a third jumper position is provided that passes no DC through the yoke. The normal position of the jumper is at the right, where both resistors are in series with the centering current.

Developing the Centering Current When the S-shaped voltage on yoke return capacitor C_{409} is less than 110 V, diode CR_{402} conducts and centering current passes through the yoke. The path for centering current is as follows: from the 110-V supply, through the series resistor(s), through diode CR_{402}, and through the yoke and the conducting horizontal output transistor.

Note: In Figure 23.30 several ferrite beads are shown as FB_{401}, FB_{403}, FB_{405}, and FB_{406}. These are low-value inductors that function as RF chokes. They suppress high-frequency transients. In some circuits, such beads are used to prevent parasitic oscillations.

23.11 PINCUSHION (PIN) DISTORTION

There is an inherent distortion of the raster shape. This is particularly noticeable in large-screen picture tubes that use wide-angle deflection (such as 110°). This distortion is most noticeable at the four sides of the raster, but actually is present to some degree throughout the entire raster. "Pincushion" (PIN) distortion is illustrated in

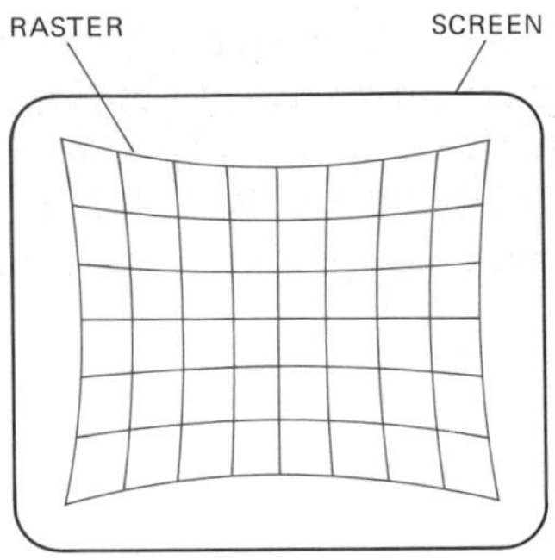

Figure 23.31 "Pincushion" distortion of the raster.

Figure 23.31. Note that all four sides bow in toward the center. (This figure shows an uncorrected condition.)

REASON FOR PINCUSHION DISTORTION

Pincushion distortion results because of deflection geometry. The tube face is not spherical with respect to the horizontal and vertical arcs swept by the picture tube electron beam(s). The face is appreciably flatter. Therefore, the distance traveled by the electron beam(s) to the screen is not uniform over the face of the picture tube. This distance increases as the beam(s) deviates from screen center and is greatest at the four corners. If uncorrected, the raster (and picture) would appear as shown in Figure 23.31.

To eliminate this problem, both horizontal (side to side) and vertical (top to bottom) corrections are required. In this chapter, we shall deal only with horizontal pincushion correction. Vertical pincushion correction is covered in Chapter 24.

HORIZONTAL CORRECTION REQUIREMENTS

Horizontal pincushion distortion can be correcting by carrying out the following two steps: (1) gradually increase horizontal width as deflection progresses from the top of the screen to the center and (2) gradually decrease horizontal width as deflection progresses from the center of the screen to the bottom.

CIRCUIT OPERATION

There are several circuits designed for horizontal pincushion correction. One such circuit (Quasar) is shown in simplified form in Figure 23.32A.

The PIN Transformer The primary (V_1–V_2) of the PIN transformer is fed from the vertical output transistor. The secondary (H_1–H_2) is connected in series with the horizontal yoke and output transformer.

Modulated Yoke Current The vertical (60-Hz) trapezoidal wave form appears across H_1–H_2 in the form of a parabola. Here it modulates the horizontal scan current. The resultant modulated horizontal scan (yoke) current is shown in Figure 23.32B. Note that the scan current is *least* at the *top* of the raster. It gradually increases to a *maximum* at the vertical *center*. Then, it gradually decreases to the *bottom* of the raster, where it is again a *minimum*. Compare this action with the sides of Figure 23.31. Note that the corrected (modulated) horizontal scan current has the opposite effect of the uncorrected pincushion raster (side-to-side only). As a result, the corrected horizontal scan current straightens the sides of the raster.

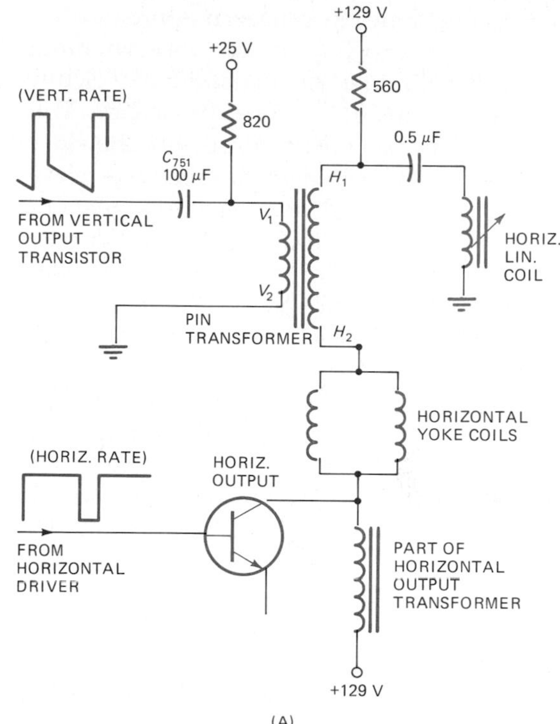

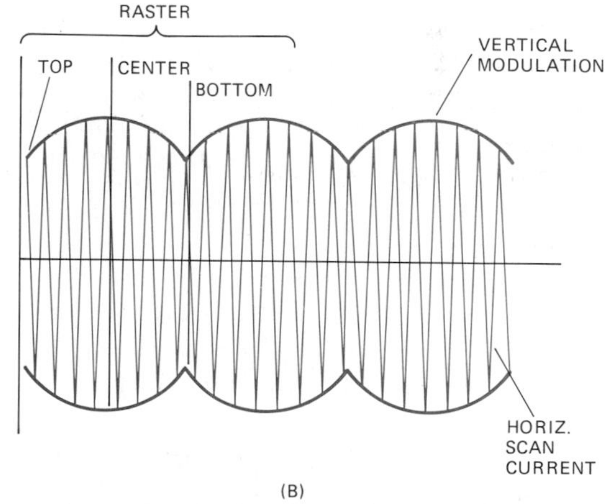

Figure 23.32A-B. Horizontal (side-to-side) pincushion (PIN) correction. **A.** Simplified schematic diagram. **B.** Horizontal yoke current across winding H_1-H_2 of the PIN transformer.

23.12 SILICON-CONTROLLED RECTIFIER (SCR) HORIZONTAL-OUTPUT CIRCUITS

The basic objective of all horizontal-deflection circuits using electromagnetic deflection is the same: to cause an approximately linear current to flow in the yoke windings in such a manner as to deflect the picture tube beam linearly across the screen. This current must be synchronized with the video information provided in the received television signal. Several manufacturers generate the desired horizontal-deflection current with circuitry that contains silicon-controlled rectifiers (SCR) and associated circuit elements. One typical circuit is described below.*

The essential components of an RCA SCR horizontal-output circuit are shown in Figure 23.33. Diode D_1 and rectifier SCR_1 provide the switching action that controls the current in the horizontal yoke windings, L_Y, during the picture tube beam-trace interval. Diode D_2 and rectifier SCR_2 control the yoke current during the retrace interval. The components L_R, C_R, C_H, and C_Y supply the necessary energy-storage and timing functions. Inductor L_{G1} supplies a charge path for C_R and C_H from B+, thereby providing a means of "recharging" the system from the power supply. Inductor L_{G2} provides a gating current for rectifier SCR_1 (L_{G1} and L_{G2} make up transformer T_{102}). Capacitor C_H controls the retrace time because it is charged up to the B+ voltage through L_R.

*Information regarding the operation of this SCR circuit is provided by permission of RCA.

To assist in the explanation of the operation of the horizontal-output circuit, the illustrations in the following discussion have been greatly simplified. In Figure 23.34 (and the illustrations following), SCR_1 and D_1 together constitute an SPST (Single Pole, Single Throw) switch labeled S_1, and SCR_2 and D_2 constitute another SPST switch labeled S_2.

TRACE TIME

Referring to Figure 23.34A, during the first half of the trace time (t_0 to t_2), switch S_1 is closed, causing a field previously produced about the yoke inductor L_Y to collapse and resulting in a current that charges capacitor C_Y. This yoke current deflects the picture tube beam to approximately the middle of the screen. The beam is at the center when this current decreases to zero at time t_2. During the second half of the trace interval, t_2 to t_5, the current in the yoke circuit reverses because capacitor C_Y now discharges back into the yoke inductor L_y, as shown in Figure 23.35B. This current causes the picture tube beam to complete its trace.

RETRACE INITIATION

However, at time t_3 a pulse from the horizontal oscillator causes switch S_2 to close, re-

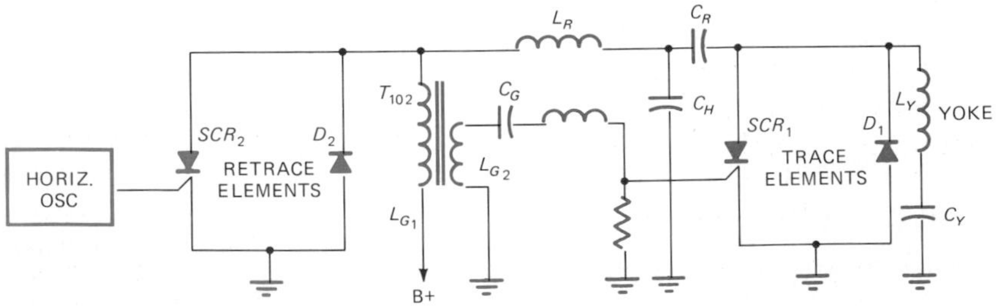

Figure 23.33 Simplified schematic of an SCR horizontal-output circuit. *(Courtesy of RCA Consumer Products.)*

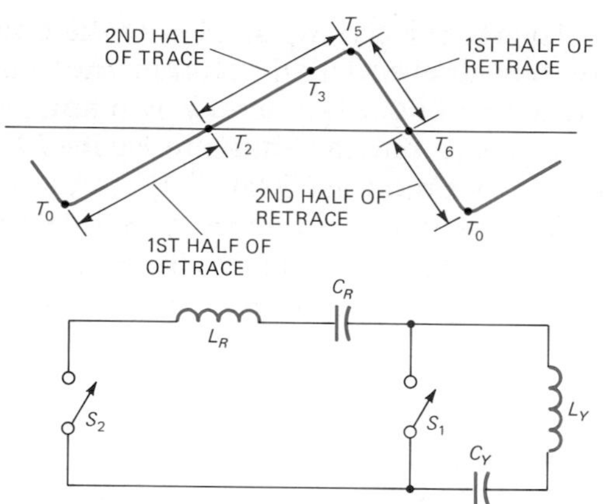

Figure 23.34 Simplified SCR output circuitry. *(Courtesy of RCA Consumer Products.)*

leasing the previously stored charge on C_R. The resulting current flows in the circuit, as illustrated in Figure 23.35C. Because of the natural resonance of L_R and C_R, this resonant current becomes equal in value to the yoke current, at t_5. At this time, S_1 opens, S_2 remains closed, and the retrace starts.

RETRACE

The simplified retrace circuit with S_2 closed and S_1 open is shown in Figure 23.35D. Basically, L_R, C_R, C_Y, and L_Y are connected in series. The natural resonant frequency of this circuit is much higher than that of the yoke circuit L_Y and C_Y because the value of capacitor C_R is very much smaller than that of C_Y. As a result, the current change through the yoke windings L_Y during retrace time is much faster than that during trace time. This, of course, causes the picture tube beam to retrace (fly back) very rapidly.

During the first half of the retrace time, t_5 to t_6, the retrace yoke current flows in the direction shown in Figure 23.35D. However, the retrace circuit current soon reverses because of resonant circuit action, and the last

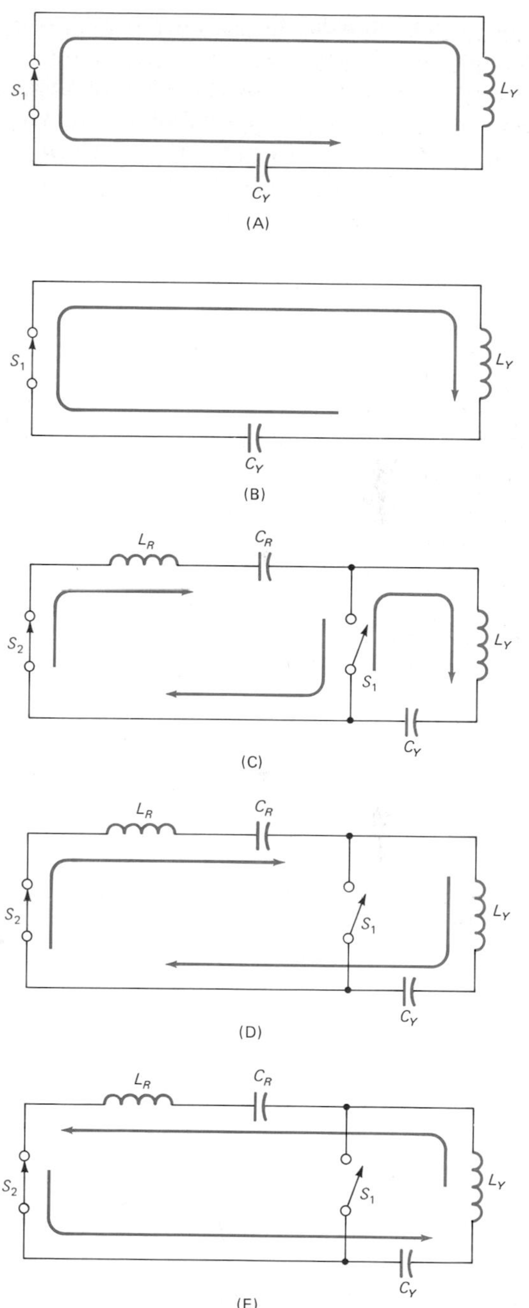

Figure 23.35A-E. Currents through portions of Figure 23.34 at different times: **A.** from T_0 to T_2, **B.** from T_2 to T_5, **C.** from T_3 to T_5, **D.** from T_5 to T_6, and **E.** from T_6 to T_0.

half of the retrace action occurs from t_6 to t_0, as shown in Figure 23.35E.

At time t_0, switch S_1 closes, and shortly thereafter S_2 opens. The field about the yoke inductor starts to collapse, and the resulting current again starts the trace interval. The trace-retrace cycle has now been completed.

SCR COLOR HIGH-VOLTAGE SYSTEM

High voltage is generated in this RCA solid-state chassis in a manner similar to most other color receivers. As shown in Figure 23.36, a vacuum tube (or solid-state) high-voltage rectifier provides a nominal 26.5 kV to the picture tube second anode. This potential varies approximately 2.5 kV over the normal range of the picture tube beam currents.

Focus and screen voltages for the picture tube are also derived from the high-voltage winding. A tap on this winding provides 1 000-V pulses, which are rectified and filtered to produce 1 000 V DC. Another winding, rectifier, and filter are used to provide −60 V for operation of the color-killer switch.

SCR HIGH-VOLTAGE REGULATION

The high voltage is regulated by controlling the amount of energy available to the horizontal-output circuitry. As previously stated, the output circuitry is supplied by the energy stored during trace time, primarily on the commutating capacitors (C_R and the auxiliary capacitor C_H), as shown in Figure 23.33. These capacitors are charged dur-

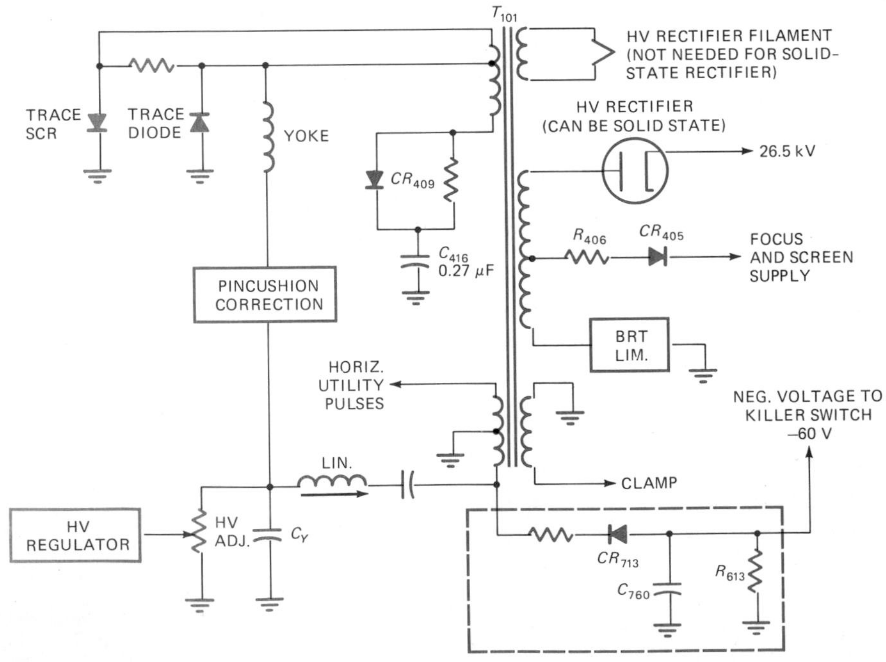

Figure 23.36 High-voltage generation circuitry for SCR horizontal-output stage. *(Courtesy of RCA Consumer Products.)*

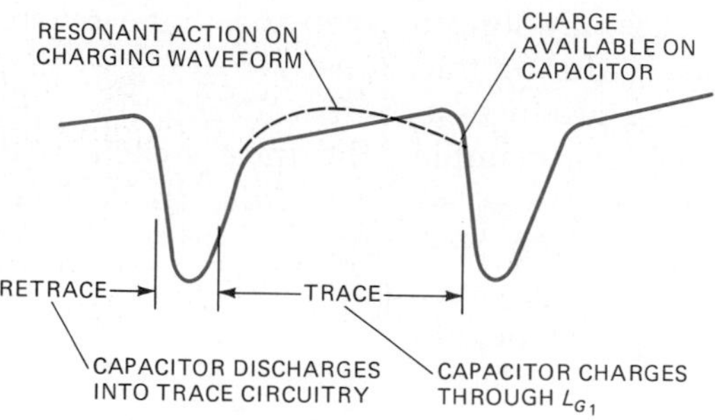

Figure 23.37 Voltage on commutating capacitor. *(Courtesy of RCA Consumer Products.)*

ing trace time through inductor L_{G1}, which is part of transformer T_{102}. In order to provide some way to control the energy on the commutating capacitors, inductor L_{G1} is designed to resonate with these capacitors at a frequency whose period approaches twice the horizontal scanning interval. The exact resonant frequency is made variable by the high-voltage regulator circuitry. Figure 23.37 illustrates the effect of this resonant action on the commutating capacitor charge. It can be seen that the shape of the wave resulting from resonant action will determine the amount of charge developed in the capacitors. More charge means a larger pulse and a larger high voltage. Likewise, if the voltage at the initiation of the retrace is lower (as shown by the dotted lines of Figure 23.37), then the high voltage will be lower.

Figure 23.38 further illustrates the relationships that exist among the elements making up the high-voltage regulating system. The resonance of L_{G1} and capacitors C_R and C_H is controlled by a saturable reactor, T_{103}, in parallel with L_{G1}. By changing the current in the reactor control windings, the total inductance represented by L_{G1} and the reactor load winding changes. The control current for the reactor is determined by the conduction of the high-voltage regulator transistor. In turn, the collector current of this transistor is controlled by the voltage across the yoke return capacitor C_Y. This latter voltage, which reflects high-voltage changes, is sampled by the high-voltage adjustment control and compared with a reference voltage provided by a zener diode. The resulting difference voltage reflects changes in the high voltage and controls the conduction of the high-voltage regulator transistor.

This regulating system is designed to

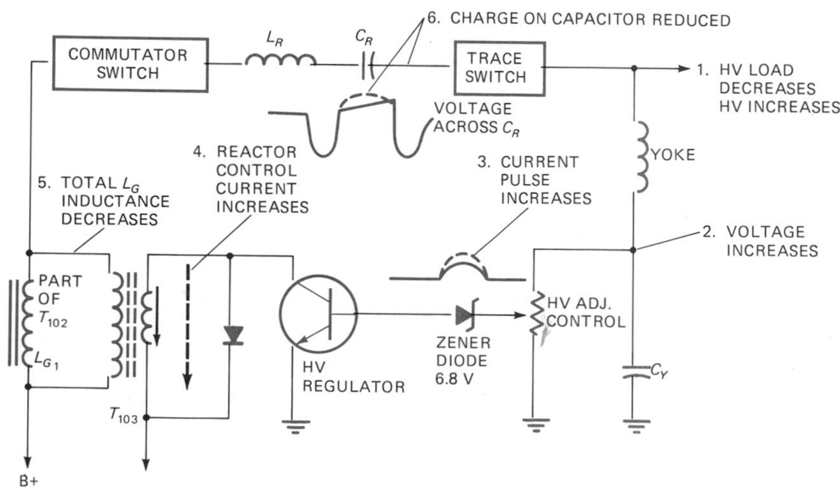

Figure 23.38 The action of the regulator circuit. *(Courtesy of RCA Consumer Products.)*

maintain high voltage substantially constant for line voltage variations ranging from 105 V AC to 130 V AC. The high voltage will drop only 2.5 kV from a nominal 26.5 kV with a picture tube beam current increase from 0 to 1.5 mA.

SCR PROTECTION CIRCUITRY

Two circuits protect the trace switch (both diode and SCR) from high currents resulting from high-voltage arcing (Figure 23.39).

One protection circuit consists of resistor R_{404} and diode CR_{409} placed in series with the primary of the high-voltage transformer. These components dampen the high ringing current that can occur under arcing conditions. This current is dissipated mainly in the resistor; the principal purpose of the diode is to allow normal initial flyback current to flow unimpeded, thereby preventing a reduction in high voltage.

The other protection circuit consists of diode CR_{403}, capacitor C_{406}, and resistor R_{405}. Diode CR_{403} conducts during the peak voltage of the retrace pulse, charging C_{406} to the peak voltage. Resistor R_{405} provides a high-resistance discharge path for the capacitor, sufficiently long to keep the diode reverse-biased during trace time. When high-voltage arcing produces a sharp voltage pulse, CR_{403} conducts, clamping the trace switch to the voltage on C_{406} and preventing the arc-pulse voltage from exceeding the breakdown voltage of the trace switch components.

A 400-megohm high-voltage bleeder resistor, R_{147}, located between the cathode of the high-voltage rectifier and ground, safely discharges the high voltage after the equipment is turned off.

23.13 HORIZONTAL-OUTPUT TRANSFORMERS AND SPECIAL ASSEMBLIES

In addition to its electrical circuitry, the horizontal-output system contains some major mechanical assemblies the reader should be familiar with.

OUTPUT TRANSFORMERS

A group of typical horizontal output transformers is shown in Figure 23.40. All of these devices may have the following terminals (or wires):

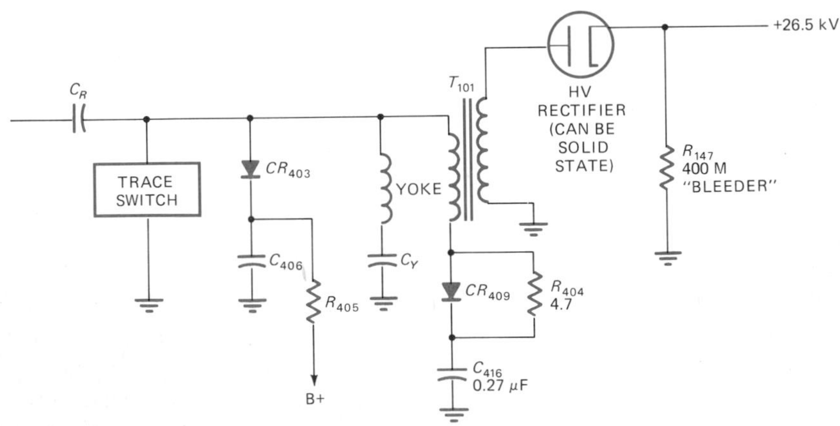

Figure 23.39 Horizontal-deflection protection circuitry. *(Courtesy of RCA Consumer Products.)*

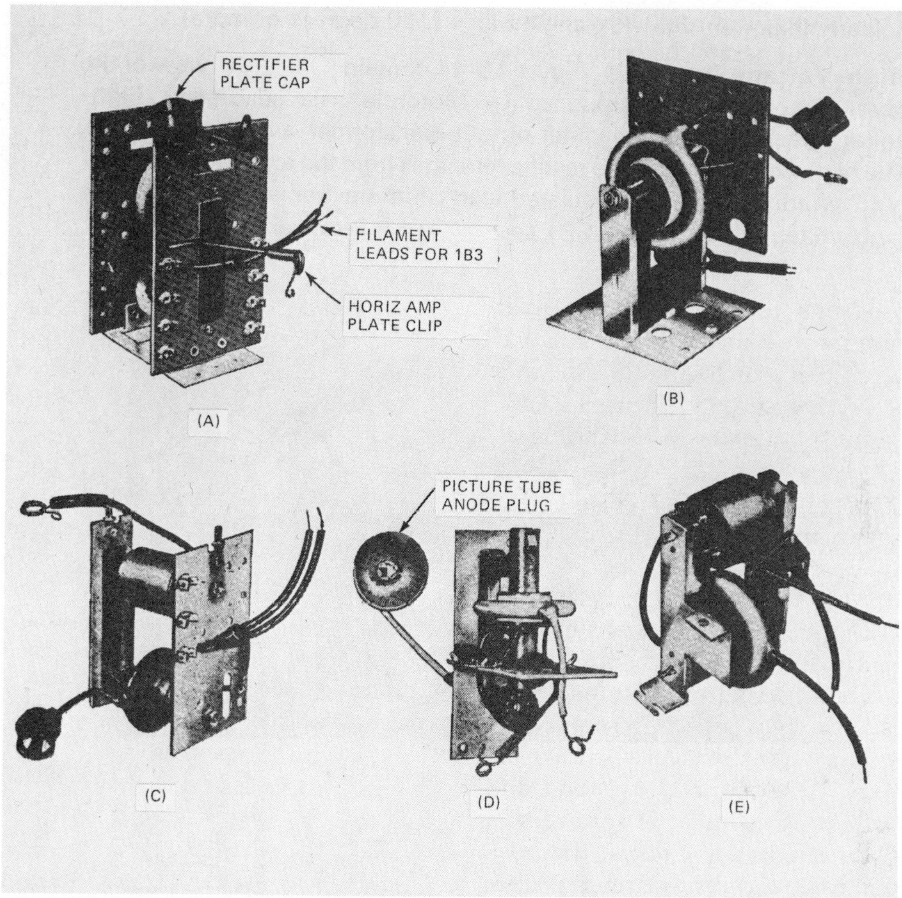

Figure 23.40A. A typical horizontal-output transformer for high-voltage flyback systems. **B.** Another view of the same transformer. **C.-E.** Other horizontal-output transformers.

1. Output tube or transistor, plate, or collector connection
2. High-voltage rectifier connection
3. Heater leads for a vacuum tube high-voltage rectifier (not present for solid-state high-voltage rectifier)
4. Terminals for the other functions: (1) yoke, (2) B++, (3) linearity, (4) width, (5) AGC, (6) horizontal AFC, (7) horizontal retrace blanking, (8) in color receivers only, focus voltage, (9) damper, and (10) centering.

Note that the above functions may not all appear in any one receiver. Also, not all of them require separate terminals. Several functions may be combined from a single pair of terminals.

THE WINDINGS

Two distinct windings are visible in Figure 23.40. One winding, the primary, carries the horizontal-output current. It is made of comparatively large size wire because the horizontal-output current is in the range of 150 to 250 mA. The other winding, the secondary, is a "doughnut" winding whose only function is to generate the high voltage. A small wire size can be used here since this winding generates from 12-to-30 kV pulses at less than 2 mA of current. This winding has many turns. A coating of special wax and/or a neoprene boot covers the winding to insulate and help the doughnut hold its shape.

The windings for the functions mentioned in Paragraph 4 above are partially visible in Figure 23.40. In Figure 23.40B, they are directly on the core. In Figures 23.40C and E, they are the upper set of windings.

COLOR TV HIGH-VOLTAGE TRANSFORMER

High-voltage transformers for color and monochrome TV receivers are very similar in appearance. Those for color sets are usually larger. This is because of the additional windings required for color sets. A typical color-TV high-voltage transformer is shown in Figure 23.41A. Transformers for tube and solid-state circuits are similar in appearance. When used for solid-state circuits, the horizontal-output tube caps shown in the photographs would not be present. Caps are generally used to connect to a solid-state rectifier.

The tube cap at the lower left is for the horizontal-output tube. This lead without the cap would go to the horizontal-output transistor. The tube cap at the upper right is for the high-voltage rectifier tube or solid-state rectifier. (See also Figures 23.45 and 23.46.) The resistor in Figure 23.41A is a 66-megohm high-voltage bleeder resistor that discharges the high voltage when the set is off. The potentiometer is for horizontal centering.

TRANSFORMER SCHEMATIC

A schematic diagram of the transformer is shown in Figure 23.41B. Inductance and

A

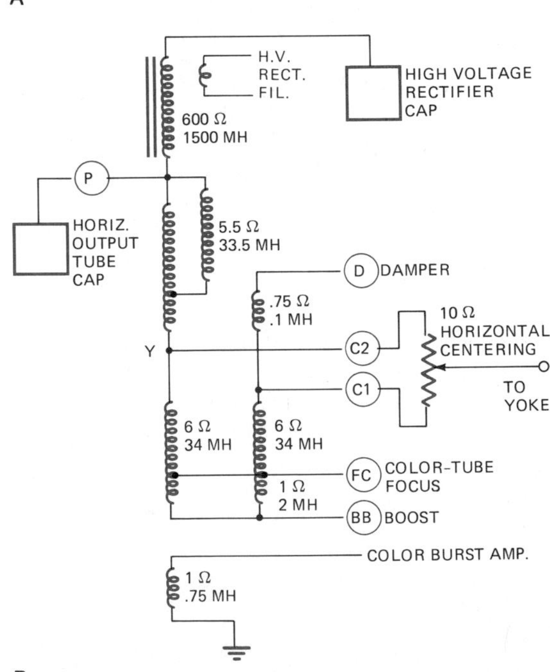

B

Figure 23.41A-B. Typical high-voltage transformer for a color receiver. **A.** Photo of the transformer. **B.** Schematic diagram of the connections and winding parameters. *(Courtesy of Triad-Utrad.)*

DC ohmic values are given for all windings. Note the windings for high-voltage, damper, centering, focus, B+ boost, and color-burst amplifier.

This particular transformer is used as a replacement in a number of manufacturer's receivers, including Magnavox, RCA, and Sylvania.

23.14 DEFLECTION YOKES

The deflection yoke consists of two sets of coils positioned at right angles to each other. These coils are mounted on the picture tube neck where the electron beam leaves the focusing electrode and travels toward the screen. Figure 23.42A shows schematically how the horizontal- and vertical-deflection windings are placed over the tube neck.

The entire assembly of deflection coils is known as a **deflection yoke.** Two typical commercial units are shown in Figure 23.42B. Note how the forward windings lap over the front edge of the yoke housing. The yoke is thus positioned right up against the flare of the tube. This way it can sweep the beam over the full screen area. This is particularly important for wide-angle tubes (110° or more). The yoke in the upper photograph of Figure 23.42B is known as a **saddle yoke** because of the manner of its construction. The yoke in the lower photo is a **toroidal yoke.**

TOROIDAL YOKES

The rear view of a toroidal yoke is shown in Figure 23.43. This yoke is used with a color picture tube. A toroidal yoke is wound on a doughnut-shaped core. This type of core confines the maximum magnetic field within itself with a minimum of external magnetic flux leakage. Thus, this yoke is

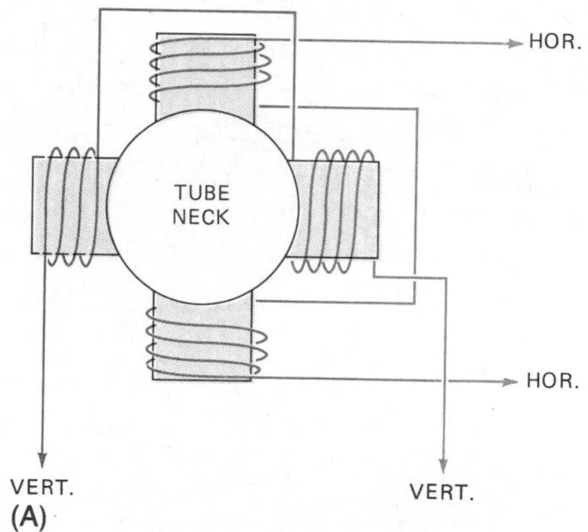

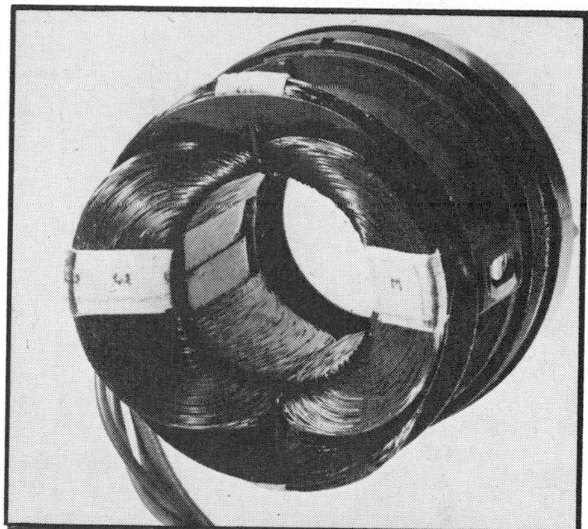

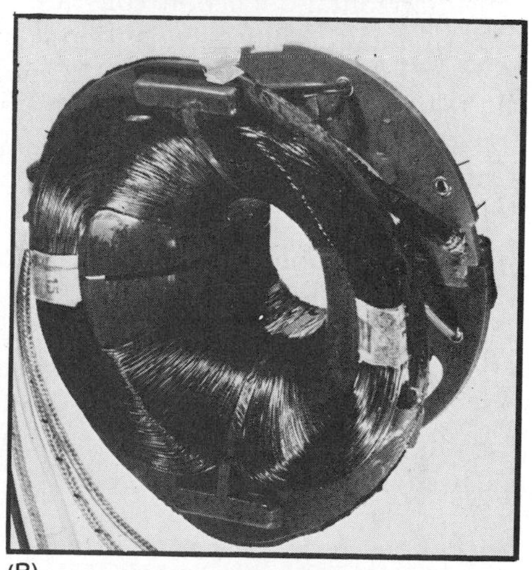

Figure 23.42A-B. The deflection yokes shown A. schematically over the CRT neck and B. pictorially.

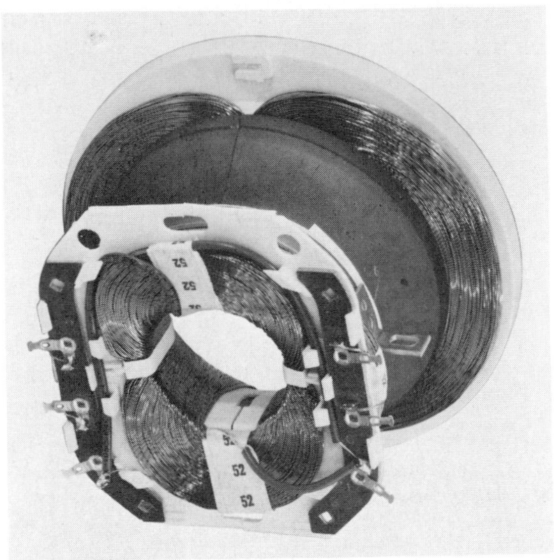

Figure 23.43 Toroidal deflection yoke used with a color picture tube. *(Courtesy of Triad-Utrad.)*

more efficient than a saddle yoke. Most yokes are now toroidal yokes. But sometimes a saddle yoke and toroidal yoke are combined in one vertical-and-horizontal deflection yoke.

The horizontal-deflection coils are connected in parallel. They have an inductance of 2.9 mH and a DC resistance of 2.5 ohms. The vertical-deflection coils are series-connected for a total inductance of 29 mH and a DC resistance of 14.5 ohms.

Toroidal Yoke Impedance Toroidal yokes have appreciably lower impedance than saddle yokes. Because of their lower impedance, toroidal yokes are used with transistor or SCR horizontal-output circuits. Such output circuits have considerably lower output impedances than vacuum tube circuits.

23.15 HORIZONTAL HIGH-VOLTAGE COMPONENTS

Figure 23.44 contains a closeup view of a high-voltage module. The solid-state high-

Figure 23.44 A high-voltage module containing the high-voltage transformer and solid-state rectifier.

voltage rectifier and horizontal-output transformer are clearly visible. The high-voltage lead to the rectifier emerges from the top of the doughnut winding. A heavily insulated lead on the extreme right routes the high-voltage DC out to the picture tube anode terminal.

SOLID-STATE HIGH-VOLTAGE RECTIFIER

A vacuum tube high-voltage rectifier and a solid-state high-voltage rectifier are shown in Figure 23.45. Note the much simpler construction of the solid-state device. Since no filament power is needed, the flyback circuit can operate more efficiently.

23.16 ADJUSTMENTS IN COLOR TV HORIZONTAL SYSTEMS

The adjustments and controls in color TV sets are nearly the same as those in monochrome TV sets, as described in Section 23.8. The major exceptions are the adjustments related to the high-voltage regulating circuits (tube sets only) and those of the high-

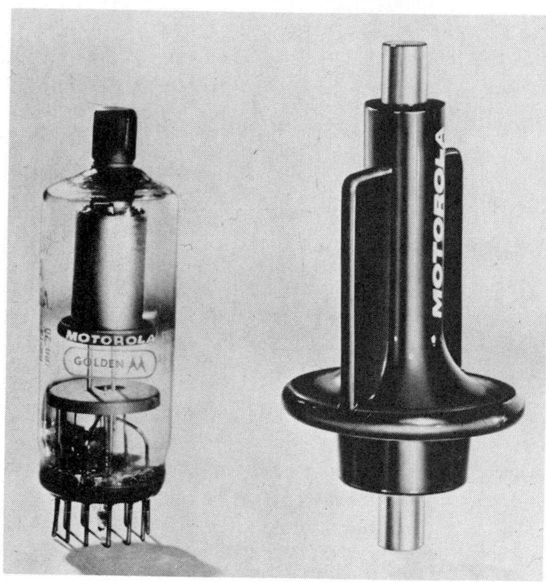

Figure 23.45 Solid-state high-voltage rectifier (right) compared to its vacuum tube counterpart.

voltage focus circuits. It must be remembered, however, that adjustment of the high voltage has an effect on the width and height of the picture. As the anode voltage is increased, the picture will get smaller; if the voltage is lowered, the picture will become larger. For this reason, adjustment of the width control may be necessary if the voltage at the anode is changed substantially.

Adjustment of the linearity, or "efficiency," coils for tube sets is described in Section 23.8. In this section, we shall discuss the linearity control in an SCR horizontal-output stage. Other types of solid-state horizontal-output stages do not require linearity adjustments.

HIGH-VOLTAGE CONTROLS[1]

Two basic high-voltage regulator circuits are used in color television sets: shunt and feedback. A typical tube shunt regulator circuit is shown in Figure 23.29. The high-voltage adjustment for that circuit is as follows:

1. Place a voltmeter across the 1-kilohm resistor in the cathode leg of the regulator and connect a high-voltage probe and VTVM to the +25-kV terminal.
2. Turn the brightness control down until the screen is *black*. No kinescope current will be flowing.
3. Adjust the 500-kilohm high-voltage control for a reading of 25 kV on the VTVM. The voltmeter across the 1-kilohm resistor should indicate a 1-V minimum. A voltage of 1 V across 1 kilohm means 1 mA of regulator current is flowing.
4. Turn the brightness control to its maximum position. If the picture "blooms," do *not* adjust the high-voltage control. The picture tube bias, screens, and drive controls must be adjusted, *not* the high-voltage control. Leave the latter control as set. Reset the picture tube bias, screens, and drive controls.
5. Notice that the voltmeter across the 1-kilohm resistor indicates 0 V (no regulator current) when the picture blooms. The important thing to remember for proper regulation is that at maximum brightness, the regulator never goes into cutoff. Cutoff means blooming and loss of regulation. Never allow the shunt regulator to be cut off by any adjustments or combination of adjustments.

FOCUS ADJUSTMENTS

A color picture tube requires up to 5 kV on its focus electrode. This must be at least ±10% adjustable to compensate for changes in the picture tube characteristics or voltages

[1]Solid-state television receivers do not require high-voltage regulators because the horizontal circuits are powered by regulated supplies.

on its various electrodes. One simple method used to obtain this voltage is to simply connect a bleeder between the 25-kV supply and ground. The 5-kV supply is tapped off with a potentiometer near the ground end. Some receivers have a separate winding on the flyback transformer. In this case, a separate rectifier and filter are required.

The procedure for the focus adjustment on a color set is quite similar to the procedure on a monochrome receiver. A strong station should be tuned in so that snow will not obscure the scan lines. The focus control is then adjusted until the scan lines are sharply visible over most of the screen. Brightness and contrast must be set at normal levels.

LINEARITY ADJUSTMENT

SCR horizontal scanning current nonlinearity, which is caused by voltage drops across inherent resistances in the trace circuitry, is corrected by two means. (1) The voltage drop resulting from the resistive effects of the trace diode and SCR is minimized by placing the trace diode at a more negative voltage than the trace SCR, as shown in Figure 23.36. This is accomplished by putting a resistor between the diode and the SCR and also by connecting the diode and the SCR to different points on the flyback transformer. (2) The remaining nonlinearity is corrected by a damped series-resonant circuit, illustrated in Figure 23.46. This circuit, consisting of L_{402}, C_{413}, and R_{419}, produces a damped sine wave current that effectively adds to or subtracts from the charge on the yoke return capacitor (C_Y). The resulting alteration in yoke current corrects for the remaining trace current nonlinearities. L_{402} is adjustable and is set to obtain the best horizontal linearity.

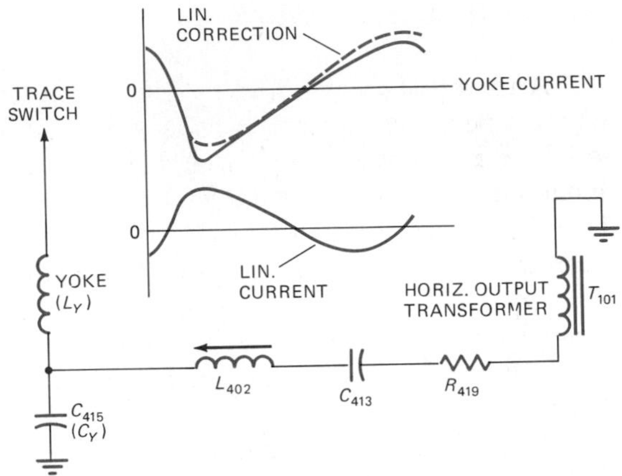

Figure 23.46 Horizontal-linearity adjustment in an SCR circuit. *(Courtesy of RCA.)*

23.17 X-RAY EMISSION

X RAYS IN COLOR TV RECEIVERS

Much publicity has been given to reports that color-TV receivers radiate x rays, which can be harmful to the viewer. Before going into the detailed explanation of this phenomenon, a few facts should be understood. The type of x rays produced by the high voltages used in color sets is called "soft" x rays and cannot penetrate skin or tissue to any appreciable depth. The most likely damage might be to the cornea of the eye. However, the levels of radiation possible from color sets are so low that, to be dangerous, it would be necessary to place one's eyes close to the source of the x rays (the high-voltage rectifier and regulator tubes). While any TV x-ray exposure is not beneficial, the amount of radiation we may receive at the dentist or during a routine chest x ray is generally far greater than we could get during years of proximity to the high-voltage section or picture tube of a color set.

NATURE OF X RAYS

An x ray is a form of radiation produced when a beam of electrons strikes some material at a relatively high speed. Generally speaking, acceleration voltages above 10 to 15 kV are required before any significant quantity of x rays is emitted. Because x rays are absorbed by glass envelopes, there is normally no significant escape of x rays from tubes until the voltages are in the range of 20 kV or higher.

When electron particles bombard material, such as a metal target, some of the energy of these electrons is converted to x rays. Unless shielded or otherwise absorbed, these x rays are emitted from the target in all directions. The energy of an x ray is proportional to the voltage accelerating the electrons. The quantity of x rays produced is very sensitive to the voltage, and a given increase in the voltage produces a *much greater* proportional increase in the amount of radiation. Therefore, it is important that the high voltage in color television receivers not be set above the specified levels.

The roentgen (R) is the international unit used in measuring x rays, and 1 milliroentgen (mR) equals 0.001 roentgen. X-ray measurements in the area of a color receiver are in the milliroentgen range.

X-ray radiation is cumulative in its effects, and any damage done to human tissue is irreparable. Even low levels are undesirable, and correctly operating TV sets do not emit any measurable amount of such radiation. It should be remembered that x-ray intensity, like any radiation, decreases in intensity as the inverse square of the distance. A 16-milliroentgen level 4 inches (10 cm) from the source will be only 4 milliroentgen 8 in (20 cm) away. With normal viewing distances of several feet from the screen, the viewer is in no danger, even if the set is not operating correctly and emits some x-ray radiation.

X rays are attenuated (reduced) by any material placed in the x-ray field. Materials used as x-ray shields include metal, glass, and ceramic, all of which are used in television receivers. The degree of attenuation is determined by the type, density, and thickness of the material and the energy of the x rays involved.

In color television receivers, x rays may be emitted from three possible sources: (1) the high-voltage regulator tube, (2) the high-voltage rectifier tube, and (3) the picture tube. (Many color picture tubes manufactured after June 1973 are specially treated to prevent excessive x-ray emission even when operated with excessive anode high voltage.)

X-RAY EMISSION FROM TUBE SETS

Emission of x rays from the regulator tube varies with voltage and tube current. Maximum x-ray radiation from the regulator tube occurs when the picture tube is dark. The emission of x rays from the rectifier tube occurs during the portion of the cycle when small reverse currents occur (not used in solid-state sets).

X-ray radiation from the picture tube depends upon beam current and voltage. Maximum emissions at rated tube voltage occur when there is a bright picture on the screen. These emissions may be further increased if the picture does not completely fill the screen.

As noted previously, excess high voltage creates even greater quantities of x-ray radiation. Conditions such as an excessive high-voltage setting, certain types of regulator tube failures, or unusually high line voltages can increase the intensity of x rays very significantly.

A color television receiver presents no x-ray hazards if it is operated at the specified power line voltage, has all the original factory-installed shields and equipment in place, and is correctly adjusted.

23.18 X-RAY PROTECTION CIRCUITS[1]

Some **x-ray protection circuits** are also called **high-voltage limit circuits** or **high-voltage hold-down circuits.** However, not all x-ray protection circuits reduce or cut off the picture tube anode voltage. Other methods in use are as follows:

1. Changing the horizontal or vertical oscillator frequency to make the picture nonviewable (the circuit for changing horizontal oscillator frequency is discussed below).
2. Preventing the video signal from reaching the picture tube.
3. Disabling the horizontal circuits, thus removing high voltage from the picture tube.
4. Lowering the anode high voltage, thus producing a dark, out-of-focus picture.

In all of these methods, the deliberately injected trouble remains until the TV-receiver defect is corrected. The end result of all protection methods is to prevent the user from viewing the picture and possibly being exposed to harmful x-ray radiation.

HORIZONTAL OSCILLATOR FREQUENCY CHANGE

In one RCA XL-100 type TV receiver, whose horizontal-output circuit is shown in Figure 23.30, method 1 is used. If excessive anode high voltage occurs, the horizontal oscillator is driven far off frequency. Thus, the picture becomes nonviewable.

Circuit Operation A simplified diagram of this RCA circuit is shown in Figure 23.47. The circuit operates by sensing the amplitude of a pulse from an auxiliary winding on the high-voltage transformer. If the pulse increases above the allowable limit, the horizontal oscillator is driven far off frequency.

With normal high voltage, the pulse is rectified by diode CR_9. The resultant DC voltage is applied to PNP transistor Q_7 as an emitter bias of +22 V, The base of Q_7 is clamped at 23.8 V by zener diode CR_{11}. Thus, during normal operation, the emitter voltage of Q_7 is less positive than the base voltage. As a result Q_7 is normally cut off and does not affect horizontal oscillator operation.

In the event of a circuit malfunction resulting in excessive anode high voltage, the applied horizontal pulse amplitude increases beyond its allowable limit. This causes an increase in the positive emitter bias of Q_7 sufficient to drive Q_7 into conduction. When Q_7 conducts (through diode CR_5), it introduces a DC error voltage into the horizontal AFC system. This drives the horizontal oscillator far off frequency until the malfunction is corrected.

23.19 TROUBLES IN HORIZONTAL-DEFLECTION CIRCUITS

Regardless of the type of horizontal-deflection system, most troubles originate in the amplifier or switching elements. Power tubes, output transistors, and SCR switches handle great amounts of energy. These generally fail first. Tubes may be substituted, and plate, screen, and grid voltages measured. Semiconductors should have element voltages measured before substitutions are made.

[1]The Bureau of Radiological Health, a federal agency, publishes a schedule of standards for color-TV x-ray radiation. After June 1970, new color sets must have a means of limiting x-ray radiation to 0.5 mR/hr. In today's sets, radiation *cannot* exceed 0.5 mR/hr, no matter what type of trouble arises in the set.

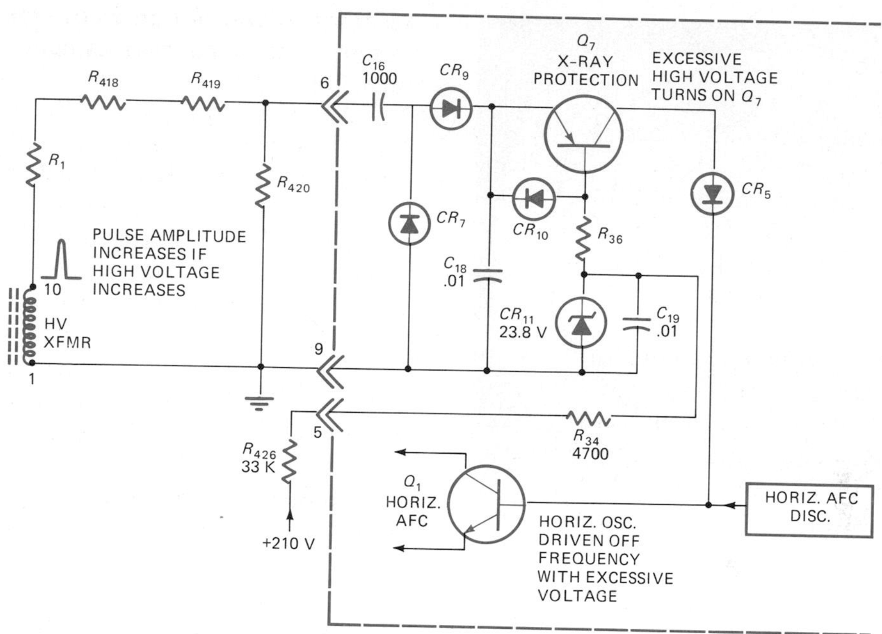

Figure 23.47 An X-ray protection circuit. Excessive anode high voltage causes the horizontal oscillator to be driven far off frequency. This makes the picture unviewable. *(Courtesy of RCA Consumer Electronics.)*

HORIZONTAL NONLINEARITY

Failure of tubes and/or flyback transformers can sometimes be traced to incorrect adjustment of the horizontal linearity, or efficiency, coil. Refer to Section 23.8 for the correct adjustment procedures. Figure 23.48 shows the appearance on the picture tube screen of a circuit defect (or adjustment) that causes poor horizontal linearity.

LOSS OF HORIZONTAL DRIVE

Replacing a horizontal oscillator tube while the set is operating will mean loss of drive to the output amplifier. In older units, without protection, this means possible destruction of the output tube. The circuit-breaker may trip if the horizontal efficiency coil is misadjusted.

In semiconductor systems, failure of the horizontal drive will cause the output transistor or SCR to turn safely off. This is just the opposite of the tube circuits. A bias check will spot this condition.

Remember that troubles on the *left half* of the screen are generally created by *damper circuit* difficulties. The *right half* of the screen is controlled by the *output amplifier* or *switch.*

WIDTH COIL MISMATCH

Width coils come in inductance values ranging from 0.05 mH up to 35 mH. Whatever the inductance, it must be matched to the transformer windings. Should any defect lead to a mismatch, such as a shorted turn in the transformer or in the width coil, it will be reflected in a picture that is either too narrow or distorted, or both. Even as little as

Figure 23.48 Horizontal nonlinearity shown on picture tube. *(Courtesy of RCA Consumer Products.)*

one shorted turn can have a marked effect, particularly in low-inductance width coils.

NARROW PICTURE HORIZONTALLY

This is also described simply as "insufficient width." This problem results from reduced horizontal yoke current and is identified by equally wide black bars on *both sides* of the raster. (Other conditions can reduce the width, but generally will show a black bar only at *one* side of the raster.)

With insufficient width, the basic problem is insufficient output from the horizontal output stage. This can be caused by a defective horizontal output tube or transistor or by a defective horizontal driver stage. Another cause is low B+ voltage to the horizontal output stage.

WIDE BLACK BAR AT LEFT SIDE

This is usually caused by defective damper operation. The picture will appear to be *folded over* from left to right. One cause of this problem is an open-circuited B+ boost capacitor.

NO RASTER, SOUND NORMAL

Chart 23.1 illustrates a practical troubleshooting procedure to be followed when there is no raster, but sound is normal.

23.20 HIGH-VOLTAGE TROUBLES

Probably the most troublesome section of the horizontal-deflection system, in both monochrome and color receivers, is the high-voltage section. Many of the defects that develop stem from a component that has failed in the high-voltage section itself. However, remember that a loss of high voltage may also be caused by a failure in some other section of the horizontal-deflection system. If the deflection wave is for some reason not reaching the output transformer, then no high voltage will develop.

The vacuum tube high-voltage rectifier is the cause of many high-voltage power supply failures. Low emission will cause picture blooming. The filament circuit may incorporate a low-value resistor. This unit can open or change the value and also cause blooming. Solid-state types are more reliable.

Corona discharge (discussed shortly) can cause a similar effect because it bleeds off energy from the high-voltage circuit. Improper "lead dress" of the high-voltage lead may cause corona problems. Replace and "redress" any such lead.

In color receivers, the addition of a regulator circuit increases the possibility of part failures. Referring to Figure 23.24, we can see how a shorted regulator tube will cause failure of the high voltage and perhaps the

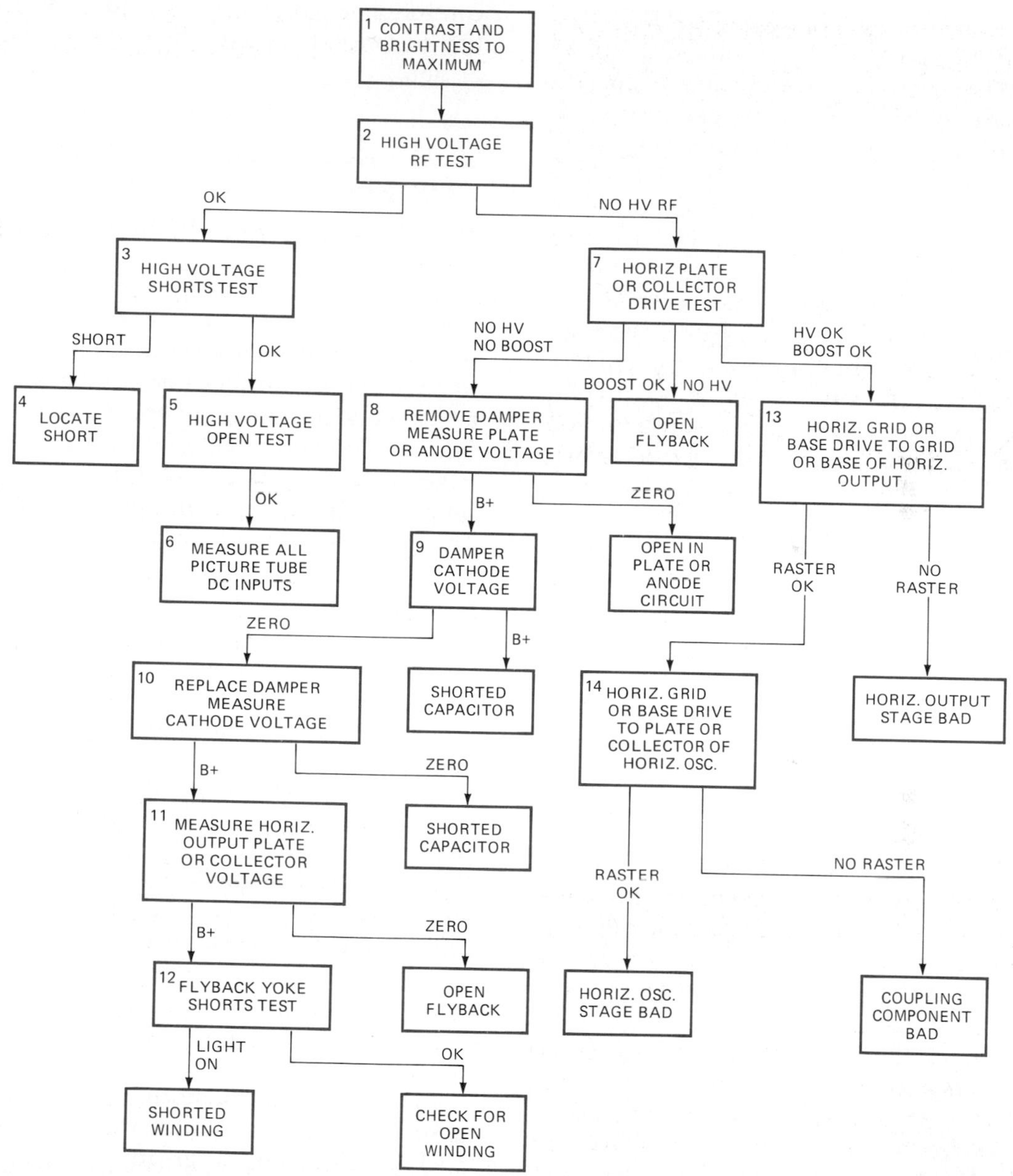

Chart 23.1 Illustrating a practical troubleshooting procedure for the condition of no raster, sound normal. *(Courtesy of B & K-Precision, Dynascan Corp.)*

high-voltage rectifier and flyback transformer as well. Misadjustment of the high-voltage control can create problems from blooming (loss of regulation) to excessive high voltage.

The focus circuits in color receivers are prone to failure. Focus coils handle high pulse voltages and are subject to breakdown.

EXCESSIVE FLYBACK CURRENT

Whenever certain horizontal circuit defects occur, the flyback may take excess current drawn in the horizontal output system and heat up the wax and/or neoprene insulation. This may cause failure. Corona may create hot spots and melt the wax to cause failure. Wax drippings in the bottom of the high-voltage cage mean that something is wrong. Be alert and perform the necessary adjustments to reduce excessive current flow in the flyback before the failure occurs. Figure 23.49 shows one possible way this problem may appear on the picture tube screen. In this case, the original short circuit is in the yoke windings.

HORIZONTAL CENTERING CONTROLS

Horizontal centering controls may carry yoke current. When the control is a carbon composition type, it is prone to pitting and failure. The usual defect shows up as loss of control or it works on one end only. In either instance, the picture will shift to the left or right.

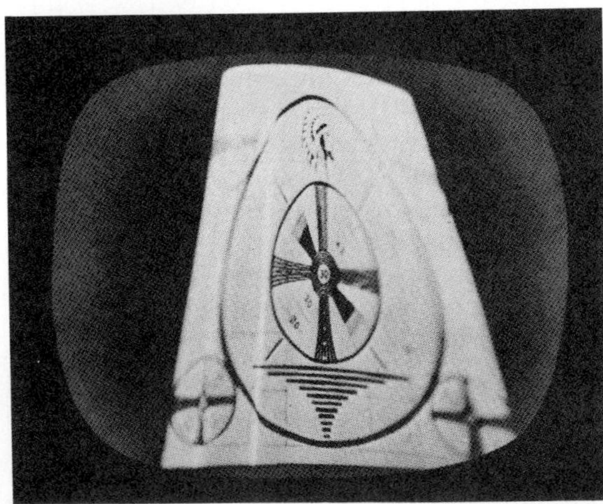

Figure 23.49 The appearance of the picture tube screen if a portion of the horizontal yoke is shorted. The effect is called horizontal "keystoning".

23.21 RELATIONSHIP BETWEEN HORIZONTAL AND HIGH-VOLTAGE PROBLEMS

Correct high-voltage generation is closely related to the proper operation of the rest of the horizontal system. A failure of the horizontal output transistor or tube will stop all delivery of energy to the flyback transformer. No energy will be forwarded to the yoke, stored, or returned during the beam retrace. High voltage will not be generated. There will be no screen brightness.

FLYBACK TRANSFORMER FAULTS

Should the lower (primary) winding of the flyback transformer open, the B+ path to the output transistor or tube will open. Again, there will be no high voltage. A short in the transformer will be most evident by the smoke and smell. A hot spot develops at the short and can be located easily. High-voltage generation will be disabled.

The doughnut high-voltage winding may open, resulting in no high voltage. A short in the doughnut, even if only one turn, can be found by its effect on the Q of the entire flyback transformer. Shorted turns absorb energy and reduce the Q of the flyback. This means a greatly reduced high voltage.

DAMPER CIRCUIT FAULTS

A horizontal output tube requires B++ in order to operate class C. Any failure in the boost filter will decrease this voltage below normal. Insufficient energy will be delivered to the flyback and the yoke. The small amount of energy returned to the circuit during the retrace will generate inadequate high voltage.

The damper charges the boost filter, and failure will either cause complete lack of high voltage or else produce a low high voltage.

FAULT IN YOKE

The yoke connects across the flyback transformer and is matched to the system. An open yoke will mean no B++ voltage. Shorted turns in the yoke are indicated by reduced high voltage and horizontal distortion (Figure 23.49). Check to see that the B+ from the TV receiver low-voltage power supply is satisfactory and does not contain too much ripple.

23.22 CORONA AND ARCING PROBLEMS

Corona and arcing are to be found in nearly all high-voltage circuits. The corona discharge is like a very fine electric spray into the air or to grounded surroundings. Areas with sharp bends or sharp points are most likely to form corona discharges.

The discharge has a very characteristic hissing or rushing sound. Placing the high-voltage section in complete darkness is an excellent way to see the faint blue glow around the discharge area. The discharge itself is not harmful, but the side effects can be disastrous to the flyback transformer. The corona discharge ionizes the air. Once ionized, the air becomes conductive and a heavy arc can form, which will burn through the insulation and wiring. At this point, a flame is nearly inevitable and the transformer is destroyed.

Arcing can occur if dust and small airborne debris are allowed to collect in the high-voltage cage or around the anode of the picture tube. Cleanliness in these areas can prevent the arcing and its crackling sound because the arc develops a path through the dust.

All wiring used in this area must be heavily insulated. Special wire with 30-kV or greater insulation property is used to conduct the high voltage to the picture tube. The caps on the leads of the high-voltage rectifier and the horizontal output tube or transistor are constructed to withstand such high voltages and prevent corona effects.

Special construction is used on a high-voltage rectifier tube socket to reduce corona effects. A corona ring is connected to the filament of the tube, as shown in Figure 23.50. This ring causes a smooth, high-voltage, equipotential surface to form around any jagged surfaces on the tube socket. High-voltage points on the tube socket can cause arcing or corona only in relation to a nearby lower-potential surface. They cannot do so when surrounded by a smooth, equipotential surface. The corona is thus minimized. Similar techniques can be used elsewhere in the high-voltage section, as applicable.

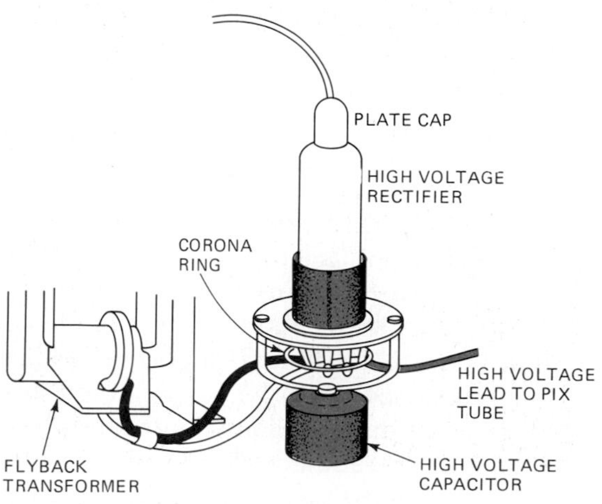

Figure 23.50 A corona ring used to reduce corona discharge.

SUMMARY OF CHAPTER HIGHLIGHTS

1. Major functions of the horizontal-output circuits are (1) supply horizontal scanning power, (2) provide picture tube anode high voltage, (3) provide focus high voltage for color tubes, (4) in many receivers, provide several B+ voltages, and (5) supply auxiliary pulses used for such functions as horizontal AFC and keyed AGC. (Figure 23.1)
2. The scanning power in the horizontal yoke must be supplied by a power device. This may be a transistor, an SCR, or a tube. (Figures 23.4 and 23.10)
3. The damper conduction provides the first portion of the forward trace. The remainder is provided by the output device (transistor or tube) conduction. (Figure 23.2)
4. Picture tube high voltage is generated during retrace (flyback) time. During this time, the output inductances oscillate at about 70 kHz. Retrace time is about 6.35-μs, roughly one-half cycle of 70 kHz. (Figures 23.5 and 23.7)
5. In question 4, the output inductances include the flyback transformer, the horizontal yoke, and any auxiliary inductances. (Figures 23.10 and 23.25)
6. The horizontal position of the electron beam(s) is dependent upon the magnitude and direction of the horizontal yoke current. (Figure 23.2)
7. The damper begins to conduct when the beam returns to the left side of the screen. At this time, it stops the 70-kHz retrace oscillation. Now the stored energy in the horizontal-output inductances is gradually dissipated, providing the initial portion of the forward trace. (Figures 23.2 and 23.5)
8. During damper conduction time, a B+ boost capacitor in the damper circuit is charged. The voltage across this capacitor is the B+ boost voltage. It may be several hundred volts higher than the B+ from the low-voltage power supply. (Figures 23.4 to 23.6, 23.25, and 23.29)
9. B+ boost voltages are not generally required in solid-state television receivers because of the relatively low operating voltages.
10. For transistor horizontal output stages, the base-drive waveform must be rectangular. The transistor acts as a switch. (Figures 23.11 to 23.13 and 23.15)
11. In Number 10, the transistor ON time is roughly equal to the OFF time. (Figure 23.15)
12. In Number 10, the damper ON time is slightly more than one-third of the forward trace time. (Figure 23.15)
13. A horizontal-output transistor must be protected against overcurrent and overvoltage. (Figures 23.9 and 23.10)
14. An S capacitor is placed in series with the horizontal yoke. This compensates for horizontal nonlinearity caused by a transistor horizontal-output circuit. (Figures 23.16 and 23.18)
15. In a solid-state horizontal-output circuit, loss of base drive will not cause output transistor damage. The output transistor will cease to conduct.
16. In vacuum tube TV sets, the drive, width, and linearity controls are very interactive. Adjustments must be repeated several times. (Figures 23.3 and 23.4)
17. In all TV receivers, it is essential to match the yoke impedance, through the flyback transformer, to the horizontal-output transistor or tube.
18. Vacuum tube color TV receivers have

anode high-voltage regulators. These are not needed in transistor receivers because their horizontal circuits are powered by regulated supplies.

19. Pincushion distortion makes the raster bow inward on all four sides. (Figure 23.32)
20. In Number 19, horizontal pincushion distortion can be corrected by modulating the horizontal scan current with a 60-Hz parabola. (Figure 23.33)
21. Horizontal-deflection yoke current can be supplied by SCR circuitry. (Figure 23.34)
22. In Number 21, the trace current is controlled by one SCR and one diode. The retrace current is controlled by a separate SCR and diode. (Figures 23.35 and 23.36)
23. The trace SCR and diode are protected from excessive currents by two circuits. (Figure 23.40)
24. Two common types of deflection yokes are the: (1) *saddle* yoke and (2) the *toroidal* yoke. The toroidal yoke is the more efficient of the two. (Figures 23.43 and 23.44)
25. Toroidal yokes have a much lower impedance than saddle yokes. Toroidal yokes are the more suitable for solid-state horizontal-output circuits. (Figures 23.43 and 23.44)
26. X-ray intensity is inversely proportional to the square of the distance between target and source. Doubling the target-to-source distance will reduce the intensity by a factor of 4.
27. In vacuum tube sets, x-ray emission can occur from the high-voltage regulator tube, high-voltage rectifier tube, and picture tube. In solid-state sets, the picture tube is the only important possible emitter of x rays.
28. Many color TV sets have x-ray protection circuits. These make the picture unviewable if the anode high voltage exceeds a predetermined value. (Figure 23.48)
29. In Number 28, the picture will remain unviewable until the malfunction is corrected.
30. In semiconductor horizontal-output systems, a failure of horizontal drive will cause the output transistor or SCR to turn safely off.
31. Low B+ power supply voltage can cause narrow picture width.
32. Picture *foldover* to the left can be caused by defective damper operation.
33. Shorted turns in a horizontal yoke can cause horizontal keystoning. (Figure 23.50)

EXAMINATION QUESTIONS

(Answers are provided at the back of the text.)

Part A Supply the missing word(s) or number(s).

1. The ______ conducts during the first portion of the trace.
2. If horizontal drive is lost, the horizontal output transistor ______.
3. Horizontal retrace occurs in about ______ to ______ μs.
4. "Ringing" at a frequency of ______ kHz begins immediately after the end of the ______ trace.
5. Boosted B+ is developed during ______ conduction.
6. Boosted B+ is generally ______ in solid-state TV receivers.
7. Picture tube anode high-voltage regula-

tors are _______ used in solid-state TV receivers.

8. In vacuum tube color TV receivers, a _______ waveform is used to drive the horizontal-output amplifier.
9. Anode high voltage is developed during the _______ portion of the horizontal scan.
10. Near the end of damper conduction, _______ conduction begins.
11. _______ cores are commonly used for horizontal-output transformers.
12. A _______ base-drive waveform is used with transistor horizontal-output amplifiers.
13. Output transistors must be given _______ and _______ protection.
14. An output transistor acts as a _______.
15. If the anode high voltage increases, the picture width _______.
16. A larger picture tube neck diameter results in the need for higher _______ current.
17. The yoke must be matched to the _______.
18. _______ to _______ kV is the approximate range of a color tube focus voltage.
19. High-voltage _______ is required in a vacuum tube color TV receiver.
20. In a transistor horizontal output stage, _______ begins at the moment the transistor is turned OFF.
21. Adjusting the DC _______ current is one method of horizontal centering.
22. _______ distortion causes bowing in all sides of the raster.
23. Modulating the horizontal scan current with a 60-Hz parabola corrects for _______.
24. The _______ solid-state horizontal output circuit does not use a transistor.
25. A _______ core is used in horizontal-output transformers.
26. A _______ yoke is wound on a doughnut-shaped core.
27. X rays are attenuated as the _______ square of the distance.
28. One type of x-ray protection circuit produces a large frequency change of the _______ or _______ deflection oscillator.

Part B Answer true (T) or false (F).

1. A trapezoidal wave of current is normally passed through the vertical-deflection coils.
2. The power output of certain deflection oscillators is adequate to drive the yoke.
3. The damper conducts only during retrace time.
4. B+ boost voltage can be as high as several hundred volts.
5. B+ boost voltage can be used for color tube focus voltage.
6. The damper permits one full cycle of oscillation to occur.
7. Output transistors need overvoltage and overcurrent protection.
8. A horizontal-output transistor will automatically turn OFF if the drive fails.
9. A rectangular base-drive waveform is required for a horizontal-output transistor.
10. A constant voltage across a pure inductance will cause linear sawtooth current to flow in it.
11. In a transistor horizontal-output stage, the S capacitor provides pincushion correction.
12. B+ boost is never used in a solid-state television receiver.
13. A variable inductor in series with the horizontal yoke can provide width control.
14. Anode high-voltage regulators are commonly found in vacuum tube color-TV receivers.
15. Horizontal centering in color TV receivers is normally accomplished by permanent magnets on the neck of the picture tube.
16. Pincushion distortion occurs only when a rectangular picture tube is used.
17. Pincushion distortion is corrected by modulating the horizontal scan currents with a 60-Hz parabolic wave.
18. In the SCR horizontal-output system, anode high voltage is provided in a manner similar to that of most other TV receivers.
19. Terminals of a horizontal-output transformer may include those for (1) output amplifier connection, (2) yoke connections, (3) AGC pulse, and (4) horizontal AFC pulse.
20. The trapezoidal yoke is commonly used for color-TV sets.
21. Most types of solid-state horizontal-output circuits do not require linearity adjustments.
22. The legal x-ray emission limit is 0.5 mR/h.
23. All x-ray protection circuits will cut off the

anode high voltage when necessary.
24. Damper circuit troubles are generally manifested at the left side of the screen.
25. A defective horizontal-output transistor can cause insufficient width.

REVIEW ESSAY QUESTIONS

1. Draw a simple block diagram of the horizontal output circuits for a solid-state color TV receiver. Label all blocks and voltages.
2. In Question 1, explain briefly, the operation of all the circuits.
3. Using Figure 23.5, explain why a trapezoidal driving voltage is applied to a vacuum tube horizontal-output stage.
4. Using Figure 23.15, explain why a rectangular driving voltage is applied to a transistor horizontal-output stage.
5. Discuss the operation of a damper stage, using Figures 23.2, 23.5, and 23.15 as aids.
6. The ringing frequency of the horizontal-output inductances and associated capacitances is about 70 kHz. What is the significance of this particular frequency?
7. In Figure 23.3, what is the function of capacitor C_3? Explain its operation.
8. In Figure 23.4, explain how the linearity control works.
9. In Question 8, where and how is the B+ boost voltage developed.
10. In Figure 23.10, what is the function of C_{118}? Of C_{107}?
11. List all the possible output connections for a solid-state color-TV receiver, horizontal-output transformer.
12. Why is it not necessary that the horizontal retrace be linear?
13. In Figure 23.10, explain the operation of the width control.
14. In Figure 23.10, in your own words, explain the operation of the current limiter.
15. Name two functions of the horizontal driver stage. (Figure 23.10.)
16. Describe briefly how the picture tube anode high voltage is developed. (Figure 23.5)
17. List four special requirements for solid-state horizontal-output amplifiers.
18. What is the basic advantage of the toroidal yoke over the saddle yoke? How is this accomplished?
19. In a transistor horizontal-output stage, what transistor current flows if the drive is lost? Explain.
20. Draw two simple circuits of width controls.
21. In Question 20, briefly explain the operation of each.
22. Draw a simple circuit of a solid-state TV receiver, horizontal centering circuit. Explain its operation.
23. In your own words, explain the reason for a pincushion raster. (Figure 23.33.)
24. In Question 23, draw and briefly explain a simple circuit (can be a block diagram) to correct for horizontal pincushioning.
25. List the components that might be responsible for x-ray radiation.
26. What factor(s) in a picture tube might be responsible for excessive x-ray radiation? Why?
27. List three methods used in x-ray protection circuits.

EXAMINATION PROBLEMS

(Selected problem answers are provided at the back of the text.)

1. The *ringing* frequency of a horizontal output system is 75 kHz. Calculate the horizontal retrace time.
2. The x-ray emission level at 5 in (12.7 cm) from a source is 25 mR. What is the level at 10 in (25.4 cm)?
3. In Figure 23.4, identify the horizontal yoke.
4. A damper diode conducts for one-third of the

forward trace. Calculate its ON time in microseconds.

5. In question 4, calculate the OFF time in microseconds of the horizontal-output transistor. (Hint: Include retrace time.)

6. In Figure 23.25, calculate the time constant of the input grid circuit of the horizontal-output tube. (Hint: The power supply points (+140 V, −70 V) can be considered to be at AC ground.)

Chapter 24

24.1 Operation of vertical-deflection circuits
24.2 Special requirements for monochrome TV transistor vertical-deflection circuits
24.3 Special requirements for color TV transistor and vacuum tube vertical output circuits
24.4 Vertical dynamic convergence
24.5 Vertical blanking
24.6 Monochrome TV vacuum tube vertical output circuits
24.7 Monochrome TV transistor vertical output circuits
24.8 Color TV transistor vertical output circuits
24.9 Vertical output transformers
24.10 Vertical yokes
24.11 Troubles in vertical output circuits
Summary

VERTICAL OUTPUT DEFLECTION CIRCUITS

INTRODUCTION

The vertical output deflection circuits provide the current needed to drive the vertical windings of the deflection yoke during the scanning process. These circuits amplify the output of the vertical oscillator to a level necessary to drive the coils in the deflection yoke. They also shape the waveform so that the current driving the yoke produces a *linear* vertical sweep. The output voltage is sometimes fed back to the output amplifier driver circuits to improve the linearity of the sawtooth wave form.

In addition to driving the yoke, the vertical output deflection circuits provide blanking pulses to blank out the picture tube during vertical retrace after each scanned field. Some circuits also provide for linearity correction and, in color receivers, for convergence of the three scanning beams.

In the early days of television, vacuum tubes were the sole means for controlling the yoke drive current. Later the transistor came into widespread use in television, and today the trend is for all solid-state sets, which includes transistors and ICs. Transistors are particularly well suited for use as current amplifiers in the present-day magnetic deflection systems. There are, however, several different vertical output deflection circuit configurations in present use. These configurations include all vacuum tube, combination vacuum tube and transistor (hybrid), and all transistor circuits.

When you have completed the reading and work assignments for Chapter 24, you should be able to:

- Define the following terms: saturable reactor, current ramp, Darlington pair, complementary-symmetry pair, vertical static convergence, vertical dynamic convergence, vertical yoke, PIN, trapezoidal wave, kickback pulse, thermistor, keystoning, parabolic wave, and vertical foldover.

- Explain the operation of television receiver vertical-deflection circuits (transistor and vacuum tube types).
- Understand the reason for vertical pincushion (PIN) and its correction.
- Describe how thermal protection is accomplished in transistor vertical output circuits.
- Describe toroidal and saddle vertical yokes and their characteristics.
- Explain how static and dynamic vertical convergence is done in a delta gun color picture tube.
- Describe the types of vertical output transformers.
- Understand some common troubles in vertical output circuits and how to isolate them.
- Describe the operation of a complementary—symmetry pair, vertical output amplifier.
- Understand three types of coupling from a vertical output amplifier to a vertical yoke.
- Draw a simple block diagram of a color TV vertical output circuit.

24.1 OPERATION OF VERTICAL-DEFLECTION CIRCUITS

VACUUM TUBE CIRCUITS

Figure 24.1 is a block diagram of a basic tube vertical output circuit. This circuit uses a single vacuum tube to amplify the signals from the vertical oscillator and a transformer to couple the amplified signals to the magnetic deflection yoke. Usually, the vertical output tube is one section of a multisection vacuum tube, the other sections being used in the vertical oscillator circuits, as discussed in Chapter 21.

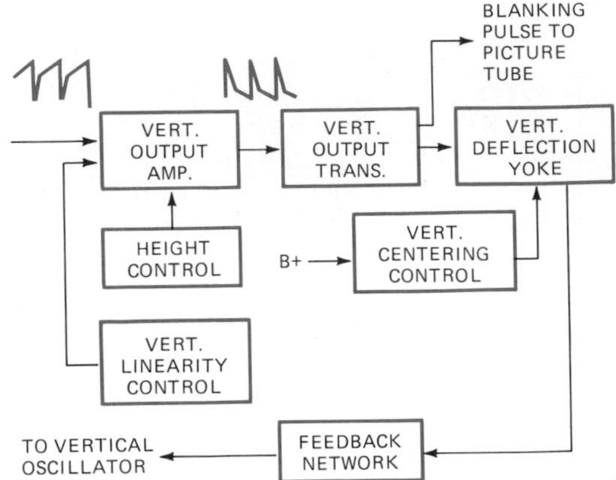

Figure 24.1 Block diagram of a basic tube-type vertical-output circuit.

The positive-going sawtooth wave form from the vertical oscillator is applied to the grid of the vertical output tube. The operating point on the characteristic curve of this tube is established by a vertical linearity control, and the output amplitude is set by a height control. These controls will be discussed later in this chapter. The output of the tube, an inverted replica of the input wave form, is impressed across the primary winding of the vertical output transformer (VOT). The main function of this transformer is to match the low impedance of the yoke to the high impedance of the vertical output tube. Also, since the vacuum tube is a high-voltage, low-current device, the VOT provides a means to step-down the voltage and thus step-up the current.

Centering In some vertical output circuits, a means for centering is provided. This portion of the circuit is simply a voltage divider that sends a DC centering current through the vertical-deflection windings of the yoke. The centering current flows through the same windings of the yoke as does the vertical driving sawtooth current. The centering current may flow in either direction, as required to move the raster up or down.

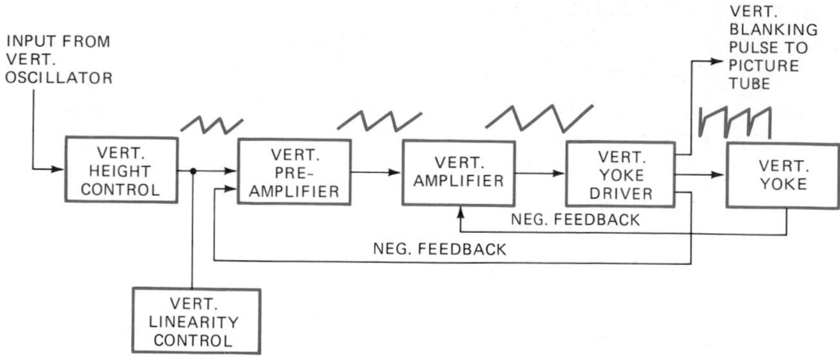

Figure 24.2 Block diagram of a transistor vertical output deflection circuit.

Feedback A feedback network, also shown in the block diagram, is used in some vertical output circuits. This circuit feeds a portion of the vertical output signal back to the input of the output tube or to a prior stage. In this way, the rise of the sawtooth current is kept uniform, providing the linear magnetic deflection field required in the yoke.

Blanking Pulses Vertical blanking pulses are produced by a network in the output stage. This network generates a pulse that is applied to the picture tube control grid, or cathode, to cut off the scanning beam during vertical retrace.

TRANSISTOR CIRCUITS

Figure 24.2 is a block diagram of a typical transistor vertical output circuit. The sawtooth wave form from the vertical oscillator is fed through a preamplifier, an amplifier, and a yoke driver stage to provide the necessary power gain for driving the vertical windings of the deflection yoke. The amplifier stages also include networks for properly shaping the wave form. Linearity adjustments are made in the biasing networks of the preamplifier stage. The height control is shown in the input circuit of the preamplifier. This control establishes the amplitude of the input signal. Linearity is corrected by feedback from the yoke and yoke driver. The driver stage also provides vertical retrace blanking pulses.

VERTICAL OUTPUT COLOR TELEVISION CIRCUITS

The block diagram of a vertical output circuit suitable for a color receiver is shown in Figure 24.3. In this circuit, there are two main features that are not present in monochrome circuits: pincushion correction and dynamic convergence. The pincushion correction network is required mainly with wide-deflection-angle picture tubes, but dynamic convergence is required in many color TV receivers. The service switch shown in the diagram disables the vertical output amplifier to facilitate gray scale adjustment.

Vertical Pincushioning Horizontal pincushioning and its correction are described in Chapter 23. Here, we are concerned only with vertical pincushioning.

Vertical pincushioning consists of the bowing of the horizontal lines in a vertical direction (Figure 24.6). To correct this condition, a parabolic voltage from the horizontal-deflection system is applied to the dynamic pincushion correction circuit. Here,

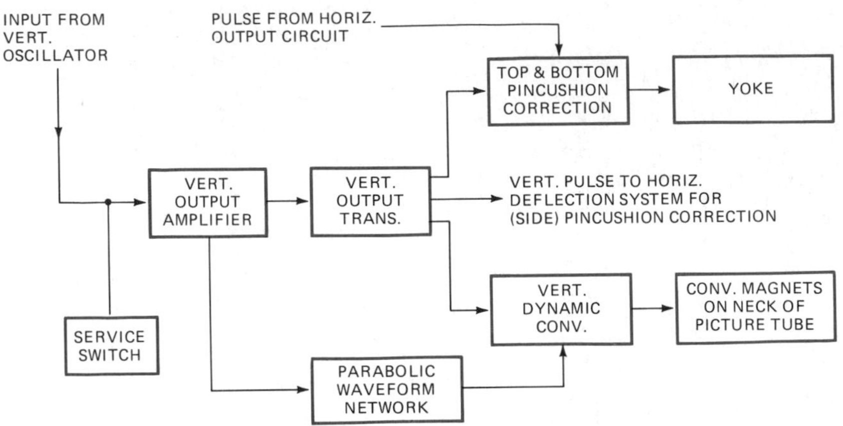

Figure 24.3 Block diagram of a vertical output circuit for a color TV receiver.

the parabolic voltages are added to the vertical output sawtooth that drives the vertical windings of the yoke. The vertical sawtooth, modified by the parabolic wave form at the horizontal scanning rate, increases vertical deflection at the center top and center bottom of the raster to compensate for the bow in the horizontal scanning lines. Service controls are provided to permit precise adjustment of the correction wave form. The circuit (for top and bottom pincushion correction) is discussed in Section 24.3.

Vertical Dynamic Convergence The purpose of the dynamic convergence circuit is to cause the three electron gun beams to converge on the face of the picture tube. As shown in Figure 24.3, a parabolic wave form is developed in a special network and is applied to the vertical dynamic convergence circuit to achieve this. Convergence windings on the VOT provide sawtooth voltages to drive the circuits. The sawtooth from the VOT is modified by the parabolic wave form and is then applied to the convergence coils mounted on the yoke housing around the neck of the picture tube. Controls are provided to adjust the amplitude and the phase of the sawtooth voltage, as well as the parabolic wave form. By properly setting these controls, the current flow through the convergence coils develops a magnetic field that converges the three beams at the edges of the raster. Dynamic convergence is discussed in greater detail in Section 24.3.

24.2 SPECIAL REQUIREMENTS FOR MONOCHROME TV TRANSISTOR VERTICAL-DEFLECTION CIRCUITS

Transistor vertical output circuits follow generally the same configuration as their vacuum tube counterpart circuits. The major difference between the two systems lies in component values, which are necessarily different. The transistor is a low-voltage, high-current device and the vacuum tube is a high-voltage, low-current device. Because of the high current flow, the transistor circuits require some form of thermal stabilization. Also, transistor operating parameters are more critical than those of the vacuum tube. Consequently, additional variable controls are required to achieve proper operation of the output circuits. This is especially true if transistors are replaced during maintenance.

LINEAR VERTICAL SCAN

To produce a linear vertical scan, the yoke current must rise at a uniform rate with no more than a 10% deviation. This requirement is accomplished by operating the output transistor as a class A_1 linear amplifier for the 60-Hz vertical-deflection sawtooth wave form. Phase shift in the amplifier is minimized by maintaining a frequency response that is flat down to 1 Hz. Wherever possible, direct coupling is used; otherwise, very large coupling and decoupling capacitors are used. Wave-shaping and negative feedback are also used extensively throughout the vertical-deflection circuits.

TRANSISTOR THERMAL PROTECTION

Figure 24.4 is a simplified diagram of a vertical output stage showing how thermal protection of the transistor is achieved. Since Q_1 is a power transistor, it operates with a substantial collector current. This high current causes appreciable heating of the collector junction. Even though heat sinks are used, this effect tends to increase collector current flow and tends also to increase the base-to-emitter current. To prevent thermal runaway of Q_1, the base-to-emitter bias current is stabilized by the thermistor shown in the base circuit of Q_1. The increasing collector current and temperature causes the base-to-emitter current to increase. Thus, more current is drawn through the thermistor, causing its temperature to increase. Increasing the temperature of the thermistor decreases its resistance. The voltage divider action of the thermistor circuit then reduces the forward bias of Q_1. This in turn prevents the collector current from increasing and stabilizes the amplifier.

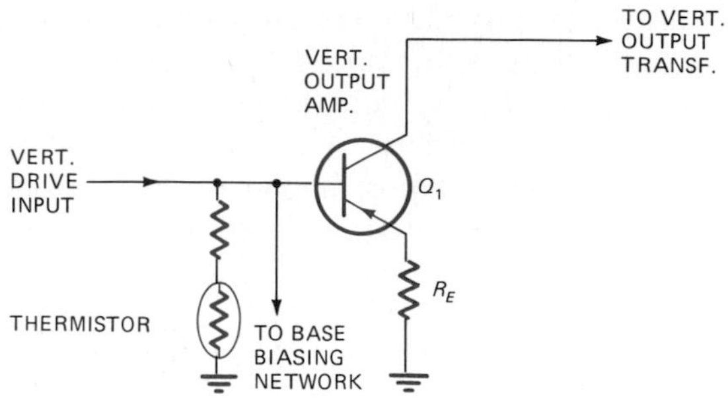

Figure 24.4 A vertical output transistor amplifier with thermal protection.

Additional thermal protection is provided by emitter resistor R_E. Any increase in collector current causes an increased voltage drop across R_E. This results in a decrease in transistor forward bias which tends to oppose the increase in collector current.

BIAS CONTROLS

Figure 24.5 is a simplified diagram showing two manual controls in the vertical output circuit. Unlike vacuum tubes, transistors used in vertical output circuits require fairly critical biasing. To achieve this biasing,

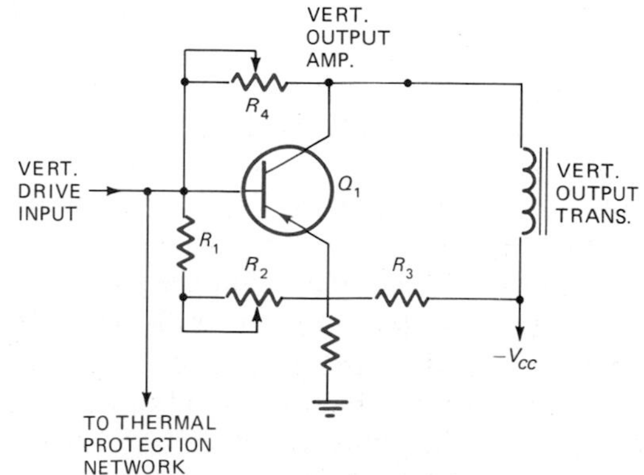

Figure 24.5 A vertical output transistor amplifier bias control circuit.

[1]Class A operation in this case means that neither the positive nor the negative peak of the output wave form is clipped.

variable voltage dividers are included in the base circuit of the transistor. In the diagram R_1, R_2, and R_3 are connected in series from the base of Q_1 to the $-V_{cc}$ supply. Adjusting R_2 establishes the proper bias for the transistor. Variable resistor R_4 is connected from the collector to the base of Q_1 to provide additional feedback and to make the tolerances of replacement transistors less critical.

24.3 SPECIAL REQUIREMENTS FOR COLOR TV TRANSISTOR AND VACUUM TUBE VERTICAL OUTPUT CIRCUITS

Vertical deflection circuits for color television receivers are virtually identical to those for monochrome receivers, except for pincushion correction and dynamic convergence. Also, since there are three electron beams in the color picture tube, each beam must be set at the precise strength to produce a monochrome picture. A service switch is provided to facilitate this setting. The circuits presented in this section operate equally well in vacuum tube or solid-state receivers.

TOP AND BOTTOM PINCUSHION CORRECTION

Wide-deflection-angle yokes used with rectangular picture tubes produce a stretching effect at the four corners of the tube face. As mentioned in Chapter 23, this stretching, which causes a bowing of the upper and lower horizontal scanning lines, is called pincushion (PIN) effect. Figure 24.6 shows the effects of pincushioning on a raster, reduced to thirteen scanning lines.

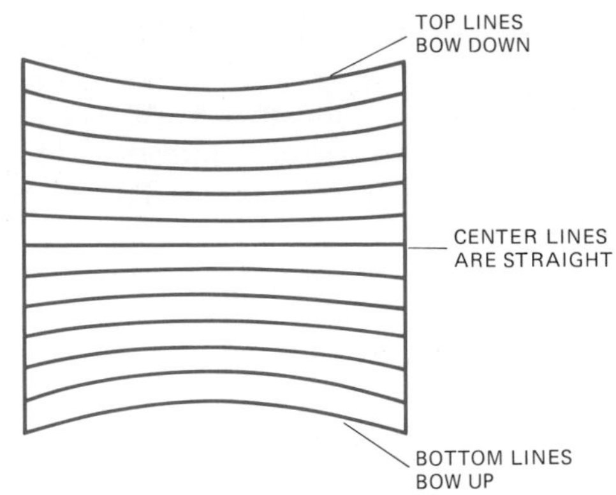

Figure 24.6 Pincushion effects on a 13-line raster.

To eliminate the pincushion distortion, and thus straighten the scanning lines, a maximum correction in a direction opposite the pincushion bow must be applied at the top and bottom areas of the raster. As vertical scanning progresses from the top of the screen, the pincushion effect decreases until, at the center of the screen, the horizontal lines are perfectly straight. Then slightly below the center, pincushioning again becomes apparent, but in the opposite direction, reaching a maximum at the bottom of the raster.

PIN Correction To correct, or bow upward, a pincushioned top-raster horizontal scanning line, the portion of the vertical-deflection sawtooth corresponding to the period of time that the horizontal line is sweeping across the screen must curve upward in a parabolic form. Figure 24.7A shows that for each horizontal line, a parabolic curve is superimposed on the vertical sawtooth. These curves counteract the pincushion effect on the horizontal lines and produce a straight-line raster.

At the center of the raster, current flow in the vertical windings of the yoke passes through zero, as shown in Figure 24.7. The horizontal scanning lines at the center of the raster are therefore perfectly straight, and correction is not needed here. As scanning progresses in the lower half of the screen, the vertical sawtooth must provide a downward-curving correction for each horizontal line. During vertical retrace, this correction

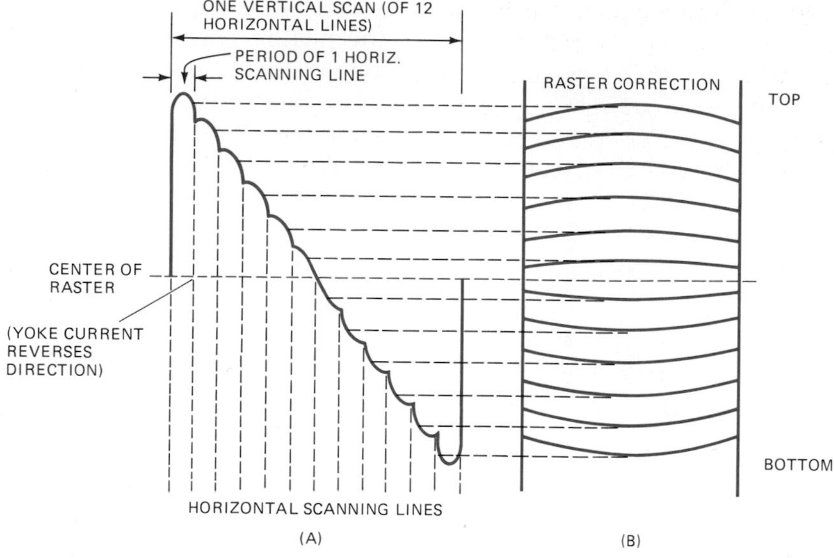

Figure 24.7 The method of top and bottom pincushion (PIN) correction.

process continues so that the pincushion correction begins at the proper place for the next vertical scan.

Since each horizontal line is corrected by a parabolic curve superimposed on the vertical sawtooth, a parabolic wave form at the horizontal scanning frequency is generated and added to the vertical-deflection current wave form. Figure 24.8 shows the circuit arrangement for accomplishing this.

THE PIN TRANSFORMER

A special transformer (T), known as a variable **saturable reactor** is used to add the horizontal frequency parabolic wave form to the vertical sawtooth. The saturable reactor has three windings, which are wound about a closed "E" core made of ferrite. A sketch of a transformer of this type is shown in Figure 24.9. The three windings are (1) the *horizontal*

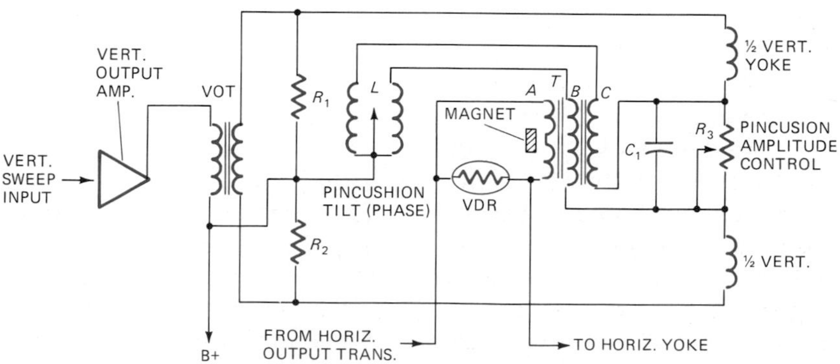

Figure 24.8 A typical top and bottom pincushion-correction circuit.

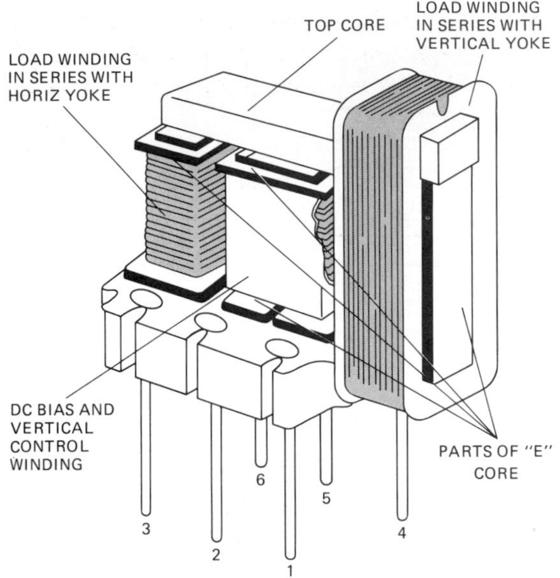

Figure 24.9 Sketch of a PIN saturable-reactor transformer.

load winding (not discussed here), (2) the *control*, or DC bias, winding, and, (3) the *vertical load* winding. In Figure 24.8, these windings are labeled A, B, and C, respectively.

Coupling the Horizontal Waves The PIN transformer couples *modified* horizontal pulses from an auxiliary winding on the horizontal output transformer into the vertical yoke. The horizontal pulses are transformed into horizontal sine waves by the resonant circuit consisting of L and C_1. Sections of a sine wave have roughly the same shape as a parabolic wave and are suitable for use here. The correction wave shape ("bowtie"), which appears across winding C of the PIN transformer in Figure 24.8, is shown in Figure 24.10. Note that there is a maximum correction at the top and bottom of the picture with no correction required at the vertical center.

Controlling Amplitude of Horizontal Waves As shown in Figures 24.7 and 24.10, the amplitude of the horizontal correction waves is maximum at the top and bottom of the picture and decreases to approximately zero at the vertical center. The amplitude of the waves in winding C (Figure 24.8) of the saturable transformer is a function of the amplitude of the vertical-deflection current flowing through the control winding B (Figure 24.8). If this current is maximum, the transformer becomes saturated and no correction voltage is coupled to winding C. At lesser amplitudes of current through winding B, coupling is accomplished in various amounts to winding C and correction waves appear across winding C. These waves have an amplitude in proportion to the current through winding B and the degree of transformer saturation produced.

Producing Bowtie Effect To obtain effect shown in Figures 24.7 and 24.10, it is necessary to pass a vertical rate current through winding B of Figure 24.8. This current is a minimum (no saturation, maximum correction) at the top and bottom of the raster and increases to a maximum (saturation, no correction) at the vertical center of the raster. This is accomplished by passing a parabolic vertical current through winding B, as illustrated in Figure 24.11.

Correction-Voltage Polarity Reversal Note in Figure 24.7 that the correction voltage has a positive direction from the top to the cen-

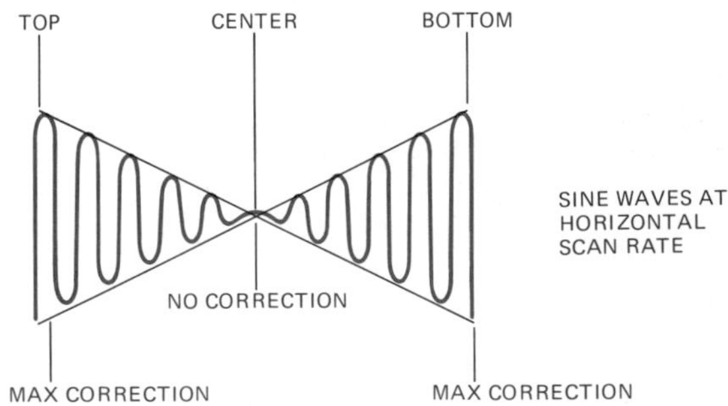

Figure 24.10 The "bowtie" PIN correction waveform appearing across winding C of Figure 24.8. (Winding C is the vertical "load" winding shown in Figure 24.9.)

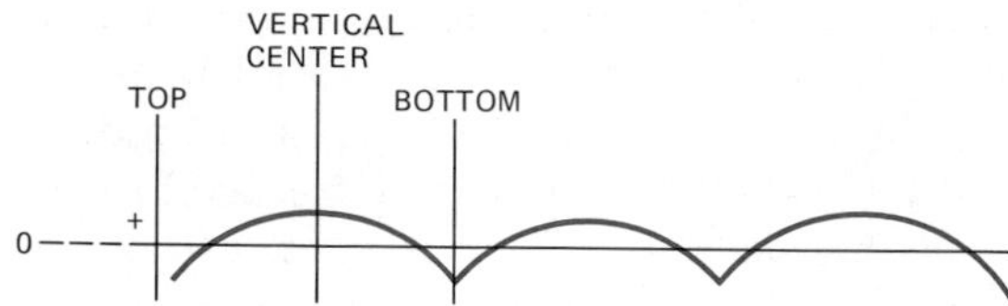

Figure 24.11 Parabolic vertical current in control winding *B* of Figure 24.8.

ter of the raster, but a negative direction from the center to the bottom. In Figures 24.8 and 24.11, we see that the parabolic current from top to center is increasing. However, from center to bottom, it is *decreasing*. Thus, we have a *reversal of current* passing through vertical center. The effect of this decreasing current reversal on the saturable reactor transformer is to reverse the output polarity of the correction voltage at winding C.

PIN Controls The slug in coil L is used to change the phase of the parabolic correction curve. With this control, we can phase-shift the curve from side to side across the picture tube screen. Coil L and amplitude control R_3 can be adjusted to produce correction curves that exactly oppose the curved horizontal scanning lines and thus cancel the pincushion effect.

In Figure 24.8, there is a permanent magnet adjacent to transformer T. In some saturable reactors, this magnet is fixed in position. It is used to provide "magnetic bias" to reduce the amount of control winding (B) current necessary to achieve core saturation. In other units, the position of the magnet is adjustable. Here, it is also used to adjust the crossover point of the polarity reversal of the correction wave. (See Figure 24.9).

24.4 VERTICAL DYNAMIC CONVERGENCE

Color picture tubes using in-line electron beams *do not* require vertical convergence. This was discussed in Chapter 18. In addition, and also covered in Chapter 18, many recent in-line color picture tubes require little or no convergence after leaving the factory.

However, delta-gun color picture tubes do require vertical dynamic convergence as well as static convergence of the three beams. The following discussion refers to such tubes.

STATIC CONVERGENCE

In delta-type, three-gun, rectangular color picture tubes using a shadow mask, converging the three beams at the center of the screen is not a particularly difficult task, as shown in Figure 24.12. This center-screen convergence is accomplished by permanent magnets mounted on the rear of the deflection yoke housing and is called static convergence.

VERTICAL DYNAMIC CONVERGENCE

When the beams are deflected to the edge areas of the screen, though, a significant problem is encountered. Figure 24.12 shows that, because the shadow mask and

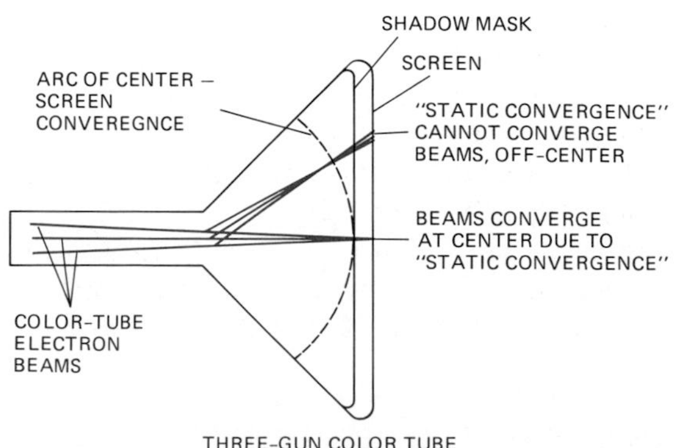

Figure 24.12 "Static convergence" is effective only at the center of the screen.

the picture tube screen are not spherical, the convergence arc curves away from the mask and screen, resulting in a loss of beam convergence. To correct this condition, dynamic convergence correction currents are generated and applied to three convergence coils 120° apart, located adjacent to the deflection yoke (see Chapter 18). Each of the convergence coils, one for each color, consists of two sets of windings, one set for vertical convergence corrections and the other for horizontal corrections, wound on the same core. In this discussion, only the vertical correction circuits will be explained.

Figure 24.13 shows what happens in the vertical center edge areas when there is incorrect vertical dynamic convergence. Note that the lines shown are the vertical bar pattern of a crosshatch generator. Corrections to converge these three lines must be provided by the vertical dynamic convergence circuits only. This condition is a result of the fact that along the vertical centerline, all horizontal corrections are zero. Away from the vertical and horizontal centerlines, in the four quadrants of the picture tube screen, dynamic convergence corrections require a

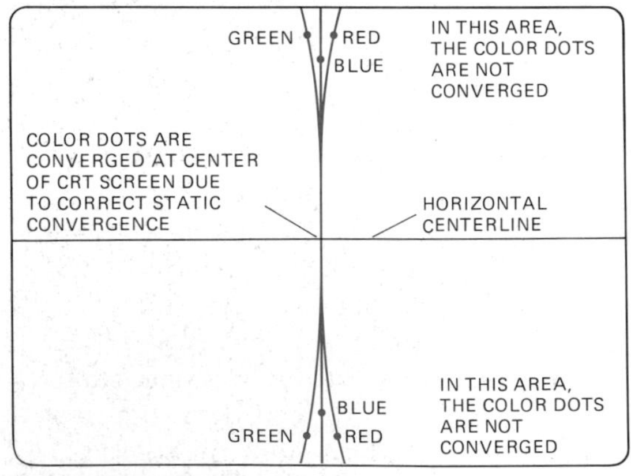

Figure 24.13 Vertical lines from a vertical-bar generator are misconverged because of poor vertical dynamic convergence. When properly converged, the three lines will be superimposed.

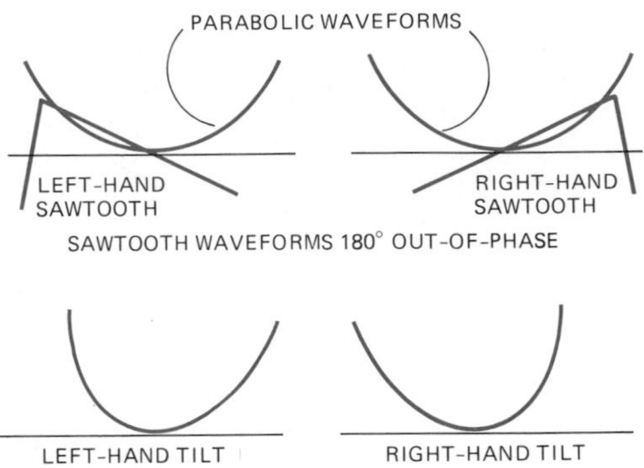

Figure 24.14 Combining parabolic and sawtooth waveforms to provide tilt.

combination of signals from both the vertical and horizontal correction circuits.

The current necessary for dynamic convergence correction is a parabolic wave form that can be set at any required amplitude and can be tilted to either side. This wave form is derived from three separate signals taken from the vertical output circuit. One of these signals, a parabolic wave, is coupled into the convergence circuits from the cathode of the vertical output tube. The other two signals are 180° out-of-phase sawtooth wave forms developed across special windings on the vertical output transformer (VOT). Combining the parabolic wave form with a right- or left-hand sawtooth produces a parabolic wave form with a corresponding tilt, as shown in Figure 24.14. The amount of tilt depends upon the amplitude of the sawtooth.

Vertical Dynamic Convergence Circuit A typical vertical dynamic convergence correction circuit is shown in Figure 24.15. This circuit is divided into three functional sections: (1) parabolic shaping network, (2) blue convergence, and (3) red and green convergence. A simplified diagram of the parabolic shaping network is presented in Figure 24.16. Voltage from the cathode of the vertical output tube is applied through C_1 to the inte-

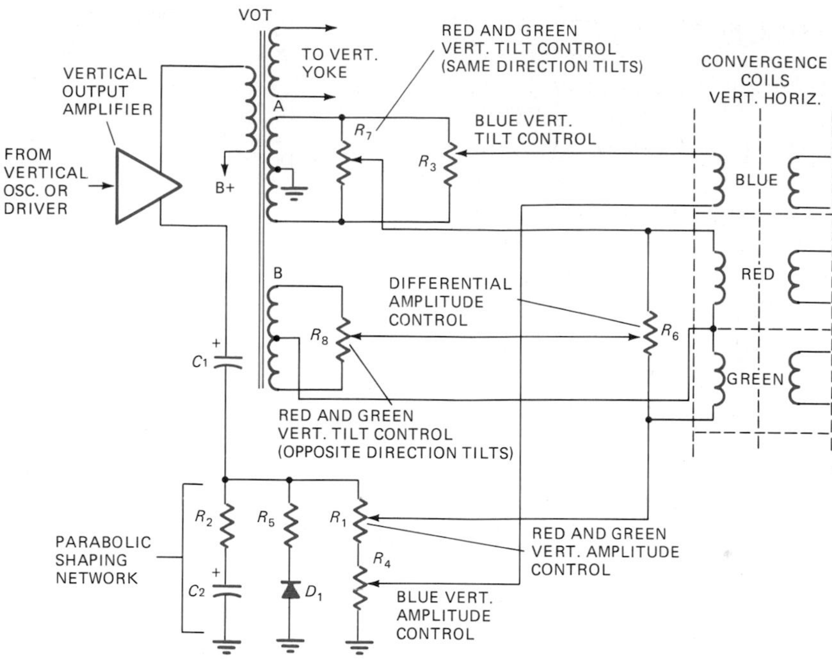

Figure 24.15 A typical vertical dynamic-convergence correction circuit.

grator-phase-shifting network consisting of R_2 and C_2. Here, the sawtooth voltage is converted into a parabolic wave form. This voltage is then impressed across the voltage divider consisting of variable resistors R_1 and R_4. Pickoffs on the two resistors provide a means for varying the input voltages fed into the blue and the red-green convergence circuits. Resistor R_5 and diode D_1 clamp the convergence correction reference at the center of the screen. This clamping is necessary to prevent any adjustment of the dynamic convergence controls from changing the static convergence. Remember that static convergence is accomplished at the center of the picture tube screen with three permanent magnets.

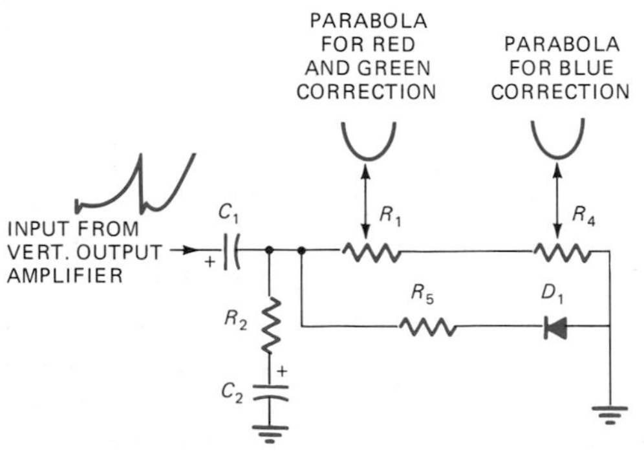

Figure 24.16 Parabolic shaping network (see also Figure 24.15).

Blue Convergence Circuit Sawtooth currents for the convergence coils are developed in two sets of center-tapped windings on the VOT, A and B in Figure 24.15. In Figure 24.17, R_3 is shown connected across winding *A* on the VOT. When the pickoff on R_3 is at the center of resistance, no sawtooth tilt current flows in the blue vertical coil. In this condition, only a symmetric parabolic current will flow in the coil. The amplitude of the current is determined by the setting of R_4. This current in the blue convergence coil causes an equal correction on both sides of the raster; the blue-beam line will remain in the center of the screen. Any magnetic field

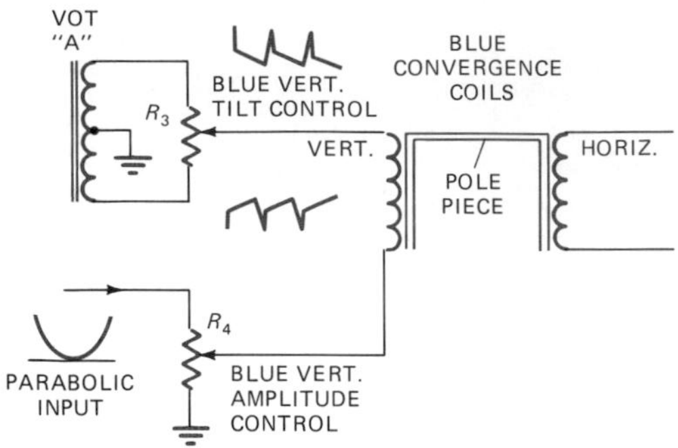

Figure 24.17 Simplified blue convergence circuit.

developed by the blue horizontal coil, wound on the same pole-piece as the vertical coil, combines with the vertical field and moves the beam accordingly.

Moving the pickoff on R_3 to either side also causes a sawtooth current to flow through the coil. Direction of this sawtooth, right-hand or left-hand, is determined by the direction of movement from the center of R_3. The sawtooth, when combined with the parabolic wave form from R_4, tilts the correction current wave form to the right or the left, depending on the direction of the sawtooth from R_3.

Red-Green Convergence Circuit Figure 24.18 is a simplified diagram of the red and green vertical convergence circuits. Here, R_1 provides the parabolic wave form for the red and green coils. Since current flows in the same direction through both coils, differential control R_6 is used to inversely vary the amplitude of the current in the two coils. In other words, changing the setting of R_6 increases the current amplitude in one coil and at the same time decreases the amplitude an equal amount in the other coil.

Sawtooth Tilt Current The sawtooth tilt current for the two coils is provided by R_8, connected across winding B on the VOT. This control, similar to the action of the blue

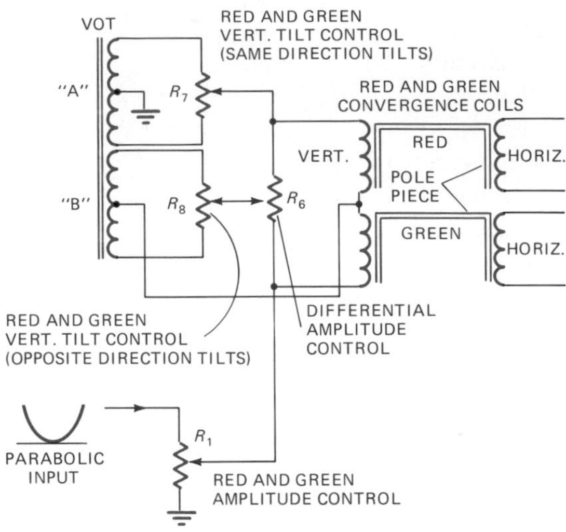

Figure 24.18 Simplified red-and-green vertical convergence circuit.

tilt control previously discussed, tilts the red and green wave forms to an equal degree, but in opposite directions, i.e., if the red coil wave form is tilted right, the green coil wave form is tilted left. Resistor R_7, connected across winding A on the VOT, provides a sawtooth current that tilts both the red and green wave forms equal amounts and in the same direction, as determined by the direction of the setting of R_7. Again, any magnetic field developed by the horizontal convergence circuits combines with the vertical magnetic field in the convergence coils. The action of both magnetic fields thus provides the means for converging the three beams at all points on the picture tube screen.

24.5 VERTICAL BLANKING

At the end of each picture tube field, the vertical-deflection system returns the electron beam in the picture tube to the top of the screen. During this return interval, the much faster horizontal-deflection system sweeps several lines across the face of the picture tube. To prevent these lines from appearing on the screen, the vertical output circuits produce blanking pulses. These pulses are then applied to either the cathode or the control grid of the picture tube to turn off the electron beam until it is at the top of the screen and in the proper position to begin the next field. The polarity of the blanking pulses is dependent upon the picture tube element to which the pulses are applied: cathode blanking requires positive pulses and control-grid blanking requires negative pulses.

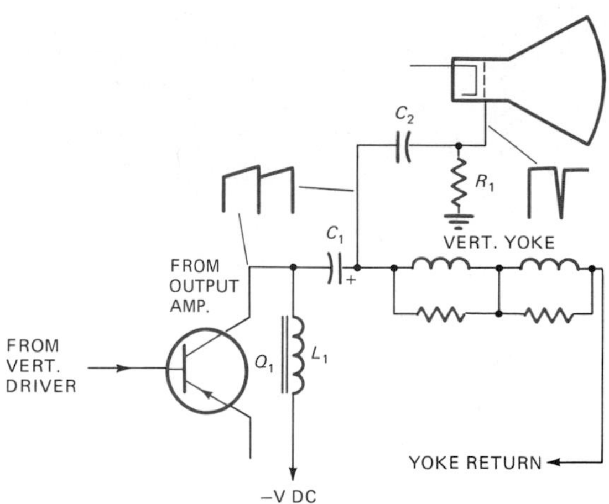

Figure 24.19 Simplified circuit for negative-pulse blanking.

NEGATIVE-PULSE BLANKING

During the vertical retrace interval, the inductive kickback from the yoke inductance generates a peaking spike on the yoke-driving sawtooth wave form. This spike extends in either a positive or a negative direction, depending upon the particular vertical output circuit configuration.

BLANKING CIRCUIT

Figure 24.19 shows a simplified blanking circuit in which negative-peaked sawtooth wave forms are tapped from the vertical yoke circuit. Even though these wave forms contain a sharp negative spike, they cannot be applied directly to the picture tube control grid because they also contain components of the sawtooth waveform. If not eliminated from the blanking pulses, these sawtooth components cause raster shading. To provide properly shaped blanking pulses, the peaked sawtooth wave forms are differentiated in the network comprising C_2 and R_1, as shown in the diagram. The output of the differentiating network (basically a high-pass filter) consists of narrow pulses built up from the higher frequencies of the sharp-peaked sawtooth wave forms. The lower frequencies that make up the ramp portion of the sawtooth wave form are attenuated and removed. In this way, only the narrow, negative pulses are applied to the control grid of the picture tube. The diagram in Figure 24.19 is typical of many solid-state TV receivers.

24.6 MONOCHROME TV VACUUM TUBE VERTICAL OUTPUT CIRCUITS

TRIODE AMPLIFIER

Figure 24.20 shows a vertical output circuit that uses a single vacuum tube, or one section of a multisection tube. This circuit is usually driven by a blocking oscillator. However, any pulse-type oscillator will also drive this circuit. Several TV-receiver manufacturers use this circuit in conjunction with a transistor oscillator stage.

The internal resistance of a vertical output tube is appreciable relative to the inductance of the vertical yoke because of the low deflection frequency (60 Hz). In addition,

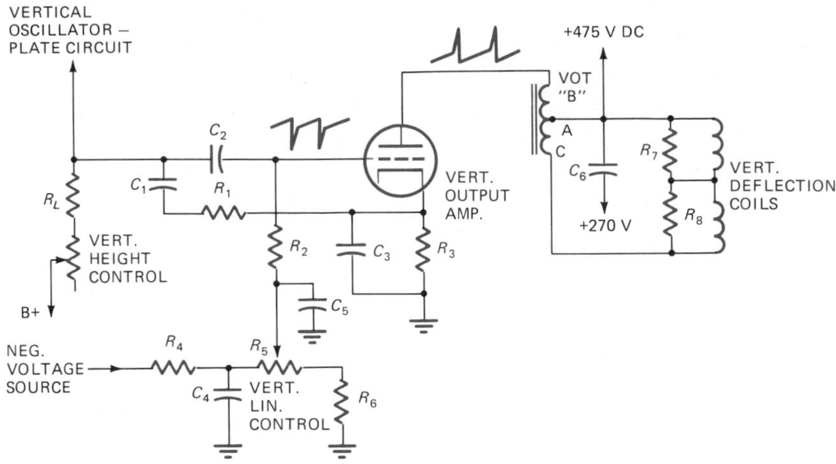

Figure 24.20 Triode amplifier vertical-output circuit for a monochrome TV receiver.

the resistance of the vertical yoke is an appreciable percentage of its 60-Hz inductance. Thus, to drive sawtooth current through this yoke, a trapezoidal voltage is applied to the grid of the output tube. The grid drive voltage in Figure 24.20 is trapezoidal. The output plate voltage of V_1 is therefore also trapezoidal, as required, to drive sawtooth current through the *R* and *L* of the yoke.

TRAPEZOIDAL PULSE AMPLITUDE

The negative-pulse portion of the grid drive trapezoidal wave form has an amplitude sufficient to drive the output amplifier far below cutoff (Figure 24.21). The requirement for this is as follows: The negative-going pulse cuts off the output tube almost instantaneously. Now the magnetic field in the VOT and vertical yoke begin collapsing, initiating the vertical retrace. However, this collapsing field induces a high positive-pulse voltage, which is applied to the plate of the output tube.

If it were not for the large negative pulse at the grid, the output tube might be forced into conduction by the high plate voltage. This could cause a longer retrace and *foldover* at the top of the picture. The large negative grid voltage that occurs during vertical retrace prevents the output tube from turning ON during retrace.

Note in Figure 24.21 that during the *trace,* a portion of the trapezoidal wave remains above cutoff. This ensures a trapezoidal wave form at the plate of V_1 in Figure 24.20, as required for the vertical yoke.

Circuit Details Vacuum tube V_1 is a high-current power triode connected in a conventional amplifier circuit. The operating point of the tube is determined by negative grid bias in conjunction with the cathode bias developed across R_3. Capacitor C_3 bypasses most of the sawtooth wave form at the V_1

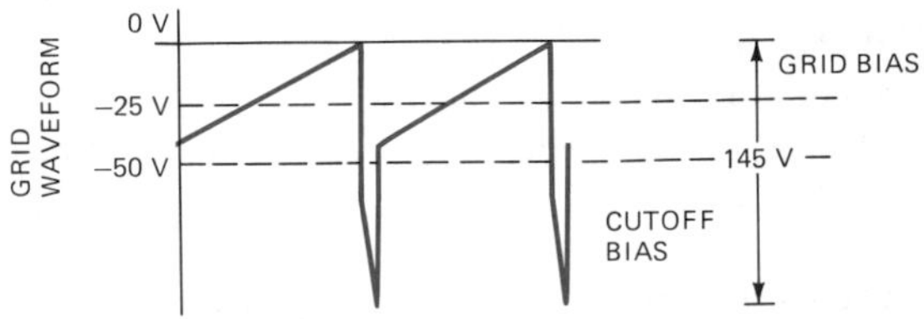

Figure 24.21 Details of the grid waveform for Figure 24.20. Note that the pulse portion drives the tube far below cutoff.

cathode to minimize degenerative losses. To provide the grid bias, voltage divider R_4, R_5, and R_6 is connected across a negative voltage source, usually the horizontal output tube grid circuit. Resistor R_5, the vertical linearity control, is a variable resistor permitting the adjustment of the V_1 grid bias and thus the establishment of the operating point of the tube on the most linear portion of its characteristic curve. Capacitors C_4 and C_5 are decoupling capacitors that prevent interaction between the V_1 grid circuit and the source of negative voltage.

Capacitor C_2 isolates the grid of V_1 from the DC plate supply of the oscillator. The height control and R_L are actually in the oscillator circuit but are shown in the figure to simplify the discussion of the C_1–R_1 action. The sawtooth voltage is developed across C_1 and is peaked by resistor R_1, which is returned to the cathode of V_1 instead of to ground. This cathode connection eliminates a small residual AC ripple that appears across cathode bypass capacitor C_3. The plate circuit of V_1 includes the vertical output transformer and the vertical windings of the deflection yoke.

In the circuit of Figure 24.20, the oscillator tube is cut off during the vertical scan interval. In this condition, C_1 is connected to B+ through R_L and the height control. Since the resistance of R_1 is very small in comparison with the resistance of R_L and the height control, C_1 charges slowly, developing a sawtooth voltage at the grid of V_1. This voltage causes V_1 to conduct. At the end of the scan interval, the oscillator conducts heavily, effectively grounding the R_L side of C_1. Capacitor C_1 then discharges rapidly through R_1 to develop the negative pulse shown in the wave form at the grid of V_1. This negative pulse cuts off V_1.

Linearity Control By varying the V_1 grid bias with the linearity control R_5, the operating point of the tube can be moved from one point on its characteristic curve to another where the curvature is less. In this way, it is possible to use the nonlinearity of the characteristic curve of V_1 to counteract any nonlinearity that may develop in the sawtooth section of the deflection wave. This can be done successfully because the curvature of the tube characteristic is in a direction opposite the curvature that develops in the deflection wave. If such correction is not made, sections of the image will crowd together. It must be remembered though that over compensation will lead to the opposite distortion and sections of the image are stretched out.

24.7 MONOCHROME TV TRANSISTOR VERTICAL OUTPUT CIRCUITS

Many of the vertical-deflection systems found in television receivers have a multivibrator circuit in which the two transistors (or tubes) serve as both the vertical oscillator and the vertical output stage. The theory of such circuits and typical examples of both transistor and tube systems are presented in detail in Chapter 21. It is advised that the reader refer to the discussions of these systems given in Chapter 21 before proceeding further.

An additional circuit of this type, one that uses transistors, is shown in Figure 24.22. In this three-stage system, Q_{302} and Q_{306} make up a multivibrator. Transistor Q_{304} is an emitter follower. Collector current flow and the feedback path of the PNP transistor circuits are also shown in Figure 24.22.

RECHARGING SAWTOOTH CAPACITORS

In operation, vertical synchronizing pulses are integrated by R_{302} and C_{300} and are then applied to the base of Q_{302} through C_{302}. At the end of each vertical scanning pe-

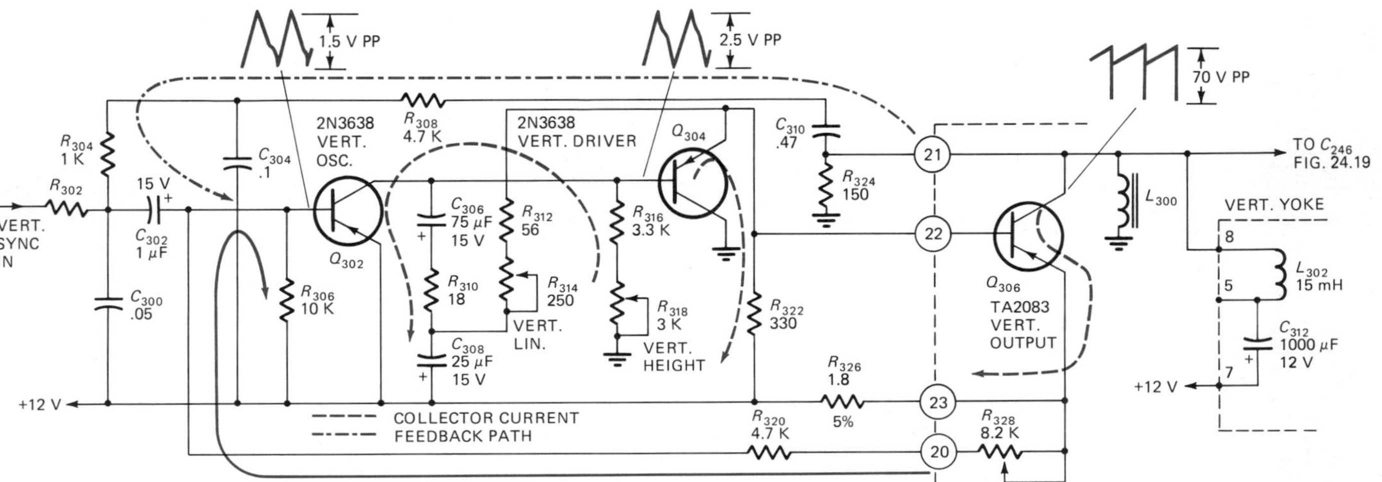

Figure 24.22 A transistor vertical-deflection system (monochrome receiver). *(Courtesy of GTE-Sylvania, Inc.)*

riod, these positive synchronizing pulses trigger Q_{302} into conduction to recharge the sawtooth-forming capacitors, C_{306} and C_{308}. A portion of the Q_{306} emitter voltage is applied through R_{328}, the vertical hold control, to the base of Q_{302}. This voltage establishes the average base bias of Q_{302} and thus determines the duration of each cycle. The positive feedback to sustain oscillations is applied to the base of Q_{302} through a shaping network consisting of R_{324}, C_{310}, R_{308}, C_{304}, and R_{304}. Capacitor C_{304} also decouples horizontal cross-talk frequencies from the feedback circuit.

VERTICAL SCANNING PERIOD

During the vertical scanning period, Q_{302} is cut off and the sawtooth-forming capacitors charge through R_{316} and R_{318}, the vertical height control. This charge current develops a positive sawtooth at the base of Q_{304}, whose collector drives the output transistor Q_{306}. The height control determines the base bias and base signal amplitude of Q_{304} and thereby establishes the amplitude of the output transistor driving voltage. The linearity of the circuit is achieved by feedback from the emitter of Q_{304} applied through R_{312} and R_{314}, the linearity control, to the junction between C_{306} and C_{308}. This voltage modifies the discharge slope of the sawtooth-forming capacitors to provide the uniformly rising current needed to drive the yoke.

VERTICAL RETRACE

The vertical windings of the yoke L_{302} are coupled to the collector of the output transistor Q_{306} across the choke L_{300}. To completely eliminate the DC component of the deflection current, the yoke is returned to +12 V DC through capacitor C_{312}. During retrace, Q_{306} is cut off and the field across L_{300} collapses very rapidly, allowing C_{312} to discharge. This action returns the scanning beam to its starting position at the top of the picture tube screen to begin the next field.

24.8 COLOR TV TRANSISTOR VERTICAL OUTPUT CIRCUITS

The color transistor vertical output circuit (Figure 24.23) described in this section consists basically of (1) a predriver Q_{601}, (2) a driver Q_{602}, (3) a top-ramp vertical output stage Q_{101}, and (4) a bottom-ramp vertical output stage Q_{102}.

The input signal to the predriver is a highly linear sawtooth wave at 59.94 Hz. This sawtooth wave is fashioned from a rectangular wave, which is the vertical output wave of a countdown IC (not shown here).

OVERVIEW OF CIRCUIT OPERATION

Two Current Ramps The schematic diagram and associated wave forms of this vertical output circuit are shown in Figure 24.23. This circuit produces two **linear current ramps** in the vertical yoke coils. The top (of screen) ramp begins with maximum yoke current in the positive direction. This current is gradually reduced to zero. At that point, the scan has reached the vertical center of the screen. Now the bottom ramp is initiated with the yoke current at zero and linearly increasing to a maximum in the opposite direction. This deflects the beams from the vertical center to the bottom of the screen.

Vertical Retrace When the beams reach screen bottom, the bottom-ramp amplifier is cut off. The collapsing field in the vertical yoke moves the beams quickly to vertical center. Now the top-ramp amplifier is driven rapidly to maximum conduction. This

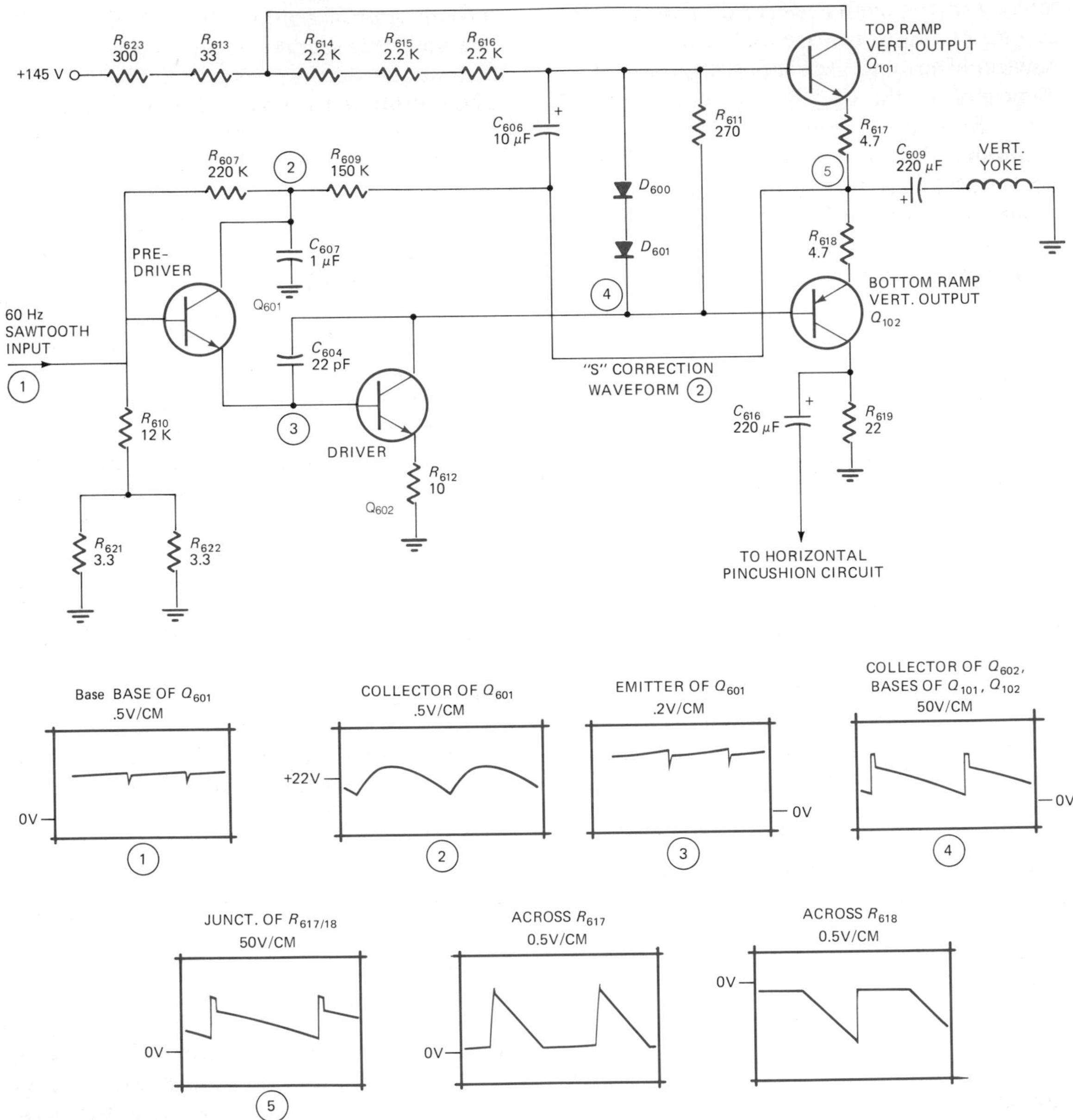

Figure 24.23 Transistor vertical-deflection system. The vertical input frequency is supplied by a digital countdown circuit. Digital circuit output is a rectangular wave that is changed to a sawtooth wave and then applied to the base of Q601.

quickly drives the beams from vertical center to screen top, thus completing the vertical retrace. The next vertical scan now commences.

Summary The vertical output circuit produces *two* linear current ramps. It must also ensure that vertical retrace begins in time with the start of the vertical synchronizing pulse received from the TV station.

CIRCUIT OPERATION

Refer to Figure 24.23 for the schematic diagram and important wave forms.

The Predriver and Driver The predriver is transistor Q_{601}. As shown in the schematic and wave form 1, its input to the base is a highly linear sawtooth with a positive rise of approximately 0.1 V. The base is forward-biased from the voltage at the junction of R_{617} and R_{618} (output amplifier circuit).

Transistors Q_{601} (predriver) and Q_{602} (driver) are a **Darlington pair.** Note that the emitter of Q_{601} is connected directly to the base of Q_{602}. Thus, the base-to-emitter current of Q_{602} is also the collector current of Q_{601}. Because of this connection, the current amplification of the pair is equal to the *product* of the gains of the individual transistors. This configuration provides a high input impedance and high amplification.

As shown in wave form 4, the output of Q_{602} is an inverted (linear) sawtooth. The sawtooth amplitude is approximately 30 V.

S Correction A parabolic wave form of voltage is developed at the collector of Q_{601} across capacitor C_{607}. This is the S correction voltage (previously mentioned) to correct for scan linearity errors introduced by the geometry of the picture tube. The parabolic wave, shown as wave form 2, has an amplitude of approximately 0.5 V. It is fed to the junction of R_{617} and R_{618}.

The Output Amplifiers The vertical output amplifiers are a **complementary-symmetry pair.** (This type of circuit is commonly found in high-fidelity audio amplifiers.) This circuit uses PNP and NPN transistors in a symmetrical arrangement. It has the characteristic of producing push-pull operation from a single input signal. It does not require any form of phase inverter to drive it. A difference of approximately 1 V is required between the two base potentials as explained presently. This voltage drop is provided by diodes D_{600} and D_{601}. Feedback capacitor C_{606} compensates for the reduction of sawtooth amplitude to the base of Q_{101}, caused by the diodes.

Top Half of Scan The negative-going sawtooth 4, goes from about +68 V to +38 V and is applied to the bases of the top-ramp amplifier Q_{101} and the bottom-ramp amplifier Q_{102}.

The initial sawtooth positive value of 68 V drives the top-ramp (NPN) amplifier instantly to maximum conduction through the vertical yoke. Now the beams are driven quickly to the top of the screen to begin a new field. As the sawtooth 4 begins its downward slope, it decreases the conduction of Q_{101} and the yoke current. Thus, the beams move down linearly toward vertical center.

When the sawtooth wave reaches about 53 V (its least positive value is 38 V), Q_{101} stops conducting because the DC emitter voltage of Q_{101} is also 53 V. (When Q_{101} stops conducting, through the vertical yoke, the beams are at vertical center.) This DC emitter voltage is determined by the charge in yoke-coupling capacitor C_{609}. The DC base bias voltages for the two output stages are determined by the base bias network, made up as follows: R_{619}, collector-to-base junction of Q_{102}, R_{611} in parallel with diodes D_{600} and

D_{601}, plus R_{616}, R_{615}, R_{614}, R_{613}, and R_{623}, to the + 145-V supply.

Bottom Half of Scan When the sawtooth voltage drops below 53 V (the emitters of Q_{101} and Q_{102} are also at 53 V), PNP vertical output transistor Q_{102} begins conducting. This starts to drive current through the vertical yoke in the opposite direction of the current from Q_{101}. The Q_{102} current builds up linearly and, passing through the vertical yoke, causes the beams to move linearly toward the bottom of the screen. When the sawtooth 4 reaches a value of 38 V, Q_{102} is conducting maximum current. At this time, the beams are at the bottom of the screen, ready for vertical retrace.

Vertical Retrace Observation of wave form 4 of Figure 24.23 shows that the lowest potential of the sawtooth is followed by a positive-going pulse that has a very short rise time. This abruptly cuts off Q_{102} because it is a PNP transistor. (Positive base voltage on a PNP transistor relative to emitter voltage will cut it off.) At the same time, the positive rise of the pulse quickly drives Q_{101} into full conduction. The net result of no current output from Q_{102} and full current conduction of Q_{102} happening very rapidly is that the beams are driven quickly from the bottom to the top of the screen.

At the conclusion of the positive-going pulse (wave 4 of Figure 24.23), the sawtooth again begins its downward course and the beams again start downward to sweep the next field.

Note: It should be noted that the positive-going pulse shown in wave 4 of Figure 24.23 is *not* part of a deliberately introduced trapezoidal wave form. A trapezoidal waveform is *not* required for this circuit, as it is for certain others. The pulse portion is actually a *"kickback"* pulse generated by rapid reversal of the vertical yoke magnetic field. The pulse is coupled to the two bases via the emitter-to-base junctions of Q_{101} and Q_{102}.

24.9 VERTICAL OUTPUT TRANSFORMERS

Vertical-deflection yokes have relatively low impedance. For sinusoidal 60-Hz currents, these range from approximately 0.4 ohms (toroidal yoke) to 20 ohms (saddle yoke). (The effective impedance for sawtooth currents is greater.)

WITH TUBE AMPLIFIERS

Vacuum tube vertical output amplifiers have much greater internal impedance than those mentioned above. Hence, when these are used, it is essential to match the tube impedance to the yoke impedance by means of a transformer. Typical transformer and autotransformer circuits appear in Figures 24.15 and 24.19.

WITH TRANSISTOR AMPLIFIERS

Some types of transistor vertical output amplifiers also use matching transformers. This is required when the amplifier impedance is appreciably higher than the vertical yoke impedance.

However, many transistor vertical output amplifiers do not require the use of matching transformers because of their inherently low impedance. Such amplifiers are frequently choke-capacitor-coupled, as shown in Figures 24.19 and 24.22. Alternatively, they may be direct-coupled, as shown in Figure 24.23.

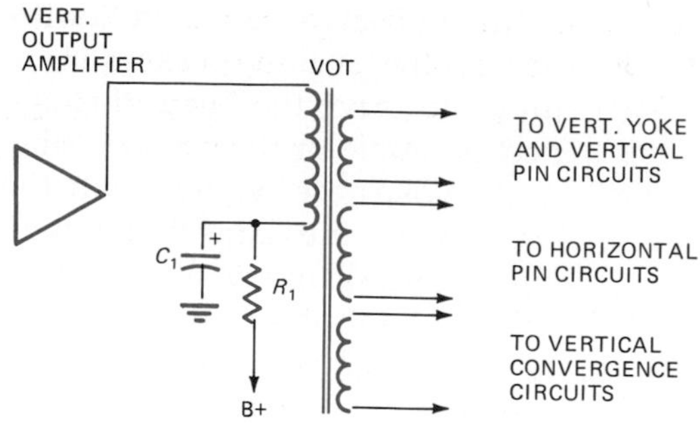

Figure 24.24 Isolation transformer used in the vertical output circuit of a color TV receiver.

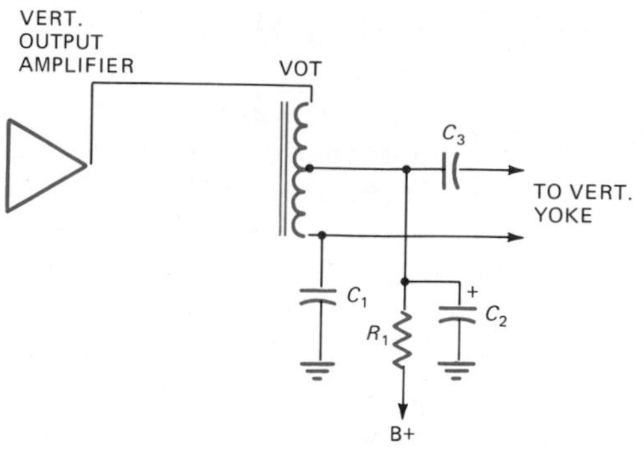

Figure 24.25 Autotransformer used in the vertical output circuit of a monochrome TV receiver.

TYPES OF TRANSFORMERS

Vertical output transformers can be classified as either isolation or autotransformer types.

Isolation Transformers Figure 24.24 is a simplified diagram of a typical isolation transformer circuit. In the diagram, C_1 and R_1 decouple the vertical output circuits from the power supply. The main disadvantage of this circuit is frequency distortion, which is inherent in iron-core transformers. Also, the inductive kickback from the yoke during retrace develops a high-voltage pulse across the primary. Transformation at the step-down ratio reduces the voltage applied to the yoke. Voltages induced across the yoke by the inductive kickback, though, are applied back across the secondary of the transformer and are therefore stepped up by the turns ratio of the transformer. For this reason, the transformer primary must be insulated to withstand the kickback voltage peaks. The convergence circuit secondary windings shown in the diagram are not used on VOTs designed for monochrome TV receivers.

Autotransformers A simplified autotransformer vertical output circuit is shown in Figure 24.25. This transformer provides the necessary step-down ratio, but the DC potential is not isolated from the secondary. In using a circuit of this type, the yoke is usually connected to the transformer through a capacitor. In the diagram, C_1 improves the frequency response of the transformer by speeding up the collapse of the inductive field during retrace. Decoupling from the power supply is accomplished by R_1 and C_2. An important advantage of the autotransformer over the isolation type is its lower cost.

24.10 VERTICAL YOKES

Vertical scanning in the picture tube is achieved by a magnetic field, whose lines of force cut horizontally through the neck of the tube. This field is developed by two coils mounted 180° apart on the neck of the picture tube. The coils are part of a yoke assembly that also includes the horizontal-deflection windings. During the vertical scanning period, the magnetic flux moves the electron beam above and below the center of the screen, depending upon the magnitude and direction of current flow in the coils. As discussed previously, a sawtooth current wave form drives the yoke.

VERTICAL SCAN CYCLE

During the vertical scanning period, the sawtooth current decreases from a maximum value in one direction to zero and then increases to the same maximum value in the opposite direction. When the current flow in the vertical coils reaches this last peak value, the electron scanning beam is at the right-hand end of the last horizontal line at the bottom of the screen. At this time, the retrace interval begins. Current flow in the coils suddenly decreases to zero and then increases to maximum once again in the opposite direction. During this retrace interval, though, the current changes very rapidly to reposition the beam at its top-screen position to begin the next field.

VERTICAL YOKE DEFLECTION SENSITIVITY

During the vertical scanning interval, the vertical amplifier delivers to the yoke windings a sawtooth current that achieves full-screen deflection of the electron beam. To accomplish this deflection efficiently, yoke inductance and resistance are kept as small as practical, for it is these two values that determine the peak-to-peak value of current required for full-screen deflection. In other words, the deflection sensitivity of the yoke depends on the peak-to-peak value of current required to drive the beam through the full deflection angle of the picture tube. For this reason, yokes are wound with the shortest wire having the largest diameter that is practical. The diameter of the neck, the anode voltage, and the length of the picture tube are varied to increase the effectiveness of the yoke.

VERTICAL YOKE INDUCTANCE

The inductance of a vertical yoke is mainly a function of the type of yoke. For saddle yokes, a typical inductance value is 45 mH, with a DC resistance of 42 ohms. For a toroidal yoke, the inductance value may be about 2 mH and the DC resistance about 1.5 ohms. These values are for the two coils connected in series.

VACUUM TUBE YOKE CIRCUIT

A simplified diagram of a typical vacuum tube vertical yoke circuit is shown in Figure 24.26. Resistors R_1 and R_2 are connected across the two yoke coils to damp out the shock-induced oscillations caused by the sudden current change during the retrace interval. The thermistor is added between the coils to equalize the yoke resistance under

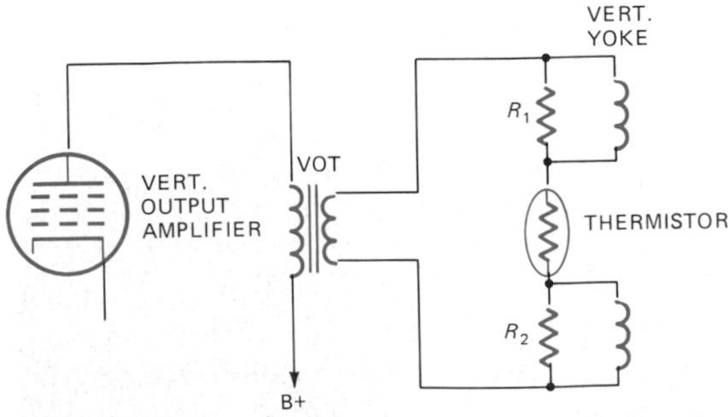

Figure 24.26 A typical vacuum tube vertical-yoke circuit, which employs thermistor compensation.

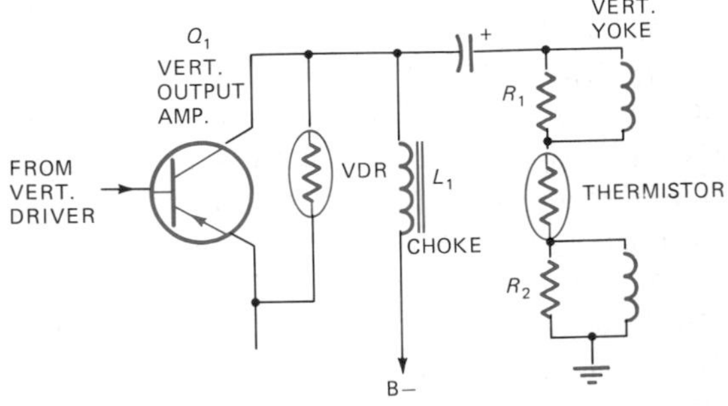

Figure 24.27 Simplified solid-state vertical-yoke circuit.

high-heat conditions. When the yoke temperature increases as a result of the heavy current flow, the resistance of the yoke increases by about 8 ohms. During this temperature rise, the thermistor is also heated and its resistance decreases sufficiently to compensate for the increase in the yoke resistance. This maintains a constant yoke current to prevent height changes.

TRANSISTOR YOKE CIRCUIT

Figure 24.27 shows a simplified diagram of a typical solid-state yoke circuit. Since a choke is used in the collector circuit, instead of an isolation transformer, the coupling capacitor is necessary to block DC from the yoke. Also, a VDR (voltage-dependent resistor) is connected in the output circuit to protect the transistor from the high kickback pulse developed across the yoke during the retrace interval.

24.11 TROUBLES IN VERTICAL OUTPUT CIRCUITS

Troubles in the vertical output circuits, in most cases, are clearly discernible on the picture tube screen. For this reason, the affected picture is by far the best place to begin a logical analysis of output circuit troubles. By carefully studying the picture tube display, the trouble can even be localized to particular circuits within the vertical-deflection system. Once this localization has been accomplished, faulty components can be isolated and replaced. In dealing with any vertical output circuit trouble, bear in mind that the primary function of the vertical-deflection system is to provide a linear scan current at sufficient amplitude to fully deflect the beam, and in color sets to provide for vertical convergence. With this function in mind, it is apparent that all troubles can be categorized in accordance with their effect on the picture: height, linearity, convergence, and retrace blanking. Except for convergence, the troubles are the same for either monochrome or color TV receivers.

HEIGHT TROUBLES

The most obvious of all vertical output circuit troubles is the complete lack of vertical deflection. This trouble is manifest on the screen by a bright horizontal line across the center of the screen (Figure 24.28). This line is similar to the condition that exists when the service switch in color output circuits is placed in the "service" position: vertical deflection is disabled. If a line is left on the screen too long, it may burn the phosphor coating inside the picture tube. This damage cannot occur when the service switch is used because the intensity of the beam is reduced.

Complete loss of vertical output may be caused by any of the following troubles:

1. No sweep input from vertical oscillator or countdown circuits
2. Loss of operating voltages in output circuits

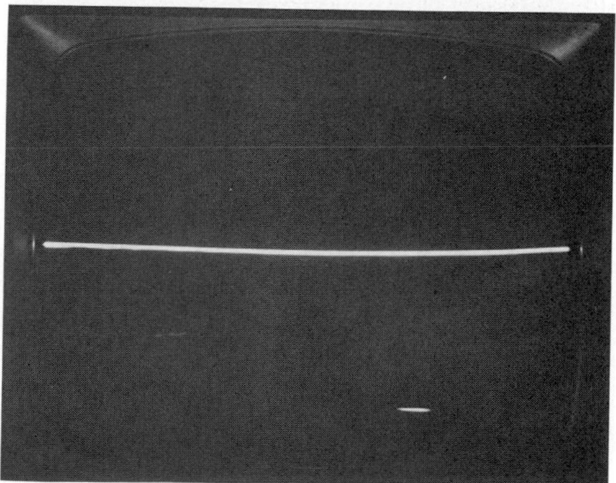

Figure 24.28 Complete loss of vertical deflection.

3. Defective output tube or transistor
4. Defective vertical output transformer
5. Open coupling capacitor
6. Open output tube cathode circuit or transistor emitter circuit
7. Open circuit in deflection coils
8. Both deflection coils or damping resistors shorted

Figure 24.30 An example of insufficient height.

Other height troubles include insufficient vertical deflection and the keystone effect. Both of these troubles produce a picture on the screen that cannot be expanded to completely fill the mask vertically. In addition, the keystone effect tapers one side of the diminished-height pattern to give a keystone shape (Figure 24.29). Even though both of these troubles affect the picture height, they are completely unrelated.

Insufficient height can be caused by increased resistance values in DC supply lines. This increase in resistance decreases the amplification of either the output or one of the drive stages. Also, weak or defective tubes and transistors fail to provide sufficient amplification (Figure 24.30). Low values of operating voltage may be eliminated as a suspect if other receiver circuits appear to operate normally.

The keystone effect is caused by a short circuit across one-half of the deflection coils. The remaining coil cannot develop a strong enough magnetic field to fully deflect the beam. Shorts of this nature may be found in the damping resistors across the coils. Shorts also develop between windings of the coils and require yoke replacement.

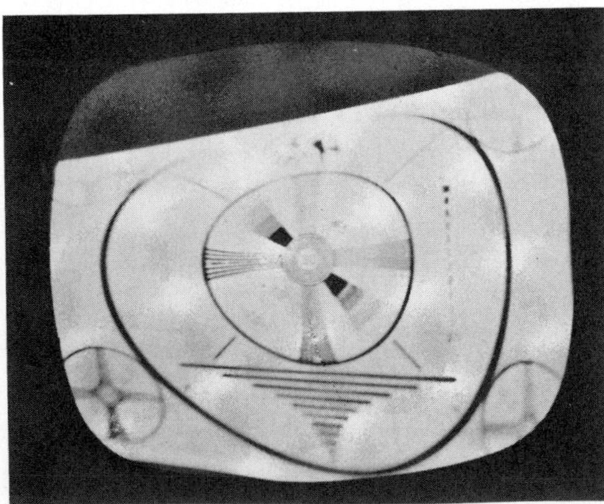

Figure 24.29 Vertical keystoning. *(Courtesy of GTE-Sylvania, Inc.)*

LINEARITY TROUBLES

Figure 24.31 shows the effect of a linear and a nonlinear scan on the picture. In Figure 24.31A, the linear scan allows equal spacing between each horizontal line. In other words, horizontal scanning progresses at a uniform rate down the screen. In Figure 24.31B, scanning progresses normally until the beam is in the lower half of the screen. Here, the magnetic field increases at a reduced rate because of the nonlinearity in the positive portion of the current wave form. This reduced vertical scanning rate allows the horizontal lines to "stack up" or crowd together at the bottom. The opposite curvature in Figure 24.31C causes crowding at the top of the screen.

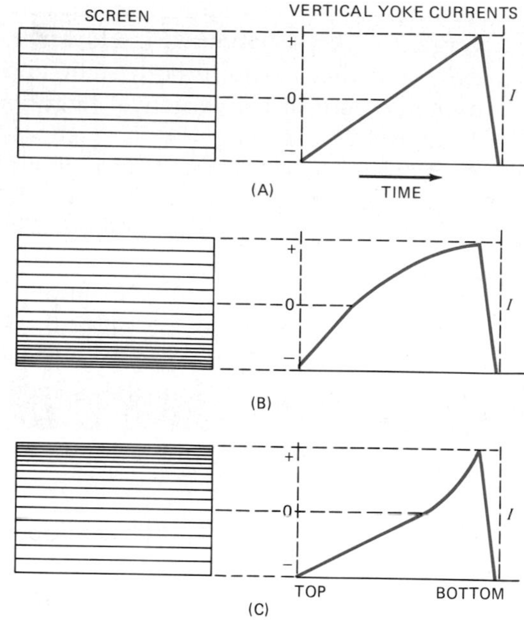

Figure 24.31 Illustrating linear and nonlinear vertical sweep and the sawtooth currents producing them.

The following troubles are the chief causes of nonlinearity in the vertical output circuits:

1. Resistance and capacitance changes in the sawtooth-forming network
2. Defects in feedback loops
3. Vacuum tube or transistor bias changes
4. Open decoupling capacitors
5. Poor filtering of horizontal cross-talk

In the sweep system, the sawtooth is formed by an RC network at a rate determined by the total time constant of the network. All resistance or capacitance changes vary this time constant and thus change the charge-discharge rate of the circuit. Also, grid or cathode bias changes in vacuum tube circuits shift the tube operation to a nonlinear point on its characteristic curve. In a similar manner, the changes in transistor biases shift the operating point of the transistor. Since many present-day circuits make extensive use of feedback for linearity correction, circuit faults could disable or distort the feedback voltage and introduce nonlinearity. These faults are usually found in leaky or shorted coupling capacitors.

SUMMARY OF CHAPTER HIGHLIGHTS

1. The vertical output circuits provide the sawtooth deflection current through the vertical yoke. (Figures 24.1, 24.2, 24.20, 24.22, and 24.23)
2. In Number 1, these circuits may also provide vertical blanking pulses and outputs to drive PIN and vertical convergence circuits. (Figures 24.3, 24.8, 24.15, and 24.19)
3. The main function of a vertical output transformer (VOT), (when used) is to match the vertical yoke impedance to the output impedance of the vertical output amplifier. (Figures 24.15, 24.20, and 24.24 to 24.26)
4. Vertical centering may be accomplished by passing DC current through the vertical yoke. (Figure 24.1)
5. Vertical blanking pulses are produced in the vertical output circuit. These pulses cut off the beam(s) during vertical retrace. (Figures 24.1, 24.2, and 24.19)
6. In a transistor vertical output circuit, a sawtooth input waveform to the base of the output transistor is required to drive a sawtooth current through the vertical yoke. If a trapezoidal input waveform is used, this is only to ensure sharp cutoff of the output transistor at the beginning of vertical retrace. However, the voltage

waveform across the vertical yoke will always appear trapezoidal because of the addition of the yoke *kickback* pulse during vertical retrace.

7. In a vacuum tube vertical output circuit, the output tube is driven by a trapezoidal wave at its grid. This is required because of the relatively high yoke-driving resistance of the output tube. (Figures 24.1, 24.20, and 24.21)
8. Vertical pincushioning (PIN) is bowing of the horizontal lines in a vertical direction. It is most evident near the top and bottom of the screen and least evident at the vertical center. (Figure 24.6)
9. Vertical PIN correction is made by adding horizontal-rate parabolic voltages to the vertical output sawtooth. (Figures 24.7 to 24.11)
10. In providing PIN correction, a special transformer known as a *saturable reactor* is used. This reactor couples the correct amplitude of correction voltages to the vertical and horizontal yokes. (Figures 24.6 to 24.11)
11. In PIN correction, horizontal-rate parabolic waves are coupled to the vertical yoke. Also, vertical-rate parabolic waves are coupled to the horizontal yoke.
12. Because of the relatively high currents (and temperatures) transistor output circuits require thermal stabilization. (Figures 24.4, 24.26, 24.27)
13. In Number 12, thermal protection may be provided by a thermistor and by an emitter resistor.
14. Vertical static and dynamic convergence causes the three electron beams to converge from top to bottom of the raster. (Figures 24.12 and 24.13)
15. Color picture tubes using in-line electron beams do not require vertical convergence (see Chapter 18). However, delta-gun tubes require both static and dynamic vertical convergence. (Figures 24.12 and 24.13)
16. A parabolic current waveform of adjustable amplitude and tilt is used to achieve vertical dynamic convergence. (Figure 24.14)
17. A vertical blanking pulse is applied to the picture tube during vertical retrace. This prevents the appearance of vertical retrace lines on the screen. (Figure 24.19)
18. One popular type of vertical output circuit used for color TV receivers uses a *complementary-symmetry* pair. (Figure 24.23, Q_{101} and Q_{102})
19. Note in Figure 24.23, that direct coupling to the vertical yoke is employed.
20. In Number 18, each output transistor produces one-half of the vertical deflection. (Figure 24.23)
21. Vertical output transformers may be classified as either isolation or autotransformer types. (Figures 24.24 and 24.25)
22. Typical parameters for a vertical toroidal yoke are $L = 2$ mH and $R = 1$ to 2 ohms. Those for a vertical saddle yoke are $L = 45$ mH and $R = 42$ ohms.
23. Transistor vertical output stages may be transformer-, choke-, or direct-coupled to the vertical yoke. (Figures 24.5, 24.15, 24.19, 24.20, and 24.22 to 24.25)
24. Vertical *keystoning* is caused by a short across one-half of the vertical yoke. (Figure 24.29)
25. Insufficient height may result from defective transistors (or tubes) or low B+ voltages. (Figure 24.30)

EXAMINATION QUESTIONS

(Answers are provided at the back of the text.)

Part A Supply the missing word(s) or number(s).

1. The ______ control in a tube vertical output circuit adjusts the output amplitude.
2. The VOT matches the ______ to the ______ stage.
3. Vertical ______ may be accomplished by passing DC current through the yoke.
4. Bowing of the horizontal scan lines in the vertical direction is known as ______.
5. Both ______ and ______ vertical convergence are required for delta-gun picture tubes.
6. A ______ is frequently used for transistor thermal protection.
7. Horizontal parabolic waves are added to the vertical ______ deflection current to correct for vertical pincushioning.
8. A PIN transformer is also known as a ______.
9. Vertical convergence is not required in color tubes using ______ electron guns.
10. A parabolic wave with tilt capability is used for vertical ______.
11. Vertical ______ pulses are generated during vertical retrace.
12. With vacuum tube vertical output stages, a ______ grid drive wave form is required.
13. In a vacuum tube vertical output stage, the ______ control varies the grid bias.
14. A vertical ______ circuit consisting of two transistors (or tubes) may serve as both vertical oscillator and output stage.
15. A complementary-symmetry circuit may serve as a direct-coupled ______ circuit.
16. Each transistor in a complementary-symmetry circuit furnishes ______ of the total vertical scan.
17. The driving wave shape of the transistor output circuit of question 15 is a ______.
18. The gain of a *Darlington pair* is equal to the ______ of the individual gains.
19. A *complementary-symmetry* circuit produces ______ operation from a single input signal.
20. A vertical yoke impedance of about 0.4 ohms would indicate a ______ yoke.
21. Vertical output transformers are classified as either ______ or ______.
22. A ______ across one-half of the vertical yoke would produce a picture *keystone* effect.

Part B Answer true (T) or false (F).

1. Vertical blanking pulses are obtained from an auxiliary winding on the horizontal output transformer.
2. Transistor vertical output circuits are generally coupled to saddle yokes.
3. A multisection vacuum tube may be used for a combination vertical oscillator, vertical output circuit.
4. Vertical centering may be accomplished by passing a DC current through the vertical yoke.
5. The input grid waveform to a vertical output tube must be rectangular.
6. Pincushion correction is required in both the horizontal and vertical directions.
7. Dynamic convergence in all color tubes is required for both vertical and horizontal directions.
8. Thermal protection of transistors is accomplished by a varistor.
9. In pincushion correction, a sine wave may be used instead of a parabola.
10. Maximum top-to-bottom pincushion correction occurs at both the top and bottom of the raster.
11. A PIN transformer couples the correction waveforms into the vertical yoke.
12. In a delta-gun tube, dynamic convergence correction requires signals from the vertical and horizontal correction circuits.
13. The vertical blanking waveform is a parabola.
14. The resistance of a vertical yoke is appreciable compared to its inductance.
15. A *Darlington pair* is frequently used as a push-pull amplifier.

16. Vertical retrace occurs at a rate determined by the 70-kHz ringing frequency.
17. A VOT is essential in connection with *top-ramp* and *bottom-ramp* vertical output amplifiers.
18. The S correction wave form helps to speed up vertical retrace.
19. During vertical retrace, a *kickback* voltage of several hundred volts is generated.
20. One result of poor vertical linearity is vertical *foldover*.

REVIEW ESSAY QUESTIONS

1. Draw a simple block diagram of a transistor vertical-deflection circuit from predriver to yoke. Label all stages properly and show appropriate simple waveforms at the input and output of each stage.
2. Draw a simple block diagram of a color-TV receiver, vertical output circuit. Include convergence and PIN circuits. Label all blocks and show all appropriate waveforms.
3. Describe two methods of thermal protection for the circuit shown in Figure 24.4.
4. Describe how the pincushion effect takes place in Figure 24.6. Use a simple sketch.
5. Explain briefly how the pincushion correction occurs in Figure 24.7.
6. Define (1) static convergence and (2) dynamic convergence.
7. Why is a trapezoidal wave form used to drive the grid of a vertical output stage?
8. What is a saturable reactor? What is its function? (Figure 24.9)
9. Explain the significance of the bowtie wave form in Figure 24.10.
10. In Figure 24.22, explain why it is possible to couple the output transistor to the yoke without using a transformer.
11. Explain how vertical pincushion correction differs from horizontal pincushion correction.
12. Draw a simple sketch of the waveform used to accomplish vertical dynamic convergence. Explain briefly why this waveform is used.
13. In Question 12, what controls are used and what do they do?
14. List all the functions performed by a color receiver vertical output circuit.
15. Explain with a simple sketch why, in an uncorrected delta-gun tube, the three beams do not converge in the vertical plane.
16. Draw a simple circuit diagram, with appropriate waveform(s), illustrating how positive-pulse vertical retrace blanking is accomplished.
17. Explain the operation of the top-ramp and bottom-ramp amplifiers (Q_{101} and Q_{102}) in Figure 24.23
18. Figure 24.32 shows an example of vertical foldover. What condition(s) might cause this problem.
19. In Figure 24.26, explain the operation of the thermistor.
20. List three possible causes of insufficient picture height. (Figure 24.30.)

Figure 24.32 An example of vertical foldover. *(Courtesy of GTE-Sylvania, Inc.)*

EXAMINATION PROBLEMS

(Selected problem answers at the back of the book)

1. A vertical yoke has an inductance of 1 mH. It has to handle the twentieth harmonic of the vertical frequency (60 Hz). What is its reactance at the twentieth harmonic?
2. A horizontal yoke has an inductance of 10 mH. At the twentieth harmonic of the monochrome horizontal frequency, what is its reactance?
3. In Figure 24.23, the voltage at the collector of Q_{102} is 1.3 V. Calculate the collector current.
4. In Figure 24.23, the emitter voltage of Q_{602} is 0.1 V. Calculate the emitter current.

Chapter 25

25.1 Review of AM and FM systems
25.2 Glossary of FM terminology
25.3 Properties of FM waves
25.4 Advantages and disadvantages of FM
25.5 Preemphasis and deemphasis
25.6 The TV receiver FM sound section
25.7 The 4.5-MHz sound-IF amplifier
25.8 Sound-IF limiting
25.9 A basic FM discriminator
25.10 The Foster-Seeley discriminator
25.11 The ratio detector
25.12 The quadrature detector
25.13 The transistor quadrature detector
25.14 The phase-locked loop (PLL) FM detector
25.15 A digital FM demodulator
25.16 The FM differential peak detector
25.17 A complete vacuum tube sound section
25.18 A complete transistor sound section
25.19 A partial IC sound section
25.20 A complete IC sound section
25.21 Complementary-symmetry audio push-pull output stage
Summary

THE FM SOUND SYSTEM

INTRODUCTION

The audio portion of all television programs is transmitted by *frequency modulation*. This choice was the result of several important factors. Of the two broadcasting systems in use today, AM and FM, the latter has proven capable of better reception under adverse conditions. First, FM reception is almost noise free, as compared to AM reception. Second, the FM system is capable of providing much better rejection of undesired signals on the same or adjacent channels. Third, for a given area coverage, the FM system is more economical and efficient, insofar as transmitter power is concerned.

As shown previously, the TV FM audio signal is transmitted within the standard 6-MHz TV channel. The FM carrier is located 0.25 MHz below the upper limit of each TV channel. Note that the maximum frequency deviation for an FM broadcast station (88- to 108-MHz) carrier is +75 kHz. However, because of bandwidth limitations in a TV channel, TV-FM deviation is limited to +25 kHz. In both systems, the transmitted audio frequency range is approximately 50 Hz to 15 000 Hz.

In this chapter, we will briefly review the principles of AM and FM. In addition, we will study the various circuits used in the FM sound section of a TV receiver. Complete TV FM sound sections will also be presented. Most modern TV receivers utilize an IC for the sound section and several of these will be covered.

When you have completed the reading and work assignments for Chapter 25, you should be able to:

- Define the following terms: amplitude modulation, frequency modulation, center frequency, frequency deviation limiter, preemphasis, deemphasis, limiting threshold, limiting knee, discriminator, ratio detector, PLL detector, quadrature detector, demodulator, complementary-symmetry circuit,

crossover distortion, digital FM detector and differential peak detector.
- Compare AM and FM transmission, giving the advantages and disadvantages of each.
- Explain why preemphasis and deemphasis are necessary and tell how they work.
- List and understand the various types of FM demodulators.
- Discuss the operation of each stage of a complete IC television receiver sound section.
- Explain various types of vacuum tube and solid-state limiters.
- Explain why preemphasis and deemphasis are necessary and tell how they work.
- Explain the operation of a complementary-symmetry audio output amplifier.
- Understand some common sound section troubles and how to isolate them.

25.1 REVIEW OF AM AND FM SYSTEMS

AMPLITUDE MODULATION

An **amplitude-modulated (AM) wave** is a carrier wave of *fixed* frequency that is varied in *amplitude.* The variations in carrier-wave amplitude are caused by the "modulation". This can be speech, music, video, digital, or some other modulating signal. The modulating signal adds to and subtracts from the carrier wave amplitude, and this produces the AM wave.

A simple example of an AM wave is shown in Figure 25.1. The equation for percentage of modulation is given in the figure (here, 50%). Note that the positive portions of the audio modulating wave increase the carrier wave amplitude, and the negative portions decrease the amplitude. The *percentage* of modulation is a function of the amplitude of the modulating wave. The greater its amplitude, the higher the percentage of modulation. To prevent distortion of the received wave, modulation is restricted to a maximum of 100% and is usually less.

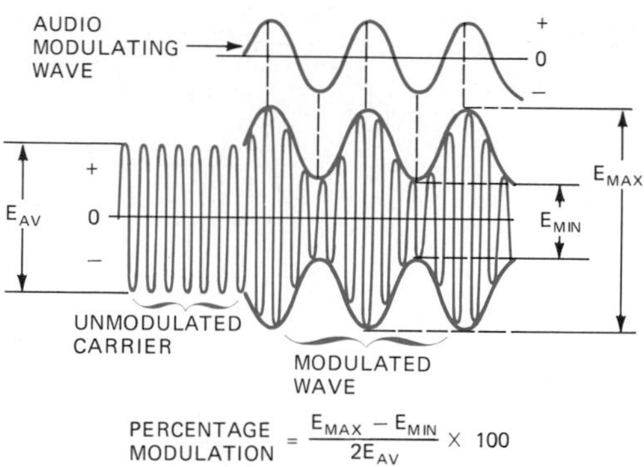

Figure 25.1 An example of amplitude modulation (AM). E_{AV} is the amplitude of the unmodulated carrier wave. E_{MIN} and E_{MAX} result from modulation. The figure illustrates a condition of 50 percent modulation.

FREQUENCY MODULATION

A **frequency-modulated (FM) wave** is a carrier wave of *fixed* amplitude that is varied in *frequency.* The carrier wave in FM transmission is also called the center-frequency wave (f_c). An example of an FM wave is given in Figure 25.2. This shows the effect of a single cycle of an audio wave that is frequency-modulating the sound FM carrier of television Channel 2 (59.75 MHz). For simplicity, only the maximum (+25 kHz) and minimum (−25 kHz) frequency deviations are shown. These are caused by modulation of the audio wave. In practice, the frequency of the transmitted wave would change smoothly in accordance with the amplitude and polarity of the audio modulating wave.

Note in Figure 25.2 that for a positive modulation signal, the transmitted fre-

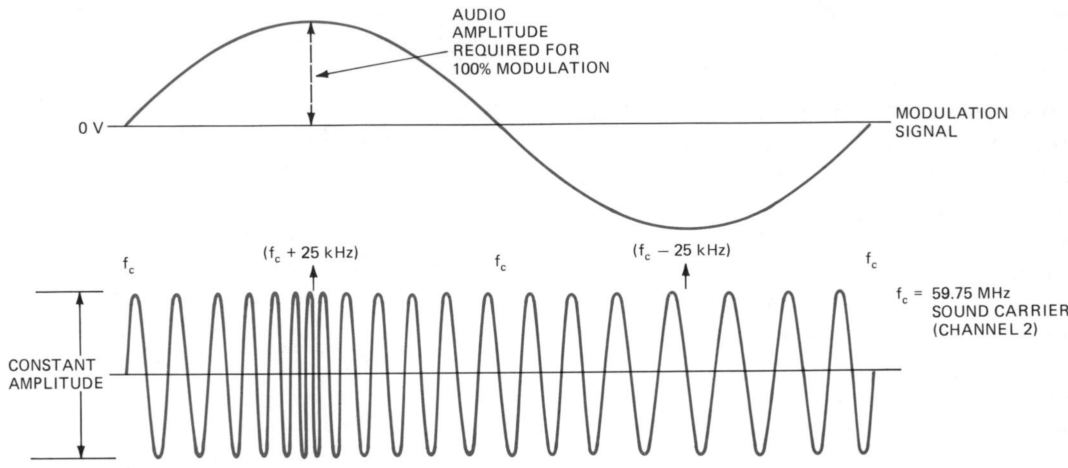

Figure 25.2 An example of frequency modulation. Note that the amplitude of the transmitted wave remains *constant* during modulation. However, its frequency varies in accordance with the amplitude and polarity of the modulation signal. Frequencies shown are for TV Channel 2.

quency *increases* from the center frequency of 59.75 MHz. For a negative modulation signal, the transmitted wave frequency *decreases* from f_c.

In FM, the *amount* of frequency deviation is proportional to the *amplitude* of the modulation signal. The *rate* of frequency deviation is the same as the frequency of the modulating signal.

25.2 GLOSSARY OF FM TERMINOLOGY

- **Center frequency** (also called *carrier* or *resting* frequency) The unmodulated frequency of an FM transmission. (In Figure 25.2, it is 59.75 MHz.)
- **Frequency deviation** The frequency difference between the center frequency and the peak frequency change due to modulation. For Channel 2, the maximum allowable frequency deviation would be 59.75 ± 25 kHz. The amount of frequency deviation is a function of the *amplitude* of the modulating signal. The higher the amplitude, the greater the frequency deviation. Thus amplitude also controls the modulation percentage (see below).
- **FM frequency swing** Twice the instantaneous frequency deviation. For television, the maximum allowable frequency swing is 2 × 25 kHz = 50 kHz.
- **FM modulation percentage** For television (and broadcast FM) audio transmission, the ratio of the actual frequency swing to that defined as 100% modulation, expressed as a percentage. For television, 100% modulation is defined as a frequency swing of 50 kHz (see "FM frequency swing"). As another example, assume a frequency swing of 40 kHz. Here the percent modulation would be $\frac{40}{50} \times 100 = 80\%$.
- **Rate of frequency deviation** This is the same as the frequency of the modulating signal. Examples are 100 Hz, 1 000 Hz, and 15 000 Hz.

25.3 PROPERTIES OF FM WAVES

As previously mentioned, an FM wave is constant in amplitude but varies in frequency. It appears as pictured in Figure 25.2. The property of constant amplitude is the characteristic that makes the FM wave so important. Most types of interference affect the amplitude of a wave much more than its frequency. For the AM signal, the interference distorts the waveform and, consequently, the information contained in the wave. FM, on the other hand, contains its information in its changing frequencies. At the FM receiver, one of the IF stages, acting as a **amplitude limiter,** smooths any irregularities in the amplitude of the incoming signal and by this process greatly reduces the interference.

FM BANDWIDTHS

The frequency bandwidth of an FM wave depends upon the strength of the impressed audio voltage. At the transmitter, the carrier frequency is fixed by a self-excited oscillator. This frequency is the mean, or center, frequency of the broadcast station. When the sounds that are to be transmitted are fed into the microphone, the mean frequency of the transmitter is varied. The louder the audio signal, the greater the deviation. For example, a frequency deviation (or change) of 25 kHz in the output might occur for a strong audio voltage, whereas only a 1-kHz change would occur if the audio voltage were weak. In the AM case, the amplitude, and not the frequency, of the wave, changes for different audio sound levels.

DIFFERENCES BETWEEN AM AND FM

AM and FM differ in many respects. This is perhaps best revealed by Table 25.1.

25.4 ADVANTAGES AND DISADVANTAGES OF FM

REDUCTION OF NOISE

Of the two broadcasting systems, AM and FM, the latter is capable of better recep-

Table 25.1 **DIFFERENCES BETWEEN AM AND FM**

FACTOR	FM	AM
Amplitude of transmitted signal	Remains constant	Varies with percentage of modulation
Audio voltage amplitude	Frequency swing of the signal is determined by amplitude of the audio modulating voltage	Determines amplitude variation of the wave
Audio frequency	Frequency of the audio modulating voltage determines rate of frequency deviation	Audio frequency controls the rate at which amplitude of the wave changes
Signal spectrum	Number of sidebands depends on amplitude of modulating signal; in television, spectrum is restricted to 25 kHz on either side of carrier	Generally limited to 5 kHz on either side of carrier frequency; determined by frequency of audio modulating wave

tion under adverse conditions. It is easier to minimize interference from other nearby stations operating on the same frequency with FM than with AM.

TRANSMITTING EFFICIENCY

Because of the arrangement of the circuits in an FM transmitter, a signal of a given wattage can be developed more economically with this equipment than with AM equipment. Specifically, the large difference in cost between the two systems lies in the audio power required to produce a signal of a given strength. With AM, the audio power is generally 50% of the carrier power, and this may entail many thousands of watts for a powerful station. In FM, on the other hand, the audio required represents only a fraction of the output power and is more easily generated.

The power relationship that exists in an AM wave between the sidebands and the carrier is in the ratio of 1:2 for 100% modulation. This is only the average power, and when the equipment is designed, it must be capable of handling much higher peak (or surge) power. Naturally, this requirement materially increases the cost of the station. In FM transmission, the power output does not increase with modulation and no additional provision for handling excess power need be made.

AUDIO FIDELITY

The matter of fidelity is not stressed because, contrary to popular opinion, just as much fidelity is possible with AM as with FM. The only problem is that, on the present crowded AM broadcast band (535 to 1 605 kHz), space is not available to permit the full 15 000-Hz audio bandwidth to be reproduced. Given sufficient spectrum space, both systems may have equal fidelity.

STATIONS ON THE SAME FREQUENCY

One definite advantage obtained with FM is due to the observed (and calculated) fact that, if two signals are being received simultaneously, the effect of the weaker signal will be eliminated almost entirely if it possesses less than half the amplitude of the other signal. This means that for one signal to completely override another at the receiver, their amplitudes need to be in the ratio of 2:1, or more. With a good antenna, it is frequently easy to tune in one station in sufficient strength so that interfering stations are eliminated entirely. No such situation exists with AM signals, where interfering stations can be heard even when there is a 100:1 relationship between the different carrier amplitudes.

Table 25.2 summarizes the advantages and disadvantages of the two transmission systems.

25.5 PREEMPHASIS AND DEEMPHASIS

In the audio frequencies fed to a transmitter, the amplitudes of the higher audio frequencies are less than those of the lower frequencies. This is due only to the natural distribution of the sound. As the sound passes through the transmitter to the receiver and then through the receiver, noise is unavoidably added to the audio signal. This noise is predominantly high-frequency audio and tends to produce a low signal-to-noise ratio at the higher audio frequencies. This condition can be overcome by the use of **preemphasis** at the transmitter and **deemphasis** at the receiver.

By this scheme, the amplitudes of the higher audio frequencies are boosted relative to the lower audio frequencies at the transmitter (preemphasis). At the receiver, the

Table 25.2 **ADVANTAGES AND DISADVANTAGES OF AM AND FM**

FM	AM
ADVANTAGES	
Better reception under adverse conditions Easier to eliminate weaker stations on the same frequency as the desired station Transmitter cheaper to build because less audio power is needed Easier to separate audio and video signals	Simpler circuitry at the receiver Alignment very easy to do
DISADVANTAGES	
Receiver more complicated Alignment more difficult than for AM receiver	Overmodulation causes distortion Most noise adds to AM signal and is heard at the speaker Difficult to eliminate weak station on the same frequency as the desired station More power needed in audio portions of the transmitter

higher audio frequencies are reduced to their original level (deemphasis). The net result of preemphasis and deemphasis is a considerable improvement in the signal-to-noise ratio at the higher audio frequencies (roughly from 1 000 Hz to 15 000 Hz).

PREEMPHASIS

The preemphasis characteristic at the transmitter is set down by the FCC and is shown in Figure 25.3A. The preemphasis characteristic may be produced by the simple circuit of Figure 25.3B, which has a 75-μs time constant. The R and L act as a voltage divider for audio signals. As the audio frequency increases, the increasing inductive reactance of L causes the audio output amplitudes to increase in the manner of Figure 25.3A. Note that the increase is almost 1 dB at 1 000 Hz, almost 3 dB at 2 000 Hz, about 8 dB at 5 000 Hz, 13.5 dB at 10 000 Hz, and about 17 dB at 15 000 Hz.

Because of the preemphasis, the higher audio frequencies will not be blanketed by the inherent high-frequency audio noise. When picked up by the television receiver, the FM signal will have the preemphasis characteristic discussed above.

DEEMPHASIS

Before the audio signal is applied to the receiver audio amplifiers, it is essential to return its relative amplitudes to those prior to preemphasis. This prevents audio amplitude distortion and simultaneously reduces the higher-frequency audio noise level.

To accomplish deemphasis, a 75-μs time constant circuit is used, as illustrated in Fig-

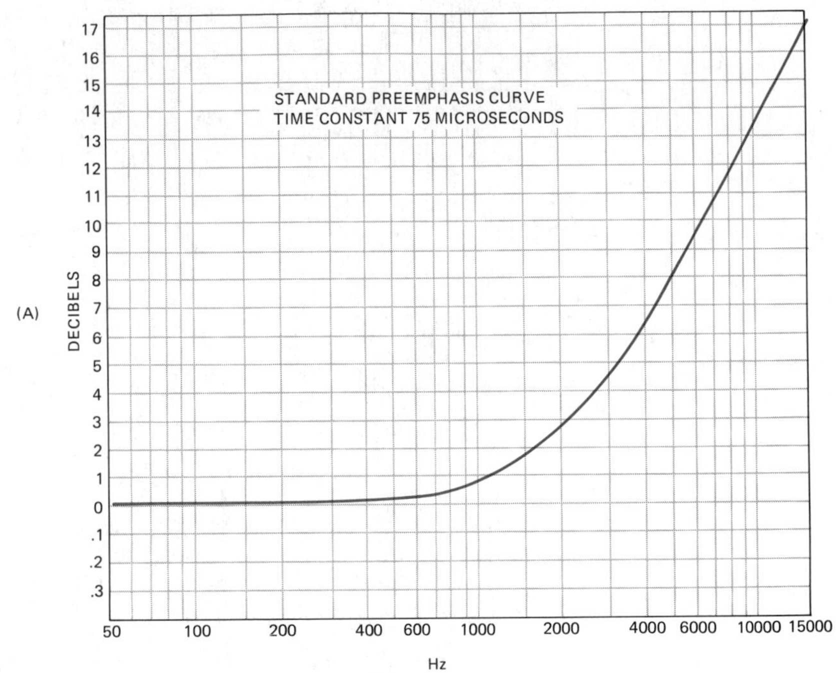

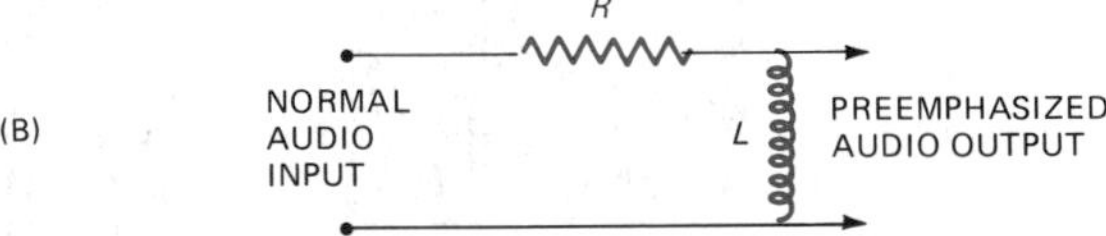

Figure 25.3 Preemphasis as employed at an FM transmitter. A. The standard FCC preemphasis curve, based on a 75-microsecond time-constant filter. B. A simple preemphasis circuit having a 75-microsecond time constant.

ure 25.4. This circuit provides a mirror-image response to that of Figure 25.3A. By means of the circuit of Figure 25.4, the higher audio frequencies are returned to their original amplitude relationship with respect to the lower frequencies.

Figure 25.4 is a voltage divider circuit. As the audio frequency increases, the reactance of C decreases and less audio voltage appears across it as an output. Since the deemphasis network must closely cancel the frequency-versus-amplitude effect of the preemphasis network, it must have the same time constant, 75 μs. As the higher audio frequencies are returned to their original amplitudes, a considerable reduction in noise occurs.

25.6 THE TV RECEIVER FM SOUND SECTION

The sound section of a television receiver is very similar to that of a broadcast FM receiver. The major differences are that television receivers have a lower IF (4.5 MHz instead of 10.7 MHz) and a narrower IF and FM detector bandwidth (50 kHz instead of 150 kHz).

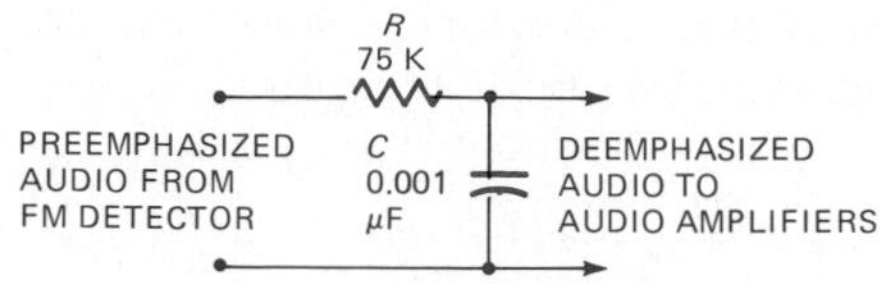

Figure 25.4 A deemphasis circuit used in the sound section of a TV receiver. The time constant is 75 microseconds. The characteristic is opposite that shown in Figure 25.3A.

Both types of FM receivers have relatively wide-band IF amplifiers, amplitude limiting to reduce noise, an FM detector, and an audio amplifier.

The TV-FM system is inherently capable of producing high-fidelity audio reproduction (50 Hz to 15 000 Hz). Although many TV receivers do not utilize this capability to its fullest extent, some of the more recent receivers are giving greater attention to improved audio quality.

BLOCK DIAGRAM

A block diagram of a typical TV sound section is shown in Figure 25.5. The section configuration for either a monochrome or a color-TV receiver is shown by means of dashed lines.

MONOCHROME TV RECEIVER INPUT

In monochrome-TV receivers, no separate sound detector is required. The 41.25-MHz sound IF and 45.75-MHz video IF are heterodyned in the video detector to produce the 4.5-MHz intercarrier sound IF. This IF is taken either from the output of the video detector or from a following video amplifier to provide a higher level sound IF signal. In either case, this IF signal is fed to the input of the 4.5-MHz sound IF amplifier.

COLOR-TV RECEIVER INPUT

In a color-TV receiver, it is important to minimize a possible 920-kHz beat frequency interference pattern on the screen. This might occur if the 4.5-MHz intercarrier sound IF signal and the 3.58-MHz color IF signal are heterodyned in the video detector. To prevent this, the sound IF at 41.25-MHz is taken off *before* the video detector and its amplitude severely reduced at the input to the video detector.[1] The 41.25-MHz sound IF

[1]An exception to this may occur in color-TV receivers that have a synchronous video detector (Figure 25.31).

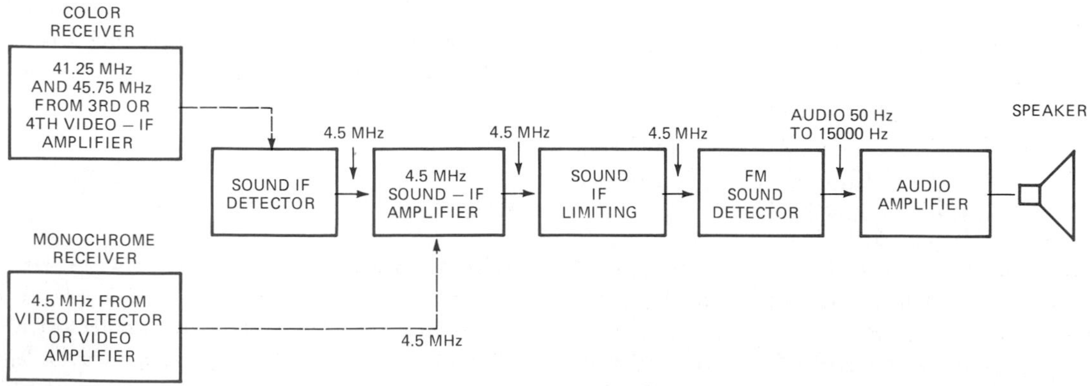

Figure 25.5 Block diagram of TV receiver sound section. Connections for either a monochrome or a color TV receiver are shown with dashed lines.

signal plus the 45.75-MHz video signal are fed to a *separate* sound-IF detector diode. The useful output of this detector is the 4.5-MHz intercarrier sound-IF signal, which is fed to the input of the 4.5-MHz sound-IF amplifier.

Note: For typical sound take off circuits, see Chapter 12.

THE 4.5-MHz SOUND IF AMPLIFIER

The 4.5-MHz sound-IF amplifier receives the FM 4.5-MHz signal either from the separate sound detector in a color TV receiver or from the video detector or video amplifier in a monochrome receiver. The basic function of this stage is to amplify the 4.5-MHz FM signal and its sidebands to a level suitable to drive a limiter or a limiting FM detector. The output amplitude of the sound-IF amplifier is about 2 to 5 V.

SOUND-IF LIMITING

The function of **sound-IF limiting** is to remove amplitude variations from the FM sound-IF signal. By so doing, most of the noise impulses are removed from the audio. Removing amplitude variations does not affect the FM portion of the sound-IF wave. Thus, audio quality is unaffected.

Amplitude limiting may be performed either by a separate *limiter* stage or by the use of an **FM demodulator** that does not respond to amplitude variations.

FM SOUND DETECTOR

The **FM sound detector** has as its input the 4.5 MHz FM IF signal. Its output is the original audio signal with a frequency range of 50 to 15 000 Hz. There are several types of FM sound detectors. These will be described in subsequent sections.

AUDIO AMPLIFIER

The **audio amplifier** may consist of one or more stages. These amplify the audio signal received from the FM sound detector to a level suitable to operate a speaker with adequate volume.

Although the FM sound detector audio output may have an audio range of 50 to 15 000 Hz, many TV receiver audio systems are not capable of reproducing this range. However, as mentioned previously, some of the higher-priced receivers may have improved audio systems to take advantage of this wide frequency range.

25.7 THE 4.5-MHz SOUND-IF AMPLIFIER

Schematically, a television sound-IF amplifier is similar to an AM broadcast receiver IF amplifier. The major differences lie in the required bandwidth and in the operating frequency. Sound-IF amplifiers for television must amplify signals with a total bandwidth of 50 kHz. This is done at a center frequency of 4.5 MHz. Compare this with an FM broadcast receiver. Here the required bandwidth is 150 kHz at a center frequency of 10.7 MHz. In contrast, the usual AM receiver has an IF bandwidth of only 10 kHz at a center frequency of 455 kHz.

Schematic Diagrams

Schematic diagrams of two typical sound-IF amplifiers are shown in Figure 25.6. Note that AGC is not used in the sound section. This is because the limiting function tends to maintain a constant signal amplitude.

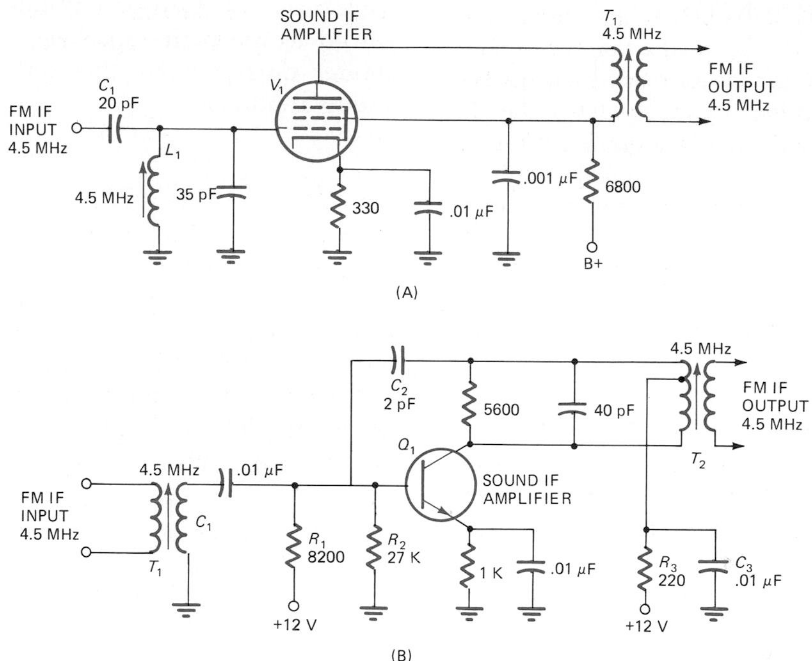

Figure 25.6 Typical 4.5-MHz sound-IF amplifiers for TV receivers. **A.** Vacuum tube sound-IF amplifier. **B.** Transistor sound-IF amplifier.

Figure 25.6A shows a typical vacuum tube sound-IF amplifier. The 4.5-MHz FM signal is applied to tuned inductor L_1 via coupling capacitor C_1. Inductor L_1 is peak-tuned to 4.5 MHz and has the required 50-kHz bandwidth. The amplified signal is coupled to the next stage (limiter or FM demodulator) by transformer T_1. T_1 is also peaked at 4.5 MHz and has the necessary bandwidth.

Figure 25.6B shows a typical transistor sound-IF amplifier. The input is via transformer T_1, which is peaked at 4.5 MHz. The signal is coupled to the base of transistor Q_1 through capacitor C_1. Forward bias for Q_1 is provided by the voltage divider action of resistors R_1 and R_2.

The output transformer T_2 has several turns on the primary winding to provide a neutralization signal through a 2-pF capacitor (C_2). This provides feedback to effectively cancel the internal base-to-collector capacitance of the transistor. Without this neutralization, the stage would tend to oscillate and might cause undesirable high-pitched sounds from the speaker.

25.8 SOUND-IF LIMITING

The first significant difference between the AM and FM super heterodynes is noted at the limiter stage or stages. Essentially, the purpose of a **limiter** is to eliminate the effects of the amplitude variations (mainly noise) in the FM signal. While it may be true that the FM signal leaves the transmitter with no amplitude variations, this is almost never true by the time the signal reaches the limiter in the receiver.

RECEIVER RESPONSE

To digress for a moment, let us see where in the receiver various parts of the FM signal could have received more amplification than other parts. An ideal response curve for a tuned circuit is shown in Figure 25.7A. With such a characteristic, each sideband frequency receives uniform amplification. Such a happy situation is seldom encountered in practice, however. The more usual state of affairs is illustrated by the curve in Figure 25.7B. Here it is apparent that the center frequencies receive more amplification than those located farther away. Hence, even if the incoming signal is perfectly uniform, amplitude variations will be present by the time it arrives at the limiter.

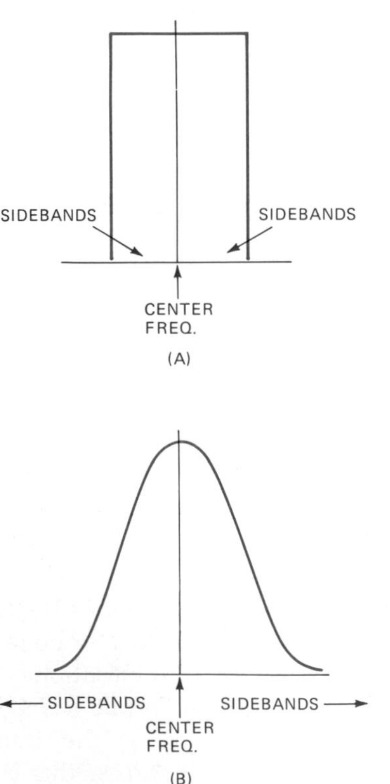

Figure 25.7 Receiver response curves: A. the ideal curve, B. a typical practical result.

The result is distortion if this wave is allowed to reach the speaker. Therefore, the limiter must remove this and other amplitude variations.

REMOVING NOISE INTERFERENCE

Most interference affects the *amplitude* of a radio signal. Thus by simply removing the signal amplitude variations without affecting its frequency eliminates a large percentage of noise interference. This is the function of sound IF limiting.

A VACUUM TUBE LIMITER

Limiters may be made with vacuum tubes, transistors, or ICs. The principle of limiting using a vacuum tube will be explained first. The basic principle remains the same for transistor or IC limiters. However, superior limiting can be obtained with high-gain ICs.

A typical vacuum tube limiter stage is shown in Figure 25.8. Except for the grid-leak method of achieving bias, its appearance is almost identical to an IF stage. Further inspection reveals that low plate and screen voltages are used. The low electrode voltages cause the tube to reach current saturation and cutoff with moderate signal levels at the grid. The use of the grid-leak bias aids in keeping the output plate current (and hence the output signal) relatively constant for different input voltage levels. With FM signals of different amplitudes arriving at the grid of the limiter, a constant output for each means the elimination of any amplitude distortion, which is exactly what is desired. With the limiter designed so that it will easily saturate and cut off, amplitude variations can be eliminated and, with them, most disturbing noises.

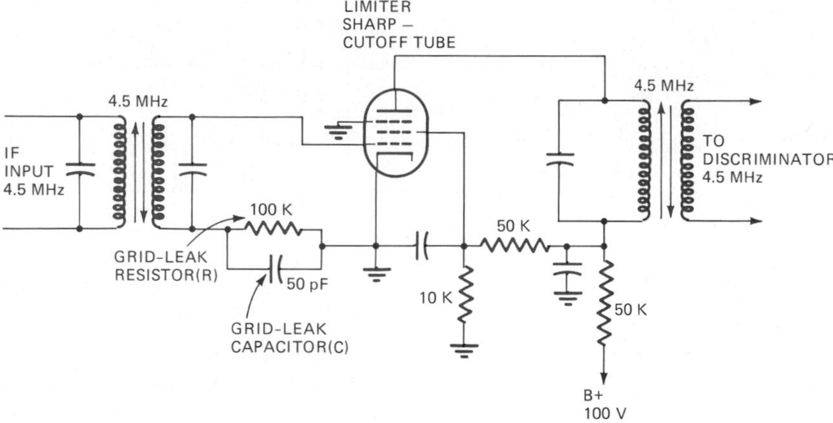

Figure 25.8 A typical vacuum tube limiter.

Grid-Leak Bias It is possible to design vacuum tube limiters on the basis of low plate and low screen voltages alone, but better results and more amplification are obtained if grid-leak bias is added to this combination (Figure 25.8). With the insertion of grid-leak bias, it is possible to raise the electrode voltages, somewhat increasing the gain.

The tube initially has zero bias with no signal at the grid. As soon as a signal acts, the grid is driven slightly positive, attracts electrons, and charges the capacitor C. This capacitor attempts to discharge through R but because of the relatively long time constant of R and C, the discharge occurs slowly. The voltage in capacitor C acts as bias, varying in value as the incoming signal varies in amplitude. In this way, it tends to keep the average plate current steady within rather wide limits of input signal voltage. A strong signal causes the grid to become more positive, resulting in a higher signal voltage in C. A larger bias is then developed. A weaker signal will develop less bias voltage, resulting in essentially the same amount of average plate current. The usual values of C range from 30 to 60 pF, and those for R are between 20 000 and 200 000 ohms.

Analysis of Limiter Action In a vacuum tube limiter, effective limiting depends upon a sharp-cutoff tube being driven to saturation and cutoff. Since grid-leak bias is used in Figure 25.8 with no input signal, the bias is zero. As the input signal amplitude increases, the bias increases proportionately.

Observe the action in Figure 25.9 for a weak signal. A low value of grid-leak bias is developed. The noise pulses on the positive signal peaks drive the tube into saturation and are thus limited. However, the noise pulses on the negative signal peaks of the weak signal do not drive the tube to cutoff. They are not eliminated.[1]

Now observe the action of the single-stage limiter of Figure 25.8 for a strong signal. Note that for a strong signal, the grid-leak bias has increased. Also, the tube is driven *beyond saturation and cutoff*, effectively eliminating noise pulses on both *positive and negative* signal peaks.

For still stronger signals, the percentage of whole signal cycles appearing in the limit-

[1]Weak-signal limiting is improved in all types of limiters if two limiter stages are used. In that case, the negative peak noise pulses on a weak signal are limited by the succeeding stage.

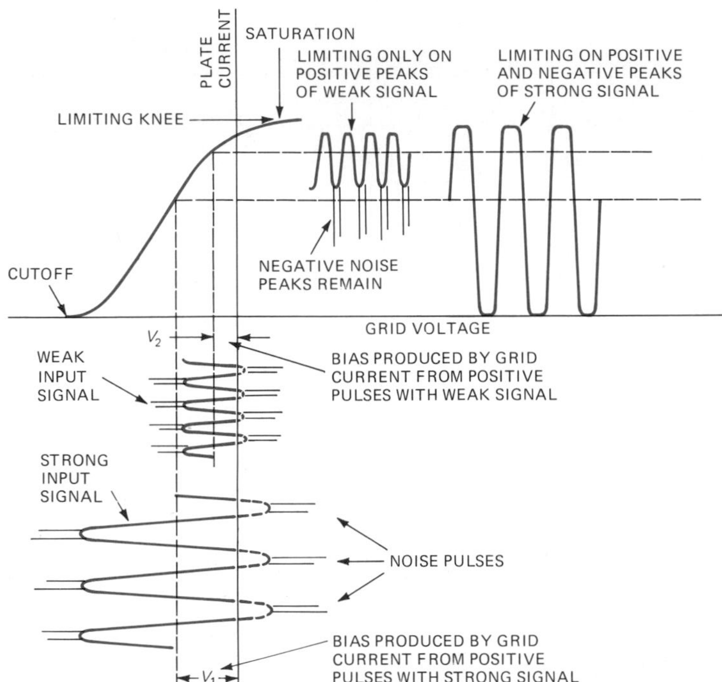

Figure 25.9 This graph illustrates how a limiter can remove amplitude variations from an FM signal. For better limiting on weak signals, two (or more) limiter stages are used.

er output current would decrease. The grid-leak bias would be further increased and the negative portion of the input signal partially or wholly cut off. In the latter case, the plate current would consist of *pulses* corresponding to the positive portions of the input signal. However, the *frequency* of the output signal is unaffected and the missing portions of each FM cycle are "replaced" by the action of the output tuned circuit.

The Limiting Knee (see Figure 25.9) The minimum input signal amplitude required to produce limiting is called the **limiting knee.** This is generally defined as the value at which plate (or collector) current ceases to increase (practically) with an increase of input signal amplitude. This point may also be called the **limiting threshold level.**

For a vacuum tube limiter (two stages), the limiting knee value is several volts. About 1 V is required for a two-stage transistor limiter. Only 100 to 200 MV is required for a high-gain IC limiter, which may contain six to eight stages.

TWO-STAGE DIODE TRANSISTOR LIMITER

Although limiting can be provided by a single stage, improved limiting can be provided by two stages. A two-stage diode transistor limiter is shown in Figure 25.10. Two stages with different input time constants are used. The first limiting stage (here, diode D_1) has a relatively short time constant so that it can limit impulse noises. The second limiting stage Q_2 has a longer time constant to limit variations of longer duration.

Transistor Limiter Operation Limiter amplifier Q_1 receives its input signal (4.5 MHz FM) from the sound-IF amplifier. The ampli-

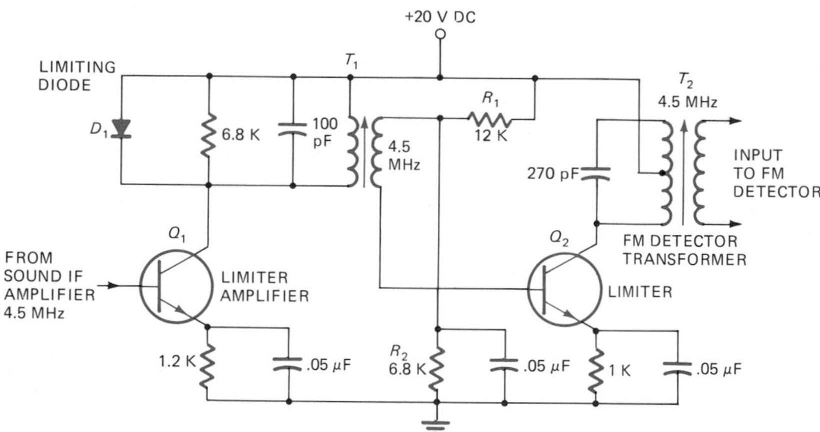

Figure 25.10 A two-stage, diode-transistor limiter.

fied 4.5-MHz FM signal appears in the collector circuit of Q_1. Here, we find limiting diode D_1 across the primary winding of the 4.5-MHz transformer T_1. Diode D_1 limits the negative portions of the signal and any AM on it. The signal polarity is inverted by transformer T_1 and is applied to the base of limiter transistor Q_2. Transistor Q_2 normally operates at saturation because of the positive base bias provided by the voltage divider R_1–R_2. What was the positive portion of the signal at the collector Q_1 becomes negative at the base of Q_2. This negative signal cuts off Q_1, limiting the second half of the input signal.

A HIGH-GAIN LINEAR IC LIMITER

A typical IC limiter is shown in Figure 25.11. This circuit has a high-gain linear IC that contains eight active stages. Because of this design, superior limiting is provided. The limiting knee of this IC is approximately 100 μV (compared with about 1 V for a two-stage transistor limiter).

25.9 A BASIC FM DISCRIMINATOR

The purpose of the second detector in an AM receiver is to obtain the audio variations from the incoming modulated signal. The same stage in an FM receiver must derive the audio variations from the different incoming frequencies. Thus, although the end product in both cases is the same, the methods used are quite different. We know that with FM a large frequency deviation from the carrier means a loud audio note, whereas a small frequency deviation means a weak audio note. The rate of deviation is the audio frequency. Hence, there must be a circuit to develop voltages proportional to the deviation of the various incoming frequencies about the FM carrier.

A simple circuit that discriminates against the various frequencies is the elementary parallel-resonant (or series-resonant) circuit. As is well known, this circuit develops maximum voltage at the resonant frequency, with the response falling off as the frequency separation increases on either side of the central, or resonant, point.

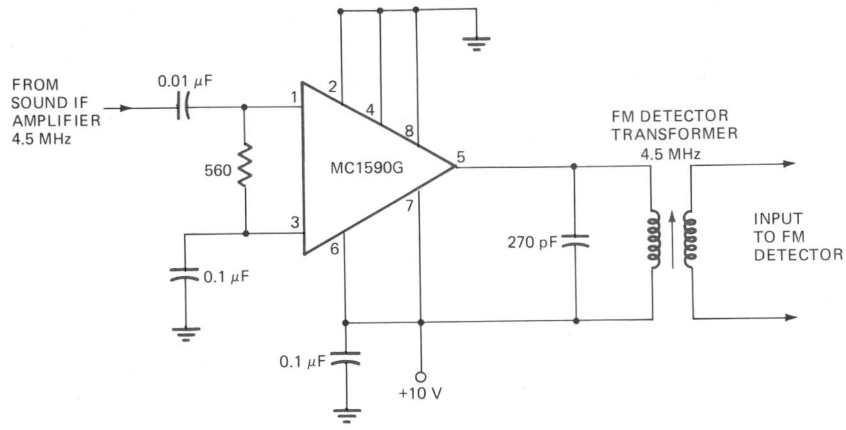

Figure 25.11 Typical limiter circuit using a high-gain linear IC. The IC contains eight active stages.

BASIC DISCRIMINATOR OPERATION

One of the first **discriminators** used in FM receivers contained two resonant circuits in an arrangement as shown in Figure 25.12. This circuit is described for background but it is not in general use today. The primary coil L_1 is inductively coupled to L_2 and L_3, each of which is connected to a diode. Each diode has its own load resistor, but the output of the discriminator is obtained from the resultant voltage across both resistors.

L_1 and L_3 Resonant Frequencies In order to determine the frequencies to which L_2 and L_3 must be tuned, it should be recalled that when an audio modulating signal alters the frequency of an FM transmitter, it varies this frequency above and below one central, or carrier, value. Thus, for a sine wave, the maximum positive portion would increase the frequency, say, by 40 kHz, while the maximum negative section would decrease the carrier frequency by the same amount. At intermediate points, less voltage would cause correspondingly less frequency deviation.

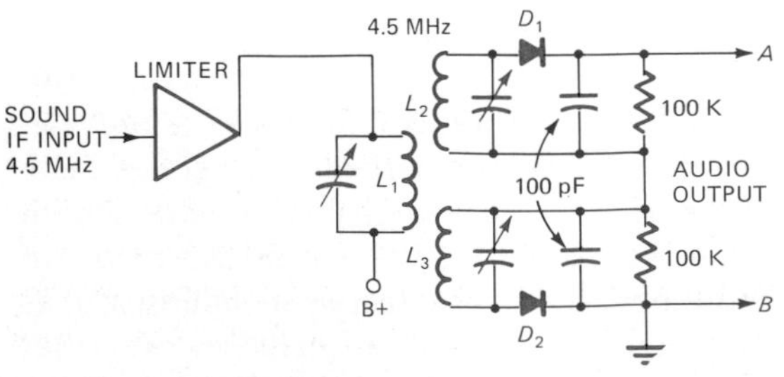

Figure 25.12 A basic discriminator circuit.

To have the discriminator function in a similar manner over the same range, L_2 and L_3 are each peaked to one of the two end points of the IF band. For example, if the IF bandspread extends from 4.25 MHz to 4.75 MHz (with 4.50 as the mean, or carrier, frequency), L_2 could be peaked to 4.25 MHz, and L_3 to 4.75 MHz. The response curves would look like those in Figure 25.13.

The Output Voltages The two curves are positioned in the manner shown because of the way the load resistors and diodes are connected in the circuit. According to the arrangement, the voltages developed across the resistors tend to oppose each other.

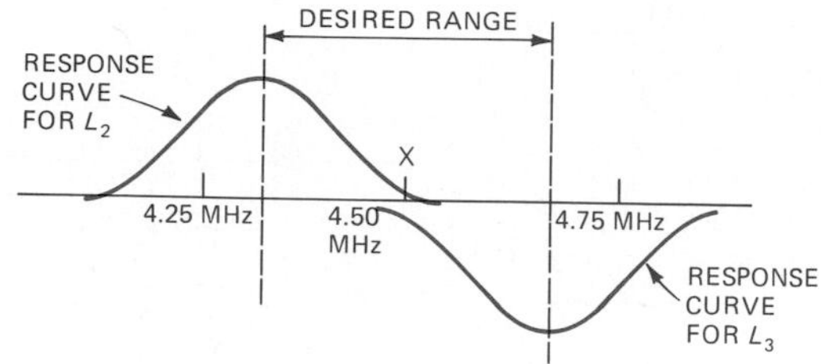

Figure 25.13 The overall response curve for the discriminator of Figure 25.12.

At the center frequency, Point X of Figure 25.13, the two voltages developed across the load resistors cancel each other and the resultant voltage is zero. Similarly, by adding the voltages at other points about the carrier frequency, we obtain the overall resultant curve shown in Figure 25.14. This is the familiar S-shaped curve of all frequency discriminators, which shows how the output voltage of the FM detector will vary as the incoming frequencies change. Specifically, suppose the signal acting at the input to the discriminator at any one instant has a frequency of 4.65 MHz. The amount of voltage developed at the output is given by Point A on the vertical axis (Figure 25.14). Then, at the next instant, if the frequency should change to 4.35 MHz, the output voltage is indicated by Point B. Notice that all frequencies below 4.50 MHz result in positive output voltages, whereas all those above 4.50 MHz give rise to negative output voltages. In this way, the audio voltages that modulated the carrier frequency at the transmitter are extracted in the receiver.

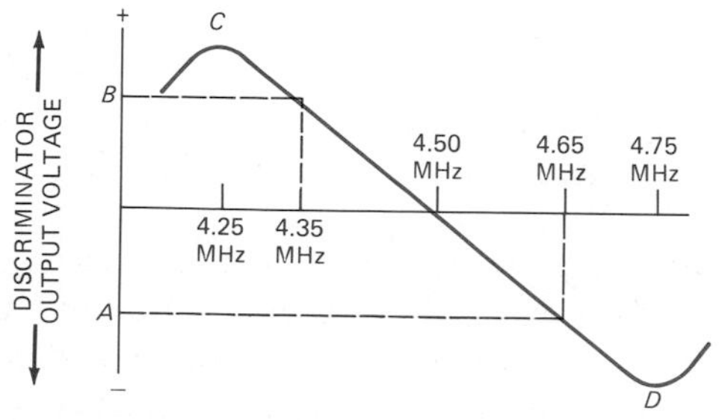

Figure 25.14 The resultant S-shaped discriminator characteristic curve obtained by adding the two separate curves of Figure 25.13.

The Linear Portion The useful segment of this characteristic curve of the discriminator is the linear portion included between the two maximum points, C and D. Any nonlinearity along this section of the curve produces amplitude distortion in the output audio signal. When discriminators are designed, Points C and D are generally set much farther apart than is required for the particular receiver. This ensures a linear curve at those frequencies that are actually used, since the response characteristic has a tendency to curve near the maximum peaks. By using a smaller range, amplitude distortion in the output signal is kept to a minimum. The sections of the curve of Figure 25.14 beyond Points C and D are completely disregarded.

The Audio Output The frequency of the audio output voltages is determined by how rapidly the frequency of the incoming IF signal varies. A large frequency deviation in the input signal gives rise to a strong output wave, and the rapidity with which this incoming frequency changes determines whether the strong output will be pitched high or low.

25.10 THE FOSTER-SEELEY DISCRIMINATOR

One may wonder why the preceding circuit was described in such detail since it is not used in modern receivers. The reason is that this circuit shows so very clearly the fundamental conversion process at the second detector of an FM receiver. Also, it is basically the same as the present-day Foster-Seeley discriminator of Figure 25.15.

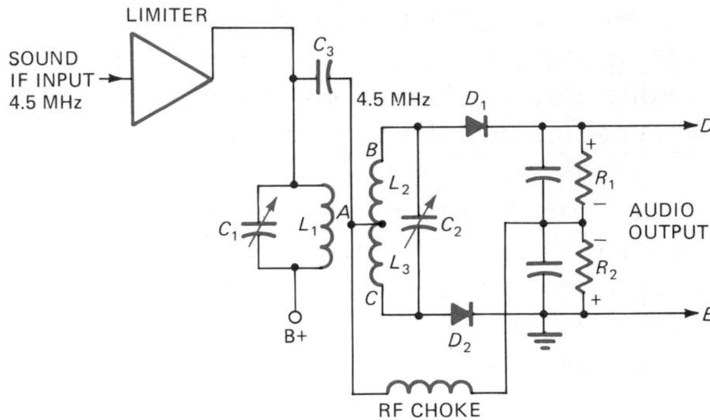

Figure 25.15 A Foster-Seeley discriminator.

Instead of having two separate tuning capacitors for the secondary circuit, only one is used. Resistors R_1 and R_2 are the load resistors, one for each diode, and the resultant output audio voltage is obtained across Points D and E. The use of one capacitor instead of two results in greater ease in aligning the circuits and economy in construction. The tap divides the secondary coil into two identical coils, L_2 and L_3. **Vector diagrams** are required to understand how the circuit responds to frequency modulation. The capacitor C_3 couples a voltage V_A at Point A. We shall consider this voltage our reference at 0° (Figure 25.16A). This is a voltage vector coming from the L_1–C_1 tank circuit. The other two vectors we require for our vector diagram come from the transformer action across L_2 and L_3. We shall call these voltages V_{BA} and V_{CA}. These voltage vectors are $\pm 90°$ out of phase with V_A when the incoming signal is at the IF center frequency. Figure 25.16A shows the vector relationship of these three voltages. The voltage V_B is simply the vector sum of V_A and V_{BA}. Similarly, $V_C = V_A + V_{CA}$ (vector addition again). These new vectors are shown in Figure 25.16A. The detected voltages across R_1 (V_{DF}) and R_2 (V_{EF}) are equal in magnitude to the peak RF voltages at Points B and C, respectively. Voltages V_{DF} and V_{EF} are equal and opposite at resonance so the voltage V_{DE} is zero.

Above and Below Resonance What happens above and below resonance? Consider the loop formed by L_2, L_3, and C_2 in Figure 25.15. Above resonance the current in this loop lags and hence voltages V_{BA} and V_{CA} lag from their mid-frequency phase. This results in the vector relationship shown in Figure 25.16B. Voltage V_B is now larger than V_C, and so the detected output, V_{DE}, is a positive voltage. Below resonance, the secondary loop current leads its mid-frequency phase. This

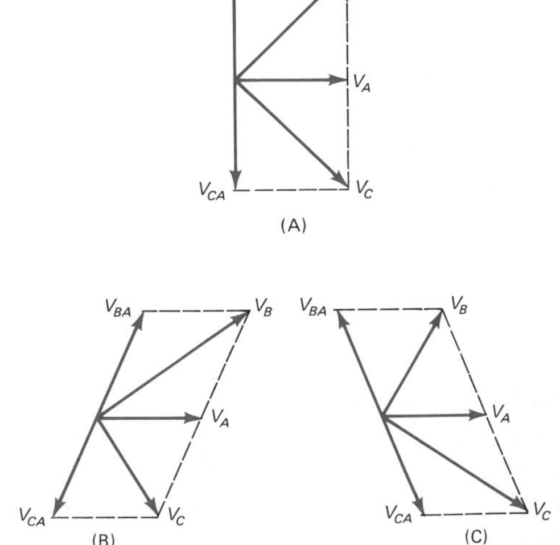

Figure 25.16 Vector diagrams for the Foster-Seeley discriminator: **A.** at resonance; **B.** above resonance; and C. below resonance.

makes V_C larger than V_B, and a negative voltage appears at V_{DE}. A characteristic curve similar to the S-shaped one in Figure 25.14 is obtained for this discriminator.

Need for Limiters Limiter stages are needed because the discriminators noted in the preceding paragraphs are sensitive to the amplitude of the incoming signal. In other words, these discriminators are not pure FM detectors. There are other FM detectors, however, that are sufficiently immune to amplitude variations to enable us to dispense with the limiter. These detectors are discussed in following sections.

25.11 THE RATIO DETECTOR

To understand why a **ratio detector** enjoys greater immunity from AM distortion in the incoming FM signal, let us compare its operation with that of the discriminator.

In the discriminator circuit of Figure 25.15, let the signal coming in develop equal voltages across R_1 and R_2. This occurs, of course, when the incoming-signal frequency is at the IF center value. Suppose that each voltage across R_1 and R_2 is 4 V. When modulation is applied, the voltage across each resistor changes, resulting in a net output voltage. Say that the voltage across R_1 increases to 6 V and the voltage across R_2 decreases to 2 V. The output voltage is then equal to the difference between these two values, or 4 V.

However, let us increase the strength of our carrier until we have 8 V each, across R_1 and R_2, at mid-frequency. With the same frequency shift as previously, but with this stronger carrier, the voltage across R_1 rises to 12 V and that across R_2 decreases to 4 V. Their difference, 8 V, is now obtained at the output of the discriminator in place of the previous 4 V. Thus the discriminator responds to both FM and AM. It is for this reason that limiters are used. The limiter clips all AM off the incoming signal, and an FM signal of constant amplitude is applied to the discriminator.

When unmodulated, the carrier produces equal voltages across R_1 and R_2. Let us call these voltages V_1 and V_2, respectively. With the weaker carrier on modulation, the ratio of V_1 to V_2 is 3:1, since V_1 becomes 6 V and V_2 drops to 2 V. With the stronger carrier on modulation, V_1 becomes 12 V and V_2 drops to 4 V. Their ratio is again 3:1, the same as with the previous weaker carrier. Thus, while the difference voltage varied in each case, the ratio remained fixed. This example demonstrates, in a very elementary manner, why a ratio detector can be unresponsive to carrier changes.

THE ELEMENTARY RATIO DETECTOR

An elementary circuit of a ratio detector is shown in Figure 25.17. In this form, the detector is similar to the detector in Figure 25.12 where each diode has a completely separate resonant circuit. One circuit is peaked slightly above the center IF value (say, T_1); the other is peaked to a frequency below the center (say, T_2). The output volt-

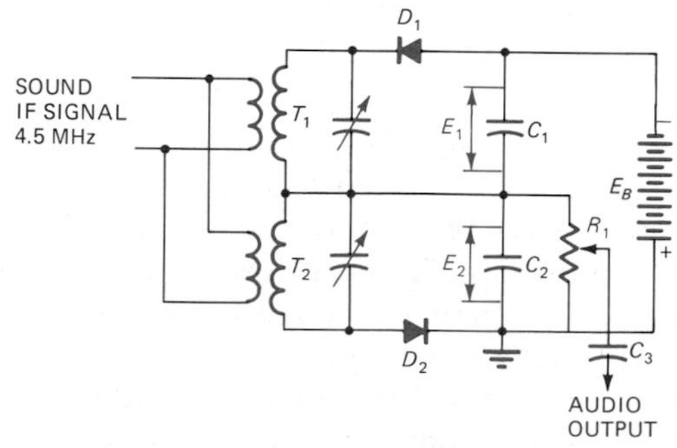

Figure 25.17 Preliminary form of the ratio detector.

age for D_1 appears across C_1, and the output voltage for D_2 is present across C_2. The battery represents a fixed voltage, E_B. Since C_1 and C_2 are in series directly across the battery, the sum of their voltages must always equal E_B. Also, because of the way the battery is connected to D_1 and D_2, no current can flow around the circuit until a signal is applied. Now, although $E_1 + E_2$ can never exceed E_B, E_1 does not have to equal E_2. In other words, the ratio of E_1 to E_2 may vary. The output voltage is obtained from a resistor connected across C_2.

When the incoming signal is at the IF center value, E_1 and E_2 are equal. This is similar to the situation in the previous discriminator. However, when the incoming signal rises in frequency, it approaches the resonant point of T_1 and the voltage across C_1 likewise rises.

Producing Audio Variations For the same frequency, the response of T_2 produces a lower voltage. As a consequence, the voltage across C_2 decreases. However, $E_1 + E_2$ is still equal to E_B. In other words, a change in frequency does not alter the total voltage, but merely the ratio of E_1 to E_2. When the signal frequency drops below the IF center point, E_2 exceeds E_1. The sum of $E_1 + E_2$ must equal E_B, however. The audio variations are obtained from the change of voltages across C_2. Capacitor C_3 prevents the rectified DC voltage in the detector from reaching the grid (or base) of the audio amplifier. Only the audio variations are desired.

Purpose of E_B The purpose of E_B in this elementary explanatory circuit is to maintain an audio output voltage that is purely a result of the FM signal. The battery voltage E_B keeps the total voltage $(E_1 + E_2)$ constant, while it permits the ratio of E_1 to E_2 to vary. So long as this condition is maintained, all amplitude variations in the input signal will be without effect.

Replacing the Battery with a Capacitor The problem of selecting a value for E_B is an important one. Consider, for example, the case where a weak signal is being received. If E_B is high, the weak signal is lost because it is not strong enough to overcome the negative polarity placed by E_B on diodes D_1 and D_2. The diodes, with a weak input voltage, cannot pass current. If the value of E_B is lowered, then powerful stations are limited in the amount of audio voltage output from the ratio detector. This is because the voltage across either capacitor C_1 or C_2 cannot exceed E_B. If E_B is small, only small audio output voltages are obtainable. To get around this restriction, it was decided to let the average value of each incoming carrier determine E_B. Momentary increases could be prevented from affecting E_B by a circuit with a relatively long time constant.

PRACTICAL RATIO DETECTORS

The practical form of the ratio detector is shown in Figure 25.18. The detector uses the phase-shifting properties of the discriminator of Figure 25.15. Resistor R and C_3 take the place of E_B, and the voltages developed across R depend on the strength of the incoming carrier. Note that D_1 and D_2 form a series circuit with R (and C_3), and any current flowing through these diodes must flow through R. However, by shunting the 8-μF electrolytic capacitor across R, we maintain a fairly constant voltage. Thus, the momentary changes in the carrier amplitude are absorbed by the capacitor. It is only when the *average* value of the carrier is altered that the voltage across R is changed. The output audio frequency voltage is still taken from across C_2 by means of the volume control.

Since the voltage across R is directly dependent upon the carrier strength, it may also be used for AGC voltage. The polarity of the voltage is indicated in Figure 25.18.

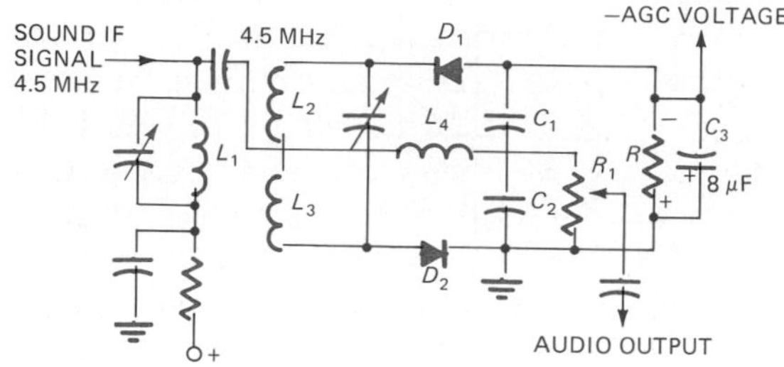

Figure 25.18 Practical form of the ratio detector.

ECONOMICAL RATIO DETECTOR

The urge to simplify circuits and thereby reduce cost is ever present among designers of television sets. Such simplification is possible with the ratio detector, as revealed by the design shown in Figure 25.19. This arrangement lacks the C_1–C_2 divider shown in Figure 25.18. In spite of the reduction, the circuit still functions satisfactorily. However, with fewer capacitors, the reader may fail to see how the difference voltage is established to provide the necessary audio output signal.

To understand how the circuit in Figure 25.19 operates, it has been drawn with lettered identification points. Current that flows through D_1 can take one of two paths. In one path, the current flows from the cathode of D_1 to the plate to Points *B, C, F, E, A,* and then back to the cathode again. The second path is cathode to anode, to Points *B, C,* then to ground, and up through C_1 to Point *D*, then to Points *E, A,* and finally back to the cathode again.

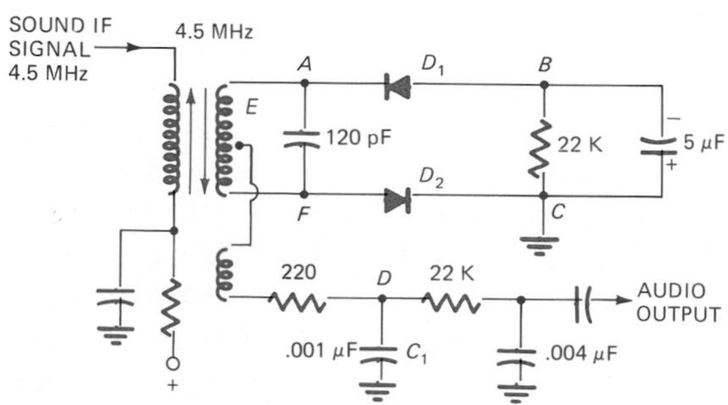

Figure 25.19 More economical form of the ratio detector circuit of Figure 25.18.

Now let us consider D_2. One path for its current is cathode to anode, to *F, E, A,* through D_1 to *B,C,* and back to the cathode again. The second path is from cathode to anode to *F, E,* and down to Point *D*, through capacitor C_1 to ground, then to Point *C*, and back to the cathode again. Note that part of the current of D_1 then flows up through C_1 while part of the current of D_2 travels down through the same capacitor. It is from these two opposing currents that the difference is established. This difference represents the audio output voltage of the detector.

BALANCED RATIO DETECTORS

The preceding ratio detector, shown in Figures 25.17 and 25.18, is an "unbalanced" circuit, so called because D_1 and D_2 are not equally balanced against ground. We can transform these circuits into a balanced ratio detector by moving the position of the ground connection, as shown in Figure 25.20. In place of resistor R, we now have two. Their function, however, remains the same.

Resistor R_1 is inserted to provide a better balance between both halves of the circuit, and R_2 limits the anode current drawn by each diode. Capacitor C_1 shunts IF voltages away from the audio output and R_3, C_2 is a deemphasis filter to equalize the audio signal back to its original form. An AGC voltage can be obtained from the negative side of the 4-μF stabilizing capacitor C_3.

25.12 THE QUADRATURE DETECTOR

Another approach to FM detection is provided by the **quadrature detector.** This

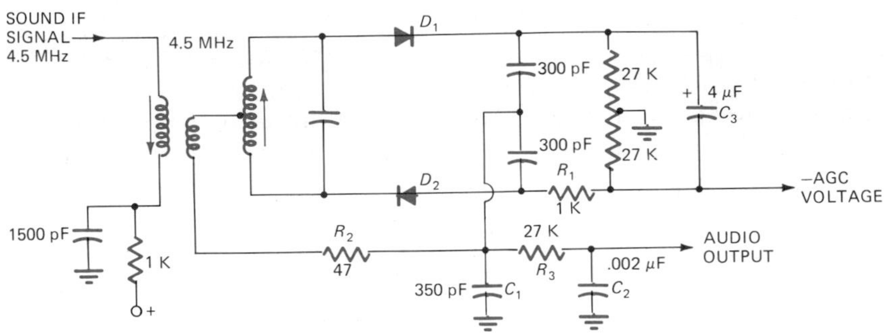

Figure 25.20 A typical form of the balanced ratio detector.

circuit acts as both a discriminator and a limiter in one stage. The original method required a special tube, such as the 6BN6. The technique has now been expanded to use more ordinary-looking tubes, such as the 6DT6. Transistor circuits are also used for quadrature detection, and ICs are frequently used also.

THE 6BN6 DETECTOR

The 6BN6 gated-beam tube was invented by Robert Adler of the Zenith Radio Corporation. It possesses a characteristic such that when the grid voltage changes from negative to positive, the plate current rises rapidly from zero to a sharply defined maximum level. This same maximum value of plate current remains no matter how positive the grid voltage is made. The current cutoff is achieved when the grid voltage is about −2 V.

Gated-Beam Tube Construction The reason for this particular behavior of the tube stems from its construction (Figure 25.21). The focus electrode and first accelerator slot together form an electron gun, which projects a thin-sheet electron stream upon Grid 1. The curved screen grid, together with the grounded lens slot and aided by the slight curvature of Grid 1, refocuses the beam and projects it through the second accelerator slot upon the second control grid. This grid and the anode which follows are enclosed in a shield box. Internally, the focus, the lens, and the shield electrodes are connected to the cathode. The accelerator and the screen grid receive the same positive voltage because they are connected to each other internally.

Grid 1 Action Electrons approaching the first grid do so head on. Hence, when Grid 1 is at zero potential or slightly positive, all approaching electrons pass through the grid. Making the grid more positive, therefore, cannot further increase the plate current.

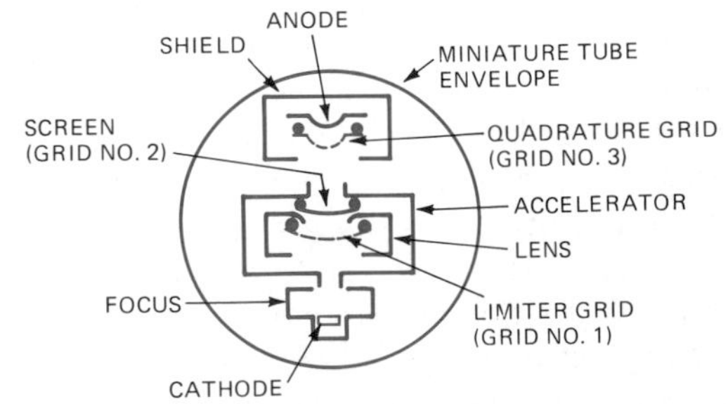

Figure 25.21 The internal construction of the gated-beam tube 6BN6.

When, however, Grid 1 is made negative, those electrons that are stopped and repelled toward the cathode do so along the same path taken in their approach to the grid. Because of the narrowness of the electron beam and its path of travel, electrons repelled by the grid form a sufficiently large space charge directly in the path of other approaching electrons, thus causing an immediate cessation of the current flow throughout the tube. In conventionally constructed tubes, the spread of the electron beam traveling from the cathode to the grid is so wide that those electrons repelled by the grid return to the cathode without exerting much influence on electrons that possess greater energy and therefore are able to overcome the negative grid voltage. It is only when the control grid voltage is made so negative that no emitted electrons possess sufficient energy to overcome it that current through the tube ceases. These differences between tubes can be compared to the difference between the flow of traffic along narrow and wide roads. On narrow roads, the failure of one car to move ahead can slow down traffic considerably; along wide roads, where there is more room, the breakdown of one car has less effect.

Grid 3 Action The electron beam in the form of a thin sheet leaves the second slot of the accelerator and approaches Grid 3. Thus, this section of the tube can also serve as a gated beam system. If this second grid is made strongly negative, the plate current of the tube is cut off no matter how positive Grid 1 is. Over a narrow range of potential in the vicinity of zero, the third grid can control the maximum amount of current flowing through the tube. However, if the third grid is made strongly positive, it also loses control over the plate current, which can never rise beyond a predetermined maximum level.

GATED-BEAM TUBE AS A LIMITER-DISCRIMINATOR

Now let us see how this tube can be made to function as a limiter-discriminator. A typical circuit is shown in Figure 25.22. It has been noted that, when FM signals reach the discriminator, they contain amplitude variations. When the 6BN6 gated beam tube is used, these signals are applied to Grid 1. If the signal receives sufficient prior amplification, it will have a peak-to-peak value of several volts. Upon application to Grid 1, current through the tube flows only during the positive part of the cycle and remains essentially constant no matter how positive the signal becomes or what amplitude variations it contains. Thus, signal limiting is achieved in this section of the tube; the electron beam is passed during the positive half periods of the applied signal, and cutoff occurs during the negative half periods. The groups of electrons that are passed then travel through the second accelerator slot and form a periodically varying space charge in front of Grid 3 (Figure 25.21). By electrostatic induction, currents are made to flow in the grid wires. A resonant circuit is connected between Grid 3 and ground, and a voltage of approximately 5 V is developed in Grid 3. The phase of this voltage is such that it slows down the input voltage in Grid 1 by 90°, assuming that the resonant circuit is tuned to the IF. (Because of this 90° difference between the grid voltages, Grid 3 is often referred to as the "quadrature grid.")

Electrostatic Induction Electrostatic induction, referred to previously, may be new to the reader. Whenever electrons approach an element in a tube, electrons already at that element will be repelled, resulting in a minute flow of current. By the same token, electrons receding from an element will permit the displaced electrons to return to their

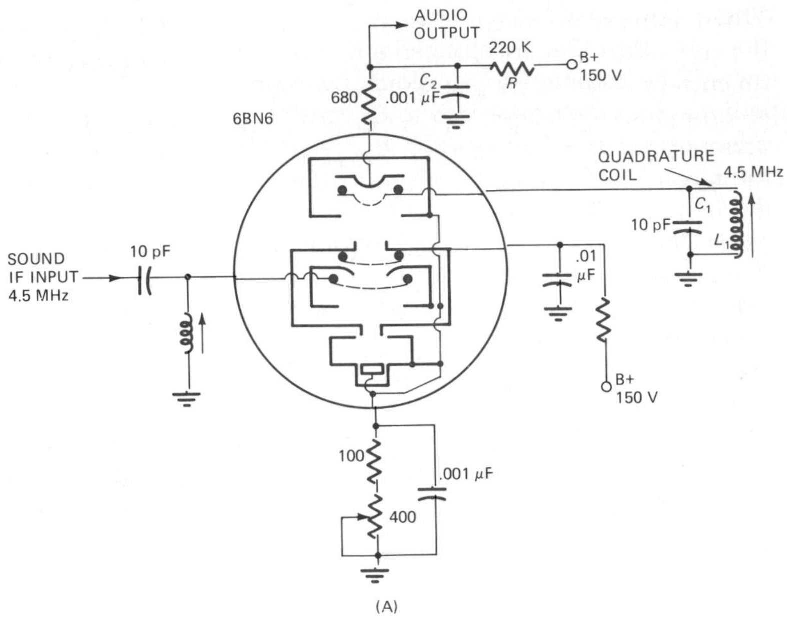

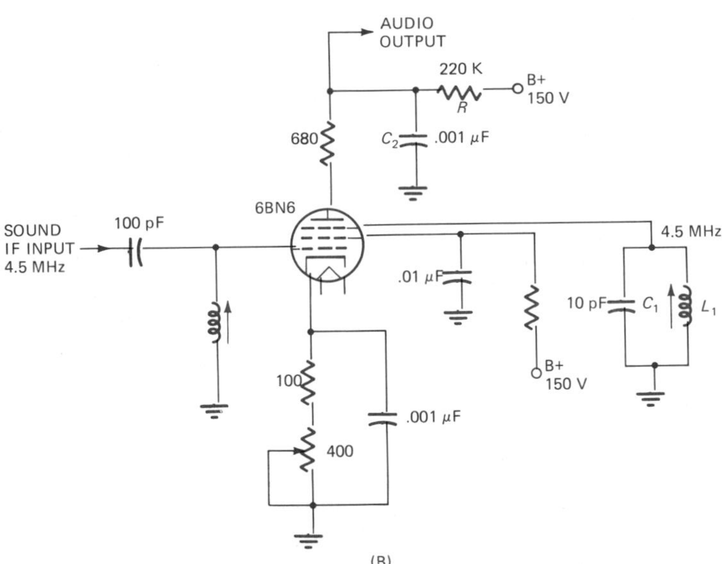

Figure 25.22 The gated-beam tube connected as a limiter-discriminator: **A.** tube shown in pictorial form; and **B.** tube drawn schematically.

previous position. Again a minute flow of current results, this time in the opposite direction. If a sufficient charge periodically approaches and recedes from an element, the induced current can achieve substantial amplitudes. This is precisely what occurs at Grid 3 in the 6BN6.

Electron Gates In the gated-beam tube, Grids 1 and 3 represent electron gates.

When both are open, current passes through the tube. When either one is closed,there is no current flow. In the present instance, the second gate lags behind the first. The plate current flow starts with a delayed opening of the second gate and ends with the closing of the first gate. Now, when the incoming signal is unmodulated and L_1 and C_1 of Figure 25.22 are resonated at the IF, the voltage on Grid 3 lags the voltage on Grid 1 by 90°. However, when the incoming signal is varying in frequency, the phase lag between the two grid voltages will likewise vary. This, in turn, varies the length of time during which plate current can flow (see Figure 25.23A).

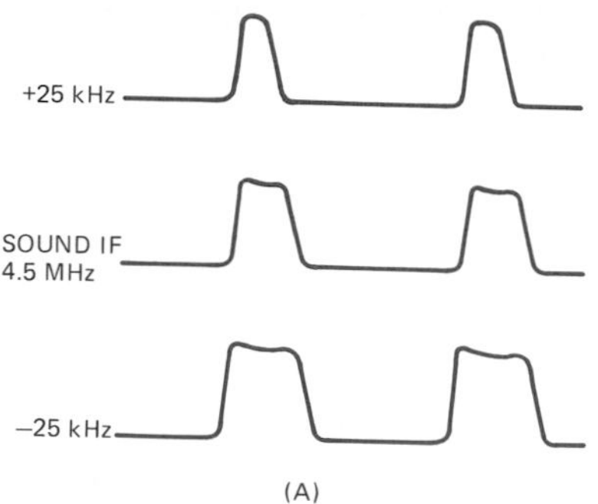

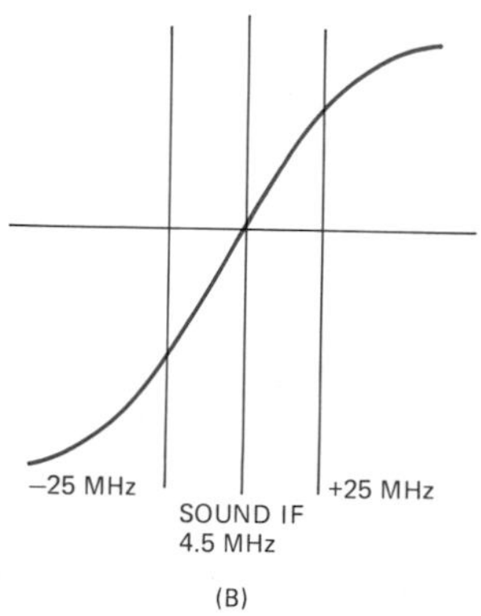

Figure 25.23A. The variation in current pulse duration with different incoming frequencies in a 6BN6. **B.** Discriminator response of the 6BN6 when connected as shown in Figure 25.22.

Thus, plate current varies with frequency. The circuit is designed so that the current varies in a linear manner. By placing the resistor in the plate lead, R of Figure 25.22, we can obtain an audio voltage to drive the audio amplifier that follows.

Typical Response Curve A typical response for a 6BN6 FM detector is shown in Figure 25.23B. Note that this curve does not have any sharp bends at frequencies beyond the range of the normal signal deviation. This makes the receiver easier to tune.

Feedback Voltage In the circuit of Figure 25.22, a 680-ohm resistor is inserted between the load R and the plate of the tube. The bypassing of the IF voltage is accomplished by C_2, but since this capacitor is placed beyond the 680-ohm resistor, a small IF voltage appears at the anode of the tube. Through the interelectrode capacitance that exists between the anode and Grid 3, the IF voltage developed across the 680-ohm resistor is coupled into L_1 and C_1. The phase relations in this circuit are such that this feedback voltage aids in driving the tuned circuit.

Cathode Bias The bias for Grids 1 and 3 is obtained by placing a resistor in the cathode leg of the tube. Since amplitude rejection, especially at low input signals near the limiting level, is a function of the correct cathode bias, the cathode resistor is made variable. Because of this adjustments can be made to compensate for tube or other component changes.

THE 6DT6 DETECTOR

The 6BN6 tube is, as we have seen, of special construction. Another tube, a 6DT6, has a similar function in the circuit, although its internal structure is more like that of an ordinary pentode. However, in the 6DT6 tube both the control grid and the suppressor grid are capable of sharply cutting off the plate current. For this reason, they resemble Grids 1 and 3 of the 6BN6. The circuit of an FM detector using a 6DT6 (or a 3DT6) (Figure 25.24) is similar to the 6BN6 circuit. So long as the incoming signal is moderately strong, quadrature grid detection takes place essentially as it does in the 6BN6 arrangement.

Locked-In Oscillations On weak signals, the 6DT6 circuit has a tendency to break into oscillation at the IF. This maintains the detected output signal constant in spite of the fact that weak signals tend to vary considerably in amplitude because of noise and fading. The oscillations arise because of the feedback that takes place between the suppressor grid and the control grid in the tube. The incoming signal at Grid 1 locks in with these oscillations and actually causes them to shift in frequency as the modulation moves the signal frequency back and forth. Normal quadrature grid detection takes place in the oscillating detector. This oscillation boosts the sensitivity of the circuit to weak signals, causing it to deliver a clearer output under adverse conditions. However, if the applied signal becomes extremely weak, the oscillator becomes unlocked, resulting in a loss of detection. Locking will occur only over a limited range of weak signal strength.

Moderate to Strong Signals When moderate or strong signals are received, the control grid draws grid current and this loads down the circuit of the input tube. This loading not only kills any tendency to oscillate but also broadens the tuning response. All this tends to limit these signals, thereby providing a certain amount of limiter action. The strong signal tends to drive the tube from plate current cutoff to plate current saturation. Thus, the current flow in the plate circuit is essentially that of a square wave and limiting action is produced. In the 6BN6, limiting is achieved by the characteristics of the tube itself.

25.13 THE TRANSISTOR QUADRATURE DETECTOR

The vacuum tube quadrature detector is easily duplicated with only three transistors. These transistors must be identical to each other and so it is desirable to include them all in one IC. The general circuit in use by a number of manufacturers is shown in Figure 25.25. The portion inside the dotted lines is

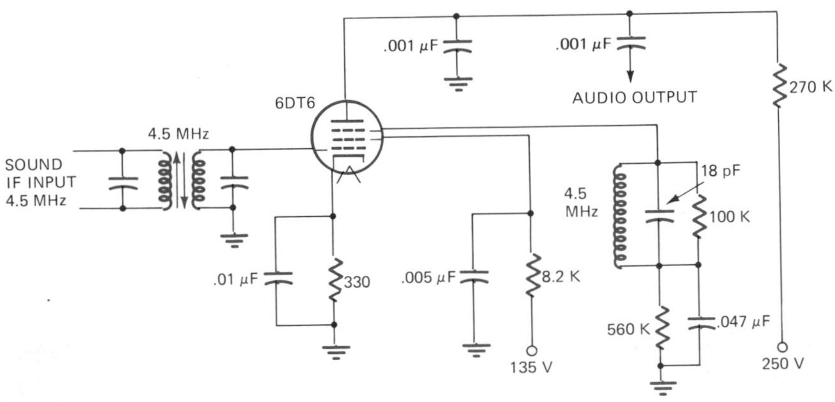

Figure 25.24 FM-detector circuit using a 6DT6.

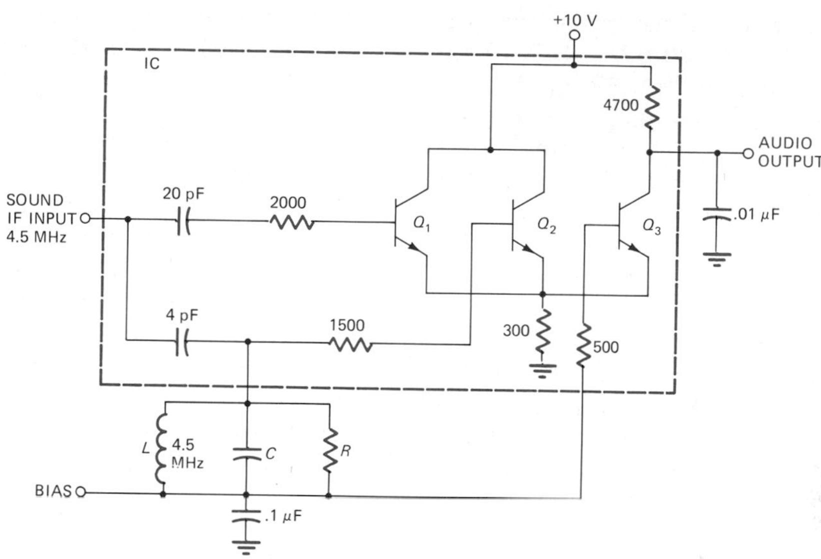

Figure 25.25 A three-transistor quadrature detector.

the part actually inside the IC. Usually the IC contains other transistors for such things as IF and audio amplifiers.

OPERATION

The principle of operation of the transistor quadrature detector is no more difficult to understand than that of the gated beam type. The IF signal has received sufficient gain before this stage so that Q_1 and Q_2 are driven very hard. The outputs on the emitters therefore have the appearance of a rectangular wave, as shown in Figure 25.23A. These two transistors perform the same function as Grids 1 and 3 in the 6BN6. The 90° phase shift is accomplished in Q_2 by driving it through a small capacitor (4 pF). The tuned circuit formed by L and C does not produce a phase shift at its resonant frequency (the 4.5-MHz IF center frequency). However, it does shift the phase of the signal linearly for small frequency changes from resonance. The net result of Q_1 and Q_2 being driven in quadrature is identical to that shown in Figure 25.23A. The only difference is that the combining of pulses is done in parallel at the emitters of Q_1 and Q_2. The gated-beam tube combines the pulses in series as they progress through the tube. The transistor version therefore produces wider pulses above resonance and narrower pulses below resonance. As seen in Figure 25.23A, the gated-beam tube does just the opposite. The audio output is identical in both methods. The final transistor (a grounded base stage), Q_3, merely amplifies the pulses and, with the help of the 0.01-μF output capacitor, removes the IF components from the audio signal.

Note: A quadrature detector incorporated as part of an IC sound section is shown later in the chapter.

25.14 THE PHASE-LOCKED LOOP (PLL) FM DETECTOR

Phase-locked loop (PLL) ICs are contained in a single, flat-pack unit. Although

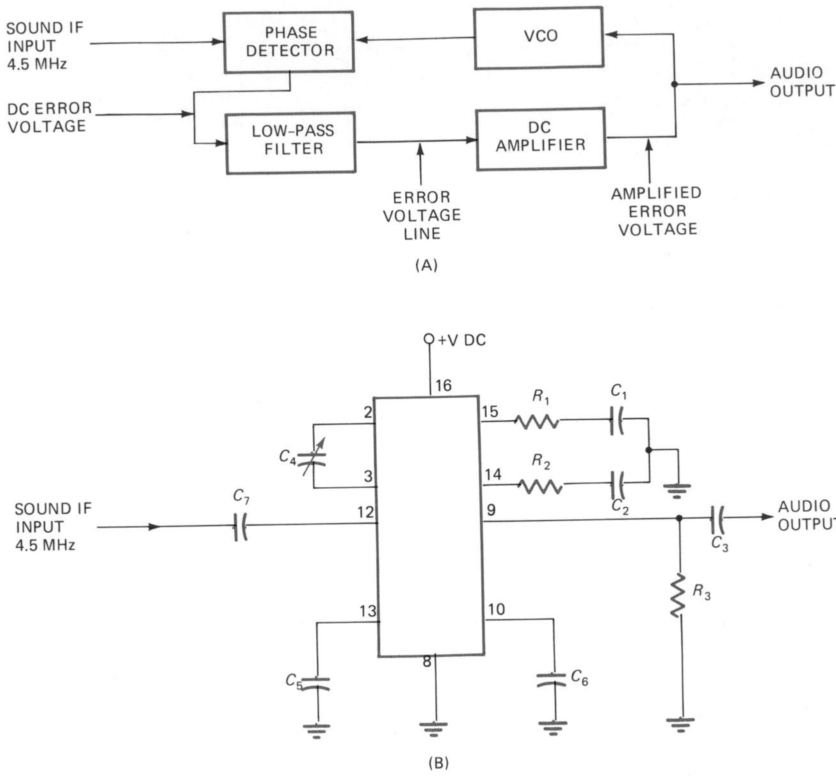

Figure 25.26 A phase-locked loop (PLL) FM detector. A. Basic block diagram. B. A typical IC schematic diagram.

PLLs are used for other purposes (frequency synthesis, for example), we are here interested only in its performance as an FM detector.

PLL DIAGRAM

The basic PLL consists of four sections. These are shown in the block diagram of Figure 25.26: (1) phase detector, (2) low-pass filter, (3) DC amplifier, and (4) voltage-controlled oscillator (VCO). A schematic diagram of a typical PLL IC used as an FM detector is shown in Figure 25.26B). Note that *no tuned circuits* are required. The sensitivity of these circuits is about 1 or 2 mV, with an upper limit of about 15 or 20 mV. Above this level, AM suppression may be lost.

THEORY OF OPERATION

The phase detector receives two inputs: (1) the sound-IF 4.5 MHz signal and (2) the signal from the VCO. The PLL operation always tries to keep the VCO frequency the same as that of the sound-IF FM signal. Since the frequency of the FM signal is constantly changing, the VCO frequency changes, too, since it is always trying to "catch up." In order to accomplish this, however, the phase detector puts out a DC error voltage through the DC amplifier to the VCO. This error voltage is an accurate copy of the original audio signal at the transmitter. The PLL therefore functions as an untuned FM detector.

The polarity of the error (audio) voltage is determined by the instantaneous fre-

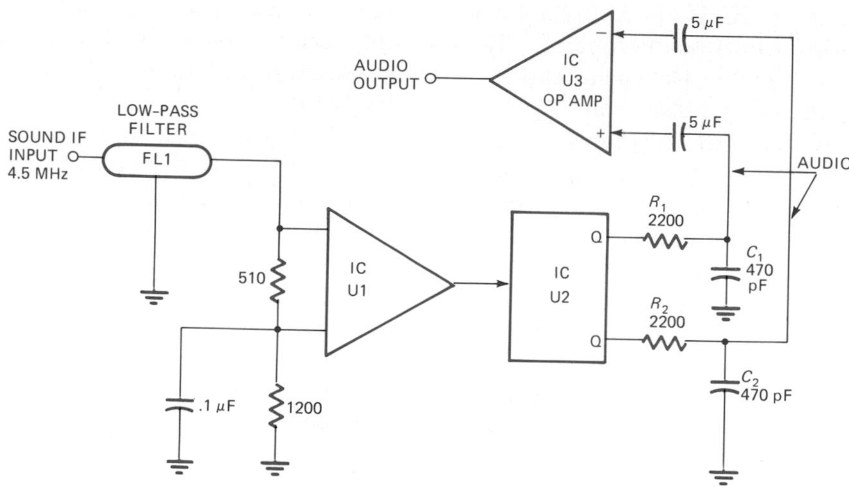

Figure 25.27 Diagram of a digital-FM detector. A monostable multivibrator (U2) and two integrator networks are used to demodulate the FM signal.

quency difference between the sound IF and the VCO. The sound IF will deviate above and below the VCO frequency.

The *polarity* of the phase detector output error voltage (audio voltage) depends upon whether the sound IF is above or below the VCO frequency. When the sound IF is at the carrier (center) frequency, the VCO is at the same frequency. At this time, the phase detector output error is zero volts. This occurs when an audio signal is passing through zero volts.

The *amplitude* of the error voltage phase detector output is a function of the instantaneous difference between the sound IF and VCO frequencies. The rate of change of IF deviation determines the error (audio) frequency.

CIRCUIT DETAILS

In Figure 25.26B, the sound-IF 4.5-MHz signal is coupled to the IC through capacitor C_7 to Terminal 12. Capacitor C_4, connected to Terminals 2 and 3, is used to set the frequency of the VCO to approximately 4.5 MHz. The minimum input signal level must be about 1 to 2 mV. The networks R_1C_1 and R_2C_2, connected to Terminals 14 and 15, form the low-pass filter shown in Figure 25.26A. The audio output is obtained from Terminal 9. Capacitor C_6 and the internal IC resistor connected to Terminal 10 are used for the 75-μs deemphasis network.

25.15 A DIGITAL FM DEMODULATOR

A diagram of a digital FM demodulator is shown in Figure 25.27. This circuit uses no coils, is relatively independent of the IF, and does not require alignment.

CIRCUIT DESCRIPTION

In Figure 25.27, the sound IF (4.5 MHz) first passes through bandpass filter FL_1. This restricts the signal to the bandwidth of the sound IF plus sidebands. Filter FL_1 may be either a ceramic or LC type of filter.

IC U_1 contains an amplifier stage and

amplitude limiters. IC U_2 is a digital-logic, monostable (one-shot) multivibrator. This circuit produces one pulse each time it is triggered by an input pulse from U_1. (The input pulses to U_2 are the AM-limited sound IF signal). The U_2 output pulses have a constant amplitude and duration. However, at any given moment their repetition rate is the same as the deviated frequency of the sound IF. (Remember that the FM signal changes frequency, or deviates, when it is modulated by audio.)

There are two complementary (opposite polarity) outputs from U_2. Each output is fed to its own integrator (R_1C_1 and R_2C_2). The integrators "average" the signals to obtain a push-pull audio signal. This audio signal is fed to the differential inputs of an operational amplifier contained in IC U_3. The output of U_3 is the audio signal, which can then be further amplified. Alternatively, the push-pull outputs of U_2 may be used to drive a push-pull preamplifier and power amplifier.

25.16 THE FM DIFFERENTIAL PEAK DETECTOR

The FM differential peak detector is used to demodulate television FM sound. A typical circuit may consist of eight transistors and a 4.5-MHz tuned circuit. This circuit is shown in Figure 25.28A. The circuit is part of an IC that serves as a complete television sound section.

CIRCUIT DESCRIPTION

The 4.5-MHz sound-IF output is applied to the resonant circuit C_2–L_1. Note that the circuit consists of two identical halves. One half includes Q_{22}, Q_{23}, and Q_{24}. The other half includes Q_{25}, Q_{26}, and Q_{27}. Also note that each end of the tuned circuit (C_2–L_1) connects to the input stage (Q_{22}, Q_{27}) of each half of the circuit.

Transistors Q_{22} and Q_{27} are emitter followers acting as buffers. They feed the **peak detectors** Q_{23} and Q_{26}. Transistors Q_{23} and Q_{26} charge capacitors C_3 and C_4, respectively, to the peak value of the voltage across tuned circuit C_2–L_1 (on alternate half cycles). The *difference* of the detected voltages (divided by R_{32} + R_{33}) produces the DC differential current at the collector of Q_{25}. This is the detected audio output signal.

DETECTOR OPERATION

At low deviation frequencies, the input voltages at the peak detectors Q_{23} and Q_{26} in Figure 25.28A are equal and the *differential* output current is zero (Point A in Figure 25.28B). As the frequency increases, the circuit C_2–L_1 resonates with capacitor C_1 to form a *series-resonant* circuit. This causes an effective short circuit at Terminal 9. The voltage at that point (V_9) is then zero (Point B in Figure 25.28B). At the same time, the voltage at Terminal 10, Q_{27} base is a maximum. The Q_{25} output current is also maximum. This produces the maximum positive differential current response at Point B.

At the sound-IF carrier frequency of 4.5 MHz, the *differential* output current returns to zero. This happens because the input voltages at the peak detectors Q_{23} and Q_{26} are again equal (Point C in Figure 25.28B).

At frequencies higher than 4.5 MHz, the tuned circuit C_2–L_1 approaches parallel resonance. Now the voltage at Terminal 9 and the base of Q_{22} approaches a maximum. Through the *differential* action of Q_{24}, Q_{25}, the Q_{25} *differential* output current approaches a maximum in the negative-going direction (Point D in Figure 25.28B).

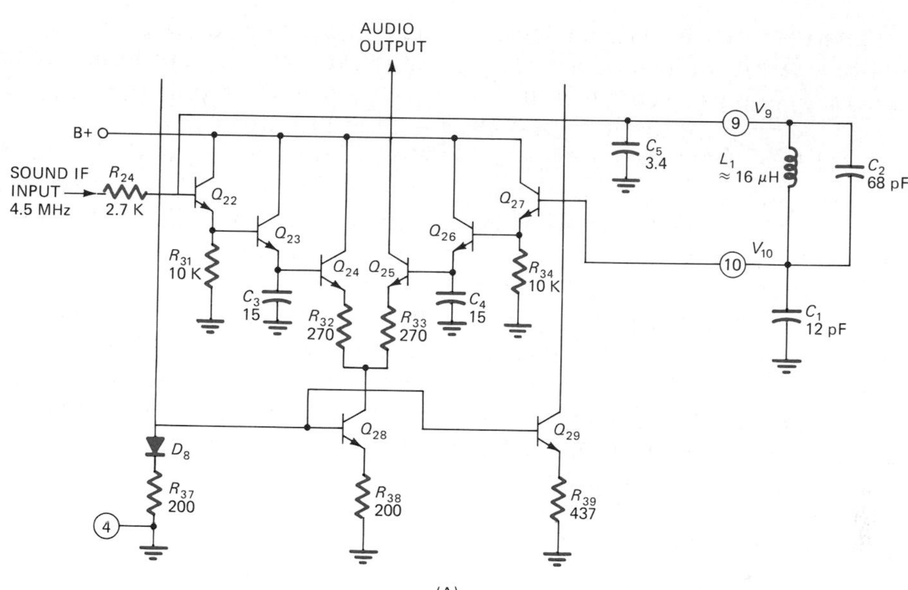

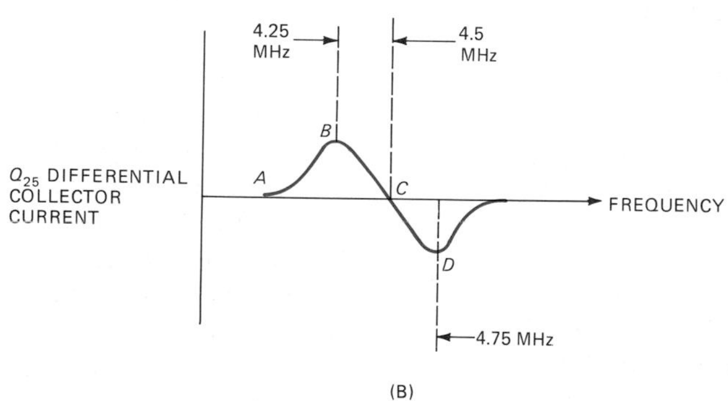

Figure 25.28 The FM differential-peak detector. **A.** The schematic diagram. **B.** The response curve. *(Courtesy of RCA Consumer Products.)*

THE S CURVE

As the FM wave deviates from a low to a higher frequency (passing through 4.5 MHz), the differential collector current of Q_{25} assumes the S-shaped curve of Figure 25.28B. This curve is basically the same as for the discriminator and ratio detector response curves required for FM detection.

Although for simplicity we have selected several specific points on the S curve, the change in detector response is smooth as

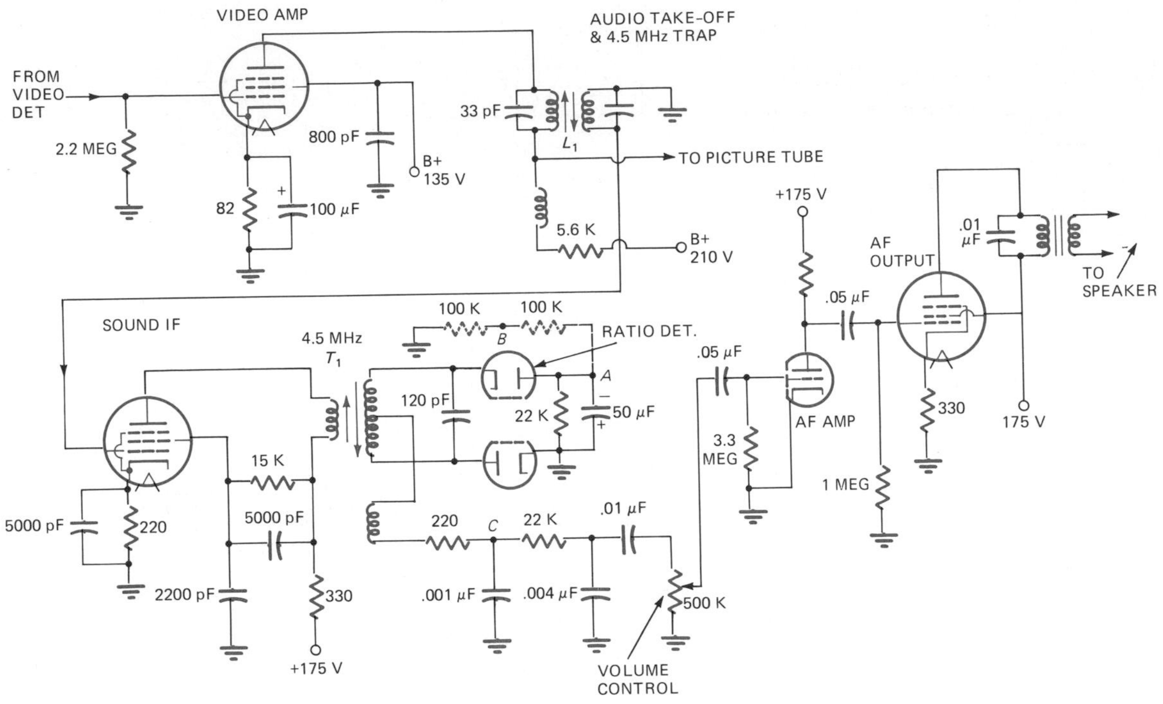

Figure 25.29 Schematic diagram of a complete vacuum tube sound section for a TV receiver. A ratio detector is used for FM detection.

the frequency deviates. Thus, the required S-curve response is produced (see also Figure 25.14).

25.17 A COMPLETE VACUUM TUBE SOUND SECTION

A complete vacuum tube sound system for a television receiver is shown schematically in Figure 25.29. One stage of sound-IF amplification (at 4.5 MHz) precedes an unbalanced ratio detector. The detected audio signal is amplified by a triode amplifier and an audio power amplifier and then fed to a loud speaker.

25.18 A COMPLETE TRANSISTOR SOUND SECTION

Figure 25.30 shows a transistor sound section for a color television receiver. It has several novel features worth mentioning. Note that there is only one transformer operating at the IF. This is a simple two-winding transformer driving an unbalanced ratio detector.

L_1–C_1 is a high-pass filter that passes only frequencies above 40 MHz. The sound IF (41.25 MHz) and picture IF (45.75 MHz) are heterodyned in the sound detector diode to produce the 4.5-MHz sound IF. Bandpass filter L_2–C_2–L_3 passes only the 4.5-MHz sound IF and its sidebands to the 4.5-MHz sound-IF amplifier Q_2. Impedance-matching network C_3–C_4 provides the proper impedance for driving the base of Q_2.

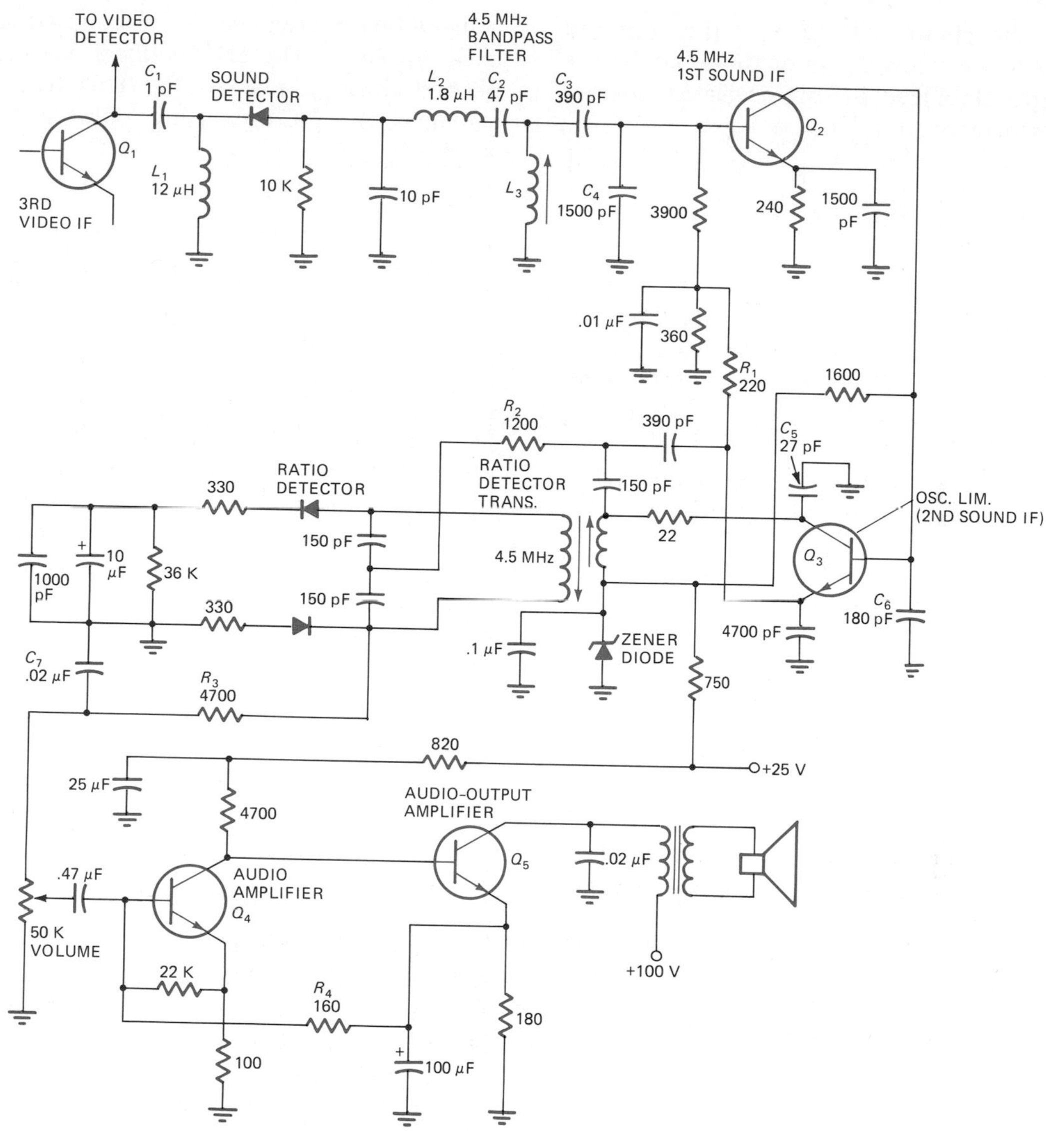

Figure 25.30 A complete transistor sound section used in a color TV receiver.

SOUND-IF STAGES

Transistors Q_2 and Q_3 form a DC-coupled two-stage sound-IF amplifier. Stabilization of the DC operating point and frequency response improvement are made possible by negative feedback through R_1. The second IF transistor, Q_3, also behaves as an oscillating limiter that has the same advantage as the oscillating 6DT6 quadrature detector, previously discussed. Weak signals are kept constant in amplitude, and the quality of the resulting audio is improved. The oscillation is sustained through the action of C_5–C_6 and the primary of the ratio detector transformer. These components make the circuit a version of the common Colpitts oscillator. A zener diode keeps the voltages stabilized in this oscillator, so that its performance remains constant.

RATIO DETECTOR

The ratio detector transformer receives its required 90° out of phase signal from R_2. This out-of-phase signal combines with the transformer output voltage in a manner similar to the Foster-Seeley discriminator (Figure 25.16). The two capacitors across the transformer secondary provide the same effect as a center tap on the secondary winding.

AUDIO AMPLIFIERS

Resistor R_3 and capacitor C_7 are the deemphasis network. Audio amplification is achieved with a two-stage, DC-coupled feedback amplifier. The DC feedback is accomplished through resistor R_4. This feedback stabilizes the DC operating point of both stages.

25.19 A PARTIAL IC SOUND SECTION

Many modern television receivers have a single IC that incorporates most or all of the sound section. Figure 25.31 shows a TV FM sound section that contains an IC and 13 other components. The two coils are single-slug tuned and simple to align. Note that the audio power amplifier and several coils, capacitors, and resistors in this sound section are connected externally to the IC. However, all the remainder of the required circuitry for the sound-IF amplifiers, limiter, quadrature detector, and low-level audio amplifier is

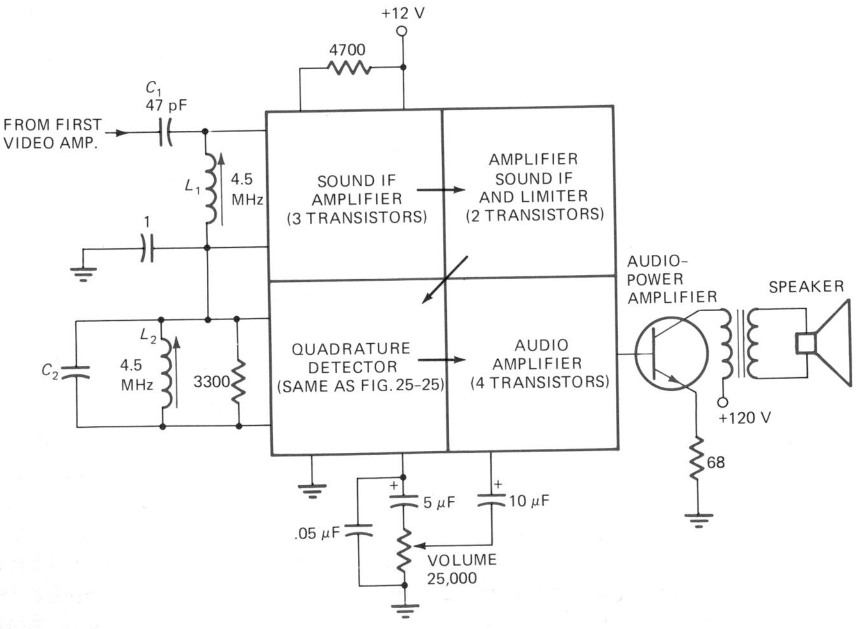

Figure 25.31 A TV sound section that uses one IC and a transistor as the only active elements. *(Courtesy of Fairchild Semiconductor Corp.)*

contained within the single IC chip. (In the IC described in the next section, the audio power amplifier as well as some additional circuits are contained inside the IC.)

AUDIO OUTPUT COMPONENTS

An external power transistor and output transformer are required for this particular IC. Some additional gain is required for the power and impedance requirements of a 4-ohm, 2-W speaker. In some other ICs (such as the one described below), no external audio power amplifier or output transformer is required.

SOUND-IF AMPLIFIERS

The two sound-IF amplifiers are not transformer-coupled or resonant circuit–coupled. Instead, adequate selectivity is obtained with just the resonance of L_1 and C_1. A gain of 2 000 at 4.5 MHz is typical. This large gain saturates the last IF amplifier and produces the square-wave signal required for the quadrature detector.

QUADRATURE DETECTOR

The quadrature detector is almost identical to the one shown in Figure 25.25. This type of FM detector is easily produced as part of an IC. Tuning merely requires making L_2 and C_2 resonant at 4.5 MHz.

AUDIO VOLTAGE AMPLIFIER

The audio voltage amplifier is a three-stage, four-transistor, DC-coupled amplifier. Direct coupling allows a good bass response and also avoids the use of capacitors. Note that, except for the speaker transformer and the two capacitors associated with the volume control, this system would have a bass response down to DC.

25.20 A COMPLETE IC SOUND SECTION

A complete IC sound section is shown in Figure 25.32. The IC contains the following stages: (1) 4.5-MHz limiter, (2) quadrature detector, (3) electronic volume attenuator, (4) audio power amplifier, (5) current limiter, (6) thermal shutdown, and (7) voltage regulator.

CIRCUIT DETAILS

The color TV receiver in which the IC shown in Figure 25.32 is used has a synchronous video detector (see Chapter 13). The output of this type of detector does not contain the various heterodyne frequencies present in the output of a diode detector. Therefore, it is possible here to tap off the 4.5-MHz sound IF at the output of a video preamplifier. This provides a relatively high-level input signal to the 4.5-MHz limiter stage of the IC.

Limiter-Detector The 4.5-MHz sound take-off coil is L_{186}. The 4.5-MHz sound IF signal is sent to the limiter via IC Terminal 10. The limiter is connected internally to a quadrature detector. The 4.5-MHz quadrature detector coil is connected to IC Terminals 14 and 15. A deemphasis and tone control network is connected to Terminal 16.

Electronic Attenuator The audio output of the quadrature detector is internally connected to an audio electronic (transistor) attenuator. (This is actually a gain-controlled audio preamplifier.) The amount of audio voltage passing through this attenuator to

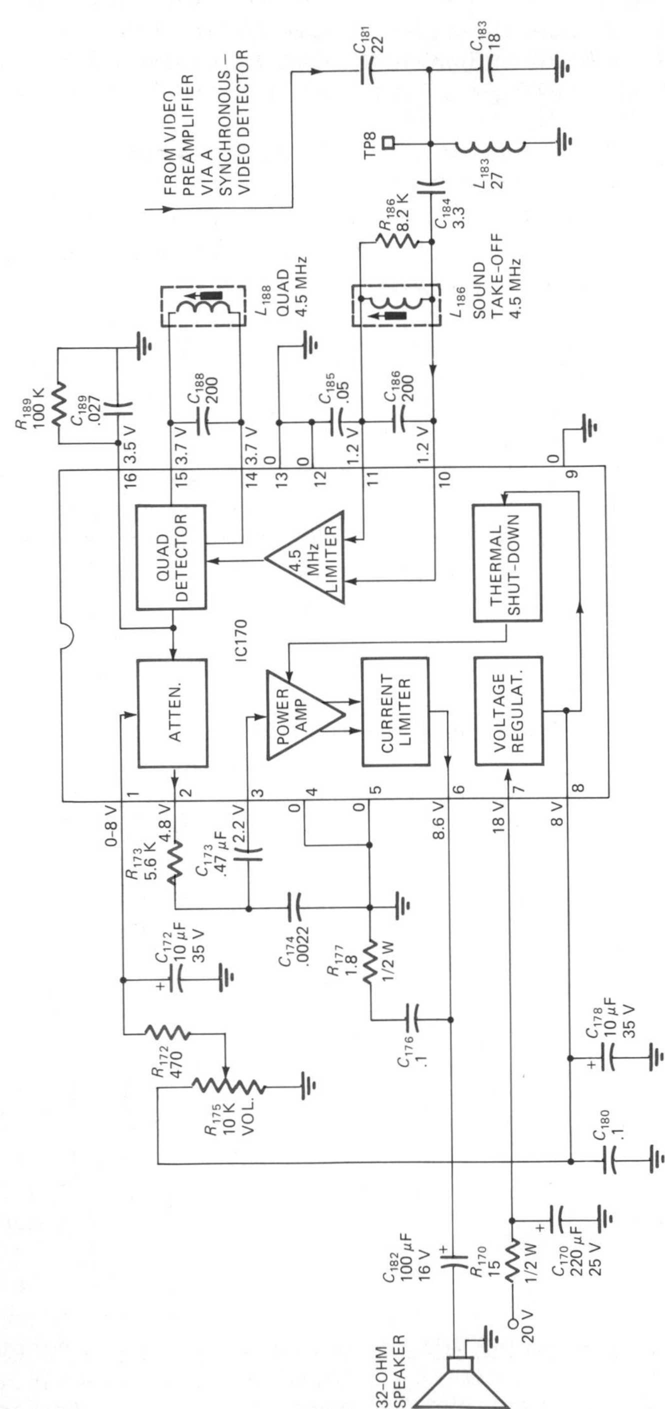

Figure 25.32 A complete sound section for a color TV receiver on a single IC. (*Courtesy of General Electric.*)

the audio power amplifier is controlled by the setting of volume control R_{175}. The advantage of this scheme of volume control is as follows: The volume control handles only DC (not audio) and can be placed at any convenient location without the use of shielded cables. There is no danger of hum pickup being introduced into the audio system.

Audio Power Amplifier The audio output of the attenuator leaves the IC on Terminal 2 and is applied to a high-frequency compensation network, R_{173} and C_{174}. It is then coupled via capacitor C_{173} to the input of the audio power amplifier at Terminal 3. The power amplifier output is through Terminal 6. From here, it is capacitively coupled through C_{182} to the loudspeaker. The speaker must be a 32-ohm type to prevent excessive power dissipation of the IC.

Overdissipation Protection The audio power amplifier is protected against overdissipation in two ways: (1) current limiting and, (2) thermal shutdown circuitry.

Voltage Regulator The voltage regulator controls the voltage for all IC circuits. Also, through Terminal 8, it supplies a stabilized voltage source for the volume control.

25.21 COMPLEMENTARY-SYMMETRY AUDIO PUSH-PULL OUTPUT STAGE

It is possible to provide push-pull operation of an audio output amplifier (or preamplifier) by using the type of transistor circuit shown in Figure 25.33. This shows a single transistor preamplifier Q_{2030} driving the

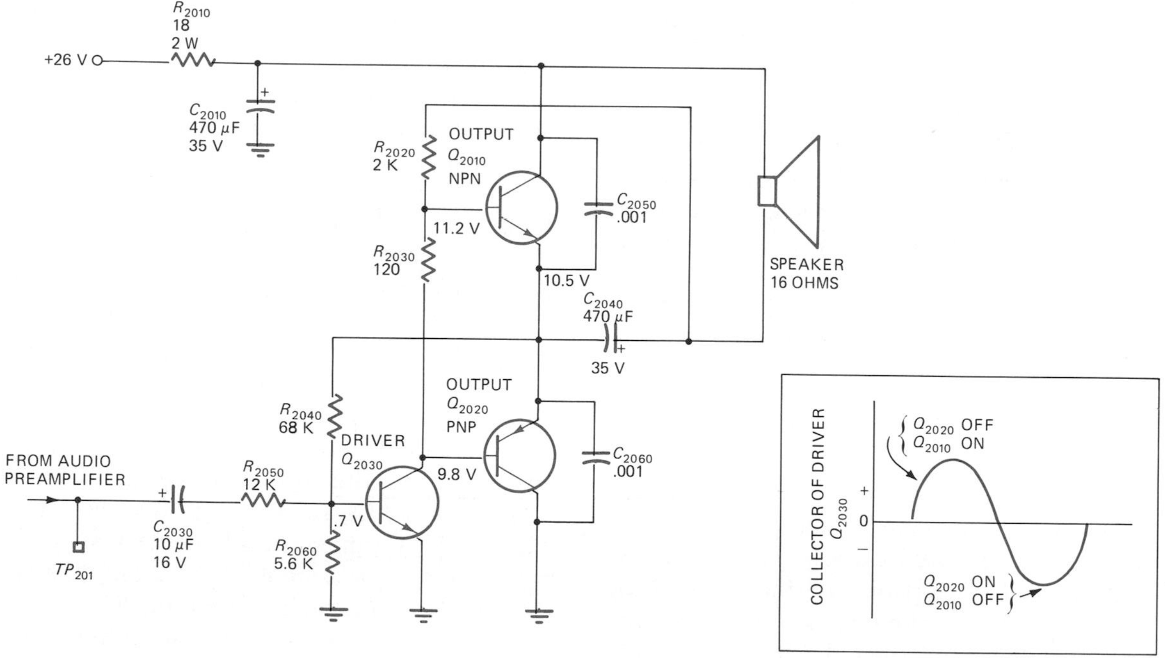

Figure 25.33 A transistor driver and audio-output section. Push-pull operation of the output amplifiers is provided by a "complementary-symmetry" circuit. No output transformer is required to drive the 16-ohm speaker. Inset shows the effect of the driver output wave on operation of the complementary-symmetry circuit. *(Courtesy of General Electric.)*

push-pull amplifier Q_{2010}–Q_{2020}. The 16-ohm speaker is connected from the common-emitter output of Q_{2010} and Q_{2020} (through capacitor C_{2040}) to the +26-V line. In some circuits, the speaker is connected from the common-emitter output through a capacitor and to ground. Note that no audio output transformer is required. The low emitter output impedance of Q_{2010} and Q_{2020} matches the 16-ohm speaker impedance reasonably well.

OUTPUT STAGES

The output circuit of Q_{2010}–Q_{2020} is known as a **complementary-symmetry** circuit.[1] By means of this arrangement, push-pull operation is obtained with a single-ended driver input. In this circuit an NPN and a PNP transistor are connected in series between ground and the power supply point. The term *complementary-symmetry amplifier* results from the use of both NPN- and PNP-type transistors.

Note: Although not shown in this book, there is also a quasi-complementary-symmetry push-pull output amplifier, in which the two output transistors are of the *same* type. However, *two* driver transistors of *opposite* types are required to develop push-pull driving voltages. Here, the drivers are connected in the complementary-symmetry manner.

CLASS OF OPERATION

For maximum power output and greatest efficiency, the complementary-symmetry amplifier is operated class B. However, the operating characteristics of the transistors become very nonlinear as cutoff is approached. If uncorrected, this condition would cause a type of audio distortion known as **crossover distortion.** To minimize crossover distortion, a small forward bias is applied to each transistor. The base of Q_{2010} (NPN) is biased positively with respect to its emitter. The base of Q_{2020} (PNP) is biased negatively with respect to its emitter.

CIRCUIT OPERATION

The circuit of Figure 25.33 is basically a class B push-pull output amplifier. This means that while one output transistor is conducting, the other is cut off. The reverse action is also true.

Assume the audio signal at the output of driver Q_{2030} is going in the positive direction (inset for Figure 25.33). This signal is applied to the base of both Q_{2010} and Q_{2020}. Since Q_{2020} is a PNP transistor, it is turned off. However, Q_{2010} is a NPN transistor and so conducts for the positive half of its base signal.

The opposite condition prevails when the driver signal goes in the negative direction. Now, Q_{2020} conducts and Q_{2010} is off. The AC currents from each transistor are coupled to the speaker through capacitor C_{2040}. In C_{2040} and the speaker, the two halves of the audio signal are combined to form the audio signal.

The power output to the speaker equals twice the maximum collector dissipation per transistor. Resistor R_{2040} at the base of driver Q_{2030} stabilizes Q_{2030} against temperature variations. It does this by applying feedback from the emitters of Q_{2010} and Q_{2020} to the base of Q_{2030}. Although not shown in Figure 25.33, some complementary-symmetry circuits have a small (1- or 2-ohm) resistor in series with each emitter leg. These stabilize the output stages against temperature variations.

Chart 25.1 illustrates a practical troubleshooting procedure for the conditin of no sound, video normal.

---

[1]See Chapter 24 for a discussion of a complementary-symmetry vertical output circuit. The operation of this audio output amplifier is similar to that circuit.

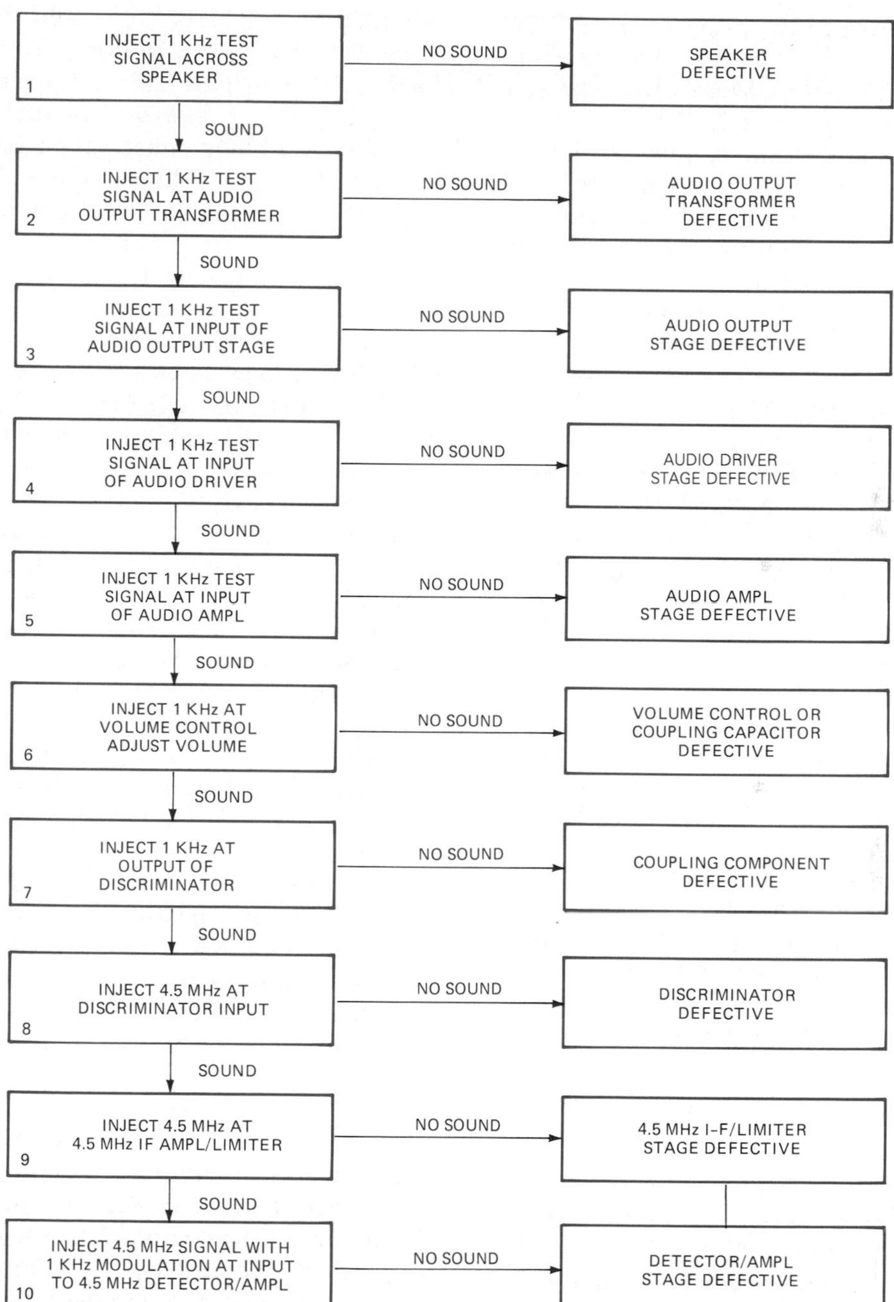

Chart 25.1 This illustrates a practical troubleshooting procedure for the condition of no sound, video normal. (Courtesy of B&K-Precision Dynascan Corp.)

SUMMARY OF CHAPTER HIGHLIGHTS

1. The audio portion of the TV signal is transmitted by FM. (Figure 25.2)
2. FM (1) is almost noise free, (2) provides better rejection of undesired signals, and (3) enables a more efficient transmitter to be used. (Figures 25.2 and 25.9 Sections 25.1 through 25.4)
3. The TV FM sound carrier is located 0.25 MHz below the upper limit of each TV channel. (Introduction.)
4. TV FM deviation is ± 25 kHz maximum. (Figures 25.2 and 25.14)
5. In FM, the amount of deviation is proportional to the amplitude of the modulating signal. The rate of deviation is the frequency of the modulating signal. (Figure 25.2)
6. Most interference is amplitude-modulated. Thus, most can be removed by AM limiting. (Figure 25.9)
7. See Table 25.1 in Section 25.3 for comparison between AM and FM. (Figures 25.1, 25.2.)
8. See Table 25.2 in Section 25.4 for summary of advantages and disadvantages of AM and FM.
9. Preemphasis is used at FM transmitters, deemphasis at FM receivers. This scheme greatly improves the audio signal-to-noise ratio at the receiver. (Figures 25.3 and 25.4)
10. A television FM sound section consists basically of (1) sound-IF detector (color sets only), (2) 4.5-MHz sound-IF amplifier, (3) limiter, (4) sound-IF detector, and (5) audio amplifier. (Figures 25.5 and 25.29 to 25.32)
11. The 4.5-MHz sound-IF amplifier increases the 4.5 MHz IF to a level suitable to operate a limiter, or limiting FM detector. (Figure 25.6)
12. This amplifier must pass a bandwidth of at least ±25 kHz. (Figure 25.6)
13. The sound-IF limiter (or limiting FM demodulator) eliminates most noise (and other AM) from the FM signal. (Figures 25.8 to 25.11)
14. Limiting is accomplished by driving the limiter beyond saturation and cutoff. (Figures 25.8 to 25.11)
15. A dual or multistage limiter is more sensitive than a single-stage limiter. It also provides better limiting. (Figures 25.10 and 25.11)
16. The minimum limiter input signal amplitude required to produce limiting is called the limiting knee. (Figure 25.9)
17. Some average limiting knee values are vacuum tube, about three volts; transistor, 1 V; and high-gain IC, 100 to 200 μV.
18. The FM detector recovers the original audio signal from the FM wave. (Figures 25.2, 25.14, 25.23, and 25.28)
19. FM detectors may be classified as (1) discriminators, (2) ratio detectors, (3) quadrature detectors, (4) phase-locked loop detectors, (5) digital FM detectors, and (6) differential peak detectors. (Figures 25.12 to 25.28)
20. A discriminator has an S-shaped response curve. The linear portion is used to recover the audio signal. (Figure 25.14)
21. In Number 20, the linear portion of the S curve for television audio should extend at least from 4.25 MHz to 4.75 MHz. (Figure 25.14)
22. The ratio detector has considerably greater immunity from AM distortion on the FM signal than a Foster-Seeley discriminator. (Figures 25.17 to 25.20)
23. In the ratio detector, a large capacitance

(4 to 8 μF) across the two diodes prevents amplitude variations of the input signal from affecting the audio output. (Figures 25.17 to 25.20)

24. A quadrature FM detector acts as both a limiter and an FM discriminator. It is seen in vacuum tube, transistor, and IC forms. (Figures 25.21 to 25.25, 25.31, and 25.32)
25. Three transistors are used in a transistor quadrature detector. This performs the same functions as a vacuum tube quadrature detector. (Figure 25.25)
26. A phase-locked loop (PLL) can act as an FM detector. It consists of (1) a phase detector, (2) a voltage-controlled oscillator (VCO), (3) a low-pass filter, and (4) a DC amplifier. (Figure 25.26)
27. In Number 26, the error (or correction) voltage output of the phase detector is an exact copy of the original audio signal. This error signal (amplified) is the audio output. (Figure 25.26)
28. The audio signal in an FM wave can be recovered by a digital FM demodulator. (Figure 25.27)
29. In a digital FM demodulator, the (limited) output pulses from the monostable (one-shot) multivibrator have a constant amplitude and duration. At any given moment, however, their repetition rate is the same as the deviated frequency of the sound IF. (Figure 25.27)
30. The FM differential peak detector can be used to recover the audio signal from an FM wave. (Figure 25.28)
31. In Number 30, the detector operation depends basically upon three factors: (1) the resonance characteristics of the tuned circuit (L_1, C_2, C_1), (2) the voltages developed by the peak detectors (Q_{23}, Q_{26}), and (3) the resultant DC differential current produced by the differential amplifiers (Q_{24}, Q_{25}). (Figure 25.27)
32. In Number 30, the detector audio output is the differential current output of Q_{25}. (Figure 25.27)
33. An oscillating limiter has the advantage of keeping even weak signals constant in amplitude (similar to an oscillating quadrature detector). (Figure 25.30)
34. One type of IC sound section includes: (1) the sound-IF amplifier, (2) the limiter, (3) a quadrature detector and, (4) the audio low-level amplifiers. (Figure 25.31)
35. A complete sound section on a single IC contains (1) limiter, (2) quadrature detector, (3) electronic volume attenuator, (4) audio power amplifier, (5) current limiter, (6) thermal shutdown, and (7) voltage regulator.
36. TV sound section troubles are classified as (1) no sound, picture normal, (2) distorted sound, (3) weak sound, and (4) 60-Hz buzz in sound, picture normal.
37. A complementary-symmetry audio output amplifier provides push-pull operation from a single-ended output stage. (Figure 25.33)
38. In Number 37, the output stage consists of one NPN and one PNP stage. No audio output transformer is required. (Figure 25.33)

EXAMINATION QUESTIONS

(Answers are provided at the back of the text.)

Part A Supply the missing word(s) or number(s).

1. The maximum FM deviation for TV is ± _______ kHz.
2. The FM system is _______ than AM, as far as transmitter power is concerned.
3. The TV FM carrier is always _______ MHz below the upper end of each channel.
4. In FM, the rate of frequency deviation is equal to the _______.
5. In FM, the *frequency swing* equals _______ the instantaneous frequency deviation.
6. Most noise interference is _______ modulated.
7. _______ is the process of increasing the amplitudes of higher audio frequencies at an FM transmitter.
8. The deemphasis network time constant is standardized at _______ μs.
9. Sound IF limiting removes the _______ from the FM signal.
10. The FM sound _______ recovers the original audio variations from the incoming FM signal.
11. The 4.5-MHz sound-IF amplifier must have a minimum bandwidth of _______ kHz.
12. Circuits for sound-IF limiting may utilize _______s, _______s, or _______s.
13. The sensitivity or *limiting knee* of a limiter improves with the number of _______.
14. A high-gain linear IC limiter may contain as many as _______ stages.
15. A *limiting knee* of about _______ μV is typical for a high-gain linear-IC limiter.
16. A discriminator does not eliminate _______ variations from an FM signal.
17. The linear portion of a television discriminator response curve goes from _______ MHz to _______ MHz.
18. The audio output of a ratio detector is a function only of the _______ of the two diode signal voltages.
19. The quadrature FM detector also performs the function of _______.
20. The 6DT6 quadrature detector will _______ on weak signals.
21. The _______ voltage of a PLL FM detector is an exact copy of the audio signal.
22. A _______ FM detector has no tuned circuits.
23. Operation of a differential peak detector depends upon the _______ characteristic of a tuned circuit.
24. A complete IC sound section may include a _______ shutdown circuit and an electronic attenuator.
25. A two-transistor audio-output stage that provides a push-pull output from a single-ended input is a _______ amplifier.

Part B Answer true (T) or false (F).

1. The maximum frequency deviation for a broadcast FM station is ±25 kHz.
2. In FM, the rate of deviation is the same as the audio modulating frequency.
3. In normal AM, the strength of the audio affects the number of sidebands.
4. For TV Channel 4, the FM signal is permitted a maximum frequency swing from 71.5 MHz to 72 MHz.
5. The unmodulated frequency of an FM transmission is called the center frequency.
6. The FM system is practically noise-free because most noise is frequency-modulated.
7. The bandwidth of an FM transmission is proportional to the frequency of the modulation signal.
8. AM is theoretically capable of providing the same audio fidelity as FM.
9. For the same effective radiated power, AM and FM stations require the same modulation power.
10. Preemphasis and deemphasis circuits are responsible for the practically noise-free reception of FM.

11. The FM sound IF for a TV receiver is 4.5 MHz with a bandwidth of ±25 kHz.
12. In a color TV receiver with a synchronous video detector, the sound takeoff point must precede the detector.
13. The function of sound IF limiting is to remove amplitude variations from the FM signal.
14. The Foster-Seeley discriminator does not require a preceding limiter.
15. The output signal amplitude of a sound IF amplifier may be about 2 to 5 V.
16. A multistage high-gain linear IC limiter has a limiting knee of about 1 to 2 V.
17. A single-stage limiter may not provide adequate weak-signal limiting.
18. The television discriminator response characteristic must be linear from 4.25 to 4.75 MHz.
19. The large (4- to 8-μF) capacitor used in a ratio detector is part of a deemphasis network.
20. As with the Foster-Seeley discriminator, quadrature detectors must be preceded by amplitude limiting.
21. On weak signals, the 6DT6 quadrature detector may oscillate.
22. The PLL FM detector requires only one 4.5-MHz tuned circuit.
23. The audio output of a PLL FM detector is taken directly from the phase detector.
24. A digital FM detector does not use any tuned circuits.
25. In a digital FM detector, the monostable multivibrator produces one pulse each time it is triggered by an input pulse.
26. In an FM differential peak detector, the two peak detector transistors also function as differential amplifiers.
27. The FM differential peak detector, like the digital FM detector, does not use any tuned circuits.
28. A complete television sound section may be contained in a single IC.
29. A quasi-complementary-symmetry amplifier uses two NPN or two PNP transistors.
30. A complementary-symmetry audio output amplifier provides push-pull voltages into a push-pull audio output transformer.

REVIEW ESSAY QUESTIONS

1. Draw a simple block diagram of a complete color TV receiver sound section. Assume that the input is from a receiver using a synchronous video detector. Label all stages, frequencies, input and output.
2. In Question 1, discuss the operation of each stage.
3. Name all the FM detectors described in this chapter.
4. In Question 3, briefly describe the operation of two FM detectors.
5. Explain briefly the effect of a modulating signal on an FM carrier.
6. Question 5, do the same for an AM carrier.
7. Define center frequency, frequency swing, frequency deviation, modulation percentage (for FM).
8. Explain the purpose of preemphasis and deemphasis. (Figures 25.3 and 25.4)
9. In Figure 25.6, what is the function of capacitor C_2? Of resistors R_1R_2? Of R_3, C_3?
10. In Figure 25.9, explain the change of bias with different input signal strengths.
11. In Question 10, why is limiting not complete for a weak signal?
12. Define limiting knee.
13. What is the approximate *limiting knee* for a high-gain linear-IC limiter? For a two-stage transistor limiter?
14. Refer to Figure 25.14. In your own words explain briefly, how the S curve shown in Figure 25.14 is produced.
15. In Question 14, what would be the result if the linear portion were less than shown?
16. In Figure 25.14, briefly explain the reason for the connection of capacitor C_3 and the RF choke.
17. In Figure 25.17, explain briefly what the effect would be if capacitor C_3 were to open circuit.
18. In Question 17 above, how is it possible to obtain audio across capacitor C_2? Could au-

dio be obtained across capacitor C_1? If so, why isn't it done here?

19. Refer to Figure 25.21. In your own words, explain briefly the function of the tuned circuit, C_1L_1.
20. In Figure 25.23, can this circuit ever break into oscillation? If so, what significance would it have?
21. Refer to Figure 25.24. Make rough sketches of the waveform outputs of the emitters of transistors Q_1 and Q_2.
22. In Question 21 above, briefly explain the function of transistors Q_1 and Q_2. What does transistor Q_3 do?
23. In Figure 25.25, what is the function of: (1) C_4, (2) R_1C_1 and R_2C_2, (3) C_6?
24. In Question 23 above, describe briefly how the PLL acts as an FM detector.
25. In Figure 25.26, is IC U_3 essential to the detection of the FM wave? Why?
26. In Question 25 above, what is the function of R_1C_1 and R_2C_2.
27. In the differential-peak detector shown in Figure 25.27, briefly discuss the following: (1) action of the tuned circuit L_1, C_2, C_1, (2) function of transistors Q_{23}, Q_{26}, and (3) function of transistors Q_{24}, Q_{25}.
28. In Question 27, how is it possible to obtain an audio output from the collector circuit of transistor Q_{25}?
29. In Figure 25.30, explain the advantage of using an oscillating limiter (Q_3).
30. In Figure 25.31, explain the function of the (1) attenuator, (2) current limiter, (3) thermal shutdown.
31. Explain the operation of transistors Q_{2010} and Q_{2020}.
32. In Question 31, why is there no audio output (matching) transformer?
33. Explain the possible causes of distorted sound and 60-Hz buzz in the sound.

EXAMINATION PROBLEMS

(Selected problem answers are provided at the back of the text.)

1. Identify the following frequencies: 4.25 MHz, 4.5 MHz, 4.75 MHz, 41.25 MHz, 59.75 MHz.
2. In Figure 25.3A, what is the preemphasis, in dB, for the following audio frequencies: (1) 200 Hz, (2) 2 000 Hz, (3) 6 000 Hz, (4) 15 000 Hz?
3. Which of the following RL combinations (similar to that shown in Figure 25.3B) could produce the preemphasis curve of Figure 25.3A? (a) R = 10 000 ohms, L = 0.75 R; (b) R = 13 300 ohms, L = 1 H; (c) R = 26 700 ohms, L = 2 H; (d) R = 6 670 ohms, L = 1.5 H.
4. In Figure 25.6, calculate the base bias of Q_1.
5. In Figure 25.6, identify (1) collector decoupling network, and (2) feedback path.
6. In Figure 25.10, (1) identify the base bias divider circuit for Q_2 and (2) calculate the base bias for Q_2.
7. In Figure 25.26, identify (1) deemphasis capacitor, (2) oscillator tuning element, and (3) low-pass filter.
8. An FM transmitter is modulated by an audio wave with an amplitude of 1 V, peak to peak. This produces a deviation of ± 10 kHz. If the peak-to-peak audio signal amplitude is changed to 2.5 V, peak to peak, what is the deviation? (Assume a linear relationship between modulation amplitude and deviation.)
9. In Figure 25.28, identify (1) peak detectors, (2) differential amplifiers, and (3) entire tuned circuit.
10. In Figure 25.33, assuming that Q_{2010} conducts at 20 mA (instantaneous) and that Q_{2020} is at cutoff, calculate the voltage at the collector of Q_{2010}.

Chapter 26

26.1 The monochrome section
26.2 The color section
26.3 Equipment for TV receiver alignment
26.4 The oscilloscope
26.5 Sweep alignment generators
26.6 RF signal generators
26.7 Marker signals
26.8 Color pattern generators
26.9 Solid-state analog and digital multimeters
26.10 Alignment using frequency sweep generator
26.11 Alignment using bar sweep generator
26.12 Horizontal AFC and anode voltage adjustments
26.13 Color picture tube adjustments
Summary

COLOR TV RECEIVER CIRCUIT ANALYSIS: TEST EQUIPMENT AND ALIGNMENT

INTRODUCTION

For our analysis of the circuits of a color TV receiver, we have chosen a GTE-Sylvania solid-state set. We shall explain receiver operation using (1) the overall schematic diagram, (2) the overall block diagram, and (3) numerous partial schematic diagrams of the individual sections.

Following the analysis of the circuits, we shall describe various types of test equipment. This will include the more recent Sencore VA48 multipurpose and bar sweep generator. Then, complete alignment and adjustment procedures for the GTE-Sylvania receiver will be given, as well as alignment procedures using the Sencore VA48.

The monochrome and color sections of the GTE-Sylvania TV receiver will be presented separately.

When you have completed the reading and work assignments for Chapter 26, you should be able to:

- Define the following terms: low-capacitance probe, demodulator probe, digital multimeter, autoranging, frequency sweep generator, bar sweep generator, color bar generator, post injection markers, gated rainbow, AFPC, AFT, marker signal generator, triggered sweep, vectorscope, gated-rainbow pattern color purity.
- Explain the use of post injection markers.
- Describe the equipment needed to check and align the circuits of a color television receiver.
- Explain the different types of oscilloscopes.
- Understand how to make color picture

tube adjustments in servicing color TV receivers.
- Discuss the use of a frequency sweep generator and a bar sweep generator in aligning a color TV video-IF section.
- Explain how to align the color circuits of a color TV receiver.
- Explain how to adjust color purity and static and dynamic convergence and how to degauss a color picture tube.
- Explain how to align the sound section of a TV receiver.
- Explain how to use a vectorscope.
- Understand the important characteristics of TV receiver test equipment.

26.1 THE MONOCHROME SECTION

In this section we shall follow the monochrome signal through the block diagram of Figure 26.1[1] Monochrome circuits are those needed to produce a monochrome signal, but many of these circuits are also used for amplifying the color signals.

TUNERS AND AFC

The UHF and VHF tuners are shown on the block diagram as being on two separate chassis. The antenna inputs for the tuners are not displayed. There are two separate AGC inputs to the tuners, and the UHF AGC is amplified. The UHF signal reduced to IF is fed through the VHF tuner to the first video IF amplifier. Four stages of video IF amplification are included.

Your attention is called to the AFC[2] closed feedback loop consisting of the tuners, IF stages, AFC drive, AFC IF, and AFC detector. This feedback loop controls the local oscillator frequency in the VHF and the UHF tuners, and prevents it from drifting. Early models of color receivers sometimes produced undesirable color changes (and sometimes a complete loss of color) whenever there was a slight drift in the local oscillator frequency. Furthermore, without the AFC circuit, a slight misadjustment of the fine-tuning control by the user often resulted in unsatisfactory color pictures and an unnecessary service call.

SOUND TAKEOFF

The sound takeoff point on the block diagram is between the fourth IF amplifier and the video detector. Note that a separate sound-IF detector diode is used, since the signal does not pass through the video detector. The 4.5-MHz amplification, limiting, and FM detection are accomplished in an IC, and the audio signal output drives a power amplifier and speaker.

Y VIDEO CIRCUITS

Let us return to the video chain in the block diagram. Following the video detector, there are three stages of Y video amplification and a video driver. The contrast control is between the first and second video amplifiers. The delay line is shown between the second and third video amplifiers. The Y video driver simultaneously feeds the red, green, and blue output amplifiers, which in turn simultaneously operate the three cathodes of the color picture tube to produce a monochrome picture.

SYNC TAKEOFF

The sync takeoff point is after the first video amplifier. The noise gate system,

[1]Figure 26.1 shows the complete block diagram, and the schematic is a foldout. This allows you to look at the schematic while you are studying the various sections of the receiver.

[2]The automatic frequency control is called AFC here. Note, however, that it is also known as AFT, for automatic fine tuning.

which inverts the noise pulses and cancels them, is fed from the video detector through a noise gate and to the sync separator. This circuit prevents noise spikes from triggering the deflection oscillators.

KEYED AGC

Keyed AGC is used in the receiver (Figure 26.4). The sync pulses for this system are taken from the first video amplifier. A pulse input from the horizontal output transformer keys this circuitry so that only the sync pulses are delivered to the AGC amplifier. The output of the AGC amplifier goes to the first video IF, and also to the VHF tuner. An additional stage of amplification is used to process the AGC voltage for the UHF tuner.

VERTICAL CIRCUITS

The sync separator delivers pulses to the vertical oscillator, which is followed by two stages of drive in the vertical output stage. Notice that vertical blanking is used to shut the picture tube beam off during retrace. Vertical blanking is accomplished by feeding a blanking pedestal to the video amplifier (Q_{212}).

HORIZONTAL CIRCUITS

The sync separator also delivers sync pulses to the horizontal oscillator. An AFC circuit is used to improve horizontal sync stability. The horizontal output stage delivers a signal to the blanker, which cuts off the horizontal retrace by sending a signal to the video amplifier (Q_{212}). This is especially important in color receivers because the color burst occurs on the back porch of the horizontal blanking pedestal. If the picture tube is not cut off during the horizontal retrace, the burst signal may produce a colored vertical line on the picture tube screen. The horizontal output stage also provides the high voltage for the picture tube. A conventional flyback circuit is used. A boost B+ voltage is also obtained. The brightness limiter and high-voltage protection circuit prevent picture tube damage from excessive voltage or drive. These circuits also limit x-ray radiation from the tube.

POWER SUPPLIES

The power supply consists of two separate systems. One produces a high-voltage B+, and the other produces a low-voltage B+.

Having discussed the block diagram in terms of the monochrome signal, we shall now look at some of the individual circuits in greater detail.

THE 180-V AND 120-V SUPPLIES

The bridge rectifier output (Figure 26.2) charges C_{500A} to the peak AC voltage, placing 185 V across voltage divider R_{516}, R_{518}, and R_{522}. A zener diode (SC_{508}) regulates the voltage drop across R_{518} and R_{522} to 135 V. Variable resistor R_{518} adjusts the forward bias to the regulator drivers Q_{504}, Q_{502}, and the B+ regulator Q_{500} and sets the emitter voltage of the regulator Q_{500}. The emitter of Q_{500} supplies the +120 V DC bus that is the regulated high B+ for the receiver.

Should the power source voltage drop, the bridge output charges C_{500A} to a lower peak voltage, and the divider network voltage is lower. However, the voltage across zener diode SC_{508} remains constant. Only the voltage drop across R_{516} decreases. This holds the voltage across R_{518} and R_{522} constant. Since the emitter voltage of Q_{500} must also remain constant, the effect of lower power source voltage is seen as a lower col-

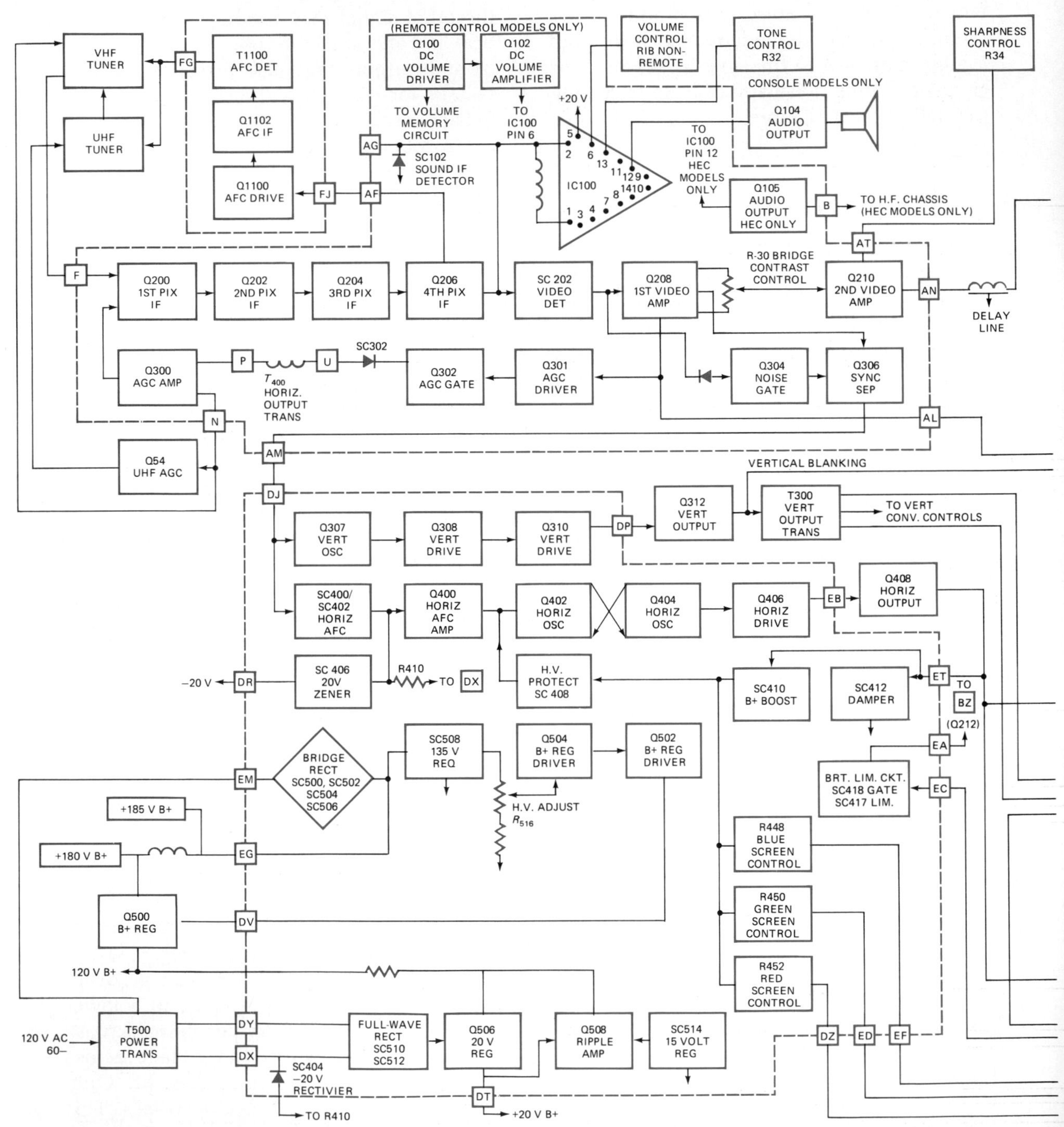

Figure 26.1 Block diagram of the Sylvania all solid-state color television receiver. *(Courtesy of GTE-Sylvania, Inc.)*

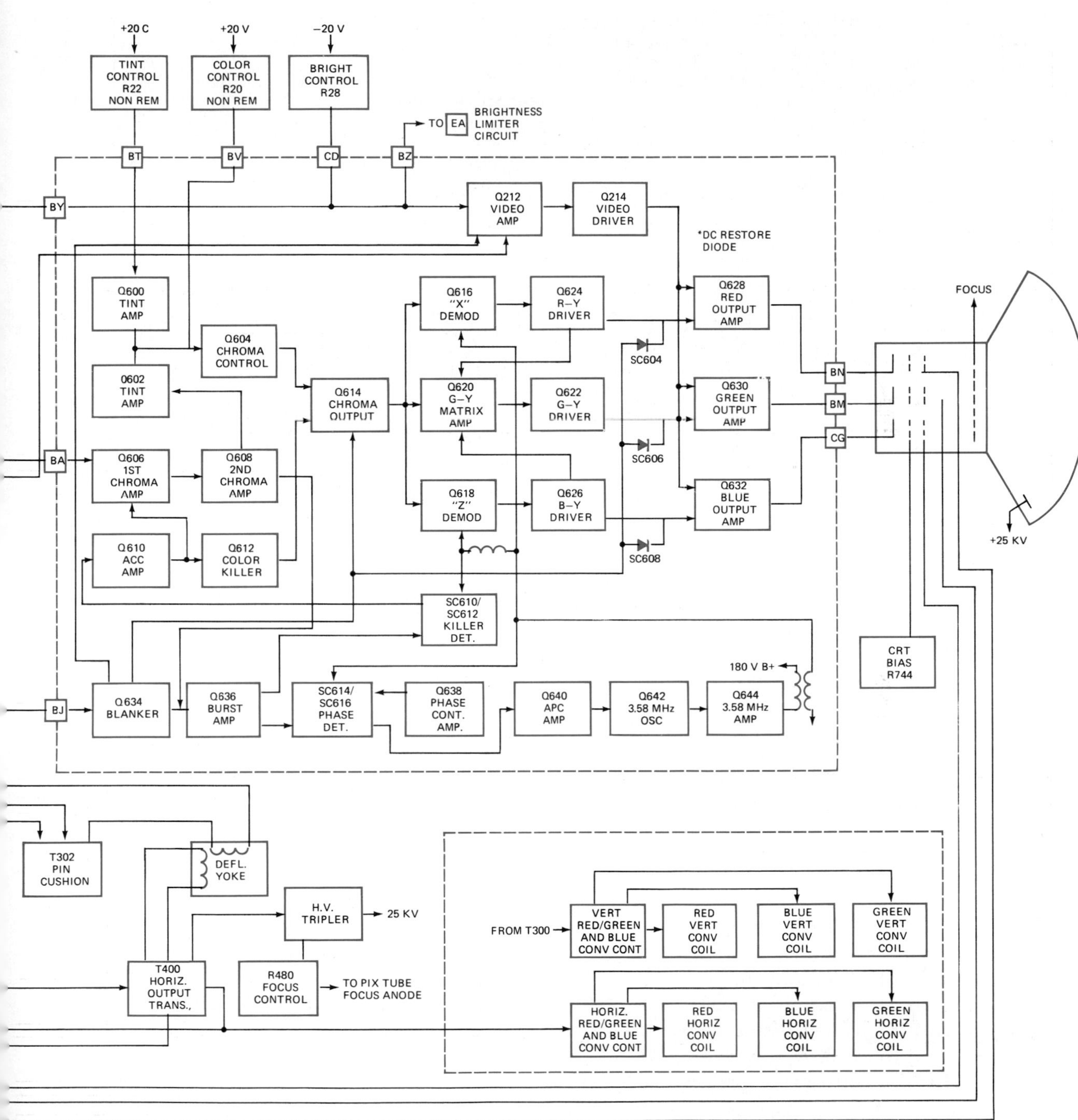

+20 C
+20 V
−20 V
TINT CONTROL R22 NON REM
COLOR CONTROL R20 NON REM
BRIGHT CONTROL R28
TO EA BRIGHTNESS LIMITER CIRCUIT
BT
BV
CD
BZ
BY
Q212 VIDEO AMP
Q214 VIDEO DRIVER
*DC RESTORE DIODE
Q600 TINT AMP
Q604 CHROMA CONTROL
0602 TINT AMP
Q616 "X" DEMOD
Q624 R−Y DRIVER
Q628 RED OUTPUT AMP
FOCUS
SC604
Q614 CHROMA OUTPUT
Q620 G−Y MATRIX AMP
Q622 G−Y DRIVER
Q630 GREEN OUTPUT AMP
BN
BM
CG
BA
Q606 1ST CHROMA AMP
Q608 2ND CHROMA AMP
SC606
Q618 "Z" DEMOD
Q626 B−Y DRIVER
Q632 BLUE OUTPUT AMP
+25 KV
Q610 ACC AMP
Q612 COLOR KILLER
SC608
SC610/ SC612 KILLER DET.
CRT BIAS R744
180 V B+
BJ
Q634 BLANKER
Q636 BURST AMP
SC614/ SC616 PHASE DET.
Q638 PHASE CONT. AMP.
Q640 APC AMP
Q642 3.58 MHz OSC
Q644 3.58 MHz AMP
T302 PIN CUSHION
DEFL. YOKE
H.V. TRIPLER
25 KV
T400 HORIZ. OUTPUT TRANS.,
R480 FOCUS CONTROL
TO PIX TUBE FOCUS ANODE
FROM T300
VERT RED/GREEN AND BLUE CONV CONT
RED VERT CONV COIL
BLUE VERT CONV COIL
GREEN VERT CONV COIL
HORIZ. RED/GREEN AND BLUE CONV CONT
RED HORIZ CONV COIL
BLUE HORIZ CONV COIL
GREEN HORIZ CONV COIL

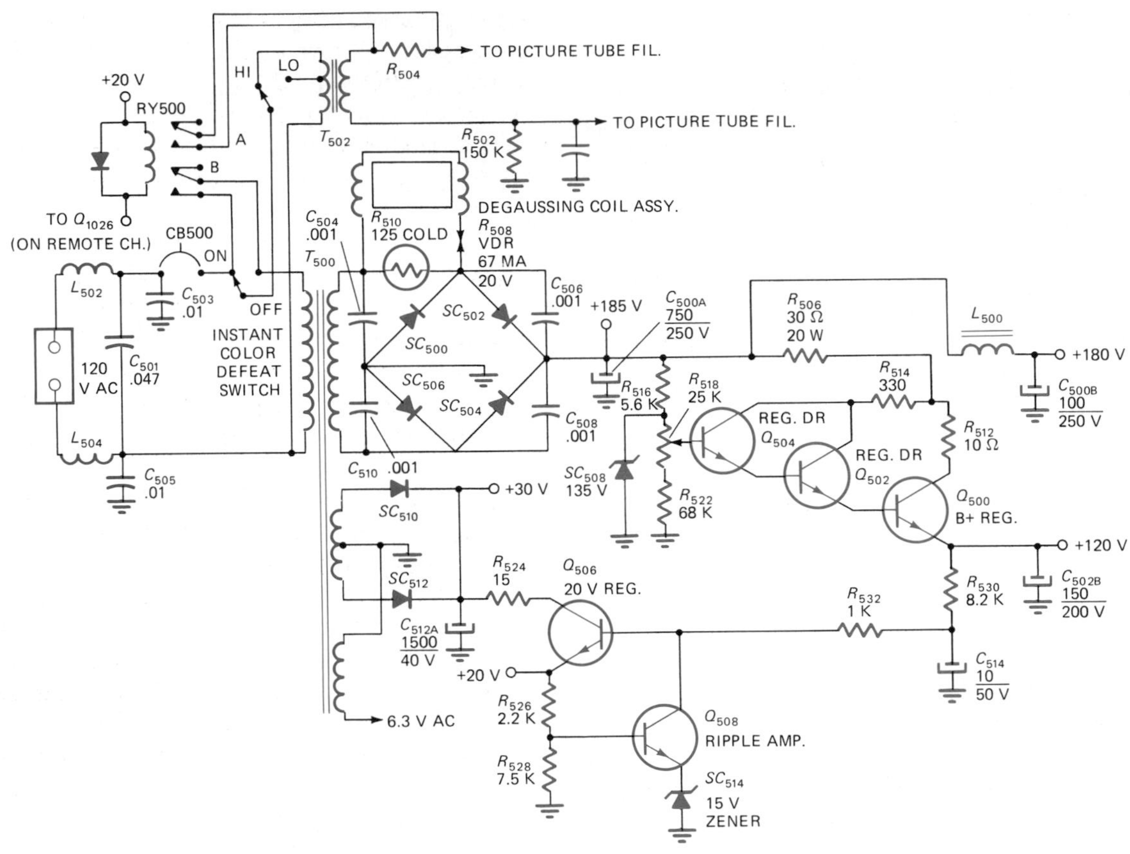

Figure 26.2 The 120-V and 180-V power-supply circuits. *(Courtesy of GTE-Sylvania, Inc.)*

lector-to-emitter voltage on Q_{500}, but the 120-V emitter voltage remains unchanged.

THE 20-V POWER SUPPLY

The low-voltage supply, illustrated in Figure 26.3, has a full-wave rectifier network consisting of SC_{510} and SC_{512} and filter capacitor C_{512A}, which is the 30-V source for Q_{506} (20-V regulator). The collector of Q_{506} is connected to the 30-V source through R_{524}.

The bias network for the base voltage of Q_{506} (which consists of R_{530}, R_{532}, Q_{508}, and zener diode SC_{514}) is connected from the regulated 120-V bus to ground. The forward bias to Q_{506} is developed across Q_{508} and SC_{514}. Transistor Q_{508} compares the voltage across SC_{514} with the +15-V DC at the junction of R_{526} and R_{528}. If the compared voltage is too high, Q_{508} conducts harder, drawing more current through R_{532}. This lowers the voltage on the base of Q_{506} and consequently lowers the voltage on the emitter of Q_{506}. Any 20-V source variations (due to load or voltage input variations) are regulated by comparing the output voltage with the zener (SC_{514}) voltage and using the error signal to control regulator Q_{506}.

VIDEO-IF SECTION AND VIDEO DETECTOR

The video-IF section is illustrated in Figure 26.4. It has four video-IF amplifier tran-

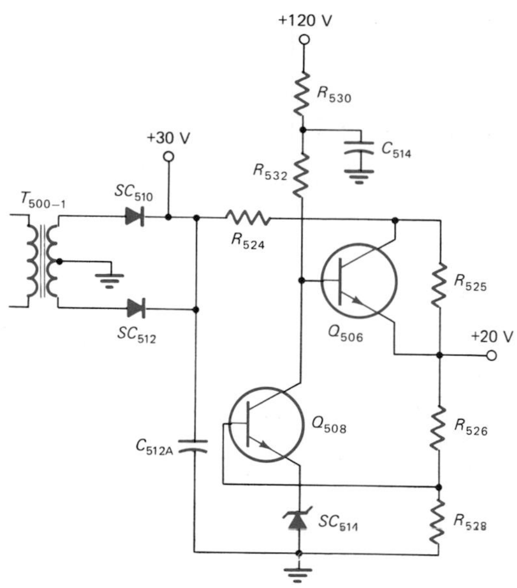

Figure 26.3 The 20-V power supply.

sistors: Q_{200}–first video IF, Q_{202}–second video IF, Q_{204}–third video IF, and Q_{206}–fourth video IF.

The RF signal from the tuner is applied to the emitter of Q_{200} through an impedance-matching coil (L_{200}), and an RC network (R_{202}, C_{202}, and C_{204}) with an adjacent sound trap coil (L_{202}). The trap coil is tuned to 47.25 MHz, which sets the upper limits of the IF bandpass. This keeps the sound and picture carrier energy of the tuned band high and the adjacent-channel sound carrier level low.

The output of Q_{206} (fourth video IF) is coupled through T_{202} (video output transformer) to SC_{202} (video detector diode). The 41.25-MHz sound carrier is trapped out by T_{202} and L_{208}. The signal is then fed into the cathode of SC_{202}, producing a video signal with negative sync pulses ($-Y$).

THE VIDEO AMPLIFIER

Figure 26.5 shows the video amplifier. Following SC_{202} (video detector), the signal inversion takes place at each successive amplifying stage (Q_{208}–first video amplifier, Q_{210}–second video amplifier, Q_{212}–third video amplifier) resulting in a positive Y signal at the output of Q_{212}. Since there is no signal inversion at Q_{214} (video driver), the result is a positive Y signal at the emitter output of Q_{214}.

Horizontal and vertical pulses are fed to the emitter of Q_{212} providing blanking signals to the picture tube. By inserting the blanking pulse in the low signal level stages, low pulse power is required. As amplification from Q_{212} and Q_{214} increases, the pulse amplitude for blanking the color picture tube beam also increases.

The input drive to Q_{210} is taken from the bridge contrast control circuit shown in Figure 26.6. The advantage of this control system in a DC video amplifier chain is its ability to control contrast without upsetting the average DC voltage level in the video system and hence the brightness level at the picture tube. The bridge circuit consists of R_{256}, R_{258}, Q_{208}, R_{270}, R_{262}, R_{264}, and the contrast control R_{30}. The contrast control has low impedance and uses the equipotential DC voltage across it to maintain the average DC level through the succeeding video amplifier stages constant. The voltage divider R_{262} and R_{264} resistance ratio is about 3:1 dividing down the 20-V B+. This places about one-fourth of 20 V across R_{262} and three-fourths of 20 V across R_{264}. The voltage at the junction of R_{262} and R_{264} is a constant voltage between 14 and 15 V positive.

The video amplifier Q_{208} is biased on by the voltage divider network in its base. This forward voltage produces a 14-V drop at the junction of R_{256} and R_{258}. Now, both ends of the contrast control are equipotential and no DC current flows through R_{30}, the bridge load (contrast control). However, R_{30} and the resistors R_{264} and R_{262} form a video voltage divider across which the video signal variations are present. When the contrast control slider is at Point *A*, maximum video drive is fed to Q_{210}. When the slider is at

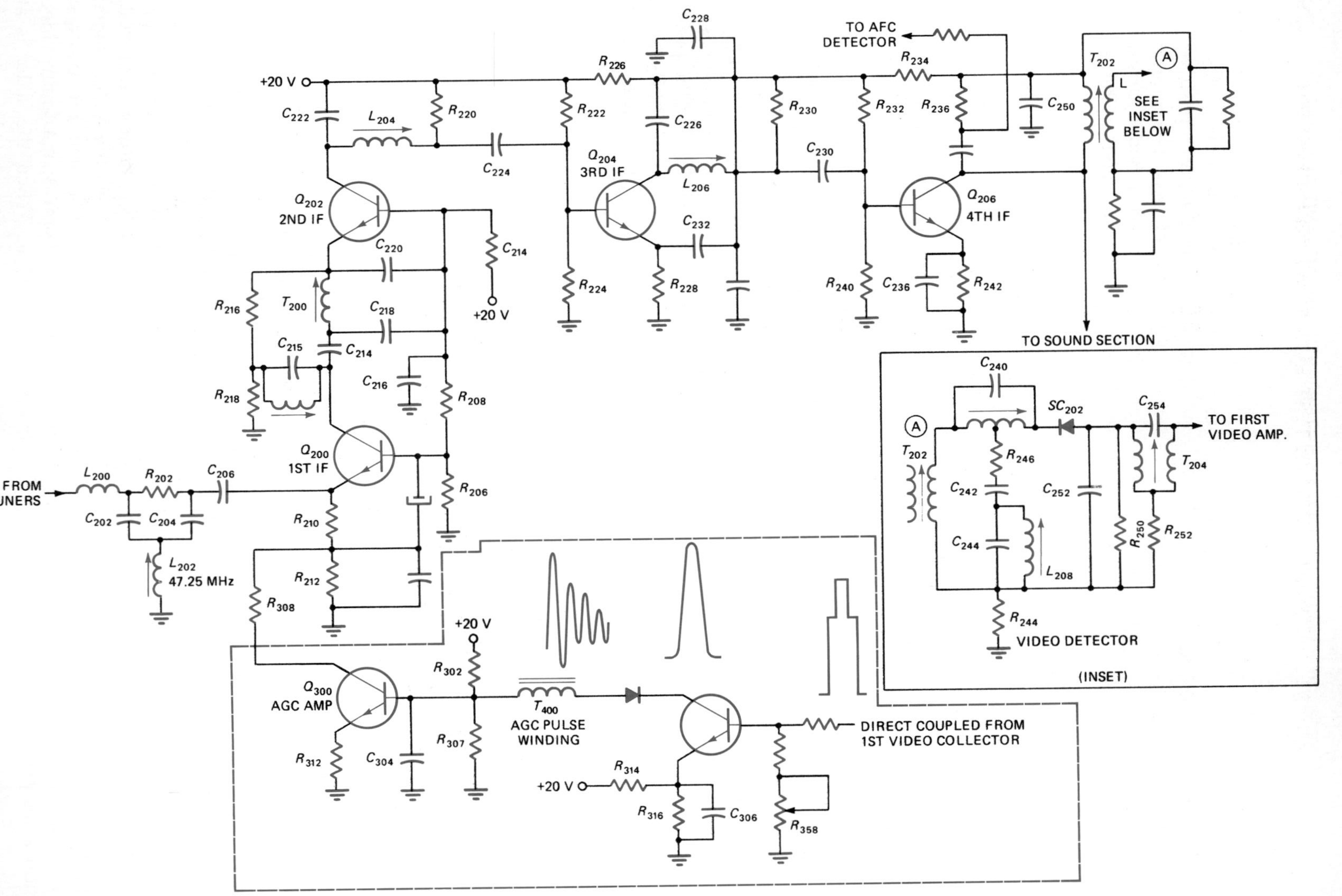

Figure 26.4 The video-IF amplifier section.

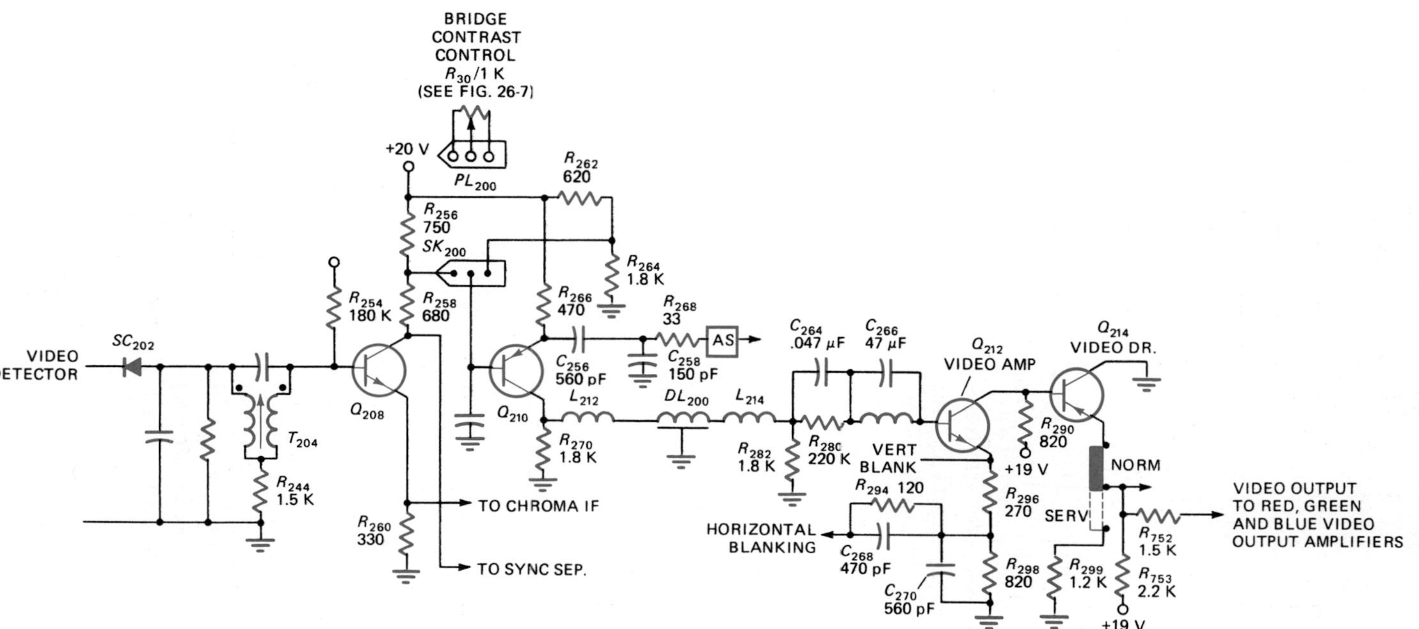

Figure 26.5 **The video amplifier.** *(Courtesy of GTE-Sylvania, Inc.)*

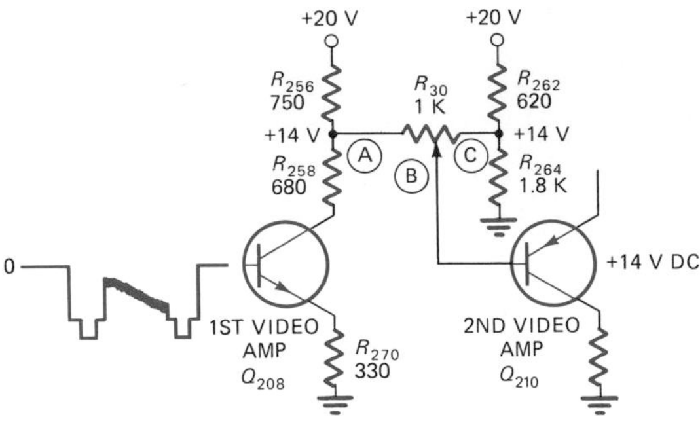

Figure 26.6 The bridge-contrast control.

Point *C*, the video drive is minimum. However, the DC voltage on the control arm does not vary. Thus, no DC shift occurs in the video amplifiers and the brightness level remains steady.

SOUND SECTION

You can follow this discussion of the sound circuit by referring to the foldout schematic inserted after pg. 804. The output of Q_{206} (fourth video IF) is applied to sound detector diode SC_{102} for 4.5-MHz sound-IF detection to avoid interference in the video detector and amplifiers. The 4.5-MHz IF is filtered through a low-pass filter (L_{102} and R_{120}) to remove all 40-MHz IF frequencies and is then impedance-coupled to the input of IC_{100}. This IC provides amplification of the 4.5-MHz IF frequencies, FM limiting, FM detection, and first stage of the audio signal voltage amplification. The resultant audio signal from IC_{100} is applied to Q_{104} (audio output) and then to the speaker system. On console models with high fidelity, the signal from IC_{100} is applied to Q_{105} (audio driver) and from there to the jack plate on the high-fidelity chassis.

THE AGC SYSTEM

The AGC circuit, illustrated in Figure 26.7, is a closed-loop system controlling RF and IF signal gain. Its purpose is to maintain relatively constant video output over a wide range of input signal levels. The system consists of a two-stage gate amplifier circuit that uses flyback pulses and sync tip voltages for regulating the AGC amplifier.

A positive horizontal flyback pulse is applied (through diode SC_{302}) to the collector of the AGC gate transistor Q_{302}. Simultaneously, the base of Q_{302} receives a positive voltage from the sync portion of the video signal at the collector of the first video amplifier. During this time, the AGC gate transistor conducts current and this results in less positive DC voltage at the filter capacitor C_{304}. The magnitude of this voltage is proportional to the current through the gate transistor and hence to the amplitude of the sync pulses at the collector of the first video amplifier, Q_{208}. This DC voltage controls the bias (and hence the gain) at the tuner and the controlled IF stages. The potentiometer R_{358}, together with R_{320} and the gate emitter bias network R_{314}, R_{316}, sets the AGC threshold voltage.

With no AGC developed, the AGC amplifier transistor Q_{300} is biased into saturation through the resistors R_{306}, R_{302}, and R_{307}. The voltage at the junction of R_{307} and R_{302} is 8 V, which gives maximum tuner gain.

With Q_{300} in saturation, the emitter current through Q_{200} and Q_{202} (series-connected) is determined by resistor R_{210}. This has a relatively low value and thus the IF stage operates at maximum current and maximum gain.

When negative AGC bias is developed and added to the existing bias at the base of Q_{300}, the current through Q_{300} is reduced. This reduces the current and the gain of Q_{200} and Q_{202}. With Q_{300} cut off by negative AGC bias, the gain of Q_{200} is low and determined by resistor R_{212} (R_{210} is low-resistance.) Transistor Q_{202} is controlled similarly by R_{218}. Further change of AGC voltage controls only the gain of the tuner (Figure 26.5).

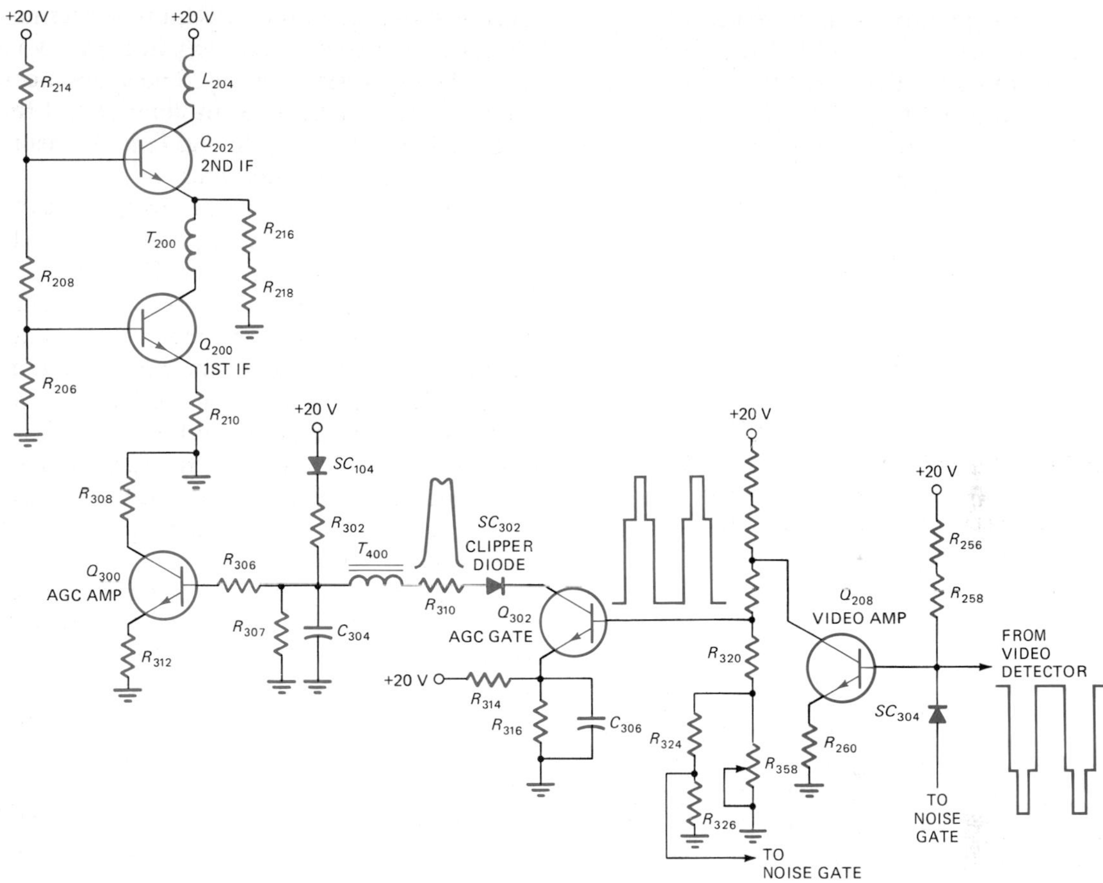

Figure 26.7 The AGC system. *(Courtesy of GTE-Sylvania, Inc.)*

AFC AND TUNING CONTROL

As shown in the receiver schematic (after pg. 804), the negative AFC circuit obtains its signal from the fourth video IF stage (Q_{206}). The tuner AFC circuit uses transistor Q_{1102} (AFC IF/detector) and Q_{1100} as the driver. The discriminator circuit is composed of T_{1100} (discriminator transformer), diodes SC_{1100} and SC_{1102}, filter circuit R_{1108}, C_{1112}, R_{1110}, C_{1114}, and the voltage matrix R_{1116} and R_{1118}. The sole purpose of the AFC circuit is to electrically control changes in oscillator frequency and IF drift by feeding correction voltage from the discriminator to the oscillator varactor located in the tuner.

The video-IF carrier, 45.75-MHz from Q_{206} (fourth video IF stage), is applied to the input of Q_{1100}, then to Q_{1102}. Transformer T_{1100} is tuned to 45.75 MHz, the video-IF carrier frequency. Should the AFC detector detect a frequency change, the matrix voltage across R_{1116}, R_{1118} changes, producing a correction voltage at their junction. This DC voltage changes the varactor diode voltage, which in turn brings the local oscillator frequency back to the correct value.

At the correct fine tuning, with the video carrier at 45.75 MHz, no output voltage is developed in the discriminator. If the RF oscillator frequency is increased for any reason, the video carrier frequency increases by the same amount. This develops negative voltage, which, when added to the tuning voltage, reduces the oscillator frequency, thus bringing the tuning back to the acceptable frequency. With the oscillator frequency reduced, positive correcting voltage is developed. This compensates for the fine-tuning error.

HORIZONTAL OSCILLATOR AND AFC

The horizontal multivibrator frequency (Figure 26.8) is controlled by the amplified output from an unbalanced comparator circuit. This circuit consists basically of diodes SC_{400} and SC_{402}, back to back and the associated wave-shaping circuits C_{412}, R_{408}, and C_{410}.

Negative horizontal sync pulses are passed through C_{402} into the unbalanced comparator from the single-ended sync separator Q_{306} (see the complete schematic). Diodes SC_{400} and SC_{402} compare the horizontal synchronizing pulse to a sawtooth voltage applied to their anodes. The sawtooth is developed by integrating a flyback pulse in the wave-shaping circuit R_{408}, C_{412}, and C_{410}.

The sawtooth wave-form is the anode voltage for each diode, and the horizontal synchronizing pulse drives the cathode of each diode. When an unbalance occurs in the comparator diodes, one diode conducts more than the other, producing an output voltage.

When the oscillator frequency is normal, the sawtooth repetition rate and the horizontal sync pulse are in sync because the pulses sit on a sawtooth AC reference. Now both diodes conduct equally, resulting in no correction voltage. However, when the oscillator frequency is lower than normal or when the oscillator phase leads the sync pulse, diode SC_{400} conducts more than SC_{402}. This unbalanced conduction results in a negative correction voltage from the comparator. Should the horizontal oscillator frequency run high or lead the horizontal sync pulse diode, SC_{402} conducts more than SC_{400}. This causes the unbalance to produce a positive voltage from the comparator circuit.

The comparator output voltage is fed to Q_{400}, a class A DC amplifier (the AFC amplifier), where the comparator output voltage either aids or opposes the AFC amplifier forward bias. When the comparator output voltage is negative, the forward bias of Q_{400} is lowered. This raises the collector voltage of Q_{400} and also the voltage drop across R_{428} and R_{432}. This increases the voltage across R_{424} and R_{426} and produces a more rapid discharge of C_{422} and C_{424}, thereby increasing the oscillator frequency.

A positive output voltage from the comparator circuit aids the forward bias at the base of Q_{400}. The AFC amplifier conduction increases, lowering the collector voltage and the voltage across R_{428} and R_{432}. This decreased voltage, appearing at frequency control R_{428}, slows down the discharge of C_{422} and C_{424}, resulting in a lower oscillator frequency.

HORIZONTAL DRIVER

Figure 26.9 illustrates the horizontal driver and output circuit. The horizontal oscillator multivibrator in the previous stage develops a square-wave pulse across resistor R_{433}. When this pulse is applied to the horizontal drive transistor (Q_{406}), it switches from cutoff to saturation. This effectively grounds the primary of T_{402} through Q_{406}. Current flows from B+ through the primary of T_{402}, and through Q_{406} to ground. A pulse is coupled tc horizontal output transistor Q_{408}.

The driver protection diode (SC_{414}) and capacitor C_{428} form a low-impedance network for reactive voltages appearing across

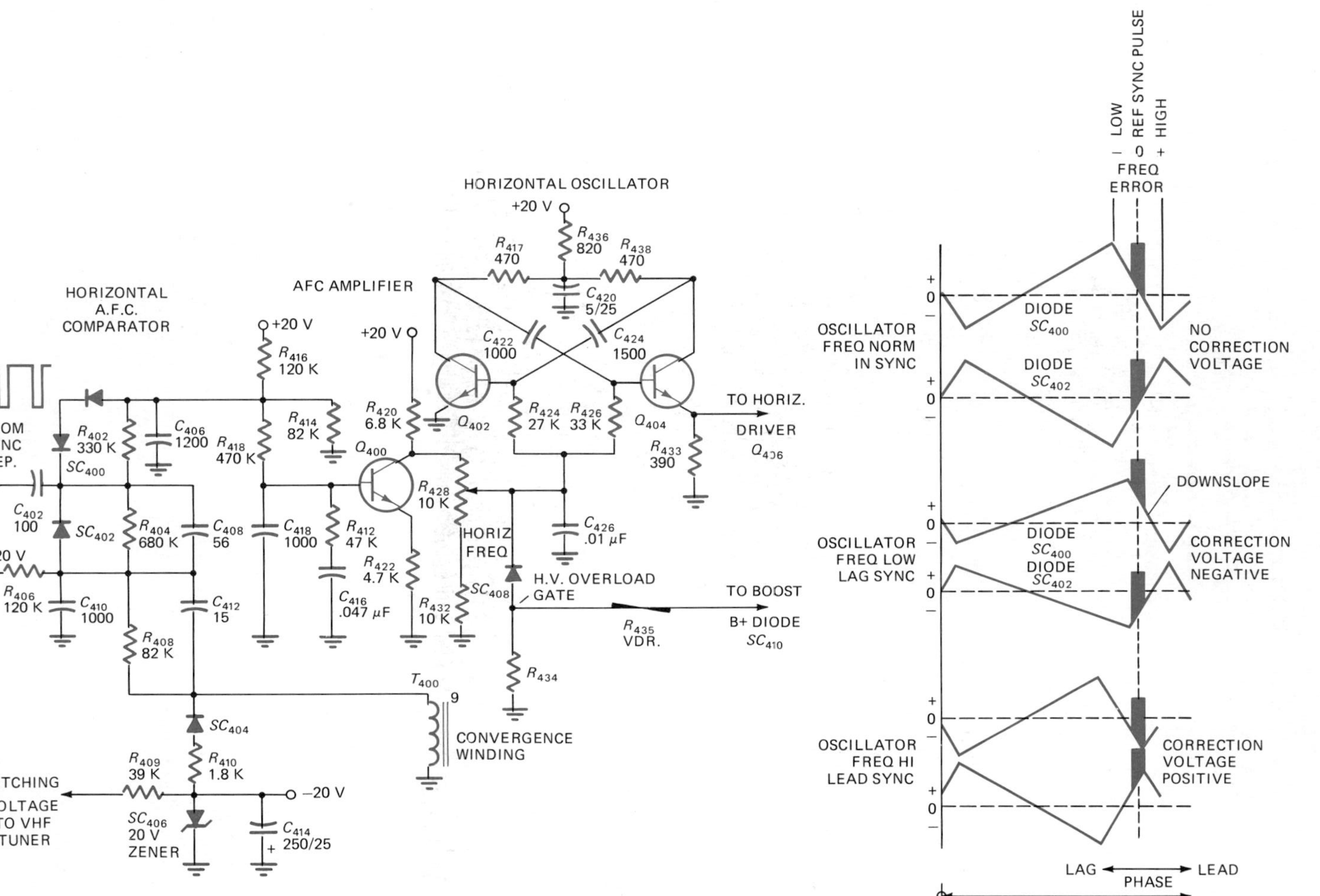

Figure 26.8 The horizontal oscillator and AFC circuit. *(Courtesy of GTE-Sylvania, Inc.)*

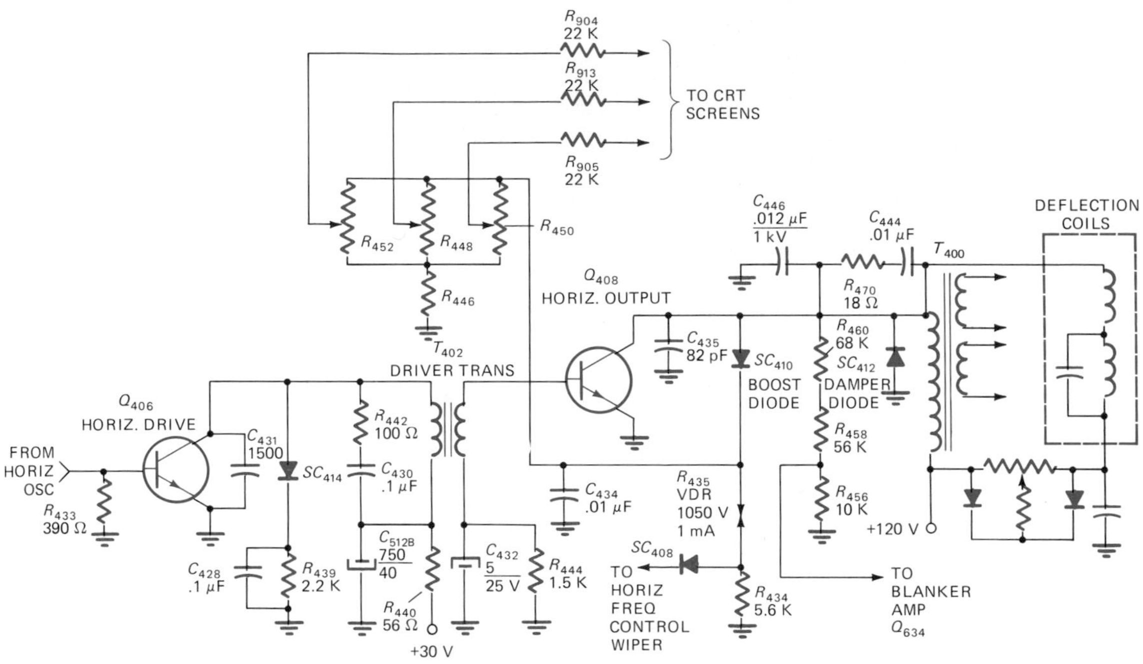

Figure 26.9 The horizontal driver, horizontal output, damper, and boost voltage circuits. *(Courtesy of GTE-Sylvania, Inc.)*

the primary of the T_{402} during the turnoff of transistor Q_{406}. As soon as the positive voltage on the collector of Q_{406} increases over the voltage on C_{428}, SC_{414} conducts and charges C_{428}. Resistor R_{439} keeps discharging C_{428} so that SC_{414} conducts again at the following turnoff of Q_{406}, thus protecting Q_{406} from voltage surges. The series *RC* circuit comprising R_{442} and C_{430} forms a damping link for all higher-frequency ringing in T_{402}. A filter network (R_{440} and C_{512B}) is used in the collector circuit of Q_{406}. This filter decouples the transistor collector and also filters the 120-Hz ripple on the 30-V B+. The voltage drop in R_{440} results in a voltage of about 13 V across C_{512B}.

HORIZONTAL OUTPUT

The driver transformer T_{402} (Figure 26.10) couples the switching signal from the driver transistor (Q_{406}) to the horizontal output transistor base (Q_{408}). This positive-going pulse turns on Q_{408}, causing it to conduct into saturation. The collector tap on transformer T_{400} is grounded when Q_{408} is saturated, and this causes current to increase through the primary winding and deflection coils. Inductive reactance opposes the current increase through the yoke coils and the T_{400} primary winding. The magnetic field develops linearly in the yoke coils, and this moves the picture tube beam from the raster center to the extreme right.

When transistor Q_{408} is turned off, the flyback transformer and the deflection coils form a resonant circuit. The decreasing yoke current flows into C_{446} until it reaches a value of zero. At this point, the electron beam has returned to the raster center. Consequently, when C_{446}, which is now charged, discharges again into the yoke, current in the yoke rises. Since this current is

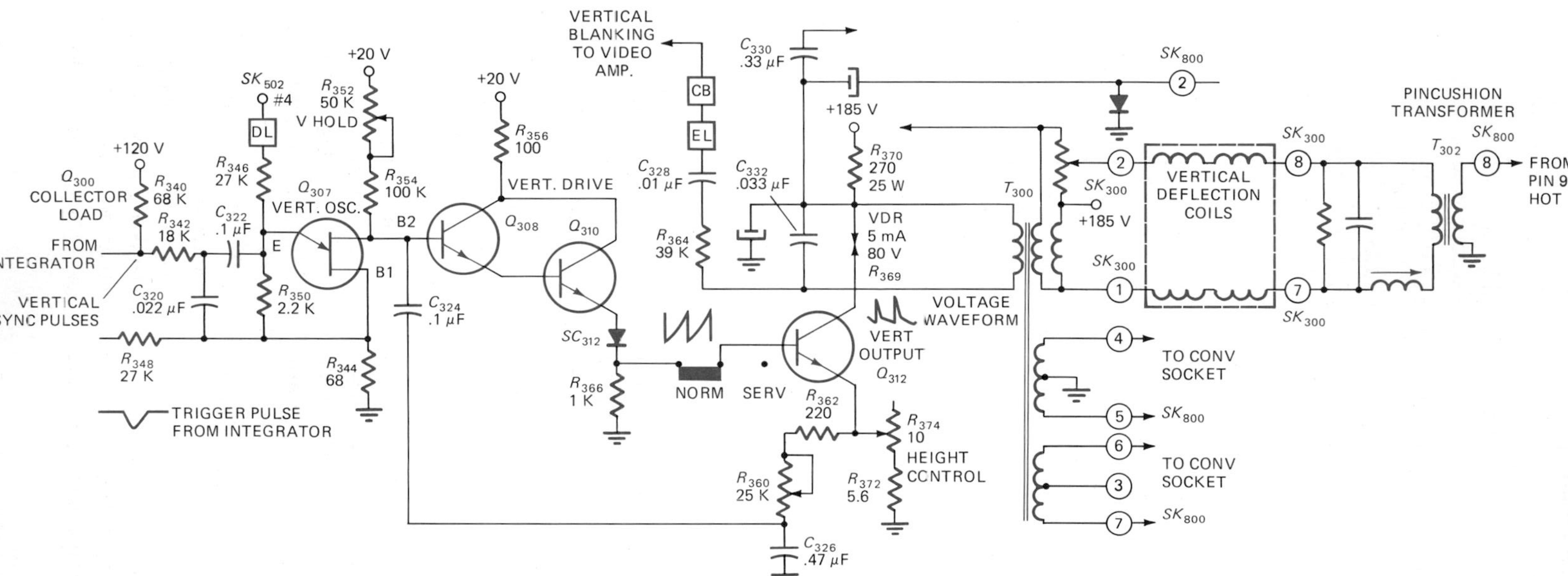

Figure 26.10 The vertical oscillator and vertical-sweep system. *(Courtesy of GTE-Sylvania, Inc.)*

now negative, it brings the beam to the extreme left of the raster.

B+ BOOST CIRCUIT

The high-energy pulse developed by T_{400} turns on boost diode SC_{410}, applying the pulse to C_{434}. This way, C_{434} becomes charged to a DC voltage that is very close to the peak collector voltage of Q_{408}, and this voltage supplies the picture tube screen controls.

DAMPER DIODE

The voltage that started rising when Q_{408} (Figure 26.9) was turned off and reached a maximum when the beam returned to the raster center is now back to zero again. The moment this voltage becomes negative, damper diode SC_{412} starts conducting, dissipating the stored energy. Thus the yoke current slowly drops to zero, bringing the beam back to the raster center.

HIGH-VOLTAGE OVERLOAD GATE

Should the voltage level across the parallel resistance network comprising R_{435}, R_{434}, R_{448}, R_{450}, R_{452}, and R_{446} increase drastically, the VDR impedance decreases, placing a higher voltage across R_{434}. This increase in voltage turns on the high-voltage overload gate diode SC_{408}, causing an increase in voltage at the horizontal control (Figure 26.10) and forcing the horizontal multivibrator into a higher running frequency. The horizontal output transistor Q_{408} is turned on for shorter periods of time, and therefore the peak currents and peak voltages in the output stage decrease. Also, the input pulse into the tripler, shown in Figure 26.2, becomes lower and reduces the second anode voltage.

VERTICAL SWEEP AND SYNC SYSTEMS

The vertical sweep oscillator, illustrated in Figure 26.10, is controlled by a negative-going pulse that is developed in the integrator circuit. This pulse is fed to the gate of the programmable unijunction transistor Q_{307} through capacitor C_{322}. The negative-going timing pulse controls the discharge of C_{324} and C_{236} through Q_{307}, thus forming the drive sawtooth fed to Q_{308}.

The programmable unijunction transistor can be better understood if it is viewed as a diode gate and a current generator. No current flows in the unijunction until the diode is forward-biased. When the anode voltage is made 0.6 V higher than the gate voltage, current flows in the diode and in the cathode. The programmable unijunction remains conducting until the anode current falls below that needed to maintain conduction.

When the vertical serrated pulse appears at the integrator circuit, the pulses are integrated, discharging C_{320} and producing a negative pulse. Capacitor C_{322} couples the negative pulse to the gate of Q_{307}, lowering the gate voltage to below the anode voltage allowing C_{324} and C_{336} to discharge through Q_{307}. When the capacitors have discharged, the unijunction no longer conducts and C_{324} and C_{326} begin to charge through R_{352} and R_{354}. The relatively long time constant of this RC circuit produces a linear voltage ramp. When the capacitors once again allow the anode to be 0.6 V higher than the gate voltage, the unijunction fires and the process repeats itself.

The vertical driver is connected in a Darlington configuration (a double emitter follower), which drives the vertical-output transistor Q_{312}. The feedback signal amplitude and shape are controlled by the adjustment of R_{360}. By combining the drive voltage developed by the charge and discharge of C_{324} with the shaped feedback voltage, both the driving signal linearity fed to Q_{308} and the current wave form from Q_{312} to the vertical transformer T_{300} are improved.

The vertical hold control adjusts the anode voltage of Q_{307}, and hence the firing voltage. Adjusting more resistance into the circuit lowers the firing point and increases the oscillator frequency. Decreasing the resistance raises the firing point, slowing the oscillator.

The height control R_{374} adjusts the emitter voltage of Q_{312}. Adjusting R_{374} to less resistance in the emitter circuit increases the current gain and also increases the deflection current and magnetic field in the yoke coils, thus producing additional vertical deflection.

26.2 THE COLOR SECTION

Before discussing the circuitry of the color section, we shall follow the color signal through the block diagram of Figure 26.1. The luminance portion of the color signal is derived from the video driver Q_{212}, which delivers its signal to the red, green, and blue cathodes simultaneously. This signal is delayed by the delay line between the second and third video amplifiers.

COLOR AMPLIFIERS

The color signal is taken from the output of the first video amplifier (Q_{208}) and delivered to the first and second color amplifiers (Q_{606} and Q_{608}). These amplifiers are also known as *color-IF* or *color bandpass* amplifiers. They amplify and pass only that portion of the signal that is related to color reproduction. Figure 26.11 shows a typical response curve for a color bandpass amplifier. The increase in the amplitude of this re-

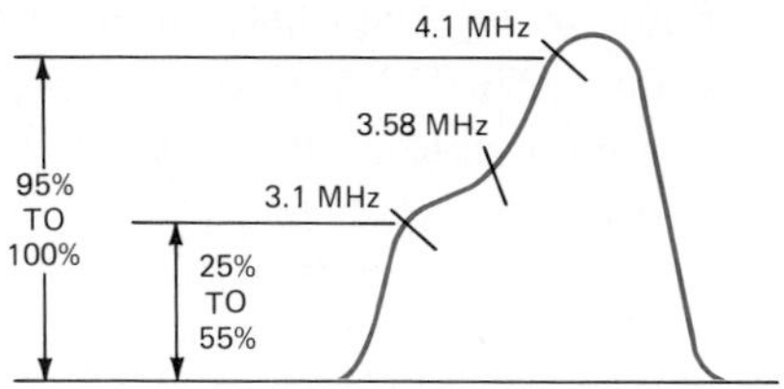

Figure 26.11 Response curve taken at the output of the second chroma amplifier. *(Courtesy of GTE-Sylvania, Inc.)*

sponse curve at the upper frequency end compensates for the drop off of the video-IF response curve at the color subcarrier point.

TINT CONTROL AMPLIFIER

Returning to the block diagram of Figure 26.1, the output of the second color amplifier passes through a tint control amplifier (Q_{602}) in which the phase of the signal is corrected for proper color reproduction. Transistor Q_{602} is a phase-shift amplifier, and the amount of phase shift is controlled by the magnitude of its collector voltage, which in turn is controlled by a tint amplifier (Q_{600}). Tint amplifier Q_{600} receives its signal either from a remote control circuit or from a variable resistor connected from B+ in sets that are not remote-controlled. In either case, the amount of DC voltage delivered to Q_{600} determines the amount of phase shift of tint amplifier Q_{602}.

COLOR CONTROL AMPLIFIER

The output of Q_{602} is delivered to a color control (Q_{604}), which determines the amount of gain of the color signal. The color control amplifier is simply a broad-band amplifier that is gain-adjustable.

COLOR OUTPUT AMPLIFIER

The color output amplifier (Q_{614}) delivers the color signal simultaneously to the X demodulator, Z demodulator, and G–Y matrix amplifier. The automatic gain control (called automatic color control or ACC when it is located in the color section) is obtained from a color-killer detector comprising SC_{610} and SC_{612}. The color control signal is amplified by Q_{610}, the automatic color control amplifier. The output of Q_{610} determines the gain of the first color amplifier and also determines whether the color killer will allow the color output amplifier (Q_{614}) to conduct. During reception of monochrome signals, Q_{614} is cut off in order to prevent noise colors on the screen of the picture tube.

3.58-MHz OSCILLATOR

The 3.58-MHz oscillator is controlled by an automatic phase control network that compares a feedback signal from the output of the 3.58-MHz amplifier (Q_{644}) with the burst from burst amplifier Q_{636}. The output signal from the oscillator is delivered to the X and Z demodulators, where it is combined with the output from the color amplifier. The output of these demodulators drives the R–Y and B–Y drivers. Note that the G–Y matrix amplifier obtains its signal from the R–Y and B–Y drivers and from the color amplifiers.

Having traced the signal through the block diagram, we shall now discuss some of the individual circuits.

TINT CONTROL

The electronic tint control uses two transistors: Q_{600} as a DC amplifier and Q_{602} as a phase-shift amplifier (Figure 26.2). The two amplifiers are emitter-follower and common-emitter configurations, stacked in series to B+, permitting Q_{600} to control the Q_{602} collector voltage.

Color tint may be changed by changing the DC voltage at the Terminal BT. The DC voltage at Terminal BT for a nonremote set is a variable-voltage source, provided by R_{22} (tint control) connected to B+ and ground. For a remote set, this DC voltage is provided by a memory module. Variable resistor R_{22} is located on the receiver control board (not shown).

COLOR CONTROL

The phase-corrected color signal is coupled from the tint amplifier to the color control (Q_{604}) (see insert). Amplifier Q_{604} gain is adjustable. This transistor is biased at 0.3 V by the forward drop of SC_{602}. This diode also compensates for gain variation in Q_{604} caused by temperature variations. With no voltage at Terminal BV, Q_{604} is biased at zero emitter current. Therefore, the transistor does not conduct and no signal appears at the collector of Q_{604}. Applying a DC voltage at Terminal BV forward biases this transistor, and we get color out at the collector. By changing the DC control voltage at BV, the emitter current can be varied. This varies the beta of the transistor and also thus varies the color output at the collector of Q_{604}.

As explained in tint operation, the DC voltage at Terminal BV is a potentiometer acting as a variable voltage source.

For a remote set, a memory module provides the color control voltage at Terminal BV. The color control, like the tint control, is located on the control board.

ACC AND COLOR KILLER

Figure 26.12 shows the ACC circuitry. The composite video signals from the emitter of Q_{208} (first video amplifier) are applied to Q_{606} (first color amplifier) through C_{616} and L_{602}. The gain of Q_{606} is automatically

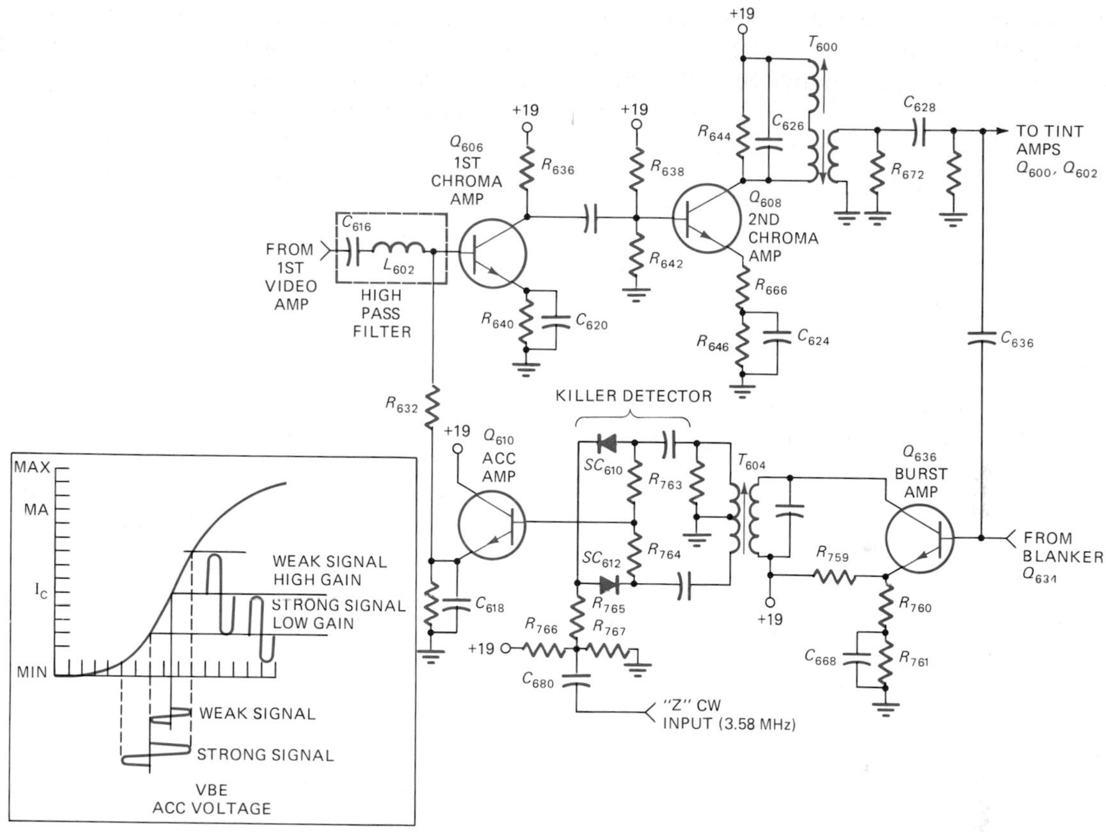

Figure 26.12 The automatic color-control circuit. *(Courtesy of GTE-Sylvania, Inc.)*

controlled by SC_{610}, SC_{612} (killer detector), and Q_{610} (ACC amplifier). The killer detector diodes, transformer T_{604}, and associated components compare the burst amplitude of the Z CW signal (3.58-MHz phase used at the Z demodulator). When no burst is present, SC_{610} and SC_{612} conduct equally and no difference voltage is developed at the R_{763}–R_{764} junction. The voltage at this junction is the sum of two voltages:

1. Voltage developed across R_{767}
2. Voltage developed due to the conduction of SC_{610}.

In the presence of burst, if the oscillator is locked, then SC_{612} does not conduct, but SC_{610} does conduct and develops a negative voltage.

Therefore, voltage at the base of Q_{610} decreases in the presence of burst. The higher the burst amplitude, the smaller the DC voltage at the base of Q_{610}. This voltage, through emitter follower Q_{610}, is applied to bias Q_{606} (first color amplifier). Reduction in this voltage means reduction in the emitter current of Q_{606}, thus reducing the beta, which decreases the gain of this stage. This way, the burst amplitude and the color amplitude are regulated.

As shown in Figure 23.5, the Q_{610} emitter voltage is also applied to the base of Q_{612} (color killer). During the presence of burst, the less-positive voltage at the emitter of

Q_{610} (ACC amplifier) acts as a turn off bias for Q_{612}. However, during the absence of burst, Q_{610} conducts harder and therefore biases Q_{612}.

X AND Z DEMODULATORS

In order to demodulate the color sidebands, the X and Z demodulators provide synchronous detection of these signals with the 3.58-MHz reference oscillator injection voltage, which is fed into the emitters of Q_{616} and Q_{618}. The phase of the 3.58-MHz signal applied to the Z demodulator is shifted approximately 90° by coil L_{608} and capacitor C_{650}. The actual shift is selected to provide the most accurate color presentation. The 3.58-MHz signal injection into the emitter of Q_{616} and Q_{618} causes pulses at a 3.58-MHz rate in their collectors.

When the color signal is applied to the bases of the demodulators, the phase and amplitude of the color signal determine the conduction levels and collector voltage changes in each demodulator. When the two signals, color and 3.58-MHz reference, are in phase across the emitter-to-base junction, the transistor conduction is lowered, raising the collector voltage toward B+. If the color signal and the 3.58-MHz reference signal are out of phase, heavy conduction occurs and this produces low collector voltage. If the two signals are 90° out of phase, the collector voltage is held at the level developed by the 3.58-MHz reference signal. These demodulator variations are direct-coupled to the bases of Q_{624} (R–Y driver) and Q_{626} (B–Y driver).

DRIVERS AND COLOR OUTPUT

Figure 26.13 shows the color drive and color output circuitry. The emitter of Q_{214} (video driver) provides a + Y signal to the emitter circuit of Q_{628}, Q_{630}, and Q_{632} (red, green, and blue output amplifiers). By using emitter drive to Q_{628}, Q_{630}, and Q_{632}, there is no signal inversion. At this point, if the color signal is absent, the + Y signal is fed to the picture tube cathode. However, when the color signal is present, color-difference signals are developed by the X demodulator and Z demodulator and applied to Q_{624} (R–Y driver) and Q_{626} (B–Y driver), respectively. Transistors Q_{624} and Q_{626} are emitter-follower configurations feeding Q_{628} (red output amplifier) and Q_{632} (blue output amplifier), their respective signals.

The G–Y signal is formed from a portion of the −R–Y and −B–Y signals. The signal across emitter resistors R_{698} and R_{702} are coupled through attenuating resistors R_{696} and R_{700} and matrixed across R_{692} and R_{694}. The resistance of R_{696} is approximately one-third of the resistance of R_{700}, and so R_{700} feeds the lesser signal to the matrix amplifier. Since R_{696} attenuates a −B–Y signal and R_{700} attenuates a −R–Y signal, the resultant voltage mix at the input of Q_{620} (G–Y matrix amplifier) is a + G–Y signal. Signal inversion occurs in Q_{620}. This results in a −G–Y signal at the input of Q_{622} (G–Y driver). Transistor Q_{622} is an emitter-follower configuration, signal polarity is maintained, and the −G–Y signal is fed to the base of Q_{630} (green output amplifier). This color-difference signal and the monochrome signal matrix across the base-emitter junction of Q_{620}.

When the +Y signal drives the Q_{630} emitter, Q_{630} conduction is controlled as though a −Y signal were being fed to its base. Therefore, this +Y signal is not inverted. When the negative G–Y signal at the base of Q_{630} is amplified, the signal is inverted to a +G–Y signal. Thus, the −Y in the +G–Y signal is cancelled by the +Y signal, and the resulting signal is the individual green signal drive to the picture tube. Identical matrix action takes place with the R and B output amplifiers.

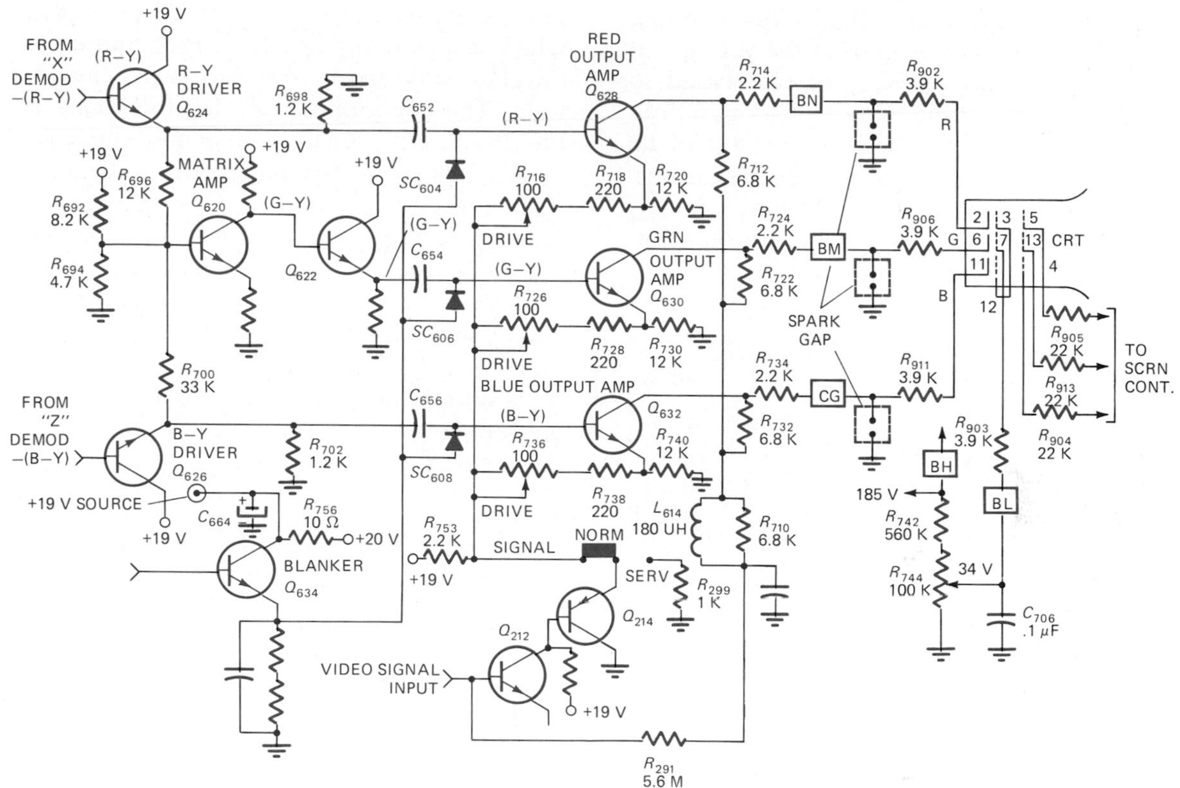

Figure 26.13 The color drive circuitry. *(Courtesy of GTE-Sylvania, Inc.)*

BLANKER CIRCUIT AND BURST

As shown in the schematic insert after pg. 804, the base of Q_{634} (blanker) receives positive pulses from a voltage divider network across Q_{408}. Operating as an emitter follower, it provides positive pulses to Q_{636} (the burst amplifier), to diodes SC_{604}, SC_{606}, and SC_{608} to establish the DC levels in Q_{628}, Q_{630}, Q_{632}, and to Q_{614} as a blanking pulse to remove the burst signal.

BRIGHTNESS LIMITER CIRCUIT

The brightness limiter circuit (see schematic) is a closed-loop regulator circuit controlling the picture tube beam current. Its purpose is to protect the picture tube by restricting its emission level, thereby increasing the picture tube life. The limiter circuit controls the forward bias of Q_{212} (video amplifier) and hence the DC level through Q_{214} (video driver) and through Q_{628}, Q_{630}, and Q_{632} (color output amplifiers) to the picture tube cathodes. Horizontal pulses are injected into the junction of SC_{418} and R_{468}. The peak-to-peak pulse across R_{468} (sampling resistor) is directly proportional to beam current. Diode SC_{418} (brightness limiter rectifier) rectifies the negative position of the pulse and develops a negative voltage across the filter network C_{436}, R_{466}. Diode SC_{417} (brightness limiter gate) is reverse-biased until the beam current reaches approximately 700 mA, which corresponds to approximately -11 V at the anode of SC_{417}, set by divider R_{462},

R_{464} off the regulated −20 V. The sampling resistor R_{468} is chosen so that the voltage on the cathode side of SC_{417} has increased to −11.5 V at this time. Diode SC_{417} goes into conduction, and the negative voltage at its anode starts to increase with beam current, thereby reducing the conduction of Q_{212} (video amplifier). From here on, the controlling action becomes continuous.

THE 3.58-MHz OSCILLATOR

The 3.58-MHz oscillator shown in Figure 26.14 is a modified Clapp configuration using 3.58-MHz crystal series resonant operation. This provides a low-impedance feedback path from collector to base and mechanically controls the oscillator by its vibrating period. In addition, the capacitors C_{694} and C_{696} form part of the feedback circuit to sustain oscillation. As long as the feedback loop gains are unity or better, the oscillator sustains operation.

Transistor Q_{640}, a Darlington amplifier, actively controls the varactor SC_{618} capacity through its forward bias and DC correction voltage from the phase comparator circuit. The oscillator load is a resonant tank circuit (L_{618} and C_{700}) tuned to 3.58 MHz. The output is capacity-coupled through C_{698} to Q_{644}, which is a 3.58-MHz buffer stage required to develop additional 3.58-MHz drive amplitude. The collector supply for this stage is the 180-V bus. Resistor R_{791} and capacitor C_{708} make up a filter that decouples the 3.58-MHz energy from the bus. Capacitor C_{704} is an additional bypass capacitor placed across C_{708} (a 30-μF 150-V electrolytic). Its purpose is to neutralize any capacitor lead inductance and assure complete 3.58-MHz bypassing. Transformer T_{606}, the 3.58-MHz output transformer, provides resonant gain to the output signals and is the coupling means to the X and Z demodulators.

THE 3.58-MHz OSCILLATOR PHASE CONTROL CIRCUIT

Phase control of the 3.58-MHz reference oscillator is accomplished by comparing the oscillator signal with the burst at the APC (automatic phase control) detector. Figure 26.15 shows the circuit. The voltage at the junction of R_{768} and R_{769} consists of two voltages:

1. The DC voltage provided by phase control amplifier Q_{638}.

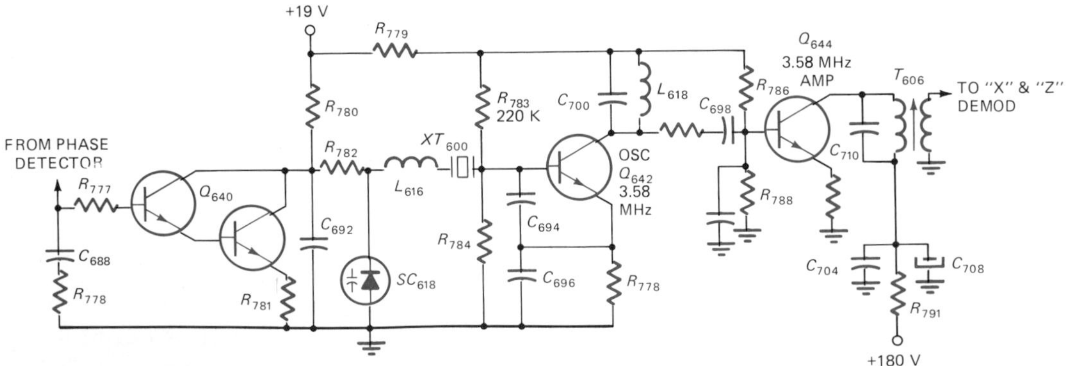

Figure 26.14 The 3.58-MHz oscillator. *(Courtesy of GTE-Sylvania, Inc.)*

2. The correction voltage provided by the APC detector, derived by comparing the phase and frequency of the 3.58-MHz oscillator to the incoming burst. With no burst or with the oscillator running in phase with burst, this voltage is zero.

The first voltage (which can be varied by R_{776}) is to bias APC amplifier Q_{640} so that with no burst (i.e., zero correction voltage), the oscillator free-running frequency is close to the burst frequency. The second voltage (i.e., the correction voltage), adds to or subtracts from the first voltage, depending on whether the oscillator is leading or lagging the burst phase, and thus decreases or increases the collector voltage of Q_{640}. The change in collector voltage of Q_{640} changes the capacity of the varactor diode SC_{618}, and this in turn changes the series resonance of the 3.58-MHz crystal, pulling the frequency and phase of the oscillator toward the burst frequency and phase.

Transistor Q_{638} stabilizes the collector voltage of Q_{640} against any temperature-caused drift.

HORIZONTAL BLANKER AMPLIFIER

The horizontal blanker amplifier circuit, illustrated in Figure 26.16, operates as an emitter follower providing positive pulses to the color amplifier, burst amplifier, color video amplifiers, and video amplifier.

The divider network consisting of R_{460}, R_{458}, and R_{456} steps down the flyback pulse to about 25 V at the base of the blanker. It is clamped to +19 V and shaped by the load consisting of R_{754}, R_{755}, and C_{662}. Inductor L_{604} and capacitor C_{642} resonate at 3.58 MHz and prevent any color signal from getting into the blanker pulse. Resistors R_{757} and R_{756} are dividers for burst suppression and gating.

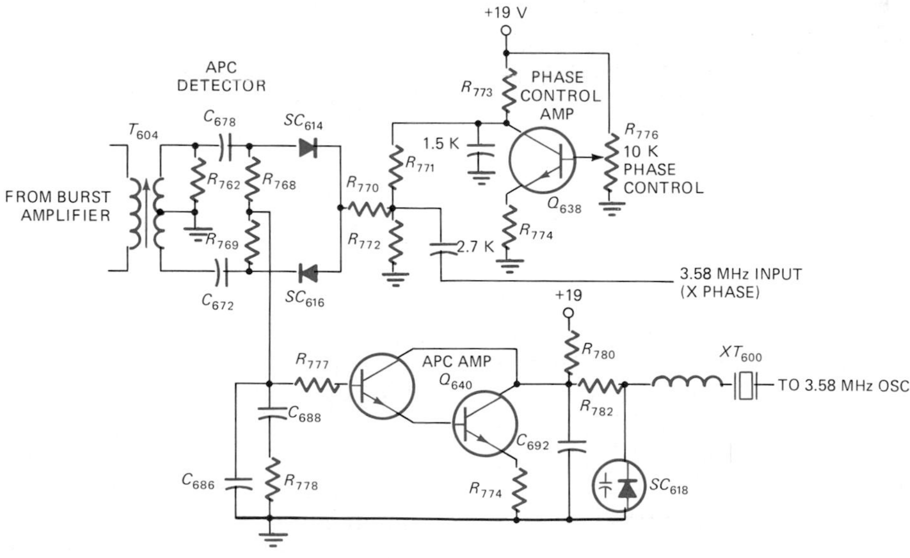

Figure 26.15 The automatic phase control circuit. *(Courtesy of GTE-Sylvania, Inc.)*

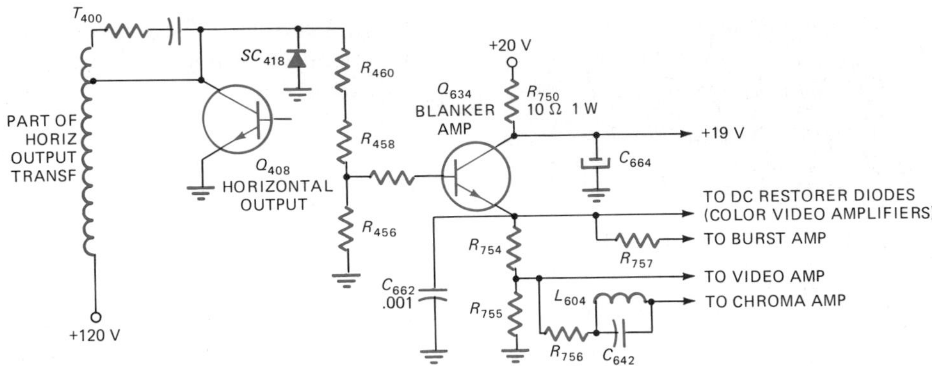

Figure 26.16 The horizontal blanker circuit. *(Courtesy of GTE-Sylvania, Inc.)*

26.3 EQUIPMENT FOR TV RECEIVER ALIGNMENT

To align a monochrome TV receiver, the following test equipment is required:

1. An oscilloscope
2. An RF frequency sweep generator or, alternatively, a bar sweep generator, such as the Sencore VA48
3. An accurately calibrated marker signal generator (frequently incorporated in the RF sweep generator). This can be a calibrated RF signal generator
4. An RF signal generator
5. A high-impedance voltmeter, such as a FET voltmeter or a digital multimeter

To align a color TV receiver, all this equipment is needed plus a color bar generator that furnishes dot and cross-hatch patterns. These functions are incorporated into the Sencore VA48 generator.

26.4 THE OSCILLOSCOPE

An oscilloscope is one of the most important test instruments used to align or troubleshoot a monochrome or color TV receiver. Some of its common uses include:

1. Observation of tuner and video IF bandpass characteristics.
2. Observation of detected video signal.
3. Observation of sync and blanking pulses.
4. Observation of the color burst signal on the back porch of the horizontal blanking pulses.
5. Observation of the color amplifier bandpass.
6. Use as a "vector scope" to study phase relationships of the color video signal.
7. Use as a DC voltmeter and a peak-to-peak AC voltmeter when measuring waveforms.

TYPICAL TV SERVICE-TYPE OSCILLOSCOPE

An oscilloscope suitable for observing and servicing monochrome or color TV equipment is shown in Figure 26.17. This is a dual-trace oscilloscope. This means that two separate waveforms can be viewed simultaneously with respect to their timing

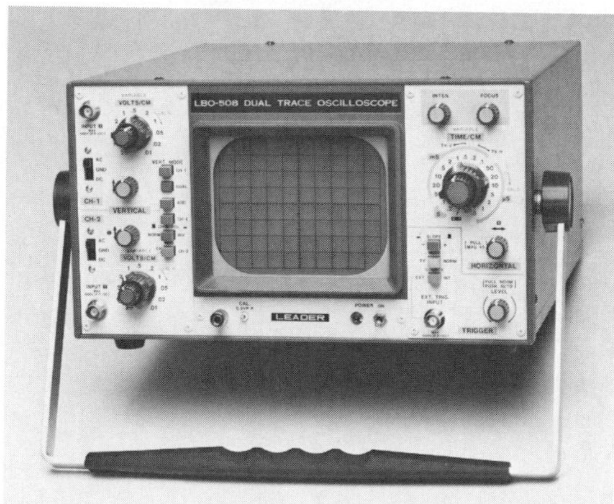

Figure 26.17 A dual-trace oscilloscope, suitable for use with monochrome or color TV equipment. A 5-in (12.7-cm) CRT is used. *(Courtesy of Leader Instruments Corp.)*

and relative amplitudes. This feature is frequently useful when checking television equipment. A few such uses are as follows:

1. Simultaneous observation of the flyback pulse and the color burst signal. These must be coincident for proper burst amplifier operation.
2. Simultaneous observation of the input and output of a circuit. This shows performance of the circuit.
3. Comparison of vertical output signal and vertical yoke current for amplitude and linearity checks.
4. Comparison of input and output waveforms of integrators and/or differentiators.

Synchronization The oscilloscope shown in Figure 26.17 features a choice of either *internal* sync or triggered sweep. With internal sync, the sweep is free-running until locked in by a portion of the wave being viewed. With triggered sweep, there is no sweep until the wave to be viewed is present. At that time, a portion of the wave (usually the leading edge) acts as a trigger to initiate the time base (sweep line). One advantage of triggered sweep is that there is no overlapping display. Only clean, whole waveforms are seen. Another advantage is the ability to magnify the sweep length (in this case, it has been magnified five times) to better examine portions of the wave. The color burst signal is a good example of a signal to be expanded and examined. With triggered sweep, the length of the time base is determined by the setting of the switch in Figure 26.17 marked "variable Time/CM."

Some Specifications Some important specifications of the oscilloscope illustrated are as follows:

1. Internal square-wave calibration. This enables vertical sensitivity calibration in V/cm. Thus, wave amplitudes can be read directly on the scope reticle.
2. Choice of single-channel or dual-channel display.
3. Front panel X–Y (horizontal-vertical) connections. This simplifies its use for alignment, phase shift analysis, and vectorscope (described below) service.
4. Vertical amplifier sensitivity, 10 mV/cm to 20 V/cm, in 11 steps.
5. Vertical amplifier bandwidth, 2 Hz to 20 MHz.
6. Vertical amplifier input impedance, 1 megohm, shunted by 35 pF.
7. Horizontal sweep speed, 0.5 μs/cm to 200 ms/cm, in 18 steps.

Accessories One important oscilloscope accessory is a **direct** and **low-capacitance probe** (Figure 26.18A). This is a dual-function, switch-operated probe. It can be used either as a direct probe or as a low-capacitance probe. As a direct probe, it does not isolate the circuit being tested. It is merely an extension of the oscilloscope connector.

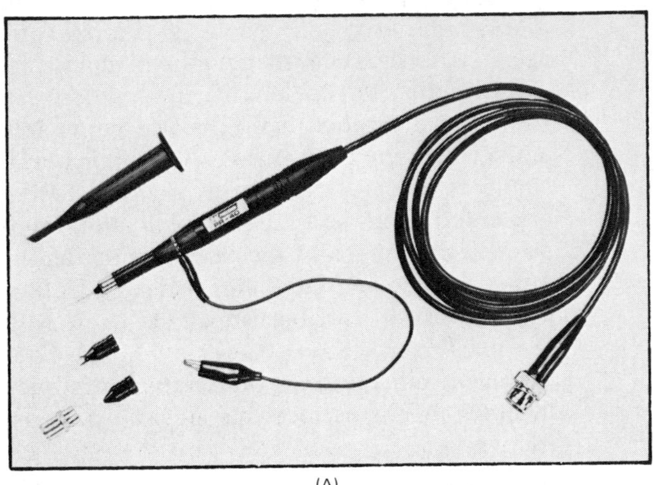
(A)

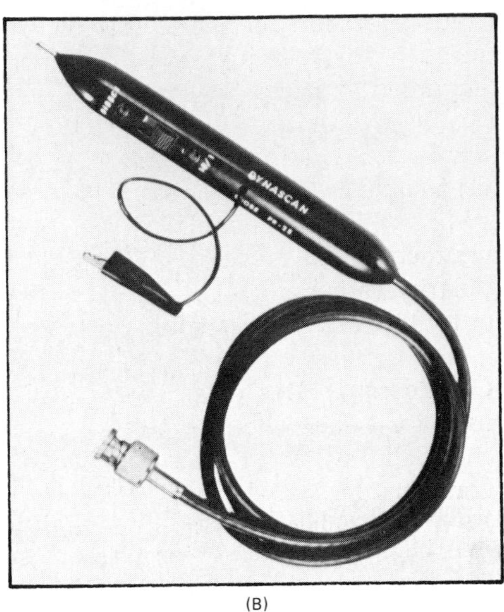

(B)

Figure 26.18 Examples of oscilloscope probes. **A.** Direct and low-capacitance probe. **B.** Demodulator probe. *(Courtesy of Dynascan Corp.)*

As such, since it is on a shielded cable, it represents an added capacitance of about 110 pF. Consequently, it can be used to measure circuits only where the added capacitance will not disturb circuit performance.

As a low-capacitance probe, it offers an isolation resistance of 9 megohms. When connected to the usual oscilloscope input of 1 megohm, it attenuates the measured signal to one tenth of the true value. Thus, when signal amplitude is being measured on the oscilloscope, the indicated amplitude must be multiplied by 10.

The probe pictured in Figure 26.18A has a frequency response up to 15 MHz. The maximum input voltage this probe can handle is 500 V, and it can be used with oscilloscopes having an input capacitance ranging from 15 to 120 pF. An adjustable compensation capacitor that shunts the 9-megohm resistor is included to match the probe to the oscilloscope. The compensation capacitor (2 to 25 pF) is adjusted while a known, accurate square wave is being watched. Proper adjustment is attained when the wave is perfectly square. The input circuit is then substantially independent of frequency (within specifications). The square wave may be the oscilloscope calibration wave.

Other low-capacity probes are useful at frequencies of up to 100 MHz or higher.

A low-capacitance probe must be used when measuring circuits dealing with high RF frequencies or nonsinusoidal wave forms. This minimizes loading the circuit under test and/or distorting the viewed wave form. A low-capacitance probe is used to measure (1) composite video, (2) color burst, (3) color video, and (4) deflection circuit wave forms.

A **demodulator probe** (Figure 26.18B) contains a diode detector. It can be used in any circuit amplifying an AM RF or AM IF signal. When so used, the modulation envelope is seen on the oscilloscope. Thus it can trace a signal through RF or IF circuits. This signal can be either from a television station or from a television-signal generator. In color television receivers, this probe is also useful in observing the performance or alignment of the color circuits.

The demodulator probe pictured in Figure 26.18B has the following characteristics: (1) frequency response up to 250 MHz (− 6 dB), (2) input impedance 30 000 ohms shunted by 45 pF, and (3) negative output polarity (other probes may have a positive output polarity).

Oscilloscope Requirements Oscilloscopes used for sweep alignment of IF stages or to check the sweep response of video amplifiers do not require a wide-band response for their vertical amplifiers. In fact, for such uses, the vertical amplifier response does not have to exceed 10 kHz. The reason for this is that we are not viewing such frequencies as 45 MHz, or even 3 MHz, directly on the scope screen. It is frequently not realized by the technician that, for the above-mentioned uses, the scope is required to reproduce not these high frequencies but rather the very low frequencies contained in the pattern repetition rate (normally 60 or 120 Hz) plus the harmonics of this fundamental rate required to reproduce the outline of the response curve. Such harmonics rarely if ever exceed about 10 kHz. This means that, particularly when servicing monochrome receivers, a relatively inexpensive oscilloscope will suffice.

Need for WideBand Oscilloscope A wide-band oscilloscope, such as one that accepts signals extending to 4 or 5 MHz, is needed only for certain special measurements. One such situation is when it is desirable to directly view the 3.58-MHz burst signal on the back porch of the horizontal blanking pulse. Another time a wide-band oscilloscope is needed is when it is desirable to examine the exact shape of the horizontal and vertical sync and blanking pulses. In this case, an oscilloscope response of somewhat less than 1 MHz will normally be adequate. The technician should carefully consider specific needs before investing an excessive amount of money in an ultrasophisticated oscilloscope that may not really be needed.

The Vectorscope The term *vectorscope* may sound rather mysterious and complex, but actually any scope can be a vectorscope and there is no need to purchase a special scope for this purpose. A vectorscope is sometimes used to view and help align the overall color circuits. It is also of considerable aid in troubleshooting some color circuits. In order to use any scope as a vectorscope, the only thing necessary is access to one vertical oscilloscope CRT deflection plate and one horizontal deflection plate. It is also necessary to be able to disable the internal sweep. The two CRT plates mentioned previously must be disconnected from any scope circuits, and there should be no input signal into the scope vertical amplifier. Now the R–Y signal from the television receiver is connected directly to the top CRT vertical plate and the B–Y signal is connected directly to the CRT horizontal right plate. The television chassis is also connected to the scope ground (Figure 26.19). When the oscilloscope and television receiver are energized, a vector pattern appears on the oscilloscope if a color bar generator is connected to the television set. Such a color bar generator (described later) is connected directly to the antenna terminals since its output is a modulated RF that can be selected to the frequencies of Channels 2 through 6.

Vectorscope Pattern Figure 26.20A shows a typical vectorscope pattern with the various phases marked for identification. The proper functioning of the color demodulators, as well as certain other color circuits, can be quickly checked with the aid of the vectorscope pattern.

An oscillogram of two lines of the rain-

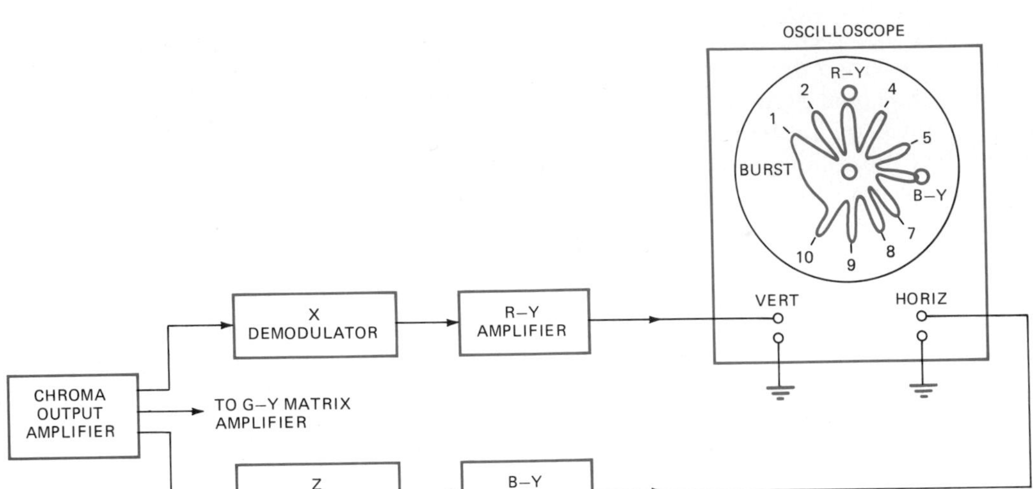

Figure 26.19 The connections for a vectorscope display. (See also Figure 26.20 for clearer display and color identification.)

bow color pattern, such as that displayed on the vectorscope in Figure 26.20A, is shown in Figure 26.20B. The colors in Part B of the figure can be correlated with those in Part A because both indicate ten colors in the same order, going clockwise in Part A.

Some manufacturers of test equipment have designed oscilloscopes that conveniently perform the dual functions of a normal oscilloscope and a vectorscope by means of simple switching operations. One such oscilloscope is illustrated in Figure 26.17.

26.5 SWEEP ALIGNMENT GENERATORS

Wideband tuned amplifiers are used in television receivers. The alignment or testing of these circuits requires the use of a signal generator, which makes it possible to view the response of the amplifier(s) on an oscilloscope. Either a frequency sweep generator or the Sencore bar-sweep generator may be used for this purpose. Both types and their uses are discussed here. Alignment procedures are given at the end of the chapter.

Frequency Sweep Generator A frequency sweep generator is one that periodically (60- or 120-Hz rate) sweeps across the desired range of frequencies. When using this generator, an oscilloscope pattern shows the actual amplifier(s) response curve. A frequency sweep generator of this type is illustrated in Figure 26.21.

The range of frequencies, markers, and added functions depends upon the particular instrument. The instrument illustrated in Figure 26.21 has the following functions and characteristics:

1. Video frequency sweep: 0 to 6 MHz
2. IF sweep: 35 to 50 MHz
3. Unmodulated and modulated crystal-controlled markers
4. Channel 4 and Channel 10 RF sweep for tuner alignment
5. 10.7-MHz sweep for FM receiver sound-IF alignment
6. RF/IF sweep widths: adjustable from 1 MHZ to greater than 10 MHz; for 10.7 MHz, from 100 kHz to 2 MHz
7. Crystal-controlled IF markers: 39.75 MHz (adjacent-channel picture trap),

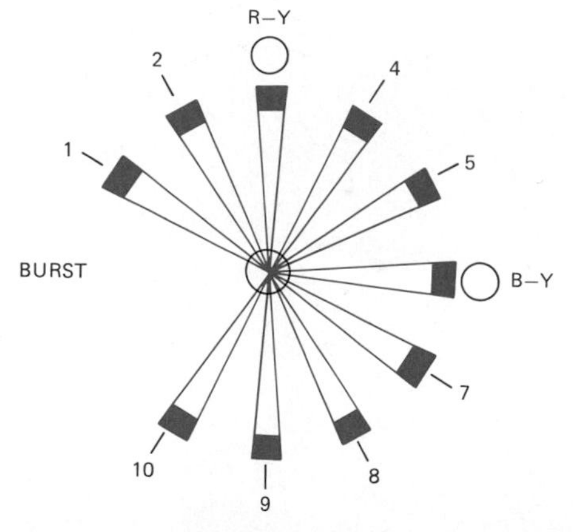

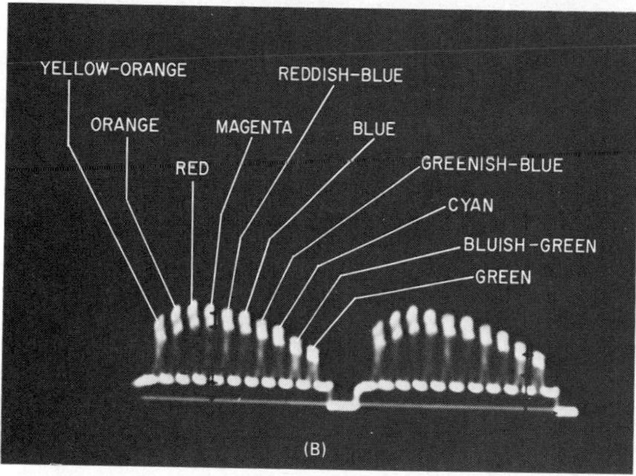

Figure 26.20A. An ideal vectorscope display. This display permits the phase relationship of the color signals to be studied. **B.** Two lines of rainbow color bars as seen on a conventional oscilloscope.

41.25 MHz (sound trap), 41.67 MHz (IF color sideband limit), 42.17 MHz (color-IF subcarrier), 42.67 MHz (IF color sideband limit), 42.75 MHz (IF tuner link), 44.00 MHz (IF center frequency), 45.00 MHz (IF response point), 45.75 MHz (picture carrier), 47.25 MHz (adjacent-channel sound trap)

8. 100 kHz continuous string of markers, used during FM or AFT alignment
9. All video IF and color markers available as RF equivalents on Channels 4 and 10
10. Three low-impedance bias power supplies included (have reversible polarity)

The Sencore Bar-Sweep Generator This generator is contained in a unit known as a video analyzer. It can perform a number of functions other than video and color-IF alignment and testing. This unit also can be used as a color TV generator. As such, it can provide color bars, dot patterns, and a crosshatch pattern. These functions are described below. (Some of the additional functions are discussed in Chapter 27.)

The Sencore video analyzer is illustrated in Figure 26.22. Its principle of operation is different from that of a frequency sweep generator. The bar-sweep system produces the standard video test patterns used by cable television companies. These patterns can be seen in the upper right section of Figure 26.22. As can be seen, the video test patterns modulate specific frequencies. These produce the *bars* on a picture tube. After detection of the video IF, or at the output of the color bandpass amplifier, the video patterns can be seen on an oscilloscope.

Note that in Figure 26.22, there are five bars corresponding to five IF alignment frequencies. Also note that there are three bars corresponding to three color alignment frequencies. When aligning either section, it is merely necessary to adjust the tuned circuits (excluding traps[1]) for one of two indications:

1. Equal amplitude of detected bar patterns on an oscilloscope
2. Equal white intensity of all bars, as seen on the picture tube screen

---

[1]Traps are separately aligned by individual crystal-controlled frequencies.

Figure 26.21 A TV frequency-sweep generator used for monochrome and color TV receivers. All required marker frequencies can be selected by individual switches. Lights on the front panel response curves indicate which markers are active. Post-injection markers are utilized. *(Courtesy of B & K-Precision, Div. Dynascan Corp.)*

We will discuss this in more detail later.

Sencore Video Analyzer Functions and Characteristics Some pertinent functions and characteristics of the Sencore video analyzer are:

1. Can be used for signal injection or signal tracing.
2. Built-in meter can be used for signal tracing.
3. Standard drive signals available for injection into any stage of television receiver.
4. Provides standard color bar patterns, including dot and cross-hatch patterns.
5. Provides *bar-sweep* patterns to align video-IF or color bandpass amplifiers.
6. Any available pattern can be injected into RF, video-IF, or audio-IF stages for troubleshooting or alignment.
7. VTR standard output for troubleshooting or aligning video tape-recorders or other video systems.
8. Yoke or flyback transformer *ringing* test.
9. Video-IF bar sweep: ten vertical bars. Six bars are RF-modulated. The remaining four bars (three on left side, one on right side) are used for black and white tracking. These latter bars are black, gray, white, and black (Figure 26.22).
10. Color bar sweep: three vertical bars

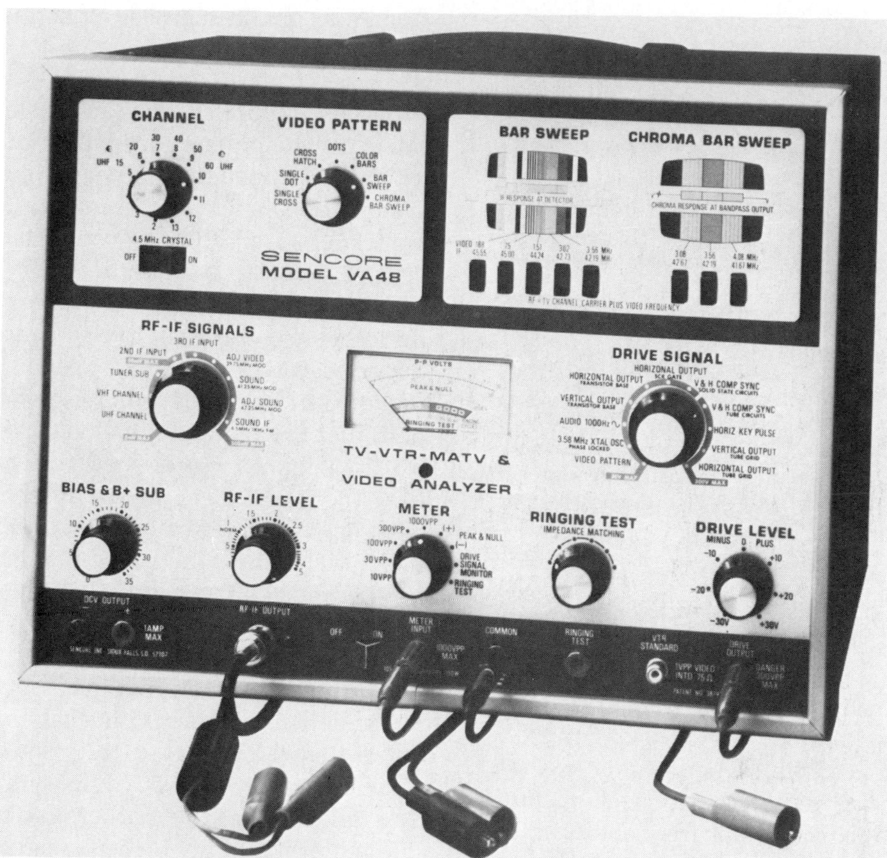

Figure 26.22 The Sencore video analyzer. This is a multipurpose instrument. *(Courtesy of Sencore, Inc.)*

for color bandpass alignment and color demodulator testing (Figure 26.22).

11. Provides crystal-controlled trap frequencies.
12. Adjustable power supply. Can be used as a bias supply or for B+ substitution. The plus or minus terminals can be chosen as the common side of the supply.

26.6 RF SIGNAL GENERATORS

When used in conjunction with a frequency sweep generator that does not furnish marker signals, an RF signal generator can supply marker signals (see below). In addition, it can be used to peak single-tuned IF circuits. Such circuits are used in the television receiver analyzed earlier in this chapter. A typical RF signal generator is pictured in Figure 26.23.

Characteristics The instrument pictured features all solid-state circuitry. It has a frequency range of 100 kHz to 100 MHz on fundamentals and up to 300 MHz on harmonics. It can be internally modulated at 1 000 Hz or externally modulated from 50 Hz to 20 kHz. For single-frequency crystal oscillator output, a crystal can be plugged into the

Figure 26.23 An RF signal generator, useful for peaking single-tuned circuits and as a marker generator. It has a range of 100 kHz to 300 MHz. *(Courtesy of Leader Instruments Corp.)*

front panel. Crystals from 1 to 15 MHz can be used.

Checking Calibration The frequency dial is accurately calibrated. However, it may be desirable (if possible), when highly accurate frequencies are required, to check the dial calibration against crystal frequencies. Such crystal frequencies should be fairly close to the desired RF signal generator frequencies. Calibration can be checked by feeding both signal sources to an oscilloscope through a diode detector. The RF signal generator is then adjusted for zero beat indication on the oscilloscope.

26.7 MARKER SIGNALS

The circuitry to produce marker signals may be incorporated in a sweep generator or supplied by an RF signal generator. It is capable of providing a single accurately calibrated signal frequency. The purpose of a marker signal is to indicate the frequency at various points in a response curve as observed on an oscilloscope screen. This aids in adjusting the tuning slugs in the resonant circuits to obtain the desired bandpass characteristics.

Video-IF Response As an example, consider the response curve of Figure 26.24, which is the response curve for the video-IF system of a television receiver. This curve would be observed if we connected an oscilloscope across the video detector load resistor. What we desire to do, once we obtain the response curve, is to determine the frequencies at various points to ensure that the curve rises and falls where it should. It is here that the marker signal is required.

If a manufacturer states in the service data that the video carrier IF is 45.75 MHz and the other end of the response occurs at 41.75 MHz, then the curve obtained should be checked for the position of these two frequencies. To obtain marker points on the oscilloscope screen, two methods are generally used. In the simplest method, the sweep signal generator contains internal crystal oscillators that superimpose their signals on the IF being swept (40 to 50 MHz) (Figure 26.21). The indication of the marker point in the visible pattern is either a slight wiggle or else a dip in the curve at this point (Figure 26.25). Note that although two marker points are indicated in Figure 26.25 when using an RF signal generator, only one is seen at a time. First the marker oscillator is set to 45.75 MHz and its position noted on the curve, and then it is set to 41.75 MHz and its posi-

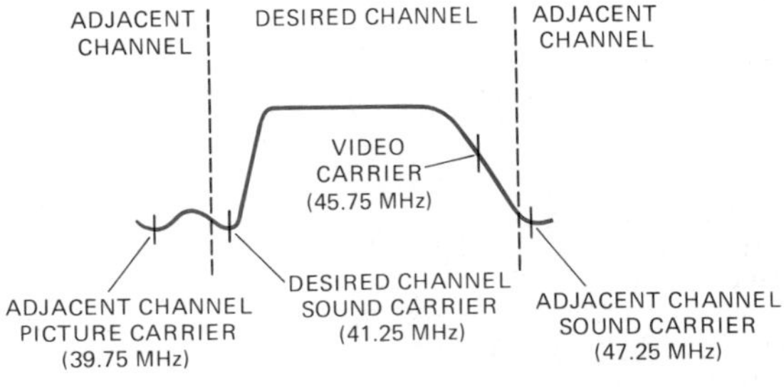

Figure 26.24 The overall video-IF response curve of a television receiver.

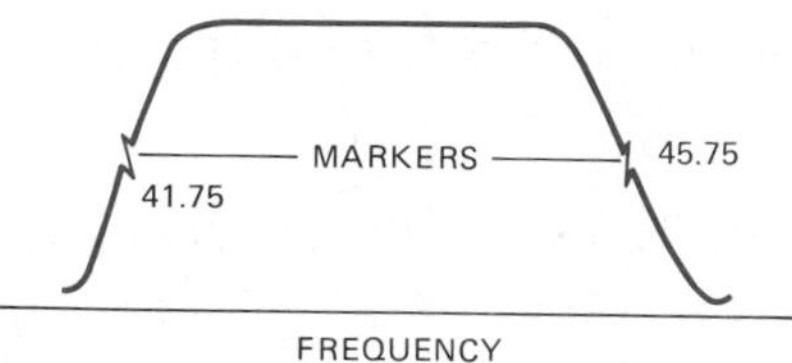

Figure 26.25 The use of marker points for indicating a frequency on a response curve.

tion checked again. In the generator shown in Figure 26.21, all markers may be displayed simultaneously if desired.

The mentioned video-IF band limits, 41.75 MHz and 45.75 MHz, represent the entire 4 MHz that can be used to transmit the details of the televised scene. Some portable receivers are designed to pass only 3 MHz in the IF amplifiers. In these instances, reference to the manufacturer's instructions will quickly indicate the band limits, and the marker frequencies can be changed accordingly.

Connecting the RF Signal Generators If the sweep generator does not contain internal circuitry for supplying the marker points, these can be obtained by the following method. Take an RF signal generator and place its output leads in parallel with those of the sweep generator, using a 50-pF isolating capacitor[1] in the signal lead of the marker generator. Set the frequency of this second generator accurately to one of the frequencies that is to be checked on the response curve, say, 45.75 MHz. With the equipment turned on, a wiggle (or dip) will appear on the overall response curve at 45.75 MHz. Note whether the response at this point is that indicated in the service manual. Now change the marker frequency to 41.75 MHz and note where this appears on the response curve. In Figure 26.24, 45.75 MHz represents the video-IF carrier and 41.25 MHz the sound-IF carrier. Other frequencies that should be checked are the trap frequencies (Figure 26.24). These include 39.75 MHz, the adjacent channel picture carrier, and 47.25 MHz, the lower adjacent-channel sound carrier. At each of these points, the response should be very low. It is advisable not to turn the amplitude of the marker signal generator too high, but to keep it as low as possible to still obtain a marker line.

Post Injection Markers The method of marking the response curve just described when using the RF signal generator has the disadvantage that the marker signal must be fed through the receiver circuitry along with the sweep signal. This same disadvantage occurs when using the sweep generator with a built-in marker system. The marker signals passing through the tuned circuits of the receiver tend to distort the response, and this often results in an unrealistic display of the overall receiver response curve on the oscilloscope. To avoid this, a system of adding markers that will not distort the response curve has been devised. This system is called the post injection marker system. (The sweep generator illustrated in Figure 26.21 uses this type of marker system.)

The post injection marker system is illustrated in the block diagram of Figure 26.26. The RF output of the sweep generator is fed to the receiver IF input and also to the input of the marker adder. The RF marker(s) (at the selected frequency or frequencies) is also fed to the input of the marker adder. In the adder, these two RF frequencies are heterodyned to produce a marker at an audio beat frequency. Since the sweep generator produces the horizontal oscilloscope sweep

[1]It may be found that attaching the marker generator directly to the sweep generator output cable causes the response curve to alter its shape. If this occurs, try inserting an isolating resistor of 10 000 ohms in the output signal cable of the marker generator, together with the 50-pF capacitor. The value of the capacitor is not critical, and values between 20 and 200 pF have been suggested by various manufacturers.

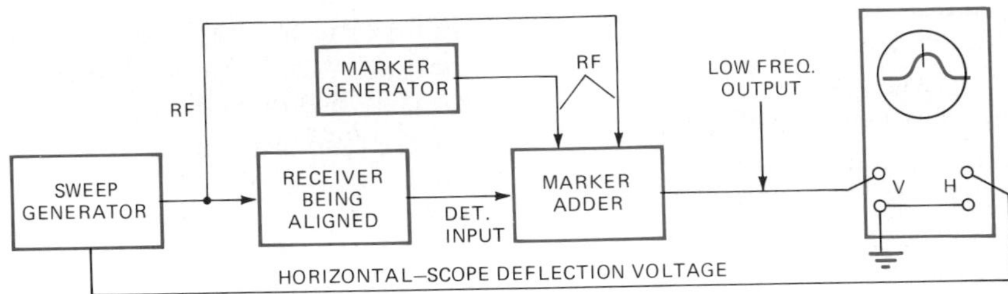

Figure 26.26 Test setup for using the post-injection marker adder. Although shown as separate blocks here, the marker circuits may be incorporated in a sweep-generator instrument.

synchronized with the RF sweep, markers appear on the oscilloscope base line even if the receiver is not connected. Note that the markers do not pass through the receiver circuits and so cannot distort the response curve no matter how great their amplitude may be. In a sweep generator, such as shown in Figure 26.21, markers may be displayed either singly or simultaneously.

Note in Figure 26.26 that the output of the receiver is fed to a detector input of the marker adder. Thus the response curve of any part of the RF or IF system can be examined directly and displayed on the oscilloscope. The detected response curve (in the marker adder) is added to the marker signals, and both are then displayed on the oscilloscope simultaneously when the receiver is connected and operating.

Vertical or Horizontal Markers If a vertical marker is displayed on the edge of a response curve slope, as shown in Figure 26.25, it may be difficult to determine exactly where the marker appears. For this reason, some sweep generators that incorporate the above marker system have a provision for displaying the markers either horizontally or vertically.

26.8 COLOR-PATTERN GENERATORS

The functions of a color-pattern generator, also known as a color bar generator, may be combined into a composite instrument such as the Sencore video analyzer pictured in Figure 26.22. It can also be a separate unit, as shown in Figure 26.27.

The color-pattern generator is an extremely useful instrument in testing or troubleshooting color television equipment. Some of its possible uses are to:

1. Check picture tube purity (using blank raster)
2. Check, align, or troubleshoot color circuits and adjust automatic frequency and phase (AFPC) circuits (using color bars)

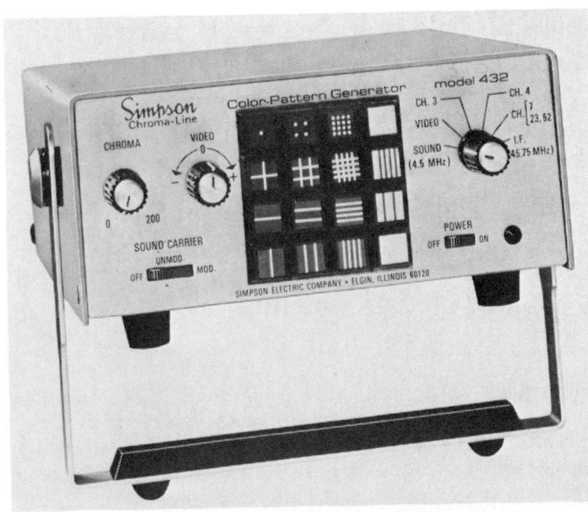

Figure 26.27 A color-pattern generator. The desired pattern is selected by any one of the 16 push-button switches. *(Courtesy of Simpson Electric Company.)*

3. Check or adjust static convergence (using dot patterns)
4. Check or adjust dynamic convergence (using cross-hatch patterns)
5. Check or adjust scan linearity (using cross-hatch patterns or individual horizontal or vertical lines)
6. Check or adjust pincushion correction (using cross-hatch pattern or individual horizontal and vertical lines)
7. Centering adjustment (cross-pattern)

Note: Delta-gun color picture tube adjustment procedures are given at the end of the chapter. Troubles in color circuits are illustrated in Chapter 27, with full color photos.

Outputs of Simpson Color-Pattern Generator For the instrument pictured in Figure 26.27, the selected pattern can modulate:

1. VHF Channel 3, 4, or 7 (switchable)
2. UHF Channels 23 and 52 (overtones of Channel 7)
3. The video-IF carrier at 45.75 MHz

Thus, a selected pattern can be fed either through the VHF or UHF antenna terminals or through the video-IF amplifier.

Other outputs of the Simpson color-pattern generator are:

1. Patterns at video frequency
2. 4.5-MHz modulated sound carrier for troubleshooting sound-IF stages

Additional Features Other features of the Simpson color-pattern generator are:

1. Adjustable RF/IF level (rear panel) and video level (front panel)
2. Horizontal or vertical oscilloscope trigger pulses (rear panel)
3. Gun killer switches, used to cut off red, green, or blue gun for purity adjustment or check
4. Color level control (front panel)
5. Nominal 75-ohm signal output through coaxial cable (balun transformer from 75 to 300 ohms can be used for receivers with 300-ohm antenna input)
6. Adjustable RF signal level on all channels (to 50 mV)
7. Composite video output (rear panel) adjustable (front panel) up to 3.25 V peak to peak (video sync can be selected to be either positive or negative by the front-panel video control)

Color-Pattern Outputs of the Sencore Video Analyzer The color-pattern outputs of the Sencore video analyzer are shown in Figure 26.28. These are:

1. Single cross for picture centering
2. Single dot for static convergence
3. Cross hatch for dynamic convergence, linearity adjustments, and pincushion adjustments
4. White dots for dynamic convergence, linearity adjustments, and pincushion adjustments
5. Color bars to check, align, or troubleshoot color circuits. Adjust automatic frequency and phase (AFPC) circuit. These are basically the same color bars as for the unit shown in Figure 26.23. However, the Sencore unit also provides a color burst signal on the back porch of the horizontal sync pulse. This provides a more stable pattern for color alignment and troubleshooting.
6. Bar sweep and color-bar sweep (previously described)

There are two basic types of color bars: (1) the gated-rainbow pattern and (2) the

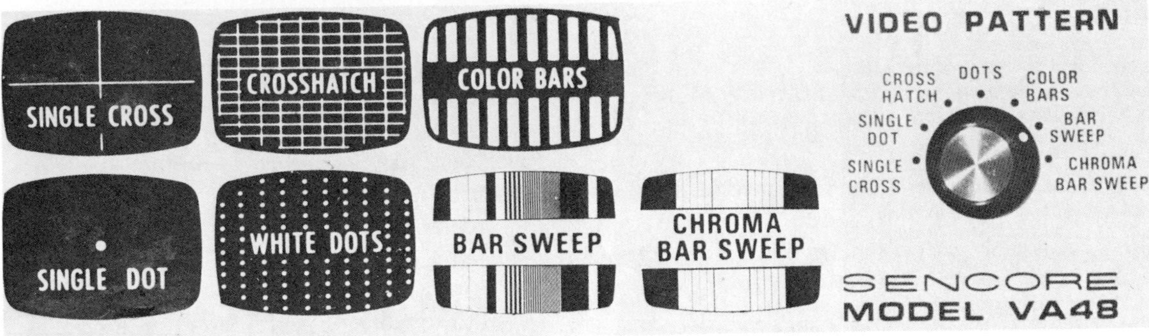

Figure 26.28 The color-pattern outputs available from the Sencore video analyzer. *(Courtesy of Sencore.)*

NTSC color bars. The gated-rainbow pattern is generally used for servicing color television receivers. The NTSC color-bar pattern is mainly used with television broadcast equipment.

Gated-Rainbow Pattern The gated-rainbow pattern is illustrated in Figure 26.29. Figure 26.29A shows the ten color bars as they appear on the screen of a color picture tube. This figure shows the colors in proper sequence. The colors are identified by number (Figure 26.20A), color, phase angle (from burst), and signal identification. Bars 1 (yellow-orange) and 7 (its complement, greenish-blue) do not have signal-identification nomenclature.

Each color bar is separated from the adjacent bar by a black vertical line. The black line is caused by gating off the signal between colors. (See Chapter 27 for color photographs of color-bar patterns.) The bars are each separated in phase by 30°, beginning with burst phase at 0° (or 360°). There is no bar corresponding to 330°.

The appearance of the gated-rainbow signal on an oscilloscope is shown in Figure 26.29B. Note the color burst signal, the ten color-bar signals, blanking between the color signals, and the horizontal sync pulses. The color burst signal is on the *back porch* of the horizontal sync as with a TV station signal.

In the gated-rainbow pattern, the subcarrier frequency is exactly 3.563 811 MHz. This differs by precisely 15 734 Hz from the 3.579 545-MHz color subcarrier frequency generated in the color TV receiver. As a result of this precise frequency difference, the phase of the color-pattern signals shifts through exactly 360° for each horizontal line. Thus the pattern repeats identically on every line, forming the vertical color bars.

The comparison between the two subcarrier signals is performed by the two color demodulators in the TV receiver. The output signals from the color demodulators and color matrix (now at the video rate) are fed to the red, green, and blue guns of the color picture tube to produce the actual color bars.

Note: There is no Y signal component associated with the gated-rainbow color bars. This color-bar pattern displays the bars with equal saturation and brightness levels, on the color picture tube. The phases and order of the various bars are shown in Figure 26.29A.

The use of a color-bar generator in adjusting the color automatic frequency and phase control circuit (including wave forms) is given later in the chapter. Color photographs of color-bar (and other) displays, helpful in troubleshooting, are given in Chapter 27.

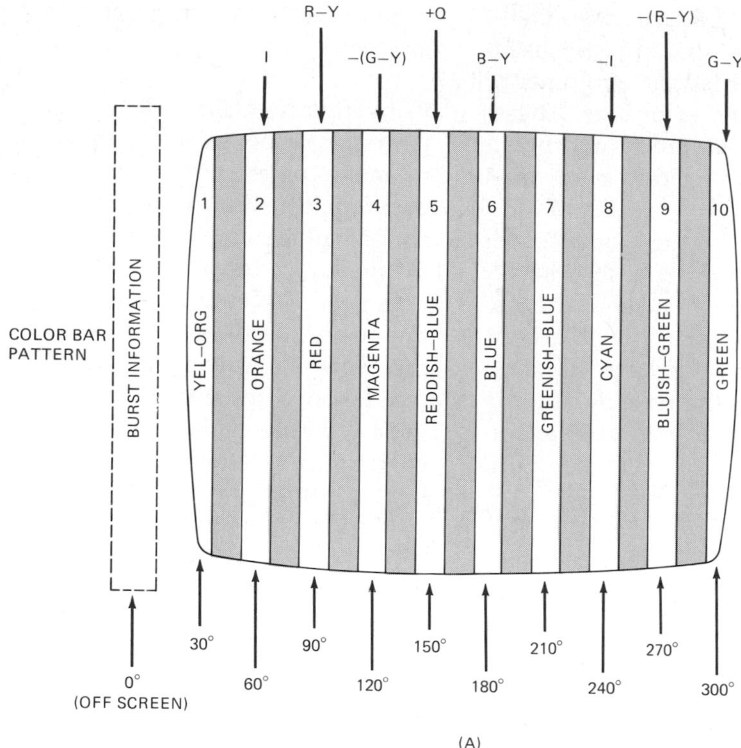

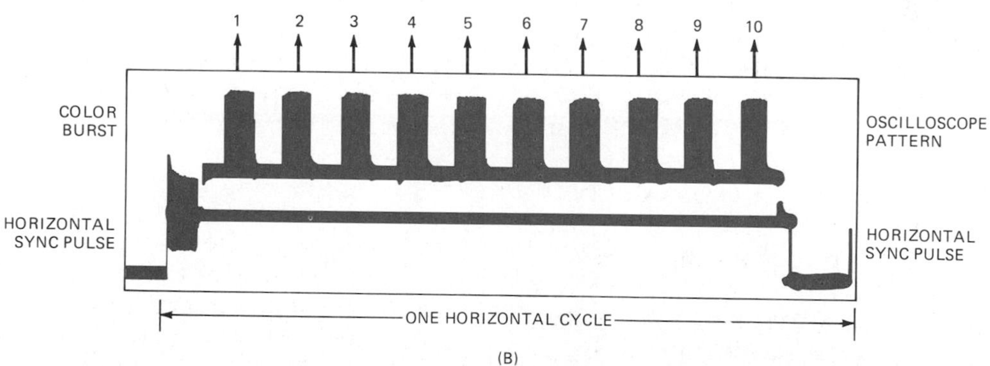

Figure 26.29 The gated-rainbow color-bar pattern. A. The 10 color bars are 30° apart in phase. See Figure 26.20A. Bar 7 is the color complement of Bar 1. All other 9 bars are identified by number, color, phase angle, and signal identification. B. The rainbow generator output waveform for the color-bar pattern, as seen on an oscilloscope. The bar numbers correspond with those of Figure 26.21A.

The NTSC Color-Bar Pattern As mentioned previously, this pattern is used mainly with television broadcast equipment. The pattern consists of six color bars. These are the primary colors, red, green, and blue, plus their complementary colors, cyan, magenta, and yellow.

Figure 26.30A shows the vectorscope presentation, and Figure 26.30B shows the oscilloscope display. Note that there are no black bars between color bars as there is with the gated rainbow pattern (Figure 26.25B). In Figure 26.30B the horizontal sync pulse and color burst are plainly visible. All colors are fully saturated (no white light dilution).

Observation of Figure 26.30B shows the color bars to be at different brightness levels. This indicates that the pattern contains a Y signal component. The effect of this component is to display the various colors in the correct shades of gray, as they would normally appear on a monochrome display.

In order to be compatible with normal broadcast signals, the burst and color-bar signals are generated at the standard color subcarrier frequency of 3.579 545 MHz. Each color-bar signal is at the correct phase angle (referenced to the color burst) to produce the six colors.

Convergence Using Dots An overall dot pattern can be used for making static and dynamic convergence adjustments. For static convergence, only the dots at the center of the screen are converged (see end of chapter for this procedure). The red, green, and blue dots near the center are converged until they merge and become single white dots.

Dots as well as a cross-hatch pattern (see below) can be used for dynamic convergence. Figure 26.31A shows a section of a color picture tube screen when all three beams are misconverged and a standard dot pattern is being fed into the receiver. The same section of this color picture tube after convergence is illustrated in Figure 26.31B. Note that the three dot groups have merged into a single dot, indicating correct convergence. For dynamic convergence, the dots over the entire area of the screen must be merged to form white dots.

Adjustments Using Cross-Hatch Pattern Dynamic convergence is frequently accomplished using a cross-hatch pattern. The pro-

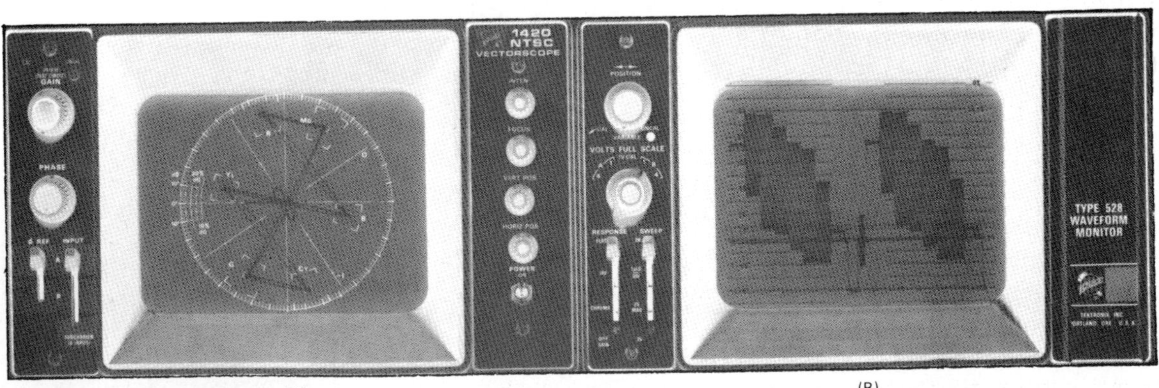

Figure 26.30 The NTSC color-bar pattern. This consists of the three primary colors, plus their three complementary colors (see text). **A.** The vectorscope pattern. **B.** The oscilloscope display. Note the "Y" signal component and the absence of black bars between color signals. (Compare with Figure 26.29B.) *(Courtesy of Tektronix, Inc.)*

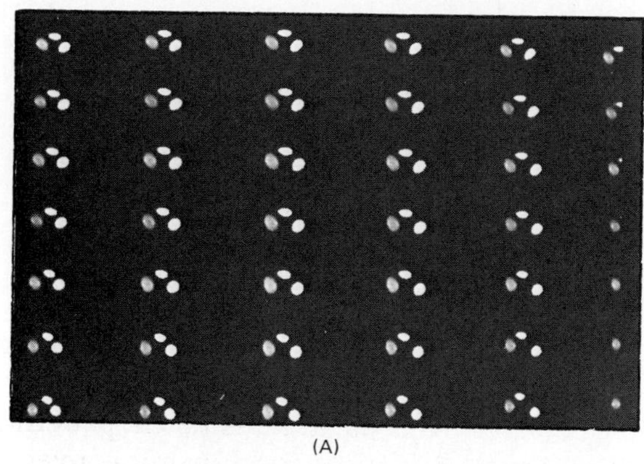
(A)

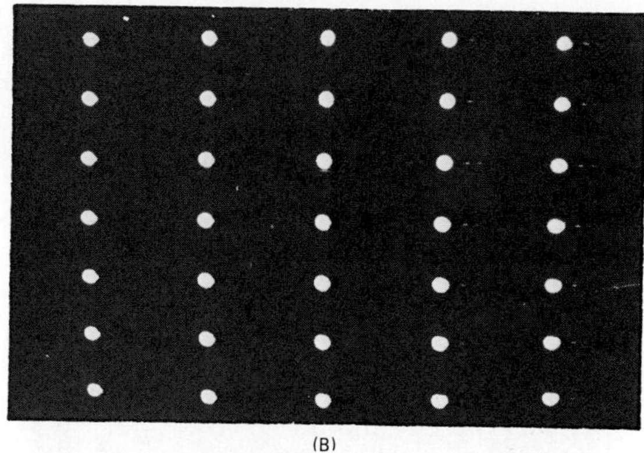
(B)

Figure 26.31A. The section of a color picture tube with all three beams misconverged is shown. **B.** The same section after convergence. Note that the three-dot groups have merged into single dots.

cedure for doing this is described at the end of the chapter. The display on a color picture tube must be adjusted for both vertical and horizontal linearity. These adjustments are the same as those made on a monochrome picture tube. For color picture tubes, there is also the need for pincushion correction. The cross-hatch pattern can be used for making all of these adjustments.

On some color generators, there is a special feature that permits a single dot or cross to be moved around the face of the screen to check dynamic convergence at the vertical and horizontal edges of the picture tube.

26.9 SOLID-STATE ANALOG AND DIGITAL MULTIMETERS

An accurate high-impedance multimeter is essential, not only in some phases of alignment, but also in general testing and troubleshooting. By definition, a **multimeter** is an instrument having a number of measuring functions. These may include (1) AC and DC voltages, (2) AC and DC currents, (3) DC resistances, and (4) decibel measurements. At the end of the chapter, in the alignment section, several uses of a high-impedance multimeter are described.

Solid-State Analog Multimeters A high-impedance solid-state, **analog multimeter** is pictured in Figure 26.32. It features an all-solid-state circuit that uses FETs, an IC, and LEDs (light-emitting diodes). The instrument has automatic polarity indication, as well as overvoltage and overcurrent protection. It can be operated from the AC line, a self-contained rechargeable cell, or a D flashlight battery.

On "LP ohms" (low power), only 30 mV is applied to the circuit under test. This provides circuit protection. Junction checks of transistors and diodes can be made using the plus or minus "diode ohms" position of the function switch. With special auxiliary probes, the meter can measure DC voltages as high as 40 000 V and RF voltages having frequencies up to several hundred megahertz.

Meter Ranges The limits of the meter ranges are as follows (in steps):

1. AC or DC voltages (on same scale): 0.03 to 300 V. 1 000 V AC and 1 000 V

Figure 26.32 A high-impedance (10 megohms), solid-state, analog multimeter. It can be operated from the AC line, a self-contained rechargeable cell, or a "D" size flashlight cell. *(Courtesy of Simpson Electric Company.)*

DC ranges available on separate jacks.

2. Input resistance: 10 megohms on all switchable AC and DC ranges; 31.9 megohms on the 1 000-V ranges.
3. AC and DC current ranges: 30 μA to 30 mA. 1-A and 10-A ranges available on separate jacks.
4. dB ranges: −50 dB to +62 dB. 0-dB reference is 0.775 V across 600 ohms (1 mW).
5. Ohms range: 0.2 ohm to 1 000 megohms.

Polarity Indication Examination of Figure 26.32 shows two LEDs. The one on the right is labeled ON and also indicates positive polarity. The LED on the left is labeled −DC and flashes if a negative voltage is being measured. Regardless of the polarity of the measured voltage, the meter needle always swings upscale.

The same type of indication is given for direct currents. For negative direct currents, the left-hand LED flashes. For positive direct currents, the right-hand LED flashes. However, in either case, the meter reads upscale.

SOLID-STATE DIGITAL MULTIMETERS

Analog meters are widely used and very practical. However, they do have several disadvantages. Some of these are (1) difficulty in locating and reading the correct scale, (2) interpolation between printed numbers, (3) limited accuracy, and (4) possible parallax errors.

Digital multimeters overcome all of these disadvantages. These instruments have large, easily read numbers and automatic placing of the decimal point. A plus or minus sign appears before the numbers. This automatically indicates the polarity of the measured voltage. The digital multimeter pictured in Figure 26.33 has an accuracy

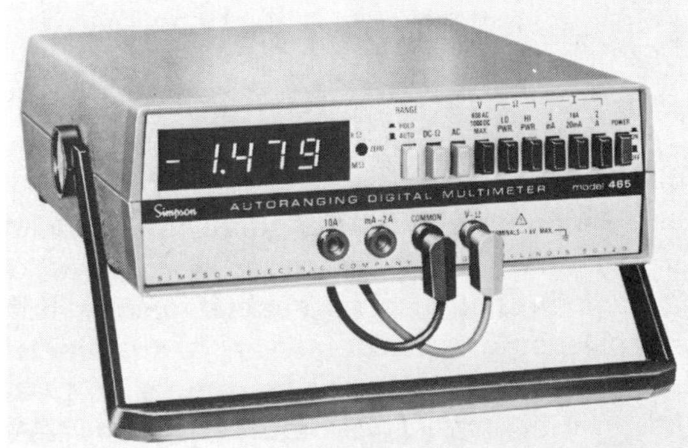

Figure 26.33 A digital multimeter. This is a $3\frac{1}{2}$ digit instrument with 0.1% DC V accuracy. It features autoranging. *(Courtesy of Simpson Electric Company.)*

of 0.1% for DC voltage measurements. Accuracies of 0.1% to 0.5% are common for such meters.

Autoranging The instrument shown in Figure 26.33 has a feature known as **autoranging.** This indicates that for all AC and DC voltage and current measurements, as well as resistance measurements, the meter automatically switches to the correct range required to perform the specific measurement. Units without autoranging have push buttons to select the desired range.

3½ Digits The meter pictured in Figure 26.33 has a display rating of 3½ digits. The meaning of the "½" digit will now be explained. With a three-digit meter, the maximum display number is 999. The decimal point is always automatically put in the right place. Other three-digit displays are 1.23, 9.99, 67.8, and so forth.

For relatively little added expense and circuitry, an additional digit can be provided to appear *in front of* the above-mentioned three digits. The added digit can only be programmed to be either unlighted (no reading) or to indicate a 1. If energized, this so-called "½" digit can practically double the reading for a particular range. Thus a 1-V range could now read 1.999 V, which for all intents and purposes is a 2-V range. (This is done in Figure 26.33, where the maximum reading on the 2-V range is 1.999 V.) Similarly, a 100-kohm resistance range could read 199.9 kohm and be called a 200-kohm range. This is also done in the meter of Figure 26.33.

Meter Ranges The various meter ranges are as follows (in steps):

1. DC voltage: 0.2 to 1 000 V, positive or negative polarity (reads from 100 μV to 1 000 V)
2. AC voltage (40 Hz to 20 kHz): 0.2 to 600 V (reads from 100 μV to 600 V)
3. Low-power resistance (max. 350 mV): 2 to 2 000 kohms (reads from 1 ohm to 2 megohms)
4. Standard-power resistance (max 2.5 V): 2 kohms to 20 megohms (reads from 1 ohm to 20 megohms)
5. DC current: 0.2 mA to 10 A (reads from 100 nA to 10 A)
6. AC current: 0.2 mA to 10 A (reads from 100 nA to 10 A)

Accessory Probes An accessory probe is available to increase the DC voltage range to 40 000 V. An additional RF probe makes it possible to measure RF voltages at frequencies up to several hundred megahertz.

26.10 ALIGNMENT USING FREQUENCY SWEEP GENERATOR

Color TV receiver alignment is presented using two methods: (1) frequency sweep generator alignment and (2) Sencore bar sweep alignment. We present the frequency sweep alignment procedure first.

In a color television receiver, a complete alignment consists of aligning the *sound-IF, video-IF, tuner,* and *color sections.* The alignment of monochrome TV receivers is essentially the same except that there is no color section. We shall discuss the alignment procedure for the GTE-Sylvania color receiver in the schematic insert after pg. 804.[1] This is a fairly standard procedure and applies with slight modifications to most color receivers. It is always a good idea before aligning any receiver to check the manufacturer's specifications, and it is especially important to check the characteristic curves shown by the manufacturer as being ideal. It is not always possible to obtain these ideal curves during

[1]Information furnished by GTE-Sylvania, and reproduced by permission.

alignment, but they will give you a better idea of what you are trying to accomplish during the alignment procedure.

VIDEO-IF ALIGNMENT AND TRAP ADJUSTMENTS

The manufacturer's recommended procedure for aligning the video-IF stages in the receiver are shown in Chart 26.1, which follows this section. To get the maximum benefit from this discussion, and also of the following alignment procedures, use the foldout schematic of the receiver to locate the various components and test points.

Step 1 Step 1 of the video-IF alignment and trap adjustment (Chart 26.1) describes the procedure for checking the combined frequency response of the mixer and the first two video-IF stages. You will note that the alignment setup notes state that the horizontal driver (Q_{406}) should be removed during this procedure. The reason for this is that Q_{406} drives the horizontal output stage (Q_{408}), which drives the flyback transformer. The keying pulse for the AGC circuit comes from this flyback transformer. Note that the winding between Pins 2 and 3 of T_{400} (see schematic) goes to Terminals U and P of the video-IF stage.

When you remove the keying pulse, the keyed AGC circuit becomes inoperative, and therefore you must apply a DC bias base voltage to make up for the lost AGC. That is the reason for applying +2.0 V to Point P on the IF board as required in Step 1.

The sweep generator cable, shown in Figure 1 of Chart 26.1, provides a 75-ohm impedance input to the UHF IF input of the VHF tuner. This simulates the normal connections. Refer to the tuner schematics of Figure 26.34 to locate the tuner connection points referred to in the following discussions.

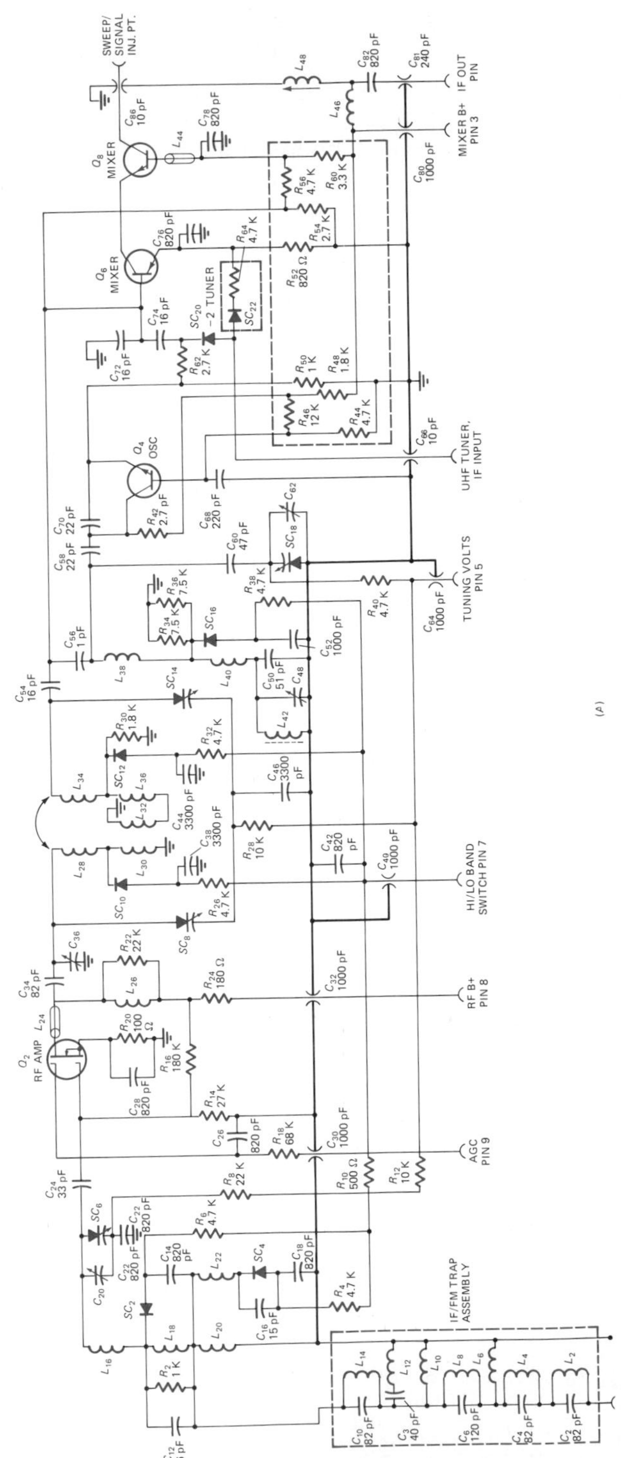

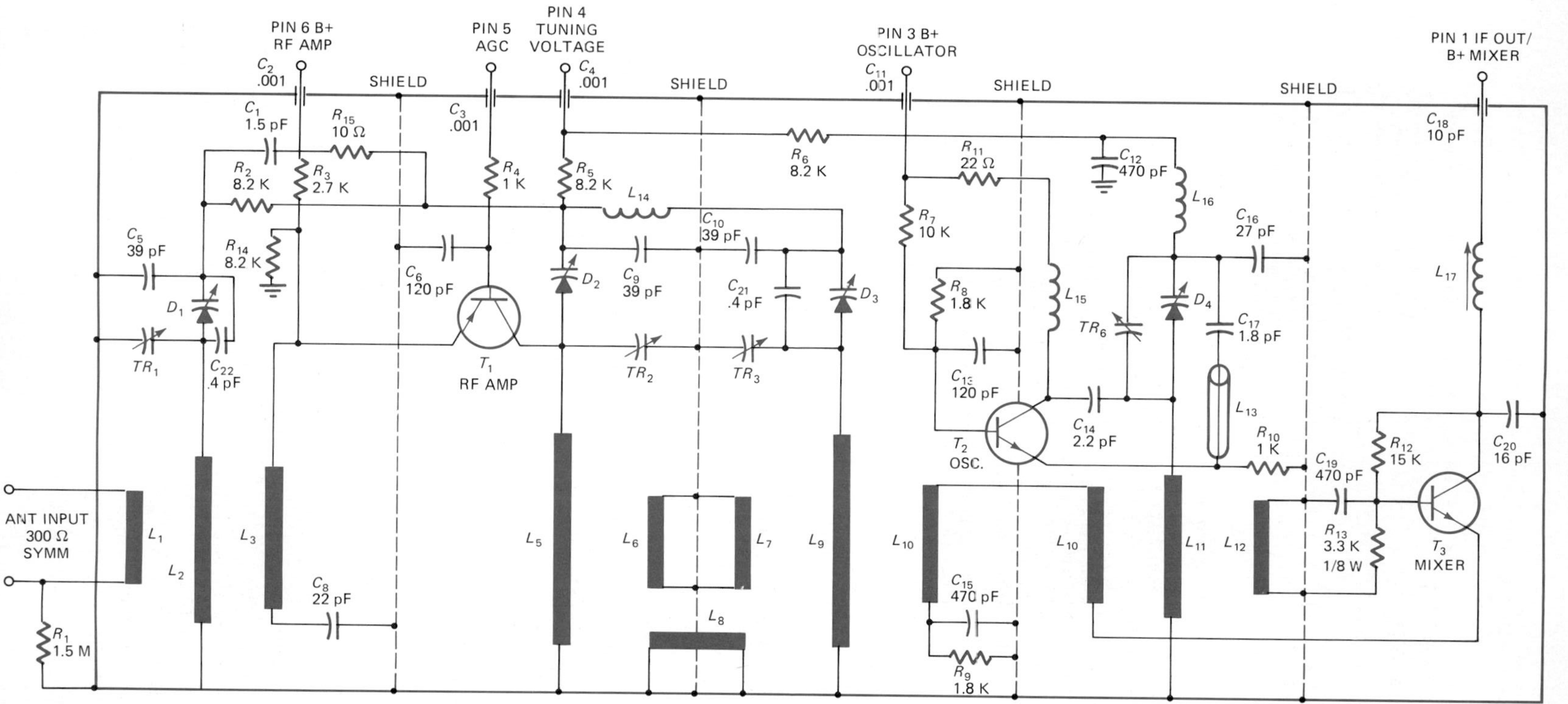

Figure 26.34 The tuners for the Sylvania all solid-state chassis. (The schematic for the complete receiver is the fold-out following pg. 804.) **A.** Sylvania solid-state VHF tuner. **B.** Sylvania solid-state UHF tuner. *(Courtesy of GTE-Sylvania, Inc.)*

ALIGNMENT PROCEDURE

PRELIMINARY INSTRUCTIONS

1. Line voltage should be maintained at 120VAC.
2. Keep marker generator coupling at a minimum to avoid distortion of the response curve.
3. Do not use tubular capacitors for coupling sweep into receiver. Disc ceramics are best.
4. For best results, solder the sweep generator ground to chasis, do not use clips.
5. Sweep generator "hot" lead must make good electrical contact at all points given under TEST EQUIPMENT HOOK-UP
6. Use scope gain for maximum peak-to-peak response curve on the scope rather than sweep generator.
7. Test equipment (tube type only) should warm up for approximately 15 minutes before alignment.
8. Bias values specified must be maintained during alignment to insure proper results.
9. To eliminate Vertical circuit interference, switch "Normal Service" switch on rear of chassis to "Service" position.
10. Markers should be kept as small as possible and should be crystal controlled.
11. Check detector response for linearity using a square wave generator at 40 MHz and 50 MHz.
12. To energize the E01-13, -14 HEC chassis when isolated from the AM/FM Amplifier system (PL104 - TV audio output plug and PL532 - TV power plug disconnected) apply 120 VAC, 60 Hz to pins 2 and 4 of PL532 - TV power plug. NOTE: There will be no audio.

VIDEO IF ALIGNMENT AND TRAP ADJUSTMENT

STEP	ALIGNMENT SET-UP NOTES	TEST EQUIPMENT HOOK-UP	ADJUST
1	Remove Horiz. Driver Q406. Apply +2.0V Bias to tie point [P] on IF board. Rotate AGC control R358 fully counterclockwise. 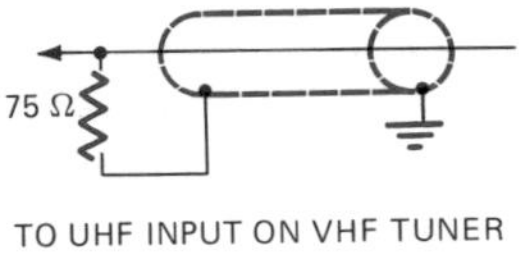 75 Ω TO UHF INPUT ON VHF TUNER Figure 1	SWEEP GENERATOR – Through a cable shown in Figure 1 to UHF IF input on VHF tuner. Set generator to 44.5 MHz with 10 MHz sweep. MARKER GENERATOR – Loosely coupled to sweep generator lead. OSCILLOSCOPE – Through network shown in Figure 2 to tie point [Z] on IF board. Calibrate oscilloscope for .01 Volt peak-to-peak output. 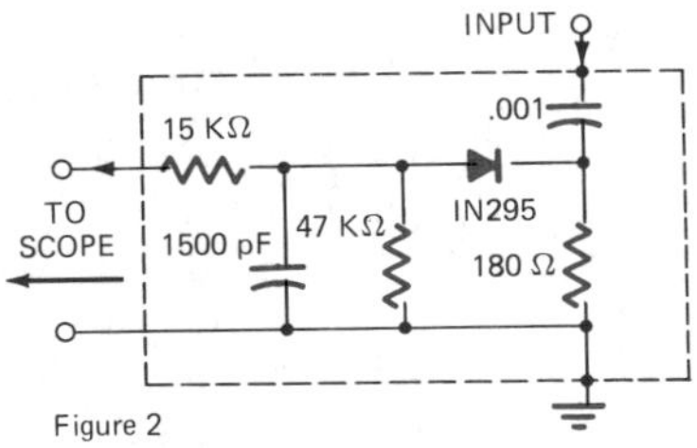 INPUT .001 15 KΩ TO SCOPE 1500 pF 47 KΩ IN295 180 Ω Figure 2	Adjust tuner IF output coil on VHF Tuner to obtain response curve shown in Figure 3. 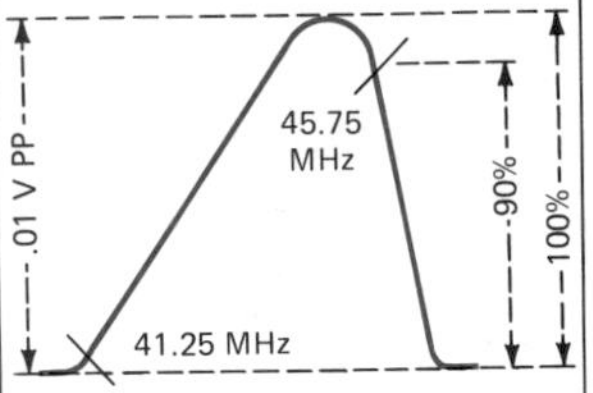 .01 V PP 45.75 MHz 90% 100% 41.25 MHz Figure 3
2	Same as step 1. 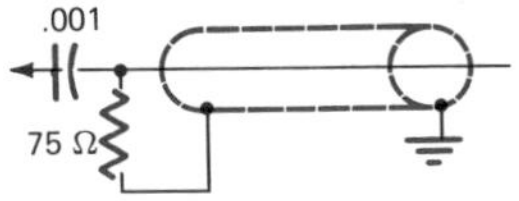 .001 75 Ω TO SWEEP/SIGNAL INJECTION POINT ON VHF TUNER (RF TEST POINT) Figure 4	Same as step 1, except: SWEEP GENERATOR – Through .001 capacitor shown in Figure 4 to SWEEP/SIGNAL INJ. PT. on VHF tuner. 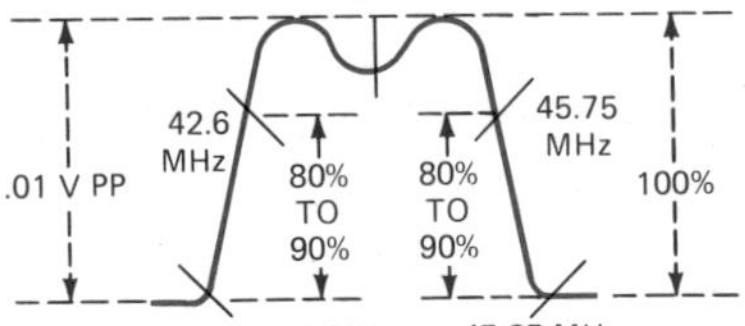 44.0 MHz .01 V PP 42.6 MHz 80% TO 90% 80% TO 90% 45.75 MHz 100% Figure 5 41.25 MHz 47.25 MHz	[L202] for minimum 47.25 MHz marker. [T200] top and bottom core to obtain response curve shown in Figure 5. Readjust tuner IF output coil if necessary. Repeat for optimum performance.
3	Same as step 2. Step 3 – Necessary only when alignment is extremely far out of adjustment.	SIGNAL GENERATOR – Through a .001 capacitor shown in Figure 4 to SWEEP/SIGNAL INJ. PT. on VHF tuner. Set output for maximum signal (RF frequencies are 44.5 MHz, 41.25 MHz and 47.25 MHz.) VTVM – Through a 33K resistor to tie point [AL] on IF bd.	[L204], [L206] and bottom core of [T202] for MINIMUM POSITIVE deflection on VTVM (44.5MHz) Set [L208] for max. inductance (core approx. center of coil). Top core of [T202], [L208] (toward top of coil) for MAXIMUM POSITIVE deflection on VTVM (41.25 MHz) Readjust [L202] slightly for MAXIMUM POSITIVE deflection on VTVM (47.25 MHz).

 Charts 26.1-26.6 reproduced with permission of GTE-Sylvania, Inc.

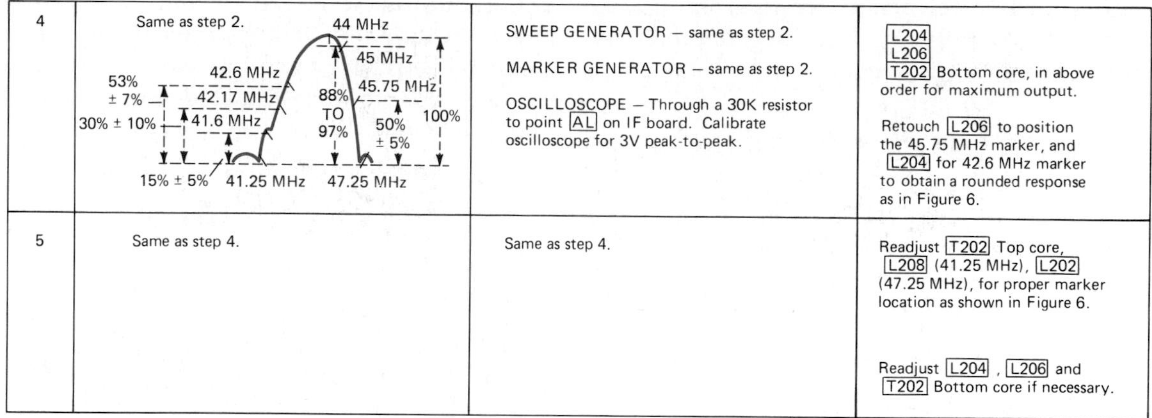

4	Same as step 2.	SWEEP GENERATOR – same as step 2. MARKER GENERATOR – same as step 2. OSCILLOSCOPE – Through a 30K resistor to point AL on IF board. Calibrate oscilloscope for 3V peak-to-peak.	L204 L206 T202 Bottom core, in above order for maximum output. Retouch L206 to position the 45.75 MHz marker, and L204 for 42.6 MHz marker to obtain a rounded response as in Figure 6.
5	Same as step 4.	Same as step 4.	Readjust T202 Top core, L208 (41.25 MHz), L202 (47.25 MHz), for proper marker location as shown in Figure 6. Readjust L204, L206 and T202 Bottom core if necessary.

Chart 26.1 cont'd.

The response curve shown in Figure 3 of Chart 26.1 is the response of the mixer stage of the tuner and the first video-IF stages.

Step 2 Step 2 of the alignment procedure in Chart 26.1 is similar to Step 1. An isolation capacitor (Figure 4 of Chart 26.1) has been added to the cabling between the sweep generator and the tuner, and the sweep generator now feeds the output point of the VHF tuner. The response curve of Figure 5 is for the first stage of IF in the receiver, but not the mixer. Note that this curve is relatively flat between 42.6 MHz and 45.75 MHz. If adjustment is necessary in this stage, then Step 1 should be repeated to make sure that the response curve has not been distorted by the adjustment.

If the response curve of Figure 5 cannot be duplicated with reasonable similarity during the alignment procedure, then it will be necessary to align the coils and traps that are used for shaping the overall response curve. This procedure is described in Step 3. An RF signal generator is used for these adjustments instead of a sweep frequency generator. The 41.25-MHz and 47.25-MHz traps are adjusted to obtain the steep sides on the skirt of the response curve of Figure 5, and the 44.5-MHz coils are adjusted to obtain the correct top shape on the curve. After these adjustments have been made, Steps 1 and 2 should be repeated.

Step 3 A VTVM[1] is used as an indicator for proper adjustment in Step 3. It is connected to the output of the first video amplifier stage. At this point the signal has passed through the detector, and only the modulation from the generator signal is available at Point AL. The modulation frequency is well within the frequency response of VTVM or FET meters.

Step 4 The overall response curve of the IF section between the tuner output and the first video amplifier is shown in Figure 6 of Step 4 in Chart 26.1. Note that the curve of Figure 6 is a typical video-IF response curve for color television receivers. The video-IF alignment procedures of Chart 26.1 are similar to those provided by other manufactur-

[1]The abbreviation *VTVM* stands for *vacuum tube voltmeter*. However, the same letters are sometimes also used for solid-state voltmeters, such as the one shown in Figure 26.32. A digital voltmeter (Figure 26.33) would also be suitable. Basically, the requirement is for a high-impedance voltmeter.

ers of color receivers. Of course, the actual adjustments will vary, but the general alignment procedure is similar.

ALIGNMENT OF THE SOUND SECTION

Chart 26.2 explains the procedure for aligning the sound section of the television receiver shown in the schematic insert.

The video carrier and the sound-IF signals heterodyne in the video detector circuit and produce a 4.5-MHz beat signal. If this signal is allowed to pass through the video amplifier section, it produces a herringbone pattern on the screen. Therefore, a trap is inserted in the video section and tuned to reject 4.5-MHz signals.

Step 1 Step 1 of Chart 26.2 indicates that a strong signal from a local station should be tuned, and that the fine-tuning knob should be adjusted until the 4.5-MHz pattern is visible in the picture.

The response of most VTVM and FET meters to a 4.5-MHz signal is very poor, and therefore the demodulation probe shown in Figure 7 must be used to measure the DC amplitude of the 4.5-MHz wave.

The sound trap (T_{204}) is adjusted for minimum voltmeter reading, thus indicating that the 4.5-MHz signal has been reduced to its lowest value. It is also a good idea to observe the herringbone pattern on the screen during this adjustment to make sure it has been minimized. However, it is not a good practice to try to make this adjustment without the aid of a VTVM or FET meter, because small changes in signal synchronization cannot be observed as easily on the screen of the television receiver as they can be with the use of a meter.

Step 2 After the 4.5-MHz trap has been properly adjusted, the next step is to apply an FM signal to the sound-IF section. This procedure is described in Step 2. With the receiver tuned off station, the FM signal generator feeds its signal to the IC and the tuned circuits are adjusted for maximum output signal.

ALIGNMENT OF COLOR BANDPASS AMPLIFIERS

Before studying the color alignment procedure, it will be helpful to review the characteristics of the color signal in the color receiver. Figure 26.35 shows a section of the receiver response curve containing the color signal. Ideally, the bandpass amplifier allows only the color signals between 3.1 and 4.1 MHz to pass through the color section. That is the purpose for aligning this section—to make sure that the color amplifiers pass color signals correctly, above and below the color subcarrier frequency.

Step 1 As shown in Step 1 of Chart 26.3, the horizontal oscillator is disabled during color bandpass alignment. Also, the 3.58-MHz oscillator is disabled. Test Point BJ is grounded to disable the blanking circuit. (The blanking circuit drives the color output amplifier Q_{614} to cutoff during the retrace period.) Test Point P is grounded on the IF strip to stabilize the AGC amplifier output during this adjustment.

The sweep generator signal is applied to the input of the second color amplifier through Terminal BC. Figure 9 of Chart 26.3 indicates the required probe. Note that the center frequency of the sweep must be 3.58 MHz (the center frequency of the bandpass amplifier) with a sweep width of only 3 MHz. As described in Step 1 of Chart 26.3, the oscilloscope is connected to the output of the second color amplifier during the adjustment. The response curve displayed is only for the second color amplifier in the color bandpass section.

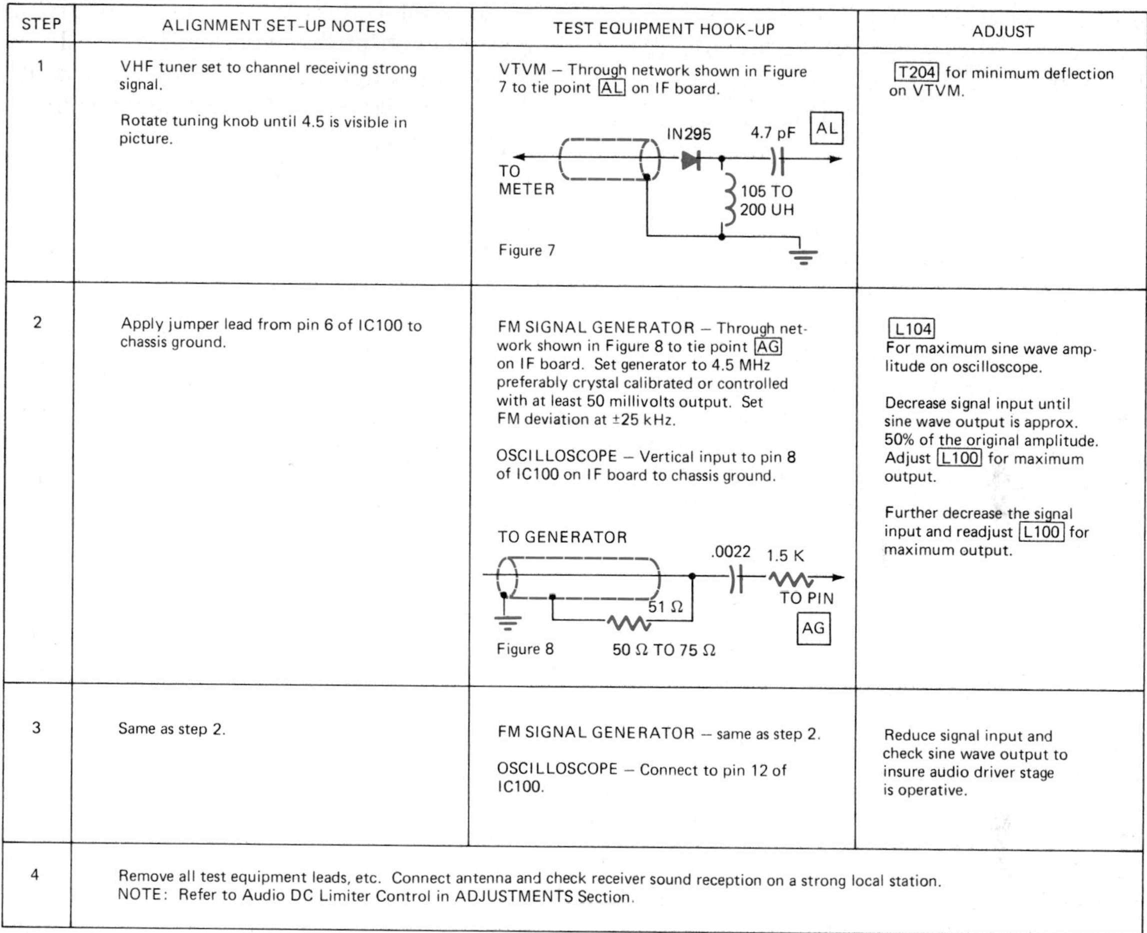

STEP	ALIGNMENT SET-UP NOTES	TEST EQUIPMENT HOOK-UP	ADJUST
1	VHF tuner set to channel receiving strong signal. Rotate tuning knob until 4.5 is visible in picture.	VTVM – Through network shown in Figure 7 to tie point AL on IF board. IN295 4.7 pF AL TO METER 105 TO 200 UH Figure 7	T204 for minimum deflection on VTVM.
2	Apply jumper lead from pin 6 of IC100 to chassis ground.	FM SIGNAL GENERATOR – Through network shown in Figure 8 to tie point AG on IF board. Set generator to 4.5 MHz preferably crystal calibrated or controlled with at least 50 millivolts output. Set FM deviation at ±25 kHz. OSCILLOSCOPE – Vertical input to pin 8 of IC100 on IF board to chassis ground. TO GENERATOR .0022 1.5 K TO PIN AG 51 Ω 50 Ω TO 75 Ω Figure 8	L104 For maximum sine wave amplitude on oscilloscope. Decrease signal input until sine wave output is approx. 50% of the original amplitude. Adjust L100 for maximum output. Further decrease the signal input and readjust L100 for maximum output.
3	Same as step 2.	FM SIGNAL GENERATOR – same as step 2. OSCILLOSCOPE – Connect to pin 12 of IC100.	Reduce signal input and check sine wave output to insure audio driver stage is operative.
4	Remove all test equipment leads, etc. Connect antenna and check receiver sound reception on a strong local station. NOTE: Refer to Audio DC Limiter Control in ADJUSTMENTS Section.		

Chart 26.2

Step 2 In Step 2 of the color alignment procedure, the sweep generator feeds its signal to the input of the video-IF section of the receiver. The oscilloscope is connected to show the overall response curve of the IF and color bandpass amplifier section with this setup. The signal at Tie Point DX is the detected sweep generator signal, which has passed through the video-IF, video detector, and color bandpass sections. Note that the inserted 45.75-MHz signal from the signal generator serves as a video-IF carrier and is essential if the resultant video frequencies are to be obtained from the output of the video detector. The video-IF carrier heterodynes against the various IF frequencies in the video detector and results in the production of the video frequencies. Without the inserted IF carrier, the output of the video detector would consist only of the very low sweep rate frequencies and the low harmonics making up the output response curve. The desired overall response curve is shown in Figure 10 of Chart 26.3.

Alternate Alignment Method An alternate method of color bandpass alignment is also shown in Chart 26.3. This method is not as

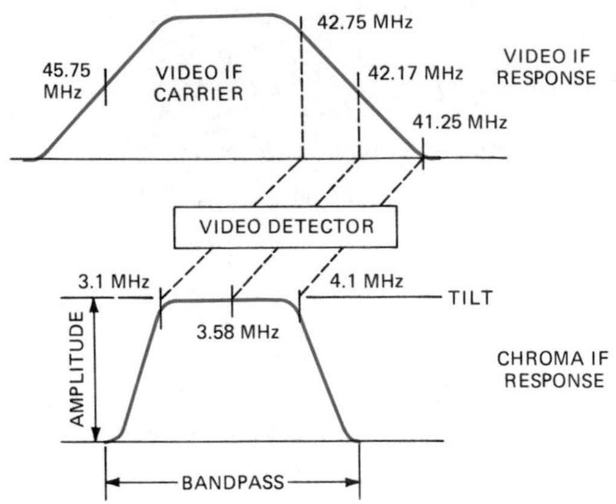

Figure 26.35 A chroma bandpass curve and its relationship to the receiver response curve. *(Courtesy of GTE-Sylvania, Inc.)*

good as the one described in Step 2. The Step 2 method enables you to view the overall video-IF and color bandpass response, whereas the alternate method views only the response of the color bandpass amplifiers. In the alternate method, a video-frequency sweep generator is used, as in Step 1, instead of an IF sweep. One advantage of the alternate method is that it makes it possible to easily align the color amplifiers in cases where the alignment is badly off. This would be difficult to do if you were also passing the sweep signal through the IF stages of the receiver. In the alternate method, the video sweep generator and markers are fed to the input of the first color amplifier. The oscilloscope remains connected to the output of the color output amplifier. Alignment is performed as shown on the chart for the response curve desired and is illustrated in Figure 14 of the chart.

COLOR AUTOMATIC FREQUENCY AND PHASE CONTROL ADJUSTMENTS

The automatic frequency and phase control circuit (AFPC) is also sometimes called the automatic phase control circuit (APC). The function of this circuit is to lock the receiver 3.58-MHz color subcarrier oscillator to the same frequency and phase as the color burst. When this is accomplished, the receiver subcarrier oscillator is exactly in step with the transmitter subcarrier oscillator. This is the condition required to properly reproduce the original colors at the transmitter. Any variation between the receiver and transmitter subcarrier oscillators results in the reproduction, at the receiver, of incorrect colors (Chart 26.4).

Step 1 In Step 1 of Chart 26.4, one of the procedures required is to ground Test Point BU. This removes the input to the burst amplifier and permits the 3.58-MHz receiver crystal oscillator to free-run without the influence of the color AFC circuit. At this time, the collector circuit of Q_{644}, the 3.58-MHz amplifier, is tuned (T_{606}) for maximum 3.58-MH output.

Step 2 In Step 2 of Chart 26.4, the ground is removed from Point BU, thus allowing the color burst to be amplified and compared with the oscillator signal in the phase detector. The adjustment of T_{604} is made to assure that it is correctly tuned and that the killer and phase detectors operate properly.

Step 3 In Step 3 of Chart 26.4, the color phase control (R_{776}) adjustment is set so that the display produced by the color-bar generator is motionless on the screen. This indicates that the 3.58-MHz oscillator phase is properly adjusted. Note that during the adjustment in Step 3, Tie Point BU is again grounded to disable the AFC. For this reason, it may not be possible to get the color bars to stand still on the screen, but if they drift at all, the drift should be very slow.

Step 4 When Tie Point BU is ungrounded, as required in Step 4, the color bars should be securely locked on the screen. The pat-

CHROMA ALIGNMENT

STEP	ALIGNMENT SET-UP NOTES	TEST EQUIPMENT HOOK-UP	ADJUST
1	A. Remove Horiz. Osc., transistor Q406. B. Remove 3.58 Osc., transistor Q642. Ground TP [BJ] on Chroma Board, & ground TP [P] on IF Board. Apply +1.7VDC bias to TP [BD] on the chroma board. Adjust Color Killer fully CW. Adjust AGC fully CW. Adjust color control for +6.5V at TP [BV] . Adjust tint control for +6.5V at TP [BT] 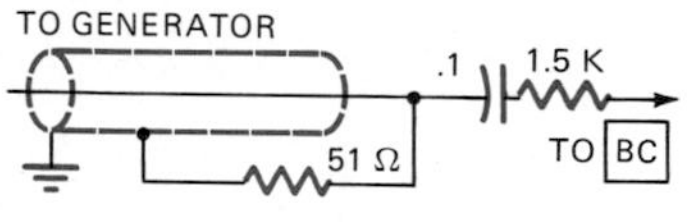 . Figure 9	SWEEP GENERATOR – Through network shown in Figure 9 to test point [BC] on Chroma Board. Set generator to 3.58MHz with 3MHz sweep. MARKER GENERATOR – Insert in parallel with sweep generator to provide frequencies of 3.1MHz, 3.58MHz, 4.1MHz. OSCILLOSCOPE – Through Video detector probe shown in Figure 11 at C628 & C636 junction on Chroma Board Calibrate scope to .1V full scale deflection. 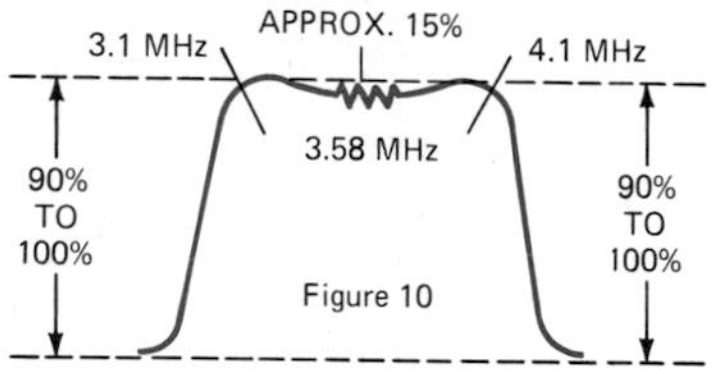 Figure 10	[T600] Top and bottom core for response curve shown in Figure 10. 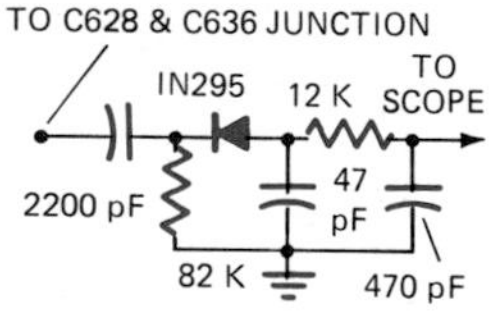 Figure 11
2	Same as step 1, except remove ground jumper and apply +2VDC bias to TP [P] 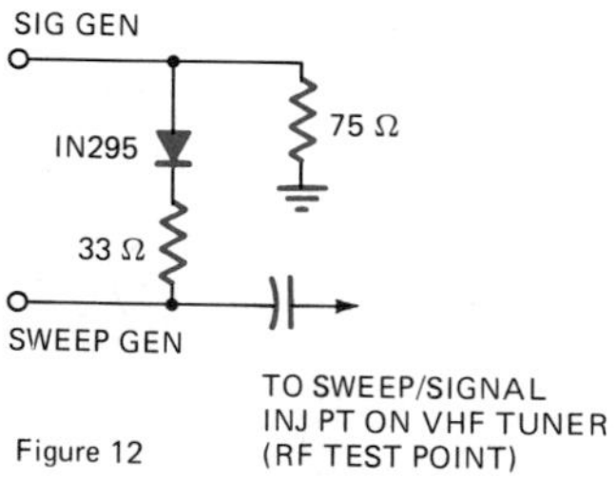 . Figure 12	SIGNAL GENERATOR – Set to 45.75MHz. SWEEP GENERATOR – Set to 3.58MHz, sweep 3MHz. The outputs of both generators through network shown in Figure 12. To SWEEP/SIGNAL INJ. PT. on VHF tuner. VTVM – Point [AL] on IF board. OSCILLOSCOPE – Through network shown in Figure 11 to tie point [BX] on Chroma Board. 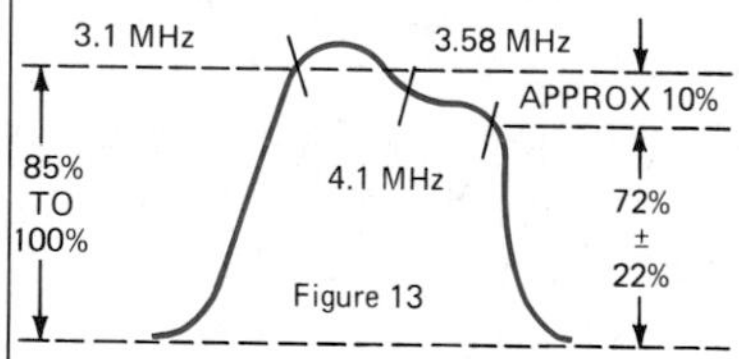 Figure 13	Set SWEEP GENERATOR to minimum for no signal output. Note DC READING ON VTVM, then increase generator output until a 1.5V reduction is indicated on VTVM. (If this 1.5V cannot be obtained, set generator to maximum and if necessary, reduce the +2V bias at tie point [P] until the voltmeter reads at least .5V DC change.) [T602] , for maximum amplitude of the 3.58MHz marker, as shown in Figure 13.

ALTERNATE CHROMA ALIGNMENT FOR STEP 2

STEP	ALIGNMENT SET-UP NOTES	TEST EQUIPMENT HOOK-UP	ADJUST
2	Same as step 1.	SWEEP GENERATOR – same as step 1, except to the base of Q208. MARKER GENERATOR – same as step 1. OSCILLOSCOPE – same as step 2. 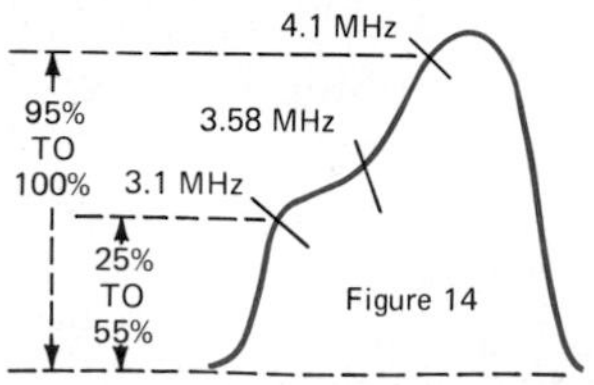 Figure 14	Adjust [T602] for maximum peak 3.58MHz marker as shown in Figure 14.

Chart 26.3

STEP	ALIGNMENT SET-UP NOTES	TEST EQUIPMENT HOOK-UP	ADJUST
1	Set Color Control for +6.5VDC at TP [BV]. Set Tint Control for +6.5VDC at TP [BT]. Set Color Killer fully clockwise. Set Phase Control [R776] at mid-range. Ground TP [BU].	COLOR BAR GENERATOR – To receiver antenna terminals. OSCILLOSCOPE – Across secondary of [T606], 3.58MHz output transformer. OR AC-VTVM – Through 4.7K resistor across secondary of [T606].	[T606] for maximum amplitude on oscilloscope. OR [T606] for maximum reading on AC VTVM.
2	Same as step 1, but remove ground from tie point [BU].	COLOR BAR GENERATOR – same as step 1. VTVM – To tie point [BD].	[T604] for minimum DC reading on VTVM. Make certain 3.58MHz oscillator is running and locked in. (1.7V or less)
3	Color Control – Same as step 1. Tint Control – Same as step 1. Killer Control – Same as step 1. Phase Control [R776] – Same as step 1. Ground tie point [BU].	COLOR BAR GENERATOR – Same as step 1.	[R776] for zero beat on picture tube (color bars stand still on screen or drift slowly).
4	Color Control – Same as step 1. Tint Control – Same as step 1. Killer Control – Same as step 1. Remove ground from [BU].	COLOR BAR GENERATOR – Same as step 1. (use low lever color signal) OSCILLOSCOPE – To test point [BN] on Chroma Board. NOTE: Very little adjustment (if any) is required. If more than 1/2 turn is required in either direction, repeat step 2 AFPC Alignment and step 2 of the Chroma Alignment.	[T604] so that when tint control is rotated from one extreme to the other, there is a minimum of + and – 30 deg. from nominal phase. Return tint control to mid-range, 6th bar on R-Y waveform should be cancelled. See Fig. 15.
5	Tint Control – Same as step 1. Killer Control – Same as step 1.	OSCILLOSCOPE – To test points [BN], [BM], [CG], respectively.	Check for proper matrixing. Waveforms and amplitudes should conform to Fig. 16.
6	Tint Control – Same as step 1. Set channel selector to channel receiving snowy raster with no color transmission.	No test equipment.	[R660] Killer threshold control clockwise until snow on screen appears colored; then rotate [R660] counterclockwise until color in the snow disappears. Check on a color program to assure setting of [R660] is not killing on color.

Chart 26.4

terns observed on the oscilloscope in Step 4 (Figure 15 of Chart 26.4) are taken at the R–Y output. Moving the tint control back and forth actually shifts the phase of the color signal. Remember that this is the color signal from the color-bar generator. The sixth bar on a color-bar display is in the blue region, and that is why it should be cancelled at the R–Y output.

Step 5 In Step 5 the waveforms and amplitudes of all three color section outputs (Fig-

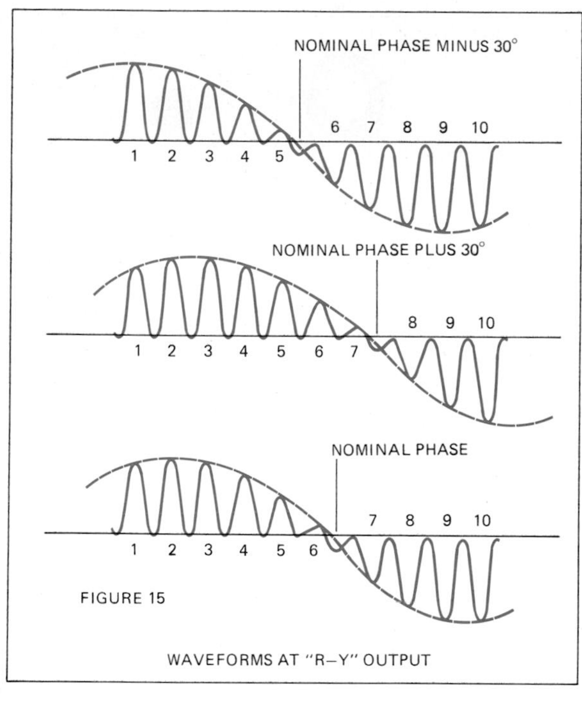

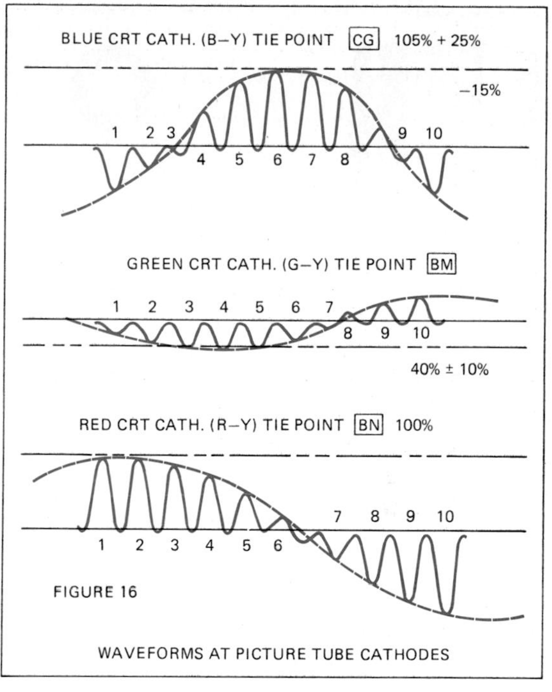

Chart 26.4 cont'd.

ure 16 of Chart 26.4) are checked with an oscilloscope. Of course, if the color oscillator is not in proper phase and frequency, these patterns cannot be obtained.

After the AFPC circuitry has been adjusted, the final step—that is, Step 6 of Chart 26.4—is to adjust the color killer threshold. Ideally, the color killer should prevent any signals from passing through the color amplifiers when a monochrome signal is being received. The reason for this is that random noise signals would cause color flecks (called confetti) on the screen of the picture tube when a monochrome picture was being displayed if the noise signals could pass through the color bandpass amplifiers. The purpose of the color killer is to automatically cut off the color bandpass amplifiers during monochrome reception. Step 6 simply explains how the color killer is adjusted to the threshold level.

26.11 ALIGNMENT USING BAR-SWEEP GENERATOR

MAKING THE SIGNAL CONNECTIONS

In the following discussion, the Sencore VA48 bar-sweep generator is referred to simply as the VA48. The connections needed when the VA48 is used for IF trap and IF-video amplifier alignment are shown in Figure 26.36.

Connecting VA48 Signal to Receiver The output signal of the VA48 should be connected to the tuner of the receiver rather than to the input of the first IF amplifier, when performing alignment. This is because the tuned circuit of the mixer is also part of the tuning for the first IF stage.

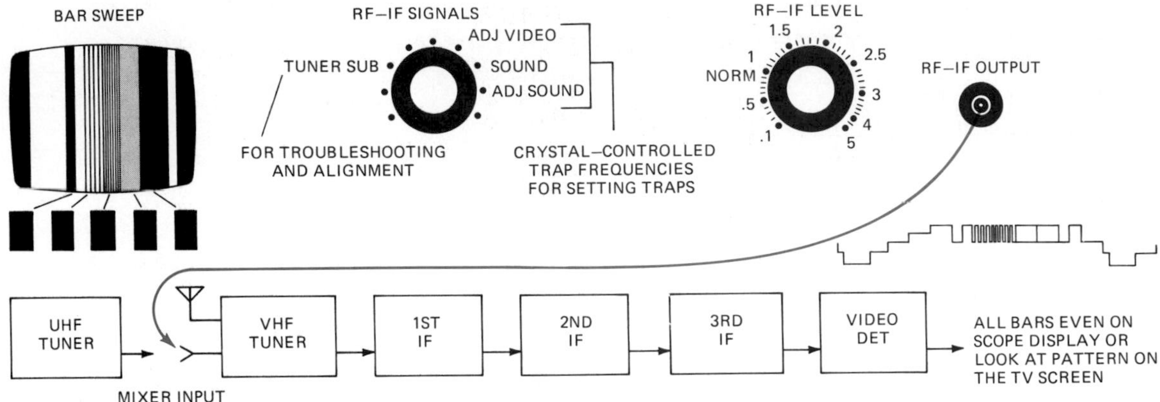

Figure 26.36 Connections of the bar-sweep generator needed to perform IF-trap and IF-video alignment. Output of video detector is connected to an oscilloscope. *(Courtesy of Sencore Corp.)*

Most receivers have separate UHF and VHF tuners with a coaxial cable connecting the two. Disconnect this cable and connect the output cable of the VA48 in its place. (Figure 26.37). Set the VHF tuner to UHF. Set the UHF tuner to an unused channel.

Note: Some of the newer, all-electronic tuners have the UHF and VHF tuners in the same package. In this case, inject the signal from the VA48 into the tuner mixer test point specified in the receiver service manual.

Connecting the Oscilloscope The oscilloscope should be connected to the output of the video detector (Figure 26.36). Do *not* connect it to the delay line. Some receivers have a 3.58-MHz trap in the first video amplifier stage, and this causes "roll-off" at some signal frequencies. Convenient test points are (1) the output of the 4.5-MHz sound trap or (2) the input to the first video amplifier stage.

IF TRAP ADJUSTMENT

The IF trap frequencies are (1) 39.75 MHz, adjacent-channel video, (2) 41.25 MHz, accompanying sound, and (3) 47.25 MHz, adjacent-channel sound. These three traps should always be adjusted before any other video-IF adjustments are made. This is because the traps affect the *overall* video-IF response.

Trap adjustment should be performed with the *lowest* signal amplitude possible. The three crystal-controlled trap frequency outputs (Figures 26.22 and 26.36) are modulated with a 1 000-Hz frequency. Thus, adjustment can be made while looking at the oscilloscope screen or the television receiver screen.

Trap Adjustment Procedure Set the VA48 to one of the three trap frequencies. Reduce the signal output until the 1 000-Hz trap pattern just disappears. Next, slightly increase the signal output until the pattern becomes visible (Figure 26.37). Now adjust the trap for the least visible pattern.

In the same manner, adjust the other traps. If two 41.25-MHz traps are used in a receiver, use the following procedure. First, detune the last trap slightly. Then adjust the first trap, followed by the second trap.

Some traps have two adjustments, either a coil and a potentiometer or a coil with

Figure 26.37 Proper trap response on screen (right) shows minimum 1000-Hz modulation. Improper response is shown at left. This indication can be on either an oscilloscope or a TV receiver screen.

two slugs. In a coil-potentiometer trap, set the coil adjustment first. In a two-slug coil, set the bottom coil first.

VIDEO-IF BANDPASS COILS ADJUSTMENT

The video-IF bandpass coils are adjusted *after* the traps are set. With the oscilloscope connected to the video detector output, set the sweep to the *horizontal line rate.* Biasing of the AGC circuit is *not* required because the bar-sweep pattern is part of a composite TV waveform.

Evaluation of the bar-sweep pattern may be made either on the screen of an oscilloscope or on the TV receiver. If a synchronous video detector is incorporated in the TV receiver, a 45.75-MHz IF carrier is required for video detection. The VA48 IF-bar signal provides this signal. If a frequency sweep and marker generator is used, the 45.75-MHz signal must be fed in separately.

Video-IF Alignment The procedure for video-IF alignment (Figure 26.38) is as follows:

1. Make connections as described for IF trap adjustment, described above. (Figure 26.36).
2. Do *not* connect an external bias supply.
3. Switch to TUNER SUB on the RF-IF SIGNAL switch. Switch to BAR SWEEP position of the VIDEO PATTERN switch. (See Figure 26.22 for location of these controls.)
4. Adjust the oscilloscope for proper triggering on the output waveform.

Note: External triggering is desirable. The composite video output from the VTR STANDARD output jack may be used for this purpose.

5. All bars on the oscilloscope should be the same height to within 10%. Also, there should be no signs of *ringing* on the sync pulse or lower frequency bars. If either symptom appears, alignment is required. Proceed as described in the following paragraphs.
6. Begin the alignment by setting the

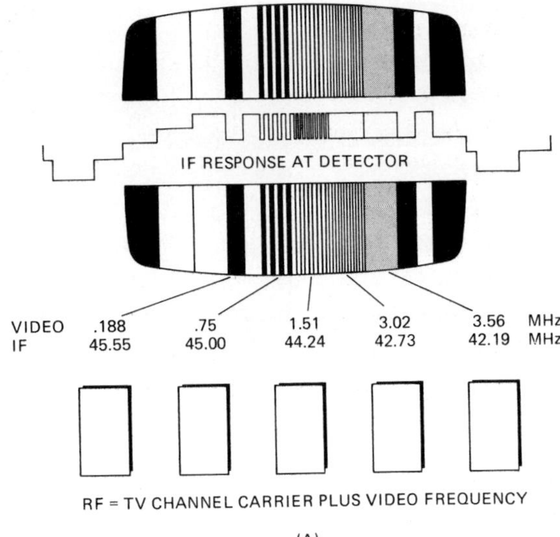

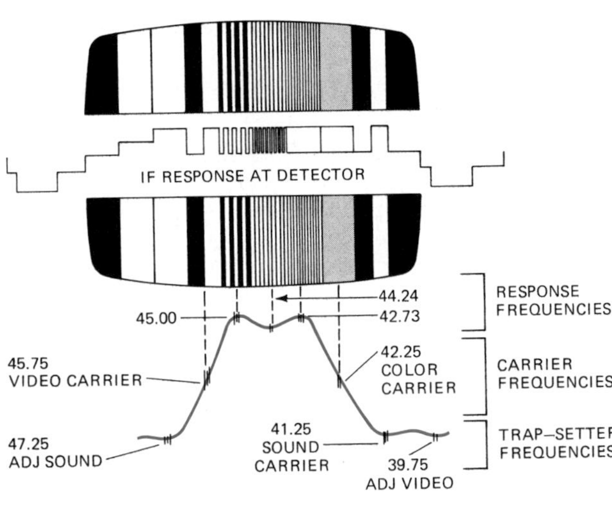

Figure 26.38 Bar-sweep alignment of video-IF stages. A. Bar-sweep buttons for the five frequency bars. Video and corresponding IF frequencies are shown. B. Video-IF response curve and spot frequencies. This curve results when each bar of the bar-sweep pattern has equal amplitude, as shown.

mixer coil (in the VHF tuner) for best sensitivity amplitude by:

a. Gradually reducing the RF IF LEVEL control until the video pattern breaks up.
b. Resetting the mixer coil for the best sensitivity.
c. Repeating Steps a and b until no further improvement is possible.

7. Increase the RF IF LEVEL control for a snow-free picture.
8. Depress each of the BAR SWEEP SELECTOR BUTTONS and observe the position of each of the frequency bars on the scope display.
   a. If the high-frequency bars are re-

duced in amplitude, and the sync, blanking, and 188-kHz pulses are rounded, the alignment is favoring the video carrier portion of the response curve.

b. The tuner mixer coil and first IF coils affect the balance between the low-frequency and high-frequency information.

c. The second IF (and third IF of a four-IF system) coil affects the medium to high-frequency information.

9. At the conclusion of balancing the IF alignment coils, recheck the traps by switching the RF IF SIGNALS switch to each of the three trap signals positions. Reset the traps if there is 1 000-Hz information present.
10. If it was necessary to reset any of the traps, move the RF IF SIGNALS switch back to the TUNER SUB position to recheck the BAR SWEEP response. Retouch the IF alignment if the trap alignment changed the IF responses.

ALIGNMENT CHECK ON TV RECEIVER SCREEN

The alignment of the video-IF stages may be checked on the TV receiver screen (Figure 26.39). If the IF- and video-amplifier system is working properly, each of the BAR SWEEP bars (except the 3.56-MHz color bar) should have the same amount of white. By reducing the brightness and contrast controls, this white level can be compared in each bar. If any of the pattern frequencies is reduced in amplitude, its white level will be reduced. If that happens, the specific bar will lose all its white content (turn black) before the others, as the brightness is reduced (see also Figure 26.40).

SUMMARY OF IF BAR-SWEEP ALIGNMENT CHECKS AND VISUAL OBSERVATIONS

Figure 26.40 gives a summary of IF bar-sweep alignment checks and results. Visual observations of the bar-sweep pattern are shown for the oscilloscope and TV screens. In addition, the effects on broadcast signals are given.

COLOR AMPLIFIER ALIGNMENT

Perform the following steps and refer also to Figures 26.41 and 26.42.

1. Connections are made as illustrated in Figure 26.41.
2. Set the VIDEO PATTERNS switch to

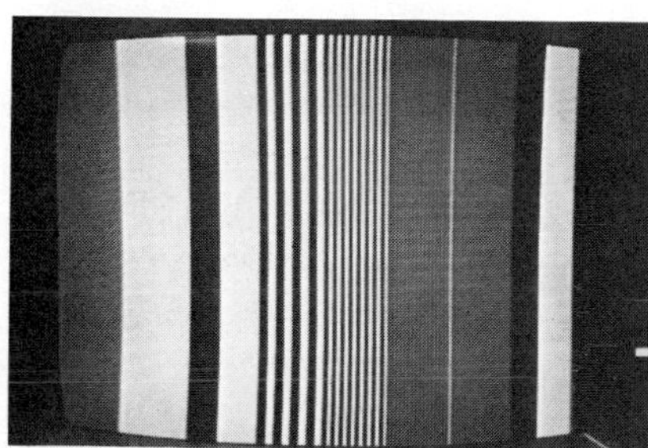

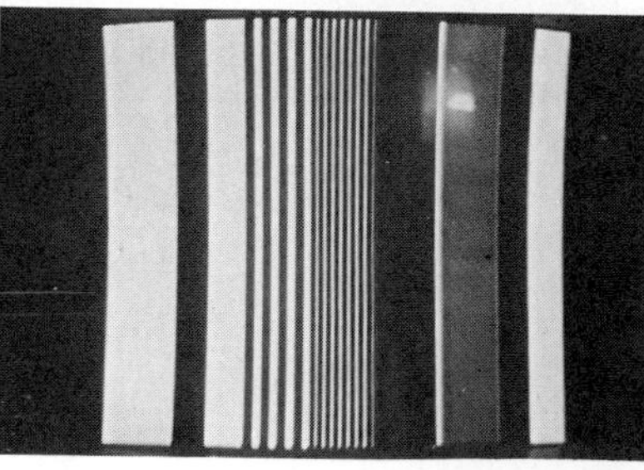

Figure 26.39 Left photo shows good bar sweep in which all bars show even white levels as brightness is reduced. Right photo shows improper pattern in which 3.02-MHz bar has dropped out before other bars. See also Figure 26.40. *(Courtesy of Sencore Corp.)*

BAR SWEEP RESPONSE ON SCREEN	BAR SWEEP RESPONSE AT VIDEO DETECTOR	RESULTING EFFECT ON BROADCAST SIGNAL
	NORMAL PICTURE	
 1. Fine tuning properly set for good resolution of 3.02 MHz bar and minimum ringing of .188 KHz and .75 MHz bars. 2. All bars become black at approximately the same time as the brightness control is reduced. a. 3.56 MHz bar goes black first. b. 3.02 MHz bar goes black second. c. Other bars disappear together.	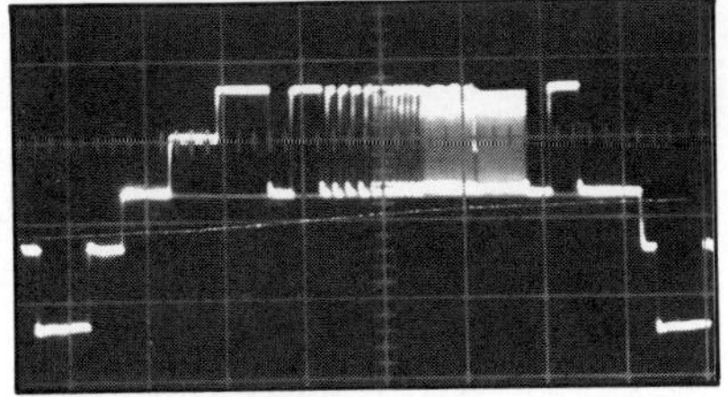 1. All bars have equal amplitudes. 2. Ringing and overshoot are minimal.	
	LOSS OF HIGH FREQUENCY RESPONSE	
 1. 3.56 MHz and 3.02 MHz bars show little contrast between white and black areas. 2. 3.56 MHz and 3.02 MHz bars disappear from screen much earlier than other bars as brightness is reduced.	 1. High frequency bars have reduced amplitude on Bar Sweep pattern. 2. IF response should be adjusted to make all bars equal amplitude. Traps should be checked for proper setting as well.	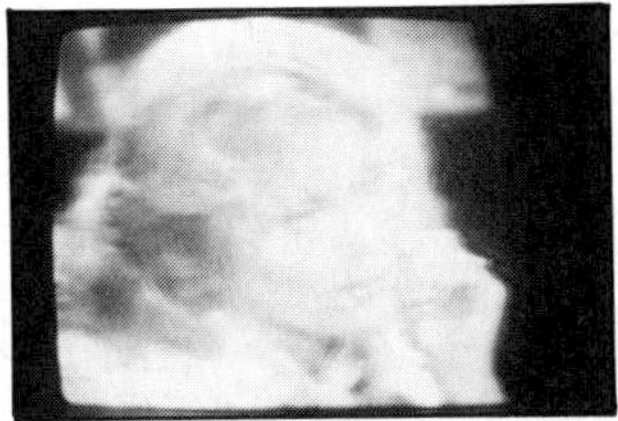 1. Low high frequency shows as loss of detail.
	LOSS OF LOW FREQUENCY RESPONSE	
 1. Picture may become unstable and show some horizontal shift in comparison to normal picture. 2. Contrast of the black and white areas of the low frequency signals in low, in comparison to the high frequency bars. 3. Ringing is present in the black areas between the low frequency bars.	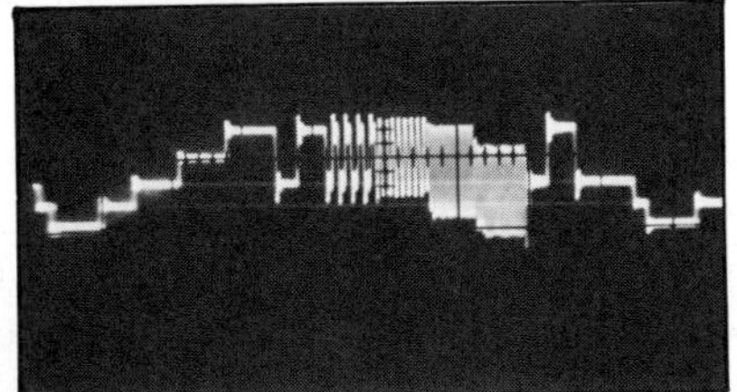 1. Sync, blanking, and grey scale levels of Bar Sweep signal are reduced in comparison to high frequency bars, indicating low video carrier amplitude. 2. Excessive overshoot is evident of square wave signals.	 1. High frequency detail is present, but large solid-color areas are overemphasized. Blacks and whites appear too strongly.

Figure 26.40 Summary of IF bar-sweep alignment checks. Results are shown on an oscilloscope and on the TV screen. *(Courtesy of Sencore Corp.)*

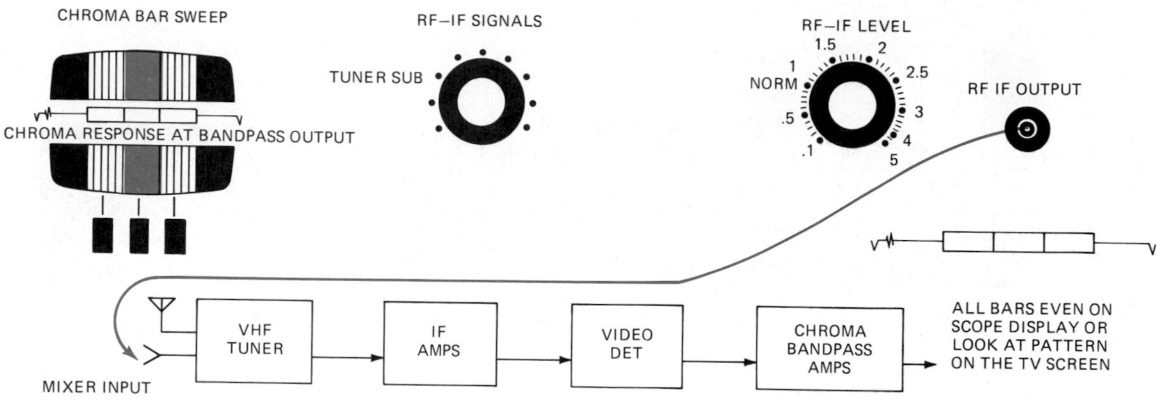

Figure 26.41 Connections of the bar-sweep generator needed to perform alignment of the chroma-bandpass amplifiers. Output of the chroma-bandpass amplifiers is connected to an oscilloscope. *(Courtesy of Sencore Corp.)*

the CHROMA BAR SWEEP position (Figure 26.22).

3. Depress all three of the CHROMA BAR SWEEP selector buttons.
4. Connect the oscilloscope to the output of the color bandpass amplifier.
5. As shown in Figure 26.42, the three bars should have equal amplitudes. If the amplitudes are not within 10% of each other, the color bandpass amplifier requires realignment.
6. As an alternate method of checking for proper alignment, using the television screen, proceed as follows: Reduce the setting of the brightness control while observing the two outer color bars (3.09 and 4.08 MHz). Reduce the control setting until either or both of these bars turn black.
   a. If both bars turn black at the same time, the bandpass amplifier is working properly.
   b. If the right-hand bar turns black first, the bandpass amplifier is favoring the high-frequency portion of the response curve.
   c. If the left-hand bar turns black first, the bandpass amplifier is favoring the low-frequency portion of the response curve.

The color-bandpass amplifier can be aligned either by observation of either the oscilloscope or the TV screen. Both methods are now described.

A. Chroma-bandpass amplifier alignment using an oscilloscope. Adjust the color-bandpass amplifier coil(s) until all three of the bars on the oscilloscope have the same amplitude. The alignment of the bandpass amplifier is now complete.

B. Chroma-bandpass amplifier alignment using the TV screen
   1. Adjust the tint control until the middle bar is bright blue.
   2. Reduce the brightness control until either the upper or lower sideband bar turns black.
   3. Adjust the color-bandpass amplifier coil(s) until both the upper or lower sideband bars have the same brightness level. The alignment of the bandpass amplifier is now complete.

SUMMARY OF COLOR BAR SWEEP ALIGNMENT CHECKS AND VISUAL OBSERVATIONS

Figure 26.43 gives a summary of color-bar-sweep alignment checks and results. Re-

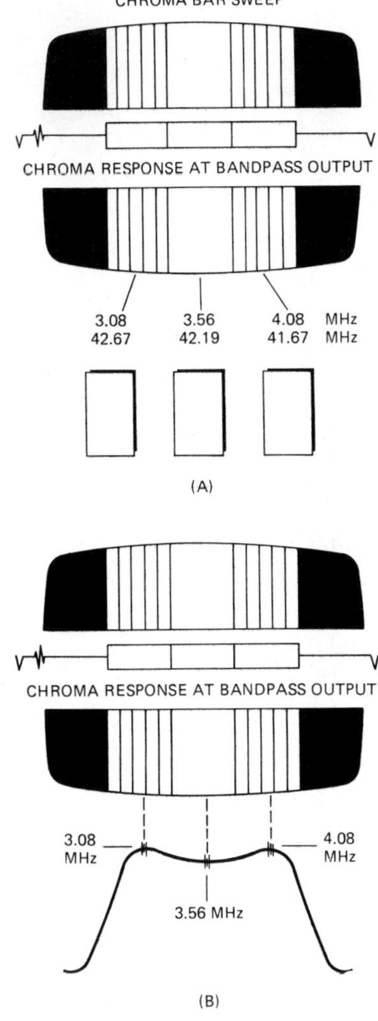

Figure 26.42 Bar-sweep alignment of the chroma-bandpass amplifier. A. Bar-sweep buttons for the three frequency bars. Video and corresponding IF frequencies are shown. B. Chroma-IF response curve and spot frequencies. This curve results when each bar of the bar-sweep pattern has equal amplitude, as shown. *(Courtesy of Sencore Corp.)*

sponses are shown on the TV and oscilloscope screens.

COLOR AFPC ADJUSTMENT

As mentioned previously, the function of the AFPC circuit is to lock the 3.58-MHz color subcarrier oscillator to the same frequency and phase as the color burst.

The BAR SWEEP pattern provides a reference for setting the local 3.58-MHz color oscillator to the proper frequency with the AFPC disabled. Sets that do not have a switch for disabling the AFPC may be set with the BAR SWEEP pattern by increasing the setting of the color killer until the 3.56-MHz bar shows a color.

To set the color oscillator in sets with an AFPC switch:

1. Feed the RF-IF OUTPUT to the VHF antenna input.
2. Set the VHF tuner and the VA48 to the same channel and fine-tune as needed.
3. Select the CHROMA BAR SWEEP position of the VIDEO PATTERN switch.
4. Move the color AFPC switch to the "service" position.
5. Adjust the 3.58-MHz adjustment for the least amount of color change in the blue bar of the CHROMA BAR SWEEP pattern. "Barber-poling" of the colors will be evident in the middle bar (Figure 26.43) when the 3.58-MHz phase is incorrect. The color will be solid across the bar when the adjustment is properly set, although the color itself may change slowly.
6. Return the color AFPC switch to the normal position.

If the receiver does not have an AFPC defeat switch, the BAR SWEEP pattern should be used. This pattern in effect disables the AFPC circuit because the pattern does not have a color burst signal to reference the local color oscillator. The color killer must be set to produce a color in the 3.56-MHz bar (Figure 26.42) before the adjustment to the local oscillator is made.

To set the color oscillator in sets without an AFPC switch:

PATTERN RESPONSE ON SCREEN		CHROMA RESPONSE AT BANDPASS OUTPUT	
NORMAL PICTURE			
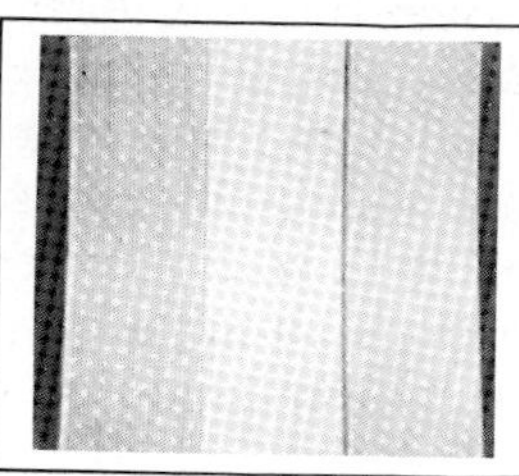	1. Center bar is bright blue, with solid color across the entire bar (no tint of other colors present at the edges). 2. All three chroma bars disappear together as the brightness level is reduced.	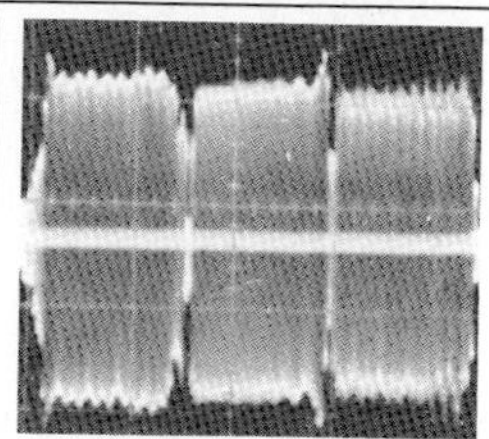	1. Amplitude of all three bars is equal.
UPPER CHROMA SIDEBAND HAS REDUCED LEVEL			
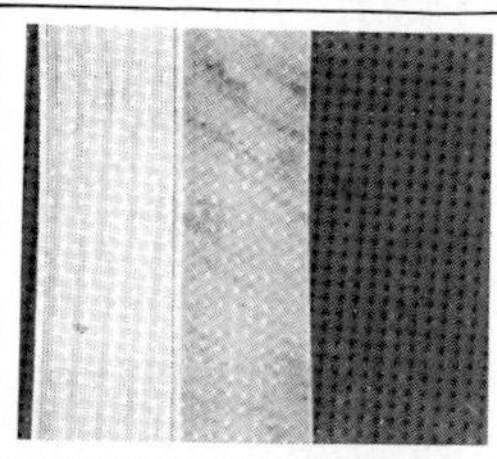	1. Upper sideband bar for 4.08 MHz disappears from screen first as brightness level is reduced.	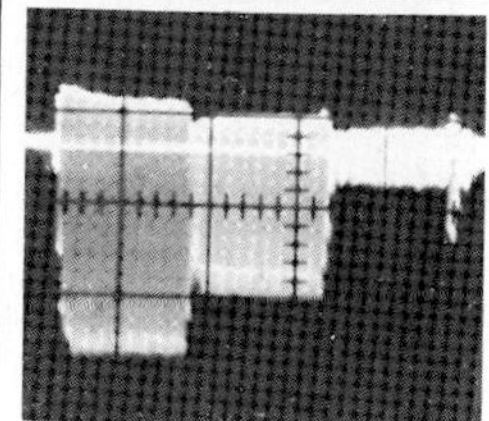	1. Amplitude of 4.08 MHz bar at the right is lower than other two bars.
LOWER CHROMA SIDEBAND HAS REDUCED LEVEL			
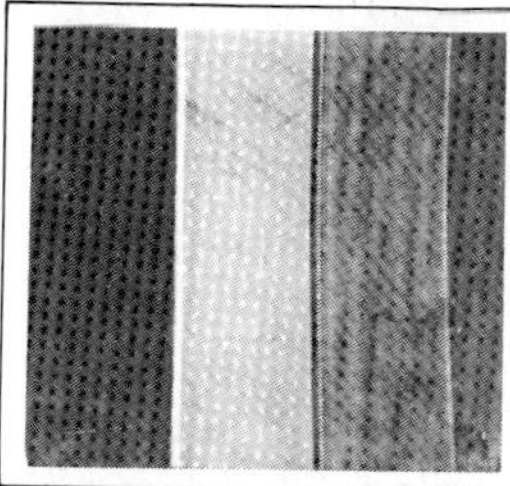	1. Lower sideband bar for 3.08 MHz disappears from screen first as brightness level is reduced.		1. Amplitude of 3.08 MHz bar at the left is lower than the other two bars.

Figure 26.43 Summary of chroma bar-sweep alignment checks. Results are shown on an oscilloscope and on the TV screen. *(Courtesy of Sencore Corp.)*

1. Feed the RF-IF OUTPUT to the VHF antenna input.
2. Set the VHF tuner and the VA48 to the same channel.
3. Select the BAR SWEEP position of the VIDEO PATTERN switch. Be sure that the 3.56-MHz (Figure 26.43) BAR SWEEP SELECTOR button is depressed.
4. Set the color killer threshold control to produce color in the 3.56-MHz bar of the BAR SWEEP pattern.
5. Adjust the 3.58-MHz adjustment for the least amount of color change in the 3.56-MHz bar. If there is a "barber-pole" effect in this bar, the adjustment should be changed until the 3.56-MHz bar produces a bar of the same color from top to bottom with the slowest amount of color change.
6. Reset the color killer threshold control for proper color killer operation.

26.12 HORIZONTAL AFC AND ANODE VOLTAGE ADJUSTMENTS

HORIZONTAL OSCILLATOR ADJUSTMENT

(See Chart 26.5) In the schematic insert, the receiver's horizontal oscillator has AFC, and this is typical of all horizontal oscillator circuits used in television receivers. As with other horizontal AFT circuits, the AFC circuit is adjusted by disabling the synchronizing pulse signal and setting the oscillator frequency so that it is approximately correct. This is the purpose of Step 1 in Chart 26.5. Note that Point DJ is now grounded and that this is the input point for the horizontal oscillator synchronizing signals. The horizontal hold control is set to approximately the center of the range, and then the horizontal frequency control (R_{428}) is set so that the free-running frequency of the horizontal oscillator is approximately correct. When the ground at Point DJ is removed, as required in Step 2, the synchronizing signal locks the oscillator.

ADJUSTING ANODE VOLTAGE

Step 2 in Chart 26.5 gives the procedure for adjusting the final anode voltage for the color picture tube. This adjustment is very important for two reasons: (1) an improperly adjusted high voltage results in poor picture quality and focus and (2) it can also produce undesirable x-rays.

26.13 COLOR PICTURE TUBE ADJUSTMENTS

CONVERGENCE ADJUSTMENTS

(See Chart 26.6.) As discussed in Chapter 18, a color picture tube must have its three beams properly converged if both monochrome and color pictures are to be correctly reproduced. (Refer to Figure 26.31 for dot patterns that indicate misconvergence and correct convergence conditions. Refer also to Chapter 18 for the discussion of convergence.)

The general convergence procedure consists of six steps, which are given at the top of Chart 26.6.

STEP	ALIGNMENT SET-UP NOTES	TEST EQUIPMENT HOOK-UP	ADJUST
1	NORM/SERV Switch in normal position. Tune receiver to a local station and synchronize picture. Short test point [DJ] on the deflection board to ground. Center Horizontal Hold Control [R28]. Increase brightness to maximum.		[R428] Horizontal Frequency Control to bring both sides of picture vertical.
2	Same as step 1, except remove jumper from test point [DJ] on the deflection board. Reduce brightness to minimum.	VTVM – Through high voltage probe to picture tube anode.	Monitor line voltage for 120 VAC. Adjust [R518], H.V. Adjust, for 25KV on voltmeter.
3	Readjust focus, height, and vertical linearity controls for proper focus and vertical size.		

Chart 26.5

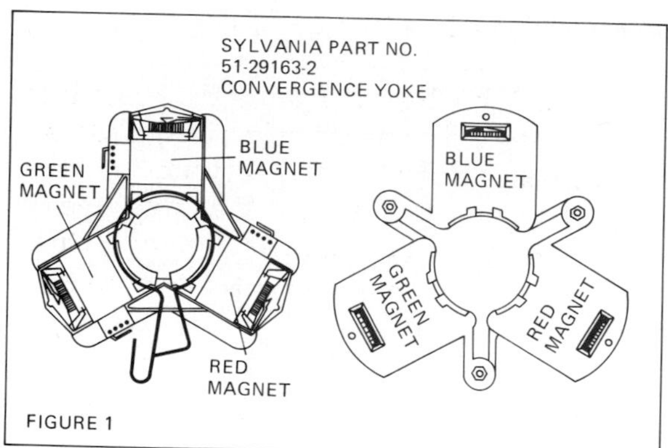

FIGURE 1

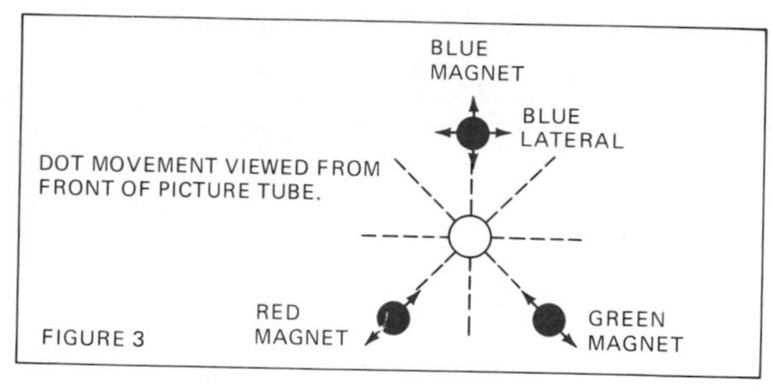

FIGURE 3

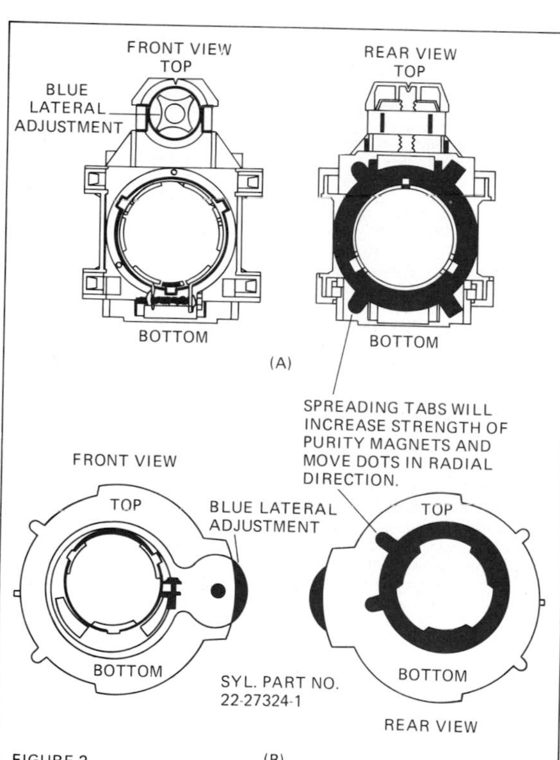

FIGURE 2

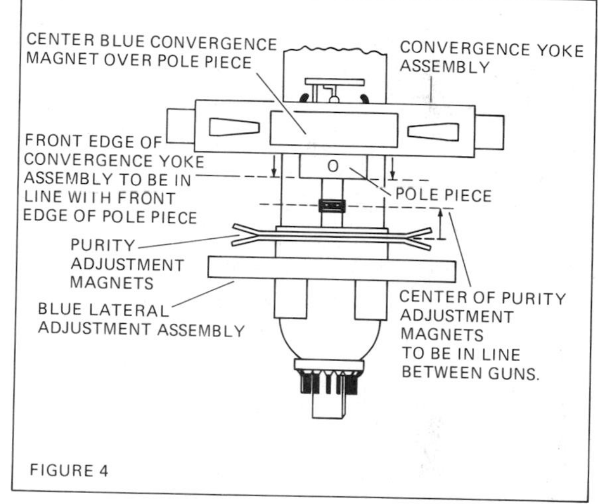

FIGURE 4

Each of the diagrams shows that portion of the screen affected by the individual controls on the convergence board. The location of the control on the convergence board is the same as the location of the diagram of its effect.

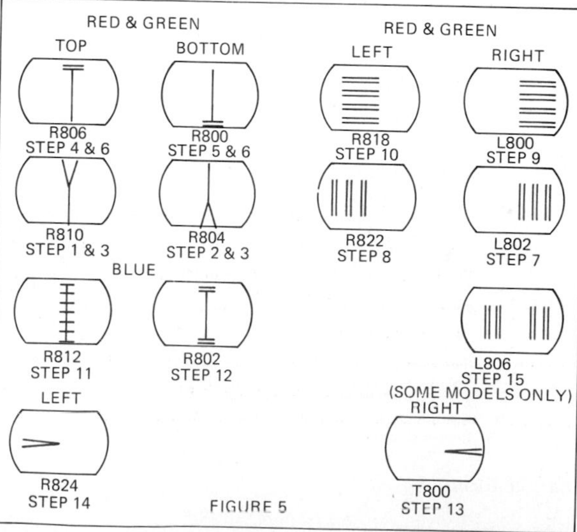

FIGURE 5

Chart 26.6

INSTALLATION OF NEW COLOR TELEVISION RECEIVER (SET-UP)

Be certain the following adjustments have been made before starting convergence procedure: H.V. Adjust, Horizontal, Focus, Height, Vertical Linearity, Width, Electrical Center and AGC.

GENERAL CONVERGENCE PROCEDURE

The following is an outline of the complete step by step Convergence procedure.

1. DEGAUSSING – Demagnetize shadow mask with degaussing coil.
2. STATIC CONVERGENCE – DC converge all three colors in center area of screen.
3. PURITY ADJUSTMENT – Adjust purity ring for red center screen area, adjust yoke position for red outer area.
4. COLOR TEMPERATURE SET-UP – Setting black and white brightness tracking.
5. DYNAMIC CONVERGENCE – Dynamically converge all three colors. Check horizontal, focus, height, vertical linearity, width, electrical centering and AGC.
6. REPEAT STEPS 2, 4, 5 for best results.

---

1. DEGAUSSING

A. Adjust receiver for a black and white picture. Observe picture for good black & white reproduction over all areas of screen.
B. If color shading is evident, demagnetizing (degaussing) of the receiver is necessary.
C. This chassis has a built-in degaussing coil and usually all that is required to degauss the receiver is to turn the receiver off for approximately 20 minutes, and then on.
D. When manually degaussing the receiver, move a degaussing coil slowly around the front faceplate of the picture tube and around the sides of the receiver; then slowly withdraw the coil to a distance of at least six (6) feet from the receiver before disconnecting the coil from the AC source. NOTE: Degaussing should be done only after receiver is placed in the position where it will be viewed and remain.

2. STATIC CONVERGENCE

A. Adjust receiver for a normal black and white picture.
B. Receiver should be in the position in which it will be operated.
C. Connect a DOT/CROSS-HATCH GENERATOR with an RF output to antenna terminals of receiver. Adjust red, green, and blue magnets (shown in Figure 1) and the blue lateral magnet (shown in Figures 2 and 4) to attain convergence of dots in the center of picture tube screen. The direction of movement of the dots using these magnets is shown in Figure 3. Lateral movement of the blue dot is accomplished by rotation of the blue lateral magnet adjustment.

3. PURITY ADJUSTMENT

NOTE: BEFORE ATTEMPTING TO ADJUST RECEIVER FOR PURITY, THE SET MUST BE TURNED ON FOR AT LEAST 15 MINUTES AND BRIGHTNESS MUST BE AT NORMAL LEVEL.

A. Tune receiver to a blank channel having a white background.
B. Adjustment for purity is most accurate while observing one color (preferably red). Decrease the Blue (R448) and Green (R450) screen controls to minimum by turning fully counterclockwise.
C. Set contrast control to minimum, brightness to normal viewing level.
D. Loosen yoke adjusting nuts and slide yoke back against convergence magnet assembly. Rotate purity ring magnets around neck of picture tube and/or spread the individual ring tabs (See Figure 2) for uniform red field area at center of screen.
E. Follow steps A, D, E, F as outlined under "COLOR TEMPERATURE SET UP" in section 4.
F. Readjust static magnets if necessary. See step 2.
G. Repeat step B under "PURITY ADJUSTMENT".
H. If the red field area at center of raster moves off center when static magnets were readjusted, the purity rings must be set again.
I. When the center of raster is red and static convergence is correct, move yoke forward and position for best overall red screen.
J. Retighten yoke adjustment nuts. Proceed with complete "COLOR TEMPERATURE SET UP", and observe all three beams.

4. COLOR TEMPERATURE SET-UP

A. Move service switch to SERVICE position.
B. Set brightness and contrast controls to mid range.
C. Set CRT Bias Control (R744) to minimum (counterclockwise) position.
D. Rotate channel selector to a free channel, rotate Contrast, Brightness, Color and the 3 screen controls to minimum (full counterclockwise) position.
E. Individually adjust screen controls so that line is just visible in the center of picture.
F. Return service switch to NORMAL position.
G. Rotate channel selector to strongest channel in the area.
H. Advance Contrast and Brightness for normal picture viewing.
I. If insufficient brightness is noticed at high brightness setting, repeat above procedure, then readjust CRT bias control (R744) for adequate brightness level after completion of screen adjustments.
J. Adjust green, red, and blue drive controls so that the bright parts of the picture are similar to, but slightly less blue, than a typical black and white receiver.

5. DYNAMIC CONVERGENCE

NOTE: MAKE CERTAIN STATIC CONVERGENCE IS COMPLETED BEFORE ATTEMPTING DYNAMIC CONVERGENCE. CHECK HORIZONTAL, FOCUS, HEIGHT, VERTICAL LINEARITY, WIDTH, ELECTRICAL CENTERING AND AGC.

A. (On some models) the Convergence board assembly is designed so that adjustment can be made from the front of the receiver. Loosen the screw(s) securing assembly to cabinet. Remove assembly and mount on top rail of cabinet with controls facing front of receiver. Secure with screws to prevent movement.
B. Adjust receiver for a normal black and white picture.
C. Connect a DOT/CROSS HATCH GENERATOR with an RF output to antenna terminals of receiver.
D. Switch to cross-hatch pattern with the following steps referring to Figure 5 while making adjustments.

6. Repeat Static Convergence Dynamic Convergence, and Color Temperature Set-Up for optimum results.

Chart 26.6 cont'd.

<table>
<tr><th>STEP</th><th>ADJUST</th><th>FOR</th></tr>
<tr><td colspan="3">DECREASE BLUE SCREEN CONTROL TO MINIMUM BY TURNING FULLY COUNTERCLOCKWISE.</td></tr>
<tr><td>1</td><td>R810 – Top R/G Vert.</td><td>For convergence of Red and Green center vertical lines at top of screen.</td></tr>
<tr><td>2</td><td>R804 – Bottom R/G Vert.</td><td>For convergence of Red and Green center vertical lines at bottom of screen.</td></tr>
<tr><td>3</td><td colspan="2">Repeat steps 1 and 2 for best convergence of Red-Green vertical center line from top to bottom.</td></tr>
<tr><td>4</td><td>R806 – Top R/G Horiz.</td><td>For convergence of Red and Green horizontal lines at top center of screen.</td></tr>
<tr><td>5</td><td>R800 – Bottom R/G Horiz.</td><td>For convergence of Red and Green horizontal lines at bottom center of screen.</td></tr>
<tr><td>6</td><td colspan="2">Repeat steps 4 and 5 for best convergence of Red and Green horizontal lines at top and bottom center of screen.</td></tr>
<tr><td>7</td><td>L802 – Right R/G Vert.</td><td>For convergence of Red and Green vertical lines on right half of screen.</td></tr>
<tr><td>8</td><td>R822 – Left R/G Vert.</td><td>For convergence of Red and Green vertical lines on left half of screen.</td></tr>
<tr><td>9</td><td>L800 – Right R/G Horiz.</td><td>For convergence of Red and Green horizontal lines on right half of screen.</td></tr>
<tr><td>10</td><td>R818 – Left R/G Horiz.</td><td>For convergence of Red and Green horizontal lines on left half of screen.</td></tr>
<tr><td colspan="3">FOLLOW STEPS A, D, E, F AS OUTLINED UNDER "COLOR TEMPERATURE SET-UP" IN SECTION 4.</td></tr>
<tr><td>11</td><td>R812 – Top Blue Horiz.</td><td>For equal displacement of horizontal blue lines lines in center of screen from top to bottom.</td></tr>
<tr><td>12</td><td>R802 – Bottom Blue Horiz.</td><td>For equal displacement of blue horizontal lines at top and bottom of screen on vertical center line.</td></tr>
<tr><td>13</td><td>T800 – Right Blue Horiz</td><td>For convergence of center blue line from middle to right side of screen.</td></tr>
<tr><td>14</td><td>R824 – Left Blue Horiz.</td><td>For convergence of center blue line from middle to left side of screen.</td></tr>
<tr><td colspan="3">Step 15 applies only to those chassis which use the TOROIDAL DEFLECTION YOKE (on some models only).</td></tr>
<tr><td>15</td><td>L806 – Left & Right Blue Vert.</td><td>For convergence of left and right blue vertical lines.</td></tr>
<tr><td>16</td><td colspan="2">Repeat all steps for best possible convergence, remove test equipment and check on air signal.</td></tr>
</table>

SLIGHT READJUSTMENT OF THE BLUE STATIC AND BLUE LATERAL MAGNET MAY BE NECESSARY. FOR OPTIMUM CONVERGENCE REPEAT THE ENTIRE PROCEDURE.

NOTE: L804 – Blue horizontal center phase is factory adjusted and MUST NOT be adjusted during the convergence alignment. Adjust only if replacement of coil becomes necessary. If coil is replaced, proceed through the convergence alignment, then connect an oscilloscope (signal lead) to test point S of convergence board, ground lead to chassis and adjust L804 until slope notch of wave shape is at 50% of amplitude. See illustration. Repeat convergence alignment for optimum performance.

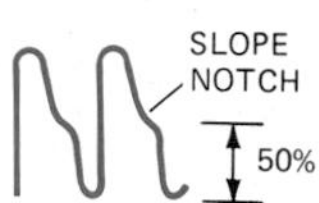

Chart 26.6 cont'd.

Degaussing The degaussing procedure is given in Step 1 of Chart 26.6 and needs no further comment here.

Static Convergence In adjusting static convergence, a dot generator is used to provide the display on the picture tube screen. Red, green, and blue convergence magnets are adjusted until the dots are superimposed and white. This condition indicates that the red, green, and blue color triads are being struck simultaneously by the proper electron beams in the center area of the screen.

The magnets around the neck of the tube that are used for making convergence adjustments are shown in Figures 1, 2, and 4 of Chart 26.6. Figure 3 of Chart 26.6 is a very important illustration because it shows in which direction the dots can be moved by each magnet. The red and green magnets move the red and green dots at an angle of 45°. The blue magnet moves the blue dots vertically and horizontally. When the set is out of convergence, dots of the three different colors are visible. As the correct adjustment of each magnet is made, the dots become superimposed to produce a pure white dot.

PURITY ADJUSTMENT

The purity adjustment is normally made with only the red gun in operation. The other two guns are turned off by adjusting the blue and green screen controls. In some receivers, switches are available to turn the individual guns on or off. Some color television generators have provisions for turning off any of the three guns. Regardless of how the blue and green guns are disabled, only the red gun should be conducting during color-purity adjustment, and the screen should display a pure red color.

COLOR TEMPERATURE AND DYNAMIC CONVERGENCE

The color temperature (gray scale) and dynamic convergence adjustments are self-explanatory and described in Chart 26.6. Keep in mind that the color temperature adjustment permits monochrome pictures to be displayed properly. Dynamic convergence adjustment assures that convergence is proper over the entire area of the screen, as well as at the center.

SUMMARY OF CHAPTER HIGHLIGHTS

1. Many so-called monochrome circuits are also used to amplify color signals. (Figures 26.1 and schematic insert)
2. The UHF tuner output signal is reduced to IF and is fed through the VHF tuner mixer to the first video IF stage. (Figures 26.1 and 26.34)
3. The tuner AFC (or AFT) circuit automatically sets the correct tuner oscillator frequency for the UHF or VHF tuner. (Figures 26.1 and 26.34)
4. The tuner oscillator frequency is particularly critical on color broadcasts in order that colors will be reproduced correctly.
5. Sound takeoff in color television receivers occurs prior to the video detector. This is to minimize the possibility of a 920-kHz beat (between 4.5 MHz and 3.58 MHz) appearing on the screen. (Figures 26.1 and 26.4)
6. The Y video driver simultaneously feeds the red, green, and blue output amplifiers to produce a monochrome

picture. (Figures 26.1 and schematic insert)

7. The noise-gate system greatly reduces or cancels noise pulses that might otherwise affect synchronization. (Figures 26.1 and schematic insert)
8. The output of the AGC amplifier (DC voltage) goes to the first picture IF and to the VHF and UHF tuners. (Figures 26.1 and 26.7)
9. Vertical blanking extinguishes the picture tube beams during vertical retrace. (Figures 26.1, schematic, and 26.10)
10. A brightness limiter and high-voltage protection circuit prevent picture tube damage and excessive x-ray radiation. (Figures 26.1 and schematic insert)
11. Horizontal blanking extinguishes the picture tube beams during horizontal retrace. (Figures 26.1, schematic, and 26.16)
12. The tint control amplifier (Q_{602}) corrects the color signal phase for proper color reproduction. (Figures 26.1 and schematic insert)
13. The first color amplifier is gain-controlled by an automatic color control (ACC) circuit similar to an AGC circuit. (Figures 26.1, schematic, and 26.12)
14. The 3.58-MHz subcarrier oscillator is frequency- and phase-controlled by an automatic phase control network. (Figures 26.1, schematic, 26.14, and 26.15)
15. The AFPC compares the 3.58-MHz oscillator output with the color burst signal.
16. The output of the 3.58-MHz subcarrier signal is fed to the X and Z demodulators. There, it is combined with the output from the color amplifier (Q_{614}). (Figures 26.1, schematic, and 26.14)
17. The output of the X demodulator drives the R–Y driver. The output of the Z demodulator drives the B–Y driver. (Figures 26.1 and schematic insert)
18. The G–Y matrix amplifier obtains its input signals from the color amplifier and from the R–Y and B–Y drivers. Its output is G–Y.
19. To align a color TV receiver, the following pieces of test equipment are required:

    1. An oscilloscope (Figure 26.17)
    2. A frequency-sweep or bar-sweep generator (Figures 26.21 and 26.22)
    3. An accurately calibrated marker-signal generator (frequently incorporated in sweep generator)
    4. An RF signal generator (Figure 26.23)
    5. A high-impedance voltmeter (Figures 26.32 and 26.33)
    6. A color-pattern generator (Figure 26.27)

20. Some important uses of an oscilloscope in checking and aligning a TV receiver are
    1. Observation of bandpass characteristics.
    2. Observation of sync and blanking pulses.
    3. Observation of the color burst signal.
    4. Used as a vectorscope. (Figures 26.19 and 26.20)
21. An oscilloscope with a dual-trace display and triggered sweep is useful for TV equipment checking or alignment. (Figure 26.17)
22. Important accessories for an oscilloscope are (1) a low-capacitance probe and (2) a demodulator probe. (Figure 26.18)
23. A vectorscope display shows the locations of the colors, with respect to their phase angle and saturation (length of bar). (Figures 26.19 and 26.20)
24. A frequency-sweep generator periodically (60 Hz or 120 Hz) sweeps across a band of frequencies. Many of these generators also provide crystal-con-

trolled marker frequencies. (Figure 26.21)

25. The Sencore bar-sweep generator (among its other functions) provides frequency *bars* for checking and aligning video-IF and color stages. It also provides crystal-controlled trap frequencies. (Markers are not required with this system.) (Figures 26.22 and 26.28)
26. An RF signal generator provides a modulated (if desired) RF signal output that is variable over a wide range (e.g., 100 kHz to 300 MHz). It can be used for peak alignment of tuned circuits or as a calibrated marker generator. (Figure 26.23)
27. The post-injection marker method does not distort the response curve. (Figure 26.26)
28. A color-pattern generator has various functions. Some are
    1. Check or align color circuits, using color bars
    2. Check or adjust static and dynamic convergence, using dot or crosshatch patterns
    3. Check or adjust picture tube purity. (Figures 26.22 and 26.27)
29. The *gated-rainbow* color-bar pattern is the one most used by TV receiver service technicians. (Figure 26.29)
30. An accurate high-impedance multimeter is essential for alignment, testing and troubleshooting. Popular types are (1) an analog solid-state meter and (2) a digital multimeter. (Figures 26.32 and 26.33)
31. In frequency-sweep alignment using markers, the actual response curve is displayed on an oscilloscope. (Chart 26.1, Figure 6, and Chart 26.3, Figure 10)
32. In bar-sweep alignment, the circuits are aligned to provide (1) equal-amplitude bars on an oscilloscope or (2) equal-brightness bars on a television screen. (Figures 26.38 to 26.40, 26.42, and 26.43)

EXAMINATION QUESTIONS

(Answers are provided at the back of the text.)

Part A Supply the missing word(s) or number(s).

1. The Sencore VA48 instrument uses the principle of ______ for alignment.
2. The AFT circuit automatically tunes the ______.
3. The sound takeoff point is prior to the ______ in a color TV receiver.
4. A noise ______ circuit is used to reduce noise pulses in the sync system.
5. A brightness ______ circuit protects the picture tube against excessive anode current.
6. AGC voltage is fed to the ______ and ______ stages.
7. The first color amplifier is gain-controlled by an ______ (abbrev.) circuit.
8. The X and Z demodulators receive the outputs of the ______ MHz ______ and the ______ amplifier.
9. The trace of ______ sweep oscilloscope is operated by the viewed signal.
10. The high-voltage overload gate (SC_{408}) (Figure 26.9) operates by causing an ______ frequency of the horizontal multivibrator.
11. The tint control amplifier Q_{602} (Section 26.2) operates by controlling the ______ of the color signal.
12. The ______ of the color signal is determined by the color control Q_{604}. (Section 26.2)
13. An ______ [or ______ (abbreviation)] circuit controls the frequency and

phase of the 3.58-MHz subcarrier oscillator.

14. The R–Y and B–Y drivers feed the ______ matrix amplifier.
15. Three color signals are produced from the outputs of ______ color demodulators.
16. Phase control of the 3.58-MHz oscillator is accomplished by comparing the ______ signal with the color ______ signal.
17. A ______ oscilloscope can be used to display two different waveforms simultaneously.
18. The bar-sweep generator uses individual ______ bars.
19. A digital multimeter is a ______ impedance instrument.
20. A reading of 1.999 V DC on a digital multimeter indicates the meter is a ______ digit instrument.
21. The modulation envelope of an RF wave can be viewed on an oscilloscope with the aid of a ______ probe.
22. A ______ is useful for checking the phases of color bars in a TV receiver.
23. A generator that periodically covers a band of frequencies is called a ______ generator.
24. Most marker generator frequencies are ______ controlled.
25. In addition to color bars, color-pattern generators develop ______ and ______ patterns.
26. Equal-amplitude detected bar patterns indicate correct circuit ______.
27. "Post-injection" refers to a system of injecting ______.
28. In performing ______ of a color-picture tube a cross-hatch pattern is useful.
29. The gated-rainbow pattern does not include a ______ signal component.
30. A high-impedance multimeter is useful in peak ______ tuned circuits.
31. Many multimeters have automatic ______ indication.
32. An ______ digital multimeter automatically selects the correct range.
33. In aligning a video-IF section, the ______ are generally set first.
34. The color-bandpass amplifier has a normal bandwidth of ______ MHz to ______ MHz.
35. The indication for bar-sweep alignment can be seen either on an ______ or on the ______.
36. Purity adjustments are generally made with the ______ color.

Part B Answer true (T) or false (F).

1. The Sencore bar-sweep generator has five frequency bars for color alignment.
2. TV receiver circuits that amplify monochrome signals do not amplify color signals.
3. Tuner AFC and AFT refer to the same function.
4. Sound takeoff in a color TV receiver is always before the video detector.
5. The Y video driver output provides the monochrome picture.
6. Three IF trap frequencies are 45.75 MHz, 47.25 MHz, and 41.25 MHz.
7. The sound intermediate frequency is 4.5 MHz.
8. The color-bandpass amplifiers have a bandwidth of about 42 to 47 MHz.
9. All color TV receivers must have three color demodulators plus a color matrix amplifier.
10. The color control can vary the color strength on the picture tube.
11. Automatic color control (ACC) is similar to AGC action.
12. X and Z demodulators are commonly used.
13. The APC (or AFPC) circuit controls the frequency and phase of the 3.58-MHz oscillator.
14. A cross-hatch pattern is essential in the adjustment of a monochrome receiver.
15. Many frequency-sweep generators have built-in crystal-controlled markers.
16. Digital multimeters are not generally suitable for alignment use because of their low input impedance.
17. RF signal generators are usually self-calibrating by means of internal crystals.
18. The Sencore bar-sweep generator also supplies dot and cross-hatch patterns.
19. A dual-trace oscilloscope with a vertical amplifier response of 2.1 MHz is adequate for the observation of the color burst signal.

20. To produce a vectorscope pattern, the R–Y and B–Y signals are connected directly to the vertical and horizontal oscilloscope CRT deflection plates, respectively.
21. Simultaneous observation of the flyback and color burst signals can be made on a dual-trace oscilloscope.
22. A low-capacitance probe should be used with an oscilloscope when viewing the composite video waveform.
23. A frequency-sweep generator periodically sweeps its range at exactly 59.94 Hz.
24. The color-IF subcarrier frequency is 42.17 MHz.
25. The Sencore bar-sweep generator produces the standard video test patterns used by cable TV companies.
26. An overall dot pattern is used only for static convergence.
27. The AFPC circuit can be adjusted with the aid of the Sencore VA48 color bars.
28. The NTSC color bar pattern is commonly used to service color TV receivers.
29. The phase separation of the color bars from a gated-rainbow generator is precisely 33°.
30. Accuracies of 0.1 to 0.5% are common for digital multimeters.
31. A frequency-sweep generator used to align a color-bandpass amplifier should sweep a range (approximately) of 3.58 ± 1.5 MHz.

REVIEW ESSAY QUESTIONS

1. Draw a simple block diagram indicating the equipment connections for frequency-sweep alignment of the video-IF section. Use post injection markers and label completely. (Section 26.10)
2. Describe the alignment of the video-IF section of a color TV receiver, including traps. Give all trap frequencies.
3. In Question 2, draw the correct overall video-IF response curve and show all important marker frequencies. (Section 26.10)
4. Draw the correct color amplifier bandpass curve, showing all important marker frequencies. (Section 26.10)
5. In Question 4, explain briefly why the bandpass is as shown.
6. In color TV receivers, explain why the sound takeoff is prior to the video detector. (Figures 26.1, schematic, 26.4 and Section 26.10)
7. Draw a simple block diagram that shows how you would align a sound-IF section using the Sencore VA48 video analyzer. Indicate control settings and required frequency output(s). Briefly describe the procedure and output indication(s). (Sections 26.10, 26.11)
8. Draw a simple block diagram to show how you would align the color bandpass amplifier with the Sencore VA48 Video Analyzer. (Section 26.11)
9. In Question 7, briefly describe the procedure and output indications.
10. In Question 8, briefly describe the procedure and two output indications.
11. List at least four uses of an oscilloscope in servicing color TV receivers. (Section 26.4)
12. Describe the difference between a single-trace and dual-trace oscilloscope. List at least two uses for a dual-trace oscilloscope. (Section 26.4)
13. Discuss the required vertical amplifier frequency response of an oscilloscope that is to be used for color TV receiver servicing.
14. List the equipment required for the alignment, testing, and troubleshooting of a color TV receiver. (Section 26.3)
15. In Question 14, describe the important characteristics of each instrument. (Section 26.3)
16. Describe, in a general way, the different techniques in video-IF alignment when using a frequency-sweep generator and a bar-sweep generator. (Sections 26.10, 26.11)
17. List all the useful marker frequencies in video IF and color alignment. Tell what each represents.
18. Tell how you would perform the following adjustments: (1) color purity, (2) static convergence, (3) dynamic convergence, and (4) degaussing. In each case, indicate what equipment (if any) is used. (Section 26.13)
19. Draw a simple block diagram that includes the following stages: (1) color output amplifier, (2) X and Z demodulators, (3) G–Y ma-

trix amplifier, (4) R–Y, G–Y, and B–Y driver, (5) red, green, and blue output amplifiers, and (6) color picture tube signal connections. Label all blocks.

20. What is a vectorscope? How is it used? (Section 26.4)
21. Explain the meaning of the rating "3½ digits" for a digital multimeter. (Section 26.9)
22. Name two common probes used with multimeters and oscilloscopes. (Section 26.4)
23. In Question 22, describe the use of each probe.
24. Explain the difference between the gated rainbow and NTSC color bars. Which has the Y component? Which is commonly used by TV receiver service technicians?
25. In the gated-rainbow display, what is the phase separation between color bars? What is between each color bar?
26. Give the sequence number and color of the ten color bars of the gated-rainbow pattern.

EXAMINATION PROBLEMS

(Selected answers are provided at the back of the text.)

1. Make up a two-column table for a color receiver. In one column list all the variable controls you can think of, (*not* alignment slugs or magnets). In the other column give the functions of each.
2. In Figure 26.4, identify the following: (1) video detector diode, (2) second 4.5-MHz trap (follows video detector), and (3) 47.25-MHz trap.
3. In Figure 26.5, identify (1) the delay line and (2) the Q_{208} signal collector load for the sync separator output.
4. In Figure 26.8, identify (1) zener diode, (2) multivibrator, (3) voltage-dependent resistor, and (4) AFC diodes.
5. For Q_{400} in Figure 26.8, I_{BE} = 1 mA and I_C = 11 mA. Calculate V_C and V_E.
6. In Figure 26.20, identify Colors 1, 4, 7, and 10.
7. In Figure 26.32, tell how positive or negative polarity is determined.
8. The maximum reading on a three-digit digital multimeter is 999 V. What would be the maximum reading if this were a $3\frac{1}{2}$-digit meter?

Chapter 27

27.1 Color-picture tube test jig
27.2 Transistor tester
27.3 Picture-tube analyzer and restorer
27.4 Picture-tube brighteners
27.5 High-voltage probes
27.6 Substitute tuner (subber)
27.7 Horizontal-output transformer, horizontal yoke and vertical-yoke testing
27.8 Analysis and isolation of TV-receiver malfunctions
27.9 Isolation by observation
27.10 Localizing color troubles by observation
27.11 Signal injection
27.12 Signal tracing
27.13 Troubleshooting the tuner section
27.14 Troubleshooting the video-IF and video-detector circuits
27.15 Troubleshooting the AGC section
27.16 Troubleshooting the video amplifiers
27.17 Troubleshooting picture tubes and associated circuits
27.18 Troubleshooting low-voltage power supplies
27.19 Troubleshooting sync-separator stages
27.20 Troubleshooting the vertical-deflection system
27.21 Troubleshooting the horizontal-deflection system
27.22 Guidepoints for troubleshooting color-TV receivers
27.23 Color-TV receiver troubles in monochrome circuits
27.24 Color-TV receiver troubles in color circuits
Summary

SERVICING AND TROUBLE-SHOOTING TELEVISION RECEIVERS

INTRODUCTION

In Chapter 26, we analyzed the various circuits of a solid-state, color-television receiver. In addition, the alignment and adjustment procedures were described. In connection with these latter items, we described some important items of test equipment and their uses.

In this chapter, we will describe briefly, additional items of test equipment that are useful in the servicing or troubleshooting of television receivers. As an example, we will describe a color-picture tube test jig, which is used to substitute for a TV receiver's picture tube and deflection components.

The remainder of this chapter will describe methods of recognizing TV-receiver faults and how to localize them to a section and a component. Many monochrome, TV screen photographs are given, as well as full-color screen photographs to illustrate various troubles in color-TV receivers.

It should be remembered that a discussion of troubles affecting the various sections of a TV receiver appear in each applicable chapter. Accompanying the text of these troubles are a number of photographs. These should be reviewed in conjunction with the troubleshooting sections of this chapter. The applicable chapter numbers are given in this chapter at each relevant troubleshooting section.

When you have completed the reading and work assignments for Chapter 27, you should be able to:

- Discuss the use and advantages of a color-picture tube test jig. Explain how proper impedance matching is accomplished.

- List at least three tests that can be performed by a picture-tube analyzer and restorer. List at least two restoring functions.
- Discuss the two basic types of picture-tube brighteners.
- Explain the function performed by an isolation picture-tube brightener.
- Explain what a "subber" is and list at least three of its functions.
- Explain how horizontal and vertical-output inductances can be tested by "ringing".
- List four common types of TV color-receiver troubles.
- Draw and label a simple block diagram of the video-IF section of a TV receiver. Discuss briefly how signal injection can be used to isolate a malfunction in any of these stages.
- Discuss the possible problems that may occur due to defects in the AGC section.
- List all the controls that can cause sync or picture defects in a color-TV receiver if they are maladjusted.
- Discuss how to reduce high- and low-frequency response in a video amplifier.
- List at least two causes of a negative picture.
- List at least three causes of the loss of anode high voltage.
- List four ways in which a color-picture tube may become defective.
- List four symptoms of defective sync operation.
- Describe horizontal "picture pulling" and list one possible cause.
- Discuss possible causes of poor horizontal linearity.
- Define horizontal foldover and describe the causes for it.
- List two types of malfunctions that cause poor (or no) horizontal sync.
- Discuss the functions of AFT, ACC, and ATC.

27.1 COLOR-PICTURE TUBE TEST JIG

A photograph of a typical test jig is shown in Figure 27.1. This is a service-bench instrument designed to facilitate color-TV receiver troubleshooting. By its use, it eliminates the need for bringing the receiver picture tube and cabinet to the service shop.

The test jig houses a color-picture tube, plus the necessary deflection components to properly operate the color-TV chassis. Vacuum-tube, hybrid and solid-state, color-TV receivers may be serviced with the jig. Yoke, convergence, and picture-tube socket adapters are used to meet the connection requirements of each individual receiver.

Horizontal and vertical-yoke matching impedances are selected by front-panel switches (see Figure 27.1). Front-panel leads are provided for ground, anode high voltage and audio (built-in speaker). For x-ray radiation safety, there is a built-in 35-kV meter.

Note: When testing a "hot" chassis TV receiver, *always* connect the chassis to the line through an *isolation* transformer.

Figure 27.1 A color-picture tube test jig. This jig can substitute for a color-TV receiver's color-picture tube and deflection components. *(Courtesy of RCA Distributor & Special Products Division.)*

27.2 TRANSISTOR TESTER

An "in-circuit," or "out-of-circuit" transistor tester is pictured in Figure 27.2. The unit pictured can test bipolar transistors, FETs, SCRs, Darlingtons and diodes. It provides automatic NPN or PNP indication. Automatic indication of silicon or germanium transistors is also provided. The unit can reliably test semiconductors in *low-impedance* circuits.

In-Circuit Tests Since most semiconductors are soldered into the circuit, it is important to be able to test them while "in-circuit." The unit pictured in Figure 27.2 (and many others), can test the part while still connected in the circuit. An "out-of-circuit" test can be made on parts not in a circuit.

With the unit of Figure 27.2, the 3 test leads can be connected in any order for in-circuit testing. Then the TEST switch is moved through its six positions. If a pulsating audio tone is heard in any position, the semiconductor is good. A tone is produced because the in-circuit transistor under test completes an audio oscillator circuit. This test is valid (with the unit pictured) for a shunt-circuit resistance as low as 10 ohms and a shunt capacity up to 15 μF. Other units may have different specifications.

Out-of-Circuit Tests For out-of-circuit tests, leakage current and loss measurements are performed. Good- or bad-test indications are shown on the panel meter. Reverse leakage can be measured from 0.1 μA to 5 mA.

27.3 PICTURE-TUBE ANALYZER AND RESTORER

A typical instrument of this type is shown in Figure 27.3. This unit can perform the following tests:

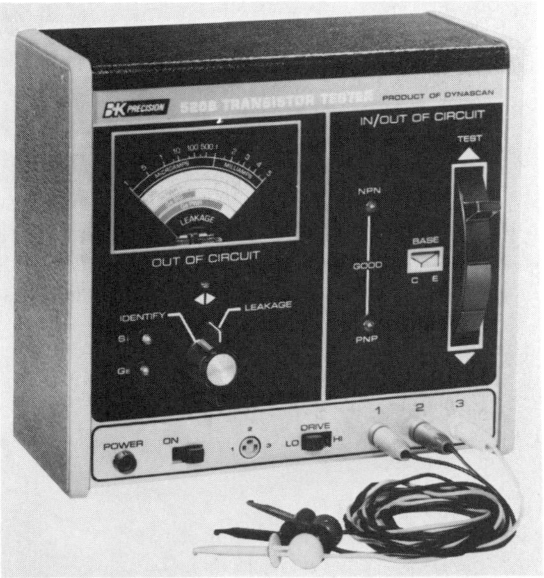

Figure 27.2 A transistor tester. This unit can test transistors either in the circuit or out of the circuit. *(Courtesy of B & K-Precision, Div. Dynascan Corp.)*

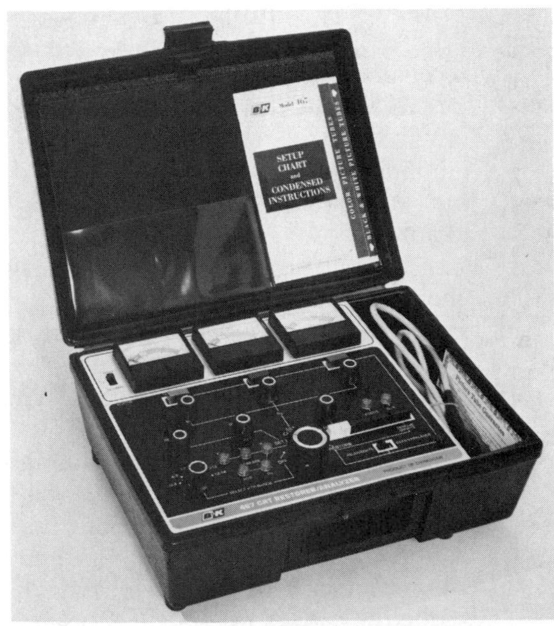

Figure 27.3 A picture-tube analyzer and restorer. This unit can test and rejuvenate picture tubes. *(Courtesy of B & K-Precision, Div. Dynascan Corp.)*

1. Cathode emission.
2. Leakage or shorts.
3. Color-tube electron-gun tracking.
4. Life.
5. Focus electrode lead continuity.

This unit can perform the following restoring functions:

1. Removal of shorts.
2. Gun cleaning and balancing.
3. Cathode rejuvenation.

Note in Figure 27.3, that there are three meters on the panel. Each of the three guns of a color tube has its performance indicated on a separate meter. All three guns of a color tube are tested simultaneously. As a result, leakage between the elements of different guns is instantly pinpointed. Cathode-to-cathode leakage is displayed, as well as cathode-to-heater and G_1-to-cathode leakage.

Note: Experience has shown that more than 95 percent of properly rejuvenated picture tubes perform as well as new tubes.

27.4 PICTURE-TUBE BRIGHTENERS

Picture-tube brighteners are made for both monochrome and color-picture tubes. In many cases the performance of worn picture tubes can be improved by the use of a brightener. Two types of brighteners are currently available, as follows:

1. Bias-network control unit (see Figure 27.4A). The unit shown is for a color-picture tube. It has individual controls for the biasing networks (G_1 to G_2 bias) of each of the three electron guns. (Picture-tube filament voltage is not boosted by this device.)
2. Filament-voltage booster unit (see Figure 27.4B). This unit is also for a color-picture tube. It works by increasing the filament voltage and thus the cathode emission. The voltage increase is in the order of 20% to 25%.

Isolation Brightener Another similar-appearing unit is called an "isolation brightener." This unit does not boost the filament voltage. It corrects the color problems that

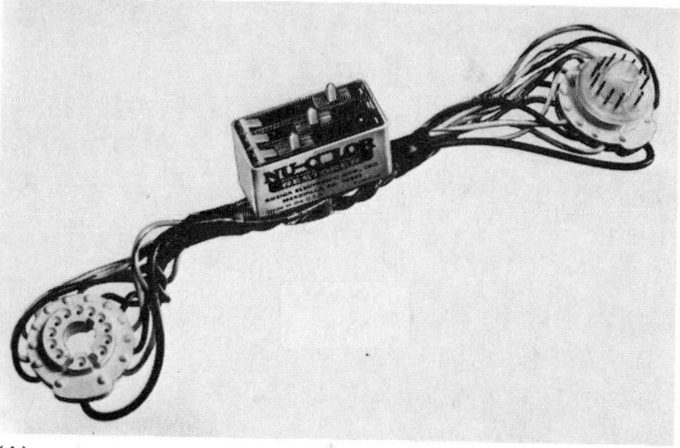

(A)

(B)

Figure 27.4 Two types of color picture-tube brighteners. **A.** This type provides increased brightness by controlling individual electron gun bias. *(Courtesy of Oneida Electronic Manufacturing Co., Inc.)* **B.** This type provides increased cathode emission by filament-voltage boost. *(Courtesy of Perma-Power Division)*

arise from a cathode-to-filament short (for one gun only). This is accomplished by the use of a transformer having separate primary and secondary windings. This isolates the picture-tube filament from ground. The cathode is thus no longer grounded and its signal functions normally.

Brightener Styles Brighteners are made with a variety of sockets to fit all picture tubes and their particular requirements. Consideration is also given to series or parallel-filament receiver connections.

27.5 HIGH-VOLTAGE PROBES

High-voltage probes can be used in conjunction with an analog or digital multimeter to extend the DC V range of the meter to a voltage in the order of 40 kV. This type of probe is shown in Figure 27.5A.

Another type of high-voltage probe has a built-in DC voltmeter. This is shown in Figure 27.5B. This device has a range up to 40 kV.

Construction Details The probes shown in Figure 27.5 are molded of Noryl thermoplastic, which has superior electrical and mechanical characteristics. Four disc barriers are used to increase the insulation and voltage-creepage distance between the handle and the probe tip. A precision high-voltage multiplier resistor (1 000 megohms) is spring suspended within the probe body.

When used with a digital multimeter

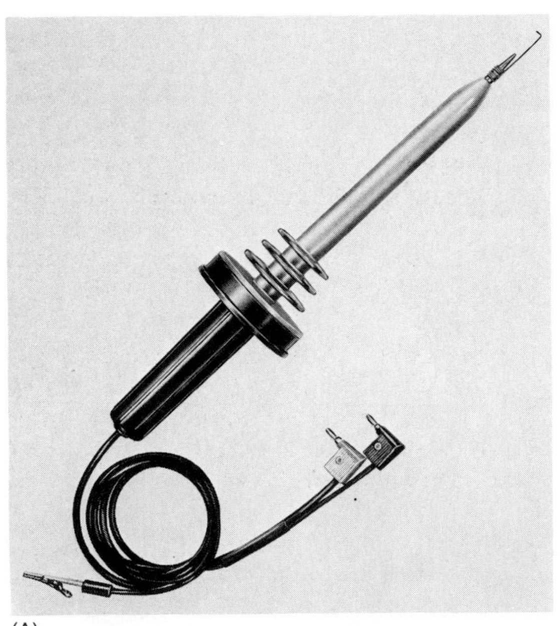

(A)

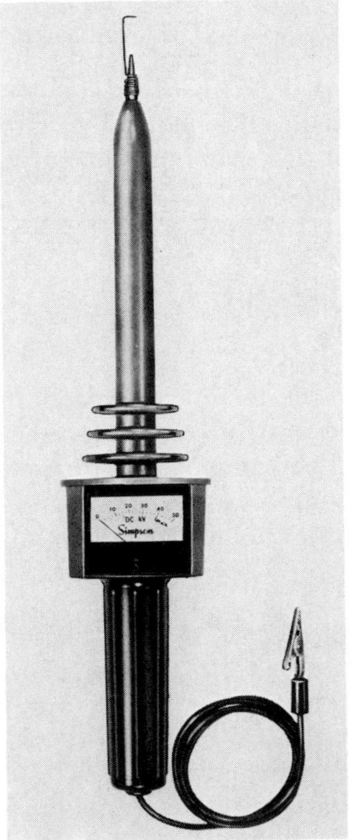

(B)

Figure 27.5 Two types of high-voltage probes. A. This probe is used in conjunction with an analog or digital multimeter. It extends the range of the meter up to 40 000 V DC. B. A high-voltage probe with a built-in DC voltmeter. Its range is also up to 40 000 V DC. *(Courtesy of Simpson Electric Co.)*

having a 10-megohm input impedance, the 200-mV (or 0.2-V) DC range is used. When used with an analog multimeter, with a 20 000 ohms-per-volt sensitivity, the 2.5-V DC range is used.

CAUTION: The above probes are to be used only by qualified personnel, trained to recognize shock hazards and trained in the precautions required to avoid possible injury. Consult the manufacturer's recommendations for using such probes.

27.6 SUBSTITUTE TUNER (SUBBER)

One way to determine if the tuner in a TV receiver is defective is to substitute an external tuner. A typical tuner "subber" is shown in Figure 27.6. The unit may be operated from the AC line, or from its two, self-contained 9V batteries. It utilizes solid-state circuits.

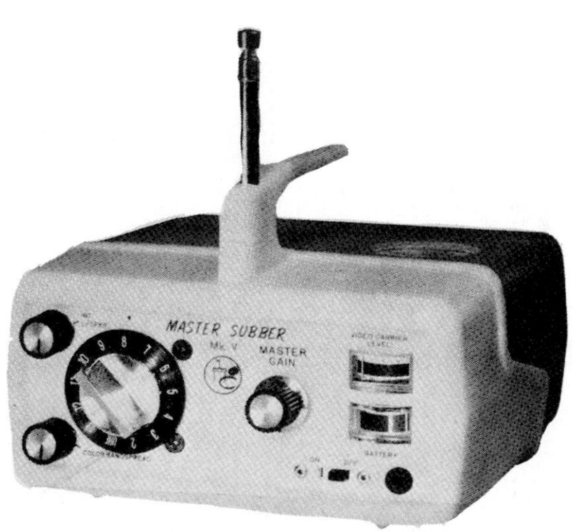

Figure 27.6 A substitute tuner (subber) used to determine whether the tuner in a TV receiver is defective. This unit also has output at video-IF, sound-IF, video, and audio frequencies. *(Courtesy of Castle Electronics.)*

Subber Specifications "Subbers" can perform other functions, in addition to substituting for the VHF tuner.

1. Substitute for VHF tuner.
2. Test UHF tuner by feeding receiver, UHF-tuner output to a 40-MHz internal amplifier. (Switch position, Channel 1.)
3. Substitute for video IF, video detector, or video-amplifier stage.
4. Substitute for sound limiter, sound detector, or audio amplifier stage.

Other Models Other models may perform fewer functions. For example, some models can substitute for both UHF and VHF tuners, but additionally, only for video-IF stages. All models can be battery operated.

SENCORE VA48 TUNER-TEST FUNCTION

(See Figure 26.23) (The basic functioning of this instrument, for bar-sweep operation, was given in Chapter 26.) In this instrument, VHF or UHF channels can be individually modulated by a video-IF bar-sweep pattern. The RF-modulated signal can be fed into individual tuner stages, or to the antenna input. By monitoring the VHF or UHF tuner output (for bar pattern), it can be determined if the tuner is operating, or which stage may be defective.

27.7 HORIZONTAL-OUTPUT TRANSFORMER, HORIZONTAL YOKE AND VERTICAL-YOKE TESTING

These units can be checked by a "ringing test," such as supplied by a separate "ringer" unit. This function is also incorporated by the Sencore VA48 unit. (See Chapter 26 and Figure 26.23) The ringing test for yokes or horizontal-output (flyback) transformers is illustrated in Figure 27.7. Connect the unit under test, in or out of the circuit, as shown. Test for GOOD or BAD on the meter. In effect, the meter "counts" the number of damped, shock-excited oscillations until they reach the 25-percent amplitude level. This test measures the "Q" (efficiency factor) of the oscillating coil. It is a reliable test of yokes and flybacks.

27.8 ANALYSIS AND ISOLATION OF TV-RECEIVER MALFUNCTIONS

A logical approach must be used to analyze a TV-receiver malfunction and to pinpoint the actual cause of the malfunction. The following approach, in general, is applicable not only to TV receivers, but also to other types of electronic equipment.

1. *Observation:* The senses of sight, hearing, and smell are called into play first. Is there a monochrome, or color picture? If so, is it abnormal and in what manner? Is the quality and volume of the sound normal? If not, how is it deficient? Use the sense of smell to determine if any part is burnt, or badly overheated. You can frequently "follow your nose" to such a part.
2. *Localizing To A Section:* This is the next step. It may be possible to localize the malfunction to a section by observation. For example, a picture out of horizontal sync only, generally indicates a fault in the horizontal deflec-

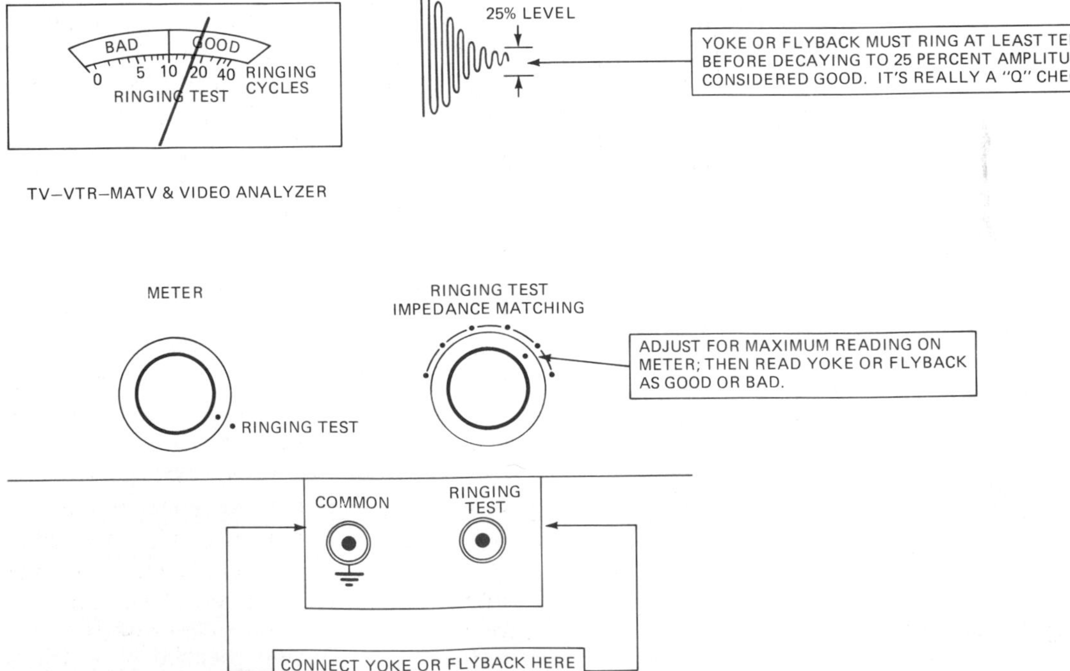

Figure 27.7 The "ringing" test ("Q" test) for yokes or horizontal-output (flyback) transformers. *(Courtesy of Sencore Corp.)*

tion section. A color picture normal except for incorrect colors, indicates a defect in the chroma section. Information on how to localize to a section is given presently.

3. *Localizing To A Stage:* Once the malfunction has been localized to a section, the next step is to localize it to a particular stage. This may be done by signal injection or signal tracing, and by voltage, and resistance measurements, as described below.
4. *Localizing To A Part:* Once the defective stage has been identified, we must find the defective part or parts. This is generally done with the aid of voltage and resistance measurements and with the use of transistor (or tube) checkers. Picture tubes can be checked with a picture-tube tester.

Notes: 1. It is vitally important *not* to simply replace parts in the hope of accidently hitting on the right one. You may actually do more harm than good this way, as well as wasting valuable time. *Always* proceed in a logical manner to find the actual part(s) at fault.

2. Be certain that the problem is not caused by a *control misadjustment*, before trying to troubleshoot the receiver. For example, a buzz in the sound may be caused by incorrect adjustment of an FM-sound detector. An incorrect AGC setting can cause a variety of symptoms including snow, overloaded picture, or poor sync.

27.9 ISOLATION BY OBSERVATION

A certain amount of common sense coupled with a good understanding of the television receiver system is needed for fast efficient troubleshooting. Refer to Figure 27.8. As an example, suppose a receiver exhibits a distorted picture accompanied by a distorted audio output. This situation immediately tells the technician that the defect is located in a circuit through which both signals pass, and it directs attention to the stages preceding the separation point. On the other hand, when only one signal appears distorted at the output, then obviously this distortion must have occurred in a circuit dealing solely with this signal. If only the audio is defective, then all attention is centered on the audio stages following the point of separation. If only the image is distorted, only the video stages beyond the separation point need be examined. With these simple facts in mind, tracking down troubles in television receivers can be simplified considerably.

In Figure 27.8, note that the sound is separated from the video after the detector. The block diagram separates the symptoms into three categories: distorted sound with normal video, distorted video with normal sound, and both sound and video distorted. The diagram is also applicable for no sound with normal video, no video with normal sound, and no sound or picture.

Further Localization Once the defect has been traced to a specific section of the receiver, the next step is to analyze that particular section with a view toward further localization. This brings us to the block diagram shown in Figure 27.9.

In the video system we have the video-IF amplifiers, the video detector, video amplifiers, and the vertical- and horizontal-sweep systems. The high-voltage power supply for the picture tube is located at the output of the horizontal-sweep system.

Once the difficulty has been traced to a particular system in the receiver, we are in a position to conduct a further analysis of our defective receiver. In the video system, for

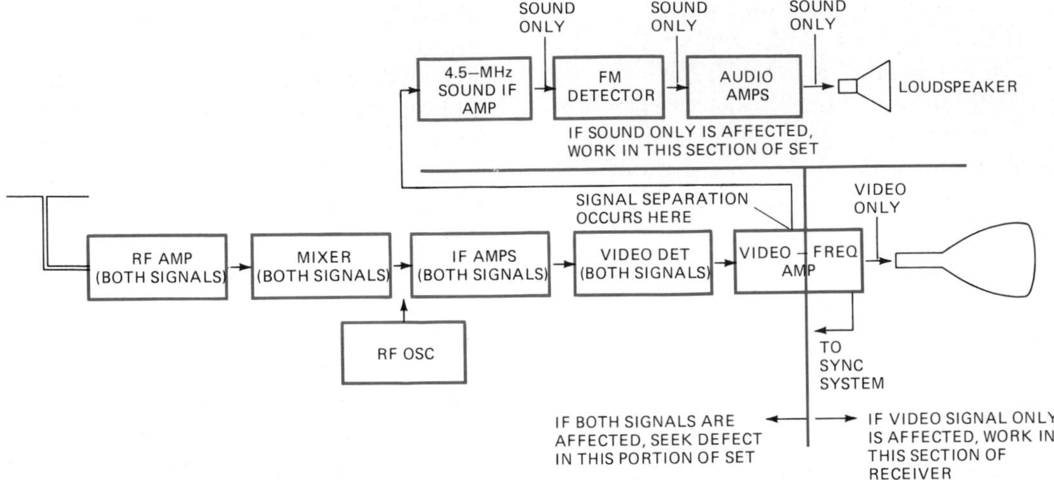

Figure 27.8 Block diagram of a monochrome-TV receiver. This diagram illustrates how observation can be used to localize a defective section.

example, breakdown of the signal path in Sections *A*, *B*, or *C* will prevent the video signal from reaching the picture tube. If the sound-separation point is prior to the breakdown, sound will be unaffected. If it is after the breakdown, sound will also be lost. In either instance, the picture-tube screen will contain a raster, but no sound.

Defective Sync Section As another example, suppose the circuit opens up in Section *D*. The sync pulses will be prevented from reaching the sweep oscillators; consequently the deflection oscillators will not be controlled by the incoming signal. On the other hand, image signals are reaching the picture tube. The visual result is a scrambled pic-

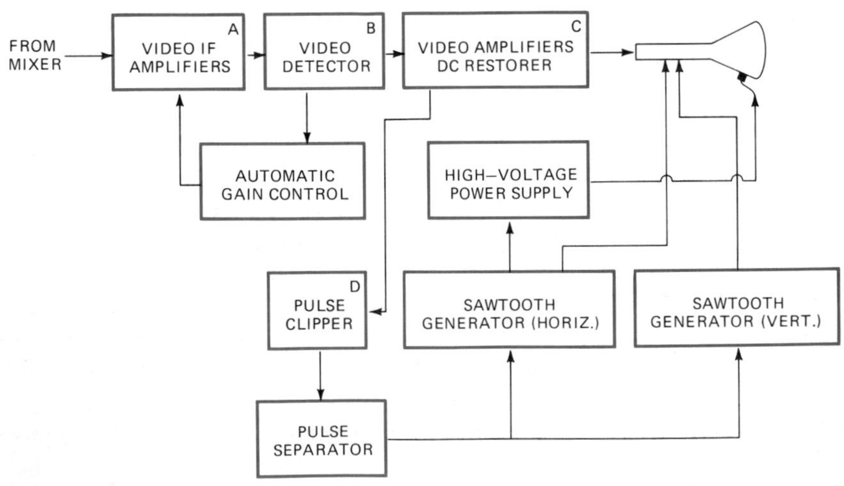

Figure 27.9 A block diagram of the video section of a monochrome television receiver.

ture. The audio section will be unaffected.

These are some of the many clues that the technician may observe. In this chapter we will undertake the analysis of many common symptoms encountered in defective television receivers. The recognition and interpretation of these symptoms will prove of great assistance in troubleshooting.

27.10 LOCALIZING COLOR TROUBLES BY OBSERVATION

Details of various color problems are described at the end of this chapter. At this time we are concerned with localizing common color problems to specific sections or stages by means of observation of the color-picture tube.

TYPES OF COLOR PROBLEMS

The types of color problems discussed here and at the end of this chapter may be grouped as follows:

1. No color picture; monochrome picture normal.
2. Weak-color intensity (strength).
3. Improper (or no) color sync.
4. Incorrect hues (tints).

For the discussions which follow, refer to the simplified-block diagram of Figure 27.10.

No Color In Otherwise Normal Picture We are looking at a normal monochrome picture. This indicates the following stages are functioning properly: video detector, video amplifier, video driver, and the red, green, and blue output amplifiers.

Since there is *no* color, as opposed to incorrect color, the trouble must be in a stage which controls *all* color signals. Thus, the trouble may be in: chroma-bandpass amplifier, color killer, burst amplifier, or the 3.58-MHz oscillator.

Weak-Color Intensity This refers to weak intensity of *all* colors, not of one specific color. This generally means the chroma-signal level fed to the X and Z demodulators and the G–Y matrix amplifier is lower than normal. Stages which may affect the chroma-signal level, are: chroma-bandpass amplifier, color killer, or burst amplifier. (Check setting of "color" control.)

Improper Color Sync In this case, the monochrome-picture sync is normal, as evidenced by a properly synced-in monochrome picture. Proper color sync depends upon the exact phase and frequency of the 3.58-MHz oscillator. Therefore, stages which may be at fault include: burst amplifier, color AFPC, or the 3.58-MHz oscillator.

Incorrect Hues This refers to *all* hues being incorrect. This problem is the result of an incorrect phase of the 3.58-MHz oscillator, with reference to the transmitted color-burst signal. [The hue (or tint) control position should be checked first.] Stages which may affect this condition are: chroma-bandpass amplifier (alignment or malfunction), burst amplifier, color AFPC, or 3.58-MHz oscillator.

27.11 SIGNAL INJECTION

Signal injection (or signal substitution) is a very useful method of isolating a TV-receiver malfunction to a section or to a stage. Signal injection is performed by injecting a test signal into a specific stage, section, or the TV-receiver antenna terminals. The performance is then checked by a monitoring device. The monitor can be an oscilloscope, the TV picture tube, or the TV speaker.

COMMON-IF SIGNAL INJECTION

An example of the technique of signal injection is given with the aid of the diagram

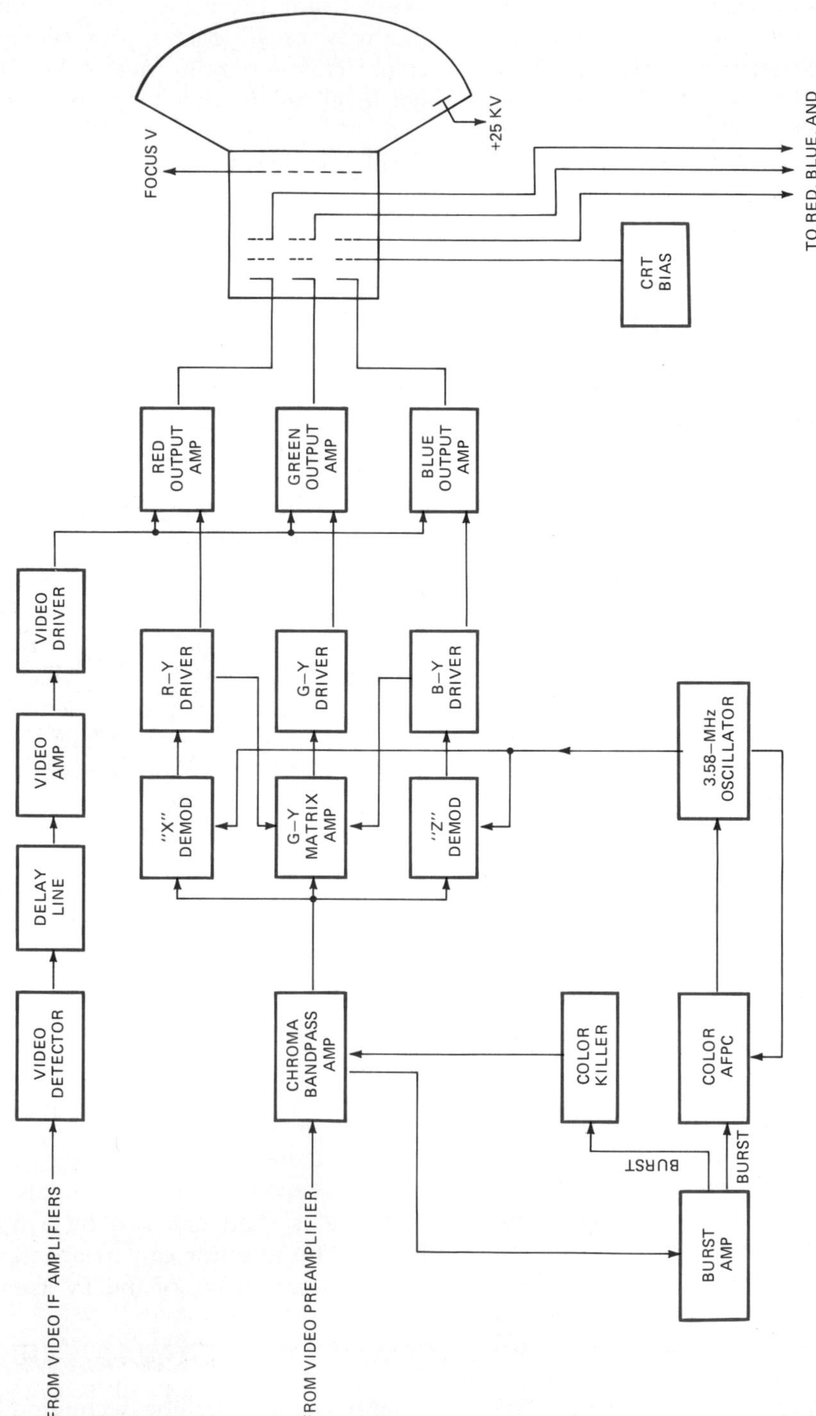

Figure 27.10 Simplified block diagram of the video detector, video amplifier, and chroma section of a color-TV receiver. This diagram will help to illustrate isolation of color troubles by observation. (See text.)

of Figure 27.11. Here we are using the Sencore VA48 video analyzer. Other types of applicable test equipment could be used as well.

For this test, an IF signal generator set to 45.75 MHz and modulated with 400 or 1 000 Hz could also be used. The output indication would be the detected 400- or 1 000-Hz frequency.

The signal output is a 45.75-MHz carrier modulated by the bar-sweep pattern (see Chapter 26). Note in Figure 27.11 that the RF-IF SIGNALS switch has three positions indicated for this particular test. Each position (including "TUNER SUB") provides a crystal-controlled output of 45.75 MHz, the picture-carrier IF. There are three positions, because each one is properly calibrated for the correct signal amplitude to be injected into a specific IF stage.

Locating A Dead-IF Stage The procedure for locating a dead-IF stage follows. Note that the indication of circuit performance can be on a scope connected at the output of the video detector. Or, it can be the indication of bars on the TV screen. (If using an IF-signal generator modulated by 400 or 1 000 Hz, the indication on an oscilloscope or the TV screen would be the audio frequency of 400 or 1 000 Hz.)

1. Set the RF-IF SIGNALS switch to the 3RD IF INPUT position.
2. Inject the 45.75 MHz from the RF-IF OUTPUT cable (through a balun transformer) at the input of the 3rd IF stage.
3. Adjust the RF-IF LEVEL control for a snowfree picture.
4. If there is no output indication, the 3rd IF stage is dead.
5. If there is an output indication from the 3rd IF stage, move the output cable to the input of the 2nd IF stage. Set the RF-IF SIGNALS switch to the 2ND IF INPUT position.
6. Now if there is no output indication, the 2nd IF stage is dead.
7. If there is an output indication from the input to the 2nd IF stage, move the output cable to the input of the 1st IF stage. Set the RF-IF SIGNALS switch to the TUNER SUB position.
8. We assume now that there is no output indication. Thus, the fault is a dead 1st IF stage.

Notes: 1. For simplicity, in the example just given, we assumed a completely dead stage. However, the same

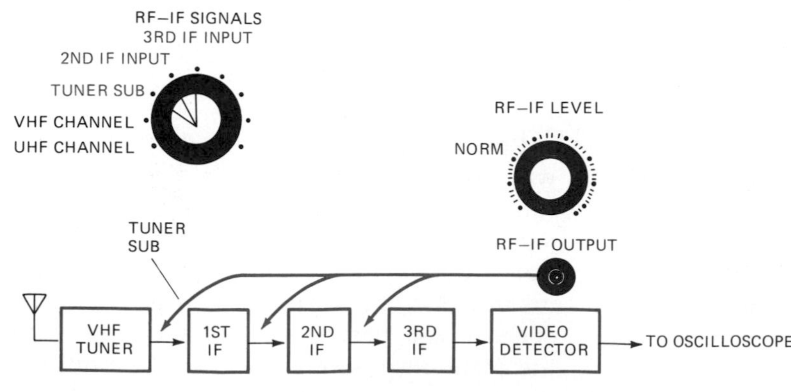

Figure 27.11 An example of the technique of signal injection (substitution). Here it is used to determine whether a faulty stage exists in the common IF section. TUNER SUB, 2ND IF INPUT and 3RD IF INPUT signals are all at 45.75 MHz. *(Courtesy of Sencore Corp.)*

technique could be used to locate a weak or misaligned stage. A weak stage would show up by lower than normal output signal-amplitude indication. A misaligned stage would be indicated by incorrect barsweep response. (See Chapter 26.)
2. A misaligned stage could be detected also, by using a frequency-sweep generator and observing the video detector output response curve.

Other defective stages in a TV receiver can be isolated by using the same signal-injection technique described above. Several examples are illustrated in Figure 27.12, using the Sencore VA48 video analyzer.

27.12 SIGNAL TRACING

In using the procedure of "signal tracing," the test signal is applied to a *fixed* circuit point. (This is unlike "signal injection," where the test signal is applied to various circuit points.) The monitoring device is then moved from point to point to check for the presence, absence, or deterioration of signal at each point. The monitoring device may be an oscilloscope, a high-impedance multimeter, or a speaker. The test signal can be the normally received TV signal, or it can be derived from a test generator.

Signal tracing and signal injection are frequently used together. For example, in Figure 27.11, the 45.75-MHz input signal could be left at the input of the 1st IF amplifier. The oscilloscope (using a demodulator probe) cable could then be moved to the output of each IF stage in turn. In this way a dead (or weak stage) can be detected. The same general principle can be applied to other sections of a TV receiver.

Notes: 1. In signal tracing, using an oscilloscope, you must use a *demodulator probe* when checking circuits prior to the video detector. When checking circuits following the video detector, use a *low-capacitance probe* with the oscilloscope.
2. In checking each section of a TV receiver by signal tracing, the correct test signal must be used for each case. For some examples of test signals, see Figures 27.11 and 27.12.

27.13 TROUBLESHOOTING THE TUNER SECTION

(See also Chapter 10.) A defect in the tuner section of a television receiver will affect both the sound and the video signals. If the TV signal is prevented from passing through completely, no sound except noise will be heard from the loudspeaker and no image will be seen on the screen. However, what will be visible is a scanning raster (see Figure 27.13).

Signal Injection Test An effective method of testing for an inoperative tuner (or one with weak output) is to use signal injection. This is illustrated in Figure 27.12A and B. In Part A, if none, or a weak signal passes through the receiver, go to Part B. If now a normal signal passes through, the tuner is defective.

A Rough Test A rough test, but one which will reveal whether or not a signal can get through is to turn the contrast control up and momentarily short across the antenna transmission line at the receiver-input terminals. Bursts of noise will be heard in the speaker and flashes of light will appear in the scanning raster. If the signal cannot get

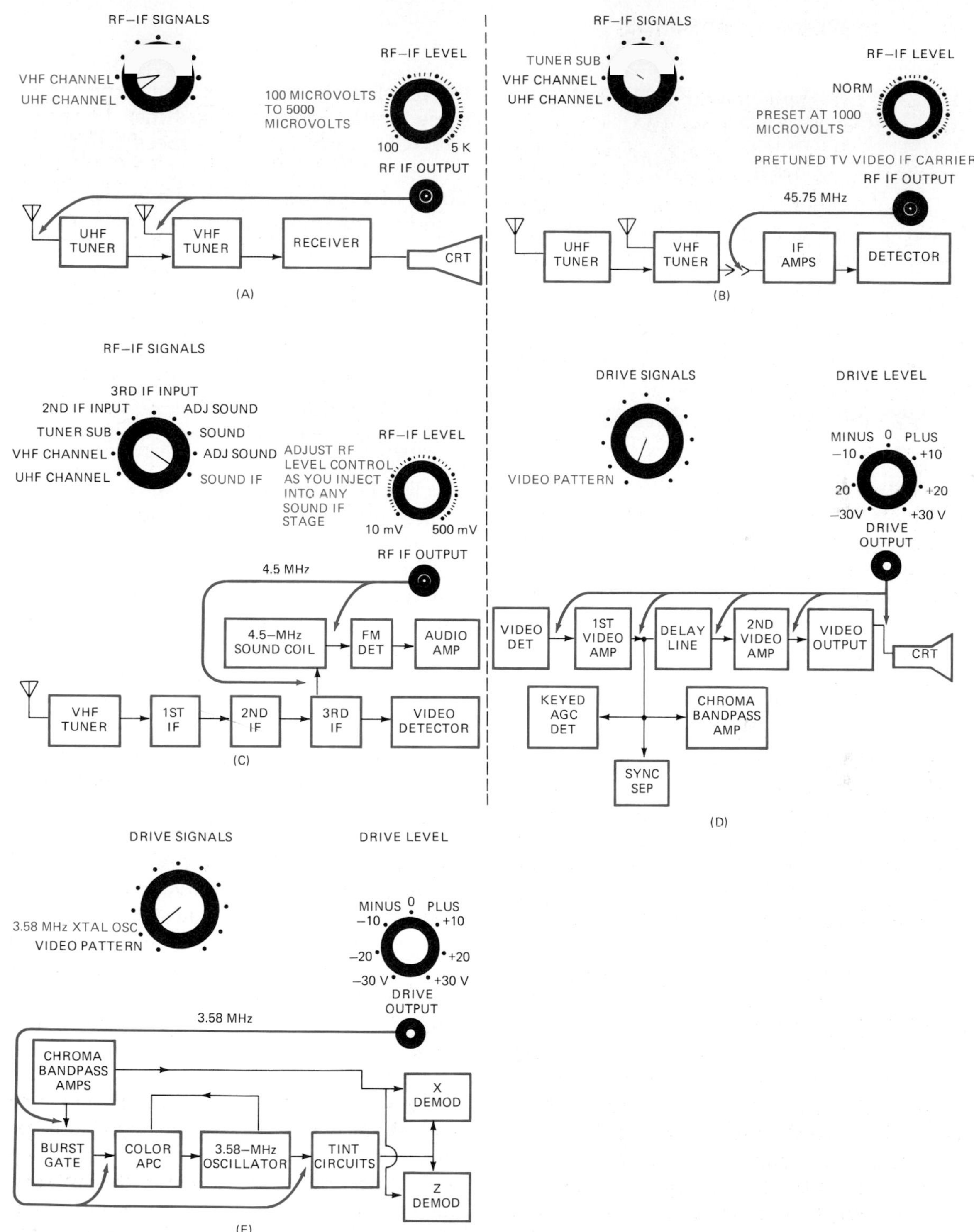

Figure 27.12 Examples of signal injection (substitution) in several sections of a color-TV receiver. *(Courtesy of Sencore Corp.)*

Figure 27.13 With a completely "dead" tuner, only a scanning raster will appear on the TV screen.

through the tuner stages, these indications will not be obtained.

Another Test Method One problem which might arise is how to distinguish between a defect in the tuner and the complete absence of any signal reaching the television receiver. The latter may occur if there is a break in the transmission line leading from the antenna to the receiver. A simple method for checking the tuner is to use an AM signal generator. Connect the instrument across the input terminals and set the generator frequency to a value about 1 MHz above the video-carrier frequency for a selected channel. Turn on the AM modulation in the generator. If dark bands appear across the screen of the picture tube, a signal is able to pass through the RF stages.

Note: The signal generator will not produce any indication on the screen if the tuner-RF oscillator is inoperative.

One Or More Channels Inoperative If only one (or several) channels are inoperative or weak, the trouble will be in the tuner. In a mechanical-type tuner, poor contacts may be the problem. In an all electronic-type tuner, some of the tuning voltages may be incorrect. (Check manufacturers specifications.)

Effect of Incorrect AGC Voltage In a vacuum-tube tuner, excessive AGC bias on the RF amplifier can cause the tuner to pass no signal, or there can be a snowy picture.

In a solid-state tuner, the RF amplifier is normally forward biased by the AGC voltage. Without forward bias, the RF amplifier is cut off. Incorrect values of forward-bias AGC voltage can cause weak or snowy pictures. (Consult manufacturer's data for correct values.)

To check for defective AGC-circuit operation, use an external-bias supply. Inject the recommended no-signal, forward-bias voltage to the RF amplifier. If normal operation is restored, the AGC system is at fault.

Additional Tests If tests show that the transistors (or tubes) are good, but the TV set is completely dead on all channels, then a voltage and a resistance check will be needed to reveal the source of trouble. Start by measuring the B+ voltages to the tuner. Check the manufacturer's specifications to make sure that these voltage readings are correct. If the receiver has an automatic fine-tuning (AFT) system, be sure to check the voltage at the tuner-AFT terminal.

27.14 TROUBLESHOOTING THE VIDEO-IF AND VIDEO-DETECTOR CIRCUITS

(See also Chapter 12 and Figure 27.11.) The next group of circuits through which the signal passes are the video circuits. These include all of the video-IF stages, the video detector, and the video-frequency amplifiers. Let us first consider the video-IF and detector circuits.

Checking Parts Failure in the video-IF system will cause the image to become distorted, lose color, or to disappear entirely from the screen. It may or may not affect the sound. After isolating the defective stage, as previously explained, the individual parts should be checked. In addition to transistors or tubes, any of the other components—resistors, capacitors, and coils–can be the cause of failure. To locate the defective unit, several methods of approach are possible. First, there is voltage checking. With a voltmeter measure the voltages of the various IF stages at the elements of the transistors or tubes. Then compare these with the manufacturer's values. This method is effective and frequently successful.

Check Power Supply Incidentally, if it is discovered that all the B+ voltages are low, trouble in the B+ power supply is indicated, and the trouble is not necessarily in the stage(s) where the voltage measurements are being made. A filter capacitor in the power supply may be leaky, the rectifier may be defective, or a shorted bypass capacitor may be at fault. In an effort to determine the one branch containing the defective component, your job now may be the tedious one of disconnecting the various branches leading off the B+ supply.

AGC Voltage Test Voltage tests are most useful in detecting open resistors, shorted capacitors, and the incorrect AGC voltages. This latter voltage is most important, because an improper AGC voltage can completely disrupt the set operation. Incorrect AGC voltages may be caused by defective transistors, tubes, diodes, or other parts.

Video-IF Stage Isolation Methods of isolating a defective video-IF stage were given in Sections 27.11 and 27.12. See also Figure 27.11. These should be reviewed now, if needed.

Signal Tracing In signal tracing the video-IF stages, with a TV-station (or equivalent TV-generator) signal, the waveform, using a scope demodulator probe, should appear as shown in Figure 27.14A and B. These waveforms should normally appear at the output of each video-IF stage. With a direct-oscilloscope probe, the same waveform should normally appear at the output of the video

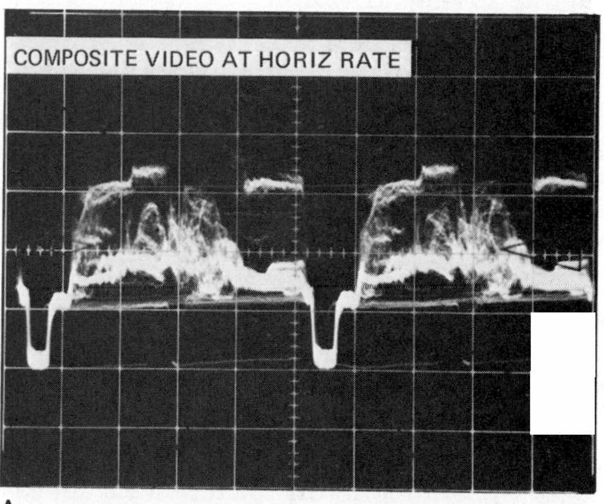

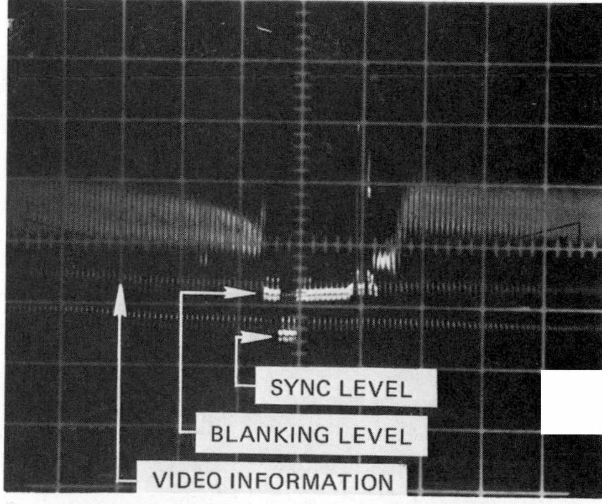

Figure 27.14 The normal composite-video signal as it should appear at the output of each video-IF amplifier and video detector. **A.** Two horizontal lines of information. **B.** Details of the horizontal sync and blanking pulse. *(Courtesy of Sencore Corp.)*

detector. In this last case, you will see the waveform only if the IF section is operating correctly.

Signal-tracing can be very effective, provided you have a good quality oscilloscope and can distinguish between a normal-appearing video signal and a distorted one. Overloading in any of the video-IF amplifiers as well as AC hum in the signal is clearly brought to light by this method (see Figures 27.15 and 27.16).

Effect of Improper AGC Voltage As in the case of the tuner-RF amplifier, the performance of the video-IF section is highly dependent upon the AGC voltage fed to it. AGC voltage is frequently fed to the 1st and 2nd video-IF stages and controls their gain. Consequently, AGC controls the signal levels of the entire video-IF section.

Improper IF-AGC voltage, which causes excessive gain can cause overloading in the third IF stage. This can cause sync compression (Figure 27.15) and resultant sync problems. In addition, the picture will appear very dark because much of the video information will be compressed toward the black level. Color information will suffer.

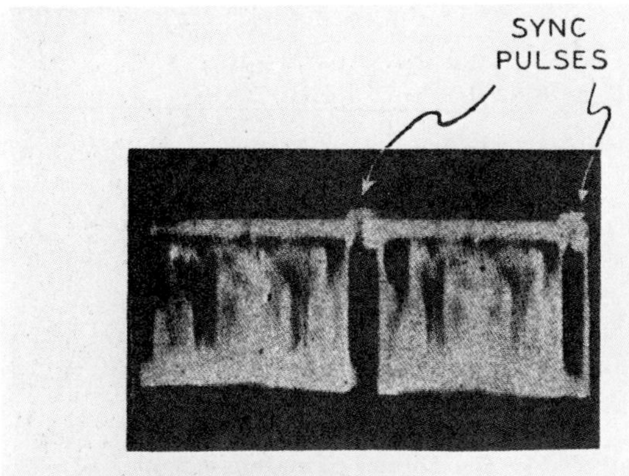

Figure 27.15 A video signal in which the sync pulses have been partly compressed. Note that the video signal extends almost up to the level of the pulses.

Figure 27.16 Result of 60-Hz hum in the video signal.

Improper AGC voltage can also cause the opposite problem, that of insufficient IF gain. A weak (or no) picture will result, with possible sync and color problems. The sound may also be lost.

If an AGC problem is suspected, it is a simple matter to check out. With an external-bias supply, connect the recommended AGC voltage to the AGC line. If the problem clears, the trouble is in the AGC section. If not, it is an IF problem.

Defective Video Detector (See also Chapter 13.) In the video detector, a completely defective diode will prevent any signal from passing through it. Defects such as a weak picture lacking in contrast, an unstable vertical or horizontal lock-in, or intermittent picture operation are other symptoms of a defective diode detector.

27.15 TROUBLESHOOTING THE AGC SECTION

(See also Chapter 14.) In the preceding two sections, we saw how incorrect AGC operation might affect the tuner-RF amplifier and the video-IF circuits. A simplified block diagram of a typical AGC section is shown in Figure 27.17. This diagram is for a solid-

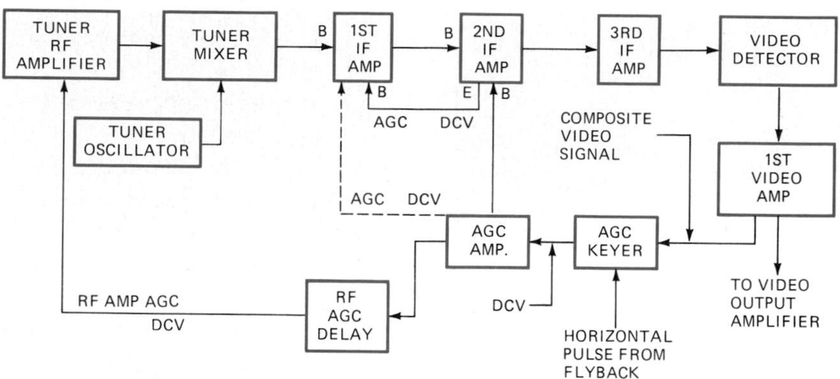

Figure 27.17 Block diagram of a typical AGC section for a solid-state TV receiver. The dashed line indicates a typical video-IF AGC connection for a vacuum-tube TV receiver. (See text.) Note: E = emitter, B = base.

state TV receiver. However, basically, except for the AGC DC V connection from the 2nd IF to the 1st IF stages, it is also applicable for vacuum-tube TV receivers. The dashed line shows a typical AGC DC V connection for vacuum-tube TV receivers (to grid circuit). For the latter, the AGC connection from the 2nd IF to the 1st IF stages would be eliminated.

In Figure 27.17, note that the AGC DC V for the first IF stage is derived from the emitter E of the second IF stage. This is known as a "totempole" arrangement.

ISOLATING AGC SECTION TROUBLES

As mentioned in a prior section, an external-bias supply can be used to isolate a possible AGC-section defect. Set the bias supply to the manufacturer's recommended voltage and connect it between the AGC line and ground. If the problem in the TV receiver clears up, the AGC section is at fault.

Note in Figure 27.17 that there are *two* AGC lines. One goes to the RF amplifier and the other to the video-IF section. Since both lines are at different potentials and serve different sections, they will have to be paralleled individually. This will partially isolate the AGC section problem. If the problem clears when the external-bias voltage is connected to the RF amplifier, the malfunction is probably in the RF, AGC-delay circuit. (See Figure 27.17.)

Troubleshooting Measures If the TV-receiver problem is cleared by the use of the external-bias supply positioned on the video-IF AGC line, the RF-AGC delay circuitry is eliminated as a suspect. (We are assuming that the 2nd and 1st IF amplifiers are normal, because of the "totempole" AGC connection.)

We proceed to track down the AGC-section malfunction as follows:

1. Check to see if there is a variable AGC control. If so, try to clear up the problem by adjusting the control.
2. With an oscilloscope, check the flyback pulse for its presence. Check the pulse amplitude against the manufacturer's specification.
3. Check for presence and amplitude of the composite video-signal input to the AGC keyer. (External bias supply must be connected for this test.)

4. Check all voltages at the AGC keyer and AGC amplifier per manufacturer's data.
5. Check the AGC keyer and AGC amplifier transistors with a transistor checker.

27.16 TROUBLESHOOTING THE VIDEO AMPLIFIERS

(Refer also to Figure 27.12D and Chapter 15.) Following the video second detector are the video amplifiers. Unless a video amplifier becomes completely inoperative, in which case no image at all is obtained on the screen, indications of other defects will be evident by their effect on the image. In a video-frequency amplifier stage, the following defects may be found:

1. A defective low-frequency compensating network.
2. A defective high-frequency compensating network.
3. Improper voltages at the transistor or tube electrodes.
4. An inoperative transistor or tube, or a defective component.

Low-Frequency Troubles When the low-frequency compensation network is defective, the background shading of the image becomes darker and the larger objects in the image "smear" (see Figure 27.18). Check bypass capacitors, load, and bias resistors. Capacitors are vulnerable components, and may be shorted or open. A fast method of checking for open capacitors is to shunt a suspected unit with another capacitor of equal value that is known to be good. Also useful are in-circuit capacitor and transistor checkers which reveal a defective unit without removing it from the circuit.

Smearing can also occur when the bias

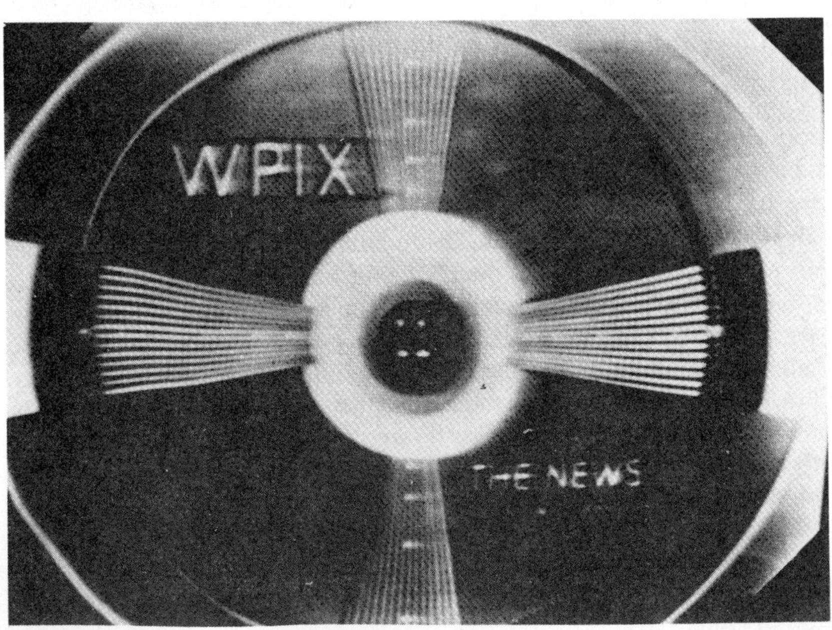

Figure 27.18 A visual indication of poor low-frequency response in a video system. *(Courtesy of RCA.)*

voltage is incorrect. Hence, measure the bias voltage at each video-amplifier stage. Leaky interstage coupling capacitors may cause incorrect-bias voltage. Low collector or plate and screen voltages produce smearing because under these conditions the transistor or tube is readily overloaded. Look, too, for a load resistor that has increased considerably in value. A transistor or tube with low gain is another cause of smearing.

High-Frequency Troubles A more difficult defect to detect is the loss of fine detail caused by poor high-frequency response of the video stages. High-frequency compensation is provided by the series and shunt-peaking coils in the video amplifiers.

The peaking coils are frequently shunted by a resistor to prevent them from sharply increasing the amplifier gain at their resonant frequency. Should this resistor increase substantially in value or perhaps open, its shunting effect would be removed and transient oscillations would develop in the coil and in its distributed capacitance whenever the signal frequency fell within this range. The oscillations would appear on the screen as ghost lines (or multiple lines) after any small object or sharply defined line or edge in the picture. Their presence is readily noticeable in test patterns. This effect, also known as "ringing" can be distinguished from ghost signals by the fact that the various lines are evenly spaced. Also, each successive line becomes progressively fainter. Ringing may also be caused by replacing a defective peaking coil with one whose value is not correct for that circuit. An "open" peaking coil shunted by a resistor would not result in a "dead" stage. However, the high-frequency response (fine details) would drop appreciably.

Quick-Stage Check If the signal is not passing through a video amplifier, as is indicated by no image on the screen (only a scanning raster), then a quick check to locate the inoperative stage is in order. Apply a 400 Hz audio signal, obtained from an audio-signal generator, across the load resistor of the video-second detector. Alternate black-and-white bars will appear across the screen if the video amplifiers are working. If the screen remains blank, move the audio generator toward the cathode-ray tube, one stage at a time, until the defective amplifier is found. Voltage, resistance, transistor and tube checks will then quickly reveal the defective component.

Signal Tracing Another approach to video-amplifier troubleshooting when no signal is reaching the picture tube is to tune in a station, then follow the video signal from the second detector to the point where the break occurs. The oscilloscope is an ideal instrument for this purpose. Place the vertical-input leads of the scope across the output-load resistor of the second detector. Set the scope-sweep frequency either to 30 Hz (to observe two fields) or to 7875 Hz (to observe two lines). Typical field and line patterns of video signals are shown in Figure 27.19. Whether the polarity of the observed signals is as shown in the figure or reversed depends upon how the detector is connected. In any event, it is the presence of the signal that is important and not its polarity.

Once the signal is observed at the output of the second detector, it can be traced to each succeeding video amplifier until it disappears or becomes distorted. Where this occurs represents the location of the defect, and voltage, resistance, and transistor or tube measurements should then bring it to light.

ADDITIONAL VIDEO-AMPLIFIER DEFECTS

Dimness in a picture, or one that lacks good contrast, can be caused by a weak video-amplifier or improper DC operating voltages.

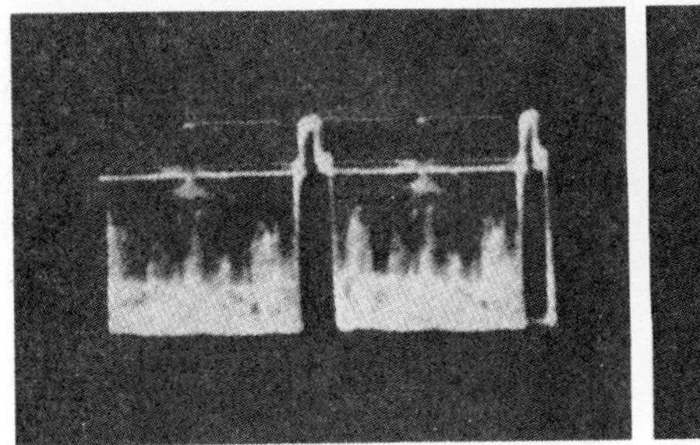
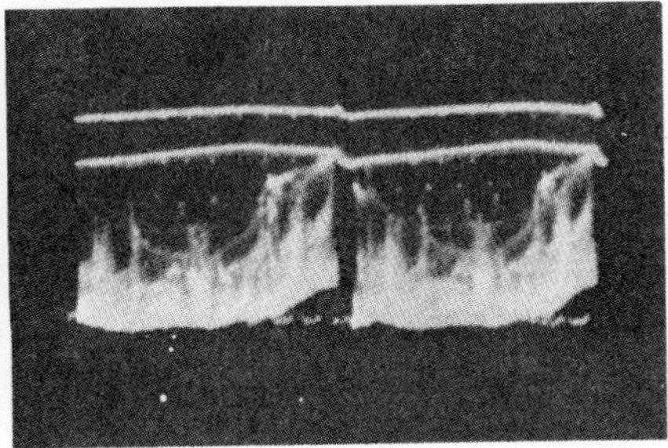

Figure 27.19 Typical line (left) and field (right) patterns of video signals.

(A dim picture may also stem from a defective picture tube or one having an excessively negative grid-bias voltage.)

Excessive Contrast On the other hand, excessive contrast—when the picture is quite dark and the grays are missing—may be caused by a defective transistor or tube, bias which is incorrect, a leaky coupling capacitor, a shorted cathode bypass capacitor, or an improperly functioning DC restorer, if one is employed. Too strong a signal from the video-second detector, due perhaps to incorrect AGC voltage, will also produce the same symptom.

Negative Picture A negative picture (Figure 27.20) represents a greater aggravation of the conditions that lead to excessive contrast. In this case, the amplifier is overdriven, resulting in the reversal of picture values, as is indicated in the illustration. Picture

Figure 27.20 A negative picture in which all the tonal values are reversed.

tubes may also cause this trouble; they may be either quite gassy or have an internal short.

4.5-MHz Beat There is usually a 4.5-MHz trap between the video detector and the picture tube. The purpose of this trap is to eliminate the 4.5-MHz beat frequency obtained by combining the video- and sound-IF frequencies in the nonlinear video detector stage. If the trap is out of adjustment, a 4.5-MHz beat pattern will appear in the picture. The beat pattern will also appear if any of the components in the trap become defective.

27.17 TROUBLESHOOTING PICTURE TUBES AND ASSOCIATED CIRCUITS

(See also Chapter 18.) The proper presentation of an image on the screen of a picture tube depends upon the following conditions: (1) a signal being present at the grids (or cathodes) of the tube; (2) the correct DC voltages being applied to the various electrodes; (3) the neck-mounted coils and magnets being properly positioned; and (4) sufficient deflection power. Failure of any one of these conditions will either distort the picture or result in its absence. Let us consider each condition separately to see what its effect will be on the picture.

NO VIDEO SIGNAL

In the absence of a signal, only a raster will be seen on the screen. The raster indicates that the high voltage plus all the other B+ voltages are operating normally. It will indicate also that the horizontal- and vertical-deflection systems are delivering the necessary deflection currents to the yoke. And from the sharpness of the raster lines and the absence of any shadows over the screen, you will know that the focus action and ion trap (not used in color tubes) are functioning properly.

LOSS OF HIGH VOLTAGE

(See also Chapter 23.) It must be kept in mind that the development of anode high voltage is dependent upon the proper functioning of the entire horizontal-deflection and high-voltage circuits. These include the following:

1. Horizontal oscillator and driver.
2. Horizontal-output stage.
3. Horizontal-output transformer and damper.
4. Horizontal yoke and associated parts.
5. High-voltage rectifier and voltage multiplier (if used).

A logical procedure (described below) is necessary to pinpoint the cause of no high voltage. The surest indication of high-voltage failure is the appearance of a perfectly blank screen. If a blank screen is accompanied by normal sound output, then we know that the low-voltage power supply is operating and we can concentrate on the high-voltage system. However, a blank screen and no sound are likely caused by a defective low-voltage supply.

Localizing Loss of High Voltage The cause of this problem must be localized between the high-voltage power supply proper and the horizontal-deflection circuits. A good place to check for this localizing is at the input of the horizontal-output amplifier. Using an oscilloscope, check the waveform (if any) at the input to the horizontal-output amplifier.

If this waveform is correct with regard to shape and amplitude, proceed to check the waveform at the emitter (or cathode). If this waveform is correct, the horizontal-output amplifier is functioning normally. The problem, most likely, is in the high-voltage power supply. Replace the high-voltage rectifier (and the damper, if required). If a voltage multiplier is used, high-voltage checks at the input and output will indicate its condition. (Use caution in any high-voltage measurements.)

No Horizontal-Output Amplifier Drive If the correct waveform appears at the input of the horizontal amplifier but not at the emitter (or cathode), this amplifier is not functioning properly. The output transistor, or flyback transformer may be faulty. Also, the B+ may be low or missing at this stage.

If the waveform at the input to the output amplifier is lacking or low or distorted, signal tracing is necessary. Using an oscilloscope, check the waveforms at the horizontal driver and oscillator stages. This will quickly isolate the defective stage.

DEFECTIVE PICTURE TUBES

(See also Chapter 18.) Color and monochrome picture-tube defects may be categorized as follows:

1. Open heater (filament).
2. Open or short circuits of picture-tube elements.
3. Low (or no) emission from one or more picture-tube cathodes.
4. Gassy tube.

Note: Rebuilt picture tubes are generally equal to new ones.

Open Heater In a monochrome-picture tube, an open heater means no cathode emission and a completely dark screen. With the heater open, you cannot see any glow at the end of the tube neck. However, this should be verified by disconnecting the socket and checking for continuity at the picture-tube plug heater terminals. This is necessary since the symptoms will be the same if there is a defect in the filament supply, the socket, or the heater wiring. If the heater is indeed open circuited, the picture tube must be replaced.

A color-picture tube has three heaters, one for each cathode. These are *internally* connected in parallel. If one heater opens up, it will mean the loss of the associated color. To restore normal color, the picture tube must be replaced.

Open or Short Circuits Picture-tube electrodes may become open or short-circuited internally. An open-circuited electrode in the tube neck may result in a dim raster, or no raster.

An open may be noted by gently tapping the tube neck and observing the screen. Momentary normal operation verifies an open connection. By the use of a picture-tube analyzer and restorer (Section 27.3) it may be possible to weld the open connection. This requires the application of a voltage of about 1 000 V between the open ends. By then gently tapping the tube neck, the open ends may weld together.

A short circuit may develop between picture-tube electrodes. In color-picture tubes, an internal short may affect only one color. The effect upon the picture will depend upon which elements are shorted together. For example, a heater to cathode short will remove the bias from that gun, increase the brightness and eliminate its picture information. Shorts between other adjacent elements will generally result in a completely dark screen.

Simple short circuits, due to the presence of foreign material, can be frequently burned out with the use of a picture-tube an-

Plate 1 Improper adjustment of the color set fine-tuning control can result in the color sub-carrier beat pattern shown here.

Plate 2 Insufficient color amplitude, which may be the result of an improperly adjusted color amplitude control or a defective band-pass amplifier stage, may result in a weak color display like the one shown here.

Plate 3 Too much color amplitude will produce a picture with oversaturated colors. An improperly adjusted color amplitude control or a defective automatic color-control circuit may cause this condition.

Plate 4 The improper adjustment of the hue control (which is also called the tint control) results in this picture with an overall green shading. This may also be caused by incorrect phasing in the color demodulation circuits. A defect in an automatic tint-control circuit can also produce this general effect. The overall color might also be red, blue, or some intermediate shade.

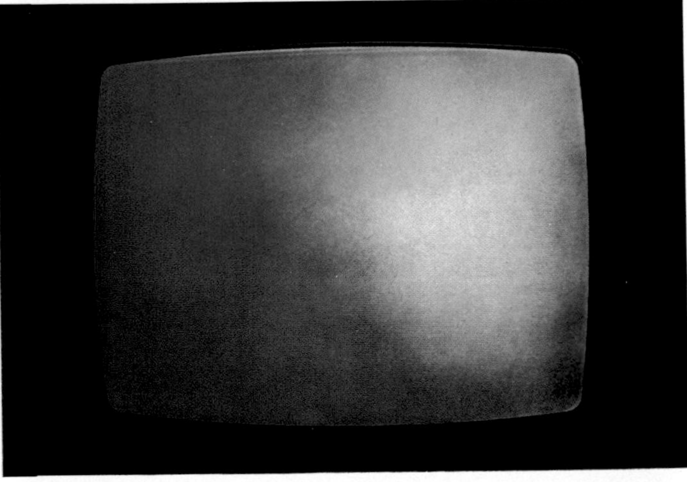

Plate 5 An example of a color tube which shows poor purity. Note the large areas of color.

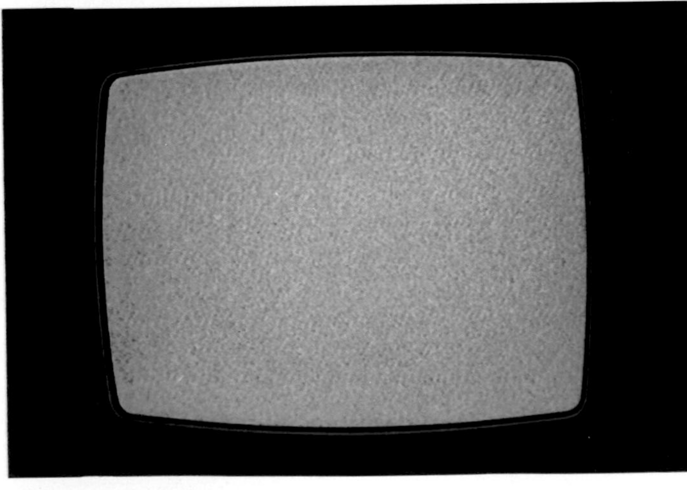

Plate 6 Colored confetti can result from low (or no) gain in the RF amplifier.

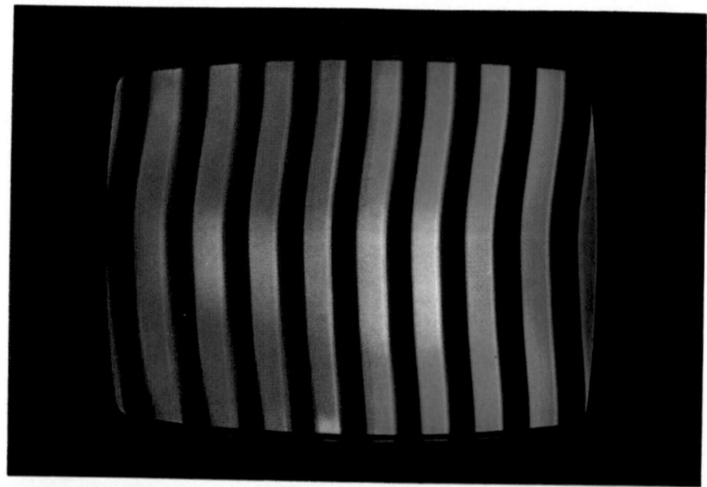

Plate 7 The appearance of the screen of a color receiver, with a color-bar generator input, when there is a 60-Hz hum in the signal circuits.

Plate 8 Ringing or close-spaced ghosts are seen here at the trailing edge of each color bar. This is a circuit-caused defect and is not due to multi-path reception.

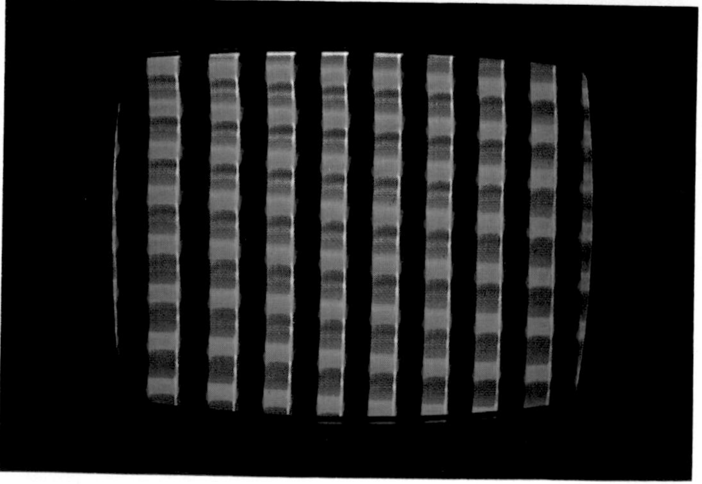

Plate 9 A color-bar generator pattern when color sync is lost. (Monochrome-horizontal sync is normal.)

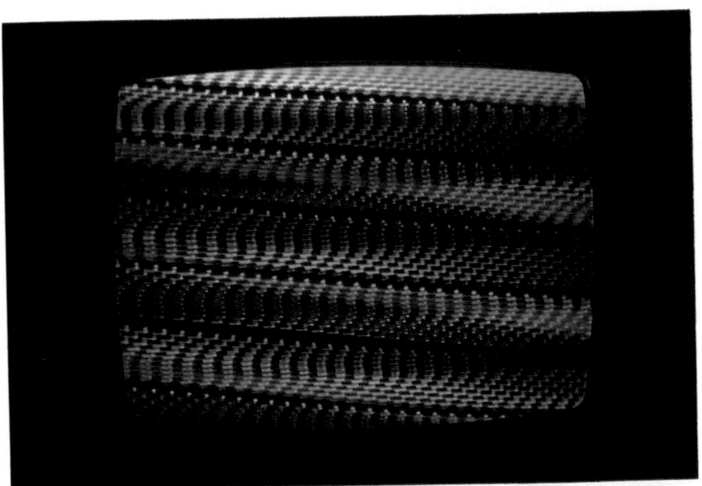

Plate 10 The effect on a color-bar pattern, when both monochrome (horizontal) sync and color sync are absent.

Plate 11 Appearance of the color-bar signal when there is no G-Y signal present. Note the complete absence of green coloring.

alyzer and restorer (Section 27.3). A high voltage (600 V or more) and current is passed through the short and may burn out the foreign material. For a heater-to-cathode short, a picture-tube isolation booster will remedy the condition. (Section 27.4).

Low Emission Low-cathode emission occurs eventually in all picture tubes. With color tubes, this symptom may occur initially for only one cathode. Here, only one color is affected. In a monochrome-picture tube, low emission results in decreased brightness. It also causes a "silvery" appearance to bright-picture areas.

A picture-tube analyzer and restorer (Section 27.3) can be used to check the emission of the cathode(s).

Low emission can frequently be remedied by the use of picture-tube brighteners (Section 27.4). It can also be remedied by "rejuvenation" using a picture-tube analyzer and restorer. (Section 27.3.)

Gassy Tube Some symptoms of a "gassy" tube are similar to those of a tube having low emission. In addition, the tube may show *negative* pictures (Figure 27.20.)

A picture tube which is somewhat gassy may sometimes cause picture-blooming, a condition which causes the picture to expand in all directions as the brightness control is turned up. Blooming is accompanied by loss of focus and sometimes by complete loss of picture and raster. The blooming or spreading out of the picture is due to a decrease in the high voltage applied to the tube. If the tube is gassy, it may be drawing an excessive amount of current, and this drain on the high-voltage supply can cause a considerable reduction in high voltage. Gas may occur inside the tube envelope due to a vacuum leak or to gas driven from a tube element. In any event, a tube with this problem is generally not repairable and must be replaced.

Effect of Ion-Trap Magnet (Monochrome-picture tubes only.) An incorrectly positioned ion-trap magnet can be responsible for reduced brightness. Owners who tamper with their sets may also bring about a reduction in brightness by moving the ion trap. Or, in moving a set about the house, the ion-trap magnet may have been jarred out of position. Or, an inexperienced technician might have had difficulty arranging the deflection yoke and centering device to obtain a shadowless picture, and might have "solved" the problem by shifting the ion-trap magnet from its optimum position.

PICTURE DISTORTIONS DUE TO DEFECTIVE PICTURE-TUBE COMPONENTS

Associated with a picture tube may be a deflection yoke, a centering magnet (monochrome only), an ion trap (monochrome only) and a focus coil. Improper placement of any of these components will have an adverse effect on the picture. By the same token, any defect within these components will also tend to distort the picture or even prevent it from appearing on the screen.

Poor Focus When a picture is not in proper focus, the cause may be: a defect in the electron gun (in electrostatically focused tubes), the application of an incorrect voltage to the focus electrode (again, in electrostatically focused tubes), an improperly adjusted focus control, a defective focus coil (in electromagnetically focused tubes), or a resistive change in the focus circuit. When a focus coil is employed, it must have an optimum amount of current flowing through it. Too much or too little current will produce defocusing. If the set uses a permanent-magnet focus unit, poor focus indicates improper placement of the magnet or possibly a weakened magnet. In Figure 27.21 the ion trap is

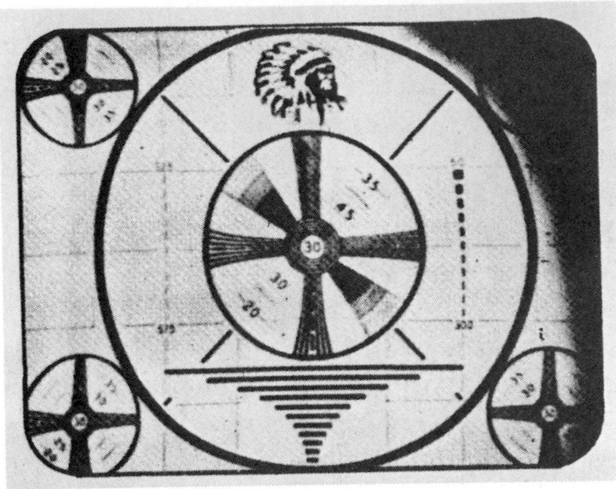

Figure 27.21 Improper placement of ion-trap magnet on the neck of a monochrome picture tube.

Figure 27.22 The shadow around the outer edge of the pattern is caused by a deflection yoke that is not as close to the cone of the tube as it will go.

out of position. In Figure 27.22 the deflection yoke is not as close to the cone (or bulb) of the picture tube as it should be.

Effects of Defective Yokes Short circuits of parts of yoke windings may produce "keystone"-picture shapes. The effect of such shorts is shown in Figure 27.23A for the horizontal yoke. Figure 27.23B shows the effect for the vertical yoke.

Effects of Other Yoke Components Thermistors (resistors that change resistance value when their temperature is changed) are used with some yokes to compensate for changes in yoke resistance when the yoke temperature changes. If the thermistor becomes defective the symptoms shown in Figures 27.23A and 27.14B may be the result.

Besides the two sets of windings in the deflection-yoke housing, there will be found damping resistors (across the vertical windings) and a small capacitor (across one of the horizontal windings). Should any of these components become defective, picture distortion will be produced. A typical appearance of the image in this case is shown in Figure 27.24 in which several ripples at the left side of the screen cause the pattern to appear wrinkled.

Note: A defective yoke can be tested by the "ringing" test shown in Figure 27.7.

27.18 TROUBLESHOOTING LOW-VOLTAGE POWER SUPPLIES

The low-voltage power supply (see also Chapter 19) generates the various B+ (and B−) voltages required by most of the stages of a monochrome or color-TV receiver. Consequently, a trouble in this supply can affect the operation of the entire receiver, or of large sections of it. This is true of color or monochrome-TV receivers.

Generally, low-voltage power supplies can be defective in the following ways: (1) no voltage output, (2) one or more voltages too low or absent, (3) insufficient filtering, (4) defective voltage regulator, (5) low or no output from a horizontal-frequency supply. (Usually associated with horizontal-deflection problems.)

No Voltage Output This condition may be caused by one of the following defects.

A

B

Figure 27.23 Distorted images that stem from the deflection yoke. **A.** Distortion as produced by a short across part of the horizontal-deflection yoke windings. **B.** Short in vertical section of yoke windings.

1. Blown B+ or AC fuse, or defective AC switch.
2. Open filter choke or series filter resistor.
3. Defective or cutoff series-pass regulator transistor.
4. Shorted filter capacitor.
5. Open rectifier in half-wave circuit.

Figure 27.24 Ripples caused by an open or wrong value of capacitor across half of the horizontal winding of the yoke.

One or More Voltages Low or Absent This condition may be caused by one of the following defects:

1. Open filter capacitor.
2. Series filter resistor open or increased in value.
3. Defective rectifier(s).
4. Open diode in a bridge rectifier, or voltage multiplier. For the bridge rectifier, this will also change the ripple frequency from 120 to 60 Hz. This will result in poorer filtering.
5. Defective voltage-regulator operation.
6. Excessive current drain in one of the circuits.
7. For a horizontal-frequency supply, if the "start-up" circuits do not work, there will be no "derived B+" and no horizontal-deflection circuit operation.
8. In Item 7, obtaining the correct voltages depends upon the proper operation of the entire horizontal-deflection system.

Insufficient Filtering This is generally caused by defective electrolytic capacitors (not shorted). In time, these may lose capacity or become open. The result is increased ripple voltage (hum). An open filter capacitor can be checked by shunting it with a known good one of approximately the same capacity. If operation becomes normal, the original capacitor is defective. A leaky capacitor must be disconnected before the "good" capacitor is substituted.

When making a visual inspection of a receiver during any servicing job, the technician should check the filter capacitors for signs of corrosion, dripping, or scaling around the base of the capacitor. The capacitor may still work perfectly, but these signs may indicate that the capacitor is about to fail and should be replaced. The failure of a filter capacitor usually produces excessive hum (in the speaker), low-volume output, and heavy black bars on the face of the screen.

Effects of Shorted or Leaky Filter Capacitors A shorted filter capacitor can cause the rectifier to burn out. It can cause the fuse to blow if the circuit is fused.

With a leaky filter capacitor, the voltage output of the power supply may be low, there is hum in the speaker, dark horizontal bars appear on the pictures, and the transformer and filter chokes may overheat.

Effects of Open-Filter Capacitors An open-filter capacitor produces most of those symptoms caused by leaky capacitors except the heating of parts. It may lower the B+ voltage. This is especially true if the capacitor in question is the input capacitor of the filter network. Hum may be heard in the speaker.

Effect on Screen One visual effect of an improperly filtered power supply is shown in Figure 27.25. The distortion in Figure 27.25 is due to a 60-Hz ripple in the power supply voltage, reaching the horizontal-deflection system. This may occur if one-half of a full-wave rectifier, or bridge rectifier, becomes inoperative. The curvature at the edge of the image represents one cycle of a 60-Hz sine wave laid on its side. If the ripple voltage is 120 Hz (open filter capacitor), then the number of "bends" is doubled. The visual effect of ripple in the vertical-deflection system is an alternate spreading and crowding of the image in the vertical direction.

Sound Bars in Picture Sound signals reaching the picture tube produce an effect which is similar in certain respects to AC ripple. Each defect causes black bars to appear across the screen, but those caused by the sound voltages are more numerous and their intensity changes in step with the amplitude of the applied audio (see Figure 27.26). The black bars produced by hum in the

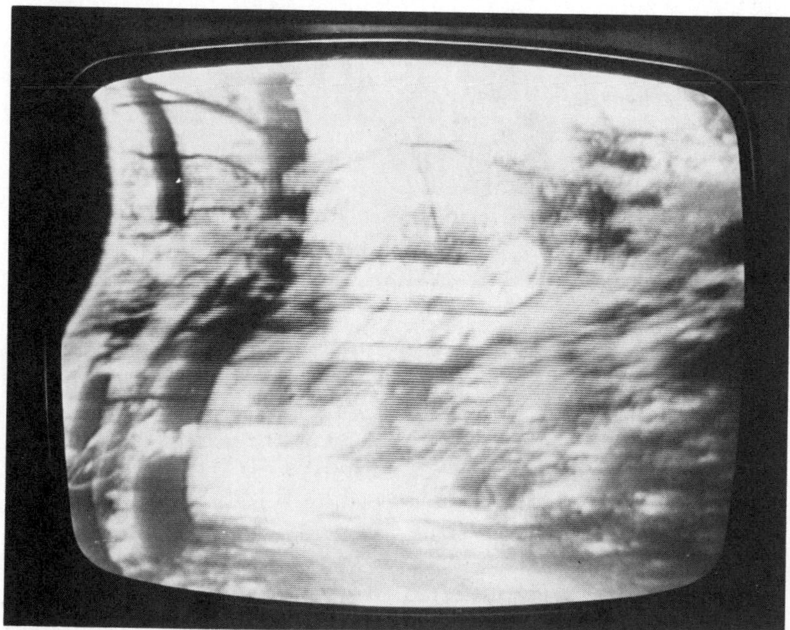

Figure 27.25 Visual effect of poorly filtered power supply. The distortion is due to 60-Hz ripple in the voltage reaching the horizontal-deflection system.

Figure 27.26 The visual effect of sound voltages in the video system.

power supply do not exceed two, are much wider than the audio bars, do not vary in intensity with the sound, and are visible on all channels.

The condition shown in Figure 27.26 can be caused by poor voltage regulation. This, in turn, can result from an open filter capacitor, or a defective voltage regulator.

Transformerless Receivers Television receivers which use semiconductor rectifiers do not always use a power transformer, which means that the technician must be careful not to ground the chassis of such receivers unless he is certain that the side of the power line to which the chassis is connected is also at ground potential. Unless this precaution is observed, it is quite possible that a fuse will be blown and any equipment connected to the receiver may be damaged. For this reason, *isolation transformers* should always be used when servicing receivers without power transformers.

27.19 TROUBLESHOOTING SYNC-SEPARATOR STAGES

(See also Chapter 20.) The sync-separator stages "clip" the horizontal and vertical-sync pulses from the composite video signal. The horizontal and vertical-sync pulses are then separated by differentiating and integrating circuits, respectively. After separation, the pulses are fed to the horizontal- and vertical-deflection circuits. There, they synchronize these circuits to be in step with the scanning at the TV transmitter. A simplified block diagram is shown in Figure 27.27.

The sync-separator stages can be defective in the following ways:

1. No horizontal and vertical sync.
2. No vertical sync only.
3. No horizontal sync only.
4. Loose vertical sync.
5. Horizontal-picture "pulling".

NO HORIZONTAL AND VERTICAL SYNC

This condition will cause the picture on the screen to roll both horizontally and vertically. This can be caused by a lack of sync pulses or by compressed sync pulses. (See Figure 27.28.)

While it may be possible to lock the picture in momentarily by adjusting the verti-

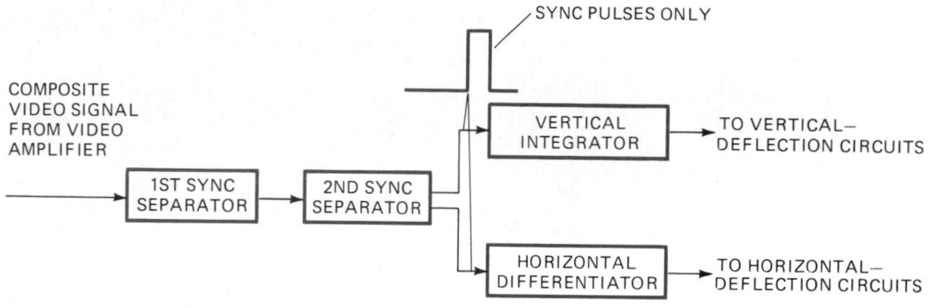

Figure 27.27 Simplified block diagram of sync-separator stages.

cal- and horizontal-hold controls, the picture will soon fall out of sync again.

Localizing Defect TV Broadcast The best way to check through a sync circuit to localize a defect, is by tuning in a TV broadcast signal or injecting a video signal from a TV generator at the output of the video detector. Then use an oscilloscope to check pulse waveforms at the input and output of each of the sync stages. The point at which to start is the input to the sync-separator section. The waveform here should be the composite-video signal (see Figure 27.29A). At the output of the sync separator the vertical and horizontal waveforms will appear as shown in Figure 27.29B. (The oscilloscope sweeping rate used to observe the vertical-sync pulses is 30 Hz; the horizontal sync pulses, 7 875 Hz.) Note how much the video signal has been suppressed, or even eliminated, while the sync pulses are being amplified.

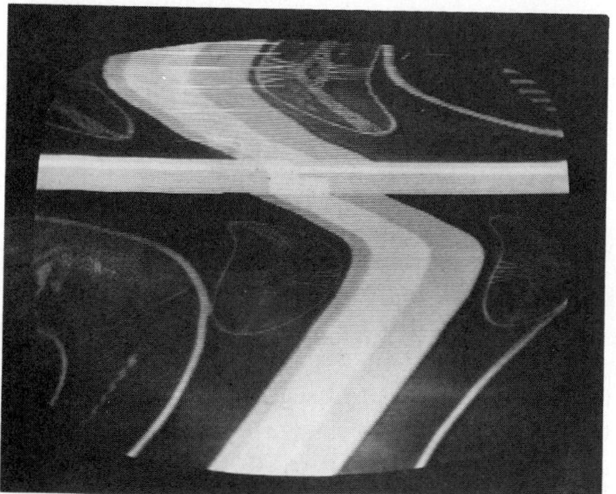

Figure 27.28 Loss of vertical and horizontal lock-in. *(Courtesy of Howard W. Sams and Co.)*

OBSERVING THE SYNC PULSES

When observing sync pulses on an oscilloscope screen, the technician will find that the horizontal pulses stand out clearer and more distinctly than the vertical pulses. One reason for this is that the horizontal pulse is simpler in structure than the serrated vertical pulse. Also, the horizontal-sync pulse occupies a greater proportion of a line than the vertical sync pulse does of a field. Hence there is more of the pulse to be observed when the scope-scanning rate is set to the proper value. These facts are borne out by the oscillograms of vertical- and horizontal-sync pulses shown in Figure 27.19.

Checking The Video Signal It is important to observe carefully the composite video signal which is applied to the input of the sync-separator section. If, for example, it is found that the sync pulses have been compressed (Figure 27.30), and it is difficult to keep the picture in sync; then it can be surmised that it is not the sync-separator stages that are at fault, but the preceding video sys-

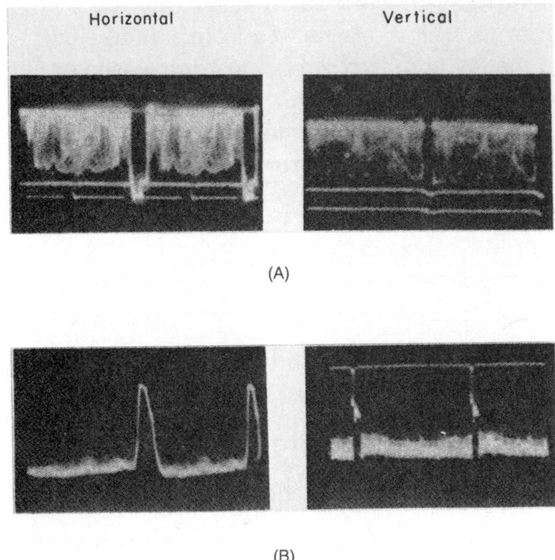

Figure 27.29 An illustration of how the sync-separator stages separate the sync pulses from the video signal. **A.** At input to sync separator system. **B.** Output of first sync-separator.

tem. It is quite possible that the defect exists in the AGC network, wherein the controlled RF and IF amplifiers are being permitted to operate at higher than normal gain because of incorrect bias. This can readily lead to overloading with subsequent sync-pulse compression.

When this signal reaches the sync separators, there is a considerable amount of video signal operating at nearly the same level as the sync pulses, and consequently it becomes impossible to effect a clear-cut separation. In the vertical system, this will show up as unstable or critical hold in. The picture will have a tendency to roll. In the horizontal system, the automatic frequency-control network may hold the picture more firmly in place. However, the hold-in range will undoubtedly be smaller than normal. Also, a bend may appear at the top of the picture. Vertical lines or objects in the picture will be found to curve to the right or to the left. In extreme cases, the top of the picture will "flag-wave," that is, will move rapidly from side to side.

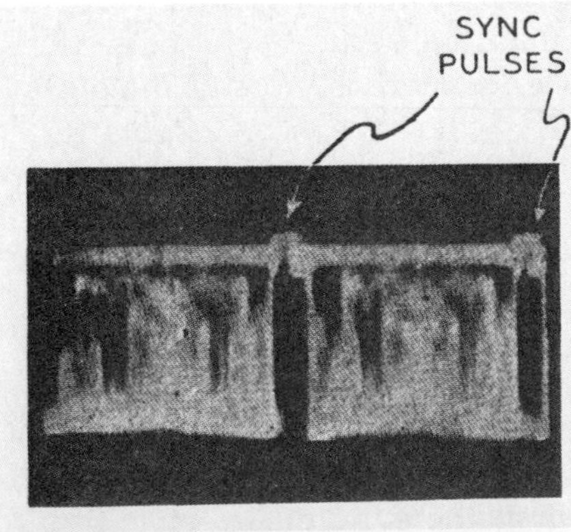

Figure 27.30 A video signal in which the sync pulses have been compressed. Note that the video signal extends almost up to the level of the pulses.

Tracing Cause of Compressed Sync Pulses

If it is found that sync pulses are being compressed or clipped entirely before they reach the sync-separator stages, then the path of the video signal should be traced back to the video-second detector with the oscilloscope. Sync-limiting or compression may occur at almost any point from the RF amplifier to the sync-takeoff point. A defective transistor or tube, a defective AGC system, a signal which is too strong, incorrect operating voltages or component values can all be responsible. Examine the signal at the video-second detector. If it looks normal, then the preceding RF and IF stages are operating normally. If the sync pulses are absent or have been compressed, the trouble exists prior to the video-second detector. Check the voltages at the video-IF transistors or tubes first. Pay particular attention to improper AGC voltages since these are a frequent source of trouble. The troubleshooting procedure should now follow along the lines previously outlined for video-IF amplifiers.

When poor sync action has been traced

to the sync-separator stages, the oscilloscope is employed first to localize the defective stage. Thereafter transistor (or tube) checks and voltage and resistance checks will be required to isolate the defective component.

NO HORIZONTAL OR VERTICAL SYNC ONLY

If only horizontal or vertical sync is lacking, the first step is to try to adjust the appropriate hold control. If only momentary hold is obtained by this method, then it is likely that sync pulses are not reaching the particular circuit. In some receivers, separate horizontal and vertical-sync amplifiers are used. These should be checked for operation, using an oscilloscope. Other possibilities are the horizontal differentiator or vertical integrator. (Figure 27.27.)

Since most horizontal-deflection circuits employ AFC for the oscillator, a malfunction of this circuit can also cause a loss of horizontal sync. Waveform and voltage tests, as well as diode and transistor tests will help to localize the problem.

Where digital-countdown circuits are used, check for the proper input waveforms and voltages. If these are correct, replace the IC.

LOOSE VERTICAL SYNC

Because of the horizontal AFC used in many receivers, horizontal sync will often hold, even with "loose" or unstable vertical sync. This is frequently caused by insufficient sync-pulse amplitude reaching the vertical oscillator. Check the sync-pulse amplitude output to the vertical oscillator with an oscilloscope. If necessary, circuit trace, moving by stages, back to the video detector. If compressed sync is found, proceed as described above.

HORIZONTAL-PICTURE PULLING

Horizontal-picture pulling must be distinguished from *raster distortion*. This is easily done by switching to a non-operating channel and observing the raster. Raster distortion is caused by a defect in the sweep circuits. However, horizontal-picture pulling is caused by a *synchronization* problem.

Horizontal-picture pulling may take one of the following forms:

1. Horizontal pulling that tends to follow picture information.
2. Horizontal picture pulling that is steady.

Varying Picture Pulling One form of horizontal picture pulling tends to change with varying picture information. This is caused by incomplete sync separation. Here, the sync separator is not completely eliminating video-signal information. Defective transistors or tubes can be the cause, as well as the bias for these units. An oscilloscope check of the sync-separator stages will quickly verify the source of this problem.

Steady Picture Pulling Here the horizontal bend does not change. This condition is usually caused by distortion of the horizontal-sync pulses fed to the horizontal-AFC circuit. The waveforms at the input of the horizontal AFC circuit should be checked first. If they are distorted, work back toward the video detector to isolate the stage producing the distortion.

Another cause may be due to inadequate filtering in the vertical-deflection circuits or the low-voltage power supply. In this case, vertical "kickback" (or flyback) pulses may be passing through the sync-separator stages, upsetting horizontal synchronization. An oscilloscope check at the input to the horizontal-AFC circuit will quickly verify if this is the case.

27.20 TROUBLESHOOTING THE VERTICAL-DEFLECTION SYSTEM

(See also Chapter 24.) Troubles in the vertical-deflection system are perhaps the easiest to analyze because the signal voltages developed here deal only with the vertical sections of the image. There are no high-voltage power supplies associated with this system, such as we find in the horizontal-deflection system. When only the vertical-deflection sytem is affected, analysis of the source of the defect is generally simple and straight-forward.

Troubles in the vertical-deflection system may be categorized as follows:

1. No vertical deflection.
2. No vertical lock (sync).
3. Jumpy picture.
4. Insufficient height.
5. Poor linearity.
6. Vertical foldover.
7. Vertical keystone raster.
8. Poor interlace.

NO VERTICAL DEFLECTION

A positive indication of complete failure of the vertical-deflection system is the appearance of a narrow, horizontal line on the cathode-ray-tube screen (see Figure 27.31). The first components to check are the vertical oscillator and amplifier transistors or tubes. The countdown circuit, if used, may be at fault. The IC then must be replaced. If these are normal, check the vertical-sweep waveforms with an oscilloscope, starting at the vertical-oscillator stage or countdown output, progressing forward to the vertical-output amplifier. Once the defective stage is located, voltage and resistance checks should reveal the defective component. Keep in mind, when making the foregoing tests, that an opening in the windings of the vertical-deflection yoke coils or vertical-output transformer is also a possibility.

NO VERTICAL LOCK

The picture may fail to lock in vertically (see Figure 27.32). Ordinarily, if the vertical-hold control is rotated, a point will be found where the frequency of the oscillator is brought close enough to the incoming vertical-synchronizing pulses to permit lock-in. The picture then becomes stable. However, if the oscillator is not functioning properly, or the vertical-synchronizing pulses are not reaching the oscillator, then lock-in will not occur. When the vertical-hold control proves ineffective, make the following tests:

1. Check the waveform at the input to the vertical-sweep oscillator. Determine whether the pulses reaching the vertical sweep oscillator are sufficiently strong to maintain control. It is quite possible that the path from the sync-pulse separator to the vertical oscillator does not transmit the full vertical sync pulse. Defective coupling capacitors, open resistors, or components which have changed value appreciably may be the causes. The transistor or tube should be checked. (If a countdown circuit is used, it may be defective. No vertical-hold control is used with a countdown circuit.)

   Check, too, the amplitude of the video signal at the point where the sync-pulse separation occurs. The image on the screen will also help determine whether sufficient signal strength is present.
2. If the foregoing test produces normal results, check the waveform at the output of the vertical-sweep oscillator. Note whether variation of the vertical-hold control has any effect on

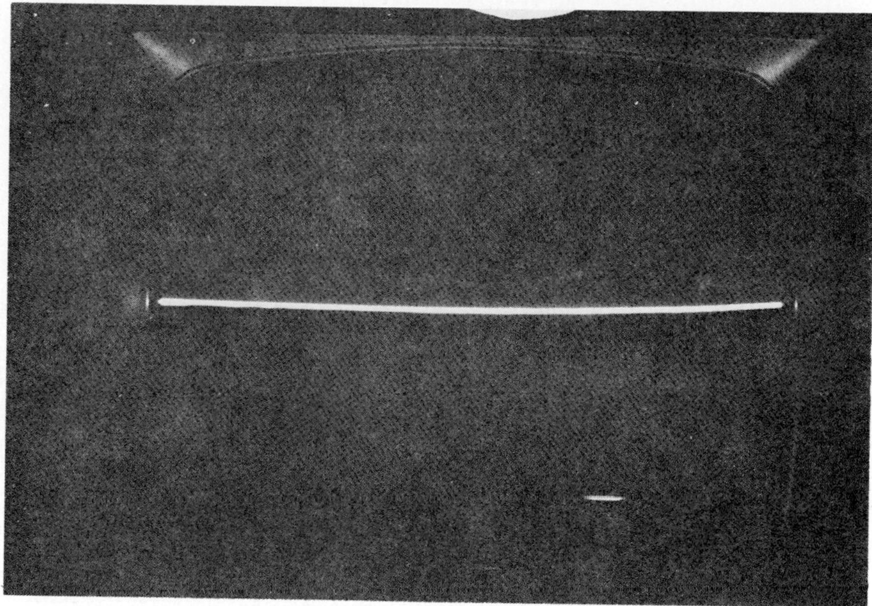

Figure 27.31 Complete loss of vertical deflection.

Figure 27.32 A picture that fails to lock-in vertically. The vertical sync, blanking and equalizing pulses create the "hammerhead" pattern between the portions of the picture.

the frequency of the observed waveform. A lack of such frequency variation may indicate an open resistor in the circuit containing the hold control. Check the vertical oscillator resistors and capacitors to find the defective component.

JUMPY PICTURE

If overloading occurs in a video-IF stage, or video amplifier, a "jumpy" picture may result. Overloading may occur due to improper AGC operation. This in turn can compress the sync pulses, resulting in erratic vertical synchronization.

DEFECTIVE VERTICAL-OUTPUT TRANSFORMER

Shorted turns in the vertical-output transformer will result in a loss of picture height. It may sometimes give the same indication that insufficient driving voltages give to the vertical-output amplifier. This type of trouble is usually difficult to detect because it reveals itself neither to normal voltage or resistance checks. If the shape and the peak-to-peak amplitude of the deflection wave are correct at the input to the vertical-output amplifier and if the amplifier appears to be operating normally, then the output transformer is a logical suspect.

INSUFFICIENT HEIGHT

The height of the picture may be insufficient. In a receiver functioning normally, adjustment of the height control will produce the proper picture height. Inability to obtain this result may be caused by one or more of the following conditions:

1. Defective vertical-output amplifier transistor or tube.
2. Incorrect voltages on vertical-oscillator and vertical-amplifier transistors or tubes.
3. Leaky capacitors.
4. Defective vertical-height control.
5. Low amplitude sawtooth waveform at any point in the circuit.
6. Defective vertical-output transformer.
7. Low line voltage.
8. Improper placement of the deflection yoke.
9. Defective vertical yoke. (Check with "ringing" test.)

POOR VERTICAL LINEARITY

This can be caused by misadjustment of the vertical linearity control or by circuit problems. The linearity control interacts with the height control. Both must be adjusted alternately to obtain the desired result. If the control adjustment is not successful there is a circuit problem.

With poor vertical linearity there may be compression of some part of the picture and expansion of another part.

The picture may be compressed at the top. This is an indication of poor linearity of the vertical-deflection voltage. The vertical-sawtooth deflection wave is developed in the output circuit of the vertical oscillator, amplified by the vertical-output amplifier, and then applied to the vertical-deflection coils. If this wave does not have the proper form, if parts of it curve or bend, the electron beam in the picture tube will not travel down at an even rate. The visual result will be a vertical bunching of lines in some sections of the image and the spreading apart of them in others. This is known as "poor linearity." When the image displays this type of distortion, the trouble is in the vertical sweep circuits. This assumes, of course, that the vertical linearity control has been properly adjusted.

CHECKING VERTICAL LINEARITY

A precise vertical linearity check can be made with an audio oscillator. Place the receiver in operation but have it tuned so that no signal is being received. This will leave the screen with a blank raster. Now connect the audio oscillator across the load resistor of the video-second detector and set the frequency to 660 Hz. As the output of the generator is turned up, a series of alternate black-and-white stripes will appear (see Figure 27.33). If the bars are evenly spaced, we know that the scanning of the screen is linear. On the other hand, a nonlinear scanning rate will cause the bars to bunch together and/or stretch at some points.

The figure 660 was suggested because it is an integral multiple of the 60-Hz vertical scanning rate. Almost any multiple of 60 may be used, although it is desirable to have at least ten or more bars on the screen.

The audio signal was injected at the video-second detector, because this same signal will also lock in the vertical-sweep system and produce a steady pattern. The sync signals are almost always taken from some point beyond the second detector, and in this way part of the injected signal will reach the vertical-sweep oscillator and lock it in.

Figure 27.33 Black-and-white horizontal bars for checking vertical linearity.

Another method, commonly used to check linearity, is by means of the crosshatch pattern obtained from a TV-pattern generator.

VERTICAL FOLDOVER

Vertical foldover in a picture might be considered as an aggravated case of nonlinearity. In form, the picture will appear somewhat as shown in Figure 27.34. The bright horizontal stripe across the bottom of the image represents the point where the scanning beam stopped moving downward. This condition generally arises from some defect between the output circuit of the vertical oscillator and the output stage. Thus, a leaky coupling capacitor between the oscillator and the output amplifier or a change in bias of the output amplifier may be responsible. It might also be wise to check the filter capacitor in the cathode leg of the output amplifier. In tube receivers, a gassy output tube is also a distinct possibility and a new tube should be tried.

Sometimes a condition will be obtained in which there is a bright horizontal line across some intermediate point in the picture. This is usually due to heater-to-cathode leakage in a vertical-output tube.

KEYSTONE RASTER

Shorted turns in the vertical-deflection yoke will produce the trapezoidal or keystone-shaped pattern shown in Figure 27.35. Which side is wider depends upon which section of the coils contains the short. This condition is called a keystone raster.

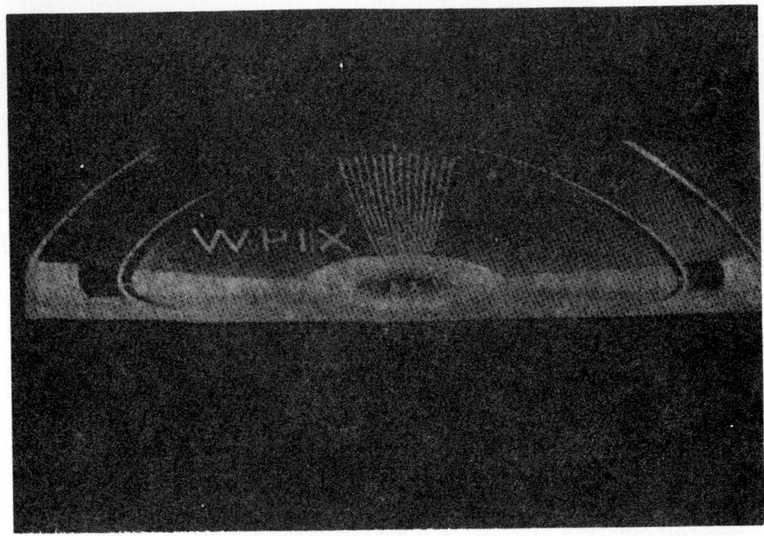

Figure 27.34 An illustration of vertical foldover. *(Courtesy of GTE-Sylvania Electric Products, Inc.)*

Associated vertical-yoke parts may also cause a keystone raster if they become defective. These parts include a thermistor or a shunt resistor (lowered value).

POOR INTERLACE

This is also called "line pairing". Common causes include

1. Horizontal pulses feeding into the vertical-deflection circuits.
2. Defective integrator network.

If horizontal pulses are in the vertical-deflection circuits, they can be seen on the vertical sweep-output waveform. On the oscilloscope, they will appear as many pulses on a single vertical sawtooth. The horizontal

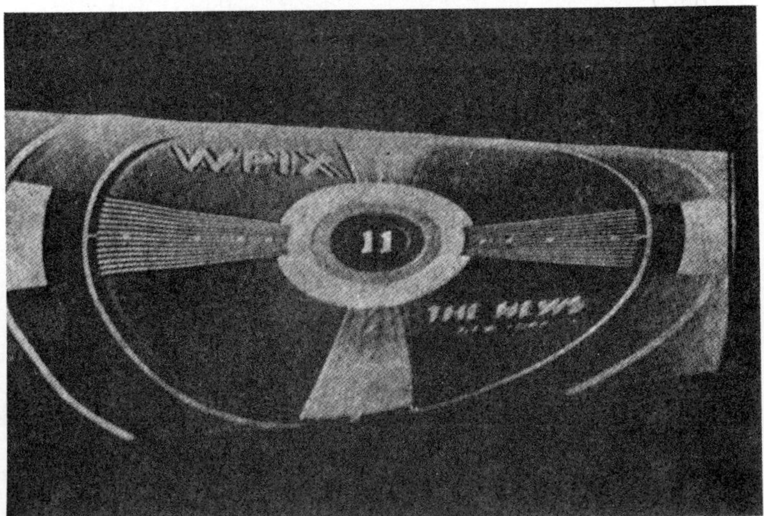

Figure 27.35 Vertical keystoning. *(Courtesy of GTE-Sylvania Electric Products, Inc.)*

pulses generally are picked up from radiation from the high-voltage circuits. Radiation may occur because of corona discharge in any portion of the high-voltage system. Radiation from the anode, high-voltage lead is another possibility.

27.21 TROUBLESHOOTING THE HORIZONTAL-DEFLECTION SYSTEM

(See also Chapter 22 & 23.) In Section 27.17, localizing the cause of no high voltage was described. For this problem, the entire horizontal-deflection system was suspect. In this section we are concerned with horizontal-deflection system troubles, other than high voltage. Horizontal "pulling," which is caused by a sync problem, was described in Section 27.19.

GENERAL TROUBLESHOOTING APPROACH

Distinctive waveforms are produced at each point in the vertical- and horizontal-deflection systems, and the reception of a signal does not alter the shape of these waves.

The best troubleshooting method is to compare the wave shapes of the voltages in the defective receiver with the corresponding waveforms given by the manufacturer in the service manual. For example, refer to the waveforms shown in Figure 27.36 (see also Chapter 26 for waveforms of an all solid-state TV receiver). These are the proper waveforms present in the vertical and horizontal circuit of this receiver when everything is operating normally. The peak-to-peak voltage values are also important in waveform checking, and these should be carefully noted. When the wave amplitudes are found to be appreciably smaller than recommended by the manufacturer, tubes and the B+ power supply voltages should be checked. On the other hand, distorted or improperly shaped waves usually indicate defective capacitors and/or resistors in the circuits. Transistor sweep circuits have waveforms similar to the ones shown in Figure 27.36 (see Chapter 26 for these).

When trouble is traced to the deflection system, check the transistors or tubes first. If these test all right, then the next job is the waveform check. Take an oscilloscope and connect its grounded vertical-input terminal to the receiver chassis.[1] Connect a test probe to the other vertical-input terminal. Then, starting at the output of the deflection oscillator, check the input and output waveforms of each amplifier, working toward the deflection coils of the cathode-ray tube. At the point where the waves disappear or are not in their proper form, voltage and component checks should be made to determine the reason for the change or disappearance of the waves. It is desirable to use the ruled plastic reticle for the oscilloscope screen and note the peak-to-peak voltage values of each of the waveforms checked. Variations of from 10 to 15 percent from the recommended values can be accepted since the adjustment of various controls and other factors can readily affect the wave amplitude by this amount.

Troubles in the horizontal-deflection system may be listed as follows:

1. No brightness (no raster).
2. Poor horizontal linearity.
3. No horizontal sync.
4. Insufficient width.
5. Horizontal keystone raster.
6. "Piecrust" or "geartooth" distortion.
7. Barkhausen oscillation and "snivets".

[1]In some TV receivers, the TV chassis is *not* the "ground" point. Be sure to check the manufacturer's literature to avoid incorrect voltage or waveform measurements.

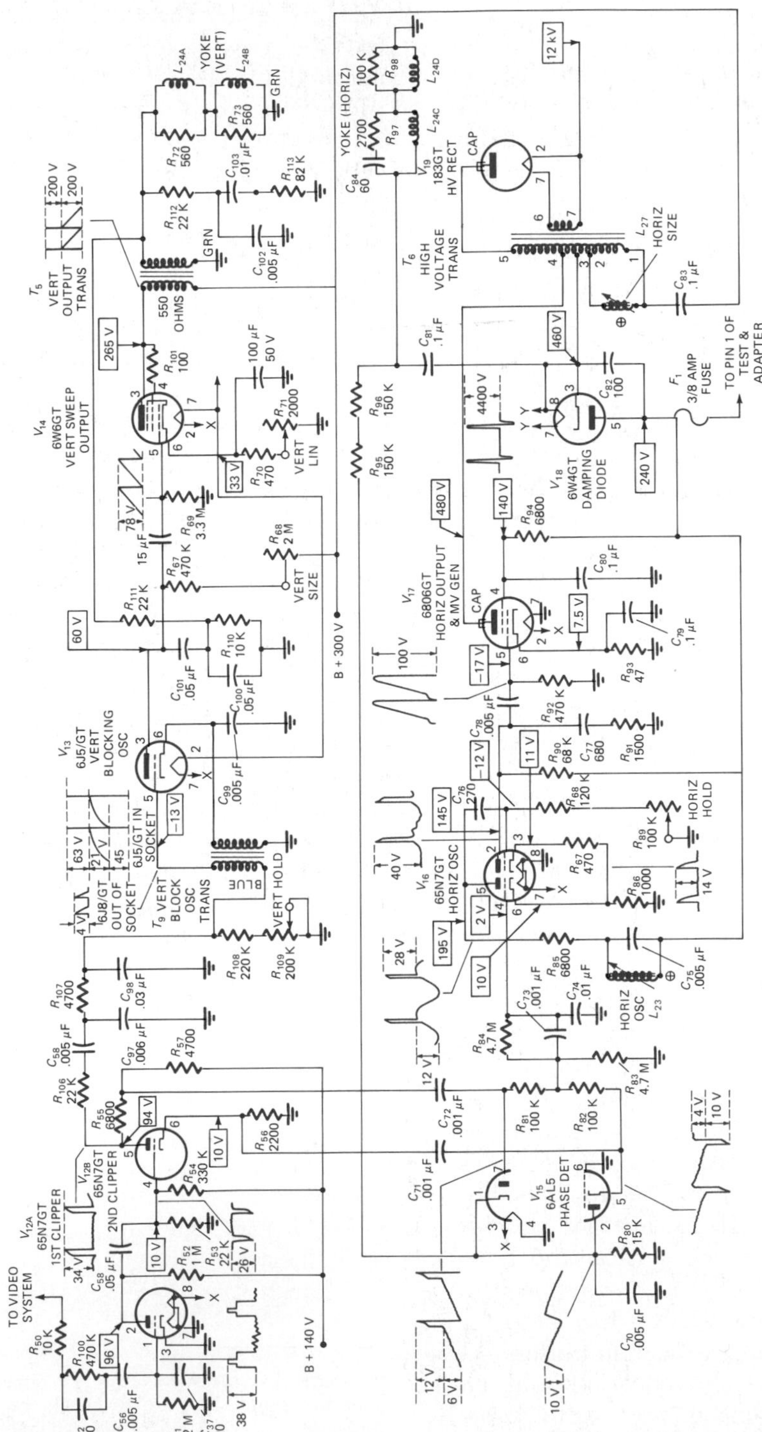

Figure 27.36 The sync separator and the vertical and horizontal sweep systems of a vacuum-tube television receiver, together with the normal waveforms found in these circuits. *(Courtesy of Quasar.)*

No Brightness

The method of troubleshooting for a condition of no brightness due to a lack of high voltage was described in Section 27.17. Refer to that section for review. As mentioned in that section, for proper operation, it is essential that all waveforms have not only the proper shape, but also the correct amplitude. This information can be obtained from manufacturers literature.

An additional trouble that can cause a condition of no brightness is a horizontal oscillator which is far off frequency. This will prevent the horizontal output circuits from functioning properly. The oscillator may be drastically off frequency because of an oscillator circuit defect. In this case, its parts and voltages should be checked.

However, the problem can also be caused by defective horizontal-AFC circuit operation. Under some conditions a defective AFC circuit can cause the horizontal oscillator to operate far off frequency. (This condition also can be caused by a defective anode-high voltage, overvoltage protection circuit, if it is the type that affects the horizontal-oscillator frequency.) To isolate this type of trouble, disconnect the AFC line from the horizontal oscillator. Now try to adjust the horizontal-hold control to obtain a single (unlocked) picture. If this is obtained, the oscillator is o.k. and the trouble is in the AFC circuit. Be sure then to check the sync pulses and the horizontal pulses at the AFC detector.

Poor Horizontal Linearity A fairly common type of picture defect is horizontal nonlinearity. The picture may be compressed or elongated at the left- or right-hand sides (see Figure 27.37).

To track down the part of the horizontal circuit most likely to contain the seat of the trouble, the technician must recall something that was learned previously. The first part of the horizontal beam travel is controlled by the damper (and its circuit), while the remaining part of beam travel across the screen is controlled by the horizontal-output amplifier.

From this it can be assumed that if the picture is impaired on the left-hand side, the defect is most likely to exist in the damper circuit. If it is the right-hand side of the picture which is distorted, the most likely place to look for the defect is in that portion of the horizontal-sweep system extending from the

Figure 27.37 An image possessing horizontal nonlinearity. *(Courtesy of RCA.)*

horizontal oscillator up to and including the horizontal-output stage, the horizontal-output transformer and the horizontal yoke.

Thus, from the foregoing reasoning, horizontal foldover caused by a defective damper or alternating, vertical light and dark stripes all quite naturally fall at the left-hand side of the picture because they arise from damper circuit defects.

The most frequent defects in the damper stage include the damper diode, any capacitors in the damper circuit, and the linearity coil (if present). In the horizontal-output stage, check the following:

1. The input signal to the output stage (shape and amplitude).
2. Operating voltages.
3. Leaky coupling capacitor at the stage input.
4. A bad horizontal-output transistor or tube.
5. Incorrect bias.

Circuit Interaction In most instances the general location of the defect, as revealed by the section of the picture which is affected, is fairly well-defined. However, because a B+ boost voltage developed by the damper stage is fed back to the horizontal-output section (and sometimes to other sections in the horizontal system as well), a certain amount of interaction between the various sections is bound to occur. For example, changes in the B+ boost filter-network components may have an effect on both sides of the picture, although the left-hand side will be the section principally affected. Or, insufficient driving voltage applied to the output stage will have its greatest effect on the right-hand side of the picture. But, since the B+ boost voltage developed depends upon the drive voltage, the left-hand side of the picture will suffer too. Thus, while the circuits directly associated with each part of the picture will have their greatest effect on that section, the close relationship between all circuits may produce disturbances in other sections of the picture as well.

Horizontal-Linearity Control

Some TV receivers incorporate a horizontal-linearity control. In the event of relatively minor linearity problems, this control should be adjusted.

Horizontal Foldover In this symptom, either the right or left sides of the picture appears to be folded back on itself for a short distance. The "foldover" actually represents an undesirable oscillation of the electron beam(s). The horizontal-output amplifier drive signal should be carefully checked first for shape and amplitude. If this is correct, then proceed as follows:

1. Foldover at left edge: this is where the damper circuit functions. (Check the damper diode and associated parts. An open B+ boost capacitor can cause this problem.
2. Foldover at right edge: this is due to malfunctioning of the horizontal-output circuits. Check the output transistor (or tube), its voltages, resistances and any other associated parts. Out of tolerance parts, including the yoke, flyback transformer or capacitors can also cause this problem. (This latter problem may cause foldover at either or both edges.)

No Horizontal Sync

(See Figure 27.38.) This can be caused by a horizontal-AFC problem or by a defective horizontal oscillator (previously discussed). The purpose of the AFC circuit in the horizontal-sweep system is to keep the horizontal oscillator locked-in with the incoming sync pulses. When this system is operating

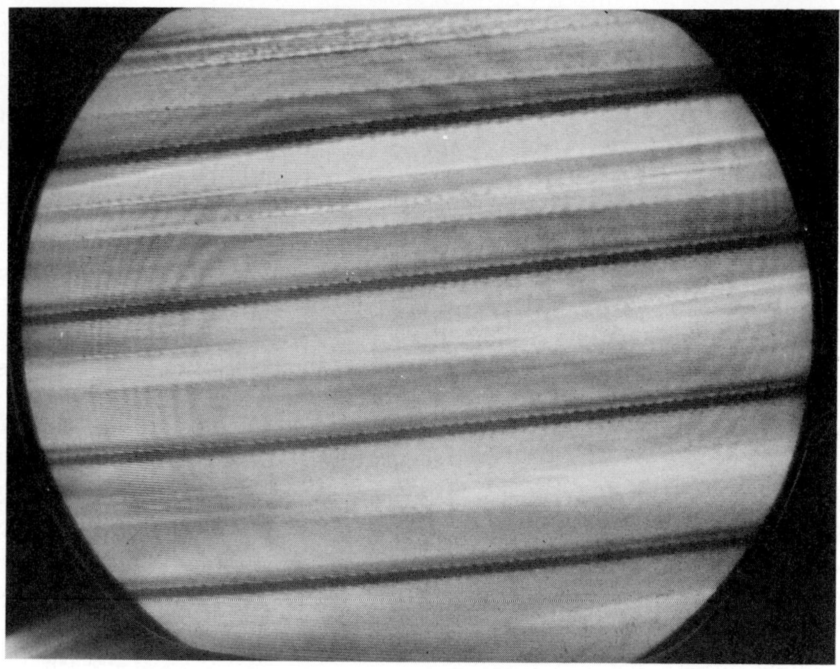

Figure 27.38 One case of no horizontal sync.

properly, random noises and momentary disturbances will have no visible effect on the horizontal-sync stability; the picture will remain securely in place. However, when the system is not operating normally, it may slip out of synchronization occasionally, or the setting of the horizontal-hold control will be critical, or the picture may not lock in at all.

Failure of the AFC system to maintain the picture in synchronization may arise from a defect in the control circuit or from a defect in some prior circuit. If the latter is true, then the sync or horizontal pulses reaching the AFC control stage are distorted in some manner or they may even be missing altogether.

Localizing AFC Fault If it is determined that the proper pulses are reaching the AFC circuit, and the oscillator operates normally, then any reason for sync instability must be caused by faulty operation of the control circuit itself. To localize the source of the trouble, several methods of approach are open. As a start, measure the voltages within the control circuit. This lends itself quite readily to the location of fairly obvious defects caused by resistances which are either open or have changed radically in value, and leaky capacitors. Transistors, diodes or tubes, whichever are used, must be checked.

Another approach to the problem is by waveform checking within the AFC circuit. To check the waveform effectively, the technician should check the correct waveforms as given in the manufacturer's literature.

Another item to check in your examination of an AFC system is whether the DC control voltage is being fed to the AFC controlled stage. Connect a high-impedance multimeter between the DC control voltage path and chassis. Then, with a signal coming in, slowly rotate the horizontal-hold control from end to end. In a properly operating system, the needle on the meter should swing back and forth in step with the hold-control variation. This indicates that the control network is developing corrective voltages to offset the changes in the horizontal-sweep frequencies produced by rotating the hold control. Failure to observe these voltage swings on the meter indicates that either no DC control voltage is being developed or that the amount developed is not reaching the controlled stage.

Changes in component values are frequently the cause of instability in AFC systems. Be especially mindful of this fact when making voltage and resistance checks.

Insufficient Width

The width control, if present, should be adjusted first. Some TV receivers have jumpers, the position of which affects picture width. If this is the case and there appears to be no obvious malfunction, try a different jumper position. The AC line voltage should be measured to insure it is within the manufacturer's recommended limits. On many recent receivers, a tolerance of 105 to 135 V AC is permitted. Some older receivers will not provide adequate width at 105 V AC line.

Insufficient width, not accompanied by obvious horizontal nonlinearity may be caused by reduced horizontal drive. This is frequently accompanied by reduced high voltage and brightness. You should carefully check the drive waveform at the base (or grid) of the horizontal-output amplifier. If the shape is correct, the amplitude may be too low. The drive amplitude is particularly important in solid-state circuits. This is because a small input-signal voltage decrease can result in a correspondingly large current change in the output transistor.

If the input waveform is normal, check the output transistor (or tube) and its voltage and resistance parameters. A partially shorted flyback transformer may also result in reduced width.

Horizontal Keystone Raster

(See Figure 27.39.) Here the raster takes the shape of a trapezoid. The top or bottom may be narrower, depending upon which half of the horizontal yoke is affected. This problem usually stems from shorted turns in one half of the horizontal yoke. A defective "balancing" capacitor connected across one half of the yoke may also produce this effect.

"Pie-Crust" or "Geartooth" Distortion

One form of horizontal picture distortion is called "piecrust", or "geartooth" distortion and is pictured in Figure 27.24. As previously mentioned, this can be caused by an open or wrong value of "balancing" capacitor across one half of the yoke. However, another cause is insufficient filtering by the RC-filter circuit between the horizontal AFC and oscillator circuits. The filter capacitor(s) may be leaky and should be checked or replaced.

Barkhausen Oscillation and Snivets

This condition can occur (if at all) only in TV receivers using vacuum-tube, horizontal-output amplifiers. The interference pattern typical of Barkhausen oscillation consists of one or more dark vertical lines at the left side of the picture. See Figure 27.40.

A high frequency oscillation may occur in some horizontal-output tubes after the tube has been cut off. If this oscillation is picked up by the signal circuits (RF or IF), it is added to the incoming signal and appears as shown in Figure 27.39.

"Snivet" interference appears as a dark, irregular vertical line (or lines) at the right side of the picture. These are caused by another form of oscillation within the horizontal-output tube. As before, the high frequency oscillation is picked up by the signal circuits. These vertical lines appear at the right side, since they are generated when

Figure 27.39 Horizontal-keystone distortion, caused by a partially shorted horizontal yoke.

the output tube exhibits a "negative resistance" characteristic. This occurs only at high values of plate current which correspond to minimum values of plate voltage.

Both of the above conditions can be eliminated by one or more of the following:

1. Change the horizontal-output tube.
2. Place a small permanent magnet (ion trap for example) near the top of the horizontal-output tube.
3. Apply +30 to +40 V to the suppressor grid of the horizontal-output tube.
4. Use ferrite beads (RF chokes) on leads of the horizontal-output tube to prevent radiation of the oscillations.

27.22 GUIDEPOINTS FOR TROUBLESHOOTING COLOR-TV RECEIVERS

Difficulties in obtaining the proper color display on the screen of a color-television receiver may be caused by one or more of the following: improper adjustment of applicable receiver controls, improper color system alignment, defective color circuits and defective color-picture tube.

IMPROPERLY-ADJUSTED CONTROLS

In early color television receivers one of the most frequent causes of poor color display was in the improper adjustment of the receiver's customer-operated controls. The manufacturers have reduced this problem considerably by the use of closed-loop feedback systems which automatically adjust the gain, phase, and frequency of various critical sections in the receiver, and thus assure the proper color display.

Fine Tuning The fine-tuning control of the receiver must be properly adjusted in order for the tuner and IF sections to be able to

Figure 27.40 The visual effect of Barkhausen oscillation.

pass the color signals. A misadjusted fine-tuning control can result in the complete loss of color display. Also, improper adjustment of the fine tuning can result in a color subcarrier beat being displayed on the screen. This latter effect is shown in Plate 1. Automatic Fine Tuning (AFT) eliminates this problem, and is used in most color-TV receivers.

Brightness-Contrast The brightness and contrast control settings are important for getting the right ratio of light to dark areas. If the contrast control setting is too low, the picture will look washed out. Most viewers prefer to adjust the brightness and contrast controls with a monochrome picture (with the color-amplitude control turned off) so they can get the proper ratio between the light and dark areas. Many color-TV receivers employ automatic circuits for those functions.

Color Amplitude If the color-amplitude control of the receiver is improperly adjusted, the colors will be either too weak as shown in Plate 2 or too saturated (strong) as shown in Plate 3. The proper setting of the color amplitude control can usually be best made after the hue (or tint) control is adjusted with a fairly low, color-amplitude control setting. Automatic Color Control (ACC) circuits help achieve the desired results.

An important preliminary test can be made with the color amplitude control to determine if there is trouble in the color-bandpass amplifier section. Rotate the color knob from its OFF position (producing a monochrome picture) to its full ON position (producing highly saturated colors). There should be no appreciable change in the picture tint when the color control is adjusted throughout its range. A change in tint is easiest to recognize by watching the flesh tones. A change in tint may cause the flesh tones to shift from their normal color toward yellow or toward red. A shift in tint with color control adjustment may indicate that the color-bandpass section should be aligned, or that a component in that section is defective.

Hue (Tint) Control The hue control varies the phase of the color signal and therefore, determines the correct coloring of the picture. Plate 4 shows an example of the picture with the improper hue control setting. The setting has turned the picture toward the green vector and thus, gives a green shade to the overall display. Remember that the hue control is properly adjusted when the flesh tones are correct. Automatic Tint Control (ATC) circuits help to maintain correct hues.

Horizontal-Vertical Hold Controls The adjustments of the horizontal and vertical hold controls of a color receiver are the same as for those in a monochrome receiver. Many of the more recent color-TV receivers do not require these controls. These receivers employ digital-countdown circuits.

IMPROPER COLOR SETUP

The term *setup* in color television refers to the process of making adjustments for purity, convergence, and gray scale (tracking). (The procedure for making the color setup was described in Chapter 26.)

If a receiver has poor purity; that is, if a purity adjustment is needed, the symptom will be a large colored area on the screen which does not change from picture to picture. This symptom is also present when degaussing of the picture tube and mounting hardware are needed. It may also indicate that automatic degaussing circuitry is not working (see Plate 5.)

Color fringes around the objects displayed on the screen generally indicate that the picture tube is out of convergence. Improper dynamic convergence is indicated by fringes in other than center areas.

If it is not possible to obtain the proper black-and-white shades on a monochrome picture, the problem is likely to be improper gray scale (color temperature) adjustment. Remember that the color section is inoperative during a monochrome picture, and improper gray scale is not a problem that can be traced to the color section.

27.23 COLOR-TV RECEIVER TROUBLES IN MONOCHROME CIRCUITS

If a receiver will not properly reproduce a color picture, the problem is not necessarily in the color section. In referring to the block diagram of Figure 26.1, you will see that the color (and monochrome) signals must pass through the tuner, IF stages, video detector, and first video amplifier stages before they actually reach the color section. Proper operation of some of the stages in the color section is also dependent upon a blanking signal coming from the blanker. This blanker receives its signal from the horizontal-sweep section of the receiver. The color portion of the color picture is superimposed upon the monochrome picture which passes through the delay line in the video amplifiers. Thus, when analyzing the trouble in a color-TV receiver, it is important to understand that the difficulty may lie in a monochrome circuit rather than in the color section. The more common problems in monochrome circuits which may affect the color portion of the picture are discussed below.

Colored Snow Pattern (Confetti) Trouble in the tuner or AGC section, can produce snow in a color picture. If the color killer adjustment is normal, the snow will pass through the color section and the display will be a colored confetti on the screen. Thus, color confetti (shown in Plate 6) is not necessarily a problem in the color section, but rather, it is due to a loss of amplification in the tuner RF section. If the RF gain is normal in the tuner and confetti occurs on the screen during reception of a monochrome picture, the trouble can be presumed to be in the color-killer section.

Normal Sound and Color Picture. No Monochrome Picture While the color elements are presumed to be normal in this condition, you must note that there can be no fine detail present in the picture, since this is ordinarily supplied by the Y signal. From the above symptoms, it should be obvious that the trouble exists in circuits which follow the sound and chroma takeoff points. Since there are different schemes for handling the Y signal in various receivers, the receiver schematic must be checked to see what circuits, following the sound and chroma takeoff points, exist in your particular receiver. In many receivers, after the takeoff points, the Y signal passes through additional video amplifier stages, including a delay line, and is applied to the three cathodes of the color tube. Thus, these stages and also the delay line would be suspect.

Hum Bars in Monochrome and Color Pictures This effect on a color-bar picture is illustrated in Plate 7. The picture shows the effect of 60-Hz voltages which have been introduced into the sync and video circuits. Note the single cycle of bend in the vertical bars caused by hum in the horizontal sync. Also note the single horizontal color bar caused by hum in the video circuits. If the hum frequency was 120 Hz, the frequency of these effects would be doubled.

In a vacuum-tube, or hybrid receiver, 60-Hz hum effects may be caused by heater-to-cathode leakage in a tube. This fault cannot cause 120-Hz effects, since the filaments are heated from a 60-Hz source. In transistor receivers, this cause for a hum problem, of course, is absent. However, in any type of receiver, hum problems can be caused by power supply defects. A failure of one of the rectifiers in a full-wave rectifier can produce 60-Hz hum, while a defective power-supply filter or bridge capacitor, may produce 120-Hz hum effects.

Raster Only; No Sound and No Monochrome or Color Picture Present The presence of a raster indicates the operation of the power supply and of the horizontal- and vertical-deflection systems. The lack of snow generally indicates that the problem follows the mixer stage of the tuner and is therefore most likely in the IF or AGC sections. Since there is neither sound nor picture, it follows that the problem exists prior to the sound takeoff point (or at it). In color sets, sound is tapped off at the last common IF stage, but always prior to the video detector. Thus, the usual source of this trouble would be found in either the IF or AGC sections. However, a completely defective mixer stage in the tuner might also be a cause for this problem and should not be overlooked.

Best Color and Monochrome Pictures Do Not Track The meaning of this trouble symptom is that the tuner fine-tuning control must be adjusted to different settings to obtain the best color and monochrome pictures. The most likely cause of this trouble is a defective or misaligned IF system, including the mixer stage of the tuner. A good way to begin the troubleshooting for this problem is to check the frequency response of the tuner and IF section, as described in Chapter 26. Any appreciable variation from normal alignment must be corrected before proceeding further. It is important to remember that IF response may be severely affected by a regenerating IF stage. A good test for the presence of regeneration is to sweep the tuner and IF stages and observe the overall response curve. While doing this, change the AGC voltage (with an external bias supply). If the response curve changes appreciably when varying the AGC bias, regeneration is present. Some causes of a regenerating IF stage are: defective neutralizing circuit, open bypass capacitor, and incorrect tuning of resonant circuits.

Ringing in Monochrome and Color Picture Ringing, or close-spaced ghosts, are illustrated in the color-bar picture of Plate 8. Note the ringing effect at the trailing edge of each color bar. This effect is easily distinguished from ghosts caused by multipath reception. This type of ringing is most easily localized by means of a sweep generator. The response of the IF and video amplifiers is checked separately with regard to their response curves. If the ringing is caused by a defect in the receiver circuitry, it usually causes serious distortion of the response curve. Some defects which may cause ringing are: regeneration of an IF stage, excessive high-frequency compensation in a video-amplifier stage, and a broken lead in the antenna transmission line or tuner input.

Color Picture Displaced Horizontally from Monochrome Picture This symptom can occur only if there is a phase difference between the color and monochrome pictures. One obvious cause of this trouble is the delay line. As discussed in Chapter 26 and elsewhere in this book, the function of the delay line (in the video (Y) amplifier section) is to delay the Y signal about 1 microsecond so that it will coincide with the chroma signals, in phasing. A defective delay line will cause the monochrome picture to either lead or lag the color picture by a small amount, depending upon the type of defect.

If the alignment is improper in either the monochrome or chroma sections, signal phase changes may occur which can cause delays or advances of the color picture with respect to the monochrome picture. If this problem is suspected, check the alignment of the monochrome and color sections, as described in Chapter 26. Remember that the alignment of any tuned circuit stages will be seriously affected by the presence of regeneration. This problem may also be caused by a defective transmission line, an incorrect impedance match between the transmission line and the receiver, or a broken lead in the transmission line.

27.24 COLOR-TV RECEIVER TROUBLES IN COLOR CIRCUITS

In connection with the following discussions, refer both to the block diagram of Figure 26.1 and to the schematic diagram inserted after pg. 804.

No Color Picture: Monochrome Picture and Sound Normal This symptom indicates that the problem exists only in the chroma circuits, since a normal monochrome picture and normal sound are present. The following are the circuits to check:

(1) *Color Killer.* The threshold control may be set incorrectly, or the circuit may be defective. The color killer (Q_{612}) will hold the Chroma Output (Q_{614}) cut off in the absence of a color picture (color burst signal), in normal operation. When a color picture is present, it should permit Q_{614} to operate normally. Note that the operation of the color killer depends upon the correct functioning of Q_{610}, the ACC amplifier, which in turn is operated by the output of the killer detector (Q_{612}). The killer detector operation requires correct signal inputs from the burst amplifier (Q_{636}) and the 3.58-MHz amplifier (Q_{644}). Thus, if there is a suspected problem of killer malfunction, all of the above mentioned circuits may have to be checked for proper operation.

(2) *3.58-MHz, Subcarrier Oscillator.* If the subcarrier oscillator signal does not reach the X and Z demodulators, color video cannot be produced from the transmitted chroma signals. In this case, a color broadcast will be

reproduced only in monochrome. In the block diagram of Figure 26.1, the subcarrier oscillator proper is shown as Q_{642}. However, note that this is followed by Q_{644}, the 3.58-MHz amplifier. Defects in either of these two stages will prevent the subcarrier from being applied to the two demodulators. The trouble may be easily localized by checking with a wide-band scope. Voltage measurements may also be useful, as well as resistance measurements in the suspected circuits.

(3) *Chroma-amplifier Section.* In Figure 26.1, note that the chroma signal is picked off from the first video amplifier (Q_{208}), as a composite video signal, and is applied to the first chroma amplifier (Q_{606}). Because of the restricted bandpass characteristics (about 3 MHz to 4 MHz) of this amplifier and the ones following, only the chroma-sideband signals are passed through Q_{606}. From this point, these signals are passed through the second chroma amplifier, (Q_{608}), the tint amplifiers (Q_{602}, Q_{600}), the chroma control (Q_{604}) and the chroma output amplifier (Q_{614}). From this last chroma amplifier, the chroma signals are applied to the X and Z demodulators and to the G − Y matrix amplifier. Since the chroma signals pass through all of the above mentioned chroma and tint amplifiers before being applied to the demodulators, it is obvious that a major defect in any of these amplifiers would prevent the passage of the chroma signal and would result in the reproduction of only a monochrome picture from a color transmission.

In localizing problems in the above mentioned circuits, the best approach would be through the use of a color-bar generator and oscilloscope. With these instruments, it would be a relatively simple matter to check the progress of the signal from stage to stage. When the defective stage has been located, then checks of the transistors (or tubes, where applicable), as well as voltage and resistance checks will make it possible to locate the defective parts.

Note: In the preceding discussions we have been assuming a complete loss of color under all signal conditions. However, it is possible that color will appear only in the presence of a very strong received signal. This indicates only a partial malfunction of the circuits previously described in numbers 1, 2, and 3 above. The best approach to localizing this type of problem is also by means of a color-bar generator and oscilloscope. However, in this case, it is particularly important to monitor the peak-to-peak values of the waveforms at each stage, since this will enable you to most easily isolate the defective stage. If you refer to the waveforms associated with the schematic of Figure 26.2, you will get a very good idea of what to expect. However, remember that these waveforms are only for the particular set shown and may be somewhat different for other receivers. Thus, it is most important to obtain the correct manufacturer's information for each receiver you will be checking. The general methods for alignment and for the connection of the test equipment to be used, however, will be very similar to that given in Chapter 26 and this may be used as a guide regardless of the receiver involved.

Abnormal Color Intensity When we speak of color intensity, we are not referring to the hue (or tint) of the color, but rather to its strength (technically, its saturation). In describing this problem as "abnormal color intensity," we mean a situation where the color intensity is too great (Plate 3) or too weak (Plate 2) and that the condition cannot be corrected by adjustment of the color-intensity control. While it is possible for other conditions in the receiver to cause this condition, we are here restricting this discussion to the possible malfunctioning of the ACC (automatic-color control) circuit. The ACC circuit controls the gain of the first chroma amplifier (Q_{606}) in a manner very similar to

AGC operation. The functioning of the ACC circuit is described in Chapter 26 and should be reviewed if necessary. The gain of the first chroma amplifier varies (normally) in proportion to the amount of ACC voltage applied to its base, which, in turn, is proportional to the amplitude of the incoming burst signal. Malfunctioning of the ACC system is best determined by checking the DC voltages throughout the system. These DC voltages should be carefully compared against the ones provided by the manufacturer. Note that the ACC system includes the killer detector, the ACC amplifier, the first chroma amplifier, the second chroma amplifier, and the burst amplifier. This is a closed-loop system and ACC problems could exist in any of the aforementioned stages.

The manufacturer will sometimes provide ACC system voltages for no-color input as well as color-input conditions from a color-bar generator. It is necessary to check the various stage voltages under both conditions, in order to localize the problem. As in troubleshooting ACC problems, it may be necessary, at times, to place a manually controlled DC voltage at the base of the first chroma amplifier and remove the voltage from the ACC amplifier. This will make it possible to isolate the problem between faults in the chroma amplifiers and faults in the remainder of the ACC loop.

Other possible symptoms of problems in the ACC loop are drifting values of color intensity and intermittent changes in color intensity. The isolation procedures here are the same as outlined previously, except that long-term monitoring is usually required.

Improper Color Sync It is possible for the color sync alone to be improper, or for both the color and monochrome sync to malfunction. Both of these conditions are discussed here:

(1) *No Color Sync: Monochrome Sync Normal.* This condition is shown in Plate 9. (The input signal is from a color-bar generator.) The first step in isolating this problem is to be certain that the defect exists solely in the chroma circuits and does not originate in monochrome circuits. This is readily determined by an oscilloscope check at the input to the chroma section (Q_{606}). With a color-bar generator providing the color signal, check the waveform and amplitude of this input signal against the manufacturer's information. Any serious deviation from specifications, indicates that the cause of the trouble precedes the chroma section.

Remember that proper color sync depends upon the exact frequency of the 3.58 MHz subcarrier oscillator, as compared to the transmitted color burst. Any circuit defects which affect the frequency of the subcarrier, or which seriously affects its amplitude will affect the color sync. Referring to the block diagram of Figure 26.1, we note that the color sync may be affected by the blanker (Q_{634}), the burst amplifier (Q_{636}), the phase detector, the phase control amplifier (Q_{638}), the APC amplifier (Q_{640}), the 3.58 MHz oscillator (Q_{642}), and the 3.58 MHz amplifier (Q_{644}). An oscilloscope is invaluable in localizing problems in the color-sync system. Defects in the operation of the blanker or burst amplifier are readily detected by means of scope waveforms. Operation of the 3.58 MHz oscillator and amplifier can also be quickly checked with a wide-band scope. The remaining color-sync circuits can best be checked by DC voltage and resistance measurements. It is important to note that the color-sync system is a closed-loop, automatic-phase control system, so that a defect which seems to be in one particular circuit may actually be in an associated circuit. However, careful waveform observations, coupled with DC voltage and resistance measurements, which are correlated with the manufacturer's specifications will make it possible to isolate the trouble. Since transistors (or tubes, where used) are a prime suspect, be sure to check these first with an in-circuit transistor tester, before spending possible needless time on other circuit checks.

(2) *No Color Sync: No Monochrome Sync.* The appearance of this problem (color-bar generator input) is shown in Plate 10. It is unlikely that both the color-sync and monochrome-sync circuits would fail simultaneously. If both kinds of sync are out together, this is usually caused by a defect in the monochrome-sync circuits. The color burst is situated on the back porch of the horizontal-blanking pulse and is gated in the color-sync circuits by a horizontal-keying pulse. Therefore, unless the monochrome-horizontal sync is normal, the color sync will also suffer. Troubleshooting for defective horizontal-sync action has been covered previously in this chapter and need not be repeated here. In the unlikely event that fixing the defect in the horizontal-sync action does not also correct the color-sync operation, refer to part 1 of this discussion.

Incorrect Hues This problem may occur in the form that all of the hues may be incorrect, or that only one or more hues may be incorrect. This problem is, of course, best checked with the aid of a color-bar generator.

(1) *All Hues Incorrect.* The correct reproduction of all color hues is dependent upon the phase of the 3.58 MHz subcarrier oscillator signal applied to the color demodulators. This phase must be synchronized with that of the transmitted color-burst signal. If the phase of the subcarrier signal is shifted, all of the colors on a color-bar display will be changed from normal and this will be readily evident on the color-tube display. Such a condition may be caused by poor alignment of the bandpass (chroma) amplifiers, as described in a prior section, and this cause will not be further pursued here, except to note that apparent poor alignment may not be caused by incorrect tuning only, but may also be the result of defective circuit components. In any case, a sweep-response curve or bar-sweep pattern will reveal the fact that a response defect exists, and it then remains to isolate the problem.

An important cue in analyzing this problem is to check and see if all the colors are presented in the correct sequence, but not in the correct position on the display. (We are of course assuming that such an obvious fault as a misadjusted hue control is not the cause of our problem.) If the sequence of colors is correct, but their position incorrect, this indicates that the subcarrier or chroma phase is wrong. In order to determine if the entire color sequence is correct, it may be necessary to adjust the hue control, since some of the color bars may be off screen. In the circuit of Figure 26.1, incorrect phasing may be caused by malfunctions of the tint amplifier (Q_{600}, Q_{602}), or in the tint control (R_{22}) circuit proper. This condition may also be caused by malfunction in the subcarrier oscillator proper (Q_{642}) or its associated circuits. These associated circuits are the automatic phase control (APC) system which includes the 3.58 MHz amplifier (Q_{644}). The automatic-phase control system includes the burst amplifier (Q_{636}), the phase detector, the phase control amplifier (Q_{638}), the APC amplifier (Q_{640}), the 3.58 MHz oscillator (Q_{642}) and the 3.58 MHz amplifier (Q_{644}). Correct alignment of all tuned circuits involved here is a must and this should be checked first if there are no immediately obvious defects in the APC system. The next step would be to carefully check all transistors for defects, which are not necessarily of a major nature. If in doubt, replace the transistor with a known good one. The next step (if the defect has not yet been found) is to make a careful check of all DC voltage and resistance values and to compare the readings against those furnished by the manufacturer. Unfortunately, an oscilloscope is not of much value generally, when looking for defects which only cause signal-phase shifts.

In Figure 26.1, the phase control amplifier (Q_{638}) depends upon the setting of a potentiometer for its proper operation (see Figure 26.2). This setting, as well as the various DC voltages associated with the stage must be carefully checked.

(2) *One or More Hues Incorrect* When only a single hue (red, green, or blue) is incorrect (missing), the trouble will naturally be found in a circuit which handles only that particular color. In Figure 26.1, this problem may be caused by a defective color-tube gun, or by a defective output amplifier (Q_{628}, Q_{630}, Q_{632}), or a defective driver (Q_{624}, Q_{622}, Q_{626}). By the use of a color-bar generator and scope, the defective channel can be readily identified and the defective stage localized. Although not used in this receiver, some color sets employ three separate color demodulators. In this event, a failure of any one of the demodulators will also produce a condition where only one of the basic colors is absent from the picture.

In the block diagram of Figure 26.1, there are two color demodulators (X and Z). The third color (G–Y) is derived from a proportion of the outputs of the two demodulators. In the event that one of the two color demodulators becomes inoperative (for example, the X demodulator), we would not only lose that particular color (red), but also the red component of chroma-video signal which is applied to the G–Y matrix amplifier (Q_{620}). Thus, the only color which would be correct would be the B–Y, or blue. This type of trouble is easily isolated by means of a color-bar generator and scope. In this case it would be found that there was no output signal from the X demodulator or the R–Y driver.

If The G–Y matrix amplifier were inoperative, the red and blue (and mixture) colors would be normally present, but there would be no green contribution to the color picture. This situation is shown in the color-bar photo of Plate 11. Note the complete absence of the color green.

SUMMARY OF CHAPTER HIGHLIGHTS

1. A color-picture tube test jig can substitute for a TV receiver's color-picture tube and its deflection and high-voltage circuits (Figure 27.1).
2. In Question 1, horizontal and vertical-yoke matching impedances are selected by front-panel switches (Figure 27.1).
3. Many types of transistor testers can test transistors, in or out of the circuit (Figure 27.2).
4. In Question 3, the unit pictured in Figure 27.2 can test bipolar transistors, FET's, SCR's, Darlington's, and diodes (Figure 27.2).
5. A picture-tube analyzer and restorer can frequently restore defective picture tubes to normal service. This instrument can remove shorts, clean and balance (track) guns and rejuvenate cathodes (Figure 27.3).
6. Picture-tube "brighteners" operate by (1) increasing heater voltage, or (2) controlling biasing networks. An "isolation" brightener corrects a cathode-to-filament short. (Figure 27.4).
7. High-voltage probes can extend the range of a multimeter, to 40 000 V (Figure 27.5).
8. A substitute tuner (subber) is used to determine if the TV-receiver tuner is defective. It may also have several additional functions, including substitut-

ing for video IF, video detector, or video-amplifier stages. (Figure 27.6).

9. A "ringing" test is useful in checking a horizontal-output transformer, horizontal yoke or vertical yoke. This test works by "counting" the number of damped oscillations until they decay to the 25% amplitude level (Figure 27.7).
10. A good procedure to use to determine the cause of a TV-receiver malfunction is: (1) observation of malfunction, (2) localizing to a section, (3) localizing to a stage, and (4) localizing to a part (Figures 27.8, 27.9, 27.10).
11. Common types of color problems may be grouped as follows: (1) no color picture, normal monochrome picture, (2) weak color picture, (3) improper color sync, and (4) incorrect hues. (Plates 1 to 11.)
12. Signal injection (or signal substitution) is a valuable method of isolating a TV-receiver malfunction. It is performed by injecting a test signal into a specific stage and monitoring circuit performance on an oscilloscope, the TV screen or the TV speaker. The test signal is moved from stage-to-stage, while the monitor remains fixed (Figures 27.11 and 27.12).
13. Signal tracing is another method of isolating a TV-receiver malfunction. In this method, a test signal is applied to a *fixed* circuit point. The monitoring device is then moved from point-to-point to determine the presence or absence of signal.
14. Signal injection and signal tracing are frequently used together.
15. In signal tracing circuits prior to the video detector with an oscilloscope, use a *demodulator probe*. After the video detector, use a *low-capacitance probe*. (Figure 26.19).
16. Testing for a defective tuner may be accomplished by signal injection. (Figures 27.12A and B).
17. To test for incorrect tuner AGC, shunt the AGC line with a bias supply set to the recommended voltage.
18. Signal injection may be used to check the video-IF circuits (Figure 27.11).
19. Signal tracing may also be used to check the video-IF circuits. The correct waveforms are shown in Figure 27.14.
20. As in Question 17, if incorrect AGC voltage at the video IF is suspected, shunt the AGC line with a properly adjusted bias supply.
21. There are two AGC lines: one to the tuner and the other to the video-IF section. These could malfunction independently (Figures 26.1, and 26.2).
22. If low-frequency compensation in a video amplifier is defective, the image background becomes darker and large objects "smear". (Figure 27.18).
23. If high-frequency compensation in a video amplifier is defective, there will be a loss of fine detail (Figure 26.2).
24. The development of anode-high voltage depends upon normal operation of the complete horizontal deflection and high-voltage circuits (Figures 26.1, 26.2).
25. Picture tube defects may be listed as follows: (1) open heater, (2) open or short circuit of internal elements, (3) low (or no) emission from one or more cathodes, (4) gassy tube.
26. A partial short in the horizontal or vertical yoke will result in a "keystone" shaped picture. (Figure 27.23).
27. Some causes of no voltage output from a low-voltage power supply are (1) blown fuse or circuit breaker, (2) defective or cutoff, series-pass regulator transistor, (3) shorted filter capacitor. (Figures 26.1, 26.2).
28. When servicing a transformerless receiver, always use an AC isolation transformer to avoid severe shock.
29. Troubles caused by defective sync-separator stages may include the following: (1) no horizontal and/or vertical sync,

(2) loose vertical sync, (3) horizontal-picture "pulling" (Figures 26.1, 26.2).

30. Compressed sync may occur due to a defect in the AGC section, or in the video-IF section (Figure 27.30).
31. Possible troubles in the vertical-deflection system may include: (1) no vertical deflection, (2) insufficient height, (3) poor linearity, and (4) no vertical lock. (Section 27.20).
32. Poor vertical interlace can be caused by (1) a defective integrator network, or (2) horizontal pulses in vertical-deflection circuits (Section 27.20).
33. Possible troubles in the horizontal-deflection system, may include (1) no brightness, (2) insufficient width, (3) poor linearity, and (4) no horizontal lock (Section 27.21).
34. Generally the horizontal damper circuit affects the left-side scan. The horizontal-output stage (and amount of drive) affects the right-side scan. (Section 27.21)
35. "Piecrust" or "geartooth" picture distortion can be caused by (1) defective horizontal-yoke "balancing" capacitor, or (2) insufficient filtering of the horizontal AFC control voltage. (Figure 27.24).
36. Improper color display may be caused by one or more of the following: (1) improper control adjustment, (2) improper color system alignment, (3) defective color circuits, or (4) defective color-picture tube. (Section 27.22)
37. Automatic circuits help to insure proper color display. Among these are (1) automatic-fine tuning (AFT), (2) automatic-color control (ACC), and automatic-tint control (ATC). (Section 27.22).
38. The term "color setup" refers to making adjustments for purity convergence and gray scale (tracking). (See Chapter 26).
39. Color fringes around objects on the screen are an indication of poor convergence. (See Chapter 26).
40. If a color-TV receiver does not properly reproduce a color picture, the problem may not be in the color section. (See Section 27.23 and Figures 26.1, 26.2)
41. Colored "snow" in a picture can be produced by a defect in the tuner of the AGC section (Plate 6).
42. Hum bars in the monochrome and color picture may be caused by a low-voltage power-supply defect (Plate 7).
43. Abnormal color intensity (saturation) can be caused by a defect in the automatic-color control (ACC) circuitry. (Plate 2).
44. Improper color sync (only) can be caused by a defect in (1) blanker, (2) burst amplifier, or (3) any part of the automatic phase and frequency-control (AFPC or ARC) circuit. (Sections 27.22, 27.23, 27.24)
45. A condition of all color hues incorrect may be caused by (1) incorrect phase of the 3.58 MHz oscillator proper, (2) poor alignment of the chroma-bandpass amplifier(s), (3) defective AFPC (or APC) circuit, (4) malfunction of the tint (hue) amplifier, or (5) misadjusted tint (hue) control. (Sections 27.22, 27.23, 27.24.)
46. If only a single hue (red, green or blue) is incorrect or missing, this may be caused by (1) defective picture tube, (2) defective color-driver stage, or (3) defective color-output stage. (Sections 27.22, 27.23, 27.24.)

EXAMINATION QUESTIONS

(Answers provided at the back of the text.)

Part A Supply the missing word(s) or number(s).

1. A color-tube test jig houses a ______voltage supply and adjustable ______ matching transformers.
2. With some "in-circuit" transistor testers, a ______ is produced when a good transistor is tested.
3. The ______ of shorts and cathode ______ can be performed by a picture-tube analyzer and restorer.
4. A high-voltage probe can be used with either an ______ or a ______ type multimeter.
5. Deflection-output inductance units can be quickly checked by a ______ test.
6. Localizing a TV receiver malfunction to a stage may be done by signal ______ or signal ______.
7. Another term for "hue" is ______.
8. No color will be seen if the ______ MHz oscillator is defective. (Normal monochrome picture.)
9. There will be no color picture ______ if the AFTC circuitry is defective.
10. If all hues of a color picture are incorrect, the ______ of the color-subcarrier oscillator is wrong.
11. A generator signal is moved from ______ to ______, in the technique of signal injection.
12. ______ black and white bars would be seen on the TV screen, when using an RF or IF generator modulated by 400 Hz.
13. The test signal is applied to a ______ point, when using signal tracing.
14. You must use a ______ probe with an oscilloscope when checking RF or IF-modulated stages.
15. A ______ is a useful instrument in determining if a tuner is defective.
16. In checking for a defective AGC circuit, shunt it with an external ______.
17. One cause of compressed sync is incorrect ______ voltage.
18. In keyed AGC, the ______ pulse provides the keying action.
19. A ______ and ______ picture can result from a defective video amplifier, low-frequency compensation network.
20. A ______ picture may result from a gassy picture tube.
21. Anode-high voltage may be lost due to a defect in any of the ______ circuits.
22. A picture-tube ______ may restore low cathode emission.
23. A defective series-pass ______ can cause incorrect (or no) B+ voltage.
24. An open, low-voltage, power supply filter capacitor will cause increased ______.
25. ______ vertical lock may be the result of compressed sync pulses.
26. Insufficient horizontal-AFC voltage ______, may result in horizontal-picture pulling.
27. An extreme case of vertical ______ may result in vertical foldover.
28. Approximately ______ horizontal black and white bars will appear on a TV screen if an audio signal at 600 Hz is fed to the video system.
29. Shorted turns in a yoke can produce a ______ shaped raster.
30. The ______ side of the raster is most affected by the operation of the horizontal damper.
31. There will be no picture-tube ______, if the horizontal oscillator is far off frequency.
32. Barkhausen lines and "snivets" can only appear in TV receivers employing ______ type horizontal-output amplifiers.
33. The ______ of the colors is automatically controlled by an ACC circuit.
34. A large, stationary colored area on a picture tube indicates poor ______.
35. Improper ______ convergence is indicated by color fringes around objects on the screen.

36. An indication of colored snow ("confetti") may result from a defective ______.
37. The ______ signal operates the color-killer circuitry.
38. If the ______ output amplifier is dead, only the color green will be missing from a color picture.
39. There will be no color(s) if the ______ amplifier is dead.
40. No color ______ is possible, with defective AFPC circuitry.
41. All hues incorrect generally results from incorrect ______ of the 3.58 MHz subcarrier oscillator.

Part B Answer true (T) or false (F).

1. A color-picture tube test jig can rejuvenate picture tubes.
2. A transistor tester will produce an audio tone for a good transistor.
3. A picture tube restorer can check leakage between gun elements.
4. One type of picture-tube brightener affects the bias network of a color-picture tube.
5. An isolation brightener eliminates picture tube G_1 to G_2 shorts.
6. A high-voltage probe is normally used with high-impedance input oscilloscopes.
7. Some "subbers" can be used to also check flybacks and yokes.
8. A "ringing" test of a yoke actually tests the "Q" of the circuit.
9. Possible control misadjustment should always be checked before actually troubleshooting a receiver (where applicable).
10. The symptom of no color picture (only) can be caused by a defective video-detector stage.
11. If the color sync is defective, the monochrome sync will be defective also.
12. In the process of signal injection, the monitoring device is moved from point to point.
13. A 45.75 MHz signal can be used to align the adjacent-channel picture trap.
14. Signal injection and signal tracing are always used separately.
15. When checking circuits following the video detector with an oscilloscope, use a low-capacitance probe.
16. A substitute tuner (subber) can completely replace the TV-receiver tuner, for test purposes.
17. If only one (or several) channel(s) operate poorly, the tuner is faulty.
18. A problem that seems to be in the video-IF section, may be caused by improper AGC operation.
19. If the shunting resistor across a video amplifier peaking coil opens, a "ringing" effect may appear in the picture.
20. Anode-high voltage cannot be lost because of a defect in the horizontal AFC circuit.
21. A picture-tube brightener can compensate for one dead filament in a color-picture tube.
22. A gassy picture tube may produce a negative picture.
23. Simple shorts in a picture tube gun can be burned out with the aid of a picture-tube restorer.
24. One open diode in a bridge rectifier will create a ripple frequency of 120 Hz.
25. In a horizontal-frequency power supply, if the "start-up" circuits do not work, only the horizontal circuits will operate.
26. No horizontal sync (only) indicates a possible defect of the integrator circuit.
27. Loose vertical lock can result from insufficient sync amplitude.
28. Poor vertical linearity can be the result of control misadjustment.
29. A vertical-keystone raster can result from shorted turns in the vertical-output transformer.
30. Insufficient width may result from low AC line voltage.
31. Horizontal-AFC circuit operation cannot affect picture brightness.
32. Horizontal foldover can be caused by a defective damper.
33. Damper operation cannot affect the anode-high voltage.
34. A digital, deflection countdown circuit eliminates the need for horizontal and vertical-hold controls.

35. Barkhausen oscillation can occur with any type of horizontal-output amplifier.
36. Misalignment of a 4.5 MHz trap can result in the appearance of a color-subcarrier beat pattern, on the TV screen.
37. It is normal to observe an appreciable tint change, when adjusting the chroma (color amplitude) control.
38. It is never possible for the color picture to be displaced horizontally from the monochrome picture.
39. Improper functioning of the color killer can prevent color from appearing.
40. The frequency, but not the phase, of the 3.58 MHz color-subcarrier oscillator is critical, to present correct colors.
41. Incorrect alignment of the chroma amplifier can affect the colors of a color picture.
42. One hue missing (red, green or blue) from a color picture, can be due to a defective picture tube.

REVIEW ESSAY QUESTIONS

1. Discuss briefly the use and advantages of a color-picture tube, test jig. How is proper impedance matching accomplished? (Figure 27.1)
2. Besides transistors, what other solid-state devices may be checked with a transistor tester? (Figure 27.2)
3. In Question 2, for an in-circuit transistor test, what indication is there for a good transistor? (Figure 27.2)
4. List at least three tests that can be performed by a picture-tube analyzer and restorer. List at least two restoring functions. (Figure 27.3)
5. Briefly discuss the two basic types of picture-tube brighteners. (Figure 27.4)
6. What function is performed by an isolation picture-tube brightener? (Figure 27.4)
7. What is a "subber"? List at least three functions that can be performed by a "subber"
8. Can the Sencore VA48 perform tuner testing? Explain briefly. (Figure 26.23)
9. In your own words explain briefly how horizontal and vertical-output inductances can be tested by "ringing". (Figure 27.7)
10. In Question 9, how does this test effectively check the "Q" of the inductance? (Figure 27.7)
11. List the major steps you would take in isolating a TV receiver malfunction. (Section 27.8)
12. In Question 11, explain briefly, how you would isolate a malfunction to a stage. (Section 27.8)
13. List four common types of TV color troubles. (Section 27.10)
14. In Question 13, briefly discuss your method of isolating two of the above troubles to a specific receiver section, or stage. (Sections 27.10, 27.11, and Figure 27.10.)
15. Draw and label a simple block diagram of the video IF, video detector and video-amplifier stages of a monochrome-TV receiver. Discuss briefly how you would isolate a malfunction in any of these stages by means of signal injection. (Section 27.11)
16. Describe briefly and in your own words a rough test (no instruments used) to detect a malfunctioning tuner. (Section 27.13).
17. Discuss briefly the possible problems that may occur due to defects in the AGC Section. (Sections 27.13, 27.14, 27.15)
18. Explain briefly how sync compression may occur. How is this condition checked? (Sections 27.14, 27.15, 27.19)
19. List all the controls that if maladjusted can cause sync or picture defects in a color-TV receiver.
20. Discuss briefly how both high and low-frequency response may be reduced in a video amplifier. (Section 27.16)
21. What are two causes of a negative picture? (Sections 27.16, 27.17)
22. List at least three causes of the loss of anode-high voltage. (Section 27.17)
23. Briefly describe a logical method to isolate the symptom of no high voltage.
24. List four ways in which a color-picture tube may become defective.
25. Can you repair an open connection in a picture gun? Explain how. (Section 27.17)
26. In a color-picture tube, if one color is weak, can this be remedied? How? (Sections 27.4, 27.17)

27. What is the effect on the picture of shorted turns in one-half of a deflection yoke? (Section 27.17)
28. In a regulated, low-voltage power supply, list at least three causes for low or no B+ voltage output. (Section 27.18)
29. In what type of low-voltage power supply(s) can a defective diode change the ripple frequency from 120 Hz to 60 Hz?
30. In Question 29, briefly explain your answer.
31. List at least four symptoms of defective sync operation. (Section 27.19)
32. What is meant by horizontal-"picture pulling"? What is one possible cause? (Section 27.19)
33. List at least five causes of insufficient picture height. Indicate which controls (if any) may affect each listed cause. (Section 27.20)
34. Briefly discuss possible causes of poor horizontal linearity. (Figure 27.37)
35. In Question 34, what part (if any) might a defective damper play?
36. What is meant by "horizontal foldover"? What might cause this condition? (Section 27.21)
37. What are two types of malfunctions which might cause poor (or no) horizontal sync? (Section 27.21)
38. What are the function(s) of the following: (1) AFT, (2) ACC, (3) ATC.
39. Explain briefly the meaning of "color setup".
40. Explain at least one cause for colored snow ("confetti"). (Plate 6)
41. How is it possible for the best monochrome and color pictures to not track? (Section 27.23)
42. What would be the symptom of an inoperative 3.58 MHz subcarrier oscillator? Explain. (Section 27.24)
43. What might be the picture symptom of a misaligned chroma-bandpass amplifier? (Section 27.24 and Chapter 26)
44. The picture displays excessive color intensity (saturation), not controllable by the *color* control. What might be the cause? (Section 27.24)
45. What is the picture symptom of an inoperative color killer? (Section 27.24)
46. Briefly explain the effect of a malfunctioning AFPC (or APC) circuit.
47. The phase of the 3.58 MHz subcarrier oscillator has shifted incorrectly by 30°. How would this affect the color picture? (Section 27.24 and Chapter 26.)
48. Only one hue in a color picture is missing. List at least two causes for this problem.

EXAMINATION PROBLEMS

(Selected problem answers are provided at the back of the text.)

1. Make up a two-column table. In one column list all the various types of test equipment (Chapters 26 and 27) used with TV receivers. In the other column, indicate their function.
2. In Figure 27.36, identify: (1) vertical integrator, (2) horizontal-frequency control(s), (3) horizontal-AFC filter.
3. Identify these numbers: (1) 72 kHz, (2) 3.58 MHz, (3) 42.17 MHz, (4) 3.1 to 4.1 MHz, (5) 55.25 MHz.
4. Refer to Figure 27.41. If C_1 shorted, what would be the collector voltage of Q_2?
5. Refer to Figure 27.41. If the base voltage of Q_1 was +4.0 V instead of +4.7 V, could there be collector current in Q_1?

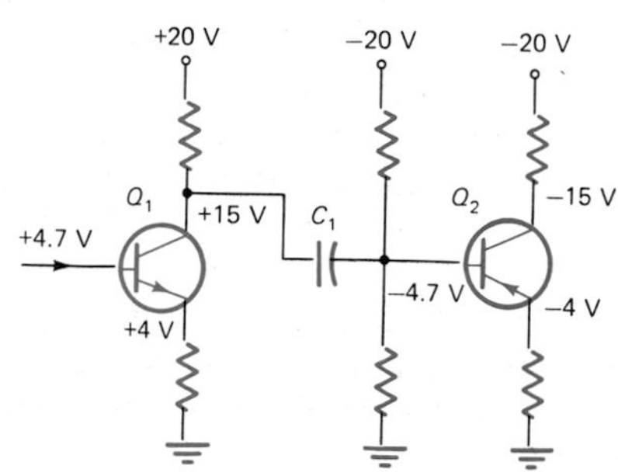

Figure 27.41 Diagram for Problem 4.

APPENDICES

Appendix I

U.S. TELEVISION CHANNELS AND RELATED FREQUENCIES (MHz)

				SOUND IF 41.25		PICTURE IF 45.75			
	CHANNEL NO.	*FREQ. LIMITS*	*CENTER FREQ.*	*PICTURE CARRIER*	*SOUND CARRIER*	*OSC. FREQ.*	*PICTURE IMAGE FREQ.*	*CHANNEL NO.*	
VHF CHANNELS	2	54–60	57	55.25	59.75	101	146.75	2	VHF CHANNELS
	3	60–66	63	61.25	65.75	107	152.75	3	
	4	66–72	69	67.25	71.75	113	158.75	4	
	5	76–82	79	77.25	81.75	123	168.75	5	
	6	82–88	85	83.25	87.75	129	174.45	6	
	7	174–180	177	175.25	179.75	221	266.75	7	
	8	180–186	183	181.25	185.75	227	272.75	8	
	9	186–192	189	187.25	191.75	233	278.75	9	
	10	192–198	195	193.25	197.75	239	284.75	10	
	11	198–204	201	199.25	203.75	245	290.75	11	
	12	204–210	207	205.25	209.75	251	296.75	12	
	13	210–216	213	211.25	215.75	257	302.75	13	
UHF CHANNELS	14	470–476	473	471.25	475.75	517	562.75	14	UHF CHANNELS
	15	476–482	479	477.25	481.75	523	568.75	15	
	16	482–488	485	483.25	487.75	529	574.75	16	
	17	488–494	491	489.25	493.75	535	580.75	17	
	18	494–500	497	495.25	499.75	541	586.75	18	
	19	500–506	503	501.25	505.75	547	592.75	19	
	20	506–512	509	507.25	511.75	553	598.75	20	
	21	512–518	515	513.25	517.75	559	604.75	21	
	22	518–524	521	519.25	523.75	565	610.75	22	
	23	524–530	527	525.25	529.75	571	616.75	23	
	24	530–536	533	531.25	535.75	577	622.75	24	
	25	536–542	539	537.25	541.75	583	628.75	25	
	26	542–548	545	543.25	547.75	589	634.75	26	
	27	548–554	551	549.25	553.75	595	640.75	27	
	28	554–560	557	555.25	559.75	601	646.75	28	
	29	560–566	563	561.25	565.75	607	652.75	29	
	30	566–572	569	567.25	571.75	613	658.75	30	
	31	572–578	575	573.25	577.75	619	664.75	31	

(Continued)

		SOUND IF 41.25			PICTURE IF 45.75		
CHANNEL NO.	*FREQ. LIMITS*	*CENTER FREQ.*	*PICTURE CARRIER*	*SOUND CARRIER*	*OSC. FREQ.*	*PICTURE IMAGE FREQ.*	*CHANNEL NO.*
32	578–584	581	579.25	583.75	625	670.75	32
33	584–590	587	585.25	589.75	631	676.75	33
34	590–596	593	591.25	595.75	637	682.75	34
35	596–602	599	597.25	601.75	643	688.75	35
36	602–608	605	603.25	607.75	649	694.75	36
37	608–614	611	609.25	613.75	655	700.75	37
38	614–620	617	615.25	619.75	701	706.75	38
39	620–626	623	621.25	625.75	707	712.75	39
40	626–632	629	627.25	631.75	713	718.75	40
41	632–638	635	633.25	637.75	719	724.75	41
42	638–644	641	639.25	643.75	725	730.75	42
43	644–650	647	645.25	649.75	731	736.75	43
44	650–656	653	651.25	655.75	737	742.75	44
45	656–662	659	657.25	661.75	743	748.75	45
46	662–668	665	663.25	667.75	749	754.75	46
47	668–674	671	669.25	673.75	755	760.75	47
48	674–680	677	675.25	679.75	761	766.75	48
49	680–686	683	681.25	685.75	767	772.75	49
50	686–692	689	687.25	691.75	773	778.75	50
51	692–698	695	693.25	697.75	779	784.75	51
52	698–704	701	699.25	703.75	785	790.75	52
53	704–710	707	705.25	709.75	791	796.75	53
54	710–716	713	711.25	715.75	797	802.75	54
55	716–722	719	717.25	721.75	803	808.75	55
56	722–728	725	723.25	727.75	809	814.75	56
57	728–734	731	729.25	733.75	815	820.75	57

			SOUND IF 41.25		PICTURE IF 45.75		
CHANNEL NO.	*FREQ. LIMITS*	*CENTER FREQ.*	*PICTURE CARRIER*	*SOUND CARRIER*	*OSC. FREQ.*	*PICTURE IMAGE FREQ.*	*CHANNEL NO.*
58	734–740	737	735.25	739.75	821	826.75	58
59	740–746	743	741.25	745.75	827	832.75	59
60	746–752	749	747.25	751.75	833	838.75	60
61	752–758	755	753.25	757.75	839	844.75	61
62	758–764	761	759.25	763.75	845	850.75	62
63	764–770	767	765.25	769.75	851	856.75	63
64	770–776	773	771.25	775.75	857	862.75	64
65	776–782	779	777.25	781.75	863	868.75	65
66	782–788	785	783.25	787.75	869	874.75	66
67	788–794	791	789.25	793.75	875	880.75	67
68	794–800	797	795.25	799.75	881	886.75	68
69	800–806	803	801.25	805.75	887	892.75	69
70	806–812	809	807.25	811.75	893	898.75	70
71	812–818	815	813.25	817.75	899	904.75	71
72	818–824	821	819.25	823.75	905	910.75	72
73	824–830	827	825.25	829.75	911	916.75	73
74	830–836	833	831.25	835.75	917	922.75	74
75	836–842	839	837.25	841.75	923	928.75	75
76	842–848	845	843.25	847.75	929	934.75	76
77	848–854	851	849.25	853.75	935	940.75	77
78	854–860	857	855.25	859.75	941	946.75	78
79	860–866	863	861.25	865.75	947	952.75	79
80	866–872	869	867.25	871.75	953	958.75	80
81	872–878	875	873.25	877.75	959	964.75	81
82	878–884	881	879.25	883.75	965	970.75	82
83	884–890	887	885.25	889.75	971	976.75	83

Appendix II

FREQUENTLY USED METRIC UNITS

QUANTITY	UNIT	SYMBOL
length	kilometer meter centimeter millimeter	km m cm mm
area	square kilometer hectare (10 000 m^2) square meter square centimeter	km^2 ha m^2 cm^2
energy, work, or heat	megajoule kilojoule joule kilowatt-hour (3.6 MJ)	MJ kJ J kWh
power or heat flow rate	kilowatt watt	kW W
electric current	ampere	A
electromotive force	volt	V
electric resistance	ohm	Ω
frequency	megahertz kilohertz hertz	MHz kHz Hz
sound level	decibel	dB
volume or capacity	cubic meter liter milliliter	m^3 L mL

QUANTITY	UNIT	SYMBOL
mass (weight)	metric ton (1000 kg)	t
	kilogram	kg
	gram	g
	milligram	mg
time	day	d
	hour	h
	minute	min
	second	s
temperature	degree Celsius	°C
speed or velocity	meter per second	m/s
	kilometer per hour	km/h
plane angle	degree	°
force	kilonewton	kN
	newton	N
pressure	kilopascal	kPa
acceleration	meter per second squared	m/s^2
rotational frequency	revolution per second*	r/s
	revolution per minute*	r/min
density	kilogram per cubic meter	kg/m^3
	gram per liter	g/L

*In expressions or operations that involve algebraic manipulation of units (unit analysis), the units of measure are 1/s or 1/min.

Appendix III

SI UNIT PREFIXES

MULTIPLICATION FACTOR	PREFIX	SYMBOL	PRONUNCIATION (USA)*	TERM (USA)	TERM (OTHER COUNTRIES)
1 000 000 000 000 000 000 = 10^{18}	exa	E	ex'a (a as in about)	one quintillion†	one trillion
1 000 000 000 000 000 = 10^{15}	peta	P	as in petal	one quadrillion†	one thousand billion
1 000 000 000 000 = 10^{12}	tera	T	as in terrace	one trillion†	one billion
1 000 000 000 = 10^{9}	giga	G	jig'a (a as in about)	one billion†	one milliard
1 000 000 = 10^{6}	mega	M	as in megaphone	one million	
1 000 = 10^{3}	kilo	k	as in kilowatt	one thousand	
100 = 10^{2}	hecto	h	heck'toe	one hundred	
10 = 10	deka	da	deck'a (a as in about)	ten	
0.1 = 10^{-1}	deci	d	as in decimal	one tenth	
0.01 = 10^{-2}	centi	c	as in sentiment	one hundredth	
0.001 = 10^{-3}	milli	m	as in military	one thousandth	
0.000 001 = 10^{-6}	micro	μ	as in microphone	one millionth	
0.000 000 001 = 10^{-9}	nano	n	nan'oh (an as in ant)	one billionth†	one milliardth
0.000 000 000 001 = 10^{-12}	pico	p	peek'oh	one trillionth†	one billionth
0.000 000 000 000 001 = 10^{-15}	femto	f	fem'toe (fem as in feminine)	one quadrillionth†	one thousand billionth
0.000 000 000 000 000 001 = 10^{-18}	atto	a	as in anatomy	one quintillionth†	one trillionth

*The first syllable of every prefix is accented to assure that the prefix will retain its identity. Therefore, the preferred pronunciation of kilometer places the accent on the first syllable, not the second.

† These terms should be avoided in technical writing because the names for denominations above one million and below one millionth are different in most other countries, as indicated in the last column.

Appendix IV

BINARY NUMBERS

Table IV.1. **A TABLE SHOWING DECIMAL NUMBERS FROM 0 TO 51 AND THEIR BINARY NUMBER EQUIVALENTS**

DECIMAL COLUMN VALUE 10 1	BINARY 32 16 8 4 2 1	DECIMAL COLUMN VALUE 10 1	BINARY 32 16 8 4 2 1
0	0	2 6	1 1 0 1 0
1	1	2 7	1 1 0 1 1
2	1 0	2 8	1 1 1 0 0
3	1 1	2 9	1 1 1 0 1
4	1 0 0	3 0	1 1 1 1 0
5	1 0 1	3 1	1 1 1 1 1
6	1 1 0	3 2	1 0 0 0 0 0
7	1 1 1	3 3	1 0 0 0 0 1
8	1 0 0 0	3 4	1 0 0 0 1 0
9	1 0 0 1	3 5	1 0 0 0 1 1
1 0	1 0 1 0	3 6	1 0 0 1 0 0
1 1	1 0 1 1	3 7	1 0 0 1 0 1
1 2	1 1 0 0	3 8	1 0 0 1 1 0
1 3	1 1 0 1	3 9	1 0 0 1 1 1
1 4	1 1 1 0	4 0	1 0 1 0 0 0
1 5	1 1 1 1	4 1	1 0 1 0 0 1
1 6	1 0 0 0 0	4 2	1 0 1 0 1 0
1 7	1 0 0 0 1	4 3	1 0 1 0 1 1
1 8	1 0 0 1 0	4 4	1 0 1 1 0 0
1 9	1 0 0 1 1	4 5	1 0 1 1 0 1
2 0	1 0 1 0 0	4 6	1 0 1 1 1 0
2 1	1 0 1 0 1	4 7	1 0 1 1 1 1
2 2	1 0 1 1 0	4 8	1 1 0 0 0 0
2 3	1 0 1 1 1	4 9	1 1 0 0 0 1
2 4	1 1 0 0 0	5 0	1 1 0 0 1 0
2 5	1 1 0 0 1	5 1	1 1 0 0 1 1

Table IV.2 **CONVERTING THE DECIMAL NUMBER 283 TO ITS BINARY NUMBER EQUIVALENT***

Division	Quotient + remainder		Binary digits
2 ⌊ 283	141 + 1	⟶	_ _ _ _ _ _ _ _ 1
2 ⌊ 141	70 + 1	⟶	_ _ _ _ _ _ _ 1 1
2 ⌊ 70	35 + 0	⟶	_ _ _ _ _ _ 0 1 1
2 ⌊ 35	17 + 1	⟶	_ _ _ _ _ 1 0 1 1
2 ⌊ 17	8 + 1	⟶	_ _ _ _ 1 1 0 1 1
2 ⌊ 8	4 + 0	⟶	_ _ _ 0 1 1 0 1 1
2 ⌊ 4	2 + 0	⟶	_ _ 0 0 1 1 0 1 1
2 ⌊ 2	1 + 0	⟶	_ 0 0 0 1 1 0 1 1
2 ⌊ 1	0 + 1	⟶	1 0 0 0 1 1 0 1 1

*Conversion from a decimal to binary by repeated division of the decimal integer. At each division the remainder becomes the next higher-order binary digit.

ANSWERS TO SELECTED QUESTIONS AND PROBLEMS

Chapter 1

Examination Questions

1. F 2. F 3. T 4. T 5. F
6. T 7. F 8. F 9. F 10. T

Chapter 2

Examination Questions

1. F 2. T 3. F 4. T 5. T
6. T 7. T 8. T 9. T 10. F
11. F 12. T 13. F 14. F 15. T
16. T 17. F 18. T 19. F

Chapter 3

Examination Questions

1. Synchronizing 2. Scanning
3. 4 to 3 4. Right to left 5. Bottom to top 6. 30; 60 7. monochrome field rate
8. odd; even 9. 525 10. $262\frac{1}{2}$
11. monochrome horizontal
12. monochrome vertical 13. 15 734.26
14. Picture 15. Vertical 16. 7
17. Horizontal sync 18. Sawtooth
19. Interlacing 20. Vertical blanking; 21
21. 16 667 22. 3.58; 8 to 11 23. Back porch 24. Horizontal 25. Horizontal and vertical 26. Blanking

Chapter 3

Examination Problems

2. $262\frac{1}{2}$ 4. 1 016 μs 5. $317\frac{1}{2}$ μs
8. 0.28 μs

Chapter 4

Examination Questions

Part A 1. (b) 2. (c) 3. (a)
4. (c) 5. (b) 6. (b) 7. (d)
Part B 1. 12.5 2. sync pulses
3. negative 4. grid 5. ± 25 6. 4.2
7. 0.6; 1.5 8. 1.25 9. detail
10. dark; light 11. saturation 12. tint; hue 13. gamma 14. luminance; chrominance 15. monochrome 16. red; blue; green 17. vertical interval test signals
18. scale

Chapter 4

Examination Problems

2. 2.76 MHz 4. (1) 30Hz to 10 kHz
(2) 10 kHz to 100 kHz (3) 100 kHz to 4.2 MHz 7. Multiburst test signal. VIR signal

Chapter 5

Examination Questions

1. F 2. F 3. F 4. T 5. T
6. F 7. F 8. T 9. T 10. T
11. T 12. F 13. F 14. T 15. F
16. T 17. T 18. F 19. T 20. F
21. F 22. F 23. F 24. T

Chapter 5

Examination Problems

1. 70%, 30% 2. 27 nA, 60 nA 5. 7.5%, 1.0%

Chapter 6

Examination Questions

1. F 2. F 3. T 4. F 5. T
6. T 7. T 8. T 9. F 10. T
11. F 12. F 13. F 14. T 15. T
16. T 17. F 18. T 19. F 20. F
21. T 22. F 23. T 24. T 25. F

Chapter 6

Examination Problems

1. 38.87 miles (62.55 km) 3. 52.67 miles (84.76 km) 4. 8 ft (2.44 m) 5. 10 dB

Chapter 7

Examination Questions

Part A 1. c 2. b 3. d 4. a
5. c 6. a 7. a 8. d 9. b
10. a
Part B 1. color signal sidebands
2. video 3. solid-state 4. IF
5. video, sound 6. video detector
7. polarity 8. AFC 9. 20 000
10. AGC 11. horizontal hold
12. printed circuit 13. modules
14. general measurements, wave form observation, signal injection 15. low voltage power supply 16. vertical deflection

Chapter 8

Examination Questions

Part A 1. b 2. b 3. a 4. d
5. b 6. c 7. b 8. a 9. c
10. a
Part B 1. white 2. yellow
3. chromaticity 4. saturated 5. 15 750
6. monochrome, color 7. 0.59, 0.30, 0, 11
8. interleaved 9. 2 10. saturation
11. 0.5 12. I, Q 13. 3.579 545
14. 920 15. frequency, phase

Chapter 8

Examination Problems

1. 30%, 59%, 11%, 89% 4. 0.58 MHz

Chapter 9

Examination Questions

1. F 2. F 3. T 4. F 5. F
6. T 7. F 8. F 9. F 10. T
11. F 12. T 13. T 14. F 15. F
16. T 17. T 18. T 19. T 20. F
21. F 22. F 23. T 24. T

Chapter 9

Examination Problems

4. 67.25 MHz 70.83 MHz 71.75 MHz.
5. damper, focus rectifier

Chapter 10

Examination Questions

Part A 1. T 2. T 3. T 4. T
5. F 6. T 7. T 8. F 9. T
10. T 11. T 12. F 13. F 14. T
15. F
Part B 1. IF 2. Front end
3. varactor 4. DC voltages
5. switching 6. random access 7. fine tuning 8. unbalanced 9. signal, noise
10. RF 11. stability 12. C_N 13. G_2
14. cascode 15. feedback 16. noise
17. MOSFET 18. AFT 19. Colpitts
20. RF amplifier 21. AFT

Chapter 10

Examination Problems

4. Antenna circuit: L_{24}, C_{66}, CR_{13}
RF circuit: L_{27}, C_{67}, CR_{14} 5. L_{28}, L_{29}, C_{68}, CR_{15} 7. 152.75 MHz, 168.75 MHz, 568.75 MHz, 976.75 MHz.

Chapter 11

Examination Questions

Part A 1. varactor 2. microcomputer 3. bistable 4. zero's and one's 5. 17 6. prescaler 7. coded decimal 8. 5 9. two signals 10. fine-tuning control 11. microcomputer 12. varactor 13. reference, tuner 14. any 15. byte 16. digital 17. programmable 18. picture IF 19. pull in, hold in 20. 34 kHz, 54 kHz 21. infrared LED's 22. pulse pattern 23. microcomputer 24. loop 25. oscilloscope

Part B 1. F 2. F 3. F 4. T 5. T 6. F 7. T 8. F 9. T 10. F 11. F 12. F 13. T 14. T 15. F 16. F 17. F 18. T 19. F 20. F 21. T 22. F

Chapter 11

Examination Problems

3. The program is 0111. 6. Tuner oscillator frequency, channel 2 = 101 MHz.

$\frac{101 \text{ MHz}}{256} = 0.39453125 \text{ MHz}. \frac{0.39453125}{4} =$

$0.98632812 \text{ MHz}. \frac{0.98632812 \text{ MHz}}{107} = 976.5625 \text{ Hz}$

8. 100 MHz to 102 MHz

Chapter 12

Examination Questions

Part A 1. picture, sound 2. sound 3. 920 4. gain, selectivity 5. section 6. before 7. reversed 8. color IF 9. 3.579545 10. vestigial 11. 39.75 12. stagger tuning 13. 39.75, 47.25, 4.5 14. series 15. intercarrier 16. picture IF, tuner 17. broadband 18. surface-acoustic wave 19. nonelectromagnetic 20. piezoelectric substrate 21. transducer 22. multistrap coupler 23. tuner, AGC 24. tuner RF 25. misaligned

Part B 1. F 2. F 3. T 4. F 5. F 6. T 7. T 8. T 9. T 10. F 11. T 12. T 13. F 14. T 15. T 16. F 17. T 18. T 19. F 20. F 21. T 22. F 23. F 24. F 25. T 26. T 27. T 28. F 29. T 30. T

Chapter 12

Examination Problems

4. RA_6, CA_3, CA_4 and LA_5. 47.25 MHz.
6. 154.75 MHz to 158.75 MHz.
7. 3.08 MHz.

Chapter 13

Examination Questions

Part A 1. diode, synchronous 2. linear 3. positive, negative 4. opposite 5. low-pass filter 6. 4.5 7. IF traps 8. gain 9. reference, modulated-IF 10. video signal 11. nonlinear 12. 45.75 13. video detector 14. loss

Part B 1. F 2. T 3. T 4. F 5. F 6. T 7. F 8. T 9. T 10. F 11. F 12. F 13. F 14. F 15. T 16. T 17. F 18. T 19. T 20. F

Chapter 13

Examination Problems

2. 5 250 kHz 4. (1) third; (2) 4.5-MHz sound; a 4.5-MHz trap is missing.

Chapter 14

Examination Questions

Part A 1. overload 2. sync-pulse level 3. peak 4. two, coincident
5. long 6. airplane flutter
7. horizontal 8. 60 9. align
10. charging a capacitor 11. delayed
12. forward, reverse 13. 0.5
14. reverse 15. sync pulses 16. weak, strong 17. horizontal 18. positive, negative, inverter 19. 500, 1000
20. coincidence gate 21. latch
22. negative
Part B 1. F 2. F 3. F 4. T
5. F 6. T 7. T 8. T 9. F
10. F 11. F 12. T 13. F 14. T
15. T 16. T 17. T 18. F 19. F
20. F 21. T 22. T 23. T 24. F
25. F 26. T

Chapter 14

Examination Problems

2. 10 μs, 190.5 μs, 5 μs, 10 μs. 3. R_1, C_2, R_2, C_3, R_3, R_4.

Chapter 15

Examination Questions

Part A 1. 2 2. 3.2 3. 4.2
4. 30 5. 100 6. smearing, low
7. compensating 8. high, sharpness
9. phase shift 10. excessive high
11. delay line 12. LDR 13. video peaking 14. interleaved 15. 3.6
16. gain 17. chroma section
18. degeneration 19. inversely, intensity
20. video peaking 21. delay line, comb
22. reverses 23. barber pole
Part B 1. F 2. F 3. T 4. T
5. F 6. T 7. F 8. F 9. F
10. F 11. F 12. T 13. T 14. F
15. T 16. T 17. F 18. T 19. F
20. F 21. T 22. F 23. F 24. T
25. T 26. F

Chapter 15

Examination Problems

1. 393 356.50 Hz 5. 54 569.91 Hz

Chapter 16

Examination Questions

Part A 1. R-C, direct 2. mutual, impedance 3. one, five 4. large
5. beta 6. alpha-cutoff frequency
7. less 8. 0.707 9. load resistor
10. 50 11. swamping resistor
12. degenerative high frequency compensation
13. low frequency 14. background illumination 15. direct coupled
16. gain-bandwidth product 17. emitter follower 18. peaking coil
19. compensation network
Part B 1. F 2. F 3. F 4. T
5. F 6. F 7. T 8. F 9. T
10. T 11. T 12. T 13. F 14. T
15. F 16. T 17. F 18. T 19. T
20. F 21. F 22. T 23. T 24. T

Chapter 16

Examination Problems

1. (1) L_1, (2) L_2, (3) R_3. 4. (c) 0.707 × 1 kHz gain. 6. 3.1 MHz. 7. 69.3 μH.

Chapter 17

Examination Questions

Part A 1. background shading
2. average value 3. darker 4. AC component 5. blanking pulse 6. DC reinsertion 7. retrace lines 8. average value 9. DC component
10. background illumination 11. DC restorer, clamper 12. shape 13. diode
14. blanking, sync 15. coupling capacitor
16. frame
Part B 1. F 2. T 3. T 4. F
5. F 6. F 7. F 8. F 9. T
10. F 11. T 12. T 13. T

Chapter 17

Examination Problems

1. R = 1 megohm 4. 35 volts
6. 0 volts

Chapter 18

Examination Questions

Part A 1. phosphor 2. dot triad, vertical stripe 3. shadow mask 4. P4
5. 6.3 6. anode 7. magnetic strength
8. 13, 14 9. lens 10. electromagnetic
11. 114 12. horizontal, vertical
13. electrostatic, electromagnetic
14. centering magnets 15. ion trap
16. implosion 17. vertical color
18. delta gun 19. delta gun 20. purity
21. black 22. in line gun 23. slotted aperture, vertical phosphor 24. one
25. one 26. focus lens
27. convergence plates 28. in line
29. vertical color stripe, aperture mask
30. 7, 12 31. implosion 32. anode
33. blooming 34. spark gaps
35. brightener
Part B 1. F 2. F 3. T 4. T
5. F 6. T 7. F 8. F 9. F
10. F 11. F 12. T 13. F 14. F
15. T 16. T 17. T 18. F 19. T
20. T 21. T 22. T 23. F 24. T
25. T 26. T

Chapter 19

Examination Questions

Part A 1. size (or weight), efficiency
2. isolation transformer 3. regulated
4. silicon 5. LC, RC 6. low
7. isolated 8. 60 9. 168 10. 504
11. solid state 12. derived 13. 105, 135 14. zener 15. ripple
16. protection 17. monolithic
18. input, output, ground 19. switching
20. switching 21. 15 734.26 22. start-up 23. voltage tripler 24. SCR
25. raster
Part B 1. F 2. F 3. T 4. F
5. T 6. T 7. F 8. T 9. F
10. T 11. T 12. F 13. T 14. T
15. F 16. F 17. F 18. T 19. T
20. T 21. T 22. T 23. F 24. T
25. T 26. T 27. F 28. T 29. T

Chapter 19

Examination Problems

2. 474.6 volts 5. (c) 50 ohms

Chapter 20

Examination Questions

Part A 1. composite video
2. equalizing 3. 31 468.8 4. vertical
5. below 6. sync separator 7. low, integrator 8. sync separator 9. noise
10. noise gate, sync separator 11. short time 12. equalizing 13. 190.5
14. 6, 6
Part B 1. F 2. F 3. F 4. F
5. T 6. T 7. F 8. T 9. T
10. F 11. F 12. T 13. T 14. F
15. T 16. T

Chapter 20

Examination Problems

1. 9H 4. (a)No 6. R_{307}, R_{308}

Chapter 21

Examination Questions

Part A 1. 59.94 2. digital 3. 10, 20 4. trapezoidal 5. free running
6. higher 7. driver 8. base bias
9. positive, negative 10. sinusoidal
11. capacitor, resistor 12. cathode (or emitter) 13. hold 14. integrator
15. 31 468 16. R-S flip flop 17. phase detector
Part B 1. F 2. T 3. T 4. F
5. T 6. F 7. T 8. T 9. F
10. F 11. F 12. F 13. T 14. T
15. F

Chapter 21

Examination Problems

1. (a) 30V 4. (a) 508.5 μs 6. C_2, R_2, R_3
9. 0.22 s

Chapter 22

Examination Questions

Part A 1. blocking oscillators, multivibrators, sinusoidal oscillators 2. IC
3. noise 4. control voltage 5. noise
6. AFC 7. sync pulse, sawtooth
8. horizontal-output transformer
9. stabilizing, noise 10. Hartley
11. frequency, phase 12. inductance
13. inductance, resistance 14. AFC, horizontal oscillator 15. base voltage
16. collector, collector, 90 17. VCO
18. horizontal, zero
Part B 1. F 2. T 3. T 4. F
5. F 6. T 7. T 8. T 9. F
10. T 11. F 12. T 13. T 14. F
15. F

Chapter 23

Examination Questions

Part A 1. damper 2. turns off
3. 6, 7 4. 70, horizontal 5. damper
6. not used 7. not 8. trapezoidal
9. retrace 10. damper 11. horizontal amplifier 12. rectangular
13. overvoltage, overcurrent 14. switch
15. decreases 16. yoke 17. output amplifier 18. 3, 5 19. regulation
20. retrace 21. yoke 22. pincushion
23. pincushion distortion 24. SCR
25. ferrite 26. toroidal 27. inverse
28. horizontal, vertical
Part B 1. F 2. F 3. F 4. T
5. F 6. F 7. T 8. T 9. T
10. T 11. F 12. F 13. T 14. T
15. F 16. F 17. T 18. T 19. T
20. F 21. T 22. T 23. F 24. T
25. T

Chapter 23

Examination Problems

1. 6.7 μs 3. L5 4. 18.8 μs (approx.)

Chapter 24

Examination Questions

Part A 1. height 2. yoke, output
3. centering 4. vertical pincushioning
5. static, dynamic 6. thermistor
7. yoke 8. saturable reactor 9. in line
10. dynamic convergence 11. blanking
12. trapezoidal 13. linearity
14. multivibrator 15. vertical output
16. one half 17. sawtooth 18. product
19. push pull 20. toroidal
21. isolation, autotransformer 22. short circuit
Part B 1. F 2. F 3. T 4. T
5. F 6. T 7. F 8. F 9. T
10. T 11. T 12. T 13. F 14. F
15. T 16. F 17. F 18. F 19. T
20. T

Chapter 24

Examination Problems

1. 7.54 ohms. 3. 59 mA

Chapter 25

Examination Questions

Part A 1. 25 2. more economical
3. 0.25 4. audio frequency 5. twice
6. amplitude 7. preemphasis 8. 75
9. amplitude variations 10. detector
11. 50 12. vacuum tube, transistor, IC
13. limiter stages 14. 8 15. 100
16. amplitude 17. 4.25, 4.75 18. ratio
19. limiting 20. oscillate 21. error
22. digital 23. response 24. thermal, volume 25. complementary symmetry

Part B 1. F 2. T 3. F 4. T
5. T 6. F 7. F 8. T 9. F
10. F 11. T 12. F 13. T 14. F
15. T 16. F 17. T 18. T 19. F
20. F 21. T 22. F 23. F 24. T
25. T 26. F 27. F 28. T 29. T
30. F

Chapter 25

Examination Problems

2. (1) 0 dB, (3) 9.5 dB 4. 9.2 V
7. (1) C_6 9. (3) L_1, C_2, C_1

Chapter 26

Examination Questions

Part A 1. bar sweep 2. tuner oscillator 3. video detector 4. gate
5. limiter 6. IF, tuner 7. ACC
8. 3.58, oscillator, chroma 9. triggered
10. increased 11. phase shift 12. gain
13. APC, (AFPC) 14. G−Y 15. two
16. oscillator, burst 17. dual trace
18. frequency 19. high 20. $3\frac{1}{2}$
21. demodulator 22. vectorscope
23. frequency sweep 24. crystal
25. dot, crosshatch 26. alignment
27. marker signals 28. dynamic convergence 29. Y 30. aligning
31. polarity 32. autoranging 33. traps
34. 3.1, 4.1 35. oscilloscope, TV screen
36. red
Part B 1. F 2. F 3. T 4. T
5. T 6. F 7. T 8. F 9. F
10. T 11. T 12. T 13. T 14. F
15. T 16. F 17. F 18. T 19. F
20. T 21. T 22. T 23. F 24. T
25. T 26. F 27. T 28. F 29. F
30. T 31. T

Chapter 26

Examination Problems

2. (1) SC_{202}, (2) C_{254}, T_{204}, R_{252}, (3) R_{202}, C_{202}, C_{204}, L_{202}. 3. (1) DL_{200}, (2) R_{256} in parallel with R_{264} and C_{201}. 7. 1 999 V

Chapter 27

Examination Questions

Part A 1. high, yoke 2. tone
3. removal, rejuvenation 4. analog, digital
5. ringing 6. injection, tracing 7. tint
8. 3.58 9. sync 10. phase
11. stage, stage 12. horizontal
13. fixed 14. demodulator 15. subber
16. bias supply 17. AGC 18. flyback
19. dark, smeary 20. negative
21. horizontal deflection 22. brightener
23. transistor 24. ripple voltage
25. loose 26. filtering 27. nonlinearity
28. 10 29. keystone 30. left
31. brightness 32. tube 33. saturation
34. purity 35. dynamic 36. tuner
37. burst 38. green 39. chroma
40. sync 41. phase
Part B 1. F 2. T 3. T 4. T
5. F 6. F 7. F 8. T 9. T
10. F 11. F 12. F 13. F 14. F
15. T 16. T 17. T 18. T
19. T 20. F 21. F 22. T 23. T
24. F 25. F 26. F 27. T 28. T
29. F 30. T 31. F 32. T 33. F
34. T 35. F 36. F 37. F 38. F
39. T 40. F 41. T 42. F

Chapter 27

Examination Problems

3. (1) Resonant frequency of horizontal-output inductances. (4) Chrome-amplifier bandwidth.
4. -20 V (No collector current in Q_2)

GLOSSARY

A

ABC Automatic brightness control.

ABL Automatic brightness limiter.

AC component The variations of a video signal.

ACC Automatic color control (amplitude).

Accumulator A computer device that stores a number, adds to it another number, and then stores the sum.

Active satellite A type of satellite that receives and retransmits a radio signal over a very wide area.

Adder (+) A circuit in which two or more signals are combined to produce an output signal that is the addition of the input signals.

AFC Automatic frequency control. Also (but rarely), automatic fine tuning.

AFPC Automatic frequency and phase control. Same as APC.

AFT Automatic fine tuning. Sometimes called AFC.

AGC Automatic gain control.

AGC comparator For the sense used in this text, a circuit that detects the difference between the AGC reference voltage and the voltage level of the synchronizing pulses. Its output is a DC voltage.

AGC reference level The tips of the synchronizing pulses.

Airplane flutter The variations of picture contrast caused by reflections from passing airplanes.

Alpha The current gain of a common-base amplifier. It is commonly between 0.91 and 0.99 and always less than unity.

Alpha cutoff frequency The frequency at which the current gain of a common-base amplifier has decreased to 0.707 of its low-frequency value.

Alphanumeric display The display of letters and numbers on a cathode-ray tube.

Ambient light Normal room light.

Antenna gain A method of comparing the ability of an antenna to capture a signal with the ability of a simple dipole to capture the same signal at the same point.

Antenna impedance The ratio of the signal voltage to the signal current at the antenna feed point (practical definition).

Antenna preamplifier A compact, solid-state amplifier mounted at the antenna.

APC Automatic phase control. Same as AFPC.

Aperture grill A thin sheet of metal placed directly behind the screen of a color picture tube. It has a large number of vertical slots and is used only with color picture tubes that use vertical color stripes. One vertical slot corresponds to each set of red, green, and blue phosphor stripes.

Aperture mask A thin sheet of perforated metal located directly behind the triad dot screen of a color picture tube. It permits the excitation of only those phosphor dots associated with the corresponding electron beams.

Aspect ratio The ratio of width to height. For the rectangular picture transmitted by a television station, the aspect ratio is 4:3.

Aural A term that is used to mean sound.

Automatic brightness control A circuit used in television receivers to automatically adjust picture tube brightness in response to changes in ambient light. Sometimes combined with automatic contrast and color control. The circuit uses a light-dependent resistor as the controlling device.

Automatic brightness limiter A circuit that prevents "blooming" on a picture tube by limiting the maximum picture tube beam current.

Automatic contrast control A circuit used in television receivers to automatically adjust picture contrast in response to changes in ambient light.

Automatic fine tuning A circuit that maintains the correct tuner oscillator frequency by compensating for drift and for moderate mistuning.

Autoranging Automatic switching of a multi-range meter to select the correct range for each specific measurement.

AVC Automatic volume control.

B

Back porch The portion of a horizontal blanking pulse that follows the trailing edge of the horizontal synchronizing pulse.

Background illumination The average brightness of a scene.

Balanced transmission line A transmission line in which both halves of the line are above ground by the same amount of voltage.

Balun A matching transformer that matches a balanced two-wire line to an unbalanced coaxial line (300 ohms to 75 ohms).

Band separator A device that accepts a composite television signal and has separate outputs for the VHF and UHF signals.

Bandpass amplifier See chroma amplifier.

Barkhausen effect Dark, vertical lines near the left side of the raster. Caused by oscillation in horizontal output tube.

Bar-sweep generator An instrument that generates frequency bars, which are used for video IF and color circuit alignment and testing.

BCD Binary-coded decimal. Basis of the digital system.

Beam blender See ion trap.

Beam deflection angle The total angle through which the electron beam can be deflected.

Beam splitter A system of mirrors and lenses that directs each primary color image to the appropriate camera tube.

Beta The current gain of a common-emitter amplifier. Common values lie between 25 and 100.

Beta cutoff frequency The frequency at which the current gain of a common-emitter amplifier has decreased to 0.707 of its low-frequency value.

Bifilar transformer A transformer in which primary and secondary windings are wound side by side to obtain extremely tight coupling.

Binary notation See binary number system.

Binary number system A number system in which only 0's and 1's are used.

Bipotential color tube A color picture tube in which two separate potentials are applied to the electrostatic focus-lens system. One of these potentials may be the high accelerating voltage.

Bistable Having two stable states (such as ON and OFF).

Bit A binary digit; usually 0 or 1.

Black surround The black coloring that surrounds each color dot or color stripe in a color picture tube to improve picture contrast.

Blanking-pulse level The reference level for video signals. The blanking pulses must be aligned at the input to the picture tube.

Blanking signal Pulses used to extinguish the scanning beam during horizontal and vertical retrace periods.

Blocking oscillator A single-stage, transformer-coupled oscillator that produces periodic output pulses.

Boosted B+ A DC voltage that results from the addition of the power-supply B+ and the average value of voltage pulses coming through the damper. Boosted B+ can be several hundred volts higher than the power-supply B+.

Booster A television signal amplifier.

Bottom ramp vertical output amplifier An amplifier that, through yoke current, drives the electron beams of a color tube from vertical center to the bottom of the screen.

Brighteners See picture-tube brightener.

Brownout A reduction of AC line voltage. Voltage may be reduced to as low as 105 V AC.

Bulk-wave reflection The wave reflection from the bottom to the top of the substrate of a surface acoustic wave filter.

Burst See color synchronizing signal.

Byte A group of bits.

C

Cable television A television system in which subscribers pay a fee. It is a combination of radio wave reception and cable distribution.

Carrier frequency See center frequency.

Cascode amplifier A two-stage amplifier in which the collector (or plate) of the input stage directly feeds the emitter (or cathode) of the output stage.

Cathode-coupled multivibrator A two-stage, tube-type, nonsinusoidal oscillator. One of the feedback paths is via a common-cathode resistor.

Cathode rejuvenation A method of restoring normal (practically) cathode emission in a picture tube electron gun.

CATV Community antenna television or cable television.

CCD Charge-coupled device.

CCTV Closed-circuit television.

CCU Camera control unit.

Center frequency The unmodulated frequency of an FM transmission.

Chip A small piece of silicon upon which all components of an integrated circuit are formed.

Chroma See chrominance.

Chroma amplifier The part of the chrominance section that receives the full color signal and amplifies it. It is used in conjunction with the 2.1- to 4.2-MHz bandpass filter.

Chroma signal See chrominance signal.

Chromaticity chart A diagram used for color mixing. The color wavelengths are plotted as coordinates of X and Y.

Chrominance The hue and saturation of a color. The modulated 3.58-MHz signal is the chrominance signal.

Chrominance section The circuits of a color TV receiver that amplify, demodulate, and matrix the color signal.

Chrominance signal The color portion of the composite video signal. The I and Q modulated color subcarrier. Phase angle of the signal represents hue, and amplitude represents saturation.

Clamping See DC reinsertion.

Closed-circuit television A television system in which the signal is transported by cables rather than by radio waves.

Coincidence circuit See coincidence gate.

Coincidence gate A circuit that produces an output only when the input pulses occur within the same time interval.

Collector-coupled multivibrator A two-stage, transistor, nonsinusoidal oscillator. Each collector feeds back to the other stage's base.

Color bars A test pattern of specific colors of vertical bars. Used to check the performance of a color television system or receiver.

Color burst signal See color synchronizing signal.

Color decoder See matrix.

Color dot triads Groups of red, green, and blue phosphor dots on the screen of a color picture tube.

Color killer The stage of the chrominance section that cuts off the chroma amplifier during monochrome broadcasts.

Color noise See confetti.

Color purity See saturation.

Color setup The process, in a color receiver, of properly adjusting purity, convergence, and gray scale.

Color sidebands The sidebands of the 3.58-MHz color subcarrier.

Color signal See chrominance signal.

Color synchronizing section The circuits of a color receiver that generate and stabilize the receiver 3.58-MHz oscillator.

Color synchronizing signal A "burst" of 8 to 11 cycles of the color subcarrier frequency (3.579545 MHz). The burst is situated on the back porch of each horizontal blanking pulse during color transmissions. It is used to synchronize the receiver's color subcarrier oscillator with that of the transmitter.

Colorplexed signal The total color video signal, including Y, I, Q, blanking, synchronizing, and burst.

Comb filter In television receivers, a filter that "combs out" the chroma signal from the composite color video signal.

Combination antenna A composite antenna designed to receive VHF and UHF TV signals, as well as FM signals.

Common-base amplifier A transistor amplifier in which the base element is common to the input and output circuits. The current gain is less than one.

Common-collector amplifier A transistor amplifier in which the collector element is common to the input and output circuits. The voltage gain is less than unity.

Common-emitter amplifier A transistor amplifier in which the emitter element is common to the input and output circuits.

Community antenna television A television antenna system in which the signal is received, amplified, and distributed to houses in a community. Provides reception in areas that cannot normally receive television signals.

Comparator An electronic circuit that compares two input signals and delivers an output signal to indicate whether or not the two input signals are in agreement.

Compatibility The capability of a monochrome TV receiver to produce a normal monochrome picture from a color broadcast and of a color TV receiver to correctly reproduce a monochrome picture broadcast.

Complementary color A color that, when added to another color in proper proportion, produces white. In color TV, a color with a phase 180° from a primary color. Thus, red, green, and blue have complementary colors of cyan, magenta, and yellow, respectively.

Complementary-symmetry amplifier An arrangement of NPN and PNP transistors that provides push-pull operation from a single input signal.

Composite video signal The complete transmitted video signal. It consists of the luminance (Y) and chrominance signals and the synchronizing and blanking pulses.

Confetti The effect of electrical noise on the picture of a color receiver.

Contrast The ratio between the dark and light portions of a television picture.

Convergence circuits The circuits that help to maintain the correct registration of the three beams in the color picture tube.

Correction voltage The DC voltage output of an AFC system.

Cosvicon A single-gun color camera tube (Panasonic).

Countdown circuit A circuit that derives the horizontal and vertical deflection frequencies by division from a stable master oscillator.

Counter See accumulator.

Coupler A device that permits two or more TV receivers to be connected to a single antenna. It does not amplify.

CPU Central processing unit.

D

Damper A diode in the horizontal output circuit. It conducts at the beginning of the trace. This stops the 70-kHz ringing and helps provide the linear initial portion of the trace.

Dark current The current flowing through a photoelectric device in the absence of irradiation.

Darlington pair A two-transistor amplifier connected so that the overall gain is equal to the product of the individual gains. It has high input impedance and high gain.

DC component The average value of the video signal. It represents the background illumination.

DC correction voltage See correction voltage.

DC reinsertion The process of aligning all blanking (and synchronizing) pulses at the same level.

DC restorer A circuit that performs DC reinsertion.

Decoder A network used to convert digital information to an analog form.

De-emphasis Reduction of the higher audio frequencies (in FM receiver) to compensate for the pre-emphasis applied to the FM transmission. With pre-emphasis, improves audio signal-to-noise ratio.

Degenerative compensation The circuit configuration in which negative feedback at the emitter (or cathode) circuit is used to widen the bandpass response of an amplifier.

Delay line An artificial delay line (0.6 to 1.0 μs) inserted into the Y channel. It equalizes the arrival time of Y and color signals at the color picture tube.

Delayed AGC The RF AGC, which does not begin to take effect until the input-signal level exceeds a predetermined amplitude; usually 500 to 1 000 μV.

Delta gun Three electron guns arranged in a triangular configuration in some color picture tubes.

Demodulator probe An auxiliary probe used in conjunction with an oscilloscope. Used to detect the envelope of the RF or video IF waves for alignment or troubleshooting.

Derived voltages Operating voltages obtained from a secondary winding of a horizontal output (flyback) transformer.

Differential amplifier An amplifier whose output is proportional to the difference between the voltages applied to its two inputs.

Differential peak detector A type of FM detector that uses peak detector and differential amplifier stages controlled by a tuned circuit.

Differentiator A high-pass filter. It provides a positive and negative sharp pulse output when the input consists of synchronizing pulses.

Digital channel switching A TV receiver channel-switching system. Digital signals are transformed into their equivalent DC voltages, which are then applied to varactors for channel selection.

Digital circuit A circuit that has only two states: ON and OFF.

Digital countdown circuit A circuit that produces output pulses of 15 734 or 59.94 Hz from a VCO operating at 31 468 Hz.

Digital FM detector A type of FM detector having no tuned circuits. It uses a digital-logic, monostable (one-shot) multivibrator. Two complementary outputs are integrated to obtain a push-pull audio signal.

Digital system In electronics, the system based on binary numbers "0" and "1" (OFF and ON conditions).

Digital tuner See digital channel switching.

Diode clamping A method of obtaining delayed AGC and DC restoration.

Dipole A resonant (tuned) antenna, cut to a half wavelength of the operating frequency.

Direct access A TV tuning system in which channels are selected by push buttons in any desired sequence.

Direct-coupled amplifier An amplifier in which the output of one stage is connected to the input of the following stage without the use of a coupling capacitor.

Direct coupling A method of coupling amplifier stages or an amplifier to a picture tube without loss of the DC component.

Discriminator A type of FM detector that derives amplitude variations in response to frequency variations.

Divider A circuit capable of dividing a frequency by a desired amount.

Domestic satellite A synchronous communications satellite, serving the continental United States, Puerto Rico, Hawaii, and Alaska.

Driven element The dipole portion of an antenna that delivers the signal to the transmission line (and ultimately to the receiver).

Driver amplifier A transistor, isolation, and amplifying stage placed between a transistor deflection oscillator and output stage.

DTL Diode transistor logic.

Dynamic convergence The process of converging the three electron beams of a color tube over the entire screen.

E

EAROM Electrically alterable read-only memory. Used in Omega digital-tuning system.

ECL Emitter coupled logic.

EIA Electronic Industries Association.

Electron-gun tracking The brightness response of each of the three electron beams in a color picture tube.

Electronic tuner A tuner that works by means of varactors. It has no moving parts.

Emitter-coupled multivibrator A two-stage, transistor, nonsinusoidal oscillator. One feedback path is via a common emitter resistor.

Emitter-follower amplifier See common-collector amplifier.

Enable To activate a circuit, such as by applying a signal or pulse of appropriate form or by removing a suppression signal.

Equalizing pulses A series of six pulses occurring before and after the serrated vertical synchronizing pulse to ensure proper interlace. The equalizing pulses occur at twice the horizontal scanning frequency.

Even field The half of a TV frame composed of the even-numbered scanning lines.

Eye resolving power The ability of the eye to distinguish between objects that are close together.

F

Ferrite bead A small bead made of ferrite powder. It is placed on a connecting wire to introduce inductance for suppressing parasitic oscillations and high-frequency transients.

FET Field-effect transistor.

Field One half of a TV frame, composed of 262.5 scanning lines. There are 60 fields per second for monochrome TV and 59.94 fields per second for color TV.

Flicker Apparent visual interruptions between successive fields of a TV picture. Occurs when the field rate is too low because then the *persistence of vision* does not give the illusion of continuous motion. With a field rate of 60 Hz, flicker is eliminated.

Flutter See airplane flutter.

Flyback The horizontal scan retrace.

Flyback capacitor A capacitor that resonates the horizontal output inductances at a frequency (about 70 kHz) to provide the desired retrace time.

Flyback transformer The horizontal output transformer.

FM modulation percent The ratio of the actual frequency swing to that defined as 100% modulation, expressed as a percentage. For TV FM, 100% modulation is expressed as a frequency swing of 50 kHz.

Footcandle A unit of illumination. Also called lumens per square foot.

Forward AGC A system of AGC in which increasing the base-to-emitter bias decreases the amplifier gain.

Foster-Seeley discriminator See discriminator.

Frame One complete TV picture. It consists of two fields and has a total of 525 scanning lines.

Free-running frequency The unsynchronized frequency of an oscillator.

Frequency deviation The frequency difference between the center frequency and the peak frequency change due to modulation. For TV FM, the maximum frequency deviation is ±25 kHz.

Frequency distortion Distortion caused by an amplifier when there is unequal amplification of all desired frequencies over the bandpass.

Frequency-sweep generator An RF, IF, or video frequency generator that periodically (60 or 120 Hz) sweeps across a desired frequency range. Used to check or align wide-band circuits.

Frequency swing Twice the instantaneous frequency deviation. For TV FM, the maximum allowable frequency swing is 2 × 25 kHz = 50 kHz.

Frequency synthesized tuning A system in which the output frequencies are derived from a single precision crystal oscillator. In this system, the basic frequency synthesizer system uses a phase-locked loop in conjunction with a digital frequency programmer.

Frequency synthesizer An electronic system that provides a number of crystal-stable frequencies derived from a single precision-crystal oscillator.

Front end See tuner.

Front porch The portion of a horizontal blanking pulse that precedes the horizontal synchronizing pulse.

FS Frequency synthesizer or frequency synthesis.

Full-wave voltage doubler A rectifier circuit whose DC output (unloaded) is equal to twice the peak value of the AC input voltage. The ripple frequency is 120 Hz.

G

Gamma A measurement that compares the contrast in the original and reproduced television pictures.

Gated-rainbow color bar pattern A test pattern of ten specific vertical color bars separated by black vertical bars. Phase separation of colors is 30°. The pattern contains no Y component and is used for color receiver servicing.

Gear tooth distortion A type of picture distortion that gives the outer perimeter of the picture the appearance of a gear tooth (or pie-crust). Can result from poor horizontal AFC voltage filtering.

Ground wave That portion of a radiated wave which travels close to the earth.

Grounded-base amplifier See common-base amplifier.

Grounded-collector amplifier See common-collector amplifier.

Grounded-emitter amplifier See common-emitter amplifier.

H

Halation The rings of light that surround the light produced by an electron beam striking a picture tube screen. This effect tends to reduce picture contrast.

Half tone A method of reproducing pictures by dividing them into small picture elements. With this method, black ink can be used to reproduce all shades of gray.

Half-wave antenna An antenna that has an electrical length equal to half the wave length of the signal being transmitted (or received).

Half-wave voltage doubler Same as above, but with common input and output connections and a ripple frequency of 60 Hz.

Hammerhead pattern A pattern on the picture tube screen when the picture is rolled halfway up. It is formed from the equalizing and vertical serrated pulses.

Head end The antennas, traps, attenuators, amplifiers, and other signal-processing equipment that precedes the splitter in an MATV or CATV system.

Height control A control that varies the deflection sawtooth amplitude.

Hertz antenna A half-wave antenna.

Hexadecimal notation A notation that uses the radix 16, with digits 0 to 9 and six more digits represented by the letters A, B, C, D, E, and F.

High The ON state; binary 1.

High-frequency compensation A scheme by which the high-frequency response of a video amplifier is extended by means of peaking coils or degeneration.

HOT Horizontal output transformer.

Hold control A deflection oscillator frequency control.

Hold-in range For an AFT system, the locked-in frequency range that the oscillator is held to.

Horizontal AFC The automatic frequency control circuitry that maintains the proper phase and frequency of the horizontal oscillator.

Horizontal blanking pulse A part of the composite video signal. It is a pedestal 10.16 μs wide, occurring between each active horizontal scanning line. It extinguishes scanning-beam current during each horizontal line retrace.

Horizontal digital countdown system A divide-by-two system that derives the 15 746-Hz horizontal frequency from a stabilized VCO operating at 31 468 Hz.

Horizontal frequency power supply A power supply whose voltages are derived from the secondary coil of a horizontal output (flyback) transformer.

Horizontal picture pulling Horizontal displacement of some portions of the picture (not raster). One cause is distortion of the horizontal synchronizing pulses.

Horizontal register Referring to the SID, the computer section that receives and temporarily stores one horizontal line of picture information at a time.

Horizontal synchronizing pulse A rectangular pulse situated above each horizontal blanking pulse. The duration is 5.08 μs. It synchronizes the horizontal scanning at the TV receiver with that at the transmitter.

Hue The name of a color corresponding to its dominant wavelength. Examples are red, green, and blue.

Hybrid tuner A tuner that has both vacuum tubes and transistors.

I

IC Integrated circuit.

IGFET Insulated gate field-effect transistor.

Image frequency An undesired frequency signal capable of passing through the IF amplifiers. Numerically, it is equal to twice the IF plus the tuned RF.

In-circuit test Testing of components without disconnection from the circuit.

Infrared transmission The transmission of information by infrared waves. The infrared spectrum lies between visible red light (about 0.75 μm) and the shortest microwaves (about 1 000 μm).

In-line gun Three electron guns arranged in a parallel configuration.

Integrated circuit A solid-state circuit in which a number of complete stages may be made on a single semiconductor wafer.

Integrator A low-pass filter. The output is the vertical synchronizing pulses.

Interdigital transducer Two interlocking comb-shaped metal-film patterns applied to a piezoelectric substrate. Used for converting RF voltages to surface acoustic waves and vice versa.

Interlaced scanning A system of TV-picture scanning. Odd-numbered scanning lines, which make up an *odd field,* are interlaced with the even-numbered lines of an *even field.* The two interlaced fields constitute one *frame.* In effect, the number of transmitted pictures is doubled, thus reducing flicker.

Interleaving See multiplexing.

Ion trap An electron gun structure in a picture tube that diverts negative ions to prevent them from burning a hole in the phosphor screen. By means of the field of an external magnet, electrons are permitted to reach the screen.

Ionosphere That portion of the earth's outer atmosphere that contains layers of free electrons and ions that affect the propagation of radio waves.

IQ system A color-TV receiver demodulation system. The I sidebands extend to 1.5 MHz and the Q sidebands to 0.5 MHz.

I signal One of the two color video signals which modulate the color subcarrier. It represents the color range from reddish orange to cyan.

Isolation brightener An isolation transformer that eliminates a heater-to-cathode short in a picture tube. It does not improve brightness.

J

JFET Junction field-effect transistor.

K

Keyed AGC An automatic gain control scheme in which the keyer device (tube or transistor) will not fire without the simultaneous application of the synchronizing pulses and the horizontal keying pulses.

Keystone raster A trapezoidal raster, which results from shorted turns in one half of a deflection yoke.

Keystoning A distortion of the raster that results in a trapezoidal raster rather than a rectangular one. There can be horizontal or vertical keystoning. Usually results from a yoke short circuit.

L

Lag The persistance of the electric-charge image in a camera tube.

Latching circuit A bistable circuit that holds a switching circuit in its "on" position.

LDR Light-dependent resistor. Its electrical resistance varies inversely with the light intensity impinging upon its light-sensitive surface.

LED Light-emitting diode.

Light-transfer characteristic A measure of camera-signal current relative to faceplate illumination.

Limiter (AM) A circuit that limits the amplitude of signals and thus reduces AM noise interference.

Limiting knee The minimum input signal amplitude required to produce limiting.

Linear detector A detector that operates linearly. In such a detector, the only output is the original modulation. The synchronous video detector is a linear detector.

Linearity control Generally, a deflection amplifier control. It can vary the operating characteristic of the amplifier to adjust the deflection sawtooth linearity.

Local oscillator The tuner oscillator.

Log periodic antenna A broad-band antenna suitable for use as a color TV antenna.

Low The OFF state; binary 0.

Low-capacity probe An auxiliary probe useful (in some types) up to several hundred MHz. It can be used with an oscilloscope or high-impedance multimeter to measure or observe high-frequency signals. It reduces the capacitive loading effect on the circuit being measured.

Low-frequency compensation A scheme used in RC-coupled video amplifiers to extend their low-frequency response and improve their phase response.

LP Low power.

LSB Least significant bit.

LSI Large-scale integration, where 100 or more interconnected circuits are on a single chip of silicon.

Luminance The amount of light intensity, which is perceived by the eye as brightness.

Luminance signal See Y signal.

M

Master antenna television A television antenna system in which the signal is received, amplified, and distributed to many receivers, as in an apartment house or a hotel.

Matrix The section of a color TV receiver that transforms the color difference signals into red, green, and blue video signals.

MATV Master antenna television.

Metal oxide varistor See voltage-dependent resistor.

Microcomputer A microprocessor combined with input/output interface devices, a type of external memory, and other elements required to form a working computer system.

Microprocessor The central processing unit of a computer, either on a single IC chip or on several chips.

Miller effect The effective increase of input capacitance of a transistor or tube. It is expressed by $C_M = C_{in}(1 + \text{stage gain})$.

Miller integrator vertical sweep system A unique vertical oscillator and output deflection system.

Milliroentgen One thousandth of a roentgen.

Module Generally, a plug-in unit containing a complete section or sections of a television receiver.

Monochrome Black and white pictures.

Monochrome signal See Y signal.

Monolithic A circuit formed within a single semiconductor (silicon) substrate.

MOSFET Metal oxide semiconductor field-effect transistor.

MOV Metal oxide varistor.

MPU Microprocessor unit.

mR Milliroentgen.

MSB Most significant bit.

Multiplexing The simultaneous transmission of two or more signals over a single RF channel. The process of interleaving the chrominance signal with the luminance signal.

Multistrap coupler With reference to a SAWF, a series of metal-film lines placed on the surface of the substrate. It greatly reduces bulk wave interference.

Multivibrator A two-stage, nonsinusoidal oscillator. In TV receiver use, it is free-running and synchronized.

N

Negative picture phase The positioning of the composite video signal so that the tip of the synchronizing pulses are at 100% amplitude. The brightest picture signals are in the negative direction, and the synchronizing pulses are in the positive direction.

Negative picture transmission The U.S. system of transmission, whereby an increase of modulation percentage of the picture carrier is caused by the picture's changing toward black.

Negative transmission The darker the television scene, the greater the AM power output.

Neutralizing circuit Within an amplifier circuit, the circuit that nullifies oscillation-producing voltage feedback from the output to the input of the amplifier.

Newvicon An ultrasensitive camera tube (Panasonic).

Noise cancellation A system whereby noise impulses are inverted and used to cancel initial noise pulses of opposite polarity.

Noise gate A transistor stage that is turned OFF by strong noise pulses. In turn, it turns OFF the synchronizing pulse separator.

Nonlinear detector A detector that operates nonlinearly, especially for weak signals. In TV receivers, it is usually a diode detector. The output of such a detector contains the sum and difference frequencies of the input carriers, as well as harmonics of these carriers. The output contains the demodulated video wave which is sent to the video amplifier.

NTSC National Television System Committee, who devised the present system of U.S. color television transmission.

NTSC color bar pattern A test pattern of six vertical color bars without separation. These are the three primary colors plus their three complementary colors. The pattern contains a Y component and is used mainly in broadcasting.

Nuvistor A miniature vacuum tube, used as the RF amplifier in some tuners.

O

Octal digit One of the symbols 0, 1, 2, 3, 4, 5, 6, and 7, when used as a digit in the number system with radix 8.

Odd field The half of a TV frame composed of the odd-numbered scanning lines.

P

Parasitic element A director or reflector of an antenna.

Peak AGC A simple AGC system that rectifies the synchronizing pulses to obtain the control voltage.

Persistence of vision The phenomenon whereby the eye retains an image for a short time after the image is no longer visible.

Phase detector A circuit that provides a DC output proportional to an oscillator frequency and a reference frequency.

Phase distortion A distortion of a signal that occurs when the phase shift of an amplifier is not proportional to frequency over the desired bandpass. Also called phase-frequency distortion.

Phase-locked loop (PLL) A closed-loop circuit used to maintain the phase and/or frequency of two signals. In TV tuners, the frequency of a VCO is compared with that of a reference-crystal oscillator. The two frequencies to be compared are first divided by digitally controlled dividers. A circuit in which an oscillator (VCO) is synchronized in phase and frequency with an input signal.

Photoconductive The property of a material in which its conductance is changed by irradiation. An increase of radiation intensity decreases the resistance.

Photodiode A semiconductor diode whose reverse current varies with the intensity of illumination.

Photoelectric effect The emission of electrons from a body, caused by radiant energy.

Picture detail The number of picture elements resolved on the picture tube screen. The greater the number, the crisper the picture will appear.

Picture overload Excessive picture contrast, or negative picture, caused by insufficient AGC voltage.

Picture tube brightener A device attached in series with a picture tube socket to restore lost brightness caused by reduced cathode emission.

Piecrust distortion See gear tooth distortion.

Piezoelectric substrate The base upon which a surface acoustic wave filter is constructed.

PIN Pincushion.

Pincushion distortion A raster outline distortion particularly prevalent with wide-angle, deflection picture tubes. It creates a raster in which all four sides are bowed inward. It is corrected in the TV receiver by the addition of parabolic voltages to the vertical and horizontal yokes.

Plate-coupled multivibrator A multivibrator where each plate is coupled to the other stage's grid.

PLL Phase-locked loop.

Plumbicon A television camera tube (N.A. Phillips).

Polarization of radio waves The direction of the electric field vector as radiated from the antenna. For example, if the electric field vector is vertical, the radio wave is said to be *vertically polarized*.

Port An input or output of a network.

Positive picture phase The positioning of the composite video signal so that the tips of the synchronizing pulses are at zero amplitude. The brightest picture signals are in the positive direction, and the synchronizing pulses are in the negative direction.

Post-injection markers A system of response-curve markers in which the markers are injected *after* the tuned circuits.

Pre-emphasis The effect of a network inserted in the audio portion of an FM transmitter. It increases the strength of the higher audio frequencies. With de-emphasis, improves the signal-to-noise ratio.

Prescaler A high-frequency divider.

Primary colors Red, green, and blue.

Processor A combination of circuits on an IC that perform a number of specific functions to obtain the required resultant output.

Programmable divider A divider whose ratio can be controlled by changing the binary number input to its ports.

Protection circuit Part of a voltage regulator. It protects against a short of the series pass element and/or excessive load current.

Pull-in range For an AFT system, when unlocked, the maximum oscillator frequency deviation from nominal that the AFT can correct.

Purity The condition of having a uniform color field individually for each gun of a color picture tube. This is obtained by adjustment of the purity magnet assembly on the neck of the tube.

Q

Q signal The second of the two color video signals that modulate the color subcarrier. It represents the color range from yellowish green to magenta.

Quadrature detector A type of FM detector in which the signal voltage on one grid or transistor lags the signal voltage on a second grid or transistor by 90° when the input signal is exactly 4.5 MHz. Input frequency deviations change the phase of the signal on the *second* grid or transistor. As a result, the output (audio) voltage varies with deviation of the input frequency.

Quasi complementary symmetry amplifier A transistor amplifier using two transistors that are both either NPN or PNP types. Must be driven by a complementary symmetry amplifier.

R

Radix The base of a number system. The radix of the decimal system is 10, that of the binary system is 2.

RAM Random access memory.

Raster The illuminated area on a TV picture tube when no received signal is present.

Rate of frequency deviation The same as the audio modulating frequency.

Ratio detector A type of FM detector in which the audio output is determined by the ratio of two IF voltages. It is relatively insensitive to amplitude variations.

RC coupling A method of coupling amplifier stages or an amplifier to a picture tube that uses a coupling capacitor and input resistor. The DC component is lost with RC coupling.

Reactance transistor A transistor circuit that appears to be an inductive or capacitive reactance.

Reference signal In general, a highly stable signal used as a standard against which other signals may be compared. As used in Chapter 13, a 45.75-MHz continuous-wave signal. This signal is one of the two inputs to a synchronous detector.

Rejuvenation See cathode rejuvenation.
Reset To restore to normal action.
Resolution See picture detail.
Resting frequency See center frequency.
Retrace The line traced by the scanning beam(s) of a picture tube as it travels from the end of one horizontal line or field to the start of the next line or field.
Reverse AGC A system of AGC in which decreasing the base-to-emitter bias decreases the amplifier gain.
RFI Radio-frequency interference.
Ringing test A test to determine the condition of deflection output transformers or yokes. It effectively checks the Q of the unit by causing shock-excited oscillations.
Ripple The AC component in the output of a DC power supply. It exists because of incomplete filtering.
Roentgen The international unit of exposure dose for X rays and gamma rays.
ROM Read only memory.
R-S flipflop A pulse generating circuit that is turned ON by a "set" pulse and OFF by a "reset" pulse.
RTL Resistor transistor logic.
Run voltage In a derived-voltage power supply, the steady-state power supply voltage.
R−Y, B−Y system A color TV receiver demodulation system. The R−Y and B−Y sidebands are each limited to 0.5 MHz.

S

Saturable reactor A PIN transformer. A transformer with an additional (third) *control* winding that carries direct current. The amount of direct current varies the core saturation and thus the output signal amplitude.
Saturation The purity or density of color. A fully saturated color has no dilution by white.
SAW Surface acoustic wave.
SAWF Surface acoustic wave filter.
Scanning The process of moving the electron beam horizontally and slightly vertically to cover the picture tube screen in successive horizontal lines.
S capacitor A capacitor placed in series with the horizontal yoke. An S-shaped wave form across the capacitor improves horizontal linearity.
S correction An S-shaped wave added to the output scanning current to correct nonlinear scanning. The nonlinearity arises from the tendency of the raster to be "stretched" at the extremes of the screen because of picture tube geometry.
SCR Silicon-controlled rectifier.
Sensor A device that senses a change in a quantity, such as light, sound, or radio waves, and converts that change into a useful signal.
Series-pass transistor regulator A voltage regulator in which a transistor in series with the supply current regulates the output voltage of a power supply.
Series peaking A method of video amplifier, high-frequency compensation in which the peaking coil is inserted in series with the coupling capacitor.
Series-shunt peaking A method of video amplifier, high-frequency compensation in which two peaking coils are used, one in series with the load resistor and the other in series with the coupling capacitor.
Serrated vertical pulse The TV vertical synchronizing pulse which is divided into six serrations. The serrated pulses occur at twice the horizontal scanning frequency.
Shadow mask See aperture mask.
Sharpness control A user-operated television receiver adjustment that controls the amount of video peaking. Also called fidelity control.
Shunt peaking A method of video amplifier, high-frequency compensation in which the peaking coil is inserted in series with the load resistor.
SID Silicon-imaging device.
Signal injection A system of troubleshooting in which the test signal is injected into various stages while the monitoring device is kept at a fixed point.
Signal tracing A system of troubleshooting (frequently used with signal injection) in which the monitoring device is moved from stage to stage while the test signal is injected at a fixed point.
Silicon-diode array A camera-tube target constructed on a thin slice of *N*-type silicon. Approximately 30 000 photodiodes are formed on the layer.
Silicon rectifier A semiconductor rectifier (diode) that can withstand high voltages, high currents, and high temperatures. It consists of a metal contact held against a piece of high-purity silicon.
Sky wave That portion of the radiated wave that travels at various angles above the horizon.

Snivets Dark vertical lines at the right side of the raster. Caused by oscillation in a horizontal output tube.

Snow Small, random, white spots produced on a television screen by noise signals.

Solid-state devices Electronic devices, such as diodes and transistors, where electrical conduction takes place through solid materials.

Solid-state tuner A tuner in which the active devices are all transistors.

Spectral response The relation between the radiant sensitivity and the wavelength of the radiation of a camera tube.

Splitter See coupler.

Stabilization coil A resonant coil used in horizontal deflection oscillator circuits to improve the oscillator noise immunity.

Stagger tuning The alignment of successive single-tuned circuits to different frequencies in order to obtain a broad-band response.

Start-up voltage In a derived-voltage power supply, a momentary voltage used to begin operation of the horizontal oscillator, buffer, and driver stages.

Static convergence The converging of the three color picture tube beams at the screen center. It is accomplished by positioning permanent magnets around the neck of the color tube.

Storage area Referring to the SID, a temporary storage site for each TV picture field. It feeds the horizontal register.

Subber See tuner subber.

Subcarrier A carrier wave that modulates another carrier wave of higher frequency. In the NTSC system, the color subcarrier at 3.58 MHz modulates the picture carrier wave of the individual channel.

Substrate The physical material on which microcircuits and other devices, for example, SAWFs, are made.

Subtractor A circuit in which the amplitude of the output signal is proportional to the difference between the amplitudes of two input signals.

Suppressed subcarrier A system whereby the subcarrier is eliminated at the transmitter but the sidebands of the subcarrier are transmitted.

Surface acoustic wave An acoustic wave traveling on the surface of the optically polished surface of a piezoelectric substrate. The velocity of the wave is only about 10^{-5} times that of electromagnetic waves. The wave has the slow velocity property of sound while retaining the frequency of its source. It can pass frequencies as high as several gigahertz.

Surface acoustic wave filter A filter that uses SAWs. It consists of a piezoelectric substrate with interdigital transducers at each end and possibly a multistrap coupler in between. The configuration of the transducers determines the bandwidth and frequency response of the SAWF.

Surface wave integrated filter Another way of describing a SAWF; for example, when it is used to establish the bandpass of a color receiver IF section.

Sweep generator See frequency sweep generator.

SWIF Surface wave integrated filter.

Switching diode A semiconductor diode that provides the same function as a mechanical switch.

Switching regulator Part of a regulated power supply. A regulator in which the series-pass element is periodically either fully ON or fully OFF. Switching frequencies range from approximately 5 to 100 kHz.

Synchronizing pulse clipper See synchronizing pulse separator.

Synchronizing pulses Pulses used for keeping the television receiver picture in step with the television transmitter picture.

Synchronizing pulse alignment Aligning the tips of the synchronizing pulses at the same DC level for correct AGC operation.

Synchronizing pulse separator The stage that removes the synchronizing pulses from the composite video signal.

Synchronous detector A device that detects only a modulated wave whose carrier is synchronized with a reference frequency.

T

Telecine The transmission of motion pictures on a television system.

Thermal protection A means of preventing thermal runaway in high-power transistors. Schemes include an added thermistor and the use of an emitter resistor.

Thermal resistor See thermistor.

Thermal runaway A condition in a power transistor in which collector current increases collector junction temperature. This allows greater collector current to flow, further increasing the heating effect. If unchecked, this action can destroy the transistor. A high ambient temperature makes the transistor more susceptible to this problem.

Thermistor A semiconductor having a high negative temperature coefficient of resistance. Its resistance is inversely proportional to temperature.

Three-terminal regulator Commonly, a monolithic power supply voltage regulator. There are only three terminals: input, output, and ground.

Time constant The product of resistance in ohms and capacitance in farads, which gives the time constant in seconds $(T = RC)$.

Tint See hue.

Top-ramp vertical output amplifier An amplifier that, through yoke current, drives the electron beams of a color tube from the top of the screen to vertical center.

Toroidal yoke A highly efficient deflection yoke that is wound around a doughnut-type core. The toroid provides for a highly concentrated magnetic field within itself, with minimum leakage of magnetic flux.

Trace On a TV screen, the movement of the electron beam from left to right. During this time, picture information is "painted" on the screen.

Transducer Any device that converts energy from one form to another.

Transformerless power supply A power supply in which the AC input voltage is applied directly to the rectifier system.

Transversal filter As used in television receivers, a tapped delay line.

Trapezoidal voltage wave See trapezoidal wave form.

Trapezoidal wave form A voltage wave consisting of the combination of a sawtooth and a rectangular wave. One use is to drive the grid of a deflection output amplifier.

Triac (triode AC semiconductor switch) A bidirectional, gate-controlled thyristor that provides full-wave control of AC power.

Triggered-sweep oscilloscope An oscilloscope in which each sweep line is initiated by the leading edge (usually) of the observed signal. It provides a stable display, capable of expansion and without overlap.

Trinicon A single color television camera tube for single tube color television cameras (Sony).

Tripotential color tube A color picture tube in which three separate potentials are applied to the electrostatic focus-lens system. This design provides smaller spot size than in a bipotential picture tube. It makes in-line electron guns practical in large picture tubes.

Tuned antenna An antenna that is cut to be resonant to a particular frequency.

Tuner That portion of a TV receiver that contains the RF amplifier, mixer, and oscillator. It selects the desired channel and rejects all others.

Tuner subber A self-contained substitute tuner. An item of test equipment that can be substituted for the TV receiver tuner to assist in isolating troubles.

Tuning voltage The DC voltage applied to the varactors of a tuner to control the resonant frequencies of the tuned circuits. A typical range may be from 3 to 28 V.

U

UHF Ultrahigh frequencies, ranging from 300 to 3 000 MHz. (Wavelengths of 1 m to 100 mm.)

Ultrasonic frequency A frequency lying above the audio frequency range, generally considered to be above about 20 kHz. For TV remote control, ultrasonic frequencies in the approximate range of 34 to 54 kHz are used.

Unbalanced transmission line A transmission line in which one half of the line is grounded. Generally, a shielded line.

V

Varactor A PN semiconductor diode whose capacitance varies with the applied voltage. It is used as a tuning element in TV tuners.

Varactor tuner A tuner in which all functions are performed by varactors. Generally, a solid-state tuner.

Varistor See voltage-dependent resistor.

VCO Voltage-controlled oscillator.

VCR Video cassette recorder.

VDR Voltage-dependent resistor or video disc recording.

Vectorscope An oscilloscope that compares both the phase and the amplitude of an applied color signal with a reference signal (3.58-MHz oscillator).

Vertical blanking pulse A positive or negative pulse developed during vertical retrace and appearing at the end of each field. It is used to blank out scanning lines during the vertical retrace interval.

Vertical foldover An extreme case of poor vertical linearity in which a portion of the picture is folded back upon itself. It may appear at the top or the bottom of the picture.

Vertical serrated pulses See serrated vertical pulses.

Vertical synchronizing pulse See serrated vertical pulse.

Vestigial sideband transmission A method of transmitting a television signal (or any AM signal) in which part of one sideband is filtered out and not radiated.

VHF Very high frequency, ranging from 30 to 300 MHz (wavelengths from 10 to 1 m).

VHS Video home system, a type of video tape recorder.

Video peaking A scheme of overamplifying the high video frequencies to about 3.2 MHz. This gives the television picture more detail and a sharper image.

Video signal That part of the composite video signal, containing the picture information.

Vidicon A television camera tube.

VIR Vertical interval reference. The VIR signal may be used at the television receiver as a reference for correct color reproduction.

VITS Vertical interval test signal. A signal that may be included during the vertical blanking interval to permit on-the-air testing of the video system of the television transmitter.

Voltage-dependent resistor A two-electrode semiconductor with a voltage-dependent, nonlinear resistance. The resistance is inversely proportional to the applied voltage.

Voltage doubler A rectifier circuit whose DC output (unloaded) is twice the peak value of the AC input voltage.

Voltage tripler A rectifier circuit whose output voltage is triple the peak AC input voltage.

V.O.T Vertical output transformer.

VTR Video tape recorder.

W

Weighted fingers Fingers in an interdigital (interlocking) transducer of a SAWF in which the amount of overlapping is nonuniform. This configuration provides specific frequency response with sharp cutoffs, as for an IF response.

Writing speed Referring to video tape recorders, the relative speed of the tape past the recording head.

X

X rays A penetrating electromagnetic radiation of short wavelengths, about 10^{-7} to 10^{-10} cm. X rays are usually generated by the collision of high-velocity electrons with a target.

X,Z system A color TV receiver demodulation system. The sidebands are limited to 0.5 MHz. Somewhat similar to the $R-Y$, $B-Y$ system.

Y

Y signal That part of the total color video signal that represents a monochrome picture. The Y signal supplies brightness and detail for a color picture.

Z

Zener diode regulator A shunt voltage regulator in which the active element is a zener diode, which operates in a reverse-current breakdown mode.

INDEX

Abnormal color intensity, 910
Absorption traps, 343–344
Accessory probe, 833
Adder, 442, 443
Adder signals, 440
Adjacent channel traps, 342
Adjustments
  in color TV horizontal systems, 706–708
  in monochrome TV receiver circuits, 686–689
Airplane flutter, 388
Alignment
  and bar sweep generator, 843
  of color bandpass amplifiers, 838–840
  equipment for, 816
  and frequency sweep generator, 833–834
  improper, in IF systems, 359
  of sound section, 838
  video IF, and trap adjustment, 834–837
Alpha, defined, 454
Alpha cutoff frequency, 455
Alpha-numeric displays, 40
"Alpha-Wrap," 31
Amplification of video signal, 370
Amplifiers. *See speceific type*
Amplitude, 74, 187, 419, 728, 760
Amplitude limiter, 753
Amplitude modulation (AM), 12
  and FM, differences between, 753, 754
Analog multimeter, 831, 832
Anik synchronous satellites, 26
Anode voltage, 499, 565, 852
Antenna(s), 117
  bandwidth, 124
  bidirectional, 127
  characteristics, 123–125
  combination, 135
  directional, 118, 124
  general-purpose, 132–135
  half-wave, 125
  impedance, 124–125
  indoor, 140–141
  installation of, guidelines for, 145
  nondirectional, 124
  placement of, 118, 122
  preamplifiers, 146–148
  rotators, 141
  stacked, 135
  tuned, 125, 127
  VHF-UHF-FM, 136, 138
Antenna directivity, 124, 127
Antenna gain, 124, 127
Antenna mast, grounding, 153
Antenna rotor control, automatic, 3
Antenna signal, 145–146
Antenna-system
  accessories, 149–150
  troubleshooting, 153–154
Aperture mask, 518–519
Aquadag coatings, 499
Arcing, 715
Arcs (flashovers), 534
Aspect ratio, 49
Attenuator, 151
Audio amplifiers, 758, 782
Audio-mute switching, 295–296
Audio system, 11
Audio voltage amplifier, 783
Automatic brightness control (ABC), 215, 436–437
Automatic brightness limiter (ABL), 215, 532, 533–534
Automatic color control (ACC), 76, 214
  and color killer, 810–811
Automatic fine tuning (AFT), 214–215, 232, 253, 296, 906
  frequency-sensitive detector, 299
  functions of, 296–300
  troubles in, 317–318
Automatic frequency control (AFC), 165, 232, 794. *See also* Automatic fine tuning (AFT)
  horizontal, 740–665
  horizontal oscillator and, 804
  and tuning control, 803–804
Automatic frequency and phase control (AFPC), 215, 840–843
  color adjustment, 850–851
Automatic gain control (AGC), 162, 163, 207, 240, 383–413, 802–803
  amplified, 386
  delayed, 350–386, 393–394
  enable and comparator, 410
  forward, 348
  functions of, 384
  gate, 398
  to IF stages, 398–399
  latch circuit
  output stages, 410–411
  processor, expanded block diagram of, 409–410
  in an RCA IC, 406–407
  reference level, 383
  to RF amplifier, 399
  in solid-state receivers, 394, 402–406
  with transistor amplifiers, 385
  troubles in, 360, 411

Automatic gain control *(cont'd)*
  types of, 385–386
  voltage, 392, 876, 878
Automatic gain control (AGC) operation, 403–406
Automatic phase control (APC), 814, 840
Automatic tint control (ATC), 76, 215
Autoranging, 833
Autotransformers, 741
  horizontal output, 676–677

Back porch, 58, 61, 828
Balun, 144, 150, 151, 228, 229–230
Band separators, 149–150
Bandpass amplifier, 210, 490
Bandpass shaping, 352, 354
Bandwidth, 417–418
  of the channel, 233
  increasing, 237
  of tuned system, 337–338
  video amplifier, 428
Bar-sweep alignment, 845–847
Bar sweep alignment checks, 847, 849–850
Barkhausen oscillation, 904
Base-collector junction capacitance, 397
Beam-alignment coil, 90
Beam-deflection angle, 500
Beat interference, 249
Beta, defined, 453
Beta cutoff frequency, 455
Bias, automatic, 580
Bias change, amount of, 395
Bias controls, 725–726
Bias-network control unit, 865
Bifilar-wound coil, 340
Binary number system, 2, 273–275
Binary numbers, converting decimal numbers to, 274, 275, 927–928
Bistable device, 273
Black level, 56, 70
Black matrix, 12
Black surround, 524
Black and white circuitry, 1
Blanker amplifier, horizontal, 815
Blanker circuit and burst, 813
Blanking, 47, 53, 723, 733
  horizontal, 53
  level of, 482–483
  negative pulse, 733
  vertical, 53, 58, 732–733, 795
Blue-lateral magnet, of color picture tube, 522
Blocking oscillator
  pulse at grid of, 608
  transistor, 609–614
  vacuum tube, 605–607
Boost circuit, B+, 808
"Boost source," 676
Boosters, 145–146, 147
"Bootstrap" operation, 676
Bow tie antenna, 130
Bowtie effect, producing, 728

Bridged-T traps, 345
Brightness, 75
  and contrast, 511–513
  factors affecting, 75
Brightness control, 167, 436
Brightness limiter circuit, 813
Brightness signal, 185
Brownouts, 561, 567
Bulk-wave reflection, 358
Burst, 811. *See also* Color burst
Burst amplifier, 813
Buzz control, monochrome TV, 167

Cable television (CATV), 1, 17–24
  antenna system, 18
  distribution-amplifier system, 18
  heterodyning down, 18
  microwaves in, 153
  system matching, 18
  trunk-amplifier system, 18
  two-way, 19
Cable television channels, frequency assignment for, 20–21
Camera control unit (CCU), 106–107
Camera fundamentals, basic, 86
Camera tube(s)
  characteristics of, 88–89, 92, 115
  efficiency of, 88
  vidicon, 89
Canadian Broadcasting Corporation (CBC), 26
Canadian Satellite System, 26
Capacitance
  semiconductor, 458
  transistor input, 458
Capacitive reactance, 660
Cascode amplifier, 238, 241, 469–470
Cathode-ray beam, 53
Cathode-ray tube, 8, 107
Centering circuit, 696
Central processing unit (CPU), 279
Channel allocations, 233–235
Channel display, 317
Channel indicator LED display circuitry, 293
Channel selector, 166
Charge-coupled device (CCD), 100
Chip, 6
Chroma alignment, 841–842
Chroma control, 75, 210
Chroma section of color-receiver, 490
Chroma signal, 185
Chromaticity chart, 181–182
Chrominance, 77, 103, 185
Chrominance amplitude, double, 442
Chrominance reference, 43
Cinematography, electronic, 95
Clapp configuration, 814
Closed-circuit television (CCTV), 1, 14, 17, 22, 89
  components, 22–23
  uses of, 23–24
Closed-loop system, 564, 802

Coincidence-gate circuit, 410
Coaxial cable(s), 1, 24, 142, 143–144
Coefficient *K*, increased, 237
Color
  vs. area, 185–186
  combining, 179–180
  elements of, 179
  purity of, 183
  saturation and hue, 183–184
  true, 179
Color amplifier(s), 809
  alignment, 847–848
Color amplitude, 906
Color bandpass, 809
  alignment, 838–840
Color bar generator, use of, 828
Color burst, 61, 192
  function of, 62–63
"Color burst" sync signal, 58
Color control, 75, 810
Color control amplifier, 809
Color field rate, 50
Color formation, 180
Color killer, 811, 909
Color mixing, 182
Color output amplifier, 809–810
Color pattern generators, 826–831
Color problems, types of, 871
Color section, circuitry of, 809–815
Color signal(s)
  cancellation, 193
  components, 77–80, 190–191
  relationship between, 212
Color subcarrier, 191–192
  frequency, 192–194
Colorplexed composite video signal, 194–196
Color television, 1, 14
  console, 6
  portable receiver, 5
  principles of, 178–199
  troubles in, 907–913
  video requirements for, 427–429
  "Y" signal and delay line, 428–429
Color television camera(s), 88
  principles, 103–105
Color television receiver(s), 211
  block diagram of, 206–207
  common troubles of, 215–216
  high-voltage circuits, 208
  horizontal- and vertical-deflection systems, 207
  I and Q system, 210
  IF response curves for, 204
  picture tube and convergence circuits, 208
  principles of, 202–218
  processing color sidebands, 208–209
  sound IF, FM detector and audio system, 205
  sync separators and AGC, 207
  video detector and amplifiers, 205–206
Color television transistor and tube vertical output circuits, 726–729
Color television transistor vertical output circuits, 737–740
Color temperature, 856
Color tracking, 436
Colpitts oscillator, 255
Comb filter, 416, 440–444
  block diagram of, 442
  factors for operating, 440–441
  frequencies rejected and passed by, 441
  luminance processing, 443
  luminance signal cancellation, 442–443
Compensating delay, 443
Complementary-symmetry amplifier, 785–787
Communications Satellite Corporation (COMSAT), 27
Community-antenna television system (CATV), 17, 152–153
Compactron, 170–171
Compatibility, 197
Composite video signal(s)
  color, 77
  monochrome, 53
  normal and special, 77
Compu-Matic circuit analysis, 285–296
  channel indicators, 293–294
  microcomputer, 289
  sensory circuitry, 291–292
  PLL, 285–289
  switching circuits, 294
Compu-Matic tuning system, troubles in, 316–317
Computer(s)
  capabilities of, 279–280
  simple program, 280
Computer-type techniques, 2–3
Confetti, 63, 907–908
Conical array antennas, 135
Continuous turners, 223
Contrast, 75
Contrast control
  automatic, 435–438
  automatic brightness and, 436
  monochrome TV, 167
  in video amplifiers, 432–435
Control voltage, use of, 309–312
  audio function, 309–310
  channel selection function, 312
  hue function, 310–311
  intensity function, 312
  volume-on function, 309
Convergence. *See also* Dynamic convergence
  adjustments, 852
  using dots, 830
Convergence circuits, 208
  blue, 731
  red-green, 732
Convergence coils, 521
Convergence current, 520–521
Corner reflectors, 130–131
Corona discharge, 715
Counters, 277
Couplers. *See* Splitters
Coupling, 673
  direct, 466–467
  interstage, 339–341
Coupling network, 450–451

Critical resolving distance, 421
Cross-coupling, 568
CRT viewfinder, 87
Current gain, 395
Cylindrical vibrating tubes, 301

Damper circuit faults, 714–715
Damper diode, 808
Dampers, solid-state, 683–685
Damping, 674–675
Damping resistor, 461
Dark current, 89
Darlington amplifier, 814
DC component of video signal, 423–424, 482–484
  loss of, 483–484
  passing, 465–466
  reinserting, 485. *See also* DC reinsertion
  shifting DC level, 482
DC reinsertion, 481–494
  with diode, 486–487
  function of, 485
  for positive-sync polarity, 487–488
  time constant, 487
DC restorer(s), 430–431, 490
DC restorer circuits, troubles in, 492
Dead-IF stage, locating, 873
Decibels (dB), 124
Decimal numbers, converting binary numbers to, 275
Deemphasis, 754, 755–756
Deflection, sawtooth, 603
Deflection coils
  horizontal, 58, 90
  vertical, 54, 90
Deflection system, 87
Deflection yokes, 705–706
Degaussing, 536, 856
Degenerative high-frequency compensation, 462–463
Degenerative traps, 344–345
Delay line, 428–429
Delta-gun tubes
  improvements in, 523–524
  operation of, 515–517
Delta-wye transformer, 345
Demodulator(s), 13
  color, 210–211
  digital FM, 777–778
  FM, 32, 758
  X and Z, 213, 812
Demodulator probe, 818–819, 874
Derived-voltage system, 560, 565–566
Diagonal deflection angle, 500
Dichroic mirrors, 104
Di-fan antenna, 130
Differential peak detector, FM, 778–780
Digital countdown system problems, 665
Digital horizontal countdown, 660–662
Digital screen display, 3–4
Digital switching, 228
Digital techniques, 2–3
Digital vertical countdown circuits, 633–634
Diode, inverting, 579–580
Diode-detector circuit, 267, 367, 368
  disadvantages of, 375
Diode phase-detector circuit, 644
Diode separators, 579–580
Diplexer, 11
Dipole antenna, 125, 128, 129
  folded, 132
  length computation, 127
  variations of, 132
Dipole array log-periodic antenna, 134
Direct-access remote control system, 312–316
Direct coupling, 488
Direct probe, 817
Direct waves, 119
Directional-coupler multitap, 18
Director element, 129
Discriminators, in FM receivers, 164, 763–765
Displays, types of, 40
Distribution amplifier, 151
Dividers, 275–276
  applications, 277
Document scanning, 95
Drift, 253
Drive controls, in vacuum tube sets, 686–688
Drive-failure protection, of output transistor, 680–681
Driven element, 128
Drivers
  and color output, 812
  horizontal, 804, 806
Dynamic convergence, 519–520, 729–732, 856
Dynamic range, 88
Dynacolor, 6

Eidophor system, 39
Eight-bay bow-tie antennas, 135
Electromagnetic deflection, 505–508
  with electrostatic focus, 508–510
Electromagnetic wave, components of, 123
Electron beam, 92
Electron beam deflection, 505–508
Electron gun(s), 13, 92, 501–505
Electronic attenuator, 783, 785
Electronic tuning, 226–228
Electrostatic focusing, 503–505, 509
Electrostatic induction, 771–772
Elements in a picture, 71
Emitter (or cathode) circuit, 464–465
Emitter-follower amplifier, 451, 468–469
Equalizing pulses, 58, 60, 574
  adding, 594
  function of, 593–594
  vertical sync and, 60–61
Even-line field, 51, 52
Eye, properties of, 77

Faceplate illumination versus output signal, 92
  linearity of, 88
Fan dipole antenna, 130–131
Federal Communications Commission (FCC), 9, 22, 40, 184
  allocation plan (1946), 130
Feedback
  degenerative, 470
  vertical output circuits, 723
Fidelity control. *See* Sharpness control
Field-effect transistors (FETs), 169, 244–249
  advantages of, 247
  types of, 245
Field rates, 49–50
Filament-voltage booster unit, 865
Films, televising, 100
Filter(s), 544–545
  color-trimming, 105
  components, values of, 545–546
Filter capacitors, checking, 888–889
Filter time constants, effect of, 643
Filtering
  IF, 371
  insufficient, 888–889
Fine tuning, 905–906
Fine-tuning control, 166
Flat pack circuit, 472
Flicker, 50
Flight simulation, 95
Fluoroscopy, TV, 95
Flyback current, excessive, 714
Flyback high voltage, 678
Flyback transformer
  faults, 714
  taps on, 672
  and yoke tester, 171
Focus
  adjustments, 707–708
  coil, 90
  poor, 885–886
Foldover
  horizontal, 902
  vertical, 897
Forward AGC, 395, 397
Foster-Seeley discriminator, 765–767
Four-way beam splitter, 104
Frames, 49–50
Frequency (ies), 41
  color video IF, 333
  intermediate, 331
  monochrome video IF, 332–333
  reversal in video IF band, 331–332
Frequency band, determining width of, 71
Frequency distortion, 422
Frequency loss, effects of, 73–74
Frequency modulation (FM), 12, 29, 750, 751–753
  and AM, differences between, 753, 754
  bandwidths, 753
  sound detector, 758
  sound section of receiver, 756–758
  stations on same frequency, 754
  terminology, 752
Frequency multiplexing, 11
Frequency response curves, 327–328, 422
Frequency sweep generator, 820
Fringe-area antennas, 139
Front end. *See* Tuner
Front porch, 56
Function discriminators, 308

Gamma, meaning of, 429–430
Gassy tube, 885
Gated AGC, 385
Gated beam tube, 770–771
  feedback voltage, 773
  as limiter-discriminator, 771–772
Gated rainbow pattern, 828
"Geartooth" distortion, 904
Generator(s)
  sawtooth, 603
  synchronizing signal, 105
Ghosts, 117
Graphic displays, 40
Gray scale, 79, 856
Grid guides, 238–239
Grid-leak bias, 761
Ground wave, 119
Grounded-base amplifier, 451
Grounded-emitter amplifiers, 243, 452–453
Grounded-grid amplifiers, 241
Grounding, 153
GTE Sylvania color receiver, 833
GTE-Sylvania solid-state receiver, 793
Guard bands, 11
Guidepoints for troubleshooting color TV receivers, 905–907

Halation, 512
Half-toned picture, 8
Hammerhead, 576, 594–595
Head end, 151, 152
Height troubles, 743–744
Hertz antenna, 125
Hewlett-Packard unit, 40
High-frequency compensation, degenerative method of, 462
High frequency gain, vacuum tube amplifiers, 455
High-frequency loss, 74
High frequency response, 455–459
  insufficient, 422
  poor, 473
High-frequency troubles, 881
High voltage
  loss of, 883
  power supply, 165
  troubles in, 712–713
High-voltage probes, 866–867
Hold-in range, 299

Home computers, 2
Home video recorder, 27, 28
Horizontal-deflection circuits, troubles in, 710–712
Horizontal-deflection oscillator, 608–609
Horizontal drive, loss of, 711
Horizontal frequency power supply, 560–566
  troubles in, 568
Horizontal and high-voltage problems, relationship between, 714
Horizontal hold control, 168
Horizontal lines (21)
  blanking, 59–60
  distribution of, 59
Horizontal output circuit
  beam scan vs. yoke current, 672
  block diagram of, 670–672
  deflection yoke and damper circuits, 671
  frequencies in, 670–671
  high voltage circuit, 671
Horizontal period, details of, 65
Horizontal sync and blanking pulses, 56–58, 574
  method of arriving at, 59–60
Horizontal synchronization, loss of, 576–577
Horizontal synchronizing pulse filter, 589–590
Hue(s), 75–76, 216, 217
  incorrect, 912–913
Hue (tint) control, 906–907
Hum, 51
  cause of, 50
Hum bars, 475, 908
Hybrid tuners, 223

Image, 7–8
  rejection, 231
Impedance coupling, 340
Impedance matching, 18
Impulse noise, 584
In-circuit tests, 864
Incremental tuners, 223
Inductive-reactance transistor, 658–660
Infrared-wave systems, troubles in, 319–320
Input-output (I/O), 279, 290
"Insufficient width," 712
Insulated-gate FET (IGFET). *See* Metal oxide semiconductor field-effect transistors (MOSFETs)
Integrated circuits (ICs), 2, 4, 6, 168–169, 329–330, 470–472
  advantages of, 6–7
  application in color TV IF system, 352–354
  detector in, 374–375
  as video driver, 471
  and video IF amplifiers, 350–354
Integrator, 591, 592
Intelsat IV, 26
Intelsat IVA, 26
Intercarrier sound system, 345
Interdigital transducers, 355
Interference filters, 150
Interference frequencies, 341–342
Interlace, poor, 898
Interlaced scanning, 51
Interleaving, 184, 186
Intermediate frequency (IF), 12, 249
  amplifier bandwidth, 334–335
  amplifiers, 163
  band interference, 231
  bandpass coils adjustment, 845
  trap, 228, 230
International Satellite Organization (INTELSAT), 27
Interphone communications, 106
Interstage circuits, 354
Ion spots, 510–511
Ion-trap magnet, effect of, 885
Isolation, video IF, 877
Isolation brightener, 865–866
Isolation by observation, 869–871
Isolation transformers, 741

Junction FET (JFET), 245–246
"Jumpy" picture, 896

Keyed-AGC system, 385–386, 387–392, 795
  block diagram of, 398
  noise impulses in, 399
  simplified, 388–389
  vacuum tube, 391–392
Keystone effect, 744
Keystone raster, 897–898, 904
Killer detector, 811
Kodak Ektalite screen, 37

Lag, 89
Large-scale integration (LSI) circuits, 37
Laser-optical videodisc recorder, 33
Latch circuits, 276, 410
"Lazy H," 132
Lead-in wires and input coil, connecting, 144
Lead monoxide, 93, 94
Level shifting, 466
Light-dependent resistor (LDR), 435–436
Light-emitting diode (LED), 107
Lightning arrestor, 153
Lightning protection, 153
Limiter(s), 164
  analysis of action, 76
  FM, 32
  high-gain linear IC, 763
  need for, 767
  purpose of, 759
  two-stage diode transister, 762–763
  vacuum tube, 760–762

Limiter-detector, 783
Limiter and filter circuitry, 655
Limiting knee, 762
Limiting threshold level, 762
Lin-clamp circuit, 631–632
Line-of-sight distances, 24, 120–122
"Line pairing," 898
Linear current ramps, 737
Linear regulator, 556
Linear vertical scan, 725
Linearity
  adjustment, 708
  poor, 896
Linearity control, 676, 688–689, 735, 902
Linearity troubles, 744–745
Local oscillator circuits, 253–256, 267
Local oscillator radiation, 232
Localization of color problems by observation, 871
Log-periodic antennas, 134
Loop bias and AGC, 352
Loss of drive protection, 686
Low-capacitance probe, 817–818, 874
Low-cathode emission, 885
Low-frequency
  compensation, 463–465
  loss, 74
  troubles, 880
Low-frequency response
  insufficient, 422
  poor, 474
Low and high frequencies, effect of loss of, 73–74, 422
Low-impedance circuits, testing semiconductors in, 864
Low-pass filter. *See* Integrator
Low-voltage power supplies, 166, 542–470
  troubles in, 566–568
  types of, 543
Lumen, defined, 88
Luminance, 77, 103
Luminance amplifiers, 416
Luminance channel, 416, 429, 431
Luminance signal, 185, 429
  converted to FM, 28

Magnavox, 37
Malfunction, analysis and isolation of receiver, 868–869
Masking voltages, 240
Marker frequencies, 335–336
Marker signals, 824
Markers, vertical or horizontal, 826
Master-antenna television (MATV) system, 17, 150–152
Matrixing, color-tube, 213
Mechanical switching, 227
  disadvantages of, 227–228
Memory module, 308–309
Mesh reflector, 130
Metal oxide semiconductor field-effect transistor (MOSFET), 169, 246–247, 308
  dual-gate, 247
  UHF RF amplifier, 249
  VHF RF amplifier, 248
Meter ranges, 833
  limits of, 831–832
Metric units, 924–925
Microcomputer, 279, 283, 289–291
Microprocessors, 2, 37, 279
Microwave links, 152
Microwave relay stations, 24–25
Microwave transmitting-receiving system, 109
Miller effect, 451, 458
Miller integrator, 627–632
Miller rundown circuit, 627
Minicam system, 109
Minimum resolving angle of the eye, 421
Mixers and mixer circuits, 249–253
  cascode, 252
  common-base, 250–251
  common-emitter, 251
  MOSFET, 252
  neutralization, 251–252
  UHF diode, 252–253
  vacuum tube, 249–250
  VHF transistor, 250
Modulating signal, 186
Modulation, infrared, 313
Monochrome circuitry, 1
  analysis of, 794–809
Monochrome portable receiver, 5
Monochrome signal in color TV signals, 197–198
Monochrome TV camera, 86–87
Monochrome TV receiver
  audio section, 164
  block diagram of, 162
  controls, 166–168
  horizontal-deflection section, 165
  ten basic sections of, 161
  troubles in, 173–174
  tuner section, 161
  vertical-deflection section, 165
Monochrome TV transistor vertical-deflection circuits, 724–726
Monochrome TV transistor vertical output circuits, 735–737
Monochrome vacuum tube vertical output circuits, 733–735
Motion picture frame rate, 50–51
Multiburst test signal, 79
Multimeter(s)
  defined, 831
  solid-state analog, 831
  solid-state digital, 832
Multiplexing, 184
Multivibrator, 614
  cathode-coupled, 619–621, 648
  emitter-coupled, 625, 626
  horizontal, 804
  stabilization, 648–649
  synchronizing, 623
  transistor, 624–625
  vacuum tube and output stage, 619
  vacuum tube plate-coupled, 614–617

Nanoamperes (nA), 88
National Television System Committee (NTSC), color TV system, 77, 178, 184–190, 830
N-channel JFETs, 245–246
Negative picture, 69, 411, 882
  transmission, 69–70
Newvicon, 22
Noise, 232, 240
Noise cancellation circuits, 399–402
Noise cancelling keyed AGC, 386
Noise-free video, 401
Noise gate, 584, 585
  diode, 399–400
Noise-inverter circuit, 400–402
Noise processor, expanded block diagram of, 407–408
Noise pulses, 408, 409
  minimizing effect of, 580–581
  negative, 401
  and sync pulses, 401
Nonlinear sweep, 685
Nonlinearity, horizontal, 711
Norelco three-tube color camera, 105
Novar, 170
Nuvistor, 170, 239, 256

Odd-line field, 51, 52
Omega system, 228
Open peaking coil, 378
Oscillations, locked-in, 774
Oscillator
  horizontal, 165, 596, 640–665
    adjustment of, 852
    troubles in, 662–664
  3.58-MHz, 810, 814, 909
  UHF transistor, 256
  vertical, 165, 596
Oscillator drift, 297
Oscillator frequency, 298, 710
  controlling, 660
  DC control of, 645–646
  sensing, 653
Oscillator stability, 232
Oscilloscope(s), 171, 816
  connecting, 844
  dual-trace, 816
  requirements, 819
  TV-service type, 816
Oscilloscope patterns, 578–579
Oscilloscope photos, 77–78
Out-of-circuit tests, 864
Output, infrared, 313–316
Output amplifiers, 165
Output circuit, 100
Output transformers, horizontal, 702–705
Overcurrent protection, of horizontal-output transistor, 679–680
Overdissipation protection, of audio power amplifier, 785
Overload gate, high-voltage, 808
Overloaded picture, 411

Parabolic current, 729
Parabolic reflectors, 131
Parallel-resonant circuit, 763
Parallel traps, 343
Parallel-wire lines, 142, 145
Parasitic element, 128
Peak AGC systems, 385, 386–387
Peak detectors, 778
Peaking control. *See* Sharpness control
Pentode video amplifier, gain of, 450–452
Pentodes, 328
Persistence of vision, 50, 193
Phase, 189
Phase angle, 187
Phase distortion, 74, 425–426
Phase-locked loop (PLL), 228, 264–265, 283, 285
  FM detector, 775–777
Phillips-MCA system, 34
Photoconductive layer, 90–94
Photodiodes, 96, 98
Pi filter, 545
Picture bending, horizontal, 596–597
Picture control, 167, 432
Picture detail, 8
Picture distortions, due to defective picture-tube components, 885–886
Picture elements, 71
Picture pulling, 893
Picture qualities, desirable, 74–76
Picture tube(s), 8–9, 13, 497–539
  bases, 500
  brightness, 865–866
  color, 208
    adjustment of, 852–856
    with delta gun and color-dot triads, 515–524
    symptoms of defective, 215–216
  defective, 884–885
  external components, 521–522
  first lens system of, 502–503
  heater volts (mA), 499
  interchangeability, 500–501
  length and neck diameter, 500
  numbering, 498–499
  protection considerations, 530–534
  safety shields, 514–515
  second-lens system of, 503–505
  specifications, 498–501
  troubles in monochrome and color, 535–536
  video signal requirements of, 417–420
Picture tube analyzer and restorer, 864–865
"Pie crust" distortion, 904
Piezoelectric effect, 302, 355
Pincushion (PIN) distortion, 696–697
Pincushion transformer, 727–729
Pincushioning, vertical, 723
Pix control, 432
Playback-chroma processing, 32
Playback circuits, 31–32
Playback preamplifier, 32
Plumbicon, 89, 93, 105, 107
  specifications, 94–95, 96
  spectral response, 94

typical operating conditions and performance, 97
Plumbicon target, 93–94
Polarity, 418
indication, 832
reversal, 646, 728–729
Polarization, 123–124
Polaroid Land camera, 40
Portable color TV camera, 107–109
Post-demodulator processing, 32
Post injection marker system, 825
Power supply(ies), 795–798
horizontal frequency, 560
hybrid color TV, 559–560
solid-state monochrome, 558–559
transformer and transformerless, 546–548
Power supply ripple, 568
Preamplifiers, 146–148
power supplies for, 147
Preemphasis, 754–755
Premodulator-processing circuits, 29
Prescalers, 275–276
Primary colors, 181
Printed circuits (PCs), 168, 329
Program-related data signal, 80–81
Programmer, 3
Projection television, 1, 14, 37–40
Pulse repetition rate (PRR), 574
Pulse width oscillator, 558
Purity adjustment, 856
Purity-magnet assembly, of color picture tube, 521
Pythagorean Theorem, 120

Q value, effect of, 235–236
Quad-in-line plastic package (QUIP), 350
Quadrature detector, 769–775, 783
Quantum efficiency, 88
Quasar Compu-matic touch-tuning system, 280–285, 312–316
Quasi-single-sideband, 72

Rabbit ears, 140
Radio frequency interference (RFI), 107
Radio-frequency waves, 1, 14
Radio wave propagation, 119–120
Radio waves, non-return of, 119–120
Radix, 273
Random access, 3, 228
Rare-earth phosphors, 524
Raster, 49, 567, 908
Ratio detector, 767–769, 782
RC differentiator, 585
RC integrator, 585
RCA color TV receiver, video-detector stage of, 374
RCA frequency-synthesized (FS) tuning, 228
RCA home TV programmer, 36–37
RCA in-line gun, color-picture tube, 524–526
RCA ministate antenna system, 139–140
RCA SelectaVision, 34
RCA XL-100, 402, 694
Reactance transistors, types of, 658
Read and write memory (RAM), 279
Read only memory (ROM), 279
Receivers, categories of construction, 160–161
Receiver video signal, 70
Record amplifier, 29
Record-chroma processing, 29–30, 32
Record-playback system, block diagram of, 28–34
Rectangular base drive waveform, 680
Rectangular screens, 513
Rectifier(s), 148, 543–544
half-wave, 547
silicon, 544
solid-state high-voltage, 706
Reference signal, 377
Reflected signal, phase shifts affecting, 122
Reflections, 512
Reflector system, 128, 129
Remote-control boxes, 18–19
Remote control devices, functions and types of, 300
Remote control system
direct-access, 312–316
electronic, 306–312
receiver, 316
ultrasonic. *See* Ultrasonic remote control system
Resetability, 226
Resistance(s), transistor input, 457–458
Resistance-capacitance (R-C) amplifiers, 449
Resistance-capacitance comparator bridge, 171, 172
Resolving power of eye, 420–421
Resonance, above and below, 766–767
Resonant circuits, 657
Resonant-stabilizing circuits, 648–649
Response curve shape, 336–337
Retrace, 48, 49, 58, 388, 737, 740
Return trace, 49
Reverse AGC, 396–397
RF amplifiers
neutralization, 243–244
transistor, 242–244
typical tube-type, 239–240
RF section. *See* Tuner
RF signal generators, 823–825
Ringing, in monochrome and color pictures, 909
Ringing test, 867–868
Ripples (hum), 888
R − Y and B − Y system, 211

S-capacitor, in solid-state horizontal-output stage, 685
S correction, 739
S curve, 779–780
Saddle yoke, 705
Satellite television relays, 1, 14
Satellites, use of, 24–26
Sawtooth capacitors, recharging, 735–737
Sawtooth current, 58, 61, 602–603, 681–682, 732
Sawtooth generator(s)
cathode-coupled multivibrator, 621–622
plate-coupled multivibrator, 617–618

Sawtooth generator(s) *(cont'd)*
vacuum tube blocking oscillator, 607–608
vertical transistor blocking oscillator and, 609–611
Sawtooth voltage, developing, 602–603
Sawtooth wave, 602, 628–629, 645, 804
Scanning, 9, 10, 47, 51, 737
frequencies, 64
introduction to, 47–48
principles, 48
Scanning lines, 9, 52
Scanning rate, 49
Schmidt optical system, 38
Screen adjustments, 197–198
Seek function, 317
Semiconductors, uses of, 6
Sencore bar sweep generator, 821
Sencore VA48 tuner-test function, 867
Sencore video analyzer, 822–823, 827
Sensory circuits, 283, 285, 291–292
Series-pass transistor regulator, 553–555
Series peaking, 460–462
Series trap, 342–343
Serrated vertical pulses, 54
Servicing television receivers, 862–913
Shadow mask, 518–519, 523
Sharpness control, 167, 438
Shield plates. *See* Grid guides
Shot noise, 232
Shunt peaking, 459–460
Shunt regulator circuit, high-voltage adjustments for, 707
Shunt-video detectors, 371–372
SI units, prefixes for, 926
Sidebands, 72–73
Signal connections, VA48, 843
Signal distribution system, 151–152
Signal electrode, 90
Signal inection, 172, 173, 871–874, 875
Signal injection test, 874
Signal loss, in IF system, 358–359
Signal substitution. *See* Signal injection
Signal tracing, 874, 877, 881
Signals, neutralizing, 243
Silicon-controlled rectifier (SCR), 532, 563–565
Silicon-controlled rectifier (SCR) horizontal output circuits, 697–702
high voltage system, 700–702
protection circuitry, 702
trace and retrace, 698–699
Silicon diode-array vidicon, 93, 95–98
Silicon imaging device (SID), 98–100
Silicon intensifier target (SIT) tube, 22
"Silicon target" vidicon, 22
Simpson color pattern generator, outputs of, 827
Single sideband transmission, 10
Single-tube color camera, 100–101
Sinusoidal oscillator, reactance-controlled, 657–660
6BN6 detector, 770
6DT6 detector, 773–774
Skipping, of channels, 312
Sky waves, 119
Snivets, 904
Snow. *See* Masking voltages
Solid state camera circuits, 87
Solid-state color TV horizontal output circuits, 694–696
Solid-state color television receiver, 3
Solid-state monochrome circuits, 678–680
Solid-state television receivers, 1, 14
Solid-state tuners, 223
Sony Corp., 100
Sound bars in picture, 889–890
Sound IF-frequency separation, 345–346
Sound IF limitings, 758, 759, 763
Sound section
IC, 782–783
transistor, complete, 780
vacuum tube, complete, 780
Sound signals, electronically generated, 303–306
Sound traps, 342, 348, 373
Special purpose television cameras, 109–111
Split-finger transducer, 356
Splitters, 149
Spurious responses in video IF band, 333–334
Square-wave response of video amplifiers, 426–427
Square wave testing, 426
Stacked amplifier, 469–470
Stagger tuning, 335, 337–339, 347–348
Staircase test signal, 79
Static convergence, 519, 729, 856
Step tuners, 223
Studio color television camera, 105–106
Studio console, 11–12
"Subber," 867
Subtractor signals, 439–440
Supply voltage changes, 232
Surface-acoustic wave (SAW), 354, 355
Surface acoustic wave filter (SAWF), 354–358
Surface-wave integrated filter (SWIF), 355
Swamping resistor, 461
Sweep generator, 426, 820–823
Switching regulator, 556–558
Sync and blanking pulse frequencies, 65
Sync pulse(s), 13, 58, 78, 369, 392, 401, 574, 644, 891–893
negative, 623
signal in DC form, 577
single-polarity, 646–648
vertical, 630–631
Sync-pulse alignment, 391
Sync-pulse separator, 164–165
Sync and video signal separation, 55–56, 577–579
Sync section, defective, 870–871
Synchronization, 48, 817, 893
effects of loss of, 575–577
loss of vertical, 634
noise reduction, to improve, 584
and phasing problems, 663
troubles, 595–596
Synchronizing circuits, 47, 574–598
Synchronizing pulse(s). *See* Sync pulse(s)
Synchronizing pulse amplifier, 581–582
Synchronizing signal, 10, 47, 61–62

Synchronous (linear) video detector, 375–378
Synchronous orbit, 26
Syncom, 26

Telecine color film camera, 110–111
Telesat Canada, 26
Telestar, 26
Television channel frequencies, U.S., 234, 921–923
Television coverage, extending, 24
Television frame rate, 51
Television receiver (s), 12–13
  four basic sections of, 12
  modern, 2, 3–6
  operation of, 12–13
  styles of, 4–6
Television standards, 40–43
Television system (s)
  applications, 16–45
  concepts, 1–16
Television transmission. *See* Transmission
Temperature-compensated slotted-aperture mask (TCM), 528
Test equipment
  for TV receiver alignment, 816
  for monochrome receivers, 171–173
Test jig, color picture tube, 863
Test pattern, normal, 472
Tetrodes, 239
Three-terminal voltage regulator, 555–556
Three-way beam splitter, 104–105
Thermal noise, 232
Time-programmable system, 3
Time-programmed channel selection, 4
Tin oxide, 93–94
Tint control, 809–810
Toroidal yokes, 705–706
Trace, 48, 49, 58
Trans-Canada Telephone System, 26
Transducers, ultrasonic, 300–301
Transformer(s)
  coupling, 236, 339–340
  horizontal output, 678
  isolation, 890
  types of, 741
Transformer power supply for tube receivers, 546–547
Transformerless receivers, 890
Transformerless TV power supply, 547–548
Transistor amplifiers, 457–458
Transistor deflection circuits, 605
Transistor horizontal output circuit, 680–683
Transistor noise-cancellation circuits, 584–585
Transister noise clipper and noise gate circuit, 585–588
Transistor synchronizing pulse separators, 582–584
Transistor tester, 171, 172, 864
Transistor-transistor logic (TTL), 2
Transistor tuner, typical, 256–257
Transistor-video amplifier, gain of, 452–455
Transistors, selection of, for video amplifiers, 468
Transmission, 10
  characteristics of, 10–11
  color, 11
  digital uses in, 3
  of picture (video) signal, 9
  of sound (audio) signal, 10
Transmission factors, automatic correction of, 80
Transmission lines, 141–142
  unbalanced, 156
Transmitter, 9–10
Transversed filter peaking system, 439
Trap adjustment, 844–845
Trap circuits, 341–345
Trapezoidal pulse amplitude, 734–735
Trapezoidal voltage, generating, 605
Trapezoidal wave forms, 603–604, 672
Triad, 517, 524
Triggered sweep, 817
Trinicon tube, 100, 101–103
Trinitron picture tube, 526–528
Triode(s), 238
  amplifiers, 240, 733–734
  synchronizing pulse separator, 580–582
  use of, 392
Tri-potential color tube, 528–530
Troubleshooting, 216, 217, 862–913
  AGC section, 878–880
  antenna system, 153–154
  horizontal deflection system, 899–905
  low-voltage power supplies, 886–890
  picture tubes and associated circuits, 883–886
  remote transmitters and receivers, 319
  sync-separator stages, 890–893
  tuner section, 874–876
  vertical-deflection system, 894–899
  video amplifiers, 880–883
  video IF and video detector circuits, 876–878
Tube deflection circuits, 604–605
Tube tester, 171
Tubes, selection of, for video amplifiers, 467
Tuned circuit, characteristics of, 235
Tuned radio frequency, (TRF), 319
Tuner(s), 161, 222, 794
  electrical characteristics of, 228
  general specifications for, 233
  interference, stability, and noise problems in, 230–233
  sources of trouble in, 266–268
  specifications, 232
  substitute, 867
  turret-type, 226
  types of, for VHF and UHF reception, 223
  UHF solid-state, 261–266
  vacuum tubes for, 223, 238

Ultra high frequencies (UHF), 24
  band switching, 295
  operation, 161–162
  television antenna, 130

Ultra high frequencies *(cont'd)*
  tuners, 223, 226, 230, 261–266, 267
Ultrasonic remote control systems, 300–302
  receivers for, 304–306
  troubles in, 318–319
Unwanted-signal paths, 122

Vacuum tube AFC, and multivibrator circuits, 644–648
Vacuum tube circuit, direct-coupled, 466
Vacuum tube color horizontal output circuits, 689–693
Vacuum tube monochrome circuit, 672–678
Vacuum tube and solid-state video IF amplifiers, comparison of, 328–330
Vacuum tube voltmeter (VTVM), 171, 172, 837
Vacuum tubes, 170, 172
Varactor(s), 222
  circuits, 259–261
  diode, 257
  tuning and tuners, 257–261, 267–268, 283
"Variable angular velocity" principle, 34
Variable-resistance transistor, 660
Vector diagrams, 766
Vectorscope, 819–820
Vertical blanking interval, 58–59
Vertical deflection oscillator system, troubles in, 634–635
Vertical digital countdown circuit, troubles in, 635
Vertical foldover, 897
Vertical and horizontal pulse separation, 589–593
Vertical Interval Reference (VIR), 3, 5, 6, 76, 80, 215
Vertical Interval Test Signals (VITS), 78
Vertical output deflection circuits, 721–749
Vertical output transformer (VOT), 722, 730
Vertical period, details of, 65
Vertical sweep and sync systems, 808
Vertical synchronization, loss of, 575–576
Vertical synchronizing pulse filter, 591–592
Vertical synchronizing pulse integration, 593
Very high frequencies (VHF), 24
  modulator circuits, 33
  operation, 162
  tuners, 223, 229
Vertical yokes, 741–743
Vestigial sideband transmission, 10, 335
Video amplifier(s), 164, 354, 416–447, 799–802
  color, 431
  design, 450–476
  monochrome, 429–431
  transistor, typical, 468–470
  troubles in, 427, 472–475
  tube and solid state, comparison of, 431–432
Video cassette recorder (VCR), 31
Video detector, 163–164, 352, 354, 367–381, 798–799
  defective, 878
  troubles in, 378
Video-disc recorders, 1, 14
Video-disc recording (VDR), 27, 33–34
Video games, 34–36
Video head and tape threading, 31
Video Home System (VHS), 28
Video IF amplifiers, 325–361
  troubles in, 358–360
Video-IF system, 204, 798–799
Video-output amplifier, 32–33
Video peaking, 438–440
Video-peaking control, monochrome TV, 167
Video signal, 72, 417–420
  DC component of, 423–424
Videotape recorder (VTR), 1, 14, 27–28
  studio color, 111–112
Vidicon
  components of, 89–90
  target operation, 90–91
  video signal, 92
Volt-ohm-milliameter (VOM), 171, 172
Voltage
  doublers, 548–550
  focus, 566
  heater, 566
  multipliers, 548–549
  ratio, gain as, 124
  screen, 566
Voltage, DC
  reduction of, 567–568
  troubles in, 566–567
Voltage-controlled oscillators (VCOs), 29, 278, 287, 377, 776, 633
Voltage-dependent resistor (VDR), 743
Voltage regulators, 551, 785
Voltage tests, 877
Voltage-variable capacitance, 458

Wafer-type incremental tuners, 223–226
Wave form(s)
  analysis, 650–652
  composite colorplexed, 194–195
  important, 196–197
  interpreting, 426–427
  observation of, 171–172
  reference voltage, 655–656
Westar Domestic Satellite, 25
Westar earth stations, 26
Western Union, 25, 26
Wide-band amplifiers, 71
Wide-band oscilloscope, need for, 819
Width coil mismatch, 711–712
Width controls
  solid-state sets, 688
  vacuum tube sets, 688

Xeron lamp, 39
X-ray emission
  in color TV receivers, 708–709

from tube sets, 709
X-ray protection circuits, 710

"Y" channel, 416
Y signal, 77, 429, 431
Y video circuits, 794
Yagi antenna, 129

Yoke(s)
effect of defective, 886
faults in, 715
Yoke-matching circuitry, 690–692

Zener diode regulator, 552–553
disadvantages of, 553